Ulrich Teubner, Hans Josef Brückner

Optical Imaging and Photography

Also of Interest

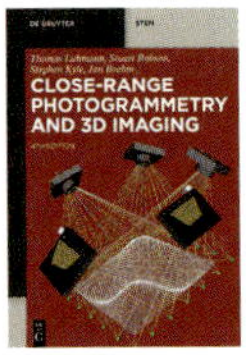

Close-Range Photogrammetry and 3D Imaging
Thomas Luhmann, Stuart Robson, Stephen Kyle, Jan Boehm, 2023
ISBN 978-3-11-102935-1, e-ISBN (PDF) 978-3-11-102967-2,
e-ISBN (EPUB) 978-3-11-102986-3

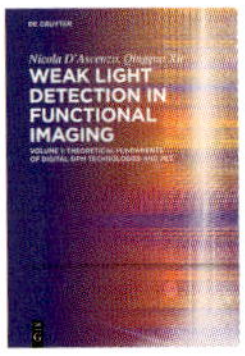

Weak Light Detection in Functional Imaging.
Volume 1: Theoretical Fundaments of Digital SiPM Technologies and PET
Nicola D'Ascenzo, Qingguo Xie, 2023
ISBN 978-3-11-060396-5, e-ISBN (PDF) 978-3-11-060577-8,
e-ISBN (EPUB) 978-3-11-060415-3

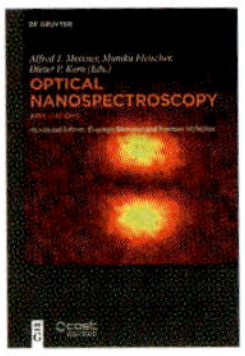

Optical Nanospectroscopy. Volume 3: Applications
Alfred J. Meixner, Monika Fleischer, Dieter P. Kern, Evgeniya Sheremet, Norman McMillan (Eds.)
ISBN 978-3-11-044289-2, e-ISBN (PDF) 978-3-11-044290-8,
e-ISBN (EPUB) 978-3-11-043498-9

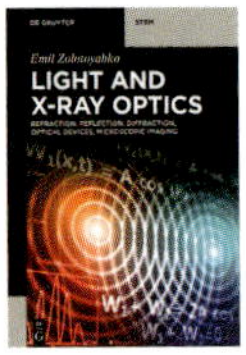

Light and X-Ray Optics.
Refraction, Reflection, Diffraction, Optical Devices, Microscopic Imaging
Emil Zolotoyabko, 2023
ISBN 978-3-11-113969-2, e-ISBN (PDF) 978-3-11-114010-0,
e-ISBN (EPUB) 978-3-11-114089-6

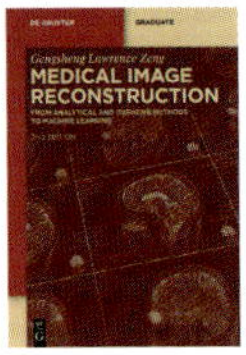

Medical Image Reconstruction.
From Analytical and Iterative Methods to Machine Learning
Gengsheng Lawrence Zeng, 2023
ISBN 978-3-11-105503-9, e-ISBN (PDF) 978-3-11-105540-4,
e-ISBN (EPUB) 978-3-11-105570-1

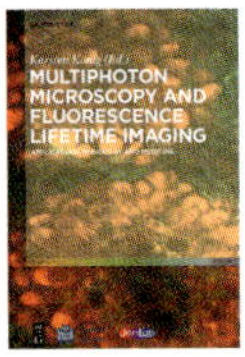

Multiphoton Microscopy and Fluorescence Lifetime Imaging.
Applications in Biology and Medicine
Karsten König (Ed.), 2018
ISBN 978-3-11-043898-7, e-ISBN (PDF) 978-3-11-042998-5,
e-ISBN (EPUB) 978-3-11-043007-3

Ulrich Teubner, Hans Josef Brückner

Optical Imaging and Photography

Imaging Optics, Sensors and Systems

2nd revised and extended edition

DE GRUYTER

Authors
Prof. Dr. habil. Ulrich Teubner
Institut für Laser und Optik
Hochschule Emden/Leer – University of Applied Science
Constantiaplatz 4
26723 Emden
Germany
ulrich.teubner@hs-emden-leer.de,
ulrich.teubner@uni-oldenburg.de

Prof. Dr. Hans Josef Brückner
Institut für Laser und Optik
Hochschule Emden/Leer – University of Applied Science
Constantiaplatz 4
26723 Emden
Germany
hans.brueckner@hs-emden-leer.de

ISBN 978-3-11-078990-4
e-ISBN (PDF) 978-3-11-078996-6
e-ISBN (EPUB) 978-3-11-078997-3

Library of Congress Control Number: 2023938348

Bibliographic information published by the Deutsche Nationalbibliothek
The Deutsche Nationalbibliothek lists this publication in the Deutsche Nationalbibliografie; detailed bibliographic data are available on the Internet at http://dnb.dnb.de.

Cover image: Harris Hawk/Wustenbussard „Paulchen“ der „Deutschen Greifenwarte Burg Guttenberg“; Author (Photograph): Ulrich Teubner; fotografiert mit freundlicher Genehmigung von Bernolph von Gemmingen
Typesetting: VTeX UAB, Lithuania
Printing and binding: CPI books GmbH, Leck

www.degruyter.com

MIX
Papier | Fördert gute Waldnutzung
FSC® C083411

This book is dedicated to our families. We would like to express special thanks, in particular, for their large amount of patience and support, to Sabine, Jan and Christine.

So eine Arbeit wird eigentlich nie fertig, man muß sie für fertig erklären, wenn man nach Zeit und Umständen das Möglichste getan hat.
(Johann Wolfgang von Goethe, Italienische Reise)

Die Entdeckung, daß es nicht so einfach ist, wie man gedacht hat, ist als Gewinn anzusehen.
(Carl Friedrich von Weizsäcker)

Foreword

With photography over the past quarter century, digital imaging has changed the way we share information and how we preserve memories in unprecedented ways. Gone are the days when you drop off 24 frames on a film roll at your local photoshop and wait a week for the prints, eventually only to find the shots are blurry or overexposed. Digital photography has created endless possibilities because the images are immediately available and transferable and can also be modified, combined and processed.

The doubts before the beginning of the millennium as to whether digital image sensors could ever achieve the resolution of high-resolution film while remaining affordable have dissipated just as quickly as questions about compact storage media that could process and store the huge amounts of data. In the 1980s and 1990s, around 20 million cameras were sold annually, ranging from the compact camera to the professional SLR camera. Today alone, 100 times as many smartphones with cameras are sold each year, i. e., almost 2 billion annually. In the 1990s, who would have thought the image quality of modern smartphone cameras that fit in a thimble, weigh just a few grams and cost less than €30 as a system component? This image quality and the linking of image data to communication networks have made digital photography so widespread today that most people around the world always carry a digital camera, constantly preserving and communicating their experiences with images. The functions of the camera have long since expanded from snapshots with friends and family to personal authentication for access to money, document scanning or data transfer via QR scan. In the future, the number of digital camera systems will continue to multiply, which will further be integrated seamlessly into our everyday lives. Billions of development funds are being invested in the development of AR glasses, autonomous driving or mobile healthcare systems.

A book on "Optical Imaging and Photography" that provides the physical fundamentals of these applications, and their technical solutions is therefore highly topical. Ulrich Teubner and Hans Josef Brückner, two professors of physics as well as enthusiastic photographers, fill a gap by comprehensively treating all aspects of digital (and analogue) photographic imaging in a single book. This is unique in the compilation. Until now, you would have had to get many different books respectively about photography and optics, optical design, Fourier optics, camera and image sensor technology and image processing. In addition, for this compilation of state-of-art technology you would have to research masses of information and data on specific camera systems, lenses, camera qualification metrology as well as the rapidly advancing developments of modern CCD or CMOS sensors from magazines, Internet sources and directly from manufacturers in the industry over many years.

Digital imaging, particularly the interplay between optics and image sensors, is the focus of the book. Nevertheless, the presentation is more fundamental: Ulrich Teubner and Hans Josef Brückner comprehensively introduce the reader to the physical basics of

https://doi.org/10.1515/9783110789966-201

all aspects of optical imaging for photography, the technical implementation of modern photo-optical systems and their performance metrics and the associated test setups from manufacturers or test institutes for the measurement of MTF curves, straylight, dynamic range, etc.

The fundamentals of "optical imaging" include the paraxial imaging equations and, for the calculation of intensity distributions, a wave-optical modeling. Wave-optical modeling, namely "Fourier optics," includes the fundamental limitation of image resolution due to diffraction as well as lens aberrations. Practical photography requires many more aspects of optical imaging, which only very rarely all occur simultaneously for industrial applications under laboratory conditions.

The imaging is essentially three-dimensional and requires lenses that can focus over a large depth range. On the other hand, there is limited depth-of-field, which depends on basic parameters such as f-number, focal length and object distance. In turn, the "blurring in depth," which the photographer calls "bokeh," exhibits special properties of the lens such as vignetting due to field diaphragms or various types of aberrations. Premium manufacturers, in addition to achieve good performance in best-focus, spend considerable effort to optimize lenses also for an appealing bokeh.

Practical photography often extends over a large dynamic range, i. e., the simultaneous presence of bright objects or light sources and dark areas. Irradiance variations of over 5 orders of magnitude or more in the scene requires image sensors as well as "HDR" imaging methods that can display these light and dark areas simultaneously. They should be displayed in such a way that fine nuances of brightness or color deviations can be distinguished by the viewer and the depiction of the scene appears natural to the human viewer at the same time. If this high dynamic range is enabled by the image sensor, then there are still the lenses, which might limit the image performance by producing disturbing ghost images or stray light haze in the image depending, e. g., on the quality of antireflective coatings.

Photographic lenses have to support the very high resolution of modern image sensors right up to "high speeds," i. e., high apertures, and this over the entire wavelength range of visible light. In addition, lenses should often be as flexible as possible, i. e., "zoomable" over a large field-of-view and focusable for a large distance range. All this is practically not compatible in one type of lens, at least not for a practical, portable system. That is why there are many different lens types such as "zoom," "macro," "tele" or "wide angle." In addition to many examples of current photo lenses on the market, Ulrich Teubner and Hans Josef Brückner also present various classic optical designs such as Double-Gauss, retrofocus type, telephoto lens, etc. and discuss the connections between the layout and optical aberrations and camera space constraints.

In addition, there are many other aspects of optical imaging for photography such as relative movements and image stabilization, environmental dependencies such as different temperatures or underwater photography, the use of comprehensive image processing including digital aberration correction, etc. This makes it clear that a representation like this one by Ulrich Teubner and Hans Josef Brückner on the optical imaging of practical photography must be very versatile, and of course, contain many other applications such as machine vision in large parts automatically.

Current digital camera systems on the market are discussed, some with interesting cross-comparisons, example.g., between smartphone cameras and large-format system cameras. For this comparison of "small" and "big" imaging systems, there are immutable physical laws: The étendue, which shrinks with the miniaturization of the optics, requires either longer exposure times or higher ISO sensitivity with a side effect on image noise. The latter reintroduces texture artifacts into the image using software-based noise reduction. Other fundamental physical effects of the miniaturization of image sensors are the increased depth of field or the limited resolution due to diffraction. These and many other complex mechanisms are described concisely with their impact on image quality. Image imperfections such as due to optics, e. g., aberrations, scattered light, ghost images, etc. or due to the image sensor such as noise, pixel sampling, rolling shutter, blooming, image lag, etc. are not only explained for themselves, but also the cause of their occurrence. At the end of the digital imaging chain, there is the representation on the display and the perceived image quality of the human observer. Accordingly, the authors also discuss perceived quality for common viewing conditions, e. g., depending on whether the images are viewed on the camera or smartphone display or enlarged on a computer screen. Sometimes their analysis concludes critically, as not all technical specifications and developments in the consumer camera market led to actual improvements of perceived image quality but were sometimes rather driven by marketing. The book is full of excellent illustrations, tabular data overviews and image comparisons for reference.

In a rapidly changing field, the blessing of application orientation presentation is always accompanied by the curse of no longer being up to date for the newest developments after a few years. Many new developments were added to the new edition, such as the multicamera systems and their multicell image sensors for smartphones, miniaturized 3D acquisition systems, the use of computational imaging or the digital correction of image errors in the optics among many other topics. However, despite all those new developments, most of the presentation will remain up to date for many years.

In the research and development of digital image systems, the entire digital image chain must be considered today when optimizing camera systems: Do all parts of the image chain, i. e., the optics, the image sensor and the image processing, harmonize? How can I improve the weakest part? Can I compensate for hardware deficits computationally by software? With what side effects? You have to ask yourself this for a huge number of camera settings, external conditions and image motifs, whether bright sunshine, twilight and darkness or high dynamic range, i. e., very bright and dark image parts at the same time, as well as different image motifs, whether finely structured, high-contrast or with fine brightness or color nuances.

So, this book is of interest to a great many, especially to those with a technical or physical education who are interested in photography. But it is also a reference book that should be on the desk of scientists and engineers of digital optical imaging systems or setups using those, whether for photographic or industrial applications.

Oberkochen, March 15, 2023 Vladan Blahnik

Preface to the second edition

After the success of the first edition, in the second edition we have complemented and deepened several of the previous discussions and corrected a couple of misspellings. But the most important change of the present book is the significantly extended content. First, this takes into account modern developments of optics, sensors and optical systems in general in the recent years. In particular, this includes advancements in sensor technology, sensors with smaller pixels and higher pixel density, curved sensors, etc. Yet we address new developments in optics, too. Here, we may mention especially the miniaturization of optics and also the application of ideas and physical concepts that have been already existing for some time and now have become sufficiently developed to be used in camera modules or cameras. Metalenses are an example. Second, the extended content includes a new chapter on cell or smartphone cameras. Here, we take into account that those devices have become the most important cameras in the sense that they are the most used cameras and that by far the most images taken in a manifold of different situations result from them. As this is relevant for miniaturized optical imaging systems, a short subchapter on imaging and focusing of Gaussian beams with ball lenses completes the second edition.

Despite of the details and practical examples, also the second edition of the textbook is intended to serve as both a tutorial that helps for deep understanding of the topics for beginners, and a useful working reference in the field of optical imaging for experienced scientists, engineers and photographers. Consequently, the main title remains the same, namely *Optical imaging and photography*. But we have changed the subtitle slightly, namely to *Imaging optics, sensors and systems*. We have kept our intention to provide a comprehensive and consistent description of imaging with a cross-link of all the involved subtopics. Furthermore, we may emphasize that it is particularly the interplay between optics and sensors that we regard as very important.

Also, the second edition is intended to concentrate on all of the relevant fundamentals, which are subject of the whole imaging process that includes the hardware (optics, sensors, etc.) as well as the physical and technical background. In that sense, it is not only relevant for photography, but for any kind of optical imaging including, e. g., machine vision. For instance, this may be seen from a typical catalogue of a manufacturer or seller of scientific and industrial optical components where, e. g., MTF curves, etc. are provided for the various kinds of optics and sensors. But, of course, even a comprehensive book such as the present one cannot cover all essentials of optical imaging. Hence, for specific topics we refer to detailed books on photogrammetry, high-speed imaging, 3D Imaging, automotive or machine vision, optical medical imaging, microscopy, video or on books on image processing, computational imaging, etc.

https://doi.org/10.1515/9783110789966-202

The authors wish to thank Dr. Vladan Blahnik for valuable discussions and comments as well as Prof. Dr. Bert Struve for his help in reading the manuscript and identifying necessary revisions.

Emden, March 2023

Ulrich Teubner
Hans-Josef Brückner

Preface to the first edition

We started the project for this book some years ago. During our laboratory work, in particular, with students of Engineering Physics in the field of lasers and optics, we encountered various situations where the imaging of objects was required at different levels of quality. Examples included the imaging of laser-produced plasmas, imaging of microstructures fabricated by laser radiation, mask projection in lithographic applications, and profiles and focal spot distributions of laser beams, to imaging large objects such as complete breadboard setups and even converting images taken with conventional light microscopes for archival storage. In all cases, optical systems were required, often in combination with modern digital electronic sensors or cameras. With the continuous development of powerful and complex digital camera systems, however, we found that basic concepts, handling and requirements for achieving the desired quality of imaging should be conveyed as well to all people working with such systems. This has to be done at least in a condensed way, incorporating all necessary steps in the imaging process chain-like imaging optics, electronic detection and image processing. Our goal was that not only students, but also experienced engineers should become capable of understanding the requirements for a given imaging problem and finding the appropriate optical system for scientific and technical applications. Moreover, being passionate amateur photographers, we discovered that also for those interested in classical photography, a physical background and technical information could be helpful.

In a more general view, imaging belongs to the most important processes in human life. The human eye is usually the reference, and images are taken in daily life and displayed on TV, computer screens, smartphone screens and so on. Thus, we can regard imaging as an important subject in general. The present book treats this subject from a technological and scientific point of view and discusses nearly all aspects of imaging in general and still imaging in particular. Here, by “still imaging,” we understand taking single images in contrast to imaging on video films. The intention is to show “what is behind” taking images and what information is contained in the images themselves.

The main title of the present book is “Optical imaging and photography” and indeed, emphasis is put on the topic of photography since photography is a topic that is demonstrative and easily accessible. This may also be of large interest for people with a bias in photography. However, imaging is treated universally, and this is indicated by the subtitle “Introduction to science and technology of optics, sensors and systems,” which shows the much broader base of the topic. This book comprises a discussion of modern image detectors as used for science and technology. Imaging and imaging technology is also essential for a lot of modern technologies such as automation, robotics and autonomous vehicles, medical applications, etc. The goal of the present book is to take those into account, too. Thus, the intention to treat the important background for imaging in a more general way relates to applications in science and technology, and in particular, for industrial purposes. Indeed, during the proofreading our manuscript, we came across the

https://doi.org/10.1515/9783110789966-203

Edmund Optics booklet "2018 Imaging Optics, A Technical Resource for Imaging Solutions," which briefly touches on a lot of the topics discussed in the present book, and thus clearly shows that the contents of the present book are well- adapted to the mentioned goal and of interest for many technical applications.

There might arise the question: what are the unique features of the present book compared to the multitude of books on related topics on the market? There are a lot of books on photography. There are standard books on optics and/or optical sensors, and for instance, on technical and industrial imaging. And there are books going very deeply into specific topics such as lens design or sensor and semiconductor technology. However, we became aware that to the best of our knowledge there is a lack of available books that cover all relevant aspects of imaging and photography in total and in compact form, comprising aspects of the optical system and the electronic sensor parts. To some extent, the recommended book "Image Sensors and Signal Processing for Digital Still Cameras" edited by J. Nakamura, may be an exception. We like it very much, but there, emphasis is put on the topics of the title and, what is more, a complete discussion, for instance of Fourier optics, is missing. In addition, the book is more than 10 years old.

On the other hand, the internet may be regarded as a good source. There are excellent websites on specific topics. A short selection of recommended links is provided at the end of this book. But also here concatenation of the relevant subtopics is often missing, which means that imaging as an integrated whole has not been available. Even more, there exists also a lot of misleading information on dubious websites. The unexperienced reader cannot discriminate. This may often lead to a smattering of information, which consequently introduces errors. There are a lot of examples; in particular, sometimes good lenses designed for analogous cameras are wrongly judged when tested in cameras with digital sensors; or the rash opinion that a larger number of camera pixels is always better; or the immature judgment of exposure corrections by ±10 EV by a specific raw converter; or the idea of an enhancement of the dynamic range of an image by usage of smaller brightness steps, and so on. Here, we would like to stimulate the reader to be very critical when reading literature and articles on popular websites, even when written by "experts."

Based on our interest in photography and maybe even more on our general experience in optics and optical imaging sensors, which we teach in lectures and which we apply in scientific experiments and technical solutions, our goal has been to write a book that closes the gap between these very special topics. We present many details and, for instance, discuss lens system constructions, lens parameters like aspheric coefficients or special and advanced imaging sensors. Of course, the latter makes sense for scientific and technical imaging. But also for everyday imaging this may be important as the example of the development of the backside illuminated CCD sensor shows. Approximately 30 years ago, we belonged to the first users of those at that time purely scientific detectors. But today this technology is implemented in standard devices such as compact and smartphone cameras.

Although the book should provide a comprehensive and consistent description of (still) imaging and be a single unit, it cannot be fully complete because this would have been out of scope for a more or less compact book. But, in particular, we emphasized cross-linking the subtopics, such as lenses, sensors, Fourier optics and so on. The book restricts to the optical, or more generally, the physical and technical background of imaging, imaging process, imaging devices, etc. In that sense, the book should also provide the essential background information for understanding the further handling of images, especially image processing. But it does not provide a workflow of image and data processing, not even partially, because there are a lot of good books on that particular topic and our intention is not to add another one. Moreover, excluded are also details of color management, which are so extensive that they would fill a separate book. Excluded is also imaging for videos, a topic that requires a lot of additional and very special discussions. Excluded is also enhanced image processing. We also do not provide much information on norms, as they are predicated on what is described in the book and as they are subject to changes. Finally, the present book is not a book of photography in the sense of being a manual on taking good pictures in the artistic sense.

Thus, as a whole, the present textbook does not only serve as a tutorial suitable for beginners and advanced learners. It may also be used as a work of reference for scientists, engineers and photographers. Photographers should be encouraged to enlarge their technical understanding that subsequently may influence their photoshooting. Even more generally, the book may be useful for those employing and assessing imaging systems including industrial or machine vision cameras and for anyone interested in imaging. Thus, we hope that the book may be of interest for a wide audience. As we concentrate mostly on the physical background, we hope that even with future progress in the field of optical imaging, the book will still be up to date for a long time given the fact that for instance sensor chips, pixel size and so on may change within the next few years, but not principal relations such as photon conversion and tone curves.

Finally, we would like to give some remarks on the book's structure. The strong concatenation of the different subtopics within the book makes it sometimes necessary to use some knowledge in anticipation of a more detailed description later on. Chapters 1 and 2 have a kind of introductory character to imaging, also with a focus to photography and examples of modern camera systems. Chapter 3 presents the basics of imaging optics that are required to understand the complexity of optical systems like modern camera lenses. Their historical evolution during the last two centuries as well as their differentiation today is given in Chapter 6, mostly based on examples of the photographic full format. Background information for sensors and detectors is given in Chapter 4 to comprehend their characteristics like noise, resolution, speed, etc., illustrated by many practical examples. We keep the discussion on electronics rather short and refer to special literature or books with a different emphasis. In Chapter 5, Fourier optics is considered in order to make out the overall transfer function of complete optical systems. Examples show how the overall quality of a system can be assessed and influenced. Chapter 8 describes some practical methods by which a multitude of different optical systems can be experimentally investigated. Some data sheets of commercially available lenses are presented, which describe their technical properties based on the investigations. In the closing Chapter 9, we dare some outlook of modern trends in optical imaging.

It is common in most textbooks that known basics are presented without detailed reference to sources. Hence, we forgo a detailed reference list. However, some selected books and articles that aid in orientation with the topics in a broader sense are compiled in the reference list at the end of the book. In cases where a singular reference is helpful, we inserted a footnote.

During the compilation of the present book, we got many valuable comments and hints directly or indirectly from various people who have influenced the progress of this book. Representatively, we would like to express our thanks to our long-time colleague, Prof. Dr. Bert Struve, for many fruitful discussions. We are also much indebted to Dr. Vladan Blahnik from the Corporate Research and Technology of Carl Zeiss AG for his kind support and valuable comments on our manuscript. We would like to thank Eberhard Dietzsch for discussions and William Claff for information on sensor data. Hartmut Jesch and the ProxiVision team, Uwe Artmann and all companies providing us with images, are given thanks. Furthermore, we are most grateful for technical support given by many staff members, our PhD students and students of our universities in Emden and Oldenburg, among which we would like to mention particularly, Volker Braun, Johannes Diekhoff, Malte Ennen, Lars Jepsen, Arno Hinrichs, Brian Holt, Gregor Indorf, Christian Menninger, James Napier, Markus Schellenberg and Sabine Tiedeken. Their help is greatly appreciated. We also extend our gratitude to Walter de Gruyter Verlag for the opportunity to present our understanding of optical imaging to an international audience. In particular, we would like to thank Konrad Kieling for the support at the initial phase of this book and Nadja Schedensack and Anett Rehner for their commitment and patience during the realization of this book.

The warmest thanks go to our families for their valuable backing and understanding during all phases of this demanding project.

Emden, September 2018

Ulrich Teubner
Hans Josef Brückner

About the authors

Ulrich Teubner studied physics at the University of Heidelberg. In 1991, he received his PH.D. degree from the University of Göttingen and in 1998 his habilitation from the University of Jena. After scientific research positions at different universities and Max-Planck-Institutes in Germany and scientific stays at the École Polytechnique in France, the Rutherford Laboratory in UK, etc., he headed the Optics Department of the Institute of Micro Technology Mainz (IMM). In 2006, he became Professor at the University of Applied Sciences Emden/Leer (Germany). He is member of the Institute of Laser and Optics in Emden and also of the Institute of Physics of the Carl von Ossietzky University of Oldenburg. His research interests are ultrashort laser pulses, high-power laser pulses, the interaction of intense laser pulses with matter, X-ray and XUV-optics, detectors and diagnostics, laser micro processing, shock waves at micro scale, ultrafast measurements with X-ray and XUV-Free Electron Lasers and optical imaging.

Hans Josef Brückner studied physics with a focus on solid-state physics and received his Ph.D. degree in 1988. After more than 10 years' experience of professional research and development in guided-wave optics and different fields of telecommunications in Germany and France, he became the Professor for Laser Applications at the University of Applied Sciences Emden/Leer (Germany) in 1999. He is a member of the Institute of Laser and Optics in Emden. His professional focus lies in the field of optoelectronics, integrated optics, optical fiber technology and optical imaging. He has been in retirement since 2020.

Both authors have been involved in teaching in the Bachelor and in the Master program of Engineering Physics, a joint program at the University of Applied Sciences Emden/Leer and the University of Oldenburg. Besides teaching of basic and advanced subjects, their teaching activities have also involved a conjoint course on optical imaging and photography.

https://doi.org/10.1515/9783110789966-204

Contents

List of symbols

Note in advance: symbols that are used locally, i. e., within a subchapter only, are not all included in the following list.

Symbols

$=$	equal to
$\approx$	approximately equal to
$\sim$	very roughly equal to
$\propto$	proportional
$\equiv$	identical to
$\neq$	not equal to

α	beam quality
α_p	power absorption coefficient
β	angle of incidence
γ	ray angle
Γ	angular magnification
δ	focal spot diameter (spot size of focus), or diameter of an image spot (δ is not strictly defined)
δ_B	spot size of focus, or diameter of an image point when this is purely diffraction limited (note that this can be related to FWHM, $1/e^2$ or, e. g., first dark strip or ring)
δ_0	spot size of focus, or diameter of an image point with respect to first dark strip or ring
δx, δy	width of an image point or spatial resolution in x-,resp., y-direction
ε	dielectric function or relative dielectric constant
ε'	real part of ε (correspondingly for other variables)
ε''	imaginary part of ε (correspondingly for other variables)
ε_0	dielectric permittivity of vacuum
η_e	external quantum efficiency (EQE, with respect to a single pixel)
η_i	internal quantum efficiency (IQE, with respect to a single pixel)
η_g	fill factor
θ, θ_i	angle of aperture or angle of diffraction
θ_0	diffraction angle for first zero position
θ_{max}	diffraction angle corresponding to $k_{x,max}$ or $k_{y,max}$; $\sin(\theta_{max}) = NA$
θ_t	telecentricity value
κ	deformation coefficient, conic parameter (in Chapter 3)
κ	geometry factor (in Chapter 5)
λ	wavelength of light
Λ	penetration depth
μ_0	magnetic permeability of vacuum
ν	frequency in general or frequency of the incident photon, in particular
ν_d, ν_e	Abbe numbers
$\Delta\nu_{ampl}$	amplifier bandwidth
Π_G, Π_P	Gaussian distribution, Poisson distribution

https://doi.org/10.1515/9783110789966-205

ρ	difference between the reciprocal curvature radii of a thin lens
ρ_E	amplitude reflection coefficient
ρ_P	power reflection factor, reflectance
σ	standard deviation
σ_{ampl}	noise of the (pre-)amplifier
σ_{dark}	dark signal noise (rms number of noise electrons)
$\sigma_{e,tot}$	total noise of a single pixel (rms number of noise electrons in total)
σ_{pe}	signal noise of the photoelectrons
$\sigma_{pe,n}$	signal noise of the photoelectrons resulting from the photon fluctuations
σ_{OE}	signal noise of the photoelectrons according to the quantum efficiency (more correctly fluctuations)
σ_{ph}	photon (or shot) noise
σ_{pix}	total noise of a single pixel (rms number in ADU; corresponds to $\sigma_{e,tot}$)
σ_{read}	read noise
τ_E	amplitude transmission coefficient
τ_f	film transmittance
τ_P	power transmission factor, transmittance
τ_{PL}	power transmission coefficient of a lens
τ_{read}	read (out) time
ϕ	phase
Φ	luminous, resp., radiant power, flux
Φ_{pix}	radiant flux or luminous flux of a pixel of the sensor
φ	angle
$\Delta\varphi$	angular resolution
Ψ	angle of view, angle of field
ω	angular frequency
Ω, $\Delta\Omega$	solid angle
$\mathbb{A}$, $\mathbb{B}$, $\mathbb{C}$, $\mathbb{D}$	elements of optical ray transfer matrix
A, A_{im}	area, area of image
A_{en}, A_{ex}, A_{eff}	area of entrance, resp., exit pupil, effective area
A_{pix}	total area of a single pixel
A'_{pix}	photosensitive area of a single pixel
a_2, a_4, a_6	aspheric coefficients
a_i, a_o	image distance, object distance
a_n, a_f, a_{hf}	near point distance, far point distance, hyperfocal distance
a_f, a_{fc}	object distance in fiber, critical length of expansion range (in Section 3.3.6)
$\vec{\mathscr{B}}$, $\mathscr{B}$	magnetic induction (often simply termed as magnetic field): 3D field vector, 1D field value or amplitude
$B_{obj}(x,y)$	brightness distribution within the object
$B_{im}(x,y)$	brightness distribution within the image
$\tilde{B}(k_x,k_y)$	Fourier spectrum of $B(x,y)$
B_{pix}	brightness on a pixel (i. e., a “pixel” within the image)
B'_{pix}	B_{pix} after image processing; units are ADU or counts later on, a screen transfers B'_{pix}, e. g., into radiant flux in W or radiant intensity in W/sr or something similar
b_r	brightness ratio
c	velocity of light in vacuum
c_f	teleconverter factor
CF	crop factor
C_j	junction capacitance of a photodiode

C_L	calibration constant for light meters
C_o, C_i	constants in object, resp., image space
C_pix	capacity within a pixel
d, d_sensor, d_FF	diagonal of image format, of sensor, of full format
d_ar	thickness of antireflection layer
dr	depth resolution
d_p	thickness of the light sensitive volume of a pixel
$\vec{\mathscr{D}}$, $\mathscr{D}$	electric displacement field: 3D field vector, 1D field value or amplitude
D	diameter
D_eff	effective usable diameter of the lens mount
D_en, D_ex	diameter of entrance pupil, diameter of exit pupil
D_im	diameter of image circle
D_p	pinhole diameter
DR	dynamic range
DS	number of steps within the signal range
DS_max	maximum number of steps within the signal range
$\vec{\mathscr{E}}$, $\mathscr{E}$	electric field: 3D field vector, 1D field value or amplitude
$\mathscr{E}_\mathrm{obj}(x,y)$	electric field distribution within the object plane
$\mathscr{E}_\mathrm{im}(x,y)$	electric field distribution within the image plane
e	elementary charge
e_Ho, e_Hi	distance from reference plane E to principal plane H
E, E_v, E_e	Illuminance, resp., irradiance
E_il	illuminance of incident light
E_o, E_i	Input, resp., output plane
E_pix	illuminance that is incident on a single pixel
e_No, e_Ni	distance from reference plane E to nodal plane N
EV	exposure value
F	fluence (identical with radiant exposure)
F_pix	fluence that is incident on a single pixel
$\overline{F}_\mathrm{pix}$	average F_pix on the pixel
FWC, N_full	number of electrons that could be accumulated at maximum within a single pixel
$f(x)$	function in general
$f_\#$	f-number ($= f/D$)
$f_{\#\mathrm{crit}}$	critical f-number
f, f_o, f_i	focal length, object, resp., image focal length
f_eq	equivalent focal length, relative to full frame format
f_Ei	back focal length
f_Eo	front focal length
f_norm, $f_\mathrm{norm,FF}$	focal length of normal lens
G	gain
G_a	amplifier gain
G_c	conversion gain
G_i	input referred conversion gain
G_out	output referred conversion gain
h	Planck's constant
h	ray elevation from optical axis, off-axis image distance
h_i	image height
$\vec{\mathscr{H}}$, $\mathscr{H}$	magnetic field: 3D field vector, 1D field value or amplitude
H, H_v, H_e	luminous, resp., radiant exposure

H_{av}, H_m	recommended average exposure, film exposure at threshold
H_{pix}	luminous exposure that is incident on a single pixel
I	intensity
I_{pe}	photo current of one pixel
I_{dark}	dark current
I_{pix}	intensity that is incident on a single pixel (prior to losses)
I'_{pix}	intensity that is incident on a single pixel after loss-correction, i. e., intensity on the photo diode surface
$\bar{I}_{pix}$	average I_{pix} on the pixel
j_{pe}	photo current density of one pixel
J_v, J_e	luminous, resp., radiant intensity
$\vec{k}$	wave vector
k	absolute value of the wave vector
k_{cut}	cut-off frequency
k_x, k_y, k_z	spatial frequencies ($2\pi/\lambda$)
$k_{x,max}$, $k_{y,max}$, k_{max}	maximum possible spatial frequency for a given optical system in x- or y-direction, resp., or in general
k_B	Boltzman's constant
K_m, K'_m	luminous efficacy
l	length
l_c	construction length (of SPC)
l_{eff}	effective available distance in camera body
l_{opt}	optical path length
l_{ot}	optical tube length
L, L_v, L_e	luminance, resp., radiance
m	integer number
m_o, m_i	slope in object, resp., image space
M_p	pupil magnification
M, M_v, M_e, M_{rel}, $M_{rel,FF}$	magnification
$\mathbf{M}_L$, $\mathbf{M}_{os}$	lens matrix, ray transfer matrix of the optical system
n, n_o, n_i, n_0, n_1, n_2, n_s, n_{ar}	refractive index
NA, NA_o, NA_i	numerical aperture, in object, resp., image space
N_e	number of electrons generated within one pixel (this includes, e. g., N_{pe} and N_{dark})
$N_{e,min}$	minimum value of N_{pe}
$N_{e,max}$	maximum value of N_{pe}
ΔN_e	uncertainty of the number of electrons generated within one pixel (usually rms value; due to noise)
N_{eff}	effective value of number of read out electrons
N_h	number of pixels in horizontal direction, e. g., of a screen or sensor
N_v	number of pixels in vertical direction, e. g., of a screen or sensor
N_{period}	number of pixels within one period of a test grating (for a PDA)
N_{pe}	number of photo-generated electrons within one pixel
N_{ph}	number of photons illuminating one pixel (prior to losses)
$N_{ph,th}$	minimum number of photons that are necessary to provide a signal beyond read noise background (prior to losses)
$N_{ph,sat}$	maximum number of photons that could be collected within one pixel to get FWC (prior to losses)
$N_{ph,18}$	number of photons (prior to losses) to achieve 18 % $\cdot N_{full}$
N'_{ph}	number of photons incident to one pixel (after loss-correction)

N_{dark}	number of charges contributing to dark current within one pixel
N_{read}	number of charges due to read noise within one pixel (rms value)
N_{reset}	number of charges due to reset noise within one pixel (rms value)
N_{SB}	value of the space bandwidth number SBN
$N_{SB,cut}$, $N_{SB,cut}^{(2D)}$	optical cutoff in 1, resp., 2 dimensions
$N_{SB,Nyquist}$	SBN according to Nyquist frequency
N_{full}	number of electrons that could be accumulated at maximum within a single pixel
OD, OD_{min}, OD_{max}	optical density
p	pixel size or pitch
P	power
P_o, P_i	points in object, resp., image space
q	(single) charge
q_{pix}	signal charge per pixel
q_{ph}	charge generated per photon
q_{full}	saturation value of q_{pix}
$\mathbf{q}$, $\mathbf{q_i}$	complex Gaussian beam parameter
Q, Q_v, Q_e	luminous, resp., radiant energy
$\mathcal{R}$	resolution
r	radius, radius of spherical lens
r_0	position of the first zero point in the image plane: distance from the optical axis
r_b	ball lens radius
R	spatial frequency ($1/\lambda$, corresponds to $k = 2\pi/\lambda$)
R_x, R_y, R_z	spatial frequencies ($1/\lambda$); where R_x etc. corresponds to $k_x = 2\pi/\lambda_x$ etc.
R_{cut}	cut-off frequency
$R_{compress}$	compression factor
R_{out}	resistance; the output voltage is measured at that resistor
R_{pix}	responsivity
R_{MTF0}	spatial frequency, where the MTF becomes zero
R_{MTF10}	spatial frequency, where the MTF becomes 10 %
R_{MTF50}	spatial frequency, where the MTF becomes 50 %
R_N	Nyquist frequency
R_S	sampling frequency
R_x, R_y, R_z	spatial frequencies (sometimes space-frequencies)
R_φ	spatial frequency with respect to the observation angle (of the eye)
s_{DOF}, s_{DOFoc}	depth of field, depth of focus
s_i, s_o	magnitude of object, resp., image distance
S_i, S_o	image, resp., object size
S	signal in general
S_{pix}	signal generated within a single pixel in ADU
S, S_{ISO}, S_{DIN}°	sensor speed, sensitivity
t	time
t_x	exposure time, resp., shutter speed
t_L	lens thickness
t_s	lens separation
T	transmission, or sometimes absolute temperature
T_F	transmission function, which describes losses of light before it is incident on the photosensitive region of the photodiode
T_{film}	film transmission
$T_\#$	T-stop number

u_o, u_i	diameter of circle of confusion in object and image space
u_p, u_d	diameter of projection blur, diffraction blur
U	voltage (in general)
U_{bi}	photodiode built-in voltage due to diffusion
U_d	photodiode voltage
U_{max}	maximum voltage at the photodiode
U_{out}	output voltage per pixel
U_{eff}	effective value of a voltage (rms value)
U_{read}	effective value of read out voltage noise (rms value)
U_{reset}	reset voltage
U_r	photodiode reverse bias voltage
V, V_i	refractive power
V_c	construction parameter (of SPC)
$v_{Ho}, v_{Hi}, v_{No}, v_{Ni}$	distance from vertices to cardinal points in a thick lens
$V_{ph}(\lambda), V'_{ph}(\lambda)$	photopic, resp., scotopic standard luminosity function
w, w_0, w_i	Gaussian beam radius
W	energy in general
W_g,	band gap energy
W_{ph}	photon energy
W_{pix}	(radiant) energy that is incident on a single pixel
W'_{pix}	energy that is incident on a single pixel (after loss-correction)
$WF_r(x,y), WF_i(x,y)$	real wave front, resp., ideal wave front
$\Delta WF(x,y)$	wave front aberration
X, Y	image width, height
X_o, Y_o	width, resp., height of an object (size in x-, resp., y-direction)
x, y, z	space coordinates
x_o, x_i	x-coordinate, in the object plane, resp., image plane
x_0	position of the first zero point in the image plane: distance from the optical axis
$X_{e,\lambda}, X_{v,\lambda}$	radiometric, resp., photometric quantity
z_a	astigmatic difference
z_R, z_{Ri}	Rayleigh length, Rayleigh length in image space
z_s	longitudinal spherical aberration

Abbreviations

1D	one dimension, one-dimensional
2D	two dimension, two-dimensional
3D	three dimension, three-dimensional
ADC	analogue-to-digital (A/D) converter
ADU	analogue digital unit; this is equal to DN (digital numbers), DV (digital values) or counts, respectively
AI	artificial intelligence
APS	active pixel sensor
APS-C	particular film-/sensor format
AR	anti reflection
BI	back illuminated
BSI	back side illumination
c. c.	complex conjugate

CCD	charge coupled device
CDS	correlated double sampling
CF	crop factor
CFA	color filter array
CI	computational imaging
CIS	CMOS Image Sensor
CMOS	complementary metal oxide semiconductor
CSF	contrast sensitivity function
DCG	dual conversion gain
DFD	depth from defocus
DN	digital number; same as ADU
DNG	digital negative
DOE	diffractive optical element
DOF, DOFoc	depth of field, depth of focus
DPD	dual photodiode
DR	dynamic range
DRI	dynamic range increase
DS	number of data steps or depth steps
DSC	digital still camera (in contrast to, e. g., a video camera)
DSLM	digital single lens mirrorless camera
DSLR	digital single lens reflex camera
DSNU	dark signal nonuniformity
DTI	deep trench isolation
DU	digital unit; same as ADU
DV	digital value; same as ADU
EBI	equivalent background illumination
EIS	electronic image stabilization
epi layer	epitaxial layer
EQE	external quantum efficiency (or overall quantum efficiency)
ESF	edge spread function
EUV	extreme ultraviolet
EV	exposure value (= aperture stop, f-stop)
FBLS	fiber ball lens system
FI	front illuminated
FF	fill factor
FFC	flat field correction
FO	fiber optics
FOV	field of view
FSI	front side illuminated
FPN	fixed pattern noise
fps	frames per second
FT	Fourier transformation
FWC	full well capacity
FWHM	full width at half maximum, measured at full width at half maximum
GS	global shutter
HCG	high conversion gain
HDR	high dynamic range
HV	high voltage
IC	integrated (electronic) circuit

iCCD	intensified CCD
iFT	inverse Fourier transformation
IPA	inverted pyramid array
IQE	internal quantum efficiency
IR	infrared
IS	image stabilization
ISP	image signal processor
LCG	low conversion gain
LDR	low dynamic range
LIDAR	light detection and ranging or light imaging, detection and ranging
lp	line pair
lp/PH	number of line pairs within the picture height
lp/PW	number of line pairs within the picture width
l/PH	number of lines within the picture height
l/PW	number of lines within the picture width
LCD	liquid crystal display
LED	light emitting diode
LSF	line spread function ($\mathrm{LSF} = \int_y \mathrm{PSF}(y)\mathrm{d}y$)
MCP	micro channel plate
MEMS	micro-electromechanical system
MLA	metalens array
MOS	metal oxide semiconductor
MOSFET	metal oxide field effect transistor
MP	mega pixel (unit used for cameras or camera sensors)
MTF	modulation transfer function
$\mathcal{MTF}$	modulation transfer function for the field
NA	numerical aperture
NIR	near infrared
NPS	noise power spectrum
OCL	on-chip lens
OD	optical density
OECF	opto-electronic conversion function
OIS	optical image stabilization
OLPF	optical low pass filter
OMA	optical microlens array
OPD	optical path difference
OTF	optical transfer function
$\mathcal{OTF}$	coherent transfer function sometimes termed as amplitude transfer function (for the field)
PC	personal computer
PCB	printed circuit board
PDA	photo diode array (this may be 1D or 2D; in a more general sense, also CCD or CMOS may be regarded as PDA)
PDAF	phase detection autofocus
PH	picture height (or height of an image or sensor)
PIA	pixel interleaved array
pixel	picture element
PLS	parasitic light sensitivity
PMT	photo multiplier tube

PRNU	photo response uniformity
PSF	point spread function
$\mathcal{PSF}$	point spread function for the field (coherent point spread function)
PTC	photon transfer curve
PTF	phase transfer function
PW	picture width (or width of an image or sensor)
QBC	Quad Bayer coding
QD	quantum dot
QE	quantum efficiency
QIS	quanta image sensor
RGB	red, green, blue
RMS, rms	root mean square
ROI	region of interest
RS	rolling shutter
SBN	space bandwidth number (term is used for both, one or two dimensions)
SBP	space bandwidth product
sCMOS	scientific CMOS
SEM	scanning electron microscope
SFR	spatial frequency response
SLR	single lens reflex camera
SLT	single lens translucent
SQF	subjective quality factor
SNR	signal-to-noise ratio
SPAD	single-photon avalanche diode
SPC	smartphone camera
SWIR	short wavelength infra-red
TBP	time bandwidth product
ToF	time of flight
TSV	through-silicon *via*
TTL	through the lens
UV	ultraviolet
vis	visible light range
VN	visual noise
WF	wavefront
WFA	wavefront aberration
XR	X-ray
XUV	extreme ultraviolet

1 Introduction to optical imaging and photography

Optical imaging has always played an important role in different ways of communication in our cultures. In modern times, with the increasing usage of electronics in various fields of daily life, new challenges for applications of optical imaging have become obvious. In private communications, simple snapshots taken by mobile phones are transmitted using quick messenger services whereas in business domains, images of various quality levels are exchanged, for instance, for analyzing, marketing or other purposes. Of course, imaging also plays an important role in many fields of science for evaluation and the presenting of results. In industry, imaging is of high importance, for instance, for monitoring processes and metrology. Consequently, issues of the present book are related to the selection and handling of scientific or technical camera systems. Besides photography, we do also include scientific and technical imaging, but not the special issues related to machine or computer vision, for instance. As for the imaging optics, we mainly restrict our consideration to lens systems in the visible spectral range. Alternative systems make uses of reflective optics such as mirrors with spherical or other shapes with examples such as Kirkpatrick–Baez mirrors or Schwarzschild objectives. Also, fiber optics or capillary optics, e. g., used in lobster eyes, may be regarded as reflective optics. All those kinds of optics and, in particular, systems for other wavelength ranges will not be covered or only little discussed.

In the present introductory chapter, we will outline our understanding of optical imaging and depict some basic general principles of it. We do not discuss the physical nature of light, like its description as electromagnetic waves, etc., namely topics that are well described in standard optics textbooks such as [Ped17, Hec16, Bor99]. We make use of such knowledge, and instead, concentrate more on the physical and technical implications for optical imaging. In the following section, we will give an idea about the main topics that are covered in this book and also topics that are not discussed or only discussed on a limited scale.

Optical imaging in our understanding comprises roughly three elements (Figure 1.1a): The real object space with a compilation of individual objects is, in general, the 3D space we live in. Images of the individual objects are located in the real image space, which in the most common case is a 2D one, for instance, a photograph. The relationship between these two spaces is established by an imaging system in conjunction with the transfer method. The transfer method is based in general on using carrier waves, for instance electromagnetic, acoustical or particle waves. From a historical point of view, the first imaging systems were optical lenses for visible light (Figure 1.1b). Thus, for simplicity we call the imaging system optical although in a strict sense this designation is only valid for the visible electromagnetic spectral range with wavelengths from around 390 nm up to 780 nm.

In a more general situation, other carrier waves, imaging systems and also imaging methods may be used. Just to name a few, acoustical waves are used for ultrasonic testing in medical and material sciences and technology, microwaves in radar imaging and

https://doi.org/10.1515/9783110789966-001

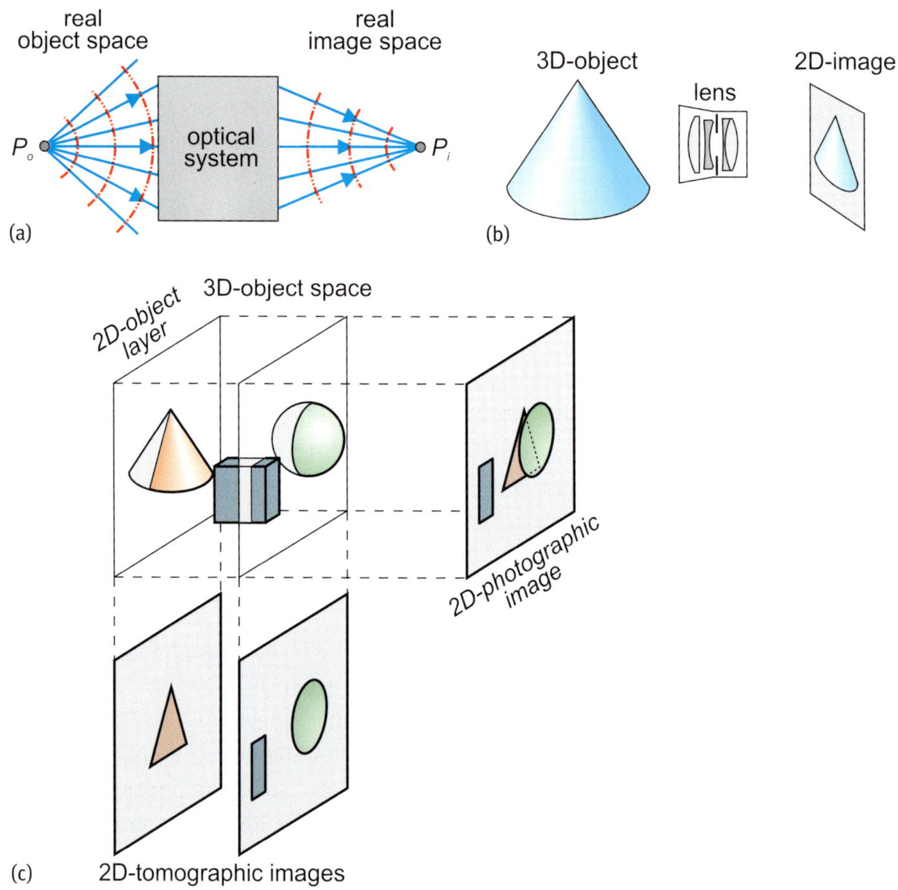

Fig. 1.1: Optical imaging based on optical systems. (a) Schematic principle; (b) example for photography with a lens as the optical system; (c) principle of tomography yielding superposition-free images, compared to photography, yielding superimposed projected images.

X-rays and particle beams like electrons for imaging objects of very small dimensions. As for the imaging methods, we can roughly differentiate between the following ones:

Holography

In general, here the complete electromagnetic field distribution of coherent waves scattered by a 3D object is stored in a hologram. This hologram is a physical 2D image of the object in a very narrow spectral range, which in the case of visible light, represents the complete 3D object structure given at any particular color. By a subsequent reproduction process, the original 3D structure can be restored.

Topography

This method is used to give 3D information about a detailed surface structure of objects. Examples are geographical applications like topographic maps. More recent applications can be found in the field of microstructures that describe the surface geometry as well as its chemical respectively physical properties using data stored in a large database.

Tomography

In contrast to topography, only 2D information is given about an object layer of a 3D object (Figure 1.1c). This method delivers sectional diagrams by doing assignment between a 2D object layer and a 2D image layer. The tomographic images have no information about the spatial depth perpendicular to the layer. However, through computational methods or image stacking, a 3D image can be reconstructed from the combination of individual 2D images.

Photography

This method can be described as a projection of illuminated objects in a 3D object space to a 2D image space (Figure 1.1b, c). Unlike in tomography, the images of different cross-sections in the object space may be superimposed. Thus, in general the information about spatial depth is lost and only limited information about it is maintained. The perspective of the image depends on the properties of the photographic setup. As with other methods, spatial information can be retrieved by a combination of multiple photographic images.

1.1 Objective of the present book

The focus of the present book is on optical imaging using photographic and related methods. Photography itself comprises many aspects that cannot all be covered within this book. We mostly limit our consideration to modern photography based on digital camera systems, but we do also include imaging systems in a more general way as used for scientific or technical applications. Thus, we do, for instance, also discuss intensifier systems. A schematic structure for the topics of interest is given in Figure 1.2. If we roughly subdivide the optical system into two sections, we have on one side the imaging optics and on the other side the image processing system including the detector or sensor. In the case of imaging optics, we can go back historically to the development of the first lens and camera systems, which started at the beginning of the 19th century. The quality of an image is determined in the first place by the lenses and their properties. Many principles of optics that have been applied to “old school” analog photography for optimizing images are still valid for modern digital photography. What is new in digital

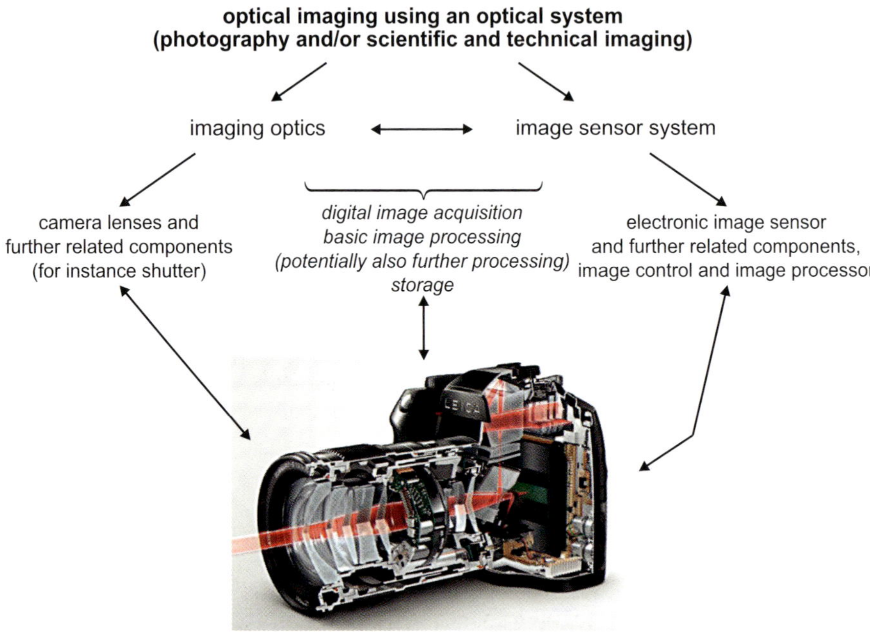

Fig. 1.2: Topics of interest as covered in the present book. The displayed camera is just an example. For scientific and technical imaging, for instance, the camera systems are usually different. In all cases, image post-processing is generally done. For scientific and technical imaging, however, post-processing may be restricted to corrections such as image field correction and noise (Inset used with kind permission of Leica Camera AG).

photography on the other side is the image processing system, which in modern systems is based on electronic image sensors. They have almost completely replaced the conventional film materials based on chemical emulsions. These films are chosen for a special photographic situation and can be easily exchanged. Image development is a detached process after image acquisition. Digital sensors, however, are integrated with the camera body, and should have a large flexibility for the exposure conditions and interact very closely with special image processors in the camera. These processors complement the remaining electronic control of exposure settings by performing further image post-processing. In some cases, also multishot techniques are used to increase the dynamic exposure range and generate high dynamic range images (HDRI). Thus, new approaches for optimizing the image quality in digital cameras are necessary.

This optimization is closely connected to the purpose of taking images. There are photographs for artistic expressions, but also for applications in metrology or documentation. For instance, for artistic expression, a certain blur or image distortion may be desired while for metrology this is not acceptable. As for a further discussion of image quality in this book, aspects of artistic expression remain mostly outside of our consideration. In case of scientific or technical applications, when photography is used for instance as a measurement tool, the quality of an image is decisive for a good metrological evaluation. In order to optimize the information that can be retrieved from an image as well as to optimize an optical system to produce good images, the following key items will be covered in detail:

- How well is the object represented in the image plane with respect to geometrical allocation? Are all proportions correct or are they distorted? Which factors influence the image resolution, its sharpness and so on? How can imaging errors be avoided in respect to the amount they are relevant? These topics are mainly related to the design of optical lenses in the system, but the sensor and image processing also influence the results.
- Which is the highest possible resolution in the image? How well is the brightness of an object reproduced? How good is the linearity of image sensors, where is the noise level or how can the signal-to-noise ratio be improved? These issues are important for radiometry, for instance to determine beam profiles, and mainly depend on the technology of image sensors.

Thus, the most important parts in the chain of the photographic imaging process covered in our textbook relate to the lens optics as well as the imaging sensors and detectors and dominate the next chapters. The sections about imaging optics and camera lenses discuss most of the topics relevant to the analog part of photography with the exception of analog films. Fundamentals are given to understand the underlying principles and help to better assess the specifications of the optics. With the advent of electronic digitization, we have seen new approaches and technologies for the transition from analog to digital signal treatment. For time-dependent signals, for instance music signals, the sampling rate of an analog signal and electronic filtering methods are decisive for the quality of the result. Likewise, in a still camera the analog image on the sensor is sampled in the 2D space. The resulting spatial frequency components, which can be manipulated or simply filtered, are decisive for the image quality like sharpness and contrast. Therefore, chapters about electronic image sensors and Fourier optics are included in our book to understand and assess a complete optical system.

We mostly limit our focus to examples in the field of photography to assess respectively select optical systems and images for their quality and usefulness. However, most of the knowledge gained can hereby be transferred to other imaging techniques, and also directly applied to general aspects of scientific imaging. Some discussion concerning that will be made as well.

A last point, which is not always obvious, but which should be kept in mind, is that the whole chain of the imaging process from the object space to the evaluation of an image also comprises in general the perception and assessment by the human eye. The process chain is schematically depicted in Figure 1.3. The first part of the overall imaging process from the object plane to the image sensor refers to the topics presented in this book. We cover only a few aspects of the second part, which is from the stored/displayed image to the retina of the human eye. However, this topic is always implicitly included

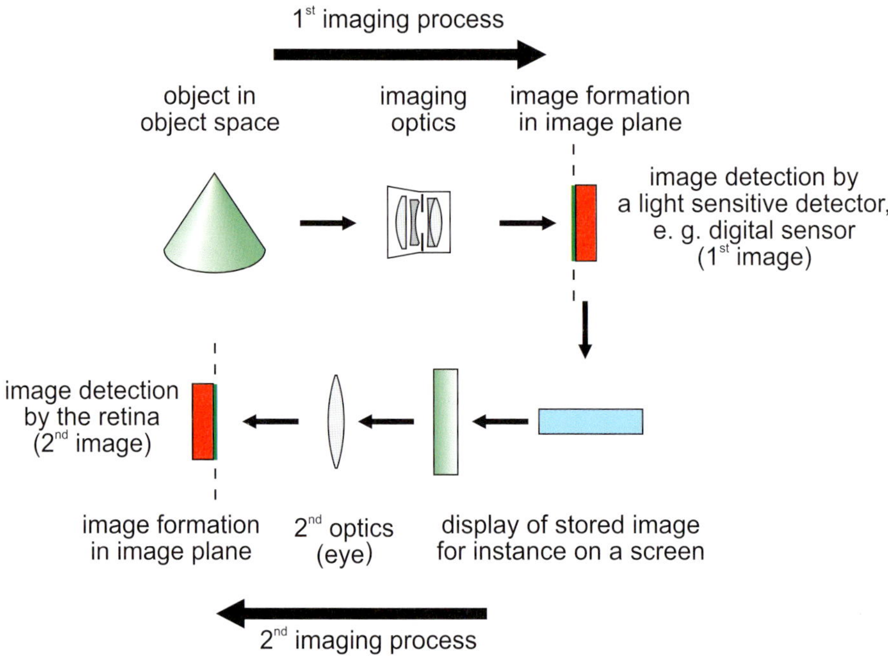

Fig. 1.3: Overall chain of process for optical imaging from the object space to the perception by the human eye.

in our consideration. For instance, the human eye or the way we view images defines the technical requirements for an optical system.

1.2 Basics of radiometry and photometry

Before starting with the main parts of the book, we will have a look at some basics related to electromagnetic radiation quantities, which are of importance in many fields of optics and will be used in our considerations. The objective of these quantities is to characterize electromagnetic sources, radiation fields and detectors. The more general description is given by *radiometric* quantities. *Radiometry* comprises measurement techniques to characterize the physical properties of electromagnetic radiation in the overall spectral range. Only a small part of that within the wavelength range of approximately 390 nm to 780 nm can be perceived by the human eye and is named as light. If we limit our description to visible light and use the physiological sensitivity of the eye to characterize the corresponding quantities, we have to deal with *photometry* and *photometric* quantities. In the following, we will first describe some radiometric quantities that are labeled with the index "e" as energetic quantities and then see the direct correspondence to the photometric ones labeled by "v" for the visual range.

1.2.1 Radiant energy, flux, fluence and intensity

The basic quantity in our consideration is the energy that is emitted by a source and may be detected at a receiver. Electromagnetic radiation, or more specifically light, can be described by waves or by using the concept of energetic particles traveling in space, named photons. In the latter case, we see the analogy to a current where its magnitude is given by particles per unit time, for instance, number of charges per unit time in the case of an electric current. Hence, we come to the idea of a photon current, which is the number of photons per unit time and is designated as *radiant flux* Φ_e. The flux is equivalent to the radiated quantity, which is the energy Q_e, emitted per time interval thus being equivalent to an optical power and measured in units of watts. We then have the relationships:

$$\Phi_e = \frac{dQ_e}{dt} \quad \text{resp.} \quad Q_e = \int \Phi_e dt \quad \text{with} \quad [\Phi_e] = \text{W} = \frac{\text{J}}{\text{s}} \quad \text{and} \quad [Q_e] = \text{J}. \tag{1.1}$$

In the strict sense, the definition of Φ_e as given in Equation (1.1) is valid only for the limiting case of Δt approaching zero and then gives the momentary value of the radiant flux. In the experimental case, however, only finite intervals are accessible. The radiant flux Φ_e then results from the energy variation ΔQ_e measured over a small finite time interval Δt yielding $\Phi_e = \Delta Q_e/\Delta t$. When substituting the differential quotient by the quotient of small differences, the resulting corresponding quantities must be interpreted as a kind of average value within the given interval.

On the other hand, if energy Q_e passes a given area A in space, we can characterize the energy density by using the term *fluence* given by

$$F_e = \frac{\Delta Q_e}{\Delta A}, \quad F_e = \frac{dQ_e}{dA}, \quad \text{respectively,} \quad Q_e = \int F_e \, dA \quad \text{with} \quad [F_e] = \frac{\text{J}}{\text{m}^2}. \tag{1.2}$$

If radiation interacts with matter, it is often necessary to qualify the situation with the flux density. Taking into consideration both time and space, then the energy per unit time and area is described by the intensity of the radiation, given by

$$I = \frac{\Delta \Phi_e}{\Delta A} \quad \text{with} \quad [I] = \frac{\text{W}}{\text{m}^2}. \tag{1.3}$$

Here, the energy is averaged over time intervals longer than some oscillation periods of time and, respectively, areas with dimensions of more than some wavelengths. Intensity is the general expression for the flow of energy per time of a radiation field in space through an area perpendicular to its flow direction. It is identical to the time-averaged Poynting vector and can be calculated from the electric and magnetic field of the electromagnetic wave as described in standard textbooks of electrodynamics and optics (see also Section 5.1).

There will be other quantities having the same physical units and also express power per area, but they are termed differently, as for instance, irradiance and exitance, respectively. The reason for it is that energy density can be used to characterize different situations, as for instance the general case of a radiation field in space. But in the specific situation of an energy density distribution on a detector surface or a surface in general, this must be characterized by the fluence or in technical terms, by the irradiance. This has to be discriminated from the situation when a source does emit a certain amount of light energy per surface element. For this situation, the appropriate quantity is the exitance, which is used to characterize the standardized energy density emitted by a source. As all these terms express the same physical situation, we often find that people in the scientific community just use the general physical terms and not the specific standardized technical quantities. In some situations in this book, we may do it likewise, however, the standardized technical quantities are given in the following section to be in line with the technical literature.

1.2.2 Solid angle and radiant intensity

Let us focus on the situation of a source emitting radiation. If the radiated power is not homogeneously distributed in space but is concentrated in a limited region, as in the case of beams exiting a flashlight, we describe this region in space using the term of *solid angle*. If we further assume that the beam has a cone shaped form with a total 2D aperture angle of 2θ then it cuts out an area A_Ω on the surface of a sphere around the source at a distance r (Figure 1.4). This area is proportional to r^2 and increases with the distance from the source as is indicated by $A_{\Omega,1}$, respectively, $A_{\Omega,2}$ in the figure. The solid angle Ω is defined to be independent from the distance achieved if the area is divided by the distance squared:

$$\Omega = \frac{A_\Omega}{r^2} = \frac{2\pi \cdot (1-\cos\theta) \cdot r^2}{r^2} = 2\pi \cdot (1-\cos\theta), \quad \Delta\Omega = \frac{\Delta A_\Omega}{r^2} \quad \text{with} \quad [\Omega] = \text{sr}. \tag{1.4}$$

2θ can also be understood as the total aperture angle under which the illuminated area A_Ω can be perceived from the source. The unit of the solid angle is steradian with the symbol sr. It is a dimensionless unit. If light from a point source is emitted into half-space, the corresponding cone has a half aperture angle $\theta = \pi$ yielding a solid angle of 2π sr. 1 sr means a cone with $\theta = 32.7°$, 4π sr characterizes the total space and is the solid angle for the radiation of a spherical wave.

The power launched into a solid angle characterizes the strength of a beam radiated by a source (Figure 1.4). This leads to a definition of the *radiant intensity* which is given by

$$J_e = \frac{\Delta\Phi_e}{\Delta\Omega} \quad \text{with} \quad [J_e] = \frac{\text{W}}{\text{sr}} \tag{1.5}$$

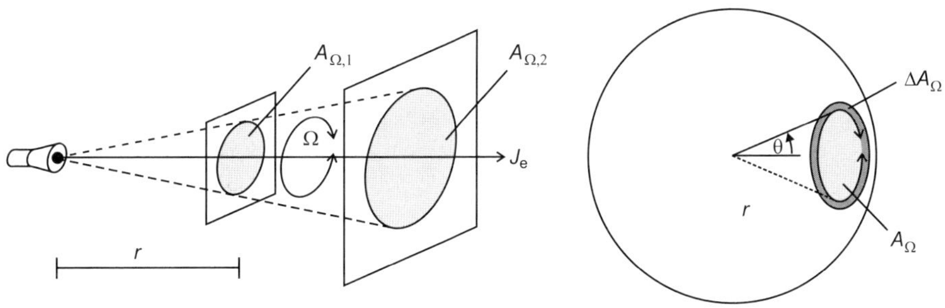

Fig. 1.4: Solid angle Ω. The surface area A_Ω cut out by a cone shaped beam with a total aperture angle of 2θ increases quadratically with the distance from the source. It should be noted that A_Ω is a spherical surface which, however, appears flat in the simple 2D representation.

The radiant intensity characterizes the source irrespectively of distance to it or any detector area used.

1.2.3 Irradiance and radiance

Many physical processes in matter, for instance the transparency change in a photographic film or the number of charged carriers generated in semiconductor material, depend on how much power respectively energy is incident to a given area. As a consequence, this effect is used to setup detectors. In order to characterize the technical situation at a detector, we use the concept of *irradiance*. If radiation from a point source hits areas at larger distances, the power per area decreases as the beam widens up. Hence, the irradiance E_e at a detector surface is defined as power per area perpendicular to the direction of radiation:

$$E_e = \frac{\Delta\Phi_e}{\Delta A_\Omega} = \frac{J_e \cdot \Delta\Omega}{\Delta A_\Omega} = \frac{J_e}{r^2} \quad \text{with} \quad [E_e] = \frac{\mathrm{W}}{\mathrm{m}^2}. \tag{1.6}$$

For a diverging beam from a point source with constant radiant intensity, we have an inverse square law for the irradiance at larger distances. As mentioned above, we have the same units as for intensity but the term irradiance is used to emphasize the situation at the detecting surface. If there is no misunderstanding, we use sometimes both terms in a synonymous way. For instance, if we measure the characteristics of an optical beam with a detector to determine its profile we describe it by its transversal intensity distribution.

Now we change our focus to the situation at a source. Real physical sources emit radiation from an extended area. We call the power per area emitted at its surface *radiant exitance* M_e, which has the same unit as irradiance and intensity. However, in a general case the source radiates into different directions and not only perpendicularly to its surface area. The radiant intensity generally decreases with increasing deviation

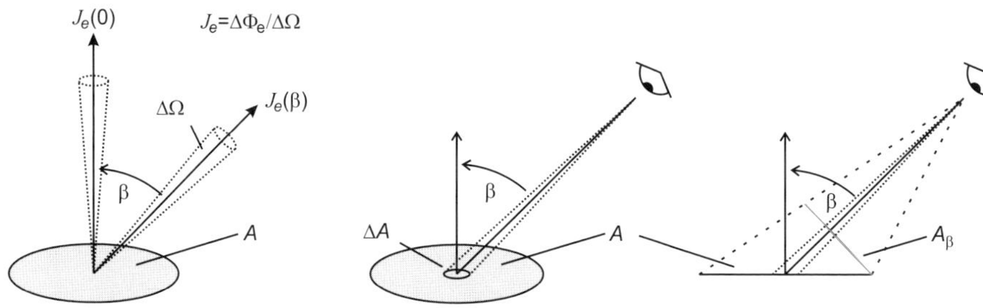

Fig. 1.5: Radiation emitted from an extended homogeneous area A. The radiant intensity J_e characterizes the strength of the beam and decreases with increasing deviation the beam direction from the surface normal. This is due to the fact that the projected area of the source decreases. By dividing the radiant intensity by this projected area, the resulting radiance L_e is a quantity that characterizes the brilliance of a source independent from the viewing direction.

β of the beam direction from the surface normal (Figure 1.5). In particular situations, for instance with a diffuse surface, it may be described in detail by a Lambertian source (see Section 1.2.4). β can also be considered the angle under which the source is perceived from far away. Thus, the best way to characterize a source is using the quantity of *radiance* L_e, which takes into account the position of an observer at a given distance. The radiance L_e is defined as the radiant intensity relative to the emitting area, which implies that L_e is larger the smaller the emitting area that the solid angle is for a given radiant intensity. In case of a very bright ray emission from a small area, as for instance a laser source, we classify the source as brilliant. Thus, L_e in some cases may also be interpreted as the brilliance of a source, which is an important quantity in optics. In a general case, the observer does not look perpendicularly to the emitting area but perceives it under the angle β. Then we get for the radiance L_e:

$$L_e = \frac{\Delta J_e}{\Delta A_\beta} = \frac{\Delta J_e}{\Delta A \cdot \cos\beta} = \frac{\Delta\Phi_e}{\Delta\Omega \cdot \Delta A \cdot \cos\beta} \quad \text{with} \quad [L_e] = \frac{\mathrm{W}}{\mathrm{sr} \cdot \mathrm{m}^2}. \tag{1.7}$$

Here, ΔA_β is the projection of the surface element ΔA in the viewing direction and becomes smaller the larger the angle β. $\Delta\Phi_e$ is the part of the source power that is radiated in $\Delta\Omega$. As we can see in the case of a Lambertian surface, this definition characterizes the source in an appropriate way independently from the viewing perspective.

1.2.4 Lambertian surface

The radiation from a surface is called diffuse if any point in the area of the source radiates uniformly in all directions. This situation is given for instance, if the area A is homogeneously illuminated by an external source and scatters diffuse light (Figure 1.5). This can also be found in the case of a semiconductor surface emitter like a light emitting

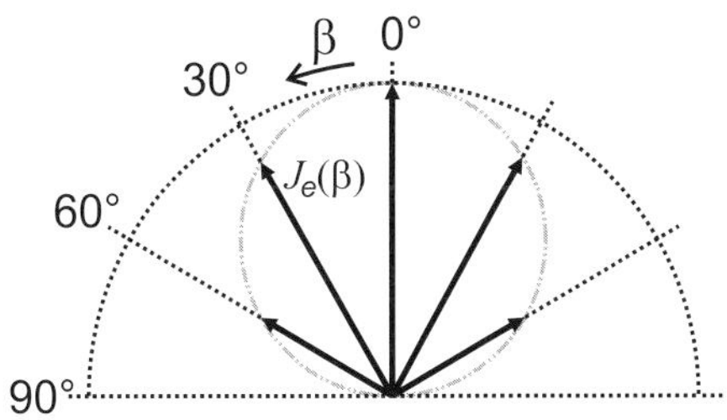

Fig. 1.6: Radiation characteristics of a Lambertian surface emitter. The radiant intensity J_e is at maximum for a radiation perpendicular to the surface ($\beta = 0°$) and approaches zero for tangential radiation ($\beta = 90°$).

diode (LED). These sources are termed Lambertian radiators, and respectively, Lambertian surfaces. Looking perpendicularly to this surface with $\beta = 0$, the radiant intensity is the highest; with increasing β the radiant intensity decreases as the effective area in the direction of radiation decreases. Under an angle of 90°, the surface can no longer be perceived, and no power is radiated into this direction. This can be described by Lambert's cosine law:

$$J_e(\beta) = J_e(0) \cdot \cos\beta. \tag{1.8}$$

It can be verified experimentally by an optical detector that is moved around the Lambertian surface at a fixed distance, thus ensuring that the solid angle remains constant. The corresponding radiation characteristic is shown in Figure 1.6. This is also the situation that can be found in the case of surface radiating LEDs that show radiation characteristics that clearly depend on the direction of emission. While the radiant intensity of a Lambertian surface decreases with increasing β, its radiance L_e remains constant for all viewing angles, implying that using L_e is a convenient way to characterize a source irrespectively of the viewing direction, respectively, distance to it:

$$L_e(\beta) = \frac{J_e(\beta)}{A \cdot \cos\beta} = \frac{J_e(0)}{A} = \text{const.} \tag{1.9}$$

1.2.5 Radiant exposure

In the classical example of taking photographs, the blackening of a film depends on the total amount of energy deposited per area in the photographic emulsion. Thus, as *radiant exposure* H_e we define the accumulated energy per area as

$$H_e = \int E_e \, dt \quad \text{with} \quad [H_e] = \frac{\text{Ws}}{\text{m}^2} = \frac{\text{J}}{\text{m}^2}. \tag{1.10}$$

The longer a detector or film is irradiated, the higher is the exposure. The definition is conceived without respect to how the accumulation process is achieved. In some cases where a sensor is irradiated by pulsed radiation, the energy density per pulsed shot is termed *fluence* and has the same units as the exposure (see also Section 1.2.1). The exposure is then the sum of all fluences after pulsed irradiation. The distinction between exposure and fluence is made in order to discriminate between the energy deposited by a single pulse, generally during a relatively short pulse time, and the accumulated energy deposited during a relatively long exposure time. It is not always possible to make a clear distinction between the terms, so we will use both of them; the meaning results from the context description.

1.2.6 Photometric quantities

All radiometric quantities discussed above have their corresponding photometric quantities and are listed in Table 1.1. They are based on the physiological sensitivity of the human eye in the visible spectral range. As this sensitivity strongly depends on the wavelength, a correction for the perceived brightness by the human eye must be done with respect to the values measured by an ideal physical detector. The relation between a pho-

Tab. 1.1: Photometric and radiometric quantities.

symbol	photometric quantity (term and units)	definition	radiometric quantity (term and units)	symbol	
Q_v	luminous energy: $[Q_v] = \mathrm{lm \cdot s}$		radiant energy: $[Q_e] = \mathrm{J} = \mathrm{W \cdot s}$	Q_e	
Φ_v	luminous flux: $[\Phi_v] = \mathrm{lm}$	$\Phi_v = \dfrac{\Delta Q_v}{\Delta t}$	radiant flux: $[\Phi_e] = \mathrm{J/s} = \mathrm{W}$	Φ_e	
F_v	luminous fluence: $[F_v] = \mathrm{lm \cdot s/m^2}$	$F_v = \dfrac{\Delta Q_v}{\Delta A}$	radiant fluence: $[F_e] = \mathrm{J/m^2}$	F_e	
J_v	luminous intensity: $[J_v] = \mathrm{cd} = \mathrm{lm/sr}$	$J_v = \dfrac{\Delta \Phi_v}{\Delta \Omega}$	radiant intensity: $[J_e] = \mathrm{W/sr}$	J_e	
M_v	luminous exitance: $[M_v] = \mathrm{lm/m^2}$	$M_v = \dfrac{\Delta \Phi_v}{\Delta A_s}$	radiant exitance: $[M_e] = \mathrm{W/m^2}$	M_e	source
L_v	luminance: $[L_v] = \mathrm{cd/m^2} = \mathrm{lm/(sr \cdot m^2)}$	$L_v = \dfrac{\Delta J_v}{\Delta A_s \cdot \cos \beta}$	radiance: $[L_e] = \mathrm{W/(sr \cdot m^2)}$	L_e	
E_v	illuminance: $[E_v] = \mathrm{lx} = \mathrm{lm/m^2}$	$E_v = \dfrac{\Delta \Phi_v}{\Delta A_d}$	irradiance: $[E_e] = \mathrm{W/m^2}$	E_e	
H_v	luminous exposure: $[H_v] = \mathrm{lx \cdot s} = \mathrm{lm \cdot s/m^2}$	$H_v = \int E_v \, dt$	radiant exposure: $[H_e] = \mathrm{J/m^2}$	H_e	detector

Note: ΔA_s is a surface element of the source, ΔA_d is a surface element of the detector
cd: candela, sr: steradian, lm: lumen, lx: lux, β: angle between beam and surface normal.
It should be further noted that the optical power density, usually measured in $\mathrm{W/m^2}$ or $\mathrm{W/cm^2}$, is designated by different terms, for instance as intensity, as irradiance or as radiant exitance, depending on the situation. For details, see the text.

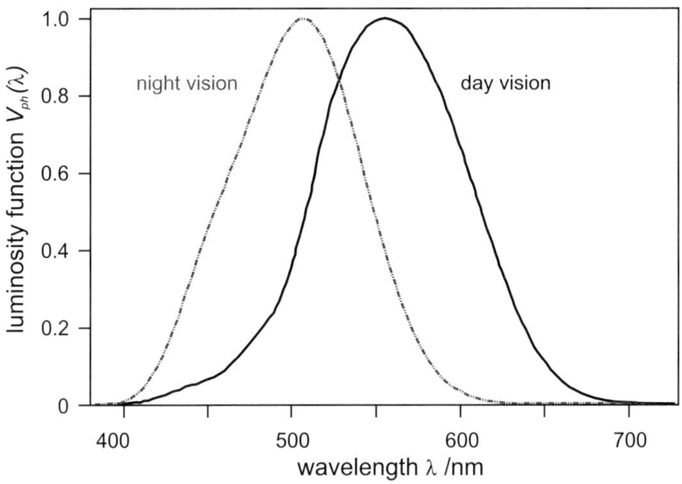

λ/nm	$V_{ph}(\lambda)$ (day vision)
400	0.002
430	0.034
450	0.067
480	0.183
500	0.355
510	0.529
555	1.000
610	0.535
630	0.282
640	0.183
670	0.033
700	0.003

Fig. 1.7: Luminosity functions of the human eye for day and, respectively, night vision; the curves are reproduced according to data from Sharpe et al.[1] and CIE.[2] Compare also Figure 4.2.

tometric quantity $X_{v,\lambda}$ at the wavelength λ and its radiometric equivalent $X_{e,\lambda}$ is given by the equation

$$\frac{X_{v,\lambda}}{X_{e,\lambda}} = K_m \cdot V_{ph}(\lambda) \quad \text{with} \quad K_m = 683\frac{\text{lm}}{\text{W}} \quad \text{(photopic: day vision).} \tag{1.11}$$

$V_{ph}(\lambda)$ is the photopic standard luminosity function of the human eye. Photopic means adaption to daytime illumination, and K_m is the luminous efficacy of radiation at 555 nm. The eye is the most sensitive at daylight in the green spectral range at 555 nm. As a consequence, the dimensionless function $V_{ph}(\lambda)$, which can be interpreted as a normalized sensitivity of the eye, has its maximum at this wavelength with $V_{ph}(555\,\text{nm}) = 1$[1] (Figure 1.7). At this wavelength, 1 W of optical power is equivalent to 683 lm of luminous flux, and thus conversely $1\,\text{lx} \cdot \text{s} \approx 0.15\,\mu\text{J/cm}^2$ under the assumption of monochromatic light. At shorter and larger wavelengths, sources emitting the same optical power are perceived to be less bright by the eye. A LED emitting an optical power of 1 W at 630 nm with $V_{ph}(630\,\text{nm}) = 0.282$ has a luminous flux of 193 lm, and thus seems to be shining much less brightly. On the other hand, a flux of 683 lm of red light at 630 nm produces the same visual impression as 683 lm of green light at 555 nm although the optical power needed at 630 nm to achieve that visual impression is more than three times as much as that for green light.

---

1 L. Sharpe, A. Stockman, W. Jagla, H. Jägle: *A luminous efficiency function, $V^*(\lambda)$, for daylight adaptation*, Journal of Vision **5** (2005) 948–968 (day vision).

2 CIE Proceedings (1951), Vol. 1, Section 4; Vol. 3, 37, Bureau Central de la CIE, Paris, 1951 (night vision).

The luminosity function strongly depends on the ambiance, for instance, the background illumination. In a low light situation where the eye is adapted to darkness, the scotopic standard luminosity function $V'_{\mathrm{ph}}(\lambda)$ must be used instead of $V_{\mathrm{ph}}(\lambda)$ in Equation (1.11) in combination with a $K'_{\mathrm{m}} = 1699\,\mathrm{lm/W}$ for night vision[3] [Ped17]. In that case, the highest sensitivity of the eye is found at 507 nm.

1.3 Basic concepts of image characterization

In this chapter, we discuss fundamental aspects and relations for image characterization. We will make use of a lot of approximations in order to highlight more clearly the important relations and to provide good enough estimates for many purposes. A deeper and more rigorous discussion of most of the discussed subjects is given later in this book. However, we would like to state that here the overview takes a more important role than in later sections of this book.

1.3.1 Imaging, "image points" and resolution

To get an introduction to imaging, we may regard a very small object that we would like to image. Ideally, this object should be a point, which means that it is infinitesimally small. In reality, it is sufficient if it is very small, particularly if it is much smaller than the wavelength of light.

Now we do apply an optical instrument or system to get an image, which then is recorded by a 2D detector and possibly stored by some additional device. For the moment we assume that the detector is an ideal one, which could detect the image with infinitesimally good quality. Of course, due to many reasons, this is not possible, but this does not play any role here. The recording device is not of interest in this section. The result of the imaging process is an "image point," which always has a finite size and shows some intensity distribution. For the moment, we must not make use of an exact definition of "intensity" as before, but it should be clear enough what we mean. Here, we should mention that with "image point" we do not really mean an infinitesimally small point, but rather an extended spot. Only for simplicity and because it is widely used, in the following we will sometimes use the expression "image point" instead of "image spot."

This distribution depends on the properties of the optical system and is not necessarily round or symmetric. Such a distribution will be discussed much more deeply in Chapter 5. Figure 1.8a shows an example of an image of a point object. We clearly see that it has some width δ, which could be measured at different positions within a line

3 CIE Proceedings (1951), Vol. 1, Section 4; Vol. 3, 37, Bureau Central de la CIE, Paris, 1951 (night vision).

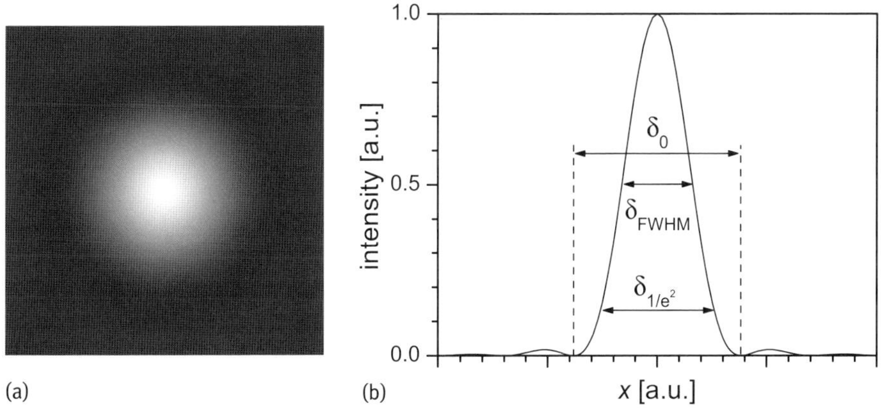

Fig. 1.8: (a) Example of an image of a point object, i. e., an “image point.” (b) Profile, e. g., measured along the horizontal or vertical line through the center or radial distribution of the light intensity, respectively. The arrows in (b) indicate the width of spot δ, measured at different positions, such as δ_{FWHM} measured at full width at half maximum (FWHM) or δ_{1/e^2} measured at $1/e^2$ of the peak or δ_0 measured between the indicated positions where the distribution becomes zero.

profile, such as δ_{FWHM}, which is measured at full width at half maximum (FWHM), etc. For the moment, it is not of much importance which of the different values of δ we take, just that one is chosen.

Now instead of taking just one simple object point we take an extended macroscopic object that can be considered to be made of many of such object points. If, for simplicity, we take each atom as one of those object points, of course, we obtain a superposition of all corresponding “image points” and all corresponding “image point” light distributions. This is the consequence of the linearity of Maxwell’s equations. Due to the huge amount of atoms within a macroscopic object, it is clear that this yields a tremendous amount of “image points” as well, and which becomes even infinite, if instead of atoms we consider infinitesimally small object points. From this, we may conclude that absolutely every position within an image of a given size is the center of an “image point” and all “image point distributions” do overlap with other ones.

Although this seems to be a hopeless situation, it is not. Let us consider a simple model (an advanced description is the subject of Chapter 5). First, we have a look at two “image points” that are well separated (see Figure 1.9a). Here, the distance d between them is much larger than their width δ, and thus they could be well identified as two different points. If the distance becomes smaller and $d \approx \delta$, then we are just at the resolution limit, i. e., we could just recognize that the two points are different ones (Figure 1.9c or Figure 1.9d). Here, with resolution we mean the lateral optical resolution (for details see Section 5.3). For even smaller distances, both points begin to merge to a single blur and could hardly be distinguished from each other (Figure 1.9c or Figure 1.9d or a situation where d is even smaller). We say those points are not resolved (Figure 1.9e).

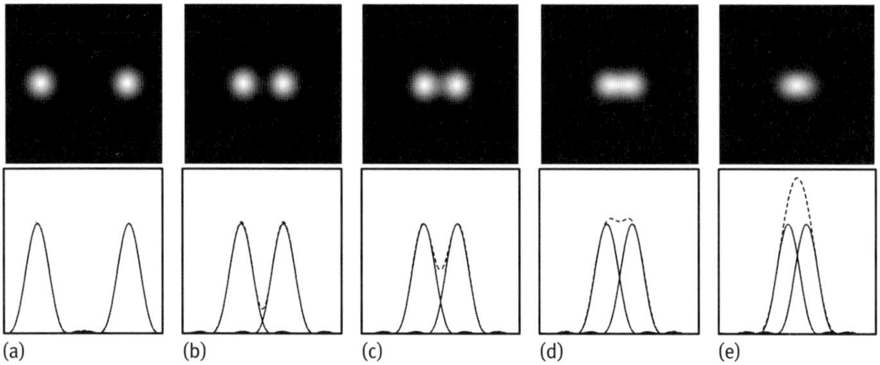

Fig. 1.9: Two image "points" located at different distances from each other. This is a demonstration of well resolved (a) and (b), just resolved (c) and (d) and hardly or not resolved dots (e), respectively. (d) corresponds to the Rayleigh criterion (see Chapter 5). The upper row shows the images, the lower one the profiles measured along the horizontal lines through the center of both points (solid lines: individual points, dashed lines and solid line in (a): superposition).

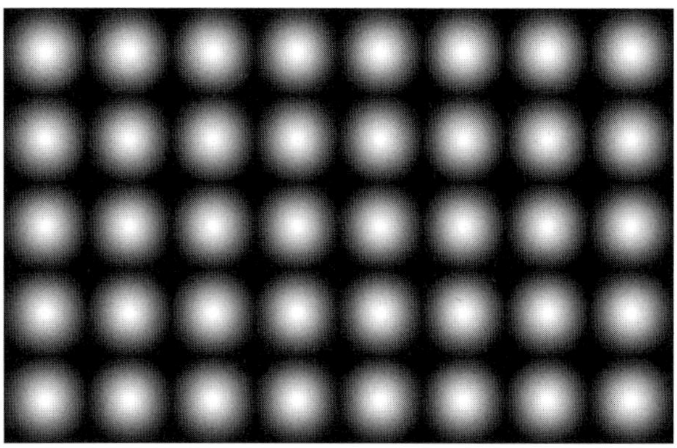

Fig. 1.10: Simple model of an image made of selected "image points."

Continuing our simple model, we will take a detector with a given size, say with a width *PW* in the horizontal direction and a height *PH* in the vertical direction, respectively. But remember, apart from its finite size, it is still an ideal one, which could display the image with infinitesimally good quality. Then we begin to fill its plane with "image points," however, not with an infinite number of points. Instead, we start to put one "image point" somewhere (e. g., in the top left corner), and then we put the next neighbor in a horizontal direction at a distance, where we can resolve those two neighbored points. Then we continue this procedure in the horizontal direction and later in the vertical direction as well, until the whole surface of the detector is filled (Figure 1.10).

Although there seems to be a straightforward similarity to newspaper pictures, which are made with a given finite number of real printer dots that are arranged in rows and columns and that may be identified as “image points” of that picture, it is very important to stress that the above discussion is related to a model that yields a finite number of *selected* “image points.” However, the physical image of any macroscopic object always still contains an infinite number of “image points.”

In the case of a nonideal detector, the situation does not change in principle. Today, usually such detectors are digital ones, which are made of a 2D array of photodiodes (a detailed discussion is subject of Chapter 4), which are named pixels, which means picture elements. If the pixel size is much smaller than δ, the situation is not much different from that with an ideal detector. If, on the other hand, δ is much smaller than the pixel size, then the pixels themselves take over the role of the imaging points. Then, of course, the resolution is worse than δ. For the moment, we may assume that 1 pixel resembles one image point, but as we will see later, resolution is 2 pixels in 1D, or 4 pixels in 2D geometry, respectively (at best, see Section 5.3). If the pixel size is the same as δ, we have a complex situation where we have to apply convolutions (see Chapter 5). However, also in such a case the basic idea of our simple model remains unchanged. The idea also remains unchanged, if we take an analogous detector such as a photographic film, where instead of regularly placed photodiodes, irregularly placed grains act as picture elements.

Furthermore, although the previous discussion is related to image observation or acquisition, it is applicable to the display of images as well. For instance, when a digital image is displayed on a digital screen, this screen is also made of pixels. In the case of a monochrome display, this is straightforward. In the case of color screens, each pixel is made of three subpixels that emit red, green and blue light, respectively. The three intensities are adjusted in a way that both the intended color and the total brightness of the pixel are reproduced correctly.

However, and although not the subject of the present book, we would briefly like to comment that for physical prints generated with digital printers the situation is somehow more complex. For photographic papers, the situation is different as well, but somehow similar to taking images on films. Again and similar to before, one may describe the resolution of a print as the number of pixels per mm or the number of pixels per inch (ppi). But here the picture elements are made of printer dots. Typically, a matrix of 16×16 dots creates one pixel. Depending on the intended grayscale value of the pixel, this value is achieved by a mixture of dots within this matrix that together form the perceived grayscale. Then one observes the reflected light from the illuminated printed image as an average of the “black” and “white” dots with low and high reflectivity, respectively. The observer should be so far away from the print that he only recognizes the matrix as one element with an average over the matrix elements. At best, the matrix is resolved as one element, but the individual dots are not resolved. This is similar to the averaging of the subpixels of a screen. For color prints, the final color of the pixel is also made of dots, but now with different colors. The background is not necessarily simple and is beyond the scope of the present work.

Altogether, the number of dots per mm or the number of dots per inch (dpi) is an important parameter for prints, and thus for the display *of available* images. But for the present book, this is not an issue. We are concentrating on imaging itself. Discussion of displaying images is restricted to basic issues. In that sense, scanner-related topics are also not a subject under discussion, as scanning is different from imaging. There, lpi, i. e., lines per inch, is an important parameter.

1.3.2 Imaging issues

The quality of an optical image depends on a lot of issues. These are, in particular, the resolution, the correct distribution of the "image points," correct color reproduction and much more. As for the correct distribution, they should be ideally located at the "correct position," which is influenced by the amount of imaging aberrations. Although most of those quality issues will be discussed within this book, here, at the beginning, we will concentrate on the resolution and the information content of an image only. We will concentrate on the very basics first, and thus neglect, for instance, bit depth, distortion and color information, which for the following discussion are not of importance. For readers who are more deeply interested in that subject, we would like to refer them, e. g., to the interesting article by Cao et al.[4] or to other ones from the same or other groups.

In the following, some very basic but also very important quantities are discussed. Some simple examples show that although here we restrict to quantities such as resolution and optical information, even a simple image characterization and, in particular, its quality, is not always straightforward.

1.3.3 The space bandwidth number

For a quite simple access to the image quality, we would like to introduce and to define the space bandwidth number (SBN). SBN is simply equivalent to the (actual) number of discriminable "image points" (not pixels) within an image (a more rigorous discussion is given in Chapter 5). If δx and δy are the lateral dimensions of an "image point" (in the x- and y-direction, resp.), and PW and PH the width and the height of the image, respectively, we define

$$N_{\text{SB}} = \frac{PW \cdot PH}{\delta x \cdot \delta y} \tag{1.12}$$

in a 2D space or

4 F. Cao, F. Guichard, H. Hornung: "*Information capacity: a measure of potential image quality of a digital camera,*" Proc. SPIE 7537, 75370F (2010); 10.1117/12.838903.

$$N_{\mathrm{SB,x}} = \frac{PW}{\delta x} \quad \text{or} \quad N_{\mathrm{SB,y}} = \frac{PH}{\delta y} \tag{1.13}$$

in one dimension (1D).

According to Equations (1.12) and (1.13), respectively, SBN in one dimension is not the same as in two dimensions. However, in this book we will not always discriminate between these, and we will just use "SBN" for both cases. Mostly, it should be clear what is meant, either because a general description is made, which holds for both cases, or it becomes clear from the context or is directly stated what is meant. In the same sense, for the moment we will not clearly discriminate between N_{SB}, $N_{\mathrm{SB,x}}$ and $N_{\mathrm{SB,y}}$ and just write N_{SB}. Later on, in Section 1.6.4, we will see that N_{SB} is conveniently given by $PH/\delta y$. In Section 5.3 we will provide more careful estimates of δx and δy, respectively.

For the moment, we will make use of this simple definition given by Equations (1.12) or (1.13), respectively, which provides a good characterization of the image quality. Please remember, here we do restrict to resolution issues only. Furthermore, for simplicity we do not discriminate between the SBN as a property of an image and the SBN as a property of an optical system. A more advanced description is the subject of Section 5.1.

The importance of SBN results also from the fact that it is independent from absolute sizes and any magnification issues, for instance, a large image with large points may be as good as a small one with small points. Moreover, if we restrict to images with only two colors, such as black and white, SBN also provides the optical information content of an image in one dimension or two dimensions, respectively. More accurately, unless stated differently (as above), with a black and white image we always mean an image with a gray level scale.

Of course, for a given image size there is a maximum possible value of SBN due to the fact that as a result of the nature of light, δx and δy could not become smaller than the minimum value. Thus, the actual value of SBN is usually smaller than its maximum possible value. Another simple example is, for instance, a detailed image displayed on a screen with a rough pixel structure. This yields an SBN of the displayed image that may be much smaller than that of the original image.

If an optical system is used to image a scene, then the SBN has to be transferred through the system. Often this leads also to a reduction of SBN during this process, which means that the image quality is reduced. But of course, at best, the final SBN obtained at the output of the optical system is the same as the initial SBN at the input side with the limitations discussed above. This is also the usual goal. An example for an exception of this goal is an image made at low light conditions, where it may be more important to achieve a large sensitivity and low noise of the optical and sensor system. In such a case, one has to accept a reduction of spatial resolution. Finally, we would like to note that only for a perfect digital camera system, the SBN is limited by the number of sensor pixels in the image plane, but otherwise it is lower. In any case, the SBN is lower than the number of pixels (see Equation (1.25)).

1.4 Resolution issues and requirements for images

Taking a photograph or taking an image in general, of course, only makes sense, if we make use of it. If, for the moment, we disregard special scientific or technical applications, usually the goal is to provide good images and to view them. To do so, it is required that the recorded image is made available by a particular output device such as a printer, a screen or a beamer and then we look upon it. Of course, this is an image process again, with the original image, i. e., the photograph, now acting as the object (see also Figure 1.3). However, in contrast to before, where the real object consists of an infinite number of object points, now the observed photograph may be *described* by a finite number of points (better: spots), which now act as new "object points" for further observation (again an imaging process; note again: this is a model, in reality there is an infinite number of overlapping image point distributions).

This "second imaging process" (second with respect to the original scene captured by the camera) is performed by the most important (optical) imaging device for humans, namely the human eye. In contrast to us, bats and dolphins, for instance, use acoustic signals for image formation. Hence, for us it makes sense to relate images and their content to the physical properties of the eye. Within this introduction chapter, we will restrict ourselves to the previously mentioned properties resolution, ascertainable information and SBN. To get an idea of what we require from an image, we need to have knowledge of the performance of the human eye. Here, in particular, we concentrate on two properties: its optical resolution and the angle of view Ψ. The relation to other properties, such as "depth resolution," will be made in later chapters.

1.4.1 Resolution and angle of view of the human eye

Unlike image formation by conventional technical systems, our human visual perception is a very complex process due to the physiological structure of the eye in combination with its behavior in viewing objects. A standard camera produces an image of a scene where in the ideal case all parts of the object space are imaged simultaneously to the image sensor. Ideally, the sensor is homogenous with respect to its resolution and sensitivity. The human eye, on the other hand, has great similarity to a camera, but the resolution on the photosensitive retina is not homogeneous, being the highest in a region called the macula, approximately 3 mm in diameter (Figure 1.11).

In a simplified model, the eye can be described as consisting of a nearly spherical vitreous body with a lens and a transparent cornea at the light entrance section. The incoming light rays are imaged to the almost spherically shaped image plane coated by the retina. The retina contains in principle two types of sensor cells termed cones and rods. The cones, sensitive to bright light and colors, have their highest density near the center of the retina in the macula. With increasing distance from the macula, their density decreases and so does the visual resolution at daylight. The rods, on the other

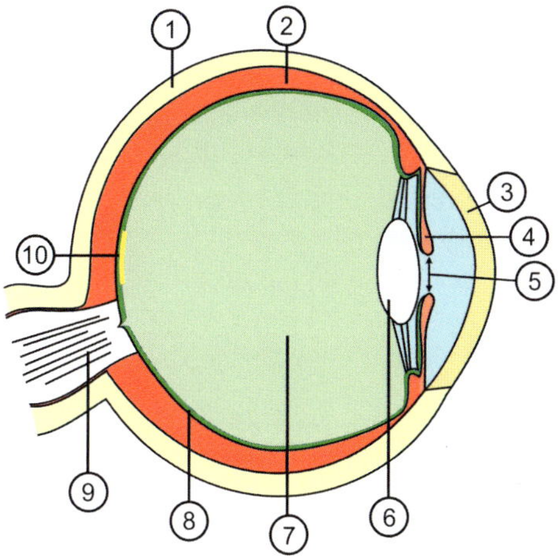

1 sclera (white of the eye)
2 choroidea
3 cornea
4 iris
5 pupil
6 lens
7 vitreous body
8 retina
9 optic nerve
10 macula with fovea

Fig. 1.11: Schematic structure of the human eye (author Talos[5]).

hand, are more sensitive to dim light than the cones and are more densely located toward the periphery of the retina. They cannot discriminate colors but only bright/dark light. Thus, with decreasing brightness the color impression and resolution degrade as the cones are only weakly sensitive at dim light, and night vision can be characterized by almost black-white imaging at lower resolution. Additionally, the spectral range for highest sensitivity of the eye shifts from the green for day vision to the blue range for night vision (see also Section 1.2.6).

The highest visual acuity of the eye is found in the center of the macula in a rod-free region of around 0.2 mm diameter, called fovea. As a consequence, in cases where it is required to see sharply over a larger region, the eye must permanently scan the area of interest and then, in conjunction with the brain, the visual impression is constructed.

As a field of view of the human eye, we understand the range of visual perception. If the eye is at rest, the part of the image that covers the macula is the center of the image and is perceived with the highest resolution and sensitivity. The visual acuity decreases with increasing distance from the image center. At the periphery, only a blurred or shadowy image impression is possible. Points in the field of view that are perceived by the same sensitivity are in general located on closed curves termed isopters. Figure 1.12a illustrates schematically some isopters (yellow lines) as a function of the angle in the field of view for the right human eye, measured by kinetic isopter perimetry.[6] The white area represents the blind spot in the eye where the optic nerve passes through the retina and where no perception is possible. The isopters are nearly concentric and asymmetric with a wider lateral extension due to shadowing by the nose. Only four isopters are shown

5 https://commons.wikimedia.org/wiki/File%3AEye_scheme_mulitlingual.svg

6 F. Grehn: *Augenheilkunde*, 31. Auflage, Springer Verlag, 2011.

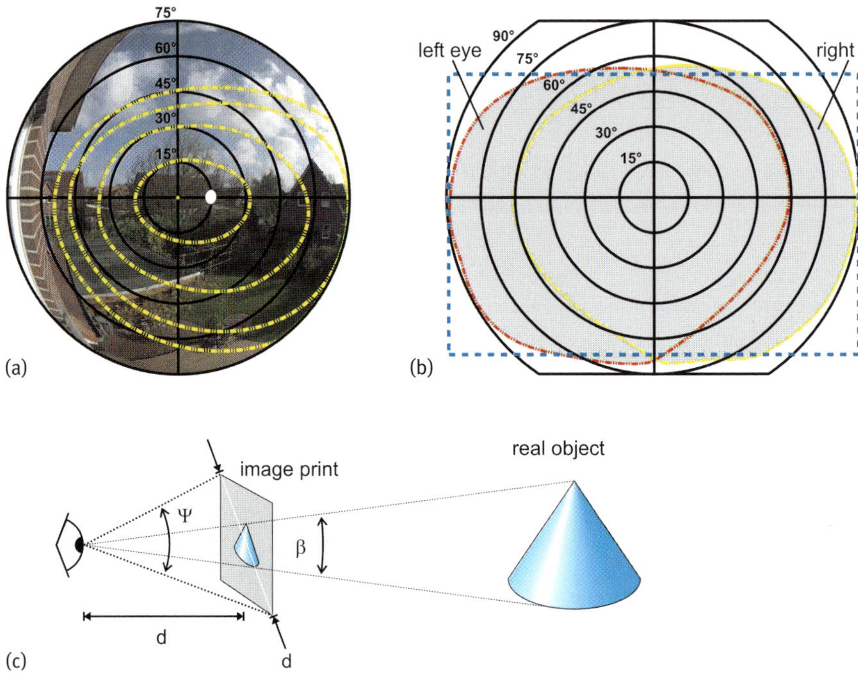

Fig. 1.12: Field of view for the human eye. (a) schematic illustration of isopters for the human right eye (yellow curves), indicating lines of constant sensitivity to light variation, measured by kinetic isopter perimetry. The white dot indicates the position of blind spot. The background image is taken by a fisheye lens of 150° of total angular field of view. (b) schematic total binocular field of view (gray area); the dotted frame is a rectangle with a 3:2 width/height aspect ratio; (c) the total angular field of view for an image viewed at the distance *d* of the image diagonal yields $\Psi \approx 53°$.

for illustration purposes, but there are also curves beyond 90° from the optical axis. The closer the isopters are to the center the higher the sensitivity of the eye. The peripheral isopter in Figure 1.12 is measured using full intensity of the test light, the subsequent inner isopters are taken for 1/3, 1/10, respectively, 1/30 of the full intensity. Seeing with two eyes for stereoscopic vision, the brain constructs a total optical impression, which is larger than that of an individual eye. As a result, the total binocular field of view extends to more than 180° in horizontal direction whereas in the vertical direction it is of more than about 120° with wider extension to the lower part. The binocular field of view is schematically depicted in Figure 1.12b. Only the central part, which is enclosed by both isopters for the right and left eye, is seen simultaneously by both eyes for stereoscopic vision. For comparison, a rectangular frame of an aspect ratio 3:2 is shown in the figure. It can be seen that image sensors or films of a similar shape and aspect ratio are well appropriate for representing visual human perception.

Moreover, Figure 1.12a also shows the image taken by a fisheye lens with a total angular field of view of about 150°. Unlike for human vision, we have nearly the same resolution all over the image field. From that consideration, it becomes difficult to define an angle of view for the human vision in the same way as it is normally done for technical optical systems (see also Section 2.2). However, in order to fix a reasonable value for the angle of view of the human eye, a different approach taking into consideration the psychology of seeing as well as the habits for observing images is necessary.

Let us assume a distinct visual range of 25 cm for reading a text or clearly observing an image. The image size must be adequately large to feel comfortable when looking at it. Empirical values show that this is the case if the image diagonal is approximately identical to that distance of 25 cm or slightly smaller. Then the eyes of the observer can comfortably scan the total image at high visual resolution. If the image diagonal d is equal to the observing distance or slightly less, the total angle Ψ is (Figure 1.12c):

$$\Psi \leq 2 \cdot \arctan(0.5) \approx 53^\circ \tag{1.14}$$

If a photo, which is observed under this condition, has been taken using a lens with the same angle of view, a very natural perspective and a plastic, nearly 3D impression is achieved. The angle β under which the imaged object on the image print is perceived is identical to the angle under which the real object is perceived by the observer when taking a photograph of it (Figure 1.12c). These considerations as well as some technical aspects to optimize lenses for a 35 mm film format with a 43 mm image diagonal may have inspired the lens designers of Leica in the years around 1920 to fix the focal length of the normal lens for that film format to 50 mm (see Section 2.2). The angle of view for this lens is nearly 47°. Thus, the angular field of view Ψ_{eye} for the human eye can be assumed to be between 47° and 53°. We will use in this book $\Psi_{eye} = 47^\circ$ as a reference value for the human eye.

As described above, the resolution of the human eye is not homogeneous across the retina. The acuity of vision depends on different parameters as, for instance, illumination, object distance and structure, as well as the symmetry of the observed object. The evaluation of the resolution is not as straightforward as for optical systems (further discussion is done in Chapter 5). Thus, a broad range of values for the angular resolution of the human eye can be found in the literature.

Two objects, for instance, points or lines, can be discriminated as being separate at an observation distance of $l = 25$ cm if their separation distance δ is between 75 µm up to 300 µm for comfortable vision (Figure 1.13a). This situation is also shown in Figure 1.9d for small dots contacting each other. The corresponding visual angular resolution $\Delta\varphi$ is given by

$$\Delta\varphi = 2 \cdot \arctan \frac{\delta}{2 \cdot l} \approx \frac{\delta}{l} \tag{1.15}$$

Using the above values for l, respectively, δ yields for $\Delta\varphi$ values between $1'$ and $4'$ of arc, respectively, 0.3 to 1.2 mrad. The minimum value of 0.3 mrad means that two human hairs close to each other can still be discriminated at a distance of 25 cm.

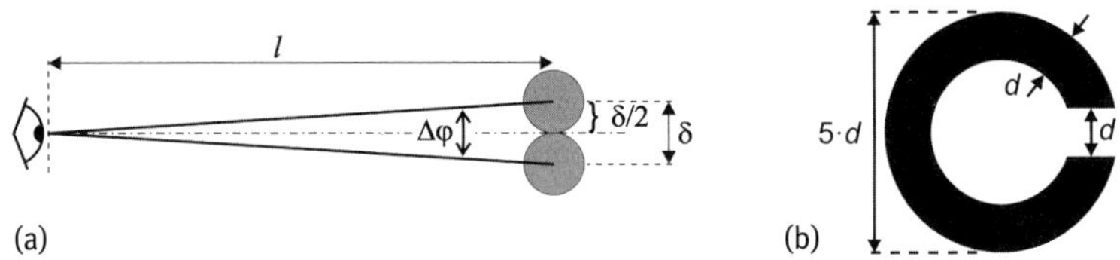

Fig. 1.13: Angular resolution of the human eye. (a) As determined by discrimination of isolated objects; (b) as determined by the perception of a Landolt-ring. The stroke width d is identical to the gap width; the diameter is $5 \cdot d$.

This is compatible with the perception of smallest apertures in a Landolt ring, which is used in ophthalmology to determine the human visual acuity (Figure 1.13b). If very narrow structures or deviations from symmetry should be detected, even values below $1'$ of arc can be found. This is the case for instance if the nonius of a vernier caliper is observed. Here, values of $5''$ of arc to $10''$ of an arc for $\Delta\varphi$ may be found. Taking all these different aspects into consideration, it is very difficult to specify the resolution of the eye as is done for technical optical systems. We think, however, that $\Delta\varphi_{\text{eye}} = 0.6\,\text{mrad}$, respectively, $2'$ of arc, is justified as a mean value for the angular resolution of the human eye in many cases. This value will be used as our reference if a comparison with optical systems is needed in this book.

In some cases, a higher resolution is required, for instance in order to ensure that aberrations of the optical system, like camera lenses, should not be perceived by the human eye. Then $1'$ of arc or is more appropriate. This is a challenging value, which lens manufacturers use for the design of high-quality lenses [Nas10]. If, for instance, we view an image print of 12 cm × 18 cm, having a diagonal of 21.6 cm, from a distance of 25 cm, then the angular resolution of $1'$ of arc corresponds to a distance of 73 µm on the print. This is 1/3000 of the image diagonal and is the limit value for details that can be perceived on the image for the described viewing condition. If the print is a 5× magnification of an original image taken by a 24 mm × 36 mm full format sensor with a diagonal of 43 mm, 1/3000 of its diagonal corresponds to 15 µm. Structure details on the sensor with dimensions below 15 µm can no longer be detected by the human eye even on 5× magnified prints observed from the distance of its diagonal. This is decisive for the allowable circle of confusion that can be tolerated by optical systems (see Section 1.5.4, Section 2.5.4 and Section 6.9).

Based on those definitions, one may approximate the maximum number of "image points" that could be distinguished within the width of an image (i. e., in one dimension), and thus we get for a consideration in one dimension:

$$N_{\text{SB}} = \frac{\Psi}{\Delta\varphi} \tag{1.16}$$

For a 2D space, neglecting differences in angle of view in horizontal and in vertical direction, SBN may be roughly estimated by the square of the value given by Equation (1.16).

According to the values given above by $\Psi_{eye} = 47°$ and $\Delta\varphi_{eye} = 2'$ of arc for the human eye, we may estimate $N_{SB,eye} \approx 1400$ or roughly 1500 (see also Section 5.2.4.5). With the more challenging resolution of $\Delta\varphi_{eye} = 1'$ of arc, we may obtain instead $N_{SB,eye} \approx 2800$ or roughly 3000 but not, for instance, 10,000, and thus at least get a feeling of the image contents that may be captured by a human eye "at once."

This means that from a typical photograph with an aspect ratio of 3:2 (or 4:3), which fully covers the human angle of view in one direction in the best case, we may estimate that we could resolve approximately 3000 "image points" in this direction and 2:3 (or 3:4) times less in the other one. Thus, in total, the maximum possible space bandwidth number for the eye is $N_{SB,eye} \approx 5 \cdot 10^6$. This is the number of perceivable "image points" in two dimensions (note: not pixels).

1.4.2 Remarks to reasonable number of "image points" and SBN in photography

From this estimation, it is clear that usually it makes no sense to increase the number of "image points" within an image, namely the SBN, much beyond the value given by $N_{SB,eye}$. Only if we are interested in large posters where high resolution at close observation distance may also be required, then images with $N_{SB} \gg N_{SB,eye}$ have to be provided. However, then we see only a fraction of the whole scene, albeit at high resolution. On the other hand, if the number of "points" is significantly fewer than in $N_{SB,eye}$, the image contains less information than what is possible at maximum, and thus image quality is worse.

Thus, an image that has the discussed optimum SBN, which is $N_{SB,eye}$ for the human eye, provides less information, i. e., a lower SBN, when observed from a closer distance. The reason is that in this case we see only a fraction of all of the "image points," which can now be perceived as larger blurs, but we cannot see more details. The limitation is given by the original "image point" size of the observed image.

On the other hand, when the original image is observed from farther away, we do have a similar situation. Due to limited resolution of the eye, we cannot resolve the original "image points." The eye only recognizes fewer "image points" within the image, which again, is equivalent to a reduced SBN of the captured image.

1.4.3 Magnified images

Depending on how magnification of an image is performed, one may discriminate different situations. To do so, let us take an image with a given fixed number of "image points," i. e., a given SBN (image 0) and consider two situations.

(I) Let us regard a simple magnification of such a *full* image ("image I"). Because both PW and δx, and PH and δy, respectively, are enlarged by the same factor, SBN does not change, and thus the information content for the magnified image is the same as before. N_{SB}, namely the number of image points," is independent of its absolute size. This means also that we cannot recognize more details in the magnified image, and thus magnification does not lead to a better resolution; only the size of the region and that of the "image points" is increased. This situation is given when printouts of different sizes are made from the same picture taken by a camera.

(II) Now let us regard a simple magnification of a given *fraction* of the full image ("image II"). In particular, we compare a selected region of interest (ROI), such as the one marked in Figure 1.14a but now enlarged to the same size as Figure 1.14a. Of course, SBN within the marked region is the same as in the expanded full image Figure 1.14b, but it is much smaller than the SBN of the full image Figure 1.14a. This is the so-called *software zoom* and leads to a loss of information content. This may be even more apparent when the ROI marked in Figure 1.14b is compared to Figure 1.14c, which was taken by a telephoto lens. Ideally, SBN in Figure 1.14a and Figure 1.14c is the same, but the quality of the image Figure 1.14b is worse.
Usually, the *software zoom* may be applied when taking photographs with a digital camera. However, unless one is interested in saving recording space, it is recommended to avoid it, because this is always coupled with loss of information. If the image should be magnified anyway, this can always be done later in a post-processing step using a computer. Then there is even the advantage of selecting different ROI more properly.

(III) For comparison to before, let us regard a true zoom or an image obtained from a fixed telephoto lens ("image III"). In this case, we do not enlarge the image discussed before, but instead we take another one by applying different optics. Such a telephoto lens has a reduced angle of view Ψ (see Chapter 2 and Chapter 6), but also a better angular resolution $\Delta\varphi$ when compared to before. At best, this telephoto lens has the same SBN as that of the optics used to capture the image 0, and thus no loss of information occurs. Figure 1.14c illustrates this situation.
This example also clearly shows the advantage of a photo camera with exchangeable or zoom lenses, respectively, when compared, e. g., to almost all mobile phone cameras. The first one does allow for high quality hardware zooms, whereas the latter ones do not. As usually the in-built lenses are wide-angle lenses with fixed focal length, any zoom is automatically a software zoom with the discussed significant disadvantages (for further discussion see Chapter 7).

▶ **Fig. 1.14:** (a) Original image, taken with an $f = 24$ mm zoom lens. The yellow marked region is enlarged to the same image size and displayed in (b). This is a software zoom. (c) The same image, but now taken with the same lens zoomed to 300 mm ("hardware zoom"). It is apparent that SBN in (b) is significantly reduced when compared to the original image (a). SBN is also much smaller than in (c), and thus image quality is much worse. However, SBN and image quality may be the same in (a) and (c) under the assumption that these two lens settings do not differ in their image quality.

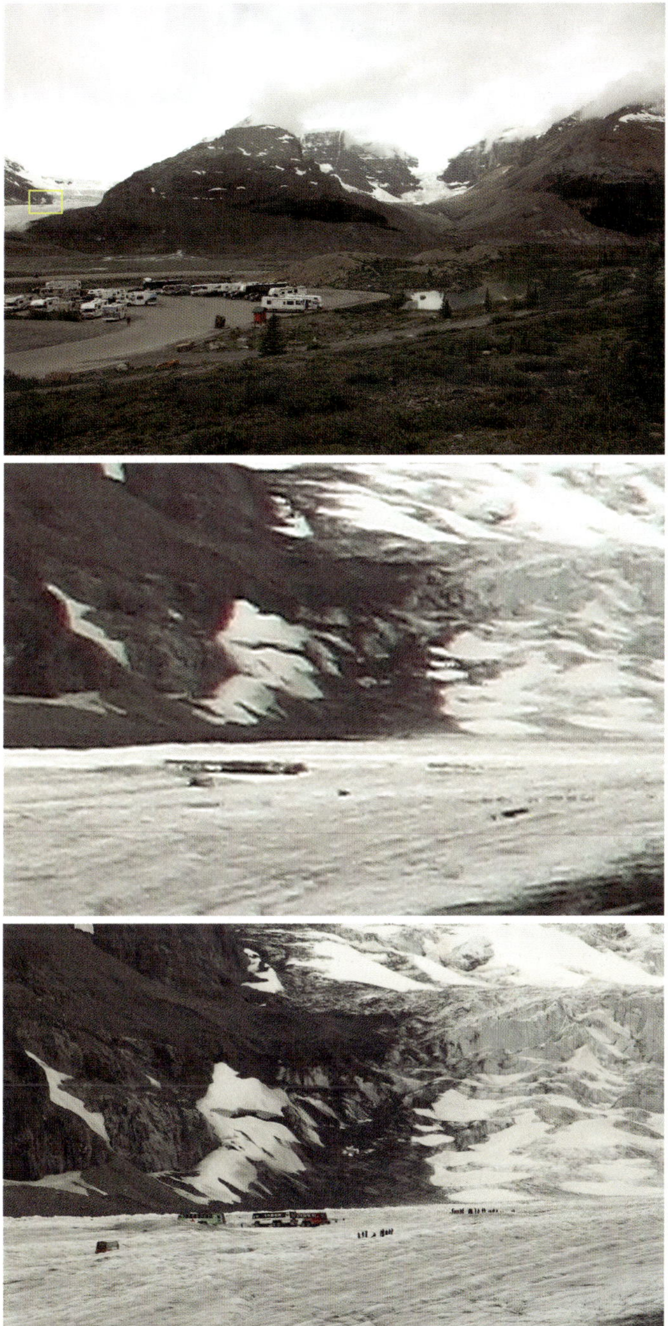

(a)

(b)

(c)

1.5 Imaging and focusing

1.5.1 Focusing and f-number

Although the present book is related to imaging, it is important to make a comparison to focusing. In this section, we restrict ourselves to the very basics and rely on a basic knowledge of geometrical optics. For deeper discussion, we refer to standard textbooks on optics (see, for instance, [Hec16, Ped17]).

First of all, we would like to note that the goals of focusing and imaging are completely different. Usually, focusing makes use of a more or less parallel beam and has the goal of hopefully concentrating all of the light within a very small spot, to achieve high fluence or intensity, for instance. Assuming such a beam, the position of the focal point F is given by the focal length f (see Figure 1.15a). The spot shows a light intensity distribution similar to that displayed in Figure 1.8. The shape of the distribution depends on the beam shape in front of the optics. The diameter of the focal spot can be obtained from diffraction theory (see standard textbooks such as [Ped17] or [Hec16]; see also Section 5.1.4 and Section 5.1.6, in particular, Equation (5.43)).

$$\delta_B = 2\kappa \cdot \lambda \cdot \frac{f}{D} \cdot \alpha \tag{1.17}$$

where D is the beam diameter, respectively, the width of the limiting aperture, for instance of a flat top profile in the near field. κ is a constant that depends on the beam shape (see Table 5.1).

In addition, it depends on the position within the profiles of the spot and the laser beam at which D and δ_B are respectively measured (cf. Figure 1.8). α is a constant that describes the beam quality. It is a measure that describes wavefront distortions within the beam, and also includes such ones that originate from aberrations from the optics. It should be noted that for Gaussian beams and perfect optics this is identical to the parameter "M^2". Table 5.1 provides some examples of focal diameters for different beam shapes (note that the values of δ_0, δ_{FWHM}, $\delta_{1/e2}$ in that table are the diameters of the PSF, but due to the same physical background they are the same as the focal spot diameters; see the remark at the end of Appendix A.8).

Equation (1.17) can be derived by assuming a source at infinite distance that emits a spherical wave, thus being equivalent to the assumption of a plane wavefront that passes an aperture with a width D and afterwards an ideal lens. The aperture introduces an "intrinsic divergence," which for instance yields to a first dark ring at the divergence angle θ_0. In the case of a slit or a circular aperture, this is further discussed in Section 5.1.4.

Equation (1.17) shows that, in particular, δ_B does depend on f/D. This ratio is the so-called f-number $f_\#$ and is an important quantity in optics in general (see Equation (2.14) and Section 3.4). It must be remarked that δ_B depends neither on f nor on D alone but only on the ratio of both. Furthermore, it can be shown that the f-number, disregarding immersion optics, cannot be smaller than 0.5 (see Section 3.4).

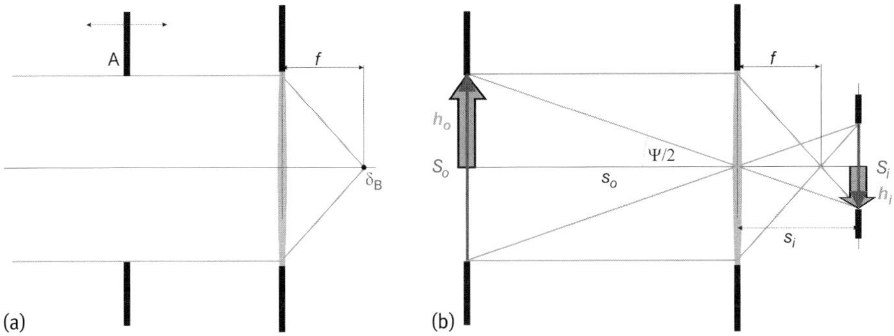

Fig. 1.15: Illustration of (a) focusing and (b) imaging. In (a) the aperture A, which limits the beam diameter may be shifted as indicated by the arrow without affecting focusing (it may even be removed; this may be a somewhat simplified consideration). The dot indicated by δ_B is the focal spot F. Within aberration free geometrical optics, it is a mathematical point which, of course, is infinitively small. However, within beam optics it has a finite size. In (b), the object may be the aperture around the solid arrow indicated with S_0, or, e. g., provided by the open arrow (compare also to Figure 1.16).

1.5.2 Imaging and imaging conditions

The goal of imaging is absolutely different. Images are taken to see what an object looks like. We would like to see its structures and hopefully also a lot of details. Thus, the image should have a light distribution that is similar to that of the object; only the absolute size may differ. This is entirely different to focusing, where all of the light is concentrated within one spot and no structure information of an object is available at all. This is also obvious from SBN: for focusing $N_{SB} = 1$ whereas imaging requires $N_{SB} \gg 1$. It should be noted that this is in sense for a number of resolvable points and information content, which for focusing could not be increased by an artificial increase of sensor size. Focusing is closely related to Fraunhofer diffraction or far field, while imaging is governed by geometrical optics and near field (although there might be severe corrections resulting from wave optics effects; see Section 5.1.2). As an example, Figure 1.15b shows the image construction within geometric optics. The position of the image can be obtained by the thin lens equation

$$\frac{1}{f} = \frac{1}{s_i} + \frac{1}{s_o} \tag{1.18}$$

where s_o and s_i are the object and the image distance, respectively (see also Sections 2.2 and 3.1).

It is obvious that the position of the image and that of the focal point are different. The size of the image S_i can be calculated from the object size S_o

$$\frac{|S_i|}{|S_o|} = \frac{|s_i|}{|s_o|} = |M| \tag{1.19}$$

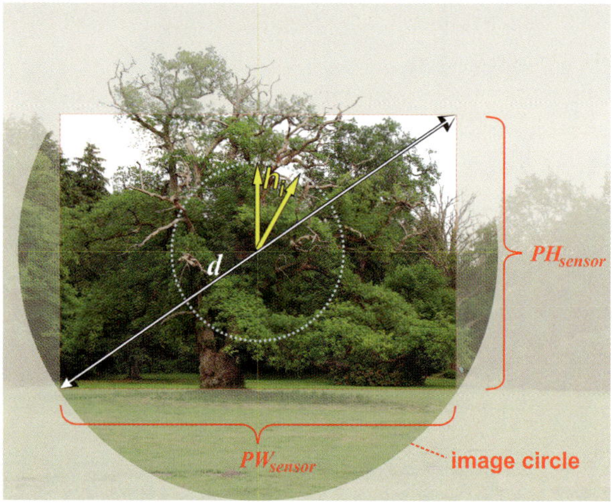

Fig. 1.16: Definitions of image parameters (see text).

where M is the transversal or linear magnification of the imaging. It should be noted here that we disregard the direction of the distances and sizes and, therefore, use only the magnitude of these quantities. In a more rigorous consideration, which is done in subsequent chapters, the direction must be taken into account where a negative M accounts for an inverted image relative to the object, as can be seen in Figure 1.15.

In order to characterize images, some key parameters are required. Figure 1.16 illustrates the situation of imaging and shows the definitions used within this book. The object is imaged into the image plane, which in principle, extends laterally to infinity. Due to restrictions on the optical system, an image is restricted to a finite area. Usually, the optics is round and symmetric, and thus, if it is well aligned, the image is restricted to the round area, termed image circle, as illustrated in the figure. The optical axis is perpendicular to the image and marks the center. If the sensor is centered as well, it cuts out a fraction of the image as shown by the rectangle in this figure. The width and the height of the sensor are designated by PW_{sensor} and PH_{sensor}, respectively. The sensor diagonal is d. The total height of the image of the scenery on the sensor is PH_{sensor} and lies within the image circle. Later on, this image is reproduced as a printout, or displayed on a screen. The corresponding width and height will be termed PW and PH, respectively.

If we consider only a part of the original scenery as the object, for instance a branch of the displayed tree, its original size is given by S_o, its size in the image plane is S_i and the ratio of both quantities yields the transversal or linear magnification M. In many cases, the term image height is used to describe the quality of the imaging process using rotationally symmetric optics. We then need to describe the transversal distance of an image point on the sensor from the optical axis. This distance is indicated in Figure 1.15 and Figure 1.16, respectively, by an arrow from the image center to any point and is designated by image height h_i. Using this notation, for symmetry reasons, there is no directional dependence of h_i. The maximum value of h_i that is possible within the captured frame depends on direction ($PH_{sensor}/2 \le h_i \le d/2$).

For the generation of real images, we need to discriminate three different conditions that result from the application of the lens Equation (1.18) and which are discussed for photography in more detail in Section 2.2:

(I) The object plane is nearly identical to the focal plane of the lens, with $s_o \approx f$. Then a high magnification is achieved, and the image plane is at a very large distance from the lens with $s_i \to \infty$ and $|M| \gg 1$. This is the typical situation for microscopy where the optical system is not specified by its f-number $f_\#$ and focal length but rather by its numerical aperture NA and magnification.

(II) The standard situation for photography and particularly for astrophotography is quite the opposite of microscopy. Here, the object plane is at a large distance from the lens and particularly, s_o is much larger than the focal length f of the system. As a consequence, the image plane nearly coincides with the focal plane in the image space and the magnification becomes very small with $|M| \ll 1$.

(III) In the intermediate range for approximately $0.1 < |M| < 1$, we have the situation of close-up photography or macro photography. For this type of imaging, the image distance is significantly larger than the focal length of the lens, which often requires special setups or lens constructions. Like for standard photography the optical system is rather described by its f-number and focal length.

This discrimination shows that the discussion about spatial optical resolution relates, for microscopy, to the object plane with the object being much smaller than the image whereas spatial resolution issues are more relevant in the image plane for photography. All spatial resolutions can also be expressed in terms of angular resolution. The angular resolution in the object space for distant objects, for instance as observed by astrophotography, is more meaningful than the spatial resolution in object space. The subsequent chapters of this book predominantly deal with the standard photographic situation, but hints to microscopy and close-up photography are also given.

1.5.3 Relations between imaging and focusing, SBN and image quality

Here two remarks are very important:

1. Although formally (real) focusing may be regarded as imaging of an object that is placed in infinity, and thus is demagnified to zero size with $M \to 0$, such a consideration makes no sense at all; even if due to wave optics, the size remains finite. This is because the goal of imaging as an information transfer process is totally missed, due to the fact that one does not get any more information of the object's structure. Again, focusing and imaging imply entirely different goals.
2. In photography, people often talk of focusing: "before an image is captured, we have to focus." However, of course, this is colloquial and not correct because as men-

tioned before, real focusing prevents any structural information, and thus the goal of imaging. What people really mean is "rendering the image sharp" by fulfilling the lens Equation (1.18) for a given object distance s_o with f given by the camera lens. However, because terms such as "one has to focus" or "focusing of the camera" are widespread, we also use this term if the meaning is unambiguous. Nevertheless, we should be always aware of the real meaning of the terms.

Nevertheless, focusing and imaging are closely related by fundamentals of optics. In particular, if aberrations are mostly absent, the size of an "image point" δ (cf. Section 1.3.1) and the size of a focal spot δ_B, both obtained with the same optics, are the same when the size is measured between the first zero positions of the spot distribution (see Figure 5.6). More generally, depending on the definition of δ and depending on the required contrast, for the resolution $\mathcal{R}$ within an image one gets, e. g., $\mathcal{R} \approx \delta_{FWHM}$ or $\mathcal{R} \approx \delta_0/2$. This is shown more rigorously in Section 5.1.6, in Section 5.3 and at the end of Appendix A.8. From Rayleigh's or Abbe's criterion, respectively, one obtains

$$\mathcal{R} = \kappa \cdot \lambda \cdot \frac{f}{D} \approx \kappa \cdot \frac{\lambda}{2 \cdot NA} \tag{1.20}$$

where NA is the numerical aperture (see standard optics textbooks such as [Ped17] or [Hec16] and also Section 3.4.3). $\kappa = 1.22$ is valid for Rayleigh's criterion whereas $\kappa = 1$ represents Abbe's criterion (see also Table 5.1). From the comparison of Equation (1.17) to (1.20), it is obvious that for $\kappa = 1.22$ we get $\delta_B = 2 \cdot \mathcal{R}$. This resolution is displayed in Figure 1.9d, where the width of each of the spots is given by $\delta_0 = \delta_B$. δ_0 is displayed in Figure 1.8.

Although the actual value of κ is not of too much importance here, at least not for the discussion of the basics as in the present chapter, we may state, that for optical imaging, $\kappa = 1.22$ is mostly used and at least a good approximation. A rigorous description is the subject of Section 5.1.

According to this discussion, $\delta = \delta_B$, which then is given by Equation (1.17) with $\kappa = 1.22$ and $\alpha = 1$ for imaging with aberration free optics. Using this knowledge, the discrimination of imaging from focusing becomes even more clear. In particular, image quality is only good if the number of "image points" within the image is large. This is the case if $PW, PH \gg \delta$ or equivalently, if $N_{SB} \gg 1$ in a 1D consideration. In two dimensions, the result is straightforward. If N_{SB} becomes smaller, the image quality becomes worse. The "worst image" and the largest "image point" size is PH (or PW; for simplicity in the following we assume $PW = PH$), neither the image size PH can become smaller than δ nor δ can become larger than PH. This is the situation of focusing and even the just used expression "worst image" should express the situation only formally. With respect to the above statement, we would like to state that talking about imaging requires at least a significant amount of "image points" within the image. There is no fixed limit; determining such one is left to the "taste" of the reader.

Here, we would like to add a further conclusive remark on that topic:

3. a) We talk about focusing when $N_{SB} \approx 1$.
   b) We talk about imaging if $N_{SB} \gg 1$. This is the situation governed by geometrical optics although there might be severe corrections by wave optics.

1.5.4 Circle of confusion

Historically, in photography the size of an "image point" is given by the so-called circle of least confusion with a diameter of u_i in the image plane. This is the limiting size of a spot that can still be perceived by the human eye. As discussed in Section 1.4.1, the size of this spot depends on the way an image is viewed. For normal viewing conditions, we assume the angular resolution of the human eye as $\Delta\varphi_{eye} = 2'$ of arc, which is equivalent to $\Delta\varphi_{eye} = 0.6\,\text{mrad}$. For high quality requirements, the resolution of the eye is assumed to be of $\Delta\varphi_{eye} = 1'$ of arc, respectively, $\Delta\varphi_{eye} = 0.3\,\text{mrad}$. As a consequence, the maximum acceptable diameter u_i for the circles of confusion, when viewing an image from the distance l, is given by

$$u_i \approx \Delta\varphi_{eye} \cdot l \tag{1.21}$$

Due to the small value, we have approximated $\tan(\Delta\varphi_{eye})$ by $\Delta\varphi_{eye}$. Reproduced images such as typical rectangular image prints are conventionally viewed at a distance that is approximately the same as their diagonal. The natural viewing angle is between 47° and 53° and the space bandwidth number of the eye is approximately $N_{SB,eye} \approx 1500$ for normal quality resolution (Section 1.4.1). This means that structures with a size of 1/1500 of the image diagonal d can still be perceived. For high quality requirements, the resolution of the eye is assumed to be of $\Delta\varphi_{eye} = 1'$ of arc, respectively, $\Delta\varphi_{eye} = 0.3\,\text{mrad}$, and thus the allowable circle of confusion must be smaller or equal to 1/3000 of the diagonal. Thus, we can define the acceptable diameter u_i of the circle of confusion with respect to the image diagonal:

$$\begin{aligned} u_i &\approx \frac{d}{1500} \quad \text{for } \Delta\varphi_{eye} = 0.6\,\text{mrad} \quad \text{(normal quality)} \\ u_i &\approx \frac{d}{3000} \quad \text{for } \Delta\varphi_{eye} = 0.3\,\text{mrad} \quad \text{(high quality)} \end{aligned} \tag{1.22}$$

For an image print-out of 12 cm × 18 cm, this means $u_i \approx 145\,\mu\text{m}$ for normal quality and, respectively, $u_i \approx 72\,\mu\text{m}$ for high quality. As the image print-out is simply the magnification of the original image on a film or digital sensor, SBN remains unchanged, yielding a circle of confusion of about $u_i \approx 30\,\mu\text{m}$ for a full format sensor in normal quality and $u_i \approx 15\,\mu\text{m}$ in high quality. If this print-out were a contact reproduction of a photographic film of the same format, the circle of confusion on the film would be identical to that of the image. As a consequence, if images are taken by cameras with different image sensors and then magnified to the same size, the requirements for acceptable circles of confusion are more demanding for smaller image sensors or films. Thus, lenses for mobile phone cameras require a much higher precision and quality than lenses for large format cameras. The relevance of the circle of confusion is discussed in the following chapters for different applications under different circumstances. The resolution of optical lenses is especially limited by lens aberrations and by the aperture stop of the lens. The resulting circle of confusion can be controlled by the f-number and is discussed in Section 2.5.4. Likewise, the depth of field and depth of focus are controlled by the f-number and the acceptable circle of confusion. This will be discussed in Sections 3.4 and 6.9.

It is important to note that the circle of confusion usually is defined for a larger value than δ_B, even in the case of small aberrations. The reason for this is that, e. g., a small value of "defocusing" is accepted, even for a high quality image. This is discussed later (see, e. g., the examples provided in Section 5.2.3 and Section 6.9.1).

So far, we only considered the circle of confusion u_i in the image plane. The associated value for the circle of confusion u_o in the object plane can be calculated by applying the image magnification according to Equation (1.19), thus $|u_o| = |u_i|/|M|$. u_o is a measure for the structure size in the object plane that is just at the limit of resolution by the optical system.

1.6 Digital input and output devices

1.6.1 Image acquisition with a photodiode array: simple man's view

Up to now, for the detector we have assumed a screen with a resolution $\mathcal{R}_{\text{screen}}$, which was either $\mathcal{R}_{\text{screen}} \to 0$ or at least $\mathcal{R}_{\text{screen}} \ll \delta$, namely some kind of ideal detector. Now we would like to discuss the particular situation, when the resolution $\mathcal{R}_{\text{det}}$ of the detector is finite. This is the case, for instance when a photographic film is used as the detector, where film represents an analogous detector with $\mathcal{R}_{\text{det}}$ given by the grain size (see Section 4.1). This is also the case for modern detectors, which today mostly are digital ones. Usually, they consist of a 2D array of individual photo detectors, usually photodiodes. A detailed discussion of such array detectors, which are mostly CCD or CMOS, is the subject of Chapter 4. The array elements are named pixels, which means picture elements.

What is the resolution of such a photo detector array (PDA)? Mistakenly, one may assume that in 1D geometry this corresponds to the size of one pixel. To get a proof, we would like to discuss this situation in one dimension by performing a simple experiment using a grating as the test object.

Such a grating is characterized by the width of its bars and gaps. Quite often the width of both is the same, and thus one talks of lines and line widths, i. e., the width of a bar or a gap, respectively. Then one period consists of the width of a bar and a gap, or a "dark line" and an adjacent "white line," and this is identical to the width of a line pair (lp).

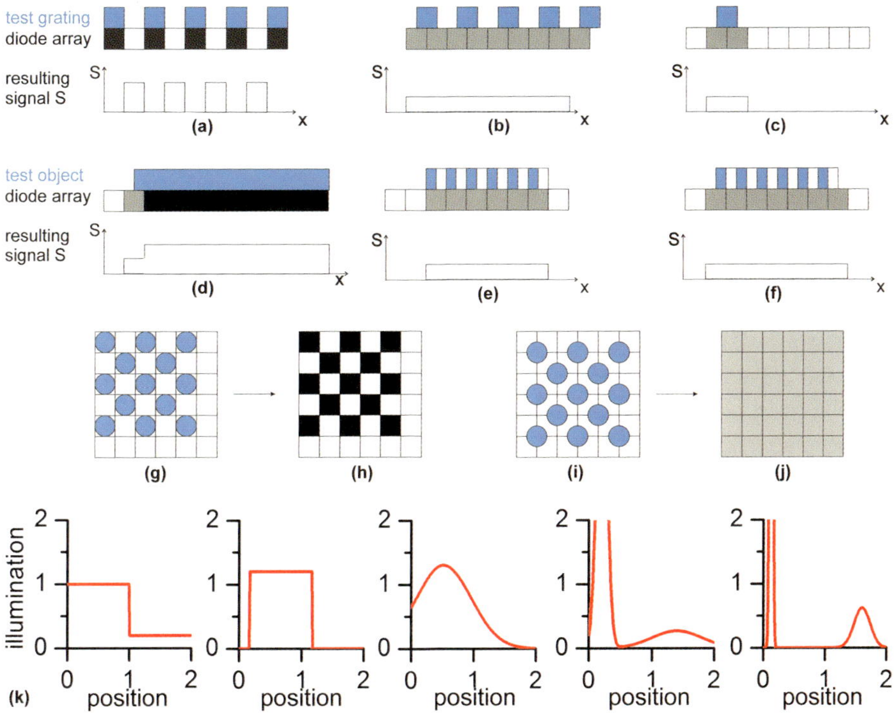

Fig. 1.17: Illustration of the resolution capabilities of a 1D and a 2D photodiode array. (a) The test object is a grating with a bar width of one pixel and a period of two pixels. The array detects the structure perfectly as illustrated by pixels that do not get illuminated because they are blocked by the grating bars, and thus are "black," and others, which do see full illumination, and thus are indicated by white color (see also corresponding signal). This situation corresponds to the "best case scenario." (b) Same test grating, but now shifted laterally as indicated. Now all pixels receive the same amount of light, namely half of the value of the "white pixels" in (a). Thus, they are displayed in gray (see also corresponding signal). This situation corresponds to the "worst case scenario." (c) The test object is a bar of one pixel width. For this bar position, this results in a signal that has a lateral width of 2 pixels. (d) test object with a sharp edge. For this position, this results in a signal that shows a extension of the sharp edge, that is distributed over 2 pixels. (e), (f) The test object is now a grating with a bar width and period that are both half of that of the grating in (a). This results in an illumination and a signal similar to that in (b). (g) and (h) show the 2D-equivalent to (a), and (i) and (j) to that of (b), respectively. (k) shows different illumination conditions on the surface of 2 neighbored pixels located between the positions 0 and 1 and 1 and 2, respectively, that all generate the same signal of the two pixels.

Here, in particular, we choose a grating with a line width of one pixel, and thus a period (or width of 1 lp) of two pixels as the test object. In Figure 1.17, this object may be either placed directly on the surface of the array and illuminated from the top with a collimated light beam or it may be regarded as the image of the object on the surface. The structure is marked in blue and would be the image of an ideal detector, even if it is shifted laterally. The size of the individual diodes (pixels) corresponds to black and white squares.

As illustrated in Figure 1.17, although at best, a test object may be well reproduced by the measurement (Figure 1.17a), in a different situation we may not get any information about its structure, although it is exactly the same object and the identical sensor (Figure 1.17b). Even for a small test object that is imaged onto the detector surface with one pixel width, the detector yields a width of 2 pixels (Figure 1.17c). Similar to that, the image of a sharp edge may yield to a sharp transition on the detector or recorded as a signal that is distributed over 2 pixels (Figure 1.17d). Moreover, if the object has an even finer structure, i. e., if the grating period is further reduced, this may also not be resolved at all (Figure 1.17e,f).

Here, it is important to state that in imaging we do not prepare situations such as displayed in Figure 1.17a to j, and thus know the real object and brightness distribution on the sensor surface. But we observe captured images without that knowledge. This is illustrated in the example displayed in Figure 1.17k: any of the illumination scenario examples (input) generates exactly the same signal for the left pixel and the same is valid for the right pixel (outputs). Thus, it cannot be discriminated which of these or even another one is the input. In other words, we do not know how the object looks like. Thus, again, we get stuck with the 2-pixel resolution limit in one dimension.

This simple man's example clearly shows that although a structure corresponding to one pixel resolution may be observed, in general, the resolution of such a digital array detector cannot be better than two pixels. In that sense, in Chapters 1 to 4 we talk about a sensor resolution of 2 (or 4) pixels. However, a more detailed analysis shows that the resolution is even slightly worse when a well-observable image is intended (see the following chapter and Section 5.3).

1.6.2 Image reproduced from a digital device, artefacts, Moiré effect

This is illustrated in another simple example using a slightly advanced consideration, for simplicity again also restricted to one dimension. As a test object, we choose a "grating" with a sinusoidal brightness structure that is imaged on the surface of the PDA (Figure 1.18). We are interested in how well this original image is reproduced by the PDA, depending on the number of pixels, namely photodiodes, per period, or in other words, the number of pixels per line pair.

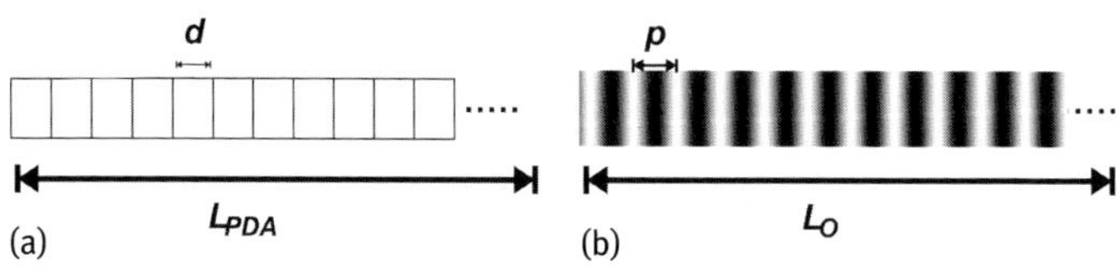

Fig. 1.18: (a) Section of a 1D PDA strip (total length L_{PDA}, width of 1 pixel, also diode, is d). (b) Section of the image of a 1D grating with a sinusoidal brightness modulation between a minimum signal that is equal to zero and a maximum signal (total length L_O, period p).

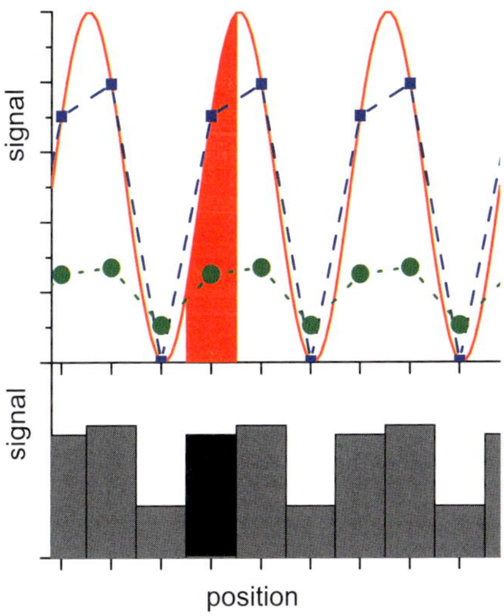

Fig. 1.19: Fluence distribution of the image of the grating on the surface of the PDA (red). The center positions of the pixels of the PDA are marked on the x-axis. The fluence value at those center positions is marked by blue squares, which are connected by dashed lines for better visibility. However, each photodiode integrates over its whole surface, which may range until its neighbors. The integral of Equation (1.24) yields the energy deposited within each diode. As an example, this is illustrated for the fourth diode as the red area. The energy signals are displayed as green dots and gray bars (for the red area, the corresponding signal bar is marked in black). Note: Red and blue lines and points correspond to fluences with the unit J/cm^2, whereas green dots and gray bars correspond to the integrated values, and thus to the energies with the unit J.

At each position x on the PDA, the amount of light illuminated on the surface may be described by the fluence $F(x)$, which is defined by the incident energy Q_e per area A,

$$F = \frac{dQ_e}{dA}. \tag{1.23}$$

This is equivalent to Equation (1.2) and shown as the red curve in Figure 1.19. However, although the fluence may change within the area of a pixel (see Figure 1.19), F is integrated over the pixel area A_{pix} and yields the total energy that is received by this pixel,

$$W_{pix} \equiv Q_{pix} = \int\limits_{pixel} F_{pix}\, dA \tag{1.24}$$

(green dots in Figure 1.19). Usually, this energy differs from pixel to pixel (see position dependence of green dots and gray bars in Figure 1.19).

It is clear that reproduction of the original image by the PDA (red lines in Figure 1.20 and Figure 1.21) depends on the number of pixels within one period. Of course, if this number is large, for instance 10 pixels per period (Figure 1.20, Figure 1.21), then the structure is well reproduced. For a smaller amount of pixels per period, i. e., if a PDA with a larger pixel size is chosen, then the structure becomes less well resolved (Figure 1.20b, Figure 1.21c).

For even fewer pixels per period, in particular for 2 pixels per period, we do not get any resolution at all (Figure 1.20c). This is similar to the situation of Figure 1.17b. The difference is only that here the grating has a sinusoidal structure, whereas in Figure 1.17, the grating has a rectangular shape. This is the resolution limit and called the Nyquist limit (see below).

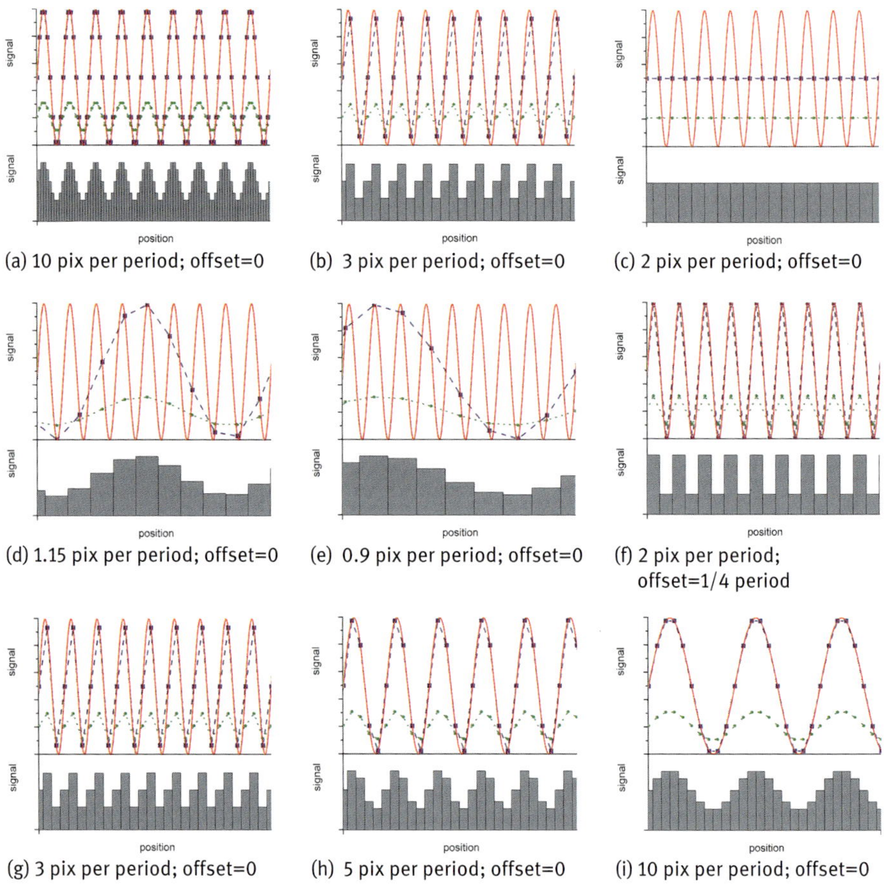

Fig. 1.20: Reproduction of a sinusoidal structure on the surface of a PDA (such a structure may be the result of the imaging of a corresponding test grating). (a) to (f): illustrate the situation when the same structure is sampled with different PDA, each of them with a different pixel width (indicated as gray bars); (g) to (i) illustrate instead the situation when the same PDA (with fixed pixel width) samples structures with different periods. The number of pixels per period and the relative shift (offset) is indicated below the diagrams. c) corresponds to Figure 1.17b, e) corresponds to a third-order spectrum; f) corresponds to Figure 1.17a. The Nyquist limit is 2 pixels per period. If it is exceeded, as in (d), (e) some kind of beat frequency occurs (blue and green curve, resp.). For further discussion, see also Chapter 5.

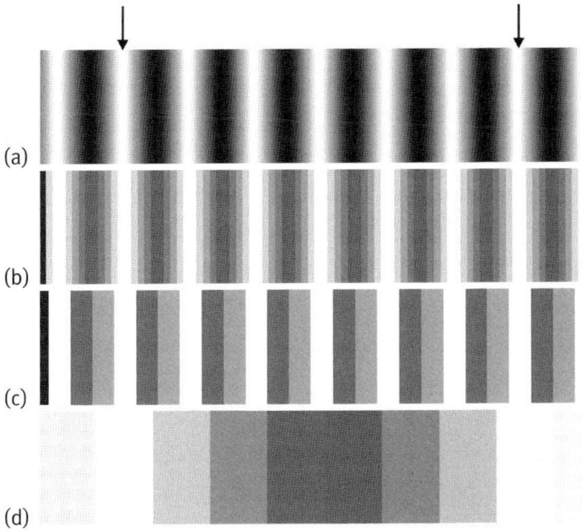

Fig. 1.21: (a) original image on PDA (equivalent with Figure 1.20); (b) image that is reproduced with a PDA with 10 pixel/period (equivalent with Figure 1.20a), this may be a reproduction of acceptable quality; (c) image that is reproduced with a PDA with 3 pixel/period (equivalent with Figure 1.20b); this still reproduces the original, however, quality is rather poor; d) 1,15 pixel per period (equivalent with Figure 1.20d); this image does not reproduce the original; it is an artefact (here we are below the Nyquist limit).

Further reduction of the number of pixels per period, i. e., further increase of the pixel size, yields that the original image is not reproduced by the PDA. Instead, the image obtained from the PDA shows an artefact, i. e., a super structure, which strongly differs from the original input (Figure 1.20d,e and Figure 1.21d).

We would like to note that only the ratio of pixel size to period is important and not the absolute values of both quantities. Hence, the above discussion yields information on both, a) how reproduction quality of a given structure could be improved, namely by increasing the number of pixels per period, and b) how a given PDA with given pixel size could resolve structures, such as test gratings, when the period is shrunken and the resolution becomes worse until resolution limit is reached.

Here, a brief comparison to signal theory may be helpful. If we identify the number of pixels per mm with the sampling rate or sampling frequency R_S, respectively, then according to this theory, somewhat simplified signal reconstruction is only possible up to the Nyquist frequency R_N, which is half of the sampling frequency. We may regard this value as a measure for the maximum information that can be transferred through a system; in other words, it sets the resolution limit, here considered for an optical system. Consequently, the Nyquist limit $R_N = R_S/2$.

For the present situation of Figure 1.20, this means that the number of periods per mm of a test grating that should be resolved cannot exceed R_N. As an example, we consider a sampling frequency of $R_S = 100$ pixels per mm. This means that we can resolve up to 50 periods per mm, which is equal to the limit $R_N = 1/2 \cdot 100$ pixels per mm. In other words, we need more than 2 pixels per period to resolve the grating structure, as shown in Figure 1.20c.

We would also like to remark that similar to the situation shown in Figure 1.17a, and in contrast to Figure 1.20c, the structure of the original image may still be reproduced roughly when the original is shifted slightly (Figure 1.20f). Hence, one might argue that lateral or spatial resolution is given by the size of 1 pixel and by no means can it be better than 1 pixel.

However, such a shift (phase shift, offset), which in the optimum case is a quarter period, cannot be guaranteed for an arbitrary image situation and, in particular, when the image is taken. In contrast to the artificial situation within our discussed well-defined experiment, in practice no information on period or shift is known. Thus, for such unknown conditions, one would expect a situation with an arbitrary shift. If then, we do have many pixels per period, the result does not depend significantly on the shift. On the other hand, if only very few, or even only 2 pixels per period are available, one might expect typically something in between the results shown in Figure 1.20c and Figure 1.20f. Thus, if one takes a lot of images with tiny varying movement of the camera or the sensor in between, one gets a series of a lot of different shifts. Then, after averaging all those reproduced images, one might expect an average reproduction with slightly improved resolution, which then may be 1.5 pixels. But such a particular image series is not the usual situation for photography.

On the other hand, this straightforward idea has been implemented as so-called pixel shift technology, for instance to increase the resolution (or the dynamic range) of cameras (including smartphone cameras). A requirement for its application is a movable image sensor, such as used for image stabilization (see Section 7.1.1), which allows for slight shifts in horizontal and vertical direction, respectively. Another requirement is a static arrangement, namely a scene, which does not change and a camera that is fixed extremely stable, for instance on a tripod. This is to avoid any movement on the micrometer level during the series of, e. g., 4 (or 16) images taken with a shift of half a pixel in between. Then within a post-process, these frames could be combined to a large single image with a potentially increased resolution. But one has to imagine not only the rather large storage space for this image series and the combined one, but, in particular, the rather artificial conditions for photography, especially for imaging with smartphone cameras. This gives more the impression that this technique should mostly demonstrate the potential of the camera. Nevertheless, we may note that the principal idea of pixel shifting is also used for the slanted edge method discussed in Section 8.3.2, but there applicable to any camera.

We would like to conclude that a reliable deduction of the structure of the original from the reproduced image for any case can only be provided if at least 2 pixels per period are available, and thus this is regarded as the resolution limit. If fewer pixels are available, or in other words, the structure is too fine to be resolved; usually one obtains artefacts (this corresponds to the so-called “alias effect” or “undersampling”).

A special effect of such artefacts is the Moiré effect. As discussed before, this occurs when the resolution of a digital device is too low. This is the result of the regular arrangement of the pixels in the PDA. For films with their irregular grainy structure, this effect does not occur. As an example, Moiré effect is also observed in TV, when, e. g., a person with cloths with a stripe structure moves over the scene, at least if the period of the stripe structure is close to the resolution limit. Figure 1.22 shows images where spatial frequencies within the scenery are higher than the Nyquist limit. For instance, the high frequencies are due to the rather small stripe structure on the guitar (Figure 1.22c), and within the inner part of the Siemens star (Figure 1.22e), respectively. They are also due to the relatively small period of the brick structure (Figure 1.22g). In those examples, the resolution of the camera was not sufficient and frequencies higher than the Nyquist limit are not suppressed, and thus the Moiré effect is seen clearly.

A particular situation arises when the structures within the image come close to the resolution limit, meaning close to the Nyquist frequency (see inner part of the Siemens star in Figure 1.22b). In that case, chromatic effects become apparent because white light emitted from the same object point is not imaged at the exact image spot (see later in this book). Consequently, the Moiré effect appears independently for different colors. Also, the Nyquist frequency is not exactly the same for all colors (usually due to the Bayer filter arrangement; see Chapter 4). As a result, the Moiré effect appears as colored stripes and areas (color Moiré effect; see Figure 1.22e).

Figure 1.22f and g illustrate that the Moiré effect can also occur as a beating effect when a high-resolution image (f) is displayed by a screen with too-low resolution (g). In this case, the photo takes the role of the object and the screen that of the camera. Here, the Nyquist frequency of the screen is lower than the high spatial frequencies of the brick structure and, therefore, we see a superlattice, which originates from frequency beating near the Nyquist frequency R_N (for details, see Section 5.2).

1.6.3 Similarity to spectroscopy

Finally, we would like to compare the situation of the artefacts discussed in the previous section to the situation in spectroscopy, because often this is more familiar. To do so, as an example, let us consider the spectrum shown in Figure 1.23a, which may be an optical spectrum or an acoustical or another one. The red line indicates the spectrum as emitted from a source and the black line that one measured by a spectrometer. However, both are different, e. g., additional peaks and other differences are apparent. These peaks do not correspond to real lines emitted from the source, but similar to the artefacts discussed above, here they originate from diffraction in second order (pure second-order spectrum here shown as the blue line). Sometimes, in addition, even higher orders are present. In case of an audio spectrum, the higher orders correspond to overtones or harmonics. Altogether, within this example, besides the pure first order spectrum, the measured spectrum includes the second order, and thus artefacts that are indicated by

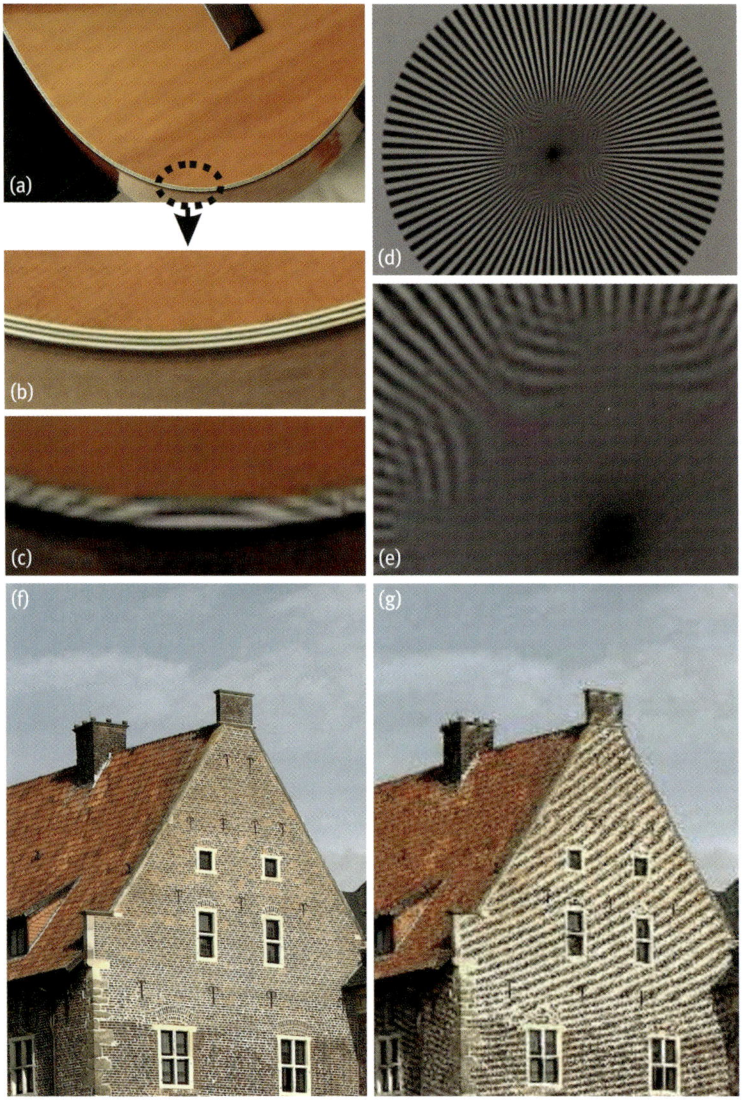

Fig. 1.22: Illustration of the Moiré effect. (a) Crop of an image of a guitar. The marked region is further enlarged and displayed as an image taken by a high-resolution camera where the image is not affected by the Moiré effect (b) and another one where it is (c). (d) Image of a Siemens star taken by a DSLR (compare Chapter 8). This image shows that the outer part is well resolved. The inner one is not and again, here the Nyquist frequency is lower than the spatial frequencies of the line structures (cf. Chapter 5). Consequently, the Moiré effect is present. (e) shows this in more detail and the Moiré effect is apparently leading to "strange" color distributions although the object is purely black-and-white. (f) and (g) illustration of the Moiré effect when a high-resolution image (f) is displayed by a screen with too-low resolution (g) (see the text).

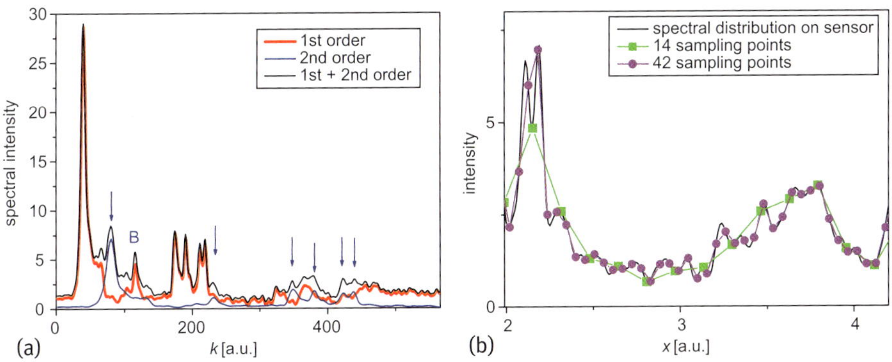

Fig. 1.23: (a) Example of a spectrum as emitted from a source (red line) and the measurement by a spectrometer (black line). Here, the variable k of the x-axis may be identified with the spatial frequency or the wave number, respectively. For further discussion, see the text. (b) Example of a brightness distribution on the sensor surface that is sampled according to the SBN of the sensor. The green symbols indicate a sensor with 14 sampling points within the displayed region (this may be the sensor width or height, resp.), the magenta ones a sensor with three times more points. The optics for both cases are the same and also supports the higher resolution. x is the spatial position on the sensor.

the blue arrows. "B" marks another artefact, namely an additional background that also originates from the second-order spectrum.

1.6.4 Space bandwidth number of digital devices

According to Equation (1.12), the SBN of a digital device, such as a PDA, is given by the sensor (or detector size, we do not discriminate) and its resolution, which may be the size of 2 pixels in one dimension. Thus, for a sensor 1024 pixels within its width or height, SBN = 512 (at best; if there is, e. g., significant charge spreading, etc., this value may be worse; see Chapter 4). Usually, SBN is determined by using test gratings such as those discussed in Section 1.6.1 and Section 1.6.2 (see also Section 5.2).

Then resolution is given by the number of lines, line pairs or periods that we can resolve with the given pixel structure. Usually, this quantity is related to the full width or height of an image or picture, and thus the resolution is given by the number of resolved lines per picture height (l/PH) or line pairs (lp) per picture height (lp/PH). Although it does not really matter if resolution is defined with respect to picture height PH or picture width PW, for photography it is more convenient to provide the resolution with respect to PH. In that case, for instance, a full format picture with aspect ratio 3:2 does not change the SBN if its width is clipped to an aspect ratio of 4:3.

The theoretical limit that we can resolve is two lines or 1 lp within 2 pixels, and thus N_V lines per PH, which is equivalent to $1/2 \cdot N_V \cdot \mathrm{lp}/PH$ (see Section 1.6.2). N_V is the number of pixels within the picture or sensor height. According to this, of course, this quantity is equivalent with SBN and hence the maximum SBN of a digital sensor is $N_{SB,max,sensor} = N_v/2$ (in 1D, in vertical direction).

Consequently $N_{SB,max,sensor} = N_v/2$ is consistent with the Nyquist limit $R_N = N_v/2$. For the situation discussed in Section 1.6.2 and, in particular, Figure 1.20, this means that the number of periods per *PH* of a test grating that should be resolved cannot exceed R_N (here in vertical direction). Correspondingly, one obtains a limit of $R_N = N_h/2$ in horizontal direction with the consequence that with respect to the full sensor and according to Equation (1.12), in 2 dimensions $N_{SB,max,sensor} = N_v N_h/4$. For this reason, any claim that the resolution of a sensor or a camera is directly given by the number of megapixels (MP) is nonsense. At best, the resolution is determined by the space bandwidth number of the sensor, which is a quarter of the sensor's pixel number. But due to further limitations by the optical and sensor system, respectively, and also because at least some contrast is needed to observe a structure (see Section 5.3), mostly it is even worse. Thus, one can formulate

$$\text{camera “resolution”} < N_{SB,max,sensor} = \frac{\#\text{MP}}{4} \tag{1.25}$$

where here #MP is the number of megapixels of the camera. Here, we have used "resolution" because this is not really a resolution, but a SBN. The correctly defined resolution has been discussed in the previous subchapters (see also Section 5.3).

Considering the example in Section 1.6.2, $R_S = 100$ pixels per mm = 2400 pixels per *PH* for a full format sensor with $PH = 24$ mm. This means that we can resolve up to 1200 periods per PH, which is equal to the limit of $R_N = 1/2 \cdot 2400$ pixels per *PH*. Again, we do need more than 2 pixels per period to resolve the grating structure.

We would like to emphasize that lp/*PH* is the most reasonable unit for digital devices such as digital camera sensors, digital screens, etc. This is in contrast to the previously-used, film-era unit lp/mm, which was often well adapted to analogous devices such as films. In particular, this was the case, if for instance the resolution of different films of the same format was compared, or the resolution of pictures taken by a full format and an APS-camera, respectively (the picture size on film is different; see Section 4.3), was compared. However, for digital images, such considerations are less helpful, because as discussed before, resolution or SBN given in lp/*PH* is independent of any absolute size or magnification, etc., but the resolution given in lp/mm is not.

This is also seen from the following simple example. Let us assume an image that should be reproduced by a PDA that consists of 14 points or pixels per *PH*. Thus, due to the 2-pixel resolution limit $N_{SB} = 7$ (then the signal is already zero, see Figure 1.20c). Now let us compare this sensor (I) with two other different ones.

Sensor (II) has the same *PH*, but 42 pixels per *PH*, which means three times larger SBN. The pixel width is reduced by a factor of three. Of course, this provides significantly better resolution (Figure 1.23b). The number of pixels per mm has also increased by a factor of three.

Sensor (III) has the same pixel width as (I), but its *PH* is a factor of three larger, and thus the total number of pixels within its *PH* is a factor of three larger as well in one dimension. Thus, again SBN is a factor of three larger than that of sensor (I), although the number of pixels per mm is the same as in (I).

The larger *PH* can then be used either to image in addition the regions on top and/or below the scenery captured by sensor (I). Obviously, this increases the information content of the image. On the other hand, the larger *PH* can then be used to enlarge the same scenery captured by sensor (I) now to the size of sensor (III). In this case, certainly, no new additional scenery is present within the image, however, many more pixels are available to record the image. This corresponds to an increased resolution, and thus an increased information content as well. Hence, in both cases this manifests in the increased SBN.

Finally, we would like to discuss three examples that clearly show that the absolute number of pixels does not necessarily determine the resolution, and, in particular, that a good camera with a sensor that has fewer pixels than a different one *may* be much superior.

Example I

We would like to compare two cameras, a typical cheap consumer camera A and a more expensive DSLR B. The relevant parameters for both cameras are tabulated in Table 1.2. Both cameras have different sensors and different lenses. The quality of the optics differs as well. The corresponding space bandwidth number of the systems $N_{SB,tot}$, resulting

Tab. 1.2: Parameters for the discussed "example I" in the text.

camera	*A*	*B*
number of pixels in total	16 MP (16.4 MP)	11 MP (10.7 MP)
sensor size (see Section 4.3.1)	approx. APS	full format
aspect ratio of *PW*/*PH*	4:3	3:2
focal length of camera lens	$f = 27$ mm (equivalent to $f = 50$ mm for full format)	$f = 50$ mm
PH	13 mm	24 mm
crop factor (see Sections 2.6.1.2 and 4.3.1) 24 mm/*PH*	1.85	1
pixel size (square shape) p	3.7 µm	9 µm
N_v	3513 pix	2666 pix
$N_{SB,sensor}$ (only according to sensor R_N)	1757 lp/*PH*	1333 lp/*PH*
$N_{SB,tot}$ in total (reduced by optics)	700 lp/*PH*	1300 lp/*PH*
this corresponds to a resolution on chip of $\mathcal{R}$	18.5 µm	18.5 µm
this corresponds to also to $N_{SB,tot}$ (in other units)	54 lp/mm	54 lp/mm
object distance s_o	8 m	15 m
image distance s_i	27.05 mm	50.17 mm
magnification M	1/295	1/299

from the combination of lens and sensor, are provided in the table. Now, we may ask which camera is better, if we restrict ourselves only to resolution.

To test the cameras, we apply a test grating with a given period $a = 0.18$ lp/mm, which means a grating period of $a^{-1} = 5.5$ mm, which is positioned in such a way that this just corresponds to the resolution of the camera under consideration. From the lens Equation (1.18) and Equation (1.19), one could deduce that camera A could just resolve this grating when it is placed in a distance of $s_0 = 8$ m of the camera. According to $N_{SB,tot}$, at this distance, it is demagnified to a period size on chip of 18.5 µm. This is identical to the resolution of the camera as 700 lp per 13 mm yields $\mathcal{R} = 18.5$ µm on the sensor.

To get a similar situation for camera B, the same test grating has to be placed in a distance of $s_0 = 15$ m. Then the test grating could again just be resolved and due to nearly the same demagnification as before, one also obtains a resolution $\mathcal{R} = 18.5$ µm on the sensor.

However, for camera A, this corresponds to five times its pixel size, namely $\mathcal{R} = 5 \cdot p$, whereas for camera B this corresponds to two times its pixel size $\mathcal{R} = 2 \cdot p$ only. This means due to the rather poor resolution of the lens (this is typical for simple consumer cameras), which is much worse than that of the sensor, the image quality of camera A is significantly less than that of camera B. Of course, this could have been seen directly by comparison of both values of $N_{SB,tot}$, but the present discussion gives a better impression of the quality of two typical cameras. Thus, indeed, although camera B is almost a factor of 2 better (resolution is 1300 lp/*PH* and 700 lp/*PH*, resp.), this cannot be seen if resolution is calculated in lp/mm (see the table).

Example II

Figure 1.24 illustrates a similar situation where a picture from a camera with more pixels yields worse image quality than that from another one with fewer pixels, but better optics. This is a typical situation for many consumer cameras that are "tuned" to a large pixel number. In this example, the image (a) is taken with a camera where the sensor consists of $2000 \cdot 3000$ pixels (6 MP) and that in (b) with another one with a 1.5 MP sensor with $1000 \cdot 1500$ pixels. (c), (d) show profiles measured along a horizontal line in the lower third of the images shown above. The lower resolution in the case of the camera with 2000 pixels (in horizontal direction; (c)) when compared to that one with 1000 pixels (d) is apparent. Here, the optics in front of both sensors is different, namely a rather poor lens in front of a sensor with more pixels (a), and a good one in front of a sensor with fewer pixels (b), respectively. But, of course, if in contrast to the situation of this example, both cameras are equipped with the same good lens, the cameras with more pixels may also yield better image quality. From those very different situations, one may conclude that a judgment of image quality based only on the pixel number is not possible at all.

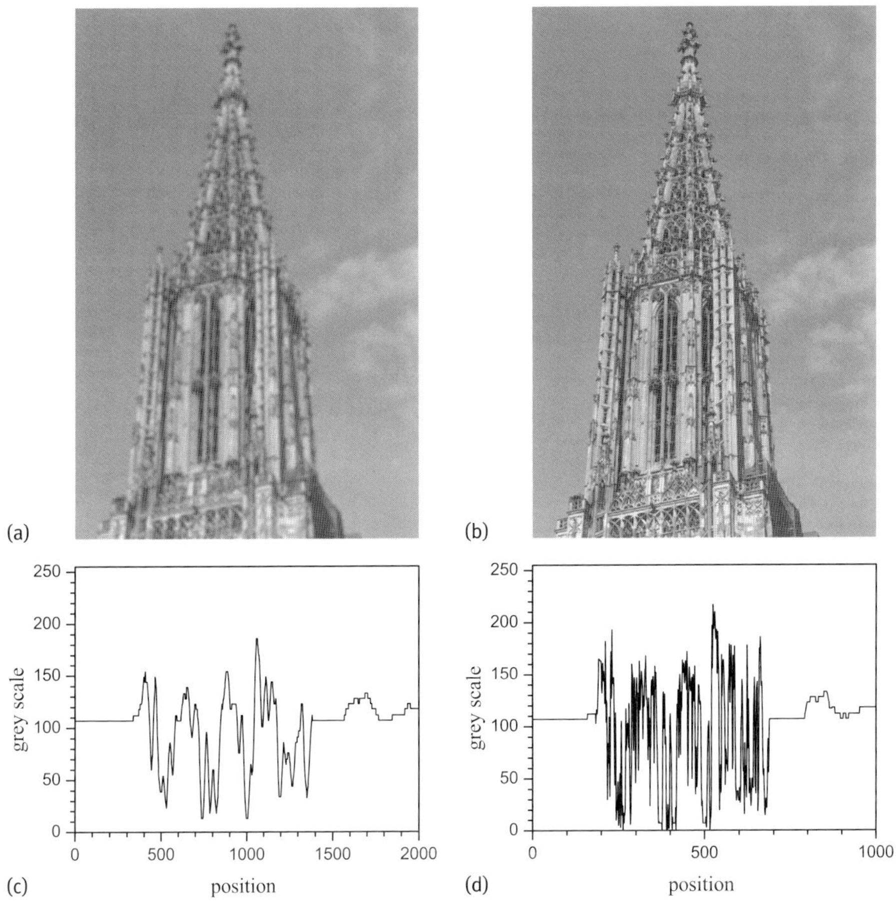

Fig. 1.24: Illustration of a typical situation for consumer cameras; more pixels do not necessarily lead to better image quality; details of the same image taken with two different cameras (see the text).

Example III

Now let us compare two cameras with different pixel sizes, namely $p = 2\,\mu\text{m}$ (camera A) and $p = 6\,\mu\text{m}$ (camera B) but both with lenses that do not lead to significant aberrations; in other words, both are nearly perfect. However, due to the nature of light, diffraction occurs. This becomes visible if the size of the diffraction pattern is too large, namely if it exceeds the size of 2 pixels. Thus, according to Equation (1.17) one may estimate that this may be observable when

$$1.22 \cdot \lambda \cdot \frac{f}{D} > 2 \cdot p \tag{1.26}$$

This situation occurs for $f/D > 6$ and $f/D > 18$ for camera A and B, respectively (here, for visible light, we assume a wavelength $\lambda \approx 550\,\text{nm}$). This also shows that even in the case of better optics for camera A than that used in example I), its $N_{SB,tot}$ may still be worse than that of camera B, which has a sensor with a larger pixel size. We may note as well that for real camera lenses usually for smaller f/D values aberrations become stronger (see Chapter 3). Again, this, in particular, affects the camera.

Finally, we will give some comments on the basis of the presented examples. The crazy drive for more megapixels, which is done by down scaling pixel size, does not necessarily lead to better camera system performance. In particular, optical quality factors such as SBN, or more general MTF, etc. (see Chapter 5) are not necessarily improved. The resolution of lenses poses a lower limit on useful pixel sizes (see also the discussion in Section 5.2). But an even more important aspect may be that smaller pixels severely reduce their sensitivity (see Chapter 4). Consequently, the camera, respectively, sensoring sensitivity and its noise, may be a more important issue than the total number of pixels.

1.6.5 Image observation from digital screens

In a previous section, we did estimate the reasonable maximum number of "image points" within an image, namely the SBN. We also discussed the SBN of sensors that record the image generated by the camera lens on its surface (input device). Now we would like to apply this knowledge to an output device, and as an example, deduce the potential of actual HDTV screens.

Current HDTV screens have $N_h = 1920$ pixels in the horizontal direction and $N_v = 1080$ pixels in the vertical direction, respectively. Thus, the aspect ratio is 16:9. If we take, for instance, a 40-inch screen, which means a screen with approximately 1 m diagonal, from Pythagoras's theorem we easily can calculate its width and height: $X = 87$ cm and $Y = 49$ cm, respectively. Thus, the width of one pixel is $D_{pix} = X/N_h = 454\,\mu m$, and the same value is obtained for its height. Again, we assume a resolution of 2 pixels. If we would like to stay at a distance L in such a way that we can resolve 2 pixels, then we get $L = 2 \cdot D_{pix}/\Delta\varphi_{eye} \approx 3\,m$.

Now the question arises: what is the width X' that corresponds to Ψ_{eye} at this distance? Due to $X'/2 = L \cdot \tan(\Psi_{eye}/2)$, we can estimate $X' \approx 2.6\,m$. Thus, in principle, we could see in the horizontal direction a scene three times larger than the screen width if we neglect that the resolution of the eye decreases at the borders (see Section 1.4.1). On the other hand, if the screen is located at a shorter distance, the observed fraction becomes larger, however, on the expense of image quality loss, because then the pixel structure of the screen becomes apparent.

Using the SBN, we could get even faster to this result. According to the 2-pixel resolution, we get SBN $= 1920/2 = 960$ for the full screen in horizontal direction, a value that is three times smaller than $N_{SB,eye}$. If we use such a screen to display images taken by usual digital still cameras, the situation becomes even worse. Such images usually have an aspect ratio of 3:2 (or 4:3). When an image in landscape format is displayed, one makes use of all the 1080 pixels in vertical direction, but because then only the $1 \cdot 3/2 = 1620$ pixels (1440 pixels, resp.) in the horizontal direction contribute, SBN $= 810$ (720) in the horizontal direction.

Consequently, if the screen should be adapted to the human eye, it must have approximately three times more pixels in its horizontal direction and the same factor in the vertical direction. Nonetheless, one should remark that this is not the typical situation how we regard pictures, because mostly pictures placed in a typical observation distance take only a fraction of our angle of view.

Finally, we would like to mention that for larger or smaller HDTV screens the situation is absolutely the same, because the number of pixels, and thus the SBN is identical to before and only all sizes scale. They all scale by the same factor, thus larger screens just require larger observation distances, but picture quality is then identical.

1.7 Optical glass

In the case of optical imaging in the visible spectral range, we mostly deal with optical lenses. Unlike in mirror optics the light traverses matter, and thus its propagation is influenced by the physical properties of the material. The lenses are usually made of glass, which is highly transparent in the visible range and is easily machinable. In the following, we give a short overview of the structure of commonly used glasses and some of their optical properties.

1.7.1 Structure of silica based glasses

Most people have a clear idea of glass from their daily use of items made of glass. From the physical and chemical point of view, glass is quite special because the term relates to a solid that has the structural properties of a liquid. This is due to the fact of its process of formation. In order to understand the amorphous glass structure, we will focus on the most important and commonly used type of glass, which is based on silicon dioxide SiO_2, also termed silica.

SiO_2 can be abundantly found in the crust of the earth in various forms, for instance, as quartz or as a major constituent of most sands. If silica is melted at very high temperatures around 2000 °C, its liquid phase contains parts of silicon, oxide and SiO_4-tetrahedrons, which are the basic building blocks of all solid configurations of silica. Silicon (Si), being a main group 4 element, has four outer electrons that each establish a covalent bond with a neighboring oxygen atom in the tetrahedron (Figures 1.25, 1.26). When the liquid melt is cooled down, different solid phases can establish depending on the thermodynamic conditions. When slowly cooling down, a crystalline solid phase is entered at a well-defined critical temperature. Figure 1.25c shows one possible example of a crystalline silica structure reflecting the strict geometry of the crystal. A long-range

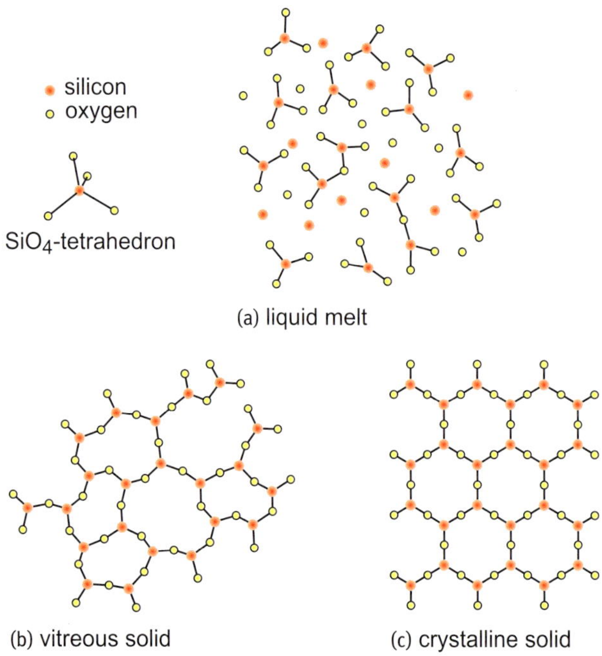

Fig. 1.25: Silicon dioxide in its different phases. (a) Liquid melt; (b) vitreous solid without long-range order; (c) crystalline solid with long-range order; in all phases only a planar representation of the network is given, which means that a fourth bond of each tetrahedron is oriented perpendicularly to the plane.

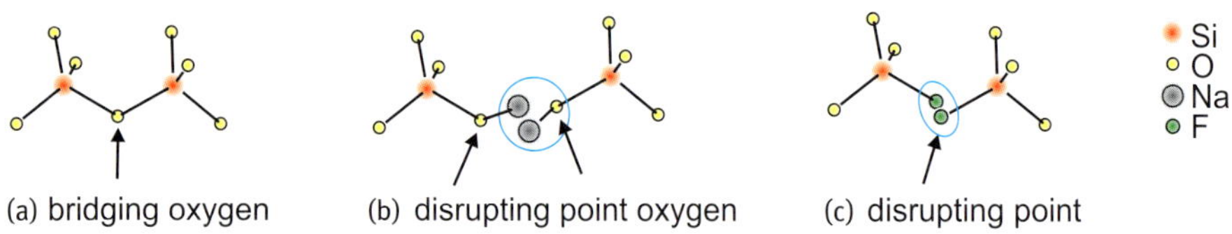

Fig. 1.26: Tetrahedron links in silica glass. (a) oxygen bridging in pure silica; (b) Na_2O network modifier creating a disrupting point; (c) F_2 creating a disrupting point.

order, which characterizes crystals, can be seen, and the distances and angles between nearest neighboring atoms are always identical throughout the whole crystal. The vitreous phase is formed when the melt is cooled down so rapidly that no equilibrium rearrangement of the atoms is possible. The liquid structure simply freezes and the viscosity increases with decreasing temperature. There is a transition range below the liquid phase down to about 1000 °C where the glass is not yet fully solidified, and a kind of glass flow is still possible. Below that transition temperature of about 1000 °C the viscosity of pure silica glass is high enough that it can be considered a real solid. In contrast to the crystallization process, there is no phase transition in the glass formation process and glass can be considered even at ambient temperature a frozen solid.

In the amorphous structure, all SiO_4-tetrahedrons form an irregular network without long-range correlation. That means that the distances between two nearest neighboring atoms is nearly constant whereas the distances and angles beyond it can no longer be predicted, unlike in crystals. It should be noted that the networks in Figure 1.25 illustrate the planar projection of a 3D structure and the fourth bond of a tetrahedron oriented perpendicularly to that plane is not shown. The oxygen linking two silicon atoms in these networks can be considered as bridging oxygens whereas the positive Si^{4+}-ions are termed network formers. There are also other possible ions that can form networks in combination with oxygen atoms. Such networks are based on oxygens like RO_2, R_2O_3 and R_2O_5 where R designates this positive ion. Possible examples are glasses of As_2O_3, B_2O_3, GeO_2 and P_2O_5. All these oxides can form glasses by themselves but can also be mixed to yield multicomponent glasses.

Besides these network formers there are also network modifiers that usually do not form glass networks by themselves but modify the structure and change physical properties like refractive index, mechanical properties and especially the melting point. Examples for network modifiers are the oxides of alkali metals and alkaline earth metals. Figure 1.26b illustrates the characteristics of Na_2O in the glass network. It is in between two SiO_4-tetrahedrons, which in this case are no longer linked by one bridging atom. If the temperature is increased the network disrupts at the Na_2O locations at much lower temperatures than in pure silica and reduces the viscosity. In soda-lime-glass mixtures, the melting point is reduced by around 1000 °C as compared to pure silica glass, thus making them well appropriate for manufacturing processes. Another property of the oxygens at the disrupting point is their higher polarizability compared to bridging oxygens. In general, polarizability is the reaction of matter to the application of an electrical field. The field displaces the electrical charges, thus a stronger response to electrical fields is expected for atoms with higher electronic density and better displacement of charges.

Doping glasses with appropriate components increases or reduces the polarizability, and consequently the refractive index of glass. Pure silica glass has a relatively low refractive index of $n = 1.46$ at the wavelength $\lambda = 600$ nm. This can be reduced furthermore by doping silica with boron trioxide or fluorine (Figure 1.27). The fluorine molecule F_2 substitutes the bridging oxygen and establishes a nonoxidic disruption point (Figure 1.26c). The disruption point oxygens of other components increase the refractive index. Examples for the index increase by doping silica glass with P_2O_5, GeO_2, Al_2O_3, TiO and ZrO_2 are given in Figure 1.27. Doping multicomponent glasses with heavier atom oxides like PbO and BaO leads to higher refractive indices but also strongly influences the dispersion characteristic of glasses. The lowest absorption of light is found in pure silica glass at $\lambda = 1.55$ µm with a value of 4.5 % power loss per km. This is a very low value compared to plastic materials for optical lenses, which are of the order of some percent loss per mm. Doping silica generally increases the absorption, which must be taken into consideration for the application.

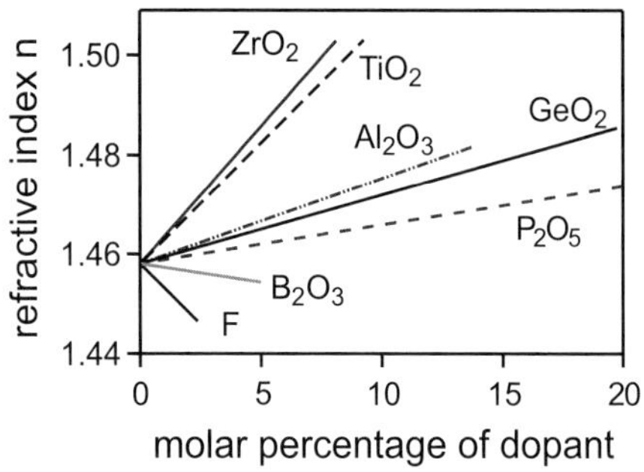

Fig. 1.27: Doping of pure silica glass by various dopants results in a refractive index change at $\lambda = 600$ nm.[7]

1.7.2 Optical dispersion in glasses

As already mentioned above, the refractive index is related to the response of matter to the electromagnetic fields of light traveling across the material. Light has a frequency of $0.5 \cdot 10^{15}$ Hz at $\lambda = 600$ nm. Consequently, all charges in matter vibrate at the same frequency, which especially affects the polarization and reorientation of molecular groups and electrons. If the frequency varies, we can see a typical resonance phenomenon, which means an increase of the response when we approach the resonance frequency from the lower frequency part. In transparent glasses, the typical resonance frequencies are above approximately 10^{15} Hz, which is the UV range below 300 nm. That implies that the refractive index is expected to increase if we approach the blue visible range from longer wavelengths in the red range. This can be seen in Figure 1.28 where pure silica glass has the lowest index increasing continuously with decreasing wavelength. The power absorption also increases when approaching the resonance. Very pure silica has the lowest power absorption and can still be used around 200 nm where the resonance frequency is not yet achieved.[8] This is virtually the shortest wavelength where pure silica lenses can be used. The quality of these lenses, however, degrades dramatically with the time of exposure in this UV range. The application of other types of glasses is recommended for spectral ranges above 250 nm.

All glasses in Figure 1.28 show a similar characteristic increase of n with decreasing wavelength, which is called normal dispersion. Changing the composition of glass and doping with special elements changes the refractive index, the resonance frequencies, and thus the overall dispersion curve. The dispersion of light in glass is exploited for instance in spectrometers to analyze the spectral components of light. Here, glass prisms are used that have different refractions at different wavelengths, and thus lead to a separation of light's spectral components. The same effect, however, is very disturbing for optical imaging of white light using simple lenses. Due to dispersion, we get multiple

7 H.-G. Unger: *Optische Nachrichtentechnik*, Hüthig Buch Verlag, Heidelberg 1990.

8 H. Scholze: *Glas*, Springer-Verlag, Berlin, Heidelberg, New York, 1977.

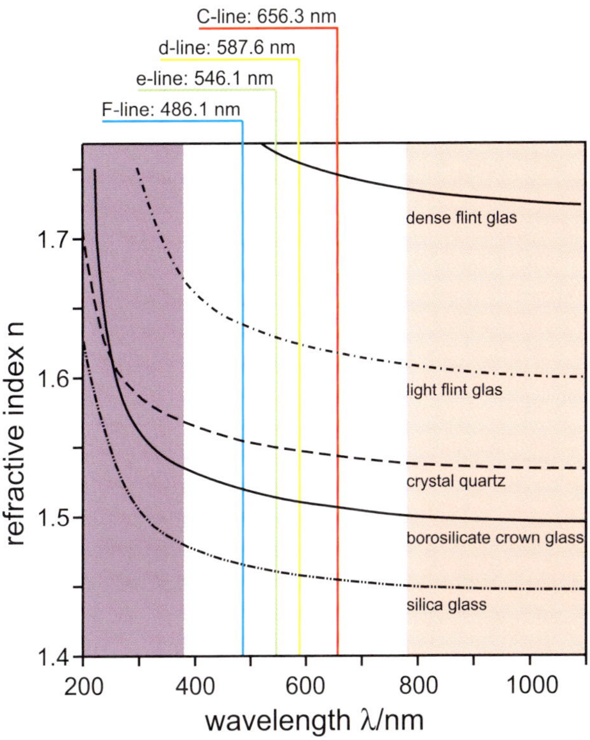

Fig. 1.28: Dispersion curves of different glass types (adapted after Hecht [Hec16]).

images of different colors and sizes as discussed with chromatic lens aberrations in Section 3.5.6. In all these cases, it is very important to characterize the dispersion characteristics of glass. The visible spectral range for the human eye is from about 380 nm up to 780 nm, which is indicated in Figure 1.28 by the unshaded area. The most sensitive range for daylight vision is around 550 nm (see Section 1.2.6) with a strong decrease of sensitivity in the red range above 640 nm and the blue-violet range below 480 nm. For many applications, it is sufficient to describe the dispersion from the blue to the red by a linear approximation. For more precise calculations, polynomial expressions are used.

In order to standardize the description of dispersion some special wavelengths are helpful. There are first of all the Fraunhofer spectral lines, the red C-line and blue F-line of hydrogen, the red C′-line and blue F′-line of cadmium as well as the yellow d-line of helium that are commonly taken as references. The green color in the middle range of the visible spectrum, where the eye is most sensible, can be conveniently represented by the spectral e-line of a mercury lamp. Their wavelengths are listed in Table 1.3 and some of their positions are indicated in Figure 1.28. The refractive index of glass at any of these spectroscopic lines is designated by a subscript, for instance, n_d, n_C, n_F, and n_e, $n_{\mathrm{C}'}$, $n_{\mathrm{F}'}$, and specified by the glass manufacturers in their specification sheets and

Tab. 1.3: Several spectroscopic lines (based on the Schott Optical Glass 2018 – Catalog[9]).

designation	wavelength	element
red Fraunhofer C-line	656.2725 nm	H
red Fraunhofer C′-line	643.8469 nm	Cd
yellow Fraunhofer d-line (also D_3)	587.5618 nm	He
green mercury e-line	546.0740 nm	Hg
blue Fraunhofer F-line	486.1327 nm	H
blue Fraunhofer F′-line	479.9914 nm	Cd

catalogs. It can be seen in Figure 1.28 that for normal dispersion n_F is always larger than n_C. Their difference $(n_\mathrm{F} - n_\mathrm{C})$ is termed principal dispersion and can be used as a measure for dispersion in the visible range. Furthermore, the refractive power of a thin lens in air is directly proportional to $(n_\mathrm{L} - 1)$, which is the difference of its refractive index relative to air as given by Equation (3.27). It is convenient to choose the index of the glass material in the middle of the spectral range at the green or yellow line. Thus, a combination of $(n_\mathrm{d} - 1)$ and $(n_\mathrm{F} - n_\mathrm{C})$ is appropriate for classifying glasses with respect to their suitability as lens material. This leads to the definition of the Abbe number, which exists in two commonly used versions, ν_d and ν_e. Historically the Abbe number ν_d was defined using the Fraunhofer d-, F- and C-lines whereas a newer version ν_e based on the spectral e-, F′ and D′-lines has been established for specifying components of optical systems:

$$\nu_\mathrm{d} = \frac{n_\mathrm{d} - 1}{n_\mathrm{F} - n_\mathrm{C}} \qquad \nu_\mathrm{e} = \frac{n_\mathrm{e} - 1}{n_{\mathrm{F}'} - n_{\mathrm{C}'}}. \tag{1.27}$$

The reciprocal value $1/\nu_\mathrm{d}$ and, respectively, $1/\nu_\mathrm{e}$ is also known as the dispersive power of glass.

Glasses of low Abbe numbers thus have high dispersion whereas low dispersion is related to having a high Abbe number like fused silica. Traditionally, glasses of high dispersion and high refractive index have been called flint glasses as some of them were historically produced by basic materials like flint containing minerals leading to these optical properties. On the other hand, crown glasses in general have a relatively low index and low dispersion, similar to fused silica. In contrast to pure silica, they have additional components that reduce the melting point and improve their chemical and physical properties so that they are better suited for industrial manufacturing at lower efforts. A typical crown glass is borosilicate, which has gained high technical importance for various applications. An overview of available glasses and their classification according to Abbe number and refractive index is given in Figure 1.29. In this scheme of

9 *Schott Optical Glass 2018 – Catalog*; http://www.us.schott.com/d/advanced_optics/ade6e884-76b0-4930-8166-f6e605e4ca10/1.5/schott-optical-glass-pocket-catalog-february-2018-us.pdf, (visited March 2018).

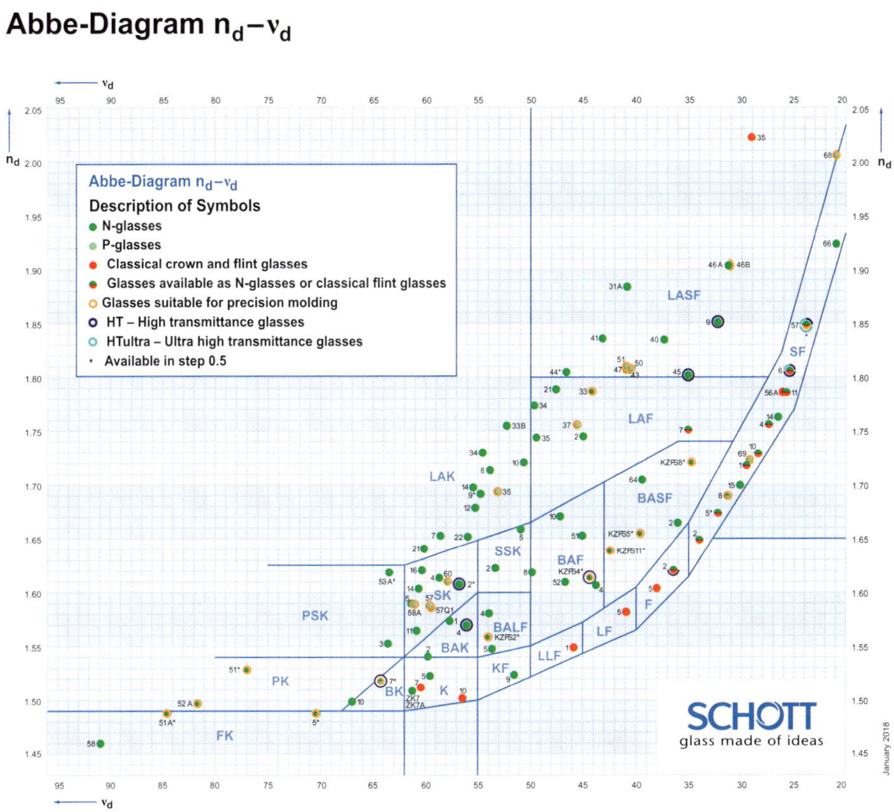

Fig. 1.29: Abbe diagram of several glasses, crown glasses (left) and flint glasses (right); with kind permission of Schott AG.

Schott, we can see that all glasses with an Abbe number $\nu_d < 50$ are flint glasses and all glasses with $\nu_d > 55$ are crown glass types. In the intermediate range, there are crown and flint glasses that can be distinguished with respect to their refractive index.

Based on the Abbe number ν_d and the refractive index n_d, an international glass code was established to identify glasses independently from the manufacturer. The code consists of six digits. The first three digits represent the rounded integer number of $100 \cdot (n_d - 1)$, the second three digits represent the rounded integer number of $10 \cdot \nu_d$. Several examples are given in Table 1.4 for glass materials from Schott as compared to pure fused silica. Some manufacturers modify or complement the code for additional information like density of mass. In general, glass manufacturers also supply their specifications for the spectral e-line for applications where the features for optical systems are required.

Tab. 1.4: Optical data of some glasses (after Pedrotti [Ped17] and Schott Optical Glass 2018 – Catalog).[10]

type of glass	glass code (international)	n_d	ν_d	ν_e	$n_F - n_C$
fused silica	458678	1.45846	67.82		0.0068
borosilicate crown (N-BK7)	517642	1.51680	64.17	63.96	0.008054
crown (N-K5)	522595	1.52249	59.48	59.22	0.008784
dense crown (N-SK15)	623580	1.62296	58.02	57.75	0.010737
barium light flint (N-BALF5)	547536	1.54739	53.63	53.36	0.010207
very dense crown (N-SSK2)	622533	1.62229	53.27	52.99	0.011681
flint (F2)	620364	1.62004	36.37	36.11	0.017050
dense flint (SF6)	805254	1.80518	25.43	25.24	0.031660

1.8 Metamaterials, metasurfaces and metalenses

The traditional methods of imaging are based on optical lenses where the refractive index for light is virtually homogeneous in the lens. The refractive index can be adjusted by appropriate doping of the base material, usually a silicate glass, as described in the previous section. Moreover, the typical types of glasses have a positive refractive index, are nonmagnetic and nonconductive, i. e., isolators in order to minimize absorption losses and to maintain a high transparency.

With the progress of micro and nanostructuring methods in the recent decades, new possibilities have shown up which allow to influence the refractive index of materials locally. A high importance is given to very thin metallic layers of subwavelength dimensions, which are deposited in a dielectric matrix or at the interface between two dielectric compounds. Due to the high conductivity, electromagnetic waves cannot be guided in metals, but in areas beneath the surface in dielectric materials. In the case of microwaves, this property is known for decades and used for guiding waves in rectangular or cylindrical waveguides with mm dimensions.[11] In integrated optics, strip-loaded waveguides of µm dimensions exploit this phenomenon for light propagation. Here, the effective refractive index is lowered below metallic stripes and light is guided in the dielectric between these sections,[12] [Sal19]. A very interesting fact is that by a periodic arrangement of very small metallic elements, i. e., of subwavelength dimensions, like rods, balls, rings, etc., in a homogeneous dielectric matrix, the refractive index can be tuned to a very broad range. This implies a variation of the electric permittivity as well as of the magnetic permeability in such a way that they can attain positive and negative

10 *Schott Optical Glass 2018 – Catalog*; http://www.us.schott.com/d/advanced_optics/ade6e884-76b0-4930-8166-f6e605e4ca10/1.5/schott-optical-glass-pocket-catalog-february-2018-us.pdf, (visited March 2018).

11 R. E. Collin: *Foundations for Microwave Engineering*, McGraw-Hill Kogakusha, Ltd., Tokyo 1966.

12 R. G. Hunsperger: *Integrated Optics*, 5th edition, Springer-Verlag Berlin Heidelberg New York, 2002.

real and imaginary parts, when described by complex quantities. A good overview of such modifications with examples is given in the textbook by Saleh and Teich [Sal19]. Unlike "normal" nonmagnetic glasses with positive refractive indexes, this variation gives rise to multiple optical applications within a relatively compact area which cannot be achieved by conventional optics. The disadvantage of such 3D metamaterials, however, is their technical realization. It is quite complicated to implement metallic elements of subwavelength dimensions for visible light as submicron resolution is required over a larger volume. A much simpler way to achieve the desired optical functionality of metamaterials is to restrict the fabrication to 2D metasurfaces where well-known technologies from planar structuring of electronic circuits can be applied. To that purpose, we will mainly focus our consideration to 2D meta structures.

Metasurfaces

Under metasurfaces, we understand metamaterials where only the interface between dielectrics is structured, and thus have a nearly vanishing thickness. Usually, the surface consists of an arrangement of metallic elements of sub-wavelength lateral dimensions having a thickness of some tens of nanometers. Conversely, thin metallic surfaces on a dielectric substrate, which are patterned in a subsequent step, yielding voids or some other structures, filled with dielectric material or simply air, can be understood as complementary metasurfaces. More advanced metasurfaces avoid metallic elements due to their relatively high absorption losses and substitute them by dielectric elements like pillars or holey dielectrics.[13] While metasurfaces with metallic elements can be very thin, significantly below 1 µm, dielectric surface elements require large aspect ratios resulting in surface thicknesses of some µm.

In order to simplify the understanding of the complex domain of metasurfaces, let us have a look to dielectrics covered with metal elements. Let us assume rectangular metal stripes which are regularly arranged on a dielectric substrate with a small distance between them. If light hits the surface in a perpendicular way, then in the areas which are not covered by metal, light is reflected and transmitted according to Snells law (Section 3.1). In the remaining parts covered by metal stripes the free electrons are nearly instantaneously displaced, leading to a plasma-like excitation (plasmons). As a consequence, electromagnetic waves, radiated by the accelerated electrons with a given amplitude and phase shift, superimpose with those traveling through the uncovered sections. In this case, the metallic elements act as tiny antennas. If the geometrical dimensions of the metal stripes as well as their separations are below the wavelength of light, then the resulting wavefront of light after the transit through the surface may be

---

13 S. W. D. Lim, M. L. Meretska, F. Capasso: *A High Aspect Ratio Inverse-Designed Holey Metalens*, Nano Letters **21**(20) (2021) 8642–8649.

locally modified and no longer homogeneous like in bulk materials. Moreover, if the metal stripes and their arrangement do not have a rotational symmetry, the wavefront modification depends on the polarization of the incident light. This results in the effect that by an appropriate choice of the geometry, size and orientation of the metallic antennas nearly any arbitrary wavefront, including polarization, can be reconstituted after traversing the metasurface. And indeed this can be done with subwavelength resolution. This is not only true for metallic antennas but in a more general way for subwavelength-spaced scatterers, may they be metallic or dielectric. As a general result for metasurfaces or complementary metasurfaces, we can state that they can be designed to fabricate optical lenses, namely metalenses, and systems with features that go beyond those of conventional and diffractive optical elements.[14] One of the current drawbacks for the exploitation of metalenses on a large scale is the mass fabrication which always requires submicron resolution. Further detailed examples for metasurfaces and metalenses are given in Section 7.5.

14 J. Hu, S. Bandyopadhyay, Y. Liu, L. Shao: *A Review on Metasurface: From Principle to Smart Metadevices*, Front. Phys. 8:586087 (2021).

2 Basic concepts of photography and still cameras

In order to understand the complexity and details of modern technical systems, it is always helpful to have a look at the historical evolution of them. The purpose of the present book is the consideration of optical systems. Typical representatives for them are still cameras. In the following, we give a short survey of them as well as some aspects of how they evolved with time. As for the term still cameras, we only focus on the fact of capturing a photograph or an image in general, and leave all aspects of movie cameras out of consideration. However, it should be noted that from a historical point of view, the driving force behind the development of still cameras and their lenses at the beginning of the 20th century was the emerging cinema market. The standard film gauge for motion picture production at that time became the 35 mm format developed by Eastman Kodak. During movie production still cameras for taking photos of scenes were needed. This led to the development of 35 mm cameras where the first commercially successful type was produced by Leica. In the following, we will give some examples of cameras with different sensors and formats. The 35 mm format, however, is of special importance and will always be the reference for comparison. The basic principle for all cases that we consider is the imaging of an object in the 3D-object space to the 2D-image plane.

2.1 Pinhole camera

The simplest type of camera is the pinhole camera. Its principle has been known since the ancient Greek world, and the "camera obscura" was used often in the Middle Ages to draw pictures of objects in the real world (Figure 2.1a). They were available in various sizes, even large enough for a painter to be inside and redraw images of projected scenery. Its name "camera obscura" is the origin of the term camera, which we use for modern imaging systems.

The principle of the pinhole camera is illustrated in Figure 2.2. It consists of a closed box without lens but with a small hole on the front side towards the object. At the rear side in the image plane, there is a photosensitive film or detector, or just a screen. All

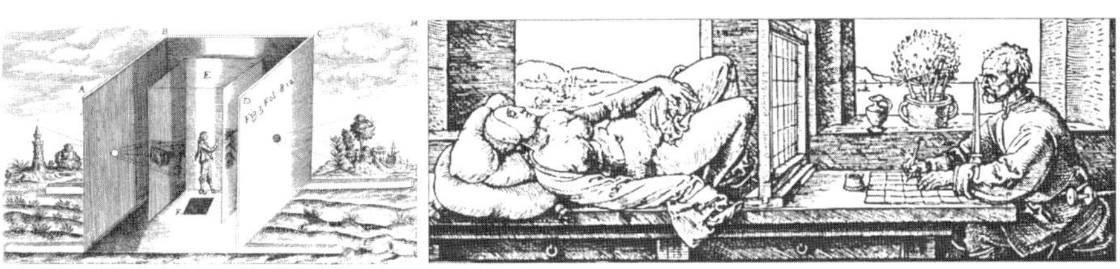

(a) (b)

Fig. 2.1: (a) Large size "camera obscura" used for painting applications, by Athanasius Kircher, 1645; (b) application of perspective drawing, woodcarving by Albrecht Dürer, approx. 1527 [Hön09] (reprinted with kind permission of Carl Zeiss AG).

https://doi.org/10.1515/9783110789966-002

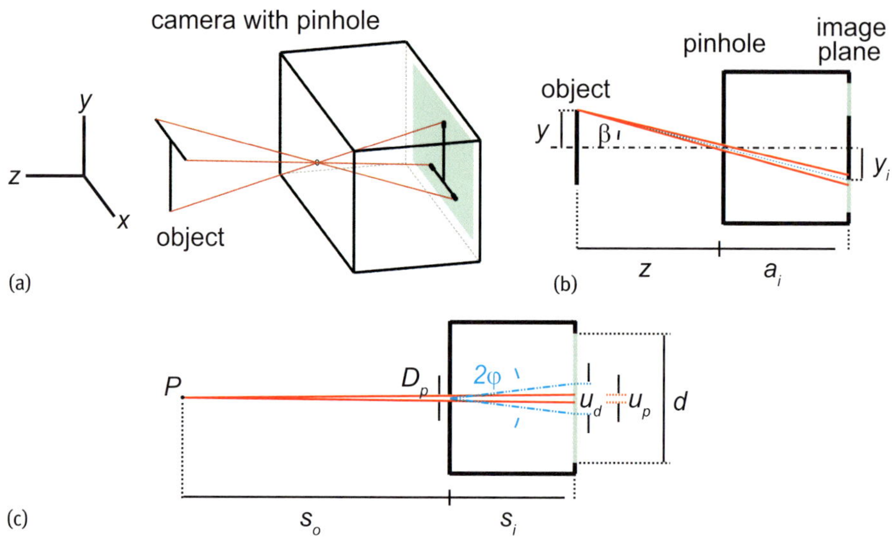

Fig. 2.2: Pinhole camera. (a) Schematic setup. The image is inverted and blurred due to the size of the pinhole; (b) projection characteristics; (c) blur due to diffraction and projection.

object rays pass through the pinhole and are projected in a straight line to the image plane, which can be arbitrarily chosen. An object point at the position x, y, z in the object space is imaged to the coordinates x_i, y_i in the image plane at a distance a_i from the entrance according to the relation:

$$x_i = -x \cdot \frac{a_i}{z} \qquad y_i = -y \cdot \frac{a_i}{z} \qquad \frac{y}{z} = \frac{y_i}{a_i} = \tan\beta \quad \Rightarrow \quad y_i \propto \tan\beta \tag{2.1}$$

This type of projection is termed a gnomonic or central projection where a 3D-space object is imaged through the center of projection to a 2D-space. The projected image is rotated by 180° with respect to the object and scaled by a factor of a_i/z. In general, the distance z is much larger than the image distance a_i, which means that we get a downsizing. Objects having straight lines will produce images that also have straight lines. That means the projected image is without distortion and we can qualify it as nearly ideal (Figure 2.3). As the imaging is from a 3D to a 2D space, information about the depth of an object is lost. As a consequence, object points at different distances from the camera located on the same ray across the pinhole will be imaged on the same point in the image plane and can no longer be distinguished.

The same type of projection is also applied when 2D images are sketched by an artist, as illustrated in Figure 2.1b. The center of projection in this case is the tip of the rod across which the artist locates the object point and its position in the image frame. The difference from the pinhole camera is that here the object and image spaces are both on the same side relative to the center of projection whereas in the pinhole camera

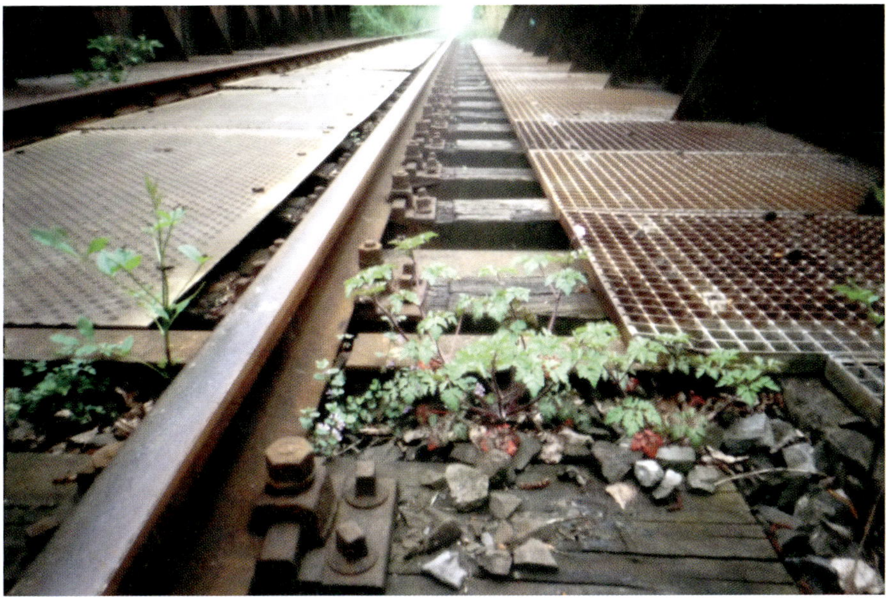

Fig. 2.3: Pinhole camera photo of a disused railway (author: Joachim K., exposure time 3 minutes[1]).

the center of projection is in between image and object space, and thus separates both spaces. The center of projection in a camera with a lens is the entrance pupil, which is described in more detail in Section 3.4.

As for the sharpness of the image in the pinhole camera, we have to take into consideration two different aspects. First of all„ due to the finite aperture D_p of the pinhole there can be more than one ray traced from a starting point P in the object space to the image plane. Consequently, this is not an unambiguous point-to-point imaging process and implies that we get a blurred spot on the image plane with a diameter u_p (see Figure 2.2c). We name this effect projection blur. If we designate the absolute values of the object and image distances by s_o, respectively, s_i, then, due to the projection characteristics, the ratios D_p/s_o and $u_p/(s_o + s_i)$ are identical and yield

$$u_p = D_p \cdot (1 + s_i/s_o) \tag{2.2}$$

The projection blur can be reduced by reducing the hole diameter D_p and also having a shorter image distance s_i. If the object is at a large distance s_o from the camera with $s_o \gg s_i$, then the projection blur u_p can be assumed to be identical to the hole diameter, thus $u_p \approx D_p$. If d is the linear extension of the sensor in the image plane, which in general is the diagonal of a rectangular format, the relative blur u_p/d is nearly independent

1 https://commons.wikimedia.org/wiki/File%3ARheda-Wiedenbr%C3%BCck%2C_stillgelegte_Eisenbahnbr%C3%BCcke%2C_Lochkamera.jpg

from the image distance and equal to D_p/d. Here and in the following, the consideration also includes a film as a sensor.

On the other hand, reducing D_p has the effect that less light enters the camera, with the image becoming darker so that longer exposure times are needed for image acquisition. What is more critical, however, is how the reduced D_p leads to an increased diffraction of light at the pinhole as described in Section 1.5. According to the Rayleigh criterion, an infinitely small light source is imaged as a diffuse Airy disk due to the limited aperture D_p (see Chapter 5). Its resulting blur diameter u_d in the image plane is given by

$$u_d = 2.44 \cdot \frac{\lambda}{D_p} \cdot s_i \tag{2.3}$$

The diffraction blur increases with the image distance s_i whereas u_p is nearly independent from it. As for the size of the pinhole, diffraction blur and projection blur act in opposite directions: Reducing the pinhole minimizes the projection blur, while widening up the hole minimizes the diffraction. The general approach for obtaining the optimum pinhole size is a mathematical procedure for minimizing the total blur. In a simpler consideration, the same result is achieved if both blur diameters are identical and nearly equal to the aperture D_p. Thus equaling Equation (2.2) with (2.3) and rearranging the equation, we get for the optimum pinhole size:

$$D_p = \sqrt{2.44 \cdot \lambda \cdot \frac{s_i \cdot s_o}{s_i + s_o}} \approx \sqrt{2.44 \cdot \lambda \cdot s_i}. \tag{2.4}$$

Now the pinhole camera is optimized for the pinhole size, but we have not yet considered if the resulting blur leads to an acceptable image sharpness. Therefore, the blur always has to be assessed in relation to the total image size. This is similar to the discussion in Section 1.3.3 where the blur takes the part of the spot size; for instance, it is equal to δx in Equation (1.12) and the image width PW is given by the image format diameter d.

As a reasonable criterion for sharpness, we chose the limiting case where the blur can no longer be perceived by the human eye if an image is regarded from a distance of its format diameter. Assuming in our consideration that d is the image format diameter then the relative blur for large object distances is approximately D_p/d and should be less than the angular resolution of the eye with 0.6 mrad. As discussed in Section 1.5.4, a normal image quality is achieved if the circle of confusion, which in this case is nearly identical to the pinhole diameter, is approximately $d/1500$. As a consequence, for a pinhole camera large image formats are favorable as the hole size is typically of the order of 100 μm, and thus for good quality, the image diagonal should be of the order of 15 cm.

The great advantage of the pinhole camera is that all objects independently from the distance to the camera are imaged without any distortions like those caused by a lens, and with nearly the same sharpness. The latter one is often expressed by qualifying the pinhole camera to have an infinitely large depth of field. A consideration of the depth of field for lenses is given in Sections 3.4.6 and 6.9.2. The biggest disadvantage of the camera is the low irradiance in the sensor plane due to the small pinhole size. This leads to long exposure times and makes the pinhole camera in general unsuitable for moving objects, except for special applications where long exposure times are required.

In spite of this disadvantage, its simplicity and the fact that it is applicable to any wavelength, makes the pinhole camera today still a very suitable instrument in science. One example of a special application is the imaging of laser-produced plasmas, for instance within laser fusion research. Here, the image is captured within the soft x-ray range where imaging in general is a very difficult task and lenses are namely not applicable. Thus, although very simple, for such kind of investigations the pinhole camera is still first choice as one of the standard optics, even in horribly expensive high-tech experiments.

Example: Pinhole camera design

A pinhole camera can be setup using a digital single lens reflex (DSLR) camera body (Figure 2.4a). The problem then is, what should be the optimum pinhole size and what is the image quality with respect to sharpness? A typical image distance between the lens mounting flange of the camera body to the focal plane of the sensor is about 45 mm; the diagonal of the 35 mm format sensor is $d = 43$ mm. If a pinhole is positioned at the mounting flange then its optimum diameter for the image distance $s_i = 45$ mm and a center wavelength of $\lambda = 0.55$ µm is calculated according to Equation (2.4) to yield $D_p = 246$ µm. This is also the size of the blur diameter, and thus the relative blur $D_p/d = 0.0057$, being equivalent to a space bandwidth number of $N_{SB} = 175$, is about eight times larger than the angular resolution of the eye. Figure 2.4b shows the image taken with an optimized pinhole aperture of 250 µm yielding a reduced image size by a factor of about 0.015 as compared to the original star target (Figure 2.4c). The blur is clearly visible and becomes more distinct with different pinhole diameters.

Choosing a larger image sensor format results in a reduced relative blur. In the case of an A4 format (210 mm × 297 mm) with a diagonal of $d = 364$ mm and an image distance

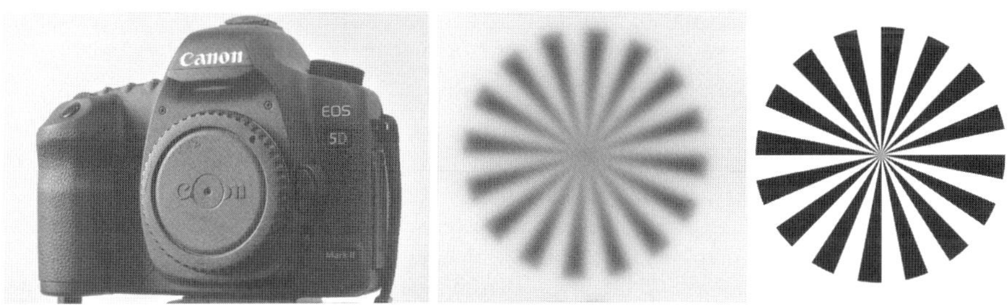

(a) (b) (c)

Fig. 2.4: Pinhole camera setup using a Digital Single Reflex camera. (a) Camera body with mounted pinhole; (b) image produced by a 250 µm pinhole size produced by excimer laser ablation on a layered substrate; (c) original star target.

being the same as the diagonal in order to maintain a natural viewing perspective when regarding the image, the optimum pinhole size is $D_p = 0.70$ mm. The relative blur now becomes $D_p/d = 0.0019$, corresponding to $N_{SB} = 526$, and is drastically reduced by a factor of three, compared with the example of the DSLR camera above. Regarding the image from a distance of about 1 m gives the impression of a sharp image as the blur is just at the resolution limit of the eye. This example, however, can only be realized using photographic film material since electronic image sensors of that size are quite expensive and not commercially available at present.

2.2 Camera with a lens

The camera with a lens can be considered a further development of the pinhole camera after glass lenses as well as photosensitive materials became available. The first cameras can be dated back to the beginning of the 19th century in Europe. The schematic drawing of a lightproof camera with a converging lens as the only entry point for light is given in Figure 2.5. The camera lens projects a real, inverted image in the 2D sensor plane if the object is at a distance a_o from a lens that is larger than the focal length of the lens. It should be noted here that we have to make a distinction between distances and lens parameters on the object side labeled by the subscript o and those on the image side labeled by the subscript i. The reason is that the quantities are measured from the lens position pointing into opposite direction thus having positive and respectively, negative values. In the case of a simple converging lens like our example in the figure, its image focal length f_i is a positive value, its focal length f_o is a negative value and both have the same magnitude f. In the present chapter, we mostly just use the magnitude f, whereas in the following chapters on more complex lens systems and calculation methods we have to be stricter with signs and directions.

In a pinhole camera, we have the ideal case of an infinitely large depth of field, and only one path of ray can be traced from any object point to its associated image point.

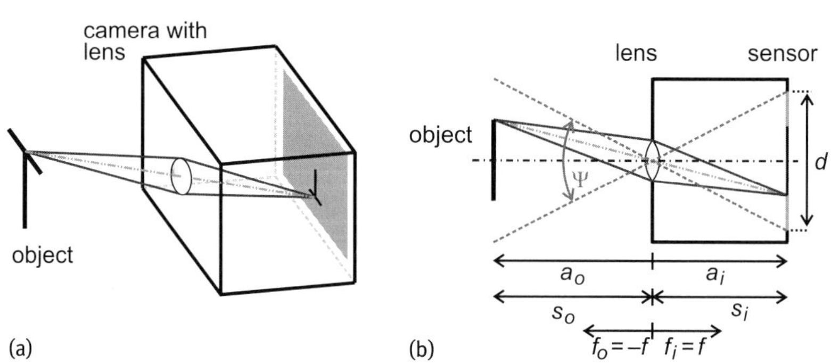

Fig. 2.5: Camera with a lens. (a) Schematic setup with inverted real image; (b) projection characteristics with path of rays.

In contrast to that situation, using a camera with a lens, a perspective and sharp imaging of an object at distance a_o from the lens occurs only at a given image distance a_i from the lens. We get a limited range of field in the object space, which means that only the object at a_o and objects within a limited range around it are sharply reproduced on the sensor at a_i (Figure 2.5b). This is due to the fact that many light rays emerging from an object point and having different light paths across the larger lens aperture are imaged to the associated image point. As a consequence of using cameras with lenses, there are brighter and sharper images possible at shorter exposure times than with pinhole cameras. The disadvantages, however, are a limited range of field for objects to be imaged, impairments of the lens quality and a necessary adjustment of the lens position to achieve the optimum image distance a_i.

The imaging properties in a simplified description are given by the thin lens formula that expresses the relationship between the focal length of the lens and the corresponding object and image distances:

$$\frac{1}{a_i} - \frac{1}{a_o} = \frac{1}{f_i} \tag{2.5a}$$

$$\frac{1}{s_i} + \frac{1}{s_o} = \frac{1}{f}. \tag{2.5b}$$

Both equations are valid for thin lenses where the thickness of the lens is neglected and all distances relative to the lens are defined respectively from the center of the lens. The more general formula is given by Equation (2.5a) where the signs for the orientation of the distances must be taken into account as mentioned above. As for our examples above the distance of the object a_o relative to the lens center is counted as negative. For photographic applications, we usually argue only with positive values. Then a more practical version is the photographic lens formula (2.5b) where s_i and s_o are the absolute values of the image, and respectively object distances, and f the absolute value of the focal length.

In the case of real images, the sizes of image and object are directly related to each other via their corresponding distances to the lens. Hence, we can define a linear or lateral magnification M for the imaging:

$$M = \frac{S_i}{S_o} = \frac{a_i}{a_o} \qquad |M| = \frac{s_i}{s_o}. \tag{2.6}$$

Here, S_i, and respectively, S_o, are the linear dimensions of image and object. A negative value means that the image is inverted as compared to the object. The magnification can also be expressed as a function of the focal length and the object distance if we substitute the image distance by rearranging Equations (2.5a) and (2.5b). We get

$$a_i = \frac{a_o \cdot f_i}{a_o + f_i} \qquad s_i = \frac{s_o \cdot f}{s_o - f} \tag{2.7}$$

and then Equation (2.6) can be rewritten as

$$M = \frac{a_i}{a_o} = \frac{f_i}{a_o + f_i} \qquad |M| = \frac{s_i}{s_o} = \frac{f}{s_o - f}. \tag{2.8}$$

This equation directly shows that a real transversal magnification is only possible if the object is at a larger distance than the focal length of the optical system and that a singularity exists for $s_o = f$. With respect to Equations (2.5b) and (2.8), we now will discriminate the following situations for taking images:

a) Images taken of objects far away from the lens: $s_o \to \infty$
For practical applications, this means that the object is at a distance s_o that is at least one order of magnitude larger than the image distance s_i, respectively, f, or $s_o \gg f, s_i$. In this case, the image is located near the focal plane of the lens with $s_i \approx f$. The difference $s_o - f$ in Equation (2.8) is approximately equal to s_o, and thus $|M| \approx f/s_o$. We get a reduced, inverted image with $|M| \ll 1$, and its size is proportional to the focal length. There is a strong implication of this proportionality for the image composition of a photographer: If we use a longer focal length, we get a higher magnification and conversely a smaller image is achieved for a shorter focal length.

b) Photographic situation: $\infty > s_o > 2f$
If the object distance is larger than twice the focal length of the lens, we get a reduced, inverted image of the object in the image plane of the camera. The image distance s_i increases when the object comes closer to the lens. If the sensor is fixed in the camera and, like in a), located nearly in the focal plane, then the position of the lens relative to the image plane must be adjusted. This is generally termed focusing and means that optimum image sharpness is achieved if the image is at the distance s_i from the center of the lens. Focusing in that case should not be confused with the situation that the image position coincides with the focal point like in a) above (see also the remark in Section 1.5.3). Objects at different distances from the lens will be rendered sharp at different positions of the lens relative to the sensor. If the position is fixed, only objects at a given distance are imaged sharp while others appear blurred. Thus, focusing can be used as a method to selectively image a given range in the object space. This range is called depth of field and will be discussed in more detail in Section 3.4.6. For the standard photographic situation, we usually have $|M| < 0.1$, which means that the object is at a distance of more than 11 times the focal length from the lens. Placing objects closer to the lens not only leads to larger image distances but also to larger image sizes. Here, we find the domain of close-up imaging where the limiting case of macrophotography is achieved when object and image distances are nearly identical. We then get $s_o = s_i = 2f$, and the image has the same size as the object, however inverted, thus resulting in a 1:1 imaging with $M = -1$.

c) Extreme close-up photography: $2f > s_o > f$
If the object distance is closer than twice the focal length, the image size in the sensor plane is life size or greater. For this type of photography, special setups are

required as it is technically very difficult to design camera lenses with extensions to more than twice its focal length. The most conventional way to achieve magnification $|M| > 1$ is using additionally auxiliary close-up lenses, extension tubes or macro bellows.

The light projected to the image plane forms a circle due to the circular geometry of the lens. It is called an image circle and is always larger than the format of the sensor. Otherwise, parts of the sensor would be shaded. One of the key parameters characterizing the image perspective is the angle of view, also termed angular field of view, indicated by the symbol Ψ (Figure 2.5b). It describes the angular extent under which the object space can be perceived through the camera lens. For a given focal length of the lens, the limitation of the field of view is due to the sensor format. Increasing the sensor format with the same lens results in a wider angle of view. Ψ can be calculated by the ratio of the largest extent of the image, which in general is its diagonal d, to the image distance s_i:

$$\Psi = 2 \cdot \arctan \frac{d/2}{s_i} = 2 \cdot \arctan \frac{d \cdot (s_o - f)}{2 \cdot s_o \cdot f} \tag{2.9}$$

In this equation, we again have substituted s_i with the relation (2.7) to get the dependency on the focal length. This exact relationship in Equation (2.9) is necessary to describe the situation if the object distance is not so much different from the focal length, as for example in close-up imaging. In most photographic situations, however, the object distance s_o is much larger than f and the image is located nearly in the focal plane. Then Equation (2.9) can be simplified, setting $s_i \approx f$:

$$\Psi = 2 \cdot \arctan \frac{d}{2 \cdot f}. \tag{2.10}$$

This simplified expression will be used for further classification of camera lenses. For cameras with a given sensor format, the angle of view becomes smaller the larger the focal length is and vice versa. Imaging using a large focal length, with a small angle of view, gives a perspective as if the scene was perceived through a telescope. If we compare this with the natural viewing perspective of the human eye when regarding a scene, its angle of view is about 47° (see Section 1.4.1). From Equation (2.10), it follows that this natural perspective is achieved for a lens with a focal length, which is slightly longer than the sensor diagonal, more precisely $f_{norm} = 1.15 \cdot d$. Therefore, a lens designed for that angle of view with a special sensor format is classified as a normal camera lens and its focal length f_{norm} is termed normal focal length. On the other hand, if a large-enough print of the image is viewed at the typical viewing distance, which is equal to the print diagonal, then the angle of view for the observer is about 53°. We then again have nearly the same natural perspective. As a consequence, the angle of view for a normal lens is conventionally agreed to be between approximately 47° and 53°.

The nearly linear relationship between the focal length of a lens and its magnification, as given by Equation (2.8) can also be referred to the normal lens. Using again the approximation that objects are fixed at s_o far away from the lens we define the relative magnification M_{rel} as the magnification with respect to that of the normal lens:

$$M_{rel} = \frac{M(f)}{M(f_{norm})} \approx \frac{f}{f_{norm}}. \tag{2.11}$$

From this consideration, we now come to the rough classification of photographic lenses. This classification makes sense only if the lens is used in combination with a well-defined sensor or film format, as mentioned above:

normal lens: a lens is termed normal lens or standard lens if its focal length is nearly equal or slightly larger than the used sensor diagonal ($f_{norm} = 1.15 \cdot d$); perspective and angle of view are similar to natural viewing ($\Psi \approx 47° \ldots 53°$)

long focus lens: the focal length is significantly longer than the sensor diagonal, and thus higher magnification than the normal lens is achieved; relative narrow angle of view ($\Psi < 47°$); the lens is also often termed telephoto lens

wide angle lens: the focal length is significantly shorter than the sensor diagonal, and thus a lower magnification than when using a normal lens is achieved; relative wide angle of view ($\Psi > 53°$)

This classification scheme is illustrated in Figure 2.6 for the example of a full format sensor and respectively, 35 mm film format (see also Chapter 4 for sensor formats). Here, the normal lens has a focal length of 50 mm and serves as a reference.

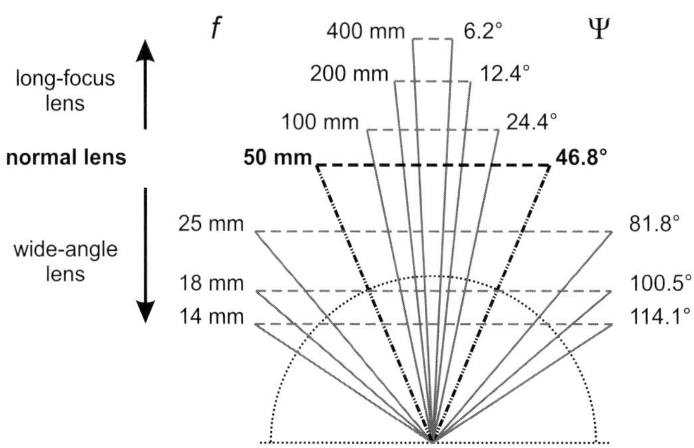

Fig. 2.6: Classification of lenses for full-format (35 mm format); the angle of view Ψ is given as a function of the focal length f.

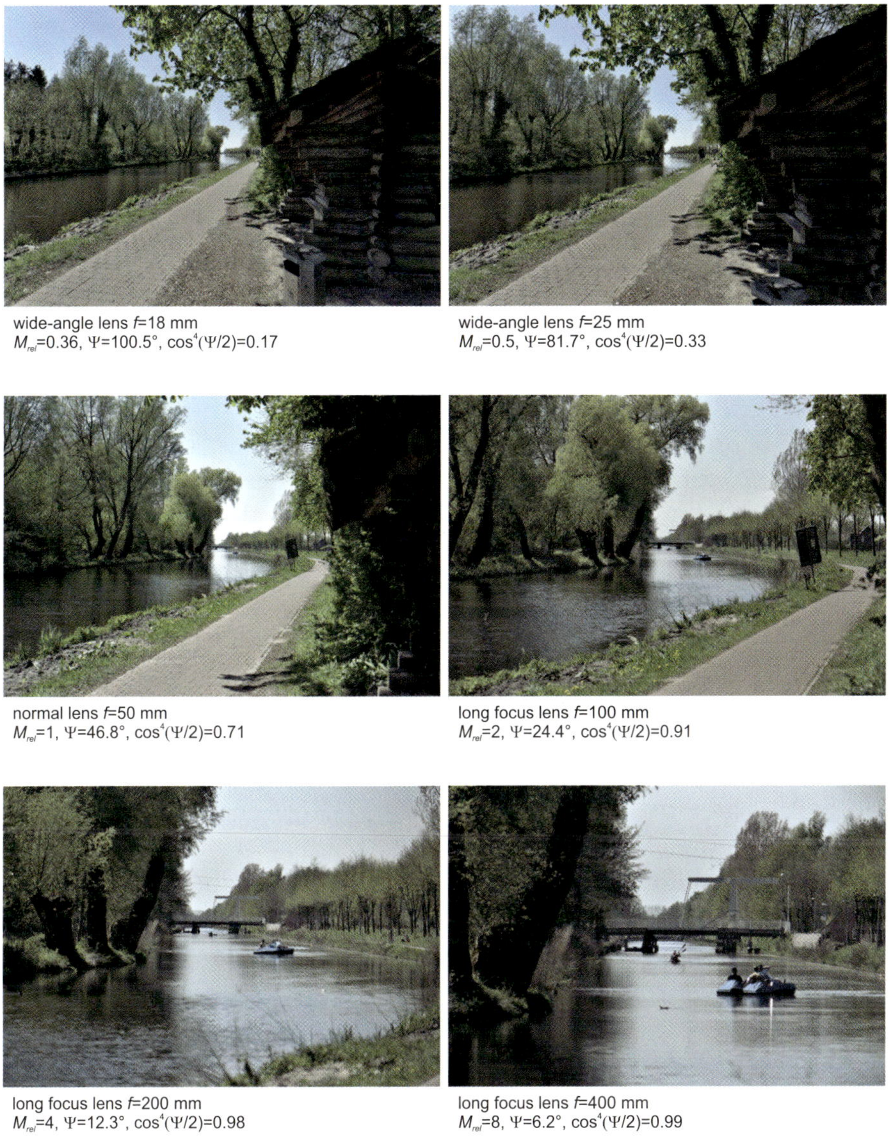

Fig. 2.7: Perspective and angle of view Ψ at different focal lengths f for full format (35 mm format).

Further examples for the perspective and angle of view of lenses are given in Figure 2.7. There, the corresponding values for the angle of view and the magnification relative to the normal lens are indicated. Moreover, the value of $\cos^4(\Psi/2)$ is calculated, which accounts for the brightness fall-off in an image due to natural vignetting. That means the larger the angle of view, the more the brightness falls off relative to the center of the image. This effect, which is discussed in more detail in Section 3.4.4, cannot be avoided and becomes more evident in cases of wide-angle lenses. A photographer chooses the appropriate camera lens with respect to the given ambient situation as well as to his individual intention for photographic design.

2.3 Illuminance and f-number

In optical systems, stops limit the transmission of light and are therefore used to control the exposure when taking images. For cameras, the most important one is the aperture stop in the lens. In most cases, the stop is continuously variable from its minimum to its maximum size and defines a nearly circular aperture. Deviations from the circular shape are due to its physical design as it generally consists of several movable blades forming a nearly circular iris diaphragm (Figure 2.8a). A circular shape is desirable as it yields a highly rotational symmetry with respect to the optical axis. The aperture stop strongly influences the depth of field (Section 3.4.6) and is a main factor for improving the imaging quality as it can minimize lens aberrations (Section 3.5). Other aperture types can be found in different types of cameras (Figure 2.8b,c). In modern automatic cameras, they are in general electrically driven, for instance, by a galvanometer.

The transmission of light is controlled by the aperture stop, which is the physical element in the optical system to limit the incoming optical power. In a more general consideration, the limiting element for incoming rays is the entrance pupil (Section 3.4), which in our consideration here is assumed to be identical to the aperture stop.

In the following, we calculate the illuminance E_i in the image plane. For simplicity, we omit the index v that designates photometric quantities that are conventionally used in the case of photography. All following considerations could be done equally well

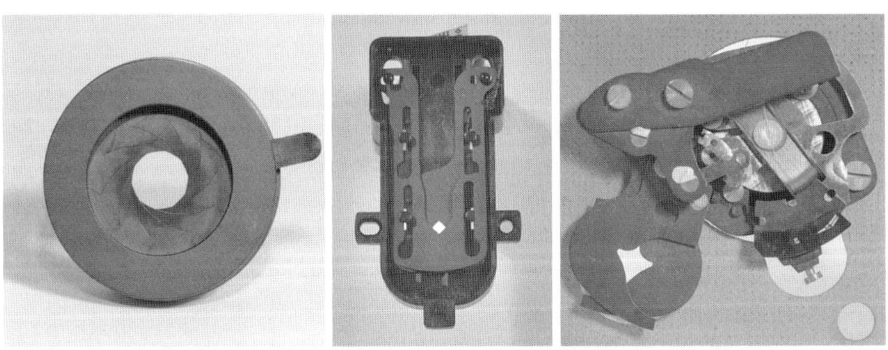

(a) (b) (c)

Fig. 2.8: (a) Iris diaphragm consisting of 10 movable blades defining a nearly circular aperture; (b) two blades driven by a galvanometer and moving in opposite directions to form a square aperture in a digital compact camera; they also act as a central shutter in the camera lens; (c) two blades driven by a galvanometer and moving in opposite directions to form an aperture in an analog film camera.

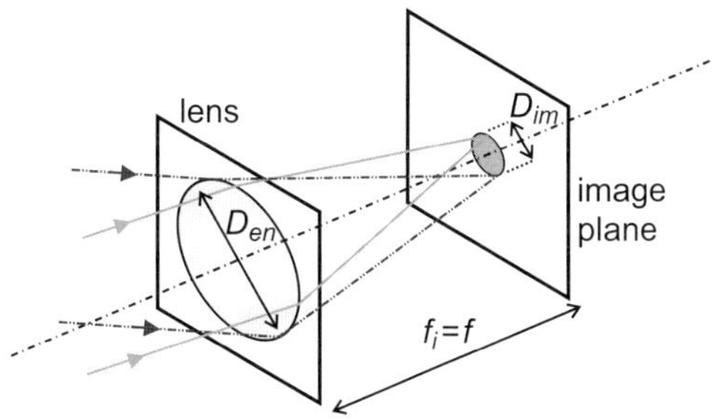

Fig. 2.9: The circular aperture stop of diameter D_{en} is the limiting element for the illumination in the image plane.

using radiometric quantities without restriction. The index i is used to characterize the situation in the image space.

In order to calculate how the aperture stop influences E_{i} we assume a circular aperture stop at the entrance of light to the camera (Figure 2.9). Its diameter is D_{en} and, in the simplest case of a camera with a single lens, D_{en} may be identical to the lens diameter. When taking an image of a nearly homogeneously radiating circular object at a large distance from the lens, light from different positions of the object passes across the entrance pupil and is imaged to a circular area of diameter D_{im} in the sensor plane.

In the case that the luminous flux respectively power of light is homogeneously distributed over the aperture stop, the total amount of luminous flux Φ entering the camera is directly proportional to the area A_{en} of the aperture stop. Assuming a circular type of stop, this flux is concentrated in the image plane over the area A_{im} of the image with a diameter D_{im}. The illuminance E_{i} is the total flux entering the camera divided by the image area, yielding

$$E_{\mathrm{i}} = \frac{\Phi}{A_{\mathrm{im}}} \propto \frac{A_{\mathrm{en}}}{A_{\mathrm{im}}} = \frac{D_{\mathrm{en}}^2}{D_{\mathrm{im}}^2} \tag{2.12}$$

As the image distance is nearly identical to the focal length, i. e., $s_{\mathrm{i}} \approx f$, the lateral image size is proportional to the focal length according to Equations (2.8) and (2.11). Thus, with image diameter $D_{\mathrm{im}} \propto f$ we get

$$E_{\mathrm{i}} \propto \left(\frac{D_{\mathrm{en}}}{f}\right)^2 \tag{2.13}$$

The illuminance in this case directly depends on the properties of the lens, namely its aperture and its focal length. This leads to the definition of the f-number $f_{\#}$, also termed f-stop for optical systems, which is the ratio of focal length to aperture diameter:

$$f_{\#} = \frac{f}{D_{\mathrm{en}}} \tag{2.14}$$

The reciprocal value of the f-number can be interpreted as a relative aperture, which implies that the illuminance in the image sensor plane is higher, the larger $1/f_\#$ is

$$E_\mathrm{i} \propto \frac{1}{f_\#^2} \tag{2.15}$$

It should be mentioned here that $f_\#$ is sometimes also termed relative aperture which, however, is not correct. The relative aperture for optical systems is indicated in the form of $1{:}f_\#$, e. g., 1/2.8, or $f/f_\#$, e. g., $f/2.8$. In the present chapter, we do not differentiate between the f-number $f_\#$, which is only defined for imaging from infinity, and the working f-number $f_{\#\mathrm{w}}$, defined for objects closer to the lens. For the usual situation in photography with objects being at large distances as compared to the focal length, there is not a great difference between them. However, for more complex optical systems and with close-up imaging, a distinction must be made (see also Section 3.4.3).

A low f-number means that the aperture D_en of a lens is large as compared to its focal length. The large relative aperture then has the consequence that the illuminance in the image plane is high, which is the case for fast lenses leading to short exposure times (see Section 2.4). A drawback in many cases is that spherical lens aberrations become more pronounced for fast lens systems due to their large diameters with wide angular apertures. Figure 2.10 shows the comparison of a fast normal camera lens ($f/1.4$) with a lens of more conventional relative aperture ($f/2.8$). In both cases, the glass elements in the lenses have approximately the same sizes but in the $f/2.8$ lens only rays closer to the center can go through, and thus the periphery, which is the origin for many aberrations, is avoided due to the smaller aperture stop. Fast lenses with f-numbers $f_\# < 1.8$ are mainly used for situations with low light or where a narrow depth of field is intended.

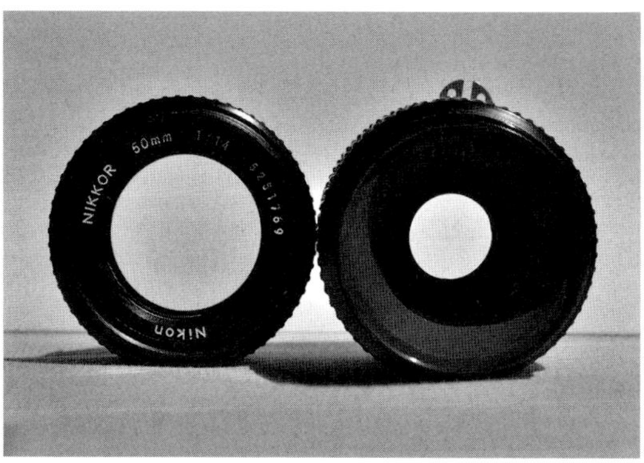

Fig. 2.10: Fast normal lens ($f/1.4$, $f = 50$ mm, left) as compared to a conventional lens ($f/2.8, f = 55$ mm, right).

2.4 Exposure

According to Equation (2.15), the illuminance in the sensor plane depends on the relative aperture of a lens. When taking images, as illustrated in Figure 2.11a for a large format camera with a film, the incoming light is projected to the sensor plane. In the light path, we find a shutter that can rapidly open and close. Its function is to allow light entrance only at the well-defined moment when taking the photo and to control the duration for the light flux needed to inscribe the information to the sensor. There are different locations possible for the shutter inside the camera. Figure 2.11 gives examples of central shutters. A central shutter generally consists of distinct metal blades that are integrated with the lens. The blades slide over each other and define in the ideal case a nearly circular aperture, which is smoother the larger the number of blades. The motion of the blades is in the way that the aperture quickly opens up to its maximum value, stays open for well-defined time and then quickly closes again. Hence, we come to the exposure time t_x during which the light flux passes through the lens while the total aperture section for the incoming flux is limited by the variable aperture stop. The total amount of energy per area that is deposited on the sensor and film is termed exposure H_i, which is directly proportional to t_x and the illuminance E_i in the sensor plane. Taking into consideration Equation (2.15), we get for the luminous exposure the relation:

$$H_i = E_i \cdot t_x \propto \frac{t_x}{f_\#^2}. \tag{2.16}$$

The photometric unit of the exposure is lx · s (lux second). The radiometric exposure expresses the accumulated energy per area and its unit is J/m^2 and respectively, Ws/m^2.

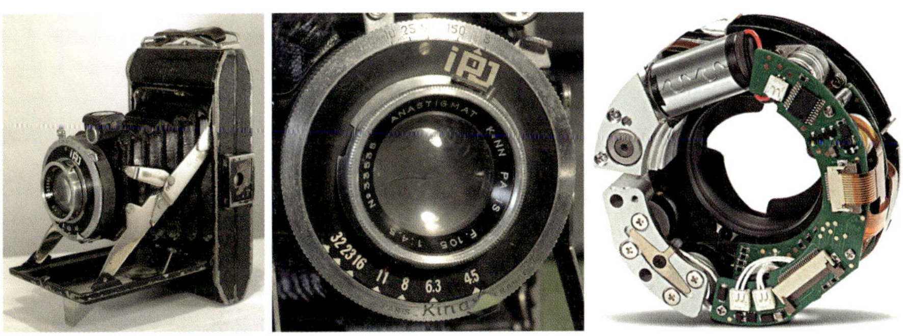

(a) (b) (c)

Fig. 2.11: (a) Historical large format camera (9 cm × 6 cm) with fixed normal lens (105 mm focal length), central shutter and adjustable diaphragm; (b) central shutter consisting of three metal blades integrated in the camera lens; (c) modern central shutter consisting of carbon fiber blades integrated in DSLR lenses (Leica-S system, with kind permission of Leica Camera AG).

The correct exposure of an image at a given object brightness can be realized by different combinations of f-number and exposure time t_x. To achieve that, the reciprocity between exposure time and relative aperture, which is expressed by the ratio $t_x/f_\#^2$ for constant exposure, must be taken into consideration. If the exposure time t_x, often misleadingly termed as shutter speed, is divided by two, for instance for faster exposure, the relative aperture $1/f_\#$ must be increased by a factor of $\sqrt{2}$. As a consequence, the f-numbers of camera lenses form a geometric series and increase by a factor of $\sqrt{2}$, namely $f_\# = 1.4, 2, 2.8, 4, 5.6, 8, 11, 16$, (see Figure 2.12b). For simplicity, the values of $f_\#$ are rounded. Its minimum value for a lens in air is theoretically 0.5 (see Section 3.4.3); for commercially available camera lenses there can be found some with $f_\# = 0.95$. As it is difficult and expensive to eliminate lens aberrations for relatively large apertures, most lenses have values of $f_\# > 1.4$.

At fixed film or sensor sensitivity, also termed sensor speed, only the combination of f-number and exposure time, namely the ratio $t_x/f_\#^2$, is decisive as described above. For example, taking images with the combination $f_\# = 8$ and $t_x = 1/30$ s results in the

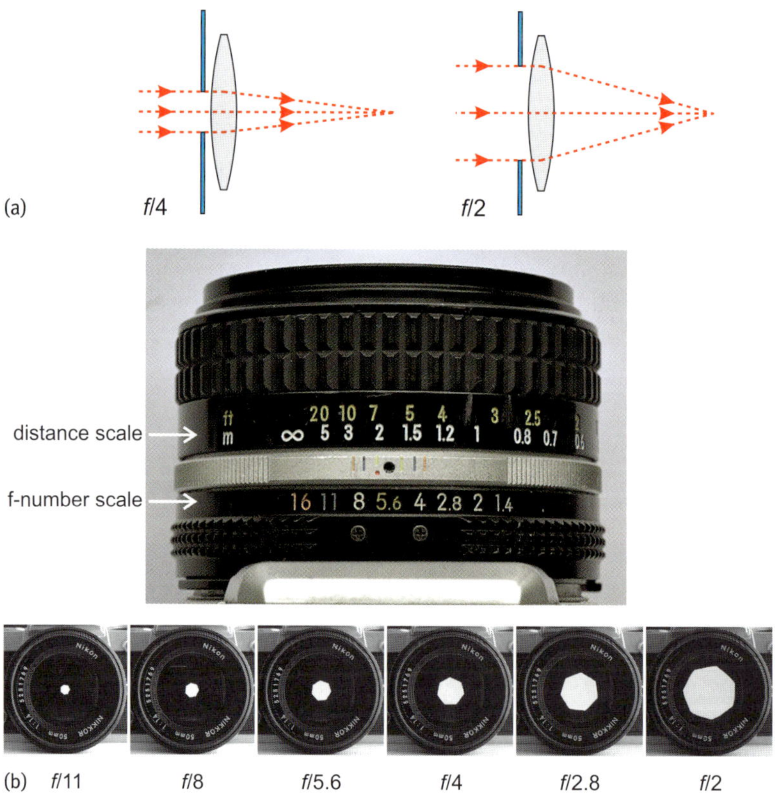

Fig. 2.12: (a) Illustration of relative apertures $f/4$ and respectively, $f/2$; (b) indication of f-numbers on a camera lens and the corresponding cross sections as seen through the lens.

same exposure as when setting $f_\# = 5.6$ and $t_x = 1/60$ s to the same lighting conditions. We therefore can attribute the same exposure value (EV) to both combinations. The definition of the parameter *EV* as well as other key parameters for photographic exposure are given in the following sections.

2.5 Key parameters for photographic exposure

Modern digital camera systems generally can take images in a fully automated mode. In that case, the best combination for exposure time, relative aperture and film sensitivity is chosen for the given situation by the camera processor. When the camera is held by the hand a relatively short exposure time is required to avoid motion blur caused by the movement of the camera or by movement of the object. This becomes more critical the longer the focal length of the lens is or with increasing image magnification. As a rough rule of thumb for the 35 mm format, we can state that the exposure time in seconds should be shorter than the reciprocal focal length of the lens in mm to avoid significant motion blur. For example, when taking images using a 50 mm normal lens, then the exposure time t_x should be shorter than 1/50 s. If the ambient situation does not change but t_x is reduced, the aperture has to be opened and $f_\#$ must be reduced according to Equation (2.16) for keeping the exposure value constant. Reducing $f_\#$ on the other hand implies less depth of focus and in some cases a deterioration of image quality due to the influence of lens errors (see Chapter 3). This becomes more obvious in low light situations and can be counteracted by choosing a higher light sensitivity of the sensor, and respectively, film. On the other hand, a higher sensitivity is related to higher noise and lower image resolution as can be seen in Figure 2.17 and also in more detail in Chapter 4.

Here, we would like to note that many modern lenses or camera systems do include an additional technical solution to reduce motion blur, namely an image stabilizer (see also Section 2.6.4 and Chapter 6). Although we will not discuss the details here, we will give a short description of the basics. Typically, there are different possibilities for stabilizing the image. The "optical image stabilizer" makes use of a particular lens or lens group within a camera lens that can be freely displaced with respect to the rest of the lens. Displacement is according to camera movement and is related to the inertia of the particular lens or lens group. If done well, this keeps the position of the image on the sensor quite stable, even if the camera is moved during exposure. Typically, image stabilization allows for increasing exposure times by a factor of two or more, when compared to the values given by the rule of thumb as discussed above. In the case of an "electronic image stabilizer" such lens movement is not possible, and thus any movement of the camera during exposure time changes the position of the image on the sensor. For this kind of image stabilizer, the real sensor size is larger than the image size offered by the camera. Thus, it can also capture regions that are somewhat outside the range provided by the camera. Now, in particular, for video cameras, the image processor can deduce the positions of all subsequent frames captured by the camera and then reattributes them so that the position of those frames within the image are always the same. This works well if the camera movement is restricted to small movements, for instance jittering. But it still allows for recording moving elements.

In the following, we will consider some key parameters for the photographic exposure as for instance, sensor speed S and exposure value EV as well as how to measure EV and some consequences of the metering.

2.5.1 Sensitivity and speed *S*

The exposure settings for an image depend on the sensitivity of the sensor. The higher the sensitivity is the less luminous exposure is required for a sensor to produce the same standard image of an illuminated object. Higher sensitivity also means shorter exposure times if the aperture remains constant. Due to this fact, films of high sensitivity are also termed as fast films and the light sensitivity as film speed. The classical photographic film shows a highly nonlinear optical density response as a function of the exposure. Only a well-defined exposure range can be used to produce images. This range is very specific for a film and is used to classify the sensitivity of this film. We will not go into details of the film properties here. Characteristics of digital sensors are discussed more extensively in Chapter 4.

The definition of the sensitivity for an electronic sensor in digital cameras is done in an analogous way as for photographic films and aims to get a correlation between the digital sensor and analog film speeds. The International Organization for Standardization (ISO) specifies four methods for determining the sensitivity of digital camera systems in standard ISO 12232:2006. These provide the sensitivity based on noise, sensitivity based on saturation, the recommended exposure index (REI) as well as the standard output sensitivity (SOS). The reason for these different methods is that due to electronic amplification and image processing in the camera, a wide range of image manipulation is possible. The standard output sensitivity S as defined by ISO for digital cameras is given by the relation:

$$S = \frac{10\,\mathrm{lx}\cdot\mathrm{s}}{H_{\mathrm{av}}} \tag{2.17}$$

H_{av} is the recommended average exposure needed to produce a standard image within a given color space or grayscale, leaving a certain exposure range for higher and lower values to guarantee a broader image contrast. Usually, H_{av} can be measured using an integral light meter. From Equation (2.17), it follows that for the standard speed of $S = 100$ or short ISO 100 an average exposure of $H_{\mathrm{av}} = 0.1\,\mathrm{lx}\cdot\mathrm{s}$ is necessary to produce the standard image. Setting the camera to higher speed value, as for instance ISO 200, only half of the exposure is required, and so on. We thus have a direct reciprocal relationship between sensor speed and exposure. The numerical value of $10\,\mathrm{lx}\cdot\mathrm{s}$ in Equation (2.17) is chosen in the way that a compatibility of the digital camera sensitivity with the traditional film-based photography is given. This means that a setting of ISO 100 in a digital camera requires the same combinations of exposure time t_x and f-number $f_\#$ for the exposure as in the case of a photographic film with the same speed. However, the exact ISO speed ratings of digital cameras depend on the exact characteristics of the sensors and image processors in the camera, and thus some manufacturer-specific variations are possible. For further discussion, see also Section 4.8.8.

The ISO speed represents the current standard, whereas definitions based on some older standards can also be found. The former ASA (American Standards Association) speed S_{ASA} was the predecessor for current ISO specifications. Here, we have a linear or arithmetic expression for the sensitivity, which means if the numerical values of S are doubled then the sensitivity is doubled in the same way. A different definition has been given by the former DIN (Deutsches Institut für Normung) standard that expresses the speed in a logarithmic way and marks the number with the symbol for degree. In that case, an increase of the value by 3° corresponds to a doubling of the sensitivity. We have the following relationship between the different standards:

$$S_{\mathrm{ASA}} = S \qquad S^{\circ}_{\mathrm{DIN}} = (1 + 10 \cdot \lg S)^{\circ}. \tag{2.18}$$

The current ISO standard S_{ISO} is a combination of the linear ASA and logarithmic DIN values, and a speed of ISO 100 is written in the form ISO 100/21° (Figure 2.13). It has become quite common that only the linear value is used and the logarithmic one is omitted. This can be seen especially in modern electronic cameras. Figure 2.13 illustrates the display of the parameter's exposure time, f-number and ISO speed on the rear display of a digital camera.

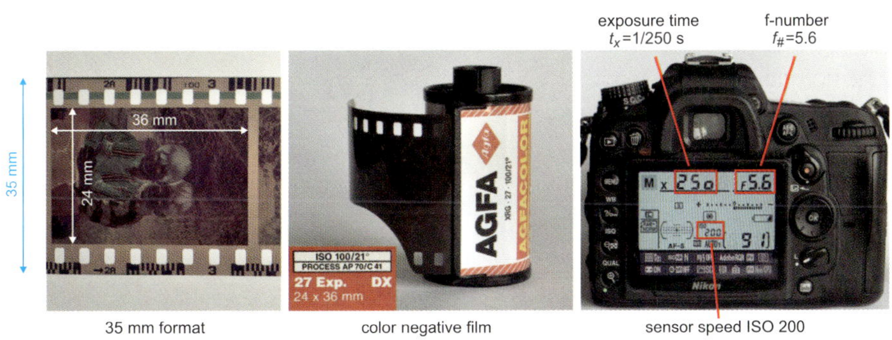

Fig. 2.13: 35 mm film for 24 mm × 36 mm image format (left); color negative film with ISO speed indication (center); exposure parameters as indicated on the back monitor display of a digital camera (right).

2.5.2 Exposure determination and exposure value

Before images are taken, normally all relevant exposure parameters are determined after light measurement by a light meter in the camera or by external meters. This is necessary to get the appropriate exposure settings $f_\#$ and t_x for a given sensor speed. As stated above, only the combination of $t_x/f_\#^2$ is decisive for an adequate exposure. This combination can be determined by a light meter. There are in principle two different methods for light meter applications: one uses a reflected light meter, the other one uses an incident light meter. Let us consider the incident light meter method, which measures the illuminance E_{il} of the ambient light incident to the scene and then displays the combination $t_x/f_\#^2$ for a given film speed. This combination is directly proportional to the recommended average exposure H_{av}. It can be seen from Equation (2.17) that H_{av} is inversely proportional to the speed S. As a consequence, if images are taken at constant incident light illumination and keeping the f-number $f_\#$ of the lens constant, then the exposure time must be reduced by a factor of two if the speed is increased by the same factor, for instance from ISO 100 to ISO 200. On the other hand, if sensor speed and f-number are kept constant but the illuminance of the incident light is increased by a factor of two, for instance from $E_{il} = 150\,\mathrm{lx}$ to $300\,\mathrm{lx}$, then the exposure time must be reduced by the same factor, as the recommended exposure is achieved in a shorter time. We thus have the following relationship between the parameter combination $t_x/f_\#^2$, sensor speed and incident light:

$$\frac{t_x}{f_\#^2} \propto \frac{1}{S \cdot E_{il}} \quad \Rightarrow \quad \frac{f_\#^2}{t_x} = \frac{S \cdot E_{il}}{C_L}. \tag{2.19}$$

The calibration constant C_L is necessary for getting a direct relationship between the exposure parameters, sensor speed and the absolute light situation. C_L is characteristic for light incident from half-space as it can be measured by the meter. It varies depending on meter manufacturer and light measurement method between $C_L \approx 240\ldots425\,\mathrm{lx}\cdot\mathrm{s}$ as recommended by ISO 2720:1974.

In order to simplify the discussion of exposure parameters, the combination of f-number and exposure time is expressed by one parameter, which is termed exposure value (EV). The perception of luminance by the human eye as well as the optical response of films and electronic sensors comprises such a large range of values from its minimum in dark areas to the maximum in bright areas that a compression of this range by using logarithmic quantities is the best way to describe it. Even more, this includes the logarithmic dependence of the perception of the human eye (see Section 4.8.5). If, for instance, we start at a given combination of f-number and exposure time and assign a given EV to it, then the next step is achieved if we increase and respectively reduce the exposure time by a factor of two and keep the f-number constant. The step can be equally achieved by increasing and respectively reducing the f-number by a factor of $\sqrt{2}$ without changing the exposure time. This is equivalent to taking the next aperture stop up and respectively, down on the f-number scale. In all cases, going one EV step up or down means that the ratio $f_{\#}^2/t_x$ increases, and respectively decreases by a factor of two thus forming a geometric series based on that factor. We therefore define the parameter for the exposure value EV using the following relation:

$$\frac{f_{\#}^2}{t_x/\mathrm{s}} = 2^{EV} \quad \Leftrightarrow \quad EV = \mathrm{ld}\left(\frac{f_{\#}^2}{t_x/\mathrm{s}}\right) = \frac{\lg\left(\frac{f_{\#}^2}{t_x/\mathrm{s}}\right)}{\lg 2} = 3.32 \cdot \lg\left(\frac{f_{\#}^2}{t_x/\mathrm{s}}\right). \tag{2.20}$$

The division of t_x by s in the equation is necessary because the exposure value is a quantity of the first dimension or, as commonly expressed, a dimensionless quantity. In general, the exposure values are rounded to integer numbers. For example, a combination of $t_x = 1/125°\mathrm{s}$ at $f_{\#} = 5.6$ gives a value of $EV = 12$ or simply EV 12. The combinations of f-number and exposure time in seconds at a given exposure value is graphically illustrated in Figure 2.14 [Sch81].

The exposure value only reflects the camera settings during the exposure and does not give any indication of the illuminance if the sensor speed is not specified. Choosing an ISO speed that is two times higher at a given illumination has the consequence that the exposure value must be reduced by one step. If a direct relation to the ambient luminosity should be established, then Equation (2.19) must be taken into consideration. In that case, the absolute value of the incident light illuminance can be determined using

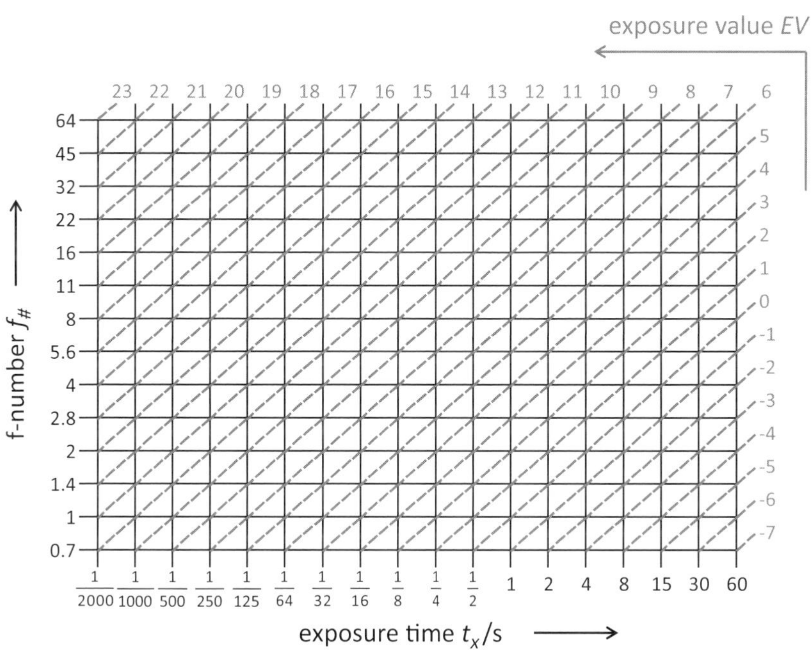

Fig. 2.14: Combinations of f-number and exposure time for a given exposure value.

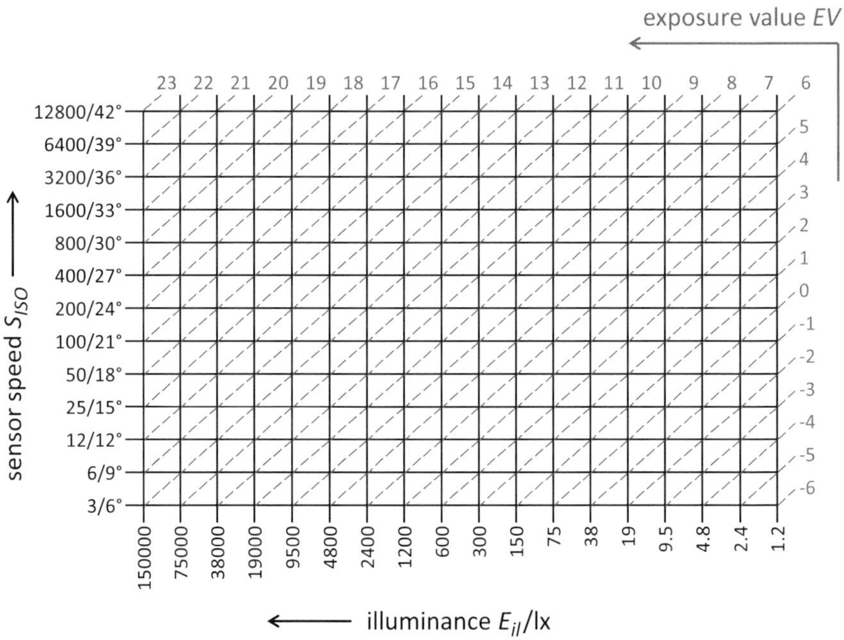

Fig. 2.15: Exposure value for given illuminance and sensor speed; calculation according to Equation (2.21) with $C_L = 240\,\text{lx}\cdot\text{s}$.

the exposure value and the sensor speed, as is done by using a light meter:

$$E_{\text{il}} = \frac{C_L \cdot 2^{EV}}{S}. \tag{2.21}$$

A graphic representation of the relationship between illuminance in lux, ISO sensor speed and exposure value is given in Figure 2.15. The diagram has been calculated for $C_L = 240\,\text{lx}\cdot\text{s}$. One of the three parameters can be determined from it if the other two are known. This principle is also incorporated in light meters to determine the exposure settings after illumination measurement.

Finally, we may remark that in general the definition of the absolute value of EV is not always used as discussed above. In particular, in Chapter 4 we also make use of differently "calibrated" values, namely we set $EV = 0$ for an illumination, which leads to the "18 % grey average" or alternatively for an illumination that just leads to sensor saturation. However, even in those cases, changes of the exposure value by one or more f-stops behave as discussed before.

Example: Exposure settings and their influence to imaging

As stated above, the recommended exposure can be realized by different combinations of exposure settings f-number, exposure time and sensor speed. The exposure value only

Tab. 2.1: Examples for combinations of exposure settings.

No.	f-number $f_\#$	exposure time t_x/s	sensor speed S_{ISO}	exposure value/EV	illuminance E_{il}/lx
1	1	1	100	0	2.4
2	8	1/60	100	12	9500
3	5.6	1/125	100	12	9500
4	8	1/125	200	13	9500
5	5.6	1/125	200	12	4900

depends on the ratio $f_\#^2/t_x$ as can be seen in Table 2.1 (no. 2, 3, 5) and is independent from the sensor speed S_{ISO}, and respectively, the illuminance E_{il} (no. 3, 5). On the other hand, the ratio $f_\#^2/(t_x \cdot S_{ISO})$ is always constant if the illumination does not change (no. 2, 3, 4). It should be noted here that we use the conventionally-rounded values for the f-number and exposure time so there will be deviations from exactly calculated results.

In many photographic situations, there will be the question for the absolute values of f-number and exposure time. Their influence is illustrated by Figure 2.16. The images, showing a watch with a moving second hand, have all been taken under the same ambient conditions. That means that E_{il} has the same value in all cases, but time and f-number are varied according to Equation (2.19). When the exposure time is increased, then the moving of the second hand becomes visible, leading to a motion blur proportional to t_x (Figures 2.16a and 2.16b from top to bottom). With increasing t_x at constant sensor speed, the relative aperture must be reduced, leading to higher f-numbers (Figure 2.16a). A consequence of this stopping down is that the depth of field increases. This can be clearly seen in Figure 2.16a where the top image has a good sharpness only in a limited area around the center. In the bottom image with reduced aperture, however, the sharpness is nearly homogeneous across the whole field.

If one wishes to have a large depth of field, stopping down is recommended. In that case, the exposure time must be increased if the sensor speed is not enhanced correspondingly. This can be seen in Figure 2.16b where only a parallel increase of the sensor speed with the f-number guarantees a short exposure time. Here, a short exposure time with a large depth of field can only be achieved at high sensor speeds.

These advantages of a higher sensor speed, however, are at the expense of increasing electronic noise, which is equivalent to higher grain sizes in photographic films. In parallel to that, information in the image is lost due to reduced optical resolution (see also Chapter 4 and Chapter 1). Figure 2.16c shows increased color and luminance noise especially in the dark areas. The detriment of the image resolution can also be seen in Figure 2.17. Details that can be seen at ISO 100 are no longer visible at ISO 25600.

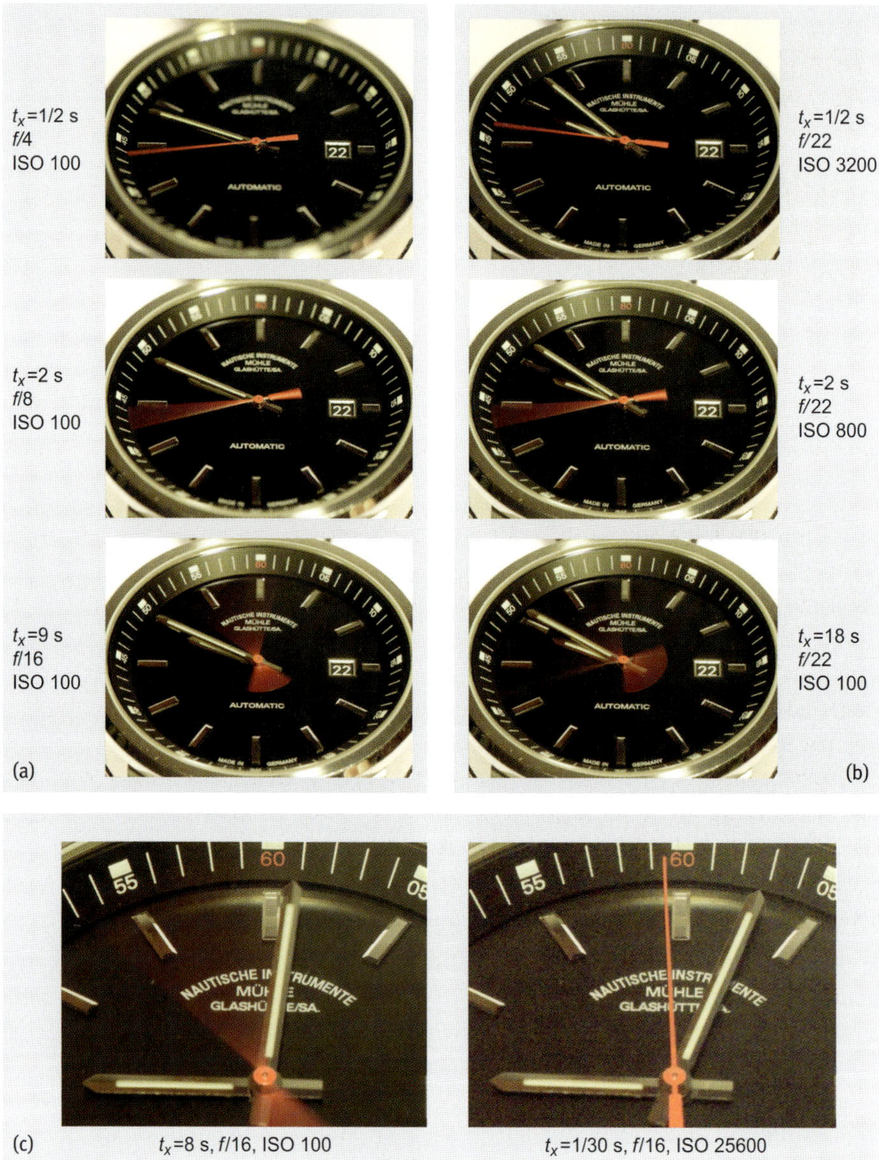

Fig. 2.16: Images shot under the same lighting conditions. (a) Images at constant ISO sensitivity and varying combinations of exposure time and f-number; (b) images at constant f-number and varying combinations of exposure time and ISO sensitivity; (c) noise influence due to increased ISO sensitivity.

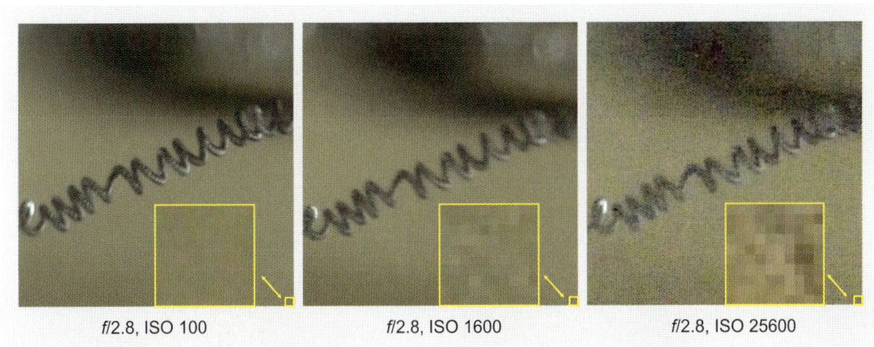

Fig. 2.17: Images of the same object captured at different sensor speeds. Increased sensor speed leads to increased optical noise and reduced image resolution (note: the large rectangular inset is a 34 × magnification of the small rectangular area in the right lower corner of 10 pixel × 10-pixel size).

2.5.3 Exposure value and relative brightness change

Important applications of imaging can be found in the domain of optical measurements. However, standard photographic cameras should be used only with care for precise measurements as the output images are strongly influenced by the way the detector data are converted to the final image data. This is discussed in more detail in Chapter 5. However, rough average measurements are possible.

If, as for instance, the light distribution over a given area should be determined, a clear relationship between the illumination and the sensor response is required. Equation (2.19) describes how the illuminance of the light incident to the sensor plane is related to the exposure value. Thus, if we choose the illuminance in the center of the sensor plane E_{ref} as a reference value, then any other value E_{il} in the plane can be expressed as a deviation ΔEV of the exposure value from the reference in the center:

$$b_r = \frac{E_{il}}{E_{ref}} = 2^{\Delta EV} \quad \Leftrightarrow \quad \Delta EV = \mathrm{ld}\, b_r = 3.32 \cdot \lg b_r. \tag{2.22}$$

b_r in Equation (2.22) is the brightness change in the image relative to the center. Let us consider the examples given in Figure 2.7. In the case of wide-angle lenses, a brightness fall-off can become very pronounced from the center to the periphery of the image plane. The theoretical fall-off due to natural vignetting increases with the angular field of view Ψ according to the factor $\cos^4(\Psi/2)$ (see Section 3.4.4). For the 18 mm wide angle lens, the relative brightness change, being equal to that factor, is $b_r = 0.17$ or expressed in exposure values, $\Delta EV = -2.6$. If the brightness in an image drops from 100 % to 50 % at the border, this can be expressed by a drop of $\Delta EV = -1$.

2.5.4 Optimum aperture and critical f-number

The aperture stop is of high importance for optical systems and has to perform different functions. As described above, it controls the light flux entering the system and influences the depth of field for imaging. Moreover, as will be detailed in Chapter 3, the overall image quality is strongly influenced by the position and size of the aperture stop. As most lenses consist of spherical elements, errors are produced due to the fact that peripheral rays through the lens are imaged to slightly different points from central rays. This can be characterized by a blurred spot due to lens aberrations having a diameter of u_{err}. The blur generally becomes more pronounced the further the image spot is off the optical axis. Lens aberrations can be minimized by reducing the aperture, which means stopping down or respectively increasing the f-number. Figure 2.18 illustrates schematically how the size of u_{err} varies as a function of the f-number for a lens of higher quality (lens 1) as well as of lower quality (lens 2) of the 35 mm format. On the other hand, stopping down increases the diffraction of light, which is more detrimental the smaller the aperture is. This diffraction blur is given by Equation (2.3) for the pinhole camera with s_i being the image distance and D_p being the diameter of the aperture. For standard situations using optical lenses the object distance s_o is in general much larger than the focal length f, and hence the image distance s_i is nearly the same as f. If we substitute s_i with f, the aperture diameter D_p with D_{en} and use the definition of the f-number $f_\# = f/D_{en}$ then Equation (2.3) can be modified to yield the blur diameter u_d, due to diffraction in a lens, for the visible range at $\lambda \approx 0.55\,\mu m$ as a function of the f-number:

$$u_d = 2.44 \cdot \frac{\lambda}{D_{en}} \cdot f = 2.44 \cdot f_\# \cdot \lambda \approx 1.34\,\mu m \cdot f_\#. \tag{2.23}$$

u_d is the diameter of the so-called Airy disk and limits the resolution of a lens that is even free of aberrations (see also Chapters 1 and 5). The diffraction blur u_d increases linearly with $f_\#$ and is illustrated in the semilogarithmic plot in Figure 2.18 by a curved line. When varying the aperture stop to optimize the image quality, we observe two competing effects acting into opposite directions: Stopping down to reduce lens aberration and stopping up to avoid diffraction blur. The optimum aperture with the least blurred

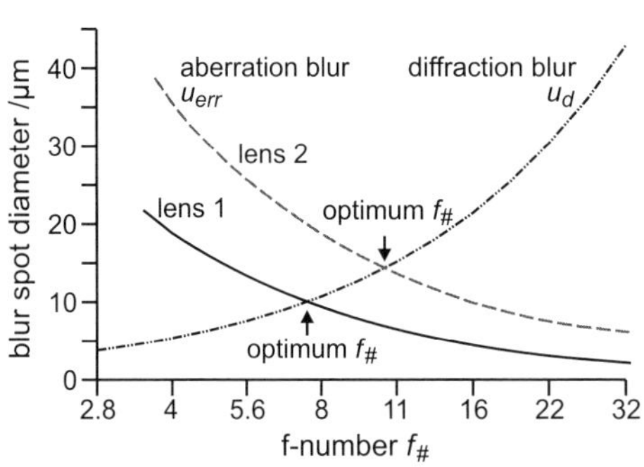

Fig. 2.18: Diffraction blur and aberration blur of different lenses. The aberration curves are tentatively shown for a lens of higher quality (lens 1), and respectively, of lower quality (lens 2) for the 35 mm format. The diffraction blur is drawn according to Equation (2.23).

spot is achieved with the f-number where the curve of u_d intersects the curve of u_{err} for a given lens. u_{err} depends on the lens design. It characterizes the quality of the lens and is usually known by measurements or detailed numerical simulations. Typical values for the optimum f-number can be roughly estimated to be between 2 to 4 stop-values above the lowest $f_\#$ of the lens. For 35 mm format lenses, typical values are found between $f_\# \approx 5.6$ and $f_\# \approx 11$. For the examples given in Figure 2.18, the optimum $f_\#$ for the higher quality lens is below 8 and smaller than that for the lower quality lens. While the diffraction blur is independent from the sensor format, the absolute values of the aberration blur u_{err} scale with the format and lens size and increase with increasing image format. Small format lenses have optimum apertures at relatively low f-numbers and require a higher precision during manufacturing. Large format lenses may have optimum sharpness at higher f-numbers.

The above consideration is only valid for a lens alone without taking into account the quality of the image sensor with which the lens is used. However, superior lenses may have such a high quality that the limiting factor for the resolution is the image sensor. Let us therefore consider a lens with low aberration errors in combination with an image sensor that has a pixel size p. The spatial resolution of this sensor is limited by the pixel size as the minimum structure size that can be resolved by the chip is always larger than $2p$ (see Chapter 1). Thus, the diffraction blur u_d is uncritical if it is smaller than $2p$ as it cannot be resolved by the sensor anymore. In order to get an image with the least diffraction blur, stopping down to a critical f-number $f_{\#crit}$ should not be exceeded, which can be calculated under the condition that u_d is equal to $2p$. We then get for the wavelength $\lambda \approx 0.55\,\mu m$ in the visible spectrum a value of $f_{\#crit}$,

$$2.44 \cdot f_{\#crit} \cdot \lambda = 2 \cdot p \quad \Rightarrow \quad f_{\#crit} = \frac{p}{1.22 \cdot \lambda} \approx \frac{p}{0.67\,\mu m} \approx \frac{1.5 \cdot p}{\mu m} \tag{2.24}$$

As an example, let us take a 35 mm sensor with 26 MP. The typical pixel size for this sensor is around $p = 5.8\,\mu m$. The critical f-number then using Equation (2.24) is roughly 1.5 times the value of the pixel size in µm, and for this example $f_{\#crit} \approx 9$. If the lens is of high quality with an optimum $f_\#$ being smaller than $f_{\#crit}$, then setting the lens to $f_{\#crit}$ yields the best sharpness. Stopping down to higher f-numbers impairs the resolution due to increased diffraction while stopping up to lower f-numbers does not lead to higher resolution due to the limitations of the sensor. At the lowest possible f-numbers of the lens, below the optimum $f_\#$, there may even the lens aberrations become manifest. Hence, it is always necessary to consider lens and image sensor as a system to achieve the optimum performance. With modern image sensors having for instance more than 30 or even 50 MP for the full-frame format, the use of superior lenses is required in order to benefit from the high sensor resolution. Also, cameras with very small pixel sizes in the order of 1–2 µm, like miniature or mobile phone cameras, have critical f-numbers between about 1.5 and 3. In some cases of low-cost cameras, the lens optics is not good enough for these small sensors and sensors with less megapixels would be more appropriate. In the following sections, we will have a look at different still cameras with different sensor chips and pixel sizes. As a key feature for these cameras, we will also have a look at their $f_{\#crit}$ and not their optimum f-numbers as it is easy to calculate $f_{\#crit}$ from the sensor specifications while the optimum f-number of a lens is only available from measurements. In most cases, it can be found that the lenses of complete systems are well adapted to the resolution of the image sensor.

2.6 Examples of camera systems

A modern camera system can be roughly subdivided into its main parts of imaging optics, the optoelectronic sensor, the image processing system and the control electronics to ensure an optimum interaction between the different process steps for taking images. If all these parts are optimized and integrated into one system, we normally classify it as a compact camera. On the other hand, system cameras can be characterized as consisting of a camera body with interchangeable modules and accessories, as for instance optical lenses, optical finders, external control, flashlight, etc. The minimum requirement normally is a camera body with an interchangeable lens. In the following, we will consider some systems and their specifications as given by the manufacturer. Furthermore, some properties can be deduced with the help of the formulas given above. We start with the most popular and most versatile system of 35 mm single lens reflex cameras and end with some recent development in the field of mobile phone cameras. A short look at scientific and industrial cameras will be given as well.

2.6.1 Single lens reflex camera

2.6.1.1 Characteristics and camera body

The main feature of a single lens reflex camera (SLR) is that the same camera lens used to expose the image on the sensor is also used to observe the scenery in combination with an optical finder before exposure. The advantage then is that the image can be observed exactly as it will be after exposure without any deviation, for instance parallaxes, as is normally the case if a separate finder mechanism is used. In order to achieve this, a hinged main mirror is positioned in the optical path of light between the lens and the image plane on the sensor (Figure 2.19). This mirror reflects the light to a focusing screen where a mirrored image is generated. The term reflex camera is due to this mirror reflection; the schematic setup for a 35 mm film format SLR is illustrated in Figure 2.19. For capturing the image, the mirror is flipped out of the light path thus allowing the same light as observed to expose the optoelectronic sensor or photographic film. In order to guarantee a reliable movement of the mirror without obstruction, a certain space between the image plane and the mounting flange of the lens is required. It depends on the mirror size, and thus the film format. For the 35 mm format, the distance from the

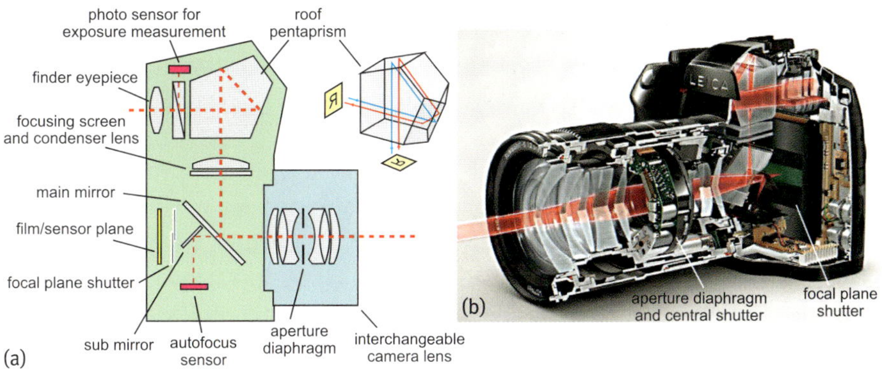

Fig. 2.19: (a) Schematic cross-section view of a SLR camera with perspective detail of roof pentaprism; (b) cutaway of a DSLR camera with a medium size format and two shutter types (Leica S, with kind permission of Leica Camera AG).

flange to the image plane varies between about 40 mm to 50 mm for different camera manufacturers.

The focusing screen for the image control before exposure is at the same distance from the mirror as the image plane. In general, the screen is a matte with etched markings on it in order to denote special areas in the field of view. It may be interchanged for manual focus cameras to other types of screens; for example, one with microprisms or a split screen indicator to facilitate lens focusing (Figure 2.20). The combination of the mirror with a roof pentaprism having three internal reflections results in an up-

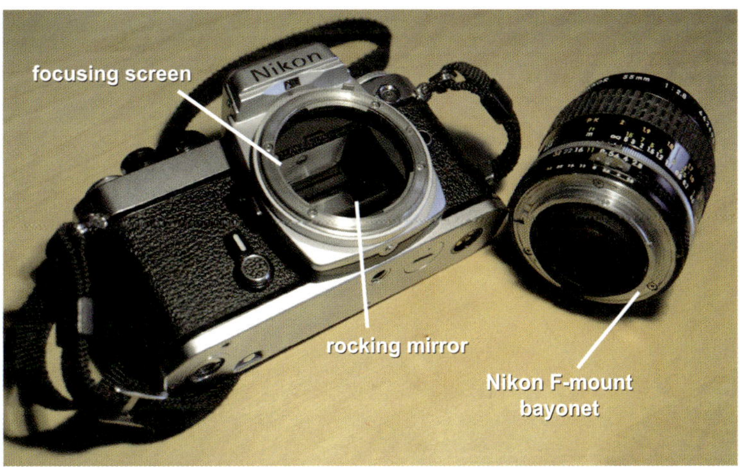

Fig. 2.20: Front view of a Nikon FE (1980) SLR body and interchangeable lens; inside the body the mirror can be identified in its rest position; the interchangeable focus screen is located below the roof prism; camera lenses with the Nikon F bayonet can be directly mounted to the body.

right and true-to-side image that can be observed by the finder eyepiece. The condenser lens is necessary to converge otherwise divergent beams, thus ensuring a bright finder image for the observer.

Light measurement prior to exposure in a SLR uses the light through the lens (TTL) and can be achieved by different methods and at different positions in the camera body. A simple method is the use of a photo sensor behind a semitransparent mirror or in combination with a beam splitter that captures a part of the TTL-light, which is illustrated in Figure 2.19 for a sensor located in the light path behind the roof pentaprism. More sophisticated sensor arrays are possible, delivering a more precise light distribution in the area that is used to compute the exposure settings in modern cameras. A typical range of exposure settings at ISO 100 is between EV1 up to EV18. As stated above, the absolute values of EV may be ambiguous due to the different definitions, but the range of the exposure settings is typically about 18 EV.

When using flashlights, light metering before exposure is not possible as the light is only available during exposure when the mirror is flipped away. Hence, the light reflected from the film respectively sensor while the shutter is open is used to control the exposure (Figure 2.21).

The exposure is done by opening the mechanical shutter. In modern film cameras, exposure times ranging from several seconds to approximately 1/4000 s are electronically controlled. Unlike the central shutter, which is integrated in the lens (Figure 2.11), the focal-plane shutter, normally used in a SLR camera, is located directly in front of the sensor and has the advantage that the exposure is a process that can be completely independent from the interchangeable lens. The focal-plane shutter is in many cases implemented as a pair of curtains, where the first curtain opens to expose the film frame and the second one closes after the correct exposure time (Figure 2.21). For very short exposure times, the shutter is not fully open to expose the frame, but the first curtain opens, and the second curtains follows at a distance behind it. This can be described as a slit moving across the image plane where the width of the slit and its traveling speed are matched to deliver the correct total exposure for the image. As a consequence, in the

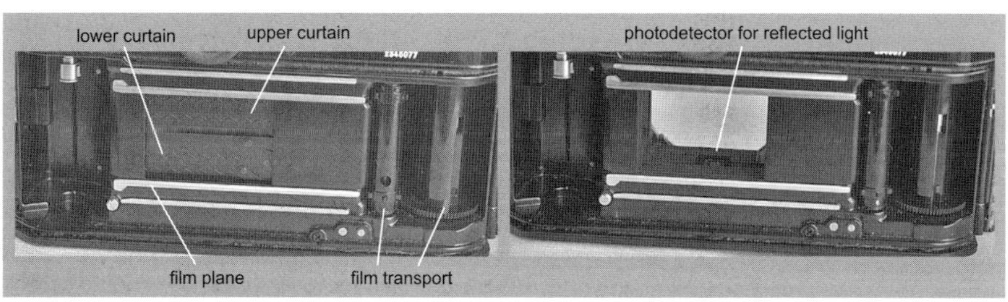

Fig. 2.21: Focal plane shutter consisting of two titanium foil shutters of a Nikon FE2 as seen from the open back (left); when the shutter is open the photodetector cell can measure light reflected from the film during exposure (right).

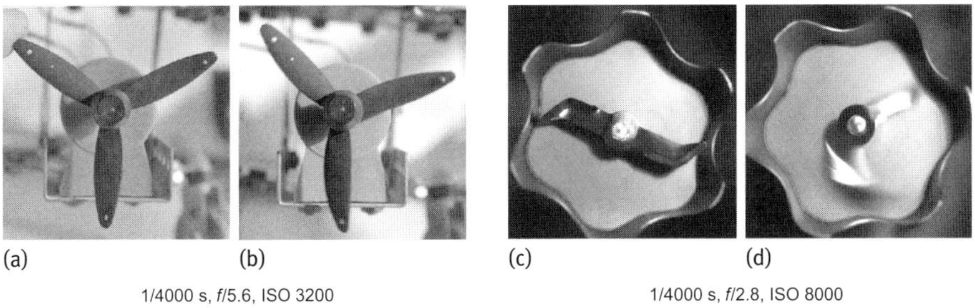

Fig. 2.22: Images taken by a focal plane shutter (Nikon D750, 1/4000 s). (a) Propeller at rest; (b) propeller rotating clockwise at moderate angular frequency; (c) chopper blade at rest; (d) chopper blade rotating clockwise at high angular frequency. For further details, see the text.

case of very short exposure times when the slit is narrow, different parts of the image are exposed at different times leading to distorted imaging of moving parts. Figure 2.22 depicts images taken by a DSLR camera where the slit moves from the top to the bottom with respect to the camera body. As the image on the sensor is inverted, the lower parts of the image are exposed before the upper parts. The resulting distortion can be easily seen with the clockwise rotating lower propeller blade in Figure 2.22b. The lower part of the propeller, taken at the beginning of the exposure with a nominal exposure time of 1/4000 s, is centered below the image center whereas the upper part is slightly rotated compared to the reference image of the propeller at rest in (a). A much stronger image distortion of a chopper blade rotating at a higher speed than the propeller can be seen in Figure 2.22d compared to the blade at rest in (c). Here, the image of the rotating blade is of no avail, hence a different exposure technique is required. If a powerful external light source with very short light pulses is used, like in the case of flashlight exposure, the moving slit exposure must be avoided. Only a small strip of the sensor would receive the light, just that part where the slit is when the light pulse is emitted. Thus, when using a flashlight, a flash synchronization is necessary that ensures that the flash is only fired when the shutter is fully open. As a consequence, the shutter speed must be limited to be longer than a flash synchronization speed, which in cases of film SLR cameras is typically of the order of 1/125 s to 1/250 s. This problem does not show up for circular central shutters, which usually are integrated in the camera lenses. Especially for medium format SRL there exist different types of camera bodies having integrated focal-plane shutters (Figure 2.19), but can also be optionally operated with lenses having built-in central shutters. It is up to the photographer then to choose which type of shutter will be used.

It should be mentioned that more and more electronic shutters based on the image readout of the electronic image sensors are realized, often in combination with mechanical shutters (see Chapter 4). However, problems may also arise there, due to sensor technology. For instance, similar image distortions like in Figure 2.22d can be observed, known as rolling shutter effects. They should be discriminated from the mechanical shutter effects although they have similar consequences.

The early generations of SLR cameras up to the 1980s could be operated only with lenses that had a manual focus. To that purpose, different focus screens in the camera body were used to facilitate a quick and reliable adjustment of the lens to get a sharp image. With the progress in electronics, the first autofocus systems showed up in the 1980s based on different sensor technologies and arrangements in the camera body. A common method in SLR cameras is to use a part of the light transmitted through the main mirror, which is then reflected by a separate submirror to an autofocus sensor system (Figure 2.19). The sensor signal is used to control the camera lens adjustment with the help of an electric motor drive. There are currently different autofocus systems depending on the manufacturer. Some camera bodies are equipped with a motor drive to actuate the lens adjustment via a mechanical shaft. In other cases, the motors are directly integrated in the lenses and optimized to the complex optical system. This is also characteristic for the most advanced systems. Modern DSLR cameras can cope with both approaches; they have integrated motors and can also control lenses with built-in drives. Quick and reliable autofocus systems are of high interest for modern camera development. The most common methods are phase-detection and contrast-detection autofocus. There are further advanced methods as well, such as Panasonic's Depth From Defocus (DFD) technology and on-chip phase-detection autofocus (PDAF, see Section 4.10.7.3). However, the topic of autofocus is quite complex and extensive. It will not be covered in more detail in the present book where the main interest is rather on optics and sensors.

2.6.1.2 Film formats and camera lenses

The most popular version of SLR cameras is the one based on a 35 mm film format. The reason for it is that film material of that format has been available in a very large variety and affordable prices as compared to larger formats. Smaller formats were never as popular due to their poorer quality. As mentioned above, the origin for this still camera format was the film material used for cinematographic production. Figure 2.13 illustrates the dimensions of a 35 mm film with its 3:2 aspect ratio (height 36 mm, width 24 mm, diagonal 43.3 mm) and shows a typical color film cartridge. For professional use there exist different larger formats of which the most prominent ones are the legacy formats 6 cm × 6 cm and 6 cm × 4.5 cm, mainly used by cameras of Hasselblad, Rollei and Mamiya, only to mention a few. But also, a new 30 mm × 45 mm format for DSLR cameras has been developed by Leica. The great advantage for the larger format is that the film or sensors have a better signal-to-no-noise ratio and higher resolution if the image properties are considered with respect to the image size, namely the space bandwidth number, but not to measurement units like mm or μm. For further discussion of sensor topics, see Chapters 4 and 5.

The larger the image, the more the ratio of focal length to image diagonal decreases, which means that lenses with the same focal length as the 35 mm format yield a different perspective for larger format images. For instance, the normal lens for a 6 cm × 6 cm format is a lens of around 80 mm focal length yielding an angle of view of around 53° while the same focal length mounted on a 35 mm format camera yields an angle of around only 30°, thus acting as a moderate telephoto lens. Due to the fact that the 35 mm format is so popular and widespread, many properties of cameras are related to this format. In order to facilitate the comparison, it has become common use to define a crop factor *CF*, which is simply the ratio of the full-frame or 35 mm format diagonal d_{FF} to the diagonal d_{sensor} of any other sensor format. As the focal lengths f_{norm} of a normal lens for a given format is virtually identical to the corresponding sensor diagonal, we get the relation, with $f_{norm,FF}$ being the normal focal length for the full-format:

$$CF = \frac{d_{FF}}{d_{sensor}} = \frac{f_{norm,FF}}{f_{norm}} \tag{2.25}$$

The practical implication of the crop factor is that the focal length of any lens used with a special format or sensor can be multiplied with the *CF* to yield the equivalent focal length of a camera lens for the 35 mm format. For our example of the 6 cm × 6 cm format, which is more precisely 56 mm × 56 mm, we have a diagonal of 79.2 mm, and thus a crop factor of 0.55. The 80 mm focal length with the 6 cm × 6 cm format has the same perspective as a 44 mm focal length for the full format, thus being a normal lens for that format.

On the other hand, the depth of field impression and the related bokeh change when a different format is used (see Section 6.9). The depth of field decreases when the image magnification increases. As a consequence, a medium format camera with its normal lens produces images with less depth of field than a full format camera with its normal lens. This is very important for image or photographic design and may help to understand why medium format cameras are very attractive for portraitists. However, cameras with larger formats need larger lenses, and thus are less easy to handle than full format cameras and are more prone to a camera shake blur.

As lenses for system cameras can be easily interchanged, there is a great variety of lenses that are optimized for special applications. Image aberrations can be generally corrected without great effort only for a limited range of focal lengths. High quality lenses are always complex combinations of different lens elements. Thus, we can roughly classify camera lenses according to their focal length, and especially for the 35 mm format (full frame) we have the following characteristics:

- The normal lens generally has a focal length of 50 mm, but also focal lengths between 40 mm and 60 mm can be considered normal lenses. Their maximum relative apertures are typically between $f/1.4$ and $f/2.8$.
- Long focus lenses above the normal lens range up to about $f = 1200$ mm (consisting of glass lenses) and $f = 2000$ mm (consisting of mirror lens elements).
- Wide angle lenses have focal lengths below the normal lens down to approximately 13 mm, while special fish-eye lenses with special projection geometry and strong

distortions can be found down to about 6 mm. Their angle of view can be as large as 180° or even slightly more.
- Zoom lenses with variable focal lengths can be found in a great variety for all ranges.
- There exist special lenses for close-up photography. Their length can be extended to achieve a large image distance. Then image sizes for 1:1 imaging are possible. Other special lenses for perspective correction (PC) can be found where whole lens groups inside the optical system can be displaced or tilted for optimum imaging.

A more detailed description of camera lenses with respect to their specifications and internal lens arrangements is given in Chapter 6. A further discussion on formats and sizes is given in Section 4.3.

2.6.2 Digital single lens reflex camera

Camera systems with digital image sensors are the consequence of advances in the production of electronic components. The principal step from the film-based camera to the digital electronic camera has only been the replacement of the film by the electronic sensor. However, the production of large area electronic image sensors is much more difficult and expensive than that of smaller ones. Thus, the first digital cameras commercially available had image sensors that were in general smaller than the 35 mm format, and also were not system cameras with interchangeable components. The commercial availability of digital single lens reflex (DSLR) cameras started in the first decade of the 2000s. In most cases, the sensors of these cameras also were smaller than the 35 mm format but the compatibility with lenses of the older SLR systems was often given. The most widespread format for DSLR cameras is currently the APS-C format with a crop factor of about 1.5. The format dimensions of about 24 mm × 16 mm vary slightly depending on the manufacturer, and thus the format is also termed differently, for instance APS-C or DX. As for larger format systems and studio cameras, interchangeable backs for films were always available, and thus a digital back with a conventional SLR was the only part necessary to realize a DSLR once the development of large electronic image sensors at high quality and affordable prices were available. Also, new larger digital formats have been developed in order to combine the flexibility of 35 mm systems with the advantages of still larger professional formats, to name only for instance, Leica (45 mm × 30 mm), Pentax and Hasselblad (44 mm × 33 mm, 53 mm × 40 mm).

2.6.2.1 Characteristics

DSLR cameras can be considered the logical further development of the traditional analog film SLR camera. They are based on the same function principle as the most advanced SLR cameras with the only difference being that the photographic film medium has been replaced by a digital electronic image sensor. This can be seen in Figure 2.19

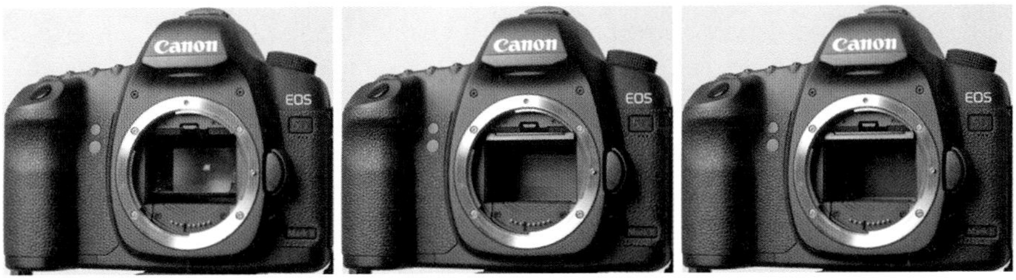

(a) (b) (c)

Fig. 2.23: Front view of a full-format DSLR camera body (Canon EOS 5D Mark II). (a) Hinged mirror at rest position; (b) hinged mirror flipped up giving view to the focal plane shutter in the back; (c) open focal plane shutter giving view to the image sensor; the slightly green color is due to the infrared stop-filter in front of the sensor.

where principal setup of a conventional SLR camera is opposed to its modern digital version of high performance. Figure 2.23 illustrates views of a full-format DSLR camera with a special look at the hinged mirror, the focal plane shutter and the image sensor, respectively. With increasing application of electronic components in the system, new features especially in combination with the image sensor and camera control processors have become available. Here, some remarkable differences should be mentioned that are available only with an electronic image sensor: images can be directly controlled before and after capturing through the use of an optical monitor display, as well as live view monitoring of a scenery allowing for video sequence recording. Also, the ISO sensitivity can be chosen individually for each image within a wide range whereas for SLR cameras that has been always fixed by the chosen film. In principle, all camera control can be done using the monitor display as a control panel.

Moreover, a purely electronic image sensor could potentially allow for an electronic shutter and the abandonment of mechanical shutters. The advantages then would be much shorter exposure times in the order of shorter than 1/10,000 s, no vibration caused by mechanical movement and a very quiet operation mode. However, there are still strong liabilities that cannot yet be overcome at the moment. The first is that reading out the information of around 10^8 photocells of a high-performance sensor in a very short time is not satisfactorily resolved, leading to image distortions like rolling shutter effects in the case of fast moving objects (see Chapter 4). The second one is that flash synchronization is very difficult or even impossible with some types of sensor technology. As a consequence, all current DSLR use a built-in mechanical focal plane shutter like the legacy SLR cameras. High-performance DSLR use complementary central shutters with specially designed lenses. There are also some manufacturers that complement the mechanical shutters with an electronic shutter for some applications, but an electronic shutter is never used as the only possibility in a DSLR.

Due to electronic advancements, exposure control can be performed by complex sensor systems measuring the light at well-defined spots in the image or distributed over a certain range. As mentioned above, a very important point is the performance of the autofocus system in a camera. The improvement of it is an ongoing process where the DSLR, due to separate light paths, still have advantages over cameras where the autofocus is controlled by the information captured by the image sensor alone. One of the disadvantages of the modern DSLR as compared to their SLR versions is their larger weight due to their complex electronics and the inevitable battery pack.

2.6.2.2 Camera lenses

Lenses for system cameras are typically optimized for a special camera design, as for instance the lens mount, the camera electronics and especially for the image sensor of the DSLR. In general, if a DSLR camera is equipped with the same lens mount as its SLR predecessor most of the legacy camera lenses may be used, however, with some restrictions. The manufacturer specific mount ensures that the focal plane is always at the same distance from the lens, whether or not the camera is a SLR or a DSLR, even with a different sensor format. For different camera and lens systems, there exist also mechanical adapters to match a lens to different mounts. However, digital image sensors work best with lenses ensuring ray paths close to a telecentric lens design on the image side due to their semiconductor properties (see Section 3.4.5). In the case of wide-angle lenses designed for film SLR, the slanted light incidence at the periphery of the image may be uncritical for analog films but may cause problems as vignetting or color aberrations when used with digital electronic sensors (see Section 4.6.1). These problems may be corrected to a certain extent after the image capture by image processing software but nevertheless this may degrade image quality. Lenses with longer focal lengths, and respectively, smaller angles of view are less prone to these aberrations, especially when used with smaller format image sensors.

Only lenses with an image circle larger than the image diagonal of the sensor should be used with a camera. Figure 2.24 illustrates the use of 50 mm lenses mounted on cameras with different sensor sizes. In the example, FX designates the 35 mm full format and DX designates the APS-C format with a crop factor of $CF = 1.5$. Figure 2.24a shows the image taken with a 50 mm FX lens designed for the full format and mounted on the corresponding sensor. We get a standard image with an angle of view of $\Psi = 47°$ as the 50 mm lens is the normal lens for full format. Mounting a 50 mm DX lens designed for the smaller crop format to the same full format camera body delivers the same image size (Figure 2.24b). However, the image is shaded at its lateral borders as the image circle of the DX lens is smaller than that of the FX lens and cannot fully illuminate the sensor. The normal lens for the smaller DX format has a focal length of around 33 mm, which can also be calculated by taking the 50 mm focal length and reducing it by the crop factor CF. As a consequence, images taken with a 50 mm lens on a DX sensor manifest a relative magnification of CF as given by Equation (2.11). This magnification of 1.5 can be seen in Figure 2.24c, and the angle of view is consequently narrowed to $\Psi = 32°$ according to Equation (2.10). This corresponds to nearly the same perspective we get when using a

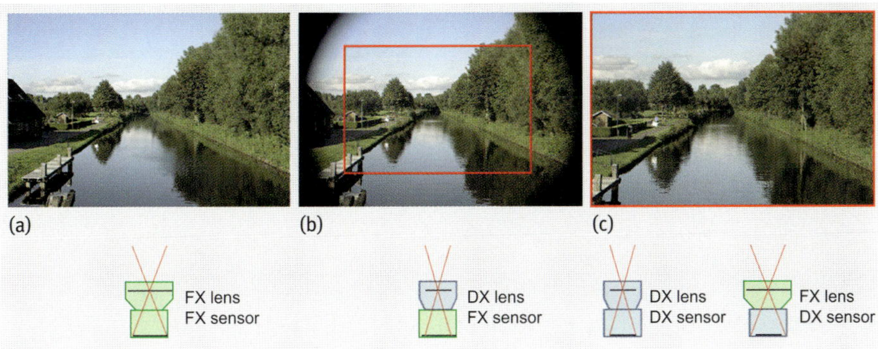

Fig. 2.24: 50 mm lenses for full format (FX) and crop format (DX), $CF = 1.5$, mounted on cameras with the corresponding sensor formats; the size of the DX sensor image is indicated by the red frame; a brightness fall-off due to mechanical vignetting is seen in b).

75 mm lens with the full format. In image c) no shading occurs because the image circles of both lenses are always larger than the DX format, and thus the sensor is fully illuminated.

In general, lenses designed for a larger format can always be used with smaller sensors resulting in a relative magnification identical to the crop factor. The overall classification of lenses for DSLR is the same as given above for SLR cameras. The normal lens for the sensor size is always the reference.

2.6.2.3 Examples for DSLR cameras

In the following, we will consider some examples for DSLR of different sensor sizes and compare their specifications. The comparison helps for understanding their performances as well as their suitability for some practical applications. But our comparison should not be regarded as a judgment on quality issues or as any recommendation. This holds also for the other examples, such as those provided in Table 2.3.

Table 2.2 lists the data for a medium format, a full format and an APS-C format camera. Their principal structure is very similar, as described above. The aspect ratio of all sensors is 3:2. They are produced in CMOS technology, and all have sensors pixel numbers between 24 MP and 38 MP. A consequence of the high pixel density is that the size of one pixel is below 6 µm. On one side, small pixels may support a high resolution; on the other side, decreasing pixel size may lead to decreasing signal to noise ratio and increasing amount of data to be processed for the image.

The assessment of a camera should always be done with respect to the intention for which the camera is used. If for instance images are taken to be reproduced and then viewed from a distance not shorter than their diagonal, a camera with a sensor of about 5 MP is sufficient due to the limited resolution of the human eye (see Section 1.4). Higher resolution is necessary if the images are intended to be largely magnified, for

Tab. 2.2: Technical specifications of some DSLR cameras with different sensor formats.

Camera	Leica S	Canon EOS 5D Mark IV	Nikon D7200
sensor format	digital medium format	digital full format	digital APS-C (Nikon DX)
aspect ratio	3:2	3:2	3:2
image sensor	CMOS	CMOS	CMOS
width/mm × height/mm	45 × 30	36 × 24	23.5 × 15.6
image size/pixel	7500 × 5000 (37.5 MP)	6720 × 4480 (30.1 MP)	6000 × 4000 (24.0 MP)
pixel pitch $p/\mu m$	6.0 µm	5.4 µm	3.9 µm
image diagonal d/mm	54.0	43.3	28.2
crop factor CF	0.80	1	1.54
ISO speed latitude	100–12,500	100–32,000 (up to 102,400)	100–25,600 (up to 102,400)
metering range/EV	1.2–20 (at ISO 100)	0–20 (at ISO 100)	0–20 (at ISO 100)
shutter	focal plane and central	Focal plane	Focal plane
exposure time	60 s–1/4000 s	30 s–1/8000 s	30 s–1/8000 s
color depth/(bit/pixel)	16	14	14
image data format	DNG (RAW), JPEG	Canon-RAW, JPEG	Nikon-RAW, JPEG
critical f-number $f_{\#crit}$	9.0	8.1	5.8
price (body, Sept. 2016)	17000 €	3500 €	1000 €

instance, to produce posters or to present details in a photo. To fully capitalize on the high resolution of a camera with sensors of more than 20 MP, the lenses must also be of a very high quality, which is often only possible if prime lenses with fixed focal lengths are used. Also, the stopping down to aperture values $f_\# > f_{\#crit}$ impairs the image quality due to diffraction and becomes more important the higher the resolution is in combination with a smaller image sensor. If for instance the number of pixels of an APS-C sensor is increased above 32 MP, then the critical f-number is below $f_{\#crit} = 5$, leaving almost no margin for a large depth of field without impairing the resolution and requiring a high lens quality at wide apertures. Moreover, any motion or shake of a camera with high resolution must be avoided. Even the mirror motion in the camera must be taken into account and should be limited. For a professional press photographer, the reliability of a camera is more important than very high resolution. Here, a rugged camera, ensuring more than 200,000 shutter cycles, and a sensor with less than 20 MP to reduce the data amount is preferred. If a camera is used in low light situations, then a larger pixel size is also more favorable as the noise performance becomes better.

A last point that should not be covered in detail here is the image data format. All high-quality cameras can store raw image data, which guarantees that no information contained in image details is lost after the capture (see Section 4.9). These details can be elaborated in a post-processing step whereas a data format like JPEG yields compressed data with loss of certain information that can no longer be restored. The advantage of a compressed data format is that much less data are necessary.

All cameras in Table 2.2 are of superior quality. For professional use, cameras with sensors equal to or larger than the full format are preferred.

2.6.3 Digital compact camera

With the advent of digital optical sensors, a variety of new digital camera types and formats has been developed. This is currently an ongoing process and a classification according to SLR and compact cameras, as has long been the case in order to differentiate with respect to high quality on one side and easy use on the other side, has become more and more difficult. The main objective of compact cameras, sometimes termed point-and-shoot-cameras, is to have an easy-to-use automatic camera with relatively small size and low weight as compared to the DSLR. As the physical dimensions in general depend on the sensor size, these cameras in most cases have optical elements and image chips smaller than those of the full format. However, with technologic advancements also high-performance compact cameras with noninterchangeable powerful zoom lenses have become available that range between the typical consumer compact cameras and the more professional DSLR system cameras. They are often termed bridge cameras. Recent developments, which are described below, present mirrorless system cameras with interchangeable lenses and relatively large sensors that can no longer be categorized according to that rough classification as DSLR for most complex and versatile systems and as compact cameras aiming rather to the consumer market. Keeping that in mind, we nevertheless go on with that rough classification scheme for the moment.

2.6.3.1 Characteristics

The typical structure of a compact camera is illustrated in Figure 2.25. The camera in principle consists of a compact lens module, which in many cases is integrated with the image sensor and an infrared filter. The camera body may have an optical finder module to inspect the scenery before capturing. It consists of an optic module by which the scenery can be separately observed, although it suffers from parallax errors as the corresponding light path differs from that across the lens. This effect is more pronounced the closer the objects are to the camera. In some modern cameras, this optical finder is replaced by an electronic viewfinder. Here, the image captured by the image sensor is electronically projected to a miniature display and observed by an eyepiece lens. The electronic viewfinder shows exactly the same image through the lens but may suffer from a certain time lag and may have a limited quality especially in critical light situations. The viewfinder can also be omitted for a more compact camera. In virtually all cases, the finder function is also carried out by the monitor display on the camera back, which is usually a liquid crystal display (LCD). The optical finder, however, is especially

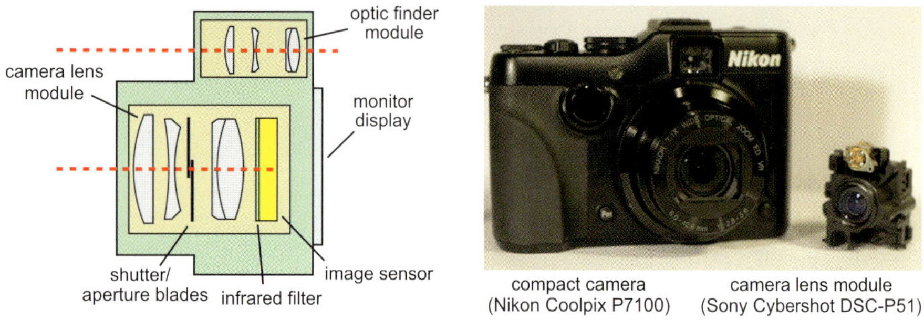

Fig. 2.25: Schematic cross section of a compact camera (left); compact camera with optical finder and a camera lens module (right).

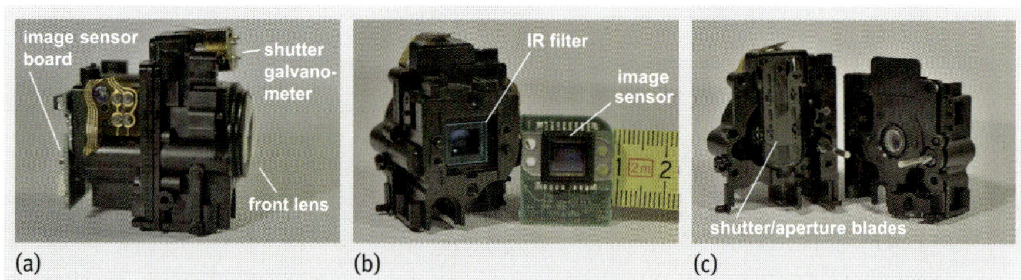

Fig. 2.26: Module of a zoom camera lens integrated with IR filter, sensor chip and shutter/aperture blades (Sony Cybershot DSC-P51, 1/2.7″ sensor with 6.72 mm diagonal).

advantageous if the outside illumination, for instance in bright sunlight, is so high that the external display visibility is impaired.

The dedicated camera lens module in the camera is in general optimized for that camera with its well-defined sensor size. An example for a lens module is given in Figure 2.25 and Figure 2.26, presenting a 1/2.7″ image sensor having a 6.72 mm diagonal. The infrared filter is used to filter out the nonvisible infrared (IR) radiation, which tends to falsify images as the silicon-based chip, unlike the human eye, is very sensitive in that spectral range. The camera lens in our example consists of four lens elements of which two can be displaced by electrical motors. One lens element is shifted in order to vary the focal length, thus performing an optical zoom function. The second element is displaced by the autofocus control to achieve optimal sharpness of the image on the sensor. This autofocus control in compact cameras is usually carried out by the image processor electronics and not by a separate autofocus sensor. As a consequence, the time delay between actuating the shutter button and image capturing may be longer than in a DSLR because all the processes for exposure metering and autofocus setting rely on sensor chip data acquisition and evaluation.

The simplest way to realize a mechanical shutter is to combine it with the aperture stop. For our example in Figure 2.26c, the aperture stop is accomplished using two metal blades moving in opposite direction thus forming a square aperture (see also Figure 2.8b). The blades are driven by a galvanometer, which ensures a very precise and quick setting of the desired aperture. During the life view function of the camera before taking the image, the aperture is fully open. For image capturing, the aperture is set to the necessary f-number while it stays at this position during the exposure time t_x. The aperture/shutter then is fully closed after the exposure during data readout to reopen again after it. For this example, the aperture stop is nearly in the center of the lens, which is beneficial for suppressing lens aberrations. Moreover, the opening and closing shutter blades constitute a central shutter for the camera. Cameras of higher quality are equipped with iris diaphragm apertures and shutters. The mechanical shutter is adapted to the electronic shutter mechanism, which is different for different sensor technologies. New generations of electronic shutters, for instance, having exposure times down to 1/32,000 s, can operate at much higher speeds than mechanical ones but also have some quality limitations and are not yet fully optimized. Most cameras make a combination of mechanical and electronic shutters.

2.6.3.2 Consequences of the compact setup

In the case of a small image sensor, the focal length of its normal lens is correspondingly shorter than for the 35 mm format. As this format in many cases serves as a reference, some manufacturers specify the lens data as equivalent values for the full format. The lens of the Nikon Coolpix P7100 camera is specified as a zoom lens 6.0–42.6 mm (Figure 2.25). The diagonal of its 1/1.7″ sensor is of 9.5 mm resulting in a crop factor of $CF = 4.56$ according to Equation (2.25). Consequently, the equivalent lens for the full format is 27.4–194.2 mm. The angle of view for that lens is from the wide-angle range of 77° down to the long focus range of 13° (Equation (2.10)). We have the same perspective with the lens of the compact camera as it would be for the equivalent lens of the full format. The depth of field, however, is different because the absolute value of the focal lengths is smaller. The image of the compact camera yields a larger depth of field for the same equivalent field of view and aperture stop than the corresponding full format image would do. This should be taken into account for image design. Moreover, due to the smaller dimensions, there are high demands for the precision of mechanics and optics during fabrication. When it is required to reduce the camera size, the number of lens elements must be kept low, and thus incorporating aspheric elements in the lens design may be a good solution. Using plastic aspheric lenses is favorable with respect to weight and production costs, but glass lenses are of better quality and long-term stability.

A critical point for compact cameras is the size of the image pixels. Small sensors and a large number of pixels require a small area for the pixels. As already discussed above with examples of some DSLR cameras, compact cameras therefore tend to be more prone to image noise. Also, the optimal apertures for these cameras are at lower f-numbers than for larger formats. This again stresses the importance of high lens quality and aspheric lenses as the errors become in general more obvious at large apertures.

Last but not least, stopping down with compact cameras to increase the depth of field is in many cases linked to diffraction blur. For the example of the 10 MP camera Coolpix P7100, the pixel pitch of 2.1 µm yields a critical f-number of $f_{\#crit} = 3.1$. This is exceeded even at full aperture in the long focus range. However, as stated above, when magnified images, taken by such a camera, are viewed at the distance of the image diagonal, it is nearly impossible to identify resolution limitations. But for image reproductions of higher quality, larger format cameras are required.

2.6.3.3 Examples for compact cameras

In the compilation below (Table 2.3), some compact cameras are presented with their technical specifications. Their great advantage is the small size and low weight combined with an overall high image quality. There is no ranking of the cameras intended, only the current status of technical realization. The examples are chosen for image sensors smaller than the full format. All cameras have definitely smaller dimensions and

Tab. 2.3: Technical specifications of some compact cameras with different sensor formats.

Camera	**Panasonic Lumix DMC-LX100**	**Sony Cybershot DSC RX-100 IV**	**Canon Powershot SX 720 HS**
sensor format	4/ 3″	1″	1/2.3″
aspect ratio	4:3 (also 3:2, 16:9, 1:1)	3:2	4:3 (also 3:2, 16:9, 1:1)
image sensor	CMOS	CMOS	CMOS
width/mm × height/mm	17.3 × 13.0 (15.8 × 10.6*)	13.2 × 8.8	6.2 × 4.6
image size/pixel	4272 × 2856* (12.2 MP*)	5496 × 3672 (20.2 MP)	5184 × 3888 (20.1 MP)
pixel pitch p	3.7 µm	2.4 µm	1.2 µm
image diagonal d	19.0 mm*	15.9 mm	7.7 mm
crop factor CF	2.3*	2.7	5.6
ISO speed latitude	100–25,600	125–25,600	80–3200
camera lens	1.7–2.8/ 10.9 mm–34 mm	1.8–2.8/ 8.8 mm-25.7 mm	3.3–6.9/ 4.3 mm–172.0 mm
35 mm equ. foc. length	25 mm–77 mm*	24 mm–69 mm	24 mm–963 mm
exposure time	60 s–1/4000 s (mech.) 1 s–1/16,000 s (electr.)	30 s–1/2000 s (mech.) 30 s–1/32,000 s (electr.)	15 s–1/3200 s (mech./electr.)
color depth	12 bit/pixel	12 bit/pixel	–
viewfinder	electronic	electronic	no
image data format	RAW, DPOF, JPEG	ARW (RAW), JPEG	JPEG
critical f-number $f_{\#crit}$	5.5	3.6	1.8
weight/g	351	298	270
price (Sept. 2016)	700 €	1000 €	380 €

* data for 4:3 format; only a part of the sensor is used

a lower weight than the standard DSLR, which has a weight of the order of 1 kg when equipped with a standard zoom lens.

It should be noted here that the Panasonic camera uses the four-thirds format (4/3″), which is slightly smaller than the APS-C sensor. The sensor is not entirely used for the camera; only 12.8 MP of the total 16.8 MP. For that reason, the image circle of the lens can be smaller, resulting in smaller and lighter lenses. Moreover, the sensor can be operated for different formats without cropping of image parts. The data for that camera listed in the table are given for the 4:3 format as indicated.

The Sony camera has a smaller image sensor and a higher number of pixels. Therefore, the pixel pitch is narrower and leads to a smaller value of the critical f-number compared to the Panasonic camera. According to the specifications, a higher resolution but a lower signal to noise behavior can be supposed. Both cameras have a 12-bit color depth, which can be exploited for image post-processing based on the RAW data format.

The camera with the smallest sensor in Table 2.3 is the most compact in this compilation. The color depth is not specified and is irrelevant as the lossy JPEG data structure is based on an 8-bit structure, and a RAW data post-processing in order to capitalize on the full quality is not possible. The pixel pitch is only of 1.2 µm, and this is the reason why the critical f-number is only of 1.8. The lens has an impressive hardware zoom factor of 40, which is the ratio between the longest to the shortest focal length. The possible f-numbers, however, are much larger than the critical f-number. In that example, a smaller number of pixels would have been more beneficial as the theoretical resolution of the small pixel size can never be achieved, due to diffraction limitation of the lens. With fewer pixels, and thus larger pixel size, the low-light performance of the camera as well as the data evaluation speed could certainly still be improved.

2.6.4 Other types of digital cameras and further developments

2.6.4.1 Mirrorless interchangeable lens camera and single lens translucent camera

Taking high-quality DSLR as one end of the still camera selection and compact cameras as the other end, there is a great variety of new developments in between that combine the advantages of different aspects. In the past few years, there has been growing interest in mirrorless interchangeable lens cameras (MILC), which typically have the compact form, like a compact camera, but use interchangeable lenses like a DSLR camera. A more convenient term for MILC is DSLM (digital single lens mirrorless), which underlines its affinity to DSLR cameras. By abstaining from a hinged mirror for an optical viewfinder as well as for autofocus and light metering sensors, the whole camera body becomes smaller as no roof prism and no electronics in parallel to the image sensor electronics are needed any more. Moreover, no mechanical noise and vibration due to mirror movement happen. As the distance from the rear vertex of the lens to the image sensor can be very short, new lenses for the compact system are developed and optimized. The short flange-to-image plane distance can be of advantage especially for wide angle lenses with short focal distances. They need a more complex retro focus design for application with DSLR, but for mirrorless cameras the design is much simpler. The short flange focal distance also allows for relative apertures larger than $f/1.0$, which is not the case for all SLR cameras. As interchangeable lenses are used, the shutter principle in DSLM is usually the same as in DSLR cameras, which is a focal plane mechanical shutter combined with an electronic shutter. The image sensors are in general the same as in the DSLR, ranging from full format to the best crop format sensors. Recently, image stabilization mechanisms have been developed by which the image sensor in the camera body is slightly shifted to compensate for the camera movement. Some methods are even combined with corresponding stabilization methods in the camera lens to reduce vibrational effects that might impair the image resolution. The technical specifications of sensors in DSLM cameras are virtually identical to that in DSLR cameras. The biggest challenge of the DSLM, however, is the speed of image data processing. Here, the image sensor is the bottleneck as all functions like autofocus and light metering rely on the data of the image processor and not on separate electronics.

A special type of digital camera termed the single lens translucent (SLT) camera has been developed by Sony. It has no hinged mirror but a fixed beam splitter at the position of the hinged mirror in a DSLR camera. The camera is of nearly the same design as a DSLR, rather than a DSLM camera. It is quite compact without an optical viewfinder but with an electronic one like most DSLM cameras. The intention of the beam splitter or "mirror" is that only a small part of the incoming light is reflected to a dedicated autofocus sensor like in a DSLR, but most of the light is reaching the image sensor and processed by it. The Sony α 99 II of 2016 has a 42 MP sensor and uses, in addition to the dedicated autofocus sensor, the data evaluated by the image sensor to adjust the lens focus position. Thus, the principle of a DSLM is combined with that of a DSLR camera. As a consequence, the camera is much faster than typical DSLM or compact cameras. There is also an image sensor stabilization function implemented. The position of the image sensor in the body is controlled by a five-axis-stabilization mechanism to counteract any camera shake detrimental to the high-resolution image sensor. This example shows that the traditional classification scheme of cameras may be abandoned in the future and a combination of different approaches may lead to a variety of high-quality camera systems. Moreover, the implementation of video functionality leads to a blend of high-quality still camera functionality with that of electronic motion picture acquisition.

In principle, all established manufacturers of DSLR cameras offer newly developed DSLM cameras of lower weight and compact bodies with their own type of new lenses. As they exist in parallel to their DSLR and have shorter flange-to-image plane distances, the manufacturers also offer special adapters to fit all traditional SLR-lenses to the DSLM cameras. Thus, there are a very large number of lenses available for them. On the other hand, DSLR cameras with their longer flange focal distance cannot use the DSLM lenses for imaging from infinity but only for a restricted shorter object distance.

In the following chapters in this book, the term DSLR is sometimes used in order to emphasize the high quality of a camera. In that case, a high-quality DSLM or SLT is implicated in this designation.

2.6.4.2 Mobile phone camera and miniature camera

In the following section, we present a short overview about mobile phone and miniature cameras. It simply serves as a quick comparison to the other camera systems presented in this chapter. A more detailed and updated discussion about smartphone camera (SPC), respectively, miniature camera systems is given in Chapter 7, which is specially dedicated to this topic.

Taking photographs has not been the prime objective of mobile phones. However, the interest of this function has continuously increased with the growing number of users in social media and networks and constitutes a key feature of smartphones. The latter can be considered as advanced versions with more functionalities than simple mobile phones. Sending online snapshots using mobile phones has become very popular and defines the requirements for their camera modules. Mobile phones have to be compact, easy to use and economical with respect to power consumption. Their thickness is between 5 mm and 11 mm with a typical value around 7 mm. Consequently, the camera lens in the optic module integrated with the image sensor must have a focal length of less than about 10 mm. The open body of a mobile phone with integrated camera module is depicted in Figure 2.27a. Figure 2.27b shows a dismounted camera module with image sensor board on the backside. The lens is as compact as possible with a minimum number of elements, which necessitates aspheric lenses to compensate for the different lens aberrations. Figure 2.27c illustrates the cross-section of a camera lens developed by Zeiss for a Nokia phone camera [Nas11a]. It consists of four aspheric lens elements and their arrangement resembles a classical Tessar lens (see Chapter 6). The resolution of this camera lens (5.2 mm focal length, f-number 2.8) is superior to the best

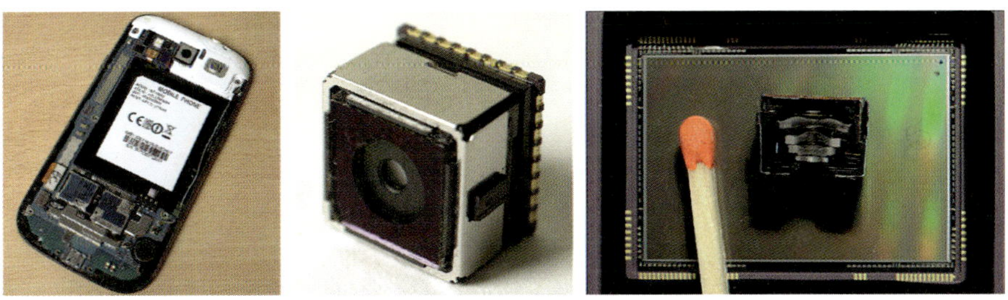

(a) (b) (c)

Fig. 2.27: (a) View to the inner parts at the backside of a mobile phone; the camera module is located at the top; (b) dismounted camera module with image sensor board of a mobile phone; (c) cross-section of a mobile phone camera module by Zeiss consisting of four aspheric lens elements; a match and a full-format digital image sensor are shown for comparison (with kind permission of Carl Zeiss AG).

full format lenses due to its very short focal length and dimensions, but only over a very small image area. Similar lens designs based on aspheric lenses can be found in other miniature cameras. Due to the compact structure of these cameras, only a fixed focal length without optical zoom is possible. Moreover, the depth of field with these short focal lengths is so wide that an adjustment of the lens position to achieve a sharp image is in most cases not necessary. Thus, we have a fix-focus lens for the majority of mobile phone cameras. For some more advanced miniature cameras, a position adjustment is done but only of the order of fractions of 1 mm. The diagonal of the image sensors is of the order of around 10 mm or less.

The disadvantage of these miniature lenses is that they are typically made of plastic materials. Hence, their durability, mechanical stability and optical properties like refractive index and dispersion can be considered inferior to those of optical glass lenses.

There is in general no space left for a variable aperture stop or even a mechanical shutter. The sizes of the lens elements already limit the opening of the camera lens and act as the physical aperture stop. Their typical f-numbers are between $f_\# = 1.8$ and $f_\# = 2.8$. Larger values like in bigger cameras and stopping down, for instance, due to variable apertures, do not make sense. This would be detrimental to the resolution given the fact that the pixel pitch in the cameras is around 1 µm. Moreover, due to the lack of a mechanical aperture stop and shutter, only electronic shutters are currently realized in the miniature cameras.

As for their detailed technical specifications, examples for some high-quality miniature cameras are listed in Table 2.4. The first two examples (Nokia Lumia, Apple iPhone) refer to cameras integrated in mobile phones. The third example (DxO one) is a miniature camera that can be connected to a mobile phone. DxO one is an autonomous camera that can be used as a standalone system, however, without a viewfinder. Once connected to a mobile phone, the screen of the phone can be used as a viewfinder and control panel for the camera. The camera has a 1″ image sensor like many compact cameras. The sensor size and pixel pitch of 2.4 µm are larger than typical sensors and pixel pitches in mobile phones. Thus, a corresponding high quality can be expected and at the same time, due to the lack of additional complexity, the size and the weight of the camera can be kept at a minimum.

The other two examples of mobile phone cameras have a pixel pitch of only 1.2 µm. Nokia Lumia 1020, using a larger chip than most of its competitors, offers impressive 38 MP resolution internally. Here, it may seem incomprehensible what the developer intended with a resolution that most DSLR camera never achieve. The aim, however, is the realization of a 3 × digital zoom factor yielding a sharp image of about 5 MP resolution. For that purpose, the images are intensively post-processed after capturing to reduce the effective number of pixels and to improve its low light performance.

The Apple iPhone 7 Plus has a smaller sensor resulting in a lower resolution which, however, is still high enough if the prime objective is not for professional photographer's use. Also, an intensive post-processing of the images is possible using multishot images captured within a short time interval with different exposure settings. The objective is to

Tab. 2.4: Technical specifications of some miniature cameras with different sensor formats.

Camera	Nokia Lumia 1020	Apple iPhone 7 Plus	DxO ONE
sensor format	2/3″	1/3″ and 1/3.6″	1″
aspect ratio	4:3	4:3	3:2
image sensor	CMOS, optical image stabilization	CMOS, optical image stabilization	CMOS
width/mm × height/mm	8.8 × 6.6	4.8 × 3.6 and 4.0 × 3.0	13.2 × 8.8
image size/pixel	7152 × 5360 (38.3 MP)	4000 × 3000 (12 MP)	5406 × 3604 (20.2 MP)
pixel pitch p	1.2 µm	1.2 µm and 1.0 µm	2.4 µm
image diagonal d	11 mm	6 mm and 5 mm	15.9 mm
crop factor CF	3.9	7.2 and 8.7	2.7
ISO sensitivity latitude	100–4000	–	100–51,200
camera lens	2.2/7.2 mm	1.8/3.9 mm and 2.8/6.4 mm	1.8/11.9 mm
35 mm equ. foc. length	28 mm	28 mm and 56 mm	32 mm
exposure time	4 s–1/16,000 s (electr.)	– (electronic)	30 s–1/20,000 s (electr.)
color depth	10 bit/pixel	–	12 bit/pixel
viewfinder	no	no	no
image data format	RAW, JPEG	RAW, JPEG	RAW (DxO), DNG, JPEG
critical f-number $f_{\#crit}$	1.8	1.8 and 1.0	3.0

compress image information over a high dynamic range (HDR) into a single image. That improves the overall visibility and is especially advantageous if the dynamic range of the scenery is much higher than that of the sensor. A special point for the Apple iPhone 7 Plus is the realization of a 2 × optical zoom factor. For that purpose, a second lens with twice the focal length of the first and a second image sensor are integrated in the phone camera. Both are fixed lenses but can be chosen at will.

As a conclusion, miniature cameras have recently progressed very much. There is, however, still the limitation due to their small physical dimensions of the sensor and its pixels. They are more prone to diffraction due to their small absolute pixel pitches and limited apertures. Small pixels gather less light than larger ones, thus especially the low light performance must be inferior to that of larger ones. Some drawbacks may be compensated by intensive image processing in the camera, for instance by multishot techniques or selectively binning neighboring pixels. However, individual image design is hardly possible due to the low versatility. The main objective here is the easy, automatic capturing of good images. For higher requirements and advanced image design, larger and more complex systems are necessary.

2.6.5 Cameras for scientific and industrial purposes

In contrast to consumer cameras, cameras used for scientific and/or technical purposes, for instance in industry, are mostly not standalone systems. Some of them are rather small (Figure 2.28a), but they require external control, typically via a PC or another suitable device. Depending on the camera, sensors with very different sizes, pixel numbers, aspect ratios, read out speed, etc., are available. Many of these can be ordered with or without color filter array, which typically is a Bayer mask, and with or without optical microlens array (see Section 4.6.3). Special camera lenses are available as well.

In general, the application purpose of using cameras for scientific and technical purposes is much more manifold than just photography. Although the following cannot be completely understood without knowledge of the following chapters, in particular Chapter 4, it is important to discriminate different situations or applications of imaging. Of course just taking pictures of objects is one intention. But if the camera should be used as a device for measurements, special care has to be taken for several reasons discussed later in this book, particularly in Chapter 4. This almost excludes consumer cameras, even expensive DSLR and DSLM, as all of them suffer from unavoidable image processing, even when used with raw data. As an example and a practical hint, if the intention is the measurement of the spatial intensity distribution of light, one should use a CCD camera with its linear response. The sensor should have a fill factor of at best 100 %, no color filter array and no optical microlens array. This avoids interpolation of data points which in the worst case is based on a guess of the interpolated signal which itself may

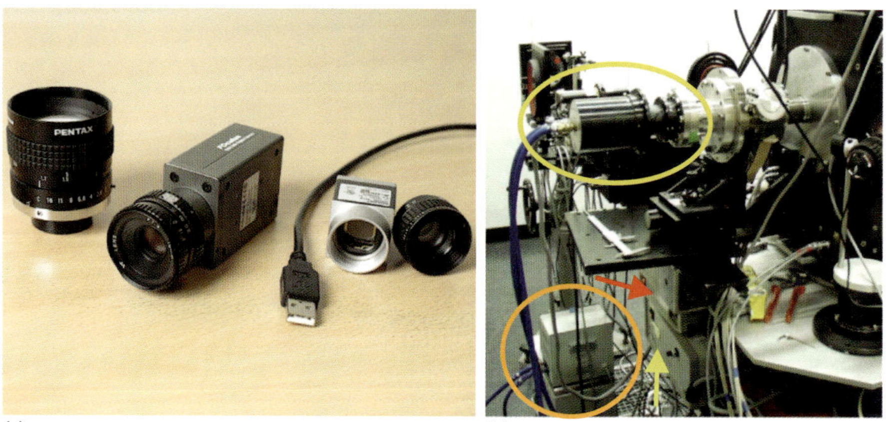

(a) (b)

Fig. 2.28: (a) Examples of typical cameras and fitting lenses usable for scientific and/or technical purposes; (b) slow-scan iCCD-camera system used for the investigation of intense laser pulse interaction with matter (see Section 4.11.4). The camera itself, marked by a yellow ellipse, is mounted to a vacuum chamber. The system is cooled to −20 °C by means of an external cooling unit (marked by an orange ellipse). The camera controller is indicated by a yellow arrow and the driver for the MCP by a red one (see Section 4.11.3). A view into the dismantled camera head is presented in Figure 4.13b.

be wrong (see Section 4.9.3). In contrast, a CMOS camera normally shows a nonlinear response, which means that one has to rely on camera calibration, if available, or this calibration has to be performed by the user himself. Moreover, CMOS sensor pixels do quite often have a complicated shape, and thus are not rectangular with a large fill factor. Shading effects may also be a disadvantage. But again, camera selection depends on application, which, for instance if high speed or high dynamic range is required but not linearity, may favor a CMOS or scientific CMOS (sCMOS) camera.

Though our intention is not to give a comprehensive representation of scientific and technological cameras, but an illustration only, we nevertheless would like to discuss briefly the setup of much advanced scientific cameras. Figure 2.28b shows a typical example. Another one can be seen in Figure 4.73c. Such systems are operated by means of an external controller, which itself is connected to a personal computer. To keep noise from dark current low (see Section 4.7.3) quite often the camera can be cooled. Depending on the application this may be done by a simple air cooler, by simple water cooling or by an external cooling cycle with special cooling liquids, which allow for temperatures much below 0 °C.

There is a vast amount of different applications and so are the types of camera systems. Just as an example, cameras are available for specific wavelength regions, ranging from long IR wavelengths down to the short wavelength X-ray region. Some of them are made for high-speed operations, namely they offer the possibility for high frame rates, other ones for slow read out do improve the signal to noise ratio (see Section 4.7.3) and even other ones for very long integration times, for instance, for astronomical imaging.

Those scientific cameras can be operated in a much more flexible way than consumer cameras. They allow to set readout for specific regions of interest (ROI), they can be operated in binning mode, background images can be captured and subtracted, and much more (see Sections 4.8 and 4.9). One part of this flexibility may be that the user has to apply his own optics. This can either be a standard camera lens, but quite often special optics have to be used, sometimes directly developed for usage with such a camera system. A particular example is imaging in the XUV- or X-ray regime, where for instance diffractive Fresnel zone plates (in XUV) or crystal optics (XR) can be used or reflective ones such as elliptical mirrors or Kirkpatrik–Baez mirrors (or pinhole cameras, see Section 2.1). Sometimes the cameras themselves are also more complicated, as discussed in Section 4.11. Other ones make use of a large array set up by a huge amount of CCD sensors.

3 Imaging optics

The preceding chapters have been introductions to the topics of the present book and gave some basic concepts for a special optical system represented by a still camera. In most cases, when the wavelength of light is much smaller than any objects considered in the imaging process, we generally use the concept of optical rays. This is the realm of geometrical optics. If the dimensions of objects become small and the wavelength of light can no longer be neglected, as for example when discussing diffraction of light or in the domain of Fourier optics, we describe the phenomena rather using the concept of waves. The objective in the following chapter is to gather an understanding of the elements that constitute a complex optical system and how their arrangement influences the performance of this system. We focus on the propagation of rays and start with geometric optics although wave optical phenomena like diffraction are not included, and thus the physical properties of optical systems are covered only within a limited extent. A very efficient approach to understand complex systems like lenses for cameras or microscopes is the use of ray propagation matrices. In the subsequent sections, we follow the classical approaches given in optics textbooks as for instance by Hecht [Hec16] or Pedrotti [Ped17]. In general, we do not present the derivations unless they are important for the understanding. We recommend the textbooks of the cited authors for more detailed considerations. As for the concept of wave optics, this will be covered later and is a prerequisite for understanding the Fourier optics described in Chapter 5.

3.1 Principles of geometrical optics

3.1.1 Huygens' principle, Helmholtz equation and rays

Let us start with the basic principle of light generation. In a simplified approach, we assume that light is generated by a point source, e. g., by atoms, and neglect for the moment any discussion about polarizations of electric and magnetic fields. A light pulse emerging from this point source propagates as a spherical electromagnetic wave with the point being the center of the sphere (Figure 3.1a). The planes of constant phases of the electromagnetic fields constitute spherical surfaces that expand with the speed of light. According to Huygens' principle, any point on that surface can be considered a secondary point source emitting itself again as a spherical wave. These secondary waves, also termed wavelets, superimpose to form the new wavefront of the propagating wave (Figure 3.1b). We use the term wavelets in the denotation used by Born and Wolf for elementary optical waves [Bor99], although this term may also be used for other optical concepts. The wavefront is the envelope of all wavelets and is thus tangent to them. The direction of propagation is always perpendicular to the wavefront, which represents a surface of constant phase. As a consequence, the local normal to the wavefront indi-

https://doi.org/10.1515/9783110789966-003

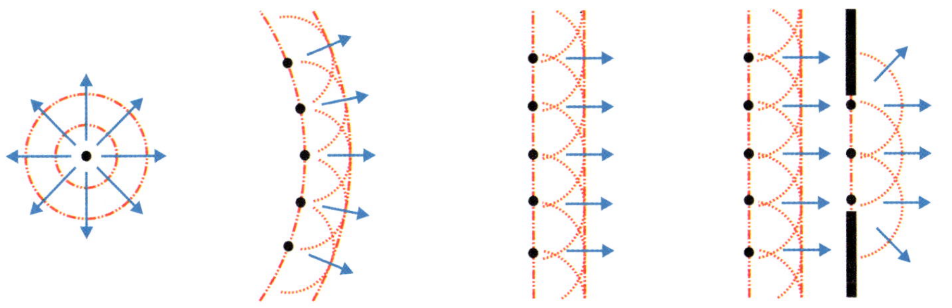

Fig. 3.1: 2D representation of light propagation; rays are indicated by arrows, wavelets by solid lines, wavefronts by dotted lines. a) Generation of a spherical wave by a point source; b) Huygens principle; c) plane wave as part of a spherical wave far away from its origin; d) obstructed wavefront causing diffraction.

cates the direction of propagation and can be interpreted as the local ray describing the propagation of light emerging from that point. In Figure 3.1, these rays are indicated by arrows.

The bend radius of a spherical wave increases with the propagation distance. The curvature is reciprocal to the bend radius, which means that the curvature of a spherical wavefront far away from its origin decreases. The wavefront becomes flat, and its curvature approaches zero at infinite distances. Thus, a plane wave can be considered the limiting case of a spherical wave at large distances where all rays perpendicular to the wavefront are in parallel. An example for that is the light coming from the sun as a point source can be described on earth using a parallel beam (Figure 3.1c). When the propagation of a parallel beam is obstructed by an aperture, the wavefront is distorted and only the central part of the wave propagates in the same direction as before. Close to the fringes, the rays locally bend away from the central propagation direction due to the restricted number of wavelets. This phenomenon is called diffraction and is described by the corresponding equations. The beam profile after transmitting the aperture has changed. In case of a circular aperture, its radial intensity profile can be expressed by an Airy pattern (Equations (5.20b) and (5.21b)). The divergence of that beam increases with decreasing aperture and can only be neglected for very large apertures. This phenomenon has been taken into consideration for the increasing diffraction blur in lenses when stopping down (Chapter 2).

A mathematical treatment of ray propagation can be done on the basis of Maxwell's equations (see Appendix A.11). This leads to the Helmholtz equation, which describes the time-independent propagation of harmonic electromagnetic waves in space:

$$\nabla^2 \vec{\mathscr{E}}(x,y,z) + \frac{4\pi^2}{\lambda^2} \cdot n^2(x,y,z) \cdot \vec{\mathscr{E}}(x,y,z) = 0 \tag{3.1}$$

This is a differential equation for monochromatic waves of the free space wavelength λ as a function of the space coordinates x, y and z. $\vec{\mathcal{E}}$ represents the complex 3D stationary electric field vector propagating in a medium, which is defined by the 3D distribution of the refractive index $n(x, y, z)$. Equation (3.1) is identical with (A.16) in Appendix A.11.

In transparent, lossless media, the refractive index is a positive, real number, and the equation is valid if the local variation of n is only slow in relation to the wavelength. An extended discussion is given in textbooks, for instance, by Saleh and Teich [Sal19].

A simple solution of the Helmholtz equation is the propagation of a spherical wave, which is depicted in Figures 3.1a and b as an illustration of Huygens' principle. Another simple solution is a plane wave propagating with a constant amplitude (see also Figure 3.1c and Section 5.1.1) in contrast to a spherical wave, of which the amplitude is decreasing on propagation.

Under the assumption of a slowly in space varying envelope, a paraxial solution of the Helmholtz equation yields a Gaussian beam, which is discussed in more detail in Section 3.1.3. Here, paraxial means that the rays, which are normal to the wavefront, are closely directed to the axis of propagation and the character of the wave is nearly that of a plane wave with a spatially modulated envelope. A Gaussian beam describes the fundamental transversal mode propagating in a laser resonator which is bounded by circular reflecting mirrors.[1] The beam exits the resonator through a circular aperture and propagates outside according to the characteristics as described in Section 3.1.3. The propagation of light, exiting from a cleaved optical single mode fiber, can be approximated in the same way.

3.1.2 Ray equation, Snell's law and reflection loss

The wavefront of propagating light can be influenced by different effects. Small particles or obstructions, as seen above, can locally distort it and lead to a deflection of rays. Another deflection of rays will happen if the refractive index n of the medium in which the wave propagates is not homogeneously distributed. n is the ratio of the speed of light c in vacuum by its speed in the medium. A higher refractive index means that the speed of light in the medium is equal to c/n and thus the light is slowed down when entering a medium of higher index. If for instance the refractive index in a glass lens is not homogenous but varies locally, then the speed of light changes correspondingly, and thus a ray in this medium is deflected from its straight path. This deflection is stronger the higher the index change per distance.

1 A. G. Fox, T. Li: *Resonant Modes in a Maser Interferometer*, The Bell System Technical Journal **40**, 453–488 (1961).

a) b)

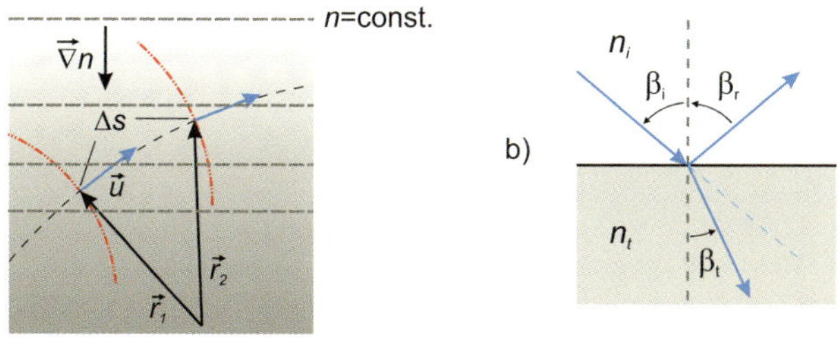

Fig. 3.2: (a) A ray propagating in a medium is deflected by the gradient of the refractive index; (b) illustration of Snell's law for refraction of light entering a medium with refractive index $n_t > n_i$.

Figure 3.2a illustrates a curved wavefront with its ray propagating in a medium with an inhomogeneous refractive index n. The position vectors $\vec{r}_1$ and $\vec{r}_2$ indicate the starting respectively end points of the wave propagation along a path Δs. $\vec{u}$ is the unit vector of the ray being normal to the wavefront and pointing into the direction of propagation. The refractive index $n(\vec{r})$ is a function of the space coordinate $\vec{r}$ with the gradient $\vec{\nabla} n$ pointing to the direction of the highest variation of n per distance. Areas of constant n are marked by broken lines. It can be shown that the deflection of the unit ray vector is directly proportional to this gradient, which is described by the ray equation:

$$\frac{\mathrm{d}}{\mathrm{d}s}(n \cdot \vec{u}) = \vec{\nabla} n \tag{3.2}$$

Thus, the direction of $\vec{u}$ is continuously deflected toward $\vec{\nabla} n$ leading to a curved ray path in the medium. This can be understood taking into account that a retardation of the wavefront in the lower part is due to the reduced speed of light in that area. If the ray enters a sector where the refractive index does no longer change with space, which means that the gradient is zero, the beam is no longer deflected as the speed of light has the same value at all positions. The beam goes on propagating in a straight line as it is known for a homogeneous medium. The consequences for practical applications are that in optical glasses of low quality, where for example the refractive index may vary slightly, deformations of the wavefront as well as deviations of the beams from their optimal light path may arise. This impairs the imaging quality and may lead to image blurring.

The principle of ray bending can also be applied for a qualitative understanding to the refraction of rays at an interface between two media having each a homogeneous refractive index. A more rigorous mathematical treatment leads to the so-called Fresnel equations, which quantitatively describe the reflection as well as the transmission of rays hitting that interface. Reflection and transmission not only depend on the angle of the beam incident to the interface and the corresponding refractive indices but are also strongly depending on the polarization of light. The continuity condition for the tangential components of the electromagnetic fields at the interface leads to Snell's law, which describes the refraction of light at that surface (Figure 3.2b). Assuming a ray incident under an angle β_i to the normal in the medium with refractive index n_i we get a beam that is reflected back to the medium under an angle β_r, which is identical to β_i. Additionally, to this reflection we also observe a beam transmitted into the medium with index n_t under an angle β_t to the normal. The relationship between the angles and the refractive indices is given by Snell's law:

$$n_i \cdot \sin \beta_i = n_t \cdot \sin \beta_t \qquad \beta_r = \beta_i. \tag{3.3}$$

The product of the refractive index with the sine of the angle is related to the numerical aperture of the interface, which will be considered in more detail in Section 3.4. For the moment, we simply remark that in general the numerical aperture is a quantity that is related to the resolving power of optical systems and indicates how much light can enter the system. Snell's law states that the numerical aperture remains constant when rays traverse the interface of two different media. Moreover, by comparing Snell's law with the ray equation it can be seen that a ray, coming from a medium with lower refractive index n_i and transmitting into one with higher index $n_t > n_i$, is bent into the medium with the higher index as the gradient of the refractive index in the interface points downwards. If the index ratio is inverted, which means that the gradient points upwards, we get a ray deflected into the corresponding direction. This is again compatible with the fact that the speed of light is slower in a medium with a higher refractive index.

If we consider a parallel beam striking the interface, then the intensity of the reflected beam strongly depends on the angle and the light polarization. The sum of the powers of the transmitted and the reflected beam equals that of the incident one. The power reflection coefficient ρ_P, also termed reflectance, is the power ratio of the reflected beam by the incident beam. In the case of perpendicular incidence, all polarizations are equivalent and then ρ_P only depends on the refractive indices according to the following Fresnel formula:

$$\rho_P = \left(\frac{n_i - n_t}{n_i + n_t} \right)^2 \qquad \tau_P = 1 - \rho_P. \tag{3.4}$$

In the case that there is no power loss, for instance, due to absorption, the power transmission coefficient, termed transmittance τ_P, is the complement of ρ_P to 1. As for practical applications, a ray in air with $n_i = 1$, striking a glass surface with $n_t = 1.5$ perpendicularly, is reflected back by a reflectance of $\rho_P = 0.04$. The same reflectance shows up for a ray coming from glass and going to air. Thus, without antireflection coating, each surface of a glass lens reflects 4 % of the incident light, which means that a simple glass lens has a characteristic reflection loss of roughly 8 % and only 92 % of the incident light is transmitted in total.

3.1.3 Gaussian beam propagation

As mentioned in Section 3.1.1, a paraxial solution of the Helmholtz equation leads to the Gaussian beam [Sal19]. In a medium with homogeneous refractive index n, it propagates along a straight line, which we consider as the optical axis and designate as the z-direction. For simplicity, we neglect any polarization and characterize the electrical field by its amplitude $\mathscr{E}$ which is rotationally symmetric around the optical axis.

Due to this symmetry, $\mathscr{E}$ can be described using the longitudinal coordinate z and the transversal coordinate h perpendicular to the optical axis. Unlike for plane waves, where the wavefront is flat, a Gaussian beam exhibits a continuously changing curvature of the wavefront during propagation. Therefore, $\mathscr{E}$ is usually represented as a complex quantity, which accounts for the wavefront bending and an additional slow phase shift along the optical axis:

$$\mathscr{E}(h,z) = \mathscr{E}_0 \cdot \frac{w_0}{w(z)} \cdot \exp\left(-\frac{h^2}{w^2(z)}\right) \cdot \exp\left(-i\frac{2\pi \cdot n}{\lambda} - i \cdot \phi_l - i \cdot \phi_t\right) \tag{3.5}$$

$$h^2 = x^2 + y^2 \tag{3.6}$$

Here, i is the imaginary unit. The imaginary argument of the complex exponential function takes into consideration the phase $2\pi \cdot n/\lambda$ in the medium, the longitudinal phase shift ϕ_l and the transversal phase shift ϕ_t. With λ being the wavelength in vacuum, we get

$$\phi_t = \frac{\pi \cdot h^2 \cdot n}{\lambda \cdot r(z)} \qquad \phi_l = \arctan \frac{z}{z_R} \tag{3.7}$$

$$r(z) = z \cdot \left[1 + \left(\frac{z_R}{z}\right)^2\right] \tag{3.8}$$

$$z_R = \frac{\pi \cdot w_0^2 \cdot n}{\lambda} \tag{3.9}$$

$r(z)$ is the changing curvature radius of the wavefront. As for the magnitude of a Gaussian beam's amplitude close to the optical axis, it suffices to consider the terms of Equation (3.5) except the complex exponential function. These are the real factors $\mathscr{E}_0$, $w_0/w(z)$ and the exponential function with the real argument. This function describes a bell-shaped curve in transversal direction (Figure 3.3b). The beam has a minimum beam radius w_0 where the wavefront is flat (Figure 3.3a). This position, at $z = 0$ in Figure 3.3a, is termed the waist with a diameter of $2 \cdot w_0$. When moving away from the waist along the optical axis, the beam widens up according to the beam radius $w(z)$:

$$w(z) = w_0 \sqrt{1 + \left(\frac{z}{z_R}\right)^2} \tag{3.10}$$

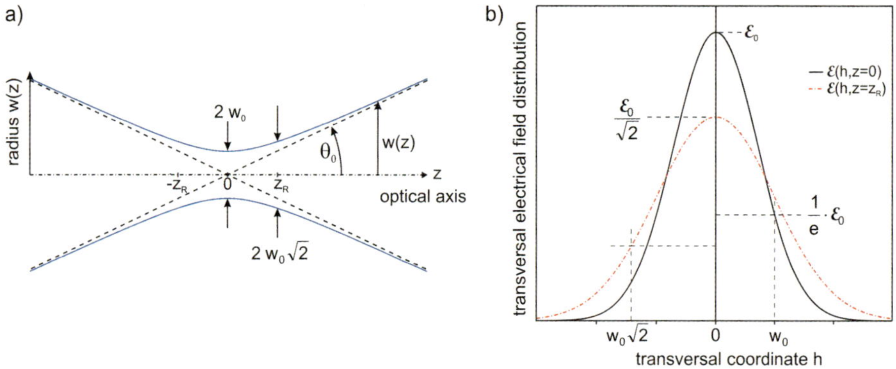

Fig. 3.3: Gaussian beam properties. a) propagation along the optical axis; b) transversal electrical field distribution in the waist and at a distance of the Rayleigh length z_R, resp.

z_R is called the Rayleigh length at which the beam radius has increased to $w_0\sqrt{2}$, thus the area of the beam or spot has doubled. It should be noted that Equation (3.10) is valid in the near or far field (see also Equation (5.27)). It can be seen that in transversal direction the amplitude of the electrical field continuously decays with increasing distance from the optical axis and approaches zero (Figure 3.3b). Hence, we define the beam or spot radius in our representation as the 1/e value of the maximum transversal electrical field value on the optical axis. This leads to w_0 for the beam radius of the waist where the beam has its maximum electrical field $\mathscr{E}_0$ in the center. When moving away from the waist along the optical axis, the beam widens up and its intensity decreases. At $z = z_R$, the beam's peak value of the transversal electrical field on the optical axis has dropped to $\mathscr{E}_0/\sqrt{2}$ while the radius has increased to $w_0\sqrt{2}$. At longer distances from the waist with $z/z_R \gg 1$, Equation (3.10) yields that the increase of the beam diameter can be approximated in a linear way. The asymptotic behavior is indicated by two straight lines having a slope angle θ_0 (Figure 3.3a):

$$\tan\theta_0 = \lim_{z\to\infty}\frac{w(z)}{z} = \frac{w_0}{z_R} = \frac{\lambda}{\pi\cdot w_0\cdot n} \cong \theta_0 \qquad (3.11)$$

θ_0 is also termed as half-angle divergence of the Gaussian beam. The smaller the waist, the larger the divergence in the far field. The product of the beam waist radius times the divergence angle $w_0\cdot\theta_0$ is an invariant quantity for small values of θ_0. This is equivalent to Equation (5.27).

As a consequence of this consideration, we can state that close to the waist and within a distance of approximately $\pm z_R$ from it, the Gaussian beam has a flat wavefront like a plane wave and its diameter varies only slightly. Thus, it is common to define this range as the depth of focus (DOFoc) of a Gaussian beam, which is equal to twice the Rayleigh length, namely $2\cdot z_R$ (see also Sections 3.3.6.4 and 3.4.6). At large distances, $z \gg z_R$, the wavefront looks like a section of an expanding sphere wave with linearly increasing curvature radius. This is compatible with the idea of a plane wave being obstructed, for instance, by an aperture (Figure 3.1d). When passing through the aperture, the light wave is diffracted and the wavefront becomes deformed. The narrower the aperture, the stronger is the effect.

The two crucial parameters that completely characterize a Gaussian beam at a given point z in space are the curvature radius $r(z)$ and the spot radius $w(z)$. Both can be combined to define the complex beam parameter $\boldsymbol{q}$ of a Gaussian beam [Sal19]:

$$\frac{1}{q} = \frac{1}{r(z)} - i\frac{\lambda}{\pi \cdot w^2(z) \cdot n} \tag{3.12}$$

Within wave optics, focusing and collimation of Gaussian beams can be described by Fourier optics, where Fourier transformation of the beam intensity profile (near field) yields the focal spot distribution (far field) and vice versa (see Chapter 5 and Appendix A.8). In a similar way, imaging can be described by a 4-f-system with two Fourier transformations (Section 5.1.3). This is different from a geometrical optics description. In some cases, however, the differences can be neglected. As typical examples for Gaussian beams, we should mention light beams from laser sources or light radiation from single-mode optical fibers. An example for imaging of Gaussian beams in comparison to geometrical optics is given in Section 3.3.6 for fiber ball lenses where the description of a Gaussian beam by the complex beam parameter is required for precise results.

3.1.4 Image formation

As already stated in Chapter 1, optical imaging is a process where light rays emerging from an object point are transferred to its image point by the use of an optical system (Figure 1.1a). The optical system generally consists of an arrangement of refracting interfaces that transform a divergent wave into a convergent wave. Here, we would like to note that we concentrate on optics based on refractive systems but not on reflective or diffractive optics. A simple example is shown in Figure 3.4 where light coming from an object point P_o is imaged to its corresponding image point P_i. The basic concept underlying imaging may be described by Fermat's principle taking into account the optical path length. The optical path length l_{opt} between two points is the geometric path length l multiplied with the refractive n in the medium, and respectively, the integral, over the path s if n is not constant in space:

$$l_{opt} = l \cdot n \qquad \text{resp.} \quad l_{opt} = \int n(s)\,\mathrm{d}s \tag{3.13}$$

Fermat's principle states that the path of light between two points, e. g., P_o and P_i, is always that one for which the traveling time l_{opt}/c is invariant against small variations of the optical path length. In a homogeneous medium without obstruction, this is always a straight line, while in a medium with locally changing refractive index, it is a

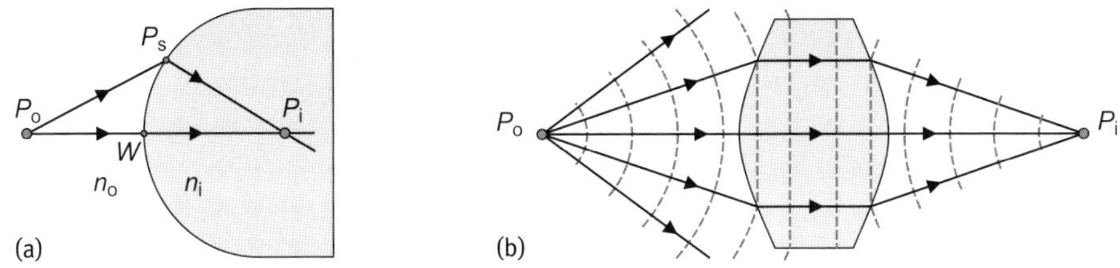

Fig. 3.4: (a) All rays from the object point to the image point are isochronous; (b) perfect imaging using a double hyperbolic lens; the lens changes the curvature of the incoming wavefront.

curved path (Figure 3.2). If we have an arrangement of refracting surfaces, for instance, as shown in Figure 3.4a in a simple case having one curved surface, all rays from P_o to P_i have the same optical path length and are isochronous, which means they need the same time. The ray taking the path $P_oP_sP_i$ has the same optical length as the ray P_oWP_i striking the surface perpendicularly. If $n_i > n_o$, then the distance P_sP_i must be shorter than WP_i as P_oP_s is longer than P_oW. It can be shown that the condition of isochronicity can be fulfilled for all rays emerging from P_o only if the surface is not spherical. However, spherical surfaces are a good approximation for rays nearly parallel to the optical axis and can be used in many cases. The problem here is that not all rays starting at P_o are imaged to P_i but are scattered around it, leading to a blurred image with limited quality as will be discussed in the subsequent chapters of lens aberrations. Figure 3.4b illustrates the situation of a 1:1 imaging using a lens with a double hyperbolic surface. Here, a perfect imaging of all points on the optical axis is possible. It can be seen that the wavefront, when entering the lens, slows down and becomes distorted as compared to its original shape. When leaving the lens, the wavefront changes its curvature again due to speeding up outside and converges to point P_i.

In order to describe the imaging properties of an optical system, let us analyze the refracted ray paths. As for the corresponding distances and orientations, conventions are necessary and will be defined as represented in Figure 3.5 for a simple optical system consisting of a thin converging lens with negligible thickness. We adopt the same convention as described in more detail in the textbook of Pedrotti [Ped08] and use this also when computing the optical matrices for a system in the next subchapters. The light propagation is assumed to be from the left to the right. The optical axis is identical to the horizontal axis and in general also represents the symmetry axis of the optical system. The object space with emerging rays is on the left side, while the image space with conjugated points is on the right side. All conjugated quantities and points are designated by the same characters but with different indices *o*, and respectively, *i* for distinction. Distances can be positive or negative depending on their directions on the optical axis. Also, distances pointing upwards in the positive vertical direction are positive and negative when pointing downwards. The angle between a ray and the optical axis is negative if the ray is rotated clockwise to be aligned with the optical axis, and positive for counterclockwise rotation. In the case of refraction at surfaces, the situation is different. Here,

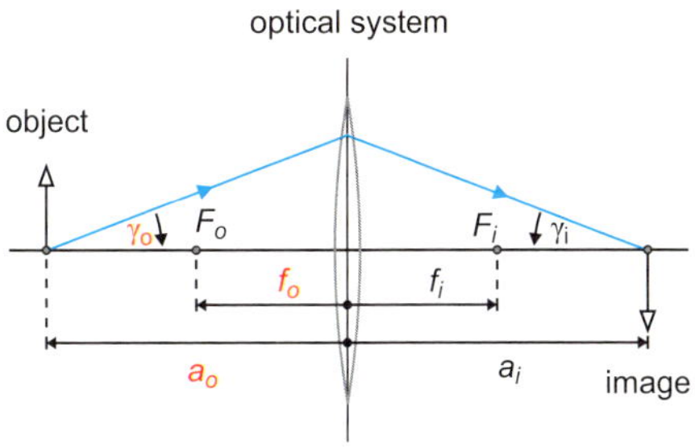

Fig. 3.5: Conventions for distances and angles; note: a_o, f_o and γ_o in the scheme are negative values.

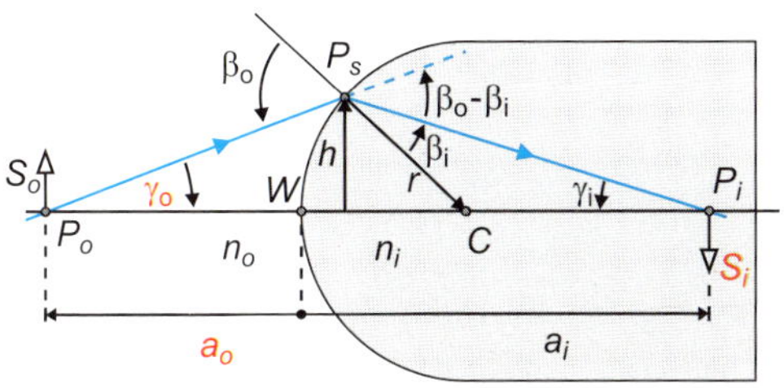

Fig. 3.6: Refraction at a spherical surface with $n_i > n_o$.

the angle between the surface normal and the incident ray is negative if the normal is rotated clockwise to become aligned with the ray, and positive for counterclockwise rotation of the normal. Convex spherical surfaces have a positive radius of curvature with the surface being to the left of its center, and concave surfaces have a negative radius. Note that the quantities a_o, f_o and γ_o in Figure 3.5 are counted negative according to the convention.

If we consider the path of two rays emerging from P_o in Figure 3.6 the image point P_i can be determined by their intersection in medium n_i. The first ray propagates along the optical axis, striking the surface at W perpendicularly and arriving at P_i in a straight line. A second ray incident to the surface at P_s under the angle β_o is refracted to the medium n_i under the angle β_i according to Snell's law (Equation (3.3)) and then intersects the optical axis at P_i. If the angle γ_o is small, then the angles β_o and β_i are also small. In these cases, using the relationship $\sin\beta \approx \tan\beta \approx \beta$, the sines and tangents of the angles can be approximated by the argument itself yielding

$$n_o \cdot \sin\beta_o = n_i \cdot \sin\beta_i \quad \Rightarrow \quad n_o \cdot \beta_o = n_i \cdot \beta_i. \tag{3.14}$$

A more detailed analysis with this paraxial or small-angle approximation leads to a general form of the refraction equation, which gives the relationship between the distances, refractive indices and the radius r of the surface curvature [Ped08]:

$$\frac{n_i}{a_i} - \frac{n_o}{a_o} = \frac{n_i - n_o}{r}. \tag{3.15}$$

If we not only consider a single point but are interested in the imaging of an extended object of size S_o, the same process applies to all points of the object from which rays can emerge. The resulting image of size S_i is inverted and has a different size when compared to the object. The linear, respectively transversal, magnification M is defined by the ratio of the image to object sizes as in Equation (2.6). But here we get additional information by taking into account the signs yielding a negative value:

$$M = \frac{S_i}{S_o} = \frac{n_o \cdot a_i}{n_i \cdot a_o}. \tag{3.16}$$

It should be noted here again that due to the small-angle approximation, the distance between the projection of P_s to the optical axis and the vertex W in Figure 3.6 is nearly zero and can be neglected.

If we consider a ray emerging from a point at a large distance from the vertex, then $1/a_o$ is nearly zero and can be neglected in Equation (3.15). The ray can be classified as a being nearly parallel to the optical axis, and its image at P_i is in the image focal point F_i with its image distance a_i becoming the image focal length f_i. In the same way, all rays that become parallel to the optical axis in the medium n_i emerge from the object focal point F_o with its distance a_o to the vertex being the object focal distance f_o. Then the focal lengths of a spherical surface in the different media can be defined by

$$f_i = r \cdot \frac{n_i}{n_i - n_o} \qquad f_o = -r \cdot \frac{n_o}{n_i - n_o}. \tag{3.17}$$

The ratio of the focal lengths is negative and proportional to the refractive indices. The sum of the focal lengths is equal to the curvature radius of the sphere:

$$\frac{f_o}{f_i} = -\frac{n_o}{n_i} \qquad f_o + f_i = r. \tag{3.18}$$

Based on this concept, the imaging properties of thin lenses can be understood if we add a second refracting surface to the right side of the first one depicted in Figure 3.6. In the simplest case, we neglect the thickness between the vertices of the surfaces. Figure 3.7 illustrates the cases of a thin converging and a thin diverging lens in air. The imaging then can be described by the thin lens formula, also known as the Gaussian lens formula, taking into account object distance, image distance and focal length of the lens as already given by Equation (2.5):

$$\frac{1}{a_i} - \frac{1}{a_o} = \frac{1}{f_i} \tag{3.19}$$

The imaging using a converging lens as well as the lateral magnification has been already described in more detail in Section 2.2, so we focus here on some principles for image formation using thin lenses.

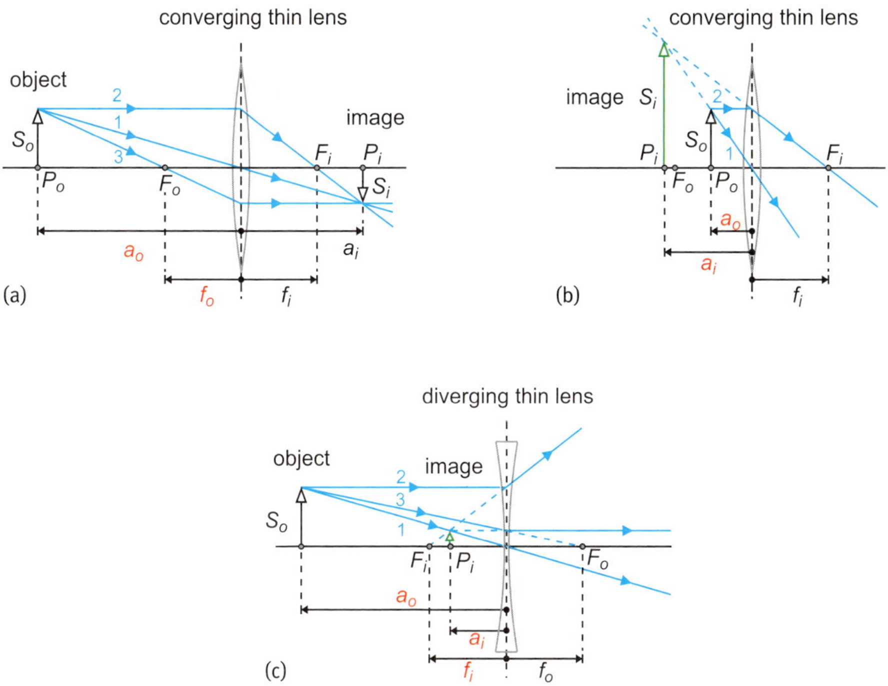

Fig. 3.7: Ray diagrams to describe the image formation for thin lenses in a homogeneous medium, for instance, in air. (a) Real image formation by a converging lens; (b) virtual image formation by a converging lens; (c) virtual image formation by a diverging lens.

Although two rays emerging from one object point are already sufficient to find its conjugated image point, let us consider three particular rays by which the image formation can be easily constructed using the graphical scheme shown in Figure 3.7. The first ray comes from the object and points to the center of the lens. In the case of thin lenses, we represent the lens as a single principal plane perpendicular to the optical axis. For a symmetric thin biconvex or biconcave lens, the principal plane is identical to the symmetry plane of the lens. Due to the negligible thickness of the lens, this center ray (1) passes through the lens without deviation from its straight path. The second incident ray (2) is parallel to the optical axis. When it strikes the principal plane, it is refracted and leaves the optical system in a straight line connecting its intersection point on the principal plane and the image focal point F_i. The third ray, entering the system in a straight line between the object point and the object focal point F_o, leaves the system as a ray parallel to the optical axis at a distance given by the intersection with the principal lens plane. It should be noted that the object focal length of a converging lens is negative, the image focal length is positive, whereas the signs are inverted for a diverging lens. As a consequence, the positions of the corresponding focal points are also inverted for converging respectively diverging lenses.

A converging lens always yields a real upside-down image with $M < 0$ if the absolute value of the object distance is larger than that of the focal length (Figure 3.7a). If the object comes closer to the focal point F_o the image becomes larger and approaches infinity if the object is nearly at F_o. If the object is between F_o and the lens vertex, no real image formation is possible as the outgoing rays diverge (Figure 3.7b). However, a virtual image can be captured by an observer or an optical system on the outgoing side. The virtual image is apparently located on the object side where the rearward extensions of the diverging rays intersect. The virtual image is upright and magnified as compared to the object with $S_i > S_o$. We then have the practical application of the converging lens as a magnifying lens or loupe as used in optical instruments.

A diverging lens never yields a real image of an object due to the diverging rays at the output, but an upright virtual image that can be made visible only by additional optics (Figure 3.7c). Here, the virtual image is smaller than the object. The same principles of image formation can be applied for thick lenses and a complex arrangement of lenses. In these cases, the spatial extension of the optical system can no longer be neglected like in thin lenses and will be discussed in the next section. We will also describe in that context how the focal length is influenced by the curvature of the refracting surfaces and the lens material.

Before dealing with spatially extended systems, let us interpret the thin lens Equation (3.19) under a different point of view. The front of a spherical wave emerging at an object point becomes more and more flat the longer it travels away from its source. This can be seen in Figures 3.1 and 3.4b. The wave entering an optical system is divergent, while the wave leaving the system has a transformed wavefront curvature that is converging. Now we can define the vergence of the incoming respectively leaving wave by the reciprocal values of their object respectively image distances. The vergence thus is a measure for the curvature of the wave. If we define the refractive power of a lens by its reciprocal image focal length $V_i = 1/f_i$, then the statement of the thin wave Equation (3.19) is that the curvature change of an incoming wave is identical to the refractive power of the optical system:

$$\frac{1}{a_i} - \frac{1}{a_o} = \frac{1}{f_i} = V_i. \tag{3.20}$$

The refractive power V_i is measured in units of diopters (dpt) with 1 dpt = 1/m. Combining two lenses at a close distance means that the refractive powers of both have to be added. As an example, a converging lens with a focal length of 100 mm has a refractive power of 10 dpt. Two lenses of this type combined yield 20 dpt and are equivalent to a lens of 50 mm focal length. If a converging and a diverging lens, which both have the same magnitude of refractive power, are combined without separation between them; then the resulting V_i is zero as the lenses have opposite signs. This system is called afocal.

3.2 Thick lenses

To neglect the distance between the vertices of a lens is only valid in some cases where exact geometrical positions are of minor importance. For a precise description of the image formation the path of a ray across the glass of a thick lens must be considered. The method to compute this path for thick lenses and lens combinations will be given in Section 3.3 dealing with the matrix method. In the following, we illustrate the image formation by thick lenses based on the cardinal planes and cardinal points. We will summarize the formulas to characterize the imaging properties of thick lenses having spherical surfaces with their centers on the optical axis. It should be stressed here once again that all results are valid within the small-angle approximation and all results assume perfect lenses without aberrations that show up in real lenses.

3.2.1 Basic lens equations for thick lenses

The ray diagrams for image formation in thick lenses is very similar to that given in Figure 3.7 for thin lenses. The main difference is that for thick lenses two principal planes and several cardinal points are necessary as opposed to one plane for thin lenses. If we focus again on three particular rays, their path across a lens is illustrated in Figure 3.8 for a converging lens:

1. A ray coming from the object focal point F_0 is refracted into the lens when striking the first surface. Then it traverses the lens in a straight line and is refracted at the second surface to leave the lens as a ray parallel to the optical axis. If we extrapolate the incoming and leaving rays to determine their apparent intersection point, we find that these intersection points are located on a plane, which is termed the first principal plane. The principal plane is perpendicular to the optical axis and intersected by it in the principal point H_0. A more detailed analysis shows that the principal plane is not a flat plane but rather a curved surface with the curvature radius of the focal length (see Section 3.4.3). However, in most cases and especially for rays close to the optical axis, the flat plane approximation is sufficient.

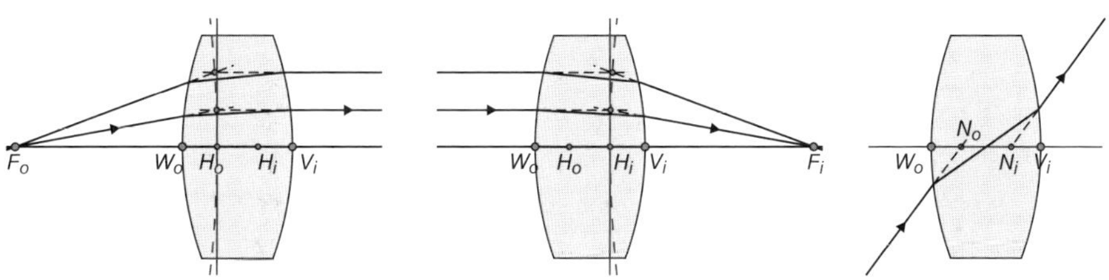

Fig. 3.8: Illustration of rays traversing a thick lens; indication of first and second principal planes at H_0, and respectively, at H_i, nodal points N_0 and N_i as well as the focal points F_0 and F_i.

2. All rays that enter the lens parallel to the optical axis are focused to the image focal point F_{i}. Again, entering and leaving rays appear to intersect in points on a plane, which this time is the second principal plane striking the optical axis at the principal point H_{i}.
3. In a thin lens ray diagram, the center ray, aiming to the center of the lens, is traversing in a straight line without deviation. In a thick lens, there is also a ray that leaves the lens parallel to its incoming path, although displaced by a certain distance due to the thickness of the lens. The extensions of the incident respectively leaving rays intersect the optical axis in the corresponding nodal points N_{o} respectively N_{i}. Thus, any ray on the object side aiming at N_{o} is refracted, displaced and refracted again to leave the lens in a parallel path, but it appears as if the leaving ray emerges from the nodal point N_{i} in the lens.

To summarize the main elements used to describe the imaging properties of a thick lens, there are six cardinal points on the optical axis: The object and image focal points F_{o} and F_{i}, the principal points H_{o} and H_{i}, and the nodal points N_{o} and N_{i}. The first and second principal planes, intersecting the optical axis at H_{o} and H_{i}, are slightly curved surfaces and are used to describe the paths of rays to the focal points respectively from them. For thin lenses, the two principal planes coincide with each other in the center of the lens, which is then also the position of the principal and nodal points.

As for the exact position of the cardinal points in thick lenses, it must be taken into consideration that all distances are directed. Their orientation is indicted in Figure 3.9, and it should be noted here that the quantities f_{o}, r_2, v_{Hi} and v_{Ni} in that scheme have negative values. The best way to define the distances is in general relative to the physical points of a lens, which can be identified in the simplest way. These are the vertex points W_{o} and W_{i} on the object or image side of the lens. The thickness t_{L} of the lens is given by its extension from the vertex on the object side to its counterpart on the image side. Then all positions of the principal and nodal points are measured with respect to the vertices, whereas the focal length is always defined as the distance from the principal point H to its corresponding focal point F.

In the following, we present formulas without derivations that allow to calculate all necessary quantities of arbitrary thick lenses with spherical surfaces as for instance

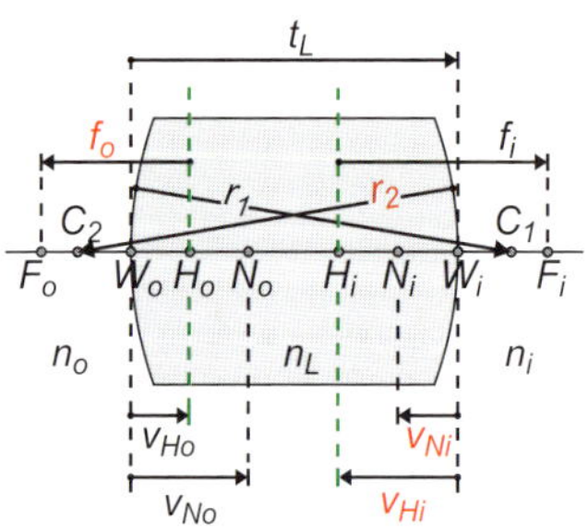

Fig. 3.9: Symbols and directed distances to describe the imaging properties of a thick lens; note that f_{o}, r_2, v_{Hi} and v_{Ni} are negative values.

convex, concave or meniscus lenses. We assume that the lens consists of a homogeneous material with refractive index n_L and can be used for image formation in media with different refractive indices. The refractive index on the object side n_o may be different from that on the image side n_i. This is the case when, for instance, images are taken from objects in water and the camera with a glass lens is submerged under water. Then the camera body is waterproof and the sensor chip in the camera is in air. Thus, all three refractive indices are different, and all quantities are different from the values of thick lenses in air.

The focal length f_o on the object side is a function of the curvature radii r_1, r_2 of the lens surfaces, the thickness t_L of the lens and the refractive indices of the media:

$$\frac{1}{f_o} = \frac{n_L - n_i}{n_o \cdot r_2} - \frac{n_L - n_o}{n_o \cdot r_1} - \frac{(n_L - n_o)(n_L - n_i)}{n_o \cdot n_L} \cdot \frac{t_L}{r_1 \cdot r_2}. \tag{3.21}$$

The focal length f_i on the image side is opposite to f_o and related to it by

$$f_i = -\frac{n_i}{n_o} \cdot f_o. \tag{3.22}$$

The focal lengths are measured from the principal points H, which for their part are measured relative to the vertices:

$$v_{Ho} = \frac{n_L - n_i}{n_L \cdot r_2} \cdot f_o \cdot t_L \qquad v_{Hi} = -\frac{n_L - n_o}{n_L \cdot r_1} \cdot f_i \cdot t_L. \tag{3.23}$$

The distances from the vertices to the nodal points are given by

$$v_{No} = \left(1 - \frac{n_i}{n_o} + \frac{n_L - n_i}{n_L \cdot r_2} \cdot t_L\right) \cdot f_o \qquad v_{Ni} = \left(1 - \frac{n_o}{n_i} - \frac{n_L - n_o}{n_L \cdot r_1} \cdot t_L\right) \cdot f_i. \tag{3.24}$$

The thin lens formula (3.19) can no longer be applied for thick lenses in media with different indices on the object and image side. The magnitude of the focal lengths f_o and f_i are proportional to the corresponding indices according to Equation (3.22). Thus, the imaging properties are expressed by the more general lens equation:

$$\frac{f_i}{a_i} + \frac{f_o}{a_o} = 1. \tag{3.25}$$

The linear magnification factor, being the ratio of image to object size, then is

$$M = \frac{S_i}{S_o} = \frac{a_i \cdot n_o}{a_o \cdot n_i}. \tag{3.26}$$

If the refractive indices on the object and image side are identical, then the principal points coincide with their corresponding nodal points and the focal lengths are equal in magnitude. Equation (3.25) then becomes identical to the thin lens Equation (3.19) with $f_i = -f_o$ and $M = a_i/a_o$. The only difference in the ray path construction scheme is that all distances in the object space are measured relative to the object principal and distances in the image space relative to the image principal plane.

3.2.2 Types of lenses and lens shapes

As given by Equations (3.21) and (3.22), the focal length of a lens depends on the curvatures of its spherical surfaces, the refractive indices and its thickness. If a lens with particular focal length or refractive power is required, there are different lens shapes and sizes possible for the same value. The main distinction for lenses is made with respect to their imaging properties: There are positive lenses or converging lenses, which enhance the convergence of exiting beams respectively reduce their divergence. This type of lenses in most cases is thicker in the middle between the vertices than at the edges (Figure 3.10). They have real focal points and a positive image focal length. By contrast, negative lenses or diverging lenses render incoming beams more divergent. In general, they are thinner between the vertices than at the edges, have a negative image focal length and produce only virtual images.

Figure 3.10 depicts different converging lenses that all have the same refractive power but different shapes (upper row). The lenses with the same focal length are also said to have a different lens bending. The lower part shows the corresponding situation

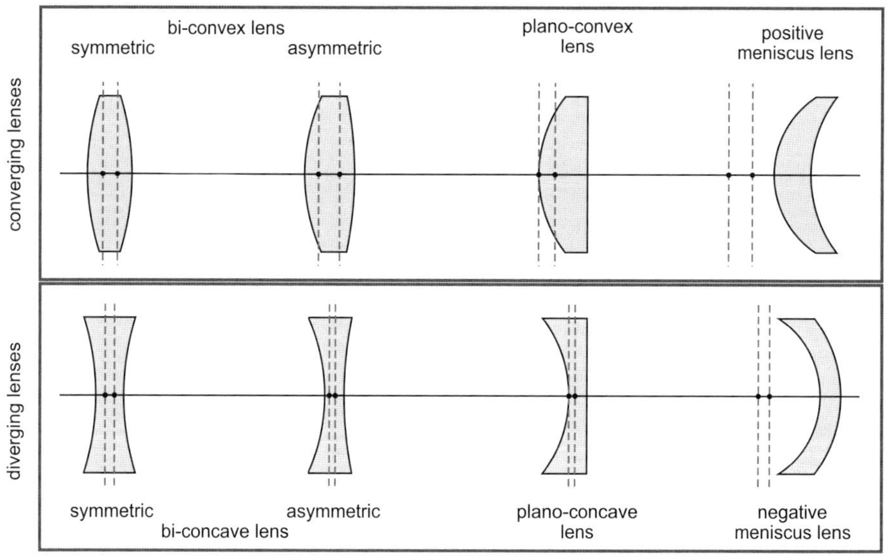

Fig. 3.10: Converging and diverging lenses of different shapes; all converging lenses have the same positive refractive power despite their different bending; likewise all diverging lenses have the same negative refractive power; the shape of the lens strongly influences the location of the principal planes, which are indicated by broken lines.

for diverging lenses. The figure also indicates the positions of the principal planes in the lenses. It can be seen that, starting with symmetric bilenses, the principal points are displaced toward the surface with the smaller magnitude of the curvature radius if the lens bending changes. In plano-convex and plano-concave lenses, one of the principal points is always located in the vertex of the curved surface. It should be noted here again that for lenses in air with $n_o = n_i = 1$ the focal lengths on the object and image side have the same magnitude. The imaging properties remain the same if the lens is reversed, which means that the entrance and exit side are interchanged. However, the position of the principal points changes with the physical position of the lens. In the following examples, the focal lengths as well as the positions of the principal points of glass lenses in air are discussed.

Example: symmetric bilenses

Let us consider a glass lens surrounded by air with $n_o = n_i = 1$. The refractive indices of glass usually have values in the range of around 1.45 for pure silica glass up to around 1.75 and above for flint glasses. For simplicity, we choose $n_L = 1.5$, which is a good approximation for crown glass and other conventional technical glass types. The refractive power and the image focal length then is respectively given according to Equations (3.21) and (3.22) by the formula:

$$V_i = \frac{1}{f_i} = (n_L - 1) \cdot \left(\frac{1}{r_1} - \frac{1}{r_2}\right) + \frac{(n_L - 1)^2}{n_L} \cdot \frac{t_L}{r_1 \cdot r_2} \approx 0.5 \cdot \left(\frac{1}{r_1} - \frac{1}{r_2}\right) + \frac{t_L}{6 \cdot r_1 \cdot r_2}. \tag{3.27}$$

For the position of the principal points as measured from the vertices, we get

$$v_{Ho} = -\frac{n_L - 1}{n_L \cdot r_2} \cdot f_i \cdot t_L \approx -\frac{f_i \cdot t_L}{3 \cdot r_2} \qquad v_{Hi} = -\frac{n_L - 1}{n_L \cdot r_1} \cdot f_i \cdot t_L \approx -\frac{f_i \cdot t_L}{3 \cdot r_1}. \tag{3.28}$$

Symmetric biconvex and biconcave lenses are bounded by spherical surfaces with identical curvature radii $r_1 = -r_2 = r$. The refractive power according to Equation (3.25) then is $V_i \approx 1/r - t_L/(6 \cdot r^2)$. In most cases, the radius of a conventional lens is much larger than its thickness, thus $V_i \approx 1/r$ and the focal length $f_i \approx r$ as for thin lenses. The principal points are located at a distance $v_{Ho} \approx t_L/3$ and $v_{Hi} \approx -t_L/3$ from the vertices and symmetrical to the center of both biconvex and biconcave lenses, as is shown in Figure 3.10. In biconvex lenses r, and thus f_i are positive, whereas in biconcave lenses they are negative.

A ball lens is the special case of a biconvex lens where the thickness can no longer be neglected. Equation (3.27) then yields $V_i \approx 2/(3 \cdot r)$ and a positive $f_i \approx 1.5 \cdot r$. The principal points are both located in the center of the ball with both $v_{Ho} = v_{Hi} \approx r$. The focal point of a glass ball lens in air is at a distance $1.5 \cdot r$ from the ball center, and thus $0.5 \cdot r$, as measured from the vertex of the ball.

Example: planar lenses

Planar lenses are not symmetric but have one flat facet whereas the other end surface is curved, for instance, $r_1 = r$ and radii $r_2 = \infty$ as illustrated in Figure 3.10. The refractive power as well as the focal length are both independent from the thickness of the lens. Using Equation (3.27), we get a refractive power $V_i \approx 1/(2 \cdot r)$, which is only half of the value of a biconvex lens, and consequently, $f_i \approx 2 \cdot r$. This can be explained by arguing that a planar lens is only one-half of a symmetrical bilens. Combining two identical lenses means that we have to add the refractive power if the thickness and the spacing between them can be neglected. One principal point in the planar lens is located in the vertex of the curved surface with $v_{Ho} = 0$ whereas the other one is at $v_{Hi} = -2 \cdot t_L/3$ from the flat side.

Example: meniscus lenses

Meniscus lenses are bounded by surfaces of which the curvature radii are directed equally (Figures 3.10 and 3.11). We assume the radii are positive, as illustrated in Figure 3.11 without loss of generality as the focal length in air is independent from the orientation. In order to see how the radii influence focal length, we will discuss different conditions:

a) $r_1, r_2 > 0$ with $r_1 < r_2$:
The outer surface has a stronger curvature than the inner one, as depicted in Figure 3.10 for a positive meniscus lens. The thickness t_L between the vertices is positive. According to Equation (3.27), we can see that the refractive power V_i as well as the image focal length f_i is always positive; hence, all these menisci are converging lenses. As for the principal points, Equation (3.27) yields negative values for v_{Ho} and v_{Hi}, which means that the principal planes are shifted to the outer side as is depicted in Figure 3.10.

b) $r_1, r_2 > 0$ with $r_1 = r_2$ (Höegh's meniscus):
Here, we have a lens where both radii are identical (Figure 3.11a). It is also termed Höegh's meniscus. We get a positive refractive power and a focal length $f_i = 6 \cdot r^2/t_L$. In order to be not afocal, this meniscus must be a thick lens. The principal planes are shifted by the same amount $v_{Ho} = v_{Hi} = -2 \cdot r$ away from the vertex outside the meniscus like in case a) for positive meniscus lenses. A special feature of Höegh's meniscus is that it has no curved image plane as will be discussed later in the section about lens aberrations.

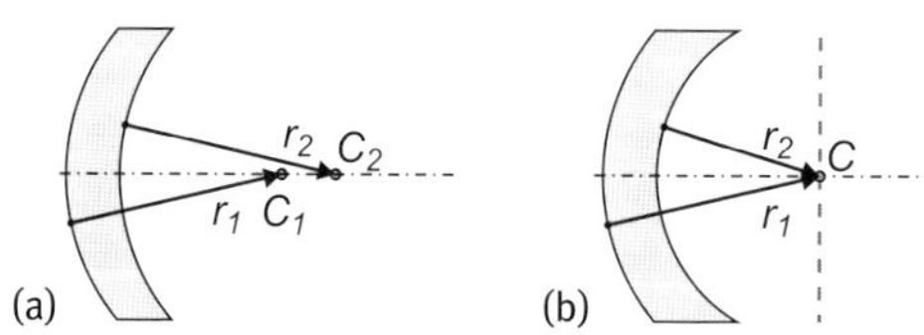

Fig. 3.11: (a) Meniscus lens with identical curvature radii (Höegh's meniscus); (b) meniscus lens with concentric surfaces.

c) $r_1, r_2 > 0$ with $r_1 > r_2$:
Unlike in a), the outer surface has a more minor curvature than the inner one. Although $1/r_1 - 1/r_2$ is negative, it is the thickness relative to the radii that determines if the overall refractive power is positive or negative (Equation (3.27)). For $t_L = 3 \cdot (r_1 - r_2)$, we even find that $V_i = 0$, and hence f_i becomes infinitely large, meaning that the lens is afocal. For larger thickness, the lens is positive. A thinner meniscus has a negative refractive power and focal length and consequently is a diverging lens. The principal planes for a diverging meniscus are located at the side of the inner surface as is illustrated in Figure 3.10.

d) $r_1, r_2 > 0$ with $t_L = r_1 - r_2$ (concentric surfaces):
This is a special case of c) where both curved surfaces have a common center point C. The lens is thinner than the afocal meniscus, and hence we have a negative focal length with $f_i = -3 \cdot r_1 \cdot r_2/t_L$. Both principal planes coincide and are located in C with $v_{Ho} = r_1$ and $v_{Hi} = r_2$ (Figure 3.11b).

3.3 Ray path calculation by the matrix method

We have summarized above the formulas for single thick lenses that were the results of the matrix method for the calculation of the path of a ray through an optical system. This method is very powerful when applied to more complex systems that consist of different lenses arranged sequentially like in microscope lenses, or camera lenses like the Tele–Tessar as displayed in Figure 3.12.

The basic principle of this method is that in a 2D description, a ray at any point in the optical system can be represented by the two parameters that indicate its position as well as its direction of propagation: there is the height h of the ray, measured perpendicularly to the optical axis, and thus characterizing the distance to it (Figure 3.13). The second parameter is the slope angle γ relative to the optical axis indicating the propagation. We would like to stress here again the sign convention as described in Section 3.1.4 where a counterclockwise rotation of the ray to be aligned with the optical axis is associated with a positive angle and otherwise negative. The propagation is illustrated in Figure 3.13 where the path of a ray emerging at point P_o on the object side can be fol-

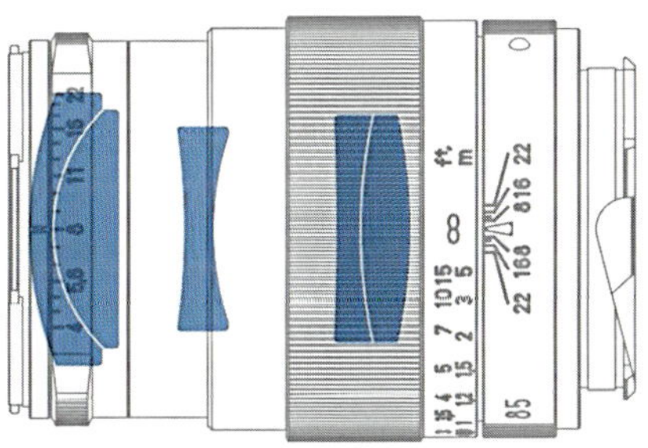

Fig. 3.12: Objective lens for 35 mm-format cameras consisting of multiple single lens elements (ZEISS Tele–Tessar T* 4/85 ZM, with kind permission of Carl Zeiss AG).

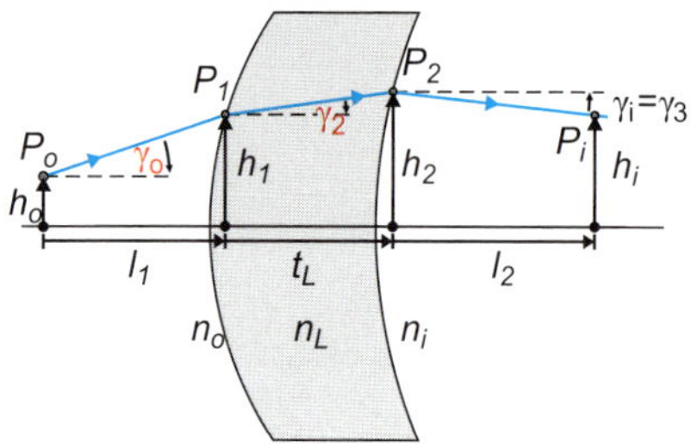

Fig. 3.13: Meniscus lens as an example for a sequence of matrix applications; γ_o and γ_2 in the diagram are negative angles.

lowed over the points P_1 and P_2 in the lens to its final destination P_i on the image side. This example shows the two principal operations that we need to calculate the beam propagation across an optical system: the translation and refraction of a beam. A third conventional operation, the reflection of a beam, is not necessary for the optical systems that we will consider in this book and will not be discussed. In the case of translation and refraction, only one parameter changes whereas the other remains invariable upon applying the operation. When the beam starts at P_o, described by the parameter pair (h_o, γ_o), it is displaced along the distance l_1 on the optical axis in a straight line and arrives at point P_1. There, due to the straight-line or "free-space" propagation, only the elevation of the beam has changed to h_1 whereas the slope angle on arrival is the same as at the starting point. In P_1 the beam is refracted. This is an operation in one point, which means that the height h_1 now remains unchanged while the slope angle γ_1 of the incoming beam in this point changes to γ_2 when leaving the point. As for the further path in the lens, the beam is again translated in a straight line and refracted when exiting the lens at the second surface. The values (h_i, γ_i) at the final point P_i result after a translation along l_2. It should be noted here again that all these considerations are based on the simplified description for paraxial rays striking inclined surfaces. Therefore, lens errors, e. g., due to large beam apertures, cannot be calculated in this way and need a more advanced method.

3.3.1 Ray translation matrix

We begin our consideration with the free space propagation. The ray starting at P_1 with its parameters (h_1, γ_1) progresses along the optical axis by the distance l (Figure 3.14). The ray parameters will be presented as 2D column vectors. The translation matrix $\mathbf{T}_{12}$

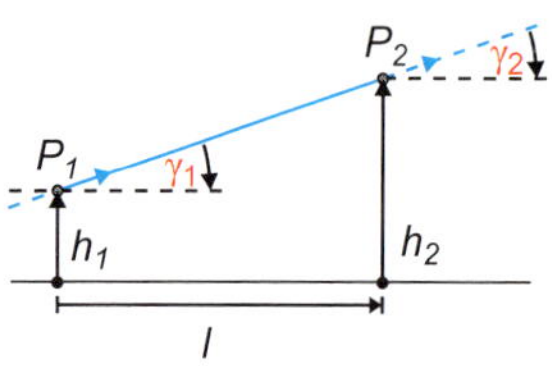

Fig. 3.14: Ray translation along a distance; γ_1 and γ_2 in the diagram are negative angles.

then is a 2 × 2 matrix operator that is applied to the input vector at P_1 and yields the output vector with the coordinates (h_2, γ_2) at P_2:

$$\begin{pmatrix} h_2 \\ \gamma_2 \end{pmatrix} = \mathbf{T}_{12} \cdot \begin{pmatrix} h_1 \\ \gamma_1 \end{pmatrix} = \begin{bmatrix} \mathbb{A} & \mathbb{B} \\ \mathbb{C} & \mathbb{D} \end{bmatrix} \cdot \begin{pmatrix} h_1 \\ \gamma_1 \end{pmatrix} \tag{3.29}$$

The matrix coefficients $\mathbb{A}$, $\mathbb{B}$, $\mathbb{C}$ and $\mathbb{D}$ result from the explicit calculation of h and γ according to Figure 3.14. During translation, the slope angle does not change, thus $\gamma_2 = \gamma_1$. The height increases linearly along l, yielding $h_2 = h_1 - l \cdot \tan \gamma_1$. In this equation, a negative sign must be associated with the slope angle as the height values are all positive while the angle according to our convention is negative. We then get two linear equations from which we can derive the relationship to the corresponding matrix elements by comparison. Using the paraxial approximation of $\gamma_1 \approx \tan \gamma_1 \approx \sin \gamma_1$, we get

$$\begin{aligned} & h_2 = h_1 - l \cdot \gamma_1 \\ & \gamma_2 = \gamma_1 \\ & h_2 = \mathbb{A} \cdot h_1 + \mathbb{B} \cdot \gamma_1 \\ & \gamma_2 = \mathbb{C} \cdot h_1 + \mathbb{D} \cdot \gamma_1 \quad \Rightarrow \quad \mathbb{A} = 1, \quad \mathbb{B} = -l, \quad \mathbb{C} = 0, \quad \mathbb{D} = 1 \end{aligned} \tag{3.30}$$

We then can write the translation matrix $\mathbf{T}$ as

$$\mathbf{T}_{12} = \begin{bmatrix} 1 & -l \\ 0 & 1 \end{bmatrix} \tag{3.31}$$

It should be noted that our considerations here are done within the frame of geometrical optics. For the presented matrix formalism, only the geometric path length l in the medium is of importance. However, in other considerations based on wave optics, a differentiation between l and the optical path length l_{opt} has to be done (see Section 3.1.4).

3.3.2 Ray refraction matrix

In the case of refraction, we consider the situation of a ray at one point. This is illustrated in Figure 3.15 where the incoming ray strikes the refracting surface at point P_1. It should

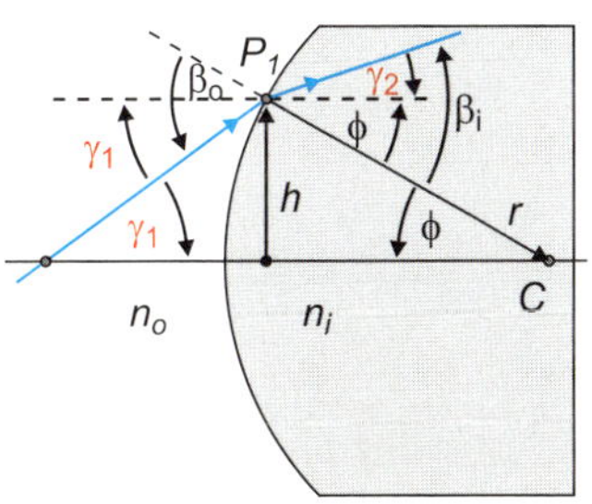

Fig. 3.15: Ray refraction at a spherical surface; γ_1 and γ_2 in the diagram are negative.

be noted that we discuss all cases within the limits of paraxial approximation. However, the angles in the figures have much larger values in order to improve the visibility and to better illustrate the principles. The ray can be characterized by its height and slope angle (h, γ_1) on arrival. The ray is refracted in P_1 according to Snell's law (Equation (3.14)) at the spherical surface bounded by the materials with different refractive indices n_o and n_i. The leaving ray then has the coordinates (h, γ_2) after refraction. The angles can be calculated following the geometry depicted in Figure 3.15 with the normal to the surface in point P_1 intersecting the optical axis in point C. Snell's law in paraxial approximation yields $n_o\beta_o = n_i\beta_i$. The angles relative to the optical axis can be calculated using $-\gamma_1 = \beta_o - \phi$ and $-\gamma_2 = \beta_i - \phi$. In paraxial approximation, we get $\phi \approx \sin\phi \approx h/r$. Hence, the equations for γ and h can be written as follows:

$$\begin{aligned} h_2 &= h = h_1 \\ \gamma_2 &= -\beta_i + \phi = -\frac{n_o}{n_i}\left(-\gamma_1 + \frac{h_1}{r}\right) + \frac{h_1}{r} = \left(1 - \frac{n_o}{n_i}\right)\frac{h_1}{r} + \frac{n_o}{n_i} \cdot \gamma_1. \end{aligned} \tag{3.32}$$

Hence, for the matrix elements we get

$$\begin{aligned} h_2 &= \mathbb{A} \cdot h_1 + \mathbb{B} \cdot \gamma_1 \\ \gamma_2 &= \mathbb{C} \cdot h_1 + \mathbb{D} \cdot \gamma_1 \quad \Rightarrow \quad \mathbb{A} = 1, \quad \mathbb{B} = 0, \quad \mathbb{C} = \frac{n_i - n_o}{n_i \cdot r}, \quad \mathbb{D} = \frac{n_o}{n_i}. \end{aligned} \tag{3.33}$$

As in the case of translation, the refraction matrix $\mathbf{R}_1$ of the spherical surface can be written as

$$\mathbf{R}_1 = \begin{bmatrix} 1 & 0 \\ \dfrac{n_i - n_o}{n_i \cdot r} & \dfrac{n_o}{n_i} \end{bmatrix} \tag{3.34}$$

3.3.3 Thick-lens and thin-lens matrix

Let us now calculate the matrix for the ray propagating across a thick lens and take as an example the meniscus lens described in Section 3.3 (Figure 3.13). As the input plane of this lens, we assume the left convex surface has a curvature radius r_1. It is virtually located in the vertex because in paraxial approximation with small angles γ the projections of all points on the input plane to the optical axis are nearly identical and in the vertex. Likewise, the output plane is assumed to be located in the vertex of the right surface with a curvature radius r_2. For all considerations, we deal with lenses consisting of glass with refractive index n_L and being surrounded by media with refractive indices n_o, respectively, n_i on the object respectively image side. Any ray striking the input plane, characterized by (h_1, γ_1) at P_1 is refracted into the lens as described above, and we get with $\gamma_1 = \gamma_o$ and $h_2' = h_1$:

$$\begin{pmatrix} h_2' \\ \gamma_2' \end{pmatrix} = \mathbf{R}_1 \cdot \begin{pmatrix} h_1 \\ \gamma_1 \end{pmatrix} \tag{3.35}$$

Then the refracted ray is translated in the lens along the distance t_L which is the thickness between the vertices. We get with $\gamma_2 = \gamma_2'$:

$$\begin{pmatrix} h_2 \\ \gamma_2 \end{pmatrix} = \mathbf{T}_{12} \cdot \begin{pmatrix} h_2' \\ \gamma_2' \end{pmatrix} \tag{3.36}$$

Finally, the ray exiting the lens has the parameters (h_3, γ_3) in the output plane after the second refraction $\mathbf{R}_2$, with $h_3 = h_2$:

$$\begin{pmatrix} h_3 \\ \gamma_3 \end{pmatrix} = \mathbf{R}_2 \cdot \begin{pmatrix} h_2 \\ \gamma_2 \end{pmatrix} \tag{3.37}$$

The subsequent application of the individual matrix operations can also be expressed by one operation $\mathbf{M}_L$, which characterizes the matrix of the thick lens that converts the input ray (h_1, γ_1) at P_1 into the exit ray (h_3, γ_3) at P_2:

$$\begin{pmatrix} h_3 \\ \gamma_3 \end{pmatrix} = \mathbf{M}_L \cdot \begin{pmatrix} h_1 \\ \gamma_1 \end{pmatrix} \quad \text{with} \quad \mathbf{M}_L = \mathbf{R}_2 \cdot \mathbf{T}_{12} \cdot \mathbf{R}_1 \tag{3.38}$$

As the matrix operation is associative but in general not commutative, the sequence of the individual matrices representing the physical circumstances of light propagation is very important. The first operation is the rightmost one in the product of the matrices, and the last operation is the leftmost one. With the definition of the matrices according to Equations (3.31) and (3.34), we get for the matrix of the thick lens:

$$\mathbf{M}_L = \mathbf{R}_2 \cdot \mathbf{T}_{12} \cdot \mathbf{R}_1 = \begin{bmatrix} 1 & 0 \\ \dfrac{n_i - n_L}{n_i \cdot r_2} & \dfrac{n_L}{n_i} \end{bmatrix} \cdot \begin{bmatrix} 1 & -t_L \\ 0 & 1 \end{bmatrix} \cdot \begin{bmatrix} 1 & 0 \\ \dfrac{n_L - n_o}{n_L \cdot r_1} & \dfrac{n_o}{n_L} \end{bmatrix} \tag{3.39}$$

This formula not only describes a meniscus lens but is universally valid for any lens of refractive index n_L with different spherical surfaces in different media at both sides as described in Section 3.2.1. If we carry out the matrix multiplication, we get the general results that have already been presented above for thick lenses. In order to understand the significance of the matrix elements and before we discuss the more general matrix of an optical system, let us consider the matrix of a thin lens submerged in a medium of refractive index n on both the object and image side. With $t_L \approx 0$, the matrix $\mathbf{L}$ describing a thin lens in medium $n = n_o = n_i$ becomes

$$\mathbf{L} = \begin{bmatrix} 1 & 0 \\ \frac{n-n_L}{n \cdot r_2} & \frac{n_L}{n} \end{bmatrix} \cdot \begin{bmatrix} 1 & 0 \\ 0 & 1 \end{bmatrix} \cdot \begin{bmatrix} 1 & 0 \\ \frac{n_L-n}{n_L \cdot r_1} & \frac{n}{n_L} \end{bmatrix} = \begin{bmatrix} 1 & 0 \\ \frac{n_L-n}{n} \cdot \left(\frac{1}{r_1} - \frac{1}{r_2}\right) & 1 \end{bmatrix} = \begin{bmatrix} 1 & 0 \\ \frac{1}{f_i} & 1 \end{bmatrix} \tag{3.40}$$

As a result, the thin-lens matrix has only one significant coefficient, which is in the first column, second row. It is identical to the refractive power and reciprocal value of the image side focal length of the thin lens:

$$V_\mathrm{i} = \frac{1}{f_\mathrm{i}} = \frac{n_\mathrm{L} - n}{n} \cdot \left(\frac{1}{r_1} - \frac{1}{r_2}\right) \tag{3.41}$$

This result is identical to Equation (3.27) in the case where we have a thin lens in air with $n = 1$. As discussed above, V_i is positive for converging, and negative for diverging lenses. The input and output planes in the case of thin lenses coincide with each other and are identical to the central lens plane.

3.3.4 Ray transfer matrix for optical systems

The design of optical systems for high-quality imaging is in general based on the combination of different lenses in order to compensate for the lens aberrations. So far, we have only considered matrices representing basic elements. When combining these elements, for instance positive and negative lenses, their exact position with respect to well-defined input and output planes must be specified. The location of the cardinal planes and points are strongly influenced by the arrangement of these elements and can be calculated from the four matrix elements of the complete optical system. The above considered matrix formalism can be generalized to calculate the transfer matrix $\mathbf{M}_\mathrm{os}$ of an optical system. The subsequent application of the matrices representing the different lens elements in the system yields the system matrix in a similar way as considered for single lens elements:

$$\begin{pmatrix} h_\mathrm{Ei} \\ \gamma_\mathrm{Ei} \end{pmatrix} = \mathbf{M}_\mathrm{os} \cdot \begin{pmatrix} h_\mathrm{Eo} \\ \gamma_\mathrm{Eo} \end{pmatrix} \quad \text{with} \quad \mathbf{M}_\mathrm{os} = \mathbf{M}_\mathrm{m} \cdot \mathbf{M}_{\mathrm{m}-1} \cdot \ldots \cdot \mathbf{M}_2 \cdot \mathbf{M}_1 \quad \text{and} \quad \mathbf{M}_\mathrm{os} = \begin{bmatrix} \mathbb{A} & \mathbb{B} \\ \mathbb{C} & \mathbb{D} \end{bmatrix} \tag{3.42}$$

In Equation (3.42), the pair $(h_\mathrm{Eo}, \gamma_\mathrm{Eo})$ represent the ray parameters at the input plane E_o on the object side of the system, and correspondingly $(h_\mathrm{Ei}, \gamma_\mathrm{Ei})$ are the parameters at the output plane E_i on the image side. The system matrix $\mathbf{M}_\mathrm{os}$ is the result after m matrix operations where the $\mathbf{M}_\mathrm{i}$ can be any of the above-described translation, refraction, thin-lens or thick-lens matrices. As for the input and output planes, which have been identical to the spherical surfaces in the examples for the lenses described above, they can be displaced by any distance, which is achieved by an additional translation matrix operation. This is sometimes required if the properties of optical systems are related to some significant planes as, for instance, the focal point of camera lenses relative to the mount flange. An example is illustrated in Figure 3.18 for a simple retrofocus camera lens setup consisting of two lenses in a body housing. The input plane is conveniently chosen as the front end of the housing, while the output plane is typically the rear end of the flange, which is used to mount the lens to the camera. The image focal F_i point

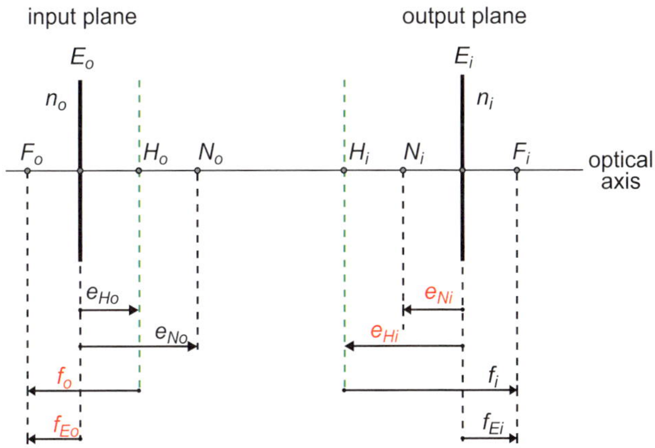

Fig. 3.16: Symbols and directed distances to describe the imaging properties of an optical system.

can be easily determined for practical purposes using its back focal distance f_{Ei} relative to the output plane.

Once the input respectively output planes are fixed the computation of $\mathbf{M}_{\mathrm{os}}$ delivers the matrix coefficients $\mathbb{A}$, $\mathbb{B}$, $\mathbb{C}$ and $\mathbb{D}$ from which all properties, and particularly the cardinal points, can be derived. The location of the principal planes and nodal points is in general measured in relation to the input and output planes as illustrated in Figure 3.16. This scheme is very similar to the one presented for thick lenses in Figure 3.9 with the fundamental difference being that the reference planes E_{o} and E_{i} can be chosen arbitrarily.

As for the focal lengths, we have to differentiate between different quantities: The front focal length f_{Eo} is the distance between the input plane E_{o}, which may be at the vertex of the first lens element or the front end of the housing, and the focal point F_{o} on the object side. The back focal length f_{Ei} is the corresponding quantity at the image side relative to the output plane. The effective focal lengths f_{o} and f_{i} are measured as usual relatively to the principal planes and yield the physical refractive power of the lens with $V_{\mathrm{i}} = 1/f_{\mathrm{i}}$.

Figure 3.17 shows a ray emerging from the object focal point F_{o} and entering the optical system at an elevation h_{Eo}, under an angle γ_{Eo}. As the ray comes from the focal point it must leave the system parallel to the optical axis with $\gamma_{\mathrm{Ei}} = 0$. Its elevation h_{Ei} is determined by the intersection with the principal plane on the object side. The matrix equation then yields

$$\begin{aligned} h_{\mathrm{Ei}} &= \mathbb{A} \cdot h_{\mathrm{Eo}} + \mathbb{B} \cdot \gamma_{\mathrm{Eo}} \\ \gamma_{\mathrm{Ei}} &= \mathbb{C} \cdot h_{\mathrm{Eo}} + \mathbb{D} \cdot \gamma_{\mathrm{Eo}} = 0 \quad \Rightarrow \quad -\frac{\mathbb{D}}{\mathbb{C}} = \frac{h_{\mathrm{Eo}}}{\gamma_{\mathrm{Eo}}} \approx f_{\mathrm{Eo}}. \end{aligned} \tag{3.43}$$

It can be seen from this result that the ratio $-\mathbb{D}/\mathbb{C}$ of the matrix coefficients is identical to the front focal length f_{Eo} if the small-angle assumption $\gamma_{\mathrm{Eo}} \approx h_{\mathrm{Eo}}/f_{\mathrm{Eo}}$ is valid (Figure 3.17).

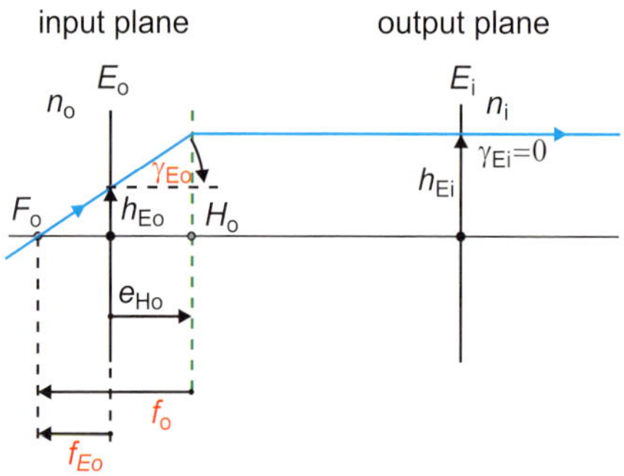

Fig. 3.17: A ray from the object focal point enters the system at height h_{Eo} with angle γ_{Eo} and leaves it at height h_{Ei}, parallel to the optical axis ($\gamma_{Ei} = 0$).

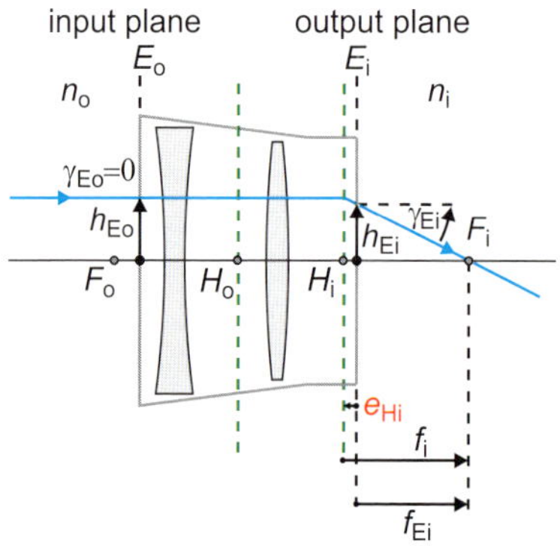

Fig. 3.18: Ray construction for a ray entering parallel to the optical axis; for illustration purposes the input respectively output planes of the simple retrofocus design are not identical with the principal lens planes but are located on the body of the camera lens.

This is the same result that we get for a thin lens in air according to Equation (3.40) with $f_o = -f_i$, $\mathbb{D} = 1$ and $\mathbb{C} = 1/f_i$. Following similar considerations for parallel incoming rays (Figure 3.18) and central rays directed to the nodal points of the system parameters, the remaining significant parameters of the optical system can be derived.

In the following, we will give a summary of these quantities without further derivation:

$$f_{Eo} = -\frac{\mathbb{D}}{\mathbb{C}} \qquad \text{front focal length (input plane } E_o \text{ to object focal point } F_o\text{)} \tag{3.44}$$

$$f_{Ei} = \frac{\mathbb{A}}{\mathbb{C}} \qquad \text{back focal length (output plane } E_i \text{ to image focal point } F_i\text{)} \tag{3.45}$$

$$e_{Ho} = \frac{n_o/n_i - \mathbb{D}}{\mathbb{C}} \qquad \text{distance from input plane } E_o \text{ to principal plane } H_o \text{ (object side)} \tag{3.46}$$

$$e_{Hi} = \frac{\mathbb{A} - 1}{\mathbb{C}} \qquad \text{distance from output plane } E_i \text{ to principal plane } H_i \text{ (image side)} \tag{3.47}$$

$$e_{No} = \frac{1 - \mathbb{D}}{\mathbb{C}} \qquad \text{distance from input plane } E_o \text{ to nodal point } N_o \text{ (object side)} \tag{3.48}$$

$$e_{\mathrm{Ni}} = \frac{\mathbb{A} - n_\mathrm{o}/n_\mathrm{i}}{\mathbb{C}} \quad \text{distance from output plane } E_\mathrm{i} \text{ to nodal point } N_\mathrm{i} \text{ (image side)} \tag{3.49}$$

$$\begin{aligned} f_\mathrm{o} &= f_{\mathrm{Eo}} - e_{\mathrm{Ho}} \\ &= -\frac{n_\mathrm{o}/n_\mathrm{i}}{\mathbb{C}} \quad \text{object focal length (principal point } H_\mathrm{o} \text{ to object focal point } F_\mathrm{o}) \end{aligned} \tag{3.50}$$

$$f_\mathrm{i} = f_{\mathrm{Ei}} - e_{\mathrm{Hi}} = \frac{1}{\mathbb{C}} \quad \text{image focal length (principal point } H_\mathrm{i} \text{ to image focal point } F_\mathrm{i}) \tag{3.51}$$

In all cases, the image focal length is the $1/\mathbb{C}$. It can be shown that the matrix determinant of the optical system is identical to the ratio of the refractive indices on the object respectively image side:

$$\det \mathbf{M}_{\mathrm{os}} = \mathbb{A} \cdot \mathbb{D} - \mathbb{B} \cdot \mathbb{C} = \frac{n_\mathrm{o}}{n_\mathrm{i}} \tag{3.52}$$

From Equations (3.46) to (3.49) follows that the distance between the principal points is the same as the distance between the nodal points.

Furthermore, optical systems surrounded by the same media on the object and image side, which means that $n_\mathrm{o}/n_\mathrm{i} = 1$, have the following characteristics:

- According Equation (3.52), the matrix determinant is equal to one, i. e., $\det \mathbf{M}_{\mathrm{os}} = 1$.
- According to the Equations (3.46) to (3.49), all nodal points coincide with their corresponding principal points, i. e., $e_{\mathrm{Ho}} = e_{\mathrm{No}}$ and $e_{\mathrm{Hi}} = e_{\mathrm{Ni}}$.
- The object and image focal length are oppositely directed but of the same magnitude, i. e., $f_\mathrm{i} = -f_\mathrm{o}$.

3.3.5 Examples of simple camera lens setups

The properties of an optical system, consisting of different individual lens elements, strongly depend on their position. Optical lens design capitalizes on that to optimize, e. g., lenses for special types of cameras. In the following, we present some lens arrangements that can be easily computed by the matrix method described above. The telephoto lens design is of particular interest as the physical length of this lens construction is remarkably shorter than its overall focal length. On the other hand, the inverted arrangement of a retrofocus design is favorable for lenses with a short focal length when a larger distance to the image plane is required.

Example 1: serial arrangement of two thin lenses separated by a distance t_s

As a first example, we consider two thin lenses L_1 and L_2 in air that are separated by a distance t_s (Figure 3.19). Their corresponding focal lengths are $f_{1\mathrm{i}}$ and $f_{2\mathrm{i}}$. The central lens

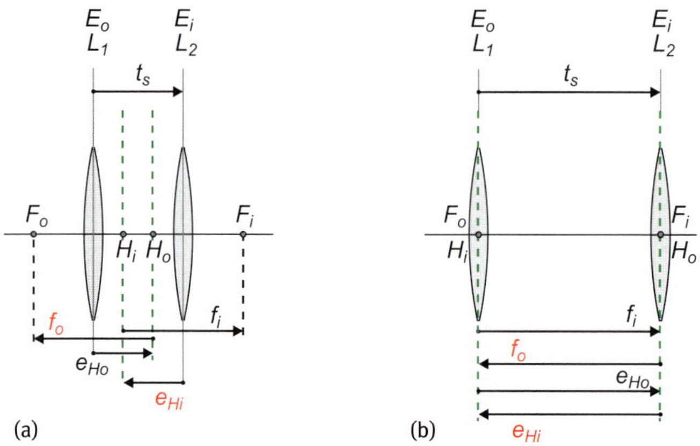

Fig. 3.19: Serial arrangement of two identical thin lenses in air, each having an image focal length of f: (a) lens separation $t_s = f/2$; the image focal length of the combination is $f_i = (2/3) \cdot f$; (b) lens separation $t_s = f$; the image focal length of the combination is identical to that of a single lens; the combination is free of chromatic aberrations (see Section 3.5.6).

planes are conveniently chosen as the input plane E_o and output plane E_i, respectively, of the optical system. The system matrix is setup with the matrix $\mathbf{L}_1$ (Equation (3.40)) for the first thin lens on the right side of the matrix multiplication followed by the translation matrix $\mathbf{T}_{12}$ (Equation (3.31)) and the second lens matrix $\mathbf{L}_2$ to the left:

$$\mathbf{M}_{os} = \mathbf{L}_2 \cdot \mathbf{T}_{12} \cdot \mathbf{L}_1 = \begin{bmatrix} 1 & 0 \\ \frac{1}{f_{2i}} & 1 \end{bmatrix} \cdot \begin{bmatrix} 1 & -t_s \\ 0 & 1 \end{bmatrix} \cdot \begin{bmatrix} 1 & 0 \\ \frac{1}{f_{1i}} & 1 \end{bmatrix} = \begin{bmatrix} 1 - \frac{t_s}{f_{1i}} & -t_s \\ \frac{1}{f_{1i}} + \frac{1}{f_{2i}} - \frac{t_s}{f_{1i} \cdot f_{2i}} & 1 - \frac{t_s}{f_{2i}} \end{bmatrix} \tag{3.53}$$

The overall refractive power of this system is given by the coefficient in the first column, second row yielding

$$V_i = \frac{1}{f_{1i}} + \frac{1}{f_{2i}} - \frac{t_s}{f_{1i} \cdot f_{2i}} \quad \text{with} \quad f_i = \frac{1}{V_i}. \tag{3.54}$$

The location of the principal points relative to the reference planes is given by Equations (3.46) and (3.47):

$$e_{Ho} = \frac{1 - \mathbb{D}}{\mathbb{C}} = t_s \cdot \frac{f_i}{f_{2i}} \qquad e_{Hi} = \frac{\mathbb{A} - 1}{\mathbb{C}} = -t_s \cdot \frac{f_i}{f_{1i}}. \tag{3.55}$$

If both lenses are diverging with $f_{1i} < 0$ and $f_{2i} < 0$, then the total system is also diverging and no real image formation can be expected using such a system. A more interesting case is found for both lenses being positive, $f_{1i} > 0$ and $f_{2i} > 0$. Now the sign of V_i depends on the size of the separation t_s and we have to distinguish three different situations:

a) If both lenses are close to each other with t_s being much smaller than any of the thin lenses focal length, V_i is positive and nearly the sum of the refractive power of both single lenses. This has been mentioned in Section 3.1.4 for a simple lens combination. Increasing the separation between the lenses leads to a decrease of V_i and an increase of the focal length. If, for instance, both lenses are identical with $f_{1i} = f_{2i} = f$ like in symmetrical lens designs we get

$$\mathbf{M}_{os} = \begin{bmatrix} 1-\frac{t_s}{f} & -t_s \\ \frac{2}{f}-\frac{t_s}{f^2} & 1-\frac{t_s}{f} \end{bmatrix} \quad \text{and} \quad \begin{matrix} f_i = \frac{f^2}{2f-t_s} & f_{Ei} = \left(1-\frac{t_s}{f}\right)\cdot f_i \\ e_{Ho} = \frac{t_s}{f}\cdot f & e_{Hi} = -e_{Ho} \end{matrix} \tag{3.56}$$

For a positive refractive power and focal length f_i, we get the range for the separation between both lenses being $t_s < 2f$. If the separation is increased from zero to $t_s = f/2$, the refractive power of the total system then drops from $V_i = 2/f$ to $V_i = 1.5/f$ and the focal length increases from $f_i = 0.5 \cdot f$ to $f_i = (2/3) \cdot f$. It should be noted that for $t_s = f/2$ we get $e_{Ho} = f/3$, which means that the object principal plane is shifted to the right while the image principal plane is shifted by the same magnitude to the left side ($e_{Hi} = -f/3$) as depicted in Figure 3.19a. Increasing the separation further to a value of $t_s = f$ leads to further shift of the principal planes and increase of the focal length. Focal points and principal planes are then located in the lens planes (Figure 3.19b). The combination of these two lenses with a separation of $t_s = f$ forms an achromatic doublet, and thus is free of chromatic aberrations (see Section 3.5.6). However, in that case the back focal length f_{Ei} is zero, which is not appropriate for practical imaging, thus $t_s < f$ is required for a symmetric camera lens combination. The overall length l of the lens combination, in this simple case with lenses of negligible thickness and without mount, extends from the first lens element to the image focal point and yields

$$l = t_s + f_{Ei} = t_s + \left(1 - \frac{t_s}{f}\right) \cdot f_i. \tag{3.57}$$

The ratio of the length l by the image focal length f_i is called the telephoto ratio for long focus lenses (see Section 6.3) and can be written after some simple calculations using Equation (3.56) as

$$\frac{l}{f_i} = \frac{t_s}{f} - \left(\frac{t_s}{f}\right)^2 + 1. \tag{3.58}$$

If both lenses are separated by not more than their focal length, we have $0 < t_s/f < 1$. For this condition, the telephoto ratio is between 1 and 1.25. This implies that a symmetric lens design always results in a lens that is longer than its image focal length.

b) If the separation is equal to the sum of both focal lengths, $t_s = f_{1i}+f_{2i}$, then V_i becomes zero. The optical system then is afocal with zero refractive power and infinitely long focal length. There is no real image formation but a beam transformation. This can be seen by analyzing the system matrix $\mathbf{M}_{os}$:

$$\mathbf{M}_{os} = \begin{bmatrix} -\frac{f_{2i}}{f_{1i}} & -(f_{1i} + f_{2i}) \\ 0 & -\frac{f_{1i}}{f_{2i}} \end{bmatrix} \quad \text{and} \quad \begin{aligned} h_{Ei} &= -\frac{f_{2i}}{f_{1i}} \cdot h_{Eo} - (f_{1i} + f_{2i}) \cdot \gamma_{Eo} \\ \gamma_{Ei} &= -\frac{f_{1i}}{f_{2i}} \cdot \gamma_{Eo} \end{aligned} \tag{3.59}$$

If the first lens has a longer focal length than the second one, $f_{1i} > f_{2i} >$ we have a telescope setup, also termed astronomic or Keplerian telescope. This results in an angle magnification Γ for rays from objects at faraway distances, having a small angle γ_{Eo} with the optical axis:

$$\Gamma = \frac{\gamma_{Ei}}{\gamma_{Eo}} = -\frac{f_{1i}}{f_{2i}} \tag{3.60}$$

Γ is equal to the ratio of the lenses' focal lengths. As we have an afocal system no real image is formed, but rather a virtual image that can be perceived by the naked eye. An inverted image from faraway scenery can be considered and appears magnified as the perceived angle is larger than without a telescope. Moreover, a beam parallel to the optical axis with $\gamma_{Eo} \approx 0$ and entering the optical system at an elevation h_{Eo} will leave the system again as a parallel beam but now at a reduced distance $h_{Ei} = -(f_{2i}/f_{1i}) \cdot h_{Eo}$. We thus have a compression of the diameter of a parallel beam entering the system. If the telescope is used in an inverted way with the first lens having a shorter focal length than the second one, the system acts as a beam expander for collimated beams and is often used in setups for handling optical beams.

c) For larger distances between the lenses with $t_s > f_{1i} + f_{2i}$, the refractive power and the overall focal length become negative. The optical system behaves like one diverging lens and no real image formation is possible.

Combining lenses is very efficient for correcting lens aberrations inherent to spherical surfaces. Moreover, short focal lengths of spherical lenses necessitate small curvature radii thus reducing the physical size of a lens. Small lens sizes on the other hand mean less light accumulation than larger sizes and may be a problem for optical imaging. A solution for overcoming this is the combination of larger lenses with lower refractive power to increase the total refractive power of the system and to accumulate more light.

As a result of the consideration above, the combination of two positive lenses can be tailored to setup a converging, diverging or afocal optical system just by choosing the appropriate separation between the lenses. This gives the lens designer much flexibility for tuning a system to the desired properties. In the next examples, we will consider how the lens combination can be used to shift the principal planes of a system relative to the physical elements as is done for special types of camera lenses.

Example 2: positive and negative lens with air gap (Gauss-type lens)

The combination of a positive and a negative lens with an air gap can be found in the Dutch telescope setup designed in 1608 by the spectacle maker Lippershey. An improved version was later used by Galileo for his astronomical observations. A modified version, consisting of two meniscus shaped lenses of different magnitudes of focal lengths and different glass materials, was proposed and calculated by Gauss (1817) in order to reduce spherical and chromatic lens aberrations in telescopes. To see how the combination of a converging and a diverging lens influences the location of the principal planes, let us consider the arrangement of two thin lenses of the same magnitude of focal length (Figure 3.20). The first one is a converging lens with $f_{1i} = f > 0$, the second one is a diverging lens with $f_{2i} = -f$, positioned at the focal point of the first lens, which means at a distance $t_s = f$ behind it. The input and output planes are also here identical with the lens planes. The matrix for the optical system in air results from Equation (3.53) by inserting the values for f_{1i}, f_{2i} and t_s:

$$\mathbf{M}_{os} = \begin{bmatrix} 0 & -f \\ \frac{1}{f} & 2 \end{bmatrix} \tag{3.61}$$

Hence, the significant quantities of the system:

$$\begin{aligned} f_i &= \frac{1}{\mathbb{C}} = f \qquad & f_{Eo} &= -\frac{\mathbb{D}}{\mathbb{C}} = -2f \qquad & f_{Ei} &= \frac{\mathbb{A}}{\mathbb{C}} = 0 \\ e_{Ho} &= \frac{1-\mathbb{D}}{\mathbb{C}} = -f \qquad & e_{Hi} &= \frac{\mathbb{A}-1}{\mathbb{C}} = -f \end{aligned} \tag{3.62}$$

We get the interesting result that the image focal length f_i of the system is identical to that of the single converging lens. As for the principal planes, they are both shifted by $-f$ to the object side as indicated by e_{Ho} and e_{Hi}. The object principal plane is outside the optical system, whereas the image principal plane coincides with the central plane of the converging thin lens. Consequently, the object focal point F_o is at a distance of $-2f$ in front of the first lens, whereas the image focal point F_i coincides with the position of

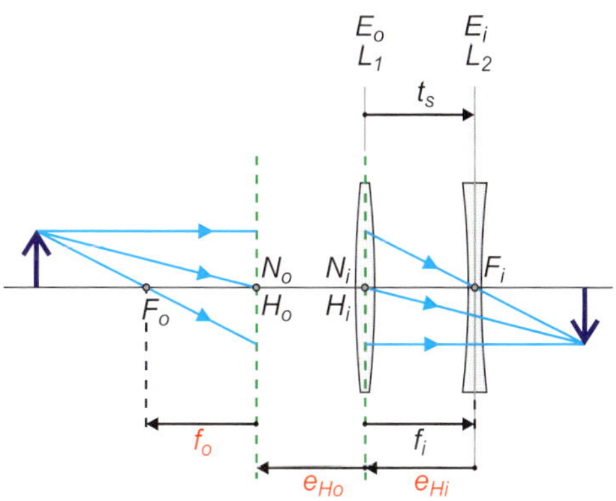

Fig. 3.20: 1:1 imaging using a lens combination consisting of two thin lenses of the same magnitude of focal length.

the diverging lens. The advantage of this setup is the correction of lens aberrations that will be described in more detail in Section 3.5. A drawback may be the fact that objects at large distances will be imaged near the image focal plane, which is identical to the second lens plane, and thus makes it impossible to locate an image sensor there. This drawback can be avoided if the separation between the two lenses is smaller than f or the magnitudes of both single focal lengths are not identical.

Example 3: telephoto lens setup

A design similar to Example 2 is used here for a simple telephoto setup as shown in Figure 3.21. The separation t_s between the two lenses, however, is smaller than the focal length of the converging lens, and thus the image focal plane is located further behind the second lens as mentioned above. A telephoto lens, which is discussed in more detail in Section 6.3, has a considerably longer focal length than a normal lens. Hence, for our example we assume a lens for the 35 mm format made up of two lenses with $f_{1i} = 300\,\text{mm}$, $f_{2i} = -f_{1i} = -300\,\text{mm}$ and a separation of $t_s = 180\,\text{mm}$ between them. With the lens planes as reference planes, the system matrix then can be written according to Equation (3.53):

$$\mathbf{M}_{os} = \begin{bmatrix} 1-\frac{t_s}{f_{1i}} & -t_s \\ \frac{t_s}{f_{1i}^2} & 1+\frac{t_s}{f_{1i}} \end{bmatrix} = \begin{bmatrix} 0.4 & -180\,\text{mm} \\ \frac{1}{500\,\text{mm}} & 1.6 \end{bmatrix} \tag{3.63}$$

The significant distances for the system are as follows:

$$f_i = \frac{1}{\mathbb{C}} = 500\,\text{mm} \qquad f_{Eo} = -\frac{\mathbb{D}}{\mathbb{C}} = -800\,\text{mm} \qquad f_{Ei} = \frac{\mathbb{A}}{\mathbb{C}} = 200\,\text{mm}$$
$$e_{Ho} = \frac{1-\mathbb{D}}{\mathbb{C}} = -300\,\text{mm} \qquad e_{Hi} = \frac{\mathbb{A}-1}{\mathbb{C}} = -300\,\text{mm} \tag{3.64}$$

The image focal length of the telephoto setup is with $f_i = 500\,\text{mm}$ longer than that of the single lens. Both principal planes are each shifted relatively to the lenses by –300 mm to the object side left (e_{Ho} and e_{Hi}). The image plane for a telephoto lens virtually intersects

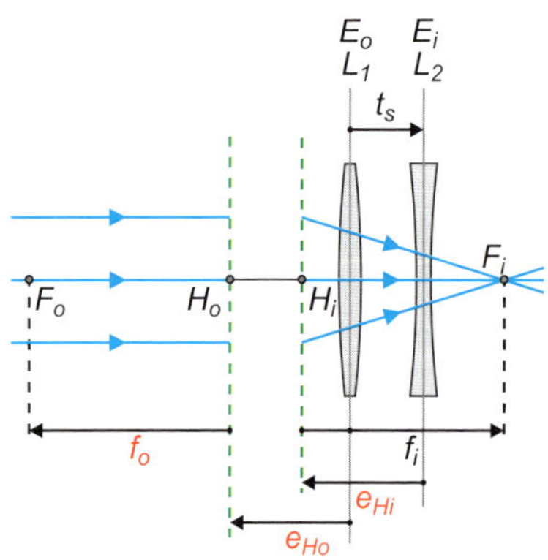

Fig. 3.21: Simple telephoto lens setup; the shift of principal planes outside the lens body results in a short design length.

the optical axis in the image focal point F_i. This one is located at the back focal distance $f_{Ei} = 200$ mm beyond the vertex of the diverging lens L_2. Thus, the overall design length of the telephoto lens from the vertex of the front lens L_1 to the image focal plane at F_i is of 380 mm. As a result, the design length is considerably shorter than its focal length of 500 mm.

For a telephoto setup, the magnitude of f_{1i} is larger than that of f_{2i} for the diverging lens. If the separation t_s is chosen as the sum of both focal lengths with $t_s = f_{1i} + f_{2i}$, then we have the Galilean telescope setup where the focal points of both lenses coincide in the image space. As the second lens has a negative focal length, the lens separation t_s is smaller than f_{1i}. The system is afocal like the Keplerian telescope described in Example 1 above and has the same angular magnification Γ given by Equation (3.60). The particular difference from the Keplerian telescope, however, is that with f_{2i} being negative the magnification Γ has a positive value. This means that the image observed by the human eye through the telescope is upright and not inverted like in the Keplerian telescope. For the nonafocal telephoto lens design, the separation between both lenses must be larger than that of the Galilean telescope but smaller than the image focal length of the converging lens and in order to get a positive back focal length. Further details about this design are discussed in Section 6.3.

Example 4: retrofocus setup for wide angle lens

As a last example, we consider a lens arrangement as illustrated in Figure 3.22.

This design is conventionally used for wide angle lenses, which typically have shorter focal lengths than normal lenses. A short focal length means that the image plane, which is located near F_i, is also close to the image principal plane at H_i. For the technical construction of interchangeable lenses, it is preferable that the image principal and focal planes should be at positions beyond the rear lens plane L_2 or even outside the camera lens body. In order to shift the principal planes to the image side, we use a lens arrangement similar to Example 3 but instead, swap its sequence. The first lens now is a diverging lens with $f_{1i} = -70$ mm, the second one a converging lens $f_{2i} = 40$ mm

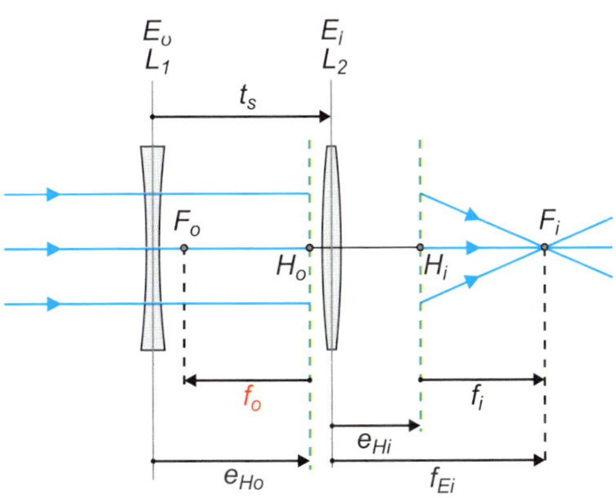

Fig. 3.22: Simple retrofocus lens setup; the shift of principal planes towards the image plane is advantageous for short focal lenses designed for system cameras.

and the lens separation is $t_S = 50\,\text{mm}$. As for the previous examples, the lenses are chosen as reference planes for input and output. We then get the system matrix:

$$\mathbf{M}_{OS} = \begin{bmatrix} 1 - \frac{t_s}{f_{1i}} & -t_S \\ \frac{1}{f_{1i}} + \frac{1}{f_{2i}} - \frac{t_s}{f_{1i} \cdot f_{2i}} & 1 - \frac{t_s}{f_{2i}} \end{bmatrix} = \begin{bmatrix} 1.71 & -50\,\text{mm} \\ \frac{1}{35\,\text{mm}} & -0.25 \end{bmatrix} \tag{3.65}$$

The significant distances and points can be calculated:

$$\begin{aligned} f_i &= \frac{1}{\mathbb{C}} = 35\,\text{mm} \qquad & f_{Eo} &= -\frac{\mathbb{D}}{\mathbb{C}} = 8.75\,\text{mm} \qquad & f_{Ei} &= \frac{\mathbb{A}}{\mathbb{C}} = 60\,\text{mm} \\ e_{Ho} &= \frac{1 - \mathbb{D}}{\mathbb{C}} = 43.75\,\text{mm} \qquad & e_{Hi} &= \frac{\mathbb{A} - 1}{\mathbb{C}} = 25\,\text{mm} \end{aligned} \tag{3.66}$$

The focal length of the system is with $f_i = 35\,\text{mm}$ shorter than the normal lens. As intended, the object side principal plane is at a distance of $e_{Ho} = 43.75\,\text{mm}$ on the right side of the diverging lens L_1, the second principal plane is at $e_{Hi} = 25\,\text{mm}$ on the right side of the second lens L_2. The distance between lens L_2 and the focal plane is $f_{Ei} = 60\,\text{mm}$ thus leaving enough free space, for instance, for a rocking mirror if the lens is mounted to a 35 mm SLR camera.

3.3.6 Ray transfer method for Gaussian beams

So far, we considered the imaging of rays within the idealized frame of geometrical optics where diffraction is neglected. Moreover, the object usually is illuminated by a noncoherent source and scatters the light into a wide solid angle. For some applications, however, like in case of coherent light exiting from an optical fiber with microscopic dimensions, light propagation using Gaussian beams is more appropriate for higher precision. Here, the Gaussian beam represents better the characteristics of the launched light than a simple bundle of geometric rays. We present in the following sections the example of a miniature ball lens for two purposes: First of all, we will see that the ray transfer method is not only valid for geometrical optics beams but also for paraxial Gaussian beams, however with some special modifications as for the beam parameters. Second, the ball lens is a good example for a miniature lens, which can be fabricated by relatively low efforts for special applications and where a mass production like for other microstructures may not be given. Special applications are for instance probes for optical coherence tomography or light transfer between optical fibers (see also Section 3.3.6.4). Here, a flexible lens design is required.

The light of a Gaussian beam is not simply refracted or translated in a straight way like a beam in geometrical optics, but its divergence changes with propagation. Here, a special feature of the Gaussian beam must be explicitly taken into account, which is that it can be considered as a parallel light bundle only over a short range. It should be noted that short is of the order of twice the Rayleigh length. In case of the ball lenses under consideration, this is of only a few mm at best, according to Equation (3.93). The beam will always diverge in the far field due to diffraction and cannot be imaged to a point of zero diameters. Hence, the propagation and imaging of a Gaussian beam through a lens deviate from that of a geometrical optical light bundle. As an example, we investigate the path of a light beam through a miniature ball lens on a microscopic scale. The beam originates from an optical singlemode fiber as a point source of about 4 µm diameter. It is imaged by a spherical lens surface, which is a section of ball lens generated at the end of the optical fiber by a fusion splicer. We first compute some characteristics based on geometrical optics, then use the matrix method for paraxial rays and finally apply the matrix method to a Gaussian beam. A comparison with experimental results validates the theoretical approaches. More details of the considerations in the following sections can be found in a separate publication of the authors elsewhere.[2]

3.3.6.1 Vortex equation for a ball lens

The problem that we want to consider is: How can we image a point source from an optical fiber core using a simple lens in a compact way? Using a separate lens, detached from the fiber, is quite expensive due to the necessarily precise alignment between fiber and lens, requiring a bulky set-up. A quite simple way is to fuse the fiber end facet and achieve a spherical surface acting as a ball lens.[3] Let us start the discussion with geometrical optics and have a look at Figure 3.23 where the cross-section of a fiber ball lens system (FBLS) is shown. Figure 3.23a schematically shows a ball lens directly attached to a singlemode fiber without any expansion range between fiber and ball lens. As a consequence, the light spot at the end of the fiber is within the focal distance of the ball lens (see also Section 3.2.2) and no real image of the spot is possible. Thus, the rays leaving the fiber end diverge. In order to achieve a bundle of parallel rays at the output, the point source must be shifted to the object focal point in glass. This can be done by splicing a section of a nocore fiber between the singlemode fiber and the ball lens in order to establish a longer expansion range (Figure 3.23b). If the expansion range is further increased, we come to the situation that the emerging beams converge and intersect the optical axis in the image point (Figure 3.23c).

The imaging characteristic can be calculated using the vertex equation (3.15), which takes into account the distances a_f, a_i the refractive indexes n_f, n_i and the curvature radius r of the spherical surface:

$$\frac{n_\mathrm{i}}{a_\mathrm{i}} - \frac{n_\mathrm{f}}{a_\mathrm{f}} = \frac{n_\mathrm{i} - n_\mathrm{f}}{r} \tag{3.67}$$

---

2 H. J. Brueckner, L. Y. Chai, S. Tiedeken, V. Braun, U. Teubner: *Design considerations and experimental investigations on fiber ball lens systems for optical metrology*, J. Phys. Photonics **5**, 035004 (2023) (https://doi.org/10.1088/2515-7647/acdba6)

3 S. Park, S. Rim, J. Kim, J. Park, I. Sohn, B. Lee: *Analysis of Design and Fabrication Parameters for Lensed Optical Fibers as Pertinent Probes for Sensing and Imaging*, Sensors (Basel, Switzerland) **18**, 4150–4163 (2018).

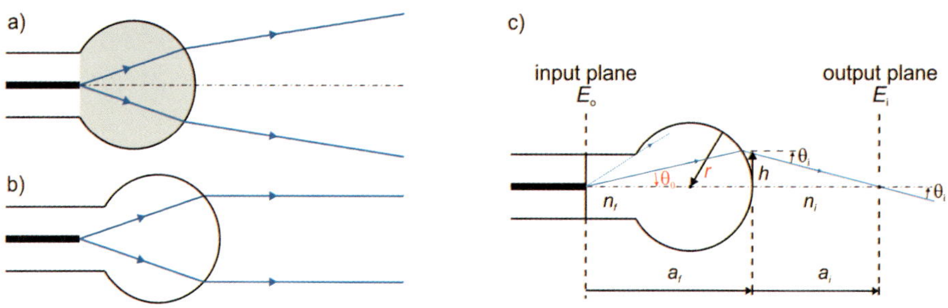

Fig. 3.23: Schematic illustration of the imaging characteristics of a FBLS. a) Ball lens directly attached to the singlemode fiber yields a diverging FBLS; b) increasing the beam expansion range can lead to parallel emerging rays; c) converging FBLS with a longer beam expansion range than b). Note: Symbols in red color indicate negative quantities.

The indices f and i refer to fiber and image space, respectively. It should be noted that according to our sign convention in this book quantities in red color are counted as negative (Section 3.1.4). Usually, the object distance a_f from the light source in the fiber to the vertex of the lens points to the negative direction. However, in order to facilitate the matrix formalism in the next section and to be aligned with the optical axis, we have redirected it, and thus it is counted positive. In the following, a_f, a_i as well as the ball lens radius r_B are positive quantities. With $r = -r_B$, we then get from Equation (3.67):

$$\frac{n_i}{a_i} + \frac{n_f}{a_f} = \frac{n_f - n_i}{r_B} \tag{3.68}$$

Rearranging that equation to get the back image distance a_i, which is the image distance measured relative to the ball lens vertex, yields

$$a_i = \frac{n_i \cdot r_B \cdot a_f}{a_f \cdot (n_f - n_i) - n_f \cdot r_B} \tag{3.69}$$

In paraxial approximation, we also calculate the output divergence angle θ_i as a function of the input divergence angle θ_0. This is valid for angles smaller than about 5°. Hence, the intersection of the starting ray with the spherical surface at the elevation h from the optical axis can be used to get the relation between the angles of divergence:

$$h \approx -\theta_0 \cdot a_f = \theta_i \cdot a_i \quad \Rightarrow \quad \theta_i = -\frac{a_f}{a_i} \cdot \theta_0 \tag{3.70}$$

Inserting Equation (3.69) in (3.70) leads to the equation:

$$\theta_i = \theta_0 \cdot \frac{a_f \cdot (n_i - n_f) + n_f \cdot r_B}{n_i \cdot r_B} \tag{3.71}$$

The denominator of Equation (3.69) determines if a_i is positive, negative or infinite as n_i, r_B and a_f all are positive quantities. In a similar way, it determines the sign of the output divergence given by Equation (3.71). We define a critical value a_{fc} for the length of the expansion range when the denominator is zero:

$$0 = a_{fc} \cdot (n_f - n_i) - n_f \cdot r_B \quad \Leftrightarrow \quad a_{fc} = \frac{n_f}{n_f - n_i} \cdot r_B \tag{3.72}$$

If the expansion range has the critical length a_{fc} then the image point is at infinity, the output divergence is zero. This is the typical feature of a collimating lens where the light source is in the object focal point and the outgoing light bundle consists of parallel rays. If we assume a negative θ_0 as depicted in Figure 3.23, larger distances $a_f > a_{fc}$ result in a positive, but finite back image distance a_i with positive divergence θ_i, which is typical for converging FBLS. Shorter distances $a_f < a_{fc}$ lead to a diverging system with negative a_i and negative θ_i. We can easily estimate the critical a_{fc} for a FBLS in air when using a typical value for glass with $n_f = 1.5$ and $n_i = 1$. We then get $a_{fc} = 3r_B$. This means that an expansion range of at least three times the radius of the ball lens is required to obtain a converging FBLS.

3.3.6.2 System matrix for a fiber ball lens system

A more flexible way to discuss the properties of a FBLS is to set-up the system matrix $\mathbf{M}_{FBLS}$, which can be used for both geometrical optical rays as well as Gaussian beams. To that purpose, let us consider Figure 3.23c where the ray starts on the optical axis at the input plane E_o under the angle θ_0. It first expands along a_f to reach the elevation h at the intersection with the spherical lens surface. The displacement is described by the translation matrix $\mathbf{T}_{er}$ in the expansion range. At the surface, the ray is refracted from the glass medium to the image space where only the angle changes and for which we use the refraction matrix **R**. Finally, in the image space the ray is translated along the distance a_i, given by the matrix $\mathbf{T}_{is}$. As described before the system matrix is established by the subsequent matrix multiplication,

$$\mathbf{M}_{FBLS} = \mathbf{T}_{is} \cdot \mathbf{R} \cdot \mathbf{T}_{er} = \begin{bmatrix} 1 & -a_i \\ 0 & 1 \end{bmatrix} \cdot \begin{bmatrix} 1 & 0 \\ \dfrac{n_i - n_f}{-n_i \cdot r_B} & \dfrac{n_f}{n_i} \end{bmatrix} \cdot \begin{bmatrix} 1 & -a_f \\ 0 & 1 \end{bmatrix} \tag{3.73}$$

It should be noted that the ray hits in the glass ball a concave surface with negative curvature radius $r = -r_B$; hence, the negative sign in the matrix element. Carrying through the matrix multiplication yields the system matrix elements:

$$\mathbf{M}_{FBLS} = \begin{bmatrix} 1 - a_i \cdot \dfrac{n_i - n_f}{-n_i \cdot r_B} & -a_f - a_i \cdot a_f \cdot \dfrac{n_i - n_f}{n_i \cdot r_B} - a_i \cdot \dfrac{n_f}{n_i} \\ \dfrac{n_i - n_f}{-n_i \cdot r_B} & a_f \cdot \dfrac{n_i - n_f}{n_i \cdot r_B} + \dfrac{n_f}{n_i} \end{bmatrix} \tag{3.74}$$

The ray at the output plane with the parameters h_{Ei} and γ_{Ei} results from the application of the system matrix to the input ray with h_{Eo} and γ_{Eo}:

$$\begin{pmatrix} h_{\mathrm{Ei}} \\ \gamma_{\mathrm{Ei}} \end{pmatrix} = \mathbf{M}_{\mathrm{FBLS}} \begin{pmatrix} h_{\mathrm{Eo}} \\ \gamma_{\mathrm{Eo}} \end{pmatrix} \tag{3.75}$$

Assigning the values $h_{\mathrm{Eo}} = 0$ and $\gamma_{\mathrm{Eo}} = \theta_0$ to the parameters of the starting ray, fixed on the optical axis, we get the following results for the outgoing ray:

$$h_{\mathrm{Ei}} = -\theta_0 \cdot \left(a_{\mathrm{f}} + a_{\mathrm{i}} \cdot a_{\mathrm{f}} \cdot \frac{n_{\mathrm{i}} - n_{\mathrm{f}}}{n_{\mathrm{i}} \cdot r_{\mathrm{B}}} + a_{\mathrm{i}} \cdot \frac{n_{\mathrm{f}}}{n_{\mathrm{i}}} \right) \tag{3.76}$$

$$\gamma_{\mathrm{Ei}} = \theta_0 \cdot \left(a_{\mathrm{f}} \cdot \frac{n_{\mathrm{i}} - n_{\mathrm{f}}}{n_{\mathrm{i}} \cdot r_{\mathrm{B}}} + \frac{n_{\mathrm{f}}}{n_{\mathrm{i}}} \right) \tag{3.77}$$

The paraxial image point is located on the optical axis with $h_{\mathrm{Ei}} = 0$. For this condition, we can calculate the back image distance a_{i} using Equation (3.76):

$$a_{\mathrm{i}} = \frac{n_{\mathrm{i}} \cdot r_{\mathrm{B}} \cdot a_{\mathrm{f}}}{a_{\mathrm{f}} \cdot (n_{\mathrm{f}} - n_{\mathrm{i}}) - n_{\mathrm{f}} \cdot r_{\mathrm{B}}} \tag{3.78}$$

When the ray leaves the ball after refraction, it propagates in a straight line and its angle is independent from its position in the image space. This can be also seen from Equation (3.77) as the only image space parameter is the refractive index, which is a constant. We thus get for the output divergence:

$$\theta_{\mathrm{i}} = \gamma_{\mathrm{Ei}} = \theta_0 \cdot \frac{a_{\mathrm{f}} \cdot (n_{\mathrm{i}} - n_{\mathrm{f}}) + n_{\mathrm{f}} \cdot r_{\mathrm{B}}}{n_{\mathrm{i}} \cdot r_{\mathrm{B}}} \tag{3.79}$$

As expected, we get the same results for image distance and output divergence as we did for the calculation based on the vortex equations above. Both considerations are based on geometrical optics.

3.3.6.3 Gaussian beam propagation

We now extend our view to the propagation of a Gaussian beam where we make profit of the system matrix $\mathbf{M}_{\mathrm{FBLS}}$ (Equation (3.74)). The complex beam parameter $\boldsymbol{q}$ of a Gaussian beam was introduced in Section 3.1.3 based on the curvature radius $r(z)$ and the spot radius $w(z)$ (Equation (3.12)). The transformation of a Gaussian beam with the parameter $\boldsymbol{q}_0 = \boldsymbol{q}(z_0)$ at the input plane into that of the output plane with the beam parameter $\boldsymbol{q}_{\mathrm{i}} = \boldsymbol{q}(z_{\mathrm{i}})$ is described in a general way by the equation [Sal19, Ped08]:

$$\boldsymbol{q}_{\mathrm{i}} = \frac{\mathbb{A} \cdot \boldsymbol{q}_0 + \mathbb{B}}{\mathbb{C} \cdot \boldsymbol{q}_0 + \mathbb{D}} \tag{3.80}$$

Here, the matrix elements of the system matrix $\mathbf{M}$ are used:

$$\mathbf{M} = \begin{bmatrix} \mathbb{A} & \mathbb{B} \\ \mathbb{C} & \mathbb{D} \end{bmatrix} \tag{3.81}$$

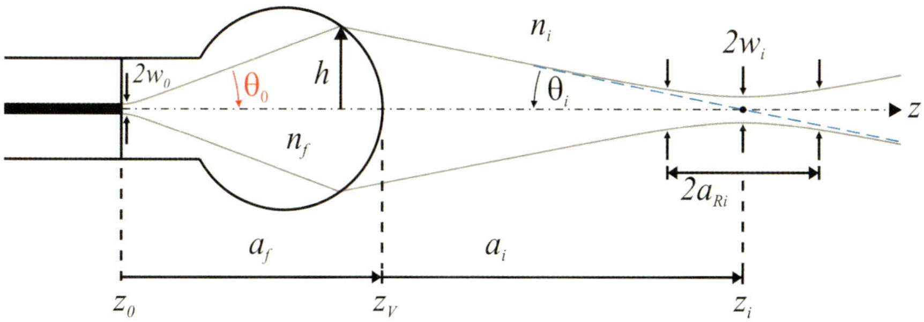

Fig. 3.24: Image formation in the fiber-ball-lens-system (Gaussian optics approximation).

For the beam propagation through our FBLS we choose the starting condition of a flat wavefront for the Gaussian beam originating at the end of the optical singlemode fiber (Figure 3.24). In this case with a reciprocal curvature radius $1/r(z_0) = 0$ and a spot radius $w(z) = w_0$, we can write

$$\frac{1}{\boldsymbol{q}_0} = \frac{1}{r(z_0)} - i\frac{\lambda}{\pi \cdot w^2(z_0) \cdot n_\mathrm{f}} = -i\frac{\lambda}{\pi \cdot w_0^2 \cdot n_\mathrm{f}} \tag{3.82}$$

Hence, we get a purely imaginary beam parameter, having a magnitude equal to the Rayleigh length z_Rf of the beam in the fiber:

$$\boldsymbol{q}_0 = i \cdot z_\mathrm{Rf} \quad \text{with} \quad z_\mathrm{Rf} = \frac{\pi \cdot w_0^2 \cdot n_\mathrm{f}}{\lambda} \tag{3.83}$$

The complex beam parameter at the output of the FBLS is given by

$$\boldsymbol{q}_\mathrm{i} = \frac{\mathbb{A} \cdot i \cdot z_\mathrm{Rf} + \mathbb{B}}{\mathbb{C} \cdot i \cdot z_\mathrm{Rf} + \mathbb{D}} \quad \text{with} \quad \begin{array}{ll} \mathbb{A} = 1 + a_\mathrm{i} \cdot \dfrac{n_\mathrm{i} - n_\mathrm{f}}{n_\mathrm{i} \cdot r_\mathrm{B}} & \mathbb{B} = -a_\mathrm{f} - a_\mathrm{i} \cdot a_\mathrm{f} \cdot \dfrac{n_\mathrm{i} - n_\mathrm{f}}{n_\mathrm{i} \cdot r_\mathrm{B}} - a_\mathrm{i} \cdot \dfrac{n_\mathrm{f}}{n_\mathrm{i}} \\ \mathbb{C} = \dfrac{n_\mathrm{i} - n_\mathrm{f}}{-n_\mathrm{i} \cdot r_\mathrm{B}} & \mathbb{D} = a_\mathrm{f} \cdot \dfrac{n_\mathrm{i} - n_\mathrm{f}}{n_\mathrm{i} \cdot r_\mathrm{B}} + \dfrac{n_\mathrm{f}}{n_\mathrm{i}} \end{array} \tag{3.84}$$

In order to further discuss the properties of the beam in the image space, we need an expression for $1/\boldsymbol{q}_\mathrm{i}$. The inversion of $\boldsymbol{q}_\mathrm{i}$ yields

$$\frac{1}{\boldsymbol{q}_\mathrm{i}} = \frac{a_\mathrm{f} \cdot \frac{n_\mathrm{i}-n_\mathrm{f}}{n_\mathrm{i} \cdot r_\mathrm{B}} + \frac{n_\mathrm{f}}{n_\mathrm{i}} - i \cdot z_\mathrm{Rf} \cdot \frac{n_\mathrm{i}-n_\mathrm{f}}{n_\mathrm{i} \cdot r_\mathrm{B}}}{-a_\mathrm{f} - a_\mathrm{i} \cdot a_\mathrm{f} \cdot \frac{n_\mathrm{i}-n_\mathrm{f}}{n_\mathrm{i} \cdot r_\mathrm{B}} - a_\mathrm{i} \cdot \frac{n_\mathrm{f}}{n_\mathrm{i}} + i \cdot z_\mathrm{Rf} \cdot \left(1 + a_\mathrm{i} \cdot \frac{n_\mathrm{i}-n_\mathrm{f}}{n_\mathrm{i} \cdot r_\mathrm{B}}\right)} \tag{3.85}$$

The real and imaginary parts of $1/\boldsymbol{q}_\mathrm{i}$ are of large importance as they directly yield the curvature radius and spot size, respectively. Separation of real and imaginary parts of

Equation (3.85) and rearranging the equation leads to the expression:

$$\Re\left(\frac{1}{\boldsymbol{q}_\mathrm{i}}\right)$$
$$= \frac{[-a_\mathrm{f}(n_\mathrm{i}-n_\mathrm{f})-n_\mathrm{f}r_\mathrm{B}]\cdot[a_\mathrm{f}n_\mathrm{i}r_\mathrm{B}+a_\mathrm{i}a_\mathrm{f}(n_\mathrm{i}-n_\mathrm{f})+a_\mathrm{i}n_\mathrm{f}r_\mathrm{B}]-z_\mathrm{Rf}^2(n_\mathrm{i}-n_\mathrm{f})\cdot[n_\mathrm{i}r_\mathrm{B}+a_\mathrm{i}(n_\mathrm{i}-n_\mathrm{f})]}{[a_\mathrm{f}\cdot n_\mathrm{i}\cdot r_\mathrm{B}+a_\mathrm{i}\cdot a_\mathrm{f}\cdot(n_\mathrm{i}-n_\mathrm{f})+a_\mathrm{i}\cdot n_\mathrm{f}\cdot r_\mathrm{B}]^2+z_\mathrm{Rf}^2\cdot[n_\mathrm{i}\cdot r_\mathrm{B}+a_\mathrm{i}\cdot(n_\mathrm{i}-n_\mathrm{f})]^2}$$
$$\Im\left(\frac{1}{\boldsymbol{q}_\mathrm{i}}\right) = \frac{-z_\mathrm{Rf}\cdot n_\mathrm{i}\cdot n_\mathrm{f}\cdot r_\mathrm{B}^2}{[a_\mathrm{f}\cdot n_\mathrm{i}\cdot r_\mathrm{B}+a_\mathrm{i}\cdot a_\mathrm{f}\cdot(n_\mathrm{i}-n_\mathrm{f})+a_\mathrm{i}\cdot n_\mathrm{f}\cdot r_\mathrm{B}]^2+z_\mathrm{Rf}^2\cdot[n_\mathrm{i}\cdot r_\mathrm{B}+a_\mathrm{i}\cdot(n_\mathrm{i}-n_\mathrm{f})]^2} \tag{3.86}$$

These equations can be used to calculate the location of the image point, the beam diameter as well as the far field divergence. Unlike for geometrical optics, where the rays intersect the optical axis in the image point, a Gaussian beam never strikes the axis but attains a minimum spot size. Hence, as the image point of the Gaussian beam we define the location of its waist in the image space. We know that at this point the beam not only has its least diameter but also a flat wavefront. This means that the reciprocal curvature radius is zero, which is equal to the real part of $1/\boldsymbol{q}_\mathrm{i}$. The back image distance a_i can be calculated from the condition:

$$\frac{1}{r(z_\mathrm{i})} = \Re\left(\frac{1}{\boldsymbol{q}_\mathrm{i}}\right) = 0 \tag{3.87}$$

and yields after some rearrangements:

$$a_\mathrm{i} = n_\mathrm{i}\cdot r_\mathrm{B}\cdot\frac{-a_\mathrm{f}^2\cdot(n_\mathrm{i}-n_\mathrm{f})-a_\mathrm{f}\cdot n_\mathrm{f}\cdot r_\mathrm{B}-z_\mathrm{Rf}^2\cdot(n_\mathrm{i}-n_\mathrm{f})}{[a_\mathrm{f}\cdot(n_\mathrm{i}-n_\mathrm{f})+n_\mathrm{f}\cdot r_\mathrm{B}]^2+z_\mathrm{Rf}^2\cdot(n_\mathrm{i}-n_\mathrm{f})^2} \tag{3.88}$$

As for the waist radius w_i, we use the relation:

$$w_\mathrm{i} = \sqrt{\frac{-\lambda}{\pi\cdot n_\mathrm{i}\cdot\Im(\frac{1}{\boldsymbol{q}_\mathrm{i}})}} = 0 \tag{3.89}$$

Inserting in this equation the expression for $\Im(1/\boldsymbol{q}_\mathrm{i})$ (Equation (3.86)) and substituting a_i by Equation (3.88), we get after some lengthy rearrangement for the waist radius w_i:

$$w_\mathrm{i} = \frac{w_0\cdot n_\mathrm{f}\cdot r_\mathrm{B}}{\sqrt{[a_\mathrm{f}\cdot(n_\mathrm{i}-n_\mathrm{f})+n_\mathrm{f}\cdot r_\mathrm{B}]^2+z_\mathrm{Rf}^2\cdot(n_\mathrm{i}-n_\mathrm{f})^2}} \tag{3.90}$$

As the divergence angle of a Gaussian beam in the far field is inversely proportional to the corresponding beam waist radius, we can directly express the output divergence for the beam using Equation (3.11):

$$\theta_\mathrm{i} = \frac{\lambda}{\pi\cdot w_\mathrm{i}\cdot n_\mathrm{i}} \tag{3.91}$$

Substituting w_i by Equation (3.90) and w_0 using Equation (3.11), we get the result:

$$\theta_i = \theta_0 \cdot \frac{\sqrt{[a_f \cdot (n_i - n_f) + n_f \cdot r_B]^2 + z_{Rf}^2 \cdot (n_i - n_f)^2}}{n_i \cdot r_B} \tag{3.92}$$

For large ball lens radii r_B and large expansion ranges a_f, as compared to the Rayleigh length z_{Rf} in the fiber, the first square term under the square root in Equation (3.92) dominates the root so that the remaining part can be omitted. In that case, we get the same result as Equation (3.79) for geometrical optical rays. We will discuss this below.

3.3.6.4 Comparison of theoretical approaches with experimental results

The design of an optical system depends on the application for which the system is used and the conditions for it. For instance, the FBLS considered above may be used for a free space propagation of light. Then a collimated beam may be required, which is emitted by a FBLS and captured by an equivalent system at the end of the free space area. The collimation has a better quality the larger the emitted beam size is as the output divergence decreases. Other applications, like sensors for optical coherence tomography (OCT), require a smaller spot size for better optical resolution, but also a long working distance may be desired.[4] In order to illustrate how the image distance a_i, the divergence θ_i and the spot size w_i of the outgoing beam are affected by the splice to vertex length a_f, let us have a look at Figure 3.25. The diagram shows these quantities as a function of a_f for a ball lens of radius 135 µm. The wavelength is 630 nm, the waist radius w_0 of the emerging beam in glass is 2.1 µm. A microscopic view of this ball lens with $a_f = 548$ µm is given Figure 3.26a.

The critical length in order to achieve a collimation lens with $r_B = 135$ µm is $a_{fc} = 430$ µm according to Equation (3.72). The geometrical optical approximation yields that in this case the image point is at infinity, which is indicated by the vertical dashed line in the back image distance of Figure 3.25. The outgoing beam is parallel to the optical axis with $\theta_i = 0$, indicated by the intersection of the dashed lines in the divergence diagram of Figure 3.25. Shorter a_f lead to diverging lenses, larger a_f to converging lenses with finite back image distances. As for the geometrical optical approach, the back image distance decreases in a hyperbolic way with increasing a_f while the divergence of the outgoing beam increases linearly.

The Gaussian approximation yields completely different results when a_f is close to its critical value. Here, the back image distance is slightly larger than zero, which means that the image position is just beyond the vertex. The image spot diameter has its maximum at a_{fc}, and the far field divergence is above zero. When the expansion area is

4 S. Park, S. Rim, J. Kim, J. Park, I. Sohn, B. Lee: *Analysis of Design and Fabrication Parameters for Lensed Optical Fibers as Pertinent Probes for Sensing and Imaging*, Sensors (Basel, Switzerland) **18**, 4150–4163 (2018).

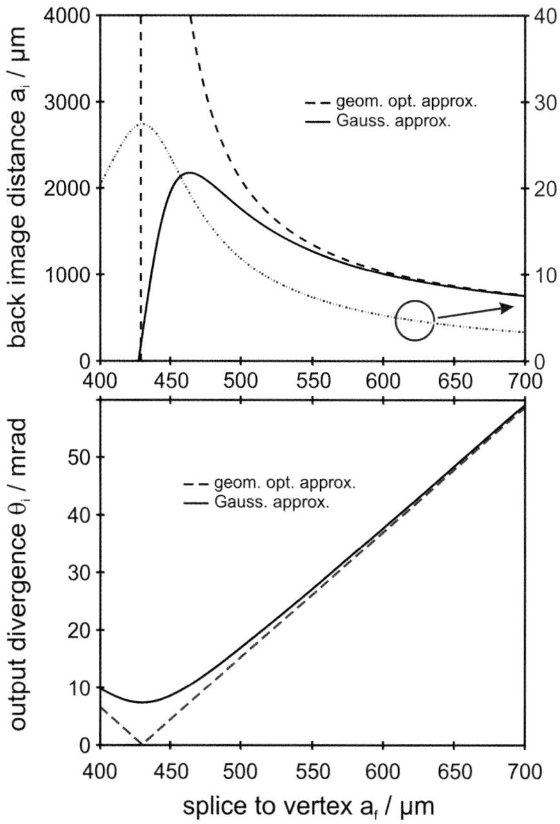

Fig. 3.25: Back image distance respectively output divergence of a FBLS with $r_B = 135$ µm, and $n_f = 1.4577$ as a function of the splice to vertex length; free space wavelength 630 nm; note that for simplicity only the magnitude of θ_i is shown in the lower diagram.

then increased, the back image distance steeply increases to achieve a maximum. Then it decreases again continuously to approach the hyperbolic curve given by the geometrical optical approximation. In both cases, the limiting value for the nearest image distance from the vertex is achieved for large a_f with $a_i = n_i \cdot r_B/(n_f - n_i)$. This is also identical to the back-image focal distance in case of geometrical optics treatment of the FBLS. For a typical refractive index in glass with $n_f \approx 1.5$, this value is equal to $2r_B$, which is the diameter of the ball lens. This is the shortest possible working distance for an FBLS in air. Also, the far field divergence of the outgoing beam approaches for large a_f the linear curve given by the geometrical optical approximation.

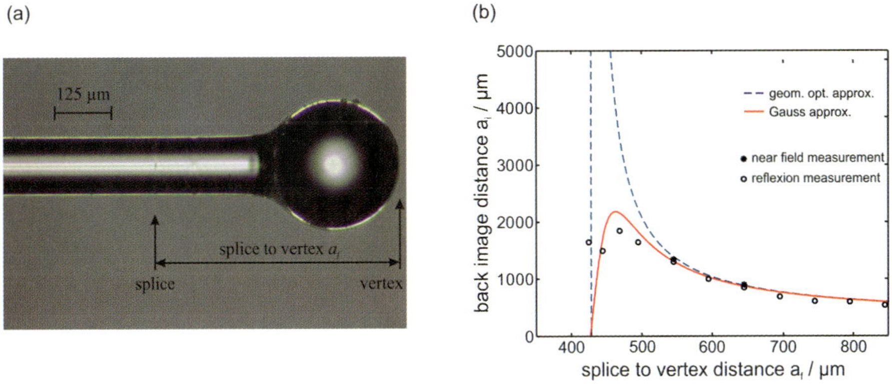

Fig. 3.26: Experimental and theoretical results for a FBLS, ball lens radius $r_B = 135$ µm. a) Microscopic view of a FBLS with a splice to vertex distance $a_f = 548$ µm; b) theoretical results based on different approaches (curves) in comparison to experimental results (circles); free space wavelength 630 nm, refractive index in fiber glass $n_f = 1.4577$.[5]

As mentioned above, the Gaussian beam spot has its largest diameter in case of the critical length at $a_f = a_{fc}$. The larger the spot size, the larger its Rayleigh length in the image space z_{Ri}:

$$z_{Ri} = \frac{\pi \cdot w_i^2 \cdot n_i}{\lambda} \tag{3.93}$$

As a consequence, the Rayleigh length is largest at a_{fc} and then decreases with increasing expansion range. As described in Section 3.1.3, within the range of $\pm z_{Ri}$ around the image distance a_i the spot radius widens up by a factor of $\sqrt{2}$. This means that light coupling to a Gaussian beam within that range leads to additional coupling losses of less than only 0.5 dB.[6] Hence, when coupling to other fibers or FBLS, like for instance for free-space light transmission, choosing $a_f = a_{fc}$ yields the range with the largest tolerances for positioning around the image distance.

On the other hand, FBLS with larger a_f are better suited for selective probes, as required, e. g., for OCT sensors, due to smaller spot sizes and restricted tolerances. The experimental results shown in Figure 3.26b are in good agreement with the theoretical approaches.

After the comparison of different theoretical approaches with experimental results, it can be concluded: For large splice to vertex distances a_f, the geometrical optical approach yields sufficiently exact results for image distance a_i and far field output diver-

---

5 H. J. Brueckner, L. Y. Chai, S. Tiedeken, V. Braun, U. Teubner: *Design considerations and experimental investigations on fiber ball lens systems for optical metrology*, J. Phys. Photonics **5**, 035004 (2023) (https://doi.org/10.1088/2515-7647/acdba6)

6 G. Grau: *Optische Nachrichtentechnik*, Springer Verlag Berlin, Heidelberg 1986.

gence θ_i. For shorter a_f and close to a_{fc}, only the Gaussian beam approximation yields precise results. In all cases when a calculation of the spot sizes is required like, e. g., for light coupling or sensor resolution, a treatment using a Gaussian beam is inevitable.

3.3.7 Software-based computational methods

The design of an optical system usually is a very complex task with tedious calculations, particularly if it is composed of many diverse lens elements. It took months for the detailed design and calculation of camera lenses around 100 years ago, and this was the reason that nearly all these types of lens inventions have been protected by patents. Today all these calculations are done using modern software-based methods. The matrix formalism presented above is a very powerful tool for understanding the characteristics of a system but has its drawbacks due to its paraxial simplifications and restrictions to 2D problems. However, its basic principles are implemented in many computational methods. As for the computer-based approaches, they can be distinguished with respect to the calculation of optical systems of different sizes: ray tracing, which follows the propagation of individual ray packages across complex optical system of macroscopic scale, and beam propagation, which simulates the light propagation in media with a slowly varying refractive index like optical waveguides or photonic devices of microscopic scale.

3.3.7.1 Ray tracing

The most commonly used method is based on analyzing the path of optical beams represented by single rays across any medium or system with an arbitrary arrangement of optical elements such as lenses. Before starting the computation, the optical system has to be defined by indicating the arrangement of lenses and their specifications as for instance shape, material, optical properties like refractive index and absorption coefficient, which usually also depend on the wavelength of light. In a next step, rays are launched into the system locally. On their path, these rays may be refracted, bent, reflected and partially absorbed, thus the physical laws of describing these phenomena have to be applied locally to determine the ray changes for the next propagation steps. Matrix calculation methods may be applied, but for more general problems finite element computational methods are used to solve the numerical mathematical equations. Details about ray tracing methods are for instance described in “Lens Design Fundamentals” by Rudolph Kingslake [Kin10]. In contrast to the simple matrix method as described above, which uses the paraxial approximation, the numerical computation incorporates the accurate formulas, thus avoiding lens errors due to a simplified approach. Also, the consideration of sagittal rays is possible whereas simple methods only deal with paraxial rays in meridional planes. The differentiation between sagittal and meridional rays, respectively, planes is discussed in more detail in Sections 3.5.2 and 3.5.3.

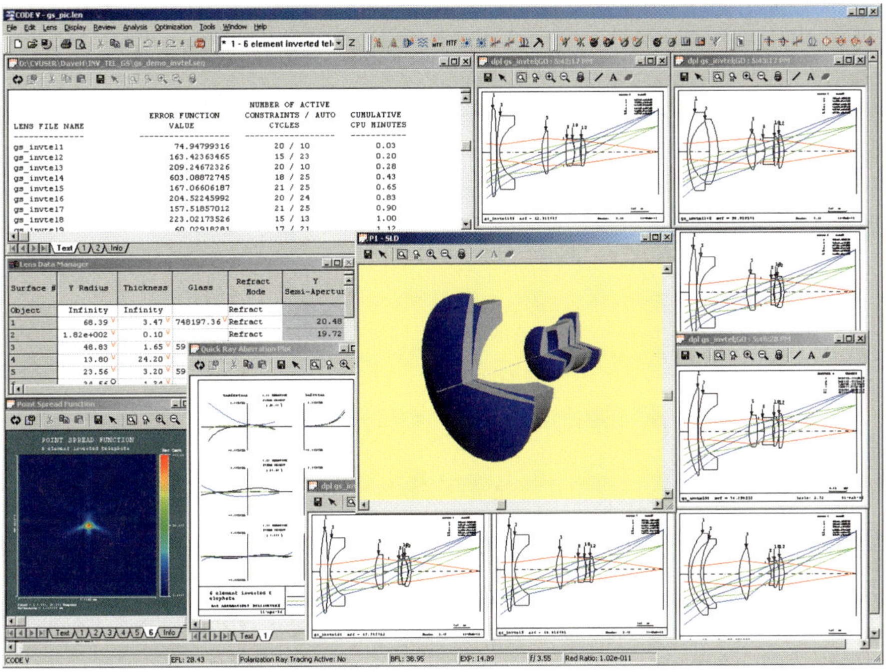

Fig. 3.27: Results of a reversed telephoto lens optimization by optical design software Code V™ (Synopsys Proprietary. Portions Copyright ©2017 Synopsys, Inc. Used with permission. All rights reserved. Synopsys & Code V are registered trademarks of Synopsys, Inc.).

A very important point using the software is the functionality to analyze any errors of the optical system and then optimize the whole setup. Optimization is not only necessary with respect to the performance of the optical system but also to its complexity, manufacturing tolerances and production costs. As the ray tracing method is only one part of the overall process of designing an optical system, a software incorporating these computational methods is also termed optical design software. Figure 3.27 illustrates the optimization of a telephoto lens using such software. All necessary parameters describing the performance and the quality of the lens can be viewed. Furthermore, multiple reflections of light beams in the lens construction are simulated to understand the imaging properties of lenses at critical situations of illumination, which for instance, lead to lens flares and ghost images (see Section 6.8).

3.3.7.2 Beam propagation

The physical dimensions of conventional optical systems are much larger than the wavelength of light. If diffraction is not an issue, light propagation can in many cases be sufficiently studied using ray optics. If, however, microscopic systems like optical waveguides or other photonic devices of integrated optical structures are considered, methods

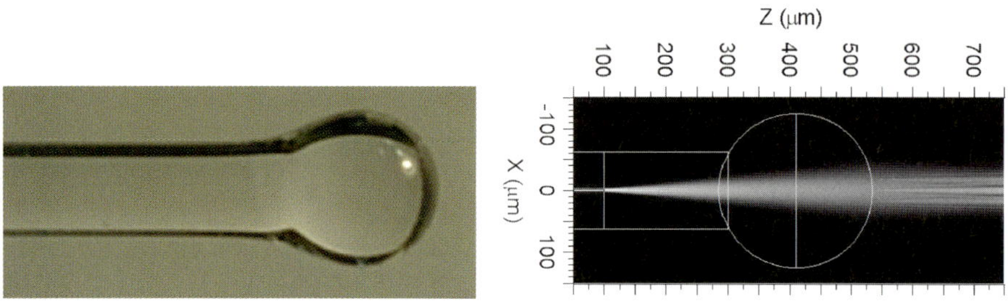

Fig. 3.28: 250 μm diameter ball lens formed at the end of a silica glass fiber; the numerical simulation for 630 nm light radiation was done using a 2D numerical beam propagation method.

based on a wave optical approach are more appropriate as they implicitly cover effects like diffraction or interference. The time-independent wave propagation in space is described by the Helmholtz equation (3.1), which is a differential equation for monochromatic waves of the wavelength λ as a function of the position vector $\vec{r}$ of which the components are the space coordinates x, y and z (see also Section 3.1.1 and Appendix A.11). In this equation, $\vec{\mathcal{E}}$ represents the 3D stationary electric field propagating in a medium, which is defined by the 3D distribution of the refractive index $n(x, y, z)$. By defining the index distribution, the optical system is fixed. After starting the simulation by launching light from arbitrary sources, for instance Gaussian beams, the spatial light distribution is calculated by numerically solving the Helmholtz equation. Also, here in general finite element methods are implemented using different approximations to reduce the mathematical complexity. The calculations can be very timeconsuming, especially for large systems with low symmetry when a full 3D treatment is necessary. Beam propagation algorithms can be implemented in more advanced optical design software packages for special applications.

Figure 3.28 shows the example of a ball lens of 250 μm diameter formed at the end of a no-core silica fiber by the use of a fusion splicer (see also Section 3.3.6). Light is fed to the system by a single-mode silica fiber for 630 nm wavelength having a core diameter of 4 μm. The simulation results were achieved using a 2D beam propagation software to optimize the no-core fiber length in order to attain a collimated output beam, which otherwise would be much more divergent without lens.

3.4 Limitations of light rays

In the previous sections, we considered the principles of image formation using some selected rays emerging from an object point. They are imaged by the optical system in the ideal case to one corresponding image point. However, due to physical limitations of the system, such as lens sizes or internal obstructions as a consequence of the lens design, only a restricted number of rays strike the image plane in a camera, and thus determine for instance the brightness of an image. On the other hand, if due to optical aberrations in the system, not all rays from one object point are imaged to the same image point; the faulty rays should be blocked in order to increase the image quality at the expense of brightness. For the discussion of these phenomena, it is necessary to introduce the concept of a chief ray with its limiting marginal rays passing a system with limitations. A classification of the limiting elements with respect to their function and impact on the imaging will be given, such as aperture stop or field stop: aperture stops control the amount of light reaching the image plane and help to improve the image quality by blocking unwanted rays. Field stops control the extent of the field of view and help to reduce vignetting. In the following consideration, we assume for simplicity circular shaped stops, if possible. It should be noted that the field stop, however, is often defined by a rectangular image frame due to the sensor geometry. But this does not restrict the principal discussions. Moreover, we limit ourselves to 2D considerations as the problems usually feature a rotational symmetry with respect to the optical axis.

3.4.1 Controlling the brightness: aperture stops and pupils

The aperture stop is the physical element in an optical system that limits the amount of light, and thus its brightness entering the system. In a camera, this stop is in general variable and often realized as an iris diaphragm (see, for instance, Figure 2.8). In the human eye, it is the iris that limits the incoming light. The aperture determines a cone of rays emerging from any point in the object plane that can enter the system. This cone can be characterized by a total aperture angle $2\theta_{en}$ (Figure 3.29). θ_{en} is defined as the angle between a central ray from a point P_o on the optical axis in the object plane and its corresponding marginal rays that still can enter the system and are not blocked by

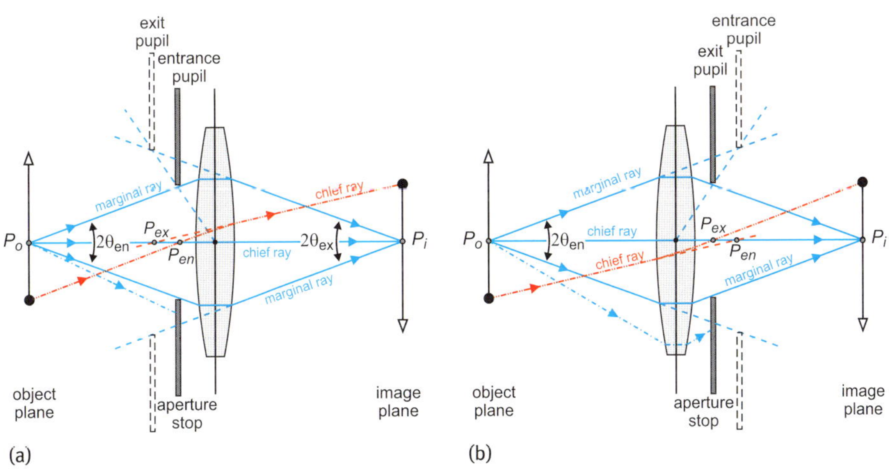

Fig. 3.29: Aperture stop in combination with a converging lens acting as a brightness limitation for incoming light. (a) Aperture stop in front of the lens; (b) aperture stop behind the lens.

the stop. Rays outside the cone are blocked. The larger the aperture stop the larger θ_{en}. Thus, the brightness of all points in the image plane is directly influenced by the size of the aperture stop, respectively, the angular aperture θ_{en}.

Figure 3.29a illustrates a 1:1 image formation, which means that the object focal point is at half distance between the object and the lens plane. The aperture stop is close to the lens and its diameter is smaller than that of the lens, so it is definitely the limiting physical element for the incoming rays. If the aperture stop opens up and is equal to or larger than the lens diameter, then the lens itself becomes the limiting physical element and acts as the aperture stop in the system. Figure 3.29b shows a similar situation like a) and the only difference here is that the aperture stop is behind the lens in the image space and not in front of it. In this case, the aperture stop is not the first physical element in the ray path but nevertheless the "bottleneck" for all rays traversing the optical system.

If one looks from the object space to the optical system, the limiting element that can be perceived by the observer is called entrance pupil. If the aperture stop is in front of the lens, as in Figure 3.29a or in Figure 3.30, it can be seen directly as it is the first component in the system. Here, the entrance pupil is identical with the aperture stop. If the stop is behind the lens, it is imaged through the lens and it is the image that is perceived by the observer. The image may be a virtual one, as shown in Figure 3.29b if the stop is within the focal distance to the lens. Then the aperture stop is viewed by the observer with the lens acting as a magnifying glass. The magnified virtual image is located beyond the lens on the same side as the stop (see also Figure 3.7b). It is also possible that the image is real, namely if the stop is at a larger distance behind the lens. Then the entrance pupil is the real image of the aperture stop in front of the lens. The concept of the entrance pupil helps to determine the angular aperture θ_{en} which is especially of

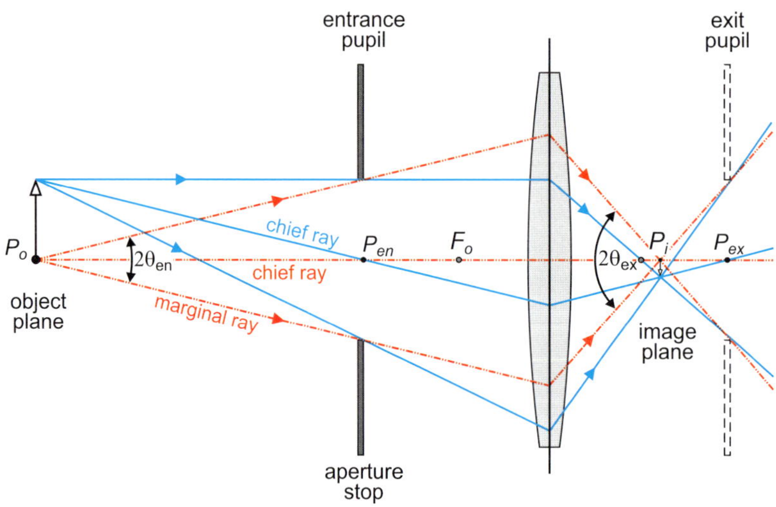

Fig. 3.30: Aperture stop in front of a converging lens resulting in a real image for the exit pupil.

avail in Figure 3.29b. The cone of entering rays can be easily constructed by drawing a straight line from P_o to the edges of the entrance pupil. The center of the entrance pupil, designated by P_{en}, is usually located on the optical axis for systems with rotational symmetry to it. If there are multiple elements in the optical system that may be limitations for ray bundles, then the entrance pupil is that element, respectively, image, which is associated with the smallest angular aperture θ_{en}.

Likewise, the exit pupil is the aperture stop or its image as perceived by an observer who looks at the optical system from an axial point P_i in the image plane. If the aperture stop is behind the lens and the last component in the system, the exit pupil is identical with the aperture stop whereas the image of the stop must be considered if the aperture stop is in front of the lens. In Figure 3.29a, the exit pupil is the virtual image of the aperture stop and located on the same side as the stop. The aperture stop in Figure 3.29b is the last element for rays leaving the system and is identical with the exit pupil as no more lenses are between it and P_i. Figure 3.30 illustrates the situation that the aperture stop is in front of the lens and its distance to the lens is larger than the focal length of the lens. Consequently, the image of the stop is a real one located even behind the lens. The center of the exit pupil is P_{ex} on the optical axis. The angular aperture θ_{ex} of the exit pupil is the angle between the straight ray from P_o through the image point P_i on the optical axis and the extreme or marginal rays coming from P_o that exit the optical system. $2\theta_{ex}$ is the total aperture angle of the cone that is formed by these marginal rays. θ_{en} and θ_{ex} in general have different values.

Usually, optical systems have more than only one lens and in many cases the aperture stop is located in between the lenses as illustrated in Figure 3.31. Then the aperture stop is neither entrance nor exit pupil. The entrance pupil in this example is the virtual image as perceived from the object space through Lens 1, which is only part of the whole system, whereas the exit pupil is the virtual image as perceived through Lens 2 from the image space. In more complex systems, a detailed analysis is necessary to determine the aperture stop, entrance pupil and exit pupil.

Based on this consideration, it can be said that the aperture stop always is the physical element, which controls the amount of light entering the optical system. The entrance pupil is the conjugate of the aperture stop and is perceived from the object space as the physical stop itself or as its image through the lens system. Likewise, the exit pupil is the conjugate of the aperture stop and perceived from the image space as the real component or its image. The exit pupil is also conjugate with the entrance pupil, which means that the exit pupil is the image of the entrance pupil and vice versa. The sizes of the entrance pupil and the exit pupil are related to each other by the pupil magnification M_p, which is given by their ratio:

$$M_p = \frac{D_{ex}}{D_{en}}. \tag{3.94}$$

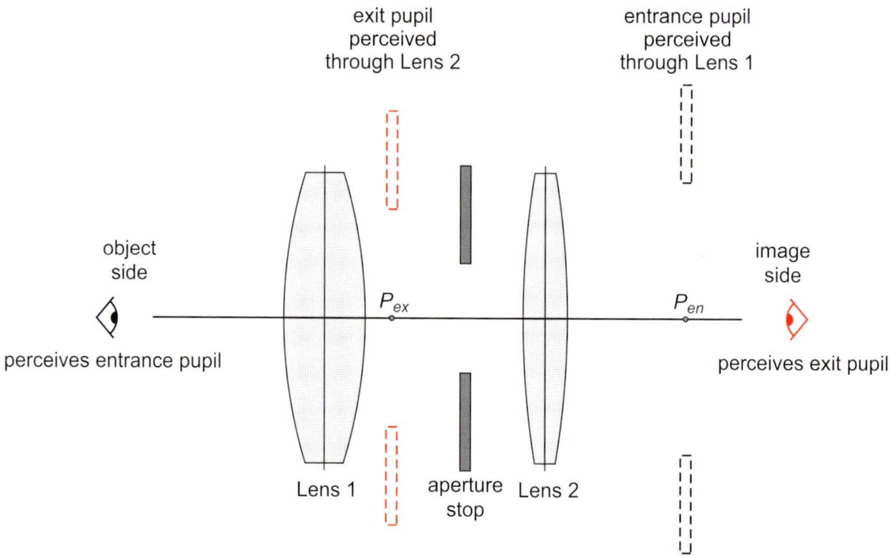

Fig. 3.31: Aperture stop in between two converging lenses.

For circular stops and pupils, D_{en} is the diameter of the entrance pupil and D_{ex} the corresponding value of the exit pupil. M_p is equal to one for lens combinations that are symmetrical with respect to the aperture stop. For other cases, the value of M_p can be calculated from the lens construction or experimentally by the inspection of the pupils from the object, respectively, image space, as illustrated in Figure 6.26.

The entrance pupil determines the angular aperture for rays in the object space, the exit pupil determines the corresponding angular aperture in the image space. All rays or their extensions that strike the edges of the entrance pupil also strike the edges of the exit pupil. This leads to the concept of chief and marginal rays, which simplify the discussion of image formation. For the construction of a ray diagram, we start with the fact that any point in the object plane is the origin for different rays entering the optical system. Among all these rays there is one significant ray, which is termed the chief ray. A chief ray emerging from any off-axis point in the object plane always passes the center of the aperture stop. For its construction, we direct the ray from the source point to the center of the entrance pupil P_{en} on the optical axis. Consequently, the chief ray leaves the optical system along a straight line between the center of the exit pupil P_{ex} on the axis and the image point that is associated with the object point (Figures 3.29 and 3.30). The chief ray can be considered as the central ray in a conical bundle of rays with the marginal rays representing the envelope of the cone. From this consideration, it can be stated furthermore that the marginal rays from the object point on the optical axis directed to the edges of the entrance pupil constitute a cone with the total aperture angle $2\theta_{en}$.

3.4.2 Controlling the field of view: field stops and windows

The field stop is the physical element in an optical system that limits the angular field of view in the object space and simultaneously also influences the extent of the image displayed in the image space. Figure 3.32 shows the image formation in a simple system consisting of only one converging lens in combination with an image format frame. This frame determines the size of the image that can be captured and is for instance in a camera with a digital optical sensor identical with the area of the active sensor pixels. In this example, the extreme points on the image that still can be registered are P_{i1} and P_{i2}. When we trace back the corresponding chief rays traversing the entrance pupil and ending at these points, we find the points P_{o1} and P_{o2} as the source points in the object space. The potential image points P_{i3} and P_{i4} are blocked by the field stop and do not contribute to the image registered by the electronic sensor or film located behind the field stop. Correspondingly, the points P_{o3} and P_{o4} in the object space are outside of the visible field as perceived by the optical system. Thus, the rays from P_{o1} and P_{o2} to the center of the entrance pupil define the angular field of view Ψ in the object space, which is equal to the total angle of aperture of a cone centered on the optical axis. All chief rays within that cone are located in the visible field of view in the object space and can be imaged on the sensor. Moreover, all rays on the surface of the cone strike the edges of the field stop in the case of a circular shape. For rectangular field stops, the field angles in horizontal and vertical directions are different. Then the maximum angular field of view measured across the diagonal of the field frame is used to specify Ψ.

Like in the case of aperture stop and pupils where entrance and exit pupils are associated with each other and control the brightness, we have here the field stop and the conjugate windows that control the field of view. Once the field stop is determined,

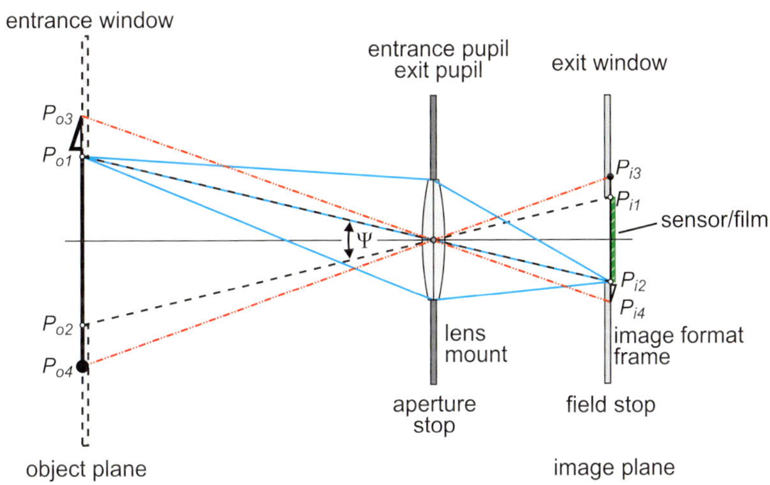

Fig. 3.32: Image format frame acting as a field stop.

the entrance window is the image of the field stop that can be perceived from the object space if there is a lens between the object plane and the field stop. This is represented in Figure 3.32 where the entrance window is the real image of the field stop. There may be cases where the field stop is in the object space and limits the field of view. Then the entrance window is identical with the field stop (see below).

Correspondingly, the exit window is the image of the field stop as perceived from the image space if there are lenses between the stop and the sensor plane. If there is no lens like in Figure 3.32, the exit window is identical to the physical field stop. Furthermore, it can be stated that the entrance window is conjugate with the exit window, which means that they are related to each other like an object and its image. In a more general definition for the angular field of view Ψ, we can say that it is the total angle formed by the rays from the center of the entrance pupil to the edges of the entrance window.

The field stop may not be necessarily located in the image plane like in the example discussed above and can be further in front of the sensor plane. This would be the case if in Figure 3.32 the image format frame was omitted, and a different element delimits the field of view. A similar situation is shown in Figure 3.30. Here, the lens mount or the size of the lens itself is the physical component that limits the field of view. Because the field stop in this case is identical to the lens mount, the conjugate entrance and exit windows are both identical to the field stop itself. As for the angular field of view in these examples, it is formed by the rays aiming from the center of the entrance pupil to the edges of the entrance window. The extensions of these rays to the object space limit the visible field in the object plane. The location of the field stop strongly influences the homogeneity of the image brightness across the image plane, which will be discussed in more detail in the next section.

3.4.3 Properties and effects of stops, pupils and windows

For a given arrangement of lenses and stops, it depends on the position of the object plane and which physical components have the function of the aperture stop or field stop. For example, an observer, and thus light from objects at large distances may "see" a different aperture stop in the system than light from a near distance. Pupils are conjugate with the aperture stop and control the brightness. Windows are conjugate with the field stop and control the field of view. Entrance pupils and exit pupils are conjugate with each other, which implies that they are related to each other like object and image. The same is valid for the entrance and exit window. In general, the positions of pupils and windows do not coincide for a given ray construction. A more general relationship between pupils and windows is sketched in Figure 3.33. The angular field of view Ψ delineates the total aperture angle of the conical bundle of chief rays that can enter the optical system. It restricts the field of view in the object space and is formed by rays from the center of the entrance pupil to the edges of the entrance window. In a

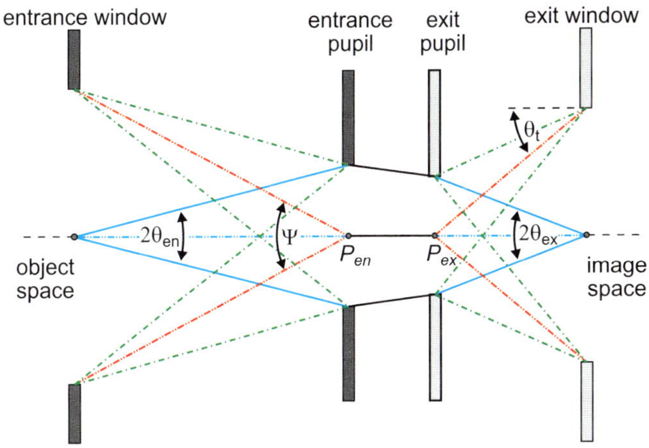

Fig. 3.33: General relationship between pupils and windows.

similar way, the total angular field of view in the image space is given by $2\theta_t$ as shown in the figure. The meaning of θ_t, also termed telecentricity, will be discussed below if the deviation from perpendicular incidence to the sensor plane is considered, as this is in some cases of more interest than the total field angle in the image space. It should be noted that the position of the windows does not necessarily coincide with the object and image planes although it is often the case and may seem from Figure 3.33.

Rays from any point in the visible object field can enter the system on different paths through the entrance pupil, thus forming conical bundles. The total angle of aperture $2\theta_{en}$ for light bundles originating from points on the optical axis is used to characterize an optical system. The angular aperture is measured between the marginal rays from a point in the object plane on the optical axis aiming at the edges of the entrance pupil. Accordingly, the angular aperture $2\theta_{ex}$ in the image space is measured between the marginal rays of an image point on the optical axis to the edges of the exit pupil. The larger θ_{en} and θ_{ex} are, the brighter the image. The influence of the entrance pupil to the brightness is also expressed in the definition of the f-number $f_\# = f/D_{en}$ as given by Equation (2.14) or the relative aperture $1/f_\#$. It should be noted here that the definition according to Equation (2.14) is only valid for rays emerging from points on the optical axis at infinite large distance in the object space. Then the incoming light beam is described by a chief ray with parallel marginal rays. The diameter of this light bundle entering the optical system is identical to that of the entrance pupil, thus allowing for more brightness with a larger entrance pupil. The bundle is imaged to a small spot at the position of the focal point in the image space. The concept of the f-number is appropriate for characterizing the brightness for imaging on the object side especially if the object is a distant one. This is the standard situation for photography where the image magnification $|M|$ is usually smaller than about 0.1. Then the image position is close to

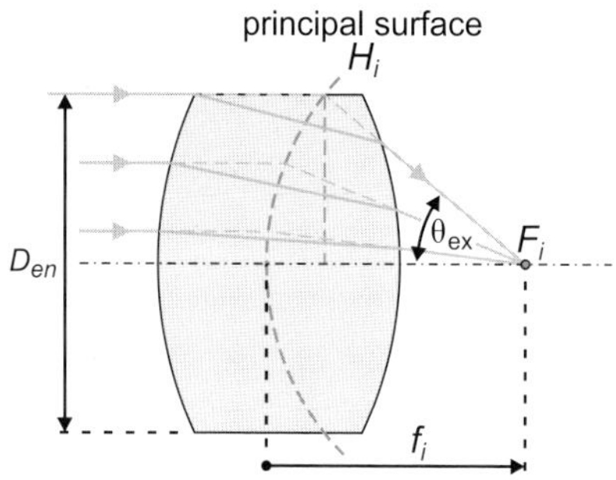

Fig. 3.34: Angular aperture θ_{ex} in the image space for parallel incident rays.

the focal plane. The brightness in the image space can be appropriately described by the image side angular aperture θ_{ex} of the light bundle (Figure 3.34).

It can be shown that in a perfect lens system fulfilling the Abbe sine condition (see Section 3.5.2) the pupils as well as the cardinal surfaces are not planes perpendicular to the optical axis but rather curved surfaces [Ber30, Bla14]. In the case of imaging objects at infinite distance, the curvature radius of the image side principal plane is identical with the image space focal length f_i. The numerical aperture NA_i in the image space in air is then related to the corresponding angular aperture θ_{ex} and the f-number with the following expression:

$$NA_i = \sin\theta_{ex} = \frac{D_{en}}{2\cdot f_i} = \frac{1}{2\cdot f_\#} \quad \Leftrightarrow \quad f_\# = \frac{1}{2\cdot NA_i}. \tag{3.95}$$

It should be stressed again that the definition of the f-number after Equation (3.95) is only valid for imaging from infinity. This is the value that is usually indicated in order to characterize a photographic lens. As the sine function is not larger than one, it follows that the minimum $f_\#$ is theoretically limited to 0.5.

The apertures θ_{en} and θ_{ex} delineate the conical bundles in the object, respectively, image space. The larger they are, the brighter the image will be. For standard photographic situations with relatively large object distances, it is sufficient to use $f_\#$ to characterize the "light gathering" ability of a lens on the object side whereas on the image side NA_i is appropriate. However, if the object plane approaches the lens, the angular aperture in the object space increases. Accordingly, the image distance increases and the corresponding angle θ_{ex} decreases. Simultaneously, the image magnification $|M|$ increases. The apertures in object and image space are linked to each other by the image magnification. Based on the Abbe sine condition, we get the following ratio between the object-side numerical aperture NA_o and the image-side numerical aperture NA_i:

$$\frac{NA_o}{NA_i} = \frac{\sin\theta_{en}}{\sin\theta_{ex}} = M. \tag{3.96}$$

Similarly, the angular field of views in object and image space are linked to each other by the pupil magnification M_p. We will not go into detail about the derivation, which can be found in textbooks:

$$\frac{\tan\frac{\Psi}{2}}{\tan\theta_t} = M_p. \tag{3.97}$$

It becomes apparent that with increasing $|M|$ due to approaching the object plane to the lens, the object-side "light gathering" ability is no more appropriately described by $f_\#$, which only characterizes the situation ideally for $|M| \approx 0$. In order to describe object-side numerical aperture in f-number terms, we use the working f-number $f_{\#w}$, which is also sometimes called the effective f-number [Smi08, Bla14]:

$$f_{\#w} = \frac{1}{2 \cdot NA_i} = \left(1 - \frac{M}{M_p}\right) \cdot \frac{f_i}{D_{en}} = \left(1 - \frac{M}{M_p}\right) \cdot f_\#. \tag{3.98}$$

For imaging from infinity, the image size goes to zero with $M = 0$. In this case, $f_{\#w}$ and $f_\#$ are identical. As the illuminance in the image plane is inversely proportional to the square of the f-number, namely $(1/f_{\#w})^2$, a variation of the object distance leads to a variation of M, and thus of the brightness in the image plane. This is of special importance for more complex lens systems like camera lenses (see Chapter 6). For thin lenses with the aperture stop in close proximity and for symmetrical lens arrangements with the aperture stop in the center between them, the entrance and exit pupils are assumed to be of the same size, and thus $M_p = 1$.

The gist of this consideration can be summarized as follows. For any arbitrary distances in the object, respectively, image space the numerical apertures in both spaces are related to each other by Equation (3.96). For typical applications in photography, the object distance is relatively large and the corresponding angular aperture θ_{en} varies with the object distance. If we disregard close-up photography, θ_{en} is a relatively small value and also the absolute value of magnification M is in general smaller than 0.1. Due to the small magnitude of M, the working f-number $f_{\#w}$ is nearly identical to $f_\#$ and thus the f-number is almost independent from the object distance and the best way to characterize the lens properties for photographic applications. On the other hand, in microscopic applications, the object to be imaged is nearly in the focal distance of the lens and the magnification M is a very large value. Then the aperture of the incoming light bundles is always virtually constant when the object distance slightly varies whereas M, and thus $f_{\#w}$ are strongly affected by small distance variations. Thus, the angular aperture and with it the numerical aperture is the best parameter to characterize the lens properties for imaging in microscopy.

Every point in the plane of the entrance window is linked by a ray to every point in the plane of the entrance pupil. The same relation holds between exit window and exit pupil. This has a consequence for the brightness distribution in the planes of the pupils, respectively, windows. If the area of the pupil planes is reduced, then the brightness of all points in the corresponding windows is also reduced. For instance, stopping down the iris diaphragm in a camera lens means a nonuniform brightness distribution in the pupil area as only the outer fringes of the aperture blocked while the center remains open. This reduces the overall image brightness but does not reduce the image area. Thus, if the brightness is not uniformly distributed over the pupil planes, then only the overall brightness in the windows is affected and does not produce any local inhomogeneity in the windows. As a consequence, small particles close to the planes of pupils that cover their effective area, as for instance dust particles or scratches on lenses close to the pupils, do in general not perturb the image but only influence the overall brightness. However, light may be scattered or diffracted at small particles, but this is out of consideration here. It is more critical if such nonuniform shading or dust occurs close to the windows as they are imaged close to the image plane and locally perturb the image.

We have seen above (Figures 3.32, 3.33) that a format frame in the image plane should be the limiting field stop that sharply restricts the visible image as well as the field of view. Here, the exit window is located in the image plane. The chief rays from points at the edges of the entrance window enter the center of the entrance pupil, leave the center of the exit pupil and are imaged to the edges of the exit window. Usually, they do not strike the sensor in the image plane perpendicularly but under the angle of which the deviation from the optical axis is measured by θ_t. This angle can be seen in different ways: θ_t is called telecentricity or telecentric value, which for a telecentric lens setup is equal to zero (see Section 3.4.5). $2\theta_t$ can also be interpreted as the total angular field of view in the image space. For semiconductor image sensors, the telecentricity should be as low as possible (see Section 4.6.1). This can be achieved by increasing the distance of the exit pupil to the image plane.

If the exit window is further away from the image plane, then the brightness distribution across the image may be more influenced than in the case above with the format frame being the exit window. This situation is illustrated in Figure 3.35 where the lens with its lens mount act as a field stop. The aperture stop is outside the focal distance in the object space and is identical with the entrance pupil. The angular field of view Ψ in the object space is fixed by the rays from the center of the entrance pupil to the edges of the entrance window. The extension of the rays to the object plane yields the visible field. The points P_{o1} and P_{o2} are clearly within that field whereas P_{o3} is just at the edge of that field and P_{o4} is outside of it (Figure 3.35b). Light bundles emerging from P_{o1} and P_{o2} have nearly the same angular aperture (Figure 3.35a). Their chief rays as well as their marginal rays traverse the system without obstruction thus producing image points P_{i1} and P_{i2} of nearly the same brightness. It should be noted here that the upper marginal ray of point P_{o2} just passes the field stop, which is no longer the case for a point with larger transversal distance to the optical axis. The light paths for points with larger distances are shown in Figure 3.35b. Light bundles from P_{o3} and P_{o4} have nearly the same aperture as from P_{o1} but roughly half of the bundle emerging from P_{o3} is not traversing the lens and blocked by the lens mount. Only the chief ray and the lower part of the bundle can pass the lens, whereas for P_{o4} even the lower marginal ray may be blocked

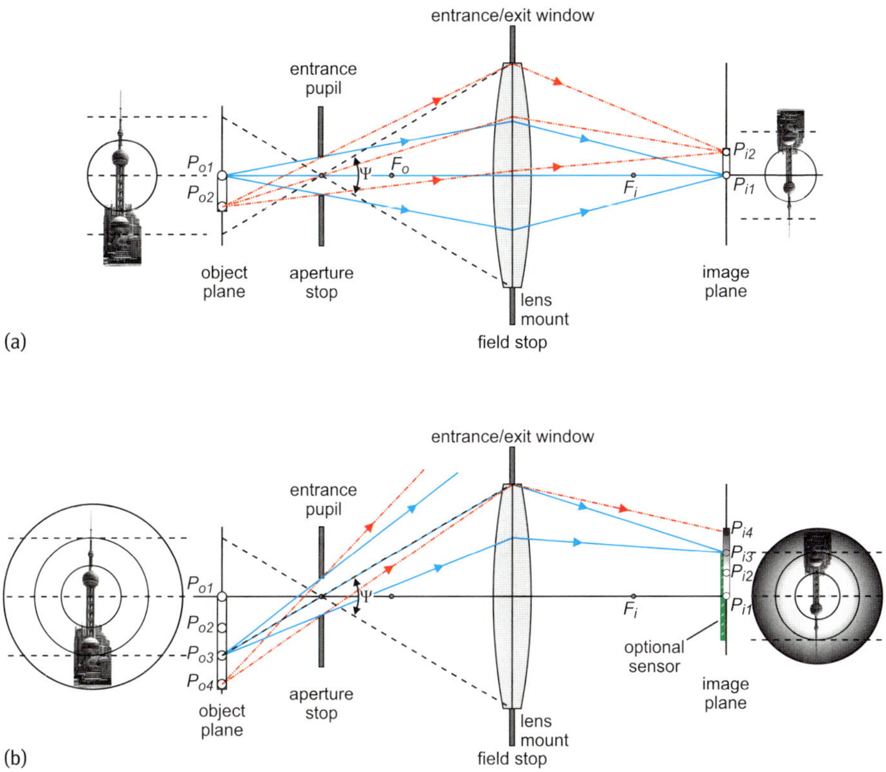

Fig. 3.35: Lens mount acting as a field stop. (a) The object points P_{o1} and P_{o2} are imaged with nearly the same brightness; (b) object points between P_{o2} and P_{o3} within the field of view are imaged to points between P_{i2} and P_{i3} with brightness fall-off; even points beyond the field of view in the range between P_{o3} and P_{o4} can be imaged in this example, although at very low brightness.

or just reach the image plane as the only ray. The consequence is that the very part of the circular image begins to dim, for which the radius is larger than that describing the circle on which P_{i2} is located. The inner circle is nearly uniformly bright, whereas beyond it the brightness continuously decreases and becomes zero for circles beyond P_{i4}. This effect is called vignetting and is due to the fact that the field stop, and thus the exit window is not close to the image plane. It should be noted that points between P_{o3} and P_{o4} in the object plane can be seen in the image although they are outside the field of view. However, they are imaged at low brightness.

3.4.4 Controlling vignetting in lens systems

We can modify the optical system by adding an optical image sensor like in Figure 3.32 to the setup (Figure 3.35b). If the diameter of the sensor format frame is about the diameter of the image circle on which P_{i3} is located, or slightly less, then this format frame becomes the limiting component for the field of view. It is now the sensor that acts as the field stop and no longer the lens mount. In this case, the exit window changes and is identical to the optional sensor, whereas the entrance window is located in the object plane. Its maximum extent is up to the point P_{o3} and its mirrored point on the other side of the optical axis. The angular field of view is the same as before as it is encompassed by the rays from the entrance pupil to the edges of the entrance window. However, the visible image does not go beyond the point P_{i3} and also the field of view does not extend beyond P_{o3}. Here, we have the situation that the field stop and exit window are both identical with the image frame but we still have vignetting in the image circle between P_{i2} and P_{i3}. Only the inner circle between P_{i1} and P_{i2} has an almost uniform brightness distribution. The only way to avoid any vignetting in this case is either to further reduce the image sensor to a diameter that is equal to that of the inner circle or to increase the diameters of all components in the optical system. The consequence of a too-large sensor for a given camera lens can also be seen in Figure 2.24. A uniform illumination in the image plane is only possible if all conical light bundles from all points in the object field can completely, that is including all marginal rays, traverse the optical system without any obstruction.

This simple consideration describes the principle of mechanical vignetting, which is the obstruction of beams due to the mechanical arrangement of lenses, mounts, stops, etc. There is a further cause for brightness fall-off at the corners of images, which is the natural vignetting. Vignetting can be quite complex, so we roughly divide it into two parts: natural and mechanical vignetting. Let us first consider the natural vignetting and then come to a synthesis to describe the whole phenomenon of vignetting [Sch14].

Natural vignetting is always existent even in perfect optical systems. In order to describe this brightness fall-off at the edges of an image, we consider the diffuse radiation of light from an extended area in the object plane. In Figure 3.36, the points P_{o1} and P_{o2} are located on small areas that can be considered as Lambertian surfaces with their diffuse light emitting characteristics. Both areas have identical luminous intensities J_o parallel to the optical axis. An observer in front of these sources of light perceives identical brightness if he is at the same distance from them in the object space. An optical system, however, for instance a camera with an entrance pupil, perceives the light from the points P_{o1} and P_{o2} under different angles. The luminous intensity emerging from P_{o1} and seen at the center P_{en} of the entrance pupil is J_o whereas the intensity from P_{o2} is seen under the angle β relative to the optical axis, and thus reduced to J_β according to the characteristics of a Lambertian surface (Equation (1.8)):

$$J_\beta = J_o \cdot \cos\beta \tag{3.99}$$

The luminous flux $\Delta\Phi_\beta$ emitted from point P_{o2} into a small solid angle $\Delta\Omega$ directed to P_{en} can be calculated using Equation (1.5):

$$\Delta\Phi_\beta = J_\beta \cdot \Delta\Omega \tag{3.100}$$

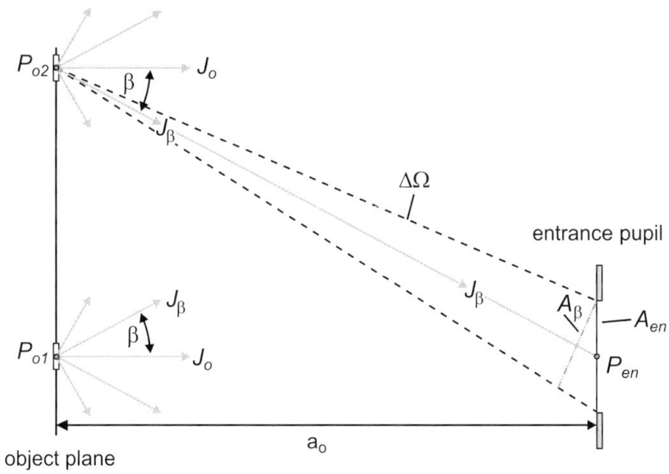

Fig. 3.36: Identical diffuse light sources in the object plane generate an angle dependent illuminance at the entrance pupil.

The solid angle under which the area of the entrance pupil can be seen from P_{o2} depends on the perceived area A_β and the distance a_β from P_{o2} to P_{en} (Equation (1.4)):

$$\Delta\Omega = \frac{A_\beta}{a_\beta^2} \tag{3.101}$$

While the projected area A_β is smaller than the area A_{en} of the entrance pupil by a factor of $\cos\beta$, the distance a_β is longer than the object distance a_o by a factor of $1/\cos\beta$. Combining the Equations (3.99) to (3.101), we find for the luminous flux entering the entrance pupil:

$$\Delta\Phi_\beta = J_\beta \cdot \frac{A_{en} \cdot \cos^3\beta}{a_o^2} = \frac{J_o \cdot A_{en}}{a_o^2} \cdot \cos^4\beta. \tag{3.102}$$

The illuminance E_β at the entrance pupil generated by the area element at P_{o2} under the angle β is equal to the flux divided by the area of the entrance pupil yielding

$$E_\beta = \frac{\Delta\Phi_\beta}{A_{en}} = \frac{J_o}{a_o^2} \cdot \cos^4\beta = E_o \cdot \cos^4\beta \tag{3.103}$$

E_o is the illuminance at the entrance pupil if the light source is perpendicular in front of the lens at a distance a_o. If this light source is shifted in the object plane in parallel to the area of the entrance pupil and is seen under an angle β from the lens, its illuminance at the lens decreases by the 4th power of the cosine of the angle. The light flux $\Delta\Phi_\beta$ entering the optical system is imaged to the sensor and leads to an illuminance $E_i(\beta_i)$ in the image plane, which also depends on the angle β (Figure 3.37). In our example,

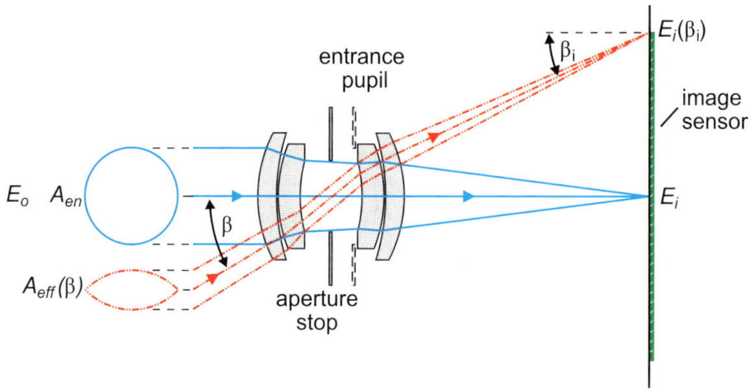

Fig. 3.37: Mechanical vignetting reducing the effective area of the entrance pupil.

we assume a symmetric lens construction with the aperture stop in the center. In this case, the pupil magnification M_p is 1 and the angle β in the object space is identical to its conjugated angle β_i in the image space. The consequence is that the image of a white extended area, which is homogeneously illuminated in the object plane and radiates diffuse light, produces an image with a brightness fall-off to the edges in the image plane. The brightness of the image has its highest value in the center and then falls off proportionally to $\cos^4 \beta_i$. In a more general consideration, even if the field angles in object and image space are not identical, namely when $M_p \neq 1$, it can be shown that it is always the field angle in the image space that determines the brightness fall-off. As mentioned above, this natural brightness fall-off can be observed even in perfect lenses. Now we will combine it with the mechanical vignetting.

Figure 3.37 illustrates a symmetrical camera lens similar to standard lenses in cameras. There are two conical light bundles coming from objects at large distances thus being represented by parallel light beams. The chief ray from a point on the optical axis is directed to the center of the entrance pupil, its parallel marginal beams aim to the edges of the pupil. The cross-section of the beam traversing the lens is identical to the area A_{en} of the entrance pupil, its conjugated value in the image space is the area of the exit pupil A_{ex}. The limitation of the light beam's brightness is due to the size of the aperture stop and its images, as well as the entrance and exit pupil. The field stop in this example is the image sensor, which limits the visible range and the field of view in the object space. The oblique chief ray in Figure 3.37 comes from the edges of the field of view. The corresponding marginal rays are parallel to the chief ray but are not the ones aiming at the edges of the entrance pupil because these latter ones are obstructed internally by the lens dimensions. As a consequence, the cross-section of the oblique beam is reduced. This can also be seen from the examples given in Figure 3.38 where a) is a nearly perpendicular look to the camera lens. We can see clearly the virtual image of the seven bladed aperture stop thus representing the entrance pupil. It is slightly larger than the dimensions of the aperture stop as the lens in front of it acts as a magnifying

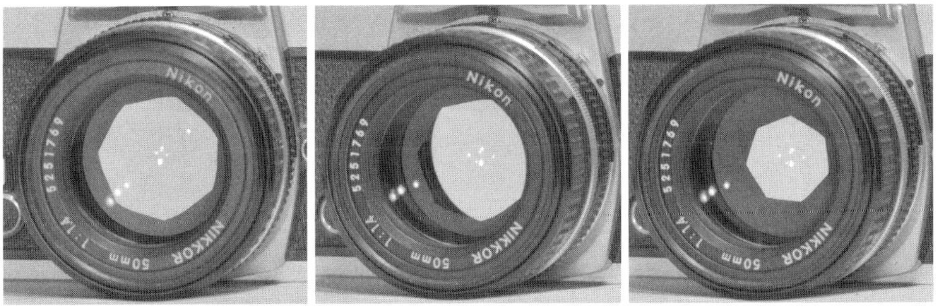

(a) (b) (c)

Fig. 3.38: Entrance pupil as seen from the object space. (a) Nearly frontal view to the virtual image of the aperture stop representing the entrance pupil; (b) at an oblique view the full area of the entrance pupil is not visible; (c) mechanical vignetting can be reduced by stopping down.

lens. The edges of the pupil can be clearly seen and define the angular aperture and with it the brightness of the incoming beam. Shifting slightly the lateral position in the object space leads to the view presented in Figure 3.38b. We can see that the effective area of the conical light bundle entering the lens decreases as the view through the lens is blocked on the left side by the rear lens mount and on the right side by the frontal mount of the lens. The image of the effective area $A_{eff}(\beta)$ in this photograph is similar to the sketched one in Figure 3.37. Its size depends on the angle β to the lens and decreases with increasing β once a limiting angle β_l is exceeded. This means that between $\beta = 0$ and $\beta = \beta_l$ there is no mechanical vignetting whereas after it the mechanical vignetting sets in. β_l depends on the f-number of the lens as well as on the construction of the lens. This mechanical vignetting leads to a reduced effective area. Besides the natural vignetting, it directly influences the illuminance $E_i(\beta_i)$ in the image plane by the ratio of $A_{eff}(\beta)/A_{en}$ and their conjugated values in the image space. Here, A_{en} is the area of the entrance pupil for frontal view, $A_{eff}(\beta)$ is the effective, angle dependent area perceived under the field angle β. Its conjugated value in the image space is the effective area of the exit pupil $A_{eff,i}(\beta_i)$, now seen under the angle (β_i). We can thus come to a more general description of the illuminance in the image plane including both parts of natural and mechanical vignetting:

$$E_i(\beta_i) = E_i \cdot \frac{A_{eff,i}(\beta_i)}{A_{ex}} \cdot \cos^4 \beta_i. \tag{3.104}$$

The larger the angle of view, the stronger the brightness fall-off in lenses. This becomes more pronounced for wide-angle lenses. The mechanical vignetting can be reduced in lenses by stopping the aperture down. This can be seen in Figure 3.37 when the aperture stop is reduced in the way that its edges just touch the marginal rays of the oblique beam. Then the bundles of the straight as well as of the oblique beam have identical cross-sections and no fall-off by mechanical vignetting exists. This is also shown in Figure 3.38c, which perceives the entrance pupil under the same angle as b). When stopping down, the full entrance pupil becomes visible having virtually the same effective area as for perpendicular view. At least in that angular range, the mechanical vignetting is then eliminated.

It should be noted here that in modern complex lens construction, namely wide-angle lenses with a retrofocus design, the exit pupil can be significantly larger than the entrance pupil and the total field angle in the image plane $2\theta_t$ is lower than Ψ in the object space. Hence, this type of lens may show a significantly lower brightness fall-off in the image plane than symmetrical lens constructions. The maximum fall-off at the corner of the image field is then proportional to $\cos^4\theta_t$ instead of $\cos^4(\Psi/2)$ (see also Sections 6.5 and 8.2). Consequently, mechanical vignetting as well as natural vignetting can be significantly reduced by an optimized lens design.

3.4.5 Telecentric lens setup

As we have seen, the position of the apertures is very important for the path of rays. Telecentric setups benefit from that by placing the aperture stop in the focal plane of the lens. We distinguish between an image-space telecentric setup and an object-space telecentric setup, which are used for different purposes. In the image-space telecentric setup, all chief rays in the image space are parallel to the optical axis, which is highly favorable in cameras with semiconductor digital image sensors in order to improve the image quality (see Section 4.6). On the other hand, in object-space telecentric setups all chief rays from points in the object space are parallel to the optical axis. This is used for a precise measurement of the object size as the image size is independent from variations of the object distance.

The image-space telecentric setup is illustrated in Figure 3.39a for a simple system consisting of a thin lens and an aperture stop. The stop is located in the object focal plane and is identical to the entrance pupil. The exit pupil is the virtual image of the stop. The image becomes infinitely large and is located at infinity, with the pupil magnification $M_p \to \infty$ after Equation (3.94). Chief rays from the object points, indicated in the figure by arrows, aim at the center of the entrance pupil, which is the focal point, and consequently leave the optical system as rays parallel to the optical axis. The marginal rays, aiming at the edges of the entrance pupil, are imaged to their corresponding image points in the image plane and determine the angular apertures in the object and image space. Stopping down the aperture results in a narrowing of the conical light bundles striking the image plane perpendicularly. An advantage of this image-space telecentric setup is that even small variations in the position of the image sensor do not lead to image distortions, as the reproduction scale remains unchanged. The telecentricity value θ_t, measuring the deviation from a parallel to the optical axis, is zero in this case. This is compatible with Equation (3.97) that for an image-space telecentric optical system the pupil magnification is infinite.

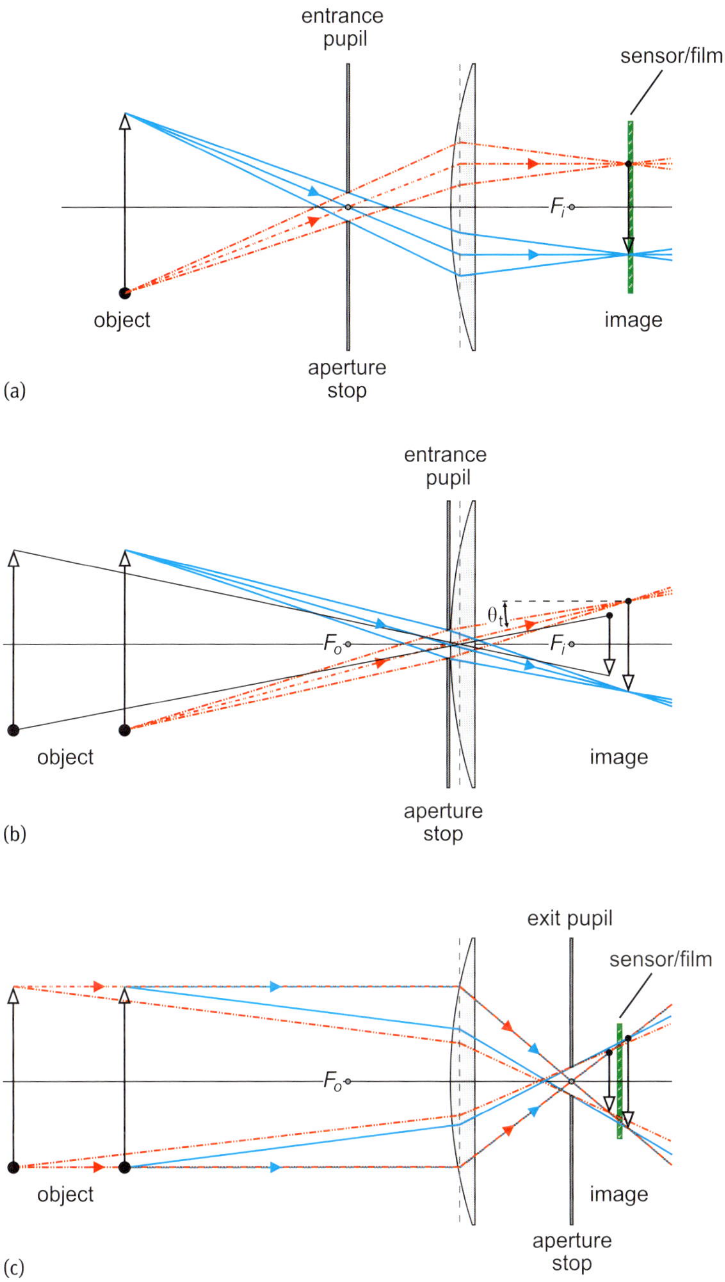

Fig. 3.39: Telecentric lens setup, chief rays are indicated by arrows. (a) Placing the aperture stop to the object focal plane leads to an image-space telecentric system with all chief rays striking the image sensor at a right angle; the exit pupil is located at infinity; (b) aperture stop close to the lens, no telecentric system with oblique chief rays, telecentricity $\theta_t > 0$; (c) object-space telecentric system as a consequence of the aperture stop being located in the image focal plane; a fixed image sensor registers the same image size from equally sized objects independently from their object distance.

Figure 3.39b shows in comparison to a) the case that the aperture stop is located close to the lens. It can be seen that unlike in a) all chief rays from points off the optical axis are oblique, in the object space as well as in the image space. Moreover, the image size depends on the object distance according to the lens equation, thus the image size decreases with the object distance. The exit pupil, being the image of the aperture stop as seen from the image side, is in that case approximately of the same size as the aperture stop and located at its position. Arguing with M_p we can state that for the stop close to the lens or in symmetrical lens constructions we get $M_\mathrm{p} = 1$, and an incoming chief ray has the same angle with the optical axis like its conjugated outgoing chief ray. If the stop is moved away from close to the lens to larger distances into the direction of the object focal point like in a), the position of the exit pupil, being the virtual image of the stop, is shifted to far distances in the object space and the angles of incidence of all chief rays to the image plane approach 90° similar to the case of the image-space telecentric setup. This can be used for the setup of symmetric lenses with the stop in the center between the lens groups to achieve a lower θ_t and thus increase the performance of digital semiconductor image sensors.

If the aperture stop is located in the image focal plane (Figure 3.39c), then the entrance pupil as the image of the aperture stop is located in the image space at infinity whereas the exit pupil is identical with the aperture stop. In this case, $M_\mathrm{p} = 0$. The chief rays in the object space, connected with the center of the entrance pupil, are all parallel to the optical axis, and thus leave the optical system in a straight line from the image focal point, which here is the center of the exit pupil. As a consequence, the chief rays of all objects, that have the same elevation from the optical axis, leave the optical system under the same angle in the image space. If we consider the images of the two objects shown in Figure 3.39c, it can be seen that they are at different positions and have different sizes. If, however, the image sensor is fixed at a position in between both images then the size of both images detected by the sensor is identical although the images may become slightly blurred. The advantage of this object-space telecentric setup is that precise optical measurements of object sizes are possible where small variations of the object distance do not influence the measured results. This is also an advantage, for instance, in laser structuring based on mask projection where ablation or material modification should only take place according to the usually demagnified structure of the mask on the sample surface. In that case, any deviations introduced, for instance, by a laser scanner, would then not lead to a degradation of the processed structure on the sample.

Both telecentric setups require that the diameter of the lens is larger than the lateral range of the parallel chief rays. This implies for the image-space telecentric setup that the sensor diameter is smaller than the lens diameter, which in general is no problem for cameras. In the case of the object-space telecentric setup, only objects smaller than the lens diameter can be measured, which means that this method is only appropriate for the measurement of small objects.

3.4.6 Depth of field and depth of focus

A very important function of the aperture stop is the control of the depth of field and simultaneously the depth of focus. Under depth of field, we understand a given range in the object space in front of well-defined object plane as well as behind it. Within that range, objects can be imaged with acceptable sharpness in the image plane. Correspondingly, there is a range in the image space that is termed depth of focus. For a given point in the object plane, we have one sharp image point. The variation of the sensor or film position around this image plane where the point can still be detected with an acceptable sharpness determines the depth of focus in the image space.

Figure 3.40 depicts the imaging using a simple thin lens with its lens mount acting as the aperture stop. The entrance pupil as well as the exit pupil are in this case identical to the stop. The aperture of a conical light bundle emerging from the object plane at P_o is determined by the diameter D_{en} of the entrance pupil. This bundle is sharply im-

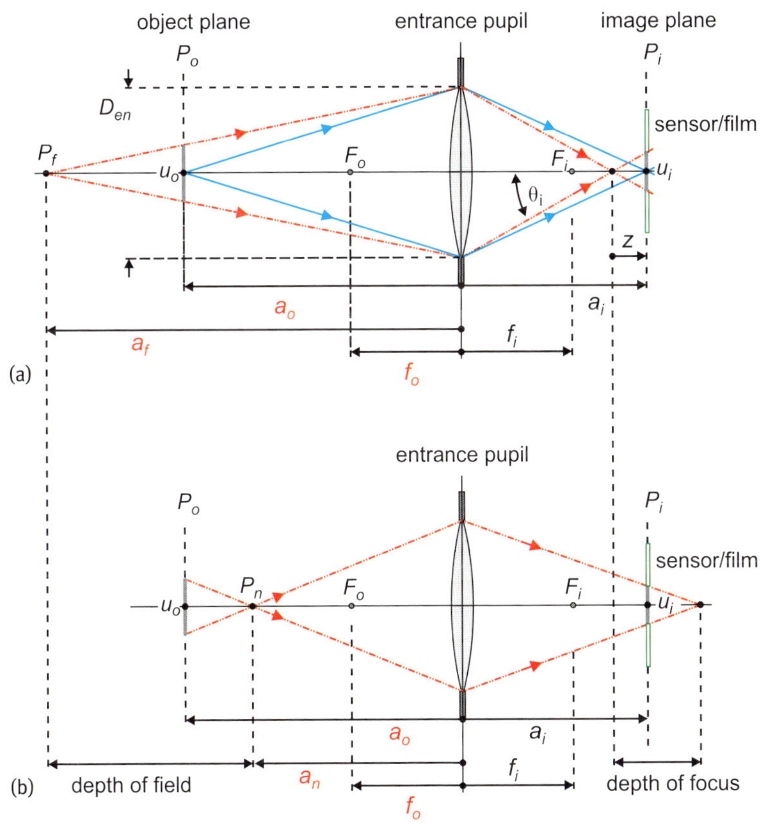

Fig. 3.40: Ray construction for depth of field and depth of focus. (a) Consideration for far point P_f; (b) consideration for near point P_n.

aged to the image plane at P_i at a point with nearly zero diameter disregarding any lens aberration or diffraction. Light coming from a far point P_f on the optical axis aiming to the entrance pupil forms a conical bundle that intersects the object plane at P_o with a sectional area of diameter u_o. As this bundle is farther away from the lens in the object space, its sharp image in the image space is closer to the focal plane at F_i and its distance to the image sensor at P_i is z. Beyond its sharp image point, the bundle widens up again and intersects the sensor plane yielding a circle of confusion with a diameter of u_i. The image of this blurry spot diameter u_i seen in the object space is the circle of confusion with diameter u_o in the object plane. Thus, u_o and u_i are related to each other like object and image with the corresponding lateral magnification $M = u_\mathrm{i}/u_\mathrm{o}$. If the blurring u_i is small enough that it cannot be perceived when viewing or analyzing the image, it is acceptable and its "blurry" image point on the sensor can still be assumed as sharp. As a consequence, the acceptable circle of confusion on the image sensor is decisive for the sharply visible range of distances in the object space as well as in the image space. For visual inspection of images, we assume an acceptable blurring diameter of $u_\mathrm{i} \approx 0.0006 \cdot d$ or more conveniently $u_\mathrm{i} \approx d/1500$ due to the limited resolution of the human eye (see Sections 1.4.1 and 1.5.4). In some cases, a higher resolution is required and $u_\mathrm{i} \approx 0.0003 \cdot d$ or roughly $u_\mathrm{i} \approx d/3000$ is more appropriate, for instance, for wide angle camera lenses. d is the diameter of the image sensor/film and in the case of a 35 mm-format sensor with $d = 43$ mm we have an acceptable blurring spot diameter of around 30 µm or 15 µm for high quality requirements. If u_i is the maximum acceptable circle of confusion, then all points in the object space from P_o to the far point P_f are imaged sharp (Figure 3.40a). Figure 3.40b shows the ray construction for light emerging from a point P_n, which is nearer to the lens than P_o. The corresponding image point lies beyond the sensor plane, thus the light bundle from the entrance pupil to the exact image point intersects the image sensor and also produces a circle of confusion on the sensor. If its diameter is identical to u_i then it is an acceptable value and the images from points between P_o and P_n on the sensor can all be considered as sharp. The depth of field in the object space now extends from the far point P_f to the near point P_n. As can be seen from Figure 3.40, the distance from the near point to the object plane is not the same as the distance from the far point to the object plane. Likewise, the depth of focus can be defined in the image space as the range between the exact image points of P_f and P_n, which is also slightly asymmetric with respect to the sensor plane.

The distances of the far, and respectively, near point can be calculated using the diagram 3.40. It should be noted that distances on the object side are counted as negative quantities. Starting with the magnification between u_i and u_o, it can be expressed using Equation (2.8):

$$M = \frac{u_\mathrm{i}}{u_\mathrm{o}} = \frac{f_\mathrm{i}}{a_\mathrm{o} + f_\mathrm{i}} \tag{3.105}$$

Assuming a positive value for u_i means that u_o is negative as the image is inverted with respect to the object. In the object space, the theorem of intersecting lines yields the relationship between the diameter of the entrance pupil to that of the circle of confusion:

$$\frac{D_{en}}{-u_o} = \frac{-a_n}{a_n - a_o}. \tag{3.106}$$

In order to get rid of the diameter of the entrance pupil in the formula, we use its relation to the f-number $D_{en} = f/f_\#$ and substitute u_o according to Equation (3.105). We find after a simple computation for the near-, and respectively, far-point distances by applying the same procedures to both a_n and a_f:

$$a_n = \frac{a_o \cdot f_i^2}{f_i^2 - u_i \cdot f_\# \cdot (a_o + f_i)} \tag{3.107}$$

$$a_f = \frac{a_o \cdot f_i^2}{f_i^2 + u_i \cdot f_\# \cdot (a_o + f_i)} \tag{3.108}$$

Adding the reciprocal values of the near-point and far-point distances leads to the following expression:

$$\frac{1}{a_n} + \frac{1}{a_f} = \frac{2}{a_o} \tag{3.109}$$

We will define the total depth of field s_{DOF} as the difference between the far-point distance and the near-point distance in a way that it becomes a positive number:

$$s_{DOF} = a_n - a_f. \tag{3.110}$$

Inserting Equations (3.107) and (3.108), then yields after some rearrangements:

$$s_{DOF} = 2 \cdot u_i \cdot f_\# \cdot \frac{a_o \cdot (a_o + f_i)}{f_i^2 - (u_i \cdot f_\# \cdot (a_o + f_i)/f_i)^2}. \tag{3.111}$$

This equation gives the depth of field as function of a combination of the object distance and the focal length. This is a relationship many photographers are familiar with. However, the depth of field is virtually independent from focal length and object distance if we do not image infinitely distant objects like in astronomy. If we use the magnification M as given by Equation (3.105) to eliminate the combination of f_i and a_o from Equation (3.111), then we get after some mathematical rearrangements:

$$s_{DOF} = 2 \cdot u_i \cdot f_\# \cdot \frac{1 - M}{M^2 - (u_i \cdot f_\#/f_i)^2}. \tag{3.112}$$

This result is virtually independent from f_i if we take into account that the circle of confusion in the image plane, expressed by u_i, in general is of the order of µm whereas the focal length is of the order of mm, even in miniature cameras like in mobile phones. Then we can neglect the corresponding part in the denominator and get a result that is especially interesting for ambient situations and especially close-up imaging:

$$s_{\mathrm{DOF}} \approx 2 \cdot u_{\mathrm{i}} \cdot f_{\#} \cdot \frac{1-M}{M^2} \tag{3.113}$$

It should be noted that for real images M is negative as we get an inverted image. For a 1:1 imaging, we find that with $M = -1$ the depth of field s_{DOF} is in the order of 1 mm when using a 35 mm-format sensor at $f_{\#} = 8$. Moreover, the equation above shows that for a given M the total depth of field varies linearly with the f-number.

In most photographic situations the focal length is much smaller than the object distance to the lens, namely $f_{\mathrm{i}} \ll |a_{\mathrm{o}}|$. Then Equations (3.107) and (3.108) can be rearranged and rewritten in simpler form:

$$\begin{aligned} \frac{1}{a_{\mathrm{n}}} &= \frac{1}{a_{\mathrm{o}}} - \frac{u_{\mathrm{i}} \cdot f_{\#}}{f_{\mathrm{i}}^2} \\ \frac{1}{a_{\mathrm{f}}} &= \frac{1}{a_{\mathrm{o}}} + \frac{u_{\mathrm{i}} \cdot f_{\#}}{f_{\mathrm{i}}^2} \end{aligned} \tag{3.114}$$

Increasing the f-number has the effect that the near-point is shifted toward the lens whereas the far-point is shifted further away. For many lenses, it is possible to choose the f-number in a way that P_{f} is shifted to infinity with $a_{\mathrm{f}} \to -\infty$. To achieve this, the denominator in Equation (3.108) must be equal to zero. The resulting value for the object distance in this case is termed hyperfocal distance a_{hf} and is given by the following formula:

$$-a_{\mathrm{hf}} = f_{\mathrm{i}} - \frac{f_{\mathrm{i}}^2}{u_{\mathrm{i}} \cdot f_{\#}}. \tag{3.115}$$

From Equation (3.109), it follows that with $a_{\mathrm{f}} \to -\infty$ the near-point distance is

$$a_{\mathrm{n}} = \frac{a_{\mathrm{hf}}}{2} \tag{3.116}$$

The hyperfocal distance a_{hf} is the nearest object distance setting of a lens for which the depth of field extends up to infinity. The total depth of field then ranges from a_{n}, which is at half distance between a_{hf} and the lens, up to infinity. For practical application with the focal length being short compared to the object distance, the hyperfocal distance can be approximated:

$$|a_{\mathrm{hf}}| \approx \frac{f_{\mathrm{i}}^2}{u_{\mathrm{i}} \cdot f_{\#}}. \tag{3.117}$$

Hence, a relatively large depth of field is achieved for wide angle lenses with a short focal length and stopping the aperture of the lens down.

In order to estimate the depth of focus in the image space, the distance z is evaluated based on Figure 3.40a. The angular aperture in the image space can be approximated under the condition of nearly paraxial rays and for a sufficiently large object distance as compared to the focal length:

$$2\theta_\mathrm{i} \approx \frac{u_\mathrm{i}}{z} \approx \frac{D_\mathrm{en}}{f_\mathrm{i}}. \tag{3.118}$$

With the definition of the f-number, we get the simple result:

$$z = u_\mathrm{i} \cdot f_\#. \tag{3.119}$$

In this simple consideration, the total depth of focus s_DOFoc is approximately

$$s_\mathrm{DOFoc} \approx 2 \cdot z = 2 \cdot u_\mathrm{i} \cdot f_\#. \tag{3.120}$$

s_DOFoc only depends on the acceptable blurring diameter u_i of the circle of confusion and the f-number. In the case of 1:1 imaging, we have a symmetric imaging situation with the same conditions in the object and image space with the magnification $M = -1$. Then Equation (3.120) must be identical with Equation (3.113). The deviation of Equation (3.120) comes from the approximation that the object distance is assumed to be much larger than the focal length, which is no longer valid. The consequence of the depth of focus for practical applications in photography with relatively large object distances is that the exact position of the sensor in the image space is less critical the larger the acceptable blurring and the larger the f-number is. For mobile phone cameras with very small sensor sizes and relative low f-numbers, the precise manufacturing becomes a more important issue than for optical systems with medium format sensors.

It should be noted that the above consideration is only valid for symmetric lens combinations where entrance and exit pupil have the same size. For asymmetric lenses, the pupil magnification factor has to be taken into account using the working f-number given by Equation (3.98) [Hor06]. The above consideration shows that the depth of field decreases the closer the object plane comes to the lens or the larger the magnification of the object is. Further implications of the depth of field for practical application, also the topic of bokeh, will be discussed in the section about camera lenses (Section 6.9) and optics of smartphone cameras (Section 7.1.1.1).

Last but not least, in all cases for DOF we should be aware that the sharpness is limited by the choice of the circle of confusion and the conditions under which the image is viewed. For instance, if the object distance is set to the hyperfocal distance, all objects from a_hf to infinity are perceived as imaged sharp. But this is only valid under the standard viewing conditions for the human eye, namely observation of the image from a distance which is identical to the diagonal of the image print (Sections 1.5 and 2.2). If the print is magnified and/or inspected by means of higher resolution than the human eye, the best sharpness may be somewhere inside the DOF. For optimum sharpness of a single object, the camera should thus be set to the true object distance. The same arguments are valid in an analogous way for the near- and far-point distances, respectively.

3.5 Lens aberrations

Optical lenses are known in our culture for several hundreds of years. The typical lens consists of glass with a homogeneous, isotropic refractive index, which means that the refraction is independent from the direction and polarization of light. The outer surface of the lens has a spherical symmetry that can be easily manufactured even by simple methods like manual grinding. But it is this simple spherical symmetry that is a critical issue for imaging when light enters an optical system under a large aperture angle and when the object points have large transversal distances to the optical axis of the system. Nearly all of the formulas that we used to describe the image formation are only valid in the small-angle or paraxial approximation. This part of the geometrical optics is also termed Gaussian optics.

The physical principle of light refracted at transparent boundaries is described by Snell's law incorporating the sine of the ray angles with the optical axis. The Gaussian optics simplifies the consideration by substituting the sine and tangent function by its argument (Equation (3.14)), which for angles β smaller than about 5° produces in general negligible errors. For larger angles, the deviations become more and more perceivable and the use of Gaussian optics in these cases to describe the results is no longer satisfactory.

In Gaussian optics, we have a clear, unambiguous imaging of an object point to its conjugate image point. A small surface in the object plane perpendicular to the optical axis is imaged to a plane that also is perpendicular to the axis in the image space. No image blur occurs in the ideal case where the imaging properties are described by the equations for thin or thick lenses or even by the matrix mechanism in the paraxial approximation. A more strict theory of imaging is based on the expansion of the trigonometric functions with a Taylor series, for instance, for the sine and tangent functions:

$$\sin\beta = \beta - \frac{\beta^3}{3!} + \frac{\beta^5}{5!} - \cdots + \cdots \qquad \tan\beta = \beta + \frac{\beta^3}{3} + \frac{2\cdot\beta^5}{15} - \cdots + \cdots. \tag{3.121}$$

The exact computation for lenses and mirrors with spherical surfaces yields deviations from Gaussian optics that increase with the angle β. For instance, the focal points for parallel rays coming from different points in the object plane are in general not identical. It also turns out that image planes in the general case are no longer flat but rather slightly curved as are also the principal planes.

When using Equation (3.121) and when the observed imaging properties can be sufficiently described by incorporating only the linear terms of β, thus neglecting all higher orders of it, we deal with Gaussian optics. If we additionally have to use the third-order terms in order to properly describe the characteristics of real lenses that are no more in

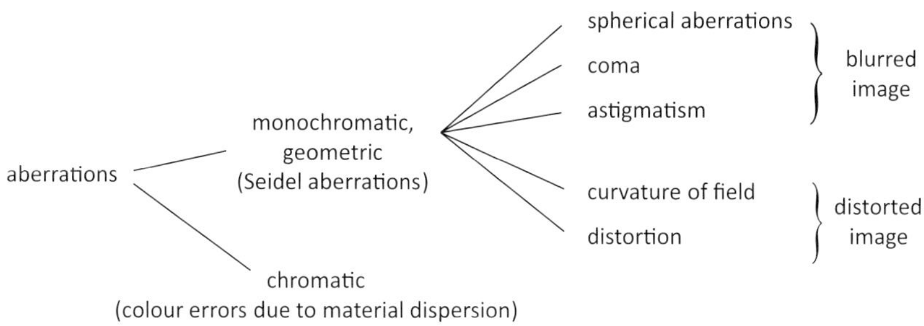

Fig. 3.41: Classification scheme for lens aberrations.

line with the Gaussian optics, the deviations from the simple first-order imaging properties are termed Seidel aberrations, referring to the scientist Seidel who classified the aberrations for monochromatic light. There are also higher-order aberrations, which we will not cover in this book. A classification scheme of the aberrations that will be discussed in the following sections is given in Figure 3.41.

We differentiate between chromatic aberrations and geometrical aberrations. Chromatic aberrations only show up when light propagates in transparent matter like glass or crystals. In these materials, the speed of light, and thus the refractive index depends on the dispersion characteristics of the material. Chromatic aberrations are also present in Gaussian optics with its paraxial approximation and superimpose the geometrical aberrations. They can be entirely avoided when no transparent materials like glass lenses are used but instead reflective imaging elements like mirrors.

Geometric aberrations are due to the spherical symmetry of the imaging elements and exist for any color of light. They are especially investigated for monochromatic light in order to characterize them separately from chromatic errors. These aberrations cause defects in image sharpness due to the fact that not all rays emerging from one object point are imaged to one single point but to a larger blurry spot. We can roughly distinguish between the situations that light enters an optical system in bundles of narrow or wide apertures. Moreover, light can originate from object points close to the optical axis or at larger transversal distance to it. This leads to the classification scheme of the monochromatic Seidel aberrations impairing the image sharpness as represented in Figure 3.42.

Image distortion and curvature of field on the other hand, may occur also with sharp images. Distortions have the effect that the image of objects in a plane no longer has the same geometry as the original, which is mainly due to the fact that the magnification is not constant across the image plane. Here, we only refer to distortions due to the deviations from the ideal Gaussian optics but not to distortions inherent to the projection from a 3D object space to a 2D image space.

In general, it is not possible to correct all aberrations in an optical system. If one special defect is minimized, then the system may be suboptimal in other respects. It is

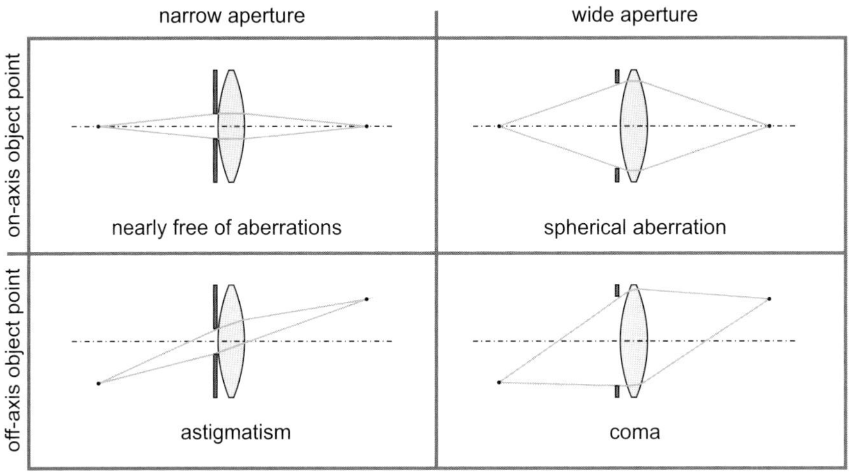

Fig. 3.42: Monochromatic Seidel aberrations that deteriorate image sharpness.

always necessary to optimize a system for a given application, and in many cases the specifications of the human eye serve as a reference. A great deal of the aberrations presented in the following sections can be minimized by the combination of different appropriate lenses. This is also the background for designing complex optical systems. Furthermore, the aperture stop is of high importance for minimizing all aberrations. As all chief rays from the object space aim to the center of the entrance pupil, which is the conjugate of the stop, its position determines the path of rays across the optical system, and thus across the lens zones. Thus, all geometrical aberrations, with the exception of the spherical aberration, are influenced by the position of the aperture stop. The magnitude of the aperture, on the other hand, can be used to control the spherical aberration, coma and the image resolution.

We start the description with the geometrical aberrations as classified by the third-order theory. In practice, it is difficult to observe them separately as they are always interrelated. In many cases, some of them are jointly minimized by the same optimizing process. However, this conventional classification scheme helps to better understand the different critical issues. After the geometrical aberrations, the chromatic aberrations, which add to all of the previously discussed, are finally described.

3.5.1 Spherical aberration

The first aberration in the above presented scheme is the spherical aberration, which can be separately discussed only for object points on the optical axis that are the origin for light bundles of large apertures entering an optical system. According to Gaussian optics described in Section 3.1.4, all rays from an object point P_0 are imaged to one single

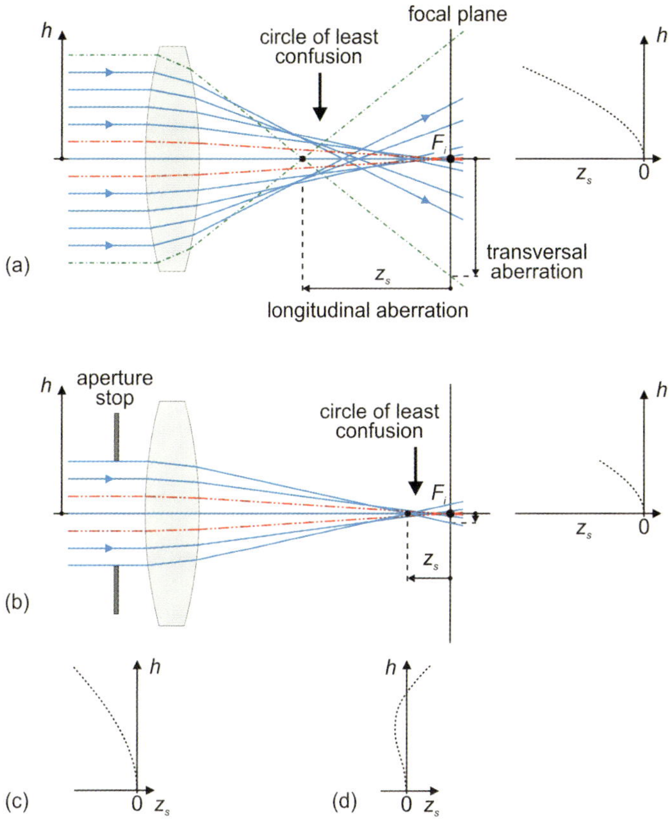

Fig. 3.43: Spherical aberration of a converging lens for parallel incident light. (a) Maximum aperture causing maximum longitudinal and transversal aberrations; (b) reducing the aberrations by stopping the aperture down; (c) schematic of a typical longitudinal aberration of a single lens; (d) corrected spherical aberration of a lens combination.

image point P_i due to refraction at a spherical surface (Figure 3.6). This assumption, however, is only valid for rays with small incident angles γ_0 and small elevations h, namely paraxial rays. All rays parallel to the optical axis in the object space are converged to the focal point in the object space. The mathematical treatment within the third-order theory comes to the result that rays from infinitely far object points on the optical axis, which means that they are virtually parallel to the axis, are intersecting the optical axis at different points if the refracting surface is a spherical section. This situation is illustrated in Figure 3.43 for a converging spherical surface.

The paraxial image focal point F_i is struck only by rays close to the optical axis. With increasing elevation h, the refraction increases. This leads to the fact that the intersection of the rays with the optical axis is shifted away from F_i toward the lens. We can formally describe this behavior by a focal length that is no longer constant but depends on the elevation h. The axial miss distance between the elevation-dependent intersection point relative to the paraxial focal point is defined as longitudinal aberration z_S. Figure 3.43 shows the schematic behavior for a converging lens. Converging lenses have negative longitudinal aberrations z_S as the marginal rays are focused closer to the lens and z_S is pointing to the negative direction of the axis. A corresponding consideration for a diverging lens yields that the focal point of marginal rays is shifted into positive directions as compared to the paraxial focal point. This implies that the longitudinal aberration z_S of a diverging lens is opposite to that of a converging lens, and thus pointing into the positive direction. A converging lens with a negative longitudinal aberration is termed spherically undercorrected while the diverging lens with a positive aberration is termed spherically overcorrected.

When a screen or sensor is placed to the focal plane perpendicular to the optical axis at F_i, then the extension of all marginal rays entering the lens at the same transversal distance from its center intersect the image plane by forming circular rings of confusion. The maximum extent of these circles of confusion defines the transversal spherical aberration of the lens. The envelope of all rays in the image space is termed caustic. It can be seen in Figure 3.43a that the waist of the caustic envelope never gets to zero for large lens apertures but has a minimum along the distance of the longitudinal aberration. When placing the screen or sensor to that position, a spot of minimum size or of least confusion is registered. The brightness distribution within the circle of confusion is not homogeneous as the rays concentration off the optical axis is higher than close to the optical axis thus generating the bright line of the caustic envelope (see also Figure 6.59).

The spherical aberration and caustics of a drinking glass filled with water and of a cylindrical mirror surface are shown in Figure 3.44 where incoming parallel sunlight is not converged to a smallest spot in the image plane but forms a diffuse image. Putting an imaginary screen in vertical direction to the center of their paraxial focal points, the images could be characterized as a spot in the center surrounded by a circular halo. A similar image is expected on a screen in the focal plane in Figure 3.43a. It should be noted that in this figure the additional chromatic aberration becomes also visible with the outer edge of the caustic envelope having a red color. This combination of spherical and chromatic aberration is termed spherochromatism and will be discussed in the section of chromatic aberrations. As a consequence, the spherical aberration in an image deteriorates its quality with respect to sharpness as well as image contrast.

As stated above, the mathematical treatment of the spherical aberration is done only for object points on the optical axis. For off-axis points, additional phenomena as described in the next sections show up. The third-order theory predicts a longitudinal spherical aberration with $z_S \propto h^2$ as well as the transversal aberration being proportional to h^3. The parabolic relationship of z_S is indicated in a diagram in Figure 3.43. It becomes evident that reducing the aperture of an incident light bundle, for instance, by an aperture stop, leads to drastic reduction of both longitudinal and transversal aberrations. The position of the stop, however, is uncritical and does not influence them. Furthermore, when the aperture is stopped down, the circle of least confusion is shifted towards the paraxial focal plane besides reducing its spot diameter.

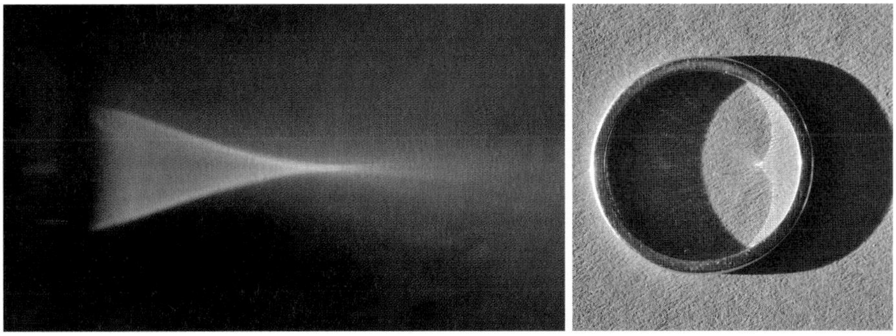

(a) (b)

Fig. 3.44: Spherical aberration and caustic in different situations. (a) Refraction of sunlight at a drinking glass filled with water acting as a cylindrical lens showing spherical and chromatic aberrations; (b) reflection of sunlight at a cylindrical surface.

The spherical aberration strongly depends on the conditions under which the imaging takes place, for instance, the object distance and the type of lenses. It is not always possible to eliminate the aberration completely, but one could minimize it substantially. For optical systems, a combination of positive and negative lenses with opposite spherical aberrations is an appropriate method for optimizing the image quality. This principle applies also to complete lens groups where the combination of a spherically undercorrected group with a spherically overcorrected group is used to minimize the total longitudinal spherical aberration. Figure 3.43d shows the corrected longitudinal spherical aberration for a lens combination compared to that of noncorrected single lens in Figure 3.43c. It can be seen that due to the correction a marginal ray at a given elevation is imaged to the same point as the chief ray on the optical axis, thus both having $z_s = 0$. All other rays in lens zones between these two rays show nonvanishing aberrations with $z_s \neq 0$. These remaining aberrations are also termed zonal aberrations.

For single lenses, the lens shape as well as the refractive index n_L of the lens material is decisive. The imaging of infinitely distant object points is achieved with best sharpness by a biconvex lens where the radii r_1 and r_2 of both spherical surfaces fulfill the following condition [Flü55]:

$$\frac{r_1}{r_2} = \frac{-2 \cdot n_L^2 + n_L + 4}{-n_L \cdot (2 \cdot n_L + 1)}. \tag{3.122}$$

A lens with these radii is called best-form lens. Assuming $n_L = 1.5$ yields $r_2 = -6 \cdot r_1$. This is a nearly plano-convex lens with the six-times more strongly curved surface pointing to the object space (Figure 3.45a). A plano-convex lens with r_2 being infinitely large, which describes the flat surface, is the best form for glass with $n_L = 1.6861$. This results from calculating n_L by setting the numerator in Equation (3.122) to zero. The spherical aberration is much stronger when the lens is oriented with the nearly flat side to the object

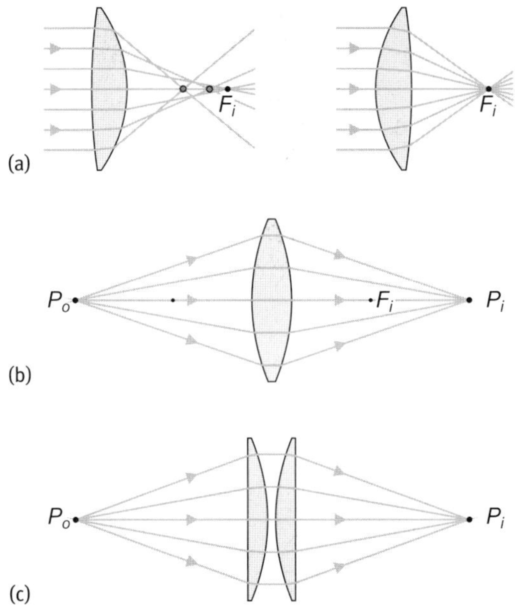

Fig. 3.45: Minimizing spherical aberrations. (a) Nearly plano-convex lenses for parallel incident light; (b) biconvex lens for 1:1 imaging; (c) combination of two best-form or nearly plano-convex lenses for 1:1 imaging.

space, thus underlining the influence of the lens form on the amount of the spherical aberration. In general, it can be said that a nearly equal distribution of the ray bending to the different refracting surfaces helps to minimize the aberration. Thus, for a symmetrical 1:1 imaging the best solution is a symmetric lens or lens arrangement. Figure 3.45b shows that situation, where a biconvex lens is the best choice for a single lens. A still better result is achieved by replacing the biconvex lens by two best-form lenses or nearly plano-convex lenses of half of the refractive power combined at close distance. Here, the total refractive power adds to the same value of the biconvex lens. The object point P_o as well as the image point P_i are each located in the focal point of the plano-convex lens. As for the optical system consisting of both lenses, however, we have a 1:1-imaging with object and image plane at twice the focal distance from the principal planes.

Furthermore, the spherical aberration is eliminated in lenses with concentric spherical surfaces, like an aplanatic meniscus (Figure 3.11b) fulfilling the sine condition as described in the next section; however, only for object points in the center. This meniscus does not produce a real image but is used with other lenses to increase the usable aperture in systems. Meniscus lenses in general are appropriate to reduce spherical aberrations in combination with other lenses.

3.5.2 Coma

A second separate aberration shows up for object points that are not located on the optical axis. It is termed the coma and increases strongly with the aperture of the light bundles emerging from an off-axis point, even at small transversal distances from the axis. A physical reason for coma can be that the principal planes in lenses are not flat planes but curved surfaces. For our consideration, we exclude the spherical aberration and assume an image plane at a fixed distance. Without spherical aberration, the position of the plane does not depend on the elevation h of a ray striking the lens. However, due to the curvature of the principal planes, the effective focal length, and thus the magnification in the image plane depend on the elevation h. The resulting image of an off-axis point is affected by comatic circles as depicted in Figure 3.46c schematically for an oblique parallel light beam of large width entering the lens. A flare like a comet tail can also be seen in the image of the sun through a lens if the sunlight does not perpendicularly strike the lens but under a given angle of inclination (Figure 3.48b, c), hence the name of the phenomenon. To describe the phenomenon, we imagine the spherical surfaces of a lens consisting of circular zones as illustrated in Figure 3.46a. The parallel, oblique light beam striking the lens consists of rays that traverse the different zones of the lens. The vertical y,z-plane intersecting the lens in the center, and thus containing the optical axis along z is called a meridional plane. The image formation in this plane is depicted in Figure 3.46b. If we only consider the rays in this plane, the ray striking the center of the lens at o_c propagates in a straight line to the image plane located in the paraxial focal point F_i. The corresponding image point i_c is at a transversal distance y_c from the optical axis. The two other rays across the extremal lens zone, striking the lens at points o_1, are refracted and converged to one image point i_1 at a distance y_e from the axis. Rays traversing the upper respectively lower part of a different zone in the meridional plane are imaged to points with distances between y_e and y_c. In the case of paraxial imaging, both points would be identical with $y_e = y_c$. In a comatic image, however, we have with $y_e > y_c$ a positive coma or a negative coma with $y_e < y_c$. If we follow the path

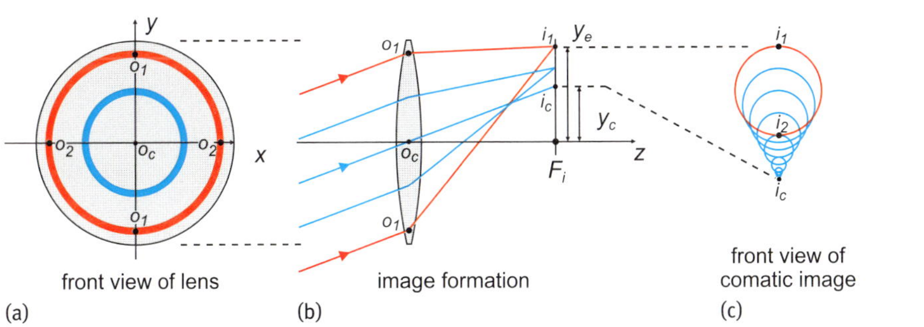

Fig. 3.46: Coma. (a) Front view of lens with circular zones; (b) image formation in a meridional plane; (c) comatic image consisting of circles each characteristic for one lens zone.

of rays through the lens zones in a horizontal cut, for instance, rays from points o_2 in the extreme zone in Figure 3.46a, then the left and right extreme rays are also imaged to one point i_2 (Figure 3.46c). This point, however, is not identical with i_1 at a distance y_e from the axis but lies on a circle in the image plane, which is constituted by the image points of all parallel rays traversing the same zone. Likewise, all rays passing the same zone produce a circular image in the image plane. The radius of these circles increases the larger the distance of their image point in the meridional plane from the image point i_c of the center ray is. The superposition of all these image circles yields the overall comatic flare. It becomes evident that by reducing the aperture the comatic aberration decreases.

The principal cause for the comatic image is that the magnification varies locally due to the curved principal planes. In the absence of spherical aberration, points on the optical axis are imaged perfectly. However, small area elements perpendicular to the optical axis must also be imaged without distortion in order to prevent coma. This means that the magnification for any ray in the imaging process must be constant and independent from the elevation or aperture angle. This leads to the Abbe sine condition, which has also been elaborated by Clausius and Helmholtz and which is fulfilled by ideal optical systems without aberrations.

In order to derive the condition, we consider an object point at S_o at a distance y_o from the optical axis (Figure 3.47). The image S_i in this example is formed by two rays, one of which strikes the refracting surface perpendicularly, and thus is not refracted. The second ray, having an angle of γ_o with the first one, is refracted at the spherical surface according to Snells' law (Equation (3.3)) and intersects the first one in the image point S_i under an angle γ_i at a distance y_i from the axis. When designating the distances

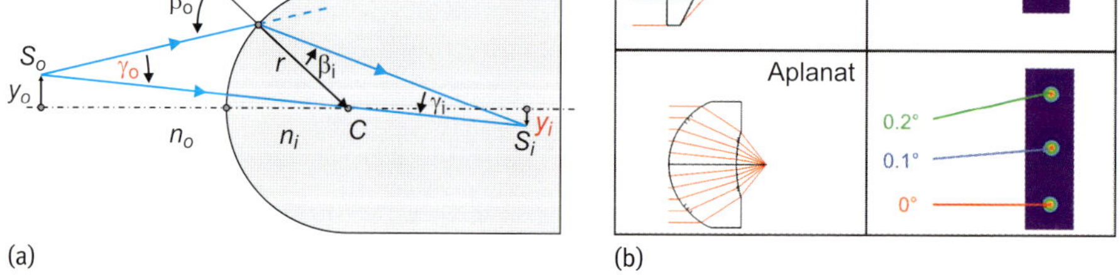

Fig. 3.47: Sine condition. (a) Derivation for a spherical surface; (b) aspherical surfaces without spherical aberration, but only the aplanat fulfilling the sine condition [Bla14] (with kind permission of Carl Zeiss AG).

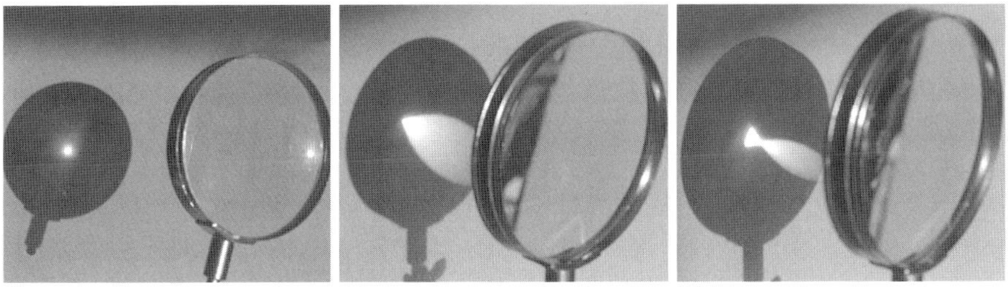

(a) (b) (c)

Fig. 3.48: Coma and astigmatism for sunlight imaged by a simple burning glass. (a) Perpendicular incidence with the image plane in the paraxial image point; (b) off-axis incidence with the image plane closer to the paraxial image point; (c) off-axis position with the image plane closer to the lens.

from S_o to C as s_o and from C to S_i as s_i, we get by applying the trigonometric sine condition to the angles γ_o and γ_i the following equation:

$$\frac{-\sin\gamma_o}{r} = \frac{\sin\beta_o}{s_o}, \quad \frac{\sin\gamma_i}{r} = \frac{\sin\beta_i}{s_i}. \tag{3.123}$$

Note that $\sin(180° - \beta_o) = \sin\beta_o$, and, according to our sign convention, γ_o is negative while γ_i, β_o, β_i, s_o, s_i are positive. The relation between β_o and β_i is given by Snells's law:

$$\frac{\sin\beta_o}{\sin\beta_i} = \frac{n_i}{n_o}. \tag{3.124}$$

With the definition of the transversal image magnification as the ratio of y_i by y_o we find the Abbe sine condition:

$$M = \frac{y_i}{y_o} = -\frac{s_i}{s_o} = \frac{n_o \cdot \sin\gamma_o}{n_i \cdot \sin\gamma_i} \quad \Leftrightarrow \quad y_o \cdot n_o \cdot \sin\gamma_o = y_i \cdot n_i \cdot \sin\gamma_i. \tag{3.125}$$

Any lens or optical system that is free of coma must fulfill the Abbe sine condition. But this may be not enough to prevent coma, and other conditions must also be fulfilled additionally. If the lens is free of spherical aberration, then fulfilling the sine condition is sufficient for getting a coma-free image. Lenses or optical systems that are both free of spherical aberration and free of coma are termed as aplanatic.

A real image without coma cannot be achieved by a single lens with spherical surfaces. An aplanatic meniscus lens like the concentric meniscus (Figure 3.11b) is an example for a lens with spherical end surfaces that is free of spherical aberration and coma. It is, however, a diverging lens thus achieving only virtual aplanatic imaging, and only for singular points. Real aplanatic imaging by a single lens is in principle only possible by aspheric lenses. Figure 3.47b gives two examples for aspheric lenses that are free of spherical aberration for object points at infinity [Bla14]. The hyperbolic lens with a flat surface in the object space and a hyperbolically shaped surface in the image space does not fulfill the Abbe sine condition, and thus is not free of coma. It can be seen that parallel incident light is imaged perfectly to a spot due to the absence of spherical aberration. However, a slight tilt of only 0.2° relative to the optical axis leads to comatic aberrations, and the coma increases with the tilt angle. If the lens is specially designed to have a particular aspheric form that fulfills the Abbe sine condition, the resulting aplanatic lens is free of coma and spherical aberration within a given tilt. Both lenses shown in Figure 3.47b have the same angular aperture of $\theta_{en} = 49°$, which is equivalent to a numerical aperture of $NA = 0.75$ in the image space. The aplanatic lens thus can be used for microscopic applications where small areas in the focal point are imaged to infinity.

In general, coma in optical systems can be prevented by the combination of lenses with positive and negative coma, which is strongly influenced by the lens shape. Like in the case of spherical aberration where lenses of different shapes and materials are appropriate for special imaging conditions, the coma also depends on lens shape and material. The plano-convex lens for reducing the spherical aberrations of object points at infinity has also relatively low comatic aberration in that case. Moreover, not only the aperture of a stop is used to reduce the aberration but also the position of the stop in the system. This can be understood after the examples in Section 3.4 where the position of a stop strongly influences the path of rays across a lens or a complex optical system. As the coma depends on the zones of a lens that are passed by the rays, the appropriate position of a stop controls the ray path, and thus can be used to reduce coma for a given situation.

3.5.3 Astigmatism

The astigmatism according to the third order theory is an aberration that appears when rays have their origin at points that are farther off the optical axis. Unlike coma, the astigmatism can be observed for light bundles even of small apertures. The narrow light bundle strikes the lens surface in an asymmetric way with respect to the optical axis, which leads to different refraction characteristics of the bundle in different planes of intersection with the lens.

In order to describe this phenomenon, we first have to differentiate between two types of significant ray planes. The meridional plane, which has already been considered in the previous section and may also be termed the tangential plane, is spanned by the optical axis and the chief ray from a point in the object space (Figures 3.49 and 3.50). Thus, this plane furthermore contains the object and image points, the center of the aperture stop as well as the center of the pupils and windows. It is usually the plane in which we construct the path of rays in an optical system. The sagittal plane is perpendicular to the meridional plane and also contains the chief ray. In general, it does not contain the optical axis but intersects it in one point. The sagittal plane may not be continuous but can have different sections if the chief ray, aiming to the center of the

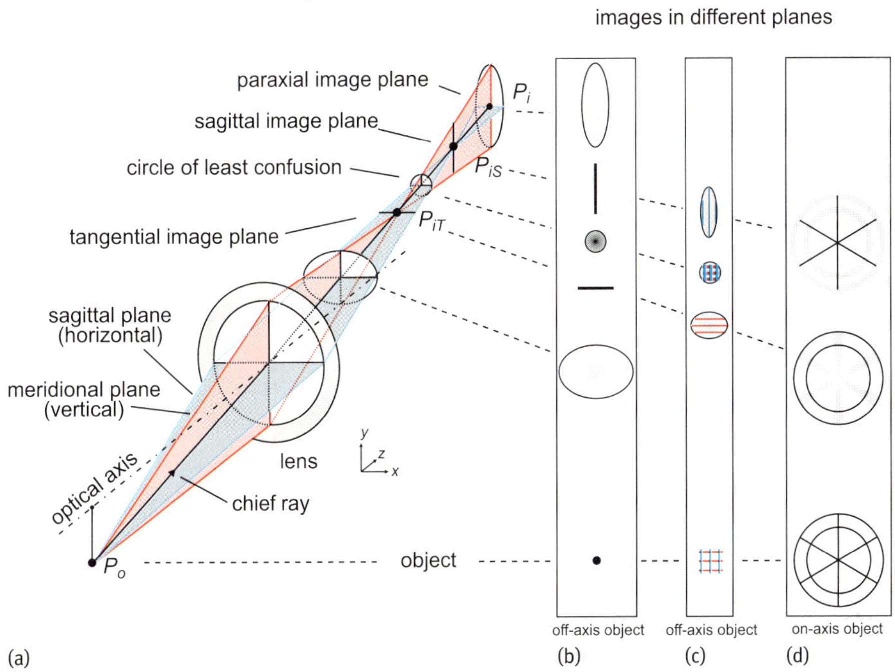

Fig. 3.49: (a) Path of rays and image formation in an astigmatic system. Images in different planes as a result of an off-axis image point (b), of a small off-axis area with orthogonal lines (c), and of a symmetrical on-axis spoke-wheel pattern (d).

entrance pupil, does not traverse the center of the lens, and thus is refracted. These different sections are inclined to each other according to the refracted chief ray. Rays in meridional planes are termed meridional or tangential rays, rays in sagittal planes are termed sagittal.

In paraxial imaging, all rays from one object point are converged to the paraxial image point no matter in which plane or sector they strike the lens. In real systems, however, small light bundles from off-axial positions do not come across the same symmetry of the surface curvature in the meridional plane like in the sagittal plane when striking the refracting surface. This is illustrated in Figure 3.50, which shows the meridional plane and the projection of the sagittal plane to the horizontal plane that includes the optical axis. In the projection of the sagittal plane (Figure 3.50b), it can be seen that the marginal ray paths in the bundle are symmetric with respect to the chief ray. This, however, is not the case for the marginal rays in the meridional plane. As a consequence of the different effective curvatures in the lens, the focal lengths as well as image distances in both planes are different. The tangential image at P_{iT} is closer to the lens than the sagittal image at P_{iS}. Both are not at the same location as the paraxial image P_{i}. The difference between the two image locations in the tangential respectively sagittal planes is defined as the astigmatic difference z_{a}. It increases with increasing inclination γ_{o} of

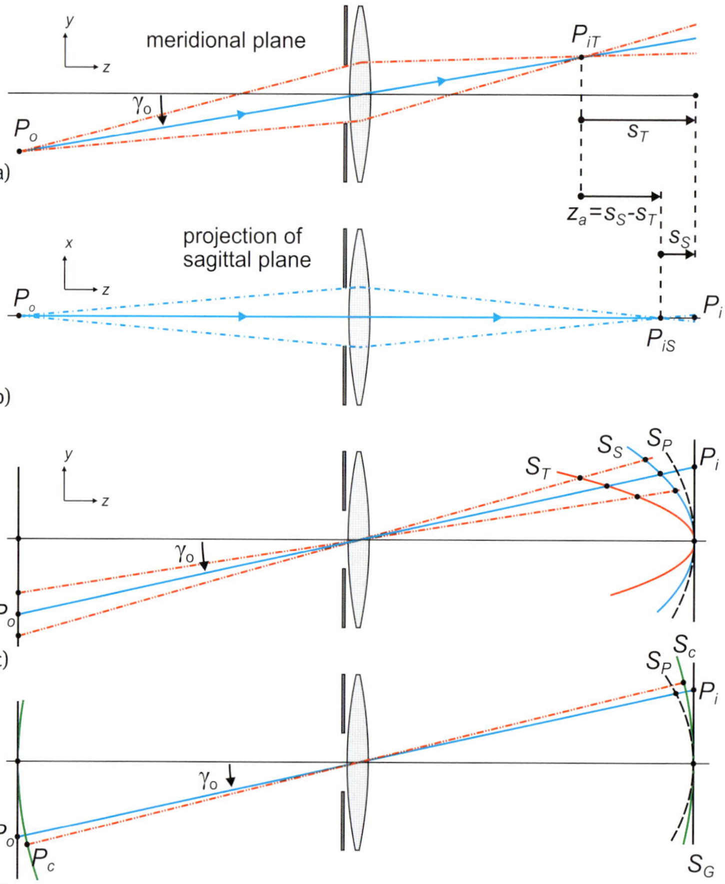

Fig. 3.50: Astigmatism and curvature of image plane. (a) Path of rays in the meridional plane for imaging by a biconvex lens; (b) path of rays in the projection of the sagittal to the horizontal plane; (c) astigmatism undercorrected ($z_a < 0$) and Petzval surface undercorrected; (d) path of rays in the meridional plane for corrected astigmatism but not-flattened curvature of field. The paraxial image point P_i is located in the paraxial image plane.

the incoming rays and with increasing refractive power of the lens. For $z_a < 0$, the astigmatism is termed undercorrected, and it is termed overcorrected for $z_a > 0$ (see Figure 3.50 and Section 3.5.4).

Let us now consider the image formation of a small off-axial object point at P_o (Figure 3.49a). A conical light bundle emerging from P_o intersects the lens in a nearly circular form. Due to astigmatism, the marginal rays in the meridional plane intersect each other in the tangential image plane at P_{iT} whereas the marginal rays in the sagittal plane are still separated. Thus, the cross-section of the bundle degenerates into a primary horizontal line image at P_{iT}. We can see that the bundle's cross-section on the way from the lens exit to P_{iT} gradually changes from nearly circular over elliptic to linear in the horizontal sagittal plane (Figure 3.49b). Following the ray path further on, the marginal rays in the tangential plane diverge while the marginal rays in the sagittal plane converge and intersect in the sagittal image plane at P_{iS}. Here again, we observe a secondary line image, but this time the line is vertical in the tangential plane. In between the line images there is a circular spot image termed the circle of least confusion. It is not a sharp image of the object point but rather a blurred structure, however, not distorted as the line images. Beyond the sagittal image plane, the cross-section of the bundle widens up and remains elliptical. It can be characterized by different divergence angles in the vertical, respectively, horizontal direction. In the paraxial image plane at P_i no sharp image is seen.

Based on the imaging characteristics of a small off-axis point, the image formation of a small area with orthogonal lines at a given distance off the optical axis in the object space can be understood (Figure 3.49c). A horizontal line can be considered as consisting of individual points, and thus will be imaged as sharp horizontal lines in the tangential image plane at P_{iT}. A narrow vertical line, however, does not lead to a sharp image in the tangential image plane but produces a blurred background. However, it yields a sharp vertical line image in the plane at P_{iS} and superimposes there the blurred background generated by the horizontal lines. As a consequence, the small line pattern of mixed orthogonal lines produces purely horizontal and purely vertical line patterns in the different image planes. At the position of the circle of least confusion, the orthogonal line pattern is imaged without distortion but blurred.

As a last example for the image formation, we consider the spoke-wheel pattern of the symmetry depicted in Figure 3.49d, which is centered as an object on the optical axis. Any of its straight lines in the object space can be regarded as an arrangement of small neighboring off-axis points in transversal direction. They are imaged as transversal lines in the sagittal plane. Due to the rotational symmetry of the pattern as well as the rotational symmetry of the lens, all spokes are imaged sharp whereas lines orthogonal to them lead to a blurred background. The circles in the object space are orthogonal to the spokes, and thus cannot be seen or only as a blurred background in the sagittal plane. With the same argument, the wheel circles of the object can be regarded as an arrangement of small transversal lines rotated around the center. Thus, the circles are sharply imaged in the tangential image plane. In between the tangential and the sagittal image planes, at the position of the circle of least confusion, the image is blurred with reduced contrast for the lines as well as for the circles.

The effect of the different image planes can also be seen when focusing the sunlight on a screen using a burning glass as shown in Figure 3.48. The best image is achieved for perpendicular light incidence (Figure 3.48a). We get a round image of the sun close to the paraxial focal plane, and only spherical aberration is expected to blur the image. When tilting the optical system including the lens, the screen and the optical axis relative to the incident beam, a strong comatic influence is seen (Figure 3.48b). However, when approaching the screen to the lens, thus reducing the image distance, the influence of the sagittal and tangential image planes to the spot image as well as the coma can be verified. Here, also the curvature of the image planes comes into play, which is often related to the astigmatism but is not corrected by the same methods. The curvature of field in the image space will be discussed as a separate aberration in the next section.

The effect demonstrated in Figure 3.48 is achieved using a simple lens where all types of aberrations superimpose. It is very difficult to display one single effect. The astigmatism of a single lens strongly depends on its shape and the refractive index of the material, and, if used with an aperture stop, also on the position of the stop. The examples presented in Figures 3.49 and 3.50a,b,c are typical for biconvex lenses. Using concave lenses or meniscus lenses results in different astigmatic differences $z_a = s_S - s_T$ that depend on the angle γ_o and can be positive, negative or zero in the ideal case of no astigmatism. Here, thick meniscus lenses with different surface curvatures and thicknesses are appropriate for correction as well as the position of stops and, like in most cases, the combination of different lenses. The difference $z_a = s_S - s_T$ is related to the curvature of the image planes and will be discussed also in the next section of curvature of field. If the tangential image point is to the left of the sagittal image point like in Figure 3.50c, the astigmatism is called undercorrected (astigmatic difference $z_a < 0$); if it is to the right side, it is called overcorrected. If both points coincide, the astigmatism is corrected.

The astigmatism as presented in this section is only due to perfect spherical lenses and related to the deviations when describing their image formation for oblique rays within the frame of paraxial optics. There is also a different type of astigmatism, which is observed in optical systems that lack of rotational symmetry. We call it axial astigmatism and it is typical for systems consisting of cylindrical or of toric lenses (Figure 3.62). They have different refractive powers, respectively, focal lengths in different planes, even for perpendicularly incident ray bundles or also for paraxial imaging. Their imaging properties and aberrations are to some extent similar to that of spherical lenses, but they are based on different physical principles and must be corrected in different ways. An optical system where axial astigmatism can be observed is the human eye where the cornea deformation leads to this type of astigmatism. It can be corrected by cylindrical lens surfaces or by toric contact lenses (see Section 3.5.7). Other optical systems that are strongly prone to axial astigmatism are electromagnetic imaging system like electron microscopes or lithographic system using charged particle beams. Here, the astigmatism can be easily corrected by calibration of the electrically tunable imaging elements like magnetic coils.

3.5.4 Curvature of field

Curvature of field is always related to the astigmatism. However, it is a separate effect in the third-order theory and will be treated in the first instance as a stand-alone aberration. It is also corrected in a different way. Unlike with spherical aberration, astigmatism and coma, all object points are imaged sharp but the image of a small area off the optical axis is imaged to a curved surface. As the image sensor is in general flat, the image on it will be sharp only in parts where the curved image surface intersects with the sensor plane. Moreover, the geometry of the recorded image will be distorted due to geometrical mismatch of image and sensor surfaces. The principles for the curvature of field are illustrated in Figure 3.50d. Assuming the absence of any lens error, and especially astigmatism, then in paraxial imaging small areas perpendicular to the optical axis with object points P_o will be imaged to P_i in the Gaussian image plane indicated by S_G in the figure. S_G is a flat plane and perpendicular to the optical axis. A closer inspection of imaging reveals that object points P_c on a spherical surface in the object space will be imaged sharp to a spherical image surface S_c, which for paraxial imaging is virtually identical with S_G. The deviation from the Gaussian ideal becomes obvious with increasing angle γ_o. The object points P_o in the vertical object plane, however, are imaged to a surface S_P, which is called Petzval surface according to the mathematician Joseph Petzval. In the case of a simple thin converging spherical lens, the Petzval surface is parabolic with its curvature toward the lens. For a single thin diverging lens, the Petzval surface is curved away from the lens. Thus, a combination of positive and negative lenses is appropriate to get a flat Petzval surface, which then may become identical to the Gaussian image plane. In a more general approach, Petzval has shown that the curvature radius r_P of the resulting S_P depends on the refracting surfaces of an optical system. Assuming m spherical surfaces, each having the curvature radius r_j as well as the refractive index n_j in front of the surface and the index n_j' behind it with respect to the orientation of the optical axis, then the following relation is valid:[7]

$$\frac{1}{r_P} = \sum_{j=1}^{m} \frac{1}{r_j} \cdot \left(\frac{1}{n_j} - \frac{1}{n_j'} \right) \tag{3.126}$$

This relation is also named the Petzval sum. The surfaces are arranged in a sequential order, which implies that $n_j' = n_{j+1}$. For a single spherical lens in air, having the radii r_1 and r_2 and the glass refractive index n_L, we get then

$$\frac{1}{r_P} = \frac{1}{r_1} \cdot \left(1 - \frac{1}{n_L} \right) + \frac{1}{r_2} \cdot \left(\frac{1}{n_L} - 1 \right) = (n_L - 1) \cdot \left(\frac{1}{r_1} - \frac{1}{r_2} \right) \cdot \frac{1}{n_L}. \tag{3.127}$$

Using the relation (3.27) for the image focal length f_i we can express r_P for a thin spherical lens in air:

$$\frac{1}{r_P} = \frac{1}{f_i \cdot n_L} \tag{3.128}$$

Thus, for a serial arrangement of m thin lenses in air the Petzval sum can be described:

7 Hans Zincken genannt Sommer, Annalen der Physik, Bd. 122, S. 563 ff., Berlin 1864.

$$\frac{1}{r_{\mathrm{P}}} = \sum_{j=1}^{m} \frac{1}{f_{\mathrm{j}} \cdot n_{\mathrm{j}}} \tag{3.129}$$

In this equation, n_{j} is the refractive index of the lens material and f_{j} its image focal length. When the Petzval sum is zero its curvature radius will be infinitely large, which is equivalent to a flat plane. Thus, the curvature of field can be prevented by the right combination of thin lenses. Moreover, Equation (3.127) shows that a single spherical thin lens, with the exception of Höegh's meniscus, is never free of curvature of image field. Höegh's meniscus (Figure 3.11a) with $r_1 = r_2$ has a flat image plane but is not free of coma and of spherical aberrations [Flü55]. Its astigmatism completely vanishes if the position of the aperture stop is adapted to the lens radius.

If astigmatism exists in an optical system, a point P_{o} on a transversal object plane is not imaged to one point on the Petzval surface but to the two different line images at P_{iS} and P_{iT} on the sagittal, respectively, tangential image surfaces S_{S} and S_{T}. As described above for astigmatism, both surfaces depend on the lens shapes, the stop position and are in general not identical. The astigmatic deviation of the image points on the corresponding planes relative to the paraxial image plane is given by s_{S} respectively s_{T} (Figure 3.50a,b). The deviation can be influenced by lens shapes, the appropriate combination of lenses and the stop position. As for the location of the different image surfaces, Figure 3.51 schematically shows examples of the sagittal and tangential surfaces S_{S} and S_{T} relative to the Petzval surface S_{P} and the ideal Gaussian image plane S_{G}, which is always flat and located in the paraxial focal point. If we consider only S_{P}, S_{S} and S_{T}, the third-order theory yields that S_{P} is always the leftmost or rightmost of these three surfaces and that the longitudinal distance from S_{P} to S_{T} is three times longer than its longitudinal distance to S_{S} (Figures 3.51a,b,c). Hence, it becomes clear that for an astigmatism-

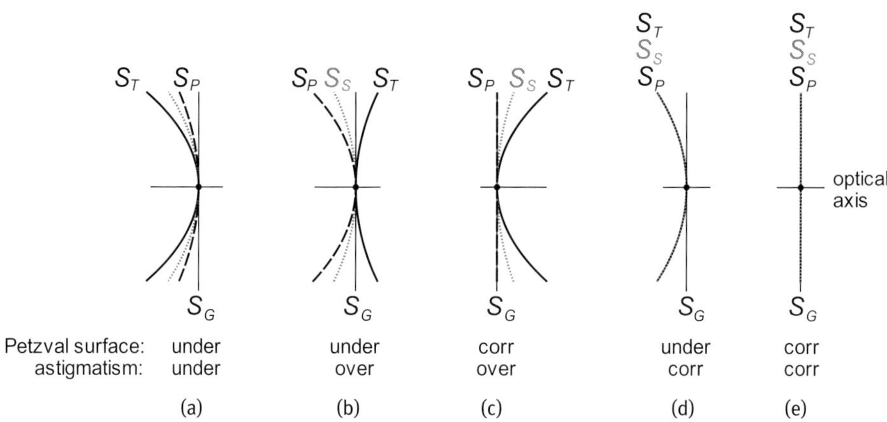

Fig. 3.51: Scheme of the location of the different image surfaces S_{G} (ideal Gaussian plane), S_{P} (Petzval surface), S_{T} (tangential image plane), S_{S} (sagittal image plane) and the corresponding corrections (under: undercorrected, corr: corrected, over: overcorrected) [Kin39].

free system, which means that $z_a = s_S - s_T = 0$, both astigmatic sagittal and tangential surfaces are identical and coincide with the Petzval surface, i. e., $S_P = S_S = S_T$ (Figure 3.51d). However, if the Petzval sum (3.129) is not nullified by the appropriate combination of lenses, the Petzval surface S_P is still curved and the image plane is not flat. If additionally the Petzval surface is corrected by bringing the Petzval sum to zero, the resulting image plane is identical with the Gaussian plane S_G (Figure 3.51e). This is the ideal case. If the astigmatism is still present, it is nevertheless possible to obtain a flattened Petzval image plane with $1/r_P = 0$ (Figure 3.51c). But this is of no practical value as the images in that plane are never sharp due to the curved astigmatic image surfaces. The best sharpness or least confusion is achieved for images on a surface in between S_S and S_T. Figure 3.51b illustrates schematically the case of a meniscus lens with the appropriate location of the aperture stop. The Petzval sum is not zero, the Petzval surface is undercorrected and the astigmatism is overcorrected with S_T being to the right of S_S. As S_S and S_T have opposite curvatures of approximately the same magnitude, the effective image plane of least confusion in between them is virtually flattened. However, the sharpness decreases with increasing distance from the center of the image field. It should be mentioned here again that the Petzval sum is only affected by the refractive index and the refractive power of a lens or lens combination, whereas the astigmatism is influenced by the type of lenses, especially its shapes, and the position of stops.

As for real anastigmats, it is nearly impossible to correct astigmatism and curvature of field completely, taking into account that the third-order aberration theory is only an approximation where higher-order terms are neglected. Figure 3.52 depicts the schematic deviations s_S respectively s_T of the astigmatic image point from the ideal Gaussian image plane as a function of the incident angle γ_0 of a ray with the optical axis. These are typical curves for specifying the quality of lenses or lens combinations. Part a) shows the typical curves of a noncorrected single converging lens where astigmatism and curvature of field are undercorrected. Part b) shows the curves of a corrected lens combination where the astigmatic difference is zero at two angular positions and nearly zero in lens zones between them. However, the curvature of field is still slightly present. Part c) illustrates the curves of an anastigmat with corrected astigmatism, which means

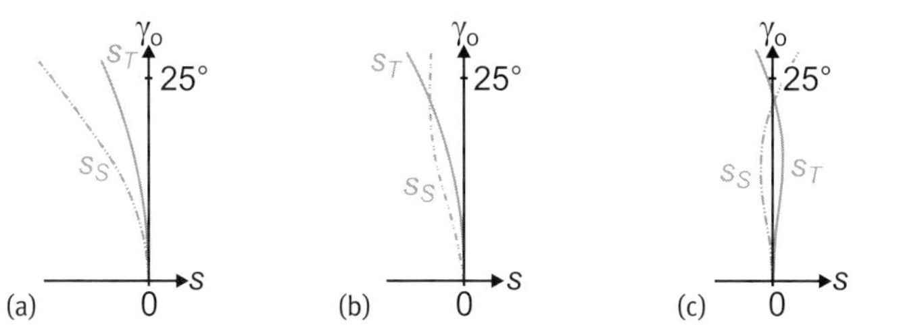

Fig. 3.52: Schematic astigmatic deviations of sagittal and tangential image points as a function of the incident angle. (a) Noncorrected converging single lens; (b) lens combination with corrected astigmatism but slightly curved image plane; (c) lens combination with corrected astigmatism and flattened image field.

small zonal aberrations, and a flattened image field. Both b) and c) are typical for good anastigmats and come close to an ideal correction.

Astigmatism and curvature of field are corrected in order to optimize imaging to the Gaussian image surface S_G. A flat image plane is due to the fact that it is technically quite expensive to produce curved image surfaces, be it on classical photographic film material or electronic image sensors. However, due to advanced technologies it has become possible to produce curved image sensors (Section 4.12). They are of high interest as the number of lens elements can be significantly reduced for curved image sensor, and the same or even better image quality can be attained (Section 5.2.7).[8] As a consequence, in many cases shorter overall lengths of complete lenses can be produced, which is of high importance for smart phone cameras (Chapter 7). But also, for other types of lenses, curved image sensors could be of advantage, as demonstrated for zoom fisheye cameras (Section 6.5.4).

3.5.5 Distortion

The last third-order aberration that we describe here is the image distortion. Like the curvature of field it does not produce blurred image points but a geometric deformation of the image. Even if all the previous aberrations are corrected, distortion becomes more obvious the larger the image size is. It increases with the transversal image distance h to the optical axis. As the underlying physical principle for distortion, we can account the transversal image magnification, which in the paraxial case is constant for all points in the object plane. In real lenses, however, the optical path of light across different lens zones results in a path dependent magnification which produces larger or smaller image sizes than in paraxial cases. Usually, thin lenses are less affected by image distortion than thick lenses. Figure 3.53 presents the examples of two virtual images produced by a ball lens and by a loupe eyeglass. Both converging lenses show an increasing image magnification with increasing distance from the lens center. The corners of a square object seem to be pulled apart in the image. The resulting effect is classified as pincushion distortion. The extreme is seen at the edges of the image formed by a ball lens. To obtain a better image quality the simplest method here is the application of a field stop in the image plane thus limiting also the usable field of view. Another type of distortion is the barrel distortion (Figure 3.54), which due to the fact that the transversal magnification decreases with the image size. This can be observed for instance with diverging lenses. Both types of distortion, however, show up in both positive and negative lenses when they are combined and especially when stops are used in combination with lenses.

8 C. Gaschet et al.: *A Methodology to Design Optical Systems With Curved Sensors*, Appl. Optics **58** (2019), 973–978.

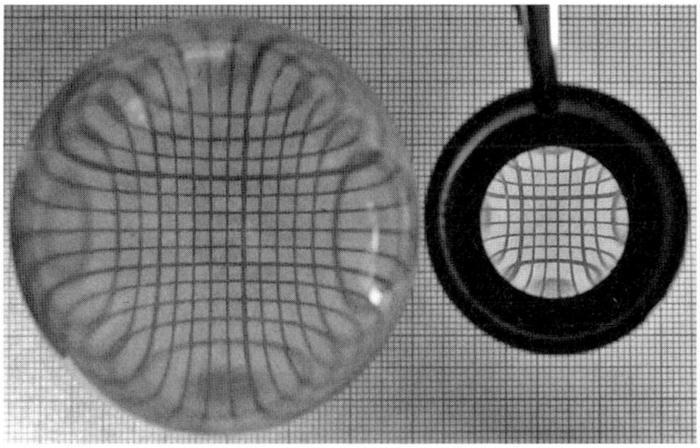

Fig. 3.53: Pincushion distortion seen in the virtual images of a ball lens (left) and of a loupe eyeglass of 10× magnification (right).

The influence of stops on the light across a lens has been already demonstrated for different applications, like for instance telecentric lens setups (Figure 3.39). Similar setups consisting of simple converging lenses with an aperture stop are given in Figure 3.54.

Figure 3.54a shows the situation where the aperture stop is located close to the thin lens. The entrance pupil is almost identical with the lens center. Thus, the chief ray from a point P_o in the object space travels in a straight line through the lens center to its image point P_i. In this case, no distortion occurs and the nondistorted image is termed orthoscopic. If the aperture stop, however, is shifted away from the lens the center of the entrance pupil is not identical with the lens center. When the stop is located on the side of the object as shown in Figure 3.54b, the chief ray of an elevated object point with its light bundle traverses the lower part of the lens. As compared to the chief ray through the lens center, which in the figure is indicated by a dotted line, the chief ray now has a longer light path to its image point than in case a). According to Equation (2.8), the magnification M is inversely proportional to the optical path length in the object space. Consequently, the absolute value of M decreases with longer paths, namely for object points off the optical axis. The image of a rectangular object is distorted in the areas that are far from the center and in this case leads to a barrel distortion. The total size of the image is smaller than the orthoscopic image. On the other hand, if the aperture stop is located in the image space (Figure 3.54c) we get the inverted situation as compared to b). Now the light path in the object space is shorter than the path across the lens center and leads to an increased transversal magnification. The image point P_i is farther away from the center than for orthoscopic imaging and is characteristic for the pincushion distortion. This consideration shows that the position of the aperture stop not only is one of the causes that distortion occurs but is also decisive for the type of distortion. Thus, the best way to minimize distortion is an aperture stop close to the lens or, in the case of lens combinations, a more symmetrical location with respect to the lens posi-

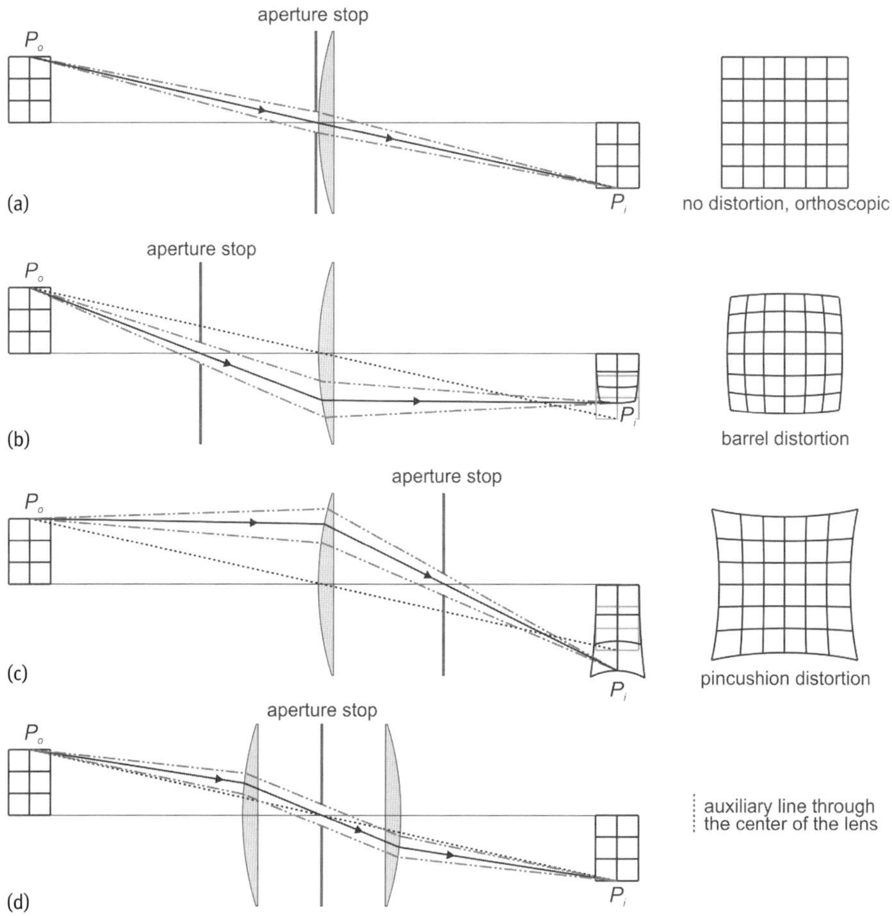

Fig. 3.54: Distortion caused by different positions of the aperture stop; chief rays are indicated by arrows. (a) Orthoscopic imaging, no distortion; (b) stop located in the object space leading to barrel distortion; (c) stop located in the image space leading to pincushion distortion; (d) nullifying distortion by a symmetrical stop position for 1:1 imaging.

tions. Figure 3.54d shows a completely symmetrical placement of the stop between two identical lenses. A similar setup is used in many camera lens constructions, especially in lenses for reproduction purposes where absence of distortion is required. In the symmetric setup, the distortion produced in one part of the lens combination is balanced by the inverted effect in the other part. A perfect elimination of the distortion is only possible in a 1:1 imaging with symmetrical conditions in the object as well as in the image space. Then also other effects like coma or chromatic aberration can be avoided.

In most practical applications, we usually have a less symmetrical situation with the image space being much smaller than the object space. Even in these cases a symmetrical setup is favorable but distortion, which depends on object distance and magnification, cannot be fully eliminated for all situations. The observed distortions in real lenses, also for well-corrected ones, may not only be of the pincushion or barrel type but also a mixture of them and is sometimes designated as wave or moustache distortion.

The image formation due to central projection, as described by Equation (2.1), yields an image height that is proportional to the tangent of the angle β under which the object is viewed from the center of projection. Using a lens, the proportionality constant between the off-axis distance h_i in the image plane from the optical axis and the tangent of the angle is the focal length f of the lens:

$$h_\mathrm{i} = f \cdot \tan\beta. \tag{3.130}$$

In the case of image distortion, as described above, the image magnification is no longer constant for fixed distances in the object and image space but varies with h_i. This can be described by an additional correction factor characterizing the relative radial distortion $K_\mathrm{rad}(h_\mathrm{i})$, which itself depends on the off-axis distance in the image plane from center of the image field:

$$h_\mathrm{i} = f \cdot [1 + K_\mathrm{rad}(h_\mathrm{i})] \cdot \tan\beta. \tag{3.131}$$

For an ideal lens without distortion, $K_\mathrm{rad}(h_\mathrm{i})$ is equal to zero all over the image field. A noticeable pincushion distortion of a 35 mm format lens is shown in Figure 3.55a where the relative radial distortion continuously increases from the center of the image field to obtain a maximum value of about 3 % in the diagonal corners of the image field [Hön09]. The radial distortion is rotationally symmetric around the center and can be measured in any direction as the deviation Δh_i of the actual off-axis distance from the ideal nondistorted value as depicted in Figure 3.55b. The relative radial distortion then is expressed by

$$K_\mathrm{rad}(h_\mathrm{i}) = \frac{\Delta h_\mathrm{i}}{h_\mathrm{i}}. \tag{3.132}$$

For the pincushion distortion the image height increases with the off-axis distance, which means that $K_\mathrm{rad}(h_\mathrm{i}) > 0$, whereas for a barrel distortion $K_\mathrm{rad}(h_\mathrm{i}) < 0$.

For the 35 mm format, the diagonal image corners are at a distance of 21.6 mm from the image center with the long image side $X = 36$ mm and the short image side $Y = 24$ mm. The example in Figure 3.55a shows that the absolute distortion is zero in the center and the largest in the diagonal corners with a maximum value of about 0.7 mm compared to the nondistorted format. At the edges straight above, respectively, below, the center the distortion is smaller than at the lateral edges.

A different type of distortion is the TV-distortion, which has to be distinguished from the radial distortion. The TV distortion K_TV simply describes the maximum image distortion in vertical direction normalized to the full vertical image extension. There ex-

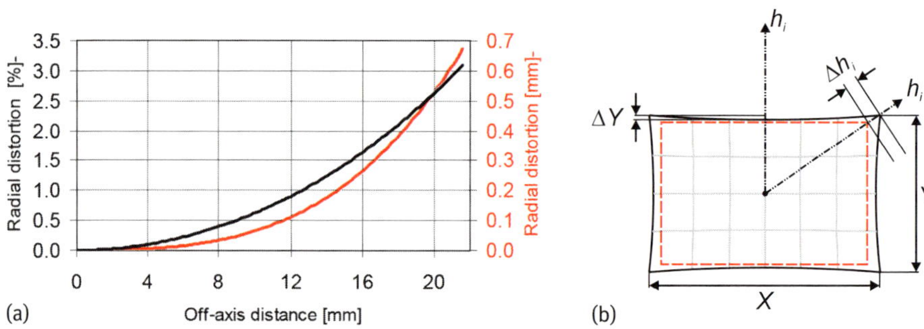

Fig. 3.55: Pincushion distortion of a 35 mm format lens with a 3:2 image ratio. (a) Relative and absolute radial distortion as a function of the off-axis distance h_i [Hön09] (with kind permission of Carl Zeiss AG); (b) parameters to determine radial as well as TV distortion.

ist different definitions for the TV distortion. Here, we use the one published by Zeiss [Hön09]:

$$K_{TV} = \frac{\Delta Y}{Y}. \tag{3.133}$$

ΔY is the vertical distortion of the long horizontal edges quantifying their bending at the corners (Figure 3.55b). Y is the short lateral format side. Other definitions specify the distortion as the total difference, namely $2 \cdot \Delta Y$ with respect to Y, resulting in a larger distortion value. Definitions with respect to the short format extension measured in the center of the image are also possible.

The TV distortion is always smaller than the radial distortion by a factor of 2 to 3 as it only takes into consideration the projected distortion on one short image side and normalizes this value to the total extension of the short side. The evaluation of the distortion generally requires test charts with a calibrated grid of geometric patterns.

3.5.6 Chromatic aberration

Chromatic aberration in lenses is due to the interaction of light with matter while traversing the lens, and thus are nearly absent in mirror optics with light traveling in air. The refractive index in glass is not constant for the wavelengths of which light is composed, and thus all ray paths of the different color components through the optical system cause images at different positions and of different sizes in the image space, depending on the color. Any color component may be affected by the Seidel aberrations as described above, and then all these colored partial images superimpose to form the overall image. Even in perfect lenses or in paraxial approximation the chromatic aberrations must be treated separately when working with white light. The generation of different colored images may be more disturbing than the geometrical aberrations but is virtually irrelevant in monochromatic imaging using light with a narrow spectrum like that of light emitting diodes or lasers.

In this section, we only focus on the normal dispersion of light in glass as considered in Section 1.7.2. It can be seen from Figure 1.28 that blue light is more strongly refracted than red light. Consequently, parallel incident white light is not focused to one single focal point but is fragmented and the colored focal spots are distributed over certain range. This is illustrated in Figure 3.56a for a thin lens. It can be seen that the image focal length of blue light $f_{\mathrm{F}'}$ is shorter than that of red light $f_{\mathrm{C}'}$. Thus, the blue focal point $F_{\mathrm{F}'}$ is located closer to the lens than the red one $F_{\mathrm{C}'}$. The indices represent the Fraunhofer spectral lines F′ and C′ in the blue and red spectral range (see Table 1.3 in Section 1.7.2). It should be noted that for the description of the dispersion in this section we use the newer specifications for optical systems. These are based on the refractive index n_e at the green *e*-line and the F′-, C′- and *e*-lines for the Abbe number ν_e. The distance between the image locations on the optical axis, which in the case for parallel incoming light is equal to the difference of the focal distances $f_{\mathrm{F}'} - f_{\mathrm{C}'}$, is a measure for the dispersion and is defined as the longitudinal chromatic aberration. If we measure that distance with $F_{\mathrm{F}'}$ in the blue range as a reference, the longitudinal aberration is negative for a positive, namely converging lens. The deviation of the image focal length f_λ at the wavelength λ relative to $f_{\mathrm{F}'}$ can be expressed by $f_{\mathrm{F}'} - f_\lambda$ and is shown in Figure 3.58a qualitatively

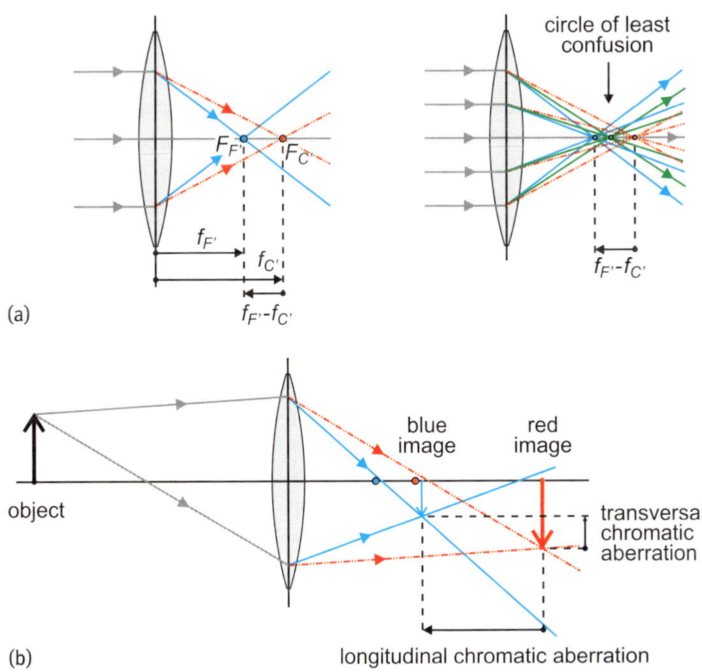

Fig. 3.56: Chromatic aberrations. (a) Longitudinal chromatic aberrations of incident light beams parallel to the optical axis; (b) longitudinal and transversal chromatic aberrations for off-axis imaging.

for a single positive lens as a function of λ. As f_λ increases with λ, its difference $f_{F'} - f_\lambda$ is negative at wavelengths above the F′-line. For a diverging lens, $f_{F'} - f_{C'}$ is positive as will be calculated below. If we consider the envelope of the refracted rays, there is no single point image with nearly zero extent like in Gaussian optics but rather a circle of least confusion between the blue and red image points. This resembles the spherical aberration but here the focal point deviation does not depend on the aperture and is also persistent for paraxial rays. Hence, stopping the aperture down does not reduce the chromatic aberration as a stand-alone effect unlike spherical aberration. The caustic always remains colored and a colored halo around the focal point of one color is seen if a vertical screen is located at this position.

The longitudinal aberration can be calculated for a thin lens in air. Starting with the Equations (3.21) and (3.22), setting $t_L = 0$ for thin lenses and inserting n_e as the refractive index of the lens material we get the lens-maker formula for a thin lens in air:

$$V_e = (n_e - 1) \cdot \left(\frac{1}{r_1} - \frac{1}{r_2}\right). \tag{3.134}$$

V_e is the refractive power in the image space at the *e*-line wavelength, and r_1 and r_2 are the curvature radii of the lens. In order to simplify the writing, we abbreviate the notation and use the symbol ρ for the difference between the reciprocal radii of the lens. Then V_e and the corresponding image focal length f_e are expressed as

$$\begin{aligned} V_e &= (n_e - 1) \cdot \rho \\ f_e &= \frac{1}{V_e} = \frac{1}{(n_e - 1)} \cdot \frac{1}{\rho}. \end{aligned} \tag{3.135}$$

The variation Δf of the focal length between the blue and red spectral range due to a variation of the refractive index can be calculated by

$$\Delta f = \frac{\partial}{\partial n} f_e \cdot \Delta n = -\frac{1}{(n_e - 1)^2} \cdot \frac{1}{\rho} \cdot \Delta n \tag{3.136}$$

If we consider the dispersion between the F′- and C′-lines centered around the *e*-line as described above, the variation of the focal length using $\Delta f = f_{F'} - f_{C'}$ and $\Delta n = n_{F'} - n_{C'}$ is given by

$$f_{F'} - f_{C'} = -\frac{1}{n_e - 1} \cdot \frac{n_{F'} - n_{C'}}{n_e - 1} \cdot \frac{1}{\rho} = -\frac{f_e}{\nu_e}. \tag{3.137}$$

Here, we resorted to the definition of the Abbe number $\nu_e = (n_e - 1)/(n_{F'} - n_{C'})$ according to Equation (1.27). It can be seen that the longitudinal chromatic aberration of a single thin lens for imaging from infinity is directly proportional to its focal length and directly proportional to the dispersive power of glass $1/\nu_e$. Moreover, the longitudinal chromatic aberration is negative for positive lenses with $f_e > 0$. It is positive for dispersive lenses with $f_e < 0$, as mentioned above. It is especially this feature, which is exploited when positive and negative lenses are combined to establish achromatic doublets or triplets where the chromatic aberration is eliminated. The magnitude of $f_{F'} - f_{C'}$ can be easily estimated for a thin lens taking into account that for many standard glasses the Abbe number is on the order of $\nu_e \approx 50$. Then the longitudinal chromatic aberration of a thin converging lens with a focal length of $f_e = 50$ mm is about –1 mm. The blue focal point is about 1 mm closer to the lens than the red one.

In a similar way, we can calculate the variation of the refractive power with the wavelength λ by using its partial derivative:

$$\frac{\partial}{\partial\lambda}V_e = \frac{\partial}{\partial\lambda}(n_e - 1)\cdot\rho = \rho\cdot\frac{\partial n_e}{\partial\lambda}. \tag{3.138}$$

If we substitute the partial derivative by the ratio of the differences, we get

$$\frac{\partial V_e}{\partial\lambda} = \rho\cdot\frac{\partial n_e}{\partial\lambda} \approx \rho\cdot\frac{n_{F'} - n_{C'}}{\lambda_{F'} - \lambda_{C'}} = \rho\cdot\frac{n_e - 1}{\lambda_{F'} - \lambda_{C'}}\cdot\frac{n_{F'} - n_{C'}}{n_e - 1} = \frac{V_e}{\lambda_{F'} - \lambda_{C'}}\cdot\frac{1}{\nu_e}. \tag{3.139}$$

Thus, the variation of the total refractive power between the F′- and C′-lines is equal to

$$\Delta V_e = \frac{V_e}{\nu_e}. \tag{3.140}$$

Like for the focal length, the variation of the refractive power of a thin lens is directly proportional to its value and directly proportional to the dispersive power of glass $1/\nu_e$..

Let us now consider the images of an object with off-axis points (Figure 3.56b). As the focal length for blue light is shorter than for red light, the image distance according to the thin lens formula (3.19) must be shorter, and thus also its image size. Not only are the images located at different positions in the image space. but their sizes are also larger the longer the wavelength is. The difference of the image sizes is called transversal chromatic aberration and can be interpreted as different magnification for different colors. Furthermore, Figure 3.56b shows that there is more longitudinal chromatic aberration, if the image distance is larger. This happens when objects are closer to the lens than at faraway distances. The chromatic aberrations strongly increase with the magnitude of the image magnification and become a serious problem especially for close-up imaging and also in telescopes.

3.5.6.1 Achromatic doublet: two thin lenses of different materials

In order to eliminate the chromatic aberrations, let us now consider a combination of two thin lenses at close distances (Figure 3.57). The minimum distance of separation t_s is achieved when the lenses are in contact to each other. The total refractive power of

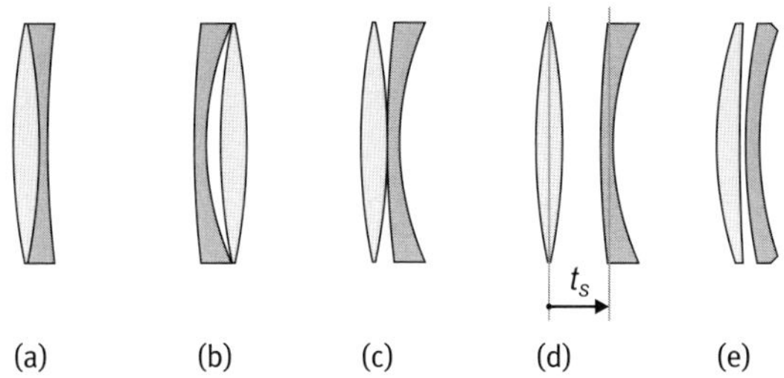

Fig. 3.57: Achromatic doublets. (a) Thin symmetric biconvex crown glass lens cemented to a negative flint glass lens; (b) doublet with edge contact; (c) doublet with center contact; (d) two thin lenses of different glass types separated by a distance t_s (dialyte); (e) two thin meniscus lenses of different glass types separated by a short distance (achromatic Gauss doublet).

the lens combination after Equation (3.54) can be expressed by the formula where the indices 1 and 2 relate to the different lenses:

$$V_e = V_{1e} + V_{2e} - t_s \cdot V_{1e} \cdot V_{2e}. \tag{3.141}$$

It should be noted that in this expression t_s is a positive quantity. For thin lenses in contact, the approximation of $t_s \approx 0$ is reasonable and their total refractive power is

$$V_e \approx V_{1e} + V_{2e} \tag{3.142}$$

The chromatic aberration is nullified if the total refractive power no longer depends on the wavelength λ. Thus, the derivative with respect to λ must be zero, and using Equation (3.139) we get

$$\frac{\partial}{\partial\lambda} V_e = \frac{\partial}{\partial\lambda} V_{1e} + \frac{\partial}{\partial\lambda} V_{2e} \approx \frac{V_{1e}}{\nu_{1e} \cdot (\lambda_{F'} - \lambda_{C'})} + \frac{V_{2e}}{\nu_{2e} \cdot (\lambda_{F'} - \lambda_{C'})} = 0. \tag{3.143}$$

This condition is fulfilled if the ratio of the lenses' refractive powers respectively focal lengths is equal to the ratio of their Abbe numbers:

$$\frac{V_{1e}}{V_{2e}} = -\frac{\nu_{1e}}{\nu_{2e}} = \frac{f_{2e}}{f_{1e}}. \tag{3.144}$$

This result shows that the ratio of the image focal lengths is negative while the Abbe numbers are always positive for crown and flint glasses. Thus, the chromatic aberration vanishes for a thin doublet consisting of a positive and a negative lens made of glass types with different Abbe numbers. Furthermore, Equation (3.144) yields that only the refractive power and the Abbe number of the lenses is decisive but not the lens shape. The first types of thin cemented achromats in the 19th century, also termed old achromats, consisted of conventional flint and crown glass types and were designed according to Equation (3.144) with an overall lens bending by which the spherical aberration was minimized. As for the curvature of fields of these old achromats, it could not be nullified because the Petzval sum could not be brought to zero using the glasses available at that time. With the production of new glass types like dense crown and light flint glasses at the end of the 19th century, achromatic doublets, termed new achromats, could be designed, which fulfilled the Petzval criterion to flatten the field curvature but still showed spherical aberrations. In order to correct chromatic, spherical and astigmatic aberrations, more complex arrangements and calculations like in camera lenses are necessary (see also Section 6.2.4). A thin lens achromat without lens separation, however, cannot be free of all of these aberrations.

As a consequence, different types of achromatic doublets are designed for different requirements. A very common type is the cemented achromat. Figure 3.57a shows this type consisting of a positive crown glass lens with curvature radii of the same magnitude. The negative lens of flint glass has one surface of the same curvature radius like the positive lens. Thus, the opposite surfaces of both lenses have a perfect match and are cemented by a very thin adhesive layer. Other types without separation are lenses that have edge contacts (Figure 3.57b) or center contacts (Figure 3.57c). The more general type of achromatic doublets is the dialyte type that consists of a positive and negative lens, both of different materials and different magnitudes of refractive powers and separated by a distance t_s (Figure 3.57d). If both lenses have meniscus shapes, they are termed Gauss doublet, since Gauss was the first to use that type of lens arrangement for telescopes (Figure 3.57e). The small distance t_s gives more flexibility in the lens design in order to reduce spherical and chromatic lens aberrations (see also Figure 3.20) and to reduce the Petzval sum of the lens. Different lens shapes like meniscus lenses and different materials are used to get a large variety of achromatic doublets.

3.5.6.2 Achromatic doublet: two thin lenses of identical materials with separation

Nullifying the chromatic aberrations is also possible using thin lenses of the same material. Returning to Equation (3.141) and inserting for the diffractive power of the individual thin lens the relation given by Equation (3.134) with n_e being the refractive index of each of the lenses we get

$$V_e = V_{1e} + V_{2e} - t_s \cdot V_{1e} \cdot V_{2e} = (n_e - 1) \cdot \rho_1 + (n_e - 1) \cdot \rho_2 - t_s \cdot (n_e - 1)^2 \cdot \rho_1 \cdot \rho_2. \quad (3.145)$$

The total refractive power is independent from the wavelength if the partial derivative of V_e with respect to λ is zero. This leads to the following expression:

$$\frac{\partial}{\partial\lambda} V_e = \rho_1 \frac{\partial n_e}{\partial\lambda} + \rho_2 \frac{\partial n_e}{\partial\lambda} - t_s \cdot \rho_1 \cdot \rho_2 \cdot 2 \cdot (n_e - 1) \cdot \frac{\partial n_e}{\partial\lambda} = 0. \quad (3.146)$$

If there is dispersion, which means $\partial n_e/\partial\lambda \neq 0$, then the wavelength independence is only fulfilled if

$$\rho_1 + \rho_2 - t_s \cdot \rho_1 \cdot \rho_2 \cdot 2 \cdot (n_e - 1) = 0. \quad (3.147)$$

If we multiply the left term with $(n_e - 1)$ and replace $(n_e - 1) \cdot \rho$ by V_e, we can write

$$V_{1e} + V_{2e} - 2 \cdot t_S \cdot V_{1e} \cdot V_{2e} = 0. \tag{3.148}$$

Rearranging this equation yields

$$t_S = \frac{1}{2} \cdot \left(\frac{1}{V_{1e}} + \frac{1}{V_{2e}} \right) = \frac{1}{2} \cdot (f_{1e} + f_{2e}). \tag{3.149}$$

The chromatic aberration vanishes for two lenses of the same material with a positive air gap of half the sum of both focal lengths in between. This is valid for all lenses, being positive or negative, and also for any lens shape. As the separation t_S is a positive quantity, the combination is achromatic only if the sum of the focal lengths yields a positive number, which is not possible using two divergent lenses. The combination of two identical converging lenses is discussed in Section 3.3.5 and illustrated in Figure 3.19b. This principle to minimize the chromatic aberrations is applied for oculars, namely the Huygens' ocular. For the particular case of a Gauss lens combination where both lenses are made of the same glass material, the air gap between them is calculated according to Equation (3.149). For the more general case of a Gauss achromatic doublet made of different glass types a detailed calculation, taking into account their Abbe numbers is necessary.

3.5.6.3 Complex achromatic systems

In the case of thick lenses or more complex optical systems, not only the focal distances vary with the wavelength but also the location of the cardinal points in the lens. This means that even if the focal lengths for different colors are identical, their focal points do not coincide since the principal planes for the different colors are at different positions, and thus also at corresponding focal points. In order to reduce the chromatic aberrations, it is not sufficient to adjust the refractive power and with it the focal length of a system for different colors but also the positions of the focal points. This is more complicated for thick lenses but uncritical for achromats consisting of thin lenses where the principal planes all coincide with the thin lens plane.

The presented principles above that minimize the chromatic aberrations using doublets generally lead to identical focal lengths only for two wavelengths, namely the red and blue color lines C′ and F′. The primary dispersion is nullified when compared to a single lens; however, a secondary color spectrum remains that leads to little chromatic aberrations in between (Figure 3.58a,b). The overall chromatic aberrations can be further reduced by combining additional lenses of special glass materials. An example is an achromatic triplet consisting of three lenses where the chromatic aberrations are zero at three wavelengths. Optical lenses or systems that have no chromatic aberrations at three wavelengths, that have no spherical aberrations and fulfill the sine condition at

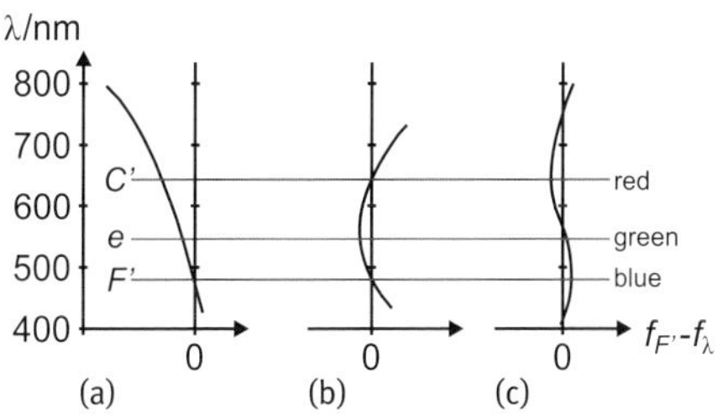

Fig. 3.58: Schematic qualitative behavior of the longitudinal chromatic aberration $f_{F'} - f_\lambda$ (in arbitrary units) as a function of the wavelength λ for different lenses. (a) Single converging lens; (b) achromatic doublet corrected for the F′- and C′-lines; (c) apochromatic lens corrected for three wavelengths.

two wavelengths are termed apochromatic. The schematic behavior of the longitudinal aberration of an apochromatic lens is illustrated in Figure 3.58c.

One special point that should mentioned here is spherochromatism. This term is used for the variation of the longitudinal spherical aberration with the wavelength, and thus a mixture of the spherical with the chromatic aberrations. Spherochromatism can be avoided by designing spherically corrected achromats, which are the result of combining lenses of different shapes, materials and thicknesses at different separations. This is a big advantage of Gauss-type lens combinations. An iterative approach to minimize both aberrations is then required where appropriate lens shapes and materials are varied [Kin10]. As the chromatic aberration increases with the focal length, the problem of spherochromatism especially shows up with long focus lenses at large aperture, which is typical for telescopes operating at night sky. However, also for compact camera modules with short focal lengths spherochromatism becomes a serious problem. Here, only few lenses can be combined due to space restrictions for miniaturization [Ste12]. Moreover, these lenses usually are combined with digital image sensors having small pixel pitches. Thus, the lenses must have wide apertures in order to avoid diffraction blur and low contrast when used with small pixel sensors. In these cases, it becomes difficult to correct both spherical and chromatic aberrations with the same approach as is done for larger optical systems based on spherical lenses. Here, often aspheric lenses are required by which many of the third order aberrations can be corrected. The chromatic aberration, however, is still a problem as in consumer cameras mostly plastic materials are used for aspheric lenses and only a limited range of Abbe numbers is available. All this has to be considered in the design of lenses of different sizes and for special applications.

3.5.7 Aspheric surfaces

As described in the preceding sections, a perfect imaging using spherical lens surfaces may only be possible for light rays propagating close to the optical axis. For instance, in the consideration of the spherical aberration, we have seen that rays parallel to the optical axis striking the surface at high elevation h from the vertex of a lens are more strongly refracted to the optical axis than predicted by the Gaussian optics. This is due to the fact that the angle between the ray and the spherical surface normal continu-

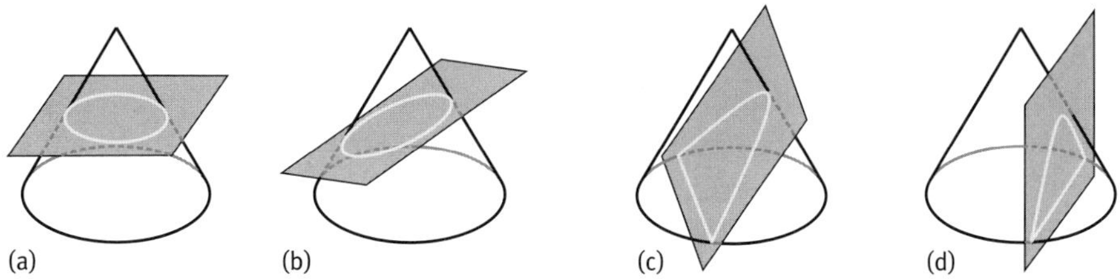

Fig. 3.59: Conic sections as a consequence of increasing angle of a plane relative to the perpendicular. (a) Circle, (b) ellipse, (c) parabola, (d) hyperbola.

ally increases with the elevation h leading to increasing aberrations. One possibility to counteract this could be if the curvature of the surface at higher elevation is reduced unlike the curvature of spherical surfaces, which is constant and independent from h. This feature can be found with surfaces that, close to the vertex, have a spherical-like geometry, which then continuously changes. A simple way to describe such a surface is to start with a conic section as the cross-section of the surface in the meridional plane and then rotate that curve around its axis of symmetry. This axis then defines the optical axis. Thus, surfaces with rotational symmetries and aspherical shapes are obtained. Such conic section curves originate from the intersection of a plane with the surface of a cone when the orientation of the plane is varied with increasing angle of the plane relative to the perpendicular (Figure 3.59).

If the plane is parallel to the base area of the cone, the conic section is a circle. A slight inclination of the plane, intersecting only the envelope of the cone, results in an elliptical conic section. In the case of stronger inclinations when the plane intersects the base area of the cone, the resulting curve will be a parabola or a hyperbola (Figure 3.59c, d). The curves that are obtained all have the common feature that their vertex section can be approximated by a circle. To describe this in a mathematical way, we start with an expression for the circular section of a spherical lens surface where the origin is centered in the lens vertex:

$$h^2 + (z - r)^2 = r^2 \quad \Leftrightarrow \quad h^2 = 2 \cdot r \cdot z - z^2. \tag{3.150}$$

Here, the spatial component z on the optical axis measures the distance behind the vertex, the elevation perpendicular to the axis is expressed by h, and the radius of the spherical lens is r. This description of a circular form necessitates only one form parameter which is the radius r of the circle. In order to describe the departure from the circular form we need a second form parameter κ modifying the influence of the second order term of z:

$$h^2 = 2 \cdot r \cdot z - (1 + \kappa) \cdot z^2. \tag{3.151}$$

κ is also termed deformation coefficient or conic constant [LdO03, Kin10]. When comparing Equations (3.151) and (3.150), it can be seen that $\kappa = 0$ describes a circle. It has become conventional to describe the cross-section of an aspherical surface by expressing z as a function of h. Rearranging Equation (3.151) thus yields

$$z = \frac{\frac{h^2}{r}}{1 + \sqrt{1 - (1 + \kappa) \cdot \frac{h^2}{r^2}}}. \tag{3.152}$$

For $\kappa = -1$, the square root in the denominator of Equation (3.152) is identical to 1 and z is a simple parabolic function of h. This can also be seen in Figure 3.60 where the normalized elevation h/r of a lens cross-section in the meridional plane is illustrated as a function of the normalized distance z/r from the vertex for different conic parameters κ. The circle and parabola are obtained in the case of the fixed values $\kappa = 0$, respectively, $\kappa = -1$. When κ is continuously varied from 0 to −1, we see a smooth transition from the circle via an ellipse to a parabola. The curve is prolate elliptic with the small end of the ellipse in the vertex with $0 < \kappa < -1$, whereas oblate elliptical curves with the long side of the ellipse toward the vertex are achieved for $\kappa > 0$. For $\kappa < -1$, we get hyperbolic curves.

As mentioned above, the aspheric surface results from the rotation of the conic section around the optical axis. The rotational surfaces constitute the refracting surfaces

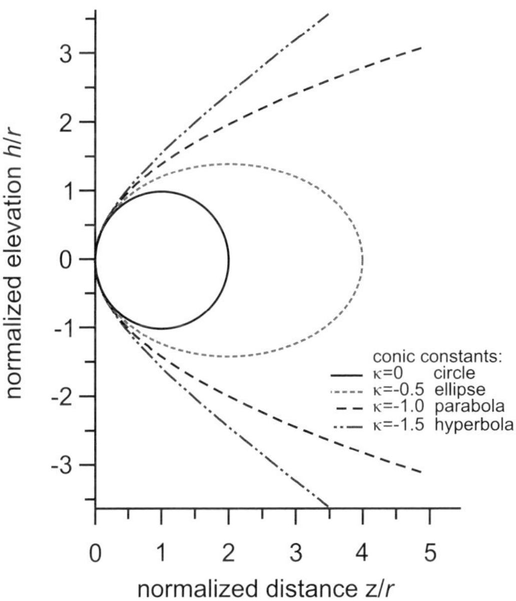

Fig. 3.60: Normalized elevation h/r of a lens cross-section as function of the normalized distance z/r from the vertex on the optical axis for different conic constants κ.

of lenses made of transparent material or the reflecting surfaces in the case of mirror imaging. Some of the rotational aspheric surfaces have the special feature to ensure exact point images without spherical aberration in optical systems for selected points on the optical axis. For instance, a rotational paraboloid, used as a reflecting surface, images incident light beams, which are parallel to the optical axis, to the focal point F_i without aberration. In air, F_i is located at $z = 1/(8 \cdot r)$ from the vertex. Another example is a hyperbolic glass lens (Figures 3.4b and 3.47b) that images parallel incident light beams to the focal point without spherical aberration. Here, the focal point is a function of the refractive index as well as of the surface curvature in the vertex. However, these special aspheric surfaces based on conic sections have special features only for a limited range of points and cannot eliminate all aberrations.

As discussed in Section 3.5.2, a hyperbolic lens can eliminate spherical aberration but not coma. Thus, a modification of the aspheric surface is necessary to improve the imaging quality. Therefore, a more general description of an aspheric surface should be used, but in all cases the spherical surface is the starting point and the departure from it is described by aspheric coefficients.

If we consider Equation (3.152) for paraxial rays, which means that h approaches zero, then the argument h in the square root can be neglected and the curve is independent from the conic constant for small values of h. The narrow paraxial segment of a spherical surface is identical to that of a parabolic or hyperbolic rotational asphere. The best way to describe the general asphere then is to expand the function of the aspherical surface with conic section as expressed by Equation (3.152) in a Taylor power series, starting in the vertex with z and h being zero. Due to the rotational symmetry with the optical axis only even-numbered powers of h are required. The Taylor series of z up to the fourth power of h and neglecting higher orders of h yields the result:

$$z \approx \frac{1}{2 \cdot r} \cdot h^2 + \frac{1+\kappa}{8 \cdot r^3} \cdot h^4. \tag{3.153}$$

When setting $\kappa = 0$, we get the result for a spherical surface. If any arbitrary rotational aspheric surface is wanted, the following power series is more appropriate:

$$z = a_2 \cdot h^2 + a_4 \cdot h^4 + a_6 \cdot h^6 + \cdots. \tag{3.154}$$

The coefficient a_2, a_4, a_6, etc. are called aspheric coefficients. If we want to describe a spherical surface by these parameters, the comparison of Equation (3.154) with (3.153) for $\kappa = 0$ yields the result for the first two significant coefficients:

$$a_2 = \frac{1}{2 \cdot r}, \quad a_4 = \frac{1}{8 \cdot r^3} \tag{3.155}$$

In order to get an impression of how well this approximation can be used to describe a spherical surface, Figure 3.61 shows different curves in comparison with a circle. Using only $a_2 = 1/(2 \cdot r)$ and setting all higher aspheric coefficients to zero results in a parabolic

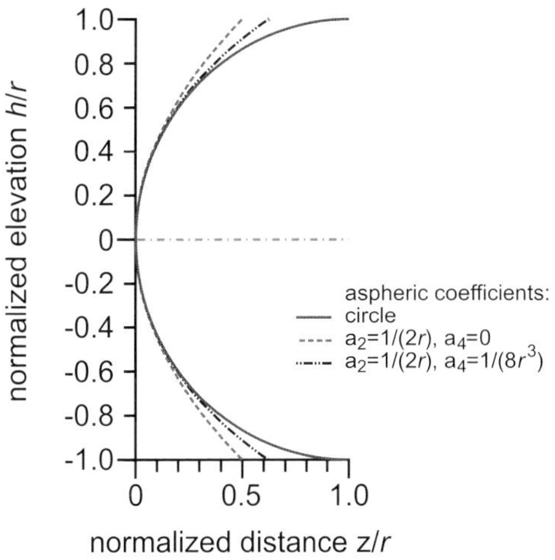

Fig. 3.61: Normalized elevation h/r as function of the normalized distance z/r for meridional sections of different surfaces.

curve that is almost indistinguishable from a circle for approximately $h < 0.2 \cdot r$. At higher elevation, a clear departure with lower curvature is seen. Setting a_2 and a_4 according to Equation (3.155) and higher coefficients to zero, the curve is identical with a circular segment up to $h \approx 0.5 \cdot r$. As for the curvature of a curve in one point, it can be illustrated by the osculating circle or sphere in that point. The radius of the osculating circle is the reciprocal value of the curvature. The osculating circle has the same first and second derivative as the curve in that point. It can be seen that the circle of radius r is always the osculating circle of the aspheric polynomial function (Equation (3.154)) if $a_2 = 1/(2{\cdot}r)$. Thus, starting the expression of any aspherical surface with that parameter means that its paraxial imaging properties are fixed by it. The calculation of the aspherical surface for larger values of h in general is done using ray tracing software where the structure is optimized in an iterative approach according to the requirements.

There exist also other mathematical forms to describe the cross-sections of aspherical surfaces. Some incorporate odd-numbered powers of h, which are necessary to define surfaces that are not rotationally symmetric with the optical axis. Other types of aspheric surfaces are, for instance, toric surfaces (Figure 3.62) or cylindrical surfaces. They have the characteristic feature that their curvature has two different values in perpendicular directions unlike spherical surfaces where the curvature is the same in all points. A cylindrical surface has a curvature only in the plane perpendicular to the cylindrical axis, and thus only rays in that plane are converged to the image plane. For rays in planes perpendicular to it, the lens is afocal. A transparent section of a general toric surface used as a lens has different focal points for rays in planes perpendicular to each other like imaging using two crossed cylindrical lenses. A section cut parallel to the rotational symmetry axis can be used as contact lens for the human eye to correct for visual astigmatism.

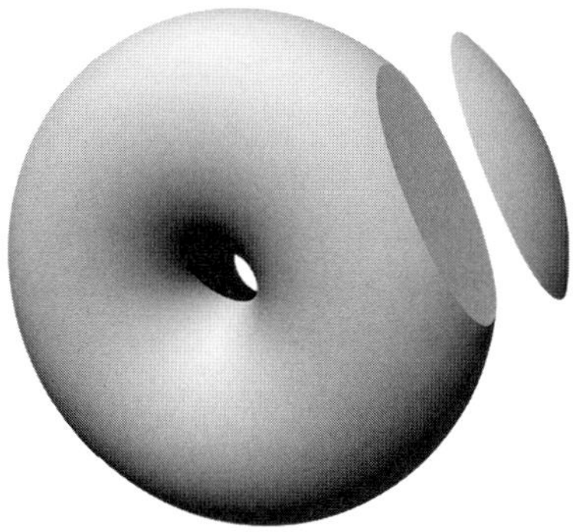

Fig. 3.62: Toric surface with a section cut parallel to the symmetry axis (author HHahn[9]).

The big advantage of an aspheric lens is that its surface can be tailored for special applications and then replace the combination of multiple lenses. The fabrication of such a lens consisting of glass, however, is much more difficult than that of a spherical lens. For instance, spherical glass lenses are easily polished and ground by automatic machines whereas for aspheric surfaces special tools are required, which must be adapted to that surface. Moreover, the surface quality is generally lower than that of spherical lenses or can only be achieved at higher costs. Therefore, aspheric glass lenses are usually reserved for special applications. On the other hand, molded aspheric plastic lenses can be produced in a much cheaper way. However, they have a problem of mechanical stability. Also, some optical properties like refractive index are less well controlled than that of glass lenses.

9 https://commons.wikimedia.org/wiki/File:Toric_lens_surface_2.png?uselang=de

4 Sensors and detectors

4.1 General, films, photodiode arrays

4.1.1 Introduction and overview of 2D detectors

In the previous chapters, we discussed the basic aspects of imaging, cameras lenses and several advanced topics related to those subjects. Although it is clear that usually an image has to be recorded by a 2D detector, e. g., a film or an electronic photo detector array (PDA, typically a matrix of photodiodes, a CCD or CMOS, see later in this chapter), in the previous chapters we mostly did not take much care how the image is recorded. However, as discussed in Chapter 1, the detector takes an important part within the imaging process, which may be not necessarily regarded as independent from the rest of the whole process (see also Figure 1.3, and the examples later in this chapter). Here, as an example, it is notified that detector noise may have a strong influence on the spatial resolution of a camera. What is more with this example, we may state that there are a lot of cameras with large pixel numbers that produce much worse images than other ones with a smaller number of pixels and less noise.

Coming back to the whole imaging process, first an image has to be generated on the surface of a suitable sensor. Then it has to be saved. To do so, first we need a sensor that is sensitive to light and that can detect the illumination pattern on its surface with spatial resolution. Usually, this image detection is performed in two dimensions. Storage of the corresponding signal may be achieved at the same time, for instance, when a film or an image plate is used. Alternatively, this can be performed in a second step, for instance, when the signal readout from a PDA is stored in a data medium.

Although all that looks straightforward, in particular, for an electronic detector system, it is not. Between image capturing within the photosensitive material of the sensor, and final storage, there are a lot of additional steps. Those are illustrated in Figure 4.1. We will not discuss those steps here. This figure just introduces the topics of the present Chapter 4. For other sensors, such as films, part of those issues is not of much relevance, but others are important as well although they may be termed differently. Thus, to some extent we will discuss that too within this chapter.

Before we continue with a more detailed discussion, we would like to discriminate between a naked detector and a detector system. For instance, a photodiode is regarded as a 0D detector and a PDA (such as a CCD or a CMOS sensor; see Section 4.4 and Section 4.5) or a film as a 2D one. Table 4.1 lists some further examples of detectors that may be used for light detection or imaging, respectively. Within this book, we define such light sensitive elements as detectors or sensors (we will not discriminate between a detector and a sensor). In addition, we define the term detector system (or sensor system, resp.) as a device consisting of the detector and additional filters and other components. This is discussed in more detail in Section 4.6.

https://doi.org/10.1515/9783110789966-004

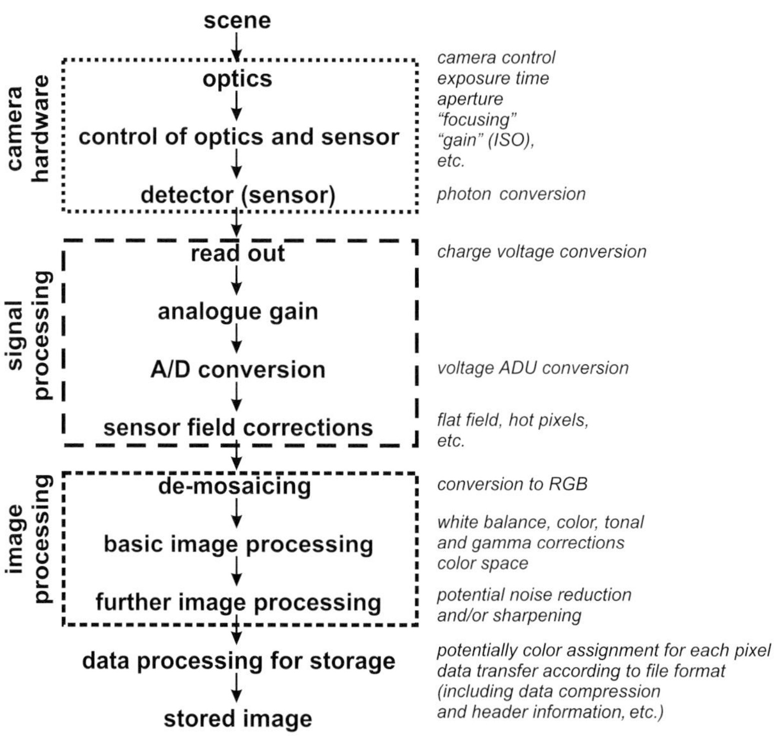

Fig. 4.1: This figure displays the major content of what has to be discussed with electronic sensors. This begins with the image generated on the sensor surface and ends when the image is stored in a data file. The stored image then can be displayed by some suitable output medium and then regarded by the eye, or it may be used for other purposes (including further image processing and/or generation of prints) as is already discussed together with Figure 1.3. For scientific and/or technical imaging, some post-processing may be done as well (e. g., corrections such as for flat field, noise reduction/smoothing, etc.).

Tab. 4.1: Examples of 2D detectors used for imaging (for details on film, CCD, CMOS, etc.; see later in this chapter).

nonelectronic detectors and such ones with electronic or laser supported readout	film image plate (photographic plate) imaging plate (based on photostimulated luminescence)
electronic detectors (digital sensors), PDA	CCD, CMOS and somehow similar sensors
image converters	photocathode, scintillator, phosphor
image intensifiers	intensifier tube, MCP, EM-CCD

Light sensitivity of the various detectors may be quite different, and thus they may operate in different spectral regions, such as the visible range, X-ray region, infrared region, or even THz range. Furthermore, the detectors may be either analogous ones or digital ones. They may have a linear or a nonlinear response when illuminated with light and they may even amplify the input signal, as done, e. g., with an MCP (microchannel plate; see Section 4.11). Although an advanced discussion of all of those detectors is beyond the scope of the present books, in the following we will treat the most important issues of the most common detectors, in particular those used within photography or scientific and technical imaging. In addition, we will discuss some special and further scientific detectors and discriminate also situations when the same detector is used in different spectral ranges, and hence shows many different characteristics.

Finally, we would like to remark that there are a lot of requirements for the detectors and detector systems. This includes several issues, particularly the goal of a good reproduction of the object or scene that should be imaged onto the detector surface (including desired resolution, potentially large contrast and low noise). Sometimes capturing with high speed may be a concern as well. In addition, there are technical aspects, such as reliability and easy handling of the system and after all, its cost as well.

4.1.2 Introduction to color reproduction

Before we continue with our main subject, we will briefly summarize some general issues related to color within optical imaging. As color information and reproduction is an extended topic on its own and fills a lot of books, within the present book we will keep that rather short and keep to the very basics.

Monochrome films or electronic PDA only render brightness information but no color information according to the visual perception of the human eye. For that purpose, a representation of the visible colors is necessary, similar to the sensibility of the color receptors in the human eye (Figure 4.2c), the cones, for red, green and blue colors, respectively. All visible colors can be described by different color models, of which we want to mention only combinations of basic colors in two ways: the first is by an additive color mixture based on the linear combination of the primary colors: red, green and blue. Like in the human eye, the additive combination is appropriate to describe colors that result from the mixture of different optical radiations from light emitting sources, hitting the eye in parallel or in fast sequences one after the other. A typical example for that mixture is the color combination from red, green and blue pixel elements on a color monitor to yield the color impression of one image pixel or the light combination in a beamer projection. It should be mentioned that the pixels are small enough so that they cannot be visually resolved as individual elements. If all three primary colors are combined in the same illuminance, a white light impression results (Figure 4.2a). Let us designate their relative illuminances by color triples (r,g,b) with decimal numbers between 0 and 1. For instance rgb (1,1,1) means maximum, equal relative illuminances for the three primary color components, leading to white color whereas rgb (0,1,1) represents the additive linear combination of green and blue with zero illuminance of red, yielding cyan. rgb (0.5,0.5,0.5) results in white color of less brightness and can be characterized as 50 % gray value whereas rgb (0,0,0) means black. Mixtures of two primary colors of

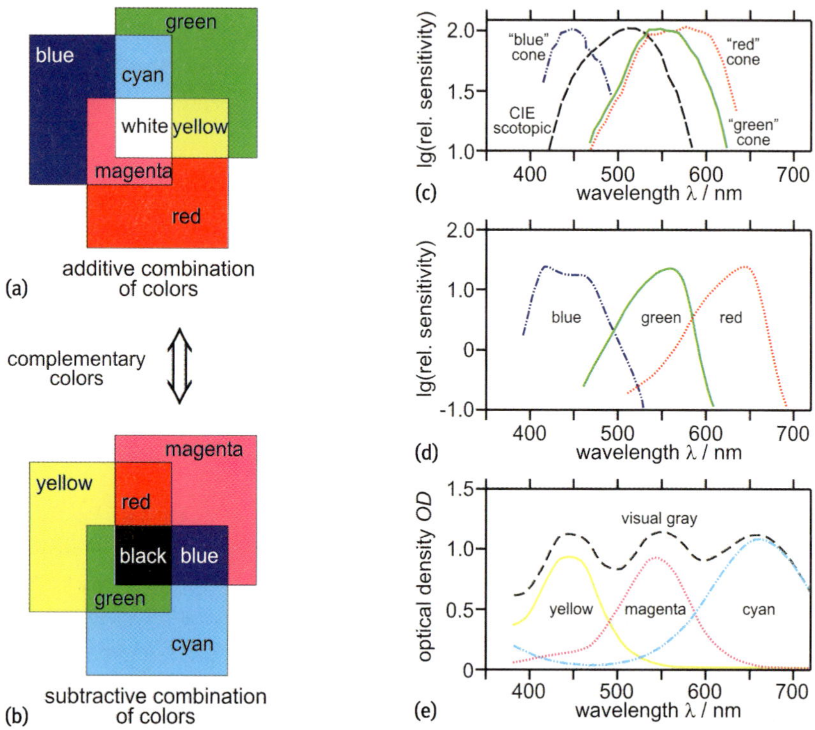

Fig. 4.2: (a) Additive combinations of primary colors. (b) subtractive combinations of their corresponding complementary colors. (c) Spectral sensitivity of the human eye (data points taken from Bowmaker and Dartnall;[1] the lines are only to guide the viewers eye). (d) Relative spectral sensitivity of the unprocessed color layers. (e) Optical density of processed layers for the standard condition of a color slide film and as perceived by the viewer.

equal illuminances leads to the colors depicted in Figure 4.2a for the additive combination, except white. A change of their absolute illuminances does not change their tone of color, if their ratio remains constant. Then only brightness is different. However, changing the ratio as well as the absolute values of their illuminances changes their tonal as well as their brightness appearance. In this way, nearly all visible colors can be realized.

Hence, it is clear that for a fixed brightness B (or luminance), there are only two independent components for the three colors (r,g,b) (remember that $r + g + b = 1$). Consequently, instead of a description of a triple of three independent absolute values $(R, G, B) = B \cdot (r, g, b)$ this can be expressed, e. g., by two colors and the absolute brightness $B \cdot (r, g, 1\text{-}r\text{-}g)$. This is the description of the 2D color space with the commonly used U-shaped chromaticity diagram. However, more closely related to the human perception is another description, which is based on the three variables luminance (or "lightness"

1 J. K. Bowmaker, H. J. A. Dartnall: *J. Physiol.* **298** (1980) 501–511.

or brightness), hue (i. e., color type, such as purple) and saturation (or "chroma"; i. e., "purity" of color). This LCH color space (lightness-chroma-hue) is related to a polar coordinate system.

Now we will briefly summarize the second way to describe visible colors. This way is a subtractive combination of colors (Figure 4.2b). Its description provides the basic knowledge necessary for the next section. Here, the primary colors are cyan, magenta and yellow, which are the corresponding complementary colors of red, green and blue, respectively. The description by a subtractive combination is of advantage if light colors are filtered out of the spectrum, for instance, by a serial arrangement of filter elements. If for instance transmission filters of the primary colors yellow, magenta and cyan are arranged in series, they block all transmission of colored light. This is exploited in light filtering layers of color films. Analogously to the designation for RGB, one can do this with CMY for the subtractive combination of the primary colors cyan, magenta and yellow, however with the decimal numbers representing filtering efficiency related to the optical density of the layer. cmy (1,1,1) means a perfect filtering of the primary colors thus yielding black and is identical to rgb (0,0,0) whereas cmy (0,1,1) yields red, which is complementary to rgb (0,1,1) describing cyan (similar to before, we use lowercase letters for the relative values between 0 and 1 and capital letters for the absolute values). It should be mentioned here that true black can never be achieved since it would imply a nearly infinite filter thickness. Thus a sequence of cyan, magenta and yellow layers of equal filter density realizes an achromatic gray tone, and the absolute values of their densities determine if it is black or rather gray. Therefore, using this simple scheme, only an approximate description is possible and exact color representation necessitates a more profound consideration of different color models also representing their hue, saturation, lightness or absorption values. Color descriptions can be transferred to different models using the appropriate relations. However, this is out of the scope of the present book and will not be discussed further.

4.1.3 Films—Principle of the photographic silver halide film imaging process

Although nowadays mostly replaced by predominantly much superior electronic detectors, we would like to state that films still have some importance. While many people regard films as very old-fashioned, they are still in use. Sometimes the decision for a film as the detector is a question of the taste of the photographer, but there are also particular cases where a film is a detector that is much superior even to the most modern ones. Here, we simply use the term film, but this should include in general photographic emulsions on different substrates like image plates based on glass, etc.

One example is that a film, in principle, could be made of large size and at a moderate cost, whereas a very large image area fully covered by electronic picture elements must be built of a vast amount of CCD or CMOS chips and becomes incredibly expensive. Nevertheless, for instance in astronomy, this is done in very exceptional cases.

Moreover, films are flexible and could be simply used as a detector that could be easily adapted to an image field, even when this is curved. Films do not need electric power supply and can be used for very long exposure times. And they hardly suffer from artefacts such as the Moiré effect. Moreover, photographic films can be considered the most advanced permanent optical storage of analog images that do not need any permanent supply or further effort. Last but not least, films have played an important role in imaging for a long time, and thus have had a strong influence on image processing, even today.

Even though image processing in a digital camera is performed in a very different way at present, the principal workflow, when compared to processing a film in a dark chamber, is basically the same. Also, terms such as the gradation curve are applied within modern image recording and processing. Even though this book concentrates on current modern electronic sensors and sensor systems, for those reasons mentioned before, in the following we will briefly discuss films as sensors or sensor systems. The very complex chemical processes are considered only in a simplified way to understand the principal behavior. A more comprehensive discussion on films may be found in the vast amount of literature about films.

Let us start with the basic primary photographic process that is inherent to almost all film-based exposures. The basic photosensitive material is crystalline silver bromide, AgBr. Here, one makes use of the uniqueness of silver, namely that it forms compounds with other elements (in particular halogenides, i. e., chlorine, bromine, iodine) quite readily (more, when compared to other noble metals such as gold or platinum) and that its compounds are less stable than those of other common metals (e. g., aluminium, magnesium). Furthermore, inimitably its compounds are photosensitive and fractured by light, yielding pure silver crystals.

AgBr features the typical cubic structure of ionic crystals with the electrons located close to the larger, negative Br^--ions. Crystals of lateral sizes between 0.2 µm for low sensitive and up to around 2 µm for high-sensitive films are embedded in a gelatin matrix of around 6 µm–15 µm thickness, depending on the film sensitivity (Figure 4.3a). These crystals may be regarded as the fundamental image particles (i. e., the smallest particles that form an image; in the case of color films those are color dye clouds; see later).

Here, we would like to remark that the discussion is to some extent controversial because in the literature (even in publications by film companies) often it is stated that the film grains act as the fundamental film particles as the image-forming element. However, it is just the statistical distribution and overlap of the small crystals that leads to an irregular aggregation of light sensitive grains. This layer, termed the emulsion layer, is the main photosensitive layer of the film. It is covered on the top by a transparent super coat as a protection to prevent mechanical damage. On the bottom, there is an anti-halo layer against undesired light reflections from the interface to the film base. All these layers are supported by a solid film base of about 100 µm–250 µm thickness, in general made of cellulose ester, or also of more advanced modern plastic materials.

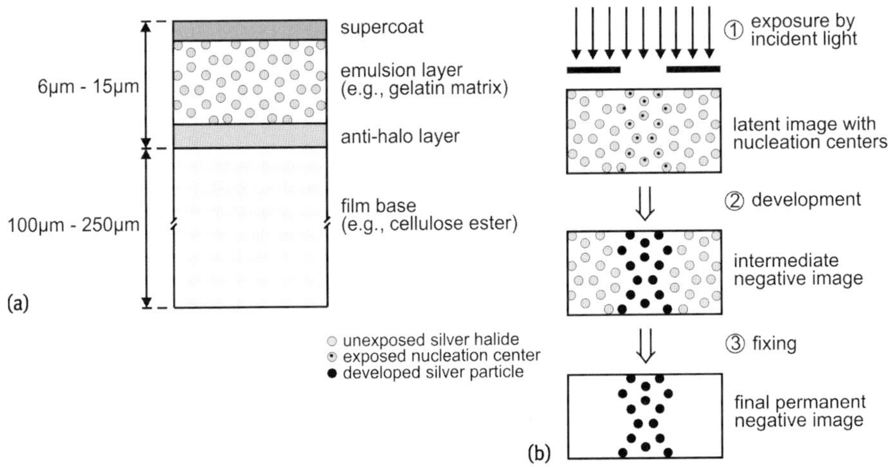

Fig. 4.3: Silver halide negative film. (a) Schematic layer structure; (b) Basic process steps: exposure, development and fixing to achieve the final image formation.

In contrast to the physical size of the crystals, film grain is a quantity perceived by the human eye together with the brain. It also results from the accumulation of the image particles located in different layers (or depths) within the film emulsion, whereas the individual particles usually are not recognized. Even more, grains observed by different magnifications or different techniques, are often perceived differently (e. g., observation with a microscope or enlarged photographic print, etc.). Grains are randomly distributed and part of them do overlap. Consequently, and in contrast to pixel-based electronic sensors (PDA), in a film the image is stored as a more or less continuous reproduction of spatial positions and gray tones. Image data stored from PDA are digitized, and thus discrete, both in spatial direction and in “intensity direction” or “data depth” (see later in this chapter).

The grain size of those crystals ranges from 1–2 µm for high-resolution films to approximately 20 µm for very sensitive films. Thus, they are approximately one order of magnitude larger than the fundamental film particles (in the case of silver crystals; in the case of a color film, the color dye clouds are approximately a factor of 5 larger, i. e., they range from 1 to 10 µm; note also that grain size depends on exposure and development). Although the optical resolution of a film (see also Section 5.2.5) and grain size are different properties, the absolute value of both quantities is often similar. Thus, films with smaller grains usually do have higher resolution but are less light sensitive (i. e., they have a smaller ISO number; see Section 2.5.1) and vice versa. This is another example where detector properties such as resolution and sensitivity are not independent from each other, and thus are related to the whole imaging chain. Table 4.2 provides some information on grain (e. g., one can recognize that grain size often is larger than the resolution of details).

Tab. 4.2: Parameters of typical films. RMS granularity is not the same as the grain size (see Section 4.7). See also Section 5.2.4. A fine grain leads to a low sensitivity of the film (a "slow film"). The grain size is rather uniform; the film shows high contrast and a short tonal range (e. g., films with ISO values up to 50). A large grain leads to a high-sensitivity film (a "fast film"). There is a variation in grain size, as the film shows lower contrast and a larger tonal range (e. g., films with ISO values of 400 and larger). (SEM is scanning electron microscope, XRD is X-ray diffraction).

	typ. size	corresponding spatial freq.	how to measure
fundamental image particles	0.2 . . . 2 µm	500–5000 lp/mm	with microscope or SEM/XRD
grain	1 . . . 30 µm	30 . . . 1000 lp/mm	with light microscope
resolution		35 to 80 (up to 170) lp/mm	R_{MTF30}-value (see, e. g., Table 5.3)
RMS granularity	7–13		microdensitometer

When light is incident to the emulsion layer, photons with energies above 2.5 eV transfer their energy to electrons of the Br^--ions. Thus, electrons are released that render the ions neutral Br, which leads to a rupture of the crystal structure at that location. The released electrons may be trapped either by Ag^+-ions, thus rendering them neutral and forming small metallic atom clusters; or they can be trapped by lattice defects in the small silver bromide crystals and constitute negatively charged nucleation centers there (Figure 4.3b). In the ideal case, the neutral bromine atoms are bonded to gelatin and become inactive. After the exposure, a latent image is formed where the exposed area is distinguished by metallic silver atoms and nucleation centers from the unexposed areas. The image, however, is not yet visible as the optical transparency in the layer has not yet changed.

Now, when light is incident on the film, according to the local light intensity distribution $I(x, y)$ (or irradiance or radiant flux density) and exposure time t_x, the film is exposed with the corresponding fluence (or radiant exposure), i. e., in physical or photometric terms

$$F(x,y) = \int_{t_x} I(x,y,t)\,dt. \tag{4.1a}$$

In radiometric expression, this is illuminance E and luminous exposure H, respectively, and thus one obtains

$$H(x,y) = \int_{t_x} E(x,y,t)\,dt \tag{4.1b}$$

This is equivalent to Equation (1.10). Note that all values, F, I, H, E are related to the image plane. Usually, both integrals simply transform to $F = I \cdot t_x$ and $H = E \cdot t_x$, respectively ("reciprocal law" or "reciprocity", i. e., for a given value H, E and t_x behave inversely proportionally, namely the same film response may be obtained if H is increased and t_x

decreased by the same factor, and vice versa). This means that the formation of the latent image as the primary imaging process in the case of time independent illumination is in the ideal case directly proportional to the amount of the exposure.

In the subsequent development process of the film, the exposed areas of the emulsion are chemically altered by a developer in an alkaline solvent. The developer solution reduces the remaining Ag^+-ions (the seeds) in the exposed silver bromide crystals to form larger metallic silver clusters, the size of which increases with the amount of exposure. This process is selective as the unexposed silver bromide crystals remain nearly unchanged. It delivers an intermediate negative image of the original scene as the exposed areas become more opaque the higher the exposure is. The development has to be performed in darkness as the remaining unexposed emulsion is still sensitive to light. Therefore, an additional fixation process in acid solution and a subsequent washing is necessary where the unexposed emulsion parts are removed. After that, the film exhibits a final negative permanent image where the transparency of the exposed areas is controlled by the density of metallic silver particles ("they look black"). It should be noted that in the development process the selectivity is not perfect and also small amounts of silver particles from nonexposed silver halide particles are deposited, but much less than in the illuminated areas. This leads to an overall slightly opaque background, which is called chemical fog.

As for the overall chain of processes, we can roughly describe the image formation using silver halide films as a three-step process consisting of the subprocesses exposure, development and fixing. The exposure is a detached process from the remaining ones and carried through by the photographer. After exposure, the film can be stored in darkness for a relative long time. The development and fixing processes are usually performed under laboratory conditions in subsequent short time intervals.

Some additional remarks to this simplified description have to be made. Although we only considered AgBr as principal light sensitive material, the emulsion may also contain other supplementary silver halide crystals such as AgCl, which is less light sensitive, or AgI, which is more sensitive than AgBr. Moreover, AgBr is only sensitive to photons with quantum energies higher than 2.5 eV or correspondingly to light of shorter wavelengths than 490 nm, which is the blue spectral range. A pure AgBr film has a limited sensitivity in the range from approximately 370 nm up to 490 nm (Figure 4.4). Therefore, complementary dye components have to be added that are sensitive to photons in the green and red spectral ranges. These compounds interact with the silver halides to generate nucleation centers in the exposed crystallites so that in the subsequent development process a silver deposition is achieved even for longer wavelengths beyond the blue spectral range. Historically, black-and-white films with sensitivity in the blue and green spectral range due to dye components were termed orthochromatic and became available after 1870. They were insensitive to red light and rendered red-colored objects white on the negative film. Moreover, these films could be developed in a red-light ambiance, which was quite comfortable for the process. After 1920, further development yielded panchromatic black-and-white films with additional dyes being sensitive

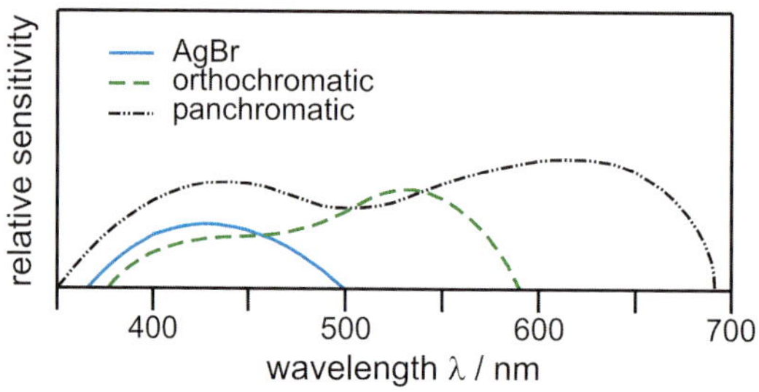

Fig. 4.4: Relative spectral sensitivity of monochromatic negative film layers. Pure AgBr layers are light sensitive only in the blue spectral range whereas panchromatic layers cover the whole range of human visible perception. Orthochromatic layers are insensitive to the red light spectrum.

in the red spectral range up to 670 nm (Figure 4.4). Now also the brightness of red colored objects could be adequately mapped to a black-and-white negative image. All current black-and-white films are in general panchromatic. Modifications of these films for special purposes can also be found, for instance with extended sensitivity in the near-infrared or even to the longer infrared range.

In the discussion above, the darkening of the film was only considered as a function of the exposure H in the image plane without further differentiation of how the exposure was achieved. We implicitly assumed the reciprocal relationship between E and t_x to yield the constant exposure $H = E \cdot t_x$. The exposure time t_x is controlled by the shutter speed in the camera whereas E in the image plane directly depends on the aperture stop. Hence, a given exposure fixed by a combination of $f_\#$ and t_x after Equation (2.16) can be realized for instance by shorter exposure time and larger aperture or vice versa.

However, it has been observed that for very short and for relatively long exposure times, the reciprocity is no longer valid. This reciprocity failure is attributed to the nature of the chemical and physical primary processes during exposure before the development and will not be discussed in detail here. Some film manufacturers like Agfa specify the range of exposure times where the reciprocity is valid with $0.1\,\text{ms} < t_x < 1\,\text{s}$ for their films. The deviation at very long times is called Schwarzschild effect.

Figure 4.5 depicts the schematic behavior of the required illuminance E_i as a function of the exposure time in order to achieve a constant amount of exposure. The true function is a smooth curve, but for illustration purposes the function is approximated by linear curve segments with a negative slope $-p$ in this log-log plot. The slope of a straight line in a log-log plot indicates a power law with the slope being identical to the exponent. Evaluating the slope, we get

$$\frac{H}{\text{lx}\cdot\text{s}} = \frac{E}{\text{lx}} \cdot \left(\frac{t_x}{\text{s}}\right)^p \tag{4.2}$$

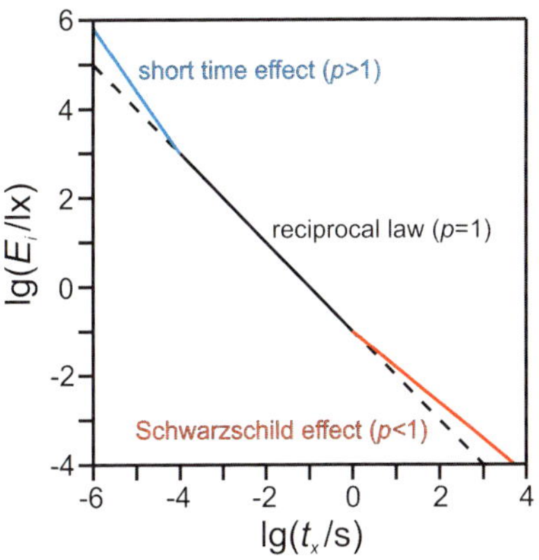

Fig. 4.5: Reciprocal law and reciprocity failures (i. e., deviations). The curve indicates (E, t_x)-combinations for the same exposure. Only in the central part of the curve the reciprocity between E and t_x is valid.

with a constant

$$p = -\frac{\lg\left(\frac{E/\mathrm{lx}}{H/(\mathrm{lx \cdot s})}\right)}{\lg(t_x/\mathrm{s})} \tag{4.3}$$

(and, of course, similar expressions for F and I).

As can be seen in Figure 4.5, the slope in the central part yields $p = 1$ and represents the usual reciprocity law for H. For the short time effect, we get a higher slope with $p > 1$ whereas the Schwarzschild effect in the long-time domain can be described using a positive $p < 1$. Both effects lead to less film sensitivity, since higher illuminations and longer exposure times are required to achieve the same exposure as estimated by the reciprocity law. In "normal" photographic situations the Schwarzschild effect can already be remarked for exposure times longer than about 1 s.

As an example, let us consider the standard situation, which requires an average exposure of $H_{av} = 0.1\,\mathrm{lx \cdot s}$. For an illumination of $E = 0.1$ lx, we need an exposure time of $t_x = 1$ s, and usually here the reciprocity is still valid. If we stop the lens down by 3 EV, the illumination in the image plane is reduced by a factor of 8 and we get $E = 0.0125$ lx. In the case, that the reciprocity would be valid with $p = 1$; we would need an exposure time that is longer by the same factor, thus $t_x = 8$ s. In the case of the Schwarzschild effect, however, assuming a typical value of $p = 0.8$, we can calculate $t_x = 13$ s after Equation (4.2). Here, a significant deviation from the conventional calculation is necessary. Another example is the recommendation of, e. g., Kodak, namely that for its KODACHROME 64 a correction of +1 EV for an exposure of $t_x = 1$ s should be applied, and similarly, the one, e. g., by Agfa for Agfachrome Professional (+1 EV for $t_x = 1$ s; +1.5 EV for $t_x = 10$ s).

Due to this fluence (or exposure), the film darkens and thus changes its transmission. For black-and-white film (these are "negatives"), the exposed areas becomes more darkened the more light is incident (this is a negative process; the corresponding tonal relation is expressed by a density curve; this response curve is not straightforward and is the subject of Section 4.8.5). As a result, according to the local light intensity distribution of the image that is projected onto the film, the image is recorded as a 2D distribution of gray tones, each following the tonal curve. Those gray tones could be measured by a microdensitometer as a local variation of the film transmission $T_{\mathrm{film}}(x, y)$ and then expressed by the negative of its logarithm, namely by the optical density

$$OD(x, y) = \lg\left(\frac{1}{T_{\mathrm{film}}(x, y)}\right) = -\lg(T_{\mathrm{film}}(x, y)). \tag{4.4}$$

4.1.4 Photographic reversal films and color films

4.1.4.1 Reversal films

As described above, the basic photographic process using silver halide film generates a negative image. For some applications, this is sufficient, like in the case of X-ray imaging for medical purposes. However, if positive images are required, a copy process must be done, which is the negative image of the first negative image. This is the standard procedure when positive photo prints on paper are generated. The principle of the second imaging process is very similar to the first one, although it is optimized to the type of photo paper material. The second negative imaging process can be avoided if a reversal photographic film is used for exposure. In this way, positive images on a transparent film base are directly produced by a modified process chain. This is required for applications such as the direct projection of motion picture films and transparent slides; the latter are also termed diapositives.

The overall process chain for the reversal film comprises roughly six steps. The two steps in the beginning are nearly identical to the steps presented in Figure 4.3b for the conventional monochrome negative imaging. Also, in the reversal film, the primary photographic process is the first exposure of silver halide crystals in the emulsion yielding the nucleation center as the latent negative image. In the 2nd step, the latent image is developed, the exposed silver halide crystals are chemically modified leading to a destruction of the exposed crystals and a deposition of metallic silver atoms. The nonexposed silver halide crystals remain nearly intact. In the 3rd step, unlike for the negative film, the negative metallic silver image is chemically removed. This removal of the dark silver particles is also called bleaching. The unexposed silver halide particles remain intact and constitute a latent positive image. In the subsequent 4th step, the remaining silver halide crystals are homogeneously exposed by diffuse lamp illumination in order to render them sensitized for the 5th step. In this step, the second development of the latent image leads to a deposit of dark silver atoms at locations that were not exposed during the primary exposure. Here, we get a real positive image showing a tonal gradation corresponding to the primary illumination. During the 6th step, a final fixing, eventually all remaining unneeded components are washed out of the film layer to render it durable.

4.1.4.2 Color negative and color slide films

Figure 4.6a depicts the typical structure of a color negative film. The top supercoat layer protects the film against mechanical damage and the UV-stop filter layer blocks short wavelengths below the visible blue range. The first light-sensitive layer in the color films is the silver halide layer, which is only sensitive to blue light. Additionally, this layer contains color couplers, which generate yellow dyes after development, the complementary color to blue. The next sequential layer is a yellow filter that only allows to pass yellow and blocks blue. This is equivalent to being transparent for green and red colors. Thus, green light can expose the green-sensitive layer, which additionally contains color couplers that generate magenta dyes after development. The next sequential filter layer is that for red light transmittance, which blocks the remaining green light. The silver halide in the red sensitive layer is only activated by red color light. The additional color couplers generate cyan dyes after development. It should be mentioned that the silver halides in the green and red sensitive layers are specially sensitized to the corresponding colors by chemical additives. As in the case of monochrome films, the lower layers are the anti-halo layer and the relative thick film base.

After the primary exposure as the first step, the silver halide particles are exposed in the layers correspondingly to their sensitivity to light. This is depicted in Figure 4.6b. For instance, white light exposes silver halide in all three light-sensitive layers and generates nucleation centers there, whereas blue, green and red light generates nucleation centers only in the correspondingly sensitive layer. In unexposed areas, there is no change. A latent image is inscribed in all three layers but is not visible. Visibility is achieved by the development in the second step. The nucleation centers from the primary exposure are now fully developed and metallic silver particles are deposited in the layers. Simultane-

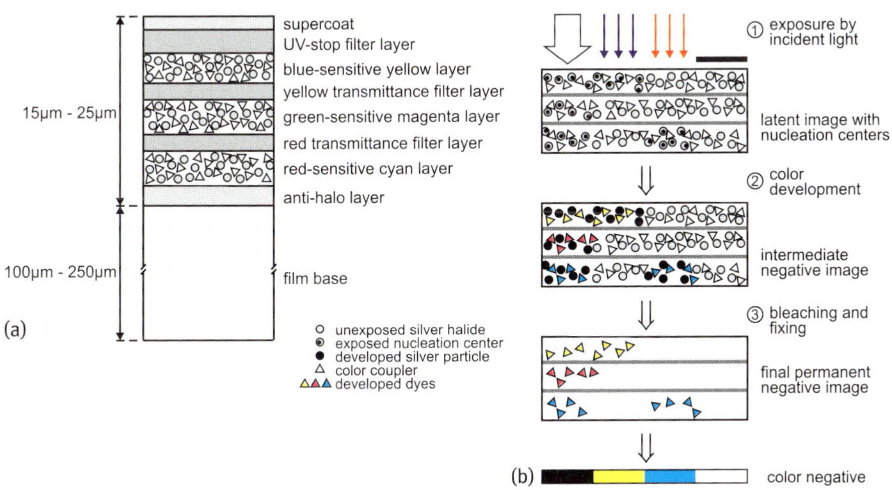

Fig. 4.6: Modern color negative film. (a) Schematic layer structure; (b) basic process steps exposure, development and bleaching to achieve the final negative image consisting of complementary colors.

ously, the couplers in the layers attached to the developed silver atoms are activated to generate the dyes in the corresponding layers. The generated dyes exhibit the complementary color of the color to which the layer is sensitive. After step 2, the intermediate image is real and consists of the complementary colors of the original incident light. As a high-dye density in a layer implies low transmittance, we have now a negative image as well for the color and also for the optical density in the layer. Because the silver atoms have no more color selective function and only attenuate the transiting light, they are washed out in a last step in parallel to fixing the permanent negative image in the film. The color negative consists of three sequential transmittance filters for the primary colors yellow, magenta and cyan. The higher their optical density is, the higher is their absorption and the less is the brightness of the light passing through the film. As can be seen from Figure 4.6, a strong white light exposure leads to black color impression, whereas blue, green and red light of low intensity leads to high transmission of their corresponding complementary colors.

The generation of a positive print is achieved by applying a second negative imaging process, similar to the first one, using color sensitive photographic negative paper to invert the first negative image. It turned out, however, that the spectral characteristics of the color dyes in the cyan, magenta and yellow layers are not perfectly complementary to red, green and blue. As can be seen from Figure 4.2d and Figure 4.2e, the magenta layer absorbs not only its central wavelength but also in the spectral range of the yellow filter, which means incident blue is more absorbed. Similarly, the cyan layer has non-negligible absorption in the ranges of the magenta and yellow filter outside its center, which affects the incident green colors. These color faults can be compensated by additional filtering in the complementary spectral ranges. This is achieved by masking the processed negative film homogeneously using a slight red and yellow dyeing, which can be seen as an orange hue of the negative film (see Figure 4.7). By this method, a better color reproduction is assured for the negative to positive copy process on color paper, where the same dyes are activated like in the film development. If additional color faults are obvious, they can be filtered out during the copy process by the corresponding color filters.

Analogously to monochrome films, a positive image can be achieved in the color film itself by additional process steps for reversal color films. After the primary exposure, only the negative silver image is developed with a special developer that only modifies the exposed silver grains without activation of the color couplers. After a second diffuse white light exposure, the remaining, previously nonexposed silver halide crystals are developed using a developer that simultaneously activates the color dyes located with them. In this way, a positive color image is generated. In a last bleaching step, all silver particles are washed out. The remaining filter layers of cyan, magenta and yellow constitute a positive color image based on the subtractive composition of the filter colors as shown in Figure 4.2. In the resulting positive image, color faults due to the nonperfect filter characteristics of the dye layers become less obvious than with the negative films and the subsequent copy process. Thus, a masking is not necessary and would also be

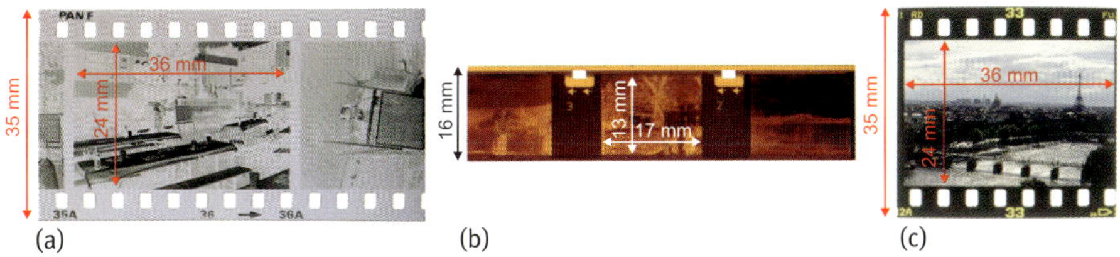

Fig. 4.7: Different processed film images on different formats. (a) Black-and-white panchromatic negative film strip, 35 mm format, with perforation parallel to the images for film transport; (b) negative color film strip, cartridge based film format 110, exhibiting an orange color masking; the perforation for film transport is between the images; pre-exposed frame lines and numbers can be seen on the strip; (c) positive color slide on a color reversal film for the 35 mm format.

detrimental. Figure 4.2 shows how the colors of the filter layers of a modern reversal-processed color slide film are perceived by a viewer under certain defined standard lighting conditions (Figure 4.2d). The spectral sensitivity of the layers is given for an unprocessed film also defined for standard conditions (Figure 4.2e). It is adapted to reproduce a natural color impression for the human visual perception.

4.2 Electronic sensors: photodiode arrays

4.2.1 Optoelectronic principles of a photodiode

Prior to a more detailed description in the following chapters, the basic principle of modern electronic image sensors, such as CCD and CMOS types are briefly discussed. Such sensors are based on a 2D array of photodiodes (PDA). Although there are differences, in particular as for the readout, the principle of light detection and charge storage is the same.

Photodiodes are light sensitive semiconductor diodes that due to the internal photo effect convert the energy of light into that of electrical charge carriers. In the ideal photodiode, one incident light photon of sufficient energy generates one electron-hole pair. One can also say that the optical power, which can be expressed as the number of photons per time interval, produces a proportional number of charge carriers per time interval that constitute a photo current. The photo current in a standard photo detector application is usually monitored as a voltage drop across a resistor with the voltage being directly proportional to the momentary incident optical power. However, in image sensors, the photodiode is typically operated in the integration mode over a certain exposure time and outputs a voltage proportional to the accumulated number of electrons in a capacitive storage well. The conversion from the optical to the electronic domain by the photodiode element in image sensors is predominantly based on the properties of a semiconductor pn-junction whereas the subsequent electronic signal treatment differs for technologies like CCD or CMOS sensors. Usually, all image sensors are realized on a silicon substrate. The reason is that this material features nearly ideal photon conversion efficiency for visible light and the integration with electronic components is very advanced.

Silicon is a main group IV element of the chemical periodic table. A pure Si-crystal is nearly an isolator at ambient temperature, which means that the valence band of the crystal is nearly fully occupied by electrons and the conduction band is nearly empty. Doping the crystal with electron donors, for instance by implanting the main group V element phosphorous, leads to n-type impurities that fill freely moving electrons to the conduction band. A p-type doping is achieved by electron acceptors, like the main group III element boron in Si, that trap electrons from the conduction band, which is equivalent to "filling holes" to this band. Both majority carriers, the holes in the p-doped crystal as well as the electrons in the n-doped crystal contribute to the overall electric conductivity. In a photodiode, the p-doped section is in contact with the n-doped side, and thus establishes a pn-junction, which is the cause for the typical diode characteristics.

Figure 4.8 illustrates the diode characteristics for a pn-junction in thermodynamic equilibrium (a) as well as in the nonequilibrium state with reverse bias in integration mode (b). The upper parts of the figure depict the energetic band structures of electrons as a function of the depth z in the diode structure. The lower parts show cross-sections of the diode. The bandgap W_g is the energetic difference between the conduction band W_c and the valence band W_v and only depends on the material. It is nearly independent from the doping, and thus is the same on the p- as well as on the n-side as the diode is made of a homogeneous Si-crystal. The Fermi-level W_F represents the chemical potential in the diode and is continuous all over when the diode is in equilibrium state. On the n-doped side, the conduction band W_c is closer to the Fermi level than the valence band. The asymmetric position increases with the dopant concentration. On the p-doped side, we have the inverse situation. Here, the valence band W_v is closer to the Fermi level. In undoped material, the Fermi level is symmetrically located in the center of the bandgap. As a consequence, we get a distorted band structure around the junction from the n-side to the p-side. The physical cause for the distortion is the diffusion of the mobile majority carriers near the junction, each to their oppositely doped side. As the remaining immobile dopant ions establish a counteractive Coulomb attraction when the mobile charges depart, diffusion is driven by the concentration gradient and is only possible over short distances of the order of 1 µm. For instance, electrons diffuse to the p-side, holes to the n-side and recombine in the area where both carrier types come in contact. Due to the recombination we get a zone that is free of mobile carriers around the junction and which is termed the depletion layer. However, the remaining immobile ions in the crystal lattice, the positive ions on the n-side and the negative ions on the p-side, establish an electrical field in between both sides directed from n to p just like the charges on isolated capacitor plates. This situation is shown in Figure 4.8a for the diode in the equilibrium state. Due to the electric field in the depletion layer and taking into account its direction, the energy of an electron on the n-side is lower than that on the p-side. This means an energy $W = e \cdot U_{bi}$ is required to move an electron

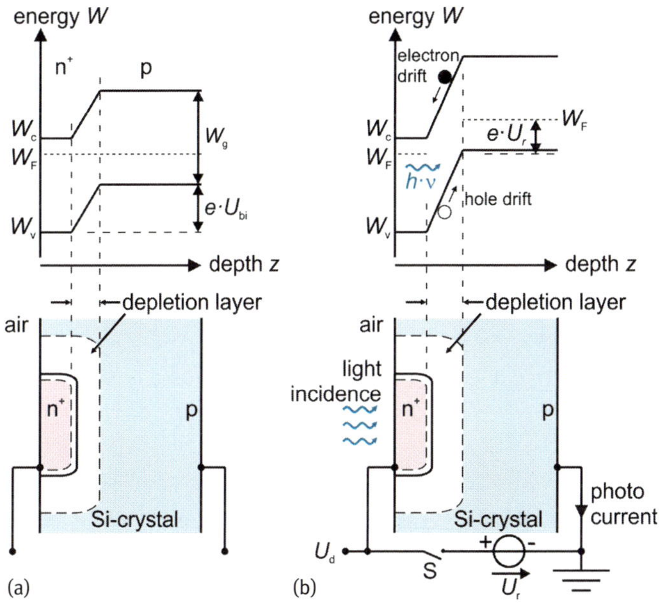

Fig. 4.8: Energy band diagram (above) and cross-sectional view (below) of a pn-photodiode. (a) Photodiode in thermodynamic equilibrium; (b) illuminated junction with reverse bias U_r in integration mode.

in the conduction band from n to p against the direction of the electric field. Here, U_{bi} is the resulting built-in voltage due to the diffusion and depends on the material, the concentration of the dopants as well as on the temperature. For Si with a dopant concentration of about 10^{16} cm^3, we get roughly $U_{bi} \approx 0.6$ V. Conversely, an electron in the conduction band located in the depletion layer is accelerated to the n-side with the lower energy level. Increasing the dopant concentration not only results in a higher diffusion voltage and a stronger electric field in the depletion layer, but an asymmetric dopant concentration shifts the location of the depletion layer toward the side with the lower concentration. Figure 4.8 shows an example for a very high-donor concentration, and consequently a high number of electrons in the conduction band, indicated by n^+. On the p-side, there is a “normal” acceptor concentration. This asymmetry causes the depletion layer to be located nearly entirely on the p-side.

When light hits the surface of the photodiode, it is absorbed in the semiconductor material if the quantum energy $h \cdot \nu$ of a photon with frequency ν is larger than the bandgap W_g of the semiconductor. This relation can be expressed as a function of the light wavelength λ taking into account the speed of light c (this is related to ν via $c = \lambda \cdot \nu$):

$$W_{ph} = h \cdot \nu = \frac{h \cdot c}{\lambda} > W_g \tag{4.5}$$

Thus, only wavelengths smaller than the cut-off wavelength λ_{co} can be converted by a photodiode with a bandgap W_g given by

$$\lambda_{co} = \frac{h \cdot c}{W_g} \approx \frac{1.24\,\text{eV} \cdot \mu\text{m}}{W_g} \tag{4.6}$$

$$(1\,\text{eV} = 1.6 \cdot 10^{-19}\,\text{C} \cdot \text{Nm/C} = 1.6 \cdot 10^{-19}\,\text{J}).$$

For a Si-based photodiode with $W_g = 1.12$ eV, we get a cut-off wavelength of $\lambda_{co} = 1.11\,\mu$m, making the diode appropriate for absorption of the visible light spectrum. The absorbed photon transfers its energy to an electron in the conduction band, and thus a transition to the valence band is made. This is equivalent to the generation of a hole in the valence band and an electron in the conduction band. Both, electron and hole are generated at the same location within the diode. If the absorption takes place in the area of the depletion layer, the electron-hole pair is immediately separated by the electric field. The electron drifts to the n-side and the hole to the p-side, giving rise to a photocurrent pulse. In order to make the consideration simple, for the moment let us assume that all photons are absorbed in the depletion layer, and we have an ideal detector meaning that each photon generates one electron hole pair. Deviations from that ideal case will be discussed later. Then the generated photocurrent is proportional to the number of photons within the incident radiant flux, which is the optical power. In order to improve the linear characteristics of the photodiode and to reduce saturation effects, the photodiode is operated with reverse bias U_r and the switch S closed in Figure 4.8b. Usually, the voltage drop of the photo current across a resistor is measured, which however, in the consideration for the integration mode is not relevant.

In the case of no illumination, there is no current at all in the circuit since the depletion layer, due to the lack of mobile charges, and represents an isolating layer. The voltage drop across the depletion layer, which can be represented as a capacitor, is identical to the applied bias voltage U_r. The corresponding capacitance of the depletion layer is also termed junction capacitance C_j. Its exact calculation is quite complex, but for our simple consideration it is sufficient to state that C_j is proportional to the cross-sectional area of the diode perpendicular to the light incidence, namely the pixel area, and inversely proportional to the depth of the depletion layer.

For the integration mode of the diode operation on the image sensor, the diode is reset in darkness by closing the switch S. Hence, the depletion layer is emptied of mobile carriers and the n^+-side is set to voltage U_r while the p-side is at the ground. Then the switch S is opened and the illumination of the diode can start. If light enters the diode, the generated carriers drift to the corresponding sides and remain on the "capacitor plates" as the external circuit is cut by the open switch. From the electric point of view, the charges constitute a photo current, which discharges the capacitor C_j and reduces the overall reset voltage U_r. When the illumination of the diode is stopped, the resulting voltage of the diode U_d can be measured. The voltage difference $U_r - U_d$ increases with the accumulated charges on C_j during the integration, respectively, exposure time, and thus is a measure for the total exposure (see Section 4.2.2). While the holes flow to the p-side which is grounded, the electrons gather on the n^+-side that can be considered as storage well for electrons. It should be mentioned here that the depth of the depletion layer increases with the applied reverse bias voltage U_r as more charges are taken out of the zone. This is indicated in Figure 4.8b where the depth of the depletion layer is somewhat larger than for the diode in the equilibrium state. As a consequence, the value of C_j can be controlled to a certain amount by setting U_r. The larger C_j is, the more electrons can gather in the storage well for the same voltage difference, which may be exploited for practical operations of image sensors (see Section 4.10.6). The dependence of C_j on the voltage, however, is one of the drawbacks for CMOS sensors where the voltage difference $U_r - U_d$ is evaluated. If the capacitance was a fixed value, this voltage difference would be directly proportional to the amount of accumulated charges, yielding a linear relationship between voltage and charges, respectively, exposure. But as C_j increases with decreasing U_d during integration, we get a nonlinear relationship, and thus nonlinear voltage-exposure characteristics for CMOS sensors (see also Section 4.8.5 and, in particular, Figure 4.46).

Conversely, for CCD sensors, it is not the voltage at the diode itself that is evaluated. The accumulated charges, which are proportional to the exposure, are transferred in steps after the exposure to storage cells where the measured voltage is directly proportional to the number of charges. Hence, the voltage at the final analog-to-digital conversion unit yields a linear relationship with the exposure (Figure 4.46).

The above simplified description must be detailed for better understanding of real sensors. Therefore, we focus on the absorption of photons in the semiconductor. The absorption is usually described by Beer's law for absorption along the propagation depth z:

$$\frac{\Phi(z)}{\Phi(z=0)} = \frac{N_{ph}(z)}{N_{ph}(z=0)} = e^{-\alpha_p \cdot z} \tag{4.7}$$

where α_p is the absorption coefficient. Here, $\Phi(z)$ is the optical power in the diode at depth z while $\Phi(z=0)$ is the radiant flux penetrating to the Si material.

Here, we disregard the reflection loss at the semiconductor air interface. For uncoated Si, this loss is typically of the order of 30 % and, therefore, an antireflection treatment is required. The topic of antireflection treatment is discussed in Section 6.8 for complex lens arrangements. The light reflection at the air-sensor interface of a complete image sensor is assumed to be of about 5 %. It follows from Equation (4.7) that the ratio of the optical power in the diode relative to the initial value at the surface is equal to the ratio of the corresponding photon numbers N_{ph} in the light flux. The number of absorbed photons is given by the difference $N_{ph}(z) - N_{ph}(z=0)$. In an ideal diode with a quantum efficiency (QE) of 1, this is converted to the same number of electron-hole pairs, whereas in real diodes the number of charge pairs is smaller, yielding efficiencies below 1 (see the next section). From this consideration, it becomes clear that the total efficiency of a diode increases with increasing $\alpha_p \cdot z$. At the bottom of the diode, ideally

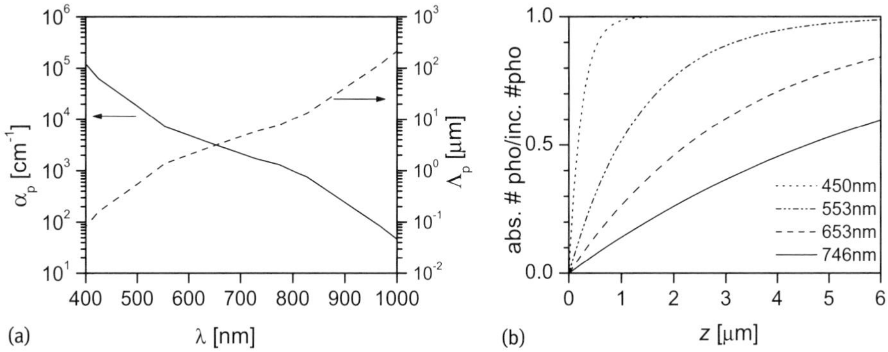

Fig. 4.9: (a) Absorption coefficient α_p and penetration depth Λ_p for bare silicon (note that there is a difference to doped Silicon). (b) Relative number of absorbed photons. Absorption data taken from Palik.[2] Note that losses due to reflection from the surface are not taken into account. Moreover, additional transmission losses may occur in the Si-layer in front of the active diode region and this adds to all additional filters in front of the PDA and this decreases $T_F(\lambda)$ (i. e., this is an additional factor in the product of all involved transmission factors; see Section 4.2.2, Section 4.6 and Section 4.10).

all photons should have been absorbed and converted, which requires a large value of $\alpha_p \cdot z$. Figure 4.9a shows the power absorption coefficient of pure Si as a function of the wavelength. Its reciprocal value $1/\alpha_p$ is defined as the penetration depth Λ_p of the radiation in the substrate, which is the depth where the radiation has dropped to $1/e \approx 37\,\%$ of its initial value at the surface. It can be seen that the absorption coefficient in Si is highest for the shortest wavelengths with a penetration depth of less than 0.1 μm around $\lambda = 400$ nm, whereas at $\lambda = 700$ nm the penetration depth is of some μm below 10 μm. The near-infrared spectrum around $\lambda = 1$ μm is relatively weakly absorbed, having a penetration depth of some 100 μm. Figure 4.9b shows the number of absorbed photons relative to the number of incident photons at different wavelengths as a function of the depth z. Blue light photons of $\lambda = 450$ nm are nearly completely absorbed after 1 μm penetration, whereas red photons of $\lambda = 653$ nm require more than 6 μm to achieve an absorption rate of 85 % (here we would like to note that, of course, a mixture of photons and wavelength is not well described in physics terms, however, it is clear what is meant and this makes discussion easier). Consequently, a total depth between 10 μm to 20 μm is sufficient for a photodiode in an image sensor for visible light whereas diodes in solar cells also exploiting the near IR range have layers of more than 200 μm depths.

As the photons in the blue wavelength range are absorbed already at very short distances, there is a risk that the generated electron-hole pairs have not yet arrived in the depletion layer where they are separated by the high-electric field in the drift zone and contribute to the overall photocurrent. Thus, the n^+-zone should be relatively shallow and of high-crystal purity in order to prevent the electron-hole pairs from recombining

2 E. D. Palik: *Handbook of optical constants of solids*, Academic Press, San Diego, 1985.

before being separated. Likewise, photons in the red spectral range may be absorbed beyond the depletion layer and the generated electron-hole pairs have to diffuse backward to the drift zone to contribute to the photocurrent. The separation of electron-hole pairs at large depth is improved by an additional p^+-zone behind the p-layer in some diode designs. In this way, an additional weak depletion layer is established at larger depths. A high-quality crystal structure in the critical zones with carrier diffusion is achieved by epitaxial layer ("epilayer") growth of the n^+, respectively, p-zones on a Si wafer substrate. In layers of lower quality with more lattice defects, electron-hole pairs may be trapped by the defects and recombine. An early recombination of the electron-hole pairs before separation impairs the overall quantum efficiency.

According to the described principles and due to the spatial light intensity distribution over the sensor surface, charges are generated and collected, namely integrated, within a 2D photodiode array (PDA; Figure 4.14). This stored pattern, which corresponds to the grayscale distribution of the image, must be readout. The method to do so is different for CCD and CMOS sensors (and there are slight differences in the semiconductor structure as well, in particular, the thickness of the epilayer). The photodiodes themselves usually have a rectangular shape on the surface, maybe quadratic, polygonal or have a different particular shape, especially when used for CMOS sensors. Typical lateral extensions, the "pixel widths," range from approximately 1 µm to more than 20 µm.

4.2.2 Charge detection and conversion

Let us now consider a single pixel that is illuminated at its local position with an intensity I_{pix} [W/cm^2]. Thus, one may calculate the corresponding power P_{pix}, fluence F_{pix}, and energy W_{pix}, on that pixel

$$P_{pix} = \int_{A_{pix}} I_{pix}\, dA \tag{4.8a}$$

$$F_{pix} = \int_{t_x} I_{pix}\, dt \tag{4.8b}$$

$$W_{pix} = \int_{t_x} \int_{A_{pix}} I_{pix}\, dA\, dt \tag{4.8c}$$

where A_{pix} is the pixel area (typically in [cm^2]) and t_x the time interval during that the pixel is illuminated, i. e., the exposure time. W_{pix} and P_{pix} are given by the radiant energy Q_e and the radiant flux Φ_e, respectively, but here related to a single pixel. Most often, all these integrals simply transform to products such as $W_{pix} = I_{pix} \cdot A_{pix} \cdot t_x$. Conversely, the average photometric quantities on the pixel, namely intensity (irradiance) and fluence (radiant exposure) and the corresponding radiometric quantities illuminance and luminous exposure are given by

$$\overline{I}_{\text{pix}} = \overline{E}_{\text{e,pix}} = \frac{W_{\text{pix}}}{A_{\text{pix}} \cdot t_{\text{x}}} = \frac{P_{\text{pix}}}{A_{\text{pix}}} \tag{4.8d}$$

$$\overline{F}_{\text{pix}} = \overline{H}_{\text{e,pix}} = \frac{W_{\text{pix}}}{A_{\text{pix}}} \tag{4.8e}$$

$$\overline{E}_{\text{pix}} = \frac{Q_{\text{pix}}}{A_{\text{pix}} \cdot t_{\text{x}}} = \frac{\Phi_{\text{pix}}}{A_{\text{pix}}} \tag{4.8f}$$

$$\overline{H}_{\text{pix}} = \frac{Q_{\text{pix}}}{A_{\text{pix}}} \tag{4.8g}$$

where Q_{pix} and Φ_{pix} are the luminous energy and flux, respectively (the related equations with the integrals are straightforward; here we skip the index “v”). Equations (4.8) describe the ideal case, namely if all photons incident on the pixel would reach that region within the semiconductor where they could be converted to photo electrons. However, as we will see later in this chapter, there are losses. This is because part of the incident photons are absorbed or reflected by the upper structures above the photodiode, which is described by a wavelength dependent transmission function $T_{\text{F}}(\lambda)$, which also includes losses in additional filters such as an IR filter (e. g., Figure 4.25c) or color filters (see Section 4.6.3). Note that the front glass may also have to be included, but usually transmission is close to 100 % in the visible with a sharp cutoff for wavelengths below approximately 350 nm. Moreover, there is an effective fill factor η_{g}, which describes geometrical losses due to the photodiode structure (see Section 4.6.1 and Equation (4.19)) so that the intensity I_{pix} is reduced to

$$I'_{\text{pix}} = T_{\text{F}}(\lambda) \cdot \eta_{\text{g}} \cdot I_{\text{pix}} \tag{4.9}$$

where all quantities with an apostrophe should indicate the related values when such losses are included (part of the losses may be compensated by an optical microlens array (OMA; see Section 4.6); note that reflectivity losses are disregarded here). Then the total number of incident photons on this pixel (accumulated during the exposure time) is given by

$$N_{\text{ph}} = \frac{W_{\text{pix}}}{W_{\text{ph}}} \tag{4.10a}$$

$$N'_{\text{ph}} = \frac{W'_{\text{pix}}}{W_{\text{ph}}}. \tag{4.10b}$$

Now N'_{ph} photons are available to be converted into photoelectrons with an efficiency η_{i} given by the internal quantum efficiency (IQE or energy related quantum yield, or

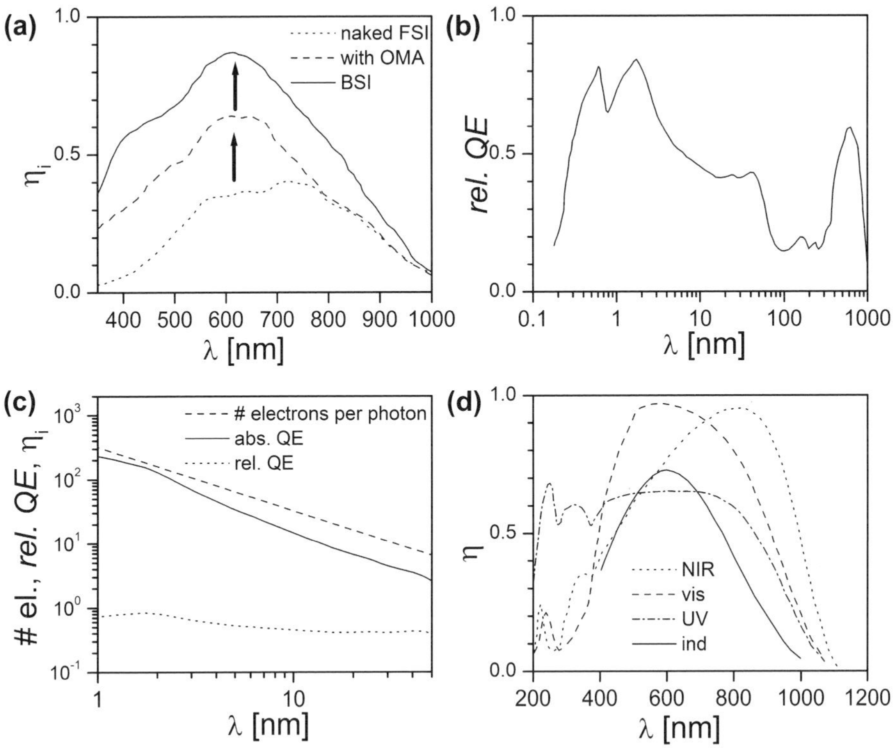

Fig. 4.10: (a) Scheme of the typical IQE of a CCD (or CMOS) monochrome sensor ranging from UV to near IR (for a color sensor, see Figure 4.24). Improvements (see text) are indicated by the arrows. (b) Relative IQE of a scientific BSI-CCD with $\eta_g = 100\,\%$ (relative IQE means that this is the probability that the photon is detected; the effective quantum efficiency is discussed in Section 4.8.7). The sensor is sensitive over a huge wavelength range. (c) Short wavelength region of the same sensor as in (b). *If* an X-ray photon is absorbed, *then* it generates a large amount of electrons. This is included in absolute IQE (e. g., at 1 keV, the relative QE $\approx$ 85 % and approximately 300 electrons are generated per photon, so thus the absolute QE is 255). (d) Typical quantum efficiency for optimized scientific sensors. "NIR": deep depletion CCD with NIR AR coating (see Section 4.10.8), "vis": BSI sensor for the visible range, "UV": BSI sensor for UV-vis region (for BSI see Section 4.10.4). "ind" indicates a CMOS sensor for industrial applications. Note that there is a change of maximum and a shift of the η-curve with temperature, which in particular, has taken into account when cameras are cooled, e. g., within scientific experiments.

sometimes termed as charge collection efficiency). IQE is a measure of how many electrons are generated per photon within the conversion region and usually is smaller than one. Figure 4.10 and Appendix A.5 provide some examples. In general, the related IQE(λ) curve and its maximum is specific to the semiconductor material and sensor/pixel design (see, e. g., Section 4.10). The external (or overall) quantum efficiency (EQE; see below) additionally includes wavelength dependent losses, e. g., due to $T_F(\lambda)$, and thus usually EQE(λ) is different from IQE(λ).

The quantum efficiency curve for different sensors may be quite different. This depends on details of the semiconductor, AR coatings, additional or absent OMA (see Section 4.6) and so on. Sensor chips may also be optimized, e. g., for the near-infrared region, for the extreme ultraviolet (XUV or EUV) or the X-ray range, respectively (see also Section 4.10 and Section 4.11). Due to the improvements made in the last decade (e. g., by usage of transparent electrodes, OMA with improved materials and improved design), today very efficient sensors are available. Special improvements have also been made by the development of back side illumination (BSI, see Section 4.10.4) technology, which originally was used for scientific detectors only but today is also applied for standard sensors.

We would like to remark that the spectral sensitivity curve of a CCD or a CMOS sensor is much broader than that of the human eye (compare Figure 4.2b, but attention, there the ordinate is on a logarithmic scale; compare also Figure 4.24), and thus the detector is sensitive in near-infrared and UV, respectively (unless the light is blocked by a window or a filter). As a consequence, if one is interested in images that include only that part of the spectrum that is also seen by the eye (namely the visible region), then the unwanted spectral components have to be removed. Usually, this is done by filters (see Section 4.6.3). On the other hand, one may take advantage of the sensitivity of the sensor outside the visible region, e. g., for surveillance cameras. The extended range is also useful, e. g., for technical or scientific applications that are not necessarily restricted to the visible region.

Figure 4.10b,c also shows that in the XUV or X-ray range a special situation occurs for the sensor because a single photon may generate quite a lot of electrons (but note that in the XUV-range only BSI sensors are suitable; see Section 4.10.4). Thus, the losses turn into gain, and hence IQE may be much larger than one (see Figure 4.10c,d and Section 4.8.7).

Consequently, in any case the number of photogenerated electrons generated within the photodiode (i. e., pixel) is given by

$$N_{\mathrm{pe}} = N'_{\mathrm{ph}} \cdot \eta_{\mathrm{i}} \tag{4.11a}$$

$$N_{\mathrm{pe}} = N_{\mathrm{ph}} \cdot \eta_{\mathrm{e}} \tag{4.11b}$$

where *EQE* is provided by

$$\eta_{\mathrm{e}}(\lambda) = T_{\mathrm{F}}(\lambda) \cdot \eta_{\mathrm{g}} \cdot \eta_{\mathrm{i}}(\lambda). \tag{4.12}$$

This includes all losses, and thus relates N_{pe} to the real number of incident photons prior to losses N_{ph} (see Figure 4.10b and Figure 4.24). Thus, it is clear that $\eta_{\mathrm{i}} > \eta_{\mathrm{e}}$ (note that a factor (1–surface-reflectivity) has to be included on the right-hand side in Equation (4.12) if reflection losses (see above) cannot be ignored). Vice versa, Equations (4.11.) may be regarded as definitions of the quantum efficiency.

The generated photoelectrons lead to a charge $e \cdot N_{\mathrm{pe}}$ and, therefore, to a photo current $I_{\mathrm{pe}} = e \cdot \dot{N}_{\mathrm{pe}}$ and after all a current density (per pixel)

$$j_{pe} = e \cdot \eta_e \cdot \frac{I_{pix}}{W_{ph}} = R_{pix} \cdot I_{pix}. \tag{4.13}$$

The photoelectron current density is directly proportional to the incident light intensity (prior to losses) with a wavelength dependent proportional constant

$$R_{pix}(\lambda) = \frac{e \cdot \lambda}{h \cdot c} \cdot \eta_e(\lambda) \tag{4.14}$$

called responsivity. The responsivity is the ratio of the photocurrent to the optical input power (in A/W) or equivalent, the amount of charges generated per photon with a given photon energy (or wavelength; in C/J). Typical values are between 0.1 and 0.5A/W (due to the only small dependence on λ, the shape of the R_{pix}-curve is not too different from that of η_e).

Integration of the photo current density over the pixel area and the exposure time, finally yields the signal charge generated per pixel

$$q_{pix} = \int_{A_{pix}} \int_{t_x} j_{pe} \, dt \, dA. \tag{4.15a}$$

This accumulated charge within each pixel (of course, $q_{pix} = e \cdot \eta_e \cdot N_{ph}$) leads to a potential change q_{pix}/C_{pix} where C_{pix} is the capacitance of the photodiode or pixel (i. e., the capacitance connected to the potential well). As a result, this charge is detected at the output amplifier (for CCD this is common for all pixels, for CMOS each pixel has an individual one; see also the previous section), and thus generates a voltage that is further amplified by a factor G_a, which is the amplifier gain

$$U_{out} = G_a \cdot \frac{q_{pix}}{C_{pix}} = G_a \cdot G_i \cdot N_{pe} = G_a \cdot G_i \cdot \eta_e \cdot N_{ph}. \tag{4.15b}$$

Here, we would like to remark that the input referred conversion gain $G_i = e/C_{pix}$ [µV/electron] is given for a single charge. The output referred conversion gain is given by $G_{out} = G_a \cdot G_i$, and thus is just the proportional constant between U_{out} and N_{pe}.

The amplified signal as a function of input signal (N_{ph}, I_{pix}) defines the response curve of the detector (see also Section 4.8.5). As long as the gain does not depend on the input, electronic detectors have a linear response up to the saturation value.

Of course, there is a maximum of charge $q_{full} = e \cdot N_{full}$ that could be accumulated within the potential well of a particular photodiode (N_{full} is the corresponding maximum number of elementary charges). This saturation value is called full well capacity (FWC). q_{full} is related to the initial and maximum voltages U_{reset} and U_{max} at the photodiode. It depends on the photodiode architecture including the layer structure and well depth and it also depends on the operation conditions:

$$e \cdot N_{\text{full}} = \int_{U_{\text{rest}}}^{U_{\max}} C_{\text{pix}}(U) \cdot \mathrm{d}U \tag{4.16}$$

N_{full} may be rather low for cheap cameras, higher for digital single lens reflex cameras (DSLR, or mirrorless ones, DSLM; see Chapter 2) and very high for scientific cameras (typically several times 10^3 up to $> 10^6$ electrons; see Table A.3 in Appendix A.5).

Straightforwardly, from Equation (4.11b), the corresponding photon number to reach FWC is given by

$$N_{\text{ph,sat}} = \frac{N_{\text{full}}}{\eta_e}. \tag{4.17}$$

Figure 4.11 shows some examples of the FWC, e. g., of typical cameras used for photography ranging from (simple) compact cameras over bridge cameras to DSLM and DSLR (black squares). In addition, examples of high-end mobile phone cameras are included (triangles) and such ones used for technical and scientific purposes, respectively (open circles). The stars mark the FWC of much advanced scientific cameras. But we would like to remark that although such cameras are used for imaging as well, they are not very suitable for photography (in particular, the camera with the star in the upper right corner is a camera made for digital radiography).

Due to the huge amount of different sensors available on the market, the displayed values represent only a very restricted selection. Some of the data can be found in Appendix A.5. As a further example, there are cameras used for scientific or industrial applications with up to 100 MP or more (not shown here) and, in particular, very special sensors are built for astronomical observations (either specially made ultralarge single sensor elements or sensors that consist, e. g., of a multisensor configuration such as a recent arrangement of fourteen CCD290-99-chips, which together yield a 1200 MP sensor). It also may be helpful to know that usually many of the sensors are available, both either with or without additional (Bayer) filters or microlens arrays (see Section 4.6).

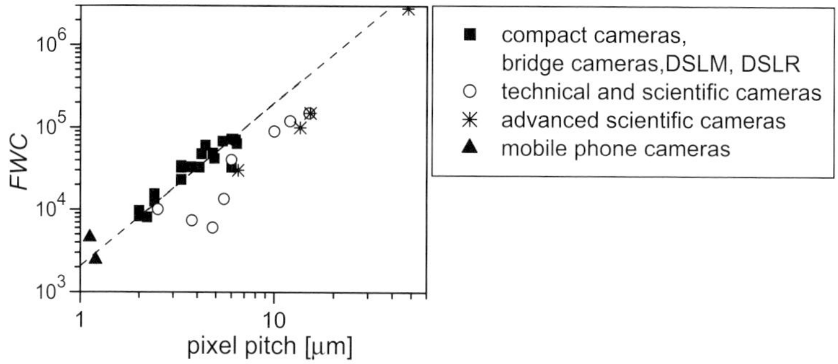

Fig. 4.11: Examples of the FWC for different cameras (see text).

The selection provides rather different pixel sizes. It may be seen that the FWC strongly increases with pixel pitch (or size; see Section 4.6.1). As we will see later in this chapter, noise, resp. SNR, is also strongly related to pixel size and so is the dynamic range (both is better for larger pixels). An interesting relation between FWC and pixel pitch may be observed as well from this figure: for most of the cameras that are used for photography and that are listed in Appendix A.5 the FWC is approximately proportional to the pixel area (and thus to the square of the pixel pitch); here, for simplicity, we assume a fill factor of $\eta_g = 1$. We may note that the provided data correspond to the physical FWC of the individual pixels, i. e., the individual photodiodes. The linear relation seems to be reasonable because FWC is related to the volume of the potential well of the photodiode. The volume itself is proportional to the cross-section of the photodiode and its depth, but its depth is limited due to the penetration depth (see Figure 4.9). Nevertheless, this is a simplified consideration, and thus the observed relation cannot be regarded as a physical law, it is just an observation and not a general rule. This can be seen as well from the figure because several of the sensor data are related to sensors used for other purposes than photography. They do not follow that relation.

4.3 Formats and sizes

4.3.1 Formats and sizes of films and digital sensors

As illustrated in Figure 4.12a, the size of a sensor is an important issue. The large open arrow should illustrate an object. Here, for simplicity, a simple lens within an aperture should indicate a more complicated lens system. Depending on the sensor size, more or less from the scenery is captured. For instance, the image field of an APS-C sensor is smaller than that of a full format sensor. If the object is too large for the image to fit on the sensor (e. g., if the open arrow is replaced by the broken line arrow), the camera has either to be tilted and shifted or if the orientation of the camera is kept constant, the sensor has to be shifted downwards (or the lens upwards). This can be done with a (tilt-)shift objective lens (an extended discussion is the subject of Section 6.7).

Figure 4.12b illustrates how much sensors of different sizes can capture. Here, it becomes clear that only a smaller fraction of a scene can be imaged when the sensor size gets smaller in the case that the optics are not replaced and adapted to the sensor. However, if the total image content of any sensor is displayed on a screen of fixed size, or if it is reproduced and printed, e. g., on a 10 cm × 15 cm paper, the absolute size of the reproduced image is the same (compare Figure 4.12c and Figure 4.12d). Thus, the image taken with the smaller sensor (e. g., an APS-C sensor, Figure 4.12d) looks like a magnified version of that taken with the full format sensor. It seems as if it were taken with a lens having a narrower angle of view and larger relative magnification compared to the full format lens (see Sections 2.2 and 2.6). Indeed, if one neglects image quality issues of the lenses and also different pixel sizes of both sensors, a full format sensor together with a

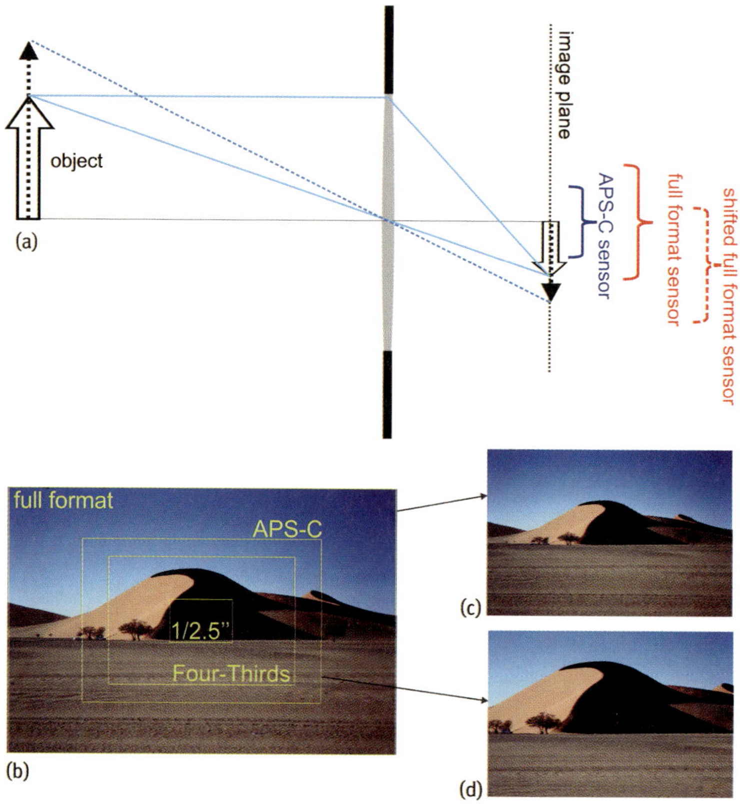

Fig. 4.12: Simple illustration of imaging with different sensor sizes (see text).

lens having a longer focal length yields the same image as a smaller sensor with shorter focal length. The ratio of both focal lengths is identical to the crop factor *CF*, which is the ratio of both sensor diagonals according to Equation (2.25).

We would like to note that within the present example the crop has the same aspect ratio (i. e., *PW*:*PH*), but there are other formats, e. g., "Four Thirds" (see below), where the aspect ratio changes. As a consequence, when displayed on a screen or if printed, usually *PH* is the same, but *PW* is different when images captured with both sensors are compared. Consequently, *PH* may be regarded the better value when image quality should be compared for different sensor sizes.

Here, we would like to comment on the sensor size. Usually, the "sensor size" is provided in inch (see, e. g., first column in Table 4.3). We have put quotation marks because this size is not the true size of the area occupied by the pixels. The provided "sensor size" just results from the historical characterisation of vacuum tubes originally used as image converters. The sensor manufacturers still keep the approximate ratio of light sensitive area to tube diameter and provide that as the "sensor size." Thus, typically a CIS with 8 mm diagonal (i. e., 0.31 inch) is sold as 1/2 inch sensor. In this book, we do not

Tab. 4.3: Typical formats for films (gray background) and electronic sensors (CCD and CMOS; light blue background). Note that sometimes the aspect ratio is provided differently, namely as height:width.

	sensor width [mm]	sensor height [mm]	sensor diagonal [mm]	aspect ratio width:height	crop factor	sensor area rel, to full format [%]	remarks
APS-film	30.2	16.7	34.5	1.8	1.3	58	compact camera
full format (35 mm film)	**36.0**	**24.0**	43.3	**3:2**	**1.00**	**100**	**SLR etc.**
4.5×6 cm^2	56.0	41.5	69.7	approx. 4:3	0.6	269	120 frame size
6×6 cm^2, medium format	56.0	56.0	79.2	1:1	0.5	363	120 frame size
6×7 cm^2	69.0	56.0	88.9	1.2	0.5	447	120 frame size
6×8 cm^2	76.0	56.0	94.4	approx. 4:3	0.5	493	120 frame size
6×9 cm^2	89.0	56.0	105.2	approx. 16:10	0.4	577	120 frame size
1/3.2 inch	4.5	3.4	5.6	approx. 4:3	7.7	1.8	
1/2.7 inch	5.4	4.0	6.7	approx. 4:3	6.4	2.5	
	5.8	4.0	7.0	3:2	6.1	3	e. g., mobile phone
1/2.5 inch	5.8	4.3	7.2	approx. 4:3	6.0	2.9	older compact cam.
1/1.8 inch	7.2	5.4	9.0	approx. 4:3	4.8	4.5	larger compact cam.,
2/3 inch	8.8	6.6	11.0	4:3	3.9	6.7	e. g., mobile phone
	13.2	8.8	15.9	3:2	2.7	13	DSLM
4/3 inch, Four-Thirds	17.3	13.0	21.6	4:3	2.0	26	
Foveon X3	20.7	13.8	24.9	3:2	1.7	33	
Live MOS	17.3	13.0	21.6	3:2	2.0	26	Panasonic, Olympus
	18.7	14.0	23.4	3:2	1.9	30	actual compact cam.
APS-C	approx. 22.3	approx. 14.9	approx. 27.1	3:2	approx. 1.6	approx. 38	not standardized
DX	23.7	15.6	28.4	3:2	1.5	43	Nikon DSLR
APS-H	29.2	20.2	35.5	approx. 3:2	1.2	68	
full format	**36.0**	**24.0**	**43.3**	**3:2**	**1.0**	**100**	**DSLR, DSLM, etc.,**
S-format	45.0	30.0	54.1	3:2	0.8	156	Leica
medium format, M-format	48.0	36.0	60.0	4:3	0.7	200	Hasselblad, Mamiya

make use of that "sensor size." Instead, we provide the real size of the sensor diagonal, width or height, which can also be obtained directly from pixel pitch and pixel numbers in horizontal and vertical directions, respectively.

For a given lens, the size of an image that could be recorded is preset by the sensor size (see also Chapter 6). Although, in principle, there are not too many restrictions, e. g., films or photographic plates could be delivered in different dimensions, in particular, for special purposes in science and medicine, practically only a couple of sizes have been established for consumer photography. For films, these are mainly the APS-film format, the 35 mm format and the medium format (see Table 4.3). Other formats, e. g., that for pocket cameras (13 mm × 17 mm) exist as well.

The most important format for photography is the 35 mm format, sometimes called "135 film" (or 35 mm film; see Figure 4.7a and c; "most important" in the sense of some kind of standard). It was introduced by Kodak in 1934 and popularized by Leica camera. This film is 35 mm wide, but has perforations for film transport, which reduces the height available for its image (see Figure 4.7). The actual frame (of an image) is 24 mm high and 36 mm wide and called full frame or full format (we also will use mostly these terms). Consequently, its diagonal is 43.3 mm and its aspect ratio (width to height) is 3:2. For electronic detectors, the situation is a little different (details of these devices are discussed here and in the following chapters). These sensors are manufactured in specific sizes predetermined by the companies, which, in particular, are the result of what present technology allows to produce at reasonable cost. This has led to a vast number of "formats" (see Table 4.3), today including also the more expensive "full format" as one of them.

Other cameras, e. g., such as ones used for scientific or industrial purposes, are equipped with other sensors. Those sensors often have different sizes when compared to those listed in Table 4.3. The pixel sizes (width and/or height) range typically from a couple of microns to more than 25 µm. The sensors have also different sizes that typically range from a few hundred pixels to several thousand pixels within sensor width (or height). The absolute sensor size can be calculated from the pixel width (or height) and the number of pixels. The sensor height is either the same as the width, but it might be also quite different, in particular, for special applications (e. g., cameras used for spectroscopic applications often have a height that is much smaller than the width).

4.3.2 Full format and crop factor

The full format has an outstanding position. The reason for using this format also with digital sensors is manifold. Many photographers are accustomed to it, and thus often it is regarded as some kind of reference.

If the same scenery should be taken with cameras of different sensor sizes, also different lenses are required since the image on the sensor should have the same field of view. For simplicity, it is assumed here that the aspect ratios are the same for the different sensors. Let us, e. g., consider the image of an object of height S_o taken with a full format camera and its normal lens of focal length $f_{norm,FF}$. If the image with size $S_{i,FF}$ covers the full height of the sensor, we have $PH_{FF} = S_{i,FF}$. If we take a photo of the same object by a different camera with a crop sensor, and the image height $S_{i,CF}$ should likewise cover the full height of the sensor, namely $PH_{CF} = S_{i,CF}$, a lens of different focal length is required as the absolute size of the image on the sensor changes. In our consideration, we need the normal lens for this format with a focal length $f_{norm,CF}$, since the viewing perspective and the field of view should be maintained in both photographs.

As stated in Chapter 2, the feature of a normal lens for any sensor format is that it has a focal length, which is approximately identical to the diagonal of the sensor format. Moreover, the crop factor CF, sometimes also called "extension factor," for a sensor format is defined as the ratio between the diagonal of the full format sensor to the diagonal of the crop sensor, respectively, $CF = PH_{FF}/PH_{CF}$. Consequently, the ratio of the focal lengths of the normal lenses must be the same as that of the diagonals, and thus be identical to the crop factor CF. This relationship is expressed by Equation (2.25). Since many photographers are used to working with the full format, it is of interest for them to know, when working with a crop format camera and a focal length f, which is the equivalent focal length f_{eq} in the full format. This equivalent value can be calculated just by multiplication with the crop factor:

$$f_{eq} = CF \cdot f \tag{4.18}$$

Conversely, using a lens of a given focal length with different formats, the field of view decreases inversely with the crop factor. This situation is illustrated in Figure 2.24 for the comparison between the FX full format and the DX crop format with $CF = 1.5$. In that example, a 50 mm lens yields nearly the same perspective with the crop sensor as a 75 mm equivalent lens with the full format. It should be noted that the consideration with respect to crop factor, equivalent focal length and perspective is only exact when the image plane is almost in the focal point of the lens, namely for imaging of infinitely distant objects, otherwise it is an approximation. This condition is quite well fulfilled for standard photographic situations where the magnification is smaller than about 0.1.

However, although the image is quite similar in both cases, there might also be differences. In particular, depth of field and depth of focus are both directly related to the sensor size since the diameter of the allowable circle of confusion scales with the sensor diagonal (see Sections 3.4.6 and 6.9). For a narrow depth of field like in portrait photography, larger format cameras are more advantageous. They additionally have a larger depth of focus in the image space, which is very favorable as for the fabrication tolerances of large format systems. Conversely, smaller format cameras have larger depth of field, which allows for more tolerances in focusing to the object plane. But they also have a smaller depth of focus, which requires a very high-manufacturing precision and becomes more challenging the smaller the sensor is.

A very important point is the noise characteristics of a sensor, especially in low light illumination. Larger sensors usually consist of larger pixels, and thus have significant better signal-to-noise ratio SNR (see Section 4.7 and Section 4.8), which yields an improved image quality.

4.4 CCD sensors

4.4.1 Basics

A particular readout arrangement for the 2D array of photodiodes was invented in 1969/1970, namely the charge-coupled device or briefly CCD, which subsequently evolved to large success, or even more, to some kind of revolution in science, technology and even daily life. The importance of this invention showed up in 2009 when W. Boyle and G. E. Smith got the Nobel prize in physics for "the invention of an imaging semiconductor circuit—the CCD sensor." An example of a CCD sensor is shown in Figure 4.13.

The readout of a CCD bases on a charge transfer of a series of photodiodes as shown in Figure 4.14. Similar to a bucket brigade, the charge generated by the incident light is collected within the potential well of the particular diode and then is passed to the

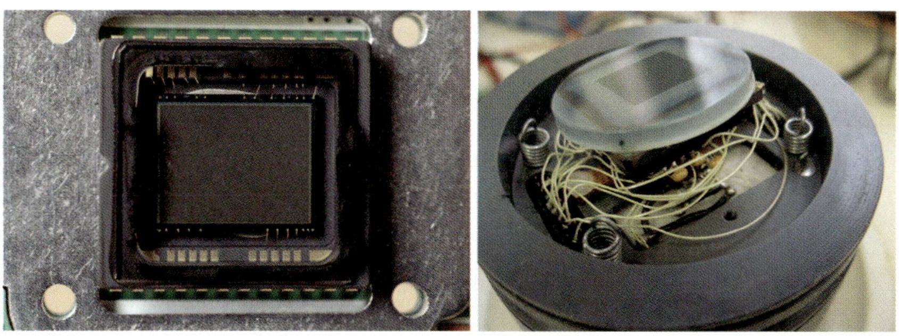

(a) (b)

Fig. 4.13: (a) CCD sensor chip of a consumer camera. (b) Image of a CCD sensor within a scientific monochrome full frame slow-scan CCD camera. The large plate on top of the sensor is a fiber optical plate (see Section 4.11). For better visibility, the detector head is dismantled.

▸ **Fig. 4.14:** Scheme of a CCD image sensor illustrated for a device consisting of an array of 4 rows and 5 columns as the photosensitive region (displayed in blue) (a). The additional transmission region, i. e., the shift (horizontal) register, is marked in green. The arrows indicate the direction of charge transfer. (a) to (d) show different readout schemes (see text). (e) Shows the realization of the transport within the semiconductor. Note that steps 4 and 7 may not be present (for illustration only). Types of CCD image sensors: (a) Full frame CCD, (b) Inter line transfer CCD, (c) Frame transfer CCD, (d) Frame-Interline-Transfer-CCD. The transfer region is displayed in gray (vertical, V-CCD) and the intermediate charge storage pixels in purple, respectively.

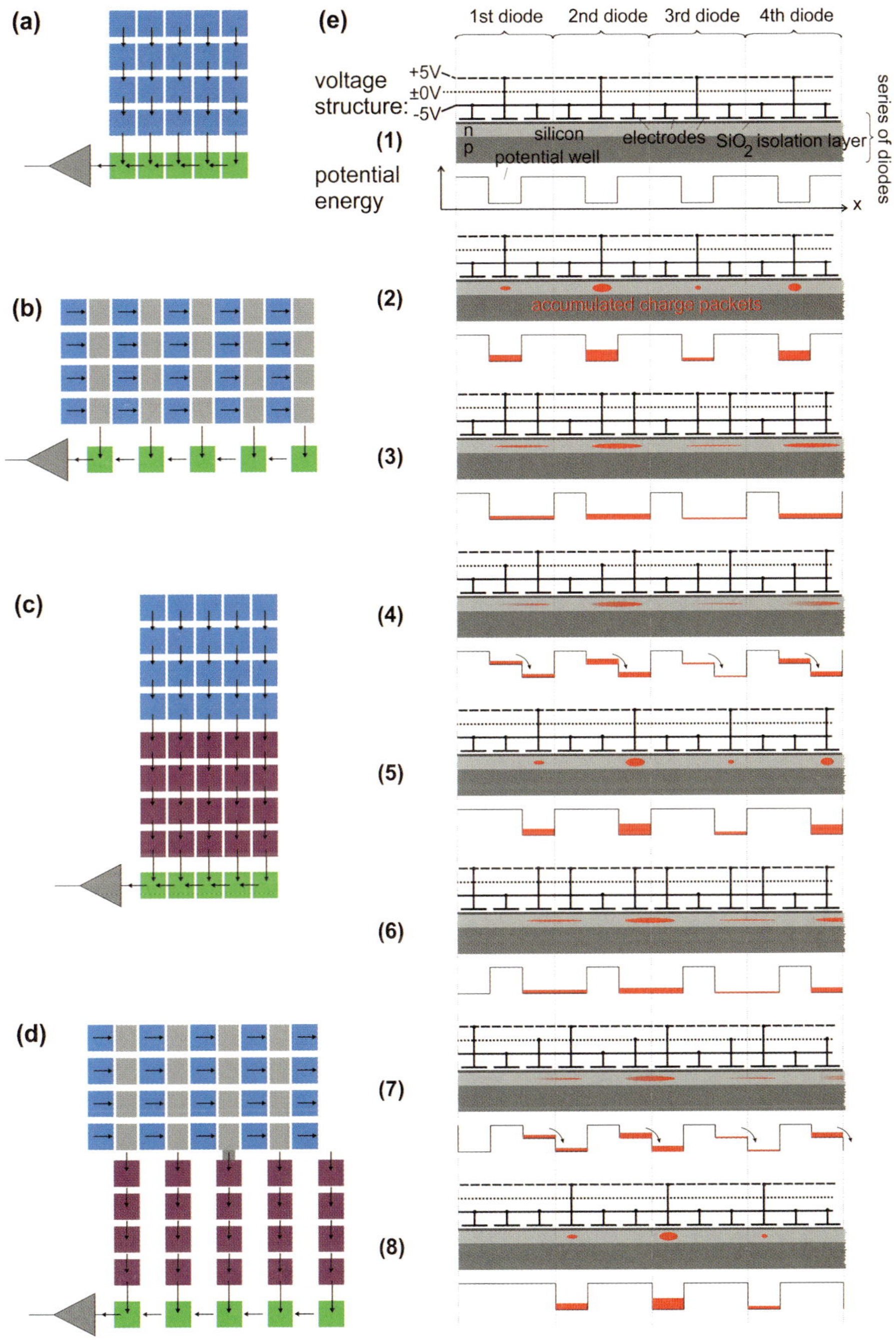
(a)
(b)
(c)
(d)
(e)
1st diode
2nd diode
3rd diode
4th diode
voltage structure:
+5V
±0V
-5V
(1)
n
p
silicon
potential well
electrodes
SiO2 isolation layer
series of diodes
potential energy
x
(2)
accumulated charge packets
(3)
(4)
(5)
(6)
(7)
(8)

neighbored one. According to that principle, this happens for all charges and all potential wells at the same time.

The charge transfer itself is realized by a succession of a selective sequence of external voltages triggered by a control circuit and applied to the device (Figure 4.14e). Figure 4.14e (1) Shows a series of neighbored photodiodes with supplied voltages prior to illumination with light (top) and the corresponding potential energies as a function of the spatial coordinate. Each diode corresponds to one pixel. The vertical dotted lines indicate the pixel boundary. Within one pixel, the optical transparent metal electrodes together with the doped silicon yield a MOS capacitor (metal oxide semiconductor) used for charge collection and transfer. (2) Shows the situation after illumination with light with a specific pattern according to the image on the sensor surface. The accumulated charges are indicated by red ellipses within the semiconductor and the red regions in the respective filled potential wells. (3) to (8) illustrate the charge transfer. For each clocking, the voltages applied to the electrodes are specified at the train lines. By this, electrons first are shifted within one pixel from one potential well to a neighbored one and then within the following clocking to the first potential well of the neighbored pixel. Then, in the same manner this continues; charges are shifted from potential well to potential well and from pixel to pixel.

Following this procedure, the charges within the first column of the 2D PDA shown in Figure 4.14a are shifted until the charge in the fourth row from the top is stored in column 1 of the (horizontal) shift register. Similar to this, the other columns of the shift registers are filled with charges, which originally have been collected within row 4 of the photosensitive region. After that, the shift register is readout pixel by pixel in the same way (following the operation principle shown in Figure 4.14e). Readout of column 1 of the shift register always means a charge transfer to the readout circuit (indicated by the triangle). In the following, this process repeats: row 4 (which was row 3 formerly) is shifted to the shift register and then readout, row 3 (which formerly was row 2) is shifted to row 4 and so on until the entire charges of all pixels are successively readout.

Readout for all “signals” occurs via the same readout circuit, which includes conversion of charge to voltage, shaping and amplification of the signal, digitization by the ADC, etc. (see also Section 4.8.6). Finally, the signals are sent to the image processor and afterwards stored in a memory and a data medium. These steps are discussed in more detail in Section 4.9. Further details, in particular, on CCD circuits, charge transfer mechanism, etc. are well described in the excellent book of Nakamura et al. [Nak06], and thus do not need to be repeated within the present book, which puts emphasis on other topics. For the same reason, the discussion in the following section is kept short.

4.4.2 CCD operation principles

In the previous chapter, the principle of the operation of a CCD sensor was discussed. However, in practice there are a lot of different types of CCD, not only differing in pixel size, geometry, number, semiconductor design and some further properties of the chip itself, but also in having different readout schemes. For instance, to boost readout, there are CCD with two readout amplifiers, CCD with a shift register for each row separately (if those are also omitted, then charge transfer becomes obsolete and one obtains an active pixel sensor (see Section 4.5)) or CCD with intermediate storage regions. Some of those types will be discussed below.

4.4.2.1 Full frame transfer CCD

The operation principle of a full frame transfer CCD (FFT-CCD) is exactly what we have discussed before (Figure 4.14a). The configuration is simple, it is easy to manufacture, and thus relatively cost-efficient. It may have the best spatial resolution. This scheme usually has the advantage of a high-dynamic range (see Section 4.8), but the disadvantage of a slow readout. Thus, application of a good shutter is essential in most situations (if light emission from the object occurs only for a time shorter than the exposure time of the camera, then, of course, an additional shutter is not necessary; an example of such a situation is imaging of an extended pulsed light source, e. g., within scientific experiments). High-quality scientific, but also good commercial cameras often use this scheme.

4.4.2.2 Interline transfer CCD

Interline transfer CCD (IT-CCD) uses additional pixels for temporal storage of charges. Storage is done in the closest position to the light sensitive ones (Figure 4.14b). In particular, directly parallel to each column of light sensitive pixels there is another one that is shielded against photons (altogether a vertical shift register). After illumination, all photon-generated charges are shifted horizontally in parallel (i. e., at the same time) to the shielded regions. From there, readout occurs conventionally, i. e., first row by row to the shift register and in the following from there to the readout circuit with the output amplifier. Due to the close neighborhood of the storage pixels, exposure could be stopped rather quickly. This allows for rather short exposure times.

Although this procedure often is called the "electronic shutter," it is not really a shutter. Nevertheless, in cheap cameras no further shutter is applied, which leads to the well-known disadvantages (see above). More expensive cameras use an additional shutter during readout time as well, which in the simplest case is just a metal plate.

The advantages of this concept are that the smear problem is reduced, at least with an improved shielding, it shows low noise and a relatively high-dynamic range. However, the disadvantage is the more complicated set up. Moreover, due to diffraction from the light sensitive pixels in the vicinity, there may be still some sensitivity to light in the shielded regions. Another disadvantage is that within the imaging area, now there are regions that are not light sensitive (i. e., the vertical shift registers) and this reduces both resolution and sensitivity (however, there could be correctives; see Section 4.6). Nevertheless, due to the advantages this CCD type is (or was) a standard sensor and mostly applied for many cameras, in particular, for compact cameras, CCD based mobile phone cameras, etc., but also for video cameras.

4.4.2.3 Frame transfer CCD

If after exposure illumination cannot be prevented, or for other reasons, another intermediate charge storage may be an option (Figure 4.14c). One possibility is prior to final readout, rather quickly to shift the entire contents of the illuminated imaging region via multichannel vertical transfer to another matrix that is light shielded. Then from there readout is done conventionally but protected against further illumination. Readout could be even done during the next exposure of the photosensitive imaging region. This scheme of the frame transfer CCD (FT-CCD) usually has the advantage that it is quite simple and allows for small pixels. But it has the disadvantage of smear (see Section 4.7.5) and needs twice the number of pixels. This increases the size and cost of this sensor. Furthermore, usually the first shift of the light sensitive to the light protected region is not fast enough to allow for very short exposure times. Hence, here again, a shutter is necessary. Usually, this type is more expensive than the interline transfer CCD.

4.4.2.4 Frame-Interline-Transfer-CCD

The Frame-Interline-Transfer-CCD (FIT-CCD) looks like a mixture of both types discussed before: it consists of an imaging/storage area as an interline transfer CCD but has a further storage area as the Frame transfer CCD (Figure 4.14d). The improvement when compared to the interline transfer type is the very low or absent smear effect, but on the other hand, it is much more complicated, much larger, more expensive and has a much higher power consumption. Even so, these sensors are often a good choice for high-speed cameras.

4.5 CMOS sensors

4.5.1 Basics

CMOS (complementary metal–oxide–semiconductor (MOS), sometimes complementary-symmetry metal–oxide–semiconductor) is a special type of a digital circuitry design implemented on integrated circuits (“complementary” refers to the design of pairs of transistors for logic functions; “MOS” refers to the manufacturing, although the materials for the MOS field effect transistors (“MOSFET”) today may have changed).

CMOS was invented in 1963. But although the first instances of manufacturing took place 5 years later, a lot of problems were present at that time and prevented fast success. Nonetheless, up through today large progress was made. Above all, this is due to the application of the highly developed manufacturing infrastructure for the production of chips by the semiconductor industry now also for image sensors. Thus, CMOS fabrication has become an industrial standard technology, some people talk about “mainstream technology." Nowadays high CMOS image sensor (CIS) performance is available. Even more, CMOS sensors have mostly replaced CCD sensors, particularly first in many low-end cameras, and later even in high-end professional cameras. Nevertheless, high-quality scientific cameras often still make use of CCD sensors (with the exception of high-speed cameras). A comparison of the advantages of CCD and CMOS sensors, respectively, is made in Section 4.5.2.3.

Similar to a CCD, a CMOS sensor consists of an array of photodiodes. In contrast to a CCD, now each pixel could be directly addressed. It has its own readout circuit including the readout amplifier. Such an array is called an active pixel sensor (APS; note: this abbreviation has nothing to do with APS format). Due to manufacturing issues, the gain and the noise of the amplifiers of the individual pixels usually differ slightly. This results in a nonuniform signal distribution even for absolute homogeneous illumination (see also Section 4.9.1). A CIS scheme is shown in Figure 4.15. Sensor examples are shown in Figure 4.16 and Figure 2.23c. For further details, we refer to the literature, in particular to special books on sensors such as [Nak06].

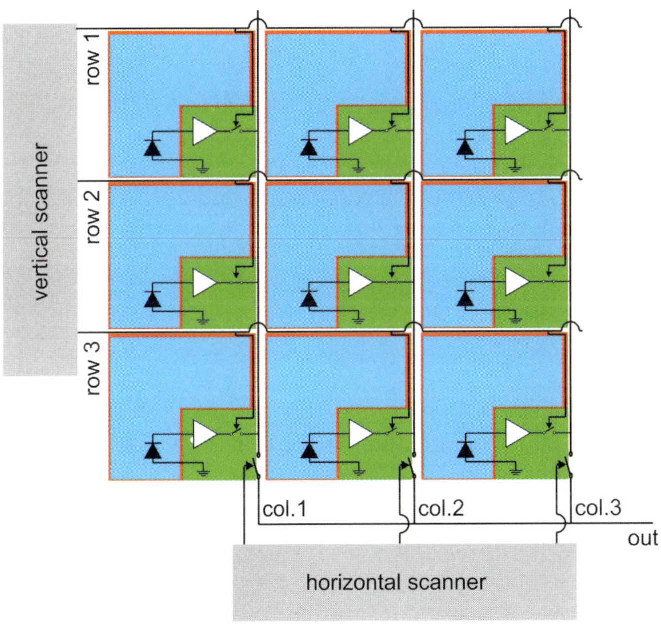

Fig. 4.15: Scheme of a CMOS image sensor. At each photodiode (pixel; marked in orange) the signal charge is converted to a voltage or a current by an active transistor inside a pixel (MOSFET). It has its own circuit (marked in green) including readout, amplifier and reset (not shown here). The light sensitive region is displayed in light blue (here L-shaped). The vertical and horizontal scanner, respectively, allows direct addressing of each pixel (XY addressing scheme). The output signal is transmitted to the ADC and later on to the image processor. In contrast to a CCD sensor, here amplifiers, ADC, clock and timing generation, bias generation oscillator, clock driver, etc. are located on the image sensor board.

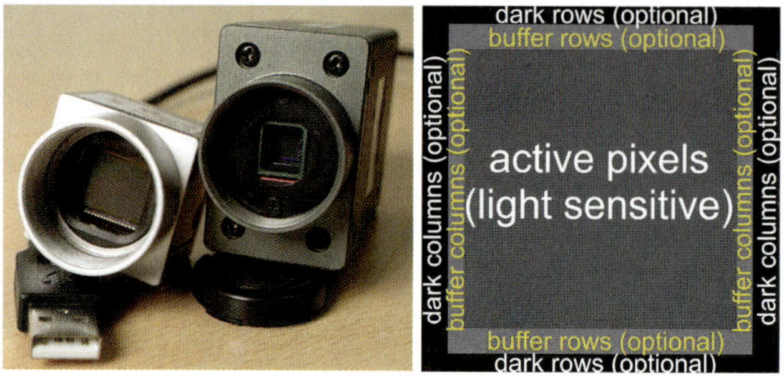

(a) (b)

Fig. 4.16: (a) Top view on the sensor surfaces of two cameras used for scientific or technical purposes, respectively: left-hand side CMOS, the right-hand side CCD (the shown cameras are equipped with much different sensor sizes; a view onto the CMOS sensor surface within a professional DSLR is shown in Figure 2.23c). (b) Scheme of the architecture of a CCD or CMOS chip. The active pixel area may be smaller than the total one. Potentially, there are additional rows and columns with pixels acting as buffer. There may be also further pixels that allow for deducing the dark signal. And there might be "barrier pixels" that should reduce interference with the currents of the neighbored circuitry.

Due to the direct access possibility, readout is quite flexible. Thus, e. g., this scheme allows for simple binning process (see later), individual signal amplification for each pixel and much more. However, due to the sensor architecture (or the individual settings), charge integration time may differ for the different active pixels. A full parallel readout of all pixels at the same time is not possible (or at least not realized in almost all cameras) because this would require a complex and rather space consuming electronics quite close to the pixel. That would reduce the fill factor; see Section 4.6. Nevertheless, there are sensor architectures, where instead of Figure 4.15, where the signals of all pixels are transmitted through the same "out" (potentially additional amplifier and ADC, see Section 4.8.6), the signals for each column have their own "out."

Usually, to control exposure time, similar to a CCD, also CMOS sensors may be equipped with additional mechanical shutters. Moreover, an additional reset scan in which the shutter pulses scan the pixel array prior to exposure is applied ("rolling shutter").

4.5.2 General issues of CCD and CMOS sensors and comparison of both sensor types

4.5.2.1 Chip architecture

The sensors of both CCD and CMOS chips do not necessarily consist of light sensitive photodiodes only.

Figure 4.16b shows that often neighbored to the light sensitive region, sensors may have additional rows and/or columns (typically up to ten) that serve as buffers. There might also be other ones that due to a special cover are made totally insensitive to light (typically up to ten or even more). These are used to determine read noise, bias, etc. (see Section 4.8 and Section 4.9). The number of rows and columns may not be equal, neither for the active pixel region, nor for the buffer region, nor for the dark pixel region.

It may be further noted that in the case of so-called electronic image stabilizers, e. g., as used in video cameras, not all active pixels contribute to the image that is recorded. The number of rows and columns, respectively, is only a fixed fraction of those of the total region of active pixels. The image or video processor compares consecutive images and if the same or nearly the same contents of both of them is just shifted slightly they are captured on slightly different regions of the chip. This forces the system to store just those corresponding regions.

For CCD sensors, most of the additional electronics is separated from the sensor chip on a separate printed circuit board (ADC, clock and timing generation, bias generation oscillator, etc.). For CMOS sensors, further around the displayed part of the sensor (Figure 4.16b), a large fraction of the electronics is located close to the pixels. This includes the analog signal processing, the ADC, the digital logic (interface, timing, output), the clock and timing control and much more.

4.5.2.2 Exposure and readout

CCD and CMOS image sensors are charge-integrating types of sensors. Consequently, prior to the capture of a new image, the signal charge on a pixel has to be removed; in other words, the pixel must be reset before a new charge integration is started. This is somewhat different for CCD and CMOS sensors, but that is not an issue here (for details, see the book of Nakamura [Nak06]).

A general problem of CCD and CMOS sensors is that light illumination during readout should be prevented. Otherwise, those pixels that are still illuminated during readout do not yield the correct reproduction of the image. Particularly, e. g., within a CCD, this becomes severe, when the illumination affects pixels with charge packets stored temporally after the rows have been shifted. This then leads to ghost images (see Appendix A.4) and smear, etc. (see Section 4.7.5). For CMOS sensors, there might be problems due to the rolling shutter effect. This effect is the electronic equivalent to the mechanical focal-plane shutter (see Section 2.6.1 and Section 2.6.2) and originates from the timing between the reset pulse and the readout pulse, which define the length of exposure in presence of continuing illumination. In videos taken with CMOS cameras, this can be well seen, e. g., from propeller planes where the rotating propeller blades look like bananas, namely similar to the example shown in Figure 2.22d.

To avoid such situations, more expensive or some scientific cameras apply additional shutters (see Section 2.6.1). Then just prior to exposure, the shutter is closed, the reset voltage (see Section 4.4.1) is applied (for CMOS cameras instead of the timing of the electronic rolling shutter, a global reset mode is applied) and then the mechanical shutter is opened for the preset time for the exposure of all pixels at the same time (all that and also the necessary “autofocus” measurement contributes to the shutter release delay). The following readout (after the shutter is closed) usually occurs sequentially. For a more detailed discussion of different readout schemes in general, we may refer again, e. g., to the book of Nakamura [Nak06].

Due to the disadvantages of the rolling shutter (RS) when used without an additional mechanical shutter, global shutter (GS) technology has been developed. In contrast to RS CIS, where each photodiode takes the role of the light sensitive part and that of the related information storage part for readout, in GS CIS the roles are separated. The light sensitive photo diode is supplemented by a fully de-coupled readout element, namely the so-called memory node, which stores the signal until readout. Both together set-up the pixel together with further electronics. Similar to the shift register of a CCD, also illumination of the memory nodes has to be prevented completely. Also, similar to the shift register, readout is made successively. However, such a GS CIS is more complicate when compared to RS sensors. There are additional transistors and, in particular, a sample-and-hold circuitry. This leads to additional noise, which especially for very small pixels becomes an issue. For those reasons, most CIS are still RS sensors. Nevertheless, there are GS CIS available, e. g., applied for automotive or machine vision applications, where image captures of fast-moving objects are required. An interesting example of a very large format CIS for industrial applications is Sony's IMX661, a CIS with a diagonal of 56.73 mm, more than 100 MP pixel and a pitch of 3.45 µm. Actually, there a lot of GS CIS with much different properties, including such ones with much larger pixels, e. g., $p = 9\,\mu\text{m}$. But it may be difficult to find GS CIS with a pitch below 2 µm.

4.5.2.3 Comparison of CCD and CMOS sensors

In general, today both CCD and CMOS technology is highly developed. In particular, CMOS technology is based on standard fabrication process in semiconductor industry. For a CCD, the linearity of the signals (with respect to the incident light, see Section 4.2.2) for each pixel usually is rather high (see also photon conversion in Section 4.8). Any further nonlinearity introduced by the image amplifier is the same for all pixels (in the case of a CCD where there is a common output amplifier for all of them) or different for all pixels (in the case of a CMOS where each of them has its own output amplifier). This is a disadvantage of CMOS sensors, although this may be compensated (at least partly) later on during post-processing (see Section 4.9). Nevertheless, for CMOS this is a source of (a spatially) nonuniform sensor response that due to the operating principle in general is fully absent in the case of CCD sensors.

The high sensitivity is another advantage of CCD sensors. Some of them are nearly sensitive to a single photon, which makes them much suitable for low light applications such as imaging in astronomy. Here, one also has the benefit of the low noise of CCD. This is another advantage of CCD. Read noise may become nearly zero, in particular, for slow scan CCD. For CMOS sensors, the electronics is located quite close to the pixels, and thus disturbances from it may couple via the substrate to the signal from the pixels. Also, with respect to dark current CCD is much superior and in contrast to CMOS, in CCD reset noise can be fully suppressed. On the other hand, progress in CMOS technology leads to advances in noise reduction, e. g., by correlated double sampling (CDS; see Section 4.7.3), which allows long time exposure times. The high sensitivity of CCD results also from the large fill factor, which may be even 100 %, whereas for CMOS sensors it is significantly smaller (see Section 4.6). This may be another issue that makes CCD preferential, in particular, for such scientific, medical and technical applications, which require a 100 % fill factor. The still significant importance of CCD sensors also shows up in current scientific journal articles, which analyze CCD sensors as important imaging devices (see, e. g.[3]).

High-dynamic ranges may be achieved for both sensor types and even more, HDR (see Section 4.9 and Section 4.10) applications become possible. An example is Fujis's fourth generation super CCD (see Section 4.10). Another one is the integration of a special pixel architecture together with the relevant hardware processing circuit within the sensor chip (HDR CMOS, HDRC; Section 4.10).

Advantages of CMOS sensors are the possible much higher speed, which makes them preferential for high-speed photography (although this is not a concern of this book) and the superior windowing, namely the flexibility to readout a preset arrangement of pixels. Conversely, CCD have the advantage that hardware binning is possible, whereas for CMOS sensors this depends on sensor architecture (see Section 4.8.3).

CCD may have blooming and smear, whereas CMOS has no smear and usually less blooming. Furthermore, biasing and clocking is easier for CMOS sensors. A special advantage for some specific scientific application is that CMOS sensor suffer less from radiation damage, e. g., by X-ray and gamma radiation.

Reliability of both sensor types is equal. Power consumption may be much lower for a CMOS when compared to a CCD on pixel level, however, taking into account the whole sensor including all the electronics that may change.

CMOS sensors are highly integrated devices, which means that in contrast to a CCD, the light sensitive part with the active pixels and the electronics is realized within a single chip. This allows for small cameras (e. g., mobile phones take advantage of this). If produced in large stock, this may be also very cost-effective. On the contrary, integration of the necessary electronics for the chip is not possible for CCD, which have to be produced in a special process that is not compatible with other standard processes in the semiconductor industry. CCD need separate circuits for signal processing, ADC, timing, etc.

On the other hand, the high integration on the chip requires longer times of chip development when compared to a CCD system. Furthermore, the high integration tailors a CMOS chip only for one or a few applications (not a problem when used for consumer cameras). Here, CCD are more flexible as they allow more easily adaption of readout, dynamic range and digitizing depth, binning, nonlinear analog processing and other

3 K. Boone et al.: *A Binary Offset Effect In CCD Readout And Its Impact On Astronomical Data*, arXiv:1802.06914 [astro-ph.IM].

customized operation modes (this may be an issue for scientific and technical applications). In particular, CCD systems may allow the replacement of the electronic part of the sensor by another one. This is not possible for CMOS sensors. Part of the flexibility of a CCD is also the possibility to setup rather large array sensors (those may also be more cost effective when compared to CMOS sensors). Current advances of CMOS sensors are discussed in Section 4.10.5.

Altogether, CMOS sensors have mostly replaced CCD sensors, mainly because they make use of a standard production technology. High volume and space constraint products take advantage of that. Image performance of the sensor may be rather high for both sensor types.

With respect to the produced number of image sensors, CCD sensors may be considered out of date. Indeed, in most cameras used for photography (and other purposes), the CCD is replaced by the CMOS sensor and, e. g., Sony has stopped to produce CCD. However, there is still a market for CCD and other manufactures continue production (some of them, e. g., e2v company, offer even replacements for Sony chips). There are still significant advantages of the CCD, in particular, for scientific and some technical applications. Thus, for specific demands CCD sensors are (much) superior to CMOS sensors. For instance, for that reason one of the currently most advanced scientific cameras for astronomy makes use of CCD (see Table A.3). And finally, we would like to remark that nevertheless there are even further developments on CCD to get rid of its main disadvantages, namely the incompatibility of its production with standard fabrication technology. Conclusively, both sensor types have their advantages.

4.6 CCD and CMOS systems

Although there are some differences, both CCD and CMOS can be regarded as pixel matrices with a lot of similarities. Hence, unless stated differently, in the following we will not always distinguish between both sensor types. In that sense, in the following we will use the terms PDA for both of them, unless particular setups are discussed that are special for one of the two types only. We will also see that the sensor itself is only part of a sensor system. A sensor system consists, e. g., of several special filters and additional optics (not to be confused with the main optics, i. e., the camera lens).

4.6.1 Fill factor and optical microlens array

Within a PDA the light sensitive elements, i. e., the photodiodes, often are not placed close by. Consequently, photodiode width and height, respectively, may differ from horizontal and vertical pitch, respectively. For definitions of pitch, etc., see Figure 4.17: The light sensitive areas are shown in white. Sometimes these are termed as “real pixels”

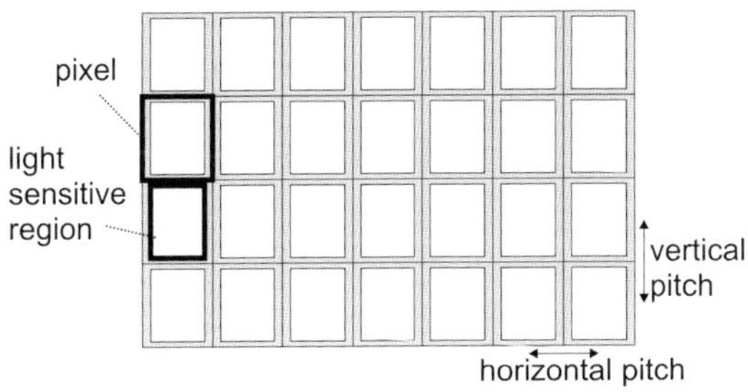

Fig. 4.17: Typical arrangement of the light sensitive elements within a PDA.

with the area A'_{pix}. However, the whole element as displayed in gray (area A_{pix}) may include a region that is not photosensitive and often all of that is termed as a pixel. Even more, the light sensitive area is not always placed centrally within a pixel and the geometry of an individual light sensitive element is not always rectangular. There may be also particular contours (see, e. g., Figure 4.15). Of course, in most cases pixels of sensors for photographic cameras do have a height equal to its width, but for other sensors that must not be the case.

One can define a fill factor (FF) η_g as the ratio of the photosensitive area A'_{pix} to the total area of a pixel A_{pix},

$$\eta_g = \frac{A'_{\text{pix}}}{A_{\text{pix}}} \tag{4.19}$$

Quite often the fill factor is smaller than one. As a result, part of the light is not detected, and hence information of the illumination conditions in the regions that are not photosensitive is lost. In general, the fill factor is larger for a CCD (here it may even be 100 % for full frame and frame transfer CCD) when compared to a CMOS sensor, where the electronic circuit is located quite close to the photosensitive part of the pixel (see, e. g., Figure 4.15).

To avoid such a situation, in many cameras (in almost all cameras used for photography), each pixel is equipped with a microlens that collects nearly all of the light in front of it, and thus increases the fill factor preferentially to a value close to 100 %. This is realized by the implementation of an optical microlens array (OMA) in front of the sensor, which then increases the QE (see Figure 4.10b). Examples are shown in Figure 4.18 and Figure 4.25.

For CCD and CMOS, the situation is little different. For CCD, the fill factor is significantly larger when compared to a CMOS. It may be even up to 100 % and in such cases the OMA may be omitted. Furthermore, due to the wiring, the light sensitive region of the CMOS has an aperture in front of it. Due to the vertical separation of light sensitive region and wiring, this results in a collimation effect and leads to shading losses (Figure 4.18). This is most severe for small pixels.

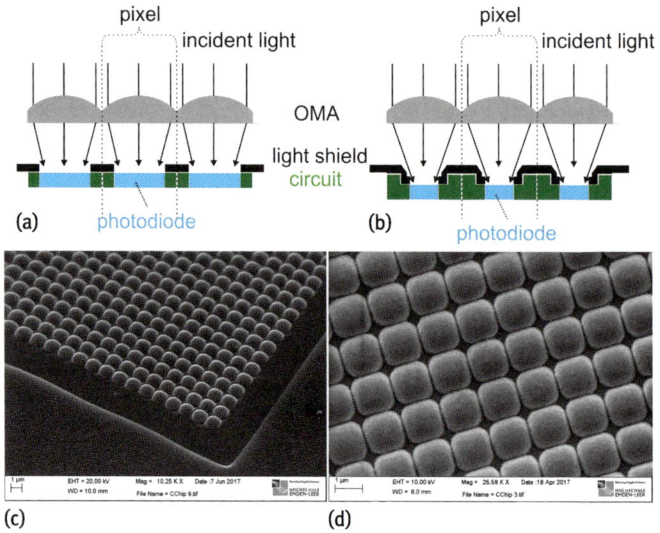

Fig. 4.18: Section of the PDA equipped with an optical microlens array (OMA). The photosensitive region may be equipped with an AR filter on top and is marked in light blue. (a) Scheme with three pixels of a CCD sensor. (b) Same for a CMOS sensor. Scanning electron microscope (SEM) images of the OMA of a CMOS sensor: (c) corner of the chip, (d) details.

Here, we may note that the task of the microlenses is light collection only. Imaging is not an issue for the OMA because the pixel itself acts as one point (but remember that the resolution in 1D is at least two pixels; see Chapter 2 and Chapter 5).

The presence of an OMA has also to be taken into account for the objective lens design. This is because in contrast to analogous cameras that accept a wide range of angles of incidence on the film as the sensor (Figure 4.19a), a microlens in front of the photodiode does not. It does require more parallel light, otherwise losses occur. As a result, cameras with digital sensors (PDA) require lens designs with a telecentricity θ_t that is preferably rather low (see Figure 4.19b and Section 3.4). Nevertheless, in the case of much oblique incidence, the microlens still may focus at least part of the light into a region outside the photodiode (Figure 4.20a). Even more, the extension of the wiring in the vertical direction acts as a collimator, which further limits the field of view for each photodiode (Figure 4.21). All that leads to shading losses, in particular, most pronounced in regions far off the sensor center where the light rays are most inclined. This special kind of vignetting adds to that one discussed in Section 3.4 and Chapter 6. But we may note that special scientific CMOS (sCMOS, see Section 4.10.5) with a special chip architecture are also available. These offer a fill factor of 100 % and are supplied without an OMA, and thus do not suffer from the shading losses, which is an issue, particularly for low light conditions.

For this reason, e. g., a high-quality objective lens designed for an analogous camera may yield quite average results when tested at a digital DSLR. Of course, this is most pro-

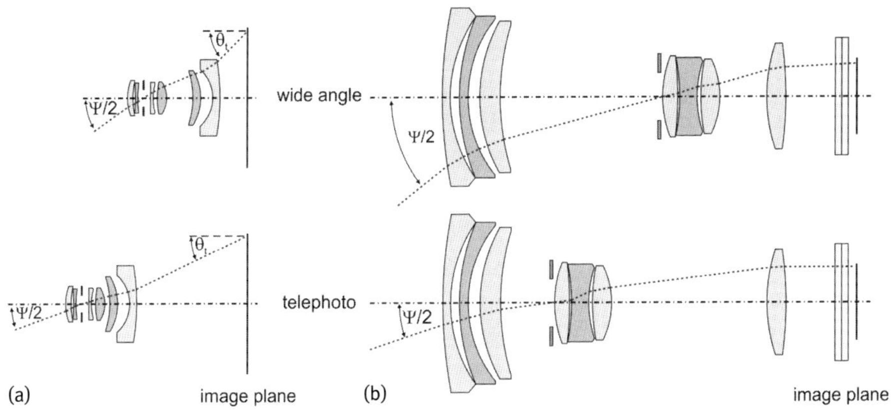

Fig. 4.19: Appropriate lens designs for analogous (a) and digital cameras (b), respectively. Lenses made for analogous cameras lead to rays that may have a strongly inclined angle, in particular at the image borders. This is most pronounced for wide-angle lenses. Lenses for digital cameras should be mostly telecentric (but usually they are not completely; see Figure 4.20; for telecentricity, see the discussion in Section 3.4 and Section 6.5.3). The telecentric angle θ_t and the half angle of view $\psi/2$ is indicated.

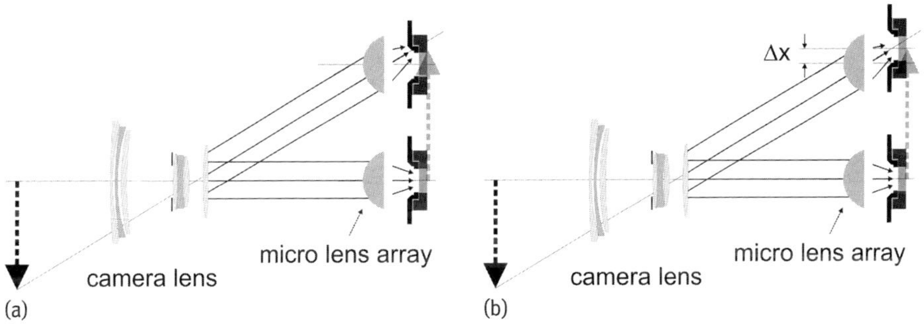

Fig. 4.20: (a) Simplified scheme of a standard microlens array consisting of a regular spaced matrix of lenses (not to scale; in reality, in particular, the object is much farther away, and the angles are much smaller). This example just shows two microlens/photodiode combinations (i. e., a pixel): one located close to the optical axis and another one located more far off the optical axis. (b) Similar scheme but now the microlens of the pixel far from the axis is shifted by a distance Δx (Δx depending on the distance to the optical axis, usually according to a linear function).

nounced for wide-angle lenses. Thus, a lot of such tests, e. g., found on the web, conclude with the statement, that one is "disappointed in the tested lens that was expected to be of high rank." However, the problem is not the lens, but the naively applied test.

To reduce shading losses for sensors of high-end DSLRs, some manufacturers apply a modified microlens arrangement. Microlenses located close to the corner are offset from their regular positions to improve coupling of obliquely incident light to the photodiode (Figure 4.20b). If designed well, an OMA significantly increases QE (see Figure 4.10).

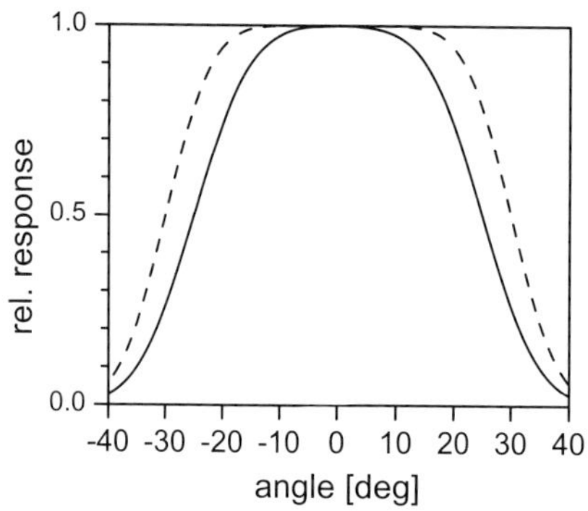

Fig. 4.21: (a) Schematic of the typical response of a microlens/photodiode combination (i. e., a pixel) with a not-shifted microlens. Depending on the actual design, the curves may have a different shape and/or smaller or wider angle of acceptance than shown in this example. Usually, the horizontal direction has a smaller angle of acceptance when compared to the vertical direction one (or vice versa; this is indicated by the solid and dashed lines, resp.).

4.6.2 Optical low pass and infrared filters

If we remember the discussions in Section 1.6 (see also Section 5.2.4), an image taken by a digital device may consist of artefacts, e. g., the Moiré effect.

According to the Nyquist–Shannon sampling theorem, all spatial image structures are unambiguously and reliably digitized as long as all sampling frequencies R are below the Nyquist limit R_N (see Figure 1.20a,b,i and also textbooks on signal processing theory). Higher frequencies, corresponding to smaller image structures on the sensor will be "folded back" into the sample range, in a similar way as an optical spectrum that consists of higher order (see Section 1.6.3), and thus generate an artefact image and not a reproduction of the original (alias effect, blue curve in Figure 1.20d, Figure 1.20e). This can be also seen in the image displayed in Figure 1.22.

In advance of Section 4.6.3, we would also like to remark here that the Moiré effect is even more present for color. As we will see, usually color images have less resolution, and thus a lower Nyquist limit and, therefore, are even more affected by too fine structures on the sensor surface. Consequently, they suffer even more on artefacts (Figure 1.22d).

To avoid such a situation, too-small structures on the sensor surface have to be avoided. To do so, an object that is imaged with a high-quality lens onto the image plane (sensor surface) with an optical resolution better than the Nyquist limit of the sensor has to be slightly decreased in its resolution. This is done with an antialiasing filter, also termed an optical low pass filter (OLPF), which reduces the contribution of too high-spatial frequencies significantly (see also Chapter 5). In other words, by a little smearing of the signal slightly to neighbored pixels as well, the smallest structures are slightly extended. Of course, this leads to somewhat reduced resolution but artefacts may be avoided that usually strongly disturb the visual impression of an image. The human eye, together with the brain, is very sensitive to such artefacts.

Although OLPF is present in most commercial cameras, in scientific ones it should be mostly avoided. Sometimes the photographer has a choice: some manufacturers offer the same camera model, one with and one without OLPF. Examples are, e. g., the Nikon

D800/D800E or also some of the current 50 MP cameras such as the Canon 5 DS/5DS R. The camera without OLPF will yield better resolution, but if, e. g., photographs are taken from structures with some periodicity or when the lens outresolves the sensor (see Chapter 5), artefacts may occur that disturb the image. Consequently, depending on the application, the photographer has to make his own decision which camera is expected to provide better images.

To realize such a filter, usually a transparent plate of a birefringent material is used. As extensively discussed in standard textbooks on optics, an image transmitted through a birefringent plate will split, and thus behind the plate, two images are generated. OLPF filters consist of a sandwich of two such plates (see Figure 4.25), where one of them is rotated by 90 degrees with respect to the other one. As a result, a small amount of the optical signal, which would have been transmitted to a specific pixel, now also contributes to four neighbored ones at the left, right, top and bottom, respectively (see also Figure 5.7, no. 8). OLPF do not simply distribute the optical information over several pixels but deliver most of the light to the central pixel and less to the neighbored ones and even less to pixels much further away. This corresponds to a smooth distribution, which avoids hard clipping, and thus avoids unwanted diffraction effects (for artefacts in MTF, see Section 5.2). It may be mentioned that reducing the alias effect but keeping resolution and sharpness as much as possible is always a challenge. On the other hand, using a lens with a resolution worse than that of the sensor would make OLPF obsolete. Even more, sometimes a further filter is added to block unwanted infrared light. This is because the sensor is sensitive to IR, but the human eye is not (see Figure 4.25).

4.6.3 Color information

Up to now only spatial light intensity distribution of the sensor has been discussed, i. e., a local intensity distribution $I_{pix}(i, j)$, where i, j correspond to row and column of the pixel within the PDA. Information on color was not an issue. In contrast to film and special sensors such as the Foveon X3 sensor (see Section 4.10.1), usually pixels of a PDA are sensitive to I_{pix} but not sensitive to color. Of course, there is a wavelength dependence due to QE, but this does not provide information on the color locally incident on that particular pixel. In other words, each pixel yields a signal, which may be regarded as monochromatic information on brightness at its local position.

But how do we then get color information, i. e., colored images? A method to generate color information is to use the superposition of several selected specific colors only, each with the appropriate intensity. In reality, these are spectral bands, e. g., such as the three shown in Figure 4.24. Such a mixture allows for generating nearly every possible color within a so-called color space (a basic discussion on that was made in Section 4.1.2; a more specific discussion is far beyond the scope of the present book).

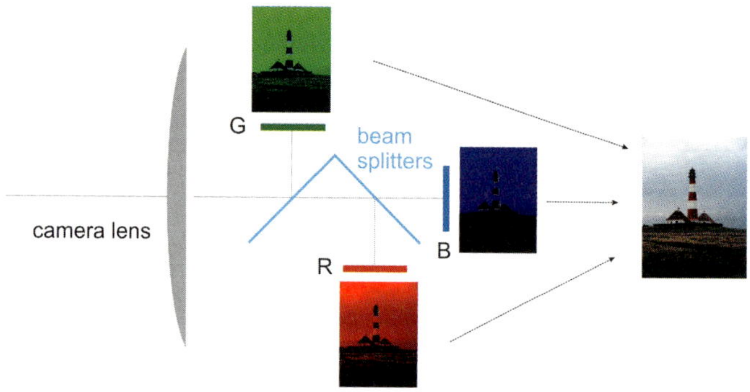

Fig. 4.22: Scheme of a three-chip camera.

One of the most common is the RGB color space. Thus, to obtain colored images, early digital cameras made use of three independent sensor systems (three "chips"; Figure 4.22). Behind the camera lens, the beam path is split into three different directions. The image is captured in the resulting three image planes by a PDA. In front of each, there is a red, green or blue color filter (marked as R, G and B, resp.). Consequently, each of these three images is taken with full resolution by its particular sensor, but only within the corresponding color range. The final image is made as a superposition of the three partial images (indicated by the broken arrows). It is important that within the final image each pixel is made exactly from the three corresponding pixels within the red, green and blue channel, respectively. This requires an exact positioning of the three sensors with respect to each other and special lens constructions incorporating beam splitters. The arrangement is rather complicated and rather expensive. Today this scheme is not used anymore for DSC.

Current cameras instead use a single sensor only. Placing a specific color filter in front of each individual pixel instead of one for the whole chip, each pixel can be made sensitive to a different color, and thus the equivalent local color information is obtained at its local position.

Generating a RGB signal by equally distributing the three colors with the rectangular pixel arrangement shown in Figure 4.17, of course, is not possible. On the other hand, this is not a strong disadvantage, because the human eye is mostly sensitive to green light, and thus more pixels within the PDA should be selected correspondingly.

An appropriate arrangement of RGB filters, which is used in most cameras is the Bayer filter (named after its inventor B. E. Bayer; see Figure 4.23a, Figure 4.26a). When this color filter array (CFA) is positioned in front of the sensor, the color can be calculated within any square consisting of four neighbored pixels within the sensor. Of course, this is an average over this area and no exact color information at the position of each individual pixel is possible, but only a guess.

Figure 4.24a,b show the spectral response of the three color channels of a typical professional DSLR and a mobile phone camera, respectively, both equipped with a Bayer

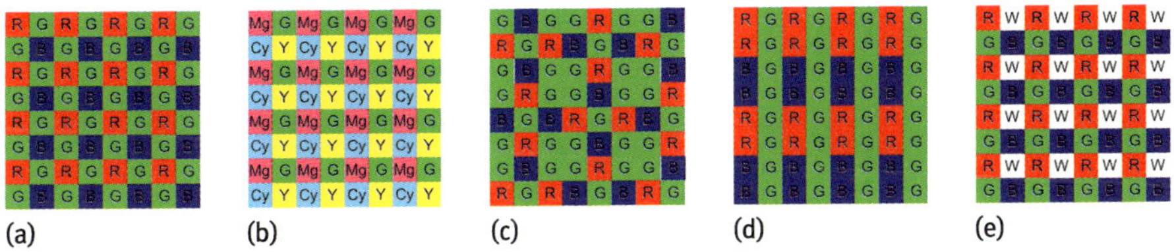

Fig. 4.23: Different color filter arrangements: (a) Bayer mask, (b) four color CYGM (Sony's emerald), (c) X-Trans (Fujifilm), (d) super-CCD EXR (Fujifilm; may also be used in inclined arrangement, see Section 4.10.3) and (e) RGBW. The symbols R, etc. mark R red, G green, B blue, Mg magenta, Cy cyan, Y yellow and W white (here no color filter, just intensity sensitive).

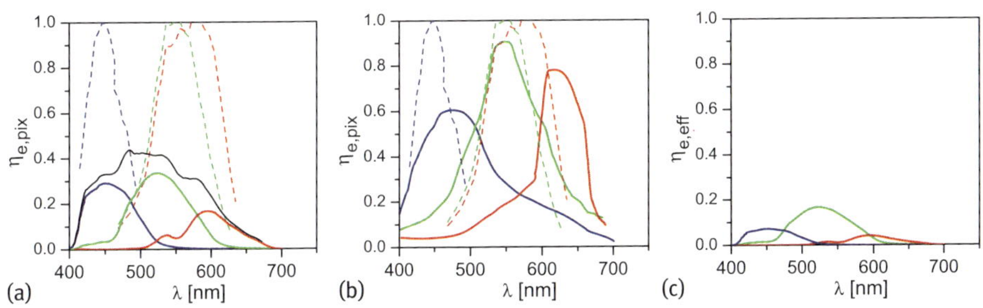

Fig. 4.24: Typical quantum efficiency as a function of wavelength for single pixels of a (a) professional DSLR (solid lines) or (b) mobile phone camera (solid lines), respectively. For comparison, the spectral response of the human eye is displayed as well (here on a linear scale too, dashed lines; compare also to films, Figure 4.2, but there on a log scale). (c) is based on the curves shown in (a) but takes into account the geometric quantum efficiency of the Bayer mask. In contrast to Figure 4.10, η is reduced by the color filter transmission.

mask filter. The colors of the curves reflect the corresponding color sensitivity of the pixel behind its filter. The black curve in (a) is the sum of the 3 RGB channels. Note that here the green channel has the same weight as the red and blue one, respectively. Here, the external or overall quantum efficiency $\eta_{e,pix}$ is considered with respect to an individual pixel, equipped with the corresponding filter of the CFA. This is a convolution of filter transmission with quantum efficiency from a curve such as that shown in Figure 4.10. With respect to an illumination of all pixels together, the situation is different. If one takes into account the Bayer mask, only a quarter of the pixels is sensitive to red light, another quarter for blue light and the rest for green light. This yields a geometric efficiency η_{geom}, which is 25 % for red and blue light, respectively and 50 % for green light. Note that $\cdot\eta_{geom}$ is not equal to the fill factor η_g. Consequently, the effective quantum efficiency $\eta_{e,eff} = \eta_{e,pix} \cdot \eta_{geom}$. As an example, Figure 4.24c shows the resulting $\eta_{e,eff}$ curves for $\eta_{e,pix}$ curves displayed in (a). For the mobile phone camera, the situation is similar, and thus the $\eta_{e,eff}$ diagram is omitted.

The comparison of the examples Figure 4.24a and Figure 4.24b also shows that cameras do not necessarily reproduce color in the same way and also color sensitivity is not necessarily similar to that of the human eye. For the present example, the DSLR may reproduce color quite well, whereas the mobile phone camera obviously shows a much stronger signal in the blue channel, which may lead to nice colors within the captured images. But on the other hand, there is significantly worse reproduction of the colors of the original scenery.

On the other hand, today effort has been made for improved color reproduction for CIS in general and smartphone cameras (SPC), in particular. Also, improvements for low light conditions are sometimes an issue. For that reason, CFA of other than Bayer type have been applied. Several examples are shown in Figure 4.23, some similar ones may use another nomenclature. Even in a different configuration the 2 green channels of the Bayer CFA are replaced by 2 yellow filters, which should improve color reproduction for SPC. But it is not our intention here to discuss if any of those CFA alternatives to the Bayer type, together with appropriate de-mosaicing procedures (see Section 4.9.3), shows a better performance. However, to the best of our knowledge, there is no CFA which can be considered to be the best choice in general.

We may remark that due to both, the quantum efficiency curves and the fact that more pixels are equipped with green light filters when compared to the other two colors (at least for Bayer CFA), the camera is most sensitive to green light. Consequently, for image capturing and processing this has to be taken into account. As an example, if illumination is done with white light, exposure should be optimized for the green channel.

In general, a photodiode is only sensitive to the incident light energy but not to the wavelength. Consequently, a photodiode behind a specific color filter within the CFA is only sensitive to the incident light that could pass the related filter and the signal is proportional to the amount of the corresponding amount of energy. This results in a sensitivity just for that particular color or wavelength range, respectively, according to one of the three spectral curves displayed in Figure 4.24. We may term the combination of the photodiode together with the filter according to the color as red, green or blue pixel, respectively. As for any of the pixels, color information is restricted to one color only, e. g., to the energy of the blue channel, the missing energies of the two other components, e. g., the green and the red channel, at this pixel position have to be *estimated*. Usually, such an estimate is made by interpolation of the red, green and blue pixels in the vicinity of that pixel position where full color should be obtained. For this procedure, a lot of different algorithms are available. Full color and brightness information at the pixel under consideration is *estimated* using one of those algorithms. As a result, for each pixel a RGB color signal is *assigned*. This procedure is called de-mosaicing. But although for usual conditions this method does reproduce color information quite well, it is important to note that the real color of the image within the local area at a particular pixel position is *not measured*, and thus will never be *known reliably*. In Section 4.9.3, we will also see, that indeed, such an estimate may be totally wrong in special situations with the consequence that the image retrieved from the sensor signal is very different from the original image on the sensor surface, which is generated with the objective lens in good quality.

There are two further consequences, in particular, when a color camera where the sensor is equipped with a CFA is compared to a pure monochrome camera without it. First, resolution is reduced. For instance, because green light is detected only by every second pixel, resolution for that color is reduced to 50 % (e. g., in horizontal direction) when compared to the monochrome camera (for red and blue light, resp., it is only 25 %). Second is that even a monochrome signal that is calculated from the color information of all pixels, due to the fact that the *calculation bases on several assumptions*, is always worse than a *measurement* performed with the monochrome camera. Thus, for most scientific applications, e. g., laser beam analysis, CFA should be avoided.

Although the Bayer filter is the most common CFA in cameras, other arrangements and/or color filters and appropriate de-mosaicing algorithms are applied as well (Figure 4.23b to e, Figure 4.26b). For instance, the arrangement Figure 4.23b used in some consumer cameras should lead to an improved color reproduction or that one of Figure 4.23c should lead to a reduced Moiré effect in line structures because every row contains all three colors. Due to that reason, the application of the OLPF is omitted there. Even other manufacturers still use the Bayer filter geometry, but they make use of yellow channels instead of the green ones. They imply a subtractive color mixing instead of an additive one. Examples are found in some Huawei smartphones. Further sensor (and camera) manufactures use even other geometries. Two of them are displayed in Figure 4.23d, and (e), respectively. Further discussion of pixel geometries is done in Section 4.10. We would also like to remark that there are CMOS sensors that are equipped with two OMA, one in front of the CFA and one behind it (not necessary for BSI-CMOS sensors; see Section 4.10.4). This should improve the collimation into the "light channel" formed by the circuit (see Figure 4.18b).

Finally, Figure 4.25 shows a typical arrangement of the different components discussed above within a sensor system. The thickness of the filter stack (here 2 mm) consisting of the IR filter and the OLPF is well seen. It may also be seen that there is a large distance between the stack and the sensor surface (2 mm). The orange color on top of the stack results from observation of the surface coating of the IR filter seen at flat angle. However, a top view clearly shows the fully transparency of the filter stack and the sensor surface below. The same sensor chip with removed filter stack is shown in Figure 4.13a.

Figure 4.26 shows microscope images of the OMA. Part of the components are coated with an antireflection (AR) film to reduce reflection losses. For instance, the index of refraction is 1.45 for SiO_2 and in between 3 and 4 for silicon. Thus, for normal incidence, according to Fresnel's equations without an AR film this would lead to a loss of approximately 3 to 40 % (compare Section 3.1.2 and Section 6.8.1).

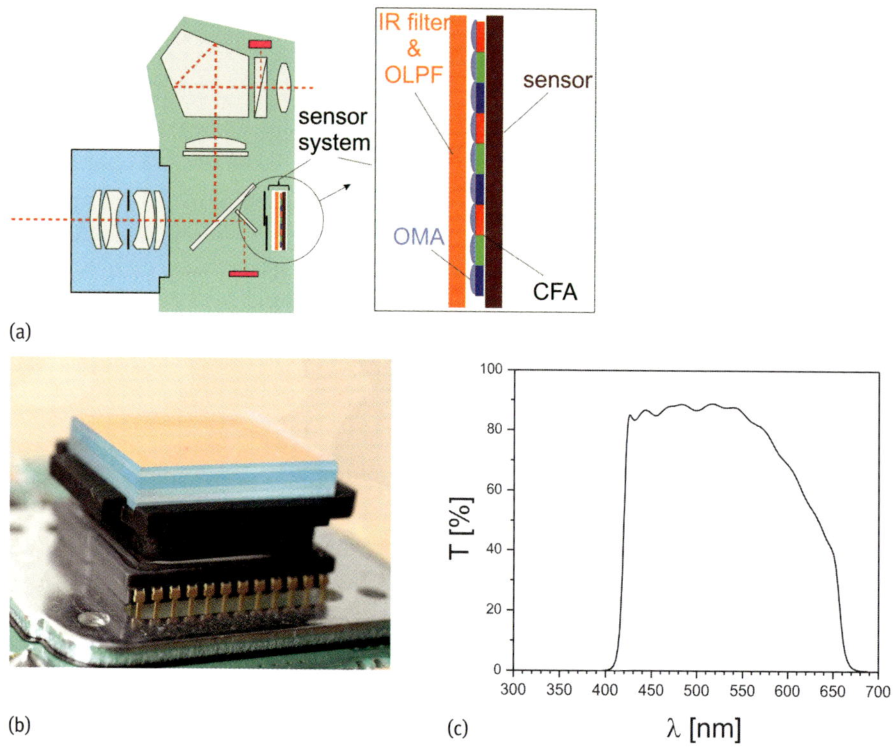

Fig. 4.25: (a) Scheme of the typical arrangement within a sensor system (b) Photography of the filter placed on the sensor chip (14.1 mm × 9.4 mm). (c) Transmission curve of the filter stack.

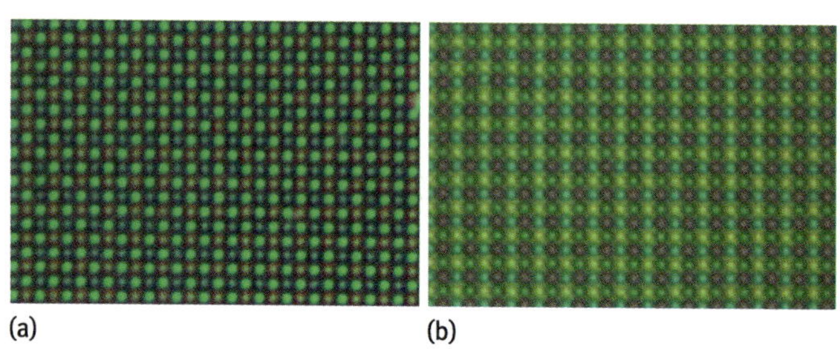

Fig. 4.26: Microscope images of a sensor with a (a) Bayer mask (photography of the surface of a mobile phone CMOS chip; 2.5 μm pixel pitch) and (b) a CYGM mask (photography of the surface of a digital camera CCD chip; 5.5 μm pixel pitch), respectively. The round shape of the microlens structure below the CFA is seen well. Compare also to Figure 4.18c and Figure 4.18d.

4.7 Noise and background

In Section 4.2.2, the maximum value signal that can be provided by a single photodiode has been discussed. And of course, for a film as the detector, the maximum value is also given by its saturation value. Now the question arises: what is the minimum signal that can be obtained? The answer is: for a photodiode or other electronic devices, this is limited by noise, for a photographic film this is limited by grain. In the following, we will discuss the most relevant aspects on noise with respect to imaging. Yet we will not give an extensive discussion on noise and related phenomena in general. For that topic, we refer to standard or special textbooks, e. g., on signal processing, and also to [Nak06]. Within this and the following Section 4.8, we will also see that noise "is everywhere" and not only limits the minimum detectable signal but affects signals in general.

4.7.1 Basics

First of all, from images taken with an analogous camera on a photographic film one may observe a grainy structure within the image that is not present in the real scenery.

As an example, Figure 4.27a shows an image where a rather homogeneous illuminated area is marked. In its enlargement and the profile measured along a horizontal line, respectively, the grain structure and the according intensity fluctuations could be well seen. Here, we will not discriminate between grain within black-and-white films and dye clouds, in the case of color films. Grain is randomly placed in the gelatin of the emulsion, and thus acts as a noise pattern. Specific grain patterns are observed for different types of films. Today, some photographers even use software that overlays specific granular patterns to digital images taken with electronic sensors to give the impression

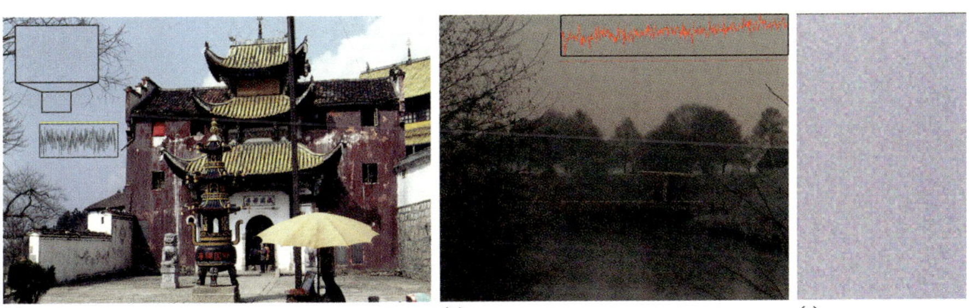
(a) (b) (c)

Fig. 4.27: (a) Example of "noise" within a slide, i. e., an image taken on a photographic film. (b) Example of an enlarged crop of a CCD image, which is strongly affected by noise (due to high ISO number). This luminance noise can also be seen as intensity fluctuations in a region, which is expected to be rather smooth. (c) color noise seen in a crop of a dark field image from a color CMOS sensor (here from dark current; see Section 4.9.1; color noise is discussed in Section 4.7.6).

of a specific familiar film. The sharpness of an image (see Section 5.2.4) is affected by grain, namely by its average size and RMS value and even its resolution.

Before we continue, we would like to make a remark on scanning on films because this is quite common nowadays. When films are scanned with a film scanner, a particular bad situation occurs sometimes, namely grain aliasing. Even in the case of rather smooth images with unpronounced grain structure (the "original"), the scan delivers a not very attractive image with large "grains." It should be noted that it cannot always be easily decided if a granular structure of a scan is similar to that of the original, unless a direct comparison is made with the original slide or the printout of a negative. The reason for grain aliasing is connected to the Moiré effect when the grain structure of the film is convoluted with the discretization of the scanner pixel (compare also, e. g., Section 1.6.2 and Chapter 5). Of course, the alias effect is most pronounced when the Nyquist frequency during scanning process is close to the spatial frequency of the grains. In such a case, the phase or offset plays a very important role (compare Figure 1.20c and Figure 1.20f) and this strongly affects the scan. But it should also be noted that a large grain structure is not the result of grain aliasing in general; sometimes it is the result of poor scanner quality or wrong scanner settings. Altogether grain aliasing is a complex effect and there are some discussions on it, e. g., on the internet.

The granularity of a photographic film (negative or slide) could be numerically described by an RMS value (root-mean-square), which is a measure of the optical density variations regions that are expected to have a homogeneous optical density and granularity is an objective measure of graininess. To determine the RMS values, a film area is exposed and developed in such a way that it has an optical density of 10, i. e., a transmission of 0.1. Then the granularity is given by the density fluctuation multiplied by 1000 within an area with 48 µm diameter given by the aperture of the microdensitometer as the measuring device. For further discussion within this chapter, more details on that subject are not important, and thus we would like to refer the interested reader to the literature.

Similar to an image from a film as the detector, images taken with cameras equipped with a CCD or a CMOS sensor are also affected by intensity fluctuations, which now are due to electronic sensor noise (see example in Figure 4.27b). In addition to the observable intensity fluctuations, colored spots may appear or even wrong colors. Thus, one may conclude that noise is a random variation of image density, visible as grain in film or fluctuations in pixel signals in digital images. We will see that noise is a key image quality factor nearly as important as sharpness. Thus, good quality images with sometimes barely visible noise may be obtained, e. g., with high-quality digital cameras, particularly DSLRs with large pixels, 4 microns wide or larger. In contrast, cheap compact or mobile phone cameras with tiny pixels may yield images that are strongly affected by noise, in particular when taken at worse conditions and/or high ISO speeds.

4.7.2 Noise distributions

Noise can be described by basic physics—the photon nature of light and the thermal energy of heat—and is always present. In many situations, the pixel signal or density variations that comprise noise can be modeled by the well-known normal or Gaussian distribution (Figure 4.28)

$$\Pi_G(x) = \frac{1}{\sqrt{2\pi\sigma^2}} \exp\left(-\frac{1}{2} \cdot \frac{(x-\overline{x})^2}{\sigma^2}\right) \tag{4.20}$$

where x is the pixel signal, e. g., given by the signal voltage (see Equation (4.15)) or the corresponding photon number N_{ph}. $\overline{x}$ is its central or mean value and σ the standard deviation, which also determines the width of the distribution. The relative number of pixels ΔN_{pix} that yield a signal within an interval ΔI (or photon number ΔN_{ph}) or infinitesimally dN_{pix}/dI (or dN_{pix}/dN_{ph}) is equal to the probability $\Pi_G(x)$. Here, this distribution is normalized to one in the sense, that the integration of $\Pi_G(x)$ from zero to infinity, i. e., the area below the curve, equals one (thus this corresponds to the fact that 100 % of the pixels yield a signal, whatever it is). As a consequence, for the Gaussian distribution, 68 % of the samples are found in an interval between $\overline{x} \pm \sigma$. This results from the integration of $\Pi_G(x)$ from $-\sigma$ to $+\sigma$.

For a set of N discrete samples x_i, one obtains

$$\sigma = \sqrt{\frac{1}{N} \cdot \sum_{i=1}^{N} (x_i - \overline{x})^2}, \tag{4.21a}$$

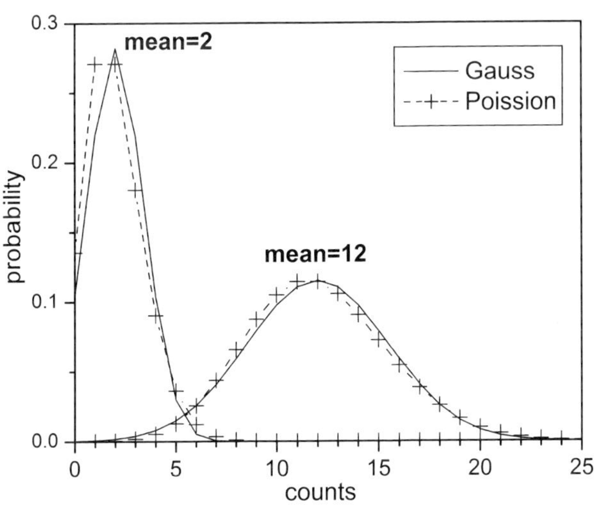

Fig. 4.28: Gaussian and Poisson distribution, each for the two different indicated mean values. Note that the dashed lines are only to guide the eye.

which is the RMS value or the square root of the variance. For continuously or quasi-continuously distributed values instead of Equation (4.21a), one obtains

$$\sigma = \sqrt{\int_{-\infty}^{\infty} (x - \bar{x})^2 f(x)\, dx} \tag{4.21b}$$

where $f(x)$ is the distribution function, e. g., $f(x) = \Pi_G(x)$ given by Equation (4.20) or by Equation (4.22).

The normal distribution does not apply for all situations. For low light levels, meaning low photon counts, in particular, the Poisson distribution Π_p instead is the appropriate one (Figure 4.28; see also standard textbooks of physics). In such a situation, the light can be described to consist of photons that are not very frequently incident on the sensor. A long-time measurement yields an average value $\overline{N}_{ph}$. For a single measurement, the probability that one measures a signal corresponding to N_{ph} (the photons correspond to independent events) is

$$\Pi_P(N_{ph}) = \frac{\overline{N}_{ph}^{N_{ph}}}{N_{ph}!} e^{-\overline{N}_{ph}}. \tag{4.22}$$

The standard deviation for the Poisson distribution simply is

$$\sigma_{ph} = \sqrt{\overline{N}_{ph}} \tag{4.23}$$

σ_{ph} is also termed "photon noise" or "shot noise." Due to $\Delta N_{ph} = \sigma$ the relative deviation is given by $\Delta N_{ph}/N_{ph} = 1/\sqrt{N_{ph}}$. It may be mentioned that for a large value of the average $\overline{N}_{ph}$ the Poisson distribution is applicable as well and then it is connatural to the Gaussian distribution.

Noise is usually also expressed as an RMS value. The average signal value corresponds to that of the original scenery, which is the signal that would have been measured if noise would have been absent. Then the noise value may be expressed by σ, which then has the same unit as the signal itself. However, as we will see later, the human eye does not respond to a light signal on a linear scale, but on a logarithmic one. Although this is not of much relevance within this section, due to its importance later on, we would like to note that for that reason it is quite common to express noise in aperture stops or exposure values EV as well.

Noise can be described by two basic types. The first one is associated with a temporal fluctuation, where the pixel signals vary randomly each time an image is captured. The second one is due to spatial variations that are caused by sensor nonuniformities. This is the so-called spatial or fixed pattern noise.

4.7.3 Temporal noise

An important noise contribution that is of the first type is photon or shot noise. As discussed before, the statistical behavior of photon arrival is described by Poisson statistics (Equation (4.22)). We will come back to this later.

Another representative of temporal noise is dark current I_{dark}, sometimes also called dark noise. In a strict sense, I_{dark} is not noise, but an unwanted signal that is affected by noise as discussed below. I_{dark} can be subtracted from the signal (see Section 4.9.1, σ_{dark} not). This originates from the fact that even without any illumination, due to the thermally activated dark current within each photodiode, charges are generated, which yield a signal,

$$N_{\text{dark}} \cdot e = I_{\text{dark}} \cdot t_{\text{x}} \tag{4.24}$$

where N_{dark} is the number of charges generated and t_{x} the time window of the measurement, usually the exposure time. I_{dark} depends on the sensor itself and yields a background signal, which depends on t_{x}. The equivalent noise can be calculated similar to Equation (4.23) and is termed "dark current shot noise," namely

$$\sigma_{\text{dark}} = \sqrt{N_{\text{dark}}}. \tag{4.25}$$

The dark current itself is due to thermally generated charges within the pn-junction of the semiconductor photodiode. Then, as a thermally activated process the dark current can be described by a generation current, which is prevailing at room temperature and below:

$$I_{\text{dark}} \propto \exp\left(-\frac{W_{\text{g}}}{2 \cdot k_{\text{B}} T}\right) \tag{4.26}$$

where T is the (absolute) temperature and k_{B} Boltzmann's constant. In addition, part of I_{dark} is due to a diffusion current as well, but its scaling with T is not much different, namely on a semilogarithmic diagram it is proportional to $-W_{\text{g}}/k_{\text{B}}T$ (see Figure 4.29). Altogether calculation of dark current is not straightforward and can be found in special literature (see also [Nak06]).

For instance, images taken with a hot camera may suffer quite a lot from dark noise, in particular, when t_{x} is long. On the contrary, the signal quality can be much improved by cooling. Thus, in particular, many scientific cameras are operated at low temperature, e. g., at −25 °C by the use of alcohols or at even lower temperatures when using liquid nitrogen, which then results in apparently $N_{\text{dark}} < 1$ even for very long exposure times (see also Section 4.10). Figure 4.29 shows some examples. The negative aspect of cooling results from the temperature dependence of the quantum efficiency, which results in a change of spectral response with temperature.

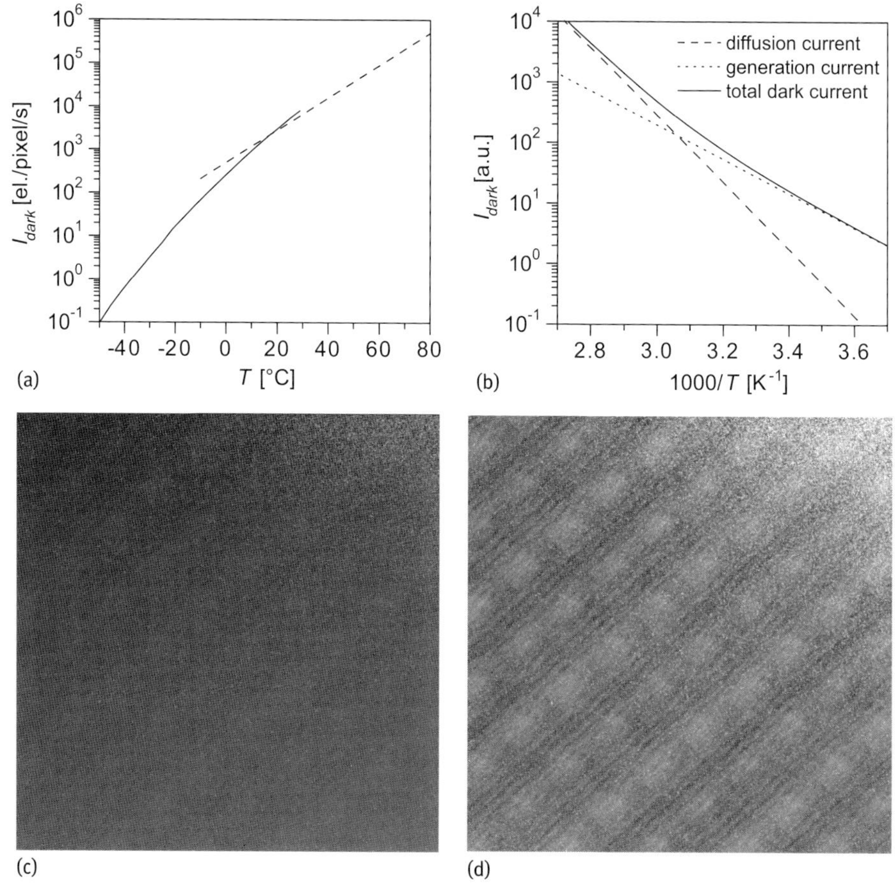

Fig. 4.29: Temporal noise. (a) shows that noise depends on sensor temperature (including the electronics). The two curves correspond to different scientific sensors. (b) shows a scheme of the different contributions to I_{dark} and the resulting total dark current as a function of inverse temperature $1/T$. As expected, I_{dark} follows a law given by Equation (4.26). Long exposure of a scientific camera equipped with a CCD at 10 °C (c) and 21 °C (d), respectively. These images have been rescaled differently for better visibility. The difference in dark current is more than a factor of 2. If cooled down to −50 °C, it is hardly possible to see noise effects.

Of course, the effect of dark noise is most pronounced when illumination is poor. Then the sum of I_{dark} and the rather small photo current I_{pe} (see Equation (4.13)) may be dominated by the first one.

As an example, a scientific camera may have a dark current of 7.5 electrons per second. By a reduction from room temperature to −40 °C, according to Figure 4.29 one may reduce this to 0.05 electrons per second per pixel. Consequently, for short exposure time this is less than the read noise of three electrons (see below) and even for a long exposure time this value may be acceptable. For imaging in astronomy, there are cameras that are cooled to approximately −100 °C and have less than one electron per hour per

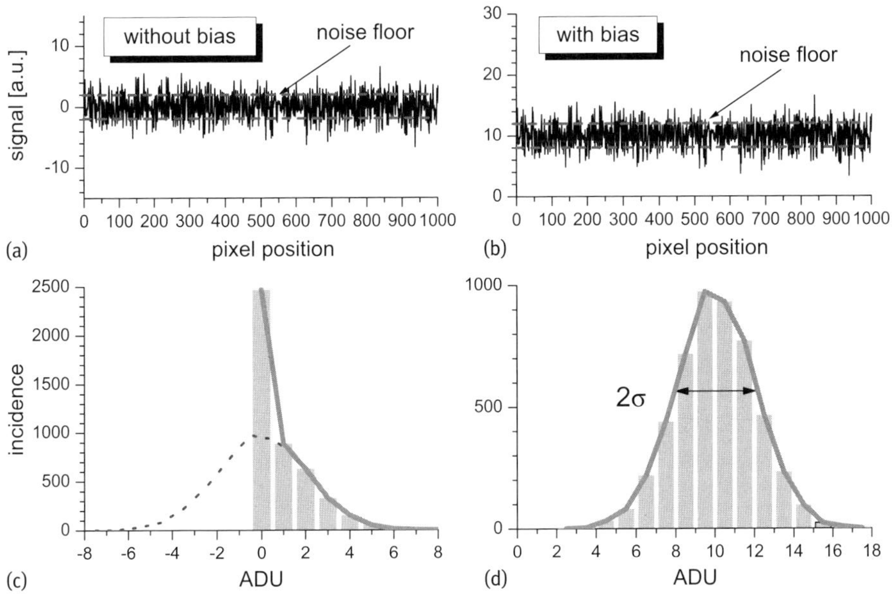

Fig. 4.30: Read noise distribution; this may be regarded as temporal fluctuations of the signal U_{out} of one pixel or, e. g., as a profile measured along a horizontal line of the sensor, which shows the fluctuations within a row of pixels (both considerations are equivalent). For discussion, see the text.

pixel. A rule of thumb is that the dark current is reduced by a factor of 2 every 5 to 7 °C. But it is important to note that if the detector is cooled below 0 °C it is essential to prevent formation of ice on its surface. This may lead to cracks, which destroy the sensor. To prevent this and to avoid thermal loads, usually the system has to be operated in a vacuum, at least the sensor surface because the full camera usually is not suitable for vacuum operation.

In addition to the noise that originates from thermally activated charges within the pixels of a PDA, noise is also generated by the readout amplifier (Johnson–Nyquist noise). This generates a voltage U_{out}, which fluctuates with positive and negative values around the average value, which usually is zero (see Figure 4.30a) or larger than zero, if an offset is applied (see Figure 4.30b). However, the corresponding noise power is given by the square of the output voltage and thus an effective voltage is measured at a resistor with resistance R_{out}, which is given by

$$U_{read} = \sqrt{\langle U_{out}^2 \rangle} = \sqrt{4 \cdot k_B T \cdot \Delta\nu_{ampl} \cdot R_{out}}. \tag{4.27}$$

This voltage depends on the temperature T and the readout amplifier bandwidth $\Delta\nu_{ampl}$. As an RMS value U_{read} it is always positive. Figure 4.30 shows the noise distribution. The fluctuations of U_{out} may be positive or negative. The average is zero; the broken lines indicate the $\pm\sigma$-values. The signal must exceed this "noise floor," but negative values will be clipped to zero upon quantization as the ADC always yields signals ≥ 0. Of course, this changes the read noise distribution (i. e., the histogram) *in the saved data files* of the image. In particular, some manufacturers, e. g., Nikon, set all negative values (indicated by broken line) to zero, and thus the resulting histogram of a dark field image does show a spike near zero ADU (Figure 4.30c) (for the discussion of histograms see Appendix A.6). It must be noted that such a diagram has to be interpreted correctly, otherwise the read noise estimated from it will be underestimated. To avoid such situations, other companies, e. g., Canon, add an offset or bias to the signal prior to quantization via ADC (Figure 4.30b). Thus, the resulting diagram reflects the noise distribution correctly (Figure 4.30d).

Furthermore, from Fourier mathematics it can be shown, that for a pulse with a given duration Δt and bandwidth $\Delta\nu$, the product $\Delta t \cdot \Delta\nu$ is always larger than a given constant, where the actual value of the constant depends on the pulse shape. This is the time bandwidth product TBP. We may also point at the similarity to the space bandwidth product SBP (Section 5.1.8) and SBN (Chapter 1).

A consequence of the TBP is the Shannon–Hartley theorem, which states that the maximum rate, which is the inverse of the readout time τ_{read}, at which information can be transmitted is proportional to the bandwidth of the transport channel.

As a result, the so-called read noise σ_{read}, which is given by U_{read} or by the corresponding electron number (see Equation (4.15); also as an RMS value) increases with a reduced readout time (of both, single pixel or full frame of the sensor):

$$\sigma_{\text{read}} = G_{\text{a}}^{-1} G_{\text{i}}^{-1} \sqrt{4 \cdot k_{\text{B}} T \cdot \Delta\nu_{\text{ampl}} \cdot R_{\text{out}}}. \tag{4.28}$$

Some examples of read noise for different cameras and sensors can be found in Appendix A.5. Typical values range from a couple of electrons per pixel up to 20 or more when the camera or the sensor chip is operated at room temperature. It may be reduced to five electrons per pixel in scientific cameras when operated at much reduced temperature. But one has to remark that read noise depends also on the gain or ISO setting of the camera (see below and also Section 4.8, in particular, Figure 4.53).

Thus, on the other hand, read noise basically can be reduced by increasing τ_{read}. This is realized in so-called slow-scan cameras (mostly scientific cameras), even though it should be noted that the term “slow” has a relative meaning, not an absolute one. It may be only really slow when compared to high-speed readouts, such as are possible with modern sCMOS sensors (see Section 4.10.5). Figure 4.31 provides examples. The clocking frequency is the inverse of the readout time τ_{read} and corresponds to $\Delta\nu_{\text{ampl}}$. According to the discussion on TBP, this is proportional to the inverse of τ_{read}. Hence, this finally results in $\sigma_{\text{read}} \propto \Delta\nu_{\text{ampl}}^{1/2} \propto \tau_{\text{read}}^{-1/2}$ (indicated by the solid line). The “error bars” of the data points (circles) should not be regarded as such ones, but instead they indicate variations of noise with operation conditions.

Examples of read noise for different commercial sensors and cameras are presented in Figure 4.32.

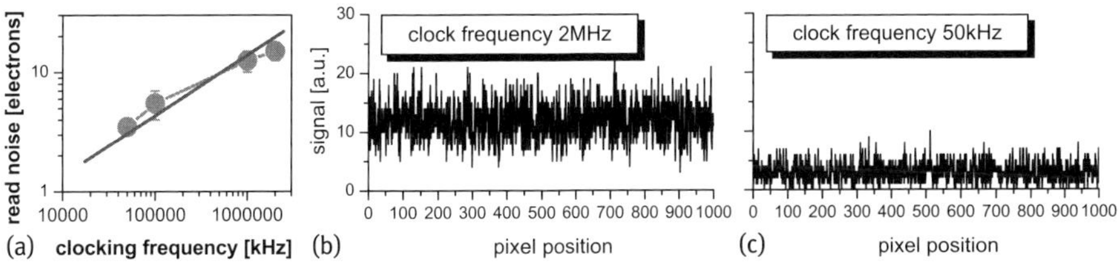

Fig. 4.31: (a) Read noise (RMS-value) dependence on clocking frequency for a high-quality scientific camera. (b) Read noise for two different clocking frequencies. Data correspond to the values shown in (a). The noise distribution follows Poisson statistics. Such a profile measured along a horizontal or vertical line of the sensor, respectively, may be obtained by taking the difference between two dark frames (background images; see Section 4.9).

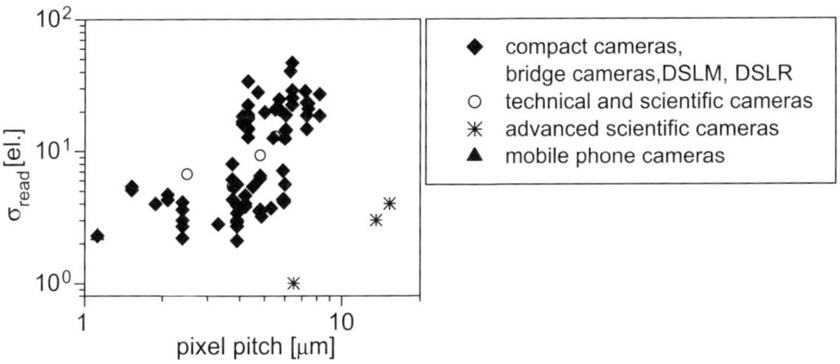

Fig. 4.32: Examples of read noise for different sensors and cameras (see also discussion in Section 4.8). The symbols correspond to those in Figure 4.11. Further examples are provided in Appendix A.5.

Finally, it should be noted that there is also a reset noise, which is due to capacitance reset prior to signal readout. It also has a thermal origin, and thus generates noise charge as well: $e \cdot N_{\text{reset}} = \sqrt{k_B T C_{\text{pix}}}$. Therefore, this is termed also "kTC noise" and is an RMS value. However, in CCDs this can be fully suppressed, in CMOS sensors it appears during reset but it can be eliminated, e. g., by a correlated double sampling method (CDS). A simple and brief explanation of CDS is that after reset, the residual charge $e \cdot N_{\text{reset}}$ can be read without affecting it. After the next exposure, the corresponding number of electrons, i. e., N_{reset}, is just subtracted from N_{pe} and read noise.

Dark current and noise, in general, depend also on the gain. Gain may be achieved by an additional output amplifier or via additional image amplifiers such as an MCP (see Section 4.11.3) that may be operated with a changeable high voltage (HV) across the plate(s), and thus operated with changeable gain.

But also simple consumer cameras or DSLR usually have an additional changeable internal "gain." This "gain" can be increased by increasing the "ISO-number." But as we will see in Section 4.8.8, this is neither a real gain nor the real ISO-number used for photographic films, respectively. This "gain" and this "ISO-number" are just equivalents that correspond to the same final effect. Nevertheless, Figure 4.33 shows the influence of those equivalents on noise.

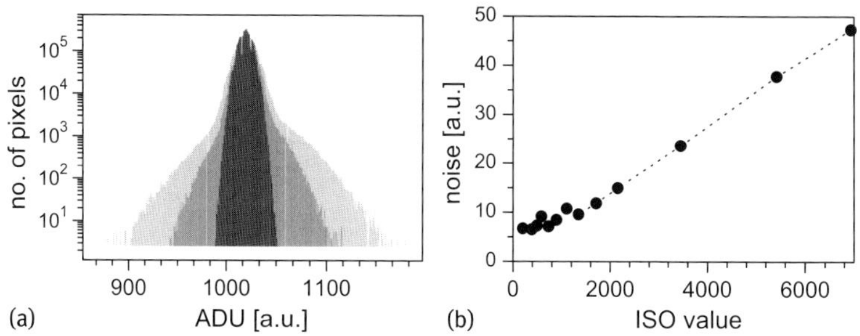

Fig. 4.33: (a) Evaluated noise from the red channel of dark images taken with a professional DSLR at different ISO settings (total noise; see Equation (4.34). The diagram shows the histograms. These correspond to the distributions shown in Figure 4.28 or Figure 4.30) for nominal ISO 100 (black), nominal 800 (dark gray) and nominal 1600 (light gray), respectively. (b) Variance values σ (i. e., the width, i. e., noise fluctuation; black circles) taken from (a) together with those deduced from further histograms (not shown here). The ISO value of the circles is according to the camera setting. For further discussion of ISO dependence of noise, see Section 4.8.8. Compare also Figure 2.16c.

4.7.4 Spatial noise

In the previous subchapter, we discussed noise associated with an individual pixel of a PDA. But although this noise *may be* almost similar for all pixels, it does not necessarily need to be exactly the same. Even more, the noise of the readout amplifier contributes as well. In the case of a CCD, due to the fact that the output amplifier is identical for all pixels, its noise is also exactly the same for all of them. However, for a CMOS, each pixel has its own one, and thus the gain may differ for different pixels. The result of all this is a fixed pattern noise (FPN). Due to the individual response of each pixel, even for a homogeneous illumination of the whole sensor, a nonuniform response occurs. This is a photo response nonuniformity (PRNU), whereas if illumination is turned off, a dark signal nonuniformity usually is present (DSNU; see also Figure 4.56). Typical values of DSNU for DSLR, compact cameras and mobile phones are 5 to 15 %. Typical values of PRNU for the same cameras (or sensors) are 0.3 to 1 % when equipped with CMOS sensors. For CCD and for further discussion, see Section 4.9.1. We would like to remark that the term FPN is not used everywhere in the literature consistently. Sometimes it is restricted to nonuniformities in the absence of illumination whereas it is also used more generally to describe the nonuniformities in cases with and without illumination. It may be interesting that pixel nonuniformity may also be used for forensic issues. Individual image sensors may be identified due to their particular unique noise patterns. On the microscopic level, pixel nonuniformity may result from irregularities of transistors, of microlenses or other ones.

Most pronounced individual responses occur for pixels that have a particular large dark current. This shows up as a large signal and as a bright (or white) spot in the image, independent of the actual illumination. Such specific pixels are called *hot pixels*.

We would like to note that bright spots occurring as single pixel events or as strong signals from several adjacent pixels may also occur at specific conditions such as camera operation in scientific experiments in the presence of hard X-rays or high-energy particle radiation. Examples include experiments with plasmas generated by very intense laser pulses, or imaging of astrophysical objects, where energetic particles or hard X-rays or cosmic rays play a role. The probability for cosmic ray events significantly increases for long exposure times. Further discussion and some examples may be found in Section 4.9.1.

4.7.5 Blooming, smear, image lag and cross-talk

Within this subchapter, several unwanted effects associated with strong light illumination, and incomplete readout are discussed.

4.7.5.1 Blooming

Blooming occurs when pixels are strongly saturated, i. e., when the photogenerated charges become larger than the FWC. Then charges spill over to neighbored pixels and/or to the shift register (see below). The result within the image is a very bright spot that has an extension significantly larger than expected from the size of the light source (Figure 4.34).

Blooming can be reduced by application of additional potential wells and overflow drains. But it must be noted that the required space reduces the fill factor. Due to their

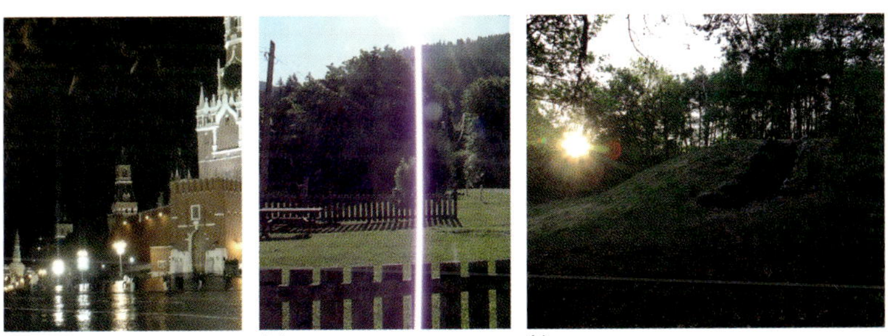

(a) (b) (c)

Fig. 4.34: (a) This image shows blooming in those regions where illumination is strong. These are the streetlamps, and in particular, the lamps in the lower center of the image. The image size of those light sources is much larger than that of the imaged object. Due to saturation, it cannot be reduced by post-processing of the image (e. g., by reduction of image brightness), it must be avoided when the image is taken. Although not too much pronounced here, smear could be seen clearly as stripes above the bright lamps in the middle of the image. More clearly, smear is seen in (b). In particular, such stripes could be present for strong intensity on the sensor (e. g., when exposure time is long). (c) Examples of blooming, flare and ghost images. (a) and (b) are crops of images taken with a CCD camera, (c) another crop, now from a CMOS camera.

sensor architecture, CMOS sensors are less affected by blooming. One has to note that antiblooming technology is of much importance if the number of pixels within a given sensor size should be increased. The reason is that this requires smaller pixels and then, due to the smaller FWC saturation, occurs much earlier.

4.7.5.2 Smear

When light penetrates deeply into the silicon bulk additional charges may be generated that then propagate to the vertical shift register of a interline transfer CCD (Section 4.4.2), which usually is light shielded. Due to a lower absorption coefficient, this is particularly the case for longer wavelengths. Moreover, there might be contributions from overflowing charges from saturated pixels and/or of strong light scattering from very strongly illuminated regions within the PDA. Subsequently, as a consequence, those unwanted charges then add to the regularly shifted charges of the individual pixels. Within the image, the result is a characteristic vertical bright strip, which commonly is called smear. This strip occurs below and/or above the bright image spot, which results from blooming (Figure 4.34). Of course, smear is expected to be more severe in systems without global shutter, when unwanted charges within the shift registers are still generated during readout process. Naturally, due to the absence of shift registers, CMOS sensors are not affected by smear. Smear may be (partially) suppressed by the photographer by reduced sensor illumination. Within advanced CCD, blooming and smear could be strongly suppressed.

Moreover, blooming may be accompanied by further disturbances such as flare and ghost images. These effects are caused by stray light and reflections within the camera system (see Section 6.8.2).

4.7.5.3 Image lag

If the charge transfer within the sensor is not complete, residual charges remain within the PDA. This results in a residual image underlying the following ones and is called image lag.

4.7.5.4 Cross talk

In some respect similar to blooming, cross talk is also due to charges that are present in regions where they should not be present. There are two contributions, namely optical and electrical cross talk. Electrical cross talk is due to signal charges that are generated within the pixel, i. e., at the correct local position, but then diffuse to a neighbored one (Figure 4.35a). There might be also a contribution from a thermal diffusion current from electrons generated by long wavelength light in the silicon bulk instead of the photodiode, that flows into the photodiodes. This also contributes to dark current noise. Optical cross talk is due to photons that enter the pixel at the correct local position, but then leave it (Figure 4.35b). One may note that due to a different layer structure CMOS may

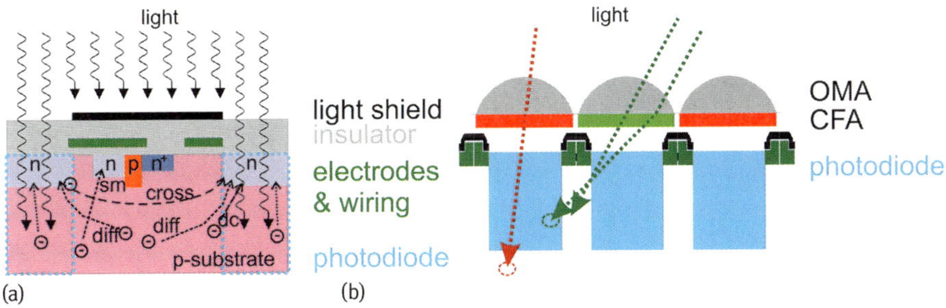

Fig. 4.35: Electrical (a) and optical (b) cross talk. As an example, (a) shows a detail of an interline CCD. "diff" denotes diffusion, "cross" cross talk, "dc" dark current and "sm" smear. The electrode of the vertical shift structure (see semiconductor structure below) and the control gate are shown in green. It can be seen that cross talk may also affect color information, e. g., photons passing CFA within a green filter may be detected by a pixel below a red filter.

be affected more strongly by cross talk than CCD. In general, also sensors with smaller pixels and larger interpixel regions are more affected by cross talk.

4.7.6 Total noise

The signal of the sensor is always affected by different noise contributions. We have discussed the most important ones. But we may comment that there are even further contributions such as dark noise drifts during readout, color noise (see below), digital artefacts, etc. However, we regard these particular contributions to be too special for the present book, and thus relate the interested reader to the literature (see, e. g., also [Nak06]). Nevertheless, we may add an interesting remark. Although it is expected that sensor noise is well characterized, this is not fully true. As an example, quite recently, namely this year (2018), a particular noise contribution of CCD sensor was recognized, which shows up as an "anomalous behavior of CCD readout electronics." With respect to a particular pixel, this effect results in an additional signal that originates from the signal of pixels that have been read previously (see[4]; this paper provides also a brief overview of sensor anomalies). For the purpose of the present book, this does not matter at all, however, for scientific applications, in particular for astronomical images; this effect may have some importance.

In general, to calculate noise for a single pixel, in principle all relevant contributions have to be added in an appropriate way. First of all, this is the photoelectron noise $\sigma_{pe,n}$ which results from the photon fluctuations (σ_{ph}; Equation (4.23)) and, in addition from fluctuations according to the quantum efficiency σ_{QE}. Due to Equations (4.11) and (4.14),

4 K. Boone et al.: *A Binary Offset Effect In CCD Readout And Its Impact On Astronomical Data*, arXiv:1802.06914 [astro-ph.IM].

one can show that

$$\sigma_{\mathrm{pe,n}} = \eta_{\mathrm{e}} \cdot \sigma_{\mathrm{ph}} = \eta_{\mathrm{e}} \cdot \sqrt{N_{\mathrm{ph}}} \tag{4.29}$$

The estimate of σ_{QE} needs a short discussion. As described in Sections 4.2.1 and 4.2.2, N_{ph} photons are converted into photoelectrons with a particular quantum efficiency η. For the moment, we will not discriminate between internal and external QE, but we will point out that this conversion efficiency follows a statistical process rather than a constant conversion factor. This means that an incident photon generates a photon electron with a specific probability p, but N_{ph} photons will not exactly generate $p \cdot N_{\mathrm{ph}}$ photon electrons. Instead, similar to the discussion in Section 4.7.2 there is a statistical distribution represented by the binomial distribution

$$p_{\mathrm{k}}^{(N)} = \begin{bmatrix} N \\ k \end{bmatrix} p^k (1-p)^{N-k} \tag{4.30}$$

where

$$\begin{bmatrix} N \\ k \end{bmatrix} = \frac{N!}{k!(N-k)!} \tag{4.31}$$

This describes that in case of N repeated measurements with one incident photon and where the measurements are independent of each other, k photons will generate $p_{\mathrm{k}}^{(N)}$ photo electrons. Similar to N repeated measurements, one can perform instead a single measurement, but with an irradiation of $N = N_{\mathrm{ph}}$ photons at the same time and $p = \eta$. This is illustrated in Figure 4.36 for $N_{\mathrm{ph}} = 10$ (a) and $N_{\mathrm{ph}} = 1000$ (b), respectively.

For this example, from Figure 4.36 (a) we see that, e. g., with a probability of $p_0^{(10)} =$ 2.8 % no electrons are generated, with $p_1^{(10)} = 12.1\,\%$ $k = 1$ electron and so on. The largest probability is found for $k = 3$ electrons, namely 26.7 %. $p_{10}^{(10)} < 10^{-3}$ and, of course,

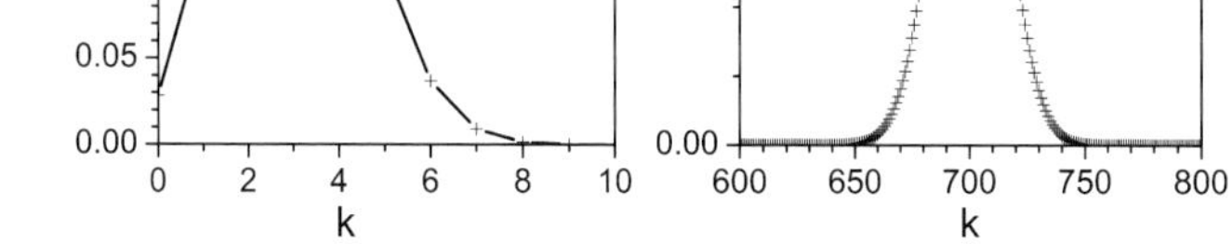

Fig. 4.36: Probability $p_{\mathrm{k}}^{(N_{\mathrm{ph}})}$ that k photo electrons are generated with a number of incident photons of (a) $N_{\mathrm{ph}} = 10$ with $\eta = 30\,\%$ and with (b) $N_{\mathrm{ph}} = 1000$ and $\eta = 70\,\%$.

the probability for the generation of more than 10 electrons is zero. The total number of generated electrons is given by the sum over $k \cdot p_k^{(N_{ph})}$, which is equal to $\eta \cdot N_{ph}$. In other words, the result of the statistical process yields (in average) the same result as Equation (4.11). Thus, it becomes clear that photo electron generation does not yield an exactly predictable value N_{pe}, but there is a distribution with a standard deviation σ_{QE}, which may be regarded as an equivalent to noise.

For large numbers of N, Equation (4.30) can be well approximated by a Gaussian distribution according to Equation (4.20) with $x = k, \overline{x} = N \cdot p, p = \eta$ and $\sigma = \sqrt{N \cdot p \cdot (1-p)}$. As illustrated in Figure 4.36 for $N = 1000$ and even for $N = 10$, this is a sufficiently good approximation. The noise equivalent resulting from the statistical photo electron generation process with a probability η and a fixed number of irradiated photons $\overline{N}_{ph}$ is given by

$$\sigma_{QE} = \sqrt{\overline{N}_{ph} \cdot \eta \cdot (1-\eta)} \tag{4.32}$$

Now, taking into account both, photon fluctuations and photo electron generation fluctuations, one can estimate the photo electron noise σ_{pe}. As both fluctuations can be well approximated by Gaussians, their convolution yields a Gaussians as well with $\sigma_{pe} = \sqrt{\sigma_{pe,n}^2 + \sigma_{QE}^2}$, and thus

$$\sigma_{pe} = \sqrt{\eta_e \cdot \overline{N}_{ph}} \tag{4.33}$$

Hence, it is important to note that, as noise is affected by the quantum efficiency, increasing η does not only enhance the sensitivity, but it also reduces noise. Figure 4.37 shows that for a rather low quantum efficiency the total photo electron noise is dominated by fluctuations during the generation process, whereas for a large value of η the most important contribution comes from photon statistics. For that reason, sensor improve-

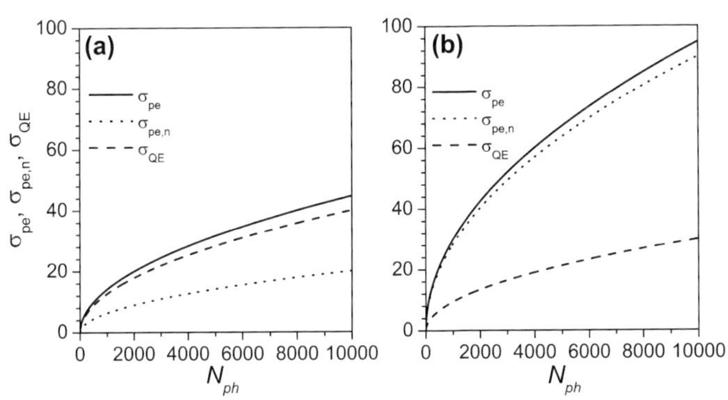

Fig. 4.37: Photo electron noise contributions as a function of incident photon number for (a) $\eta = 0.2$ and (b) $\eta = 0.9$.

ments take profit from improvement of semiconductor physics related to increasing QE (see Section 4.10).

Together with the dark signal and its fluctuations (σ_{dark}; Equation (4.25)), and the read noise σ_{read} this results in a total noise (in electrons, for a single pixel)

$$\sigma_{e,tot} = \sqrt{\sigma_{pe}^2 + \sigma_{dark}^2 + \sigma_{read}^2} \tag{4.34}$$

For noncorrelated contributions, the squares have to be added. For further discussion, see also Section 8.4. Equation (4.34) is valid for a CCD. For a CMOS sensor, one has to include additional terms which describe PRNU and DSNU. But one has to be aware that taking into account such spatial noise (see Section 4.7.4) changes $\sigma e,tot$ from being the total noise for a single pixel to that of the sensor or a comparison for different pixels that are irradiated nominally with the same photon signal. The noise resulting from PRNU and DSNU then can be derived from the equations provided in Section 4.9.1 (see also Section 8.4). Care has to be taken for present correlations, such as that of $G(i)$ in Equation (4.49) with the incident photon signal $W(i, t_x)$, which may not allow to add the corresponding additional noise as a square under the square root. Moreover, for any digital sensor there is noise due to data quantization (see Section 4.8.6; a detailed discussion can be found in [Eas10]). An extended discussion of sensor noise can be found in the EMVA1288 standard [EMVA1288].

For strong illumination, σ_{pix} is dominated by shot noise (Equation (4.23)). Quite the opposite, i. e., for low light conditions σ_{pix} is dominated by the last two terms. As an example, for a DSLR and not too large t_x, $\overline{N}_{dark} < 1$ electron and $\sigma_{read} \approx 3 \ldots 20$ electrons. Typically, the number of electrons in the dark current is of the order of 0.1 electron per second, which means that for a typical exposure time $t_x < 1\,\text{s}$ it does not contribute much. However, for long time exposure this may become different.

We would like to point out that the effect of noise on image quality is not the same for physical or technical measurements and for photography. For the former, this is quite clear and strongly related to the previous discussion. For the latter, additionally one has to take into account that human beings perceive noise differently in many ways. Perception not only depends on the scenery itself but it also depends on spatial structure size. A simple example for the first case is that noise is perceived as less annoying in sceneries taken at night or when fog is present. Also, within images of landscapes noise is often less disturbing when compared to images of sceneries with regular and smooth structures as present in architecture. Consequently, consideration of noise should not be restricted to the pixel level. A somewhat more general description of perception of noise leads, e. g., to the concept of visual noise (VN), which is briefly discussed in Section 8.4.

Up to now, we restricted to pure intensity fluctuations, namely to “luminance noise,” But noise also appears as a “color noise” or “chroma noise.” As noise is a statistical effect that does affect all pixels independently, a color signal that is the result of superposing the signal of several pixels behind the CFA (“RGB signal”) is obviously strongly affected by signal fluctuations. Quite simply, if due to a fluctuation within a region of 2×2 pixels behind a Bayer CFA, e. g., the signal of the "blue pixel" becomes lower than its average value and the "red pixel" provides a signal that exceeds the average value, there is a resulting red shift of the color at the corresponding position within the image. A consequence of chroma noise is also that an image of a homogeneous plane with only gray tones and taken with a color camera does not result in a black-and-white image, which means that it does not consist of gray tones only, but it consists of tiny color speckles (see example displayed in Figure 4.27c and Figure 2.16c).

4.8 Dynamic range, signal-to-noise ratio and detector response

4.8.1 Dynamic range

Usually, the dynamic range *DR* describes the ratio between the largest and the smallest possible power, fluence or intensity values that could be measured with a detector. But there are several definitions, and thus the "definition" of the dynamic range is not always strict. *DR* may be related to signals such as the detectable number of electrons or the charge within the pixel N_{pe} that is generated by an optical signal or the related voltage U_{out}, which is proportional to N_{pe} (see Section 4.2.2). Such definitions are related to the sensor, and thus are called output-referred. In a similar way, the input-referred dynamic range is related to the ratio of the brightest and darkest region within the scenery.

In the previous chapter, the minimum and maximum signal that can be provided by a single photodiode (pixel) has been discussed. Thus, on a pixel level, the dynamic range *DR* can be defined as

$$DR = \frac{N_{e,max}}{N_{e,min}} \tag{4.35a}$$

where $N_{e,min}$ and $N_{e,max}$ are the minimum and maximum values of N_e, respectively. We would like to note that here and in the following, there is no clip due to data quantisation by the ADC (see Section 4.8.6, otherwise this would set the maximum (digital) signal). We further would like to note that sometimes this ratio is termed contrast instead of *DR* and that more exactly, $DR = (N_{e,max} - N_{e,min})/N_{e,min}$. However, because $N_{e,max} \gg N_{e,min}$, Equation (4.35a) is a very good approximation.

In such a way, this characterizes the ability of the detector to sense dark and bright regions within a scene. However, instead of the direct ratio in Equation (4.35a), more often the dynamic range is expressed in logarithmic units, namely

$$DR = \log\left(\frac{N_{e,max}}{N_{e,min}}\right) \tag{4.35b}$$

where the logarithm "log" in principle may be with respect to any base (for further definitions of the logarithmic function, such as $\lg(x)$ and $\mathrm{ld}(x)$ see Appendix A.1). Within optical imaging, it is convenient to characterize exposure in exposure values EV, and thus it is reasonable to use the logarithm with a base of 2. Hence,

$$DR = \log_2\left(\frac{N_{e,max}}{N_{e,min}}\right) \equiv \mathrm{ld}\left(\frac{N_{e,max}}{N_{e,min}}\right) \tag{4.35c}$$

provided with “EV” or f-stops or bit, respectively. Please be aware that the values of DR according to Equation (4.35a), Equation (4.35b) and Equation (4.35c) are not equal. Thus, e. g., in Equation (4.35c), it may have been better to write $\mathrm{ld}(DR)$; however, this is not convenient. Usually, it is clear what is meant and thus we have avoided writing $\mathrm{ld}(DR)$ or defining a separate term such as “DR_{EV},” etc. Also, in general, mostly it becomes clear from the context if a given value of DR is in absolute numbers or given as a logarithm. In the first case, there is no attribute, and in the second, one EV or bit is added.

As another example, in the case of a black-and-white film with a locally varying transmission $T(x)$, it is reasonable to use the logarithm with a base of 10, and thus DR would correspond to the density range. This range is related to the optical density $OD(x) = -|\lg(T(x))|$ (see Equation (4.4)). Again, for simplicity here, we restrict to a 1D description, and in general one obtains $T(x, y)$.

According to the above definition of the dynamic range, DR describes the optical dynamic range ($N_{pe} \propto N_{ph} \propto W_{pix}$; see Equations (4.8), (4.11)), which is related to the optical signal $DR = W_{pix,max}/W_{pix,min}$. Due to Equation (4.15b), this is identical to the electric dynamic range of the voltages (i. e., amplitudes), namely $DR = U_{out,max}/U_{out,min}$, and thus in logarithmic units this yields $\lg(U_{out,max}/U_{out,min})$.

On the other hand, if related to electrical power, this is associated with the square of the voltages, e. g., as done in physics or in electrical engineering. Then DR is defined as

$$DR = \log_{10}\left(\frac{U_{max}}{U_{min}}\right)^2 bel = 20 \cdot \lg\left(\frac{U_{max}}{U_{min}}\right) \mathrm{dB}. \tag{4.35d}$$

If DR has the “unit” dB, it is indicated that, in contrast to above, the dynamic range is calculated by Equation (4.35d) (see also the example below). But we would like to comment that although usage of dynamic range in dB is quite common, it is not so well related to the incident light as the dynamic range in absolute numbers or in EV because U_{out} is proportional to N_{ph} or the intensity of the incident light, but U_{out}^2 is not.

For photography, $N_{e,max}$ is given by FWC and usually $N_{e,min}$ by σ_{read}. The signal has to exceed read noise or dark noise for long exposure time (see the previous chapter and Figure 4.38). However, e. g., if the detector has a certain threshold for signal detection, then $N_{e,min}$ is determined by this value. Hence, usually

$$DR = \frac{N_{full}}{\sigma_{read}} \tag{4.36}$$

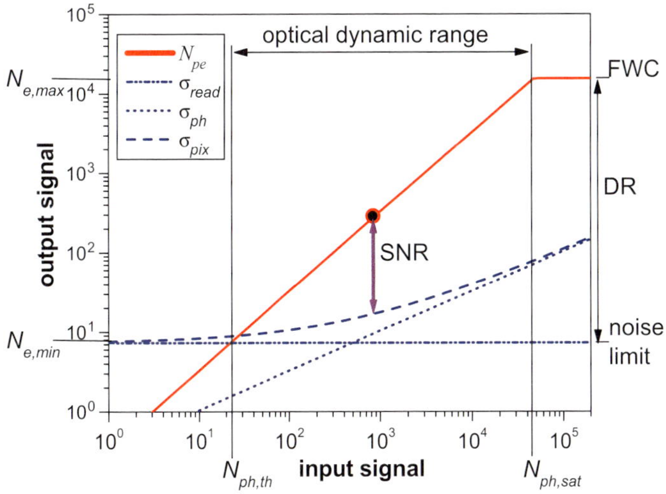

Fig. 4.38: Scheme of the photon conversion characteristics (detector response curve) and definition of dynamic range *DR*.

or *DR* is provided by the corresponding logarithmic values (see Equation (4.35b), etc.). Of course, for a more accurate calculation σ_{read} has to be replaced by σ_{pix}.

As an example based on these definitions, we regard a sensor with FWC = 15,000 electrons and σ_{read} = 10 electrons. Then we obtain the optical dynamic range $DR = 1500$ or in logarithmic units $DR = 10.5$ EV (i. e., 3.2 orders of magnitude) or $DR = 20 \cdot \lg(1500)\text{dB} = 64$ dB. Such values are usually tabulated in data sheets for sensors.

The corresponding minimum number of photons that are necessary to provide a signal beyond noise background, $N_{ph,th}$ ("threshold value"), and the maximum number of photons that could be collected within one pixel to get FWC, $N_{ph,sat}$, respectively, are given by

$$N_{ph,th} = \frac{\sigma_{read}}{\eta_e} \tag{4.37a}$$

$$N_{ph,sat} = \frac{N_{full}}{\eta_e}. \tag{4.37b}$$

We note that more correctly $N_{ph,th}$ is given by σ_{pix}, but close to the threshold this may be approximated by σ_{read}. A typical example for photon conversion characteristics (for detector response, see Section 4.2.2) is shown in Figure 4.38. More specific examples are presented later. Here, the input signal may be given by the number of photons prior to losses N_{ph} illuminating one pixel, by the fluence, the exposure or any similar related quantity. The output signal may be U_{out} or the number of electrons generated within one pixel, N_e. The lower limit of the input signal is given by one photon, while the lower limit of the output signal, in principle, is given by one electron, however more realistically the limit is given by the noise floor, i. e., $N_{ph,th}$ and threshold exposure, etc. The upper limit is set by saturation $N_{ph,sat}$. This corresponds to the saturation exposure. The corresponding output signals are $N_{e,min}$ and $N_{e,max}$, respectively, if the output signal is N_e. Otherwise, it is just the minimum and maximum detectable signal. As an example, the dot on the curve indicates a specific signal for which the SNR is shown as an arrow (see the following chapter).

Up to now, the discussion of the dynamic range has been restricted to that of a single pixel. However, for the full sensor or a whole system consisting of the optics and the sensor, that may be different. In Section 4.7.5, we have discussed that even when it is expected that a particular pixel is not illuminated besides, e. g., read noise of that pixel, further noise generation may occur due to blooming, cross talk, etc. But spurious signals do not only result from the interaction within the sensor system itself, but they may also result from an unwanted illumination, e. g., due to stray light and (veiling) flares (see Section 6.8 and also the discussion of macro contrast in [Nas08]).

For an explanation, let us consider the following situation (see also Section 8.2.2). The object consists just of a very small but rather bright source within a fully black area. Thus, it is expected that the image is also very small so that potentially only very few pixels are illuminated but all the other ones do not receive any photons. However, due to internal reflections within the lens system and/or stray light there might be the generation of a background signal $N_{e,min,opt}$ on the whole CIS or part of it, namely outside the image region (see also discussion at begin of Section 6.8). Such a situation may even occur, when the image of the object would be located outside the sensor region. In any case, this sets a lower limit for $N_{e,min}$ in Equation (4.35). Moreover, due to photon statistics this background signal is not just a constant, which may be subtracted as an offset. As $N_{e,max}$ is not affected, and if $N_{e,min,opt}$ becomes significant, the dynamic range of the optics is given by $DR_{opt} = N_{e,max}/N_{e,min,opt}$. On the other hand, as long as $N_{e,min,opt}$ or according to Equation (4.23) $\sqrt{N_{e,min,opt}}$ is significantly smaller than σ_{read}, the dynamic range of the system is not affected by the optics. Otherwise, it becomes lower than expected from Equation (4.36). This may be illustrated by a simple example of astronomical exposure, where an image of a dim star in the neighborhood of a bright star should be taken. In that case, the stray light haze of the bright star may prevent that the image of the dim star becomes visible. Moreover, the situation may become further critical in the presence of flares as discussed in Section 6.8.2.2 (see also Figure 6.52). In particular, from Figure 6.54 it can be recognized that depending on the locations of both stars, the dim star may have to be rather bright to become observable. We would like to mention as well that even not necessarily such problems can be avoided or reduced if the bright light source is located outside the object field that is imaged onto the sensor. Indeed, it is a challenge for lens designers to minimize those unwanted effects. The present example also shows that although in standard situations there might be not much restrictions due to DR_{opt}, it is not straightforward to provide the dynamic range of the optics, and thus of the whole system in general and for all photographic situations, in particular, for a system with low quality lenses.

4.8.2 Signal-to-noise ratio

Considering noise on its own is not very meaningful. It is only meaningful in relationship to a signal. Similar to the dynamic range, the signal-to-noise ratio SNR is defined now as the ratio of an actual signal N_e instead of its maximum value N_{full} to its minimum value (see example shown in Figure 4.38)

$$\mathrm{SNR} = \frac{N_e}{N_{e,min}} \tag{4.38a}$$

or in logarithmic units, e. g.,

$$\mathrm{SNR} = \mathrm{ld}\left(\frac{N_e}{N_{e,min}}\right). \tag{4.38b}$$

A typical example of SNR for conditions presented before, is shown in Figure 4.39. The diagram also shows the limiting values, i. e., the dominance of read noise at low input levels and that of photon noise at large ones. According to Equation (4.38a) in the first case, the SNR is given by

$$\mathrm{SNR} = \frac{N_e}{\sigma_{read}} \tag{4.39a}$$

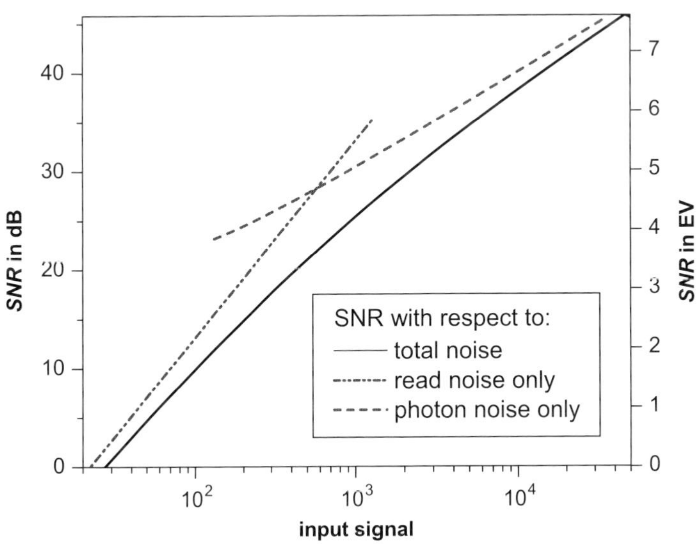

Fig. 4.39: Scheme of the SNR calculated for the conditions shown in Figure 4.38. Note: Of course, on a log scale, negative SNR values do not make sense because then the signal would be below noise level.

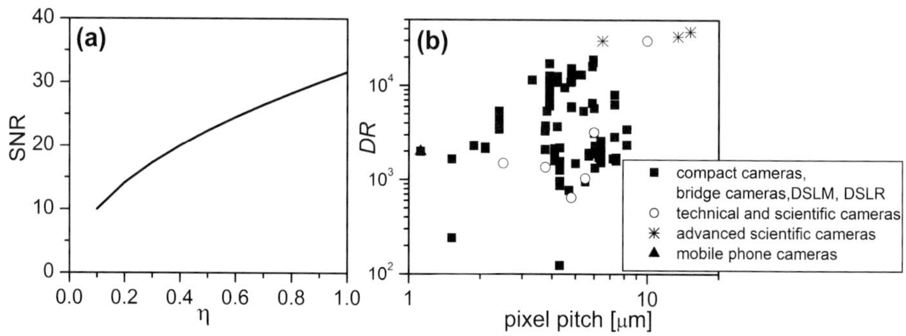

Fig. 4.40: (a) Example how an improvement of η leads to an enhancement of the SNR, here for $N_{ph} = 1000$ incident photons. In this diagram, SNR is restricted to photon noise only (b) Examples of DR of different sensors and cameras. The symbols correspond to those in Figure 4.11.

whereas in the second case,

$$\mathrm{SNR} = \frac{\eta_e \cdot N_{ph}}{\sqrt{\eta_e \cdot N_{ph}}} = \sqrt{\eta_e \cdot N_{ph}} \tag{4.39b}$$

Thus, it may be seen that an enhanced quantum efficiency leads to an increase of SNR (Figure 4.40a).

But one has to be aware that for colored pictures due to the usually strong wavelength dependence of η_e and N_{ph} there is not a simple common *SNR* value (see also Figures 4.10. and 4.24). From Figure 4.38 and Figure 4.39, one may also recognize that although *DR* covers 11 EV, the SNR is lower. The reason for that is nature, in particular, the limitation by photon noise when a large number of photons is incident on the sensor.

An important issue is that larger pixels are more sensitive and less affected by noise. Consequently, even if two cameras have sensors with the same pixel number, but different pixel size, they may have the same SBN (and SBP) but image quality may differ strongly.

For low light conditions when read noise is important, better image quality shows up as a smoother and more detailed image, because due to Equation (4.15a), Equation (4.15b) and Equation (4.39a) the signal, and thus SNR usually becomes larger for pixels with larger A_{pix}. Therefore, as both, signal current and dark current scale linearly with pixel area, according to Equation (4.25) read noise only scales with the square root of A_{pix}, which may be given by the square of the pixel pitch. As a consequence, SNR scales linearly with the pixel pitch. We may remark that this discussion is somewhat simplified, e. g., see Figure 4.32, as both dark current and capacity influence noise.

If more light is present, photon noise may dominate, but according to Equation (4.23) noise again only scales with the square root of A_{pix}. Then due to Equation (4.39b) the signal, and thus SNR become also larger for pixels with larger A_{pix} (linear scaling according to Equation (4.39b): SNR $\propto N_{pe}^{1/2} \propto A_{pix}^{1/2} \propto$ pixel pitch; see also Section 8.2).

In a similar way, the dynamic range also scales on the pixel pitch. Dividing the values of Figure 4.11 by those of Figure 4.32, one obtains Figure 4.40b. Hence, in total, Figure 4.40b may also be regarded as a typical comparison of a compact camera or a simple DSLR with a professional full format DSLR (remember also the brief discussion in Section 1.6.4). Finally, one has to be aware that SNR is more important than the absolute value of the noise and that for strong illumination this is dominated by the natural law of photon statistics rather than by the sensor characteristics.

4.8.3 Binning

To reduce noise and/or to increase readout, scientific or technical cameras often have the possibility of coupling several user-selected neighbored pixels together as some kind of macropixel prior to readout and digitizing. This method is called *binning* (see Figure 4.41). Binning is done as well in a few consumer cameras, including mobile phone cameras and, in particular, in a few DSLRs. But in those cases, the user cannot influence the results as this is done automatically in some the photographic modi or particularly in video modus.

Of course, due to its larger size, the charges of the participating pixels of a macropixel are accumulated. Thus, the signal is more strongly increased than the noise. The explanation for this is based on the same arguments as in the previous chapter. This may be illustrated in the following example. If one pixel receives a signal S then, according to Equation (4.23), the noise is proportional to $\sqrt{S}$. If four pixels are binned, the signal is $4 \cdot S$ and noise is $\sqrt{4 \cdot S}$, and thus SNR is increased by a factor of $\sqrt{4}$. Because binning is performed prior to readout, readout noise for a single and a macropixel is the same. Even though image performance with respect to noise is improved, usually this may be with the expense of a reduced spatial resolution, at least for good illumination conditions. However, for very faint illumination and single pixel readout the resolution may

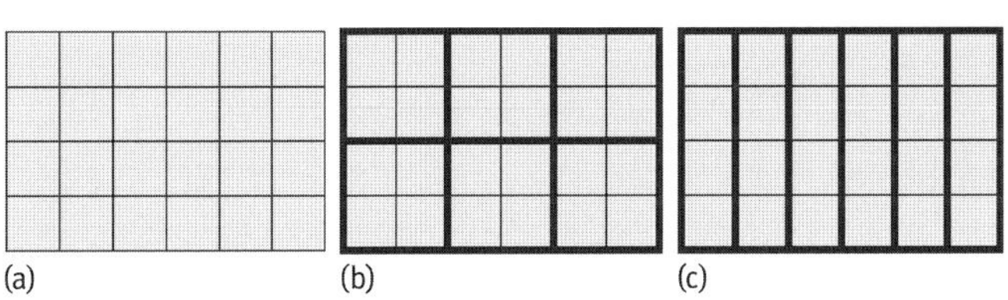

Fig. 4.41: Simple sketch of a PDA with (a) no binning, (b) 4×4 binning (indicated by the thick regions) and (c) binning of stripes (typically used for applications such as spectroscopy).

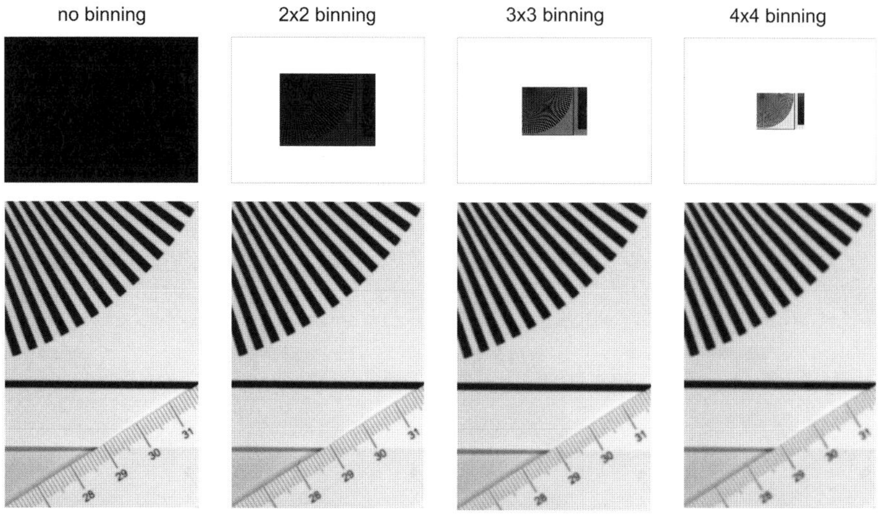

Fig. 4.42: Examples of binning. All images are captured at the same illumination conditions and exposure settings. Top row: from left to right (here for each of the images, the width of each macropixel is displayed on the same width as that of the original pixel): due to binning, the number of pixels in width and height, respectively, is reduced by a factor of 1 (i. e., original image without binning), 2, 3 and 4. Consequently, the total image becomes smaller. At the same time, the amount of light within the macropixels is increased by a factor of 1, 4, 9 and 16, respectively. The lower row shows a set of similar images, but now magnified to the same absolute size. Furthermore, the image brightness is changed to the same value so that the effect of a reduction of resolution becomes apparent. Note that 2×2 binning corresponds to the situation displayed in Figure 4.41b.

be destructed, and thus not necessarily better than when binning is applied. Figure 4.42 shows an example.

It may be remarked that binning is performed in hardware, namely within the sensor. This is superior to a later combination of pixels within the recorded image as some kind of post-processing, which does not improve SNR.

A further positive effect of binning is that due to the fact that the number of the macropixels is lower than that of the real pixels, readout is performed faster even with the same clocking frequency. A negative effect may be that FWC of the macropixel is the same like that of a single pixel, and thus saturation of the accumulated charges occurs earlier. An exception may be CMOS sensor binning in average mode or a CCD with special transfer pixel design with larger FWC. We would like to remark that the potential wells of the pixels of the readout register of some scientific CCD are larger than those of the light sensitive pixels, which then may avoid saturation effects unless binning of too many pixels with too much signal is performed. However, this plays only a role at high light level, i. e., where binning anyway usually makes no sense. Due to the readout principle, hardware binning is straightforward for CCD, whereas for CMOS sensors due to the independent pixels with own readout, it depends on sensor architecture whether hardware binning is possible or not.

4.8.4 Requirements

To judge the necessary dynamic range of a detector, it is obligatory to specify the requirements. For scientific imaging, the judgement simply depends on the requirements for the actual experiment and similar, this is for camera sensors used for industrial applications. For scientific or industrial purposes, images taken with high-dynamic range (i. e., exceeding 8 EV, see below) may be analyzed directly, and thus one may take profit of the large *DR* of an appropriate camera. This is straightforward and does not need further discussion.

For photographic imaging, the demand is related to the performance of the human eye as well as, here, in particular, its dynamic range. The demand may be related to the scenery, too, which the photographer would like to image. In the following part of this section, we will restrict ourselves to that subject.

As shown in Table 4.4, in principle, and in particular, when restricted to direct light only, the dynamic range that can be found in nature can be huge, namely eight orders of magnitude, i. e., more than 26 EV, respectively f-stops. It may be even larger, when one takes into account that photographers frequently make use of stray light as well. Although in a typical scenery the dynamic range is much lower, it is still very large and usually covers many orders of magnitude. Here, and in the following, of course, *DR* describes the range of brightness that can be recorded or displayed; color information is not a subject here. Consequences for colored images will be discussed later.

Now the question may arise, how large is the dynamic range of the human eye? The answer is not straightforward. The aperture of the eye is automatically adapted for different brightness regions, and thus covers a range of more than seven orders of magnitude, i. e., more than 24 f-stops. The pupil opens and closes, and thus the eye adjusts similar to automatically adjusted cameras, e. g., video cameras. On the other hand, if the

Tab. 4.4: Light conditions in nature. These may also be compared to the illuminance during a cloudy night without moon and without additional light (this leads to an illuminance of the order of 10^{-4} lx) and a bright day or a very brightly illuminated room (of the order of 10^5 lx). The upper end of the scale has to be extended if direct light of even brighter sources is included as well.

luminance [cd/m²]	10^{-3}	10^{-2}	10^{-1}	10^0	10^1	10^2	10^3	10^4	10^5
condition	clear star light	crescent	full	street light	sunset sunrise	overcast	overcast	hazy	direct
		moon light					sun light		

Tab. 4.5: Accessible *DR* of typical films, cameras (with CCD or CMOs sensor), scanners, etc. ("input devices"). The DR of output devices such as photographic prints and screens and also that of the eye is shown for comparison. See also the discussion in the text.

device	*DR*/EV
negative films	10–14
special films	>30
slide films	>10
mobile phone, compact camera	8
bridge camera	10–12
DSLR, DSLM camera	10–14
scientific and industrial cameras	8–16
scanners	10–16
paper prints	<8 (typ. 6)
typical screens and beamers	8–10
human eye (instantaneous)	10–14

adjustment is fixed, namely adapted to a specific region within the scenery, the "instantaneous dynamic range" is approximately between 10–14 f-stops only. But we would like to point to the logarithmic response characteristics discussed in the following section.

This may be compared to other devices (see Table 4.5). Film negatives usually cover 10 to 14 EV (3 to > 4 density ranges), and slide films may exceed 10 EV. Today such dynamic ranges are also reached by professional DSLR, some of them even get to $DR = 14\,\mathrm{EV}$ (or more, see Section 4.10) and scientific cameras (or industrial ones) may even surpass 16 EV. In the case of such a large DR, the image data are stored in special raw data files or, e. g., in 16-bit TIF files. On the other hand, the *DR* of the sensor chip of simpler cameras such as mobile phone cameras or compact cameras may also surpass 8 EV. Nevertheless, for those, the common output is a JPG file, and thus the *DR* of the image is limited to 8 EV.

However, image acquisition is only part of the whole process (see Figure 1.3). The photography is displayed and intended to be observed by human eyes. This subprocess mostly suffers strongly from the large restrictions of the available output devices. Here, we restrict to *DR* and also, in advance to Section 4.8.6, to depth resolution *dr*, whereas lateral resolution is the subject of Chapter 5. For instance, a classical photo print has a dynamic range limited to approximately only six f-stops because the black pigments still reflect typically 2 % of the incoming light. Thus, the *DR* of the displayed image, i. e., the print, is even smaller than that of any analogous or digital camera, respectively. Also, the *DR* of usual TV screens or other screens or beamers is smaller than 8 EV. For simplicity, in the following we will just write "screen" for all of these devices. Some modern screens may have $DR \sim 10$ bit and special high-quality professional screens even higher *DR*. It is important to note that we always refer to the "static contrast" since the usually accentuated "dynamic contrast" does not make sense for still images because this is related to the comparison of *different* frames. But to make use of a range larger than 8 bits differently processed and stored images are required (compare Section 4.9), otherwise again, the limit is set to 8 bits by the usually displayed JPG files. On contrary, the dynamic range of slides when imaged with an analog projector is mostly limited by the $DR \sim 10$ EV of the slide film. This is also the case for DS, which also exceeds 8 bits.

To conclude this situation, one may recognize that there are many advantages of the large *DR* of many detectors (see later also) but the displayed images are strongly restricted by the *DR* of the output devices. Thus, even today only professional image presentations with appropriate hardware may achieve the same *DR* as a high-quality slide film presentation or even surpass it (and possibly *dr* as well). All sensor–output device combinations require image processing (see Section 4.9) so that at least the *perceived* image shows a good result. Here, image processing of > 8 EV raw data can take profit (see later).

Although intrinsically obvious, finally we would like to remark that not only the *DR* of the input and the output devices influence the dynamic range of a viewed photography, but the surrounding conditions such as ambient light do as well. This is, e. g., because even an original black with a light signal of zero or quite close to zero from a screen or print may not be observed as fully black but as a dark gray. Regarding images from a screen in a really dark room improves the situation so that one may come close to 8 EV.

4.8.5 Detector response

Although the dynamic range is an important issue, both for scientific imaging (and industrial applications) and for photographs and of course also for the human eye, one has to also take into account the full response curve.

For instance, the natural photoreceptors of eyes have logarithmic response characteristics with the consequence that even small brightness differences can be discriminated in shadow or dark regions, whereas this is not possible in bright regions. In other words, the depth resolution is better for low light conditions than for strong light conditions. This corresponds to the Weber–Fechner law. Films behave similarly, and thus are somehow well suited to provide a good reproduction of the *perceived* light intensity distribution of a scenery. In contrast to that, many electronic detectors usually have more linear response behavior (see also Section 4.2.2), which then makes them well suitable as measuring devices in principle in science and for industrial applications. Thus, to reproduce the *perceived* light intensity distribution of the eye, the image taken with a CCD or CMOS camera usually has to be processed prior to observation (see also Section 4.9).

4.8.5.1 Response curves of films

Before we focus on the response curves of modern electronic detectors, first we would like to discuss the response curves of films because later image processing makes use of

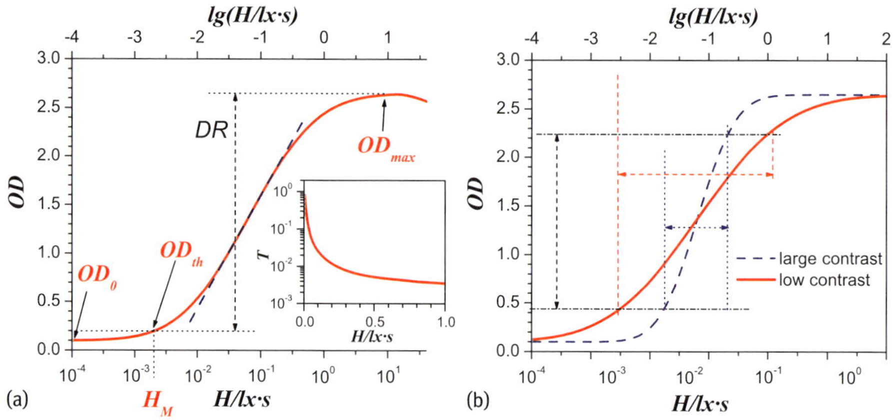

Fig. 4.43: (a) Density curve (tonal curve) of a film. The dotted line indicates $OD_{th} = OD_0 + 0.1$, namely the speed point. The insert shows the corresponding curve for the film transmission T, but now with H on a linear scale. (b) Two density curves with different γ-slopes, and thus different contrast. Remember that an approximate relation for monochromatic light at 550 nm is: $1\,\text{lx} \cdot \text{s} \approx 0.15\,\mu\text{J/cm}^2$.

procedures and the definitions related to them. Furthermore, as shown in Section 4.1, films still have some importance as detectors.

Figure 4.43 shows a typical response curve of a film. The optical density OD is plotted as a function of the luminous exposure H or instead of fluence F on a logarithmic scale. This curve is roughly s-shaped and strongly nonlinear and usually has a linear section in its middle. It is important to comment that "linearly" is only meant in the sense of the shown double logarithmic plot only. The curve starts with a residual density OD_0, which characterizes the fog. At the threshold value OD_{th}, exposure (0.1 above fog) begins, but the film may be still in the region of underexposure. The slope at the point of inflection is shown as a broken blue line. In the nearly "linear" middle section the film is exposed correctly and then transforms into a region where the film is overexposed until it reaches its maximum density at the saturation with OD_{max}. The range between OD_{th} and OD_{max} yields the dynamic range DR of the film. For even stronger illumination, solarization may occur, which means density reversal, namely a further exposure of the film does not increase the optical density but conversely results in higher film transmittance. This is not very pronounced for modern films. For further details on this and the following topics related to films, the reader may havc a look for special books on films or [All11].

The middle range is characterized by the slope γ of the "linear" section

$$\gamma = \frac{\Delta OD}{\Delta \lg(H/(\text{lx} \cdot \text{s}))} \tag{4.40}$$

and is termed as Gamma value. ΔOD denotes the difference between two different OD-values selected in the linear region and $\Delta \lg(H/(\text{lx·s}))$ the difference of the corresponding logarithms of $H/(\text{lx} \cdot \text{s})$. OD is defined by Equation (4.4). The value of γ does not only depend on the emulsion of the film, but also on the process of development. This is a similarity to electronic detectors as well.

Also, the ISO value of a film can be calculated from the density curve. First, one has to determine the "speed point," which is the minimum luminous exposure H_m measured at a value 0.1 above OD_0. Dividing $0.8\,\text{lx} \cdot \text{s}$ by this value yields the ISO number when the film is developed according to $\gamma = 0.8/1.3$. The arithmetic film speed S_ASA and the corresponding logarithmic speed S_DIN are then defined by

$$S_\text{ASA} = \frac{0.8\,\text{lx} \cdot \text{s}}{H_\text{m}} \qquad S^\circ_\text{DIN} = \left(10 \cdot \lg \frac{1.0\,\text{lx} \cdot \text{s}}{H_\text{m}}\right)^\circ \approx (1 + 10 \cdot \lg S_\text{ASA})^\circ \tag{4.41}$$

where H_m is the speed point, which has a value of 0.1 above the minimum value OD_0. Consequently, for instance, for $S_\text{ASA} = 100$, $H_\text{m} = 0.8\,\text{lx} \cdot \text{s}/100 = 8\,\text{mlx} \cdot \text{s}$ ($\approx 1.2\,\text{nJ}\,\text{cm}^2 \approx 33\,\text{photons}/\,\mu\text{m}^2$ for monochromatic light at 550 nm; see Figure 4.43a). This is compatible with the recommendation that an average exposure of $H_\text{av} = 0.1\,\text{lx} \cdot \text{s}$ is required for a standard image at a film speed of ISO100. The average exposure usually is a value that can be measured using an integral light meter. The film speed as defined by Equation (4.41) is based on the low-level signal similar to a speed based on noise for a digital system. This differs from the definition of speed based on saturation as discussed in Section 4.8.8 (in particular, Equation (4.46)).

As described in Section 2.5.1, the film speed S_ISO is a combination of both values but usually only the arithmetic film speed is indicated. For the example shown in Figure 4.43, the speed point is at approximately $10^{-2.7}\,\text{lx}\cdot\text{s}$, and thus the ISO number in the presented diagram is roughly 400.

If the density curve is strongly nonlinear, then the definition of γ may not be practicable. In such a case, the average gradient $\overline{G}$ of the middle section is more suitable. To estimate $\overline{G}$ as a first point on the density curve, one chooses the speed point and as a second one a point, which has an exposure that is $10^{1.5}$ larger. Although this leads to different results, one has to note that in both cases people talk about the gamma value.

Because the irradiance, and respectively, illuminance, on the film is proportional to the radiant intensity I_e respectively luminous intensity I_v emitted from or reflected by the object, from Equation (4.40) it becomes clear that due to detector response the ratio of two different values of I_e originating from different parts of a scenery representing the object contrast, is changed by the detector response (here the film):

$$\frac{T_\text{film 2}}{T_\text{film 1}} = \left(\frac{I_\text{e,1}}{I_\text{e,2}}\right)^\gamma = \left(\frac{I_\text{v,1}}{I_\text{v,2}}\right)^\gamma = \left(\frac{I_1}{I_2}\right)^\gamma . \tag{4.42}$$

We take account of the Weber–Fechner law in the same way a "γ-correction" is applied to display images or videos on screens.

Due to limitations of the image contrast $T_\text{film 1}/T_\text{film 2}$ and OD_1/OD_2 (y-axis), respectively, which originate from film fog and saturation, the reproduced object contrast

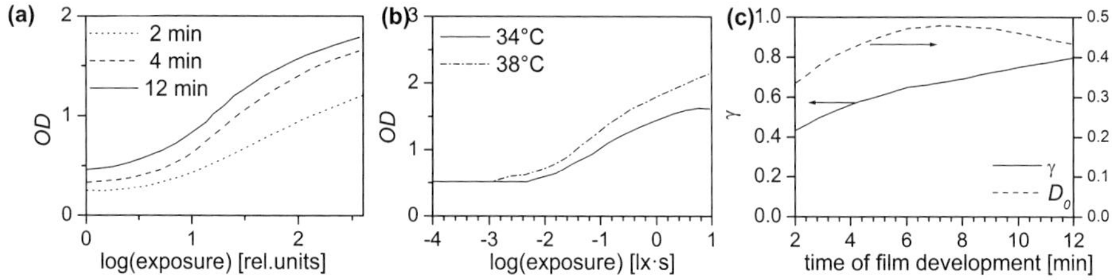

Fig. 4.44: Dependence of the tonal curves of the negative on development time (a), (c) or temperature (b). ((a), (b): Data taken from Agfa data sheets of Agfa XRS Professional; the development at the lower temperature yields a flatter purple curve and smaller grain; (c) Data taken from [Sch81]).

(x-axis) is limited as well. However, in principle, it is possible to adjust the output contrast with respect to the input by making usage of the γ-value. This is shown in Figure 4.43b where the density curves of two films with different γ-slopes are displayed. For a given object contrast of a scenery indicated by the vertical arrow, the latitude for both films is very different, indicated by the horizontal arrows of the corresponding color: one film offers a large tolerance of exposure for the photographer, e. g., if the exposure is not very well done (small slope). The other one provides more details in faint differences in object brightness (large slope). We would like to note that $OD = 0$ corresponds to maximum transmission (brightest signal) and a large OD to an opaque (i. e., dark gray or black) film. For reversal films, of course, a print then will yield the complement, i. e., dark print for low light, etc.

The photographer has the choice of γ-value by selection of the film. Moreover, the γ-value of a given film depends on the development time of the film and also for that for the prints (see the example in Figure 4.44). Furthermore, it may depend on the exposure time as well (see Section 4.1.2).

For color films, each of the different light sensitive layers (see Section 4.1.2) may have slightly different tonal curves. In particular situations, due to the Schwarzschild effect this may also lead to color shifts, and thus in addition a correction of exposure time may be necessary, the application of an additional color filter is recommended. Figure 4.45 shows tonal curves of different films. It is clearly seen that the dynamic range of films used for photography is approximately 10^3, which is approximately 10 EV.

4.8.5.2 Response curves of electronic detectors

The response curve of electronic detectors is very different from that of films. In particular, CCDs usually have a very linear characteristic curve (see Figure 4.46 and Figure 4.51). Under the assumption of a linear behavior of the output referred conversion gain G_{out} and according to Equation (4.15b) and Equation (4.11b) describing the inherently linear relation of photon to electron conversion, the signal is proportional to the fluence and exposure, respectively, with a proportional factor that could easily be deduced from the characteristic curve.

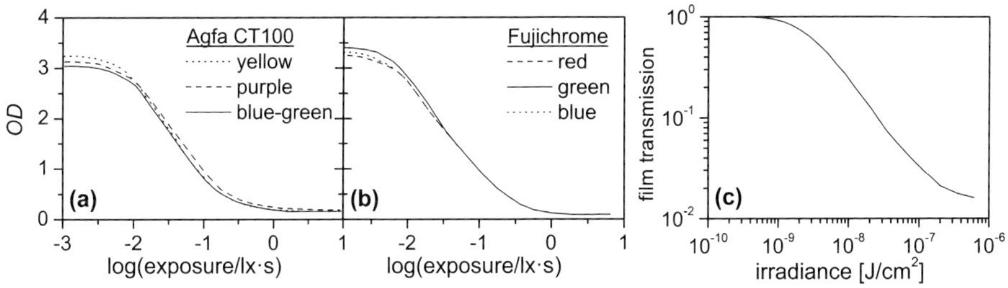

Fig. 4.45: (a) Density curves (tonal curves) of a Agfa CF100 slide film (a) and a Fujichrome Provia 100 F color reversal film (b). Data taken from corresponding data sheets. For comparison (c) shows the corresponding curve for a quite special (and hugely expensive) film without supercoat or surface protection layer, namely the Kodak 101-01. This film has been used, e. g., for measurements of the laser-plasma emission (scientific application), e. g., by one of the authors. If calibrated, it can be used to deduce the amount of emission. The measurements of the exposed film are performed with a microdensitometer and yield, e. g., an image of the plasma or the spectra in the soft X-ray and XUV range.

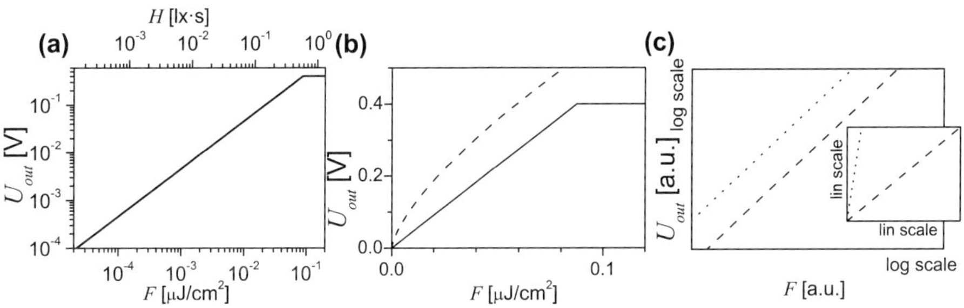

Fig. 4.46: (a) Example of a photon conversion characteristic of a CCD ($U_{\text{out}}(F)$). (b) Same plot but on a linear scale (solid line). (Data taken from [Nak06]). For comparison, a photon conversion curve of a CMOS sensor is shown as dashed line (Note: For better visibility, a different FWC than the CCD has been chosen). (c) Example of the photo response curve for a linear sensor with a high sensitivity (dotted line) and a sensor with a low sensitivity (dashed line).

On the other hand, usually the situation for CMOS sensors is different (see Section 4.2.1). Often the response is nonlinear and the output signal scales with a power of fluence F,

$$U_{\text{out}} = a \cdot F^b \tag{4.43}$$

where the coefficients a and b may have to be determined experimentally (for CMOS $b \leq 1$, for CCD, $b = 1$). This becomes even more difficult because those coefficients are slightly different for each of the individual active pixels. Equation (4.43) arises from nonlinearities within the pixel including MOSFET, signal processing, etc.

Other CMOS sensors may provide a logarithmic response. Of course, also in that case the photo current increases with the amount of incident light, but the output circuit may have an exponential voltage-current relation and with an ADC adapted to that situation (see Section 4.8.6). In any case, this requires a detector calibration. This is done within the image processor, which then finally yields a rather linear response (see Figure 4.51 and also Section 4.9). For scientific applications, this is even more essential. Here, proof of linearity should be made by the scientist and, if necessary, calibration as well. For high-image quality, there are even further corrections necessary (see later).

The sensitivity slope of an electronic detector can be easily extracted from characteristic curves such as presented in Figure 4.38, Figure 4.46 and Figure 4.51, respectively, which are based on Equation (4.15b) and Equation (4.11b). For a CCD this is straightforward, for a CMOS, usually this is measured at the half between the minimum and the maximum signal.

Figure 4.46a provides an example of typical output voltages as a function of fluence for $\lambda = 550\,\text{nm}$ light for a CCD with the following parameters: $A_{\text{pix}} = (5\,\mu\text{m})^2$, $G_{\text{out}} = 40\,\mu\text{V/electron}$, $\eta_e = 0.5$, $N_{\text{read}} = 12$, $N_{\text{full}} = 20{,}000$. At that wavelength, $1\,\text{lx} \cdot \text{s}$ corresponds approximately to $0.15\,\mu\text{J/cm}^2$. U_{out} is measured at the charge detection node and displayed on a double logarithmic scale. From those data, one may obtain a sensitivity slope of 300 mV per 67 nJ/cm^2, which corresponds roughly to 4.5 V per $\mu\text{J/cm}^2$ or 670 mV per lx ·s or 3.9 µV per photon on a pixel.

Figure 4.46c shows two curves of different sensitivity. The sensor with high sensitivity, in principle, does allow us to discriminate between smaller changes in the input signal, and thus allows better depth resolution (see below). As discussed before and also in Section 4.10.6 (see Figure 4.77c), the high-sensitivity curve may correspond to a sensor with a lower FWC (but this is not a must) and the other one may correspond to a larger FWC. The difference in FWC may, e. g., be due to sensor size, due to applied voltage (both see Section 4.10.6) or due to other reasons. The insert in (c) shows the same curves on a lin-lin scale.

In addition, there is a dependence on "ISO gain" (see Section 4.8.8) and due to the wavelength dependence of η_e a further dependence on wavelength of the sensitivity. As briefly discussed in Section 4.2.1, this is much pronounced for the X-ray range. There, e. g., for Si-photodiodes, one electron hole pair is generated per 3.62 eV photon energy, and thus one 500 eV photon generates approximately 138 photoelectrons, one 4 keV photon approximately 1100 photoelectrons and one 8 keV photon approximately 2200 photoelectrons. Even more complicated, a mixture, e. g., of the same or a different number of photons of different energy generate a number of photoelectrons, which may not be interpreted reasonably, and thus makes it difficult to interpret the whole image.

To avoid this situation, there are several possibilities. The first is just to use monochromatic radiation. Second, in the case of spectroscopic applications, wavelength selection may be done within the spectrometer and because then at each spatial position on the detector the signal is produced by photons of a single known photon energy only (in case that higher orders are absent). Consequently, during post-processing it is possible to correct the signal properly.

On the other hand, when operated in single photon counting mode, it is possible to use a detector such as a CCD for a direct measurement of an x-ray spectrum. This is rather convenient, e. g., for spectroscopy of laser-produced plasmas (see, e. g.,[5]). To do so, the distance between detector and source has to be large enough so that just by statistics it is very unlikely that within t_X any pixel is hit by more than one photon. Then such a single photon, which hits a particular pixel, generates a pixel-signal strength that expresses its energy. All pixels together then express the contribution of all photons with all energies. Sorting them by signal strength (i. e., by photon energy or wavelength, resp.), a histogram is generated and this is identical with the spectrum.

4.8.5.3 Comparison of the response curves of electronic detectors and those of films

In spite of the advantages of the linear response curve of electronic detectors, which is apparent for CCD and achieved for CMOS after correction, there is also a disadvantage, when compared to films, namely the abrupt cut-off at the saturation value unlike a smooth saturation value of films. This leads to clipping effects that are not present in images taken with films.

Figure 4.47 shows that the response function of a digital camera may show some variations, but it may be somehow similar to that of a film. Nevertheless, noise may be mostly lower for the digital camera, in particular at low light conditions and the dynamic range higher.

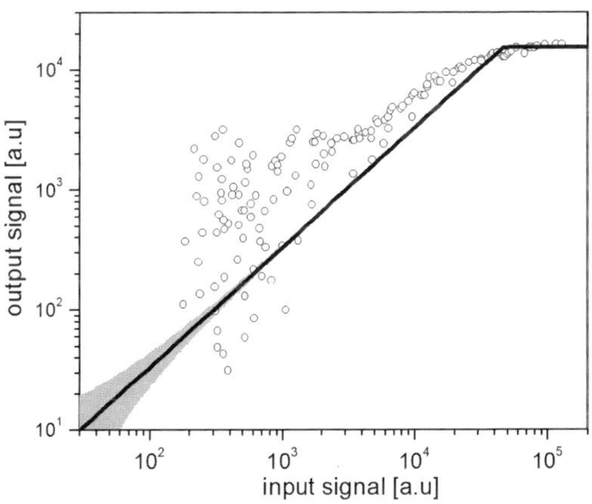

Fig. 4.47: Comparison of the photon conversion characteristics of a typical reversal film (scattered circles) and response curve of an electronic detector (solid line; that curve is the same as that displayed in Figure 4.38). The gray shaded area illustrates the signal uncertainty according to the total noise (broken curve in Figure 4.38).

5 W. Lu et al.: *Optimized Kα x-ray flashes from femtosecond-laser-irradiated foils*, Phys. Rev. E **80** (2009) 026404.

4.8.6 Data quantization and depth resolution

It is clear that a scenery of a given object contrast may consist of a smooth gray tonal distribution with faint differences. For simplicity, for the moment we assume that all gray values are occurring with an equal distribution, but even if this not the case, this does not change the principle of the following discussion. Now, the goal is to reproduce this as well as possible.

For an ideal digital system, this can be done by a linear "quantization" into as many steps or channels as possible. The quantization is done by an analog-to-digital (A/D) converter (ADC), which converts the analog output voltage (see Equation (4.15b)) into a digital signal $S_{pix} \propto B_{pix}$, which has the unit ADU (analog digital unit, sometimes also called digital number DN or digital unit DU or just "counts").

B_{pix} is called brightness, or sometimes "intensity," i. e., the signal of a pixel. Although, for convenience, partly we will use also the expression "intensity" for B_{pix}, it is clear that this is not the fully correct physical term. But we would like to note as well that the term "brightness" cannot be considered a well-defined expression. Sometimes brightness is related to the strength of the visual perception of an "image point" from a screen or a printout, respectively, but elsewhere it is regarded to be more or less the same as luminance. Within this book, we do not specify this very strictly because the meaning of B_{pix} comes clearly from the context. For instance, when used as described above, B_{pix} is directly proportional to N_e or the corresponding voltage and thus corresponds to the luminance. On the other hand, it may be modified to get better adaption of this photometric, value with respect to the response of the eye (see Section 4.9).

If we neglect background, noise, etc., S_{pix} and B_{pix}, both are proportional to U_{out}, which itself is proportional to N_{pe} (see Equation (4.15b)). If we put all proportional constants into a single factor, namely the conversion gain G_c, which includes G_a and G_i and has the unit counts per electron or ADU per electron, one obtains

$$S_{pix} = G_c \cdot N_{pe} \tag{4.44}$$

A real ADC has a limited number of channels available between the minimum and maximum value accepted by the ADC. The signals may be given by voltages or by real numbers or floating point values and converted into digital ones, which are integer values within the given limits. As an example, this is shown in Figure 4.48 for an ADC, which allows a gradation within 16 channels for 4 bit and eight channels for 3 bit, respectively. Here, we call the number of data steps or depth steps or number of channels *DS* which, of course, determines the depth resolution dr (= DR/DS).

Although the dynamic range *DR* of a detector system may be high, the *DS* may be not. In general, it is important to discriminate between those two quantities. From Figure 4.49a and Figure 4.49b, this is obvious.

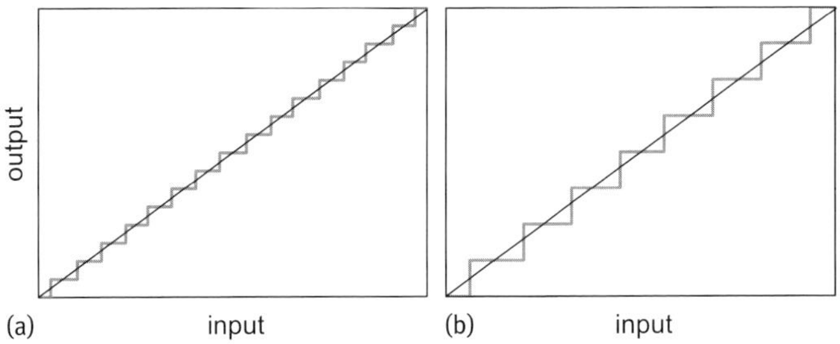

Fig. 4.48: Conversion of an analog signal (proportional to N_e, input, straight black line) to a digital one (S_{pix}, output, "step line") for two different numbers of bits n: (a) $2^n = 16$ channels, $n = 4$, (b) $2^n = 8$ channels, $n = 3$. The number of steps is $2^n - 1$.

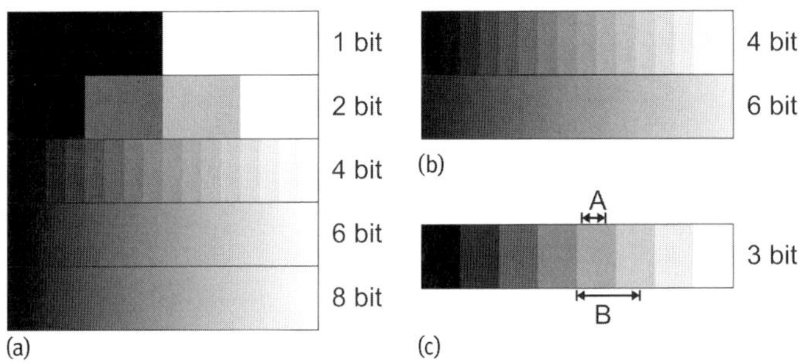

Fig. 4.49: Illustration of dynamic range *DR* and number of steps *DS*.

Figure 4.49a shows an artificial diagram with signals generated by five different sensors. All of them have the same dynamic range (i. e., the minimum and maximum signal that could be provided is identical for all sensors; here we have chosen $DR = 255$ and one count as the smallest value, which may result from noise). We may remind that prior to quantization, *DR* must not be given by an integer value. In this example, depth resolution of the sensors is different. *DS* is indicated at the side. All sensors are illuminated with exactly the same smooth linear tonal gradation. However, for some reason, the first sensor can only discriminate between two different brightness values (i. e., 1 bit), the next one between four (i. e., 2 bit), etc., and the last one is able to reproduce tonal differences within 256 different channels and 255 different steps. Of course, for sensors with larger depth resolution, e. g., corresponding to 14bit, faint tonal differences are even better resolved, but remember, to display that, e. g., within a printout or this book, one needs an appropriate output device that is able to reproduce those sensor signals.

Figure 4.49b shows that a sensor could have a larger *DR* but smaller *DS* and vice versa. In this example, the upper diagram displays the output of a 4-bit sensor, which has a dynamic range that extends from 1 to 255 (i. e., $DR = 255$) but only with 4-bit resolution (16-channel gradation, $DS = 16$). The image of the 6-bit sensor below has a smaller dynamic range, which extends from 20 to 220 only (i. e., $DR = 11$), but due to $DS = 6$ bit it can much better resolve faint tonal differences.

Before we continue, we would like to comment on the DS of usual consumer cameras. In the case of a monochromatic camera, the signal of each pixel is just the brightness value, which is stored as an 8-bit signal (or a 16-bit raw data signal for advanced cameras), i. e., as a gray tone on a scale of 256 different channels. For a color camera, the principle is the same. The only difference is that now in front of each pixel there is a color filter (due to the CFA). Consequently, the data are obviously still 8- (or 16-) bit and *could* be saved correspondingly. Only after image processing, in particular, "de-mosaicing," each pixel is *assigned additional* color information (see Section 4.9.3).

Hence, with respect to the data depth of the detected signals, color cameras with CFA are 8-bit cameras as well, although nearly all cameras manufacturers *advertise* that they are 3 times 8, i. e., 24-bit cameras. Although strictly not wrong, such declarations mostly lead to a totally wrong impression because prior to storage, due to image processing within the camera, full color information for all pixels is estimated. This additional information then requires the storage of all three-color channels for each pixel, which corresponds to $3 \cdot 8$ bit, but maximum DS of brightness remains unchanged. It is still 8 bit, i. e., $0 \dots 255$ different brightness levels and not 2^{24} different ones. A look for bit depth when the color image is converted to grayscales easily shows this as well. To conclude: these $3 \cdot 8$ bit do not correspond to 24-bit depth and also $3 \cdot 16$ bit for larger data depth of advanced cameras do not correspond to 48-bit depth. For the latter ones, data are saved as 16-bit values for each pixel in raw data files and de-mosaicing is then done within post-image processing. Here, the manufacturers use only 16 bits for each pixel and do not "waste" data capacity by saving $3 \cdot 16$ bit. Consequently, within this book we will talk about an 8-bit camera when its $DS = 256$, i. e., 8 bits and a 16-bit camera when its depth resolution corresponds to 16 bits.

Coming back to the general discussion of data depth, the question now is how many bits per pixel are necessary to get a good reproduction of captured image. From Section 4.7, it is known that due to noise, the uncertainty of the signal strength of each pixel is given by noise. Hence, it makes no sense to resolve the signal better than this; in other words, the corresponding channel width of the ADC should not be smaller. This is illustrated in Figure 4.49c for a 3-bit sensor: if the uncertainty due to noise is given by the arrow marked as "A" (if this is dr), then the displayed depth resolution may be well adapted. However, if the uncertainty due to noise is given by the arrow marked as "B" (if this is dr), then it is well seen that this exceeds the channel width, which does not provide a better *reproduction* of faint tonal differences of the scenery at all; it just leads to better resolution of noise.

This is also illustrated in Figure 4.50: from (a), one can see that if one regards a specific channel, which well shows the corresponding average analogue signal, due to strong noise fluctuations, the actual signal may be so strong or weak that it is sampled in the next channel or that one before instead. On the other hand, if the increments are large enough, so that a strong fluctuation is still mostly correctly recorded in the same

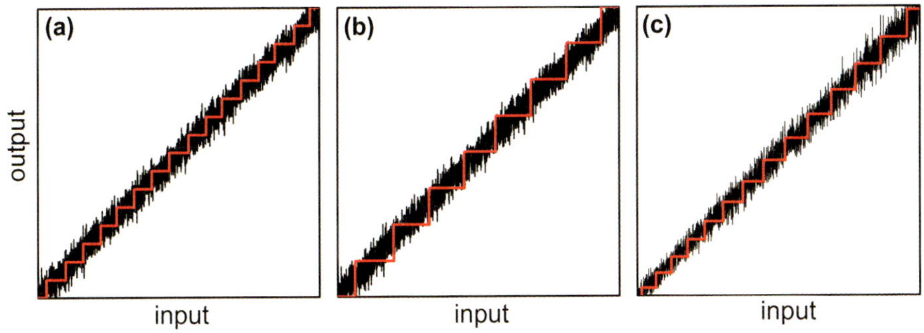

Fig. 4.50: Same as Figure 4.48, but now with an analog signal that is the subject to noise (black line). A 16-bit A/D conversion makes no sense (a), whereas an 8-bit A/D conversion may be appropriate (b). (c) shows a nonlinear conversion with an adaptive width.

channel as the corresponding signal without noise, then the A/D conversion is much better adapted (Figure 4.50b). Of course, the step size should not be significantly larger than the RMS value of the output voltage to keep depth resolution as high as possible. Of course, due to the statistical character of noise it cannot be fully avoided that the analog signal is sampled in the "wrong channel." This can be regarded as an additional noise. This quantization error is always at least 1 bit and also the quantization noise cannot be described in a straightforwardly simple way.

In summary, the gradation may be optimal when the step size is appropriate to the RMS value of noise, and hence the reasonable maximum number of steps is

$$DS_{\max} = \frac{N_{e,\max} - N_{e,\min}}{\Delta N_{pe}} \tag{4.45}$$

Here, the signal is given by N_{pe} with an uncertainty of ΔN_{pe}. As before, $N_{e,\min}$ and $N_{e,\max}$ indicate the minimum and maximum values of the signal. As ΔN_{pe} is given by noise (i. e., $\Delta N_{pe} = \sigma_{pix}$), usually it depends on the input signal (see Figure 4.38). Hence, ΔN_{pe} may be chosen in the range between σ_{read} and σ_{pe}, but of course to discriminate even signals with the smallest noise, it is preferable to define ΔN_{pe} by read noise and then accept some kind of overshooting at large signals (in the following subchapter, this becomes more clear). Even smaller steps just lead to a better "resolution of noise," but not of the signal itself, while larger steps would decrease depth resolution. Here, we would like to note that usually it is expected that 70 % of the N_{pe} values are within $N_{pe} \pm \Delta N_{pe}$. However, in principle, one can choose another confidence interval and of course, this then will affect $DS_{\max}$.

In the preceding discussion, $DS_{\max}$ has the same numerical value as DR. More generally, there might be a signal offset, however, if this is subtracted; again both values have the same numerical value. However, we would like to note that the meaning is somewhat different: DR provides the ratio of the signal range to the minimum signal, whereas $DS_{\max}$ yields the maximum number of steps within the signal range.

As a consequence, statements that a larger bit depth of an ADC leads to a better reproduction of subtle tonalities from a scenery becomes wrong when the decrease in step size (due to an increase in bit depth) is surpassed by noise. For instance, there are a lot of professional DSLRs equipped with 14-bit sensors, but due to the noise limit, the real bit depth i. e., DS_{max} is much below, e. g., 11 bits only. These DSLRs still may be of high quality, but they are far beyond the stated specified depth resolution or DS, which of course, is easily justified by measurements. Consequently, one should be aware of wrong statements on bit depth resolution.

A 14-bit ADC, of course, has a larger bit depth, and thus can discriminate more shades in principle, but this does not necessarily provide a better tonal *reproduction* of the analogous signal at all when compared to a 11-bit one. The only advantage of larger bit depth, i. e., an ADC with more channels, is that the image may contain more tonal grades, and thus it may look smoother. This advantage may be important to avoid posterization and banding. These kinds of artefacts result when, e. g., the object has a surface with a smooth continuous gradation of the colors. If this cannot be reproduced smoothly as well, at some positions the right “transition color” or its corresponding lightness or brightness between neighbored ones is missing (see, e. g., the example in Figure A.10). This results in clearly seen step-like color bands. At this point, the perceived image quality depends more on smoothness than on a correct reproduction of the colors within the gradient across the surface. Note that posterization and banding may also result from poor quality image post-processing or poor data compression. In particular, when the image is processed within an 8-bit depth instead of, e. g., a 32-bit floating point space, subsequent rounding may result in missing tonal values, which often becomes apparent.

Again, one has to be aware that when the bit depth of the ADC exceeds DS_{max}, not necessarily faint differences in gray tones observed in the image reproduce real tonal differences. One should also be aware of nature laws, such as photon statistics, which yield significant noise within the bright regions of an image. The capture of that noise cannot be avoided at all within the imaging process including the sensor, it can only be reduced artificially by post-processing.

According to that, we would like remark that the definition of the dynamic range is according to Equation (4.35), which may be larger or smaller than the “dynamic range of the ADC.” But it is clear that if it is larger, then the dynamic range is finally limited by the ADC or the range of the data output, i. e., the file format. Appendix A.5 shows examples of sensors and cameras that have $DR > 8$ EV, but the DR of the device is limited to 8 EV.

Finally, we would like to note that today it is common that the digitalisation follows the procedure described before (Figure 4.50a and b): all discretized brightness values differ according to the same noise interval. Usually, this corresponds to a constant step width and height (or channel) according to $\Delta N_{pe} = \sigma_{read}$, and as a result one accepts overshooting for large signals. Consequently, this leads a large number of channels and a large amount of data per pixel, especially for high-quality sensors, which have a low read noise and a large dynamic range. Then typically a 14- or 16-bit depth of data capacity is required.

It is obvious that in principle the number of bits could be reduced if the step width and height, respectively, could be reduced in an adaptive way as shown in Figure 4.50c. Instead of an equal or linear distribution of the brightness values within a given dynamic range, the distribution can be made nonlinearly within the same range and, in particular, it can take into account that photon noise increases with photon number (see Equation (4.23)). Thus, if, e. g., one assumes a dynamic range of 16 bit and for simplicity that photon noise is dominant, one can show that approximately 256 different gray values are sufficient to represent the signal within the full 16-bit range. This means that only 8-bit storage capacity is necessary (cf. Figure 4.50c), whereas a quantization with a fixed step width that is equal to that of the lowest channel would need 16-bit storage space for the up to 65336 different gray values. One has to note that, of course, this quantization has nothing to do with HDR, etc. (see Section 4.9.5). It also does not affect the display of the image, e. g., on a screen, where usually the dynamic range of that output device is 8 bit at best. Indeed, based on error analysis, detailed theoretical investigations and measurements have confirmed that the method of an adapted step height is feasible ("quasi lossless image data compression")[6,7]. Such a nonlinear data quantization could be realized (compare also "gamma curve" and Section 4.9.4) and, if well done, the number of necessary bits could be restricted approximately to a value that is given by the maximum SNR of the sensor. As a consequence, the application of the discussed method could save a large amount of data storage without leading to a significant loss in image quality. This is important because future sensors may have an even lower noise level and/or a larger FWC. Thus, without (lossless) data compression, data capacity will strongly increase and related data handling as well (interfaces, processing).

4.8.7 Examples of photon conversion characteristics

First, we would like to note that here and in the following two sections a linear detector response is assumed not only for CCD sensors, but also CMOS sensors. The latter is not straightforward for the sensor system discussed above. However, at least after possible post-processing by the image processor, the response curve usually is linear, too (see Section 4.9). This is displayed in Figure 4.51a. In this example, the measured photon conversion is provided for the red, green and blue channel, respectively, of a pixel behind the

6 B. Widrow, I. Kollar: *Quantization Noise – Roundoff Error in Digital Computation, Signal Processing, Control, and Communications*, Cambridge Univ Press, URL www.cambridge.org/9780521886710, 2008.

7 B. Jähne, M. Schwarzbauer: *Noise equalisation and quasi lossless image data compression – or how many bits needs an image sensor?*, tm – Technisches Messen 2016; **83** (2016) 16–24.

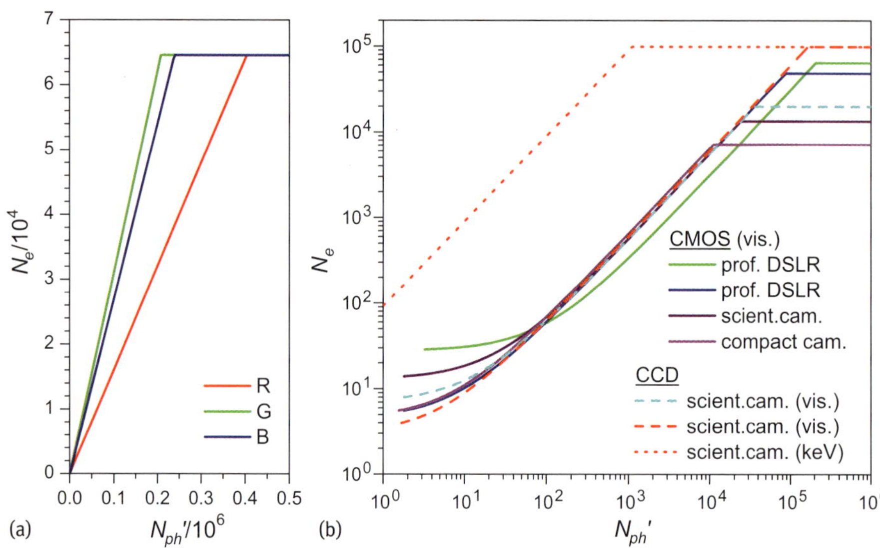

Fig. 4.51: Photon conversion curves for cameras used by photographers and such used for scientific or technical measurements. Some are equipped with CCD, others with CMOS sensors (data taken from Appendix A.5). (a) Measured data of a professional DSLR on a lin-lin scale. (b) Examples for different cameras. In contrast to (a), here the scale is log-log (hence, in principle, the curves in (a) and (b) are similar).

corresponding filter of the Bayer mask. Due to the quantum efficiency displayed by the solid lines in Figure 4.24a, here the signal for red light is significant lower than that for green and blue light, respectively. The green curve in (a) and (b) displays the same data and shows the FWC value of $N_{full} = 15{,}500$ and the noise level of $\sigma_{read} = 7.3$ electrons, which corresponds to 44 photons. The thermal noise at room temperature is less than 0.12 electrons per second, and thus negligible for $t_x < 1$ s. Hence, read noise dominates. Due to $\Delta N_{pe} = \sigma_{read}$, and the preceding discussion, $DS_{max} = N_{full}/\sigma_{read} = 2100$.

On the other hand, noise near saturation is given by photon statistics (Equation (4.22)). Hence, for $\eta_e = 33\,\%$, $N_{ph,sat} = 47{,}000$ (see Equation (4.17) and Equation (4.37b)) and due to Equation (4.23) and Equation (4.11b) this corresponds to ΔN_{pe} of approximately 70 electrons. Therefore, $DS_{max} = 320$ steps (or 8.3 bit) would be sufficient. But due to the fact, that a common step size in depth should not exceed even the lowest uncertainty range, the distribution into 2100 steps corresponding to 11 bits is reasonable although this leads to an "overshooting" when light illumination is not weak. This value is confirmed by independent OECF measurements (see Section 8.4). "Overshooting" is even more distinct, when compared to the camera specification of 14 bits (see the discussion at the end of the previous subchapter).

Appendix A.5 provides further opto-electronic properties of selected cameras. Photon conversion curves of some of those cameras including the previous example are displayed in Figure 4.51b where the input signal is given by the number of incident photons

N_{ph} and the output signal by the number of electrons N_e. N_e is the sum of photoelectrons N_{pe} and electrons due to noise. This is in contrast to Figure 4.38, where just N_{pe} is plotted as a function of N_{ph}. Here and in similar diagrams, N'_{ph} should indicate the number of photons on the sensor surface. It makes not too much sense to consider the number of photons incident on the sensor system including CFA, OMA, etc., i. e., the photon number N_{ph} prior to geometrical losses and losses resulting from additional filters in front of the sensor. Nonetheless, the principle of the example remains unchanged if one considers N_{ph} instead of N'_{ph}, only the numbers in the example change. This corresponds to a horizontal shift of the curves.

As another example, Figure 4.51b shows that the photon conversion not only depends on the camera but also on its usage. Here, in particular, for the same scientific camera with a backside illuminated CCD sensor, the detector response curve, DR and DS_{max}, are much different when used for the visible (red dashed curve) and keV range (red dotted curve), respectively. In this particular example, $N_{full} = 10^5$ and $\sigma_{read} = 3$. The thermal noise at the operation temperature of −40 °C is less than 0.05 electrons per second, and thus negligible. In the visible range, η_e is approximately between 0.1 and 0.7, whereas at 1 keV corresponding to $\lambda = 0.1$ nm, $\eta_e = 90$ (see Section 4.2.2). Consequently, $DR = DS_{max} = 10^5/3 = 33{,}000$ (i. e., 15 bit) in the visible range. However, at 1 keV the minimum signal is given by 1 photon, and hence $N_{e,min} = 90$ and the maximum signal still by $N_{e,max} = N_{full}$. As a result, $DR = 10^5/90 \approx 1100$ (i. e., 10 bit) which only shows that even very expensive very high-end cameras cannot beat natural laws. Due to the low value of σ_{read}, noise is dominated by photon noise, which ranges from $\sigma_{ph} = 1$ for 1 incident photon to $\sqrt{1100}$ at the saturation value. Thus, the appropriate step depth is $\Delta N_{pe} = \sigma_{ph} \cdot \eta_e = 90$, and hence again DR and DS_{max} have the same value.

4.8.8 "ISO-gain" for digital sensors

The sensitivity of a film as the detector is clearly described by its ISO value (see Section 2.5.1 and Section 4.8.5). Hence, a change of detector sensitivity means to exchange the film to another one with a different ISO value. For electronic detectors, the situation is different. CCD and CMOS sensors have an intrinsic sensitivity that is determined by the silicon structure and the quantum efficiency of the sensor itself. In addition, the transmission of the optics and filters of the camera system is fixed, which altogether yields a sensitivity that cannot be changed. For that reason, the "ISO gain" of a camera is not a real gain but an artificial one. For image intensifiers such as MCP, this is even more different (see Section 4.11). Thus, the ISO value that cameras usually allow to set has to be regarded differently, in particular, as a parameter used for data readout, etc.

The principle of the application of an "ISO-gain" is as follows. Let us regard the potential well of a single pixel. In principle, this could be filled with electrons up to FWC, which typically is the situation for ISO 100 (note that the maximum value of the

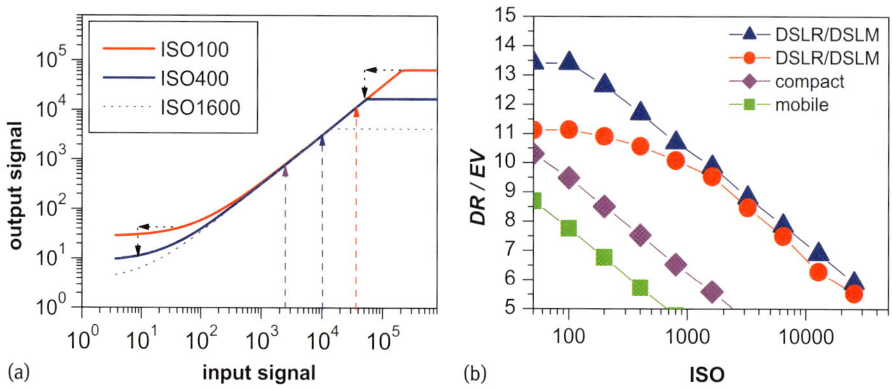

Fig. 4.52: (a) Principle of the application of an "ISO-gain." (b) Dynamic range as a function of ISO number (for the values given by the camera settings) for two different professional DSLR from different manufacturers, a compact camera and a high-end mobile phone camera.

ADC is set to FWC). When this is changed to a number, which is a factor of 2, 4, 8, etc. and higher, then the charge accumulation within the pixel is stopped as soon as 1/2, 1/4, 1/8, etc. of the FWC is attained. Thus, the pixel well is "regarded to be full" at only a fraction of its FWC (this fraction is the new maximum value of the ADC), and hence a smaller amount of light is recorded. Subsequently, during the readout, this signal is amplified electronically. This introduces some additional noise, which may be rather small for smaller ISO values such as 200 or 400, but significant for high ones. The trick within this procedure is that the amplification is applied before analog/digital conversion, which leads to a reduction of quantization errors.

The effect of this kind of "ISO-gain" is illustrated in Figure 4.52a, where the gain curve of a professional DSLR at ISO 100 is shown in red. Applying an "ISO-gain" of 400 leads to the blue curve. It can be clearly seen that this corresponds to a reduction of FWC by a factor of 4 and also σ_{read} is almost reduced by this factor. Thus, the blue curve is approximately obtained by a shift of the red curve as indicated by the dotted arrows. For ISO 1600 (broken curve in magenta), FWC is reduced nearly by a factor of 16, but σ_{read} only by a factor of 12, which then reduces *DR* to 1200 or a 10.2 bit only. It may be noted that the ISO values are those given by the camera settings; the real ISO values are 72, 285, 1093. The dashed arrows indicate the points where the signal has reached 18 % of its maximum value ("18 % gray"; see Section 4.8.9).

Figure 4.52b shows that for higher ISO values there is a loss in dynamic range. Here, *DR* is calculated according to Equation (4.36), which corresponds to a SNR = 1 for the lower limit of the range. On the other hand, it may be more reasonable to use a value for the lower limit that is clearly above the noise level, e. g., SNR = 3.

Of course, to get approximately the same amount of counts or ADU when images of identical scenery and identical illumination are taken at different ISO settings, the conversion factor in Equation (4.44) has to be changed accordingly. This is shown in

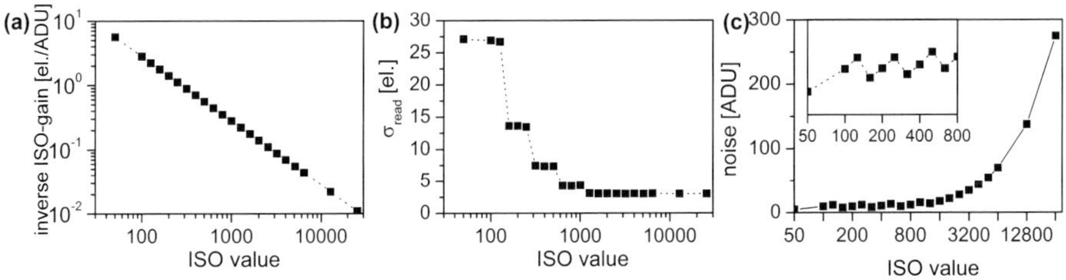

Fig. 4.53: (a) "Inverse gain" (G_c^{-1}; note that doubling the ISO number corresponds also to the application of half of the "gain" so that absolute values of images taken by different ISO values are the same) and (b) read noise as a function of the ISO value. If curve (b) is divided by curve (a), one gets plot (c). Note that the so-called "unity gain ISO" discussed by some people can be obtained from (a): it is that ISO value where the inverse ISO gain is equal to 1, i. e., where one electron leads to one ADU (it is common to assume a 12-bit ADC to get comparable values). As in the log-log plot, the dependence of the inverse ISO-gain on the ISO value is linear, the "unity gain" may be used to characterize this curve.

Figure 4.53a. The procedure of application of an "ISO-gain" described above also affects the read noise. Let us, e. g., compare the read noise $\sigma_{\text{read}}(\text{ISO } 400) = 7.4$ with $\sigma_{\text{read}}(\text{ISO } 100) = 27$ (see Figure 4.53b). In the ideal case, the SNR of the camera is not changed when the ISO setting is different. Thus, both noise values have to be the same, in other words this should be ISO invariant, and consequently the image quality should not be affected (for the two examples in Figure 4.52b, this is the case for low ISO values; note the different behavior of the camera sensors). But in real cases, noise even increases with the ISO number (see Figure 4.53c) and the dynamic range decreases (see Figure 4.52b; sometimes the ISO value where this occurs is called the "ISO invariant at ISO"). This can be seen if we multiply the σ_{read} values with the appropriate inverse gain values G_c^{-1} (see Equation (4.44)), namely 1/(2.8 electrons per ADU) $\cdot\sigma_{\text{read}}(\text{ISO } 100) = 9.6$ ADU and 1/(0.7 electrons per ADU) $\cdot\sigma_{\text{read}}(\text{ISO } 400) = 10.6$ ADU. One can see that they are approximately equal. At larger ISO numbers, the situation becomes worse (see Figure 4.53c), which is the reason for the reduced dynamic range (Figure 4.52b).

We would like to comment that beside a constant factor, plot (c) is the same as Figure 4.33b. The data points in (c) and Figure 4.33 are obtained from different measurements, but the main difference here is that the data points here[8] are corrected to get the right ISO value (see Figure 4.54), those in Figure 4.33 not yet (here for tutorial reasons). It is apparent how noise increases with the ISO number. Note that the "oscillations" in the insert in (c) are real; they originate from the processing (hardware and software) within a specific camera (see below). Consequently, noise can be minimized slightly by avoiding specific ISO settings.

The reason for the increase of noise at larger ISO numbers is that any noise added from the sensor readout prior to amplification is further amplified. Within a simple

8 Data taken from photonstophotos website by W. Claff: www.photonstophotos.net

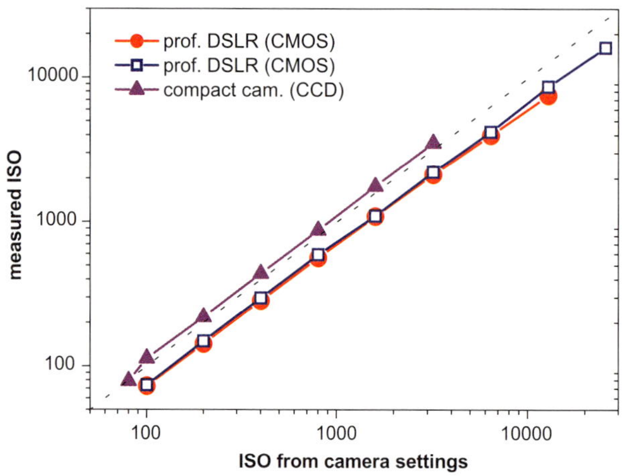

Fig. 4.54: Deviation of the real ISO values (measured) from the values stated by the manufacturers as the camera settings for three different cameras. The black dotted line corresponds to camera settings that would agree with the real ISO values. As may be seen, the ISO numbers of some cameras are lower than the real ones, but there are also examples where they exceed the real values.

model, one may discriminate between noise that originates before and after the amplifier, and as usual, one must add them quadratic. Note that the noise that originates before amplification has to be multiplied by the gain before it becomes squared. Then, at low ISO numbers, the total read noise is dominated by the read noise that originates after the amplifier and at large ones by the other term. Some manufacturers may make use of a two-amplifier scheme, which somewhat complicates our noise discussion.[9] That may also be the reason for the "oscillations" in Figure 4.53c.

The real "ISO gain" per definition is given by

$$\mathrm{ISO} = \frac{78}{H_{\mathrm{sat}}/(\mathrm{lx}\cdot\mathrm{s})} \tag{4.46}$$

where H_{sat} is the luminous exposure at the saturation value at FWC. This is the usual definition of "speed" for electronic sensors and termed "saturation-based ISO". It differs from the definition for the speed of films given in Equation (4.41), which is based on the low-level signal similar to the noise-based speed.

For example, the red curve of Figure 4.52a corresponds to FWC of $N_{\mathrm{full}} = 61{,}100$, and thus the corresponding number of photons is $N_{\mathrm{ph,sat}} = 1.8 \cdot 10^5$. Thus, for 6.4 µm pixels illuminated with green light, one obtains approximately $H_{\mathrm{sat}} \approx 1.09\,\mathrm{lx}\cdot\mathrm{s}$ ($F_{\mathrm{sat}} \approx 0.16\,\mu\mathrm{J/cm^2}$), and hence ISO = 72.

9 Web information by E. Martinec, University of Chicago; see http://theory.uchicago.edu

The real ISO values according to its definition (Equation (4.46)) in comparison to those stated by the manufacturers as the camera settings are displayed in Figure 4.54 for two camera examples shown in Figure 4.52b together with another one. These are typical examples, because for most cameras such deviations are usual. The reason for this is that sometimes manufacturers "calibrate" the cameras to a different gray level to "improve exposure conditions." Finally, we would like to remark that for some DSLR cameras, ISO 100 is the "natural" value and the ISO-50 value leads to nothing else than the ISO-100 setting together with an exposure correction of +1 EV.

4.8.9 The "universal" curve

Usually, photographers do not work directly with diagrams such as shown in Figure 4.38 or Figure 4.51. Instead, they are accustomed to work with EV as the input signals and with an output signal that is displayed linearly between 0 and 1. To obtain such a curve, one as to take the logarithm of the values of the x-axis, in particular $\mathrm{ld}(N_{\mathrm{ph}})$, which yields the x-axis in EV. Note that +1 EV corresponds to a factor of two more incident photons. But the absolute number of photons is not of interest for photographers. Thus, a calibration should be applied, namely the 18 % gray level standard, per the definition that means that the number of photons necessary to achieve 18 % of FWC ($N_{\mathrm{ph},18}$) is set to $EV = 0$ as a reference value. Under the assumption that $N_{\mathrm{full}} \gg \sigma_{\mathrm{pix}}$, such a "calibrated" exposure value can be easily calculated from the

$$EV = \mathrm{ld}\left(\frac{x}{x_{18}}\right) \tag{4.47}$$

$$N_{\mathrm{ph},18} = \frac{0.18 \cdot N_{\mathrm{full}}}{\eta_{\mathrm{e}}} \tag{4.48}$$

where x may be equal to N_{ph}, F or H and x_{18} to the corresponding value at the 18 % gray level (e. g., $N_{\mathrm{ph},18}$). This value has been chosen because usually illumination calibration of cameras is made with respect to this "average gray." Under the assumption of a typical scenery, the image is assumed to be exposed correctly, when the average of its tonal distribution is set to 18 % of the maximum signal of the camera, i. e., when the sensor becomes saturated. The value of 18 % is due to the fact that the average reflection in a typical room is 18 % of the incident light.

Thus, for the examples shown in Figure 4.52a, $EV = 0$ corresponds to $N_{\mathrm{ph},18} = 37510$, 9800 and 2438 photons for ISO 100, 400 and 1600, respectively. Although those values differ significantly, if the output signal is normalized to its maximum and displayed as a function of calibrated EV, the curves of all sensors and cameras do not differ when displayed on a linear scale. Thus, this curve may be regarded to be somehow universal.

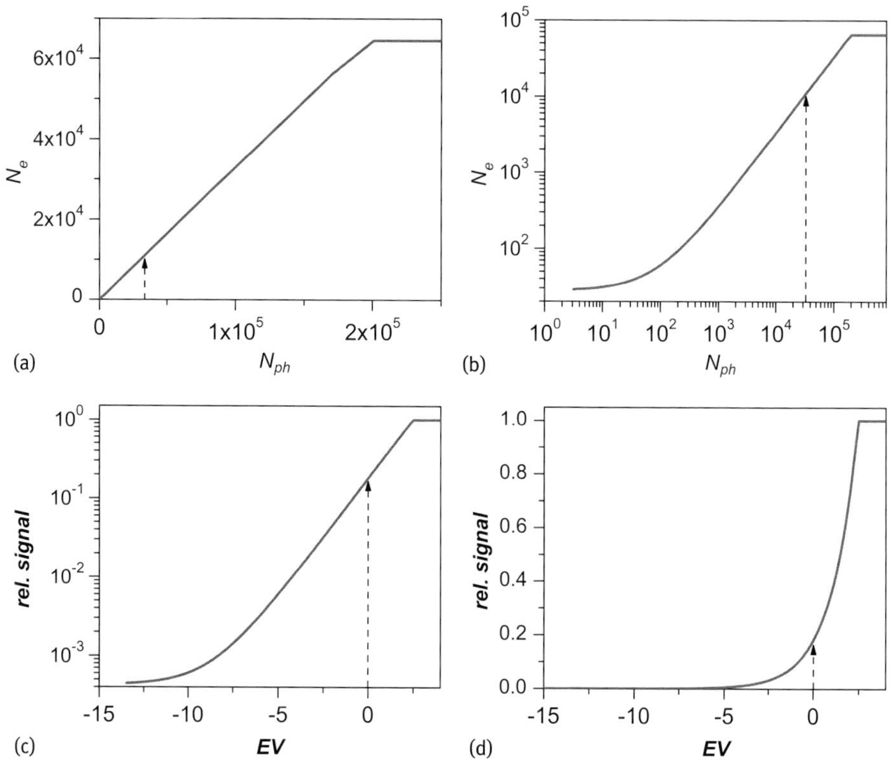

Fig. 4.55: Photon conversion curves. (a) to (d) are obtained from the same signals.

This can be seen from Figure 4.55: (a) shows the photon conversion curve $N_e(N_{ph})$ on a lin-lin scale (compare Figure 4.51b). (b) shows the same curve as (a), but now on a log-log scale (compare Figure 4.51a). In (c), N_{ph} is recalculated into EV according to Equation (4.47). Note: Because EV is a logarithmic value, this axis is still a log axis with respect to the input signal, but it is a linear one with respect to EV. In comparison to (b), now the ordinate is normalized to the maximum, i. e., FWC. (d) shows the same curve as (c), but now the ordinate is displayed on a linear scale (as it is usually shown in image processing software for the related tone curves; see Section 4.9). (c) and, in particular, (d) may be regarded as "universal curves." For good and bad sensors, the upper part of the universal curve is absolutely identical. Although the lower part may differ (see Figure 4.51), this cannot be seen in plots such as (d). Hence, although the upper part of the curve, and in particular, the maximum is the same, the minimum exposure value is given by $\mathrm{ld}(N_{ph,th}/N_{ph,18})$. For the example shown in Figure 4.51a, this minimum is $EV = -8.5$. Together with the maximum $EV = +2.5$ (see below), $DR = 2.5 - (-8.5) = 11$, as expected for that camera. The 18 % gray value is marked by an arrow in all diagrams.

The absolute value of the radiant and luminous exposures, respectively, for the maximum value may strongly depend on the specific camera and its ISO setting. For the minimum, this was just discussed. But in any case, due to its definition, the maximum cannot exceed 100 %, which is 5.56 times the average value of a typical scenery. Thus, application of an $EV > +2.5 = \mathrm{ld}(5.6)$ leads to saturation for that average, and only for less bright regions in the scenery it may make sense to increase the brightness of the whole image during post-processing by a larger value, but then usually most of the image becomes saturated (see Section 4.9).

4.9 Basics of image processing and modification

For films, basic image processing is performed in the dark room. Thus, for instance as briefly shown in Section 4.8.5, influence on the gradation curve may be taken by proper adjustment of the development process. However, although post-processing of films will not be considered further within this book, we will concentrate on the post-processing of images of digital sensors. This may be somewhat regarded as a digital darkroom. But, as stated previously, we will restrict ourselves to the basics of image processing and will not provide a comprehensive discussion, e. g., including the various possibilities of image manipulation by modern programs. Nonetheless, we also discuss some image corrections that are important, in particular, for scientific applications.

The goal of this chapter (and the whole book) is not to provide the whole workflow of raw data and image processing, even not in part because there are a lot of good books on that particular topic, and our intention is not to add another one. But we would like to provide some essential background information on that topic.

4.9.1 Sensor field corrections

Let us consider a PDA that is illuminated by light. Each pixel yields a signal $S(c, r)$ where (just here) c may indicate the column and r the row within the "pixel matrix," i. e., the coordinates. Now, for simplicity, we will renumber those signals beginning with $i = 1, 2, \ldots$ until the end of the first row, then we continue with the first column of the second row, etc. The resulting signal is termed $S(i)$. The signals usually are provided as ADU or counts (see Section 4.8.6) and are proportional to the output voltages. Now, according to the discussion in Section 4.7 and Section 4.8, the signal of each pixel consists of different contributions,

$$S(i, t_x) = S_{\mathrm{noise}}(i, t_x) + G(i) \cdot W_{\mathrm{pix}}(i, t_x) \tag{4.49}$$

where $S_{\mathrm{noise}}(i)$ results from σ_{pix} and is the total noise and $G(i)$ is the gain. Both quantities are related to a particular pixel (see Section 4.2.1). A consideration of a pure noise fluctuation that is not related to an individual pixel, such as photon noise, is not necessary here, because this is not of relevance for correction of the sensor signal. $W_{\text{pix}}(i)$ is the amount of light energy just on that pixel.

$S_{\text{noise}}(i)$ results from the temperature dependent dark noise and is even present without illumination, namely for $W_{\text{pix}}(i) = 0$ (see Section 4.7.3). Usually, $S_{\text{noise}}(i)$ is not constant for all the pixels. It depends on i, and hence leads to a so-called *dark signal nonuniformity* (DSNU), which is a fixed pattern noise (FPN). This also depends on exposure time and is made up from two contributions:

$$S_{\text{noise}}(i, t_{\text{x}}) = b(i) + a(i) \cdot t_{\text{x}}. \tag{4.50}$$

The first term results from the electronics and is regarded as the bias. The second term results from charge accumulation. $a(i)$ can be identified with the dark current including some constants. As dark current varies from pixel to pixel, it is some kind of FPN, which is overlaid by the thermal fluctuations. Figure 4.56 shows an example.

Pixels with a large dark current are called *hot pixels*, sometimes also *white pixels*. They always yield a bright signal much brighter than that of the pixels in its vicinity. They occur even in dark frames. In contrast to the signal of “warm pixels,” that of hot pixels does not scale with illumination. Hot pixels originate from leakages in the sensor and are located at fixed positions. There might also be *black pixels*, i. e., pixels that due to damage do not respond to light.

The second term in Equation (4.49) may suffer from gain variations between all the different pixels and, therefore, cause the *photo response nonuniformity* (PRNU). Of course, this is important for CMOS sensors, where each photodiode has its own output amplifier. Differences may originate from the fabrication process. Conversely, for CCDs, all photodiode signals are shifted to the same output amplifier, and thus $G(i)$ is a constant for all of them. PRNU leads to a *fixed pattern noise* (FPN), which is impressed to the real signal distribution $W_{\text{pix}}(i)$. However, FPN can be compensated for in the following way:

First, one has to measure the bias $b(i)$, which can be done by measuring the signal with blocked light illumination, i. e., closed shutter, for a very short time with $t_{\text{x}} \to 0$. For this condition, due to Equation (4.50) and Equation (4.49), the bias frame is given by $b(i) = S(i)$. Mostly, it is not constant but may contain some structure. It is usual to take several bias frames and average them. For astronomical images, it may be possible to neglect individual $b(i)$-values within the averaging process, in particular, when for the same pixel i, one of the $b(i)$-values strongly differs to the others. This is because for such a case, a cosmic ray event most probably occurs only once, but not for the same i in more than one frame.

Second, a dark frame has to be measured, again without light, but with the same value of t_{x} that is used for the exposed image, which should be corrected. This yields a dark frame signal,

$$D(i, t_{\text{x}}) = b(i) + a(i) \cdot t_{\text{x}}. \tag{4.51}$$

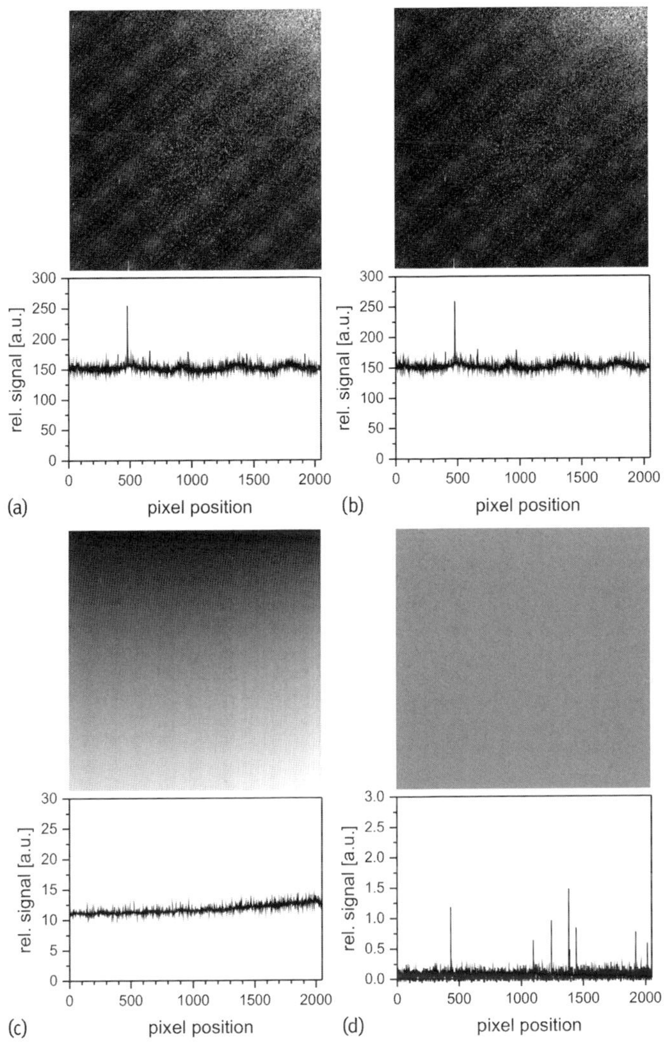

Fig. 4.56: Illustration of flat field correction for a scientific camera with a CCD sensor. (a) Single dark frame. (b) Average of 10 dark frames (not including that shown in (a), but same conditions). For comparison, (c) shows a bias frame. The dark field is identical to (b) and includes the bias. If the dark field is subtracted from an image, the image is "flat field corrected." (d) Same as (a) after FFC.

This is a background that should be subtracted from that image that should be corrected. As a practical note, we would like to remark that due to fluctuations of dark current, etc., it is advantageous to take several dark frames and average them prior to dark frame subtraction, which then yields $\overline{D}(i, t_x)$. We would like to remind that I_{dark} can be subtracted from the signal, but not σ_{dark} (see Section 4.7.3). Dark frame corrections are essential for long exposure times, such as are used in many scientific experiments or in astronomy.

Third, the sensor should be illuminated with a uniform light distribution, i. e., $W_{\text{pix}}(i, t_x) = W_{\text{uni}}(t_x)$ is the same for all pixels. Medium to strong illumination is preferred, e. g., close to saturation. Homogeneity can be achieved, e. g., by taking an image of a homogeneous illuminated area, may be with strong defocusing in addition to getting a smoother image. This yields a white or flat field frame

$$F(i, t_x) = b(i) + a(i) \cdot t_x + G(i) \cdot W_{\text{uni}}(t_x). \tag{4.52}$$

Similar to before, averaging over several frames yields $\overline{F}(i, t_x)$. Now, the correction factor $g(i)$ for each pixel signal can obtained from the following relations:

$$a(i) = \frac{\overline{D}(i, t_x) - b(i)}{t_x} \tag{4.53}$$

(this is not really necessary) and

$$\overline{F}(i, t_x) - \overline{D}(i, t_x) = G(i) \cdot W_{\text{uni}}(t_x) \tag{4.54}$$

$$\langle \overline{F}(i, t_x) - \overline{D}(i, t_x) \rangle_i = \langle G(i) \rangle_i \cdot W_{\text{uni}}(t_x) \tag{4.55}$$

where $\langle f(i) \rangle_i$ denotes the average of the function f with respect to i (i. e., the average over all pixels i within the same frame). Dividing Equation (4.55) by Equation (4.54) yields the correction factor

$$g(i) = \frac{\langle G(i) \rangle_i}{G(i)}, \tag{4.56}$$

with which the dark frame corrected uncalibrated image S has to be multiplied to get the corrected one

$$S_{\text{corr}}(i, t_x) = (S(i, t_x) - \overline{D}(i, t_x)) \cdot g(i) = G \cdot W_{\text{pix}}(i, t_x). \tag{4.57}$$

Here, it is important to note that all those images (namely S, D and F) should be captured at exactly the same exposure conditions such as temperature, etc. S_{corr} may be identified with B_{pix} of the corresponding pixel.

A *flat field correction* (FFC) according to this equation is shown in Figure 4.56 and another more simple FFC is shown in Figure 4.57. In Figure 4.56, for better demonstration the camera has not been cooled and the exposure time is set rather long ($T = 19\,°\text{C}$, $t_x = 6$ min, readout 500 kHz). The dark field is identical to (b) and includes the bias. If the dark field is subtracted from an image, the image is "flat field corrected." In (d), this is illustrated for the single dark frame shown in (a). The result is a very homogeneous image. The very tiny fluctuations result from noise, which becomes very low if the camera is cooled (see Figure 4.29). All this can be well seen from the sum of the profiles measured along the last 171 horizontal lines of the sensor shown below the images. Note the different scales in (c) and (d) when compared to (a) and (b). Note as well that the

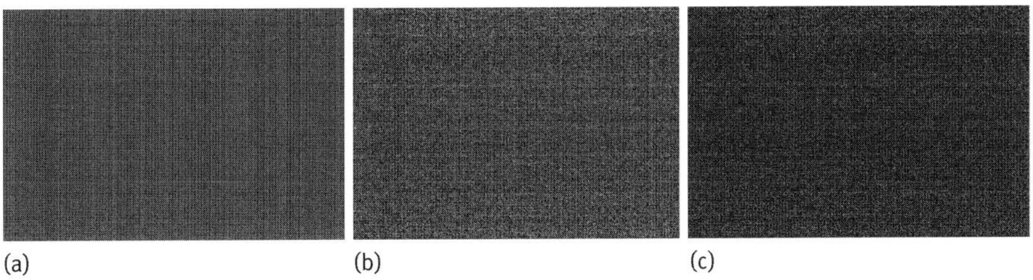

Fig. 4.57: (a) Average of seven bias frames of a DSLR with a CMOS sensor. (b) Another bias frame of the same camera. (c) Subtraction of (a) from (b) leads to a rather homogeneous frame. Note that in this simple example correction has been restricted to bias; correction for $g(i)$ is not made. More generally, the whole dark frame has to be included.

brightness of the images (a) and (b) is on the same scale but that of (c) and (d) has been increased artificially for better visibility. The stripes in the images (a) and (b) are called dark columns. They originate from traps that block or reduce vertical charge transfer. The traps originate from defects within the semiconductor; they lead to "dead or hot columns" or to a charge transfer inefficiency.

The last term in Equation (4.57) shows that as expected from Equation (4.15b), for all the individual pixel signals of a sensor with a flat field or of a sensor after flat field correction, a common gain should be applied. Thus, this gain given by $G = \langle G(i)\rangle_{\mathrm{i}} \equiv g(i) \cdot G(i)$ is independent of i.

FFC is a standard calibration method, and if carefully done, it usually leads to improved image quality. For CCD sensors, usually dark frame subtraction is sufficient.

In the case of scientific cameras, this procedure is a must. It has to be done by the user after saving the images separately, namely $S^{(s)}$, $D^{(d)}$ and $F^{(f)}$, where here the indices $s = 1, \ldots, d = 1, \ldots$ and $f = 1, \ldots$, respectively, indicate a series of the corresponding frames. For instance, a series of different images taken at the same exposure conditions could all be calibrated in the same way using the values $\overline{D}$ and $\overline{F}$, respectively, that are averaged over several frames. For commercial cameras used for photography, such kinds of image processing are performed only in part. As an example, a correction of dust on the sensor can be regarded as a kind of FFC.

Finally, we would like to remark that as any subtraction, etc. that leads to a flat field does sum up noise, again for scientific purposes, CCD may be preferential when compared to CMOS.

4.9.2 Basic image corrections

4.9.2.1 Image processors and raw converters

In the following, we would like to discuss image corrections for photography that are made to improve the quality of the *perceived* image. Usually, this is not done for scientific applications and, in particular, not when the camera is used as a measurement device.

The most basic corrections are made within the camera's image processor. Further ones are done there as well or later on by the user during the raw post-processing as described below. The very first processing steps performed within the camera are some electronic corrections such as discussed in Section 4.5.2 and Section 4.9.1. Then the data are converted and saved either into a raw data file with the full data depth of the ADC within the storage medium of the camera. Alternatively, or in parallel, they are further post-processed and only afterwards saved into a $3 \cdot 8$ bit signal JPEG data file or a TIF data file with either $3 \cdot 8$- or $3 \cdot 16$-bit data depth: note that according to Section 4.8.6 we avoid writing 24 and 48 bit for equitable reasons. It should be remarked that data saving into raw data is not possible for all cameras, but quite usual for advanced ones and the same holds for the generation of TIF files. Today there are also smartphones that are able to save raw data or DNG; see below. In the case of generation of JPG data or TIF data, data processing is done within the camera, however, with more restrictions when compared to external raw data processing using special software programs, namely raw converters.

In particular, even though most cameras allow a lot of different settings, and many of them yield quite good results, using a raw converter for post-processing offers the advantage of optimizing the final image in a much better and more desired way than can be done by the camera itself. Furthermore, most raw converters work nondestructively, i. e., processing steps can be withdrawn without any loss or image degradation.

Examples of raw converters are proprietary ones such as *Digital Photo Professional* provided by Canon for their cameras or ones that are implemented in commercial software such as *Adobe Camera Raw*. The latter is not a standalone software but used in many of the well-known Adobe programs. There are also ones that are raw converters such as *DXO Optics Pro* or other ones that are available as freeware, such as *RawTherapee*. A special position takes the open-source program *DCRaw* developed by D. Coffin because it supports raw data file formats of almost every camera on the market and also allows much more direct access to the data than nearly all other raw converters. It has for instance the exceptional possibility for data readout without application of tone mapping, etc. Moreover, as an open-source software, it may also be regarded as some kind of guarantee, if in the future existing raw data files are not more supported by commercial raw converters. It may be remarked that *DCRaw* is also implemented in many other programs such as *Gimp* and even a lot of commercial software makes use of it. Furthermore, there are programs such as the freeware *ImageJ*, which is a powerful program for image processing in general. Using special plugins, it can also read many raw data formats. Of course, the listed raw converters should be regarded as examples without particular recommendation with the exception of the special programs *DCRaw* and *ImageJ*, which, in our opinion, are really recommended.

To make the following discussion easier, we will concentrate on the tasks and procedures performed by raw converters. If data processing is done within the camera, it is obvious that the corresponding procedures are done there.

4.9.2.2 Raw data

Raw data include specific information related to some of the issues discussed in the previous chapters and how the data are pre-processed. The data contain even a lot more of information, potentially even more than expected for simple image processing. Raw converters make use of many of those data, but do not give access to that information directly. However, that information can be extracted by special programs such as *RawDigger*.

Although raw data have to be post-processed further, due to the early processing within the camera, they can only be regarded as nearly "raw" in the sense that they do almost, but not fully, reflect the original sensor data. How much the raw data differ from the original sensor data depends on the manufacturer and is unknown in most cases. One example of processing prior to storage of different manufacturers has been discussed with Figure 4.30, where the manufacturers do save different information in their files. Moreover, even for the same manufacturer, for different camera settings, differences in raw data preprocessing have been observed. Another example is the photon conversion curve of CMOS sensors. Although for CMOS sensors this curve is not linear (see Equation (4.43) where $b < 1$), within the raw data file it is saved as a linear curve up to the saturation limit. Consequently, linearization is always done by the image processor, which again shows that raw data are not fully "raw."

In addition to the ability to properly read the raw files, raw converters need additional information for accurate image corrections. In particular, this is the calibration data that result from measurements performed by the manufacturer or the raw converter developer for all supported camera models. The measurements are carried out at different illumination conditions to get proper data for white balance and color correction. Other measurements deduce signal properties such as noise. Sometimes converters include information on camera lenses as well, and thus are able to get also good corrections with respect to the optics, for instance, aberrations. Such information can also rely on theoretical data only; more advanced data are obtained from real measurements of all camera lens combinations. For a good correction, it is essential to consider the exact combination. Consideration of camera and lens separately is not sufficient. Here, in particular, camera manufactures with their own converters take advantage of the perfect knowledge of their own cameras and lenses. But, e. g., also the well-known *DXO Labs* have a huge database of carefully measured results, and thus allow for very precise corrections and high-quality images.

Raw data are always raw in the sense that they must be post-processed to obtain the intended image. Even though raw data "images" are "positives," they are somehow similar to film "negatives" in the sense that both are originals and have to be processed in the dark room or by the digital equivalent, namely the raw converter, respectively, before they can be regarded as a hopefully nice image on a screen or used to generate a printout. In other words, both raw data (sometimes they are called "digital negatives") and film "negatives" contain "full image information" but are not directly usable as the picture.

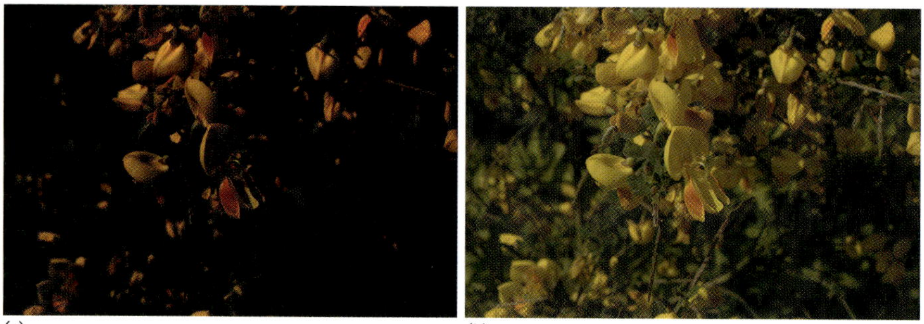

(a) (b)

Fig. 4.58: (a) Example of an image directly displayed as raw data. (b) Shows the same image, when a scaling based on a gamma curve has been applied (here $\gamma = 0.45$).

This can be seen from Figure 4.58a, which shows a direct display of the raw data, here however, after de-mosaicing (see Section 4.9.3). The original data are stored as 16-bit values, but due to the fact that 16-bit values cannot be displayed within the printed pages of this book, they are linearly scaled to 8 bits, for instance, by *DCRaw*. Linear scaling is shown in Figure 4.58a and printed as an 8-bit image. This still reflects the linear photo response curve. Figure 4.58b shows a "tone-mapped" version of the same image (see Section 4.9.4), which may be regarded as the image that has been intended.

Based on the issues discussed above and using the camera and lens specific calibration, now raw data processing could be performed. This involves the issues listed below. We have to note that that some of them are only reasonable when the mentioned data are available and when the corresponding camera or camera lens combination is supported. Usually, the user can decide if and how much possible corrections are applied, at least for some of them.

- De-mosaicing
- Color correction including color space adaption for screen, printer, etc.; this is necessary because electronic detectors do not "see" as the human eye; mostly color correction is done to improve the perceived image, not necessarily to reproduce the colors correctly (there are only few cases in photography where the intention is a strong demand)
- Tonal corrections (gradation, γ-curve, brightness, contrast)
- White balance (by tweaking the tonal curves of R,G,B accordingly)
- Dark current and other noise corrections (such as CDS)
- Correction for pixel responsitivity including correction for defect pixels
- Correction for chromatic aberration
- Stray light correction
- Shading and distortion corrections or more general, aberrations
- Application of sharpening
- Data conversion into JPEG or TIF format and storage

Prior to discussion of more details, we would like to mention that the application of those corrections on the raw files directly is not straightforward and requires a lot of knowledge. For that reason, nearly all raw converters at first apply some standard settings and then display the image. These settings then may be changed by the user in the desired way. In the following, we will concentrate on sensor related image post-processing; optical corrections are not subject of the present subchapter and will be discussed later.

4.9.2.3 Digital Negatives

Raw data storage usually occurs in a not documented very special and sometimes even encrypted proprietary format, which is specific to the manufacturer. To get rid of such proprietary solutions, a special well-documented format, the so-called digital negative (DNG) has been developed by Adobe Inc. The goal has been to establish this as an open standard with the advantage that DNG files should be accessible for long time. Some manufacturers make use of DNG in their cameras instead of generating their own raw file formats. In particular, this is the case for small camera manufacturers, which cannot expect that an own proprietary development of such a format will be incorporated in the raw converters on the market.

However, the contents of DNG files are not equal to raw data. The first ones are more extended TIF files and do not include the same amount of information as mentioned in the previous chapter. Conversion of raw data into DNG is always accompanied with data processing. Due to the fact that the proprietary raw data are not documented for third parties, any processing related to DNG has to interpret the data. Even if this would be well done, conversion is always accompanied with more or less loss. It is impossible that a processed data set includes more information than the original one.

One example of such processing is de-mosaicing (see the following section). When we follow our discussion above, we remember that there is no common procedure for this process, and moreover, improvement of de-mosaicing algorithms is still continuing. Hence, improved de-mosaicing of available raw data may be possible in future, but not for already pre-processed DNG data. And even DNG generators have been subject to changes of the de-mosaicing process. Thus, as de-mosaicing is one key element of raw data processing, this disadvantage of pre-processed DNG is obvious even if the DNG itself is of high quality.

Although there are advantages of DNG, it must be mentioned that there are even more disadvantages as well. First of all, in contrast to the claim, DNG is not a standard. It is neither a DIN nor an ISO norm. For this reason, spread of usage is restricted. For instance, there are important high quality and rather frequently used raw converters that are able to work with most camera files but not with DNG files at all, or at least not with all DNG versions. Second, the intended long-term preservation is not guaranteed. The Deutsche Forschungsgemeinschaft (DFG; German National Science Foundation) does not classify DNG to be suitable for digital preservation. This is quite similar to other proprietary formats. There is always a risk that DNG may be discontinued (or the related company may disappear from the market; this would not be the first case; note that also important software that once was expected to have a long lifetime does not exist anymore). There is no reason why in that sense DNG should be regarded to be superior to other proprietary data formats (of course also vice versa). Third, although DNG is an open format, this is only in the sense that the format is documented and the documentation is public and free of cost. But DNG is patented and cannot be regarded as a free format or creative commons. It is proprietary as well. Thus, e. g., application of legal rules may lead to severe restrictions such as prevention of offering DNG file viewers or processors and so on. Even more for this reason, users working with DNG files may be tethered to a single company with its goodwill, with its products only and potentially rather expensive rental solutions and restrictive instructions (monopole position; see also the discontinued support of several older DNG versions). Thus, not to be critical would be naïve and risky. Nevertheless, weighting of advantages and disadvantages and the decision for or against usage of DNG should be made by the user. But it must be always clear that only raw data do contain the (nearly) full information about what is captured by the sensor.

4.9.3 De-mosaicing

One of the very first processing steps performed by a raw converter is de-mosaicing. As discussed in Section 4.6.3, color information is obtained from pixels located behind the color filter array, usually the Bayer mask. Here, we will mostly restrict ourselves to that geometry. For other geometries, discussion is straightforward and monochrome or other types of color sensors, such as the Foveon X3 sensor, do not need this procedure, and thus for those the present chapter is irrelevant.

Within the CFA, each pixel records the intensity of a single color only and only for that position where it is located. Indeed, a single color has a certain spectral bandwidth (see Figure 4.24), but one attributes one of the three colors red, green or blue to each pixel. On the other hand, one is interested in obtaining full color information for the position of each pixel, which due to the lack of the missing measurement, is not available.

Thus, one has to find a method that allows us to *estimate* the two missing colors at each pixel position. This method is called *de-mosaicing*. De-mosaicing is based on the use of advanced methods and algorithms to calculate the missing information under specific assumptions. Most simply missing color information of a pixel is interpolated from the information from its neighbors. It also takes into account the full structure of the sensor system, e. g., the presence of OLPF. The method is challenging. Each company has its own “tricks” and similar to a cook, its own recipes including specific tone curves for the different colors to achieve a nice color reproduction for the specific camera sensor setup. By this way, the camera manufacturers claim to get optimum color reproduction for specific situations (see picture presets in Section 4.9.4), and so do the software companies of the raw converters.

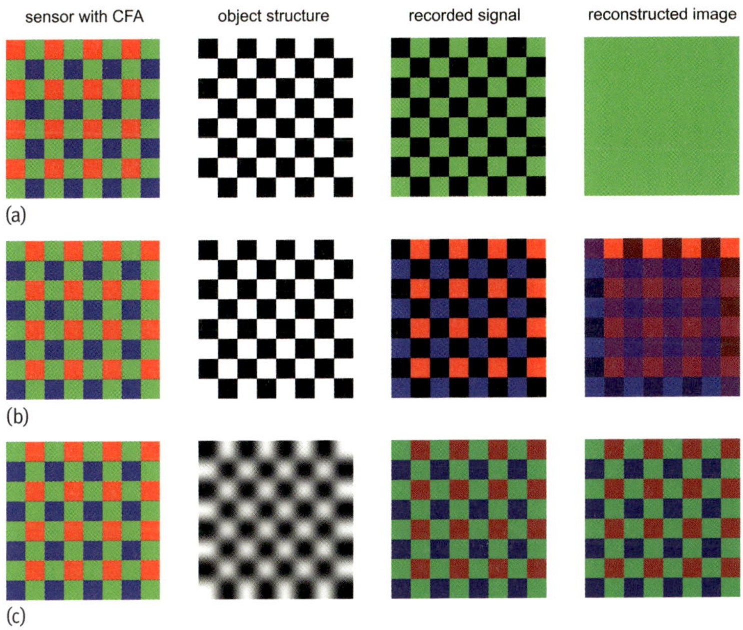

Fig. 4.59: Examples of de-mosaicing for sensor with a Bayer mask (see first column).

But as missing information can never be calculated, the results of de-mosaicing have to be regarded as estimates. Though de-mosaicing in most cases yields good results, there might also be situations where it may yield totally wrong results. Figure 4.59 shows some examples, where the sensor displayed in the first column is illuminated with a structure shown in the second column. Due to the CFA, the chip records the signal shown in the third column. From that, a simple de-mosaicing procedure may yield an image as shown in the last column. Obviously, this is not at all consistent with the object and its image on the chip (second column). (a), (b) and (c) show several examples. The difference between (a) and (b) is just a shift of the object structure by one pixel. (c) shows how an OLPF improves the situation by slightly blurring the image on the sensor, which results at least in a reproduction of the object structure information.

Further details on methods of de-mosaicing and advanced algorithms are not the subject of the basic discussion here.

4.9.4 Tone mapping

Conversion curves, in particular, photon conversion curves, of an optical input to a signal output were introduced in Section 4.8. After the first corrections, e. g., FFC and also potential linearization in the case of CMOS sensors, the signal of a single pixel shows a linear relation between the output and the input signal, i. e., its illumination, respectively. Here, and in the following, we restrict to the linear range between the noise floor

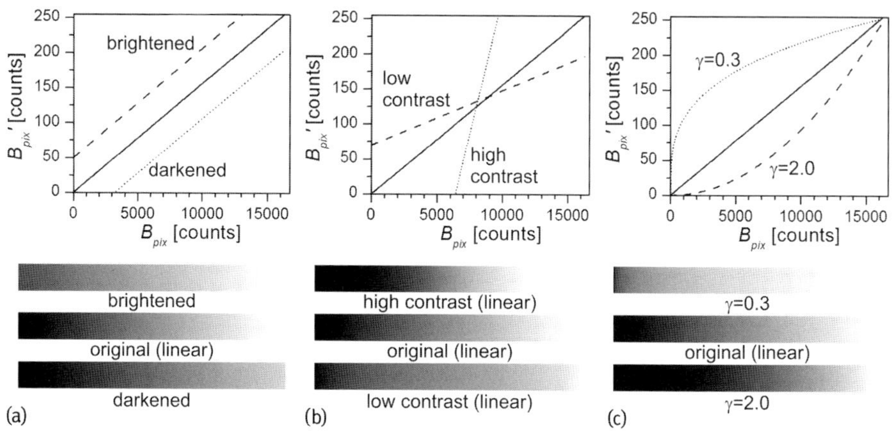

Fig. 4.60: Illustration of the rescaling of the brightness distribution. But note, if, e. g., $\gamma = 0.45$ (*for the data*) and the gamma value of a screen is set to the typical value of 2.2, then the *displayed* curve is linear again (see also Section 4.9.5.2). Then, of course, for other tone curves this is not the case.

and FWC (see, e. g., Figure 4.46, Figure 4.51). The dynamic range may be rather high and extend over several decades. Furthermore, as discussed in Section 4.8.6, due to ADC conversion the data are converted into a largely extended range, e. g., with 14 or 16 bits, which we will use as an example in the following (whether this kind of conversion is reasonable or not, is not an issue here).

This situation is displayed in Figure 4.60, where, similar to Figure 4.51a, the input range may extend over several decades, e. g., from zero to several hundred thousand photons. Similar to that, the output scale also covers a lot of orders of magnitudes (see Figure 4.51a), and via ADC this range may now extend from $B_{\mathrm{pix}} = 0$ to 16,384 counts, i. e., 14 bits.

At this point, if there were a high-performance output device such as a screen, printer, beamer, etc., allowed to display such a large range, then at least observation would be more or less straightforward. However, such devices are currently not available, and thus a rescaling of, e. g., the 14-bit data to the range of the output device is essential. If we disregard special output devices, which are in fact able to display a brightness distribution on a scale even larger than 10 bits, nearly all other output devices are related to the standard data depth of 8 bits, as it is used for the JPG data format as well (see Section 4.8.4). As a consequence, the 14-bit signals of our example have to be squeezed to an 8-bit scale.

Although this can be done with a simple linear transformation just by rescaling the range (see Figure 4.60a), of course, this causes a loss in information, in particular, of depth resolution and *DS*. At this point, even more important is the fact that then the image would not look good (see Figure 4.58a). The reason for this is the logarithmic response of the human eye as discussed at the beginning of Section 4.8.5; even small brightness differences can be well discriminated in shadow or dark regions with relatively high-depth resolution (i. e., small *dr*), whereas in bright regions, depth resolution is much worse.

Most simply this problem can be solved transferring the linear output scale to the curved scale of a gamma curve, which means that the output that is the final pixel brightness B'_{pix}, which is observed in the final image, is given by the input B_{pix} according to a power law B^{γ}_{pix}, with a gamma value that can be adapted to the specific situation (Figure 4.60b and Figure 4.60c). γ is defined in Section 4.8.5 and Equation (4.40). More generally, $B'_{pix} = c \cdot (B_{pix})^{\gamma} + b$, where in addition, contrast c and background b may be adapted as well. This improves the image quality (see Figure 4.58b), but due to the large amount of possible mixtures of the three parameters γ, c and b, even such a quite simple rescaling leads to a large amount of differently looking images with results that cannot be easily pre-seen. As a consequence, image optimization is not necessarily straightforward.

Figure 4.60 provides some examples. The digitized signals B_{pix}, e. g., as saved in a raw data file, from the photon conversion curves provide the input, i. e., $B_{pix} \propto N_e(N_{ph})$ for the rescaling; the rescaled values B'_{pix} are provided as the output here and usually are saved in a JPG or TIF file, respectively. (a) shows a simple "direct" linear transformation (solid line) and linear transformation with positive or negative background, respectively (dashed and dotted lines). (b) illustrates the application of an increased or decreased contrast. (c) shows a transformation using a gamma curve ($\gamma = 0.3$ (dotted line) and 2.0 (dashed line), respectively). The patches below the diagrams illustrate how a linear grayscale wedge may be changed when the brightness or contrast of its intensity distribution is changed or when its linear scale is turned to a gamma curve.

When a film is used as the detector, its characteristic curve, namely the density curve or tonal curve, is strongly nonlinear and well adapted to the human eye's response curve. However, most generally this is not the case for electronic detectors. Thus, for the reproduction of a good visual *impression*, one has to apply an appropriate transformation, e. g., by rescaling to a gamma curve.

This can be done also by multiplying a suitable transfer function (Figure 4.61b) to the photo response curve (Figure 4.61a), which may be equivalent to the above described (non)linear rescaling of the output values. The result is the tone curve, sometimes also called the gradation curve (Figure 4.61c), which describes the final output signal or an intermediate one, if further image processing is made later on, but now on an 8-bit scale, and still as a function of illumination (e. g., N_{ph}). For comparison, Figure 4.61c also shows a gamma curve, which may be regarded as a simple version of the tonal curve used by raw converters.

Although in such diagrams usually the abscissa is displayed on a logarithmic scale, for clearer discussion and to see linearities much better, here linear plots have been preferred. Exposure is given by F_{pix}, H_{pix} or N_{ph}. Due to the fact that in the linear range of the sensor $B_{pix} \propto N_e \propto N_{ph} \propto$ exposure, Figure 4.61c and Figure 4.60c are equivalent although Figure 4.60 describes the relation of two signals, namely B'_{pix} as a function of B_{pix}, whereas Figure 4.61c relates B'_{pix} directly to the original illumination.

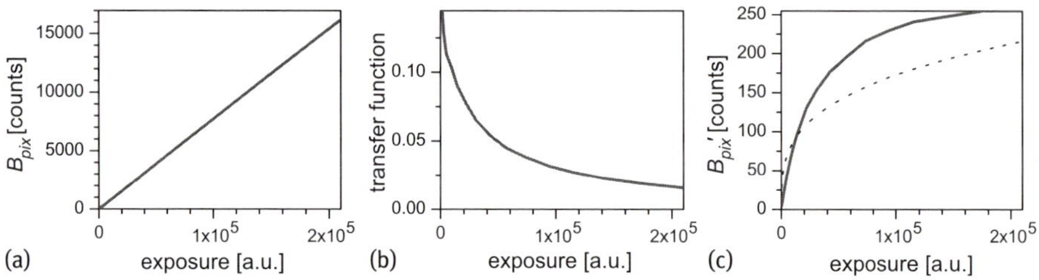

Fig. 4.61: Illustration of the procedure of tonal mapping (see the text). (a) Corresponds to the photo response curve (linear region only); (b) Nonlinear transfer function $B'_{\text{pix}}/B_{\text{pix}}$; (c) Tone curve, i. e., the dependence of the brightness of an image point (i. e., a pixel within the image) on the original illumination. For comparison, a gamma curve with $\gamma = 0.3$ is shown as well (dotted line).

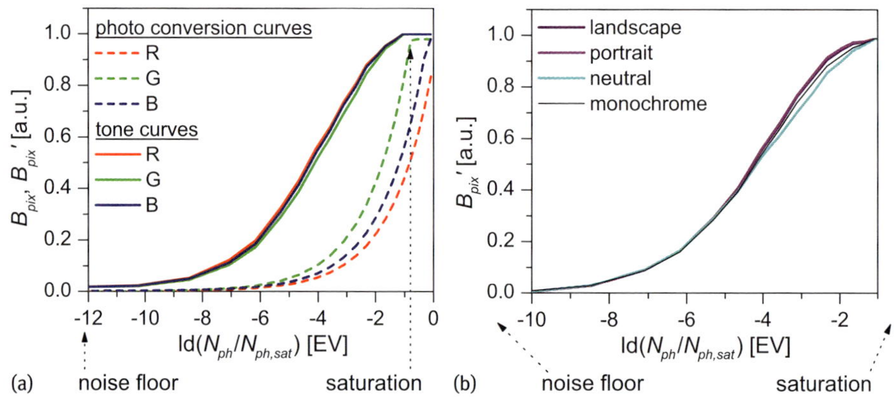

Fig. 4.62: Normalized tonal curves of a professional DSLR. (a) Example of the experimentally deduced tone curves (B'_{pix}) for the red, green and blue channel, respectively (solid lines). The photon conversion curves (B_{pix}, also normalized) that are displayed for comparison (dashed lines). (b) Different tone curves of the same camera/raw converter as could be used for different situations (picture presets or styles as indicated in the insert). In (b), the arrows point to regions outside the displayed region.

It is clearly seen that such a tonal curve now allows for discriminating small differences in dark regions much better than before. Even small changes in the input signal yield even more significant differences in the output signal. On the other hand, differences in bright regions of an image have to be quite large before they can be recognized similar to the response curve of the eye for which one assumes $\gamma \approx 0.3 \ldots 0.5$. This curve has a similarity to that of a film, which is much better seen in Figure 4.62, which shows exactly the same curve as in Figure 4.61c, but now with a logarithmic abscissa (see discussion below). This procedure is called *tone mapping*, sometimes also tone reproduction.

The curves in Figure 4.62 are similar to those in Figure 4.61c, however, as usual, now the abscissa is plotted on a logarithmic scale ($EV = \text{ld(exposure)}$), and thus linearity of the additionally displayed photo response curves is not well seen. The photo response curves are identical to those shown in Figure 4.51a, but the scaling of the axis in Figure 4.51a corresponds to that in Figure 4.55a and that of the present figures to that of Figure 4.55d. But note that here $EV = 0$ is set at saturation of the green channel, whereas in Figure 4.55d $EV = 0$ is set at the 18 % gray value. Nevertheless, this is not of much relevance here. Within Figure 4.62, the exposure is given by F_{pix}, H_{pix} or N_{ph} and normalized to the corresponding value when the B_{pix} becomes saturated. Usually, B'_{pix} is given by 0 to 255 counts (8-bit scale), but here it is normalized to its maximum value at saturation.

For rational reasons, the range of the displayed abscissa may be regarded as limited, with its minimum given by the exposure corresponding to the noise floor and its maximum to the exposure that leads to saturation. For better illustration only, in (a) this range is somewhat extended. From (a), one can recognize that the curves result from an 11- to 12-bit camera.

Also, in similarity to films, where the emulsions are well designed not only for physical and chemical reasons, but also to yield a brightness and color reproduction that looks quite good for the photographer, transfer functions or tonal curves, respectively, are designed as well to get a high-quality perceived image. In general, there are many different "designs," and some of them even try to reproduce film curves. However, manufacturers of raw converters try to generate tone curves with respect to specific situations (e. g., landscape, portrait) and also to the taste of the photographer. In other words, the goal is that the photographer sees the image of a landscape as he *expects* or *wishes* to see it. Examples of such tone curves usually offered for selection by raw converters are shown in Figure 4.62b. These curves are made for the same camera as in Figure 4.62a. For clearer visibility of the curves, the range in this diagram shows only a fraction of the full range. For supplementary discussion of tone curves, see Appendix A.7.

The results, when these different tone curves are applied to the raw image displayed in Figure 4.58a are shown in Figure 4.63. It may be seen that although the differences between the curves presented in Figure 4.62b are subtle, the resulting effect when applied to real images is apparent (see Figure 4.63). This is quite general; even small changes in brightness, contrast or tone curve may lead to observably effects within the image.

Depending on the preset, such tone curves are not necessarily the same for the different channels (i. e., red, green, blue) and they are also different for other ISO settings. They differ also for different cameras of the same company and different raw converters yield different results as well even for an identical raw data file. Thus, for instance, even turning colored images into monochrome ones may lead to significantly different pictures. This is somehow an observation that is similar to pictures taken with different black-and-white films.

We would also like to make the reader aware that a tone curve of a color image that leads to a clip in the highlight region may lead to color changes in that region. In particular, this is the case when one of the three RGB curves comes to saturation and the others have not yet. Usually, such effects become apparent in images including a bright sky.

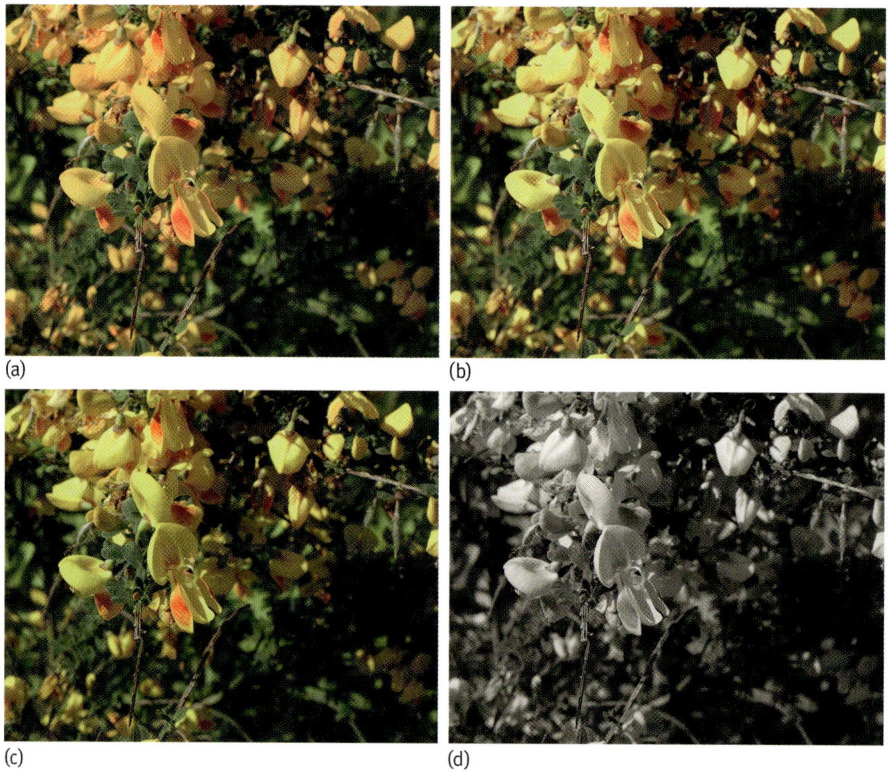

Fig. 4.63: Same image as in Figure 4.58a after the application of the tone curves displayed in Figure 4.62b: "landscape" (a), "portrait" (b), "neutral" (c) and monochrome (d). Note: Here it is not an issue whether it makes sense to apply, e. g., the preset "portrait," to a flower; it is just to see effects. In addition to different tone curves, it is expected that further corrections to the image are performed as well.

More generally, for color images, different tone curves may also lead to different colors. Sometimes this causes problems, in particular, when colors should be reproduced well. As an example, if one demands that the natural green color of grass is at least nearly correctly reproduced, one will recognize that this is a challenge. Testing different raw converters and different advanced settings quite often it is hardly possible to get an acceptable reproduction when one compares the image displayed on a calibrated screen with the original scene, which may be just outside the window. For instance, one of the most distributed raw converter software fails totally in this discipline and only a few other raw converters yield results that are satisfying. This is an example that obviously shows that commercial raw converters are made to improve the perceived image quality according to a specific taste. This is not well-defined and as an example it may be noticed that even a preset such as "portrait" may be differently perceived by people from different continents because tone curves have been adapted according to the local likings.

All this mainly excludes raw converters and usually also many cameras made for photography from being scientific devices for measurements (see the discussion in Appendix A.7, and there, in particular, Section III). This is even more the case because when raw converters are used, it cannot be avoided at all that a transfer function is applied. An exception is *DCRaw*, which allows one to do so. But we may note that even with film negatives, there is processing of the printouts in the dark room. An exception to see really raw data are slide films, but of course those raw data depend on film material and film development as well. Even when 16-bit raw data are directly saved into 16-bit TIF files without any further image manipulation (i. e., by putting all settings such as for brightness, contrast, etc., to zero), the raw converter automatically applies a tone curve to the raw data. The only choice of the user is to change the picture presets, but none of them allows for obtaining a linear tone curve such as shown in Figure 4.61a. Even turning on the so-called neutral is strongly nonlinear.

This can be seen easily from the histograms Figure 4.64 (for the discussion of histograms, see Appendix A.6). The abscissa is scaled linearly and may just be considered to display the signal brightness on a range between 0 and maximum brightness of the image on a linear scale. The ordinate, which shows the number of pixels that have been recorded with the corresponding brightness, is displayed on a linear axis as well. The differences are obvious, from the histograms and from the pictures themselves. Not only the absolute values, but, in particular, the shapes are different. This clearly shows that the images displayed in Figure 4.63 are strongly manipulated with respect to the original one, which is necessary to obtain a nice look. Note: Often the abscissa is provided in *EV*, i. e., it is a log axis. But then the ordinate has to be rescaled as well (Appendix A.6).

More quantitatively, the differences between raw data or its conversion into a linear 16-bit TIF file and any file delivered from a raw converter can be obtained by taking a photograph from a calibrated grayscale target (e. g., or an OECF target; see Section 8.4) and the use of an image-analyzing software. Here, from the raw data image one can

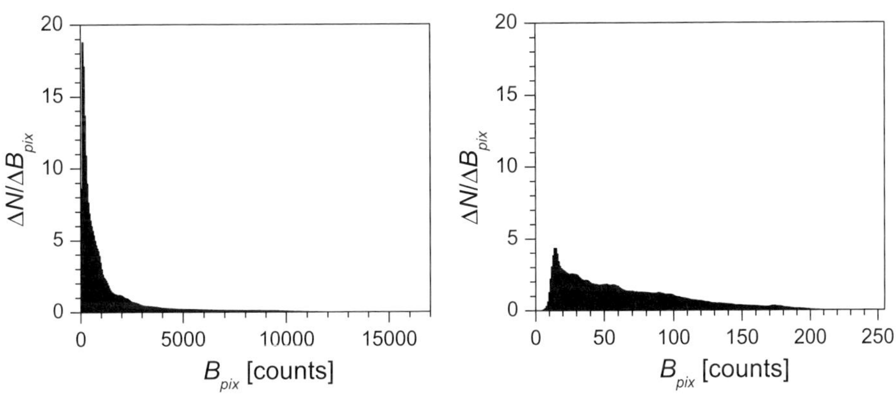

Fig. 4.64: (a) Histogram of the 16-bit image of the linear raw data (Figure 4.58a) and the 8-bit image of the tone mapped data (Figure 4.63, image with preset "neutral").

extract (e. g., by *DCRaw*) the linear photon conversion curves, whereas any file stored by the raw converter yields tone curves for the same grayscale target. This is displayed in Figure 4.62a (see also Appendix A.7).

Finally, it may be remarked that for the same reason discussed before, namely the adaption to the eye's response, tone mapping commonly is also applied within 8-bit color consumer cameras. Even those cameras transfer an 8-bit linear range to an 8-bit range nonlinear one (see Figure 4.60c or Figure 4.61c, but then there with an 8-bit scale for the input as well.

4.9.5 Further tone mapping, HDR and final remarks

The described tone mapping according to a selected preset may be only one part of a more comprehensive tone mapping procedure as part of the "workflow." Further tone curve manipulation by the user is possible, in particular, changes of contrast, brightness and so on. To show this, and even more to show that the previous discussion is the basis of raw converters, the following example serves as an illustration, however, without a substantial discussion of image processing in general. Here, we restrict ourselves to rather simple additional changes to a preset tone curve and its effects. But by no means is our intention to optimize tone mapping for the images of our examples.

As the example, Figure 4.65 shows a raw data image with a range of exposure values that in the original raw data exceeds 11 EV (see histograms). However, after the application of tone mapping using a standard tone curve according to a picture preset of the raw converter, an image is generated, which may be regarded to be exposed somehow correctly (Figure 4.66a). Nevertheless, due to this tone mapping, now clipping at highlights occurs within this restricted range of only 8 EV, although the raw data are not severely clipped. Thus, the brightest parts of the image have been saturated. This is also well seen from the histogram, which is clipped at approximately 2.5 EV.

To improve this situation, the standard tone curve (dotted line) can be shifted (by +2 EV; black solid line in Figure 4.66b), which darkens the image (subtraction of 2 EV from the original brightness as indicated in the image label). This also leads to a shift of the histogram to the left-hand side. Consequently, here clipping does not occur any longer (or is only a minor effect). But due to the limited range of 8 EV, only, of course, now there is hardly discrimination of gray levels within the dark regions.

On the other hand, to see more details in the dark region, a picture could be simply made brighter by using the large dynamic range of the camera. To do so, the shift of the tone curve can be done in the opposite direction. Consequently, this results in a shift of the histogram as well, because then more or less pixel brightness occurs in a region where it has not been before (Note: The spatial position, of course remains, unchanged). The resulting image now clearly shows structures (and even colors) in those regions, but the image as a whole is overexposed (Figure 4.66d).

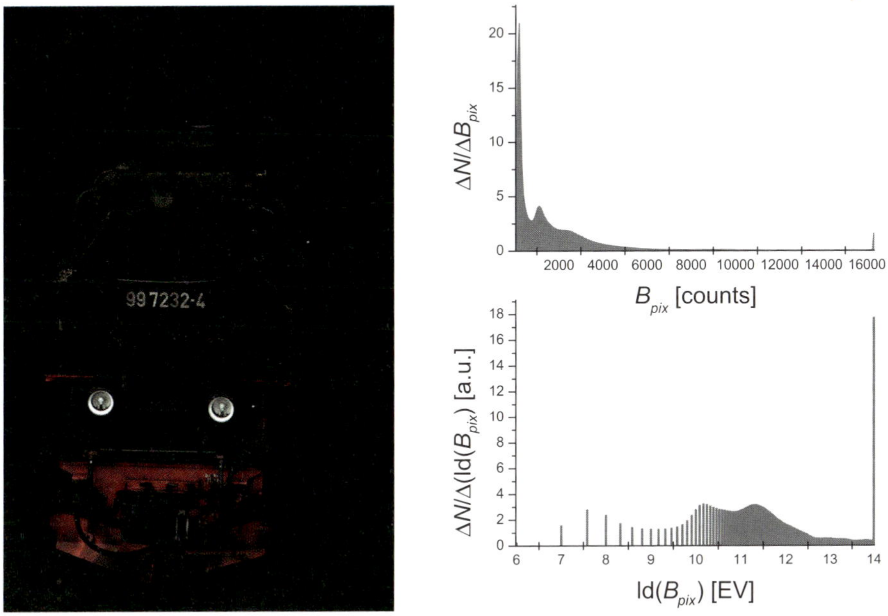

Fig. 4.65: Raw data image and histogram of the raw data of the image. Both histograms are the same, but when displayed on a log x-axis, the y-values have to be rescaled (in the lower histogram this leads to bins that have the correct height; see Appendix A.6 but are not equidistant anymore as in the upper one). Here, the abscissa is not calibrated, i. e., $EV = 0$ is set at an arbitrary position.

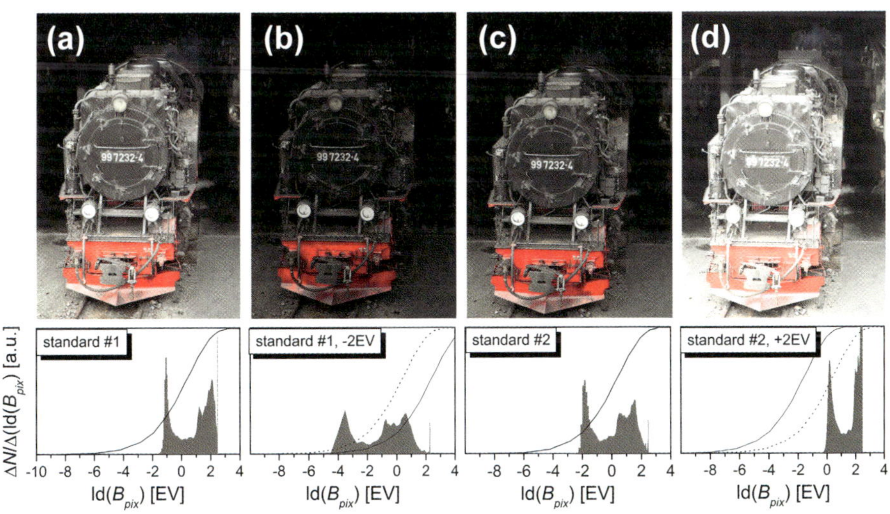

Fig. 4.66: Examples for further tone mapping. (a) and (c) show the images after the application of a standard tone curve and (b) and (d) the same ones after application of tone mapping with a shifted tone curve (see the text and Appendix A.7). Usually, the position of $EV = 0$ is set at the saturation point or the 18 % gray level. The latter calibration is used here for the histograms. However, for the tonal curve $EV = 0$ is set at the position where the tone curve has dropped to 50 % (this is a typical calibration used by some raw converters).

Although optimized tone curves could improve the situation, for images such as discussed here, usually it is difficult or even impossible to apply a tone mapping that fulfils the discussed desires. This is the subject of the next chapter. Supplementary discussion of tone curves and other examples are the subject of Appendix A.7.

4.9.5.1 Increase of dynamic range: HDR and DRI

In most cases, the dynamic range of a DSLR with a dynamic range of at least 11 bits is sufficient and this is the case as well for standard compact cameras that cover 8 bits only, i. e., $DS = 8$ bit. However, as seen in the previous example, sometimes a particular effort is necessary to produce nice-looking images. We have to note that we did not take care for that in the above examples as we just wanted to see and discuss effects. But there are also situations where even strong efforts are not successful. Hence, it is a challenge to produce good-looking 8-bit images from scenery with a very large dynamic range. Just this is the problem, namely the lack of commonly available output devices with large enough bit depth.

An example of a scenery with a contrast ratio within the scenery that obviously exceeds the possibilities of standard image processing is that one discussed in the previous example. This is again shown in Figure 4.67a. As before, this image is the result of "normal" image processing using 14-bit raw data from a DSLR with $DR = 11$ EV. The tone curve is optimized insofar that one gets a well-exposed image together with the best possible enhancement in the dark regions and still not overexposed bright regions. Of course, the restrictions are severe, because in the low light regions hardly any details can be observed. However, increasing brightness would brighten the whole image very much, and this would not be acceptable for a good picture (compare Figure 4.66d).

Due to the larger dynamic range of the combination image capturing/image displaying device, taking the picture with an analogous camera with a slide film would solve that particular problem. This would be also the case if the image were taken with a DSLR with a large dynamic range and if the image were processed properly and stored in a 16-bit TIF-file, *and* if this image then could be displayed on a screen that would have the appropriate dynamic range and depth resolution. But today that is very unusual.

For this unfortunate reason only, one may solve the present problem using the "High Dynamic Range" method (HDR). In the following, HDR is briefly explained. An extended discussion is the subject of special books related to this topic. Usually, HDR is made from a series of images from the same scenery taken at different EV. In such a "bracketed series," it is recommended to keep $f_{\#}$ constant. The method of HDR is based on an analysis of image brightness content and locally adapted image processing, in particular, instead of the usage of a global tone curve. The image information of the mentioned single high-dynamic range raw data image or the information of all images of the bracketed series is taken into account. Data processing is usually made with 32-bit floating point data instead of the standard 8-bit integer data, but finally, as usual, results in a "low dynamic range" (LDR) image. Data processing is quite tricky and, as an example, makes

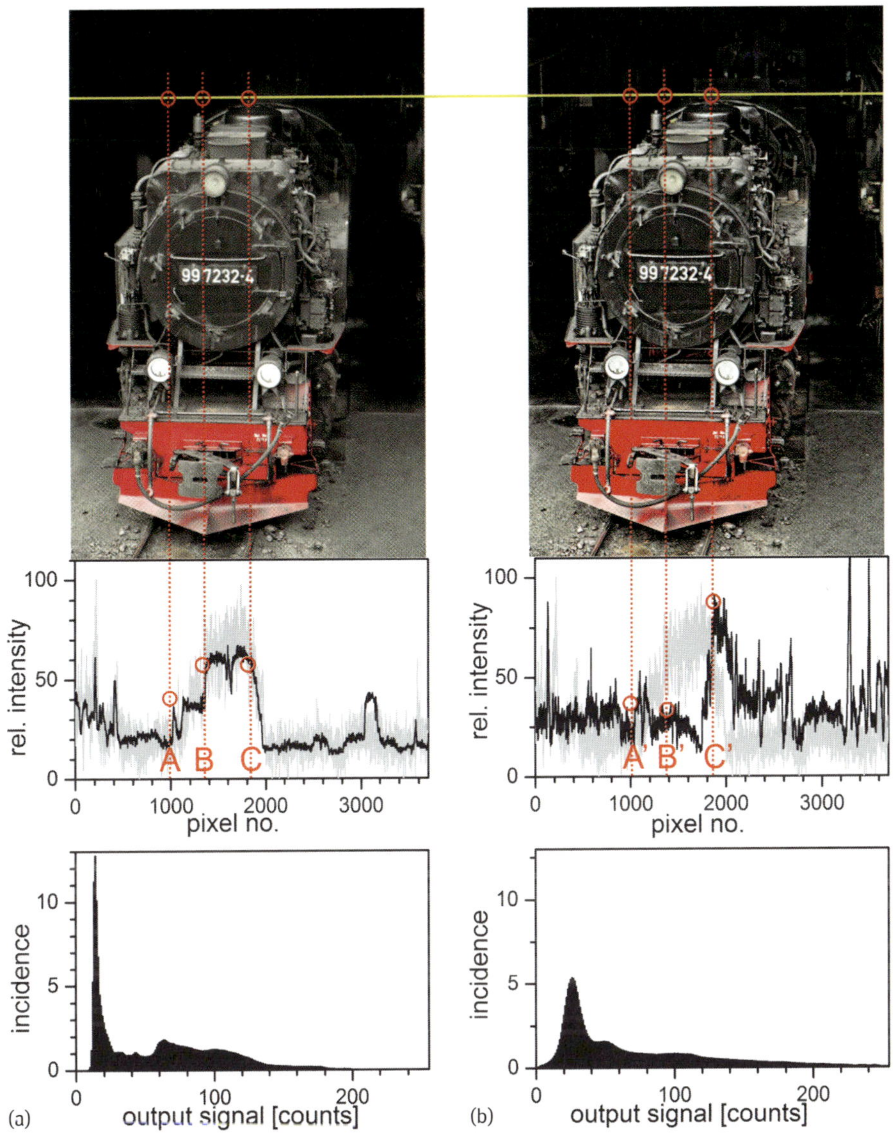

Fig. 4.67: An example of an image of a scenery with a very large input dynamic range. (a) Image that results from "normal" image processing using a common tone curve for the whole image. (b) Image that results from an HDR image processing. The yellow line marks where the line profile across the image is measured. This profile is shown below the pictures. The profile in (a) (black line) may be compared to that (light gray line) of the raw data image (Figure 4.65) and fits quite well. This is not the case for the HDR image (b) where strong differences can be seen.

use of brightness perception, which depends on the vicinity of the region under consideration. Altogether advanced image processing algorithms are applied. Special HDR programs are available on the market, but there are also implementations to some raw converters. Quite often the user has the possibility of selecting different methods for HDR calculations.

The result of an HDR image is seen in Figure 4.67b. The image looks good, it is not overexposed and it shows details in the dark regions (may be not seen clearly on the paper print of this book). Now even a blue color may be recognized (above the yellow line). The result allows the conclusion that HDR may produce good-looking pictures. However, those images are heavily manipulated. This is demonstrated by the profiles measured along the horizontal lines shown in Figure 4.67. From the profiles of the original image (Figure 4.67a), the relative intensities of three exemplarily selected points within the image, which may be pixels, or potentially an average over a small area around them are marked. Those have a direct relation to the corresponding object points within the scenery because the same tone curve is applied to the brightness of all pixels.

Now we may compare the line profiles of the raw data image shown in Figure 4.65, which are displayed as the gray lines in the diagram in Figure 4.67, to those of the tone mapped image (Figure 4.67a, black line in the diagram). From this comparison, it may be seen that the ratios have changed, however, points with originally larger intensity are still larger. Although not of importance here, we may note that in addition to tone mapping, further image processing, such as noise reduction, has made the black curve smoother than the gray one. As a result of the common tone curve, this holds for all points or pixels within the image. However, from the line profile of the HDR image (Figure 4.67b), it becomes clear that intensities within the image are somehow "arbitrarily" changed. Their relative values do not reflect at all the ratios observed in the original scene. This can be well seen from the three example points marked in the image: the line profiles clearly show that point A is less bright than B and C has the same brightness as B. The same points in the HDR image, namely A', B' and C' show a totally different brightness behavior. Moreover, the related histograms displayed at the bottom are also significantly different.

The reason for this is that now there is not one common tone curve for the whole image, but different tone curves are adapted locally to different parts of the image to enhance the so-called micro contrast. Of course, the accountable HDR procedure is not really arbitrary, but it leads to an unpredictable brightness distribution within the image.

A tone mapped image cannot be considered a measurement. But a back transformation of the intensity distribution within the image to that within the object, in principle is possible in most cases, even though in reality, due to irreversible transformations, this mostly fails. Nevertheless, in principle or as a rough approximation, this is possible, at least for a simple tone mapping procedure. However, an HDR image is always totally away from a measurement, and back transformation to the original scenery (i. e., the object) is not possible at all.

Here, we would emphasize that these statements should not be regarded as a criticism of HDR and HDR images (HDRI) in general. They may look nice, and thus HDR fulfils the goal. Thus, today many cameras, including those implemented in mobile phones, offer the possibility for HDR even within the camera and could be automatically set for a bracketed sequence. The advantages are apparent. But HDR should never be regarded as presenting a general solution for displaying images with huge contrast. In particular, professional press photographers should avoid as much as possible image manipulations, in particular, severe ones. This is also fixed by the rules of news agencies, in particular, after the heavy discussion on world press photography in 2013. Also, camera manufactures have reacted on that issue by using image data verification systems, not because of HDR, but to make any image manipulation verifiable.

An alternative to HDR may be image fusion or blending. In that case, also a series of images has to be taken at the same illumination conditions and with exactly the same geometry, but again with different exposure values. Thus, within the image with a rather long exposure time, small differences in brightness can be seen, but the highlight regions become overexposed. For short exposure times, the situation is the reverse: within the highlighted regions the intensity becomes well resolved, but the shadow regions become almost black. Then, later on, when post-processed, the different images are fused to a single one. In contrast to HDR, the bit depth of the original images remains unchanged during the whole processing procedure (usually 8 bit). For each pixel, the brightness value is calculated from the brightness values of the same pixel of the bracketed sequence. This calculation corresponds to an averaging process, where the pixel values of that series are weighted. The weighting factors are functions of the pixel position, i. e., they depend on the brightness distribution of the vicinity of the pixels, namely if these are located in a dark region or a brighter one, etc. There are variants of this method, which sometimes are called exposure fusion, exposure blending, which works somewhat differently, dynamic range increase (DRI), or pseudo-HDR. Special software is available on the market as well. The most simple and straightforward alternative to all those methods is averaging of several images. Again, a series of images have to be taken at the same illumination conditions etc., but now with the same exposure values. When photon noise dominates, then noise decreases with the square root of the number of images. Consequently, e. g., a series of four images corresponds to a factor of two more photons in total, which is similar to the reduced noise of a photograph taken with +1 EV. As usual, these methods and also HDR all may have their advantages, and these even depend on the particular scenery. It is the decision of the user to apply any of these techniques. Further discussion of HDR is the subject of Section 7.4.3.

4.9.5.2 Additional and final remarks

In contrast to data from digital images taken by cameras, for digital data resulting from scans from films, advanced tone mapping is not necessarily required and not even demosaicing. But, of course, tone mapping is nevertheless sometimes helpful. This is because the tonal curve of the film (and, in particular, its sigmoid-type shape) is imprinted in the film, and thus in the linear data of the scan. Hence, tonal mapping may be only necessary to compress the larger dynamic range of the film into the common 8-bit depth (according to Figure 4.60a).

We also would like to remark that in general, gamma curves also play a role when linear signals that are displayed on a screen should be perceived as linear. Due to the fact that a gamma curve with $\gamma \approx 0.3 \ldots 0.5$ is a good assumption for the human eye, the screen has to change the signal intensity according to the reciprocal value of this, namely to $\gamma \approx 3.3 \ldots 2$. In that case, both gamma curves cancel and the signal is perceived to be linear.

To conclude this chapter, we would like to emphasize that scientific images that should be used as measurements should have a linear relation between the input and output signal. Thus, if, e. g., a DSLR is used, the image data should be extracted as linear ones from the raw data. Nonlinear detector response together with detector calibrations may be an alternative.

On the other hand, photographic images cannot be considered measurements. Such pictures are made to look nice, and thus the data received by the sensor have to be manipulated in a suitable way. Raw converters make use of the advanced possibilities of image processing; in particular, they give access to the larger dynamic range of the lossless 12-, 14- or 16-bit raw data, when compared to the lossy 8-bit JPG data. But even for an identical raw data file, different raw converters nearly always yield different results for the processed images even if the user puts much effort to apply the same modifications. Neither de-mosaicing, nor application of tonal curves is comprehensible or transferable from one raw converter to the next. Likewise, due to different algorithms and different likings, white balance, color reproduction or noise are predominantly treated in a different way. In addition, unless special raw converters are used as most likely to be found as open source in the web, photometric reproduction of images even from raw data files is nearly impossible. Moreover, with a raw converter it is not possible to get an image that is not processed at all. *DCraw* may be an exception and self-developed programs by the user may be possible as well. Also, programs such as *RawDigger* or the *SilverFast HDR-Studio* scanner software allow for reading raw data and saving them as TIF files without further change of data.

Here, we would like to stop that discussion because tone mapping and standard image processing, and in particular, the practical application, is well described in nearly every book on photography. Thus, such discussion will not be repeated here. Even more on that topic can be found in the literature specialized on that particular matter or in related web articles. Nevertheless, we will continue the discussion on image processing with special issues related to smart phone cameras in Sections 7.3 and 7.4.

4.10 Advanced and special sensors and sensor systems

Although the present book concentrates on the fundamentals of optical imaging and the present Chapter 4 on the basics of sensors, especially the most common ones, we would like to extend the discussion somewhat. In particular, we will implicate modern developments, introduce some special image detectors and we would like to draw the reader's attention, e. g., to the availability of image converters and intensifiers. But although some of those devices are common within science and technology, we also would like to note as well that part of the related topics is rather special. Thus, some of the details are much beyond the scope of the present book. For further reading, we refer to the special literature and further information on the web from the related institutions and companies. On the other hand, there are developments in sensor technology that today have been implemented in modern cameras, including SPC modules, or might be implemented in future. In the following, we will just concentrate on introducing the particular devices and give some remarks on them. We will begin with special and advanced sensors for photographic issues.

4.10.1 Sensor with stacked color information

Usually, color sensitive sensors are based on placing a particular color filter arrangement in front of the pixel array, where the most common is the Bayer mask (see Section 4.6.3). The disadvantage is that each individual pixel receives only limited information, in particular, information on the color contribution of that part of the light spectrum that is not blocked by the filter (Figure 4.68a). One consequence of this is also the reduced spatial resolution when compared to the same camera without CFA (see Section 4.6.3 and Section 5.2.6). Another one is that the most likely color at each pixel position has to be *estimated*, which may lead to color errors and artefacts such as color Moiré, etc. (see, e. g., Section 4.6.3, Figure 1.22e).

This situation is totally different with color films. At every position on the film, full color information is obtained directly, and one can expect the same spatial resolution as with a black-and-white film if the grain size would be the same for both films. The reason for this is that within the film, color information is recorded in different depths at the same (x, y) position (see Figure 4.6).

An alternative sensor that uses a similar approach is the Foveon X3 sensor. Within this sensor, each photo site is made of a layer stack, where the different layers are sensitive to different spectral components as shown in Figure 4.68b. According to the difference in attenuation length in silicon (see Figure 4.9), the thickness of the three layers is different with a total thickness of 5 µm. This arrangement may be regarded as equivalent to a matrix of "three-component photodiodes," where the blue, green and red light sensitive photodiode components, respectively, are placed on top of each other.

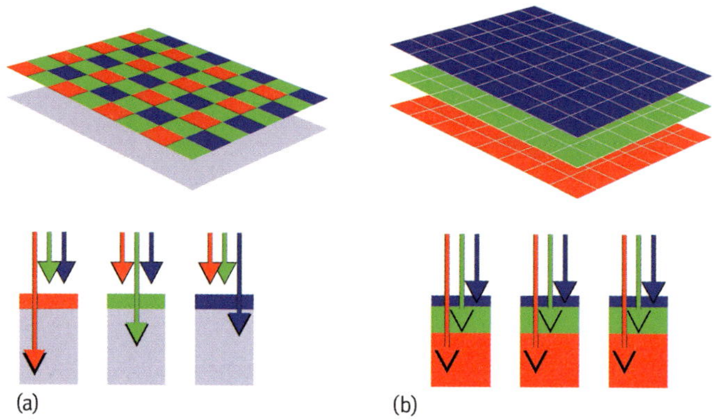

Fig. 4.68: (a) Standard sensor using a Bayer mask. The colored regions correspond to the matching color filter. The light blue region marks the CCD or CMOS pixels. (b) Scheme of the Foveon X3 sensor. In contrast to (a), here the colored regions correspond to the light sensitive region with respect to that color. In (a) and (b), the arrows indicate light of the corresponding color. Note that the upper row only illustrates the arrangement of filters and light sensitive regions. The displayed layer thickness is not at scale.

Cameras equipped with such CMOS sensors are, e. g., Sigma's compact cameras of the dp Quattro series, its DSLM of the ds Quattro series and its SD1 DSLR camera (4800×3200 pixels times three layers).

Of course, the Foveon sensor with a given number N_h and N_v, respectively, of pixels within one color plane has a significant better spatial resolution when compared, e. g., to a standard sensor with the same values of N_h and N_v. But we would like to comment that the improvement is not a factor of 3 and it is not serious to claim the effective number of pixels is the sum of all pixels within all three layers as some marketing people do. This may also be seen from the stored images, which similar to those of a comparable Bayer sensor camera, have the same pixel numbers, each of them with a color information. For JPG files, each site consists of a value with a depth of 8 bits for each of the RGB channels. A realistic comparison of spatial resolution and judgement of superiority may only be the result of a carefully performed measurement of the MTF.

Although any sensor may suffer from Moiré artefacts if the Nyquist limit is exceeded, an obvious advantage of the Foveon sensor is the absence of the color Moiré effect. Consequently, the requirements on the OLPF may be regarded to be less severe so that it may be even omitted (this also leads to an increase of resolution; see the discussion in Section 4.6.2).

A disadvantage may be the color reproduction. It is not our task to judge that quality, but we would like to point out that it is discussed that color reproduction may not be as good as expected and even not as good as that obtained with conventional CFA technology. Moreover, due to the absorption properties of the top and the intermediate layer, there is less light below, and thus noise is increased. Spectral clipping effects may be present as well. On the other hand, the next generation of Foveon sensor is currently under development. But again, a reliable comparison of the quality of images obtained with the Foveon sensor to other ones can only be made on the basis of well-done experiments. A personal judgement may be made by the reader.

More recently, there is a new development by a team of Empa and ETH Zürich (both Switzerland), which have produced a pixel that does not suffer from the above-described disadvantages and, in addition, is much smaller ("$MAPbX_3$-detector"[10]). In contrast to the Foveon sensor, where the layers are made of silicon, now they are made on a base of lead halide perovskites; the top layer contains chlorine perovskite and absorbs blue light, the intermediate layer bromine perovskite and absorbs blue and green light and the iodine perovskite bottom layer is opaque. The absorption coefficient in green is an order of magnitude larger when compared to silicon, and in the red range it is larger by two orders of magnitude. For blue light, it is approximately the same.

Finally, we would like to remark that there are approaches similar to the described sensor schemes by other companies and institutes. Fuji's three-color sensor is an example. Sony's "Three-layer Stacked Color Image Sensor With 2.0- µm Pixel Size Using Organic Photoconductive Film" is another one[11] and the developers of an even other "vertical color sensor" claim a better color recognition when compared to the Foveon sensor.[12] Moreover, there are ideas of splitting color information of each pixel (see next chapter). For instance, there is a patent of Panasonic, where instead of each single filter within the CFA there is a splitter that directs the colored light at each of these positions to three RGB-channels.

4.10.2 Sensor with a color-sorting metalens array

The usual color sensitive sensors, namely other ones than the just discussed sensors with stacked color information, suffer from the losses according to the CFA. Concretely, each color filter in front of a specific pixel suppresses all wavelengths outside the transmission band of that particular filter. This is a significant loss, which becomes well apparent, when we compare Figure 4.24a and c. We may be reminded of the discussion in Section 4.6.3, where for a Bayer mask it is concluded that the geometric efficiency ηgeom is 25 % for red and blue light, respectively, and 50 % for green light. Thus, the overall quantum efficiency is rather low. To overcome this problem, it would be preferable to replace absorptive filters by lossless arrangements that just redistribute the different

10 S. Yakunin et al.: *Nondissipative internal optical filtering with solution grown perovskite single crystals for full-color imaging*, NPG Asia Materials (2017) 9, e431.

11 H. Togashi et al.: *Three-layer Stacked Color Image Sensor With 2.0-µm Pixel Size Using Organic Photoconductive Film*, 2019 IEEE Int. Electron Devices Meeting (IEDM), San Francisco, CA, USA, 2019, pp. 16.6.1–16.6.4

12 N. Li et al.: *van der Waals Semiconductor Empowered Vertical Color Sensor*, ACS Nano **16** (2022) 8619–8629.

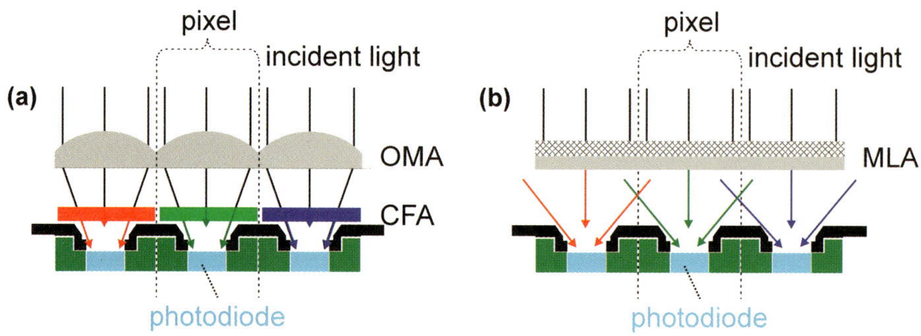

Fig. 4.69: (a) Scheme of color selection within a conventional OMA/CFA Bayer scheme as shown in Figure 4.18 and Figure 4.35, respectively. (b) Scheme of color distribution by means of MLA where color-splitting geometry may also be arranged in a way equivalent to the Bayer scheme. Black arrows indicate white light, colored arrows the corresponding color.

wavelength regions to the corresponding pixels. In that way, the number of photons per pixel could be increased by up to 2 to 3 times. A consequence of the improved (photon collection) efficiency is that the sensor sensitivity becomes larger and the SNR will be increased as well (see, e. g., Figure 4.40a).

Several arrangements to achieve this situation have been proposed. But many of them suffer from still low efficiency and/or have problems with an oblique angle of incidence (see Section 4.6.1) and/or depend on the incident light polarization and/or are not compatible with the CMOS technology of CIS fabrication and so on. However, based on recent works, one may have to overcome these problems (see, e. g.,[13,14]), in particular, for CIS made of very small pixels as used for smartphone camera modules. This approach of a "full-color sorting lens array" makes use of metasurface lenses made of nano-structures on the subpixel scale as shown in Figure 4.69 (metalens array, MLA; for basics, see Section 7.5). Now, in contrast to the scheme of Figure 4.68a, Figure 4.69 shows that all the light in front of one pixel is either collected and directly focused onto its surface (similar to an OMA), or it is directed to the appropriate neighbored pixels. The larger efficiency of light distribution is evident.

The single elements of the MLA, which may replace the microlens and the color filter, consist of nanostructures, which generate locally varying phases for the incident light (see Section 7.5). This requires a sophisticated design of the phase front affecting the nanosurface profile to achieve the desired wavelength-dependent directional focusing. In some sense, that may be regarded as the introduction of a specific wavelength dependent wavefront modification, similar to the unwanted WFA in Figure 5.24.

---

13 J. S. T. Smalley et al.: *Subwavelength pixelated CMOS color sensors based on anti-Hermitian metasurface*, Nature Communications **11** (2020) 3916, https://doi.org/10.1038/s41467-020-17743-y

14 M. Miyata, N. Nemoto, K. Shikama, F. Kobayashi, T. Hashimoto: *Full-color-sorting metalenses for high-sensitivity image sensors*, Optica **8** (2021) 1596.

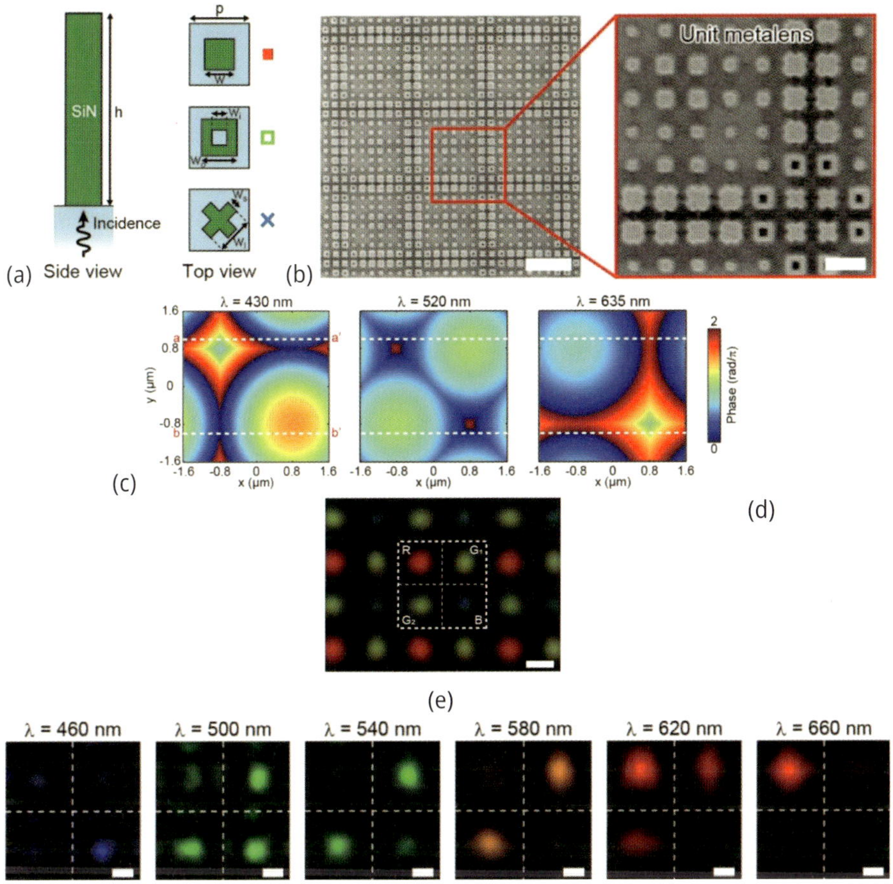

Fig. 4.70: Example of a color sorting MLA. (a) Schematic of the geometry of the nanoposts. The posts are made of silicon nitride on a fused silica substrate. The typical post height is $h = 1250$ nm. (b) SEM images of the MLA. (c) Phase profiles for wavelength sorting and focusing according to the RGB Bayer geometry. The pixel size and focal length are 1.6 and 4.0 μm, respectively. (d) Crop of the measured focal plane intensity profile of the MLA when illuminated with white light (RGB Bayer geometry). (e) Measured focal plane intensity profiles within a Bayer cell. Results are shown for different wavelengths of the collimated incident light as indicated. The scale bars in (b) are 2 μm (left) and 500 nm (right), that in (d) 1 μm and those in (e) 500 nm. Images are taken from Figures 2 and 3 (both in part) from[15]. Reprinted in with permission from M. Miyata et al., Optica **8** (2021) 1596 ©2021 Optica Publishing Group.

Within the discussed work of Miyata and coworkers, by applying a standard fabrication process, the nanosurface has been made of dielectric nanoposts of different shapes and arrangements (see Figure 4.70). The suitable arrangement of the posts has led to

15 M. Miyata, N. Nemoto, K. Shikama, F. Kobayashi, T. Hashimoto: *Full-color-sorting metalenses for high-sensitivity image sensors*, Optica **8** (2021) 1596.

a phase, which depends on the wavelength but is independent of polarization (Note: Usually, effort is made to obtain achromatic properties, but for the MLA the chromatic behavior is applied for wavelength splitting; for these and other details, see[16]). As a result, the reported predicted transmittances are 64 %, 75 % and 97 % for the blue, green and red light, respectively, for 4 µm pixel pitch. Although the experimental values are somewhat lower, there is still a large improvement in QE, which again may be compared to Figure 4.24c. In total, it was claimed that the signal was increased by a factor of 2.8 in comparison to a CFA-based experiment with an angular response of 16.5° (compare Figure 4.21).

In conclusion, it is expected that MLA technology with efficient color-splitting elements will improve the performance of CIS significantly. The enhanced QE will lead to larger sensor efficiency and an enhanced SNR, even for a pixel pitch of less than a micron. This, in particular, is a key issue for high performance SPC. Indeed, based on simulations, Miyata et al. have shown that their modified MLA even works well for $p = 0.84$ µm. The NA was up to 0.284, and thus compatible with SPC sensors. Furthermore, it is claimed that spatial resolution may be improved as well due to "additional space-color information" and improved de-mosaicing procedures (Section 4.9.3). However, there are still the restrictions of color-specific pixels (see Section 4.6.3) and the resulting resolution limits so that this claim has to be proved. Moreover, albeit all that, the main restriction of the smaller FWC of small pixels still cannot be overcome by MLA and this still restricts the SNR, and thus the sensor and SPC performance, in particular, when compared to DSLR with large pixels.

4.10.3 Pixel interleaved array CCD

Although Fuji's super CCD is no more up to date, nevertheless it is interesting to discuss the basic idea. Usually, pixel arrangement is done in a rectangular matrix as illustrated in Figure 4.17 and Figure 4.71a and d. An interesting alternative arrangement is the "pixel interleaved array CCD" (PIA-CCD) shown in Figure 4.71b and c. Most simply, this corresponds to a pixel arrangement that is tilted with respect to the sensor aperture. As illustrated, this leads to a decrease of the pixel pitch in the horizontal and vertical direction, respectively (by a factor $2^{1/2}$). The pitch at 45° directions has not changed.

The advantage of that scheme is that it fits much better the resolution of the human eye, because it has been observed that the eye is most sensitive to horizontal and vertical lines and less to inclined directions. Indeed, it has been found that the spatial resolution of the human eye is a function of the tilt angle of a test grid. In particular, the relative

16 M. Miyata, N. Nemoto, K. Shikama, F. Kobayashi, T. Hashimoto: *Full-color-sorting metalenses for high-sensitivity image sensors*, Optica **8** (2021) 1596.

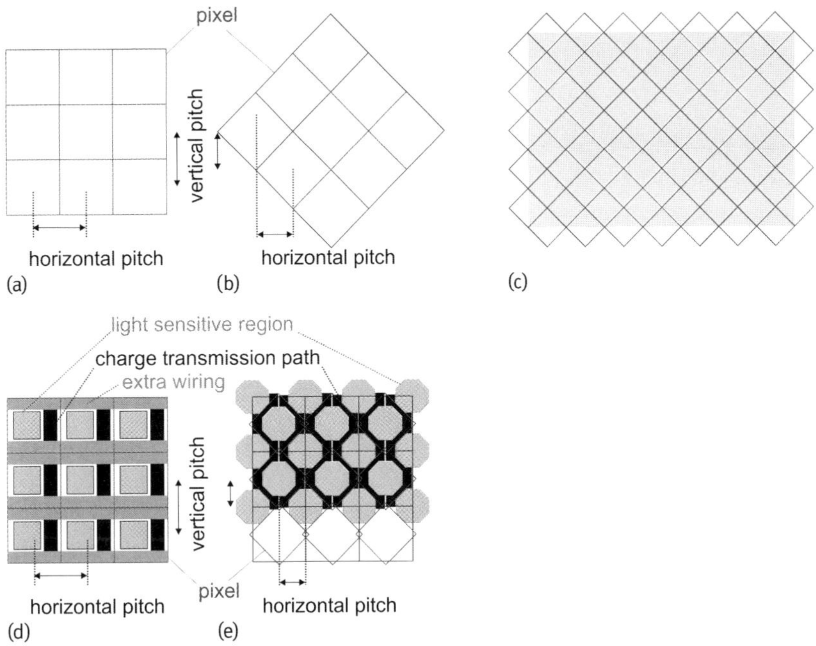

Fig. 4.71: (a) Standard layout of a CCD or CMOS sensor with a pixel arrangement in a rectangular matrix, which is oriented as displayed. (b) Simple sketch of a PIA-CCD. (c) Orientation of the pixels with respect to the rectangular sensor aperture (indicated by light gray). (d) A more detailed arrangement of a standard IT-CCD. (e) "Super CCD" according to PIA-CCD arrangement (note that this example slightly differs from (b) because the pixels are somewhat smaller than those in (d)).

contrast sensitivity oscillates between 0 and –6 dB, with the maxima for grid orientations in horizontal and vertical direction, respectively (see, e. g., [Nak06] and also the cited literature in that book; for resolution measurements with test grids, see Chapter 5 and Chapter 8). Consequently, the PIA-CCD geometry leads to a significant increase of the effective resolution in the corresponding directions and due to the tilt dependence of the eye not necessarily to a loss in any other direction.

Of course, standard output devices always make use of a square geometry oriented as shown in Figure 4.71a. Consequently, during image processing within a camera with a PIA-CCD, recalculation of the signal is done in a way that the signals first are calculated on a grid that has half of the period of the original via an appropriate interpolation. This corresponds to a sensor with twice as many pixels in the horizontal and vertical direction, respectively, and consequently a higher Nyquist frequency by a factor of 1.4. Note that physically there are not more pixels; the signals are just estimates at the corresponding positions (see Section 4.9). The final output of the camera still may not consist of more pixels as it has physically, but the MTF may be better or expressed quite simply: the "two-pixel-resolution" discussed in Section 1.6 has been somehow beaten, but only in a particular direction.

A special development of this kind of sensor is Fuji's so-called "super CCD sensor." Figure 4.71e shows an example with light sensitive regions of octagonal shape. In comparison to an interline CCD, another advantage of this sensor type is that the fill factor may be larger due to up to 30 % larger area of the light sensitive region, which leads to an increased FWC, and thus a better SNR and dynamic range. Furthermore, the octagonal shape matches the OMA in a better way. Newer developments make use of somewhat modified arrangements, some use CMOS instead of CCD and also split pixel and/or BSI technology (for both see below). PIA-CCD have been developed as image sensors usable for progressive scans and can be produced by standard technology. A discussion on details of the semiconductor layout, the charge transfer and the sensor layout in general, can be found, e. g., in [Nak06].

4.10.4 BSI CCD and BSI CMOS

In the 1990s, CCD detector systems became quite useful in many fields of science and technology. However, such systems could not be used for short wavelengths. First of all, detection in blue was poor due to the absorption in the polysilicon front gate electrons. Later on, that problem was reduced by application of electrodes with better transmission (see Section 4.2.2). Absorption at shorter wavelengths, namely in UV, could be easily prevented by removal of the front glass window of the detector. Usually, this entrance window is placed for protection of the sensor surface. Although there was strong demand to replace films (see, e. g., Figure 4.45c) and other detectors by modern CCD systems also for even shorter wavelength ranges, i. e., soft X-ray and XUV, at that time this was impossible.

Figure 4.9, in particular, indicates that absorption strongly increases with decreasing wavelength. Indeed, in XUV, the imaginary part of the complex index of refraction, for instance, at $\lambda = 10$ nm, is $n'' = 2 \cdot 10^{-2}$, which leads to a penetration depth of 40 nm. In general, strong or even huge absorption is present for λ between 1 and 100 nm in any kind of matter, and thus prevents detection with standard semiconductor detectors (Figure 4.72), because the radiation becomes fully absorbed in the silicon quite close to the surface before it gets converted in the photosensitive layer.

To avoid this situation, light has to be coupled most directly into the photosensitive layer without passing material before. Due to the gate structure, this is not possible from the front side. However, by removal of semiconductor material from the rear side, the sensor can be thinned down so much, namely to approximately 10 to 20 μm thickness, so that absorption is restricted to the region where the electron hole pairs are generated if the sensor is illuminated from that side. This is the idea of back side illuminated (BSI) CCD detectors.

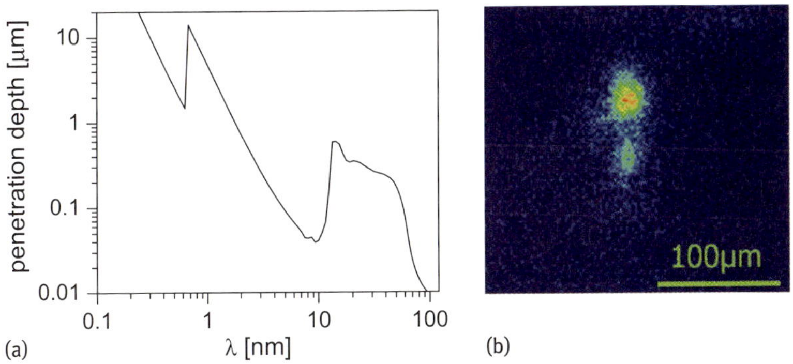

Fig. 4.72: (a) Penetration depth for pure silicon in the XUV and soft X-ray region (data taken from[17]). As a result, transmission through a 10 µm thick silicon layer is nearly zero and becomes only measurable for a thickness below 1 nm. (b) Example of an image of a laser-produced plasma taken with a Fresnel zone-plate as the optics and a BSI CCD detector (false color reproduction; bright region on top; the less bright spot below is a reflection; data taken from[18]). The wavelength is 3.4 nm.

In these "early times," thinning of the sensor chip was difficult. Particularly, there is a strong requirement of precision with respect to homogeneity and semiconductor thickness, in particular, for those fragile chips of about 10 to 20 µm thickness (Figure 4.73). Otherwise, shading effects, loss in sensitivity and nonuniform response of the sensor area will occur. Formerly, there was a lot of waste, and thus BSI CCD systems were very expensive. Nevertheless, due to their high sensitivity (see also Section 4.8.5), they became successful tools, e. g., for measurements of the plasma emission in the soft X-ray and XUV-range. An actual example of such a camera system has been discussed in Section 2.6.5. A good review article is still [Gru02].

Since that time, technology has become much advanced so that nowadays BSI sensors can be fabricated more easily. BSI CMOS have also become common, too, for the visible spectral range, and thus they are used in many digital cameras, including mobile phone cameras. In comparison to the standard front side illumination, BSI has the advantage of higher quantum efficiency. Even values above 90 % are possible (see also Section 4.2.2). For CMOS sensors, in particular, this is much pronounced because for front side illumination there is a significant shielding yielding a small fill factor and a shading effect due to the wiring at the front side (see Figure 4.18b). Hence, illumination from the back side improves the situation, e. g., by a factor of 4. Altogether, the sensor becomes much more sensitive. For a CCD, BSI does not lead to those discussed *additional*

17 B. L. Henke et al.: *Low energy X-ray interaction coefficients: photoabsorption, scattering and reflection*, Atomic Data Nucl. Data Tables **27** (1982) 1–144.

18 U. Teubner et al.: *High Brightness X-Radiation and Plasma Frequency Emission from Femtosecond Laser Plasmas, Applic. of High Field and Short Wavelength Sources VIII*, OSA Tech. Digest (Opt. Soc. Am., Washington DC, 1999), 94/WA3-1-96/WA3-3.

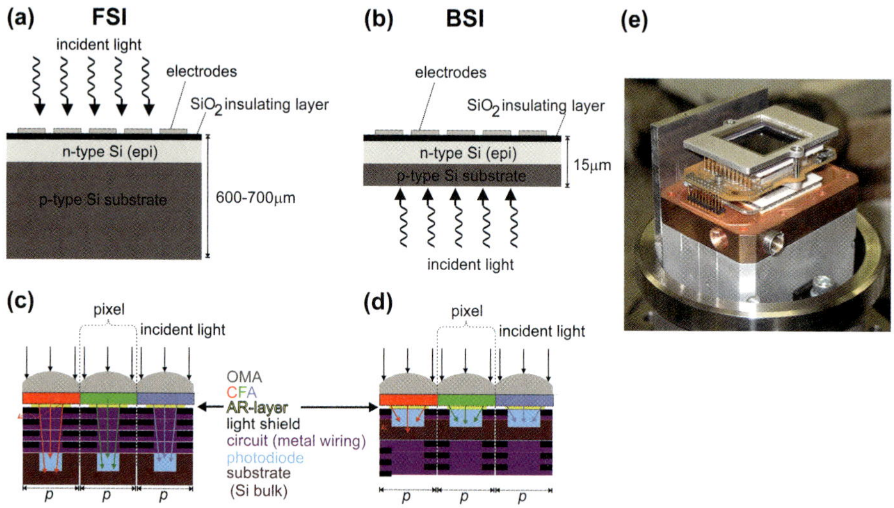

Fig. 4.73: (a) Standard CCD (or CMOS) chip, front side illuminated and (b) BSI-CCD sensor. (c) and (d) show FSI and BSI CIS, respectively. (e) View of the camera head of a scientific camera with a BSI-CCD sensor. "epi" is the epitaxial layer. Note that the sensor orientation with respect to the incident light in (b) is rotated by 180° with respect to (a), (c) and (d).

advantages because 100 % fill factor is often standard and the discussed shading effects resulting from collimation are absent.

It may be mentioned that the thinning process of the CMOS chip may result in small defects on the rear side that lead to an increase of dark current, and thus of noise. Cross talk may become a problem as well. On the other hand, current technology including improvements in design have reduced these problems.

4.10.5 Advances in CMOS technology

Image sensors based on CMOS technology have tremendously been developed and today they take a large fraction of the total optoelectronics market. In particular, behind integrated circuits and together with lamps, which today are primarily LED, CIS take the second largest market share. Among many other applications for CIS, this is driven by image sensors for machine vision, industrial cameras, automotive and mobile applications, especially SPC. However, there are also significant improvements beside the mass market, e. g., an evolution of advanced CMOS sensors.

4.10.5.1 Scientific CMOS sensors

As has been discussed in Section 4.5.2, both CCD and CMOS sensors have important advantages. To take advantage of both of its properties, years ago hybrid CCD/CMOS image sensors were developed. These devices make use of CMOS readout circuits bump-bonded to a CCD sensor substrate. These devices are rather complicated, and thus are expensive and today have been surpassed by a "second generation CMOS" called scientific CMOS, abbreviated as sCMOS (some companies even claim a second and a third generation of sCMOS).

This new technology should offer advanced performance, in particular, for demanding scientific applications, such as microscopy for biological applications, and fast 3D measurements. Significant improvements comprise improved quantum efficiency, exceptional low read noise, high resolution, large dynamic range, large SNR and all that at a high-frame rate. In comparison to other sensor technologies, it may be expected that for rather low light signals, typically < 50 photons per pixel, iCCD or EM-CCD systems (see Section 4.11) are superior, but for larger signals, sCMOS may show at least the same performance. Both kinds of sensors are expected to be superior to the widely and commonly used scientific interline CCD systems at least below approximately 200 photons per pixel. However, a general statement about which sensor is the best, e. g., the EM-CCD or a sCMOS, is not easily answered, as it may strongly depend on the application and the operation conditions, such as the amount of sensor cooling (see also brief discussions in Section 4.11.4 and Section 4.11.5). Some sensor parameters are listed in Appendix A.5.

4.10.5.2 Advances for small pixels: textures, depth increase and deep trench isolation

For image sensors in general, a large quantum efficiency is important (EQE). This includes low losses in front of the light sensitive volume, a large absorption within the volume and a high IQE. Within the last years, a lot of camera modules have become smaller and the pixel number has increased. In particular, this is the case for SPC. The pixel pitch has been decreased to 0.7 µm and below, with actually the smallest values even below 0.6 µm (2022). This has required smaller CIS with larger pixel density, which consequently has set new demands, such as a necessary improvement of pixel isolation. Also, the fraction of active area to die area of the CIS has grown (from approximately 60 % in year 2014 up to more than 80 % in 2022). More functions, such as dual gain or direct modification of the signal within a pixel (see below), have been included or have been intended to.

However, as discussed before and also later in Chapter 7, there are a lot of problems for CIS with a small pixel pitch. These are the restricted photon absorption, due to small pixel area and thickness, the relatively low FWC and sensitivity, which lead to restrictions in dynamic range and SNR, the cross talk and so on. To overcome these problems, novel technologies have been developed and applied for the current CIS generation. Albeit the small pixel pitch, the goal is to increase the sensor sensitivity, FWC and SNR. There are several strategies, which will be discussed in the following.

Around the years 2017 and 2018, a pixel pitch of $p \sim 1$ µm and a thickness of the light sensitive volume $d_p \sim 3$ µm was rather common. However, as the penetration depth Λ_p is significantly larger than d_p (see Figure 4.9), a significant fraction of light is lost by transmission through the photodiode. For the reduction of losses, structuring of the front or rear side of the light sensitive volume has been discussed. Examples have been surface textures, such as inverted pyramids (IPA, Figure 4.74a) at the side where light is incident and/or suitable cavities at the other side, which should lead to a reflection of the otherwise transmitted light. Those so-called “front inner lenses” must have a much lower index of refraction than the adjacent layer to allow for the back reflection. Various arrangements have been proposed, such as uniform or nonuniform IPA, nanorods, nanoholes or other nanostructures. All those should augment light trapping and absorption. As an example for the infrared region, where due to the large value of Λp the losses are more pronounced and shown in Figure 4.74b and c. Due to the diffraction at the structures and total reflection at the pixel boundary with applied IPA (or for instance appropriate nanoholes), absorption and efficiency improvements of several ten percents have been reported for the NIR region (see, e. g.,[19]) when compared to CIS without those additional structures.

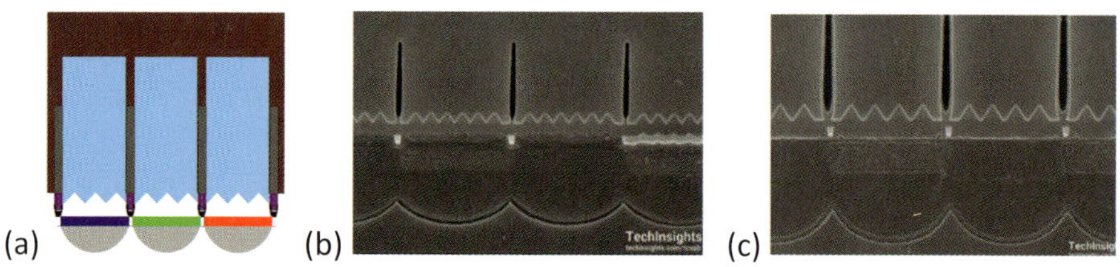

Fig. 4.74: Examples of a “back-surface” equipped with IPA (note that “back-surface” is related to BSI and means the side, which is illuminated (lower side in Figure 4.73b and top side in Figure 4.73d)). (a) Scheme for a CIS with a Bayer CFA for the visible range. The white region is made of SiO_2 or another material with a (much) lower index of refraction than the silicon of the photodiode. The pyramid dimensions comprise only a fraction of the pixel pitch, with p in the range of 0.7 to 1.4 µm. SEM images of CIS made for NIR: (b) Sony, (c) OmniVision. Note that (a) to (c) have the same orientation with respect to the incident light as Figure 4.73b but are rotated by 180° when compared to Figure 4.73d and Figure 4.75. (b), (c) images taken from[20], courtesy of Techinsights.

Actually, the thickness of the active silicon layer where the photons get absorbed has been further increased. Nowadays, $d_p \sim 4$ µm in the visible range, which leads to an aspect ratio as high as 4 to 6 (thickness/pitch) and $d_p \sim 7$ µm in NIR. Due to the absorption coefficients, a further increase will not make too much sense. In the blue wavelength region, the full amount light is already absorbed and in the red wavelength region there is not much potential left (see Figure 4.9). Nevertheless, an increased absorption is obtained. The difficulty now is that part of the absorption and photoelectron generation

19 S. Yokogawa et al.: *IR sensitivity enhancement of CMOS Image Sensor with diffractive light trapping pixels*, Sci. Rep. 7 (2017) 3832.

20 Z. Shukri: *The State of the Art of CMOS Image Sensors*, TechInsights Inc. Inc. https://www.techinsights.com/ebook/ebook-latest-development-trends-cmos-image-sensors (visited January 2023).

occurs in deeper layers, and this makes optical (color) and electronic cross talk much more a problem. To get rid of that problem (and other ones such as interference) or at least to reduce it, the individual regions of the neighbored pixels have to be separated much better within the semiconductor. This is accomplished by deep trench isolation (DTI) together with an optical scattering layer that lengthens the photon path (see above and Figure 4.75). Additionally, an optical separation in front of the semiconductor region with light separation in between the CFA may be adapted. Samsung's ISOCELL technology and subsequent ones makes use of this method. Similar technologies are also applied by other CIS manufacturers, e. g., by Sony, and used in SPC of these or other companies. Current developments work on further improvements, e. g., by the application of additional plasmonic diffraction structure (see, e. g.,[21]).

There are also CIS made of dual pixels with in-pixel DTI, which are intended, e. g., for PDAF (see Section 4.10.6.2). Moreover, there is the possibility of optical structures in front of and/or behind the absorption region to improve light management.

A similar development, now related to an increase of the response in NIR, in particular, between 850 and 940 nm, is OmniVision's Nyxel technology and similar ones. This makes use of three steps. The first one is the increase of the silicon thickness, currently to a thickness exceeding 7 µm. For light entering close to normal incidence, this does

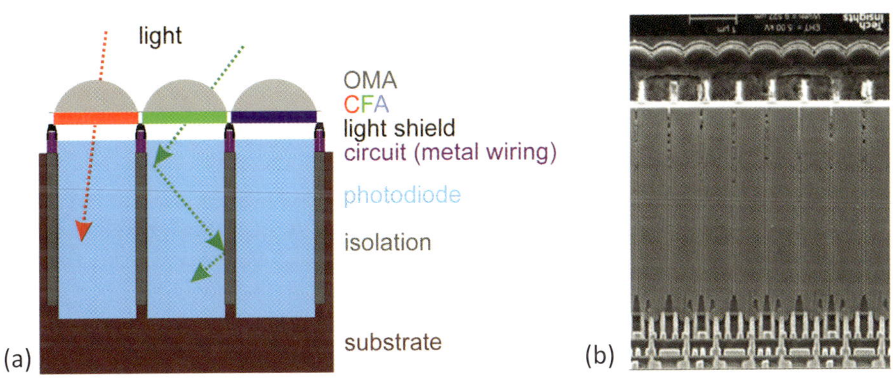

Fig. 4.75: DTI scheme. The scheme is the same as in Figure 4.74 but here without IPA and orientated as Figure 4.73d. (a) Scheme of pixels with DTI (not in scale). In reality, the isolation is significantly thinner than displayed and may also extend to full depth as shown in (b). There are different generations of CIS, which differ in silicon thickness, trench isolation width and depth, isolation material and potentially a rear side metal oxide grid. (b) SEM cross-section of the Samsung S5KGW3 (courtesy of Techinsights,[22]), which is a 64 MP CIS with $p = 0.7$ µm and an epi-layer thickness of 4.1 µm. Note that (b) is rotated by 180° with respect to (a). (b) image taken from[22], courtesy of Techinsights.

---

21 A. Ono et al.: *Near-infrared sensitivity improvement by plasmonic diffraction for a silicon image sensor with deep trench isolation filled with highly reflective metal*, Opt. Express **29** (2021) 21313.

22 Z. Shukri: *The State of the Art of CMOS Image Sensors*, TechInsights Inc. Inc. https://www.techinsights.com/ebook/ebook-latest-development-trends-cmos-image-sensors (visited January 2023).

not cause problems, however, for inclined light rays this could lead to cross talk. Consequently, this is suppressed by DTI (second step). A third step is involved by an optical scattering layer that lengthens the photon path and improves image quality. Altogether, this improves the sensor performance significantly, in particular, for low light conditions. Though the performance in the visible range is almost unchanged, the quantum efficiency is claimed to be enhanced by a factor of 3.

The technologies discussed in this section have led to the rather high FWC (in comparison to years before, but not when compared to DSLR), for instance, to the CIS of the high-end SPC with the data given in Table 7.7. Nevertheless, still significant electrical cross talk may occur, which then still leads to increased noise, degraded spatial resolution and reduced contrast (MTF).

4.10.5.3 CMOS stacking technology

Another advance in CIS chip production is CMOS stacking technology (Sony, Omni Vision, Samsung, etc.). Within this technology, the layer of consumer CMOS sensors is separated so that the electronics is shifted to an additional layer (see Figure 4.76b) and there is even a development of a tri-layer stacked CIS with an added DRAM in the third layer (Figure 4.76c). The separate layer(s) is connected to the layer with the photo sensitive region (glued, soldered, wafer bond, interconnection with through-silicon via (TSV), etc.) and also serves for stability of the whole chip. We would like to note again the fragility of the very thin BSI layer.

Discussed advantages of stacked sensors are the further increased fill factor, and thus higher sensor sensitivity, less noise, larger dynamics, the mechanical stability, the increased storage and computing power, a decrease of power consumption, shorter exposure intervals, lower cost and there may be further ones. A faster readout allows the reduction of the rolling shutter effect when shooting fast-moving subjects without global shutter. Under discussion is also that in principle every pixel can be controlled individually. In such a way, e. g., the ISO-value may be set individually for each pixel, and thus change across the sensor area. Or, even more, in the future each pixel may be equipped completely with its own processor, which then would offer a huge potential for much more extended image processing (some people talk about a nanocomputer for each pixel; if this makes sense is another question). A quite recent example for this is

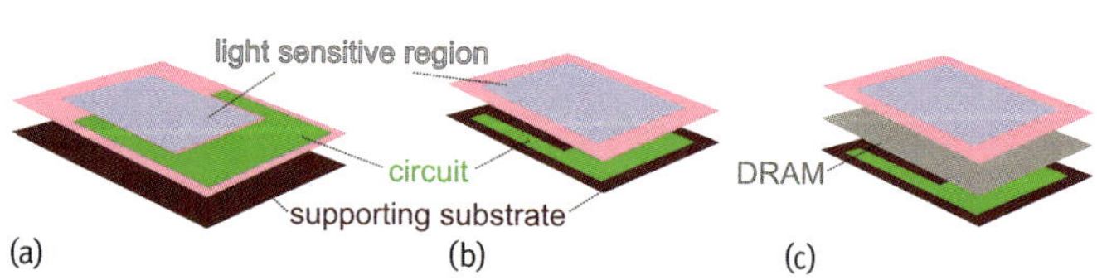

Fig. 4.76: Scheme of a (a) conventional BSI CMOS sensor, (b) stacked BSI CMOS sensor, (c) "tri-stacked" BSI CMOS sensor.

the development of CMOS imaging sensors, where all pixels could be readout in parallel. This corresponds to an electronic global shutter (see also Section 4.4.2) instead of a rolling shutter.

Altogether, such an advanced technology together with improved material quality and technologies such as DTI has led to a significant improvement of CIS characteristics. Stacked CIS are mostly applied to SPC, but much less to other cameras. In 2020, the market of SPC image sensors is covered approximately 90 % by stacked BSI CIS, partly tri-stacked, but mostly bi-stacked. The use of monolithic CIS has decreased significantly for SPC but is still important for other camera sensors.

4.10.6 Hardware technologies for dynamic range extension

It is obvious that for many applications a high-dynamic range (HDR) of the signals is desirable. But conventional sensors may not provide this; especially, sensors to be used for automotive cameras or SPC modules are examples. Where automotive cameras can be regarded as examples that indicate the requirement of say $DR = 16$ to 20 EV or 100–120 dB (see Equation (4.35)), respectively, to cover the brightness of traffic signs and at the same time that of persons in the dark, SPC modules illustrate the problem of a small pixel size and the resulting limit in FWC, which lead to strong limitations of DR (see Chapter 7). The desired large dynamic range has led to the development and the implementation of several techniques that will be discussed in the following sections. Here, we restrict the topic to light detection. Post-processing, and in particular, tone mapping and compression is not an issue of the next sections as this is subject of Section 4.9.5 and Section 8.4. Schemes of extension to a pretty high-dynamic range (HDR) make use of different operation modes where readout time and/or gain is changed, possibly even during exposure. Alternatively, they make use of a special sensor or pixel design with, e. g., split or dual pixels, which will be operated in different modes. We may note that depending on the realization of a particular HDR scheme, a specific hardware configuration is required. There are pixel circuits, which result in linear and such ones that result in nonlinear response. Sometimes several of the mentioned techniques are coupled, even linear and nonlinear ones. For an overview and some more details on HDR CIS for automotive applications, we may refer, e. g., to[23]. Details on HDR CIS for SPC are discussed below and in Chapter 7.

4.10.6.1 Staggered HDR technology

One of the most simple methods to get an HDR image is to take a series of images from the same scenery at different EV as described in detail in Section 4.9.5.1. But this is not

23 I. Takayanagi, R. Kuroda: *HDR CMOS Image Sensors for Automotive Applications*, IEEE Transactions On Electron Devices, **69** (2022) 2815.

a hardware-based technology. On the other hand, for CIS with a rolling shutter (see Section 4.5.2.2), one can make use of that to obtain a HDR image. In that scheme, several images are exposed with different times t_x within a readout cycle of the whole sensor. After a first exposure of a row of pixels with $t_{x,1}$ and its readout, the signals are stored in a line buffer. Then a reset pulse initiates a new exposure, now with $t_{x,2} < t_{x,1}$, and a new readout pulse initiates reading of this row with reduced exposure and combination. Thus, the two exposures occur immediately one after the other and within a larger time window in which exposure and readout of other rows still continues simultaneously until all signals of the CIS are readout twice. Finally, both frames with the different exposure times are combined. For linear photon conversion (see, e. g., Figure 4.55), the dynamic range is extended by a factor given by $t_{x,1}/t_{x,2}$. Besides this basic just recently commercialized scheme of staggered HDR, there are also more complicated similar ones. However, analogously to the dual gain scheme discussed later (see Section 4.10.6.5), there is the disadvantage of the difference of SNR when the frames with $t_{x,1}$ and $t_{x,2}$, respectively, are compared. This, in particular, is the case when photon noise dominates. As a result, this may be recognized as a visible artefact.

4.10.6.2 Split pixel and subpixel technology

To increase spatial resolution of a sensor even more, the number of pixels has to be increased. However, this requires the shrinkage of the pixel size with all the disadvantages discussed before, in particular, the reduction of the dynamic range. For CCD, one solution is that each pixel is split into two sections, which both consist of a photodiode (Figure 4.77a). For CIS, this is similar. One of the sections may have a rather large area and a high sensitivity, which is necessary for low and medium light conditions and good depth resolution (see Section 4.8.5 and Section 4.8.6). The other one has a small area and a low sensitivity and is used to extend the dynamic range, when the first diode becomes saturated. The reduced depth resolution of that diode is not so much an issue, because noise is much larger for bright light illumination (see Section 4.7 and discussion in Section 4.8.6). The full photodiode consisting of the two subpixels are operated with a different gains ((Figure 4.77 c); see also Section 4.10.6.3). The final signal at each pixel position is then calculated by the image processor on the base of the signals of both partial(?) diodes, which are readout individually.

An alternative is to place assigned smaller pixels with smaller microlenses in between the larger ones (Figure 4.77b). The operation mode is similar to before. Such technology is applied, e. g., for automotive applications where traffic scenes should be well captured even in the presence of strong light sources such as lamps, and thus a very large dynamic range is required. Even other sensors with split pixels are available as well. Some of them make use of the same or a similar idea to increase the dynamic range.

Even other sensors (and cameras) also make use of split pixels, but for other reasons. As an example, there are DSLR cameras such as the Canon 5D Mark IV equipped with a

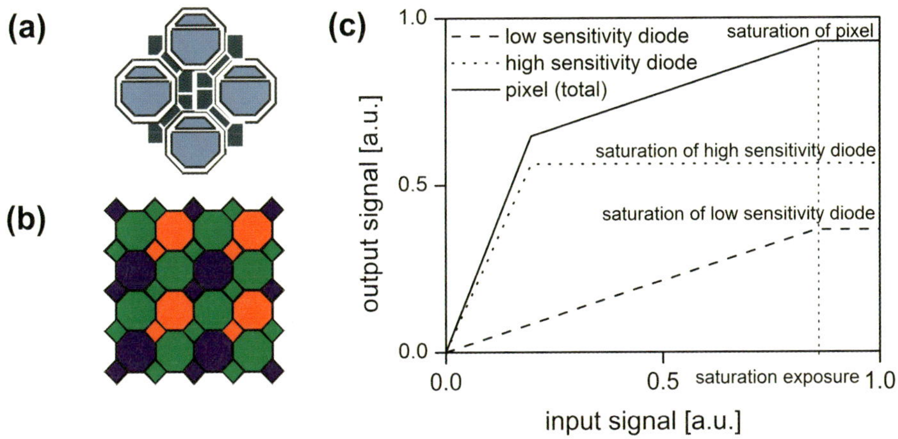

Fig. 4.77: (a) Details of an advanced "super-CCD" sensor for extended dynamic range with "split pixels." Note that due to the spatial disparity and the resulting different angular response (see Figure 4.21), this scheme may be well adapted to cameras with fixed lenses. But not necessarily it is suitable for cameras with interchangeable lenses. (b) CMOS sensor with a Bayer-mask-like CFA with smaller pixels in between larger ones. (c) Scheme of the increase of the dynamic range according to the application of different gain modes. Here, both axis are on linear scale. The curve "pixel (total)" in (c) means both parts of a split pixel or the large pixel together with the assigned smaller one.

"dual pixel sensor," where here the splitting ratio of the pixel is 1:1. Those are based on the standard rectangular geometry shown in Figure 4.71a and are equipped with a Bayer filter mask. The "dual pixel sensor" relies on the idea that images are generated, for each half of the (split) pixel. It is expected that due to the slightly different optical geometry both images are slightly different as well (slightly different perspective, change of bokeh, ghost and flare reduction). There are also some similarities to the plenoptic camera (see Chapter 9), because the direction where the light comes from differs somewhat for both parts of the pixel. Nevertheless, although this should offer the opportunity to some "post-capture refocusing" of the stored image, in comparison to the method applied for a real plenoptic camera, this opportunity is expected to be really small.

Moreover, there are special sensors where the splitting of the pixels is used for an even other purpose, namely the phase detection autofocus (PDAF; see Section 4.10.7.3). This should lead to improvements in accuracy and measurement time for a DSLR camera. Also, mobile phones cameras make use of that.

4.10.6.3 Dual or multiple conversion gain

A further route to reduce disadvantages of a small pixel size, in particular, for low light conditions has been the introduction of photodiodes with dual conversion gain (DCG). It makes also use of DTI and stacking technology. DCG pixels have a circuitry that allows for a readout with low and high-conversion gain, respectively (high-conversion gain, HCG and low conversion gain, LCG; see Figure 4.78). Without going too much into details,

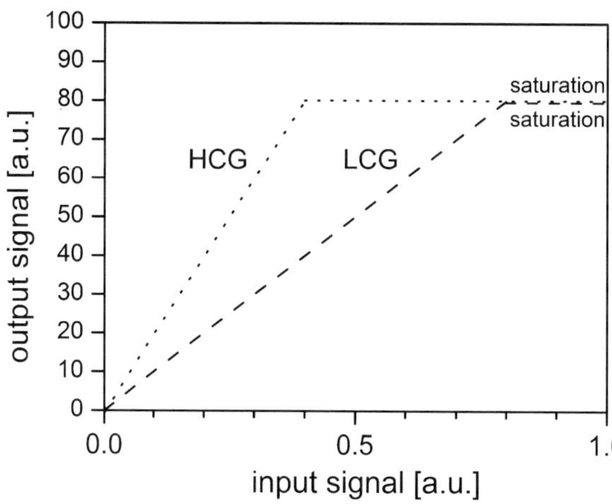

Fig. 4.78: Scheme of the two different modes in a pixel with DCG.

this results in two different photon conversion curves, which could be adapted to the illumination conditions. Switching the special operation mode of the pixel from LCG to HCG depends on the camera. Disadvantage of this method is the space requirement for the additional electronics. But one can make use of the additional space obtained from stacking technology, so that this is not an issue.

With LCG higher FWC is achieved, so that the dynamic range is extended. This has to be paid for with a larger noise level from electronics (not seen in the linear plot of Figure 4.78) but also that is not really an issue for bright illumination. The extension of FWC to a virtual larger value can be seen from the fact that the exposure in LCG mode has to be larger to get saturation. The maximum output signal of the pixel is the same as the maximum in HCG mode, and thus still can be processed by the ADC which may have its limit just at this signal. However, the image signal processor will interpret the digitized signal correctly according to the exposure and conversion gain and setup the "combined" image properly. If carefully applied, usage of DCG may complicate image processing, however, due to the availability of high-performance signal processing, this is not a restriction today. It is even possible to make use of more than two different gains, and thus to set up the combined image on base of multiple gains with different slopes (cf. Figure 4.78). It is possible to combine a series of images from the same scenery taken at different (ISO)-gains instead of different EV (however, then one has to be aware of the disadvantage with respect to SNR artefacts as discussed below).

With the possibility to switch the operation mode, and to make use of the virtual increase of FWC, the maximum SNR increased. It may be mentioned that this scheme has some similarity to the split or dual pixel technology. However, those always provide the signals of the two fractions of the pixel, whereas the signal from single DCG pixels depends on conversion mode. But when combined with multipixel cell arrangements, the similarity is quite more close (also with respect to PDAF, both of that is discussed in Section 4.10.7).

Nevertheless, there may be also problems resulting in the use of DCG. At the crossover between LCG and HCG, the signal may be observed quite differently, e. g., the SNR of the signals, and visible artefacts may result. But there are technologies, such as Teledynes's LACERA technology for which the company claims an unsurpassed dynamic range and a highly accurate signal measurement within their scientific sCMOS cameras.

4.10.6.4 Full well adjusting method, skimming HDR

CMOS technology is quite flexible and allows for one or more signal resets during the charge integration process. Note that in the case of a CCD the pixels could not be individually addressed, and thus this process is not possible. During integration, a specific voltage is applied to the gate transistor. If this is reduced just at the beginning of the next integration period, then FWC is virtually increased ("skimming HDR"). This increases the dynamic range for bright light, which finally leads to a sensor response curve, similar to that displayed in Figure 4.77c. On the other hand, there are disadvantages for which we refer to the special literature. Some CMOS sensors available on the market do allow such a dual slope, a triple slope or even more, in general, a multiple slope integration.

4.10.6.5 Sensor with complementary carrier collection

Although not suitable for photography, an interesting architecture developed for automotive applications is a 3.2 µm BSI-pixel with a capacitive deep trench isolation (DTI, see Section 4.10.5.2) based on standard architecture. It has a geometry that allows to collect the generated charge carriers, namely electrons and holes in parallel.[24] The first ones are detected in one channel and yield the low-level signal (FWC = 33000 charge carriers) and the second ones the high-level signal (FWC = 750000 charge carriers). For the high-gain mode, the noise floor is 1.2 electrons which, in total, yields a dynamic range of 116 dB, i. e., $10^{116/20} = 630000$ or 19 EV.

4.10.6.6 Logarithmic high-dynamic range CMOS sensor

As discussed many times before (see, e. g., Section 4.8.5), the human eye has logarithmic response characteristics (Weber–Fechner law; see Figure 4.79) whereas a linear detector such as a CCD does not, and thus a tonal curve has to be applied to get a similar response (see Section 4.9 and, in particular, Figure 4.62).

24 F. Lalanne et al.: *A 750 K Photocharge Linear Full Well in a 3.2 µm HDR Pixel with Complementary Carrier Collection*, Sensors **2018**, 18, 305 (www.mdpi.com/journal/sensors).

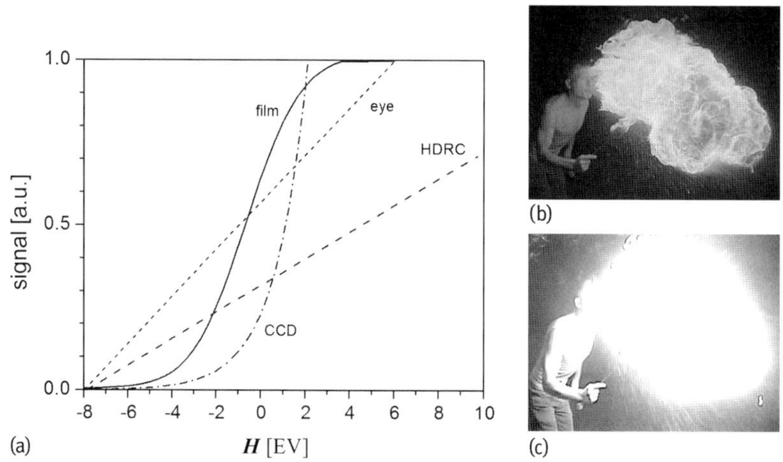

Fig. 4.79: (a) Response curves of different detectors and the human eye. Note that within this half-logarithmic plot, a logarithmic curve is a straight line. (b) High-dynamic scenery captured with a HDRC sensor, in comparison to a conventional CCD sensor (c). Source of (b), (c): Institut für Mikroelektronik Stuttgart.

On the other hand, special CMOS sensors also allow for a logarithmic curve. An example is the so-called high-dynamic range CMOS sensor (HDRC) developed by the Institut für Mikroelektronik Stuttgart, Germany. This sensor generates an output voltage, which is not linear to the incident light as discussed in Section 4.1.3 (see Equation (4.15b)), but proportional to the logarithm of the irradiance or illuminance, respectively (see Figure 4.79). Furthermore, the ADC cannot work with equal step sizes as described in Section 4.8.6 but has to be adapted properly. The claimed dynamic range is 26 EV.

4.10.7 Sensors with large pixel number and/or special pixels

4.10.7.1 Multipixel cell technology

The third route to reduce the disadvantages of the small pixel size for low light conditions is binning. As already discussed in Section 4.8.3, this leads to an increase of the number of collected photons and also to an increase of SNR. This strategy is used in Tetracell technology (Samsung), Quad-Bayer (Sony), 4-Cell (OmniVision) and other configurations. In the following, we will term that as multipixel cell technology.

Figure 4.80a shows the configuration of a standard CIS using a Bayer mask as CFA. All pixels are operated as discussed in the previous sections and each of them has its own microlens. The cell structure to cover RGB color is given by four pixels, e. g., by those within the square made of the thick line (this and the other thick line squares are for illustration only). Figure 4.80b presents another CFA arrangement, but in principle this is still the same as before. A group of four neighbored pixels is equipped with a common color filter and 4 of those groups compose a cell similar to Figure 4.80a. Thus,

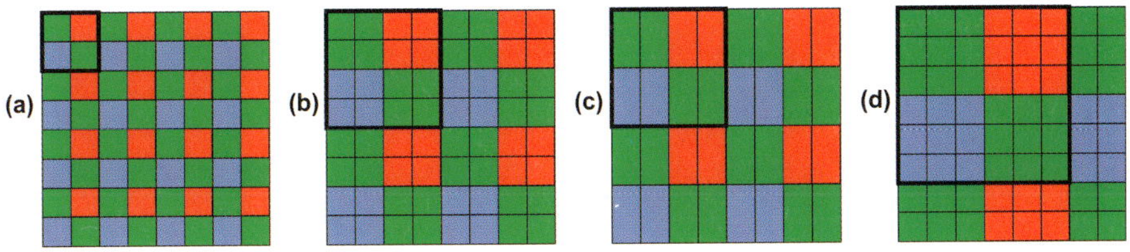

Fig. 4.80: Scheme of a standard CFA arrangement (a) and multipixel cell arrangements (b) to (d). This illustrates the binning possibilities (see the text).

the cell structure to cover RGB color now is given by 16 pixels, e. g., by those within the square made of the thick line. Again, each pixel may have its own microlens (Sony calls that Quad Bayer Coding, QBC) or alternatively, each group of 2x2 pixels may be equipped with a single common microlens for all of them (Sony calls that QBC 2x2 on-chip lens, OCL). There are further arrangements with two rectangular shaped pixels below a common microlens instead of four square-shaped ones (i. e., 2x1, "split pixels," "dual pixels"; Figure 4.80c). Although CIS with such arrangements have been used in SPC as well (e. g., in iPhone 11), this is not much an issue here as the technology and the advantages are covered by the following discussion also. A further development is the Nonacell configuration by Samsung (Figure 4.80d), which can be understood straightforwardly from the discussion of Figure 4.80b, but now with a group of 3x3 pixels and a cell again made of four of such groups. Introduced in year 2020, it is applied in their 108 MP sensors. In a very similar way, the even more recent 200 MP sensor development makes use of that. Here, a group of 4x4 0.64 µm pixels forms a macro pixel with a pitch of 2.56 µm.

The advantage of such configurations is that they can be used in different ways. Let us discuss that on base of Figure 4.80b. If a lot of light is available, each pixel is operated independently as usual in the standard Bayer mask configuration. The only difference is the other configuration of the CFA, which has to be taken into account when de-mosaicing by the image signal processor takes place. But we may note that although there might be the advantage of the potentially large resolution (but only if supported by the optics), a worse color reproduction may be expected because now the guess of the right color at each pixel becomes more difficult when most pixel neighbors have the same color filter (see Section 4.6.3).

In case of less light or even low light, each group of 2x2 pixel is binned. Accordingly, one may regard such a group as a macro pixel and with the consequence that the full sensor consists of such macro pixels placed according to the original Bayer arrangement. Of course, the original number of pixels then is reduced by a factor of 4 effectively, which makes a 48 MP CIS to a 12 MP sensor (and a 108 MP also to 12 MP, and a 200 MP also to 12 MP). Such a sensor with effectively 12 MP may be even superior for low light conditions (see also Section 8.3; note as well that approximately 12 MP is regarded as some kind of standard for SPC). This corresponds to an adaption of pixel size to illumination conditions. However, although binning can yield to an increase of FWC in principle, due to other limitations this may not be usable in general. Such a limit may be the saturation of the ADC. But one can overbear this problem by combining multipixel cell technology with DCG technology, which then effectively increases FWC as discussed above. If four submicron pixels are grouped, in that way FWC may be increased, e. g., from 6000 electrons to 24,000 electrons, and thus SNR increases as well, here by a factor of approximately 2 (see, e. g., Figure 7.13). Moreover, in principle, this scheme also allows for different exposure time settings within a macro pixel structure (e. g., the multicell), which then gives access to an additional widening of the dynamic range. Another possibility of multipixel cell technology usage is discussed in the following chapters.

A rather new development of multipixel cell technology may also be Leica's "triple resolution technology". Here beside full 60 MP images, binning allows for fast image capture with 36 MP or extremely low noise images with 18 MP (although already the 60 MP images are of very low noise; see also Table A.3, no. 7). At present, further knowledge has not been made available.

4.10.7.2 Polarized sensors

A totally different sensor, however, with some match to the discussed multipixel cell technology is a particular "polarized sensor," which implements polarization filters of the camera system into the sensor itself. From basic optics and basics of imaging, it is known that depending on the illumination conditions a suppression of light polarized into a specific direction may help to get a more clear image. In the same way, for many applications, especially for industrial ones where visual inspection of low contrast or highly reflective samples is an issue, observation of light at different polarization directions may be helpful because this allows for more clear images and a better optical characterization.

For such applications, special polarized sensors are available, where instead of a CFA, below the OMA, an array of polarizers made of nanowire grids with an AR coating is placed with an air-gap above the CIS. As an example, within Sony's polarization technology, the array is made of an arrangement similar to Figure 4.80a. However, each of the four pixels, which form a macro pixel (marked by thick square) have a wire grid with a different orientation in front of them so that linearly polarization could be detected simultaneously at 0°, 45°, 90° and 135°, respectively.

4.10.7.3 Phase detection autofocus

Although not directly related to sensor imaging issues as the subject of Chapter 4, there is an opportunity for phase detection autofocus (PDAF) on basis of the just discussed sensor technology. Thus, we will discuss that briefly (for autofocus in general, see also, e. g., Sections 2.6, 6.3 and 6.4). Originally, PDAF was used with split or dual pixels, both in SPC and DSLR, and later on with groups of four pixels (Section 4.10.7.1). PDAF makes use of potentially different path lengths from an object point to determine the right focus position. Briefly, this is illustrated in Figure 4.81.

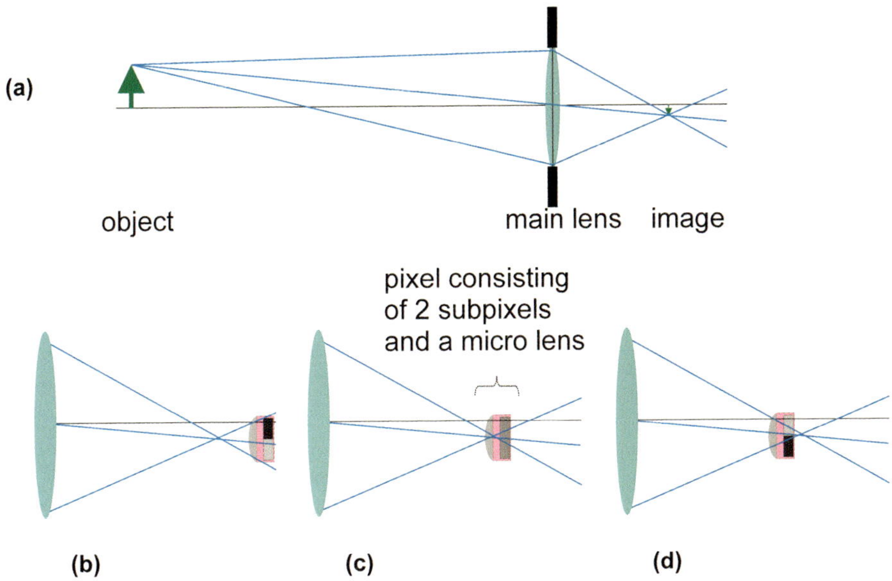

Fig. 4.81: Scheme for illustration of PDAF (see the text).

Figure 4.81 (a) shows how an object (green arrow) is imaged within geometrical optics. Here, we concentrate on the arrow head and its position in the image plane. A single pixel consisting of two subpixels and a common microlens is positioned in focus (c) or on its right-hand side (b) or on its left-hand side (d). If placed in "focus" (i. e., at the position where the image is the sharpest), both subpixels receive the same signal (c), here shown as same gray level. In (b) or (d), one or the other subpixel gets more light (shown as light gray subpixel; the other one is shown as a dark gray subpixel). Thus, the signal ratio changes with the sensor position, and hence allows camera "focusing," which usually is done automatically ("autofocus"). This should be sufficient to understand that focus measurements that are necessary for generating sharp images are possible with PDAF. Further, this then allows to control the autofocus of the main lens, e. g., of a SPC. However, PDAF sets also requirements to the main lens design (in particular, telecentricity), and thus again, this is an example that the sensor system and main lens should not be regarded separately. We may also point out that PDAF requires good isolation of subpixels from each other as discussed before. A further understanding of PDAF is not of importance for this book.

Nevertheless, we will continue the discussion with the realization of PDAF within CIS. In DSLR, PDAF is commonly used in a separate detector array located behind the flipping mirror. In SPC, the CIS or better say, part of its pixels, is used within one of three common solutions of an autofocus system. It should be noted that contrast measurements for the autofocus as applied in the past have been replaced. The first solution makes use of masked pixels and was introduced by Sony in 2014. Particular photodi-

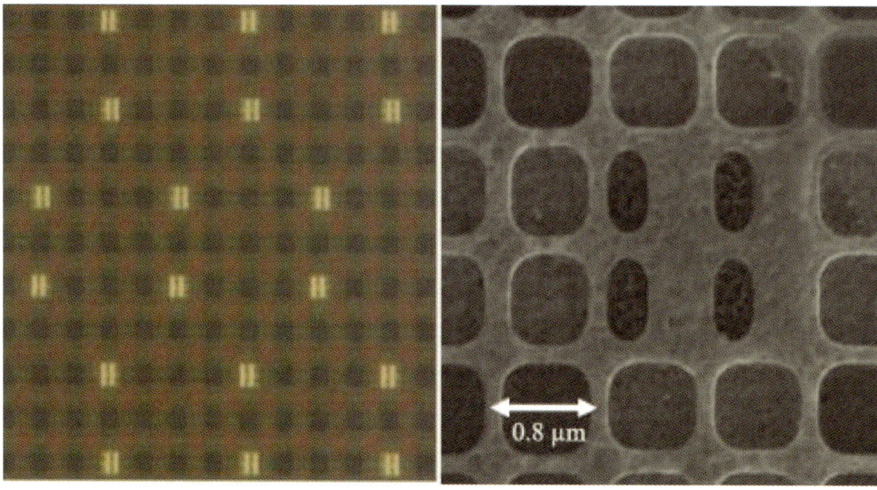

Fig. 4.82: Left-hand side: View on a CIS with a CFA with masked pixels showing PDAF pattern. Right-hand side: SEM image of 2x2 PDAF pixel group with color filters removed. The CIS is a GD1 from Samsung with Tetracell (see Section 4.10.7.1) and a clear channel to enhance PDAF output. Images taken from[25], courtesy of Techinsights.

odes behind their microlenses are covered (masked) partly by an absorbing metal layer (Figure 4.82). According to the principal situation shown in Figure 4.81, this allows for detection of at least some signal dependence on incident direction. If then one compares the signal of two different masked photodiodes, one can deduce information with respect to the correct sensor position on the optical axis. However, a key demand is a low PRNU. There is also a disadvantage. Although PDAF by usage of selected masked pixels within the CIS matrix works well, it has to be noted that those are not available for image capture. This leads to tiny blind spots on the sensor.

Consequently, in general, PDAF may affect the images. Although this must not necessarily be always noticed, this potentially occurs from time to time. Then even in images captured with high-end DSLM and DSLR equipped with PDAF, some image degradation may be present. Known effects are PDAF striping and PDAF banding. We will not go deeper into that topic. But we may mention that the related artifacts rather often may be fixed with particular software tools applied during the image post-processing so that there is not too much loss in image quality.

In a similar way as done with dual pixels, the second solution makes use of selected neighbored pixels, namely a photodiode twin, with a common microlens with double width (Figure 4.83).

25 Z. Shukri: *The State of the Art of CMOS Image Sensors*, TechInsights Inc. Inc. https://www.techinsights.com/ebook/ebook-latest-development-trends-cmos-image-sensors (visited January 2023).

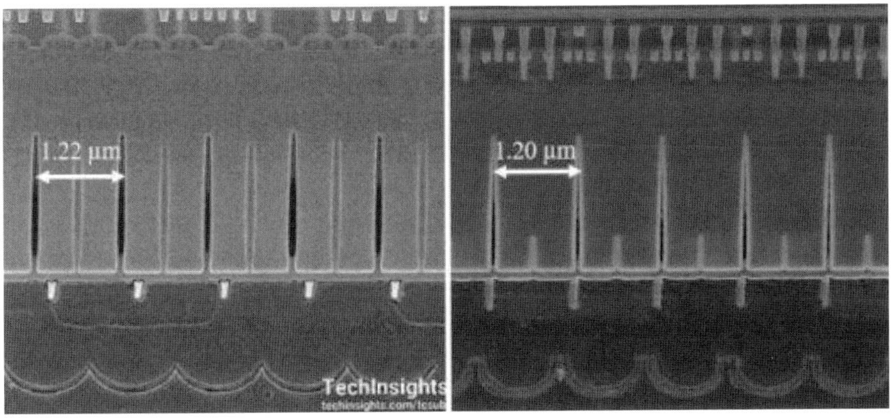

Fig. 4.83: Dual PD, full array PDAF with in-pixel trench isolation. Left-hand side: Sony IMX100 with Octa PD (1.22 µm pitch, 50 MP Quad-Bayer CIS), right-hand side: Samsung S5KGN1dual PD (1.4 µm pitch, 50 MP). Images taken from[26], courtesy of Techinsights.

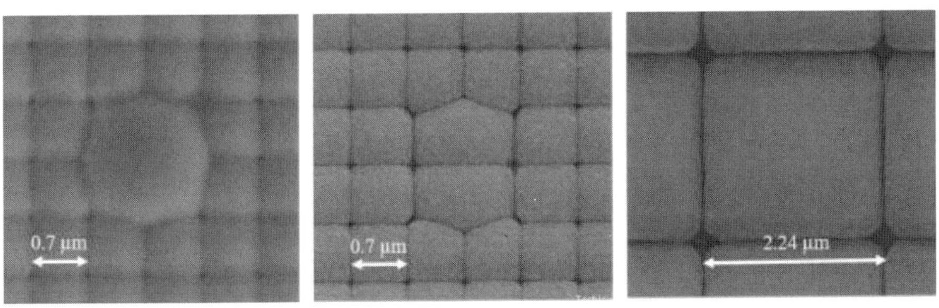

Fig. 4.84: SEM images of effective 2x2 OCL PDAF by (a) Omnivision (0.7 µm pixel of 64 MP CIS OV64B), (b) Samsung (0.7 µm pixel of 108 MP CIS HM2) and (c) full array 2x2 OCL PDAF by Sony (1.1 µm pixel of 48 MP CIS IMX689). A full array autofocus has no PDAF-dedicated pixels and, therefore, all pixels can be used for data acquisition. Images taken from[26], courtesy of Techinsights.

The third solution is quite similar and bases on a 2x1 on-chip lenses. Due to the fact that all pixels can be used for imaging, and as well as for PDAF, it does not suffer from blind spots. PDAF with OCL can be used for pixel pitches of 0.7 and below and developments are made to change from 2x1 to 2x2 pixels to allow for horizontal and vertical PDAF, respectively. We may add that there are also CIS with particularly designed pixels, namely "horizontal-vertical dual pixels" with a slanted in-pixel DTI. Note that red and blue channels in a OCL PDAF cell may be replaced by green channels. As an example, Canon makes use of such a technology (Dual-Pixel-CMOS-AF, CDAF).

26 Z. Shukri: *The State of the Art of CMOS Image Sensors*, TechInsights Inc. Inc. https://www.techinsights.com/ebook/ebook-latest-development-trends-cmos-image-sensors (visited January 2023).

In conclusion, multipixel cell technology combines high resolution, possibility of achieving HDR via image processing and PDAF in one element, namely such a CIS.

4.10.7.4 Time of flight sensors

Modern camera systems, used for industrial or automotive applications or such ones used in SPC, today often consist of several cameras. Those cameras may have different optics, e. g., to capture images at wide angles or to operate as telephoto cameras. If there are many of them as, e. g., Nokia N9, in particular cases they allow for stereoscopic measurements and from triangulation object distance can be determined (see, e. g., [Bla21],[27]).

On the other hand, there may be one (or more) special cameras that just provide additional data rather than that those data lead to a direct contribution to the imaging process. Depth sensing is an example. Although not subject of the present book, we will give a brief description because, in particular, several SPC make use of such data for image generation. Consciously, we write "generation," because the smartphone image processor uses this information, e. g., for portrait photography to calculate an artificial image (see Section 7.4.4). Potential other issues are not a subject here and we will also restrict to small depth sensing modules. For 3D sensing and imaging in general and for detailed information, we refer, e. g., to special books such as [Luh19] and to a lot of scientific and technical papers. A short overview may be found in[27,28,29].

There are several methods of depth sensing related to optical imaging. Here, we concentrate on time-of-flight (ToF) measurements of optical signals. Another example is briefly discussed in Section 7.1. Those signals may be obtained, e. g., by scanning the scene with a laser beam. This is made by a LIDAR system (light detection and ranging), which is the optical equivalent to the well-known RADAR (radio detection and ranging). Depending on the LIDAR system, it is possible to generate very large 3D-maps, e. g., of the molecular distribution of the atmosphere or of the air traffic or it is possible to generate 3D-profiles of very small regions, e. g., with a resolution in the µm range or below. Currently, of much interest are scanning LIDAR systems for autonomous driving, which operate in a range of intermediate distances, pretty similar to photographic imaging of not too far objects. To a certain extent similar to this, Apple iPhone 12 and 13 implement this method based on a system by Sony, which consists of a vertical cavity surface emitting laser (VCEL) with near IR emission and single-photon avalanche diodes (SPAD; see Section 4.11.5) for detection. In another example the interested reader may find some

27 *Time of flight white paper*, Texas Instruments 2014; https://www.ti.com/lit/wp/sloa190b/sloa190b.pdf

28 I. Gyongy, N. A. W. Dutton, R. K. Henderson: *Direct Time-of-Flight Single-Photon Imaging*, IEEE Trans. Electr. Devices **69** (2022) 2794.

29 C. Bamji, J. Godbaz, M. Oh, S. Mehta, A. Payne, S. Ortiz, S. Nagaraja, T. Perry, B. Thompson: *A Review of Indirect Time-of-Flight Technologies*, IEEE Trans. Electr. Devices **69** (2022) 2779.

details on a MEMS-based LIDAR imaging sensor with electronic scanning without mechanical moving parts.[30]

Alternatively, to that, depth information can be obtained by an illumination of the whole scene with a nanosecond IR-light pulse of a special LED or a laser diode. Then the optical part of the ToF system captures the reflected light and images it through an optical filter, which suppresses all light outside the wavelength range of the emitter source onto a special image sensor. Each sensor pixel measures the time interval between pulse emission and recording. From this and the velocity of light, the distance to different object points of the scene is calculated. Such a time measurement requires more electronics and a more complicate pixel design which in total leads to larger pixels, when compared to CIS. This limits also the total size of ToF sensors, their pixel number and the spatial resolution of the ToF camera. Currently, a typical example is Sony's IMX316 chip with $240 \cdot 180$ pixels of 10 µm pitch, which is implemented, e. g., in Sony Xperia 1 III and Huawei P30 Pro smart phones. Another example is Infineon's IRS2875C ($240 \cdot 180$ pixels) of REAL3 family also used for computational photography (Bokeh, see Section 7.4; also usable for 3D video) or the underdisplay ToF camera development of Infineon together with partners that should be used for face-recognition within a smartphone security system. Even other ToF sensors, e. g., such ones used for industrial applications may have more pixels and potentially smaller ones. Nevertheless, the pixel number is still much smaller when compared to CIS (e. g., $640 \cdot 480$ or $1024 \cdot 1024$ pixels with 5 or 3.5 µm pitch, resp.). We may note that some of those ToF cameras make use of the measurement of phase differences rather than direct time measurements.

4.10.8 Advancements for the IR region: deep depletion CCD

A demanding spectral range is the near-infrared (NIR) region. Here, in particular, light signals are quite often rather low, which is a challenge for the sensitivity and noise performance of the sensor. However, usual sensors show a low quantum efficiency in that wavelength region (s. Figure 4.10a). Moreover, due to the relatively long wavelength and the rather thin "BSI-photodiode layer," etaloning is a large problem.

Etaloning occurs in a thin transparent layer with high reflectivity at its upper and lower interface. For BSI-sensors, the high reflectivity results from the large difference in the index of refraction at the interfaces, which leads to multiple reflections of the incident light and to interferences with maxima and minima. This can be observed as a fringing effect, in particular, when wavelength is varied. This can be seen in the spectrum and in the wavelength dependence of the quantum efficiency yielding a 30 to 50 %

30 X. Zhang, K. Kwon, J. Henriksson, J. Luo, M. C. Wu: *A large-scale microelectromechanicalsystems-based silicon photonics LiDAR*, Nature **603** (2022) 253.

modulation, mostly pronounced for wavelengths larger than 800 nm. In addition, spatial etaloning occurs due to the fact that the layer does not have a fully constant thickness.

To solve this problem, specially designed CCD have been developed, namely the deep depletion CCD (DD-CCD). DD-CCD are covered with a special NIR antireflection layer on its surface, i. e., the top layer, and have a roughened bottom layer. Both lead to a suppression of etaloning. Although DD-CCD are also illuminated from the rear side, they have a thickness approximately twice that of a thinned BSI-CCD. Moreover, further improvements such as usage of high-resistance silicon with highly doped substrate are applied. The quantum efficiency has been increased to more than 90 %. For example, Princeton Instruments offers cameras for astronomical imaging with a quantum efficiency between 90 and 98 % over the whole wavelength range between ~420 nm to ~900 nm and, e. g., to >30 % at 1000 nm (see also Figure 4.10d). The disadvantage may be a larger dark current, which requires stronger cooling.

4.11 Image converters and image intensifiers

In the following, we will discuss some basics of image converters and image intensifiers. To do so, we have to discuss some general basics beforehand. But then we will mostly concentrate the discussion on the subject of the present book, namely the application of those devices for the capture of "still images," as used for scientific and technological issues. This is related to the operation of converters and intensifiers in analog mode. The alternative photon counting mode is not an issue here and with the exception of the conversion of an amplified electron distribution into visible light, we will not discuss their potential as particle detectors for electrons or ions. A good overview of most of the following topics can be found in the excellent review article of Gruner, Taken and Eikenberry [Gru02]. Although today this article may be regarded as out of date, it addresses nearly all relevant aspects of importance. For those readers, who have deeper interests in that special subject we refer them to related textbooks and articles.

4.11.1 Image converters

Although direct imaging on a suitable detector is preferential, there are situations where this is not possible:

(I) The first reason may be that the detector is not sensitive to the particular wavelength. Thus, e. g., IR-, XUV- or X-ray radiation has to be converted to visible light prior to detection by a standard sensor such as a CCD. Usually, this is done with scintillators and phosphors, which due to fluorescence and/or phosphorescence convert the incident "light" into visible light. Here, we extend the definition of light from visible down to the X-ray range. We would also like to note that the definitions of fluorescence and phosphorescence are not consistent within the literature.

An example is the P20 phosphor ((Zn,Cd)S:Ag), which converts UV light (250–300 nm) into the green spectral region. Other examples are the P45 phosphor (Y_2O_2S:Tb), which converts XUV- and soft X-rays quite efficiently to visible green light and the P43 phosphor (Gd_2O_2S:Tb), which is suitable for harder X-rays and also emits in the green. The quantum efficiency depends strongly on photon energy and may vary between 10 and > 90 %. For instance, the P43 phosphor peaks with > 90 % at approximately 10 keV. Other suitable converters are made of activated Thallium, namely CsI:Tl, Na:Tl and so on. Phosphors, in principle, can be used for both conversion of UV or shorter wavelength electromagnetic radiation to visible light or for the conversion of particle radiation to visible light, but not all phosphors are suitable for all conversions. Although decay time plays a role for some applications, usually it is not an issue for taking still images.

The absorption coefficient determines the overall efficiency. It depends strongly on the absolute thickness of the phosphor, and in particular, due to the large penetration depth for hard X-rays, for such kind of radiation the phosphor has to be chosen rather thickly (Figure 4.85). In contrast, phosphors suitable for XUV-radiation detection have to be rather thin (Figure 4.87a). Consequently, phosphor screens also allow for imaging this kind of radiation as well. Phosphor efficiency depends on incident wavelength or photon energy, respectively (and also on particle energy as phosphors are usable for particle detection as well). Depending on the phosphor, the number of generated photons in the visible may be proportional to the incident energy (photons or electrons, resp.). This a necessity for their application as part of an intensifier system, as discussed below. But note that this is not a general rule. For specific conditions nonlinearities, saturation and damage may occur. Altogether image converters and phosphors are a subject of its own and well described in the literature. A good and rather comprehensive overview together with the application of phosphors together with CCD sensors is given by [Gru02].

(II) The second reason for the usage of an image converter is the application of an image intensifier such as an MCP (see Section 4.11.3) because this is not sensitive to

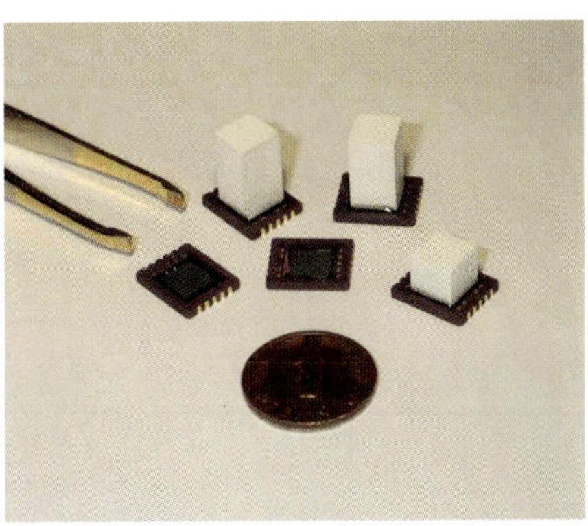

Fig. 4.85: Examples of X-ray sensitive phosphors of different thickness. Here, the CsJ(TI) phosphor is covered with white varnish for protection against humidity. The phosphors are mounted on a silicon diode (two naked ones are placed in front) that detect the visible photons generated within the phosphor.

the incoming "light," as it is "solar blind." In such a case, the 2D light distribution has to be converted to an equivalent distribution of electrons. In most cases, this is performed by means of a photocathode. The incident light generates photoelectrons via the photoelectric effect, and thus transforms the light intensity $I(x,y)$ into a number of photoelectrons $N_{\text{pho-el}}(x,y)$, which is proportional to $I(x,y)$. According to the quantum efficiency of that process, for photocathodes used for MCP, one obtains typically 0.1 to 0.5 electrons per photon (see, e. g., Figure 4.90a). Typical currents per incident light power are 0.02 to 0.16 mA/W.

After electron amplification within the intensifier as described in the following sections, the electron distribution has to be converted back to a light distribution. This is done with a phosphor. Although, in principle, the same phosphors can be used as discussed before (see reason (I)), here they are not used to convert electromagnetic radiation from one wavelength to another one, but for the conversion of particle radiation, namely electrons, to visible light. Of course, it is not the electrons themselves that are converted, but their distribution $N_{\text{el}}(x,y)$. Furthermore, to avoid confusion, for discrimination we term the first ones (I) still to phosphors and the second ones to luminous screens (II). The conversion efficiency of the luminous screens significantly depends on the electron energy (Figure 4.90c).

(III) A third reason to apply image converters is the adaption of a given or intended image size in the image plane to the size of an existing or intended sensor. If, for instance the image is large, but the sensor is small this can be solved by the application of a relay optics (Figure 4.86a) or a suitable fiber optics (FO; Figure 4.86b). A fiber optical taper or plate is an array of a huge amount of glass fibers, each typically with 6 μm diameter, which are either in parallel (Figure 4.87b) or, e. g., shaped (Figure 4.86b). Light guiding is based on the total reflection on the inner side of the fibers.

For cameras used for photography this is not an issue. If, e. g., the sensor is rather small, then a suitable objective lens is chosen, which differs from that used for a

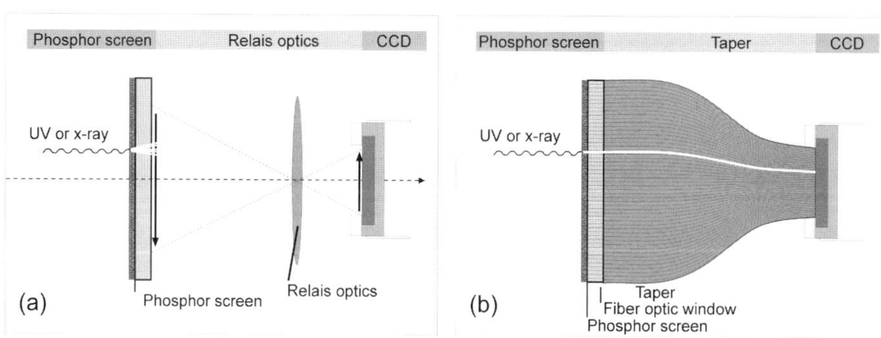

Fig. 4.86: Optical relay, here adaption of a smaller sensor to a larger sensitive image area: (a) performed by an optical imaging system, (b) via a fiber optical taper. Source: ProxiVision GmbH, Bensheim, Germany.

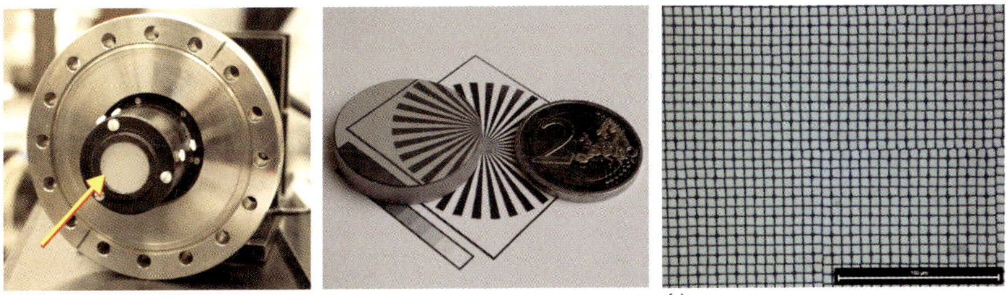

(a) (b) (c)

Fig. 4.87: Fiber optical taper (a) Thin Gd_2O_2S:Tb phosphor on top of a commercial FO taper. This phosphor is suitable for the investigation of the XUV emission of a laser-produced plasma. (sedimented "coffee cup mixture" by the author's group). The phosphor is part of a iCCD system (the whole camera is shown in Figure 2.28b), and the scheme is shown in Figure 4.92b. (b) Same FO taper as in (a) but here without phosphor coating. (c) Transmission microscope image of the taper shown in (b). The horizontal bar in the right bottom corner corresponds to a length of 100 µm, and consequently, a single fiber has a diameter of 6 µm.

larger sensor (see the example in Section 4.3.2). However, for technical and scientific purposes such arrangements are often applied.

Further potential reasons for the application of converters are discussed below (see Section 4.11.4).

4.11.2 Basics of light signal intensifiers

When the photon signal is rather weak and/or, in particular, is below the detection threshold of the detector in use, it has to be intensified before or alternatively a light signal intensifier may replace this detector. This is the case if, for instance, the quantum efficiency of the photodiode is too low or if the signal is below the noise limit.

A typical example of a 0D detector that makes use of light intensification is a photomultiplier or photomultiplier tube (PMT; Figure 4.88). Photomultipliers are standard detectors, and thus well described in the literature including standard textbooks of physics, optics, etc. Thus, here we will restrict ourselves to a very brief description.

The principle of a PMT (see Figure 4.88a) is as follows. The incident light hits a photocathode, e. g., made of an alkaline metal (for other types see Section 4.11.3), which results in emission of photoelectrons due to the photo electric effect. These electrons then enter the secondary electron multiplier, which is made of a series of cascaded electrodes, the dynodes. The dynodes are made of metals with large secondary emission coefficients with a typical emission of 3 to 5. The voltage between the entrance and the exit of the PMT is set to high voltage U_B, typically 2 to 3 kV, with a series of resistors R as voltage dividers in between. The individual dynodes are connected to the resulting electric potentials, so that there is a voltage of typically 100 to 200 V between two neighbored ones

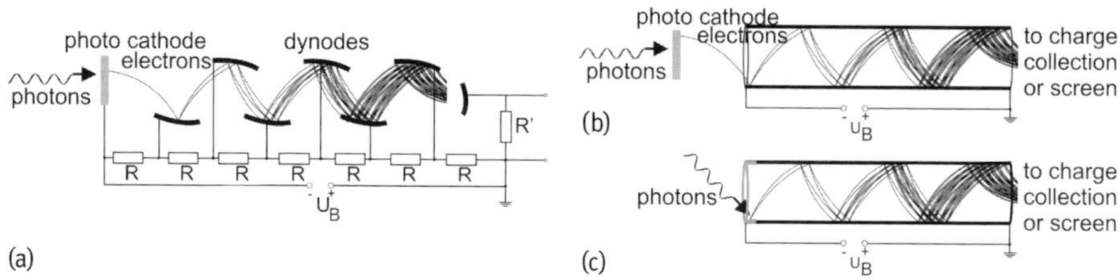

Fig. 4.88: Principal scheme of a photomultiplier (a) and a channeltron (b), (c). In (a), the PMT is encapsulated. Often channeltrons are encapsulated as well and also have a photocathode (b). But there are open channeltrons as well (those have to be operated in a vacuum) (c). These devices have a coating at its entrance (e. g., CsI:Tl; see "first reason" in Section 4.11.1) where the electrons are generated by the incident light (marked in gray). Note that there is also a high voltage between the channeltron and the screen (not shown here).

so that the typical kinetic energy of the electrons is 100 to 200 eV. This scheme leads to an avalanche-like increase of the electron number.

The gain G_{PMT} is determined by the dynode collection efficiency η_{PMT}, the applied voltage U_{B} and the gain coefficient g_{PMT}. This coefficient is given by the product of a geometry and material dependent constant and the number of dynodes, which all together leads to $G_{\mathrm{PMT}} = \eta_{\mathrm{PMT}} U_{\mathrm{B}}^{g_{\mathrm{PMT}}}$. Depending on the number of dynodes, the selected dynode materials and shape, etc., the total gain may reach 10^3 to 10^8. The resulting large number of electrons is detected at the exit of the tube as a charge, current or voltage. Instead, by application of a further potential difference, the electrons may be directed onto a phosphor where as a consequence a huge number of photons is generated. In that sense, the PMT intensifies the light. PMT may be rather linear devices. We may note that for visible light the PMT itself is placed in an evacuated glass housing with an entrance window in order to protect the photocathode and the dynode coatings against air and humidity. For the detection of short wavelength radiation, the entrance window, which otherwise would block the radiation, has to be removed. Although this destroys the special coatings, this does not cause problems, because for high enough photon energy, the light electric effect works well.

A rather similar but much smaller detector is a channeltron. The principle of the device is the same, but now the discrete set of dynodes located along the high-voltage line is replaced by a single quasi-continuous electrode (Figure 4.88b,c). This electrode has the shape of a pipe or capillary. Similar to the PMT, there is a high voltage applied between the entrance and exit of the tube. Usually, the inner wall of this hollow is covered with a high-ohmic lead glass on a ceramic substrate, which leads to a reduction of the electric potential along its extension. As a consequence, there is a potential difference between different sections of the device, which leads to acceleration of the electrons and secondary electrons. The shape of the channel must not necessarily be straight, and indeed, is often curved to avoid ion back propagation.

4.11.3 Microchannel plate intensifiers

The generations and "nomenclature" of image intensifiers are shown in Table 4.6. Typically, modern image intensifiers make use of the method of light signal intensification discussed in Section 4.11.2. With some similarity to photodiodes, channeltrons can be used to set up a 2D array that then acts as part of an image sensor system.

Arranging photodiodes in an array without further intensification leads to a simple photodiode array, or to a CCD or to CMOS sensor. An array of channeltrons comprises the principle of a microchannel plate (MCP; Figure 4.89, Figure 4.91). Although arrays of PMT are available, too, due to the large PMT cathode diameter of 5 to 13 mm or even larger, an array of PMT for imaging issues makes no sense.

Tab. 4.6: Generations of image intensifiers and their "nomenclature." Note that near focus intensifiers and inverters are not necessarily Gen 0 and Gen 1. Originally, the development was mostly driven by the military (it is still an important driver with large interest in "night vision"), but today there is general and wide usage in science and technology as well. This is not limited to "seeing life images" at low light conditions. Also, our purpose, namely the capture of "still images" is an issue. EM-CCD is not included here as it is discussed separately (see the section below). Note that with today's advanced intensifiers, the word "generation" is not used anymore to discriminate how old the system is, but instead is used to discriminate between the different types. Advances in photocathode technology become apparent from Figure 4.90.

Gen 0 and Gen 1: since approximately 1940; Gen 0 is not useful anymore today realization as near focus intensifiers; basics: photocathode and luminescent screen with a high voltage in between (e. g., 16 kV across a distance of 1 mm; hence, consequently there has to be a vacuum in between); conversion of IR to visible light; generation of multiple electrons (hardly a gain) since approximately 1950 (first demonstration by Philips in the 1930 s; usage by German and US military); Inverter: additional electron-optics which allows to demagnify the image, which is top side down; also cascade tubes, which lead to some gain (approx. $10^{2...4}$)
Gen 2 (including Gen "2 plus" and "2 super plus"): since approximately 1960 additionally to Gen 0: MCP in between the cathode and the screen (this was a much significant advance); proximity focus on screen; gain $10^{4...7}$ (highest for 3-stage)
Gen 3: since approximately 1970 further development of Gen 2; in particular, photocathode based on GaAs (or variants such as InGaAs or GaAsP); this has lead to a much increased sensitivity, in particular, in IR, much smaller devices and many other important improvements; gain > 40,000; compact proximity focus MCP intensifiers (without further electron-optics, and thus exact 1:1 image magnifications without distortion); immunity against electrical, magnetic and electromagnetic fields
Gen 3 filmless intensifiers: further development; previously needed ion barrier film (to prevent ion feedback and to extend lifetime for, e. g., GaAs photocathodes) are not necessary for the newly applied GaAsP and GaAs photocathodes; advantage: higher sensitivity and faster gate speeds (similar to Gen 2)
Gen 4 and "4G": since approximately 1990s further development of Gen 3; open performance specification ("4G" by European image tube manufacturer PHOTONIS); note: instead of Gen 4, these plates are often still called Gen 3

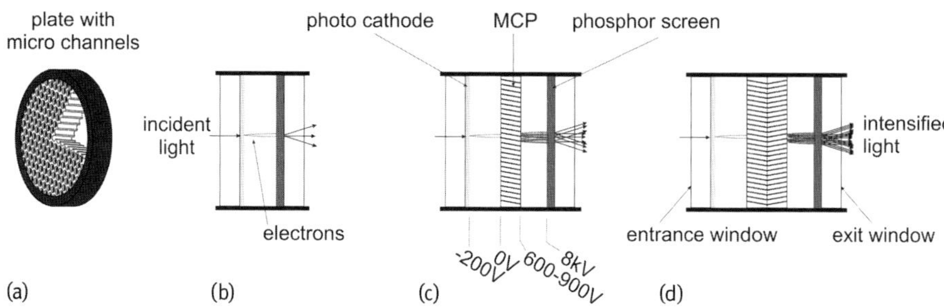

Fig. 4.89: Scheme of (a) the plate with the microchannels, (b) a proximity focus image intensifier, (c) a single stage MCP and (d) a "chevron" or V-stack MCP. Here, the MCP is proximity coupled to the screen. A fiber optical coupling similar to Figure 4.86b behind the screen is possible as well. Examples of the applied voltages in between are shown for the illustration in (c).

The three main components of a complete MCP intensifier are the photocathode, the microchannel plate itself and the phosphor screen (Figure 4.89). A complete system further needs a readout of the phosphor screen, which can be done, e. g., by a CCD coupled directly or via a fiber optical taper. Some MCP have a thin film on top of the channel plate surface to prevent ion feedback more efficiently (approximately 80 % transmission for the electrons). We may note that depending on the situation "MCP" may denote the plate only (Figure 4.89a) or the whole intensifier system (Figure 4.89c and Figure 4.89d).

In the case of an encapsulated MCP, there is first a suitable entrance window, e. g., Quartz or MgF_2 if the device should be operated in UV. This corresponds to the situation illustrated in Figure 4.88b (see also Figure 4.91b). Subsequently, just behind the window, follows the photocathode, which today often is made of GaAs or from variants such as InGaAs or GaAsP. Typical quantum efficiencies are displayed in Figure 4.90a. A voltage of typically 150 to 200 V directs the electrons more or less perpendicular to the entrance of the channels, each with a typical diameter in the range of 6 to 100 µm.

In the case of open channels, this system has to be operated in an ultrahigh vacuum. This corresponds to the situation illustrated in Figure 4.88c, Figure 4.91a. Then instead of the usage of a separate cathode for the generation of photoelectrons, the electrons are produced, e. g., at the CsI- or NaI-coating or made of another suitable composition at the entrance of the channels. Open MCP are used, e. g., for the XUV and soft X-ray range. The microchannels within the MCP are of circular shape and straight, with a typical length diameter ratio of 40 to 100. Exceptionally, there are also channels with a square cross section. The channels are set at a tilt angle of typically 5° to 15° with respect to the surface normal to avoid ion feedback along the channel, which would lead to an unwanted emission of further electrons (Figure 4.89). For the same reason as for channeltrons, special MCP with curved channels are available as well.

After amplification, i. e., boosting the number of electrons a high voltage (HV) between MCP and luminescent screen (see the example in Figure 4.90c) directs the electrons again, now to the luminescent screen close to the exit window. Typical values for

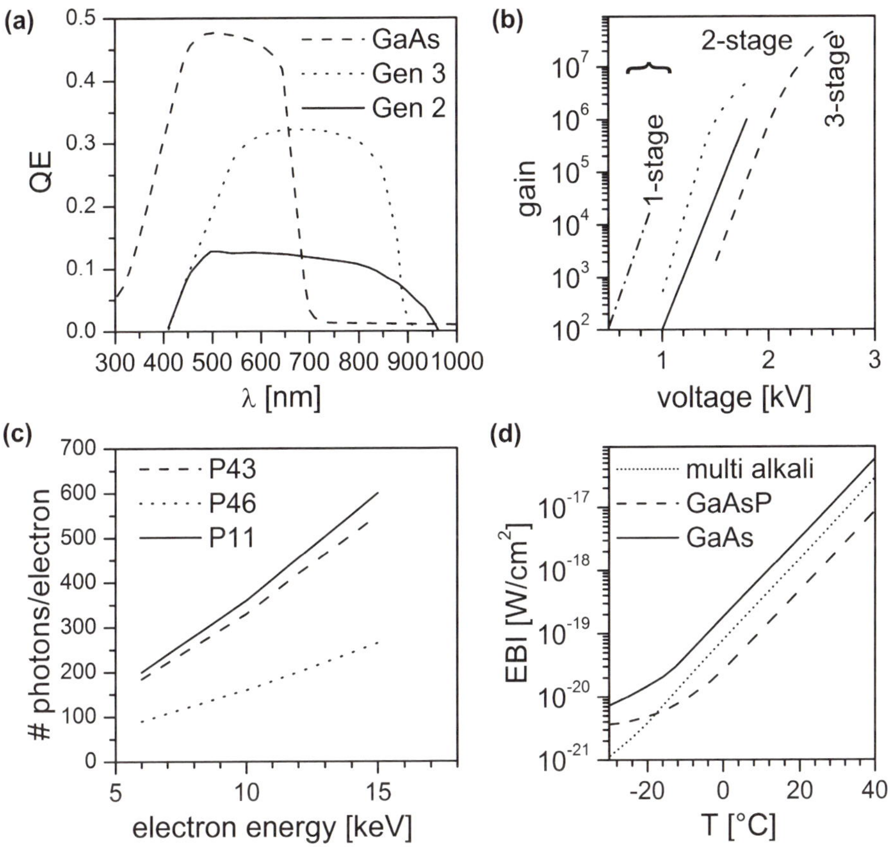

Fig. 4.90: (a) Efficiency of intensifier photocathodes (b) Examples for the gain of 1-stage (dashed-dotted line), 2-stage (dotted and solid line) and 3-stage (dashed line) MCP (data taken from ProxiVision GmbH (solid line) and Hamamatsu (other lines)). Note that the displayed "gain" of both suppliers is defined differently: the solid curve provides the physical gain of the amplifications stage only, whereas the other lines are referred to the gain of the whole system (see the text), which of course is larger. (c) Examples of the dependence of the "conversion of electrons into photons" by luminous screens (see the text). The phosphors for the present example have a grain size of approximately 1 µm and a total thickness of 4–5 µm. They are sedimented on a fiber optical taper (on plain glass the efficiency is expected to be larger by 40 %). (Data taken from ProxiVision GmbH for proximity focus image intensifier diodes). (d) EBI-dependence on temperature (data taken from Hamamatsu).

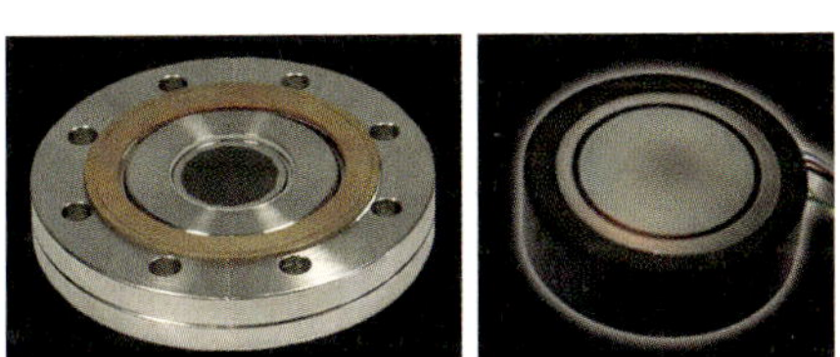

(a) (b)

Fig. 4.91: Images of an MCP: (a) open MCP, (b) encapsulated MCP. Both systems have 25 mm diameter active area. Source: ProxiVision GmbH, Bensheim, Germany.

the HV across the plate(s) could be taken from Figure 4.90. As can be seen from Figure 4.90c, the conversion efficiency depends on electron energy, and thus a high voltage is advantageous. Remember that an electron accelerated by a voltage of x volts gains an energy of x eV. The HV also results in electron propagation perpendicular to both surfaces, namely MCP and screen, on a rather straight path. Although not much common, alternatively the screen can be omitted, and the electrons may directly bombard an EB-CCD (see Section 4.11.4). Note that 1 W light power corresponds to approximately $2.7 \cdot 10^{18}$ photons per second (for $\lambda = 550$ nm) and 1 mA to approximately $6 \cdot 10^{15}$ electrons per second. Consequently, 100 photons per electron correspond to approximately to 0.2 W/mA. For calculation into photometric values, see Section 1.2.5.

We would like to note that there might be some electron signal cross talk at the interfaces and there might be a cross talk between a channel and its neighbors as well. Moreover, there might be an influence of a potentially included FO (see Figure 4.87c). There are MCP that consist of one, two or three plates that act as intensifier stages (Figure 4.89). In that case, the plates are directly coupled in a geometry as shown in Figure 4.89d. There are also configurations with a sub-mm gap in between. A two-stage MCP with this geometry is called a V-stack or "chevron," a three stage MCP a "Z-stack."

In total, the number of the capillaries within the array of a single MCP plate is of the order of 10^4 to 10^7. The entrance aperture of all channels together covers approximately 60 % of the total effective area of the MCP, which usually is circular though there are rectangular versions available as well. Special versions with funnel type channels are available too. Such devices may have up to 90 % coverage. MCP are available with diameters of the effective area of approximately 15 to 40 mm or more, either as an open or encapsulated version. Examples of channel diameter/pitch/channel length/tilt angle (in µm and degree, resp.) for typical MCP are as follows: 12/15/480/8, 4/5/200/12, 12/15/480/12, 25/31/1000/8. More details can be taken from the data sheets supplied by the manufacturers.

MCP are applicable from the NIR to X-ray range (and even for particle detection). Using fast high-voltage pulsers, MCP could be gated, which means that the exposure time (time when gain is present and the MCP yields a signal) could be limited down to a couple of ns (special MCP could be even gated with several ps). This makes MCP well suitable for high-speed imaging.

The goal of the application of an MCP within an imaging system is imaging at low light conditions, where usual imaging as discussed before fails or at least is difficult. More details on that are discussed below in Section 4.11.4.

The gain G_{mcp} is determined by the length to diameter ratio of the channels (typically 40–60) and the secondary emission coefficient. G_{mcp} scales with a power of the applied HV U_{mcp},

$$G_{\mathrm{mcp}} \propto (U_{\mathrm{mcp}})^{g_{\mathrm{mcp}}} \tag{4.58}$$

until it becomes saturated. Before saturation, the gain coefficient g_{mcp} can be evaluated from diagrams such as Figure 4.90b. Typical values can be deduced from a log-log plot similar to Figure 4.90b and result, e. g., in $g_{mcp} \sim 9, 18, 21$ for a 1-stage, 2-stage and 3-stage MCP, respectively. Some theoretical background and further details can be found in the "classical paper" on MCP detectors of Wiza.[31] The typical gain per stage or plate is 1000 and very uniform across the area. Multistage plates have higher gain (see Figure 4.90). At large photo or electron current, in particular at large HV, respectively, saturation effects occur. Then additionally, due to rejection of the highly charged electron cloud, spatial resolution becomes limited (see Chapter 5).

The total gain provided by an MCP may be estimated from values such as those displayed in Figure 4.90. In the case of a proximity focus image intensifier (Figure 4.89b), it is just the efficiency of typically 0.1 to 0.5 electrons per photon times the reconversion of electrons to photons that may lead in total to a couple of hundred electrons. This is useful for many applications. For a MCP system, in addition this has to be multiplied by the gain of the amplifier plate itself.

Before we continue, we would like to comment on "gain." Most simply one can denote gain just as the ratio of the number of electrons after amplification to those entering the MCP, i. e., the gain of the amplification stage. On the other hand, gain can be related to the total amount of light emitted from the screen with respect to the amount of incident light on the photocathode. This is the gain of the whole system. This definition can be done with respect to photon numbers (photon gain) or with respect to radiant exitance from the screen to the irradiance on the photocathode, where both quantities in the numerator and denominator, respectively, are provided in W/cm^2 (radiant emittance gain). Both of those radiometric definitions are appropriate for physical measurements. Of course, if more appropriate, this gain can be calculated as well from energy densities (J/cm^2) instead of power densities (W/cm^2). Alternatively, the gain can be expressed in photometric quantities as a luminous gain as luminous exitance from the screen/illuminance on cathode in units ($(lm/m^2)/lx$).

The low light "limit" of the illuminance is approximately 10^{-5} lx, which corresponds to the order of the irradiance of 10^{-13} to 10^{-12} W/cm^2 or a couple of photons on the area covered by a single channel. This is just the result of the demand that there must be at least a signal of one electron at the entrance of the amplifier, normally a channel of the MCP. There is another limit that is given by a signal $B_{EBI} = B_{pix} = g \cdot I_{pc}$ measured as the MCP output, or system when there is no MCP, in absence of illumination but with HV switched on. g is the gain generated by the MCP including the luminescent screen. This signal results from dark emission from the photocathode, namely electrons, cosmic rays and other ones that produce an intensity I_{pc} at the MCP entrance. To generate the same output signal, now if B_{EBI} is disregarded, an irradiance or illuminance, which is called EBI, has to be irradiated onto the MCP. We may note that sometimes the requirement for

31 J. L. Wiza: *Microchannel plate detectors*, Nucl. Instrum. Methods **162** (1979) 587–601.

EBI is twice the value of B_{EBI}, and thus (1 to 2) $\cdot B_{\mathrm{EBI}} = g \cdot \mathrm{EBI}$. This is the so-called equivalent background illumination (EBI) and is independent of gain g although, of course, different values of g lead to different values of B_{EBI}. Typical values are shown in Figure 4.90d.

We will not go to further details but comment that the above values result from rough estimates only. Altogether the relations for gain, "noise" or EBI SNR and dynamic range are not straightforward. Note as well that often photon statistics plays the important role unless the HV is quite large. Moreover, an MCP is hardly used on its own but it is always a part of an optical imaging system. In consequence, it is the dynamic range and the SNR of the system that is important. As our intention is not an extended description of sensor technology, we avoid a deeper discussion. Instead, we follow the intention of our book and concentrate on the entire system. This will be continued in the next chapter.

4.11.4 Intensified CCD and CMOS cameras

An intensified CCD (iCCD) simply consists of an image intensifier coupled to a CCD. Today, there are intensified CMOS and sCMOS cameras as well, but for simplicity, in the following we just write iCCD which, in particular, is reasonable due to the advantages of the CCD for scientific applications even though sCMOS catch up (see "Comparison of CCD and CMOS sensors in Section 4.5.2). A complete iCCD camera consists of an optical imaging system, the camera lens, the iCCD and peripheral devices such as the camera controller. In the case of special applications, such as XUV or X-Ray imaging experiments, the "camera lens" is setup separately as it is rather special, for instance, Fresnel zone plates, mirrors used at gracing incidence, Bragg reflectors, etc. Thus, this optics is not part of the iCCD camera in such a system.

Quite often, the intensifier is an MCP. There are other intensifiers used instead, such as proximity focus image intensifiers, which today are advanced systems too. But basically most simply speaking they are photocathodes coupled to a luminescent screen with a high voltage in between and with a vacuum between both planes.

A typical and quite common iCCD camera is shown in Figure 4.92a. Similar to a "normal" camera, the object is imaged with a camera lens (CL) onto the surface of the detector. Here, the detector consists of an MCP or a proximity focus image intensifier, for instance, one of the versions displayed in Figure 4.89b–d. The output image on the luminescent screen (S) can be regarded as an intermediate image, which then via a relay optics (RO) is further imaged onto a CCD sensor. We have to note that lens coupling can provide high-image quality but has low coupling efficiency. This, in particular, results from the emission angle of the screen, which covers half of the full solid angle behind the phosphor. According to the NA or f-number of the optics, only a small fraction of this solid angle can be collected. A straightforward estimate shows that even for a $f_{\#} = 1$ lens with 100 % transmission the transfer efficiency is 12 % for a magnification $M = 1$. Note

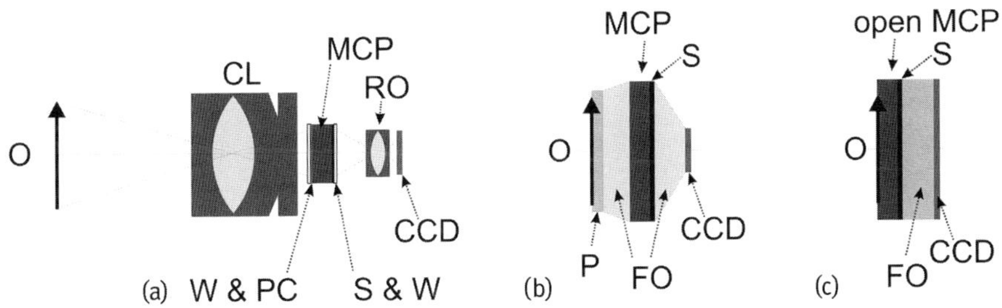

Fig. 4.92: Examples of iCCD configurations. O object, CL camera lens, RO relay optics, W window (entrance and exit), S luminescent screen, FO fiber optics, PC photocathode, P phosphor.

that from geometry, the dependence of the efficiency on M can be easily calculated (see, e. g., [Gru02]). Note as well that even specially developed relay optics, e. g., with $f_{\#} = 0.8$, are available.

The configurations displayed in Figure 4.92b and c are typically used for imaging in the UV, XUV or X-ray-range. In both cases UV-, XUV- or X-ray-optics, not shown in this figure, image the object onto the surface of a suitable phosphor such as P45 (b) or onto the surface of an open MCP, e. g., with a CsI:Tl converter at its entrance (c). Note that in (c) this image is indicated by “O.”

In (b), the phosphor is fiber optically coupled (FO), namely directly attached with a suitable index matching oil between the surfaces to the MCP and at the same time the image is enlarged. For demagnifying arrangements, FO couplings are usually, but not always, more efficient than a coupling via relay optics. An estimate is straightforward, but see also, e. g., the discussion and examples in [Gru02]. Furthermore, optical distortions can be rather low and the systems are compact. But also for FO, due to limitations of the NA and also due to the finite transmission of the fibers, loss of light is present as well.

In the present example, enlargement is done because the SBN of the MCP is assumed to be smaller than that in the object plane. If, for instance, the resolution in the object plane is 20 µm but the resolution of the MCP due to a large microchannel diameter and possibly some cross talk is 40 µm, then a magnification by a factor of 2 or more would avoid a reduction of the SBN if the total size of the MCP is large enough. For instance, if the phosphor has a diameter of 20 mm, then SBN = 1000 and it is the same for an MCP with 40 mm diameter. In the case of a high-resolution CCD, the intermediate image on the luminescent screen then may be transferred, again by means of a FO, to the CCD surface. In this example, even though there is a demagnification behind the MCP, which should lead to a matching of its output size to the CCD sensor size, there is no further change of the SBN, i. e., no loss in resolution, if we disregard some minor loss due to coupling effects. In the arrangement of Figure 4.92c, there may be no need of the adaption of different sizes so that a straight FO taper can be applied. For those examples, we do have to remark that the magnification and demagnification in Figure 4.92b is not the same as that in the discussed example; it is just an illustration.

Any other configuration than that discussed, i. e., replacement of FO tapers by relay optics or vice versa and/or change of magnification and/or substitution of encapsulated MCP by an open MCP or a proximity focus image intensifier and so on, in principle is possible. But, of course, one has to consider, e. g., if the optics (including FO) are suitable for the wavelength range intended to be used, e. g., conventional glass optics is not applicable for UV light.

Although today advanced CCD, CMOS or EM-CCD cameras (see Section 4.11.5) can be very efficient and detect, in principle even nearly single electrons, the advantage of the additional intensifier is that amplification is done prior to detection by the CCD, etc., with the result that now many photons are available for detection. Consequently, a consideration of noise from the final sensor may be of minor importance. Nevertheless, of course, photon noise is always an issue. Then due to the intensifier properties, CCD noise and dark current usually do not play a large role even if the CCD sensor is not cooled, though cooling still may lead to some improvements, in particular, when intensification is small.

Light signal intensification makes sense in the regime where sensor noise plays an important role. This, in particular, is the case below the linear regime displayed in Figure 4.51b typically between 10 and 100 photons, but not in the linear regime itself. For illustration, we will discuss an example. Let us assume an iCCD consisting of an image intensifier with an adjustable photo gain g. This effective total gain should include losses due to the efficiency of the photocathode and any coupling losses, efficiency of the relay optics etc., but also the "gain" of the luminescent screen (see Figure 4.90c). However, losses may be compensated by an increase of the MCP gain. Furthermore, the iCCD should have an $EBI = 1$ photon per pixel per second. The number of the corresponding photoelectrons generated by the photocathode then is determined by its efficiency. The intensified light signal is captured by a CCD, which should have a read noise of $\sigma_{\text{read}} = 5$ electrons (RMS), a quantum efficiency $\eta_{\text{ccd}} = 0.8$ and $\text{FWC} = 10^5$. For simplicity, in the following we will regard all estimates on the basis of 1 pixel and $t_x = 1\,\text{s}$. Hence, neither the diameter of the microchannels nor the pixel size of the CCD has to be known. We would like to remark that the knowledge of the corresponding intensities in W/cm², illuminance in lx and so on is not necessary, but they can be easily estimated for given microchannel diameters, CCD pixel size, etc.

If we apply $g = 1$, and do not illuminate the iCCD, according to EBI one would generate 0.8 electron in the CCD. This is much below σ_{read}, and thus read noise of the CCD dominates. In this case, the dynamic range is given by $DR = \text{FWC}/\sigma_{\text{read}} = 20{,}000$, i. e., 14 EV or 86 dB. Note that this dynamic range is larger than that of a scientific CCD operated in keV regime (see Figure 4.51b). If one increases the gain, DR will still have approximately the same value as long as g is small enough so that EBI does not lead to a corresponding number of electrons in the CCD that exceeds σ_{read}.

However, if now the gain is set, e. g., to $g = 100$, EBI would yield approximately 80 electrons, still within $t_x = 1\,\text{s}$. In that case, EBI dominates and $DR = 10^5/80 = 1250$ yielding 10 EV or 62 dB. For even higher gains, DR is further reduced accordingly. But quite often the exposure time used for imaging with iCCD-cameras is much shorter, in particular, if one takes advantage of the short gating times that are possible as discussed above. Then the dynamic range is still $DR \approx \mathrm{FWC}/\sigma_{\mathrm{read}}$ even when a higher gain is applied.

Another important subject is the signal-to-noise ratio. Now let us assume $t_x = 1\,\mathrm{ms}$, which is not a very short exposure time for an iCCD. Let us further assume an input signal of $N_{\mathrm{ph}} = 5$ photons now within this shorter t_x. EBI would yield 10^{-3} photons, and thus could be totally neglected. After intensification with $g = 10$, this would yield 40 electrons in the CCD. A naïve estimate would yield $\mathrm{SNR} = 40/\sigma_{\mathrm{read}} = 8$. However, the major role now takes photon noise. For an intensifier system, this is given by

$$\sigma_{\mathrm{ph}} = g \cdot F \cdot \sqrt{N_{\mathrm{ph}}} \tag{4.59}$$

where F is a noise factor that results from a detailed description of the amplification process itself. F is typically in the range between 1.3 and 2. For further discussion, we refer to[32] (compare also Equation (4.23)). Including η_{ccd}, this results in a noise of approximately 33 electrons. Thus, σ_{read} can be neglected, and consequently $\mathrm{SNR} \approx 1.5$ or more directly $\mathrm{SNR} = N_{\mathrm{ph}}^{1/2}/F$, which is independent of gain if the intensified photon signal is large enough.

On the other hand, this may be compared to a system that makes use of the assumed high-quality CCD only, namely without the image intensifier. In that case again, $N_{\mathrm{ph}} = 5$ photons within t_x would yield four electrons, which is even lower than the read noise especially if we take into account the additional shot noise from the photon statistics. Consequently, $\mathrm{SNR} < 1$, and thus no signal could be detected. Only an ideal CCD, i. e., a CCD with $\eta_{\mathrm{ccd}} = 1$ and $\sigma_{\mathrm{read}} = 0$, and $\sigma_{\mathrm{tot}} = \sigma_{\mathrm{ph}}$ would yield $\mathrm{SNR} \approx 2.2$, and thus a detectable signal.

If now, similar to before, the input is increased to $N_{\mathrm{ph}} = 50$, SNR is approximately the same for the iCCD and the CCD only. The actual value for the iCCD depends on the actual value of F. For even larger inputs, the pure CCD system becomes superior. Consequently, today with very low read noise BSI-CCD systems available, previous usage of iCCD systems has mostly been overtaken by them. Nevertheless, for particular situations such as when the photon signal is very low (see above), if very short gating times are required (see below) or if for some reason image converters should be used (e. g., when the object area is much larger than that of the CCD or if radiation damage during hard X-ray exposure should be avoided), iCCD systems still may be the best choice. Anyway, high-quality systems of both types are rather expensive.

An extended discussion on SNR and, in particular, a comparison with respect to the discussion of noise for both system can be found, e. g., in Dussault and Hoess.[32]

32 D. Dussault, P. Hoess: *Noise performance comparison of ICCD with CCD and EMCCD*, SPIE **5563** (2004) 195D.

A similar estimate shows that for long or very long exposure times, signals of iCCD become worse when compared to those of a high-quality CCD. In such cases, noise is dominated by dark current and EBI, respectively. Because for a good CCD the former is always much smaller than the latter, iCCD is much more strongly affected by noise. On the other hand, if readout time is an issue, a CMOS-based system or an appropriate MCP/CCD combination may be superior to a CCD that is optimized for low noise signals. This is an important issue in cell biology. Moreover, extremely short exposure times can only be achieved with an iCCD system, which is an issue in other fields of science. A so-called gated CCD allows t_x~ns or even approximately 10 ps for a special scientific system, e. g., a "framing camera." Among special electronics, a very fast pulser is achieved by application of a very short positive voltage pulse to the photocathode, which prevents electron propagation to the microchannels.

Finally, we would like to hint on a special variant of the arrangement shown in Figure 4.89a. In an electron-bombarded CCD (EB-CCD), the luminescent screen is replaced by a CCD. This provides some gain, which has been useful for special applications such as fluorescence microscopy.

4.11.5 Electron-multiplying CCD

Simply speaking, iCCD are based on an array of channeltrons and electron-multiplying CCD (EM-CCD) on an array of diodes similar to avalanche photodiodes. The latter may be regarded as the semiconductor equivalent to photo multipliers. There are avalanche diode arrays as well, e. g., such ones made of single-photon avalanche diodes (SPAD). Those are used for ToF-sensors or LIDAR-sensors which, in particular, are of large importance for automotive applications (see Section 4.10.7.4; for further reading see, e. g., [Luh19]). EM-CCD make use of an on-chip amplification scheme that is based on impact ionization. Again, for the same reason discussed in the previous chapter, amplification is done prior to readout, which allows signals to be increased above read noise level.

The well-known impact ionization process in atomic and plasma physics occurs when an electron has high enough kinetic energy to ionize an atom. Within semiconductor physics, similar to that, electrons that have high enough kinetic energy may generate a further electron-hole pair. This is the case, when, in particular, an external electric field applied by setting a sufficiently large voltage to the device accelerates the free electrons. This leads to impact ionization within the valence band of the lattice. Thus, electrons are transferred to the conduction band. If the energy of the primary and secondary electrons is high enough, this process can continue, which then results in an avalanche effect. This is similar to the breakdown that occurs when a high-intensity laser pulse interacts with a dielectric and generates a plasma.

In an EM-CCD, this happens within the additionally added cells located behind the normal shift register (Figure 4.93a). Due to an applied voltage, e. g., in the range displayed in Figure 4.93b, there is always a small probability of $p_{\mathrm{ionis}} = 1$ to 2 % for the ion-

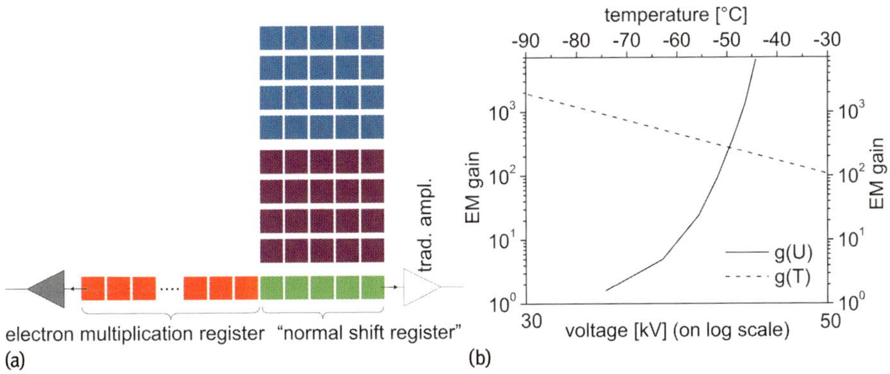

Fig. 4.93: (a) The principle of an EM-CCD is the same as that of the frame transfer CCD illustrated in Figure 4.14c, but with an additional multiplication register consisting of several hundred amplifier cells (orange). The amplifier shown by dotted lines and marked as "trad. Ampl." indicates a second amplifier when dual amplifier mode is available (see text). (b) Illustration of the dependence of the gain (solid curve) on the applied voltage and temperature (dashed line), respectively. Note that even small changes in the applied voltage lead to a large change in gain.

ization process, where p_{ionis} depends on the applied voltage and the temperature (see Figure 4.93b). Typical values are 0.01 to 0.016. However, due to large number of stages, typically several hundred, EM gains $g_{\text{em}} > 1000$ can be achieved.

The amplification may be estimated simply by

$$g_{\text{em}} = (1 + p_{\text{ionis}})^s \tag{4.60}$$

where s is the number of stages. Thus, for instance, a device with 500 stages and $p_{\text{ionis}} = 0.012$ yields a gain of approximately $g_{\text{em}} = 400$. For that reason, usually the potential well of the pixels within the amplifier stages is larger than that of the light sensitive pixels.

Although most of the other parameters and properties of an EM-CCD are not much different from a normal scientific CCD, due to the nature of the stochastic amplification process, an additional noise factor similar to that introduced in the previous chapter (see Equation (4.59)) occurs. A typical value for an EM-CCD is $F = 1.3$. As common to intensifier systems, when gain is large, there might be a reduction of dynamic range with increasing gain. In contrast to iCCD, for EM-CCD *DR* typically increases with g_{em} before it flattens and afterwards decreases. This has been discussed in Section 4.11.4, and thus we omit further discussion here. EM-CCD do not suffer from EBI, but dark current signal is amplified as well so that cooling becomes important.

With some similarity to iCCD, EM-CCD are also well suitable to fast readout. Thus, many cameras have amplifiers specially designed for that purpose. As a consequence, read noise is increased, but this is not an issue with the signal amplification prior to readout and SNR may be still large. Nevertheless, the dynamic range decreases with

Tab. 4.7: Typical parameters of an EM-CCD. Note that IQE depends on wavelength (compare Figure 4.10). Here, the provided value corresponds to the peak.

IQE [%]	pixel pitch [µm]	no. pix. horiz.	no. pix. vert.	gain (2 examples)	FWC [el.]	σ_{read} [el.] (rms)	I_{dark} [el/pix/s] at −80 °C
up to >90 %	16	512	512	4 together with fast readout	$4 \cdot 10^5$ (EM readout) $1 \cdot 10^5$ (normal readout)	30	10^{-4}
up to >90 %	16	512	512	1200 together with slow readout	$4 \cdot 10^5$ (EM readout) $1 \cdot 10^5$ (normal readout)	1	10^{-4}

gain. However, to allow for high-dynamic range measurements as well, sometimes a second traditional amplifier is introduced, which allows slow readout. Readout of different ROI at the same time may be possible, each of them with another amplifier (see Figure 4.93a). This is called dual readout. Thus, the EM-CCD can be operated as a normal CCD as well.

For illustration Table 4.7 shows typical parameters of an EM-CCD.

A comparison of the performance of iCCD and EM-CCD is not easy. Moreover, there are even cameras which combine both systems to an EMi-CCD (see, e. g., PI-MAX 4: 1024 EMB from Teledyne, Princton Instruments). The manufacturer claims several benefits, in particular, the possibility of very fast gating (down to < 500 ps). Performance any of those systems depends on the application, and thus a general statement does not make sense. There is a lot of discussion in the literature and here, again as an example, we may refer to[33]. But in any case, we would like to remark that a serious comparison requires a careful consideration of all parameters. An example is that for such a comparison the pixel size of the sensors within the compared systems has to be more or less the same. Alternatively, one may make use of binning. As an example a sensor with 10 µm used in 4×4 binning mode can be compared to another one with 20 µm pixels.

4.12 Curved sensors

Within optical imaging, the surface geometry of the image field takes an important role as has been discussed in Chapter 3. Usually, in contrast to the sensor surface, the image field is not flat. Distortions and image degradation may become significant, when the sensor surface deviates strongly from the curved Petzval surface. To avoid such a situation, a lot of care has to be taken with the design of the optical system, and, in particular, Petzval's condition has to be fulfilled (s. Section 3.5.4) for a flat image surface. For high-performance lenses, such designs are rather sophisticated. The camera objective

33 D. Dussault, P. Hoess: *Noise performance comparison of ICCD with CCD and EMCCD*, SPIE **5563** (2004) 195D.

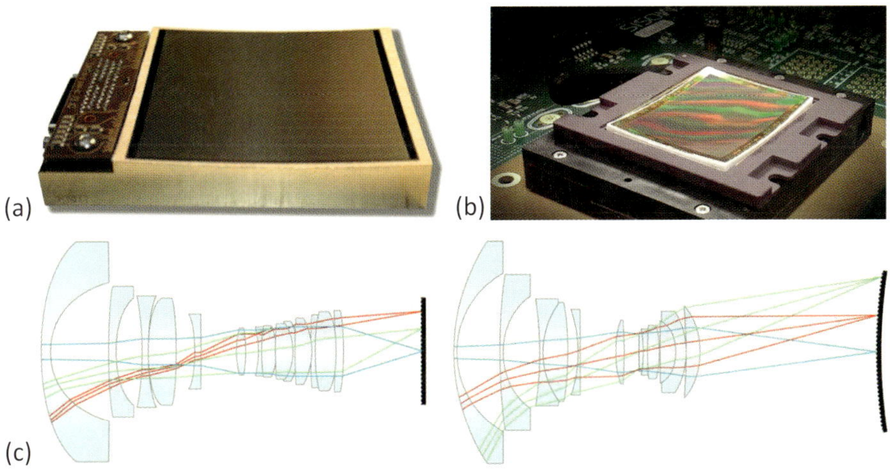

Fig. 4.94: Curved sensors: (a) CCD, (b) CIS. (c) Application of curved CIS: Fish-eye lens camera with a flat and curved sensor, respectively. (a) Courtesy Andanta GmbH, Olching, Germany. (b), (c) Courtesy Curved CMOS sensor—company CURVE SAS—contact@curve-one.com—Curving all CMOS and IR type sensors.

construction is made of many individual lenses to achieve a rather flat image field. For all such situations, image sensors with a proper surface curvature (e. g., see Figure 4.94) would be advantageous and allow for a less complex lens design.[34] To emphasize this, we may cite from [Bla21]:

> A strictly concentric system delivers consistent image quality from a spherically curved object surface to a spherically curved image surface. For a curved image sensor with simple monocentric lenses, excellent image performance is feasible [50]. For SPC lenses, it has been shown that curved image sensors can result in lenses with one f-stop superior f-number and comparable aberration performance [51]. In particular, extreme wide-angle lenses would benefit significantly from curved image fields.

However, the creation of curved sensors is a major challenge.

Sensors may not only be developed and adapted with respect to parameters directly relevant to them, such as the specific wavelength range, noise issues, FWC and so on, but also the shape of the sensor surface may have to be adapted to a particular lens design. For the optical system of many animals and humans, namely the eye, this is obviously made by nature in so far as the retina acts as a curved image sensor.

Also, for scientific applications the idea of a curved sensor is not new. Here, e. g., films mounted on a curved surface in an appropriate holder have been commonly used for very long time. Also, e. g., curved MCP have been used as imaging detectors for experiments performed at short wavelengths for more than two decades, although due to

34 D. Shafer: *Lens designs with extreme image quality features*, Adv. Opt. Techn. **2** (2013) 53–62.

the very high costs the usage has been very rare. This includes the application in X-ray spectrometers, where the curved image plane is located on the Rowland circle (see standard textbooks on X-ray optics, such as[35]; see also, e. g.,[36]). Another example is the use of CCD and CMOS sensors in astronomy, where a curved sensor leads to advantages in optical designs with improved image quality such as homogeneity, reduced monochromatic and chromatic aberrations, etc.[37] (see also references therein).

In the same way, curved sensors may revolutionize optical systems of cameras, e. g., for the consumer market, in particular, for miniaturized ones. For high-end cameras and DSLR and DSLM and their lenses, a lot of effort is taken to correct the image field so that it becomes rather flat. Conversely, for SPC and a lot of today's compact cameras this effort is hardly taken and corrections are made as a post-process via image processing, mostly within the camera. However, as we will see in Section 7.4, even advanced image processing cannot avoid image quality degradation. Consequently, there is an interest in curved CIS which then, in principle, allow for a simpler lens design. In particular, this is an issue for small pixel pitch. Here, additional aberrations should be as low as possible and, therefore, curved image sensors might be mostly helpful. Also vignetting (see Sections 3.4 and 4.6 and Chapters 6, 7 and 8) may be reduced and illumination uniformity much increased. Note that less light loss at the edges and in the corners does not only improve the straightforwardly visible image quality. But more light also means less noise and this is an additional improvement (note an improvement of 1 to 2 EV is equivalent with 2 to 4 times more photons, and thus according to Equation (4.39b) a factor 1.4 to 2 times better SNR). Figure 4.95 shows an example of a meaningful result,[38] where vignetting is almost absent. With a well-designed lens, a further advantage is the reduced inclination of the light incident on pixels far off the center (see the discussion and figures in Section 4.6.1). Finally, the advantages include a less complex lens fabrication process, less volume consumption, less weight and less cost. Furthermore, the fewer optical components needed may allow other lens designs so that also the actual discontinued optical zoom for SPC may find a resumption. For those reasons, current research and development on curved image sensors has attracted attention again. In particular, this is concerning curved CIS with a potential usage for the mass market of SPC or, e. g., security applications.

---

35 D. Attwood, A. Sakdinawat: *X-Rays and Extreme Ultraviolet Radiation* (2nd ed.), Cambridge University Press, 2016/2017; see also D. Attwood: *Soft X-Rays and Extreme Ultraviolet Radiation*, Cambridge University Press, 1999.

36 E. Förster, K. Gäbel, I. Uschmann: *X-ray microscopy of laser-produced plasmas with the use of bent crystals*, Laser Part. Beams **9** (1991) 135.

37 S. Lombardo, et al.: *Curved detectors for astronomical applications: characterization results on different samples*, Appl. Opt. **58** (2019) 2174–2182 and references therein.

38 B. Guenter et al.: *Highly curved image sensors: a practical approach for improved optical performance*, Opt. Expr. **25** (2017) 13010.

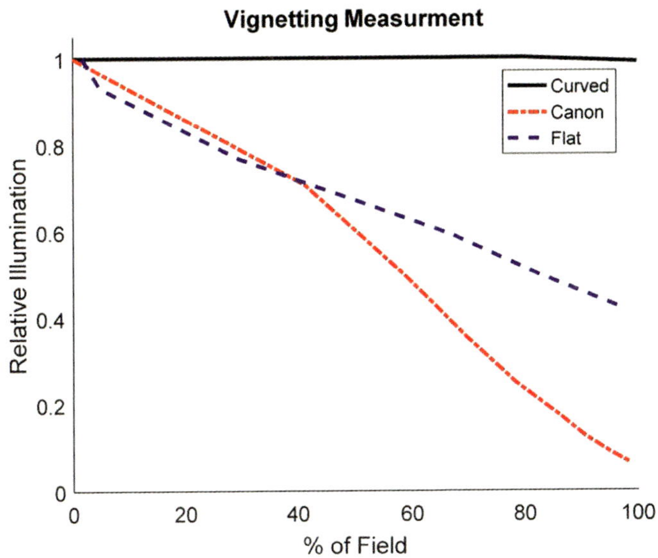

Fig. 4.95: Measured relative illumination performance of a curved CIS (Aptina AR1820HS with applied curvature; black line) and comparison to a commercial full frame DSLR equipped with an $f_{\#} = 1.2$ optics (Canon 1DS mark III with a 50 mm lens; red line) and a camera consisting of an Edmund Optics 6 mm lens mounted on a flat Aptina AR1820HS CIS (blue dotted line). Image taken as part of Figure 8 from[39]. Reprinted with permission from B. Guenter et al., Opt. Expr. **25** (2017) 13010, https://doi.org/10.1364/OE.25.013010; #286903 Journal ©2017 The Optical Society, Optica Publishing Group.

There are several possibilities to create curved sensors. Direct generation, e. g., via lithography, would be one possibility. However, this is not yet common. Thus, today the usual way is to use available flat CIS and apply precisely a well-defined deformation. Similar to bending of very thin flat crystals for imaging issues in the X-ray range, also flat CMOS sensors may be bend. But in contrast to a "passive" crystal, an electronic sensor is an "active device." The challenge of curving an existing sensor is to avoid a negative influence on its electronic and electro-optical properties arising from the induced mechanical strain and stress during the bending process (see, e. g.,[39,40]). Currently, (2022) companies such as Microsoft, Sony and other ones have developed the curved CIS, which now is available to the consumer market. Companies such as Curve-One and Silina have been among the first who have developed the necessary automation for curving a whole batch of sensors. This is in contrast to previous attempts where curving has been applied to a single sensor only. Now it is intended to transfer the technology to large sen-

39 B. Guenter et al.: *Highly curved image sensors: a practical approach for improved optical performance*, Opt. Expr. **25** (2017) 13010.

40 S. Lombardo, et al.: *Curved detectors for astronomical applications: characterization results on different samples*, Appl. Opt. **58** (2019) 2174–2182 and references therein.

sor manufacturers, and thus to reach high-volume markets. Hence, next generation SPC modules, cameras for automotive and security may make a profit.

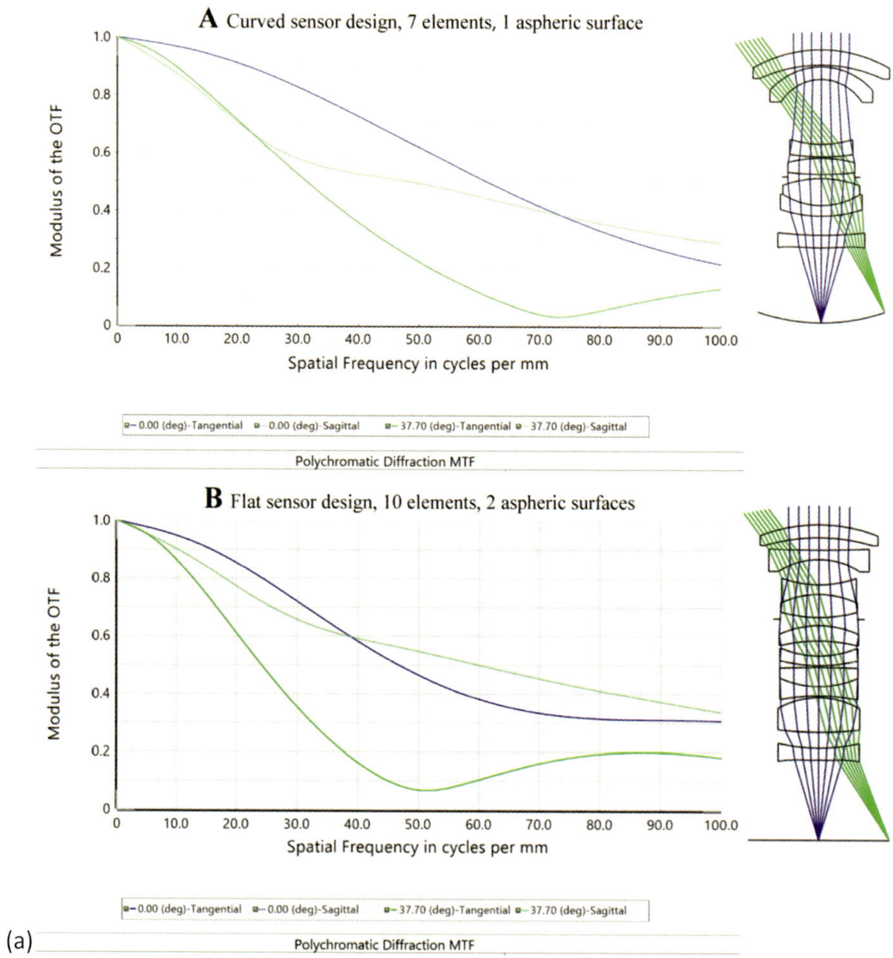

Fig. 4.96: Optical performance of cameras consisting of a flat CIS and its curved version, respectively,[41]. Both variants are equipped with appropriately designed lenses for an image circle of 43 mm. (a) lens design and simulated MTF of camera with a 28 mm wide angle lens and $f_{\#} = 1.7$. (b) the same, but for camera with a 80 mm lens and $f_{\#} = 1.5$. A and B indicate curved and flat sensors, respectively. For a discussion of MTF, see Chapter 5. Cycles per mm is equal to lp/mm. (a) Images are taken as Figure 10 from[41]. (b) Images are taken as Figure 11 from[41]. (a) and (b) Reprinted with permission from B. Guenter et al., Opt. Expr. **25** (2017) 13010, https://doi.org/10.1364/OE.25.013010; #286903 Journal ©2017 The Optical Society, Optica Publishing Group.

41 B. Guenter et al.: *Highly curved image sensors: a practical approach for improved optical performance*, Opt. Expr. **25** (2017) 13010.

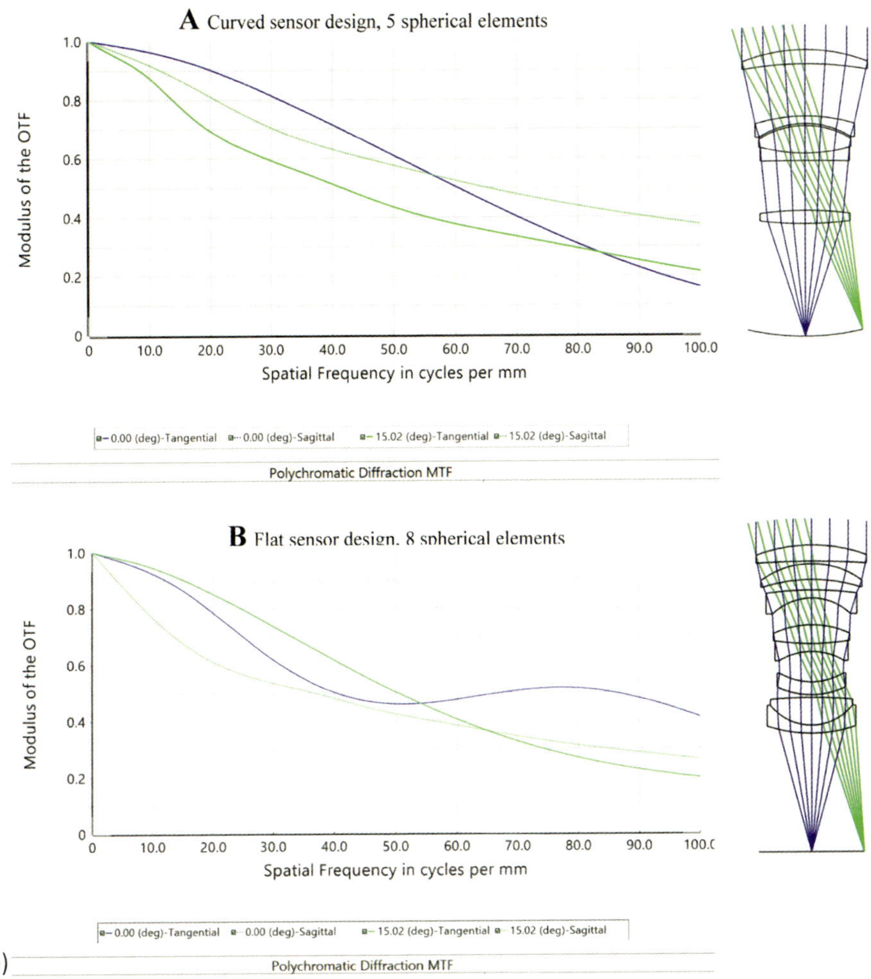

(b)

Fig. 4.96: (continued)

Figure 4.94 shows an example of a commercially available curved image sensor together with an appropriate lens design. It can be seen that the whole system, consisting of lens and sensor, is smaller and less complex when compared to a corresponding system with a flat sensor. Figure 4.96 (and Figure 5.36 in Section 5.2.7) shows that rather similar image quality may be obtained with a significant less complex lens design when a curved sensor is applied (here we restrict to the quality of the MTF, see Chapter 5). The system with the curved sensor may be even superior although less lens elements and less aspheric surfaces are applied. However, it must be noted, that in general, such comparisons must be made for cameras equipped with flat and curved sensors respectively, together with their appropriate lenses, which do not much differ in all their other properties such as sensor size, pixel pitch, read noise, f-number and so on. Due to different lens designs, it is not straightforward that the lens quality is well comparable. Consequently, one has to be careful with such comparisons. A further example is presented in Section 6.5.4.

Finally, we may add that new developments of image sensors on thin flexible foils may further boost the application of image sensors with a well-adapted surface. But in any way, curved sensors have to match the lens design and vice versa. Accordingly, patented lens designs are available. Again, this is an example that a high-performance camera requires matching of the optical and the sensor system.

5 Fourier optics

5.1 Fundamentals

5.1.1 Basics, electric field, amplitude and phase and remarks in advance

The access to optical imaging may be manifold. Optical imaging may be described within geometrical optics and with some wave-optical extensions. This has been extensively discussed in Chapter 3. Fourier optics provides another access, which is not easily understandable and not so straightforward, e. g., for camera lens design, when compared to the methods described. However, this modern wave optical approach provides a very fundamental description of imaging (and much more) with incoherent and coherent light and allows for deep analysis of optical systems.

Generally, this requires also an extended discussion with an extended explanation why the field or the intensity distribution in the object plane, the Fourier plane and the image plane are related to each other in a particular way. Such a discussion would only give rise at least to several long chapters and that would be beyond the limit of the present book. Nevertheless, with simple physical arguments we will provide the essential background quite clearly in a way that is at least sufficient for the present book. A deeper study can then be done on the basis of standard textbooks on the fundamentals of optics such as [Bor99, Hec16, Sal19]. Even more, a good comprehensive discussion of the whole topic of Fourier optics can be found, e. g., in the textbooks of Goodman [Goo17] and Easton [Eas10].

Before we continue, we would like to make several brief remarks. First, of course, geometrical optics do not include wave optical phenomena, and thus in a physically fully correct discussion, terms of geometrical optics should not be mixed with those of wave optics. However, to make the discussion somewhat easier and more clearly understandable, we will nevertheless do so. But although we will make use of terms such as rays together with phenomena such as diffraction, the physical idea behind all that is still correct. This is because optical rays are just the paths of light propagation with a direction given by the local wave vectors. Consequently, the rays are always perpendicular to the wavefronts. In the following, we disregard effects such as birefringence. Moreover, and also due to a simplification of the discussion, we will often restrict our discussion to 1D geometry, as the extension to 2D is straightforward.

Second, a physically correct Fourier optical description of a complicated objective lens would require to consider considering principal planes and limitations of the ray bundle by apertures such as the entrance pupil and the exit pupil and the corresponding pupil functions. However, due to the restriction to the goal of the present book, such an extended discussion is avoided, and the interested reader may be referred to special literature or textbooks such as, e. g., [Goo17] and [Smi08]. Thus again, for simplicity, we strongly restrict ourselves to the fundamental relations, which means also that we mostly assume that the optics can be represented by a simple thin lens or two of them

https://doi.org/10.1515/9783110789966-005

when we discuss the equivalent 4-f-system. As a result, we do not have to discriminate between entrance pupil and exit pupil here but restrict to a single relevant aperture with a diameter or width D only. Nevertheless, such a discussion provides a good illustration of the relevant physics.

Third, in principle, the discussion of diffraction, imaging and so on can be made for coherent light, for instance, emerging from a laser and incoherent light, which is the "normal" light used to describe the situation in photography. As usual and for easier understanding, we would like to begin our discussion with electromagnetic waves, which are simply speaking, all in phase. This corresponds to coherent light. Here, we restrict to spatial (or transversal) coherence. In that case, the related Fourier transformations have to be applied to the fields. This is mostly the case in Section 5.1.1 to 5.1.4. However, as imaging is typically based on incoherent light, the discussion in the following chapters is mostly restricted to that particular case. In Section 5.1.7, imaging with coherent and incoherent light is compared.

After these comments, we would like to begin our discussion at the very beginning and quickly introduce the most important relations. Wave optics is based on the common description of light by electromagnetic waves. These waves are the solution of the wave equation that itself results from Maxwell's equations (see Appendix A.11). The particular solution, of course, depends on the boundary conditions. The most simple solution is a plane monochromatic light wave that travels in the positive or negative z-direction. Its electric field is described by a pure sine wave

$$\vec{\mathcal{E}}(z,t) = \vec{\mathcal{E}}_0 \cdot \sin(kz \pm \omega t) \tag{5.1}$$

with angular frequency ω and wave number k. $\mathcal{E}_0$ is its amplitude. The argument of the sine function is its phase

$$\phi = kz \pm \omega t. \tag{5.2}$$

If $\mathcal{E}_0$ is time dependent as well, the wave still may be quasimonochromatic. The sign in front of the second term of the phase ("+" or "−") defines the direction of the wave propagation to the left or right, respectively. In a similar way, one may describe the magnetic field component of the electromagnetic wave, but in the nonrelativistic case such as it is the case in "normal" imaging, it is sufficient to restrict the calculation to the electric field component. This is quite basic and extensively discussed in standard textbooks of electrodynamics and optics.

Another description is the complex representation of a light wave, namely

$$\mathcal{E}_c(z,t) = \mathcal{E}_0 \cdot \mathrm{e}^{i(kz \pm \omega t)} \tag{5.3}$$

where we one makes use of

$$
\begin{aligned}
e^{i\phi} &= \cos(\phi) + i \cdot \sin(\phi) \\
\sin(\phi) &= \frac{1}{2i}\left(e^{i\phi} - e^{-i\phi}\right) \\
\cos(\phi) &= \frac{1}{2}\left(e^{i\phi} + e^{-i\phi}\right).
\end{aligned}
\tag{5.4}
$$

It may be important to note that any monochromatic electromagnetic wave can be fully described by its amplitude and phase. In a similar way, any light pulse or wave packet can also be fully described by its amplitude and phase. However, then both $\mathscr{E}_0$ and ϕ may be functions of t and z as well in the temporal domain. In the spectral domain, the pulse is fully characterized by its spectral amplitude and spectral phase. We may also note that for the present book, usually the term $\exp(i\omega t)$ is often not of much importance in the calculations as all time-dependent phenomena oscillate with the same frequency of the monochromatic wave. Thus, we can restrict phase terms to $\exp(\pm ikz)$.

The electric field in Equation (5.3) is a complex function, and thus the fields in Equations (5.1) and (5.3) are not the same. But as usual, functions that represent real quantities, namely physical quantities that can be measured, such as the electric field, have to be real as well. We may comment that it is not possible to measure the electric field of light $\mathscr{E}(t)$ directly, but this is just due to the ultrashort oscillation period, which cannot be measured directly because there is no detector that has high enough temporal resolution. Nevertheless, this is not a principal restriction and indirect methods give access to $\mathscr{E}(t)$, insofar that it has to be regarded as a measurable physical quantity. In that case, it is clear that one has to either take the real part of the related complex function or just add its conjugate complex. As a result, $\mathscr{E}(z,t) = \mathfrak{Re}(\mathscr{E}_c(z,t)) = (\mathscr{E}_c(z,t) + \text{c.c.})/2$, where c. c. denotes the complex conjugate.

Within this book, we concentrate on the complex description. Furthermore, to avoid extended equations, in the following, we omit the subscript "c", e. g., we simply write $\mathscr{E}$ instead of $\mathscr{E}_c$. Then calculations can be performed more easily and at the end of all calculations, if necessary, the real value can be calculated as just explained.

The physical property accessible to a measurement is the intensity I of the electromagnetic wave, which is identical to the absolute value of the time-averaged Poynting vector. The average is made typically over a couple of field periods (see the standard textbooks on electrodynamics and optics). Thus, in general, one obtains

$$
I(z,t) = \frac{\varepsilon_0 c}{2}\left|\mathscr{E}_0(z,t)\right|^2 \tag{5.5a}
$$

or

$$
I(x,y,z,t) = \frac{\varepsilon_0 c}{2}\left|\mathscr{E}_0(x,y,z,t)\right|^2 \tag{5.5b}
$$

if we include the lateral distribution as well. The brightness distribution within the object B_{obj} and image B_{im}, respectively, is given by this intensity or the corresponding fields

$\mathscr{E}_{obj}$ and $\mathscr{E}_{im}$ and field amplitudes $\mathscr{E}_{obj,0}$ and $\mathscr{E}_{im,0}$, respectively. Using exact physical expressions at this point might be somewhat difficult because light incident on the detector in the image plane has to be described as intensity I_{obj}. However, when the image is observed, it has to be described as brightness B_{obj}. But if we disregard detector response and post-image processing, both quantities describe the same image in nearly the same way. Thus, for symmetry reasons in the following we use B_{obj} for any case. But if requested, all B_{obj} in the following expressions may be replaced by I_{obj} and transferred to B_{obj} by taking into account detector response and post-image processing.

Moreover, we would like to remark that in the exact description of the object and image, respectively, one has to discriminate between the coordinates in the object plane, e. g., (x_o, y_o) and those in the image plane, e. g., (x_i, y_i). Consequently, the corresponding distributions are $B_{obj}(x_o, y_o)$ and $B_{im}(x_i, y_i)$, respectively. On the other hand, we can avoid the introduction of the involved additional symbols when we agree that (x, y) as arguments of B_{obj} are the coordinates in the object plane and when the same ones are used for B_{im} they are the coordinates in the image plane.

We would further like to remark that within this chapter we rather concentrate on the essential relations and dependencies. Those are mostly the shapes or structures of light field distributions. Thus, for simplicity, we mostly omit prefactors such as $\varepsilon_0 c/2$ and also factors such as $(2\pi)^{\pm 1}$, $(2\pi)^{\pm 1/2}$, etc. and in that sense we may simply write, e. g.,

$$B_{obj}(x, t) = \left|\mathscr{E}_{obj}(x, t)\right|^2 \tag{5.6a}$$

$$B_{im}(x, t) = \left|\mathscr{E}_{im}(x, t)\right|^2 . \tag{5.6b}$$

Of course, it might have been easier and more correct to write the relation as $B_{obj} \propto |\mathscr{E}_{obj}|^2$ instead; however, we feel that an equation such as Equation (5.6) looks more clear.

The omittance of the prefactors is not a drawback, because they can always be easily calculated when the brightness distribution within the object is known. Then an integration over its area yields the total power emitted from the object or the total energy, when a time integration is additionally performed. This leads directly to the necessary factor in between the absolute square of the electric field emerging of the object $\mathscr{E}_{obj}(x, y)$ and $B_{obj}(x, y)$. If one furthermore includes the solid angle of acceptance of the optical system, namely the camera, and also the losses (e. g., of the included lenses, filters, etc.), then one also obtains the corresponding factor in the image plane, namely that in between $\mathscr{E}_{im}$ and B_{im}.

5.1.2 Background of Fourier optics, diffraction with coherent light

For the moment, we would like to restrict our discussion to coherent light that is pure monochromatic. The discussion of incoherent light, which is most important later on, will follow. In addition, we would like to remark again that within this book, we always mean spatial coherence. Temporal coherence is not an issue for the purpose of the present book as we consider virtually stationary states at a fixed point in time.

Fourier optics with coherent light relies on the fact that the diffraction pattern of a light field that is the far field is equal to the Fourier transformation of the light field distribution emerging from the object such as, e. g., a slit or grating or any other object. This is the near field. This relation is valid within the approximation of Fraunhofer diffraction, which is the most important situation for diffraction. Some basics of Fourier transformation are explained in Appendix A.2. We will not derive and prove this statement of the fundamentals of optics here, but we will explain the necessary details and discuss them and, of course, make use of it for the application of optical imaging. Within this book, unless stated otherwise, we will always restrict to cases where the requirements for Fraunhofer diffraction are fulfilled. For the basics of diffraction in general, we refer readers to standard textbooks of optics.

A very basic situation is sketched in Figure 5.1. Within this 1D example, a slit is illuminated by collimated light, in particular, a plane wave with a constant amplitude, e. g., say $\mathscr{E}_0$. The slit serves as an amplitude object and transmits the light with a transmission function $T(x)$ that is equal to 1 when x is within its opening, which has a width D, and 0 otherwise. In principle this may be a phase object or a mixture of an amplitude and a phase object, but for simplicity we avoid that discussion here. This situation is displayed in Figure 5.1, where

$$T(x) = \mathrm{rect}\left(\frac{x}{D}\right) \tag{5.7}$$

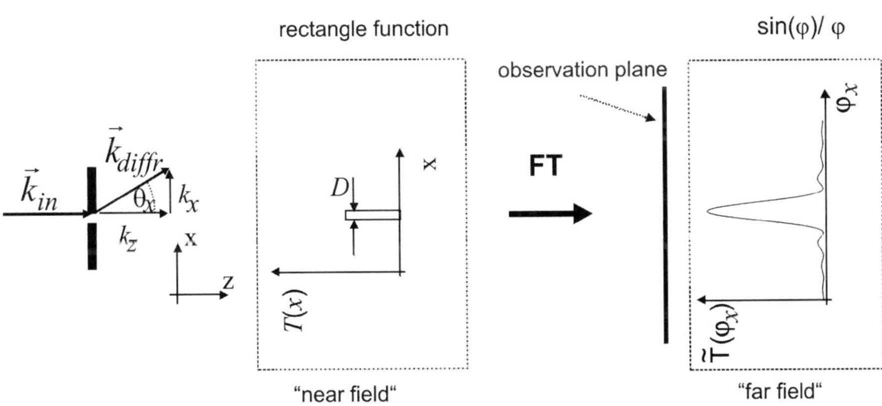

Fig. 5.1: Scheme of the diffraction at a 1D slit illuminated with a plane wavefront propagating in the z-direction (or a nearly plane wavefront from a collimated light beam). This scheme shows a basic diffraction experiment in one dimension. The opening of the slit is located in the x-direction and its width is given by D. In a more general case, diffraction may occur in the y-direction as well, where y is perpendicular to the paper of this book. As usual, the x-, y- and z-axis, respectively, define the common coordinate system.

is sketched as a “box” extending from $x = -D/2$ to $x = +D/2$ (“rectangle function”, see Appendix A.1). On a screen far away from the slit, this leads to the well-known diffraction pattern displayed behind the observation plane in Figure 5.1. In order to understand the following, at least the very basics of diffraction have to be known.

Of course, the diffraction pattern can be calculated by a rigorous integration, but knowledge in optics allows a more easy and modern discussion. Mathematically, this situation corresponds to a Fourier transformation of the near field distribution, which then yields the far field distribution. The near field is identical to the product of the electric field of the incident light and the transmission function of the slit. Consequently, here the structure of the near field is given by $T(x)$. According to the above statement, the far field distribution is just given by

$$\widetilde{T}(\varphi_x) = \mathrm{FT}[T(x)], \tag{5.8}$$

which defines the Fourier spectrum $\widetilde{T}(\varphi_x)$, namely the diffraction pattern. Inserting the rectangle function for $T(x)$ and taking its Fourier transformation results in

$$\widetilde{T}(\varphi_x) = D \cdot \frac{\sin(\varphi_x)}{\varphi_x} = D \cdot \mathrm{sinc}(\varphi_x) \tag{5.9}$$

where k_x or φ_x is the conjugated variable to x. φ_x is directly related to the diffraction parameters via the following equations:

$$\varphi_x = k_x \frac{D}{2} \tag{5.10}$$

$$k_x = k \cdot \sin(\theta_x) \tag{5.11}$$

$$k = \frac{2\pi}{\lambda} \tag{5.12}$$

k_x is the so-called spatial frequency in x-direction. It depends on the diffraction angle θ_x (in the x-direction). k and λ are the absolute value of the wave vector and the wavelength of the incident light, respectively.

Without absorption, energy and momentum are conserved. Thus, the wavelength of the incident and the diffracted light are the same. Via the deBroglie relation, the momentum is given by $\hbar\vec{k}$. Hence, the conservation of momentum demands that the wave vector of the incident and that of the diffracted beam have the same absolute value, namely $|\vec{k}_{\mathrm{diffr}}| = |\vec{k}_{\mathrm{in}}| = k$. However, as can be seen from Figure 5.1, the wave vector changes direction and a new x-component is introduced (see Equation (5.11)). Note that the x-component of the incident wave is zero. More generally, diffraction in the y-direction may occur as well so that in total the vector of the diffracted light is given by

$$\vec{k} = (k_x, k_y, k_z). \tag{5.13a}$$

Its components are the spatial frequencies, which are given by (see Figure 5.1)

$$k_x = \sin\theta_x \cdot |\vec{k}| = \sin\theta_x \cdot \frac{2\pi}{\lambda} \tag{5.14a}$$

$$k_y = \sin\theta_y \cdot |\vec{k}| = \sin\theta_y \cdot \frac{2\pi}{\lambda}. \tag{5.14b}$$

In technical optics and photography, it is typical to relate the spatial frequencies to wavelength only, i. e., without the factor 2π,

$$\vec{R} = (R_x, R_y, R_z) = \frac{\vec{k}}{2\pi}, \quad |\vec{R}| = \frac{1}{\lambda}. \tag{5.13b}$$

In the space and spatial frequency domain, the spatial coordinate $\vec{r} = (x, y, z)$ is related to its conjugated variable $\vec{k} = (k_x, k_y, k_z)$ in the same way as time t and angular frequency ω are in the time and frequency domain. In that sense, one might term $\vec{k}$ angular spatial frequency, but this is not common. And again, we would like to emphasize that in the near field the distribution is given as a function of the spatial coordinate, whereas diffraction is a phenomenon that is related to angles, not to distances (i. e., the far field is related to φ_x, θ_x or k_x, respectively, etc.). We may note that this is even the case when we observe the diffraction pattern on a screen at a large distance L of the object and measure the signal brightness at a distance x_1 with respect to the optical axis. In this case, $\tan(\theta_x) = x_1/L$ and because diffraction for the conditions within this book usually is small, $\theta_x \approx x_1/L$.

We would also like to remark that the above equations represent the diffraction formulae as well. This can be easily seen, e. g., for the first minimum, which requires that $\sin(\varphi_x) = 0$ (see Equation (5.9)). This is fulfilled for $\varphi_x = \pi$, and hence from Equation (5.11) one obtains $\sin(\varphi_x) = 0$, and this is equivalent to $\sin(\theta_x) = \lambda/D$ for the first diffraction order.

In general, the situation is similar. Then instead of a slit, the object may be represented by a slide that has a gray tone transmission $T(x,y)$ that represents the object structure. This slide, which in principle may also change the phase locally, then is illuminated by a light field $\mathscr{E}_{in}(x,y)$. Obviously, the near field then is given by $\mathscr{E}_{obj}(x,y) = \mathscr{E}_{in}(x,y) \cdot T(x,y)$. The far field, which is the spectrum in the Fourier plane, which is identical to the diffraction pattern and the spatial frequency spectrum, is given by

$$\widetilde{\mathscr{E}}_{obj}(k_x, k_y) = \mathrm{FT}\left[\mathscr{E}_{obj}(x,y)\right]. \tag{5.15a}$$

Its intensity or brightness distribution is obtained from the so-called “power spectrum” (in the sense of Equation (5.6b))

$$\widetilde{B}_{obj}(k_x, k_y) = \left|\widetilde{\mathscr{E}}_{obj}(k_x, k_y)\right|^2. \tag{5.15b}$$

Before we continue, we would like to remember that here and in the following we have still restricted ourselves to coherent light with the consequence that Fourier transformations have to be applied to the fields and afterwards they have to be squared to obtain the observed brightness distributions. We also would like to remark, that the "slide object" must not be necessarily used in transmission geometry. Instead, the principle of our discussion does not change for reflection geometry.

5.1.3 "4-f-system"

For better understanding the fundamentals of imaging, we regard a simple model system. In particular, similar to before, we illuminate a slide as the object with collimated light as sketched in Figure 5.2a. For the moment, for simplicity we choose an optical grating as the test object and restrict ourselves again to 1D geometry. The collimated light is given by a plane wave. However, the idea of the following discussion does not change with other diffracting slides or real objects and also not for 2D geometry.

As usual, the diffraction pattern can be observed very far away on a screen, i. e., in the far field, or at a finite distance, when a lens is used as a "transformer" because a lens just transforms the angular distribution to a distance distribution within the image plane. In the present 1D example, this distance is just measured perpendicularly to the optical axis, with the optical axis as the zero position. Here, we place the lens at a distance equal to its focal length f, and the screen also at the focal distance behind the lens. This shifts the diffraction pattern to the plane of the screen. This plane is called the Fourier plane and again contains the far field. For the moment, we disregard that the diameter of the lens is finite, and we assume that no aberrations are present.

Then one could complete the setup with another arrangement that is symmetric to the first one (see Figure 5.2b). In that way, one obviously generates a light distribution in the image plane that is absolutely equivalent to that in the object plane. There is only one difference, namely that the light distribution is upside down, when compared to

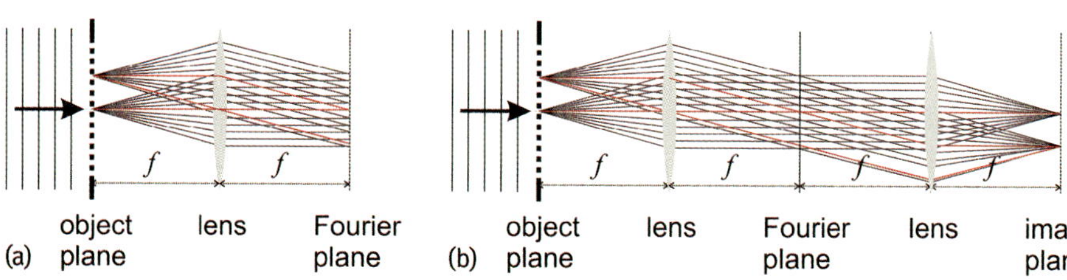

Fig. 5.2: 2-f-setup (a) and 4-f-setup (b) as an illustration of imaging. The grating in the object plane is illuminated by a plane wave (the wavefronts are indicated in front of it). One (a) or two (b) lenses (both with a focal length f) are positioned in a distance f with respect to the object and Fourier plane, respectively. The red and gray rays indicate a light path construction according to geometrical optics. For clearness, the diagrams display some selected rays only.

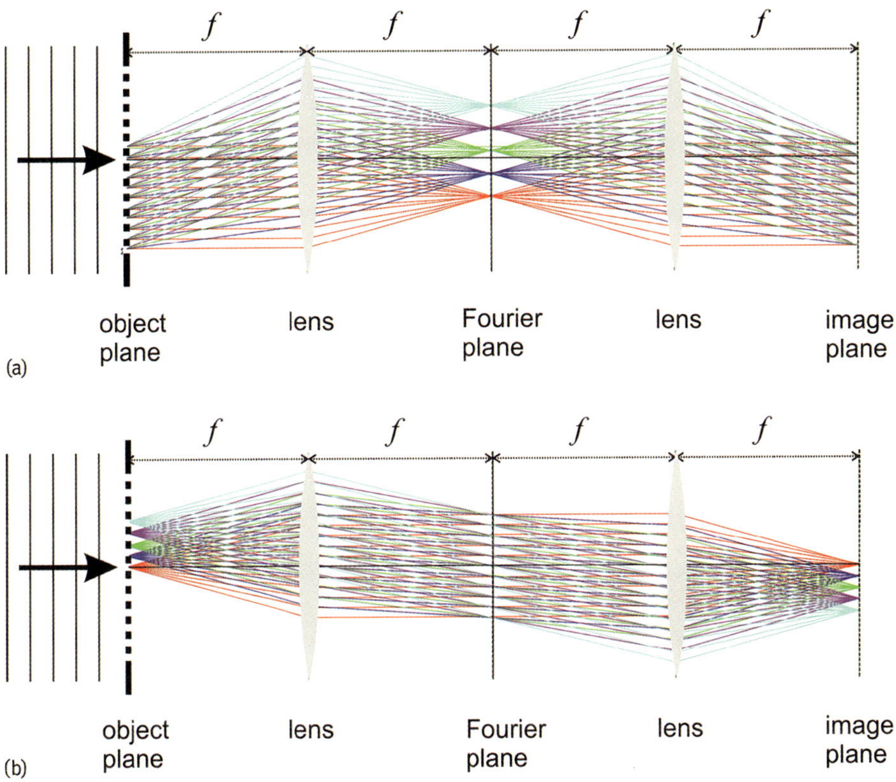

Fig. 5.3: 4-f-setup as an illustration of imaging. For clearness, this diagram displays some selected rays only. The rays in (a) and (b) are absolutely identical (but not always all rays are shown) but displayed in different colors. In (a), all rays that are parallel before the first lens are given the same color (different colors correspond to different angles). It is easily seen that rays emerging at different (diffraction) angles are focused to different points in the Fourier plane. In (b), all rays emerging from the same point are displayed in the same color. Note that (b) shows positive and negative diffraction angles, whereas (a) for clearness shows positive angles only.

that in the object plane. Consequently, we may conclude that at those ideal conditions this so-called 4-f-setup provides ideal imaging.

Before we continue with the actual discussion, we would like to show more clearly what happens. Figure 5.3 again shows a 4-f-setup. The grating as the object is illuminated by collimated light indicated by the plane wavefronts. The diffracted light is represented by the corresponding rays. From Figure 5.3a, it can be easily seen that all rays that are emitted from the object in the same direction are focused on the same points in the Fourier plane and later become parallel again in the image plane (in a more physical expression one would argue that all Huygens wavefronts, i. e., secondary waves, propagate in the same direction). On the contrary, all rays that originate from the same point within the object plane become parallel in the Fourier plane and coincide again in the image plane (Figure 5.3b). In any case, the different angles correspond to different diffraction maxima, minima or whatever in between or to different diffraction orders. Just as an example, in Figure 5.3a red color may correspond to the zeroth order, blue to first order, green to second order and so on. The negative diffraction orders are not shown in Figure 5.3a.

In terms of Fourier mathematics, the sketches of Figure 5.3 correspond to Fourier transformations: first, a Fourier transformation of the object field distribution yields the field distribution in the Fourier plane (see Equation (5.14)). Then an inverse Fourier transformation, *would* result exactly in $\mathscr{E}_{\mathrm{obj}}(x,y)$, i. e., this *would* be identical to the original

$$\mathscr{E}_{\mathrm{obj}}(x,y) = \mathrm{iFT}\left[\widetilde{\mathscr{E}}_{\mathrm{obj}}(k_{\mathrm{x}},k_{\mathrm{y}})\right]. \tag{5.16a}$$

However, in optics there is a slight difference, namely that the second transformation is a forward Fourier transformation as well. This is also the reason why images are seen as upside down in the image plane. If we disregard this reversed orientation as we will do in the following, because this is not of importance for the discussion, we also get back the original, namely the field and the brightness distribution within the image. Again, those are identical to that of the object

$$\mathscr{E}_{\mathrm{im}}(x,y) = \mathrm{FT}\left[\widetilde{\mathscr{E}}_{\mathrm{obj}}(k_{\mathrm{x}},k_{\mathrm{y}})\right] \tag{5.16b}$$

and $B_{\mathrm{im}}(x,y) = \lfloor \mathscr{E}_{\mathrm{im}}(x,y) \rfloor^2$.

From this, we may conclude that the 4-f-setup is representative of 1:1 imaging using a single lens or a more complicated optical system, namely an objective lens. It is also representative for imaging using curved mirrors. This is shown again in Figure 5.4a, which shows the imaging process within geometrical optics and in Figure 5.4b, which shows the equivalent for ideal imaging using the equivalent 4-f-setup. Note that for the moment we still assume that the diameter of the lens is infinite, and no aberrations are present; an extensive discussion of Fourier transformation by a lens, e. g., is given in [Goo17]. Examples for other magnifications are presented in Figure 5.4c and Figure 5.4d, respectively. But this will not be regarded further in the following because the subsequent discussion of 1:1 imaging can be straightly applied to other magnifications as well. Furthermore, in most cases the magnification is already included automatically (see the remark at the end of Appendix A.8).

Theoretically, one could consider a 4-f-system without any constraints at all. In this case, the setup would consist of lenses with infinite diameter and without any aberrations, even if this is not realistic. This is not only impossible due to the "huge" diameter of the lens, but also because its focal length f and its diameter D are strongly related to each other. In the discussed situation, the Fourier spectrum would consist of an infinite number of diffraction orders, i. e., one would have an ideal and complete spectrum.

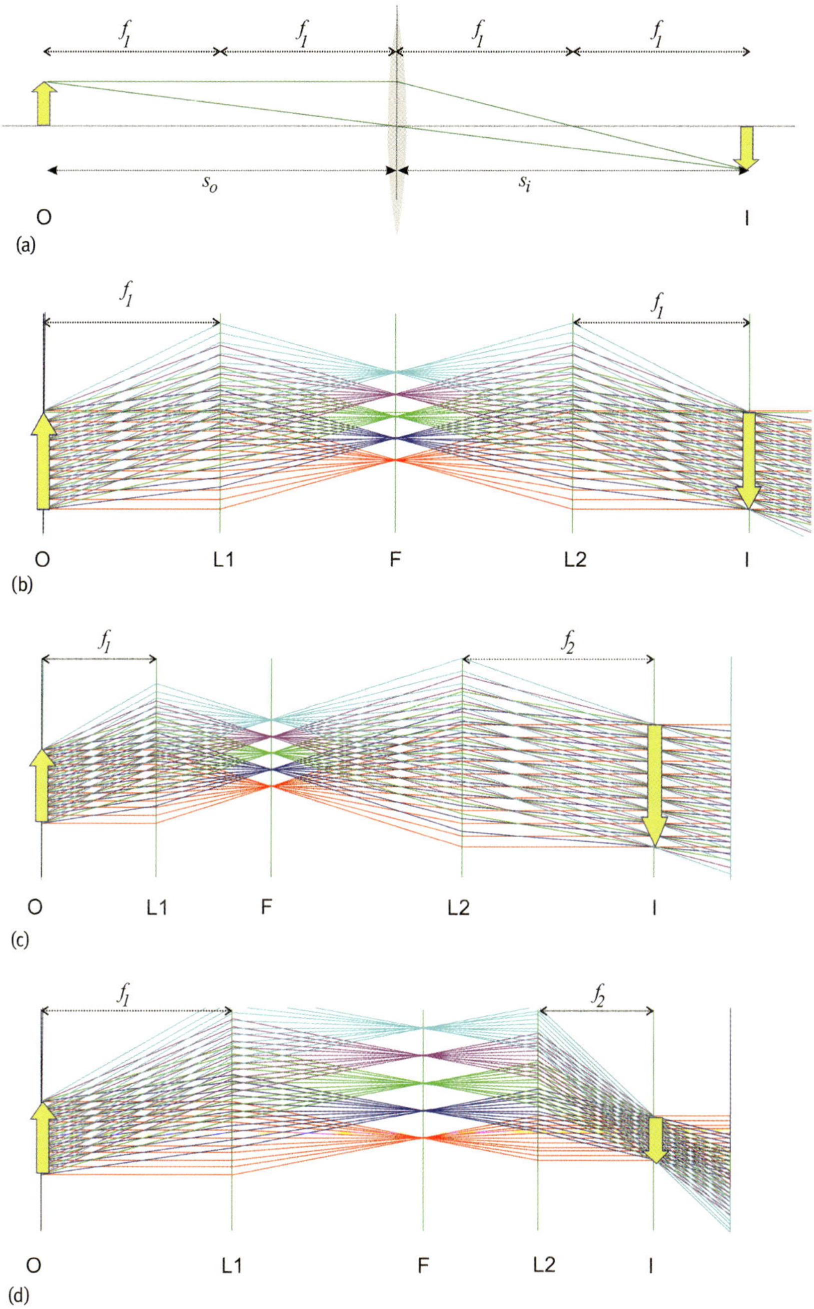

Fig. 5.4: Sketch of $M = 1{:}1$ imaging (a) and its equivalent (b). (c) and (d) show the equivalents to $M = 5{:}3$ and $M = 3{:}5$ imaging, respectively. In those equivalents shown in (b) and (c), respectively, the focal lengths of the two lenses differ. O denotes the object located in the object plane, I the image located in the image plane, F the Fourier plane and L1 and L2 the two lenses of the equivalent. The object and the image are shown as yellow arrows. In the real setup, the lens is located at the position of the Fourier plane of the representative.

Consequently, in the image plane, the image would be an ideal representation of the object.

However, in reality the spatial frequency spectrum deviates from the ideal one, and thus the image differs more or less from the ideal representation of the object. An extended discussion of this matter is the subject of textbooks on optics and/or Fourier optics, respectively. Nonetheless, we will make use of this model in the next section by having a look at the background of the theory of image formation. As a final remark of this section, one has to emphasize that the Fourier plane need not to be the focal plane of the lens performing the transform. Rather the Fourier transform always appears in the plane where the source is imaged (see [Goo17]).

5.1.4 Imaging and point spread function

5.1.4.1 Point spread function (PSF)

In the following, we will discuss the basic concept of optical imaging, which is based on Abbe's theory and Fourier optics. Although we will go in detail through it, it still may be somewhat confusing, and thus the following chapters are summarized in Appendix A.8.

Let us take a singular point or a very, very small spot as the object, the point source, and take an image of it with an ideal optical system that does not suffer from aberrations. Here, we restrict to illumination with nearly chromatic or quasimonochromatic light, namely a single wavelength only or a very narrow spectral range.

As discussed before, this system can be represented by a 4-f-system. Again, we restrict our discussion to 1D geometry, as an extension to 2D is straightforward. The object may be located at x_0 and its brightness may be characterized by a delta function in the ideal case (here and in the following the prefactor "const" is not of importance as only the field distribution is of interest)

$$\mathscr{E}_{\mathrm{obj}}(x) = \mathrm{const} \cdot \delta(x - x_0). \tag{5.17}$$

Again, here we do not take care on absolute values of the amplitude and appropriate proportional constants, and thus normalization is not necessary. For further simplicity, we assume $x_0 = 0$. Then, from Equation (5.15), one can calculate the distribution within the Fourier plane:

$$\widetilde{\mathscr{E}}_{\mathrm{obj}}(k_x) = \mathrm{FT}[\delta(x)] - 1. \tag{5.18}$$

In the case that the point is located somewhere else, i. e., $x_0 \neq 0$, the discussion would not change significantly; the result of the Fourier transformation would just show up as a phase function (see Appendix A.2). Thus Equation (5.18) shows that the Fourier spectrum of a point source imaged by an ideal optical system is just a constant, which of course leads to a homogeneous signal in the Fourier plane that extends to ± infinity.

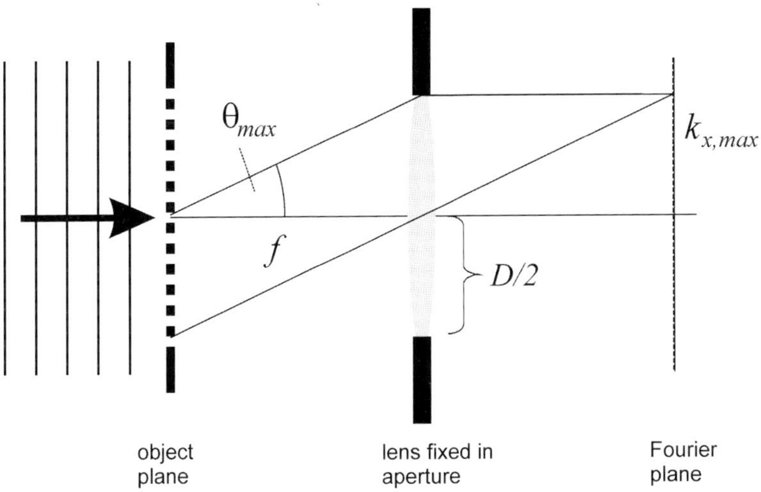

Fig. 5.5: Illustration of clipping of the spatial frequency range in the Fourier plane. Here again, this is shown for a 4-f-setup (similar to Figure 5.2a, here only the left part of the 4-f-system of Figure 5.2b is shown). The lens is mounted within an aperture that fully blocks the light outside its diameter D. Due to the finite size of D, diffraction angles are accepted only up to a maximum value θ_{max} (for simplicity, we write θ_{max} instead of $\theta_{x,max}$). This limits k_x to the range up to $k_{x,max}$ (and also $-k_{x,max}$, not shown here), larger values can be regarded to be clipped.

Now we compare this ideal system to a real optical system, which for simplicity, may be realized by a simple lens, which for the moment, should not introduce any aberrations. Such a "real system," of course, does not have an infinite, but a finite diameter D. For that reason, clipping occurs at a specific θ_{max} as illustrated in Figure 5.5. Consequently, this corresponds to a clipping in the Fourier plane as well. As a result, the range of the spatial frequencies k_x is limited, and the maximum and minimum possible value is $\pm k_{x,max}$. Of course, in addition this introduces losses, i. e., the amplitude is reduced, and thus we may expect an influence on the quality of the image as well.

For the calculation of the image, most simply this can be taken into account by multiplying the Fourier spectrum $\widetilde{\mathcal{E}}_{obj}(k_x)$ with a rectangle function, which is 1 between $+k_{x,max}$ and $-k_{x,max}$, and otherwise zero, i. e., $\mathrm{rect}(k_x/(2 \cdot k_{x,max}))$ (see the definition of $\mathrm{rect}(x)$ in Appendix A.1). From this, the field distribution within the image can be easily calculated

$$\mathcal{E}_{im}(x) = \mathrm{FT}\left[\widetilde{\mathcal{E}}_{obj}(k_x) \cdot \mathrm{rect}\left(\frac{k_x}{2k_{x,max}}\right)\right] = \mathrm{FT}\left[1 \cdot \mathrm{rect}\left(\frac{k_x}{2k_{x,max}}\right)\right] \tag{5.19}$$

where the "prefactor" "1" on the right-hand side results from the calculation of $\widetilde{\mathcal{E}}_{obj}$ in Equation (5.18). In case of $x_0 \neq 0$, one would have to write e^{ikx_0} instead of 1. From Equation (5.19), we get

$$\mathcal{E}_{\text{im}}(x) = 2k_{\text{x,max}} \frac{\sin(k_{\text{x,max}}x)}{k_{\text{x,max}}x} = 2k_{\text{x,max}} \cdot \operatorname{sinc}(k_{\text{x,max}}x), \tag{5.20a}$$

which is a *sinc*-function (see Appendix A.1). Of course, $k_{\text{x,max}}$ depends on D, and for a given lens consequently on its f-number because $k_{\text{x,max}}$ increases with D, and also the brightness of the image becomes larger if $f_\#$ is reduced. The exact relation of $k_{\text{x,max}}$ on $f_\#$ will be discussed in Section 5.2.4 (see Equation (5.61)), but here for the moment details are not important. In two dimensions, the discussion is similar.

From Equation (5.20) and Equation (5.5), the brightness distribution of the image as seen by a sensor can be calculated in 1D:

$$B_{\text{im}}(x) = |\operatorname{sinc}(k_{\text{x,max}} \cdot x)|^2. \tag{5.21a}$$

Again, here and in the following, we ignore the prefactor $2k_{\text{x,max}}$ in Equation (5.20a) because this is just a constant, which does not affect the light distribution itself. In the case of a circular aperture (2D), the radial distribution of the field is given by

$$\mathcal{E}_{\text{im}}(x) = \frac{J_1(k_{\text{x,max}} \cdot x)}{(k_{\text{x,max}} \cdot x)}. \tag{5.20b}$$

Here, x is the radial coordinate or the linear position within, e. g., a profile measured along a horizontal line. In similarity to the *sinc*-function in Equation (5.20a), this function is called the *jinc*-function. $J_1(x)$ is the Bessel function of the first order (see Appendix A.1). Now the image is given by the following function (Airy pattern) instead of the square of the sinc function (see standard textbooks of optics)

$$B_{\text{im}}(x) = \left|\frac{J_1(k_{\text{x,max}} \cdot x)}{(k_{\text{x,max}} \cdot x)}\right|^2. \tag{5.21b}$$

The function in Equation (5.20a) and Equation (5.20b), respectively, is called the point spread function for the field $\mathcal{PSF}$. Here, $\mathcal{PSF}$ is given for the described two cases of a 1D slit and 2D circular aperture, respectively, in presence of an aberration-free optics that is dominated by diffraction only. $\mathcal{PSF}$ is also called coherent point spread function, because a point source is spatially coherent per se.

The function in Equation (5.21a) and Equation (5.21b), respectively, is related to the intensity or brightness and termed as point spread function (PSF), in other words, the PSF is the image of a point source object. The point spread function is also called impulse response because it describes the response of the optical system to a point source object.

We would like to note that those point spread functions usually are normalized. Of course, due to energy conservation, the normalization factor can be calculated easily. Integration over the PSF must be identical to the amount of light emerging from the point

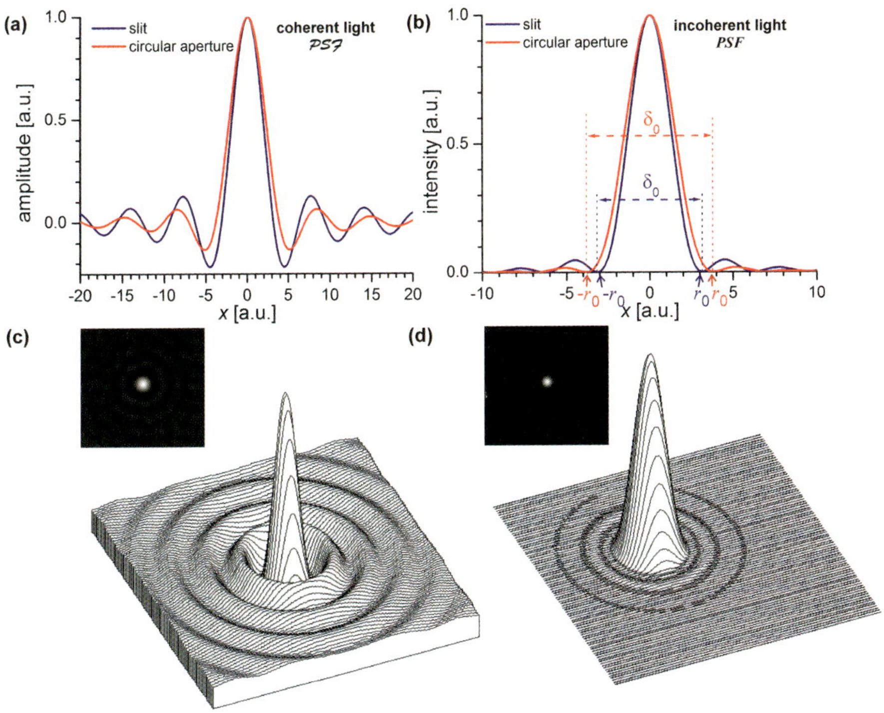

Fig. 5.6: (a) $\mathcal{PSF}$ and (b) PSF of monochromatic light for an ideal optical system that does not suffer from aberrations. Examples are provided for a slit aperture (i. e., 1D geometry, blue lines) and a circular aperture (1D profile measured along a line through the center of the 2D distribution displayed in (c) and (d), resp., red lines). The width of the PSF is given according to the notation in Figure 1.8. (c) and (d) show the according field distribution and intensity 2D distribution, respectively, of the circular aperture in the image plane.

source object corrected by the amount of angular acceptance and losses introduced by the optical system.

Figure 5.6 shows the point spread function for an ideal optical system. In a real optical system, aberrations are present in addition and this influences the shape, and thus also the "size" of the PSF and the $\mathcal{PSF}$. (see, e. g., Figure 5.10 and Appendix A.9). One has to note that Figure 5.6 displays the point spread function for monochromatic or quasimonochromatic light. One may define light as quasimonochromatic if its spectral width is much smaller than its wavelength, i. e., $\Delta\lambda \ll \lambda$. For polychromatic light, for each wavelength corresponding plots are superimposed independently. But according to Equation (5.24), the width of the PSF is different for each wavelength. Consequently, if the detector is not wavelength sensitive, the minima are smeared out, otherwise colored rings appear at different positions. This corresponds to a chromatic "error," but this has nothing to do with the chromatic aberration of the lens.

Although this discussion within this subchapter has been related to coherent light it can be regarded as general, because any light source that is a point source can be regarded automatically as fully coherent spatial coherence. We will make use of that in the following section.

5.1.4.2 Width of the point spread function and invariants

To deduce the width of the point spread function, it is convenient, e. g., to deduce the first zero points $x = \pm r_0$ of it. In principle, one can deduce also the FWHM or the $1/e^2$-width, but that does not change the principle of the following discussion. For the example of Equation (5.20a) or Equation (5.21a), respectively, the condition that the numerator becomes zero outside the central maximum is $k_{x,max} r_0 = \pm\pi$. With Equation (5.14), we get $k_{x,max} = k \cdot \sin(\theta_{max})$. For better readability, we write θ_{max} instead of $\theta_{x,max}$. Hence, together with the numerical aperture of the optical system NA, which is defined as (see Figure 5.1),

$$NA \equiv \sin(\theta_{max}) = \frac{k_{x,max}}{k} = \frac{R_{x,max}}{R} = R_{x,max}\lambda \tag{5.22}$$

one obtains $k \cdot NA \cdot x_0 = \pm\pi$. When we define $\delta_0 = r_0 + |-r_0| = 2r_0$, then with Equation (5.12) finally this results in

$$\delta_0 = 2\frac{\lambda}{2 \cdot NA} = \frac{\lambda}{NA} \tag{5.23a}$$

for a slit. The corresponding result for a circular aperture is

$$\delta_0 = 1.22\frac{\lambda}{NA} \tag{5.23b}$$

and in general

$$\delta_0 = 2\kappa \cdot \lambda \frac{1}{2 \cdot NA} \cdot \alpha \tag{5.24}$$

with a constant κ that describes the geometry (see Table 5.1) and a constant α, which is a measure of wavefront distortions introduced by the optical system. If such distortions are absent, $\alpha = 1$, otherwise α is larger than one. This is equivalent to Equation (1.17).

Equation (5.24) is equivalent to

$$\delta_0 \cdot \sin(\theta_{max}) = \lambda, \tag{5.25}$$

which may be compared to the well-known diffraction formula (see the textbooks):

$$D \cdot \sin(\theta_0) = \lambda. \tag{5.26}$$

Tab. 5.1: Factor 2κ for the calculation of the diameter of the PSF or the focus for different functions $T(x)$ where x is the lateral (in 1D) or radial (in 2D) coordinate, respectively. The notation of the subscripts of δ is according to Figure 1.8. Note κ does not depend on geometry only, but, of course, depends also on where the width is measured (e. g., between the "first zeros" or at FWHM, etc.). The 2κ-values provided for the Gaussian correspond to δ and D, both measured as FWHM or both measured as $1/e^2$ width, respectively (see also, e. g., Equations (5.26), (5.43), (5.44)).

geometry, i. e., $T(x)$	**2κ for δ_0**	**2κ for δ_{FWHM}**	**2κ for $\delta_{1/e2}$**
"box" (rect, 1D geometry)	2	0.89	1.4
circular aperture (2D geometry)	2.44	1	1.65
Gaussian (1D geometry)	–	0.44	$4/\pi = 1.27$

Here, θ_0 is the diffraction angle for first-order diffraction at a slit with a width D. This points out that there is a general relation that states that the product of the relevant aperture D_{diffr} times the (sine of the) corresponding diffraction angle θ_{diffr} for a given optical system is an invariant; in other words, it is a constant. This very important relation may be written as

$$D_{diffr} \cdot \sin(\theta_{diffr}) = 2 \cdot \kappa \cdot \lambda \cdot \alpha = \text{const} \tag{5.27}$$

This product can be propagated through an optical system and at best stays constant if no further wavefront distortions are involved. With poor optics, this product may increase during light propagation through the system. As an example, an ideal beam expander would increase the beam width from $D_{diffr,1}$ to $D_{diffr,2}$ and at the same time the "intrinsic" beam divergence $\theta_{diffr,2}$ will be reduced: $D_{diffr,1} \cdot \sin(\theta_{diffr,1}) = D_{diffr,2} \cdot \sin(\theta_{diffr,2})$.

Consequently, this relation relates also the near field with the given aperture of the optical system to the far field (see Equation (5.26) and Equation (5.25)). Here, we would like to point out that Equation (5.26) and Equation (5.27) are also equivalent to the beam parameter product known from laser physics, to Abbe's sine condition and also to the SBN (see Chapter 1) and SBP (Section 5.1.8). The 2D equivalent to $D{\cdot}\sin(\theta_{diffr})$, or to $D{\cdot}\theta_{diffr}$, because usually the "intrinsic" beam divergence is small, is the product $A \cdot \Delta\Omega$, which is called the etendue. The etendue plays an important role, e. g., in lithography. Here, A is the cross-sectional area of the beam and $\Delta\Omega$ the solid angle of intrinsic divergence. The etendue is constant, which is the 2D equivalent to Equation (5.27).

In contrast to the discussion before, usually an optical system such as a camera is affected by aberrations. These aberrations are not necessarily simple ones such as a spherical aberration or coma, but often a rather complex mixture. Indeed, the so-called third-order aberration, i. e., Seidel's aberrations, are not independent from each other. As a result, the PSF may also be more complicated. This is illustrated in Figure 5.7, which shows the image of a single point object (or point-like object, such as a star) taken with different camera lenses. Other examples are shown in Appendix A.9.

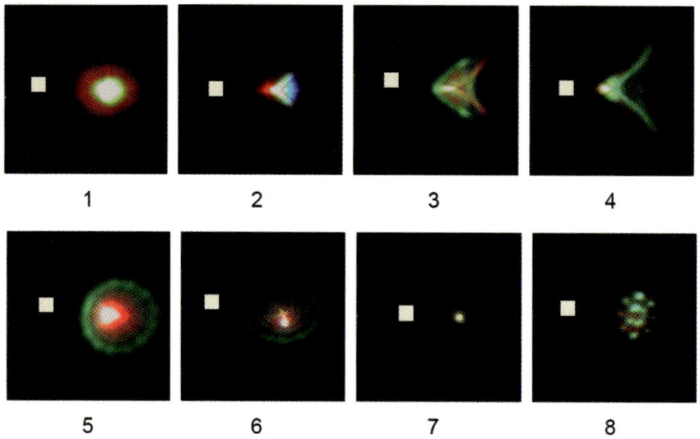

Fig. 5.7: Examples of the PSF of camera lenses. The images are observed and captured with a high-quality microscope located in the image plane. Images 1 to 6 are typical for moderate image quality, which is obtained by typical lenses with large f-numbers at full aperture, wide-angle lenses off from the image center, or if the image is slightly defocused. A small white square has been pasted onto each image for a size comparison. It represents an 8.5 µm pixel like those of a 12 MP, 35 mm full format camera. All these point spreads are thus considerably larger than this (relatively large) pixel area. The PSF in image no. 7 is an example of outstanding imaging performance, although it should be noted that a digital sensor generally does not see such small PSF. Image no. 8 shows the same PSF behind a low pass filter (OLPF, s. Section 4.6.2). The image quality is therefore artificially deteriorated as the OLPF increases the PSF considerably. These images and the related information and text are taken from [Nas08].

Taking an image of a point-like object is not the typical situation in photography although this is typical for astrophotography, but this provides direct access to the PSF, which cannot be seen in nearly all other images, because the PSF of the "image points" all overlap (see above).

5.1.5 Optical transfer function, modulation transfer function and phase transfer function

5.1.5.1 Convolution and optical transfer function OTF

As discussed at the end of Section 5.1.2, a real object may be considered as a slide acting as a diffraction structure that is illuminated from behind. This object may be considered to consist of an infinite number of object points that may be regarded as mathematical points or infinitesimal area elements.

The individual object points may be identified by their coordinates (x,y) and for a real nontrivial object, they differ in the amount of light $A(x,y)$ emerging from them. Although $A(x,y)$ differs for the different points, all of them have the same PSF in the sense that the shape of the distribution is the same. This is because the PSF is related to the optics, not to the object.

The optical system then transfers each of the object points to the corresponding spot within the image plane and the overlapping infinite number of image spots yields the image. The absolute light distribution of an individual spot is given by the PSF multiplied by its corresponding $A(x,y)$. For simplicity, for the moment we do not discriminate between PSF for the intensity and $\mathcal{PSF}$ for the field. This will be done correctly later. Within this subchapter, we would like to indicate this product by PSF′.

PSF itself depends on the properties of the optical system. Again, if aberrations can be neglected, in a standard optical system where the effective aperture is circular, the PSF is given by Equation (5.21b). Similarly, this holds for $\mathcal{PSF}$; however, this is straightforward from the discussion related to PSF.

To get the overall light distribution within the image plane, this procedure has to be done with all the infinitive number of object points, which is equivalent to an integration over the infinitive number of image spots, namely all the PSF′. Mathematically, this corresponds to a convolution of $A(x,y)$ with PSF: $A(x,y) \otimes \mathrm{PSF}(x,y)$ yields the distribution within the image (Appendix A.3 explains this convolution process in more detail).

Here, we may recall Section 5.1.1, that more strictly the coordinates in the object plane and image plane, respectively, differ, and thus should be indicated by different symbols. However, it is clear that from the distributions in the corresponding planes that they can be clearly discriminated, so that we use the same symbols for them in both planes.

Now we have to discriminate between coherent light and incoherent light. For fully coherent light, the emission of each individual point may be additionally characterized by its corresponding phase, which may lead to interference effects as well. Consequently, the phase has to also be included in the light distribution of the object, and thus $A(x,y)$ is given by $\mathcal{E}_{\mathrm{obj}}(x,y)$. If the phase is the same for all object points, then, of course, it is possible to restrict to the amplitude term of $\mathcal{E}_{\mathrm{obj}}(x,y)$.

Then integration over the image spots is equivalent to the convolution of $\mathcal{E}_{\mathrm{obj}}(x,y)$ with $\mathcal{PSF}$. In other words,

$$\mathcal{E}_{\mathrm{im}}(x,y) = \mathcal{E}_{\mathrm{obj}}(x,y) \otimes \mathcal{PSF}(x,y). \tag{5.28}$$

To simplify discussion, in the following again we mostly restrict to 1D geometry. If we assume an optical system that is dominated by diffraction only, here represented by a slit aperture with a width D, $\mathcal{PSF}(x) = \mathcal{E}_{\mathrm{im}}(x)$ given by Equation (5.20a). In 2D geometry with a circular aperture $\mathcal{PSF}(x,y)$ is given by Equation (5.20b). If aberrations are present, the appropriate $\mathcal{PSF}$ has to be taken. Examples are given in Figure 5.10 and Appendix A.9.

To make it more clear, we would like to remark that in contrast to before, here $\mathcal{E}_{\mathrm{obj}}$ is given by “many object points” whereas in Section 5.1.4 we discussed a single object point only. This means that *if* we would restrict to a single object point here as well, *then* $\mathcal{E}_{\mathrm{obj}}(x) = \delta(x)$ and as a result from the convolution, *then* we would simply obtain that $\mathcal{E}_{\mathrm{im}}(x)$ is identical to $\mathcal{PSF}(x)$ as it is by definition.

Calculation of the convolution in Equation (5.28) in a general case is possible, but this may be a difficult task. More easily, the calculation can be performed by application of the convolution theorem (see Table A.1 in Appendix A.2):

$$\widetilde{\mathcal{E}}_{\mathrm{im}}(k_x) = \mathrm{FT}\{\mathcal{E}_{\mathrm{obj}}(x) \otimes \mathcal{PSF}(x)\} = \mathrm{FT}\{\mathcal{E}_{\mathrm{obj}}(x)\} \cdot \mathrm{FT}\{\mathcal{PSF}(x)\} \tag{5.29}$$

where $\widetilde{\mathcal{E}}_{\mathrm{im}}(k_x)$ is the Fourier transformation of $\mathcal{E}_{\mathrm{im}}(x)$. Remember also that we restrict the discussion to shapes and structures only, and thus always omit factors such as 2π. The electric field distribution within the image plane then is obtained from the (inverse) Fourier transformation (remember also the remark in Section 5.1.3):

$$\mathcal{E}_{\mathrm{im}}(x) = \mathrm{FT}[\mathrm{FT}\{\mathcal{E}_{\mathrm{obj}}(x)\} \cdot \mathrm{FT}\{\mathcal{PSF}(x)\}] \tag{5.30}$$

The Fourier transformation of the $\mathcal{PSF}$ defines the transfer function of the field, which we define as

$$\mathcal{OTF}(k_x) = \mathrm{FT}\{\mathcal{PSF}(x)\}. \tag{5.31}$$

Thus, Equation (5.30) can be rewritten as

$$\mathcal{E}_{\mathrm{im}}(x) = \mathrm{FT}[\widetilde{\mathcal{E}}_{\mathrm{obj}}(k_x) \cdot \mathcal{OTF}(k_x)]. \tag{5.32}$$

Equation (5.31) yields the relevant transfer function for coherent light, which is closely related to OTF for incoherent light (see below). Sometimes this is called the coherent transfer function. $\mathcal{OTF}$ is related to the field and OTF to the intensity. We would also like to remark that $\mathcal{OTF}$ is sometimes termed amplitude transfer function, in particular, when phase effects are not an issue. Equation (5.30) and Equation (5.32) follow directly from rigorous diffraction theory (see, e. g., the textbook [Bor99]).

Our example with the 1D optical system can be described by an ideal cylindrical lens with a slit aperture. When we assume that this system is dominated by diffraction, $\mathcal{PSF}$ is given by Equation (5.20a) and the brightness distribution within the image plane then is obtained from Equation (5.32) by $B_{\mathrm{im}}(x) = |\mathcal{E}_{\mathrm{im}}(x)|^2$. Then the corresponding $\mathcal{OTF}$ is given by the Fourier transformation of the *sinc*-function in Equation (5.20a), which, of course, is a rectangle function, namely

$$\mathcal{OTF}(k_x) = \mathrm{rect}\left(\frac{k_x}{2k_{x,\max}}\right). \tag{5.33}$$

The definition of rect(x) yields 1 between $x = -1/2$ and $x = +1/2$ (see Appendix A.1). Consequently, here, for coherent light, the cut-off is at $\pm k_{x,\max}$. This can be seen from Figure 5.8a. For a 2D square aperture, the maximum transferable spatial frequency is given by $\pm k_{x,\max}$ and $\pm k_{y,\max}$, respectively.

For fully incoherent light, the phase of the individual object point is not an issue. Briefly and somehow simplified, one might argue that each individual point corresponds

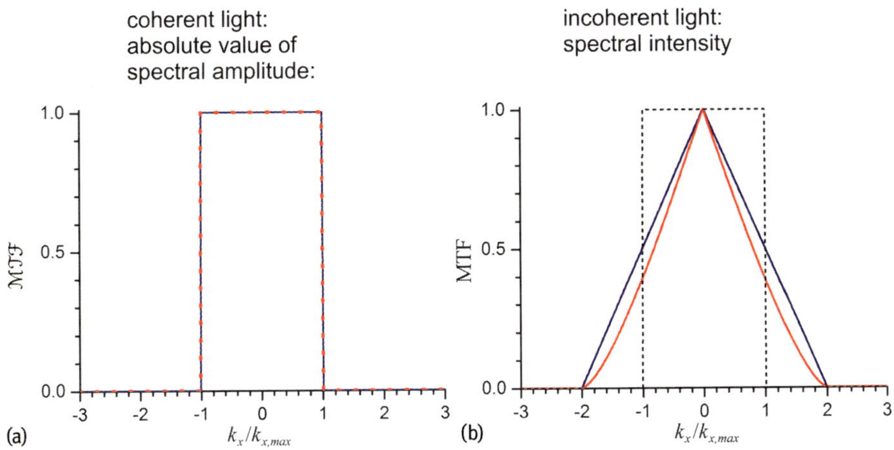

Fig. 5.8: (a) Coherent light: normalized $\mathcal{OTF}$ for the $\mathcal{PSF}$ in Figure 5.6a. (b) Incoherent light: normalized OTF for the PSF in Figure 5.6b. Here, we do not discriminate between $\mathcal{OTF}$ and $\mathcal{MTF} = |\mathcal{OTF}|$, and OTF and MTF $=$ |OTF|, respectively. The blue and red curves, respectively, correspond to the curves with the same color in Figure 5.6 (blue: 1D slit aperture or 1D profile measured along the horizontal line of the 2D MTF for a 2D square aperture; red: profile measured along a line through the center of the 2D MTF for a circular aperture or circular lens in circular aperture). For comparison, the dashed black line is again the intensity distribution for coherent light (identical to distribution in (a)). With respect to the $k_x/k_{x,max}$ value when the OTF becomes zero, we refer to the remark in the text. (a), (b) show the line profiles in k_x-direction at the position $k_y = 0$. For partially coherent light, the curves are located in between the solid and dashed curve in (b), respectively. A plot and discussion can be found, e. g., in the textbook [Smi08].

to an ensemble of point emitters that differ in phase only within the ensemble. Each of the individual points of the ensemble may be treated as discussed in Section 5.1.4; however, for all of them together, i. e., each ensemble that forms one object point with the corresponding $A(x, y)$, the phase is a "mixture," and consequently no phase effects occur between the different ensembles, namely the different object points. As a result, $A(x, y)$ is given by $B_{obj}(x, y)$. More easily, we may argue that it is common knowledge that for coherent light the light field has to be considered and for incoherent light the intensity or brightness. Hence, for incoherent light Equations (5.29) to (5.31) change to

$$\widetilde{B}_{im}(k_x) = \mathrm{FT}\{B_{obj}(x) \otimes \mathrm{PSF}(x)\} = \mathrm{FT}\{B_{obj}(x)\} \cdot \mathrm{FT}\{\mathrm{PSF}(x)\} \tag{5.34}$$

$$B_{im}(x) = \mathrm{FT}[\mathrm{FT}\{B_{obj}(x)\} \cdot \mathrm{FT}\{\mathrm{PSF}(x)\}] \tag{5.35}$$

$$B_{im}(x) = \mathrm{FT}[\widetilde{B}_{obj}(k_x) \cdot \mathrm{OTF}(k_x)] \tag{5.36}$$

with the optical transfer function

$$\mathrm{OTF}(k_x) = \mathrm{FT}\{\mathrm{PSF}(x)\}. \tag{5.37}$$

5.1.5.2 OTF of a cylindrical and a spherical lens

We have to note that PSF $\neq \mathcal{PSF}$ and OTF $\neq \mathcal{OTF}$ but we would like to remark that OTF is the normalized autocorrelation function of $\mathcal{OTF}$. For PSF, the inequality can be easily seen from our example with the 1D aberration-free diffraction dominated optical system (slit aperture): PSF is given by Equation (5.21a) and $\mathcal{PSF}$ by Equation (5.20a). Consequently, $\mathrm{OTF}(k_x) = \mathrm{FT}[|\mathrm{sinc}(k_{x,\max} \cdot x)|^2]$. Although not of importance in the following, we may comment that the right-hand sight of that equation is equal to the convolution of two rectangle functions. From this or more easily from a direct calculation of the Fourier transformation, we obtain (see also Appendix A.1 and Appendix A.2)

$$\mathrm{OTF}(k_x) = \mathrm{triang}\left(\frac{|k_x|}{2k_{x,\max}}\right) = \begin{cases} 1 - \dfrac{|k_x|}{2k_{x,\max}} & \text{for } |k_x| < 2k_{x,\max} \\ 0 & \text{otherwise} \end{cases} . \tag{5.38}$$

Obviously, this is different from $\mathrm{FT}[\mathrm{sinc}(k_{x,\max} \cdot x)] = \mathrm{rect}(k_x/(2 \cdot k_{x,\max}))$. Equation (5.39) is the OTF of a cylindrical lens without additional aberrations. The exact relation of $k_{x,\max}$ on $f_\#$ will be discussed in Section 5.1.6.

Note, in contrast to Equation (5.33), due to the definition of $\mathrm{rect}(x)$ and $\mathrm{triang}(x)$, for incoherent light, the cut-off is at $\pm 2 \cdot k_{x,\max}$, which means that the spectrum is twice as large as that for coherent light. But this is reasonable, because if the same lens with same $f_\#$ and all the other conditions identical, except the degree of coherence, the amount of light transferred through the system must be the same. In other words, the energy transmitted through the system is independent of the degree of coherence of the light. As usual, the energy can be easily calculated by integration of $\mathcal{OTF}$ and OTF, respectively, over all spatial frequencies from $-\infty$ to $+\infty$, i. e., the integration over the rectangle and triangle functions in Figure 5.8. For a spherical lens (see below), energy conservation is fulfilled in the same way, but this includes 2D integration (!) over the cylinder and the function displayed in Figure 5.9b, respectively. Of course, energy must not be conserved, if different lenses, different $f_\#$, etc. are used and compared.

For 2D apertures, in general Fourier transformations have to be 2D as well and cannot be restricted to the 1D line profiles. Only for special cases such as a 2D rectangular aperture, x and y can be separately considered so that a 1D Fourier transformation works well. But for a circular aperture, this is not the case. Figure 5.9 shows two examples for incoherent light. Nevertheless, even for a 2D optical system with a circular aperture it is possible to get an analytic solution for a 1D line profile through the center of the OTF where $k_x = k_y = 0$. The OTF can be regarded as the convolution of two "circle functions," see, e. g., [Goo17], which yields

$$\mathrm{OTF}(k_r) = \frac{2}{\pi}\left(\arccos\left(\frac{k_r}{2k_{r,\max}}\right) - \frac{k_r}{2k_{r,\max}}\sqrt{1 - \left(\frac{k_r}{2k_{r,\max}}\right)^2}\right) \quad \text{for} \quad |k_r| < 2k_{r,\max}. \tag{5.39}$$

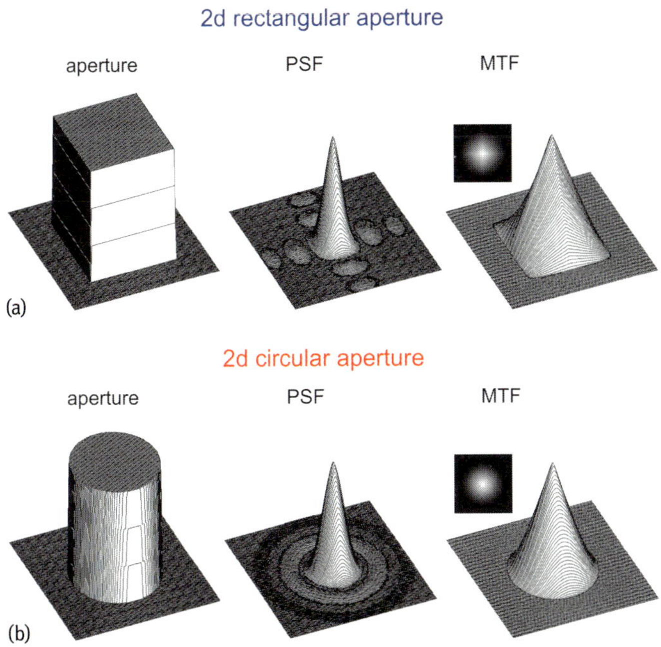

Fig. 5.9: (a) 3D plots for a 2D square aperture illuminated by incoherent light: aperture function (left), PSF (incoherent light, middle), MTF (right). (b) same for a 2D circular aperture. For the values when the functions become zero, we refer to Figure 5.8 and the remark in the text.

We would like to remind that this is not obtained from a 1D Fourier transformation of Equation (5.21b), but from a 2D Fourier transformation of the corresponding 2D distribution, which can be either performed in Cartesian coordinates or more easily in polar coordinates:

$$\begin{aligned} \mathrm{OTF}(k_x, k_y) &= \mathrm{FT}_{2D}\{B_{im}(x,y)\} = \mathrm{FT}_{2D}\{B_{im}(r(x,y), \phi(x,y))\} \\ \mathrm{OTF}(k_r) &= \mathrm{FT}_{2D}\{B_{im}(r)\} = \mathrm{FT}_{2D}\left\{\left|\frac{J_1(k_{r,max}\cdot r)}{(k_{r,max}\cdot r)}\right|^2\right\}. \end{aligned} \tag{5.40}$$

Of course, in the final result k_r may be identified with k_x and k_y when the OTF or MTF is displayed in the horizontal or vertical direction, respectively. This is displayed as the red curve in Figure 5.8b. Alternatively, one can make use of the line spread function LSF and its 1D Fourier transformation, which is discussed in Appendix A.8.

5.1.5.3 Cut-off frequency

We would like to draw to attention that there are different specific definitions of the cut-off frequency k_{cutoff} and $R_{cutoff} = k_{cutoff}/(2\pi)$. Here, we define $k_{x,max}$ and $R_{x,max}$, respectively, according to Section 5.1.6. We would like to note that this definition is also used by

[Goo17], but in some textbooks such as that of [Ped08], $k_{x,max}$ is set equal to k_{cutoff}, i. e., the functions $\mathrm{OTF}(k_x/k_{x,max})$ in Equation (5.38) and in Equation (5.39) become zero for $k_x/k_{x,max} = 1$, whereas within the present book for incoherent light $k_{cutoff} = 2 \cdot k_{x,max}$, and thus zero position is at $k_x/k_{x,max} = 2$. But in any case, k_{cutoff} corresponds to the Nyquist frequency given by the optics. There might be other cut-off frequencies as well, particularly that originating from the sensor (see Chapter 1).

As can be seen from Figure 5.8b, the maximum transferable spatial frequency for incoherent light is twice that for the coherent case, namely $2 \cdot k_{x,max}$. Hence, from the extension of k_x to larger values one could assume that this implies also a better resolution for incoherent light when compared to coherent light under the assumption of the same optics used in both cases. However, this is not so simple as we will see below. A more extended discussion can be found, e. g., in the textbook [Goo17]. In addition, we would like to remark that the relevant range of spectral frequencies is not related to the same physical value, namely the spectral field amplitude and spectral intensity, respectively, and thus of course, simple statements on image quality are not straightforward.

5.1.5.4 OTF, MTF, PTF

One can summarize that for usual imaging the Fourier transformation of the brightness distribution within the image plane is given by the product of the Fourier transformation of the brightness distribution within the object plane and the OTF for incoherent light (see Equation (5.36)). Thus,

$$\mathrm{OTF}(k_x) = \frac{\widetilde{B}_{\mathrm{im}}(k_x)}{\widetilde{B}_{\mathrm{obj}}(k_x)}, \tag{5.41}$$

which explains also the name of this transfer function. The OTF usually is a 2D complex function that can be written in terms of amplitude and phase where its amplitude is termed modulation transfer function $\mathrm{MTF} = |\mathrm{OTF}|$ and its phase, phase transfer function $\mathrm{PTF} = \arg(\mathrm{OTF})$:

$$\mathrm{OTF}(k_x) = \mathrm{MTF}(k_x) \cdot \exp(i \cdot \mathrm{PTF}(k_x)). \tag{5.42}$$

The relation for $\mathcal{OTF}$, $\mathcal{MTF}$ and $\mathcal{PTF}$ is similar. We would like to comment that the OTF or $\mathcal{OTF}$ for any optical system, now potentially including aberrations, has to be chosen in such a way that it is appropriate to the specific situation, a multiplication in the Fourier space with $B_{\mathrm{obj}}(x,y)$ or $\mathscr{E}_{\mathrm{obj}}(x,y)$, respectively, must yield the PSF or, respectively, $\mathcal{PSF}$ or the observable $|\mathcal{PSF}|^2$ in the image plane (see also Appendix A.8).

The complex OTF characterizes an optical system and has a spatial and directional dependence. A good optical system transfers a large range of spatial frequencies, and in particular this includes high frequencies that are of much relevance for fine details within the image. Then one says that the MTF is of high quality. The PTF describes a shift

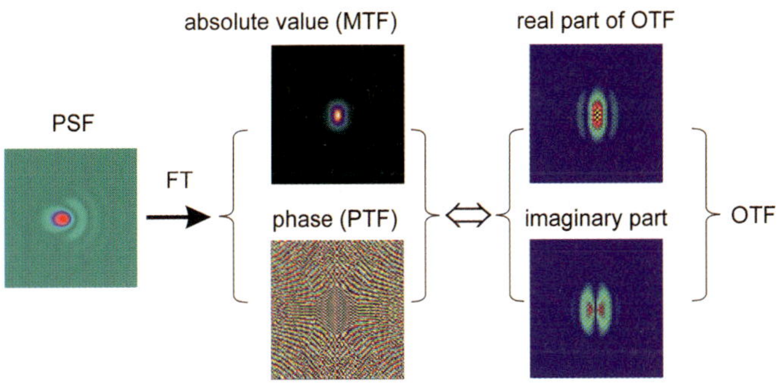

Fig. 5.10: Calculated PSF of an optical system that exhibits coma (illumination with incoherent light) and the corresponding OTF. Calculation is made for monochromatic light. For better visibility of the faint structures, all plots are displayed in false colors. The real part and the imaginary part of the OTF contain positive and negative values. The blue background is the zero signal. The absolute value of the OTF is the MTF and its phase the PTF. 3D plots and 1D line profiles of this example are presented in Appendix A.9. The phase oscillates between $-\pi$ and $+\pi$ and has been "backfolded" when its value exceeds that range.

(of a part) of the image with respect to the object when compared to a perfect imaging system.

Appendix A.2 summarizes the Fourier optics relations. Figure 5.10 shows an example of the OTF of an image from an optical system that suffers from aberrations. Other examples are provided in Appendix A.9.

Usually, an optical system such as a camera may have a more complicated OTF or MTF than that discussed above (see the examples in Figure 5.7). This is the issue of Section 5.2 and Section 8.3. But to conclude this chapter, we would also like to remark that in the most general case, the light distribution in the object plane, namely $B_{\mathrm{obj}}(x,y)$ is also quite complicated, and consequently its Fourier transformation $\widetilde{B}_{\mathrm{obj}}(k_{\mathrm{x}}, k_{\mathrm{y}})$, namely its diffraction pattern, is as well. As an example, Figure 5.11 shows an image and its spatial power spectrum, which is the product $\widetilde{B}_{\mathrm{obj}}(k_{\mathrm{x}}, k_{\mathrm{y}}) \cdot \mathrm{MTF}(k_{\mathrm{x}}, k_{\mathrm{y}})$, in other words it is the diffraction pattern of the object modulated by the MTF. For a good optical system, the MTF could be a rather smooth function (see the examples later in Chapter 5), and thus at least the central part of the spatial power spectrum may well represent the diffraction pattern that would have been observed if the object were a slide with infinitesimal resolution.

For color images, the discussion is the same, but for each color separately. This then leads to a linear independent superposition of the diffraction patterns or Fourier spectra of all involved colors. This is a consequence of the linearity of Maxwell's equations in linear optics.

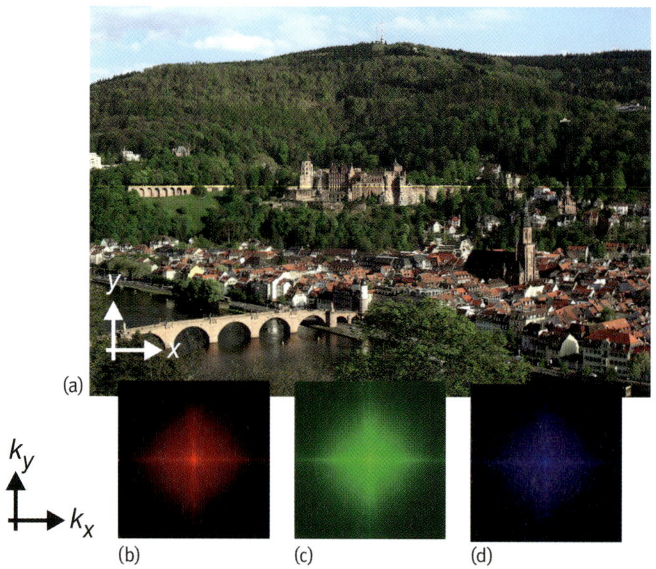

Fig. 5.11: Image $B_{im}(x, y)$ (a) and its spatial frequency spectrum $\tilde{B}_{im}(k_x, k_y) = \tilde{B}_{obj}(k_x, k_y) \cdot \mathrm{MTF}(k_x, k_y)$ (here shown as the "power spectra" for the three different color channels) in k_x and k_y-direction, respectively, (b) to (d). In (a), the abscissa provides the x-coordinate and the ordinate the y-coordinate. In (b) to (d), the abscissa provides the spatial frequency k_x and the ordinate the spatial frequency k_y.

5.1.6 Resolution, maximum frequency and contrast

5.1.6.1 Maximum frequency

As discussed in the previous sections, a real optical system cannot transfer all spatial frequencies up to infinity, but only up to a certain value $\pm k_{x,max}$. This leads to a finite size of the point-spread function as discussed in Section 5.1.4 and consequently to a limited resolution.

From Figure 5.12, one can see the typical situation of imaging. The image plane is located at a distance s_i behind the lens. Within the image plane, the first minimum of the PSF is located at a distance $\pm r_0 = \pm\delta_0/2$. Usually, in standard photography, the object distance s_o is much larger than s_i and, therefore, $s_i \approx f$ as deduced from the lens equation. Also, usually the diffraction angle for the first zero, i. e., θ_0, is rather small because the entrance pupil (aperture) $D \equiv D_{en} \gg \lambda$ (see Section 3.4.3). Hence, $\sin(\theta_0)$, $\tan(\theta_0)$ and θ_0 are almost identical. Thus, together with $\theta_0 \approx 0.5 \cdot \delta_0/s_i$ (see Figure 5.12) and Equation (5.26), or more generally Equation (5.27), this yields the width between the first zero positions or the diameter of the first dark ring:

$$\delta_0 = 2\kappa \cdot \lambda \frac{f}{D} \cdot \alpha. \tag{5.43}$$

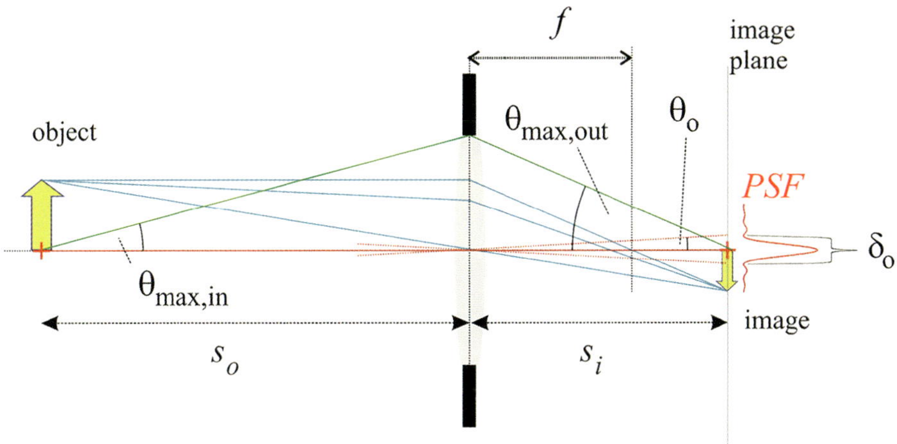

Fig. 5.12: Typical situation in photography. Note usually $s_o \gg s_i \approx f$, but a plot with such dimensions would be difficult to read. The large yellow arrow, the object, is imaged. Here, we disregard aberrations. We consider a singular point within this object, marked by a red cross on the optical axis. Within the image this point yields a spot provided as the PSF, its center is also marked by a red cross. The first zero positions are provided by the diffraction angle θ_0 and the distance in between $+r_0$ and $-r_0$ determines δ_0.

Comparison with Equation (5.24) and using the definition for $f_\#$ yields (see also Section 3.4.3)

$$f_\# \equiv \frac{f}{D} = \frac{1}{2NA} \tag{5.44}$$

where $NA \equiv NA_i = \sin(\theta_{ex})$ is related to the exit pupil (aperture; see Equation (3.95)).

Before we continue, we would briefly refer Section 3.4.3 and give two related comments. First, wavefronts usually are only flat in near or far field, respectively. Principal planes, the entrance and exit pupils as well as the cardinal surfaces are not planes perpendicular to the optical axis but rather curved surfaces (see Section 3.4 and [Ber30, Bla14]). This often leads to confusion and, in particular, to the mistaken application of the tangens function instead of the sine function [Bla14]. Second, we would like to remark that Equation (5.44) is only valid for $s_0 \to \infty$ or at least large s_0. For finite values of s_0, this is an approximation only and Equation (3.98) has to be used instead. In particular, this is the case for macro photography. Third, we note that Equation (5.44) is exactly valid for aplanatic optics (sinc-condition fulfilled). For ideal focusing optics, such as exactly aligned parabolic mirrors or hyperbolic lenses, this is approximately valid only (because $NA = \sin(\theta_{out})$ and $\tan(\theta_{out}) = 0.5 \cdot D/f$).

Now, for a given aperture D, $f_\#$ or NA, respectively, $k_{x,max} = k \cdot \sin(\theta_{x,max}) = k \cdot NA$ and $R_{x,max} = k_{x,max}/(2\pi)$ can be calculated (see also Equation (5.11)).

According to the geometry displayed in Figure 5.12 with $\sin(\theta_{x,max}) = (D/2)/s_i$ and the assumption that the object is far enough away so that we can approximate the image distance by the focal length, we get

$$k_{x,max} = \frac{k}{2f_{\#}} = \frac{\pi}{\lambda f_{\#}} \quad (5.45)$$

$$R_{x,max} = \frac{R}{2f_{\#}} = \frac{1}{2\lambda f_{\#}}. \quad (5.46)$$

From the previous relations, it is straightforward to express

$$\delta_0 = \frac{\kappa \cdot \alpha}{R_{x,max}}. \quad (5.47)$$

Consequently, for $\kappa \cdot \alpha = 1$, $\delta_0 = 1/(R_{cutoff}/2)$, which is equivalent to the statement that the resolution is given by half of the Nyquist frequency (see Section 1.6), here with respect to the optics only (for sensors see later). This is the limit where the MTF becomes zero. However, instead we may consider the spatial frequency where the MTF equals 10 % as the limit, then resolution is reduced, e. g., to 90 % of that value, namely from $k_x/k_{x,max} = 2$ to $k_x/k_{x,max} = 1.8$ (see the blue line in Figure 5.8b).

Equation (5.48) may also be used for discussion in presence of aberrations. Here, for simplicity, aberrations may be expressed by the factor α, which increases with the amount of wavefront distortion (see Section 5.1.4). A more detailed discussion is made in Section 5.2.3. This leads to a decrease of $R_{x,max}$ to a spatial frequency $R_{x,max,distort} = R_{x,max}/\alpha$. As a result, $\delta_0 = \kappa/R_{x,max,disort}$ is increased.

All those relations do not provide additional physics; they just show how the different quantities can be related to each other in different ways. All that is straightforward.

5.1.6.2 Resolution and contrast

For the determination of the resolution, we may simply regard two object points and their image spots given by the PSF in the image plane. A more extended discussion is given in Section 5.1.7. As long as one can clearly discriminate both points within the image, they are well resolved (see, e. g., Figure 1.9). According to Rayleigh's criterion, they are regarded as just resolved, when the maximum of the PSF of one of these points is located at the first minimum of the other one as is illustrated in Figure 5.13. In this case, the superposition of both PSF describing the observable brightness clearly shows a dip. Thus, the resolution is $\delta_0/2$, which according to Equation (5.23b) is given by

$$\mathcal{R} = \frac{\delta_0}{2} = \frac{0.61 \cdot \lambda}{NA}. \quad (5.48)$$

We may comment that the graph displayed in Figure 5.13 is somewhat artificial because it just shows a profile along a single line of a 2D distribution. A good judgment of resolution has to be made on the full 2D distribution as displayed, e. g., in Figure 1.9 or, e. g., an integration of all horizontal line profiles in the vertical direction.

In the case of a 1D distribution, such as obtained from thin vertical line objects, profiles along a single horizontal line through the PSF are sufficient (see, e. g., Appendix A.8).

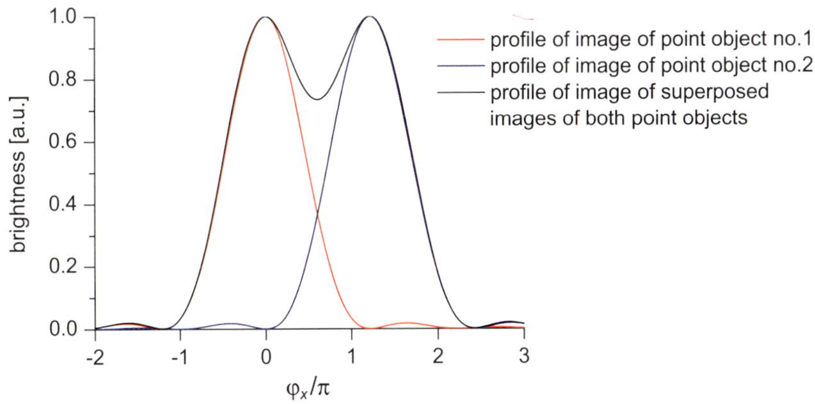

Fig. 5.13: Illustration of the resolution according to Rayleigh's criterion. The optics are affected by its finite circular aperture only (same situation as in Figure 5.6). Line profiles through the center of the PSF in the horizontal direction of the images obtained from two-point objects (red and blue line, resp.). Note that the red line could only be measured when the image of object no. 2 is blocked and vice versa. Otherwise only the brightness distribution resulting from the superposition of both images is observable (black solid line). Here, the images of object nos. 1 and 2, respectively, are separated by a distance of $\mathcal{R} = \delta_0/2 = 1.22 \cdot \lambda \cdot f_\#$. Here, the maximum value $B_{max} = 1$ and the dip in between has a value of $B_{min} = 0.735$.

Similar to this, the resolution of the optical system determined by Abbe's criterion according to Equation (5.23a) can be written as

$$\mathcal{R} = \frac{\delta_0}{2} = \frac{\lambda}{2 \cdot NA}. \tag{5.49}$$

From diagrams such as Figure 5.13 and Figure 5.14, the contrast can be obtained as

$$K = \frac{B_{max} - B_{min}}{B_{max} + B_{min}}. \tag{5.50}$$

For the present example displayed in Figure 5.13, we get $K = 15.3\,\%$ for the circular aperture and 10.5 % for the slit. This corresponds to MTF values of approximately 15 % and 10 %, respectively (see Section 5.2.2). Although those contrast values may be regarded to be close to what the human eye could still recognize, in practice, those levels may be too low to be observable, in particular, when significant noise is present.

5.1.7 Differences for imaging with coherent and incoherent light

To judge image quality in general when imaging is performed either with incoherent or coherent light, respectively, but with the same optics is not so straightforward as it may be assumed naively. First, there are no general criteria on image quality. If furthermore, one takes into account the rather complex response of the human eye in combination with the response of the brain, then *a general judgment* is nearly impossible. Even a quite simple comparison, namely the comparison of the cut-off frequency, which for illumination with incoherent light, is twice that for coherent light, is not helpful because one cut-off frequency is with respect to the field amplitude and the other one with respect to the intensity (see Section 5.1.5). Consequently, here it makes no sense to discuss the many possible aspects. This is done in many scientific journals and also in some textbooks such as [Goo17]. Nevertheless, the following brief discussion provides a simple and quite instructive example for illustration.

Let us regard again a single point object that is imaged with an optics that is affected by its finite aperture only, but not by aberrations. Imaging is performed either with incoherent or coherent light, respectively, but with the same optics.

Again, we regard this in 2D and 1D geometry, respectively. The 1D situation may instead also be regarded as 2D when we assume a thin straight line as the object, although one has to be aware of the comment to Figure 5.13 given in the previous chapter. Independently of the degree of coherence, in the image plane one would observe the brightness distribution given by the red line in Figure 5.14a for 2D or in Figure 5.14c for

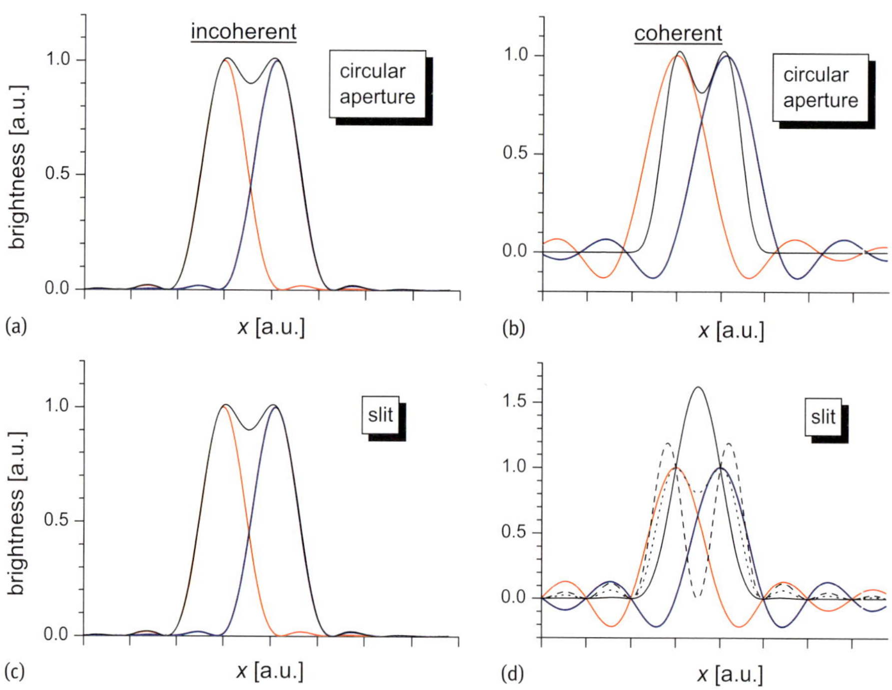

Fig. 5.14: Profiles measured along horizontal lines of images (black solid lines) obtained from two point or line objects separated by some distance. The situation is considered within the 4-f-model system of an optics that is affected by its finite aperture only (2D geometry with circular aperture in (a), (b) and 1D slit aperture in (c), (d)). The distance between the objects in (a) to (d) is always the same.

1D. If instead the object is located at another position, one would observe, e. g., the distribution shown as a blue line. In the Fourier plane, the power spectrum of both points is absolutely the same, but according to the shift theorem the spectral phase differs (for the definition of the power spectrum, see the textbooks of Fourier mathematics).

However, if now the object consists of both point objects or line objects together, coherent and incoherent light do affect the imaging process differently. This can be seen from Figure 5.14.

For each point, a Fourier spectrum that extends to $\pm\infty$ is generated. However, this spectrum is clipped as discussed previously. In the image plane, this leads to the shown brightness distributions of both points for the incoherent light ((a),(c)) and to the field distributions for the coherent ((b),(d)) light (shown as red and blue solid lines, resp.). For incoherent light, the brightness distributions, namely the PSF (see Equation (5.21a) and Equation (5.21b), resp.) of both points have to be added and the observed image in the image plane is shown as a black solid line. For coherent light, the field distributions of both points (see Equation (5.20a) and Equation (5.20b), resp.) have to be added. The observed brightness distribution, namely the image, is the absolute square of the resulting field (black solid line in (d)).

From this example, it may be seen that in 1D geometry the two points or lines separated by the same distance would lead to an image that allows us to see that the image consists of two features. One may say that the two objects in the image obtained with incoherent light (i. e., the black line) displayed in Figure 5.14c are just resolved. However, this is not the case for the image obtained with coherent light (i. e., the black line displayed in Figure 5.14d). Here, the image just shows one spot only and obviously it is impossible to recognize that the original object consists of two well-separated points or lines.

As a practical application of this observation, one may regard a simple lithographic process for computer chip production. Typically, the electronic circuit contains a lot of horizontal and vertical "lines" that have to be imaged. In such a case, coherent light is not preferable because that may lead to less resolution and due to pronounced "ringing" sharp edges are not reproduced well (see Figure 5.15).

However, if instead of homogeneous illumination still with coherent light, for some reason the light of the second point source is phase shifted, pretty good resolution may be observed. This is illustrated by the dashed and dotted lines in Figure 5.14d. They are similar to the solid line but a phase term of $e^{i\pi}$ and $e^{i\pi/2}$, respectively, has been multiplied to the field function of the blue curve. The effect on resolution is apparent. But that is a quite unusual situation and rather subject to special experiments. Also, in 2D geometry the situation may be different (see Figure 5.14a and Figure 5.14b) and even other effects such as speckles may play a role in the case of coherent light.

In spite of all these interesting aspects, in the following we restrict to incoherent light, which is the common case for imaging. Nevertheless, as the first of two final remarks of this section we would like to note that within astrophysical photography one may have to consider partial coherence. This is the case for imaging of stars that are "far

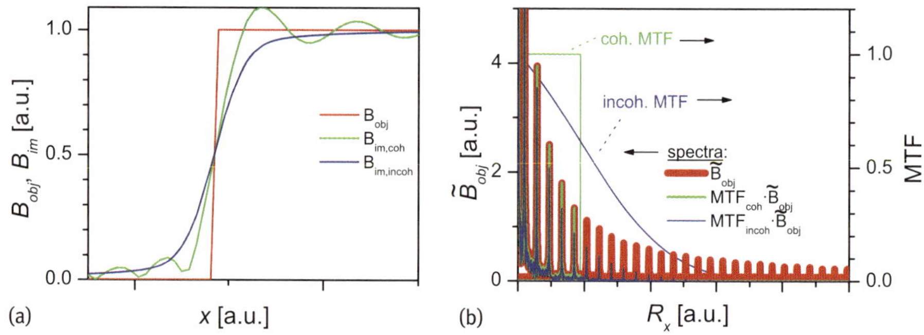

Fig. 5.15: Imaging of a sharp edge (red curve in (a)). (a) images obtained with coherent (green curve) and incoherent (blue curve) light, respectively. (b) shows the corresponding spectra and MTF curves, respectively. The MTF curves are the same as displayed in Figure 5.8.

enough" away because then the emitted spherical waves become nearly plane waves as the observer on the earth. We would like to remind that a point source object at infinite distance always emits spatially coherent light (see Section 5.1.4). Indeed, one can make also use of that to determine the distance to the star by a measurement of the degree of coherence of its light. This is exploited in stellar interferometry (see, e. g., textbooks such as [Bor99] or [Hec16]).

The second remark is that light is never absolutely fully coherent, not even in the "best" laser, and there is even no absolutely fully incoherent light. Even more, during propagation through an optical system, the degree of coherence may change, and it even may increase; see, e. g., [Sal19].

5.1.8 Space bandwidth product

Although the following also does not provide new physics when compared to Section 5.1.4 and Section 5.1.5, we would like to have a look at the same subject from another point of view, in particular, because it is very important. If one rewrites, e. g., Equation (5.47) or, e. g., Equation (5.25), then one obtains the space bandwidth product (SBP)

$$\boxed{\delta \cdot R_{x,\max} = \kappa \cdot \alpha \geq \kappa = \text{const}} \tag{5.51a}$$

or

$$\boxed{r \cdot R_{x,\text{cut}} = \kappa \cdot \alpha \geq \kappa = \text{const}} \tag{5.51b}$$

with the geometry factor κ, e. g., from Table 5.1, which also includes where the width of the spot is measured, e. g., as FWHM or at the first zero position ($R_{x,\text{cut}} = 2 \cdot R_{x,\max}$;

see Figure 5.8 and Equation (5.61)). If, e. g., measured at the first zero position $r = r_0$, $\delta = \delta_0$ and $\kappa = 1$ or 1.22 for 1D or 2D geometry, respectively. This is nothing else as the generally important relation in optics Equation (5.27). But here one may interpret δ as the uncertainty or spatial spread of the brightness and $R_{x,max}$ (Equation (5.46)) or $R_{x,cut}$ (Equation (5.61)) as the bandwidth of the system. In that sense, Equation (5.51a) expresses that a limitation of the spatial frequency range leads to a limitation in the exact "knowledge" of the spatial position of the image point. This also does not allow an accurate measurement of angles below the diffraction limit. Thus, in total, Equation (5.51a) corresponds to Heisenberg's uncertainty principle $\Delta x \cdot \Delta p_x \approx h$ where Δx is given by δ_0 and the uncertainty of the momentum by $\Delta p_x = \hbar \Delta k_x$. Heisenberg's uncertainty principle is equivalent to the uncertainty principle from Fourier mathematics but includes an important physical interpretation.

Again, SBP can be propagated through an optical system. At best, there is no loss, i. e., the SBP before the optical system and after transmission through the system is the same. Any wavefront distortion introduced by the optics will lead to an increase of SBP. In the optimal case, $\delta_0 \cdot R_{x,max} = \kappa$, any wavefront distortion may be described by an additional factor $\alpha > 1$ (here δ_0 is measured between the first zero positions and κ discriminates the 1D case related to Abbe's criterion and the 2D case related to Rayleigh's criterion, resp.).

We may mention that analogous to Equation (5.51a), in the time-frequency range there is a similar relation, namely the time bandwidth product (TBP)

$$\tau \cdot \Delta\nu \geq \text{const} \tag{5.51c}$$

where τ is the pulse duration and $\Delta\nu$ the spectral bandwidth of the pulse. This corresponds to Heisenberg's uncertainty principle $\tau \cdot \Delta E \approx h$. In that case, for a bandwidth limited pulse, $\tau \cdot \Delta\nu$ is equal to a constant. In the case of a pulse front distortion, $\tau \cdot \Delta\nu$ becomes larger than the constant. For a given bandwidth, this leads to a longer pulse. This subject is very important for laser physics with ultrashort pulses.

Equation (5.51a) states that δ_0 is a function of the optical system, and consequently, also $r_0 = \delta_0/2$ divided by the picture height PH and also its inverse. Thus, one may estimate the number of selected "points" within the image field, i. e., within $PW \cdot PH$ and follow the procedure discussed in Sections 1.3.1 and 1.3.3. When we restrict to one dimension and choose a distance $\delta y = r_0$ (see Equation (1.13)), then according to Abbe's or Rayleigh's criterion and with Equation (5.47), we obtain

$$N_{SB}^{(1D)} = \frac{PH}{r_0} = \frac{R_{y,cut} \cdot PH}{\kappa\alpha} \leq \frac{R_{y,cut} \cdot PH}{\kappa} \tag{5.52a}$$

For two dimensions, the discussion is straightforward:

$$N_{SB}^{(2D)} = \frac{PH}{r_0} \cdot \frac{PW}{r_0} = \frac{R_{x,cut} \cdot PH}{\kappa\alpha} \cdot \frac{R_{y,cut} \cdot PW}{\kappa\alpha} \tag{5.52b}$$

This is the (maximum) space bandwidth number SBN. Obviously, the presence of aberrations or wavefront distortions, which may be characterized by a value of $\alpha > 1$ (Section 5.1.4.2), leads to a lower value of the SBN. However, we may remark that although N_{SB} from Equation (5.52) is a good measure for the SBN, it slightly overestimates the number of resolvable image "points." Nevertheless, this is not much, and thus for the moment this estimate is sufficient. A more detailed discussion with a better applicable SBN is made in Section 5.3.

Physically, Equation (5.52) describes the same situation as Equations (5.51a) and (5.51b). A poor optical system leads to an increase of SBP and to a decrease of SBN. Again, this is the number of different "image points" that can be discriminated from each other. This is the information content of the image (see Section 1.3.3). Of course, propagating that number through the optical system, would lead at best to a transfer of full information (namely all data points are still available), but quite often loss of information happens, which reduces the number of data points. But in no case is additional information of the object obtained. This describes the situation within an optical system qualitatively.

For image processing or manipulation as a chain of process steps after image capture, this is equivalent. Each step may either keep the SBN or lead to a reduction, which most simply may be described by a parameter α, similar to that in Equation (5.52). It is important to hint that any loss of information from a previous step cannot be compensated by any later process step. Consequently, by no means and by whatever advanced image processing software, a missing signal or information can be recovered. In other words, neither today nor in future computations and computational imaging (see Section 8.4) can shift physical limits.

Here, one has to note that although SBN and SBP are quite helpful quantities to get a first judgment for the performance of an optical system, at this point it is important to keep in mind that these quantities provide just a simple number, which is not sufficient for a more careful judgment. This and a direct relation to the MTF is discussed in more detail in Section 5.2.2 and in the following. A closer look on the information content of an image shows that this is related to conservation of energy and momentum. This leads to a little modification of the above statement without changing the basic idea. In the physical sense, the total energy within an image is given by the 2D-integral over $B_{obj}(x, y)$ in the x-y-space. Due to energy conservation or due to Parseval's theorem, this is identical with the 2D-integral over $\widetilde{B}_{obj}(k_x, k_y)$ in the k_x-k_y-space. Though to the apparatus function of the system, namely its MTF, introduces a loss and even further losses result from image processing such as noise reduction or contrast enhancement which, in principle, can also be described by a MTF (a discussion of the MTF of a system is subject of Section 5.2.3 and those which follow). In other words, the 2D-integral over $\widetilde{B}_{im}(k_x, k_y) = \widetilde{B}_{obj}(k_x, k_y) \cdot MTF(k_x, k_y)$ is smaller than that over $\widetilde{B}_{obj}(k_x, k_y)$. Yet a redistribution of the image content, in principle, may be possible without introducing severe losses. For instance, contrast enhancement or noise reduction may "improve" the MTF at intermediate spatial frequencies even though this has to be paid for with a decline at large frequencies. Nevertheless, this may improve the *perceived image quality*.

5.1.9 Image manipulation

As we have seen, in the ideal case the spatial frequency spectrum would be fully transmitted through the optical system, but in a real system this is not the case. The influence of the optical system then may be described as a change on the spatial frequency spectrum, which then affects the image. In the same way, any further changes applied to the spectrum will result in particular changes to the image. Changing the spectrum is usually termed filtering. This also gives access to a directed manipulation of the image.

5.1.9.1 Low and high-pass filters

The easiest way is the application of a filter in the Fourier plane. This can be done via both experiment and computing. As an example, Figure 5.16 shows some of the most simple operations. Figure 5.16b and Figure 5.16c show a typical low pass filter that only allows the central part of the spectrum to pass the filter so that only the low frequencies remain. In an experiment, such a situation is arranged by putting an appropriate small aperture at the right place in the Fourier plane, e. g., in a 4-f-setup. This is also a very simple example of optical computing; in spite of the simplicity, in this way a rather complex mathematical calculation, namely a convolution, is performed very quickly and much faster than could be done by a digital computer.

In computations, low pass filtering (Figure 5.16b) is realized first by a Fourier transformation of the image, second a multiplication of the Fourier power spectrum with the corresponding filter function and third an (inverse) Fourier transformation. As described in Section 5.1.5, this procedure corresponds to a convolution process, which in the case of a low pass filter results in a blurred or smoothed image. In comparison to Figure 5.16b, this is even more pronounced in Figure 5.16c, where the central part is restricted very close to the zeroth order. Here, the mask is so small that it can hardly be seen. If the diameter of the filter is even more decreased so that the filter removes all higher diffraction orders so that only the zeroth order can pass, then structure information is totally lost. Usually, this is bad for imaging, but very important, e. g., in laser physics, because this “cleans” the beam by spatial filtering.

Figure 5.16d and Figure 5.16e also provide low pass filtering but now in one direction only. A slit restricts the spectrum in the k_y- or k_x-direction, respectively. Consequently, smoothing occurs in one direction only. As an example, Figure 5.16d shows that pillars that are exactly oriented in the vertical direction are mostly removed and vertical edges are smoothed by vertical slit filters. In contrast to that, horizontal features are hardly affected. In Figure 5.16e, this is the reverse.

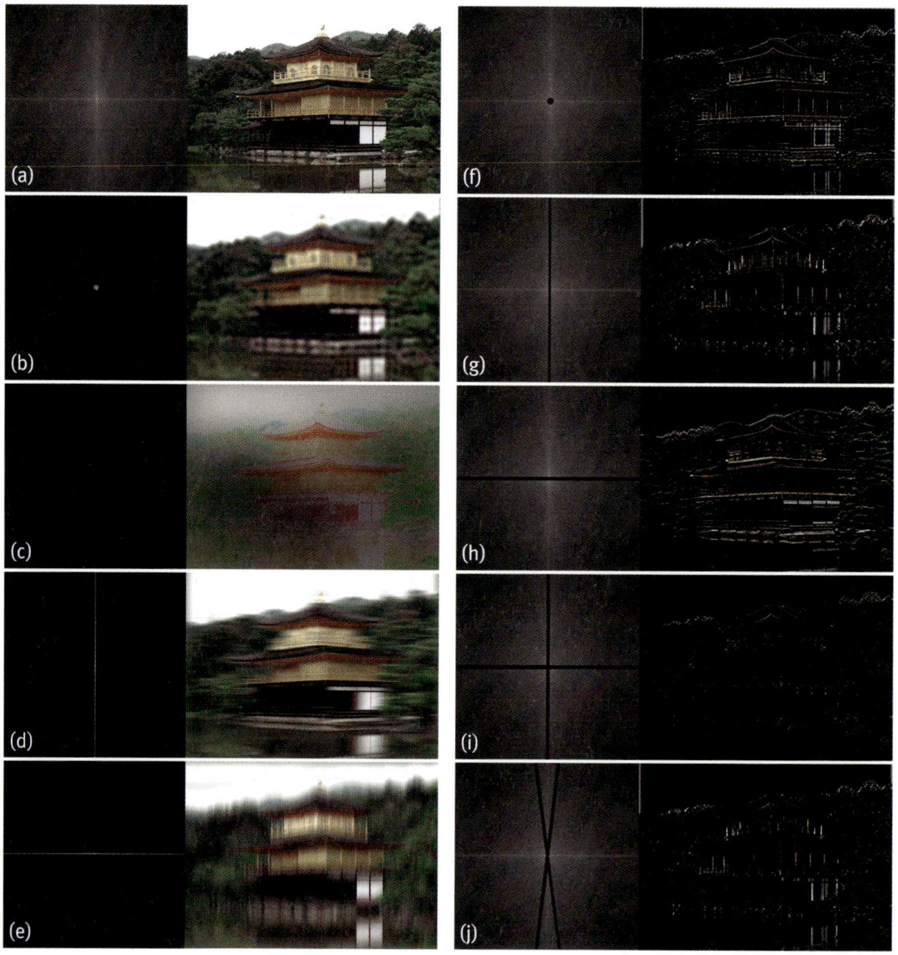

Fig. 5.16: Examples of image manipulation. (a) shows the original image and its Fourier spectrum. The other images are all filtered with a mask that allows only part of the spectrum to pass. The left column in each figure shows the filtered spectra. The resulting "reconstructed" images after Fourier transformation of the filtered spectra are shown in the column on the right-hand side. (b), (c) illustrates low pass filters. (d) and (e) also provide low pass filtering but in one direction only. (f) to (j) high-pass filters (the originals are very dark; thus, for better viewing, they are brightened).

Figure 5.16f corresponds to a high-pass filter. This filter blocks all low frequencies. Due to the fact that this includes also the zeroth order, which is responsible for the whole brightness within the image, images that are filtered in such a way are usually relatively dark. The effect of high-pass filters is that the images mostly show edges or regions with high-contrast changes. Such filters yield to a contrast enhancement. In particular, high-diffraction orders are responsible for the details within an image. Blocking of high-frequency components results in smoothing (see above).

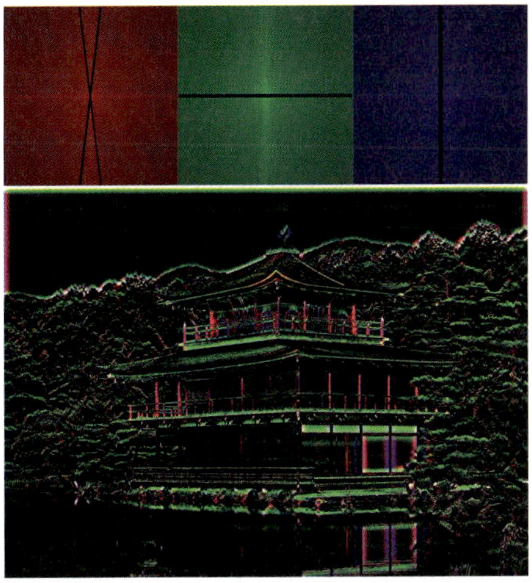

Fig. 5.17: Application of different filters for the Fourier spectra of the three RGB colors, as shown in the upper row. The original is very dark; thus, for better viewing it has been brightened.

In the same way as before, one can restrict the visibility of details and contrast changes to vertical (Figure 5.16g) or horizontal (Figure 5.16e) features, respectively. Figure 5.16i illustrates that one can restrict also to contrast changes that are in neither the vertical nor horizontal region. Figure 5.16j provides another example. Even more complicated and specially constructed filters may lead also to removal of unwanted features such as dust on images or scratches or horizontal stripes in earlier TV images.

Figure 5.17 shows that one may also affect the color composition of the image. This can also be seen in Figure 5.16c, where not all colors are affected in the same way.

Such manipulations are also the basis of image processing, image recognition, etc. This is implemented as well in image manipulation software such as *Photoshop*®, *Gimp*, etc. Of course, such software involves much more complex algorithms. We also refer to special literature on image processing.

As we have seen, the smoothing of an image can easily be made by application of a low-pass filter, where high-pass filters, in principle, lead to a sharper image. Further discussion on "sharpness," etc. will be mentioned in Section 5.2.8. However, Figure 5.16 clearly shows that high-pass filtering only does not lead to an improved image quality.

5.1.9.2 Unsharp masking

Although further discussion on image processing is not the intention within this book, we would like to finish this section with a short illustration of an important standard process of image quality improvement, which is quite instructive. This is image sharpening via so-called "unsharp masking" (USM).

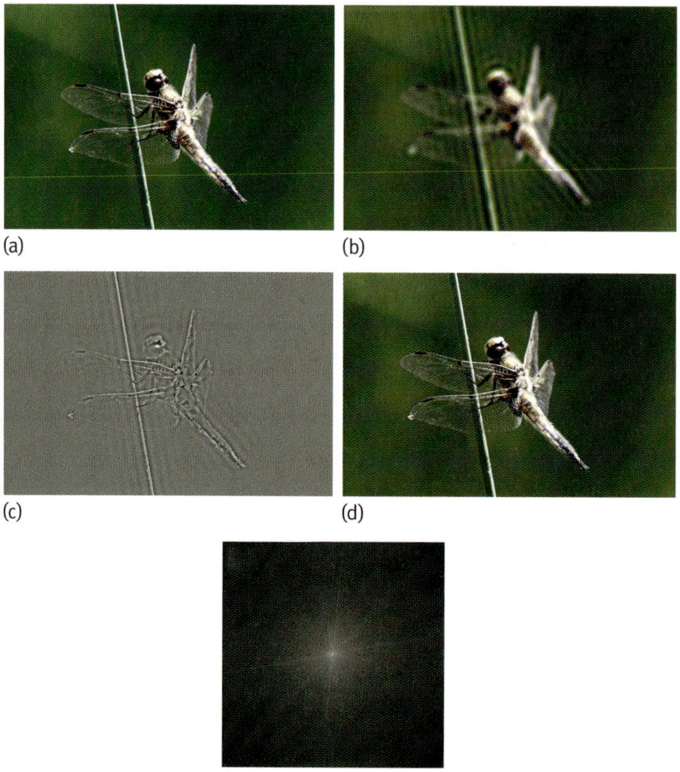

(a) (b) (c) (d) (e)

Fig. 5.18: Principle of image sharpening via USM. (a) original image, (b) unsharp copy, (c) unsharp mask, (d) sharpened image and (e) Fourier transformation of the original. Rather small reproductions, e. g., in the present book, do not allow us to see the effect quite well, however, the reader may do his own test with large scale reproduction, e. g., on a screen. Here, it is interesting that due to the sharp border of the blade of grass, a strong diffraction spectrum in perpendicular direction is observable. Even more, due to a lot of further sharp edges, clearly visible diffraction patterns (i. e., the spatial frequency spectra) extending in different directions are apparent.

The principle of this process is illustrated in Figure 5.18. First, the original image (Figure 5.18a) is blurred. The result is an unsharp copy (Figure 5.18b). This blurred copy is multiplied by a factor a (between 0 and 1), which determines the strength of the whole process and then is subtracted from the original image. The result is an “unsharp mask” (Figure 5.18c). Using this unsharp mask as a filter for the original leads to a sharpened image (Figure 5.18d).

However, it is important to mention that this method is not always applied at once to the whole image. Sometimes this is done subsequently for small regions of the image only. As a result, there is a local contrast enhancement (in contrast to a global one as shown, e. g., in Figure 5.16f; see also see Section 5.2.7).

Usually blurring is obtained by application of a low-pass filter with the principle described above (see, e. g., Figure 5.16b). The low-pass filter mask (see left column in Figure 5.16) here is usually a rectangular mask with a size of 5 pixel × 5 pixel or 9 pixel × 9 pixel. Other masks are also possible. We have to note that other filters or procedures of image manipulation may be used as well. It should be stressed that in no case can resolution within the image be increased. Processes such as USM may lead to an increased perceived sharpness as the local contrast between differently bright regions is enhanced, although sometimes resolution is even lost. Terms such as sharpness and perceived sharpness will be discussed in Section 5.2.7

5.2 Discussion of the MTF

5.2.1 Test objects, MTF, contrast, spatial frequency units

The "natural" function (or test object) that is directly related to the MTF is a sine function or sine grating, respectively (because $B_{obj}(x)$ corresponds to an intensity, the sine function may be represented by $[\sin(k_0 \cdot x) + 1]/2$; see Figure 5.19a and also the relations in Appendix A.2). Even more, as shown in Appendix A.2, any structure can by synthesized from sine functions as a linear superposition, so that the following discussion leads to quite general statements on the MTF.

As an example, Figure 5.19a shows such a sine grating ($k_0 = 2\pi/20\,\mu\text{m}$ or $R_0 = 50\,\text{lp/mm}$; $B_{obj}(x) = (\sin(x) + 1)/2$). For a pure sine grating, there are only two frequency components (see Appendix A.2), namely at $\pm k_0$ or $\pm R_0 = \pm k_0/(2\pi) = 50$ periods/mm. This corresponds to diffraction in the ± first order; higher orders are not present. Consequently, the spatial frequency has the unit cycles per mm or if we identify the areas around the minimum and the maximum, respectively, with a "dark line" and an adjacent "white line" and both together as a line pair (lp; see Section 1.6.1), then 1 period/mm ≡ 1 cycle/mm ≡ 1 lp/mm which, for a full format sensor ($PH = 24$ mm), is identical to 24 lp/PH. We would like to remind that it is much more useful to discuss resolution and MTF, respectively, in relative terms of lp/PH instead of absolute terms of lp/mm because this then allows a direct comparison of cameras with different sensor sizes. In other words, one should take into account the size of the sensor. Sometimes in the literature the unit l/mm, i. e., lines per mm, is also used, where 1 lp/mm = 2 l/mm (two lines form one line pair), but we do not make use of that.

In the case of the MTF of a digital sensor, the number of pixels within one period of such a test grating may be N_{period} and thus the corresponding SBN is half the number of pixels in the vertical direction. The factor 1/2 arises from the Nyquist limit, which is given by 50 % of the maximum pixel number (see Section 1.6.2) divided by N_{period}, namely $\text{SBN} = 0.5 \cdot N_v/N_{period} \cdot \text{lp}/PH$. This is equivalent to the consideration that one line *pair* corresponds to two pixels, and thus the maximum R_x is given half of the number of pixels within the picture height. This is also the maximum SBN of the system.

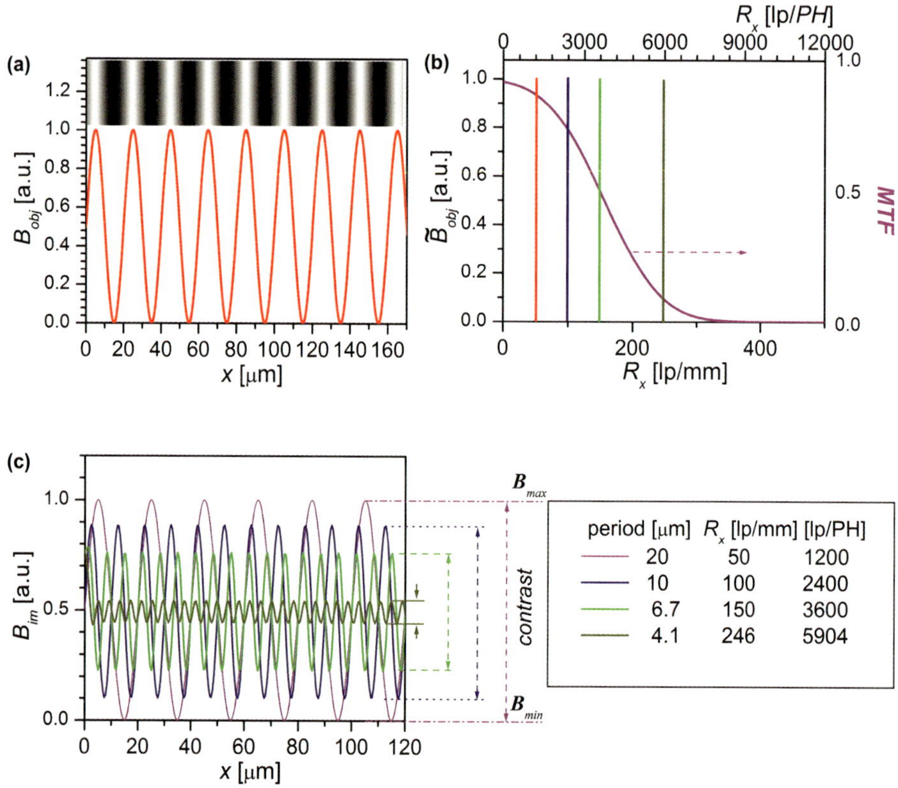

Fig. 5.19: (a) sine grating with a period of 20 μm as the object. (b) Examples of the spatial frequencies of four different sinusoidal test gratings (only the positive frequency spectrum is displayed; there is the negative one as well but this is symmetric). The thick solid line in magenta shows the MTF (just an example), which limits the transfer of different spectral components through the system. (c) Images of the corresponding test gratings in the presence of the MTF shown in (b).

In the present example, Figure 5.19b displays spectra of four different sine gratings, each spectrum in a different color. Here, the periods are 20, 10, 6.7 and 4.1 μm, respectively, yielding $R_0 = 50, 100, 150$ and 246 lp/mm, respectively. It is important to note that here the spectral amplitudes of all gratings are chosen to be the same with a value of 1 (relative units).

After multiplication with the MTF, an example is displayed as the curve in magenta, and a Fourier transformation, the corresponding images are obtained (see Figure 5.19c). The colors correspond to the same ones with the spatial frequencies displayed in (b). Their relative contribution, i. e., the amplitude, is given by the MTF. The legend provides the period and the spatial frequency.

It is clearly seen that with a reduced value of the MTF, image quality becomes worse, in particular,, the contrast is decreased, and thus resolution as well, until there is no structure information at all at high R_0 values. This is equivalent to the situation with the two-point objects in Figure 1.9. Of course, this happens when R_0 exceeds the cut-off frequency. However, this usually occurs much before the MTF value becomes nearly zero because typically even contrast values below 5 to 10 % cannot be resolved by the human eye. This corresponds to the Rayleigh limit (olive curve with 246 lp/mm in Figure 5.19). Furthermore, it becomes worse if additional noise is involved during the imaging process. As discussed in Section 4.8.1, noise influences the dynamic range, and thus the observable contrast as well (see, e. g., Section 7.3 and Figure 7.24). Then, of course, reduced contrast reduces the MTF.

Due to the well-known relations for Fourier transformation (see Appendix A.2), it is straightforward that the sine function samples the MTF. In other words, the contrast value (see the definition in Equation (5.50)) that is obtained from $B_{\text{im}}(x)$ is equivalent to the value of the MTF of the corresponding sine grating (see the examples in Figure 5.19c). In other words, the MTF reflects the relative amplitude (see the examples in Figure 5.19b) and $\text{MTF}(R_x)$ is equal to the contrast of the image of a sine grating with period R_x^{-1}.

Up to now, this has been discussed for a test grating with full contrast, i. e., contrast = 1 for the object. If the object contrast is reduced by some factor, the image contrast is reduced as well by the same factor, even for perfect imaging. Again, due to the relations for Fourier transformation, it is then clear that $\text{MTF}(R_x)$ is the contrast at R_x *relative* to the contrast of the object, which here is the sine grating.

Because arbitrary objects could be described by a superposition of sine functions, in general the MTF at a given value R_x provides the contrast for that R_x *relative* to the contrast of the same spatial frequency in the object pattern. In case of a fully modulated sine function as the object, more correctly $(\sin(x) + 1)/2$ with contrast $K = 1$ (see Equation (5.50)), the MTF is equal to the contrast: $\text{MTF}(R_x) = K(R_x)$ (see Figure 5.19).

5.2.1.1 Bar gratings

The previous can be seen from another easy and illustrative example, where the test object now is a rectangular or bar grating with a bar width that is equal to its gap width. As briefly explained in Section 8.3.1, the corresponding grating function can be described by a series of sine functions, and consequently, its frequency spectrum, i. e., the MTF, consists of the fundamental frequency, i. e., first order and higher orders that are odd orders only for this special symmetry. The relative contributions of the fundamental frequency and the higher orders can be seen from Equation (8.1) (see also the general equation for the Fourier transformation of a bar grating in Appendix A.2). The grating of the present example is shown in Figure 5.20a. The fundamental or major frequency, respectively, is the same value of R_x when compared to the corresponding sine function with the same period (see Figure 5.19).

As a result, the contrast value given by the contrast transfer function of a bar grating that is obtained from $B_{\text{im}}(x)$ is not equivalent to the value of the MTF of the corresponding sine grating. At best, it is a moderate approximation. For further discussion of the bar grating and other test objects and their relation to MTF, see Section 8.3.1.

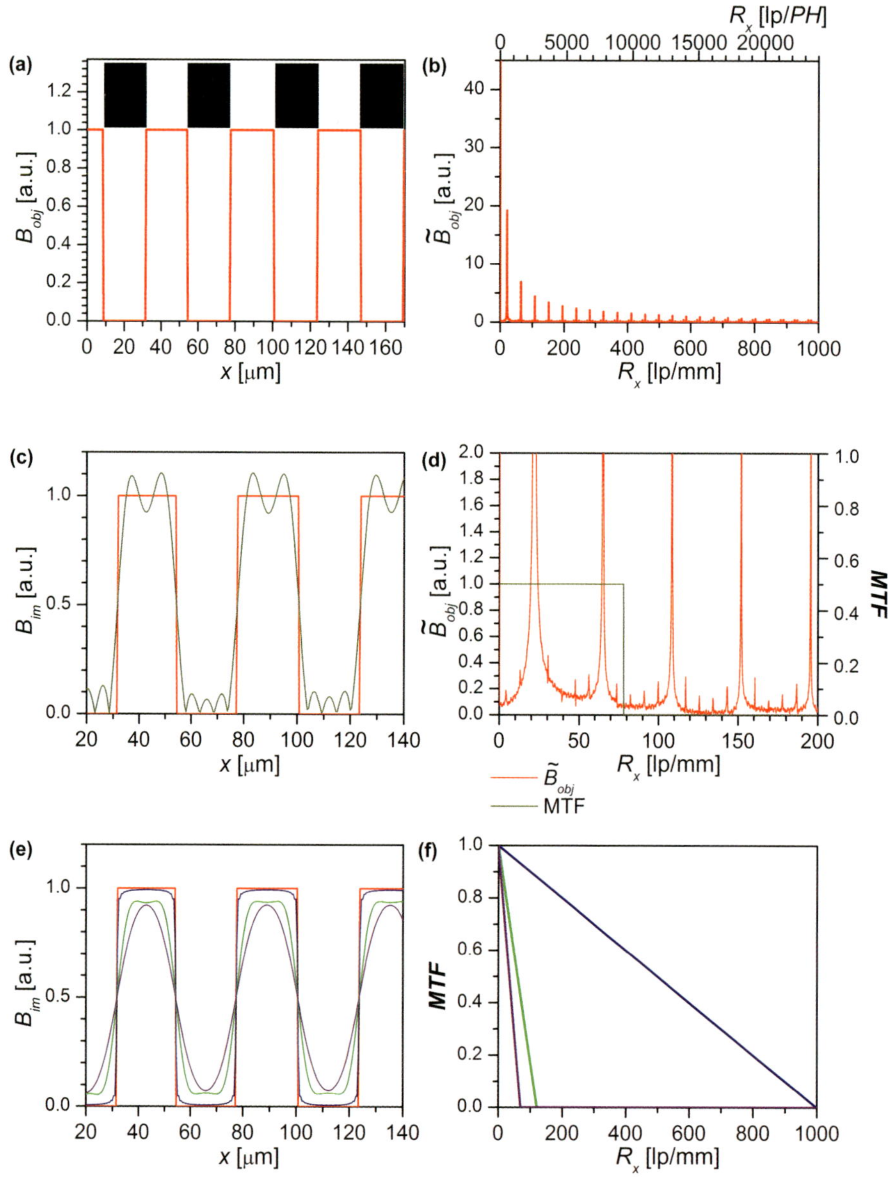

Fig. 5.20: (a) Test object $B_{\mathrm{obj}}(x)$ with a period of 46 µm and (b) and its Fourier transformation $\widetilde{B}_{\mathrm{obj}}(R_x)$. The fundamental frequency is 1/period, i. e., 22 periods or cycles or line pairs per mm. The grating itself is shown in (a) on top of the diagram. (d) shows details of (b) together with an artificial MTF curve. (c) shows the images obtained with the MTF curve displayed in (d): red curve original (with constant MTF without any cut-off), dark yellow curve image with the MTF displayed in (d). (e) shows the images obtained with the MTF curves displayed in (f).

Now we would like to illustrate the effect of different MTF curves on the image quality of this test grating. To do so, first we make use of a MTF curve that is purely artificial to illustrate a quite crude cut-off provided at the Nyquist frequency of the sensor (see the dark yellow curve in Figure 5.20d). Note that such hard clips lead to diffraction effects and ringing, in experiments and in simulations (see also Figure 5.15) but note as well that a realistic sensor MTF usually looks differently (see Section 5.2.4 and Section 5.2.5). This curve also looks like the MTF for coherent illumination, but this should not be an issue here.

Figure 5.20d shows a detail of the spectrum (Figure 5.20b) and the artificial MTF curve. In Figure 5.20c, the resulting image is plotted together with the original object structure. It can be clearly seen that the original object is not well reproduced because there is a significant suppression of higher orders, or more generally, the high frequencies that are necessary to describe the sharp edges of the bars are absent. This corresponds somehow to a low-pass filter (see Section 5.1.9). Another example is shown in Figure 5.20e and Figure 5.20f. Different MTF characteristics affect the image differently, but again, the more the higher spatial frequency components are reduced, the worse the image becomes. For instance, such curves may be the result of different $f_{\#}$ (see Figure 5.27). In particular, it may be seen that the rather modest "blue MTF" leads to a rather high-quality image, the "purple MTF" mostly restricts the spectrum to the component of the fundamental, and thus the image is more or less the same as that from a sine grating as the object. Here, the image cannot reproduce the sharp edges of the bar grating at all. But it should also be pointed out that at least this fundamental frequency has to be transmitted to obtain at least some structure information of the object. On the other hand, the larger the cut-off, the better the reproduction of the test structure (see Figure 5.20e). We may note again that noise also influences the contrast. For a "noisy sensor," this effect may be implemented in the MTF, which then results in poorer performance of the system.

5.2.1.2 More realistic MTF curves

Now we would like to continue this illustration with an extension to more realistic MTF curves and to the effect on the contrast and on the image quality. This is shown in Figure 5.21 and Figure 5.22. Again, the original object structure is displayed in red, and the lines of the image curves in Figure 5.21a correspond to the MTF curves of the same color in Figure 5.21b. The corresponding grayscale images are displayed and commented in Figure 5.21c. The results are obvious and do not need further discussion. But one has to emphasize that one has to discriminate between resolution and contrast (in the image). This is also seen in realistic images such as those shown in Figure 5.22.

Figure 5.22 clearly shows that depending on the MTF, one can obtain either an image with high resolution or with large contrast. One can also obtain both at the same time or even none of that, but it is not necessarily the case that high resolution comes together with large contrast (or vice versa).

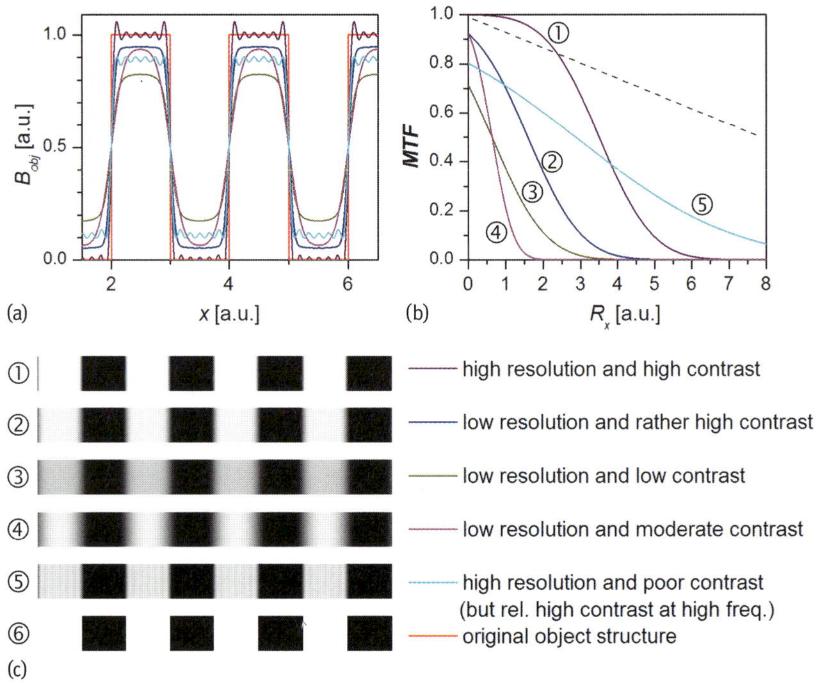

Fig. 5.21: Original structure and images (c) of a bar grating obtained for different MTF curves (b). (a) shows profiles measured along a horizontal line of the images displayed in (c). The broken line in (b) may indicate the optical resolution limit of a 1D aberration-free optics. The total MTF of the system usually is worse and may never exceed this limit. The purple curve is somewhat artificial at lower frequencies, but on the other hand, such a curve is typical when image processing takes significant influence. Of course, in such a case resolution cannot be improved, only sharpness (see the discussion in Section 5.2.7).

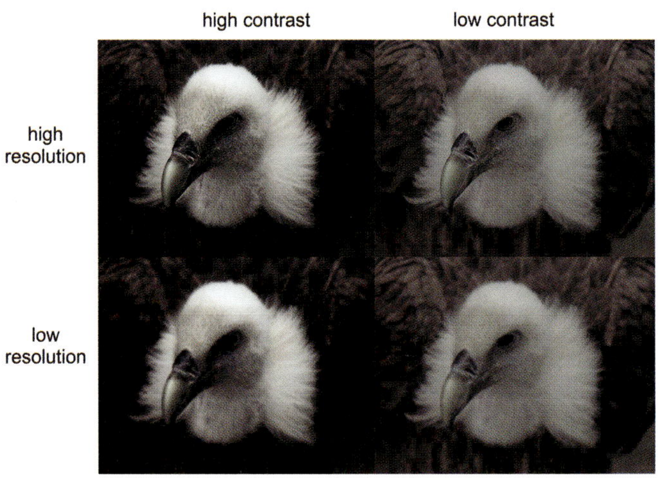

Fig. 5.22: Resolution and contrast in images of a real scenery. Left column: high contrast, right column: low contrast; upper row: high resolution, lower row: low resolution. Note that this serves for illustration only, as none of the images is optimized.

5.2.2 Image quality characterization by means of a single MTF value

Although diagrams such as those displayed in Figure 5.21 (or, e. g., Figure 5.33 and Figure 5.34) are quite useful and do characterize the MTF of a camera system well, the full MTF curve is not always easily usable, in particular, when different systems should be compared or if the dependence of the MTF on a specific parameter such as $f_\#$ should be investigated. In such a case, it makes sense to restrict the comparison to a single number and, in particular, to compare specific values of the MTF curve only. Such a value is, e. g., the spatial frequency where the MTF has dropped, e. g., to 50 % of its maximum, to 20 % or, e. g., to 10 %. We term these values MTF50, MTF20 and MTF10, etc., respectively. The corresponding spatial frequencies where these limits are reached are R_{MTF50}, R_{MTF20}, R_{MTF10}, etc. The spatial frequency values at which the MTF becomes 0.1, namely R_{MTF10} often is regarded as the minimum contrast that humans need so that structures become visible. This also corresponds to the resolution according to Rayleigh's criterion. R_{MTF10} may also correspond to the SBN (see Section 1.3 to Section 1.5 and Section 5.1.8), but of course, SBN could also be defined with respect to, e. g., R_{MTF50} or any other value.

The restriction to a single value yields only limited information, nevertheless, it is helpful in a similar sense as the PSF could also by characterized by a single value, namely its width. This can be measured, e. g., as its FWHM, its $1/e^2$-width, or the width between the first zero positions (see, e. g., to Figure 5.6). Altogether this is similar to the characterization of a focal spot by its diameter only, instead of the full profile or by the characterization of the temporal distribution of a pulse by its duration only.

As an example, we may compare the dependence of the MTF50 on the actual f-number. As we will see in Section 5.2.3, the MTF depends on $f_\#$ and there is a trade-off between diffraction effects (see Figure 5.27) and aberration effects, which both depend on $f_\#$ (see also Section 2.5.4). As a consequence, there is an optimum $f_\#$ where the lens shows the best resolution. But one has to be careful and not necessarily regard this point as the best for taking a photograph. There is also an $f_\#$, sometimes called "sweet spot," where also the depth of field in combination with the overall sharpness impression is taken into account. The optimum depends on the optics, or more exactly, on the actual system that is regarded. Figure 5.23 shows an example.

5.2.3 OTF and MTF of a system

Usually, an optical system such as a camera, consists of many elements that influence its images. Such an element is the camera lens (the optics) and another one is the sensor (or the sensor system with optical microlens array OMA, optical low pass filter OLP, etc.), and there might be even more elements such as filters, additional apertures, etc. We would like to note that filters may not be restricted to homogeneous gray or specific color filters, but also special filters with spatially varying tonal grades, softening, star gratings, etc. may be involved. All those components have to be included in a rather

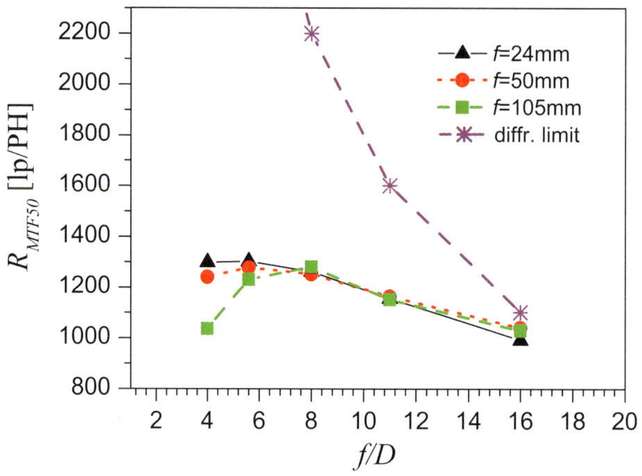

Fig. 5.23: Dependence of the R_{MTF50} of a DSLR camera with a zoom lens operated at $f = 24$, 50 and 105 mm, respectively. For the theoretical (diffraction) limit, see below (Figure 5.27).

complicated convolution process. However, similar to before, performing the calculations in the Fourier plane by application of the convolution theorem makes it quite easy to do so. In particular, then the OTF of the system consists of different contributions, which appear in a simple product,

$$\mathrm{OTF}_{\mathrm{total}}(x) = \mathrm{OTF}_{\mathrm{optics}}(x) \cdot \mathrm{OTF}_{\mathrm{filter}}(x) \cdot \mathrm{OTF}_{\mathrm{sensor}}(x) \cdot \ldots . \tag{5.53}$$

The different contributions are indicated by their indices and potentially even those may be written as products themselves. Equation (5.53) corresponds to a linear operation and thus requires that all involved processes are linear as well. For instance, if the detector response is nonlinear, the calculation cannot be performed as easily, and convolution may have to be calculated explicitly.

Although the OTF provides full information on the optical system, in most cases within optical imaging it is sufficient to restrict to the MTF, and consequently, in the following we will do so as well:

$$\widetilde{B}_{\mathrm{im}}(k_x) = \widetilde{B}_{\mathrm{obj}}(k_x) \cdot \mathrm{MTF}(k_x). \tag{5.54}$$

Similar to OTF, the MTF can also be written as a product of the different contributions:

$$\mathrm{MTF}_{\mathrm{total}}(x) = \mathrm{MTF}_{\mathrm{optics}}(x) \cdot \mathrm{MTF}_{\mathrm{filter}}(x) \cdot \mathrm{MTF}_{\mathrm{sensor}}(x) \cdot \ldots , \tag{5.55}$$

where all of them can be calculated or measured separately with the same restrictions as discussed above. But one has to pay attention. Separation in such a product is only possible when phase is lost during light propagation from one component to another one. Usually, this is the case for separate components such as the optics and the sensor.

However, this is not the case within a component itself. Thus, e. g., separation in an MTF for diffraction and another one for aberrations is not possible. In the following chapters, we will discuss some aspects of these individual terms and later on we will regard the system in total. In any case, we will restrict to imaging with incoherent light.

5.2.4 MTF of lenses, objectives and the human eye

In the following, we will consider particular MTF curves that are obtained in the presence of aberrations, where the aberrations originate from the optics or, e. g., are the result of defocusing. Those MTF curves may be regarded as practical examples.

5.2.4.1 Wavefront aberrations

Section 5.1.5 discusses the MTF of a simple cylindrical or spherical lens, respectively, for the case that aberrations are absent. In that case, the MTF is determined by diffraction only and not much detail of the lens has been taken into account. The lens has just provided the operations necessary for Fourier optics transformations.

On the other hand, a rigorous wave optical treatment regarding a single lens as a phase object element can be performed. In addition, this may be extended to a system of lenses, including apertures, in particular, the entrance and the exit pupil. All this includes calculation of PSF and MTF or OTF in the presence of aberrations. However, it is clear that this requires a rather extended discussion as it can be found in specialized books (e. g., [Goo17, Smi08]). It is not necessary to repeat such a detailed discussion here. Nevertheless, we would like to illustrate some basic background and, in particular, discuss the reason and the effect of aberrations in principle and provide some examples. Again, for simplicity, we use the word “lens” either for a single lens only or a lens combination like a camera lens. For the same reason, only a simple lens is used in the figures.

Aberrations lead to deviations of the real wavefront $\mathscr{E}_0 \cdot \exp(\mathrm{i} \cdot k \cdot WF_\mathrm{r}(x,y))$ from the ideal one $\mathscr{E}_0 \cdot \exp(i \cdot k \cdot WF_\mathrm{i}(x,y))$. This can be described by the wavefront aberration:

$$\Delta WF(x,y) = WF_\mathrm{i}(x,y) - WF_\mathrm{r}(x,y), \tag{5.56}$$

which then determines the optical path difference (OPD).

As discussed in Section 3.5 and [Kin10, Smi08] and standard optics textbooks, lens aberrations may be described by third- and higher-order aberration theory (Seidel aberrations). Thus, $\Delta WF(x,y)$ may be expressed by

$$\Delta WF(x,y) = A\cdot(x^2+y^2)^2+B\cdot y\cdot(x^2+y^2)+C\cdot(x^2+3y^2)+D\cdot(x^2+y^2)+E\cdot y+F\cdot x+G \tag{5.57}$$

where the coefficients A to G describe spherical aberration, coma, astigmatism, defocusing, tilt about the x-axis, tilt about the y-axis and a constant term, respectively.

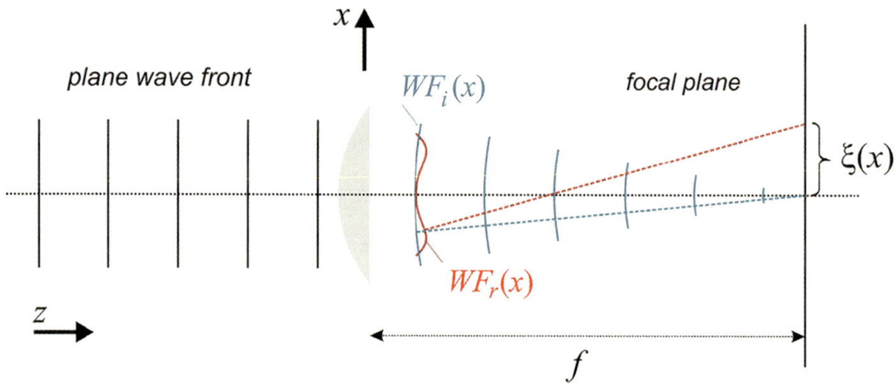

Fig. 5.24: Focusing of a wave with plane wavefront incident on a lens (black solid line). The wavefront behind the lens without aberrations and in the presence of wavefront aberrations (WFA) is shown as a blue solid line and red solid line, respectively. Two corresponding rays are plotted as dashed lines.

It is well known (see Section 5.1.1) that the local direction of the wave vector (the "ray") is perpendicular to the (real) local wavefront, i. e., to the plane of constant phase $WF_{\mathrm{r}}(x,y)$,

$$\vec{k}(x,y) = \left(\frac{\partial WF_{\mathrm{r}}(x,y)}{\partial x}, \frac{\partial WF_{\mathrm{r}}(x,y)}{\partial y} \right). \tag{5.58}$$

This is shown in Figure 5.24. Instead of a perfectly converging ray bundle, which does intersect the focal plane on the optical axis, now each ray may have a deviation. The distance between the ideal and the real ray position in the focal plane is given by $\xi(x,y)$, or $\xi(x)$ in 1D geometry. $\xi(x,y)$ depends on the focal length f and $\Delta WF(x,y)$.

Similar to focusing, within imaging process aberrations may also lead to a shift of the position of the "image points" with respect to the ideal case. Of course, this reduces the image quality. Although there is no clear general criterion for image quality, it is possible to use $\xi(x,y)$ and take its mean square as a measure for quality characterization of such deviations. In addition, due to the diffraction effects discussed before, there is another contribution. Both contributions are included in the PSF and the corresponding OTF. Figure 5.10 has shown an example.

5.2.4.2 Defocusing

Another example is defocusing. Figure 5.25 shows the OTF for a focusing error in a system with a square pupil. In 1D, this corresponds to a slit. For the more appropriate 2D situation of a circular aperture, the calculation is not straightforward and requires integration. However, differences of the results when compared to those of the square pupil, are not too severe, and thus for illustration the present discussion is sufficient. An ana-

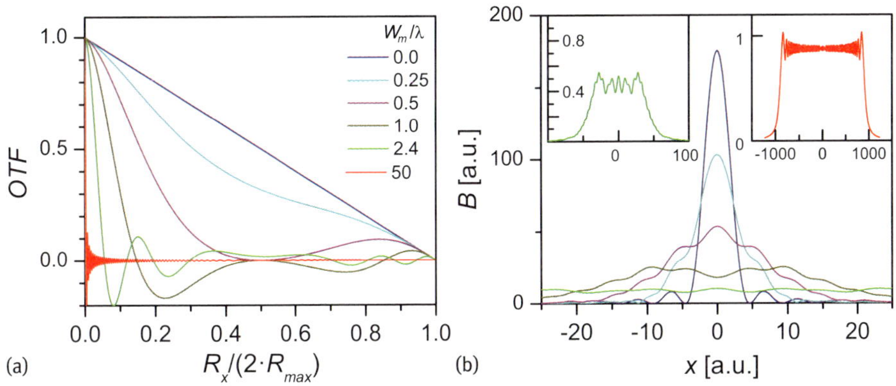

Fig. 5.25: (a) OTF in the case of "defocusing." The different curves are calculated from Equation (5.59) and correspond to different aberrations, which are indicated by the corresponding values W_m/λ. The additional dependence on $f_\#$ cannot be seen here because the abscissa is provided in normalized values (this similar to Figure 5.8) (b) Corresponding PSF, here given by the brightness distribution $B(x)$. The curves and colors in (b) correspond to those in (a) (see legend). The blue curve without aberration (i. e., $W_m/\lambda = 0$) is identical to the blue one in Figure 5.8b and Figure 5.6b, respectively, i. e., that of a slit illuminated by incoherent light. The inserts show extended ranges. A decreased resolution is indicated by a fast drop of the OTF (a) and a large width of the PSF (b).

lytic expression for the OTF in the presence of wavefront aberrations due to defocusing can be found in the book of Goodman [Goo17]:

$$\mathrm{OTF}(\hat{k}_x, \hat{k}_y) = \mathrm{triang}(\hat{k}_x)\mathrm{triang}(\hat{k}_y) \cdot \mathrm{sinc}\left(8\pi\frac{W_m}{\lambda}\hat{k}_x(1-|\hat{k}_x|)\right)$$
$$\mathrm{sinc}\left(8\pi\frac{W_m}{\lambda}\hat{k}_y(1-|\hat{k}_y|)\right), \tag{5.59}$$

with $\hat{k}_x = k_x/(2k_0)$ where $k_0 = (D/2)/(\lambda s_i)$. The values at the curves in Figure 5.25 indicate the amount of defocusing aberration given by a parameter,

$$W_m = -\frac{1}{2}\left(\frac{1}{s_{det}} - \frac{1}{s_i}\right)\cdot\left(\frac{D}{2}\right)^2 \tag{5.60}$$

where s_{det} indicates the actual position of the detector plane measured as the distance from the lens, which for a well-"focused" system would be identical to s_i. For a defocused camera, s_{det} deviates from s_i. Thus, inserting the actual values of s_{det} and s_i into Equation (5.60) and then W_m in Equation (5.59) allows the calculation of the OTF as displayed in Figure 5.25a.

Typically, for a small aperture even a value of W_m of the order of one does correspond to a deviation $|s_{det} - s_i|$ of much less than a mm. As a simple example, we may assume $f = 50\,\mathrm{mm}$, $f_\# = 4$, $s_o = 5\,\mathrm{m}$ and $\lambda = 550\,\mathrm{nm}$. Then, from the lens equation, we obtain $s_o = 50.505\,\mathrm{mm}$. Then, e. g., for a defocusing of $|s_{det} - s_i| = 18\,\mu\mathrm{m}$, from Equation (5.60) we get $W_m/\lambda = 0.25$ and a defocusing of 75 µm results in $W_m/\lambda = 1$. Figure 5.25 shows some examples. To get some insight, these will be discussed in the following.

From Figure 5.25, it may be seen that the resolution is strongly decreased for poor "focusing," and thus the whole image becomes blurred. This blur can be described by the Strehl ratio, which is a simple number of quality measure of image formation. The Strehl ratio is defined by the ratio of the PSF peaks of the image with aberration to that without aberration, where the latter is given by the diffraction limit only. From the example in Figure 5.25b, we obtain 0.6 for $W_m/\lambda = 0.25$ and 0.3 for $W_m/\lambda = 0.5$, respectively. Usually, a Strehl ratio larger than 0.8 is considered as (mostly) diffraction limited.

Even more, an interesting effect may be observed. If, e. g., we have a look at the green curve with $W_m/\lambda = 2.4$ and begin our discussion at $R_x = 0$, then as usual, with an increase of the frequency, the OTF becomes smaller until it reaches zero. Here, this is the case at approximately at $R_x/(2R_{max}) \approx 0.06$. At that point, resolution is totally lost. However, if we continue, the OTF becomes negative, and later on the related MTF, which is the absolute value of the OTF increases again. Consequently, again there is resolution for those frequencies. But due to the negative value of the OTF, now there is a contrast reversal. Thus, e. g., when the object is a grating with a period in that frequency range (here $0.06 < R_x/(2R_{max}) < 0.12$), within the image an originally black line becomes a white one and vice versa. The grating is resolved with a contrast given by the MTF. This resolution is called "spurious resolution" and only beyond the next zero point, at $R_x/(2R_{max}) \approx 0.12$ the OTF becomes positive again and resolution with correct "assignment" of black and white lines, respectively, is recovered. Nonetheless, one must be careful because an isolated contrast or resolution measurement at a frequency beyond the first zero may indicate an image quality that is not real.

Figure 5.25 also shows that for very strong defocusing, the PSF is determined by geometrical optics only. Remember that geometrical effects are convoluted with diffraction effects. Thus, it is trivial that the latter do not much contribute when they play a minor role. Therefore, the OTF in Equation (5.59) is provided by the sine function only. This yields the PSF, which is the geometrical projection of the exit pupil onto the image plane (compare also Section 3.4.6). Here, in 1D this is a slit described by a rect function (see insert in Figure 5.25b). Nonetheless, we may remark that the profile in the detector plane does not show the perfect shape of the rect function. There is ringing close to the edges, which is caused by diffraction. This means that although not much pronounced, diffraction cannot be fully neglected. A good detailed "3D discussion" of defocusing and its effects on MTF can be found in the article by Nasse [Nas08]. Some examples for other common aberrations are shown in Figure 5.10 and in Appendix A.9.

In the following, we will have a more detailed look at three typical camera lenses and potentially the related cameras. In particular, for illustration and as another example of defocusing we would like to estimate the MTF for the circle of confusion (Section 3.4.6) for the three camera lenses discussed in Section 6.9.1, namely I) a high-quality 35 mm wide-angle lens with $f_\# = 1.4$, II) a more relaxed 50 mm angle lens with $f_\# = 2.8$

Tab. 5.2: Parameters for the discussed example in text. The wavelength is $\lambda = 550$ nm and the object distance $s_o = 20$ m. Given values are shown with the gray background.

parameter	example I: 35 mm wide-angle lens	example II: 50 mm normal lens	example III: mobile phone camera ($CF = 7.2$)
f	35 mm	50 mm	3.9 mm
$f_\#$	1.4	2.8	1.8
$D = f/f_\#$	25 mm	18 mm	2.2 mm
PH	24 mm (full format)	24 mm (full format)	3.6 mm
u_i	15 µm	30 µm	4 µm
$s_{DOFoc} = 2 \cdot \lvert s_{det} - s_i \rvert \approx u_i \cdot f_\#$	42 µm	168 µm	14 µm
image distance s_i	35.061 mm	50.125 mm	3.901 mm
W_m/λ (Equation (5.60))	2.4	2.4	0.5
curve color in Figure 5.25	green	green	magenta
R_{MTF10} (read from curve)	$R_x/(2 \cdot R_{max}) = 0.049$	$R_x/(2 \cdot R_{max}) = 0.049$	$R_x/(2 \cdot R_{max}) = 0.319$
$2 \cdot R_{max}$. (from Equation (5.46))	1300 lp/mm	650 lp/mm	1000 lp/mm
$2 \cdot R_{max}$. (from above values)	31,000 lp/PH	15,550 lp/PH	3640 lp/PH
R_{MTF10} (from above values)	64 lp/mm	32 lp/mm	320 lp/mm
R_{MTF10} (from above values)	1530 lp/PH	764 lp/PH	1174 lp/PH
number of pixels per PH, N_v	3200	1600	3000
Nyquist limit of $R_N = N_v/2 \cdot$ lp/PH	1600 lp/PH	800 lp/PH	1500 lp/PH
$1/u_i$	1600 lp/PH	800 lp/PH	975 lp/PH

and III) a lens of a high-quality mobile phone camera with $f_\# = 1.8$. The corresponding diameters of the circle of confusion u_i and the total depths of focus s_{DOFoc} will be discussed in Section 6.9.1. Here, we will only make use of those values taken from that chapter. The parameters of the three examples are listed in Table 5.2. We would like to remark that within this discussion we neglect other aberrations, namely those discussed in Section 3.5.

If we assume an object distance of $s_o = 20$ m, then for the given focal length f, from the lens equation one can calculate the image distance s_i and subsequently from s_i and s_{DOFoc} one gets s_{det}. With these values and the aperture D, from Equation (5.60) we may estimate W_m/λ. Next, using Equation (5.59) for examples I and II, one obtains the green curve in Figure 5.25 (see the legend: $W_m/\lambda = 2.4$) and for example III the magenta curve (see the legend: $W_m/\lambda = 0.5$). From these curves, one can read, e. g., the value $R_{MTF10}/(2 \cdot R_{max})$, namely the spatial frequency $R_x/(2 \cdot R_{max})$ where the OTF has dropped to 0.1. To get the absolute value for that spatial frequency, one has to calculate $2 \cdot R_{max}$ (see Equation (5.46)).

The obtained absolute values of R_{MTF10} may be compared to typical Nyquist limits. As examples, we may assume the number of pixels per PH given in Table 5.2. For example, I and II this may be a DSLR with a 3:2 aspect ratio, and thus we have $N_h \cdot N_v = 1.5 \cdot N_v^2 = 15$ MP and 3.8 MP, respectively. The camera of example III may have a 4:3 ratio, which results in 12 MP. The corresponding Nyquist limits are listed, too.

Now we would like to give some comments on those three examples that have been chosen to illustrate several aspects. Example I shows that a high-quality DSLR equipped with a high-quality lens may yield superior results, even if slightly defocused. Particularly, the R_{MTF10} gives excellent results with > 1500 lp/PH. If the camera has a larger number of pixels, even smaller values of u_i can be supported and the resolution increased. This is the typical situation for a DSLR with 20 MP or more. For the present situation, the number of pixels and the circle of confusion fit well with each other.

Example II may illustrate a good analog SLR camera-lens combination with u_i = 30 µm from the film era. However, even moderate lenses for DSLR support smaller u_i, and thus better MTF curves, and consequently, may lead to better image quality, at least if s_{DOFoc} is similar to Example I. However, for the selected number of pixels given in Table 5.2, N_v and u_i also fit well with each other. For a film SLR camera, the reader may compare the $1/u_i$ value with the MTF of the film as discussed in Section 5.2.4.

Example III shows that a mobile phone camera may also yield high-quality images in the sense of the MTF curve. Other issues, such as noise, good color reproduction, etc. are not an issue here. However, to achieve the good results listed in Table 5.2, a well-manufactured lens and good environmental conditions are necessary. This, in particular, is important to mention because even for a high-quality design as discussed in [Ste12], the design results from the theoretical calculations may not be achieved during mobile phone manufacturing and in particular, not for all devices. Poor environmental conditions may influence the camera lens, and thus the MTF because usually those camera lenses are made from plastic materials, which are much more influenced, e. g., by temperature, than optics made of glass. Moreover, in contrast to DSLR, mobile phone cameras are more sensitive to good light conditions. Besides all that, it is important to note that the good MTF values of Example III are obtained only for images taken directly with the mounted camera lens, which usually is not a zoom lens. Consequently, any zoom is a software zoom, namely a crop of the full image, and that significantly reduces image quality (see Section 1.4.3; exceptions are the very rare SPC with real zoom lenses, see Section 7.3.1.3). As an example, the lens with f = 3.9 mm with a crop factor CF = 7.2 corresponds to a focal length of 28 mm. Then, if the intention is, e. g., a picture with a "normal lens," i. e., a focal length that corresponds to a 50 mm lens for full format, then a crop of roughly 50 % has to be taken. This would reduce the Nyquist frequency to 50 % as well, and consequently, this leads to a significantly degraded MTF curve. Finally, one has to keep in mind that for mobile phone cameras with fixed focus, defocused imaging may occur quite often. We would like to note that for the present example, theoretically the number of pixels would even support a smaller value of u_i. However, then s_{DOFoc} would become smaller, which may not be practical and again, one has to take into account that, e. g., the mobile phone camera under consideration is not manufactured perfectly. A deeper discussion, also how noise effects degrades the MTF is the subject of Sections 7.2 and 7.3.

5.2.4.3 Apodization

Aberrations and risk of defocusing effects could be strongly reduced if the f-number is increased. In that case, diffraction effects may become more and more important, and finally they will dominate. This situation was discussed in Section 2.5.4 (see also Section 5.1.6). In particular situations, this may lead to problems. An example for such a situation is astronomical photography with imaging of two stars of significantly different brightness, which are quite close to each other. The image consists of a superposition of both PSF. If then the maximum of the PSF of the weaker star is at the position of the first bright side lobe of the more intense star, then the weaker star may not be seen. Note that typically stars are so far away that they may be regarded as virtual point sources.

To avoid such a situation, the side lobes (see Figure 5.6) that originate from diffraction at the aperture of the optical system have to be suppressed. This can be realized by an appropriate filter in front of the lens. This filter must have a smooth radial distribution of transmission, such as given by the Gaussian function. This means full transmission at its center and then a decrease of the transmission with increased radial distance. The effect of this filter can be described by a MTF, which here is the Fourier transformation of the Gaussian function. This is a Gaussian function as well (see Appendix A.2). Thus, instead of clipping by a "hard aperture" described by a rectangular function, which yields to the side lobes, this "soft aperture" leads to a smooth decrease of the MTF with increasing spatial frequency. If its width is not too small, ringing is avoided (compare Figure 5.21). Hence, if the width of the filter function, i. e., the Gaussian function, is chosen properly, then the side lobes of the diffraction are removed and the faint image of the weak star becomes visible. Of course, overall optical resolution is somewhat decreased, but if all this is well done, still an image with good quality may be obtained. The width of the Gaussian must be smaller than that of the hard aperture, but still broad enough so that the bandwidth is not too much restricted. This method is called apodization.

5.2.4.4 Dependence on wavelength and f-number and cut-off frequency

Although it is not always recognized, it is clear that the MTF also depends on wavelength and $f_{\#}$. For instance, from Figure 5.8 this cannot always be seen because the abscissa is provided in normalized values. This is due to the dependence of k_{max} on those quantities (see Equation (5.38) and Equation (5.39)). The dependence is displayed in Figure 5.26 and Figure 5.27, respectively. The MTF [lp/PH] calculated for full format (i. e., the SBN) is displayed on the upper and right axis in the figures, respectively.

The MTF of a slit and a circular aperture of a diffraction limited system is determined by $2k_{max}$ (see Equation (5.45)). Hence, the cut-off frequency of the aberration-free

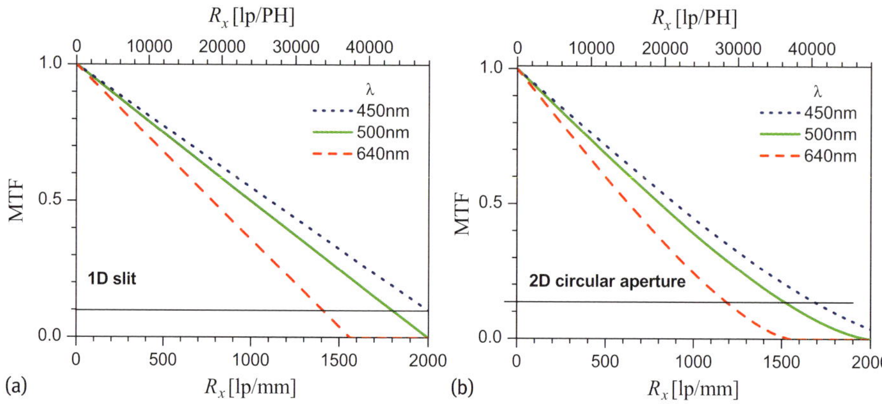

Fig. 5.26: Wavelength dependence of the MTF for a 1D slit (a) and a circular aperture (b), respectively, for an aberration-free optical system (for λ = 450, 500 and 640 nm, respectively; see the legend). Here, $f_\# = 1$. For green light, the present curve is the same as the blue and red one in Figure 5.8b, respectively. In (b), R_x denotes the radial coordinate. The upper axis is calculated for a full frame camera. The solid horizontal line indicates that value, where the MTF is 10 %.

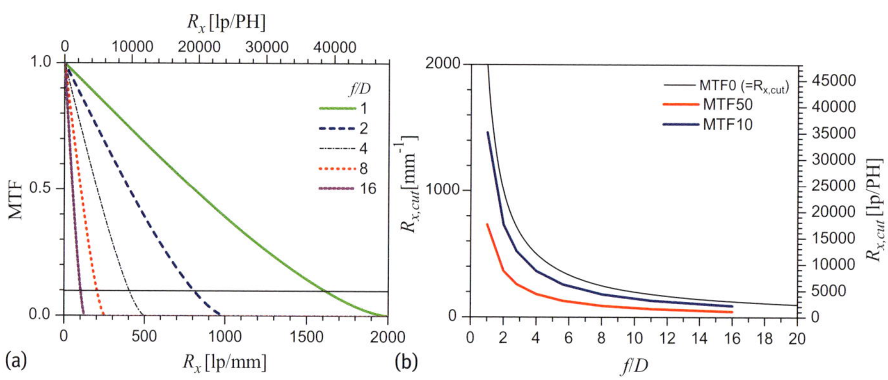

Fig. 5.27: MTF for a circular aperture for five different f-numbers (see legend) for an aberration-free optical system illuminated with 500 nm light (see Equation (5.39)). (b) Dependence of the cut-off frequency and R_{MTF10} and R_{MTF50}, respectively, of the optics on $f_\#$ (from Equation (5.61b)). The upper axis in (a) is calculated for a full frame camera.

optics is given by

$$k_{\text{x,cut}} = 2k_{\text{x,max}} = 2 \cdot 2\pi \frac{\frac{D}{2}}{\lambda s_{\text{i}}} \approx \frac{2\pi}{\lambda f_\#} \tag{5.61a}$$

$$R_{\text{x,cut}} = 2R_{\text{x,max}} = 2 \frac{\frac{D}{2}}{\lambda s_{\text{i}}} \approx \frac{1}{\lambda f_\#}. \tag{5.61b}$$

The right-hand side is an approximation for an object that is far enough away so that we can assume that the image distance is given roughly by the focal length. We would like to comment that Figure 5.27a shows that a smaller $f_\#$ leads to larger $R_{x,max}$, and consequently, to better resolution. However, for real systems there is a trade-off between this effect and the usual increase of aberrations when $f_\#$ is decreased (see, e. g., Figure 5.23). An exception are lenses of extreme high quality where aberrations are still rather low at small $f_\#$ values. In such case, there is an optimum $f_\#$, which depends on the system.

5.2.4.5 MTF of the human eye

Here, it might be interesting to discuss briefly the MTF of the human eye. From Figure 5.28, one can recognize that for large pupil diameters, the MTF is strongly affected by aberrations, whereas for small ones, e. g., $D = 1$ mm (red curve), the eye is not too far from the diffraction limit. This is illustrated in Figure 5.28a where, due to normalization of the abscissa, the black dashed curve is the diffraction limit for all pupil diameters D. R_φ is the spatial frequency in units, cycles/degree, namely lp/degree.

Although not of too much relevance for the following, we may provide an approximate relation between R_φ and R_x. This relation may be obtained on the basis of the so-called "reduced schematic eye" (according to Gullstrand), where R_x is referred to the lateral direction on the sensor, i. e., the retina. When not accommodated, namely when

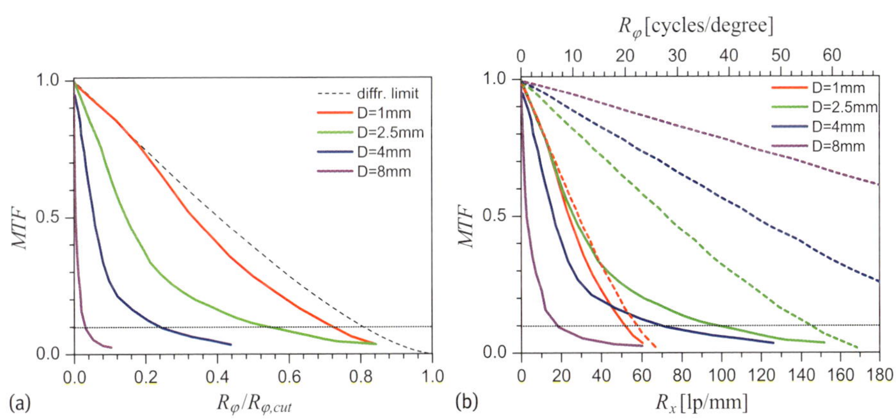

Fig. 5.28: (a) MTF for the human eye for four different pupil diameters (solid lines) and measured at $\lambda = 632$ nm (data taken from[1]). The black dashed line shows the diffraction limit for all D. The abscissa is normalized. (b) Same MTF curves as in (a) (solid lines), but on an absolute scale of the x-axis and for an object that is far enough away so that we can assume that the $s_i = f_{eye}$. The dashed lines provide the diffraction limit according to Equation (5.61b). The solid dotted horizontal line indicates that value, where the MTF is 10 % and may correspond to the resolution limit.

1 P. Artal, R. Navarro: *Monochromatic modulation transfer function of the human eye for different pupil diameters: an analytical expression*, JOSA Comm. (1994) 246–249.

the object is at infinity, the image side focal length of the optical system for the reduced schematic eye is approximately $f_{eye} \approx 22$ mm and the object side focal length is approximately 17 mm. From f_{eye}, the corresponding f-number $f_\# = f_{eye}/D$ can be calculated for each curve in Figure 5.28. Furthermore, for the accommodated reduced schematic eye the distance from the lens to the retina is given roughly by f_{eye}. Consequently, 1 lp/degree takes a lateral distance of approximately $f_{eye} \cdot 1$ degree ≈ 384 µm on the retina. This value corresponds to a frequency of $(384\,\mu m)^{-1} \approx 2.6$ lp/mm. With Equation (5.61b), which yields the cut-off frequency in lateral direction on the retina $R_{x,cut} \approx (\lambda f_\#)^{-1}$ in units lp/mm, we obtain the corresponding $R_{\varphi,cut}$.

However, the absolute value of the spatial frequency in lp/mm can be calculated more easily from the normalized values provided, e. g., at the abscissa in Figure 5.28a, namely from $R_\varphi/(R_{\varphi,cut})$, which is identical to $R_x/(R_{x,cut})$. Therefore, we get $R_x = R_\varphi/(R_{\varphi,cut}) \cdot R_{x,cut}$ and together with Equation (5.61b), $R_x \approx R_\varphi/(R_{\varphi,cut})/(\lambda f_\#) \approx D \cdot R_\varphi/(R_{\varphi,cut})/(\lambda f_{eye})$. Making use of this, Figure 5.28b shows the corresponding curves, now in absolute values.

As an example, we may have a look on the curve with $D = 2.5$ mm (green curve). The resolution can be estimated as that value of the spatial frequency, where the MTF is 10 %, namely R_{MTF10}. For the present case from the diagram, we can read $R_{MTF10} \approx 100$ lp/mm. As 1 lp/degree is equivalent to 2.6 lp/mm (see above), R_{MTF10} is equivalent to 38 lp/degree or in other words 1 lp per 0.45 mrad: This value may be somewhat better if we accept MTF values somewhat below 10 %. This result is equivalent to the resolution of 1 lp/$\Delta\varphi$, where $\Delta\varphi$ is the visual angular resolution from Section 1.4.1. According to Equation (1.16), for the present example we obtain $N_{SB,eye} \approx 1800$, again in good agreement with Section 1.4.1.

We would like to note that here we do not have the intention of a detailed discussion of the properties of the eye. Rather we would like to provide a rough approximation for comparison with camera lenses and systems and allow for some judgments. Albeit in the above discussion, one has also to pay attention that the displayed MTF does not resemble what human beings perceive, it just resembles the physical property of the "eye" as a system similar to the combination of a camera lens and its sensor. Nevertheless, although there is a large difference between the MTF displayed in Figure 5.28 and the CSF displayed in Figure 5.42a at low frequencies, comparison in the high-frequency region shows that the cut-offs do not differ significantly.

5.2.5 MTF of sensors

Prior to the discussion of the MTF of sensors and cameras within this chapter and Sections 5.2.6 and 5.2.7, we would like to note that here the effect of noise will be included only in part. Further discussion of image degradation due to noise is subject of Chapter 7.

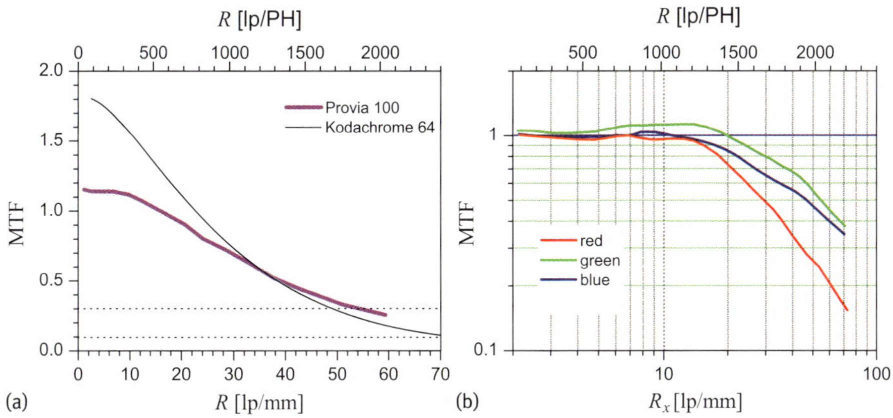

Fig. 5.29: Examples for the MTF of films (full format). (a) FUJICHROME PROVIA 100F Professional and KODACHROME 64 (both daylight-type color reversal films with ISO 100 and 64, resp.) (b) MTF for different spectral regions (red, green and blue, resp.; PROFESSIONAL PORTA 160 film). Data are taken from Kodak and Fuji data sheets, respectively. The dotted lines in (a) do indicate the MTF30 and MTF10, respectively. Note that MTF(R_x = 0) = 1 per definition [Nas09]. The humps and MTF values larger than 1 may be due to edge contrast enhancement or chemical diffusion.

5.2.5.1 Films

All spatial detectors have a certain spatial resolution. Photographic films are such detectors. For films, resolution depends on the film material and the development process. This may be characterized by the MTF. Figure 5.29a shows some examples in a plot that allows easy comparison with the corresponding plots for other detectors and the human eye, respectively. The double logarithmic plot in Figure 5.29b is more typical in the sense, that this is displayed in the data sheets by the manufacturers. This plot also shows that there is a wavelength dependence of the MTF (compare the discussion above).

In general, typical full format films used for photography support spatial frequencies up to between 40 and 150 lp/mm, i. e., 960 . . . 3600 lp/PH. Excellent black-and-white films support more than 1000 lp/mm, i. e., 24,000 lp/PH. Special films together with special processing used for text documentation and special films for purposes such as holography may support several thousand lp/mm. The actual value depends on the emulsion, in particular, on grain size, emulsion thickness, etc. Furthermore, films single value resolution characterization often is expressed by the R_{MTF30}-value. Some examples are given in Table 5.3. In practice, MTF is further reduced if the film is not perfectly flat within the image plane. Even a deviation of 0.1 mm may result in a strongly reduced contrast.

5.2.5.2 Digital sensors

As we have seen in Section 1.6, also digital sensors are affected by resolution. Similar to Section 5.2.1, this can be explained by the usage of test gratings and related contrast function measurements, where the contrast function is identical to the MTF. Although

Tab. 5.3: Examples of R_{MTF30}-values of several 35 mm films. Here, the resolution is the inverse of R_{MTF30}.

film	R_{MTF30}-value [lp/mm]	R_{MTF30}-value [lp/*PH*]	corresponding resolution [µm]
Kodak Ektachrome 160	35	840	29
Fuji Provia 100F RDP	55	1320	18
Fuji Velvia RVP	80	1920	13
Kodak Technical Pan	>140	>3360	7
Kodak Pantomic-X	170	4080	6

there is a difference for sine gratings and bar gratings, for the following discussion of the principle of the MTF of a PDA, this is not important.

Similar to Section 1.6.1, first we will make use of a simple man's view to illustrate detector resolution by placing the test gratings directly on the detector surface. Illumination is made with a collimated light beam from the top. All that is quite the same as shown in Figure 1.17, and thus needs no further explanation. The only difference is that now we will apply gratings of different periods. This is somehow similar to Figure 1.20, but here we will have a closer look.

Figure 5.30a shows four different situations, where the signal strength of the individual photodiodes is characterized by their gray value. In situation (1), the structure is well reproduced, namely there is no light behind the grating bars and only the edges are not perfectly resolved. The contrast is nearly 1 because the maxima are provided by the fully illuminated pixels (shown as pure white) and the minima by those where the signal

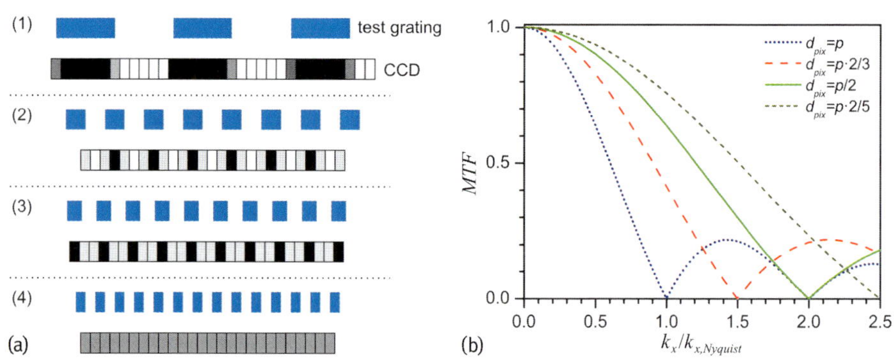

Fig. 5.30: (a) Illustration how objects with different structure sizes are reproduced by a PDA with a fixed given pixel size. The grating is illuminated from top. The transmitted light is detected by the PDA (e. g., a CCD). (1) to (4) represent test gratings with different periods, namely with different spatial frequencies. Similar to Figure 1.17, each of the four different situations illustrates the resolution capabilities of a 1D photodiode array or a row of a PDA. The explanation of the sketches is identical to that given in the caption of Figure 1.17. (b) MTF_{geom} of a PDA or 1D sensor system including OMA, etc. with pixels of rectangular shape and uniform responsivity. The pixel period, namely the pitch, is p and the width of the active area is d_{pix}. Curves are shown for different values d/p (see the legend).

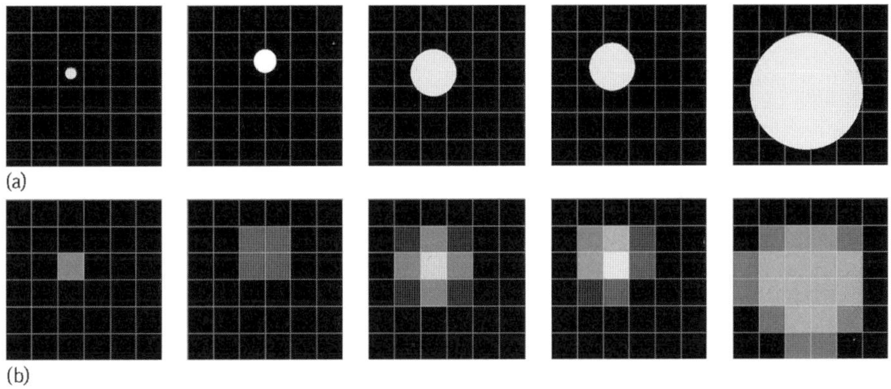

Fig. 5.31: (a) Illustration how images produced on the sensor surface (upper row) are reproduced by a PDA ((b) lower row; the grayscale reflects intensity or brightness level). These diagrams may be regarded as a 2D representation to the situation illustrated in Figure 1.17 and Figure 1.20 (the pixels yield a signal according to the amount of light they are illuminated with (charge integration)).

is zero or at least rather low (shown as pure black). In situation (2), this becomes worse, namely only the pixels behind the center of the bars are not illuminated and only the pixels just in the middle of two adjacent bars may get full light. All other pixels are partly illuminated (shown as gray). Thus, the structure becomes less resolved. In situation (3), there are no pixels that get full light and in situation (4) contrast is fully lost because all pixels show the same gray level. The period of that grating corresponds to the Nyquist frequency.

Thus, in total we see that the finer the structure, i. e., higher k_x, the worse becomes the contrast (compare Section 5.2.1). This is also shown by the gradient of the first bow in Figure 5.30b but remember also the dependence on phase (see Figure 1.17 and Figure 1.20). At higher frequencies, higher orders are present, e. g., represented by the second and the third bow in Figure 5.30b (see also Section 1.6.2). Another example of a sensor MTF is displayed in Figure 8.16b.

In two dimensions (Figure 5.31), the situation, in principle, is the same and again dependence on phase has to be taken into account, which here is illustrated again by shifting the object by half a pixel with respect to the sensor (compare the first two images and the third and fourth image, resp.).

Similar to Equation (5.55), the MTF of an image sensor made of such discrete elements, i. e., the pixels, consists of several contributions and may be written as $\mathrm{MTF}_{\mathrm{sensor}} = \mathrm{MTF}_{\mathrm{geom}} \cdot \mathrm{MTF}_{\mathrm{electronics}}$ and where the MTF due to the electronics $\mathrm{MTF}_{\mathrm{electronics}} = \mathrm{MTF}_{\mathrm{diff}} \cdot \mathrm{MTF}_{\mathrm{CFE}}$ arises from a diffusion term and another one that comes from the charge transfer efficiency.

$\mathrm{MTF}_{\mathrm{geom}}$, sometimes also called $\mathrm{MTF}_{\mathrm{integral}}$, is just due to geometry and includes the pixel pitch p, the width of the active photodiode d_{pix} and geometrical shape (cf. also Figure 4.17). It further may include effects of the optical microlens array (OMA) and if present, further apertures, filters, in particular the optical low-pass filter (OLPF), etc. in front of the pixels (see Section 4.6). This part is a common function for all pixel-based imaging sensors with the same value d_{pix} and shape. MTF_{geom} is provided by the magnitude of the Fourier transformation of the responsivity

$$\mathrm{MTF}_{geom}(k_x, k_y) = \left|\mathrm{FT}\{R_{pix}(x, y)\}\right| \tag{5.62}$$

In the case of rectangular pixels, this is given by a product of the rectangular functions in the k_x and k_y direction, respectively, and the related comb functions in the Fourier space. This comes from the convolution of the Fourier transformation of the pixels PSF and the comb functions. The pixels PSF are related to the finite sensor widths $N_h \cdot p$ and $N_v \cdot p$, respectively, and result in the rectangular functions. For a not-rectangular shaped pixel, see, e. g., the example in Figure 8.16b. The comb functions are the result of the periodicity of the sensor pixels with their periods given by d_{pix} (a derivation can be found, e. g., in [Goo17]). As a result, for this simple geometry one obtains

$$\mathrm{MTF}_{geom}(k_x, k_y) = \left|\mathrm{sinc}(k_x p)\right| \cdot \left|\mathrm{sinc}(k_y p)\right| \tag{5.63a}$$

or more generally,

$$\mathrm{OTF}_{geom}(k_x, k_y) = \mathrm{sinc}(k_x p) \cdot \mathrm{sinc}(k_y p) \tag{5.63b}$$

Equation (5.63) takes into account that OTF_{geom} and MTF_{geom} become zero at the Nyquist frequency, which corresponds to the resolution limit of 2 pixels. This is similar to the diffraction at a slit with an opening of $2p$ ($2p$ instead of $1p$ due to the sampling theorem) with its Fourier transformation $\mathrm{sinc}(k_x \cdot (2p)/2)$. There is also similarity to the diffraction function of an optical grating, where the pattern curve becomes zero when the gap to period ratio equals multiples of 1/2 or 1/3, etc. The corresponding behavior for the MTF is seen in Figure 5.30b (see legend) where one can see that the ratio of pixel width to period takes influence on the location of the zero points.

The OTF of the sensor is given by Equation (5.63b) when we assume that the OTF of the electronics plays a minor role only. It includes the spectral phase information (PTF), which may be necessary for the correct Fourier transformation that relates to the PSF of the sensor.

Similar to the diffraction function of an optical grating where the pattern curve becomes zero when the gap to period ratio equals multiples of 1/2 or 1/3, etc., this is also the case here for the MTF (see the legend; note that the sampling frequency is the inverse of 2π times the period p and the Nyquist frequency is half of that value). This is displayed in Figure 5.30b. Here, we can see that, again similar to a grating, the ratio of pixel width to period influences the location of the zero points.

Also similar to the grating, where the free spectral range is limited by its first diffraction order, the Nyquist frequency, also called critical frequency, is given by half of the sampling frequency, which yields

$$R_N = \frac{1}{2p} \tag{5.64a}$$

in units of lp/mm or

$$R_N = \frac{N_v}{2} \tag{5.64b}$$

in units of lp/PH (see the discussion in Section 5.2.1). Here again, we would like to remind (see Section 1.6.2 and Section 4.6.2) that spatial frequencies larger than R_N lead to aliasing effects (see second and third row in Figure 5.30b). Consequently, the Nyquist frequency R_N is regarded as the cut-off frequency of the sensor.

$MTF_{electronics}$ strongly depends on the sensor electronics and includes a term MTF_{diff}, which depends on charge diffusion including cross talk. This involves diffusion parameters, particle densities and the photon absorption coefficient. In principle, it can be obtained from the diffusion equation in the detector substrate (see Section 4.7 and Figure 5.32a). $MTF_{electronics}$ depends also on another term, namely the MTF of the charge transfer efficiency MTF_{CFE}, which accounts for the modulation loss due to an incomplete charge transfer. It has to be noted that the diffusion term does also depend on wavelength because the penetration depth of the light depends on λ. This is the case, in particular, for oblique incidence and shows up in cross talk by light propagation from one to a neighbored pixel that is affected by the penetration depth (see Section 4.7.5 and Figure 4.35 and Figure 5.32a). BSI reduces such and similar effects so that BSI sensors (CCD or CMOS) have an improved $MTF_{electronics}$, and thus an improved MTF_{sensor}. The interested reader may find a more detailed discussion on this subject in special literature related to sensor electronics and semiconductor physics.

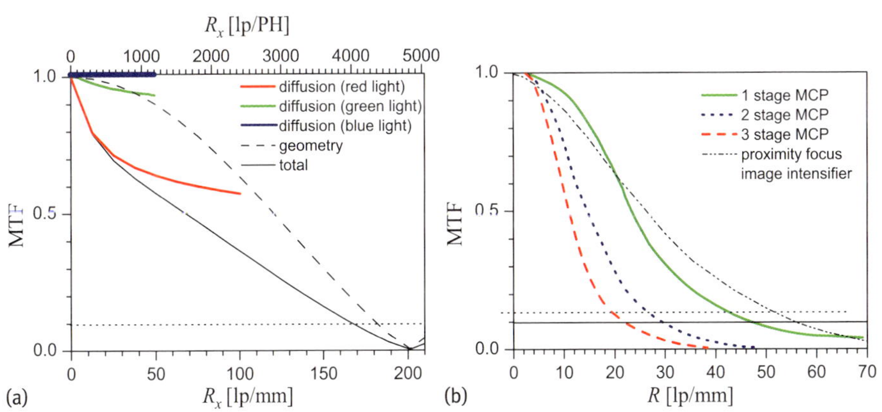

Fig. 5.32: (a) Contributions to the MTF of a scientific sensor. MTF_{geom} (broken black line), MTF_{diff} (for red, green and blue light, resp., solid lines in corresponding color), and resulting overall MTF (i. e., MTF_{sensor}; black line; here for red light channel; for green light MTF_{sensor} is only a little affected by MTF_{diff} and for blue light MTF_{sensor} is nearly the same as MTF_{geom}). These are representative curves for a CCD or CMOS sensor, respectively. (b) MTF of image intensifiers.

We may also note that the individual contributions to the sensor MTF are not always separable. In particular, for CMOS sensors this might be difficult. Furthermore, although all the previous discussion seems to be straightforward, it is important to remark that in general it is not possible to define a complete and mathematically correct $\mathrm{MTF}_{\mathrm{sensor}}$, not even in theory. This is due to nonlinearities within the semiconductor and the electronics (remember, MTF is a linear function). Furthermore, integrated data processing such as flat field correction may also influence the MTF in a nonlinear way and lead to a reduction of the MTF. Nevertheless, an approximate or pseudo-MTF may be used and in that sense we will regard $\mathrm{MTF}_{\mathrm{sensor}}$ in the following.

Similar to CCD or CMOS sensors, image intensifiers and also the other components of an iCCD (see Section 4.11.4) have an MTF. An example is the photocathode. Its resolution depends mainly on the acceleration voltage between the photocathode and the microchannel plate (MCP). A typical value for the MTF10 is 140 lp/mm. An example for the MTF curve of MCP is shown in Figure 5.32b. The limiting value for a one-stage MCP is approximately 45 lp/mm or 1100 lp/diameter and 1800 lp/diameter for plates with a diameter of 25 mm and 40 mm, respectively. Typically, for MCP the limiting values usually are provided for MTF values between 3 and 5 %.

MTF curves become also slightly better with increased voltage. There is an increase of MTF with irradiance as well, but this is not straightforward. First, in the region below approximately 10^{-5} lx the MTF limit increases approximately with the square root of the input signal. In this regime, there is so little light that the "image" is not made by a continuous photon flux but rather results from single photon events. This leads to a very irregular light distribution that only slightly reflects the object distribution. Details cannot be observed until irradiance is significantly increased. This is the photon-counting or low light level regime. If, on the other hand, sufficient light is present, a constant value for the MTF limit is reached, e. g., the 45 lp/mm mentioned above. This is the photon noise or high-light regime.

The third element of an MCP system is the luminescent screen, i. e., a phosphor. Here, the limiting resolution is given by the voltage between the microchannel plate and the screen, by the thickness of the phosphor layer and by the grain structure. To maintain high resolution, it is also essential to use an index matching oil, if, e. g., the phosphor is connected to a fiber optical taper (for details on PSF and MTF of phosphors see, e. g., [Gru02]). A rough estimate of the 3 %-MTF-limit in lp/mm is $500/d$ where d is the phosphor thickness in µm with an upper limit of approximately 100 µm. Because for an MCP with large gain, a large d may not be an issue, the phosphor may not influence the MCP system too much.

For an iCCD camera, the relay optics, either a lens objective or a fiber optics, may also play a role. A commercial photo lens would lead to a significant reduction of the MTF, even if stopped down, e. g., to $f_\# = 8$. We may mention that this large $f_\#$ would increase the coupling loss even more (see Section 4.11.4). Consequently, it is common to make use of specially developed relay optics. For such optics, the MTF usually is much better than that of the MCP over the whole image field (compare to Section 5.2.8), so that its relative importance within the camera system may be rather low.

5.2.6 MTF of a camera system and its components

Most of the MTF curves shown up to now have either been related to optics only or to sensors only. Now, we will have a look at a system. Regarding Equation (5.55), the MTF of a system is the product of the MTF of its components. An example is an iCCD camera where all elements have their own MTF, and even coupling in between them may play a role. Consequently, all components are not necessarily independent from each other and thus not all of them can occur as a term within the product in Equation (5.55). Often one of the components has a cut-off frequency that is significantly lower than that of all the other components, and thus dominates the MTF of the system $MTF_{system} \equiv MTF_{total}$. Such situations are shown in Figure 5.33 where different cameras with different combinations of low- and high-quality lenses and sensors, respectively, are compared.

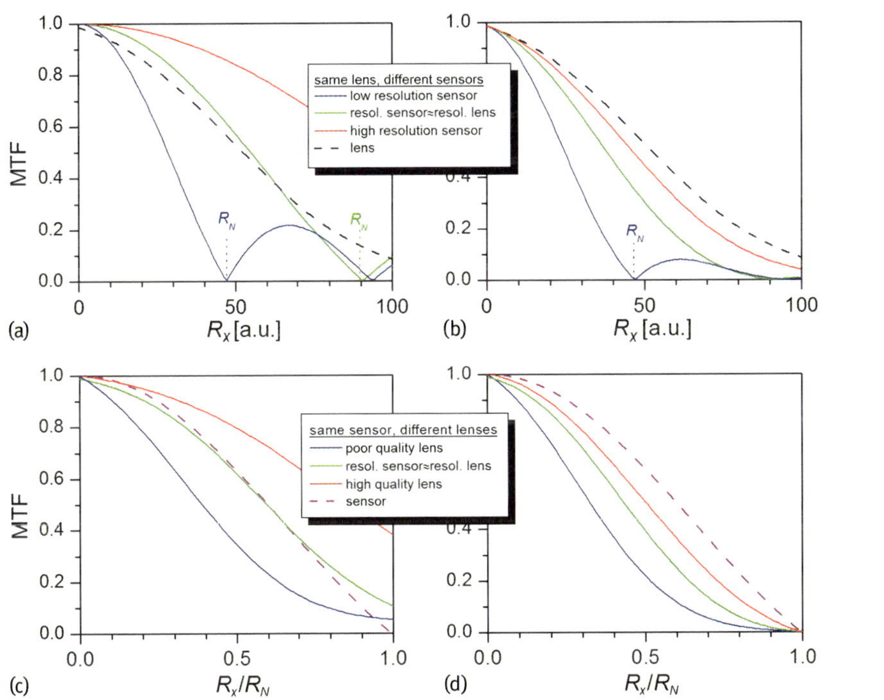

Fig. 5.33: Artificial curves of the MTF of a system, i. e., a camera. (a) MTF of the components (solid lines): a given lens (MTF_{lens} shown as black broken line) is combined with three different sensors (MTF_{sensor} shown as red, green and blue line, resp.). (b) resulting MTF of the system MTF_{system} (solid lines) and for comparison again MTF_{lens} (broken line). (c) MTF of the components (solid lines): a given sensor (MTF_{sensor} shown as a magenta broken line) is combined with three different lenses (MTF_{lens} shown as red, green and blue line, resp.). (d) resulting MTF of the system MTF_{system} (solid lines) and for comparison again MTF_{sensor} (broken line).

Figure 5.33a,b show the situation (I) that a high resolution sensor is significantly better than the chosen lens, and thus the corresponding system is dominated by the lens. In situation (II), a poor resolution sensor is significantly worse than the chosen lens and, therefore, the corresponding system is dominated by the sensor. Note that the useful range of spectral frequencies of the camera is limited by the Nyquist frequency R_N.

Figure 5.33c,d shows the situation (I) that a high-quality lens is significantly better than the sensor, and consequently, the corresponding system is dominated by the sensor. In situation (II), a poor-quality lens is significantly worse than the sensor, and hence the corresponding system is dominated by the lens. This corresponds to the example of Section 1.6.4. This "outresolving" of the lens may also happen if a camera body with a sensor of, say 50 MP, is used with just a standard lens of the manufacturer. To take profit of such a body, of course, one has to use very high-quality lenses.

From this discussion, one can also conclude that one has to be careful to judge the quality of a lens on the basis of a MTF measurement of the whole system and vice versa for the sensor. But, of course, specific combinations may be compared quite well with different ones.

5.2.7 MTF curves of cameras

Figure 5.34 shows MTF examples of real cameras, where the presented curves indicate typical examples of several camera lens combinations to show what is available. Here, we place emphasis mainly on the essentials rather than MTF curves of the latest and most advanced models. This means that if the following discussion is understood well in addition to the other issues related to this topic (see Chapter 8), then understanding and judgment of MTF curves should be straightforward, in principle. For a more detailed understanding and a more advanced judgment, we refer readers to the extended discussion of MTF curves in the excellent articles from Nasse [Nas08, Nas09]. But here we also refer to the continuation of the present discussion in Section 8.3 where we will see that MTF characterization is not always straightforward. This is because there is no single curve for one system, since MTF may change over the image field, it depends on $f_{\#}$, ISO-value, and so on.

The MTF curves in Figure 5.34 are those from a compact camera, a professional DSLR camera with a good lens, a medium format camera system and a high-performance monochrome camera. The parameters are listed in Table 5.4. We may note that an MTF of a mobile phone camera does not make too much sense because this will be dominated by the unavoidable image processing that may dominate the result. The analysis of the principal capabilities of such a system would require much more effort and a rigorous investigation of the system and its components with exclusion of the influence of the image processor.

Figure 5.34a shows the MTF as a function of R_x in mm^{-1}. In this case, of course, a system with smaller pixels allows for more lp/mm (blue curve), however, as discussed in

Tab. 5.4: Parameters for the camera systems of which MTF curves are displayed in Figure 5.34. Note that here $f_{\#}$ is roughly the same for all camera systems. The colors indicate the relation to the MTF curves in Figure 5.34.

camera	sensor size [mm × mm]	sensor size (pixels)	pixel width [μm]	Nyquist limit [lp/*PH*]	crop factor	lens: *f* [mm] (full format equivalent)	lens: $f_{\#}$ in diagram
compact camera	5.875×4.4	3456×2592	1.7	1296	6	approx. 8.3 (approx. 50)	
DSLR	36×24	5616×3744	6.4	1877	1	50 (50)	2.8
medium format	45×30	7500×5000	6.0	2500	0.8	70 (56)	2.5
monochrome	35.8×23.9	5952×3968	6.0	1984	1	90 (90)	2.5
industrial camera	36×24	12,024×8016	3.0	4008	1		not shown
diffraction limit	–	–	–	–	–	–	2.8

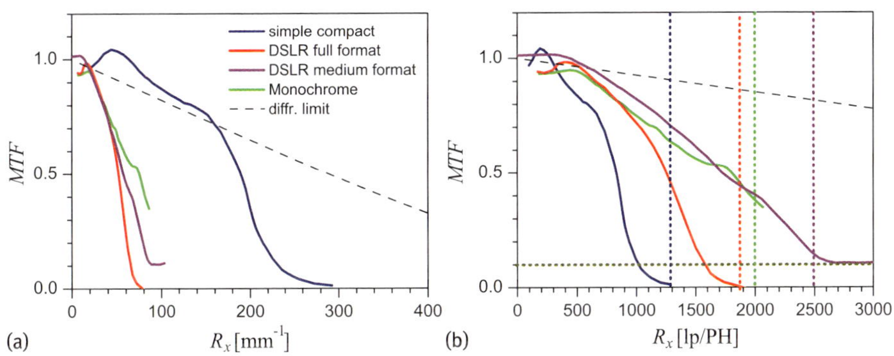

Fig. 5.34: Experimentally deduced MTF of different camera systems (camera lens combinations; see Table 5.4). The curves and the legend for (b) are the same as in (a). The broken line curve in (a) and (b) indicates the diffraction limit for an aberration free optics with $f_{\#} = 2.8$, which is the same or close to that used during the measurements (see Table 5.4). The Nyquist limits for the different camera systems are indicated by the vertical dotted lines in the corresponding color (see Equation (5.64b)). The horizontal dark yellow dotted line in (b) indicates the resolution limit according to R_{MTF10}. We would like to note that not all displayed curves seem to replicate the "real" MTF carefully. This is discussed in the text.

Section 1.6.4, a larger value of lp/mm does not necessarily lead to a better image. Better images require a larger number of lp/*PH* or lp/*PW*. Consequently, the same curves, but now displayed with changed units at the abscissa, allow better comparison of image quality (Figure 5.34b).

R_x in lp/*PH* or lp/*PW* is equivalent to the SBN, which is a measure of the information content within the image (for full format $\mathrm{lp}/PW = 1.5 \cdot \mathrm{lp}/PH$). The reader may compare these curves with the corresponding curve of the eye (Figure 5.28) and its R_{MTF10} or SBN value.

From those curves, a large difference in image quality is expected, but we would like to note that, of course, MTF is only one issue out of several (see, e. g., Chapter 8). Indeed, this is the case as can be seen from Figure 5.35. From that figure, it is obvious that for all compared camera systems, the compact camera provides the poorest image quality (blue curve in Figure 5.34). The difference from the other cameras is huge. This is even the case if we rescale the image taken with the DSLR to the same pixel number (see Figure 5.35b).

Before we continue our discussion, we would like to make some comments on Figure 5.35. The displayed frequency range for the DSLR in (c1) is twice as large as that of the compact camera (a1) in each dimension and that of the downscaled image of the DSLR (b1), respectively (R_N of the DSLR is a factor of 1.45 larger when compared to the compact camera). The range of both, k_x and k_y ranges from $-2 \cdot R_N$ to $+2 \cdot R_N$. However, due to the aspect ratio of 3 for the present image (all those are crops), here the resolution in the x-direction, namely given in lp/PW, is a factor of 3 better, when compared to the y-direction, where this is given in lp/PH. The distributions are displayed in false colors according to the bar shown on the right-hand side. Due to the same scenery, $\widetilde{B}_{obj}$ is almost the same for all three images. Therefore, the differences indicate the differences of the MTF, and thus the quality of the image. This can also be seen from the 1D MTF line profiles obtained from other measurements that are shown in Figure 5.34 as the blue and red curve, respectively. The diagrams Figure 5.35a1 to c1 serve for illustration. They clearly show the strongly restricted frequency range of the image (a), when compared to that obtained with the DSLR. This is even the case for the image (b). This also indicates that at the same spatial frequencies, the MTF values of the compact camera are significantly lower than those of the DSLR.

Now we would like to come back to the discussion of Figure 5.34. The DSLR system, the medium format system and the monochrome system have a Nyquist frequency not far from that of the eye or even exceeding it. Thus, they provide excellent image quality. However, although the Nyquist frequency is approximately the same for both systems, the monochrome one is much superior because at high-spatial frequencies (even close to the Nyquist frequency) the MTF indicates high contrast (this also shows up in the much larger R_{MTF10} value). It is also evident that its lens is much better than the sensor of the camera. But it should also be noted that the sensor of the monochrome camera is still superior to that of the DSLR with the Bayer mask sensor: with a Bayer mask, only half of the pixels "are green" and only one-fourth are sensitive to red or blue light, respectively. As a result for green light, $R_{N,green} = R_{N,monochrome}/2$ and that for red or blue light $R_{N,red} = R_{N,blue} = R_{N,monochrome}/4$. Of course, this affects the MTF as well, as can be seen, e. g., from the example displayed in Figure 8.16a. Consequently, if for a specific application, color is not an issue, a camera without a Bayer mask, namely a monochrome camera, is preferable when compared to one with Bayer filters (but there might be other reasons as well to choose a monochrome camera). In particular, this is often the case for scientific applications. An example is the measurement of a laser beam profile or a laser focus. Here, it is very advantageous to use a monochrome camera. But we would

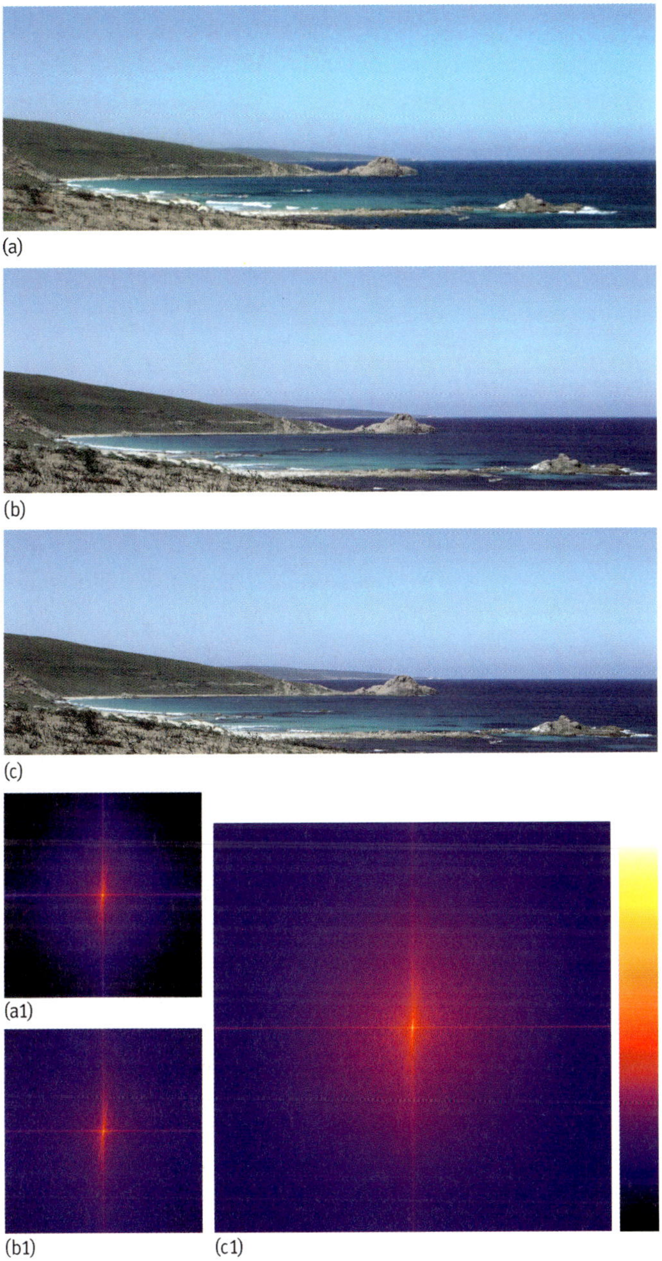

Fig. 5.35: Image examples of the same scenery captured with two different cameras, (a) a simple compact camera (2.5-inch sensor; blue curve in Figure 5.34) and (c) a full format DSLR (red curve in Figure 5.34). (b) For additional comparison we rescaled the image displayed in (c) to get the same pixel number as the simple compact camera. (a1) to (c1) show the 2D distributions $\widetilde{B}_{\mathrm{im}}(k_x, k_y) = \widetilde{B}_{\mathrm{obj}}(k_x, k_y) \cdot \mathrm{MTF}(k_x, k_y)$ of the corresponding images displayed above.

like to mention as well that this disadvantage of color cameras is not present for a sensor with stacked color information (see Section 4.10.1) because the color information is not distributed into neighbored pixels of the sensor surface, but always located at the same (x, y) coordinate where it is obtained at different depths within the semiconductor. Hence, if the number of pixels is the same in the sensor plane, such a sensor may have the same resolution as the monochrome sensor. However, there may be other effects that lead to differences, e. g., charge diffusion effects (see discussion in Section 4.10.1).

Coming back to the presented example, the medium format system yields the best performance of the MTF of the camera systems displayed in Figure 5.34. Here, performance of the lens is roughly similar to that of the monochrome system, but the sensor has a larger Nyquist limit. Table 5.4 gives also an example of a 12-bit camera that is used for industrial applications. This 96 MP camera has a Nyquist limit of 4008 lp/PH, and thus if equipped with an appropriate lens, should have a superior optical performance.

Here, we have to add an essential comment. Nearly all of the displayed curves seem to not fully replicate the "real" MTF. As seen above in this chapter, "real" MTF steadily decreases with R_x, but the displayed curves that result from a measurement do not. In particular, some of them show a "hump" at intermediate frequencies. Such a hump usually is the result of image processing. This is done automatically within the camera, or it results from an unavoidable process applied by the raw converter (see Section 5.1.9 and Section 5.2.7). In both cases, this usually results from contrast enhancement.

It is also clear that the total MTF cannot exceed that of the individual contributions and $\text{MTF}_{\text{optics}}$ cannot exceed the diffraction limit (see broken line in Figure 5.34). Thus, it seems likely that the displayed MTF of the medium format system (magenta curve) has just been shifted slightly upwards to give the impression of a better performance. If this possible shift were compensated, it may be speculated that it would not be too different from the MTF displayed by the green curve (which is still excellent).

It is unambiguous, that for a realistic judgment of the image quality that could be achieved with a camera system, namely a judgment restricted to the "hardware," that additional image processing should be avoided as much as possible. Otherwise, one would judge the image processing capability of the camera, which is not useful, because this also can be performed as post-processing in a computer. Within simple or compact cameras or within mobile phone cameras, image processing cannot be avoided. Therefore, for such systems, simple judgments on the basis of MTF curves are at least partially questionable. This may also be one reason why MTF curves of mobile phone cameras usually are not published.

5.2.7.1 MTF curves of cameras with curved image sensors

Up to now, we have discussed the MTF of cameras with image sensors that are flat. In the following, we provide an example of such ones obtained with curved CIS as discussed in Section 4.12. This is important because one may expect that in the future curved image

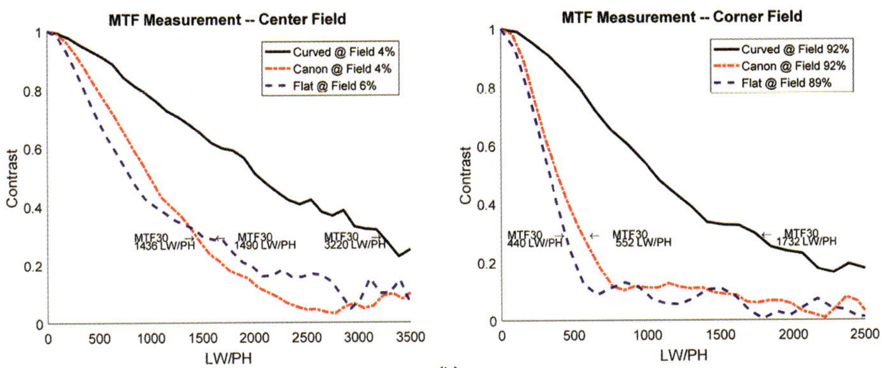

(a) (b)

Fig. 5.36: Measured MTF of a curved CIS (Aptina AR1820HS with a subsequent bending; black line) and comparison to a commercial full frame DSLR equipped with an $f_\#$ = 1.2 optics (Canon 1DS mark III with a 50 mm lens; red line) and a camera consisting of an Edmund Optics 6 mm lens mounted on a flat Aptina AR1820HS sensor (blue dotted line). (a) measured in the image center and (b) in the corner. The Nyquist limit is 3680 LW/PH for the Aptina sensor and 3750 LW/PH for the DSLR. Note that here the unit of the abscissa is LW/PH = line width per PH, which is twice of the usual lp/PH. Image taken as part of Figure 8 from[2]. Reprinted with permission from B. Guenter et al., Opt. Expr. **25** (2017) 13010, https://doi.org/10.1364/OE.25.013010; #286903 Journal ©2017 The Optical Society, Optica Publishing Group.

sensors may become an important role, in particular, as first cameras with such sensors have been introduced into the consumer market. Figure 5.36 shows the measured MTF of two combinations of a lens with a flat sensor in comparison to a camera that uses a curved CIS (another example has been shown in Figure 4.96 in Section 4.12). The field of view and the f-number is rather similar in all cases. It can be seen that within this example the camera with the curved sensor has a MTF30, which is twice that of the cameras with the flat sensors. The improvement of the image quality is obvious. However, this needs a brief discussion. The present example is made for a large aperture, namely $f_\#$ close to 1, where it is rather difficult to design and manufacture lenses with very high-image quality. Nonetheless, such lenses exist, in particular, the Otus 1.4/55 by Zeiss company, which is claimed to be the best standard objective lens worldwide. There are other lenses quite close to it, such as lenses of the Milvus series offered by the same company.

If one compares, e. g., the MTF55 of the Otus (mounted to an appropriate DSLR body), which is above 960 lp/PH, with the MTF55 value taken from Figure 5.36 for the curved sensor that is slightly below 960 lp/PH, or if one does a similar comparison at other positions of the image field or at other MTF values, then one can clearly see that a curved sensor with an appropriate lens may not be superior with respect to the MTF. Nevertheless, vignetting of the camera with the curved sensor may be smaller. But there is a large

2 B. Guenter et al.: *Highly curved image sensors: a practical approach for improved optical performance*, Opt. Expr. **25** (2017) 13010.

difference in the lenses. Especially when compared to the lens connected to the curved CIS, the Otus lens, which is made for flat sensors is a rather complex, bulky and very expensive lens. But this is an extreme situation. Most common lenses do not allow for such a large aperture and if available, usually they are not operated at, e. g., $f_\# \leq 1.4$. In that way, for the present example, it can be expected that, similar to other high-quality lenses operated at, e. g., $f_\# = 2$, the DSLR with the same lens will also lead to a high-quality image similar to the camera with the curved sensor.

This discussion follows our comment at the end of Section 4.12, where we have pointed out that attention has to be paid for comparison of cameras equipped with flat and curved sensors, respectively. If optimized for both cameras with a given $f_\#$, the corresponding optics may be significantly superior to the sensor and then due to Equation (5.55) the MTF is mostly determined by $\mathrm{MTF}_{\mathrm{sensor}}$ (see also Figure 5.33). Here, we have assumed a camera with not more than say 20 to 30 MP, and thus disregard sensors, which have a MTF much superior than that of the optics.

On the other hand, when the MTF of the sensor is much superior to that of the optics the situation changes (see also Section 7.3.1). In that case, and when the sensors are mostly the same and only differ in their radius of curvature (for the flat one this becomes infinity), the Nyquist frequency is the same as well. As a consequence, the MTF becomes zero at the same spatial frequency value and it is also expected that $\mathrm{MTF}(R)$ may not differ strongly, too. The exact course of the curve depends on the sensor design, including the shape of its surface. Thus, a camera with a curved sensor may not show a better image quality. However, as discussed above and in Section 4.12, the design of the appropriate lens may be simplified, which together leads to camera with the advantages discussed there. This, in particular, is important for future application of curved CIS for SPC and other miniaturized cameras. They may take profit of potentially simpler lens design, smaller building depth, etc.

5.2.7.2 Megapixel delusion?

The examples discussed before may also be used for a brief discussion if it makes sense to use cameras with more pixels, namely sensors with higher Nyquist frequency, in particular, ones with 50 MP or more (recent developments has even led to sensors with more than 200 MP). Of course, there are several disadvantages of these cameras. They are very expensive and they require very expensive lenses that support the large cut-off frequency of the sensor. For DSLR, there are not many lenses of that quality. One example, which even is expected to support 100 MP, is the Otus series of the company Zeiss. For SPC, sensors with more than ~50 MP have become more common also for reasons other than resolution (Section 4.10.7). Further discussion will follow in Chapter 7.

Images captured with cameras with more than ~ 50 MP need a huge amount of data capacity on the storage device. In addition, data processing of such images requires advanced hardware and relatively long processing time. Even more, if the sensor consists of an enormous number of pixels, the pixel size is rather small and this might be accompanied with higher noise. Then, although MTF of the sensor is huge, in principle image quality may be significantly reduced by noise (see Sections 7.2 and 7.3).

Furthermore, small sensor pixels require a very stable camera setup and objects that are not moving much during exposure time of the photographs. This is because lateral shifts of the order of the pixel size have to be avoided. This may be hard to fulfil when the pixels become significantly smaller so that shake during t_X may become unavoidable. Moreover, if one would like to make use of the high resolution, "focusing" of the camera has to be very accurate (see also Figure 5.25). This sets a high requirement for the usual (auto)focusing system.

On the other hand, if these "problems" become solved (by low noise sensors, stable setup and so on), image quality may be really improved. At first glance, this seems not to be the case, because shifting the Nyquist limit and, e. g., R_{MTF10} to values beyond the resolution limit of the eye may seem to not make much sense. As discussed in Chapter 1, there is an optimum distance at which a picture with given SBN, e. g., provided by R_{MTF10}, can be viewed at best resolution. Larger values of SBN then only may be useful if one is interested in making high-resolution crops, i. e., "software zoom" or if the goal is large posters that are not viewed in total.

However, this first estimate is only partly true. Of course, one cannot beat the resolution limit of the human eye. But if one compares the contrast of say a 20 MP camera, which may have a contrast of 10 to 20 % close to its Nyquist frequency $R_{\mathrm{Nyquist,\,20\,MP}}$, the contrast of a 50 MP within the same spatial frequency region may be much larger, e. g., 50 %, although larger frequencies, i. e., $R > R_{\mathrm{Nyquist,\,20\,MP}}$, cannot be resolved. The result of this larger contrast is better image quality. This was discussed above and can easily be seen by comparison of the green and the red curve in Figure 5.34 in the range above approximately 1500 lp/PH. This is similar to HiFi technology, where an increase of the cut-off frequency of the acoustic amplifier improves the audio quality, even if it is shifted far beyond the cut-off frequency of the human ear. The present discussion is a good example that in terms of resolution and contrast issues, it may make sense to use a system that has a limit that is much larger than that of the eye, e. g., the R_{MTF10}. Practically, this is also shown in Figure 5.42 and the related discussion.

Concerning multi-ten MP sensors with small pixels, we have to remark that the related discussions on MP delusion in the literature and the web are somewhat strange. Often those discussions are related to the small pixel size. Somehow similar to what we did above, they address the problem of noise. However, there are already related sensors, which may be regarded as a crop of those large sensors, e. g., 50 MP or larger, namely the APS-C sensors. But those have not been criticized within the same discussions. Even if their noise is larger when compared to, e. g., 20 MP sensors, they work quite well. Of course, upscaling of an APS-C sensor to a full frame, just by "arranging together several smaller sensors" is really a challenge and properties do not always scale in the same way. But this is not the issue here. Again, the issue from the user point of view is the application of a camera lens that supports the small pixel size and this over the whole full frame field. Furthermore, it is ridiculous to restrict discussion simply to absolute numbers. Pixel and sensor size have to be included for both. This becomes clear, e. g., for medium format 80 MP cameras, which have undoubtedly high performance or for the huge compound sensors used within astrophysical imaging. But, of course, the megapixel hype for compact cameras and mobile phone is obviously not reasonable.

5.2.8 Sharpness, perceived sharpness, acutance and noise reduction

Sharpness is an important parameter for judging the quality of an image. However, what does sharpness mean? It is straightforward that sharpness has to do with the detail information of the image. If too many details are missing, there is some lack of image information and usually such an image is regarded as "unsharp." If, for a moment, we restrict to black-and-white images where only brightness information on a grayscale is present, then the information content of an image may be described by the SBN. On the other hand, SBN is related to SBP, and thus to the MTF and also to resolution. Other important physical parameters that influence detail information as well are the brightness of the different colors within the image and the color itself. The dynamic range and the depth resolution are other ones. Consequently, there are a lot of parameters that may contribute to sharpness.

Restriction to such parameters may be important if one is interested in scientific applications. However, in photography those physical parameters and, in particular, resolution, are only part of the sharpness that is *perceived* by human beings as a subjective impression, because on one hand, one has to take into account the sensitivity of the eye, and on the other hand, the image processing by the human brain. Examples for the first include wavelength sensitivity (see, e. g., Figure 4.24). Another one is the particular response to a specific spatial frequency range. Although $R_{x,max}$ and $R_{y,max}$ play an important role, the fact is that the eye is most sensitive to the intermediate spatial frequency range (see Figure 5.42 and the discussion below). A *measurement* of sharpness may yield a different result (see also Chapter 8).

It is important to note that humans are also subject to optical illusions. For instance, a particular property of humans is to see faces even in "arbitrary structures," e. g., by recognizing a face in the moon. Another example for the importance of the image processing by the human brain is displayed in Figure 5.37 This illusion shows that the perceived luminance of a part on an image depends on the surrounding one. This and other illusions are considered in image manipulation, especially in HDR (Section 4.9.5).

Consequently, for photography, image optimization with respect to the *perceived* sharpness becomes important, and thus, instead of the spatial frequency response (SFR), which here may be regarded to be equivalent to the MTF, the subjective quality factor (SQF) or acutance plays the major role (in a more strict sense MTF is well-defined as the transfer function of the system or part of it; SFR may include more, e. g., additional

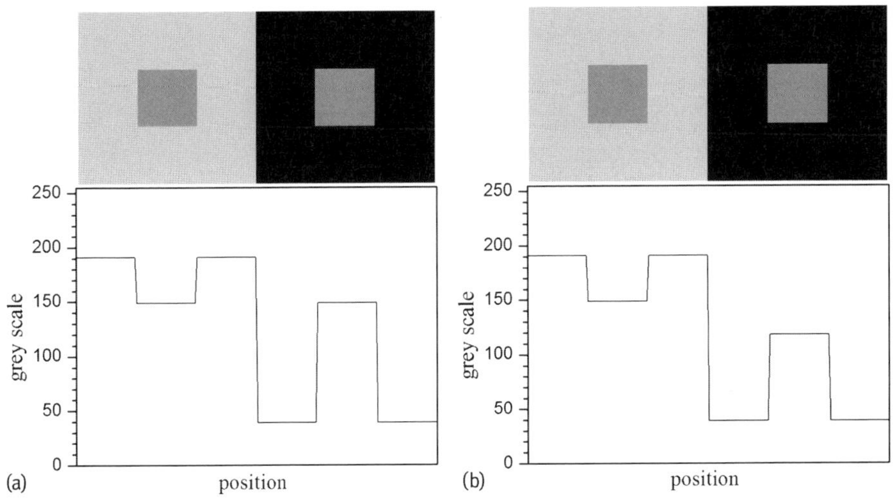

Fig. 5.37: Example of a simple illusion that also affects the perceived image: (a) although the radiance (physical quantity) of the inner dark gray squares are the same (see line profiles below the images), the impression is that they are not (same radiance, but different luminance, i. e., perceived radiance; this is the light technical property). (b) If one changes the radiance of the gray value of the square within the black surrounding, the "perceived radiance" becomes the same, although the physical values are different (i. e., the real ones, see line profiles). But note that for science and technical applications only the real or measured values are acceptable.

influence of image manipulation and potentially noise; see Section 8.3). Acutance is associated with the change of a sharp edge on a spatial coordinate, and thus is given by the gradient of the brightness in the vicinity of the edge structure. It is important to note that this should not be confused with resolution. One may mention again that resolution depends on the camera system. Acutance depends on that as well, but also on the post-processing of the image.

Acutance is a local property. Therefore, sometimes this is called "microcontrast," which has to be discriminated from the global contrast within the image: contrary to a change of the contrast for the whole image at once, mostly this can be adjusted by a parameter within image processing. Image sharpening relies on a sharpening process in small regions of the image, which is successively repeated for the whole area. This process is controlled by other parameters.

Together with resolution and other issues, acutance may influence the perceived sharpness. But note that too much microcontrast often degrades the image quality (see below and, e. g., Figure 5.40 and Figure 8.11). Even physical sharp images may not necessarily look good enough so that sharpening becomes important. This is illustrated in Figure 5.38. Here we may comment that although the images (a) and, in particular, (c) "look better" than (b) which is the original, the latter is the sharper one. Profiles measured along a horizontal line in the vicinity of the center of the vertical beam are displayed

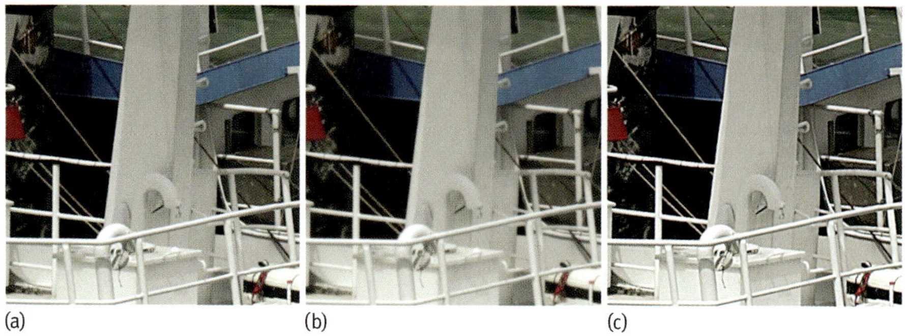

Fig. 5.38: Images that illustrate acutance. (a) is a sharpened version of the original displayed in (b). It has high acutance and high resolution. (b) original image (from raw data). It has low acutance but high resolution. The image (c) has been blurred at first, which definitively reduces resolution. Afterwards, it was strongly sharpened. This leads to high acutance, but resolution remains low. Due to the small size of the image displayed in this book, sharpening has been made stronger than it would be reasonable for a larger print. This is necessary to make the sharpening effect at least somehow visible.

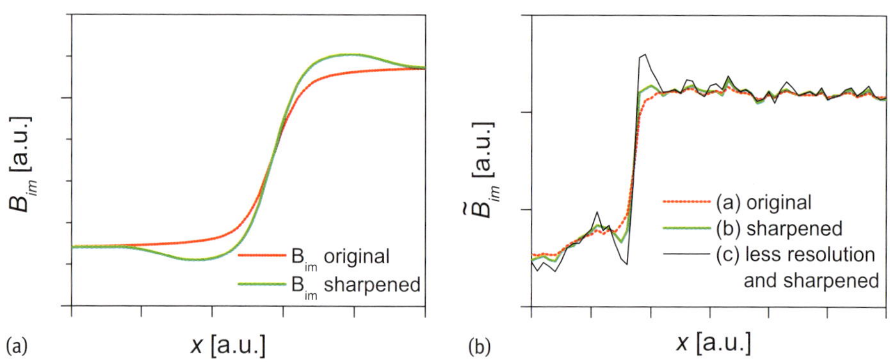

Fig. 5.39: Sharpening of the image of an edge. (a) The red curve is the profile measured along a horizontal line of the edge from the original image, the green one is that of a sharpened edge according to Equation (5.65). (b). Details of line profiles of the real images displayed in Figure 5.38. The overshooting of the sharpening effect is clearly seen (green and black curves). In particular, strong exaggeration is present in the curve of the image with less resolution for which a quite strong sharpening has applied (see the strong fluctuations below and above the edge). Note again, none of the sharpened images has a better resolution than the original; this is still best for the original.

in Figure 5.39b ((a) to (c) in the legend of Figure 5.39b indicate the corresponding pictures in Figure 5.38). The difference from Figure 5.22 is that there the global contrast is changed, whereas here, it is the local one. As discussed, this leads to sharpening.

It is essential that sharpening is done very carefully and generation of artefacts is avoided. Once generated, such artefacts cannot be removed later on, unless strong smoothing is applied. But then this has to be paid by a strongly decreased image quality. We would like to remark that it is much preferable to apply sharpening as a post-processing of the captured image, e. g., on a computer; otherwise, one has to rely on automatic image processing within the camera, which may be acceptable or not.

Although not all photographs intend to reproduce sharp images, in general the perceived sharpness is one of the most important parameters for image quality. Examples where perceived sharpness may not be intended may be found in portrait photography or in sports photography when there is the goal to show the dynamics of a fast-moving person or object through blurry representation.

A simple method that can lead to edge enhancement makes use of a subtraction of a proportion of the brightness values of the neighboring pixels from that of the pixel that is actually processed. In 1D, these are two or more neighbored pixels. The proportion and the radius around the actual pixel offer two parameters that can be adjusted (parameters C and V in Equation (5.65)). This procedure can be described by a simple multiplication in the Fourier space with the following MTF:

$$\mathrm{MTF}_{\mathrm{sharp}}(R_{\mathrm{x}}) = \frac{1 - C \cdot \cos(2\pi \cdot \frac{R_{\mathrm{x}}}{R_{\mathrm{x,max}}} \cdot V)}{1 - C}. \tag{5.65}$$

Multiplication of this function with $\widetilde{B}_{\mathrm{im}}(R_{\mathrm{x}})$ and afterwards, a Fourier transformation then leads to edge enhancement, i. e., a sharpened edge. Figure 5.39a shows an example.

An edge as that one displayed in this figure, is a representation of a boundary within the image. Here, sharpening has led to an exaggeration of the edge, and in particular there is an overshooting at the upper and lower side of the edge, respectively. This significantly contributes to the perceived sharpness. It may be recognized that the perceived sharpness is enhanced, but it has to be remarked as well that if too much sharpening is applied, halos become visible, and which then significantly degrade the image quality (examples are displayed in Figure 5.40).

Here, we may comment that although the perceived sharpness is an important issue of the image quality, noise (or grain in case of films) is so as well. Too much noise may degrade image quality. But within image processing, unfortunately sharpening and noise reduction are just opponents: either the image becomes sharpened, then noise (or grain) or other small artefacts (such as, e. g., visible dust particles) become exaggerated (see Figure 5.40, Figure 5.41), or their visibility is reduced by smoothing, which is just the opposite of sharpening.

It is very difficult and a real challenge, to improve image quality with respect to both issues. Nevertheless, large progress in image processing and, in particular, in noise reduction ("denoising") has been made. Today advanced methods and "artificial intelligence" (AI) are applied. For instance, modern noise reduction is not restricted to the application of a common low pass filter for the whole image or to a smoothing of the image, e. g., by weighted averaging of the pixel signals. But it may be applied locally, i. e., noise reduction can be made differently in different image regions. Moreover, denoising may take into account the specific noise profile of a particular electronic sensor (see Sections 4.7.4 and 4.9) or the specific grain structure of a particular film. We will not

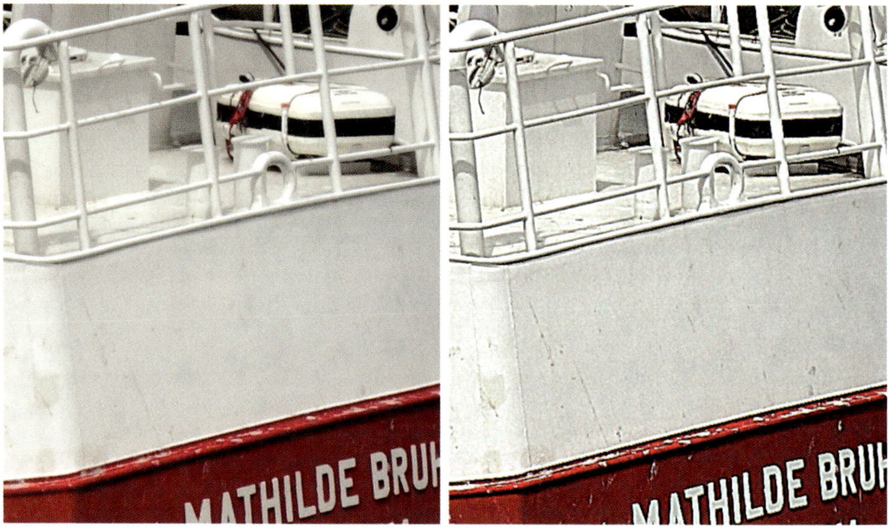

(a) (b)

Fig. 5.40: Example of an image of a digital camera that is processed with too much sharpness (b) and with well-adapted sharpness (a). In (b), for demonstration purpose, the strength of the sharpening process has been intended very large so that it does generate artefacts and leads to an unnatural appearance of the picture. Another example is presented in Figure 8.11.

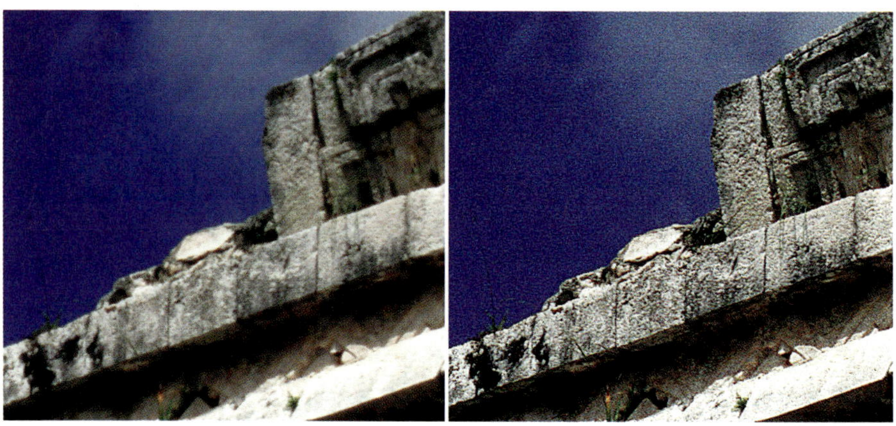

(a) (b)

Fig. 5.41: Similar example as presented in Figure 5.40, but now for a slide film image that is scanned and afterwards processed with too much sharpness (b) and with well-adapted sharpness (a). Here again for demonstration, the strength of the sharpening process is chosen to be very large so that it clearly exaggerates grain. The effect on noise in a full digital image is somehow similar.

discuss all this in more detail and refer to Section 7.4.2 and to the special literature and to software, such as *NeatImage* or *DeNoise AI*.

There are a lot of other methods for image sharpening. Some of them are rather complicated, but simple high pass filtering is not adequate because then the image is strongly darkened (see, e. g., Figure 5.16). One of those methods, namely "unsharp masking" (USM), has been briefly discussed in Section 5.1.9.2.

Again, we would like to emphasize that sharpening may improve the perceived image quality but there is absolutely no way to improve resolution within an image after it has been captured. This means that the physical resolution of an image could be at best the same as that before sharpening or image processing in general.

5.2.9 Judgment of MTF curves

Judgment of MTF curves is not so simple as it seems unless the situation is as clear as in the previous examples presented in Figure 5.34 and Figure 5.35, respectively. Of course, MTF curves with high contrast over a broad frequency range indicate a high-performance system. For scientific or pure technical purposes and, in particular, if the camera system is used for measurements, MTF may be used for judgment quite well. But if the goal is not a measurement, but photography, which relies on the perceived image, the MTF curve and the absolute MTF values alone are not a sufficient criterion for predicting the subjectively perceived image quality in any case. The curves must be assessed appropriately and the viewing conditions in each case must be taken into account. Remember also those very simple examples on the role of viewing conditions that have been discussed in Section 1.4 and Section 1.6.4. And even more, of course, a simple number such as the R_{MTF50} is not sufficient at all for reasonable judgments on image quality, in particular, when different camera systems are compared. For instance, this can be seen from Figure 5.43 where all three curves have the same R_{MTF60}, but obviously the curves are much different and so is the camera performance. Such numbers just allow, e. g., an easy comparison such as in Section 5.2.2 and Figure 5.23.

On the other hand, one can learn quite a lot from the results of a careful advanced measurement of MTF curves generated for different conditions. Although even in that case one cannot get fully rid of subjective factors with respect to image quality, nevertheless a lot of quality issues may be clearly addressed. This is also subject of Section 8.3.

To get at least some idea of "reading" MTF curves, we will discuss two of those displayed in Figure 5.43. For further reading, we refer to the excellent articles by Nasse [Nas08, Nas09]. In his articles, Nasse presents a lot of examples of real images and the related MTF curves together with an interpretation and a detailed discussion.

First of all, based on our previous notes and also following Nasses's remarks, we may note that appraisal on image quality depends on a lot of factors, such as motive, viewing conditions like illumination, viewing angle, and distance, etc. and maybe others. It is also not sufficient to consider MTF curves only (see Chapter 8). Nevertheless, if assessed appropriately, they provide a hint for a reasonable judgment. But assessment

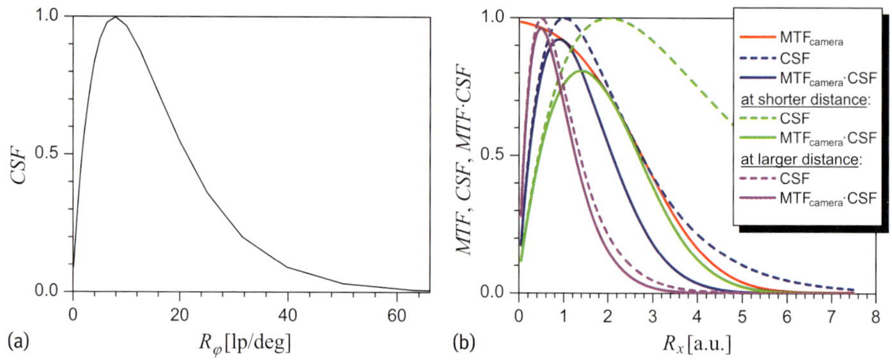

Fig. 5.42: (a) Contrast sensitivity function of the human eye (CSF). (b) Example of a "resolution plot" for a given MTF_{camera} from a camera system (artificial, red curve).

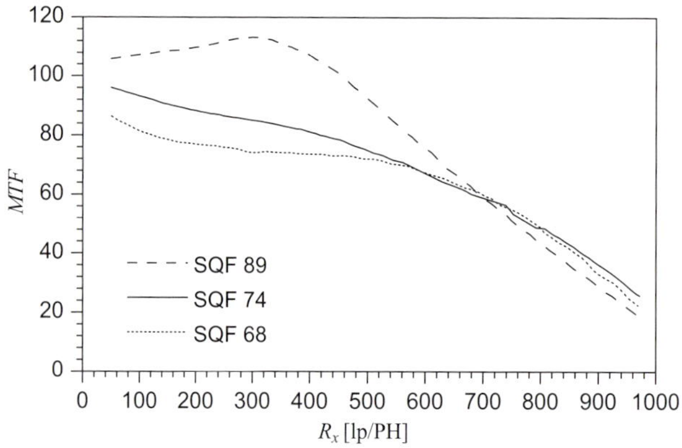

Fig. 5.43: Example of MTF curves of camera systems (data taken from [Nas09]).

also means that one has to take into account the contrast sensitivity function of the eye (CSF) (Figure 5.42). Consequently, we will introduce this function next.

Figure 5.42a shows that the eye is most sensitive within a specific spatial frequency range. Here, the spatial frequency R_φ is given in cycles or lp per degree. As a consequence, resolution with respect to the object depends on the viewing distance and even in a more general way, the perceived image is influenced by the viewing conditions of the object (see Figure 5.42b and Appendix A.12). There is a good approximation formula for the CSF (Figure 5.42a)[3]

$$\mathrm{CSF}(R_\varphi) = (0.0192 + 0.114 \cdot R_\varphi \cdot \mathrm{deg/lp})(\mathrm{e}^{-0.114 \cdot R_\varphi \cdot \mathrm{deg/lp}})^{1.1} . \tag{5.66}$$

3 J. L. Mannos, D. J. Sakrison: *The Effects of a Visual Fidelity Criterion on the Encoding of Images*, IEEE Transactions on Information Theory **20** (1974) 525–535.

We have to remark that the CSF of the above equation is in arbitrary units and not yet normalized.

The dependence on the viewing conditions, in particular on the object distance, is displayed in Figure 5.42b. Here, we may regard a test object that is an image of a perfect test grating with linear period variation such as displayed in Figure 8.6b. Let us assume that the original test grating has full contrast, i. e., the brightness extends from 0 to 1, but the photography has not ("first imaging process" in Figure 1.3). Depending on the distance between the bars where each of these distances corresponds to a particular R_X, there is a particular contrast within the photography, which is given by the contrast $K(R_X) = \mathrm{MTF}_{\mathrm{camera}}(R_X)$. This original photography is now magnified and then serves as the object which should be observed. This means imaging by the eye, i. e., the "second imaging process" in Figure 1.3.

As an example, a full format photography, which serves as the new object, is magnified by a factor M. As a result, the original values of the grating bar distances, and hence also the new spatial frequency values within the new object (for the "second imaging process" in Figure 1.3) are changed to $R'_X = R_X/M$. For observation with the eye, one has to also take into account the CSF. For a specific viewing condition, such as a specific distance, the effective MTF then may be obtained from the product of the resulting CSF and the MTF of the system or camera (see Figure 5.42b). However, CSF is a function of R_φ in lp/deg, and consequently for a particular observation distance this has to be transferred to a particular value of R'_X in lp/mm. R'_X is a function of R_φ and distance. For instance, for 50 lp/deg and a distance of 3 m, one obtains 1 lp/mm.

Now we may discuss three different viewing distances. First, we chose the distance in such a way, that for the highest frequencies $\mathrm{MTF}_{\mathrm{camera}}$ (red curve in Figure 5.42b) and CSF (blue dashed curve) do not much differ. Here, ideal conditions are assumed, i. e., an ideal printout or screen with identical resolution, contrast, etc. as the image captured by the camera. The resulting "contrast observation" is given by the product of both curves and shown as a blue solid line.

Of course, if the same image is observed from farther away, this is our second example; less details could be resolved (magenta lines).

Vice versa in the third case, observation from closer distance allows to see more details (green curves). Even from the simple Equation (1.15), it can be seen that the number of lp/mm, which could be just resolved at one distance, changes when the distance is changed.

Altogether, the magenta, blue and green curves, respectively, clearly show that depending on the observation distance, the eye has its optimum resolution range, namely the highest $\mathrm{CSF} \cdot \mathrm{MTF}_{\mathrm{camera}}$ or contrast values, in a particular frequency range. The reader may check that by a simple experiment using the test chart in Appendix A.12.

One further step to judge image quality under consideration of the viewing distance is the usage of the so-called subjective quality factor (SQF). As Nasse states in his article, *"It has been shown in many experiments with test subjects and many different images that there is a fairly useful correlation between the subjective quality assessment and the area under the MTF curve.*" This area can be calculated by the SQF, which is equal to the integral of $\mathrm{CSF}(R_x)\cdot \mathrm{MTF}_{\mathrm{camera}}(R_x)$ over $\mathrm{d}\ln(R_x)$. We may remark that $\mathrm{d}\ln(R_x) = \mathrm{d}R_x/R_x$ and R_x is a function of R_φ and the observation distance.

We will not continue with details on that within the present book but summarize that a reasonable judgment on the perceived image quality also requires that one takes into account the viewing conditions. Additionally, one may remark that for a more advanced evaluation, even for incoherent light, the PTF cannot be fully neglected and has to be included [Nas08].

However, in spite of the discussed problems with a reasonable judgment on image quality and camera performance on the basis of MTF curves, it is possible to learn quite a lot of those curves for photography. Moreover, for scientific and technical applications, problems with viewing conditions are absent, and thus MTF curves give more direct access to a camera system performance.

To provide a final example for such a judgment, the reader may have a look at Figure 5.43, which is taken from Nasse's article. Following the discussion of the related section in this article, first it is recognized that the resolution (R_{MTF20} or R_{MTF10}) is expected to be quite similar for all three curves. An inspection of all three related images (real pictures, no test charts; but those are not shown here), points out that not the image corresponding to seemingly the best MTF, namely the dashed line curve (SQF 89) is the best perceived one. The reason for that is that the acutance, which is equal to contour definition, plays a major role and high acutance requires flat curves. But it is clear that it is advantageous if these flat curves at the same time have high MTF values. In particular, only then the image is free of blooming and flares (see Section 4.7.5 and Section 6.8) and allows the observation of very fine structures even in bright regions of the image, namely well-reproduced structures with good contrast. The bend of the dashed line curve (SQF 89) within the most important frequency range (see Figure 5.42) reduces the acutance when compared to the flatter solid line curve (SQF 74). This diagram also shows that the SQF is not always a reliable quantity. Although SQF is highest for the dashed line curve (see legend), as discussed, the optical system with this curve is not the best of the three displayed examples.

We may summarize this chapter with a couple of comments. Good optical systems are not necessarily characterized only by the highest possible resolution, although this may be one preferable parameter, and quite large MTF values in the intermediate frequency range. But they are characterized by a well-balanced curve with respect to both, spatial frequency distribution over the whole image field (see Section 8.3). But we would like to remind the reader that MTF is only one parameter; other ones are, e. g., color behavior, distortion, and noise or grain resulting from the digital or analogous sensor, etc.

Altogether, MTF curves have been found to be very successful for lens and sensor characterization. Introduced by Zeiss in the 1930s (by H. Frieser), MTF data have been provided for their lenses quite early and many decades later other companies have followed. However, to the best of our knowledge, nearly all other manufacturers provide calculated MTF curves only. The Zeiss company provides *measured* ones, and these, in particular, for the sold camera lenses. This is of much importance as well. This difference cannot be considered at all minor. On the contrary, there might be a huge difference between design values and the ones of real lenses even in the case of high-quality manufacturing. Moreover, there might be an enormous variation in the batch for rather cheap lenses as used in simple compact or mobile phone cameras. In addition, further changes result from environmental conditions, in particular, for cheap plastic lenses.

Beside those MTF curves supplied by manufacturers, there are tests and investigations by independent laboratories that deliver also measured curves, mostly for complete systems. This is also the subject of Chapter 8. In any case, we may alert that although we expect that many of the published curves are reliable, some of them are not. In particular, one has to be careful, when curves exceed limits that are physically preset. And lastly we would like to note that different manufacturers tune their cameras and lenses in different ways. As an example, a large manufacturer puts emphasis on a larger contrast at high frequency whereas another one may prefer more smooth images. But picture quality may be excellent in both cases. And again, evaluation of MTF curves in general is not straightforward and forejudged statements may be totally wrong.

5.3 Resolution, SBN, MTF and PSF

Although within Chapters 1 and 5 and Appendix A.8 the physics of resolution, SBN and MTF has been discussed completely, this chapter intends to provide a practical illustration based on those discussions and the related equations (the most important ones are summarized in Table 5.5). For that purpose, the following illustrations are made for the usual imaging conditions. Thus, they are restricted to incoherent light, although one has to remark, that imaging of, e. g., stars as (virtual) point objects, may be considered as imaging with coherent light (see Section 5.1.7). For the sensor, we assume square pixels with a fill factor of 1 and a pitch p.

5.3.1 Resolution in general

In a very general sense, the image of an object may be regarded as resolved, when at least a minimum information of its structure can be recognized. This defines “resolution.” In this way, the definition is not strict. It is obvious that it depends on the object, the observer, the illumination conditions and there may be more. As we will see shortly, the resolution differs for 2 points and a grating structure. It depends also on the contrast demand of the observer. And, of course, on the amount of available light, where in the worst case, it is nearly or fully dark and then one can recognize nothing. But one can define resolution more strictly by well-defined methods of measurement and analysis.

In the following, we restrict to 1D geometry with the resolution limit given by Abbe’s criterion. The analysis for the full 2D geometry with the resolution limit given

Tab. 5.5: Compilation of formulas that are relevant for resolution and SBN, for a 1D geometry with a slit aperture $\kappa = 1$, for a 2D geometry with a circular aperture $\kappa = 1.22$ (see Table 5.1). For a system free of aberrations $\alpha = 1$, otherwise $\alpha > 1$ (α is measure of wavefront distortions or aberrations, see Section 5.1.4.2). Note that $f_\# \equiv f/D$ and $R_{x,0}$ is the spatial frequency that corresponds to the first minimum of the sinc function or the Airy pattern. Note that due to $k_x/k = R_x/R = \sin(\theta_{diffr})$ Equations (5.27) and (5.51) are equivalent (CONST = const $\cdot\,\lambda$; D_{diffr} may be set equal either to D or to δ). The most important basic relation is that for the SBP, which is marked by a gray background.

Formula	Equations or Figures
$\delta \cdot R_{x,max} = \kappa \cdot \alpha \geq \kappa = \text{const}$	Equation (5.51a) (far field)
$r_0 \cdot R_{x,cut} = \kappa \cdot \alpha \geq \kappa = \text{const}$	Equation (5.51b) (far field)
$D \cdot R_{x,0} = \kappa \cdot \alpha \geq \kappa = \text{const}$	Equation (5.51c) (near field)
$D_{diffr} \cdot \sin(\theta_{diffr}) = \kappa \cdot \lambda \cdot \alpha \geq \kappa \cdot \lambda = \text{CONST}$	Equation (5.27) (general)
$\delta_0 = 2 \cdot \kappa \cdot \lambda \cdot \frac{f}{D} \cdot \alpha = 2 \cdot \kappa \cdot \frac{\lambda}{2 \cdot NA} \cdot \alpha$	Equations (1.17), (5.23), (5.24), (5.43)
$\delta_0 = \frac{\kappa \cdot \alpha}{R_{x,max}} \Rightarrow r_0 = \frac{\kappa \cdot \alpha}{R_{x,cut}}$	Equation (5.47)
$\delta_0 = 2 \cdot r_0$	Figure 5.6 and Section 5.1.4.2
$k_{x,max} = 2\pi \cdot \frac{\frac{D}{2}}{\lambda s_i} \approx \frac{2\pi}{2 \cdot \lambda \cdot f/D}$	Equation (5.45)
$R_{x,max} = \frac{\frac{D}{2}}{\lambda s_i} \approx \frac{1}{2 \cdot \lambda \cdot f/D}$	Equation (5.46)
$k_{x,cut} = 2k_{x,max} = 2\pi \cdot 2 \cdot \frac{\frac{D}{2}}{\lambda s_i} \approx \frac{2\pi}{\lambda \cdot f/D} = 2\pi \frac{\kappa \cdot \alpha}{r_0}$	Equation (5.61a)
$R_{x,cut} = 2R_{x,max} = 2\frac{\frac{D}{2}}{\lambda s_i} \approx \frac{1}{\lambda \cdot f/D} = \frac{\kappa \cdot \alpha}{r_0}$	Equation (5.61b)
$N_{SB,y} = \frac{PH}{r_0} = \frac{R_{y,max} \cdot PH}{\kappa\alpha} \leq \frac{R_{y,max} \cdot PH}{\kappa}$	Equation (5.52a)
$N_{SB,x} = \frac{PW}{\delta x}; N_{SB,y} = \frac{PH}{\delta y}$	Equation (1.13)
$R_N = \frac{1}{2p}$ (in lp/mm); $R_N = \frac{PH}{2p}$ (in lp/PH)	Equation (5.64)

by Rayleighs's criterion then is straightforward. But we would also like to note that not necessarily a simple expansion of the 1D to the 2D geometry gives more insight, because for an image detector made of a rectangular matrix of rectangular pixels the measurement of the resolution in any direction other than the horizontal or the vertical will be significantly affected by the pixel geometry, which altogether then significantly reduces the resolution (see also 7.3.1.1).

This may also affect the MTF as the product of the transfer functions of the optics and sensor, respectively. $\mathrm{MTF}_{\mathrm{sensor}}(k_x)$ is given by Equation (5.63) and we see that the terms with k_x and k_y separate. Thus, a simple multiplication of the $\mathrm{MTF}_{\mathrm{sensor}}(k_x)$ with $\mathrm{MTF}_{\mathrm{optics}}(k_x)$ (Equation (5.38)) is straightforward. But this is not the case for the product of $\mathrm{MTF}_{\mathrm{sensor}}(k_x)$ with $\mathrm{MTF}_{\mathrm{optics}}(k_r)$ (Equation (5.39)). k_r depends on k_x and on k_y and the terms cannot be separated. Nevertheless, the product $\mathrm{MTF}_{\mathrm{sensor}}(k_x) \cdot \mathrm{MTF}_{\mathrm{optics}}(k_x)$ (instead of $\mathrm{MTF}_{\mathrm{optics}}(k_r)$) is correct when we restrict to the horizontal or vertical direction, respectively. More generally, some effort has to be taken into account to treat this situation properly.

5.3.2 Relations between resolution, SBN and MTF with respect to optics only

We begin this illustration with an image of a single point as the object, where the point may be a mathematical point described by a δ-function (or correctly δ-distribution; see the note in Appendix A.1). An experimental realization may be a star, which can be considered as a virtual point object (see, e. g., Appendix A.8 or Section 5.1.4). The image, i. e., the PSF, is displayed in Figure 5.44a (dashed line; see also Figure 5.6 and related discussion). Its width δ_0 may be described by the first zero values, which are located at $x = \pm r_0$, and thus $\delta_0 = 2 \cdot r_0$. In case of a diffraction limited system as used for the present illustration, δ_0 is given by Equations (1.17), (5.47), etc. (see Table 5.5), and the PSF is shown as the dashed line in Figure 5.44a. Here, i. e., in Section 5.3.2 and Section 5.3.3, $\kappa = 1$ and $\alpha = 1$ (see Table 5.5). More general, the optical system may suffer from aberrations and it also includes the apparatus function of the sensor, namely its MTF. Thus, the PSF may differ from that displayed in Figure 5.44a. Namely, it may have a different shape and a different value of δ_0 and r_0, respectively.

To determine the resolution, one can make use of 2-point objects (Sections 5.1.6 and 5.1.7). According to Abbe's criterion, they are regarded as just resolved with a contrast $K = 10\,\%$ when the maximum of the image of one point is located at the first minimum of the other one, i. e., when the distance between both maxima is $r_0 = \delta_0/2$. This situation is shown as the solid line in Figure 5.44a.

Naively, one may expect that adding further images of point objects at distances of multiples of r_0 will yield the same contrast. However, for such a periodic structure of a "10-point grating" K is much lower and for 50 or even more points the resolution is almost lost (Figure 5.44b to d). This is also described by the related SBN (Equation (5.52)). To obtain the same contrast as for 2 points, and thus to obtain a resolution according to Abbe's criterion, the distance between the points has to be increased to $\delta x = \delta y = 1.06{\cdot}r_0$.

Consequently, filling the image field with image spots (see Section 1.3.1) with inter-point-spacings δx and δy, respectively, yields the maximum SBN with respect to spots that can be just resolved (see Figure 1.10 and Section 1.3.3). For the present illustration

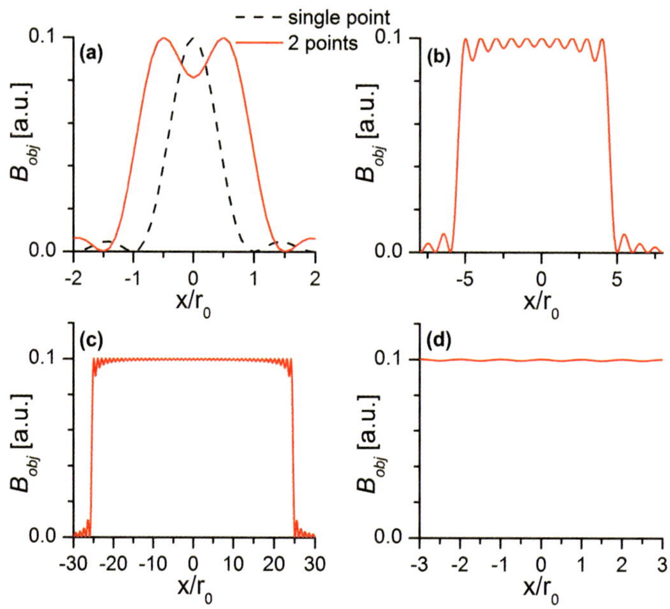

Fig. 5.44: (a) Image of a single point object (dashed line; same as Figure 5.6b) and of 2-point objects (solid line; same as Figure 5.14c), respectively. (b) Image of 10-point objects and (c) of 50-point objects. (d) shows details of (c). Here, as an example we assumed a system that consists of diffraction limited optics only. In general, of course, the PSF has to be taken for the whole camera system.

in one dimension, this yields

$$N_{\mathrm{SB,points}} = \frac{PH}{1.06 \cdot r_0} \approx \frac{PH}{1.1 \cdot r_0} \tag{5.67}$$

which is obviously slightly smaller than $N_{\mathrm{SB}} = PH/r_0$ from Equation (5.52).

However, in general the characterization of the resolution and the SBN based on a "grating" of point objects as described above is possible but not very practical. The better method is the usage of sine gratings or bar gratings, respectively (Section 5.2.1). Bar gratings seem to be more straightforward and are more easily to realize, but due to Fourier mathematics, sine gratings are the most appropriate structures as any real object structure can by constructed by an infinite set of sine functions.

Consequently, the resolution can be characterized by contrast measurements of sine gratings and this is equivalent to the analysis of the related MTF (Section 5.2.1; grating described by $(\sin(x) + 1)/2$). If again we consider a contrast $K = 10\,\%$ as the limit, where the image of a sine grating can be just resolved, then this is equivalent to the $\mathrm{MTF10}^{(\mathrm{optics})} \equiv \mathrm{MTF}(R_{\mathrm{MTF10}}^{(\mathrm{optics})}) = 0.1$ (see related discussion and Section 5.2.2). For the example of a diffraction limited system, the MTF is shown in Figure 5.8 and it is given by Equation (5.38). $R_{\mathrm{x,max}} = 1/\delta_0$ and $R_{\mathrm{x,cut}} = 2 \cdot R_{\mathrm{x,max}} = 1/(\delta_0/2) = 1/r_0$. With Equa-

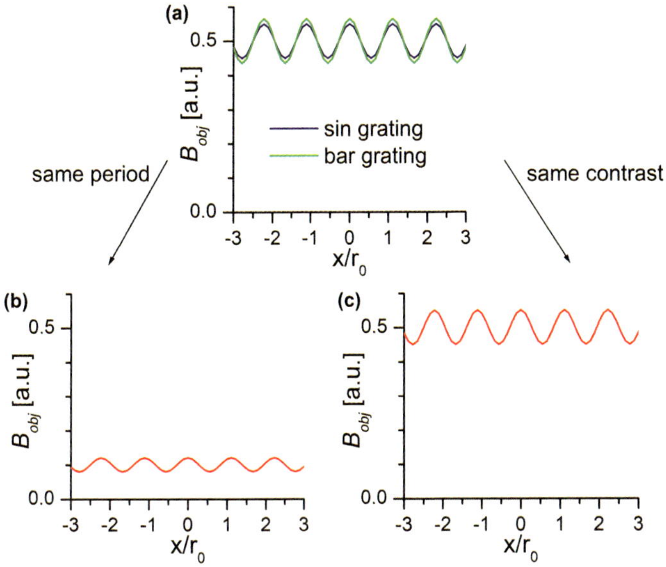

Fig. 5.45: (a) Images of a sine grating and a bar grating with a period of $1.11 \cdot r_0$. (b) Image of 50-point objects located at a distance of $1.11 \cdot r_0$ ($K = 20\,\%$) and (c) $1.06 \cdot r_0$ ($K = 10\,\%$) from each other. Here, again, as an example we assumed a system that consists of diffraction limited optics only. In general, of course, the PSF has to be taken for the whole camera system.

tion (5.38), one gets $R_{\text{MTF10}}^{(\text{optics})} = R_{\text{MTF10}}^{(\text{optics,Abbe})} = 0.9/r_0$, which means that the image of a sine grating with a period $a = 1.11 \cdot r_0$ has a contrast of $K = 10\,\%$ (see also Figure 5.45c).

If for this more common method, we would identify the maxima of the sine-grating image as the image points instead of considering the maxima of the image in Figure 5.44b and Figure 5.44c as the points, then

$$N_{\text{SB,MTF10}}^{(\text{slit})} = \frac{PH}{1.11 \cdot r_0} \approx \frac{PH}{1.1 \cdot r_0} \tag{5.68a}$$

i. e., the SBN is almost the same as before.

For those reasons, we may conclude that the maximum resolvable number of “image points” is

$$N_{\text{SB,MTF10}} = \frac{PH}{1.1 \cdot r_0} \tag{5.68b}$$

To get some numbers, as an example we may regard a system that is only diffraction limited, which also means that $\text{MTF}_{\text{system}} \approx \text{MTF}_{\text{optics}}$. For simplicity, we assume $f_\# = 1$ and a wavelength of $\lambda = 0.5\,\mu\text{m}$. We consider the picture height of a full format camera, i. e., $PH = 24\,\text{mm}$. Then the resolution limit is $\delta_0/2 = r_0 = \lambda \cdot f_\# = 0.5\,\mu\text{m}$ (Figure 5.44a) and $R_{\text{x,cut}} = 2000\,\text{lp/mm}$. Hence, one can calculate $PH/r_0 = 48000$. But because only 2 points are present, this is not a SBN (nevertheless this quotient is the same as N_{SB} from

Equation (5.52)). From Equation (5.67), we get $N_{\text{SB,points}} = 45283\,\text{lp/PH}$ (Figure 5.45c) and from Equation (5.68a), we get $N^{(\text{optics})}_{\text{SB,MTF10}} = 43200\,\text{lp/PH}$ (Figure 5.45a). Both describe the resolution limit according to Abbe's criterion, but the interpoint-spacing δy is slightly different. If we would take the same $\delta y = a = 1.11 \cdot r_0$ for the "point grating" as for the sine-grating ($\delta y = a$), then, of course, the SBN would be the same, but the contrast of the "point grating" would be $K \approx 20\,\%$ (Figure 5.45b). Then this would correspond to a more relaxed observation of the "image points." This again, justifies that the SBN may be estimated from Equation (5.68b): $N_{\text{SB,MTF10}} \approx 44000\,\text{lp/PH}$ or be identified with $R^{(\text{optics})}_{\text{MTF10}}$ in units of lp/PH.

From a similar discussion with Equation (5.39) for a circular aperture, one gets $R^{(\text{optics,circ})}_{\text{MTF10}} = 0.8/r_0$ (see also Figure 5.47c) and

$$N^{(\text{circ})}_{\text{SB,MTF10}} = \frac{PH}{1.25 \cdot r_0}. \tag{5.68c}$$

The limiting case where the resolution is fully lost (i. e., MTF0) is given by R_{cut} from Equation (5.61) (in vertical direction), i. e., by

$$N_{\text{SB,cut}} = R_{\text{cut}} \cdot PH, \tag{5.69a}$$

which means that a sine grating with a period of R^{-1}_{cut} cannot be resolved at all. We may note that this is independent of the geometry, and thus of κ. In particular, Equation (5.69) is the same for the diffraction limited 1D geometry (slit aperture) and the diffraction limited 2D geometry (spherical aperture). Furthermore, it does not depend on aberrations. We may also note that, in general, any SBN as a dimensionless number is given by the related value of R multiplied with PH or PW, respectively,

$$N_{\text{SB},x} = R \cdot PW \tag{5.69b}$$

$$N_{\text{SB},y} = R \cdot PH. \tag{5.69c}$$

It is trivial that if R is given in lp/PH that then its numerical value is identical with N_{SB}. For instance, for full format and $R = 70\,\text{lp/mm} = (70 \cdot 24)\text{lp}/(24\,\text{mm}) = 1680\,\text{lp/PH}$, and thus $N_{\text{SB}} = 1680$.

We would like to add that not necessarily the best value of $\text{MTF10}^{(\text{optics})}$ describes the system, which is judged by humans to have best optical performance. Instead, e. g., MTF50 or MTF60 could be a better measure (see Section 5.2.9 and also the discussion by Nasse [Nas08, Nas09]). For our example, that would correspond to $\text{MTF50} = \text{MTF}(R_{\text{MTF50}}) = 0.5$ with the SBN given by $R_{\text{MTF50}} = R_{\text{x,max}} = 24000\,\text{lp/PH}$ this would be $N_{\text{SB,MTF50}} = 24000$.

As the above description of the different situations may have led to some confusion, Table 5.6 provides a summary.

Finally, we would like to note that in contrast to the MTF, which is exactly defined, we would regard the SBN as an estimate of the resolution-related image quality close to

Tab. 5.6: Summary of SBN with respect to PH (mostly) and with respect to different demands on "limiting" (or perceivable) contrast K.

symbol	equation	object	contrast	remarks
$N_{SB} = \frac{PH}{r_0}$	(5.52)	2 points or lines (1D)	$K = 0.1$	spots with width δ_0 at a center-to-center distance of r_0; resolution limit according to Abbe's criterion
$N_{SB} = \frac{PH}{r_0}$	(5.52)	many points or lines (1D)	$K \approx 0.01$	spots with width δ_0 at a center-to-center distance of r_0; resolution almost lost; nevertheless, this yields a simple estimate of the SBN
$N_{SB,points} = \frac{PH}{1.06 \cdot r_0} \approx \frac{PH}{1.1 \cdot r_0}$	(5.67)	many points or lines (1D)	$K = 0.1$	spots at a center-to-center distance of $1.06 \cdot r_0$
$N_{SB,MTF10} = \frac{PH}{1.1 \cdot r_0}$	(5.68a)	grating lines (1D)	$K = 0.1$	bar/sine grating with a period of $1.11 \cdot r_0$; common criterion for resolution
$N_{SB,MTF10} = \frac{PH}{1.25 \cdot r_0}$	(5.68c)	(2D)	$K = 0.1$	same as before but for 2D; cf. both green curves in Figure 5.47b
$N_{SB,cut} = 1.0 \cdot \frac{PH}{r_0}$	(5.69)	(1D)	$K = 0$	limiting case; vanishing contrast ($\kappa = 1.0$, $\alpha = 1$)
$N_{SB,cut} = 1.22 \cdot \frac{PH}{r_0}$	(5.69)	(2D)	$K = 0$	limiting case; vanishing contrast ($\kappa = 1.22$, $\alpha = 1$)
$N_{SB,MTF50}$			$K = 0.5$	particular importance with respect to human perception
$N_{SB,x} = R \cdot PW$	(5.69b)	(1D)	any K	this is a general relation for horizontal direction; valid for any chosen K
$N_{SB,y} = R \cdot PH$	(5.69c)	(1D)	any K	this is a general relation for vertical direction; valid for any chosen K
$N_{SB}^{(2D)} = N_{SB,x} \cdot N_{SB,y}$		(2D)		

the maximum possible number of observable "image points" only (more correctly may be image spots). For that reason, it is sufficient to make use of Equation (5.68) or just take $R_{MTF10}^{(optics)}$ (or $R_{MTF10}^{(system)}$, see below) in units of lp/PH. Alternatively, one may regard the SBN as an estimate of the resolution-related image quality close to the "easily observable" number of "image points," which does not much differ from R_{MTF50} in units of lp/PH. Here, "estimate" instead of an exact definition takes into account that SBN may be also closely related to humans with their different perceptions and also illumination and observation conditions play a role (an “estimate” is not a must; of course, SBN can also be well-defined). And one has to be aware that beside the resolution-related image quality there are other issues of image quality as well.

5.3.3 Relations with respect to the sensor and the whole camera system

Now we would like to add a pixel sensor to the optics, and thus consider a full system. The sensor should be simply described by Equation (5.63) but still we restrict to 1D geometry. Further, we neglect effects such as cross talk or diffusion, which would degrade $\mathrm{MTF}_{\mathrm{sensor}}$. Hence, the sensor-MTF is given by the blue curve in Figure 5.30. Similar to the optics, $R_{\mathrm{MTF10}}^{(\mathrm{sensor})}$ is the frequency for which $K = 10\,\%$. This is lower than the Nyquist frequency, which defines the limit when the MTF becomes zero (note that then due $K = 0$ full structure loss is just attained). From Equation (5.63) and $\mathrm{sinc}(0.904 \cdot 2\pi) = 0.1$, one obtains $R_{\mathrm{MTF10}}^{(\mathrm{sensor})} \approx 0.9 \cdot R_{\mathrm{N}}$:

$$N_{\mathrm{SB,MTF10}}^{(\mathrm{sensor})} = 0.9 \cdot \frac{PH}{2p} \tag{5.70}$$

This is slightly worse when compared to the “2-pixel resolution” of the Nyquist limit

$$N_{\mathrm{SB,MTF0}}^{(\mathrm{sensor})} = \frac{PH}{2p} \tag{5.71}$$

The MTF of the system is $\mathrm{MTF}_{\mathrm{system}} = \mathrm{MTF}_{\mathrm{optics}} \cdot \mathrm{MTF}_{\mathrm{sensor}}$. (Equation (5.53)). From $\mathrm{MTF}_{\mathrm{system}}$, one can determine the spatial frequency where the contrast becomes $K = 10\,\%$. This is $R_{\mathrm{MTF10}}^{(\mathrm{system})}$. For a given wavelength and a given optics, $R_{\mathrm{MTF10}}^{(\mathrm{system})}$ is only a function of p.

This is shown in Figure 5.46. The dotted green line describes $R_{\mathrm{MTF10}}^{(\mathrm{optics})}$, which does not depend on p but just on r_0 (from Equation (5.38), one gets $R_{\mathrm{MTF10}}^{(\mathrm{optics})} = 0.9/r_0$; see Section 5.3.1). It dominates the behavior of the complete system for very small pixels. Here, the sensor is much better than the optics. The dashed blue line describes $R_{\mathrm{MTF10}}^{(\mathrm{sensor})}$. In principle, it does not depend on r_0, but here we have taken the optical resolution as the reference, so $R_{\mathrm{MTF10}}^{(\mathrm{sensor})} \cdot r_0$ does depend on r_0. For large pixels, this is the dominant term for the whole system. With $R_{\mathrm{N}} = 1/(2p)$, one gets $R_{\mathrm{MTF10}}^{(\mathrm{sensor})} \approx 1/(2.2 \cdot p)$ (dashed blue line). The solid red line is $R_{\mathrm{MTF10}}^{(\mathrm{system})}$, which includes both, optics and sensor and approaches $1/(2.2 \cdot p)$ for large pixels, and thus becomes independent of r_0 (when not normalized).

In Figure 5.46a, the resolution is described by the maximum spatial frequency that is transmitted by the system and that still allows to observe a contrast of at least 10 %. Here, this frequency is normalized to that for the optical part only, namely to $R_{\mathrm{cut}} = 1/r_0$ (remember, here $\kappa = 1$ and $\alpha = 1$). The pixel pitch is also given as a relative value. As any value of $r_0 = \lambda \cdot f_{\#} \cdot \alpha$ can be selected, this plot can be directly applied to obtain $R_{\mathrm{MTF10}}^{(\mathrm{system})}$ in absolute values for any value of λ or $f_{\#}$. It can even be applied for a case

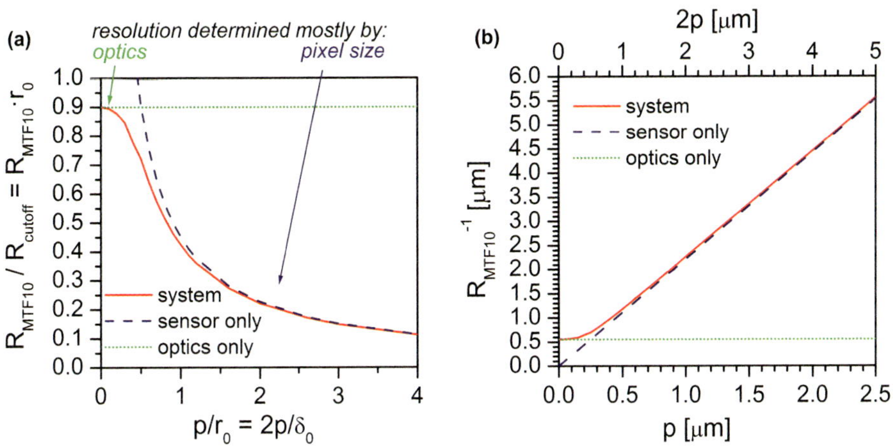

Fig. 5.46: 10 %-resolution limit described by R_{MTF10} for the system and the optics and the sensor, respectively. (a) General relation. The curve scales with $r_0 = \lambda \cdot f_\#$; this may even include aberrations when one sets $r_0 = \lambda \cdot f_\# \cdot \alpha$. For a given p/r_0, the image of a sin-grating with a spatial frequency of R_{MTF10} would have a contrast of 10 %, and thus be just resolvable. (b) Same relation, but now for particular values: 500 nm wavelength, $f_\# = 1$, no aberrations (i. e., $\alpha = 1$). For a given pixel width p, the image of a sin-grating with period $= R_{\mathrm{MTF10}}^{-1}$ displayed at the ordinate would have a contrast of 10 %.

with aberration if the aberration can be described by a simple value of α (see note in Table 5.5). As an example, Figure 5.46b is obtained from Figure 5.46a by multiplication of the x-axis values with $(\lambda \cdot f_\#)/\,\mu\mathrm{m} = 0.5$. The new values of the ordinate are obtained by dividing $(\lambda \cdot f_\#)/\,\mu\mathrm{m}$ by the ordinate values of Figure 5.46a.

An approximation for the $K = 10\,\%$ resolution limit for both, Abbe's criterion and Rayleigh's criterion is

$$R_{MTF10}^{(\mathrm{system})} = \sqrt{R_{MTF10}^{(\mathrm{sensor})^2} + R_{MTF10}^{(\mathrm{optics})^2}}\,. \tag{5.72}$$

Although not very accurate, this approximation may be regarded to be still sufficiently good for most estimates. Anyhow, to get a better result it is recommended to perform a simulation based on the real MTF of the sensors and the optics.

Resolution is fully lost when the spatial frequency exceeds $R_{\mathrm{cut}} = 1/r_0$ for the optics or $R_{\mathrm{N}} = 1/(2p)$ for the sensor, respectively. The smaller of both values sets the upper limit for the system. Figure 5.47 shows an example of a system where $R_{\mathrm{N}} = R_{\mathrm{cut}}$, which consequently is also equal to R_{MTF0} of the system.

5.3.4 System-PSF and integrated pixel signals

Without providing new information, finally we would like to relate the MTF and the PSF of the system with the signal measured by the pixels in a simple conclusive way.

Again, we restrict to a 1D discussion in x-direction because extension two 2 dimensions is obvious.

First, we would like to recapitulate that in general the PSF is the impulse response of the apparatus. This yields a fluence $F_{pix}(x, y)$ on the sensor surface (Equations (1.23) and (4.8)), which is the energy of the incident light per pixel area in the image plane. In the 1D consideration, we may replace the pixel area with the pixel width p and integrate over its height or we make use of a "1D fluence" $F'_{pix} = dW_{pix}/dx$. This is straightforward and needs no further discussion.

PSF_{optics} (green curve in Figure 5.47a,b) may be obtained in case of an ideal sensor in the sense of a sensor with $p \to 0$, i. e., if PSF_{sensor} would be a δ-function. PSF_{sensor} may be obtained for an optical system, which is ideal in the sense, that it would reproduce a point object as an image point with a diameter $r_0 \to 0$, i. e., if PSF_{optics} would be a δ-function. For real systems, both functions are finite.

For an ideal optics in the sense that it is limited by the diffraction at the aperture only (diffraction limit, no aberrations), the width of PSF_{optics} may be characterized by r_0. For an ideal sensor in the sense that it is only limited by the finite pixel size (no additional effects such as cross talk, charge diffusion, etc.), the width of PSF_{sensor} is given by $2p$ as has been discussed many times in Chapters 2 and 5. In general, PSF_{system} is the convolution $PSF_{optics} \otimes PSF_{sensor}$ (red curve in Figure 5.47a,b). This is equivalent to the product $MTF_{optics} \cdot MTF_{sensor}$ in the Fourier plane (same curve colors).

On the other hand, the pixels integrate $F'_{pix}(x)$ over their widths as described previously (Sections 1.6.2 and 4.2.2, Equations (1.24) and (4.8), Figure 1.19). For real imaging, integration is made over $B_{im}(x)$ but here we restrict to the integration of $F'_{pix}(x_i)$ over $PSF_{optics}(x)$ for each of the pixels. x_i defines the begin of the i-th pixel. This yields a curve of data consisting of the sampling points $x_i^{(0)}$. We may repeat this procedure, at slightly shifted pixel positions $x_i^{(1)} = x_i^{(0)} + \Delta x$ where Δx takes a small fraction of p. This adds additional sampling points. Further continuation until $x_i^{(n)} = x_i^{(0)} + p$ increases the number of sampling points and leads to a smooth curve. In case of small Δx, $PSF_{optics}(x)$ is reproduced. But note that this would be the result of really a lot of images, which then would have to be averaged. All of them have to be taken at exactly the same conditions. Effects that are unknown or hardly characterized or not stably reproducible (noise, statistical charge spreading, etc.) will prevent proper reconstruction. In reality, a single image is taken, or a series with not well-defined conditions. Such a series is hardly made carefully with the well-defined tiny slightly shifted pixel positions and identical image conditions. To some extent pixel shift technology may be an exception, see Section 1.6, but even that may be restricted to one direction only (i. e., 1D).

Examples of captured images and integration of the $F'_{pix}(x_i)$ are shown in Section 1.6. Another one is displayed in Figure 5.47 where the resulting pixel signals are displayed as the dashed bars with the bar area proportional to the energy W_{pix} accumulated by the corresponding pixel (Equation(1.24)). Figure 5.47a and Figure 5.47b show two situations of pixel locations, which differ by their relative positions, i. e., a "phase shift" of

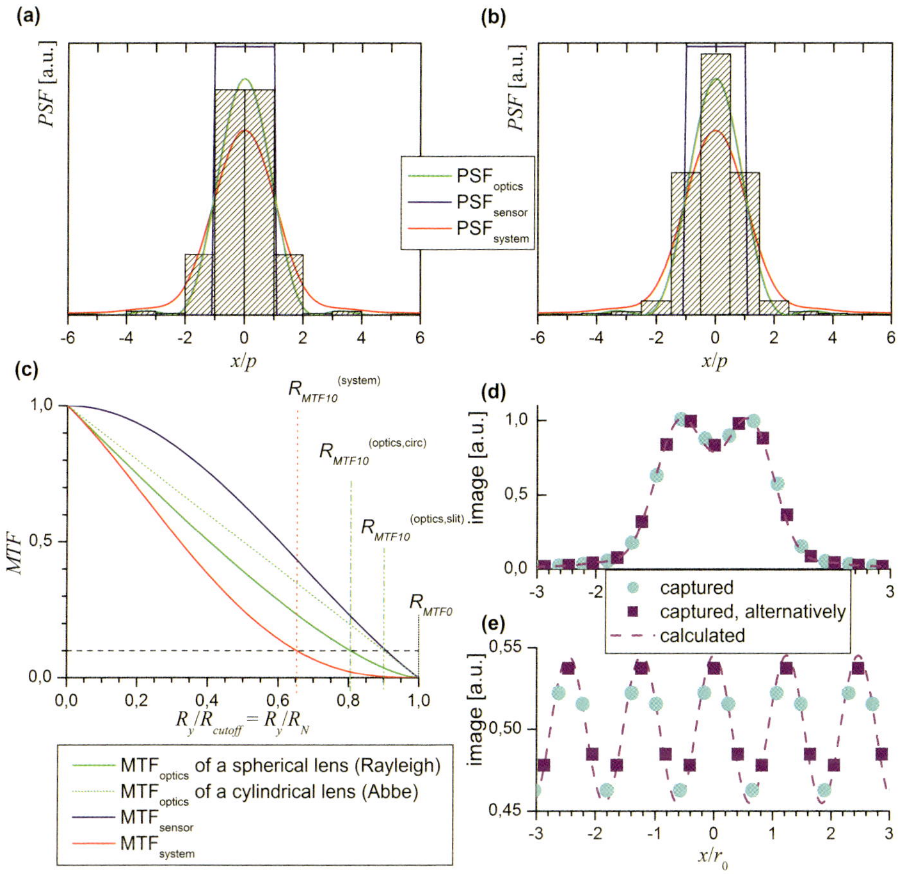

Fig. 5.47: (a), (b) PSF, sensor pixel signals and (c) MTF for a system that consists of an aberration free spherical optics and a sensor with an MTF that is provided by Equation (5.63). R_{MTF0} has been chosen that it is the same for the optics and the sensor. The bars indicate the position and width of the pixels with respect to the captures $\mathrm{PSF}_{\mathrm{optics}}$. Their area is a measure of the integral over $\mathrm{PSF}_{\mathrm{optics}}$ within their width and corresponds to the pixel signal. Although here $\mathrm{MTF}_{\mathrm{system}}$ is calculated for a diffraction limited optics with a circular aperture, for comparison the MTF of a cylindrical lens with a slit aperture as used for a 1D discussion with Abbe's criterion is included as well. (d) Image of 2-point objects located at a distance slightly larger than $1.1 \cdot r_0$, which corresponds approximately to the $K = 10\,\%$ resolution limit of the system. (e) Image of a sine grating with a similar period as the object that also leads to $K \approx 10\,\%$ (this is similar to Figure 1.19). The points in (d) and (e) correspond to the signals of the pixels located at the indicated positions.

$p/2$ (see also discussion in Section 1.6.2). But again, we would like to hint that an image is captured without knowledge of the relative phases or pixel shifts during the time of irradiation. In real photography, one cannot arrange the condition displayed in Figure 1.17a or Figure 5.47b and avoid that of Figure 1.17b or Figure 5.47a. Most likely, the relative pixel positions are even different from those displayed in these figures.

This affects the resolution. Yet resolution is not determined by a reconstruction of PSF_{optics} only but by the complete apparatus function of the system PSF_{system}. Consequently, although $PSF_{optics}(x)$ is sampled with the frequency $1/p$, which is twice the Nyquist frequency, the resolution is evidently lower (Section 1.6). According to the Nyquist–Shannon theorem, the Nyquist frequency R_N is half of the sampling frequency $R_S = 2R_N = 1/p$. Thus, taking into account this resolution limit, sampling means that at any position x the measured signal is provided by PSF_{system},which is the convolution of $PSF_{optics}(x)$ with $PSF_{sensor}(x)$ given by a rectangle function (see Appendix A.1) with a width of $2p$ (see Figure 5.47; for convolution see, e. g., Appendix A.3). This is equivalent to the Nyquist limit of MTF_{sensor} and OTF_{sensor}, respectively, (Equation (5.63)), which is related to PSF_{sensor} by Fourier transformation. For further details of such kind of signal transfer, the reader is referred to books on signal theory.

Here, we would like to note that although integration of $F'_{pix}(x)$ over one pixel is possible (and applied by the sensor) and although Fourier transformation of this rectangle function with a width of p in the Fourier plane leads to a sinc-function with the argument $k_x \cdot p/2$, the latter is not equal to MTF_{sensor} or OTF_{sensor} where the argument of the sinc-function is $k_x \cdot p$. A single pixel does not yield any resolution. In a general sense and may be more relaxed than in the discussion at the very beginning of Section 5.3, resolution is related to structures. The absolute minimum to set up a structure may be 2 pixels, even if these can only discriminate 2 states such as on/off, or high/low, etc. This sets the limit. This view does not change when we image, e. g., a bar grating as discussed together with Figure 1.17. Then again we may consider sampling. Sampling is based on the signals of many pixels, and it is also restricted to the 2-pixel limit.

Figure 5.47d and e show the image of 2-point objects located in a distance slightly larger than $1.1 \cdot r_0$ in the image plane and the image of a sine grating with a period slightly larger than $1.1 \cdot r_0$, respectively. This corresponds approximately to the MTF10 resolution limit of the system (Figure 5.47c; we would like to remind that there is a small difference in the contrast values of both images, see discussion in Section 5.3.1). The dashed magenta curves result from a calculation with Equation (5.36) with the transfer functions displayed in Figure 5.47c. The data points result from the sampling of the image generated by the optics on the sensor surface with the sampling frequency RS according to the pixel pitch. In practice, this is the integration performed by the pixels as described above (or in Section 1.6.2), but now for the image instead of PSF_{optics}. The squares and circles show examples of 2 different "measurements" (of course, here as simulated data) at arbitrary camera positions. This means that there is a slight shift of the sensor and pixel positions, respectively, for the two images. It is well seen that the structure is reproduced by a single data set of sampling points, and it is resolvable. But here, close to the resolution limit, reproduction of the object structure is limited and the "measured" contrast is little worse, when compared to the theoretical curve. Even so, an average of lot of images taken at least for quite the same conditions is expected to reproduce the theoretical one just by statistics.

As a final remark for resolution, we would like discuss the situation shown in Figure 5.47c. Here, the cutoff frequency of the sensor, $R_N = 1/(2p)$, and that of the optics with a circular aperture, $R_{cut}^{(optics)} = 1.22/r_0 = 2.44/\delta_0$, are identical. Equating both quantities yield

$$p = \frac{\delta_0}{4.88} \approx \frac{\delta_0}{5}. \tag{5.73}$$

The pixel pitch is then about one fifth of the Airy disk diameter. For full format cameras, lenses of high quality are usually designed to have a circle of least confusion that is equal to 1/3000 of the image sensors' diagonal (see Sections 1.4 and 1.5 and [Nas10]). In the ideal case of diffraction limited optics, the circle of least confusion is equal to δ_0, and with $d_{sensor} = 43.4$ mm for a full frame sensor we get $\delta_0 \approx 15$ µm, which corresponds to an optical cutoff at 3900 lp/PH. If the sensor cutoff should be identical to the optical one, a pixel pitch of about $p = 3$ µm is required, corresponding to a 96 MP sensor for the full frame format. Most consumer full format cameras have significantly less pixels and a lower Nyquist limit. For a typical consumer camera with a 24 MP sensor, we get $p = 6$ µm, $R_N = 2000$ lp/PH, and thus the overall resolution of these cameras is mainly dominated by MTF_{sensor} when operated with high-quality lenses. If the lens should not become the bottleneck of these cameras, its allowable circle of confusion should not exceed 30 µm. This is the minimum requirement for lenses of standard quality for the full frame format. Nevertheless, 2000 lp/PH is still such a good resolution that it cannot be resolved by the human eye under standard viewing conditions (see also Sections 1.4.1 and 5.2.4.5).

The situation becomes a bit different in the case of smartphone cameras (SPC). If we take a typical sensor diagonal of $d_{sensor} = 6$ mm, then the circle of least confusion for high-quality lenses, as 1/3000 of the diagonal, is of about 2 µm. This is fulfilled by most high-quality SPC for their standard lenses, which speaks in their favor (see Chapter 7). If the sensor cutoff was to match that of these high-quality optics, the pixel pitch should be of about 0.4 µm whereas the typical minimum pixel pitch is of about 0.8 µm. Hence, for standard lenses in SPC, usually the MTF_{optics} is superior to that of MTF_{sensor}. Nevertheless, the system resolution quality of the standard modules in SPC is still quite high. On the other hand, especially for long focus modules, the optical cutoff may become relatively low due to diffraction that the optics are the bottleneck for the overall image resolution. Further details are discussed in Chapter 7.

6 Camera lenses

After the principles of geometrical optics, Fourier optics and optical sensors have been covered in the preceding chapters, and we now consider more details about complete camera lenses and their features for special applications. In order to understand the current state of camera lenses, a look upon their historical evolution in the domain of photography is very helpful. This will be given after understanding the requirements of lenses and will be followed by the presentation of different lens constructions. We mainly focus on lenses for the standard 35 mm format in photography. Practical impacts of lenses to control depth of focus, like bokeh, will be discussed as well as the importance of high-quality antireflection coating for lenses. Despite our focus to camera lenses, many considerations can be transferred to other applications of imaging.

6.1 Requirements for camera lenses

The assessment of optical lenses is only reasonable with respect to their image field on the sensor for which they have been designed and optimized. A classification scheme for photographic camera lenses has already been presented in the introductory Chapter 2. The changing perspective for the different type of lenses like long focus, wide angle and normal lenses is illustrated in Figure 2.7. The term normal lens for a camera means that the focal length of the lens is approximately the same as the diagonal of the sensor format with which it is used. Long focus and wide-angle lenses are the terms for large, respectively short, focal lengths in relation to the image diagonal, and thus have an impact on the angular field of view. A good lens should have a high resolution combined with a high-image contrast over the whole sensor area and it should be free of distortion. Resolution and contrast may be conveniently described by the modulation transfer function (MTF). Basics of the MTF are discussed in Chapter 5. Measured specifications of camera lenses and their interpretation to assess the quality of optical systems are given in more detail in Chapter 8.

A high transparency of the lens, also conveyed by the term fast lens, is required when images should be taken under low light conditions and the exposure time is limited to short values. The main feature of a fast lens is the high-relative aperture, which is indicated by a low f-number. This, however, necessitates more complex constructions to correct aberrations that become especially apparent at low f-numbers.

With the development of digital, semiconductor-based image sensors, a lens design should also take into consideration a ray path not too far away from an image side telecentric ray path. Especially wide-angle lenses designed for cameras with photographic films may show a poorer quality with digital sensors and may be used only with restrictions by digital cameras. It should be mentioned, however, that modern lenses for digital sensors never have a true image space telecentric design as this would be too expensive for consumer cameras and result in very large and heavy lenses (see Section 3.4.5). If

https://doi.org/10.1515/9783110789966-006

special lenses are used for dedicated scientific and technical application, this may be different. Usually, the center of the exit pupil of the lens is far enough from the image plane that the angles of the chief rays with the normal vector of the image plane are small enough and do not lead to strong shading effects. Moreover, residual shading can be mathematically compensated if the lens data are stored in the camera and can be used by the image processor. Image processing steps after the exposure nevertheless should be applied with caution as they tend to reduce the original resolution and to impair the principal image quality (see Chapter 5).

In order to achieve a high resolution, all lens aberrations preventing stigmatic imaging should be corrected. This is usually done by a combination of many different lenses of high-surface quality and requires a high-precision production technique for rugged complex optical systems. The aperture of the lenses should be as large as possible because diffraction, which may impair the resolution, increases with decreasing aperture. On the other hand, larger apertures always require more efficient corrections of aberrations leading to sophisticated lens designs. At the beginning of the 20th century, ideas of very powerful lens designs were already available. However, they could not be effectively realized due to the lack of antireflection treatment of lens surfaces and lack of available glass materials. The technical implementation of the antireflection coating in the lens production process turned out to be one of the key factors of high-performance lens developments. All glass surfaces have to be coated by precise methods after complex calculations. The design of powerful lenses has always been influenced by the development and availability of glasses with high-refractive index and low dispersion. The higher the index, the lower the curvature of the lens surface can be for a given refractive power, which then reduces the aberrations and complexity of lens systems. This leads to more compact camera lenses and facilitates the process of miniaturization optics.

Since the beginning of the 21st century, the development of semiconductor image sensors with pixel pitches in the order of 1 μm has necessitated new approaches to lens design. The correction of lens aberrations by the conventional combination of multiple lenses is partly or, in the case of mobile phone camera lenses, completely substituted by aspheric lenses. The mass production of plastic aspheric lenses can be done at low costs. The fabrication of aspheric glass lenses by classical polishing methods, however, is still much more expensive than that of spherical ones, and their quality is also not yet at the same level. The availability of high-quality aspheric lenses at low prices as well as the development of new sensors will certainly be a driver for the further development of camera lenses.

The minimization of lens aberrations as discussed in Chapter 3 is only possible for a limited range of image magnification. Thus, it is reasonable to discriminate photographic camera lenses roughly with respect to the above-mentioned scheme of normal, wide-angle and telephoto lens, but also special purpose lenses like perspective control or macrophoto lenses. Zoom lenses in many cases represent a good compromise between versatility, quality and prize. This is especially true for compact consumer cameras where it is not possible to change the lens. Here, a complex high quality zoom lens may be designed for one special camera. For system cameras with interchangeable lenses and camera bodies with or without mirrors, however, there are more restrictions and there exist also different approaches for the manufacturer. Unlike microscope lenses or macrophoto lenses, photographic lenses are usually corrected for imaging large distant objects. Most of the modern lenses are anastigmats where the aberrations are corrected to a large amount and distortion is less than about 5 %.

In all cases, we should not forget the purpose of using lenses and cameras. The megapixel race, which was seen in the first decade of our century seems to have come to an end with the current resolution of more than 20 MP for a consumer camera. They are largely sufficient if the printed images are to be viewed by the human eye. For scientific, professional or industrial application, however, requirements beyond the consumer domain will still increase. If the images are to be simply considered on small displays like those of typical smart phones, the number of pixels usually is of minor importance.

6.2 Short history of photographic lenses

The development of complex camera lenses started roughly at the beginning of the 19th century. The background for the technological progress was the increasing demand for microscopes of higher quality as well as for better photographic lenses with the availability of storing images on film materials. For the development of photographic lenses, different phases could be identified, which lay out the framework for the systematic presentation of the lenses in this chapter. Relatively simple lenses can be found before the end of the 19th century. After 1886, new types of glasses could be successfully manufactured by Schott, Abbe and Zeiss to design novel types of achromats leading to more powerful lenses. However, the number of reflecting surfaces had to be kept at a minimum until efficient antireflection coating after 1930 became possible. This technology led to more complex camera lenses after 1930.

The amount of computation for optimum lens design increased dramatically with the number of refracting surfaces. Up to that time, all computations had to be performed manually, and thus only few variations of a lens were possible. New numerical computational methods were required and became very effective with the emergence of modern computer systems. In particular, roughly after 1950 there was a progressive improvement of camera lenses due to numerical variations of lens parameters. At around the same time, lanthanum doped glasses became available that had an additional impact on the lens design. Moreover, the growing interest for SLR cameras with hinged mirrors required new lens constructions leading to retrofocus lenses. Even if the principal layout of some lenses has not changed, the optimization of lens parameters and the improvement of manufacturing methods led to much better performances of modern lenses.

In the following, we present a brief overview about the development of photographic lenses in order to understand how the variation of lens designs was intended to correct the aberrations and eventually led to the current state of modern lenses. The examples that we present are chosen for illustration purposes only and imply neither ranking of lenses or manufacturer or advertising for them. The compilation is not complete, neither is it intended to be complete. More detailed presentations of the topic can be found in various excellent books of which we would like to mention only "A History of the Photographic Lens" by Rudolph Kingslake [Kin89], one of the pioneers in the field of optical design.

6.2.1 Simple photographic lenses

The simplest method for photographic imaging is using one single lens. However, a single lens exhibits many types of aberrations, which in general deteriorate the modulation transfer function MTF as well as the point spread function (PSF). Examples for their impact on MTF and PSF are illustrated in the Appendix (Appendix A.9).

When using a single lens, the best sharpness for objects at a large distance is achieved by a biconvex lens with the ratio of the curvature radii given by Equation (3.122) or approximately by a planoconvex lens with the curved surface oriented toward the object space. This lens, however, is not free of astigmatism and has a curved image plane. Unlike on-axis points, object points off the optical axis are not imaged sharply. For photographic purposes, especially landscape photography, lenses with a more unique sharpness across the whole image area, even at the cost of the center sharpness, are more favorable. An appropriately shaped meniscus lens in combination with an aperture stop has been proposed by Wollaston [Kin39] at the beginning of the 19th century and was used in many low-priced box cameras in the first half of the 20th century. The lens shape and the position of the stop for this fixed-focus lens were calculated to minimize the astigmatism and coma. Two versions were used in general, one with the stop in front of the concave side of the meniscus lens, and the other one with the convex side oriented to the object space with a stop behind the lens (Figure 6.1a,b).

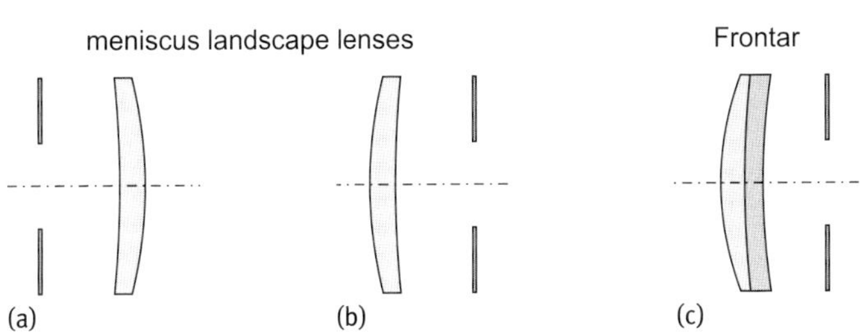

Fig. 6.1: Simple camera lenses. (a) Wollaston meniscus landscape lens (1812) with front stop; (b) meniscus lens with rear stop; (c) Goerz Frontar (1948), achromatic doublet with rear stop, converging meniscus made of crown glass, diverging meniscus made of flint glass.

As these lenses had typical relative apertures of less than about $f/11$, spherical aberration as well as coma were strongly reduced by the small aperture, and a good depth of field was achieved. The inconvenience with these "slow" lenses was that, due to the lack of very photosensitive material in the 19th century, relatively long exposure times were necessary, which made these lenses less appropriate for portrait photography but rather for landscape photography. Moreover, as the film formats in the 19th century were usually relatively large with diagonals on the order of 10 cm or more, the focal lengths of the normal lenses were of the same order.

The longitudinal chromatic aberration is proportional to f/ν, as expressed by Equation (3.137). It becomes more critical for long focal lengths and cannot be corrected by the aperture. Even in black-and-white photography the chromatic aberration must be avoided as the image contrast and sharpness are strongly reduced. As a consequence, different achromatic doublet lenses like the achromatic landscape lens of Chevalier (1821) or Grubb (1857) with a front stop were introduced. A similar lens with a rear stop, corrected chromatic and spherical aberration as well as of acceptable astigmatism was the Frontar lens of Goerz that could still be found in box cameras of the 20th century (Figure 6.1c).

6.2.2 Petzval portrait lens

With the availability of more sensitive film materials in the 19th century, there was an increasing demand for portrait lenses of relatively larger apertures. In 1840, Petzval from Vienna designed a more complex photographic lens based on mathematical considerations and achieved a high relative aperture of $f/3.2$, which was a record value at that time. The lens features a cemented achromatic doublet consisting of crown and flint glass. It is followed by a diverging meniscus lens of flint glass and a biconvex crown glass converging lens (Figure 6.2), which form a Gaussian achromatic doublet with an air gap. The six glass-air surfaces are still acceptable with respect to the light transparency of the lens. The aperture stop is in the center of the lens behind the cemented doublet thus reducing distortion. The sharpness in the center of the image is remarkably high and the lens is well corrected for spherical aberrations, chromatic aberrations, coma and distortion. However, there is still a considerable astigmatism and the image plane is

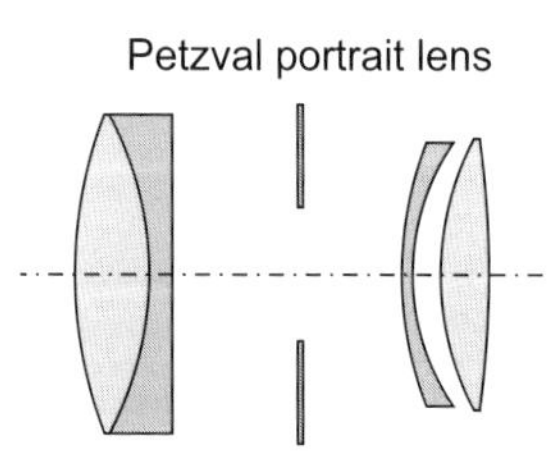

Fig. 6.2: Schematic design of the Petzval portrait lens (1840).

not flat, which is mainly due to the fact that appropriate glass types were not yet available at the time of its design. As a consequence, the sharpness suffers at the outer parts of the image field which, however, may be favorable for portrait photography and yields a nice, blurred environment. The design of the lens has been improved by other lens developers to achieve a larger relative aperture as well as a less curved image plane. This lens is reproduced as of 2014 as a portrait lens for modern DSLR cameras and capitalizes on the special bokeh and blurred off-center image parts.

6.2.3 Early symmetric lenses

In order to overcome aberrations of existing lenses, a symmetric lens design was chosen by Steinheil for the Periskop in 1865 under the influence of Seidel. According to Seidel's third-order theory, distortion, coma and transversal chromatic aberration can be minimized by a symmetrical lens setup. The simple setup of the Periskop consists of only two symmetrically arranged meniscus lenses (Figure 6.3) with the aperture stop in the center. The lens is nearly free of these aberrations for a standard imaging of objects, being at a large distance compared to the focal length and is free of them for 1:1 imaging. However, there still exist the remaining spherical aberration and longitudinal chromatic aberration if they are not corrected for the individual elements of the symmetrical camera lens. Astigmatism and curvature of field are not corrected by this symmetrical arrangement.

A further improvement was done by Steinheil and independently by Dallmeyer at nearly the same time. By using spherically corrected achromats they were able to design lenses with larger relative apertures, which were named Aplanat by Steinheil and Rapid Rectilinear by Dallmeyer. The latter name indicated that the lens has a large relative aperture and is free of distortion. Moreover, Steinheil and Dallmeyer discovered the importance of the aperture stop position to reduce the astigmatism and were able to reduce even that type of aberration. However, there was still the problem of a curved image field leading to distortion-free but not homogeneously sharp image all over the

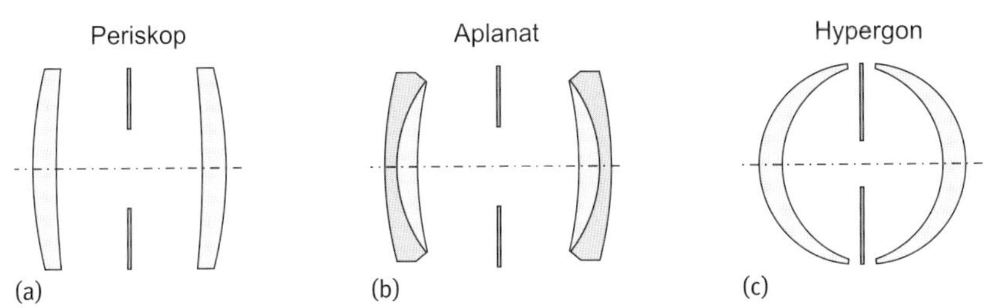

Fig. 6.3: Schematic design of symmetric objective lenses. (a) Periskop by Steinheil (1865); (b) Aplanat by Steinheil and Rapid Rectilinear by Dallmeyer (1866) with diverging meniscus of the cemented achromats made of flint glass; (c) Hypergon by Goerz (1900).

image plane. The term aplanat is still used to characterize lenses that are free of spherical aberration and free of coma.

6.2.4 Early anastigmats consisting of new and old achromats

All lenses considered so far still had the problem that a curvature of image field existed and the Petzval sum could not be reduced. There are different possibilities to minimize this sum: if a single lens is used then only a thick meniscus lens with identical curvature radii, a Höegh's meniscus, yields a flat image plane with the Petzval sum being zero. This is a necessary condition but not sufficient to achieve an effective flat image plane because the astigmatism must be corrected as well (see Section 3.5.4). Both of these features can be found with the Hypergon lens (Figure 6.3c) of the year 1900 where the radii of the two symmetrical Höegh's meniscus lenses are matched to the stop position in the center. Then the lens is free of astigmatism and has a flat image plane. However, the longitudinal chromatic aberration is still existent. Moreover, large spherical aberration as well as coma requires a small relative aperture not larger than $f/22$ for an angular field of view of about 135°.

To avoid chromatic aberrations, achromatic doublets are necessary, which can be designed as Gauss achromatic doublets with an air gap or as cemented achromats. The latter ones have the advantage of fewer glass-air interfaces, which was highly desirable at times when antireflection coating was not yet available. Thin cemented achromatic doublets before the end of the 19th century, however, still had the problem of a curved image field as there were no appropriate types of glass available to reduce chromatic aberrations and the Petzval sum simultaneously. The condition for achromatism of two thin lenses at close distances, thus forming an achromatic doublet, is given by Equation (3.144). It states that the ratio of the refractive powers V_1 and V_2 of the two lenses is equal to the negative ratio of their Abbe numbers ν_1 and ν_2 yielding $V_1/V_2 = -\nu_1/\nu_2$. Bringing the Petzval sum to zero according to Equation (3.129) implies that the ratio of the refractive powers is equal to the negative ratio of the refractive indices $V_1/V_2 = -n_1/n_2$. If both conditions for achromatism and flat Petzval surface should be fulfilled simultaneously, then we get the following relationships:

$$-\frac{V_1}{V_2} = \frac{n_1}{n_2} = \frac{\nu_1}{\nu_2} \Leftrightarrow \frac{\nu_1}{n_1} = \frac{\nu_2}{n_2}. \tag{6.1}$$

The typical glass materials known in the 19th century were crown glasses that had a low refractive index and at the same time a low dispersion, which is equivalent to a high Abbe number. The known flint glasses, on the other hand, had larger refractive indices and simultaneously larger dispersion, meaning a lower Abbe number. Refractive index and Abbe numbers of some "old" glass types are compiled in Table 6.1 on the left side. It can be seen that with increasing index n_d the Abbe number ν_d continuously decreases.

Tab. 6.1: Index and dispersion of some "old" and "new" glasses [Kin39].

"old" types of glass				"new" glass pairs			
type of glass	**index n_d**	**ν_d**	**ν_d/n_d**	**type of glass**	**index n_d**	**ν_d**	**ν_d/n_d**
hard crown	1.5157	60.5	39.9	barium flint	1.6530	46.2	27.9
extra-light flint	1.5290	51.6	33.7	light flint	1.5674	43.8	27.9
light flint	1.5746	41.4	26.3	dense barium crown	1.6098	53.3	33.1
dense flint	1.6041	37.8	23.6	extra-light flint	1.5290	51.6	33.7
extra-dense flint	1.7402	28.4	16.3	dense barium crown	1.6016	59.9	37.4
				telescope flint	1.5151	56.4	37.2

Thus, the ratio of ν_d/n_d decreases correspondingly with the crown glass having the highest value due to the low dispersion.

It was only after 1886 that the production of "new" glass types by Schott, who added components like barium or boron to the glass compositions, provided the lens designers with glasses of new properties. Now glasses with higher index and lower dispersion became available so that the condition (6.1) could be fulfilled by pairs of "new" glass materials. The right side of Table 6.1 lists some of these "new" glass pairs that have nearly identical ratios of ν_d/n_d. Unlike old achromats, the "new" achromatic doublets now additionally fulfilled the Petzval condition but had other drawbacks.

A cemented achromat, which should be converging to get a real image, must have a total positive refractive power. Thus, the magnitude of the refractive power of the converging lens is larger than that of the diverging lens. The condition of achromatism requires for the converging lens that its Abbe number is larger than that of the diverging lens. Therefore, the refractive index of the converging lens in old achromats was made of crown glass with a higher ν_d and lower refractive index. This influences the bending of the shapes of both individual lenses in the doublet that could be chosen in a way that their total spherical aberrations are nearly eliminated. As for new achromats with positive refractive power, we get the situation that due to Equation (6.1), not only the Abbe number of the doublets' converging lens but also its refractive index has a higher value than that of the diverging lens. As a consequence, differently bended lens shapes compared to old achromats are required. However, it is not possible to design the lens bendings in a thin new achromat in a way that the overall spherical aberration vanishes. "New" achromatic doublets are therefore in general spherically undercorrected and not free of spherical aberrations but have a flat image field.

The first anastigmatic achromatic lens was the Ross Concentric lens of 1888, which had a symmetrical setup of two meniscus-like shaped new achromats (Figure 6.4a). The stop position in the center was calculated to achieve a lens without astigmatism or curvature of field. As the spherical aberration was still remarkable at large apertures, the lens could be used only at values smaller than about $f/20$, but then with a high quality all over the image field with a total field of view of about 60°.

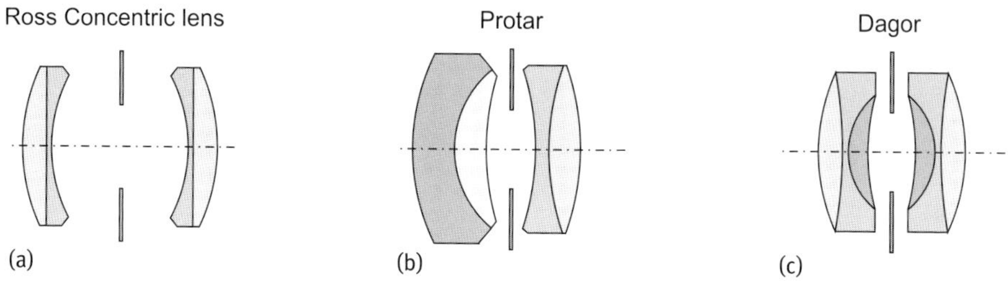

Fig. 6.4: Early anastigmats. (a) Ross Concentric lens (1888), symmetrical design using new achromats, no spherical correction; (b) asymmetric combination of an old achromat with a new achromat in the Zeiss Protar (1890); (c) Goerz Dagor (1893) using a symmetrical arrangement of corrected triplets.

The problem of spherical aberration could be overcome by combining a spherically undercorrected new achromat with an old achromat that was intentionally overcorrected. These newly emerging anastigmats, which were free of spherical and chromatic aberration, free of coma, astigmatism and curvature of field, evolved principally in two phases. In the first phase, we find an asymmetric arrangement of an old achromat with a new achromat, separated by the aperture stop in between them. An example for this design is the Zeiss Protar from 1890 with a cemented "old" achromatic doublet in front of the stop and a "new" one behind it (Figure 6.4b). This Protar type was nearly completely corrected but was limited to a relative aperture of $f/8$.

This could be improved subsequently by converting the "new" achromatic doublet on the right side into a triplet. Thus, the crown glass converging lens on the right side was replaced by two positive halves inclosing the negative flint glass element. In the second phase, a symmetrical design was preferred where both old and new achromats were combined and this combination was placed on either side of the aperture stop. This led to a modification of the Protar, which was nearly identical with the Goerz Dagor from 1893 (Figure 6.4c). The four elements of both achromatic doublets were merged into triplets where the positive first lens on the left side formed a new achromat with the negative second lens and the negative second lens formed an old achromat with third positive lens. The overall shape of the triplets was that of a meniscus with the new achromats being on the outer convex side of the meniscus.

Further modifications of these types of camera lenses included conversions of the triplets into quadruplets and even into quintuplets. Also, small air gaps between the usually cemented lenses are found to yield better results with meniscus lenses in some cases, which eventually led to the double-Gauss anastigmats as discussed below. The principle of these early anastigmats can be described that they consisted of two identical lens groups in air and the symmetrical setup led to a reduction of all transversal aberrations like distortion, coma and transversal chromatic aberrations. As for the remaining aberrations, the individual groups, where the lenses were in general cemented, had such a high degree of correction that they could be used as standalone camera lenses of lower quality or combined with other similar types of different focal lengths. Thus, sets of interchangeable lenses could be formed to allow for higher flexibility. All these anastigmats had relatively large angular field of views of up to 90°; their relative aperture did not in general exceed $f/4$. They were expensive and not very appropriate for portrait photography.

6.2.5 Anastigmats consisting of three lens groups

In parallel to the early anastigmats, which in most cases consisted of two cemented groups, different approaches were pursued simultaneously in order to simplify the lens design and to satisfy the demand for a better portrait lens. The only useful portrait lens around 1900 was the Petzval lens with a large relative aperture. However, it suffered from a curved field and astigmatism. Hence, there was a demand for a better lens with a narrower field of view unlike the early anastigmats that were more appropriate for landscape photography with larger angular fields. Unfortunately, the early astigmats eventually were not of a simple design. Their complexity was due to the cemented groups consisting of up to five elements. However, these groups have the advantage of only two glass-air interfaces, and thus yield low reflection losses and better contrast. But there are constraints as the radii of the individual positive and negative elements must have the same values and the overall refractive power is just the sum of the two lenses' values.

There are more degrees of freedom to reduce the overall aberrations when positive and negative lenses with air gap between the elements are used. The Petzval sum of a lens combination only depends on the refractive power and the refractive index of the individual lenses, but not on the lens position and the lens bending. If the two lenses of a cemented doublet are separated, the Petzval sum does not change but the overall refractive power (see Section 3.3.5). Additionally, with the lenses being no longer cemented, their bending can be adapted separately to correct for spherical aberrations as well as to influence the astigmatism. The Cooke Triplet, designed by Taylor, was the consequence of these considerations.

Taylor started with the basic idea of an achromatic lens doublet with a zero Petzval sum in 1893. The simple separation of the positive and negative lens increases the overall refractive power but also generates considerable transversal aberrations. Therefore, the converging lens of the doublet, consisting of barium crown glass, was split into two parts. One part was mounted in front of the negative biconcave flint glass element and the other one behind it (Figure 6.5a). The use of new glasses with higher refractive index permitted the use of lenses with larger curvature radii thus reducing the amount of aberrations. The aperture stop was located behind the negative lens element in the rear air gap and could also reduce the overall sensitivity of the lens to light reflections at the six air-glass interfaces. The original Cooke Triplet had a relative aperture of $f/3.5$ and an angular field of view of 55°.

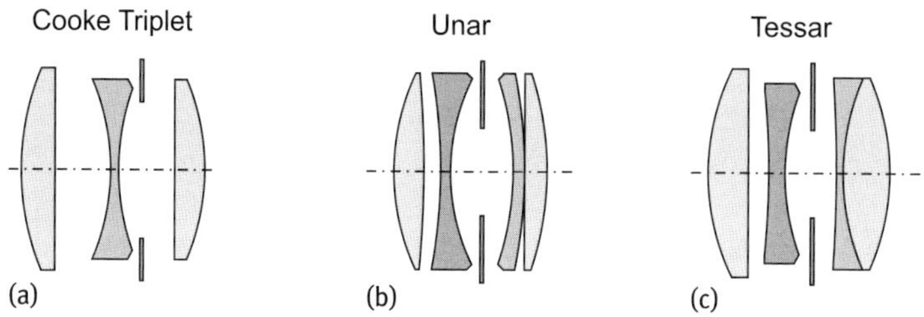

Fig. 6.5: Anastigmats. (a) Triplet by Cooke (1893); (b) Unar by Zeiss (1900); (c) Tessar by Zeiss (1902).

A different approach to a triplet lens structure resulted in the design of the Zeiss Tessar. As mentioned above, attempts to reduce the aberrations with less complexity than the early anastigmats led to the replacement of the cemented lens groups by positive–negative lens couples with air gap. The designer of the Zeiss Protar, Paul Rudolph, developed the Zeiss Unar lens in 1900, which also consisted of four lenses but with small air gaps between them (Figure 6.5b). Rudolph realized that the frontal dialyte, a crown glass positive and a flint glass negative lens couple, has fewer zonal aberrations than a cemented couple and also gives more flexibility to the whole design. The rear doublet is a Gauss-type achromat of different glasses. A big disadvantage of the Unar, however, at the time of its development was that the eight glass-to-air interfaces produced more internal reflections than the typical conventional anastigmats. Since antireflection coatings were not yet available, the internal reflections should be kept at a minimum. They are a main cause of stray light in a lens and cause a strong reduction of the image contrast. More than six glass-to-air interfaces at that time were hardly tolerable. As a consequence, Rudolph combined the dialyte front couple of the Unar with the cemented new achromat of the Protar to create the Tessar in 1902 (Figure 6.5c). Its name makes reference to the Greek numeral of four. The Tessar consists of four lenses in three groups thus having only six glass-to-air interfaces. The Tessar resembles the Cooke triplet but its design comes from a completely different consideration. The frontal Gauss-type couple has a low refractive power with large radii of curvature. It has low zonal aberrations and is mainly intended to correct the remaining aberrations of the rear strong new achromat like in the Protar. The consequence has been a lens of a relative aperture of $f/6.3$ initially, which could be increased up to $f/2.8$ and field angles ranging from approximately 45° up to 75° by modifications in later years. The Tessar features a simple lens design and a high image quality at moderate relative apertures and angular field of view values. This makes the lens universal for many applications and is the reason that the Tessar type with its various modifications has become one of the most successful camera lenses. The impressive sharpness in the central part of the image is also expressed by its marketing as the "eagle eye" of the camera by the Zeiss company.

Due to its simple design, the triplet lens was modified by many manufacturers with the intention of further reducing the aberrations and to increase the relative aperture. One or all three elements of the triplet can be replaced in principle by more complex lens

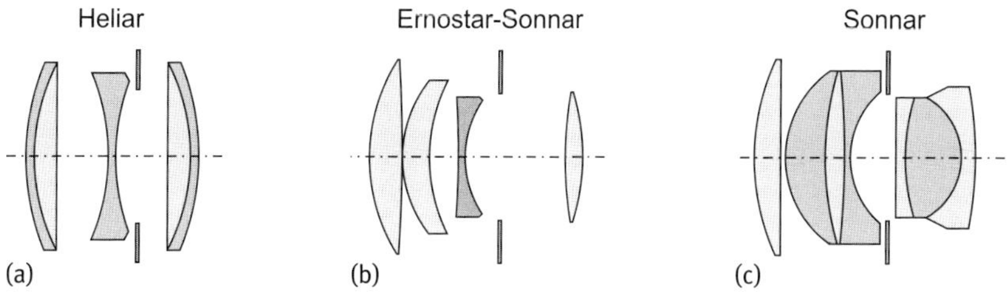

Fig. 6.6: Modifications of the triplet design. (a) Heliar by Voigtländer (1900); (b) Ernostar–Sonnar by Zeiss–Ikon (1928); (c) Sonnar by Zeiss (1932).

combinations, which lead for instance, to the Heliar lens of Voigtländer (Figure 6.6a). This lens of high quality had a symmetric design, which reduced further all transversal aberrations that have been always present with the Cooke Triplet. In the Heliar design, the positive front and end lenses were substituted by cemented achromats, which helped to reduce the remaining aberrations.

Lens designers realized in the following years that the insertion of a positive meniscus lens in the front airspace of the triplet was very favorable in order to reduce the spherical aberration, and thus to increase the relative aperture of the objective lens [Kin89]. Ludwig Bertele, the designer of Ernostar lenses in the years after around 1920, implemented that approach for the Ernostar–Sonnar type of Zeiss–Ikon in 1928 and achieved a $f/2.0$ camera lens of 100 mm focal length (Figure 6.6b) by a relatively simple lens arrangement. Further modifications of the triplet design by Bertele eventually resulted in the Zeiss Sonnar lens almost 30 years after the original Tessar. In the Sonnar lens, the negative center element, now integrated with the positive meniscus of the Ernostar–Sonnar, as well as the positive rear element are realized as cemented triplets (Figure 6.6c). The Sonnar, representing a typical three lens group design, featured a very good correction of aberrations up to a relative aperture of around $f/1.5$ with an angular field of view of less than about 45°.

6.2.6 Double-Gauss anastigmats consisting of four lens groups or more

As described in the preceding section, there have been different approaches to designing the individual achromatic lens combinations that make up the complete camera lens. Whereas the cemented achromats have the advantage of only two glass-to-air surfaces, they are less flexible in the lens design to reduce the aberrations due to their constraints for identical lens radii of the cemented lens elements and the very short distance between them. It turned out that using a dialyte achromat, consisting of two air-spaced lenses of arbitrary shape, or a Gauss achromat, consisting of two air-spaced meniscus lenses, led to better overall results in the lens design than using cemented achromats. The idea of Taylor for the design of the Cooke Triplet, which was using one positive–

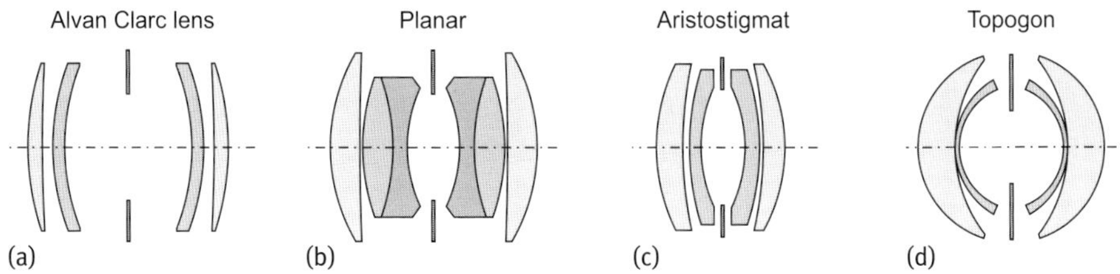

Fig. 6.7: Examples of symmetric double Gauss anastigmats. (a) Alvan Clarc lens (1888); (b) Zeiss Planar (1896); (c) Goerz Aristostigmat (1902); (d) Zeiss Topogon (1933).

negative lens couple with a lens separation and modifying it, can be extended to have two identical couples with air gaps. They are arranged symmetrically around a central aperture stop to minimize the oblique aberrations. This concept using biconvex and biconcave lenses was implemented for instance in the Goerz Celor lens in 1898. The combination of a dialyte achromat on the front side of the lens with a Gauss type on the rear end can be found in the Zeiss Unar (Figure 6.5b). However, it was not very advantageous and was replaced by more successful lenses like the Tessar. The most versatile and widespread types of camera lenses consist of two Gauss type achromats with the aperture stop in between them. They are classified as double-Gauss lenses and exist in symmetrical as well as asymmetrical versions. From the beginning- they already featured high imaging qualities at relatively large apertures. However, due to at least eight glass-to-air surfaces they had problems with internal reflections impairing the contrast and producing "ghost images." They became much more successful after the invention of antireflective lens coatings around 1935 so that almost all modern lenses of high aperture can be regarded as a further development of this double-Gauss principle [Bla16].

Among the first to have implemented the double-Gauss design were Alvan Clark in 1888 and Rudolph from Zeiss in 1897 for the Planar lens (Figure 6.7a, b). Both lenses have symmetric lens arrangement with the aperture stop in the center. The Clarc lens had a relative aperture of about $f/8$ and was not very successful. Rudolph realized that a much better quality could be achieved by a similar design with thicker lenses and a smaller air gap between the positive and negative elements on either side of the objective lens. This led to the development of the Zeiss Planar (Figure 6.7b) where the negative meniscus was composed of two different types of glasses with the same refractive index but different dispersion characteristics. As a consequence, the chromatic aberrations could be corrected at will without influencing the remaining aberrations. The oblique aberrations could be minimized simply by the strict symmetrical lens arrangement. The name Planar comes from the characteristic of the lens having an image field of very low curvature thus being nearly flat. The Planar had a superior image quality with a large relative aperture of $f/4.5$ for the time of its invention. It became a very successful lens in later times after the availability of antireflective lens treatment and after some design modifications leading to the Biotar (Figure 6.8a). Due to its initial sensitivity to incident bright light, the Planar had to step back the simpler Tessar, which was designed only some

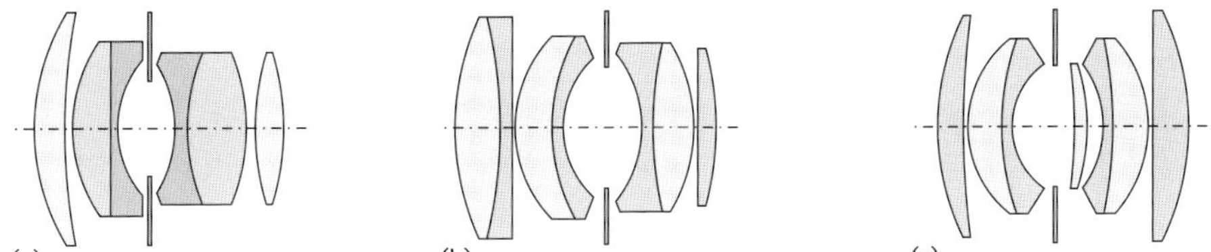

Fig. 6.8: Examples of asymmetric double Gauss anastigmats. (a) Zeiss Biotar (1927); (b) Leitz Summitar (1939); (c) Leitz Summilux (1960).

years after the Planar. In 1902, the Aristostigmat of Goerz appeared, which was based on the same principles as the Planar but had only four lenses made of newer glass types (Figure 6.7c). The Aristostigmat could be realized in a strict symmetrical setup but was also still modified to be slightly asymmetrical to reduce the remaining aberrations for the typical nonsymmetric photographic situation. All aberrations including astigmatism could be reduced a lot by only four lenses. An extreme angular field of view achieved by the four-lens double-Gauss design could be found with the Zeiss Topogon (Figure 6.7d). It was computed by Richter in 1933 to have an overall field angle of 100° at a relative aperture of $f/6.3$. The Topogon was a highly corrected anastigmat with very low chromatic aberration. It featured low distortion even at large angles and for a long time was the standard lens for aerial photography and aerial metrology. However, vignetting at large angles was a problem that could be counteracted by special graded density filters to reduce the central illuminance relative to the lower light situation in the peripheral parts of the image field. A smaller version for 35 mm format Contax cameras was the $f/4$ lens with a focal length of 25 mm.

Design improvements to increase the relative aperture of the double-Gauss type required a more or less asymmetrical setup. Merté from Zeiss modified the Planar lens to develop the Biotar in 1927 (Figure 6.8a). Compared to the Planar, it had different curvature radii, lens dimensions and also different glass in front and behind the aperture stop. By these asymmetric corrections, it was possible to reduce the disturbing reflections in the lens and to enhance the relative aperture up to $f/1.4$ with low aberrations. After World War II, Zeiss split up into two independent enterprises in the western and eastern part of Germany. Therefore, the asymmetric Biotar type was merchandised then under the brand names of Planar and Biotar for political reasons. Many other modifications based on the initial double-Gauss have been made after 1930 and especially after the development of the antireflective coating technology for lens surfaces. A similar design was chosen by Leitz for the 50 mm Summitar $f/2$ (1939) where the frontal positive meniscus of the Biotar was replaced by a cemented doublet and the remaining lenses were made of different glasses and had different bendings (Figure 6.8b). Other variations of the Summitar have the rear positive lens split up into two individual meniscus lenses of different bendings. Unlike the classical double-Gauss type, the Leitz Summilux

35 mm, $f/1.4$ of 1960 incorporates an additional positive meniscus lens in the rear half (Figure 6.8c).

As already mentioned above, most modern objective lenses of high aperture, especially normal lenses or lenses of focal lengths nearby, can be considered as modifications of the Biotar design. Further improvements were achieved in the years after 1950 when lanthanum crown glass became available. The higher refractive index made it possible to increase the curvature radii of spherical lenses, and thus reducing the corresponding aberrations. Moreover, by implementing one or more aspheric lens surfaces the overall performance could still be enhanced as was the case for the 50 mm Leica Noctilux $f/1.2$ from 1966 featuring two handmade aspheric surfaces. At that time, the manufacturing of aspheric lenses was very costly due to missing adequate automated production technology. The camera lenses with the largest relative apertures were manufactured and commercially available are the 50 mm Canon $f/0.95$ from the years after 1960 for rangefinder cameras and the 50 mm Leica Noctilux $f/0.95$ from 2008 (Figure 6.17). Both lenses represent modifications of the double-Gauss design. The Leica Noctilux has a modern design optimized for digital image sensors.

The examples given in this section about the historical development of photographic lenses were chosen to understand the steps toward achieving the high-quality lenses that we have today. This step-by-step development was necessary as the computational methods, usually done by hand or slide rule, were very arduous before the availability of digital computers. Today the optimization of objective lenses by computers is a standard procedure yielding results that often deviate from the typical classification schemes presented here.

6.3 Long focus lenses

According to the classification scheme presented in Chapter 2, a long focus lens designates a camera lens of which the focal length is significantly longer than the sensor diagonal with which it is used. If the distance to the object is kept constant, the magnification increases with the focal length and the angle of view is narrower than that of a natural viewing perspective. Total angle of views for portrait lenses typically range between 40° and 20° across the image diagonal, which means that the focal length of the lens is in between about 60 mm and 120 mm for the 35 mm format. Telephoto lenses have still longer focal lengths and narrower angles of view. Possible applications for long focus lenses are in the area of nature and sports photography where it is necessary to focus on image details. Another important application is portrait photography where for instance the face of a person should be sharply imaged whereas the background should become intentionally blurred. For that purpose, it is necessary to create a nice bokeh, which depends on the focal length as well on the relative aperture of the lens. Long focus lenses usually are very sensitive to motion blur due to shaking. Therefore, large relative apertures are usually required in order to achieve short exposure times and to control the desired bokeh.

If a large angular magnification with a narrow field of view is to be attained, generally large lenses are required. Especially fast lenses with a large relative aperture tend to be very heavy. Portrait lenses with moderate long focal length and relative magnification of about two compared to normal lenses are often designed as double-Gaussian anastigmats or sophisticated triplets of Sonnar type with good correction of all types of aberrations. These designs, however, are not appropriate for easy-to-manage lenses of still longer focal lengths, particularly super telelenses. The weight of super telephoto lenses for sports and nature photography usually exceeds that of the camera body. Thus, a different approach for relatively short and light lenses is necessary and has been implemented in the telephoto lens design. This design can be virtually found for all lenses with longer focal lengths than about 135 mm for the 35 mm format.

6.3.1 Telephoto principle

The telephoto principle is based on the lens arrangement of the Galilean telescope. It consists of a positive lens L_1 at the entrance of the optical system with a negative lens L_2 at a given distance t_s behind it. Figure 6.9a shows the telescope setup (see also Example 3 in Section 3.3.5). The negative lens L_2 is located at the position where its object focal point F_{2o} coincides with the image focal point F_{1i} of lens L_1. If we designate the image focal lengths of both lenses by f_1 and f_2, respectively, then the separation t_s between the lenses can be written as the sum of both focal lengths $t_s = f_1 + f_2$. This separation is shorter than f_1 as the image focal length of the diverging lens is counted negative. Incoming parallel light is converged to F_{1i} by the first lens, diverged by the second lens and leaves the system parallel to the optical axis. The focal length of the telescope is infinite, the system is afocal. The angular magnification Γ of the telescope is given by the ratio of both focal lengths (Equation (3.60)), which in this example is $\Gamma = -f_1/f_2$. This is a positive value as the focal lengths have opposite signs. It should be mentioned that the magnitude of f_1 must be larger than f_2 for a true image enlargement with $\Gamma > 1$. In terms of refractive power that means that the diverging lens is more strongly refracting than the converging lens.

In order to get a positive, finite focal length as well as an appropriate back focal length, the setup has to be modified. The image focal length f of the lens combination and the back focal length f_{Ei} can be calculated after Equations (3.45), (3.47) and (3.53), yielding

$$f = \frac{f_1 \cdot f_2}{f_1 + f_2 - t_s} \qquad f_{Ei} = f \cdot \frac{f_1 - t_s}{f_1}. \tag{6.2}$$

In order to get a positive finite focal length f, while f_2 is negative, it follows from Equation (6.2) that the separation t_s must be larger than $f_1 + f_2$, and thus larger than for the

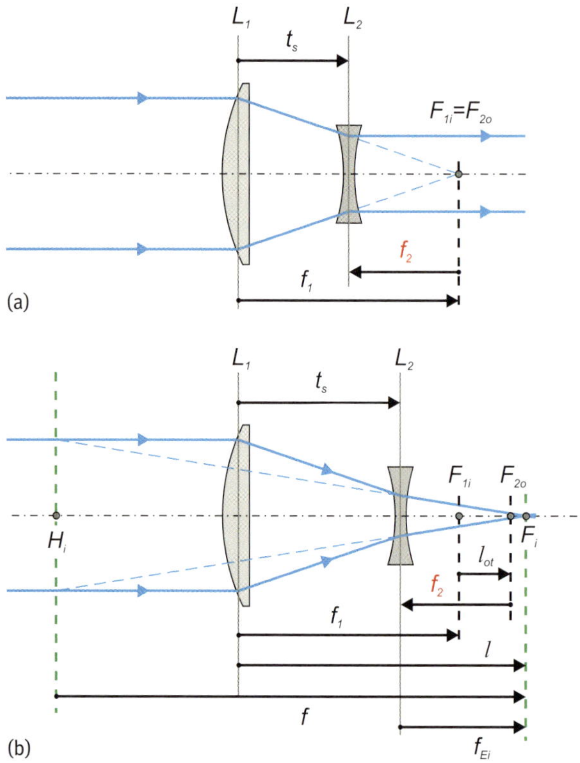

Fig. 6.9: Combination of a positive and a negative lens, with $\Gamma = 2$. (a) Galilean telescope, $t_s = 0.5 \cdot f_1$; (b) schematic telephoto lens design, $t_s = 0.73 \cdot f_1$, telephoto ratio 0.61.

telescope. However, for the back focal length f_{Ei} to become positive, t_s must not exceed f_1. This setup is termed telephoto lens design and illustrated in Figure 6.9b for $\Gamma = 2$ and $t_s = 0.73 \cdot f_1$. For comparison, the telescope in part a) has a shorter separation of only $t_s = 0.5 \cdot f_1$.

The back focal length f_{Ei} is measured from the second lens to the image focal point F_i. The overall length l of the telephoto camera lens extends from the vertex of the first lens to F_i, which is equal to the sum of t_s and f_{Ei}:

$$l = t_s + f_{\mathrm{Ei}} = f + t_s \cdot \left(1 - \frac{f}{f_1}\right). \tag{6.3}$$

The length in the telephoto design is always shorter than the focal length as the bracket in Equation (6.3) becomes negative and t_s is positive. This is due to the fact that the focal length f of the system is also larger than f_1 as can be seen from Equation (6.5) below. This is a main feature of this telephoto design and allows for shorter and less heavy lens constructions. This cannot be achieved by a symmetric lens design as is shown by

Example 1 in Section 3.3.5. The ratio l/f is termed the telephoto ratio and is always smaller than 1 for this telephoto design.

The distance between the two lenses in the afocal Galilean telescope is called tube length and is equal to $f_1 + f_2$. In the telephoto design t_s is larger than for the telescope, and its exceeding difference compared to the telescope is termed optical tube length l_{ot}:

$$l_{ot} = t_s - (f_1 + f_2). \tag{6.4}$$

In a more general consideration, the optical tube length l_{ot} of a two-lens combination is defined as the distance from the image focal plane of the first lens to the object focal plane of the second lens. Thus, for an afocal telescope setup the optical tube length must be zero as both focal planes coincide. l_{ot} is positive for the telephoto design but should not be larger than the magnitude of f_2 lest the back focal length f_{Ei} becomes negative. In that case, the negative lens would be located beyond F_i, thus we have the restriction $0 < l_{ot} < -f_2$. The focal length as well as the back focal length of the optical system can then be expressed using the optical tube length:

$$f = -\frac{f_1 \cdot f_2}{l_{ot}} \qquad f_{Ei} = -f\,\frac{l_{ot} + f_2}{f_1}. \tag{6.5}$$

Here, we can see that the focal length f of the lens combination is always larger than f_1 as f_2 is negative and l_{ot} is positive and smaller than $-f_2$. It can be understood from this consideration that by mounting a strongly refracting negative lens behind the positive lens the overall focal length of the systems can be enhanced whereas the overall length of the system is shorter than the focal length. This is also expressed by the fact that the image principal plane at H_i is shifted to the left side and is even located outside the objective lens due to the negative lens (see examples in Section 3.3.5 for shifting the principal planes).

Using the angular magnification Γ, the focal length can be written as

$$f = \Gamma \cdot \frac{f_2^2}{l_{ot}}. \tag{6.6}$$

For the relative magnification f/f_1 and for the relative back focal length f_{Ei}/f_1 we get after some rearrangement:

$$\frac{f}{f_1} = -\frac{f_2}{l_{ot}} = \frac{1}{\Gamma} \cdot \frac{f_1}{l_{ot}} \qquad \frac{f_{Ei}}{f_1} = \frac{1}{\Gamma}\left(\frac{f}{f_1} - 1\right). \tag{6.7}$$

These equations can be used for the design of a telephoto lens based on a positive lens with a given f_1. A high relative magnification is achieved if it is combined with a negative lens of weak refractive power, namely large magnitude of f_2, at a short optical tube length. A large magnitude of f_2 relative to f_1 is equivalent to small Γ. Thus, reducing Γ increases the relative magnification but also the back focal length, and it eventually increases the length as well as the weight of the telephoto lens. On the other hand, increasing Γ for a shorter lens design requires a more strongly refracting negative lens with stronger surface curvature, which implies more lens aberrations. Therefore, Γ is usually not larger than about 3 [Flü55]. As for the telephoto ratios, many values can be found in the range between 0.5 and 0.8.

Critical points for the telephoto design are longitudinal chromatic aberration, curvature of field and distortion. In modern lenses, the telephoto principle is implemented by composite positive and negative lens groups with the ratio of their refractive powers given by Γ. Virtually, all lens aberrations are corrected for each group as well as possible. Longitudinal chromatic aberrations are directly proportional to the focal length and become more pronounced in telephoto lenses. Thus, a special effort for their corrections has to be done here, particularly for the remaining secondary spectrum of the achromats. Therefore, special glass types with very low chromatic dispersions are required for high-quality lenses of this type. Some manufacturers use calcium fluoride-based glasses or modifications, sometimes designated as extreme–low or ultralow dispersion glasses. If Γ is in the order of 2–3, it is not possible to satisfy the Petzval condition for a flat image field as the difference in the refractive indexes is not sufficient to compensate for the large imbalance of the focal lengths. Thus, it is difficult to maintain the same image sharpness all over the image field. In some cases, additional field flattener lenses can be found at the end to achieve a better homogeneity of sharpness over the image field.

Due to the asymmetric lens arrangement relative to the aperture stop, the telephoto lens is prone to distortion. A true telephoto lens tends to exhibit a pincushion distortion whereas the more symmetric design types are nearly free of aberrations. Additional lens elements, in modern systems for instance aspherical lenses, are located mainly at the rear end of the asymmetric constructions to correct the remaining distortions. Another critical point for lenses, especially for the asymmetric telephoto construction with a positive front and a negative rear group, is the change of lens parameters upon focusing, which will be discussed in the next section.

6.3.2 Focusing by moving lens groups

In order to get a sharp image on the sensor of the camera, object and image distance have to be adapted according to the lens equation. In the common situation of photography where the image plane is fixed in the camera, the position of the lens is shifted relative to the camera to satisfy the conditions. This is usually called focusing. It should be stressed again that here, as well as in the whole Chapter 6, focusing means a method to achieve a sharp image of a wanted object in the sensor plane. There is also another meaning of focusing which is discussed in detail in Section 1.5 (see also Figure 1.15).

The simplest way of focusing is to displace the entire lens barrel with the individual lens groups in the barrel remaining at fixed positions relative to each other. This situation is illustrated in Figure 6.10a where a large part of the lens has to be moved in axial

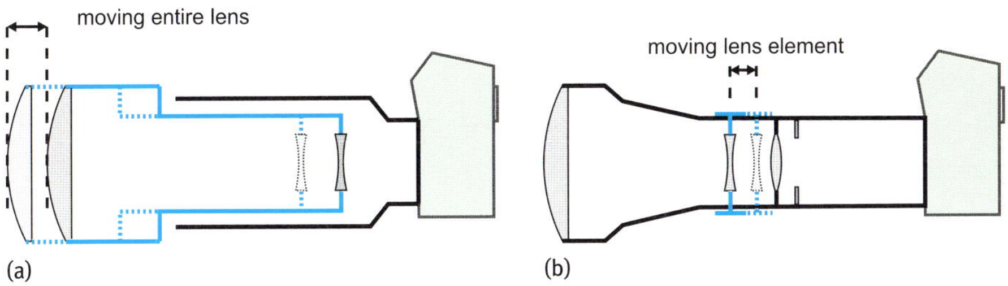

Fig. 6.10: Focusing methods. (a) Unit focusing by moving the entire lens; (b) internal focusing by moving a lens element. Note: Focusing here means to achieve a sharp image in the image plane.

direction. The total extension of the lens changes accordingly. This focusing is called unit focusing. For moving the entire lens barrel, a rugged mechanical setup between the moving parts is required, which makes the whole lens heavy and not easy to handle. For large and heavy camera lenses, this may be quite uncomfortable. An alternative to the unit focusing method is the axial movement of only one or two lens groups within the lens barrel and leaving the barrel fixed with the camera (Figure 6.10b). Compared to the scheme in part a), the negative rear element is split up and only one element of this group is displaced. The barrel neither moves nor rotates and its extension remains unchanged. This method is called internal focusing and was introduced in the early 1970s by Nikon for its telephoto lenses. The method allows usually for much quicker focusing and necessitates less shift distance. As there are only minor moving parts, the whole construction becomes less heavy and easier to handle.

There is a significant difference between these two methods: when the entire lens is moved, the individual lenses remain fixed to each other, and the overall optical lens properties do not change. All corrections of aberrations remain optimized. Particularly, the sizes of the entrance and exit pupils, thus the pupil magnification M_p and the focal length do not change. By adjusting the lens to a different object distance, however, the magnification changes, which also changes the working f-number according to Equation (3.98). The image brightness changes correspondingly, which is not an issue for still cameras where the exposure is controlled by the exposure time. For movie cameras with a fixed exposure time, however, this can become noticeable if the changes are significant and, therefore, has to be avoided. Telephoto lenses in general have $M_p < 1$ and feature much stronger variations of the illumination in the image plane upon focusing than wide-angle lenses with a retrofocus design and typically with $M_p > 1$ [Bla14]. Moreover, focusing may change the size of the visible image. This effect is termed “breathing.” All these effects can be minimized by a different focusing mechanism. On the other hand, when an individual lens element is moved, the overall focal length as well as other lens parameters change. Thus, all lens optimizations must be performed with respect to this lens movement. Further focusing methods can be differentiated where for instance only the front lens group, or only the rear lens group, or both groups are moved.

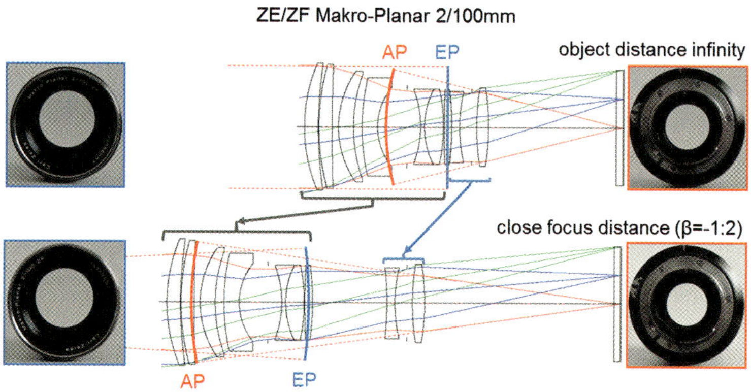

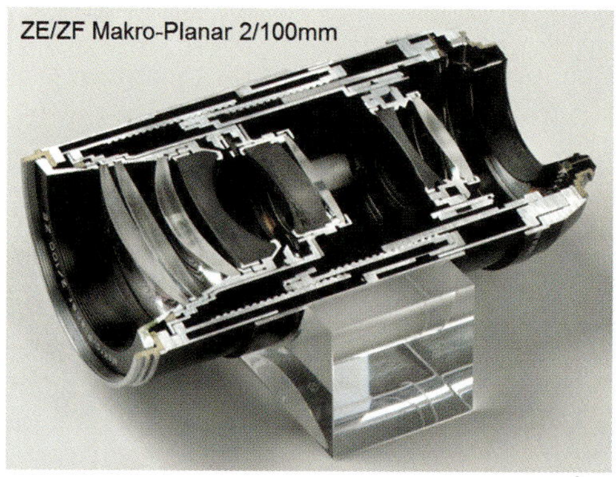

Fig. 6.11: Focusing by floating elements in the Zeiss Makro Planar 2/100 mm. Two lens groups are moved to maintain the imaging quality at different image magnifications; EP: entrance pupil, AP: exit pupil; β in the figure designates the overall image magnification; note that EP and AP slightly change upon focusing [Bla14] (with kind permission of Carl Zeiss AG).

Figure 6.11 illustrates the focusing mechanism by moving two lens groups for the Zeiss Makro Planar 2/100 mm [Bla14]. In this floating elements design, a high-imaging quality independently from the object distance is maintained. Especially for lenses with low f-numbers and for close-up imaging with relatively large image distance ranges, a compensation of the lens parameter variation is required. The moving of several lens groups becomes more sophisticated in zoom lenses where the focal length changes over significant ranges and the object and image distances in the lens have to be adapted correspondingly (see Section 6.6). It should be noted that the floating lens design is done in order to guarantee a high-optical quality even when lens groups are moved. Unlike the internal focusing, however, the overall length of the lens may change.

6.3.3 Examples of modern long focus lenses

In the following, we consider some examples of long focus lenses for 35 mm format cameras, with a sensor diagonal of 43 mm. If the lenses are used with SLR cameras, a minimum back focal length of the distance from the lens mounting flange to the image plane is required in order to not obstruct the moving mirror in these cameras. This distance is for various 35 mm SLR cameras of the order of 40–50 mm, and thus comparable to the focal length of a normal lens. As a consequence, the lens design for moderate long focus lenses can be expected to be similar to that of a normal lens, particularly the lenses between 85 mm and 135 mm lens, which are well suited for portrait photography. Figure 6.12 shows the cross-sections of some modern Nikon long focus lenses for DSLR cameras between 85 mm and 200 mm focal length. The 85 mm and 135 mm lenses feature a double-Gauss anastigmat design with asymmetric groups and the aperture stop in the center. This design ensures a high-image quality with good correction of all lens aberrations. The lengths of both lenses including their flange focal distances are longer than their focal lengths. There is a special feature of the 135 mm lens where through a patented control mechanism, one lens in the rear side of the lens can be slightly moved in order to decrease or increase the spherical aberration. The spherical aberration in an image depends on the object distance. Once the object at a given distance is sharply imaged to the sensor plane by the optimized lens arrangement, then the blur in the foreground or the background can be intentionally varied by the slight element shift to get the desired bokeh (see Section 6.9.3). By contrast, the 180 mm lens no longer has the double-Gauss design but rather the telephoto design with a telephoto ratio of 0.8. It also disposes of the internal focusing mechanism as described above and which can be found with telephoto lenses of significantly longer focal lengths (Figure 6.13). In order to reduce the overall chromatic aberration, one lens element made of special glass with anomalous partial dispersion is implemented.

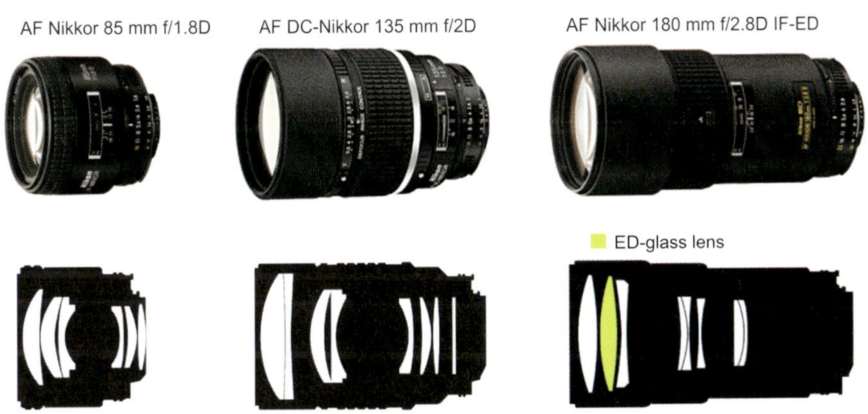

Fig. 6.12: Nikon long focus lenses and their constructions (with kind permission of Nikon).

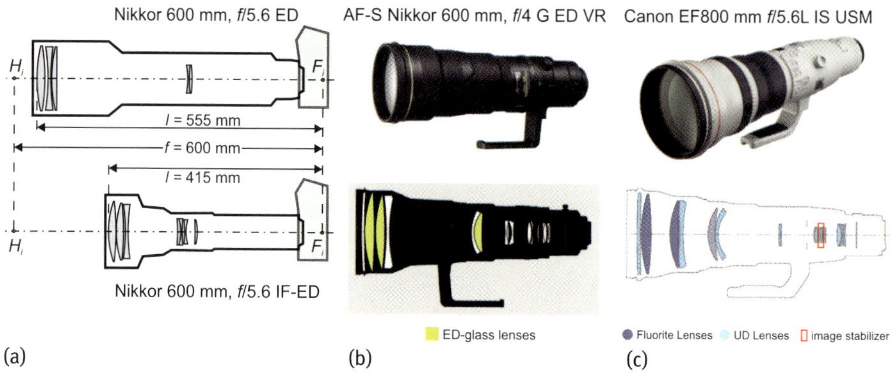

Fig. 6.13: Telephoto lenses. (a) Nikkor 600 mm, $f/5.6$ ED (before 1974) and Nikkor 600 mm, $f/5.6$ IF-ED (before 1977); (b) AF-S Nikkor 600 mm, $f/4$ G ED VR, internal focusing, vibration reduction (with kind permission of Nikon); (c) Canon EF 800 mm $f/5.6$L IS USM, internal focusing, image stabilization (with kind permission of Canon).

The evolution of a telephoto design can be verified in Figure 6.13a,b) for the Nikkor 600 mm lens. Part a) shows the schematic setup of the classical telephoto design (Nikkor 600 mm, $f/5.6$ ED, before 1974) with a positive front group and a negative rear group. The telephoto lens head was combined with a focusing unit to establish a lens of 555 mm length including the back focal distance and a telephoto ratio of 0.9. A redesign of the lens led to a significantly shorter construction with the Nikkor 600 mm, $f/5.6$ IF-ED having a telephoto ratio of about 0.7. The negative rear group was split, and the central lens group could be moved forth and back to realize the internal focusing mechanism as described above. The most recent version has the same basic principle (AF-S Nikkor 600 mm $f/4$ G ED VR, Figure 6.13b) but has been complemented by additional lens groups to implement features such as vibrational reduction (VR) and autofocus as well as to increase the relative aperture of the lens. It incorporates lenses of special glass for low chromatic errors and has a telephoto ratio of around 0.8.

A similar basic construction principle can be found with the Canon EF 800 mm f/5.6L IS USM (Figure 6.13c). Autofocusing as well as manual internal focusing is achieved by the movement of some lens elements, and also image stabilization by slight lateral lens shifts. Special glass elements based on fluoride and ultralow dispersion glass lenses are implemented. The telephoto ratio is of about only 0.6.

As for moderate long focus lenses, some other lens designs should be mentioned here. Figure 6.14a shows the Leica Apo–Summicron-M 2/90 mm. The Leica M-system was traditionally conceived for rangefinder film cameras with the 35 mm format and was then adopted by the mirrorless digital cameras as the successors of the film cameras. As there is no hinged mirror in the camera body, there are different possibilities for lens designs with shorter back focal lengths. The lens design can be characterized as an asymmetric double-Gauss setup consisting of only five lenses. One lens has an aspheric surface, two lenses are highly refracting, and two lenses are made of special low dis-

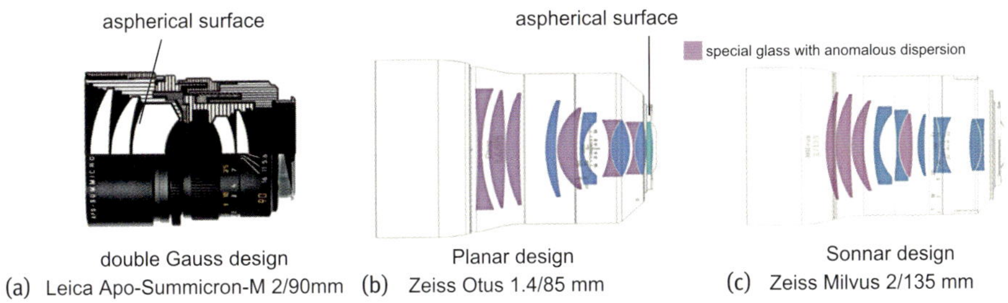

Fig. 6.14: Examples of moderate long focus lenses. (a) Leica lens (with kind permission of Leica Camera AG); (b) and (c) Zeiss lenses (with kind permission of Carl Zeiss AG).

persion glass. Having only five lens elements, the camera lens has a low weight and its length including the flange focal distance is less than 20 % longer than its focal length. Although of apparently simple design, the lens is classified as apochromatic meaning that it has a superior chromatic correction.

Also based on the asymmetric double-Gauss design is the Zeiss Otus 1.4/85 mm (Figure 6.14b). The early asymmetric double-Gauss by Zeiss was termed Biotar. It is the base model for all lenses manufactured after World War II by Zeiss Jena, termed Biotar, and by Zeiss Oberkochen, termed Planar. The Zeiss Otus is a successor in this tradition. As a modification of the original Biotar/Planar design, it features more lenses in both front and rear groups, some made of special glasses, and an aspheric lens as the last element. By that an apochromatic quality in combination with a nearly constant superior image quality all over the image plane like for medium format cameras has been achieved. As a consequence, the lens design must be more complex than the original Biotar and the lens is relatively large with a correspondingly high weight. Focusing is manually achieved by moving individual lenses or lens groups termed floating lens elements by Zeiss.

As a last example for long focus lenses, we consider the Zeiss Milvus 2/135 mm (Figure 6.14c). This focal length is nearly the limit for portrait lenses of the 35 mm format. It neither has the asymmetric double-Gauss nor the telephoto setup. It is rather based on a triplet structure following the Sonnar design, which was known for achieving large relative apertures with angles of view that are narrower than with normal lenses. Like the Otus lens, it is apochromatic with a manual focusing mechanism based on floating elements. It is made of spherical lenses, partially of special glasses. Its length including the flange focal distance is less than 20 % longer than its focal length but not as long as the Zeiss Otus of shorter focal length.

As for the apertures of long focus lenses beyond 135 mm, there can virtually no lens be found having a relative aperture larger than $f/1.8$. The reason for it can be understood in the following consideration: Well-corrected lenses for high-quality imaging require small f-numbers in order to reduce diffraction. This necessitates large entrance pupils respectively apertures. Especially in the case of long focus lenses, this leads to much larger and heavier lenses, when compared to normal lenses and wide-angle lenses, respectively. Furthermore, this incorporates a lot of expensive glass, and also optical manufacturing becomes more elaborate. In total, that results in high costs. All this is obvious from the following numerical example, when we compare three high-quality diffraction limited lenses with $f_\# = 1.8$. For a normal lens with $f = 50\,\mathrm{mm}$, $D_{en} \approx 28\,\mathrm{mm}$, for a wide-angle lens with $f = 28\,\mathrm{mm}$, $D_{en} \approx 16\,\mathrm{mm}$, but for a long focal lens with $f = 200\,\mathrm{mm}$, $D_{en} \approx 111\,\mathrm{mm}$. Vice versa, if such a large D_{en} is not possible, the image quality is reduced. Hence, fast long focus lenses of high quality become very bulky, with large front lenses, and expensive.

6.3.4 Teleconverters

Although consisting of at least two groups of opposite refractive power, modern telephoto lenses are optimized as one complete unit. This is unlike early telephoto lenses from the end of the 19th century, which were made of two separately optimized positive and negative groups and where the positive group could be even used as a standalone objective lens. The disadvantage of these telephoto lenses has been that the negative rear end group magnified all remaining aberrations of the positive first group thus requiring a very good correction of both separate units. This was not very successful at that time. This principle of combining a fully corrected objective lens with an optimized negative lens group, acting as one single lens, was picked up again for SLR camera lenses in the 20th century to extend their focal lengths. The negative lens group is contained in a teleconverter unit, which is mounted between the lens and the camera body. Figure 6.15 shows the example of a 2× teleconverter, which is used to double the focal lengths of lenses longer than normal lenses. It can be seen in Figure 6.15b that the teleconverter represents a negative lens producing a demagnified virtual image. The internal design of a modern teleconverter by Nikon is shown in Figure 6.15c. There are seven lenses in several groups, one having an aspherical surface. As stated above, it is of high importance

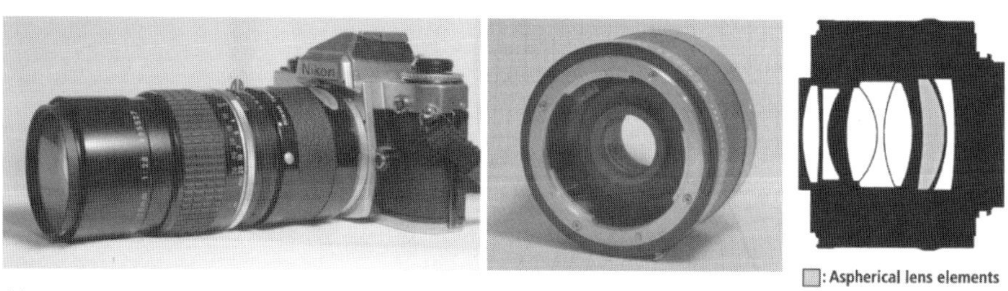

(a) (b) (c)

Fig. 6.15: 2× teleconverter. (a) Nikon Teleconverter TC-200 mounted between SLR body and long focus lens Nikon 2.8/135 mm; (b) Nikon Teleconverter TC-200; (c) lens design of Nikkor AF-S Teleconverter TC-20E III with aspherical lens element (with kind permission of Nikon).

to achieve a very good correction of all aberrations for the teleconverter in order not to deteriorate the overall quality. Teleconverters exist in versions with different extension factors. Usually, they have factors of 1.4× and 2× but also 1.7× and 3× can be found by some manufacturers. Like all lenses, they are specially designed for various types of lens mounts and adapted to the special features of the corresponding lens systems.

The fact that the teleconverter is mounted behind a fully optimized lens with the image focal length f has the consequence that the entrance pupil D_{en} of the lens combination remains the same as that of the first lens. On the other hand, the focal length increases proportionally to the teleconverter factor c_{f} and, therefore, the f-number of the combination $f_{\#\mathrm{c}}$ increases accordingly:

$$f_{\#\mathrm{c}} = \frac{c_{\mathrm{f}} \cdot f}{D_{\mathrm{en}}} = c_{\mathrm{f}} \cdot f_{\#}. \tag{6.8}$$

Thus, a 2× teleconverter doubles the focal length but also reduces the relative aperture $1/f_{\#}$ by a factor of 2. This means that the brightness in the image plane changes by 2 EV, which is equivalent to slowing down the aperture by two stops. The general relationship is given by the following consideration. The illuminance E_{ic} in the image plane for the lens-teleconverter combination can be written according to Equation (2.15):

$$E_{\mathrm{ic}} \propto \frac{1}{f_{\#\mathrm{c}}^2} = \frac{1}{c_{\mathrm{f}}^2 \cdot f_{\#}^2}. \tag{6.9}$$

The brightness change b_{r} by using the teleconverter relative to the lens without a converter then yields

$$b_{\mathrm{r}} = \frac{E_{\mathrm{ic}}}{E_{\mathrm{i}}} \propto \frac{f_{\#}^2}{c_{\mathrm{f}}^2 \cdot f_{\#}^2} = \frac{1}{c_{\mathrm{f}}^2}. \tag{6.10}$$

The change in exposure value ΔEV can be calculated using Equation (2.22). We then get

$$\Delta EV = \mathrm{ld}\, b_{\mathrm{r}} = 3.32 \cdot \log_{10} b_{\mathrm{r}} = -6.64 \cdot \log_{10} c_{\mathrm{f}}. \tag{6.11}$$

Thus, a 1.4× teleconverter slows the aperture of the lens combination down by 1 EV, a 1.7× converter by 1.5 EV, a 2× converter by 2 EV, and a 3× converter by 3.2 EV.

The use of a perfectly corrected teleconverter changes the resolution of the lens combination by increasing the diffraction blur due to decreasing the relative aperture. In the case of a very good prime lens of large aperture, however, this may hardly become visible (see Section 2.5.4 for optimal f-number). Moreover, the aberrations of the prime lens are magnified. Thus, it depends on the quality of the prime lens and the image sensor if the resolution is impaired. Definitely, the use of a teleconverter increases the necessary exposure time if the prime lens aperture remains unchanged, which is a critical point for long focus lenses.

6.4 Normal lenses

Normal lenses, also called standard lenses, are designed for given sensor formats and render images that have approximately the perspective of natural viewing when used with these sensors. This corresponds to an angle of view in the range of about 47° to 53° and is achieved if the focal length is nearly the same or approximately 20 % larger than the image diagonal of the sensor. Thus, for the 35 mm format the normal lenses have focal lengths between about 43 mm and 50 mm but also lenses up to 60 mm are still considered as normal lenses. Due to the image perspective, they can be classified also as universal objective lenses without any special requirements. In general, they have a nearly symmetric lens design. Therefore, we find the fastest lenses among the normal lenses since the high relative aperture requires a high amount of correction, which can be best done with that type of lens arrangement. Commercially available lenses with f-numbers as low as 0.95 are offered only by a few lens producers. The famous Zeiss Planar 50 mm $f/0.7$ was manufactured only in a limited number for special customers. As for the maximum relative aperture, there are limits given by the construction of the camera body and the lens mount for which the lens is designed. Figure 6.16 illustrates the body of a DSLR compared to that of a mirrorless camera for interchangeable lenses. Each lens mount is specified with respect to its circular diameter and the flange focal distance (FFD) from its camera body contact to the sensor plane. In the DSLR body, the FFD must be larger than in a mirrorless camera to guarantee that the hinged mirror is not obstructed by protrusions of the lens. The angular aperture in the image space θ_{ex} is limited by the effective available distance l_{eff} from the last lens element to the sensor as well as the effective usable diameter of the lens mount D_{eff}. Both quantities are smaller than the values specified for the mount due to construction details like material thickness or thread pitches. The maximum angular aperture then is given by

$$\theta_{ex} = \arctan \frac{D_{eff}}{2 \cdot l_{eff}}. \tag{6.12}$$

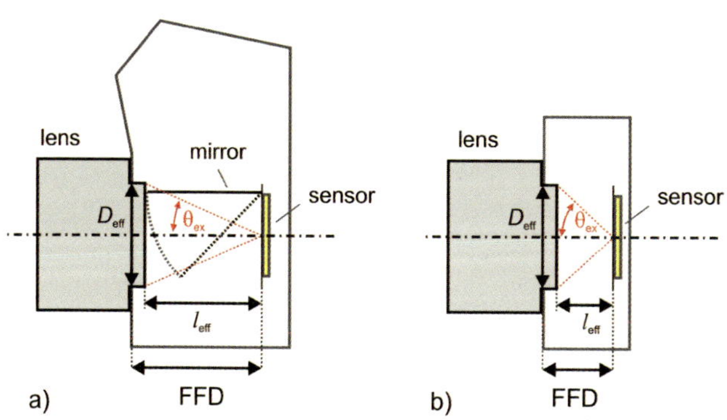

Fig. 6.16: Schematic drawing of camera body with interchangeable lens. (a) DSLR camera; (b) mirrorless camera; (FFD: flange focal distance).

Thus, the minimum f-number for the lens can be calculated according to Equation (3.95):

$$f_{\#} = \frac{1}{2 \cdot \sin \theta_{ex}}. \tag{6.13}$$

Typical dimensions of SLR cameras for the 35 mm format are about 38 mm for l_{eff} and about 36 mm for D_{eff}. For these values, a maximum relative aperture of $f/1.2$ is possible like for Nikon F-mount, M42, etc. However, it would be of no avail to mount lenses with lower f-number to these systems. Canon EF mount has a larger diameter, which is compatible with a lens of $f/1$. Mirrorless camera bodies have similar mount diameters like the SLR bodies but significantly shorter distances l_{eff}, thus allowing the use of lenses with larger apertures. For instance, for the Leica M mount, originally conceived for the classical range finder 35 mm film cameras and still used for modern digital full format cameras, values of $D_{eff} < 38$ mm and $l_{eff} < 22$ mm can be estimated. There is no problem to use lenses with $f/0.95$ or even lower f-number. Modern developments of camera bodies without mirror and having lens mounts like Leica SL, Sony E, Canon EF-M, Nikon 1-mount, just to mention a few, have still shorter FFD and can potentially be adapted to use lenses with $f_{\#} < 1$. It is always possible to mount lenses with larger FFD to cameras with shorter FFD by the use of a corresponding adapter. Conversely, lenses with a shorter FFD can only be used with cameras bodies of longer FFD within a limited object distance close to the lens. In this case, focusing at objects at infinity is not possible.

In the following, we will consider the design of some modern normal lenses for digital sensors of 35 mm format. Figure 6.17 shows a compilation of Leica lenses of different relative apertures, where all M-lenses are manually focused and are designed for the relative short FFD of rangefinder cameras. The comparison of the lenses allows for understanding the complexity, which is necessary to obtain the desired results. The Summarit M 50 mm $f/2.4$ (Figure 6.17a) is a very small and compact lens featuring a classical, slightly asymmetric Gauss anastigmat design. It resembles its predecessors Summar $f/2$ from 1933 and Summitar $f/2$ from 1939 (see Figure 6.8b). The relatively high symme-

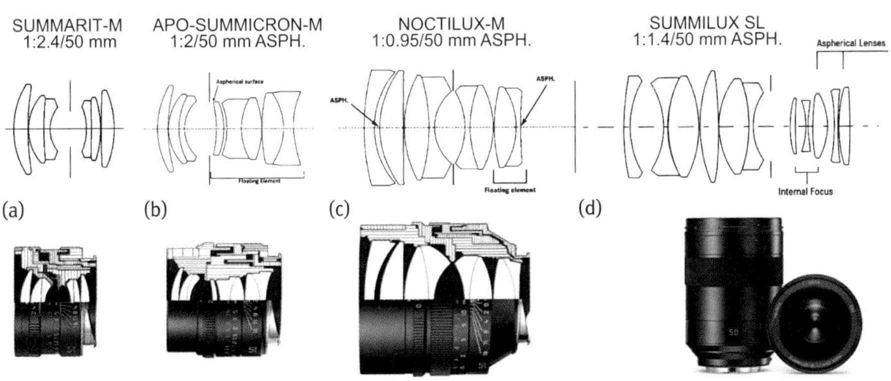

Fig. 6.17: Leica 50 mm lenses for the 35 mm format (with kind permission of Leica).

try with a central stop guarantees low transversal aberrations. The Apo–Summicron M 50 mm $f/2$ ASPH. (Figure 6.17b) is an apochromat where the zonal chromatic aberrations are nearly completely abolished. It also shows a double-Gaussian design, however with larger asymmetry and more lenses. The rear lens group incorporates an aspherical element and is moved relative to the first group during manual focusing. The Noctilux-M (Figure 6.17c) is one the fastest lenses commercially available. Like the Apo–Summicron, it has an aspheric element and the last cemented lens is a floating element, which is moved during focusing. Due to its very large relative aperture, it is of large size and weight. Like the other lenses, it has a basic double-Gauss design, however, with more modification in order to correct the additional aberrations due to the large aperture. The last lens is the Summilux SL50 mm $f/1.4$ (Figure 6.17d), which is the most recent lens in this compilation. It is a development for the mirrorless SL system with a fully automated autofocus mechanism and internal focusing, thus the lens extensions remain always constant. It contains two aspheric elements. The overall design of the two lens groups strongly deviates from the classical design of the Summarit M 50 mm and is also more complex than the Summilux 35 mm from 1960 (Figure 6.8c).

Figure 6.18 illustrates a comparison of different Canon 50 mm normal lenses. These lenses are all designed for use in DSLR cameras with the Canon EF mount. They can be operated optionally in autofocus or manual focus mode. The EF50 mm $f/1.8$ STM (Figure 6.18a), having the smallest aperture, again is the most compact one and also features the classic slightly asymmetric double-Gauss design. The EF50 mm $f/1.2$L USM lens (Figure 6.18b), being larger and having a more than one stop larger aperture, has an

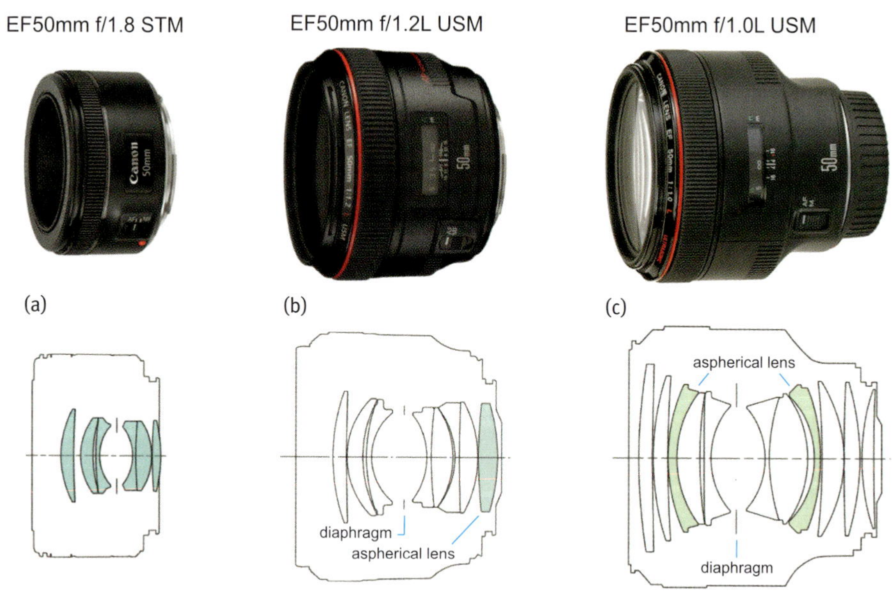

Fig. 6.18: Canon 50 mm lenses for the 35 mm format (with kind permission of Canon).

additional aspheric lens element of highly refracting glass. This is required for additional correction due to the larger aperture. In the EF50 mm $f/1.0$L USM (Figure 6.18c), the relative aperture is additionally increased by 0.5 stop, which leads to much higher complexity of the lens arrangement with two aspherical lens elements. The overall design principle is still based on the classical slightly asymmetric Gauss anastigmat, but it is obvious that additional meniscus lenses are added on both sides of the symmetrically located aperture stop. Splitting up the refractive power of one meniscus lens into two closely arranged lenses means that the curvature of one meniscus lens can be reduced, and thus less spherical aberration shows up which is a critical part for fast lenses. This principle can also be seen at other lenses of large apertures.

Normal lenses of Zeiss produced for the use with DSLR cameras having the Canon EF mount or Nikon F mount are illustrated in Figure 6.19a, b. The Planar T* 1.4/50 is a modern modification of its classical predecessors Planar (1896) and Biotar (1927). Compared to them, it has additional meniscus lenses in the front and rear groups in order to achieve lower aberrations at the larger relative aperture. The lens is focused manually by unit focusing. A completely different design is found with the newly developed Otus 1.4/55 (Figure 6.19b). It features a modified Distagon design, which is based on the retrofocus arrangement of wide-angle lenses with a negative lens group in the front and

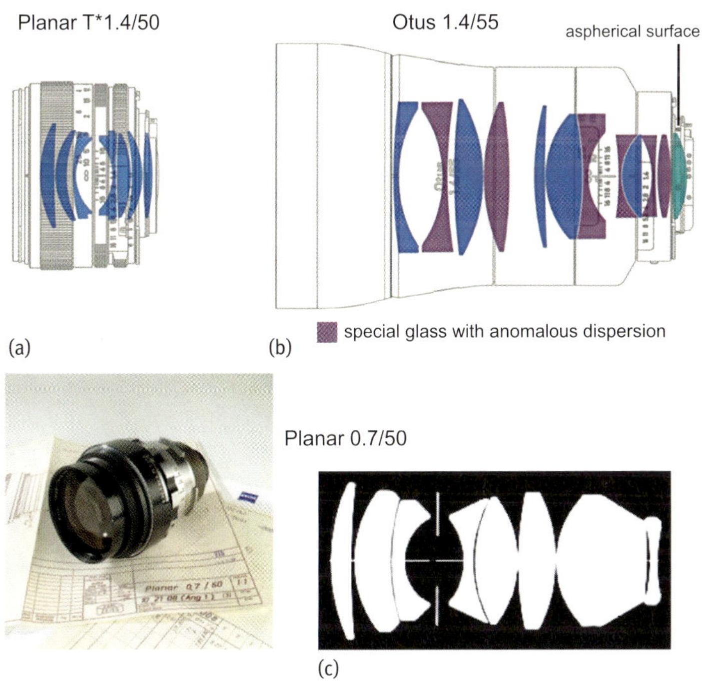

Fig. 6.19: Zeiss normal lenses for 35 mm format, (a) and (b), compared to Planar 0.7/50 (c) (with kind permission of Carl Zeiss AG).

a positive group behind it (see Section 6.5). By that arrangement, a very homogeneous quality can be maintained over the whole image field. Its superior quality is due to a quite complex combination of special glass lenses with a rear aspheric element. The lens is focused manually using floating elements. Figure 6.19c illustrates for comparison the famous Zeiss Planar 0.7/50, which was developed in the 1960s on request of NASA for taking pictures of the dark side of the moon [Nas11b]. It also features the Planar design, but the rear group is very dissimilar from the front group and has a long extension. The back focal length is only of 5.3 mm, the image circle of 27 mm is comparable to that of an APS-C sensor and the whole lens is quite large and heavy. It cannot be considered as a normal lens for the 35 mm format due to the small image circle, rather a long focus lens for the APS-C format. Nevertheless, it was fitted with a central shutter and precisely mounted to a modified large-format Hasselblad camera body where the mirror has been removed.

As last examples for normal lenses, we consider two 50 mm lenses of Nikon (Figure 6.20). Both lenses have the same relative aperture and are operated in manual or autofocus mode. They show the same principal slightly asymmetric double-Gauss anastigmat construction like most normal lenses. The G-lens can be considered a further development of the D-lens with modification of the individual elements and one additional lens.

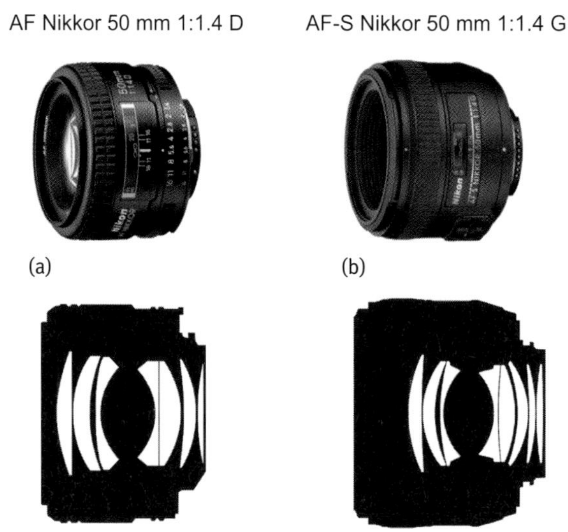

Fig. 6.20: Nikon 50 mm lenses for the 35 mm format (with kind permission of Nikon).

6.5 Wide-angle lenses

Lenses with focal lengths that are more than about 20 % shorter than the diagonal of the sensor format, for which they are developed, can be considered wide-angle lenses. Thus, for the standard 35 mm format these are lenses with a focal length of less than roughly 35 mm and the corresponding total angle of view is larger than 60°. Applications of these lenses are for instance in the fields of landscape, architectural or aerial photography. The angular range from about 60°–75° is moderate and of particular interest for street photography and for documentation. Above angles of 80°, the lenses may be termed as a super-wide angle. A special perspective distortion can be found with the category of fisheye lenses that can have total angular fields exceeding 180°. As the angular field of view is much larger than for normal lenses, a homogeneous brightness distribution and the absence of distortion over the whole image field are critical points for the development of these lenses. Moreover, with the lower image magnification due to the short focal length, a high sharpness is required in order not to lose detail information. In order to meet the different requirements with respect to the applications, there are basically three groups of wide-angle lenses reflecting their construction principle: the retrofocus design, the nearly symmetrical achromat design and the fisheye design, which is an extreme version of the retrofocus design.

6.5.1 Retrofocus design

The retrofocus principle has been known a long time and has been applied for the projection of slides in order to increase the image size for a given distance from the slide to the screen [Kin89]. The principle of this projection method is illustrated in Figure 6.21. An illuminated slide object at a distance between the focal distance f and $2 \cdot f$ from the lens is magnified to a screen by a single converging lens (Figure 6.21a). The image can be further magnified without changing the distance from object to image if a diverging lens is inserted into the ray path between the converging lens and the image (Figure 6.21b). However, the object and image distances have to be adjusted properly. This lens can be considered an amplifier lens for the image.

Light paths are reversible. Thus, when the lens combination is inverted, we have the situation that objects with larger extensions in the object space can be imaged to the sensor plane than by using only the single converging lens. This is the opposite of the telephoto principle where the focal length is increased by adding a negative lens behind the positive lens. Here, in the reversed telephoto setup, the focal length is reduced by the diverging lens in front of the positive lens. The construction is sketched in Figure 6.22a for parallel light, which enters a system of two thin lenses. It can be seen that the focal length f is shorter than the back focal distance f_{Ei}. The diverging lens defines the input plane, the converging lens the exit plane from which the back focal distance f_{Ei} is measured to the image focal point F_i of the system. This arrangement is similar to Example 4

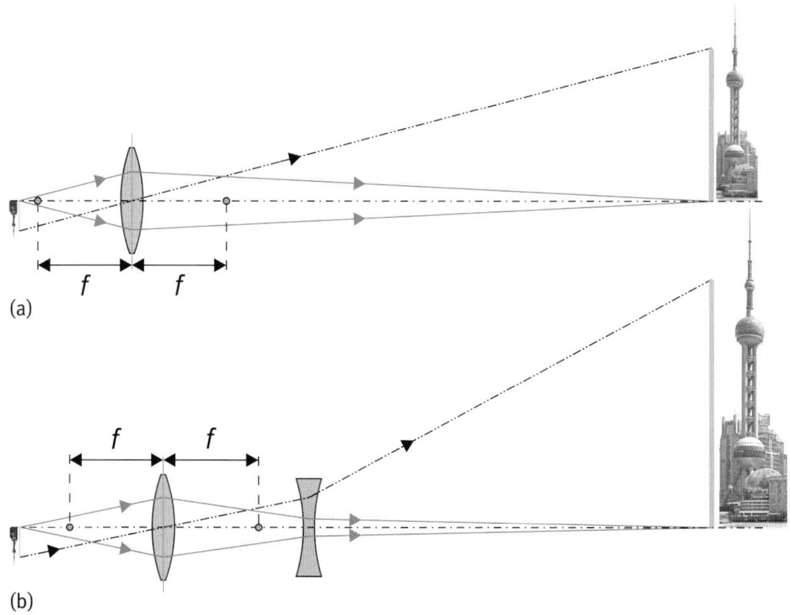

Fig. 6.21: Slide projection. (a) Using a single converging lens; (b) reversed telescope setup to achieve a larger image magnification by an additional amplifier lens.

in Section 3.3.5. The back focal length f_{Ei} can be calculated using Equation (3.66). If we assume the image focal length of the diverging lens being the negative quantity f_1, and that of the converging lens being the positive f_2, we get for the back focal length of the combination:

$$f_{Ei} = \left(1 - \frac{t_s}{f_1}\right) \cdot f. \tag{6.14}$$

In this equation, f is the image focal length of the lens combination. We can see from Equation (6.14) that the back focal length f_{Ei} for that setup is always larger than f as f_1 is negative and the distance t_s between the two lenses is counted as positive. The reason for f_{Ei} being larger than f is due to the fact that the negative lens in the front shifts the principal plane at H_i beyond the vertex of the second lens into the image space. This is very advantageous for SLR cameras where more space is needed for the mirror movement. Moreover, the image focal length of the lens combination is shorter than f_2 of the individual converging lens alone if the lens separation is sufficiently large. This can be seen from considering the refractive power $V_i = 1/f$ of the lens arrangement after Equation (3.66) with the negative $V_1 = 1/f_1$ and the positive $V_2 = 1/f_2$:

$$V_i = V_1 + V_2 - t_s \cdot V_1 \cdot V_2 = V_1 \cdot (1 - t_s \cdot V_2) + V_2. \tag{6.15}$$

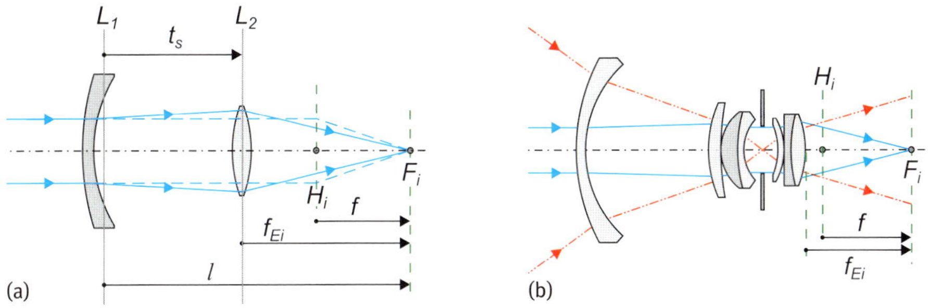

Fig. 6.22: Retrofocus construction. (a) Principal scheme for thin lenses; (b) lens construction of Flektogon 2.8/35 mm by Carl Zeiss, Jena, 1950.

The condition that the total focal length is positive and shorter than that of the converging lens can be expressed as $V_i > V_2$. As V_1 is negative this inequality can only be fulfilled if the bracket term in Equation (6.15) is negative, yielding

$$1 - t_s \cdot V_2 < 0 \quad \Leftrightarrow \quad t_s > \frac{1}{V_2} = f_2. \tag{6.16}$$

As a consequence, a wide-angle lens with a reduction of the focal length compared to the converging lens alone is achieved if the lenses are separated by more than the focal length of the converging lens. This is the case for the examples shown in Figures 6.21 and 6.22.

The principle of the reversed telephoto setup was first implemented for wide angle SLR camera lenses by Angenieux (1950) and Zeiss Jena (1950). The main intention was to shift the principal planes to attain more space in the back focal area of the camera in order to not obstruct the motion of the hinged mirror as the focal length of these lenses was below 35 mm. Angenieux used the term retrofocus as a trademark for their lenses but it was soon adopted as a generic name for this type of lens design. The first retrofocus lens of Zeiss was termed Flektogon 2.8/35 mm. The layout of this lens shows clearly the principle of a negative meniscus lens largely separated from a positive, slightly asymmetrical double Gauss anastigmat (Figure 6.22) with the aperture stop in its center. It had a moderate angle of view of 63° as it was designed with a focal length of 35 mm for the 35 mm format. One of the drawbacks of the retrofocus design was that due to the asymmetry and the large distance between the negative front and positive rear groups the transversal aberrations are much stronger than for symmetric constructions. Furthermore, due to this asymmetric distribution of refractive power, retrofocus lenses tend to exhibit a barrel-type distortion, which is opposite to the type of distortion seen with telephoto lenses. The correction of these aberrations requires a much higher effort in the lens combinations, and thus leads to quite complex, large and heavy lenses, if high relative apertures are to be attained. The first retrofocus lenses were still far from the quality of modern lenses and, therefore, had only moderate apertures to guarantee acceptable performance.

6.5.2 Symmetric lens design–Biogon type

The problem of transversal aberrations is almost not present in nearly symmetric constructions like the Planar or similar lenses. In principal, they could be downscaled for smaller focal lengths and dimensions. A further advantage of small lenses is that longitudinal spherical aberration, which is proportional to the square of the entrance pupil diameter, is less pronounced than in bigger lenses. Therefore, larger relative apertures are easier to achieve than in big lenses. However, the angular field of view in constructions like the Planar is also limited, mainly by the increasing vignetting. The problems that show up are especially visible in the periphery of the image field and corrections for oblique rays are quite expensive. A very powerful layout to solve the problems for wide-angle lenses was given by the Russian mathematician Rusinov in 1946. The proposal is very similar to the retrofocus design, but the difference is that it seems as if two retrofocus arrangements are symmetrically combined with the positive lenses oriented to each other in the center and the negative menisci at the outer extremes (Figure 6.23). The lens is slightly asymmetric with the aperture stop located in the center. Due to the lens arrangement the aperture is still fairly large at large oblique angles, thus reducing vignetting effects. The overall angular field of view was about 133° but the maximum relative aperture was only $f/18$.

A similar, but more advanced concept was independently developed by the German designer Bertele and produced as Wild Aviogon for medium format cameras and as Zeiss Biogon 4.5/21 mm for the 35 mm format in 1952 (Figure 6.23b). The focal length of 21 mm is nearly identical to half of the 35 mm format diagonal and, therefore, the total angular field of view is of 90°. The Biogon was excellently corrected with high sharpness even in the corners of the image field. Due to its high symmetry, the lens distortion is less than 0.1 % and is virtually not perceived [Nas11c]. All lenses of the rather symmetric

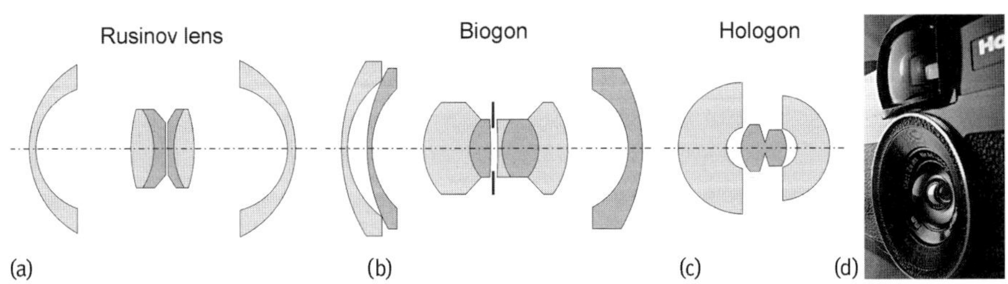

Fig. 6.23: Retrofocus construction. (a) Rusinov lens (1946); (b) Zeiss Biogon 4.5/21 mm (1952); (c) Zeiss Hologon 8/15 mm (1966); (d) Zeiss Hologon camera (with kind permission of Carl Zeiss AG).

construction type, also the more recent ones, are still termed Biogon. An extreme version of the symmetric type is the Zeiss Hologon from 1966 with a focal length of 15 mm and a corresponding angular field of view of 110° (Figure 6.23c). The basic layout is a simple triplet although its production and assembly require extreme precision. It consists of two meniscus halves with a nearly ball lens in the center. The ball lens is notched in the center to establish a fixed relative aperture of $f/8$ thus being a central stop. It is virtually a fixed-focus lens as the depth of field and the sharpness are extremely high. Its brightness can be controlled by using filters. For instance, graded filters are necessary to compensate for the natural vignetting, which cannot be influenced by stopping down. Due to its very short back focal distance of only 4.5 mm, it did not fit to conventional cameras and was adapted to a special camera body (Figure 6.23d). A revised modern version for Leica rangefinder cameras has been developed by Zeiss. The short back focal length as well as the large deviation from a telecentric ray path in the image space makes this type of lenses very difficult to use with modern digital camera systems. The consequences and a comparison with the retrofocus design are given in the next section.

6.5.3 Properties and examples of modern wide-angle lenses

The main driver for the retrofocus construction was the popularity of SLR cameras after around 1950, which required more space for the hinged mirror. By the negative front element in an asymmetric lens, the back focal distance is increased, the principal planes are correspondingly shifted and the overall focal length can be reduced. The transversal aberrations can be very well corrected, but a much larger effort than for symmetric lenses is necessary. Thus, the lens tends to become more complex, larger and also heavier. The retrofocus construction of lenses developed by Zeiss is termed Distagon compared to the more symmetric Biogon. The size difference between Distagon and Biogon lenses can be seen in Figure 6.24a by the comparison of two Zeiss lenses developed for the 35 mm format Contarex film SLR camera. The Distagon lens is much larger than the Biogon and can be used with the camera in normal operation mode of the hinged mirror due to its large back focal length. The Biogon lens has almost the same length but due to its extension into the camera body and its short back focal length the hinged mirror must be fixed in its upper position and cannot be used. As for the image quality, it took several decades until the Distagon lenses achieved the same quality as the more symmetric Biogon types [Nas11c]. The image contrast in the retrofocus Distagon lenses decreases especially at the corners of the image field, which is often mainly due to the stronger chromatic aberration. Figure 6.25 illustrates the chromatic dispersion by a microscopic view of white tangential line images at an image height of 10 mm, which is roughly at half of the distance from the center of the image to its diagonal corners. The left image is captured for the Zeiss Distagon 4/24, which was designed in the 1950s. The center image comes from the Distagon T*2.8/21 developed for Contax SLR cameras in 1992, the right image of the Biogon T* 2.8/21 ZM developed for the M-mount rangefinder camera system

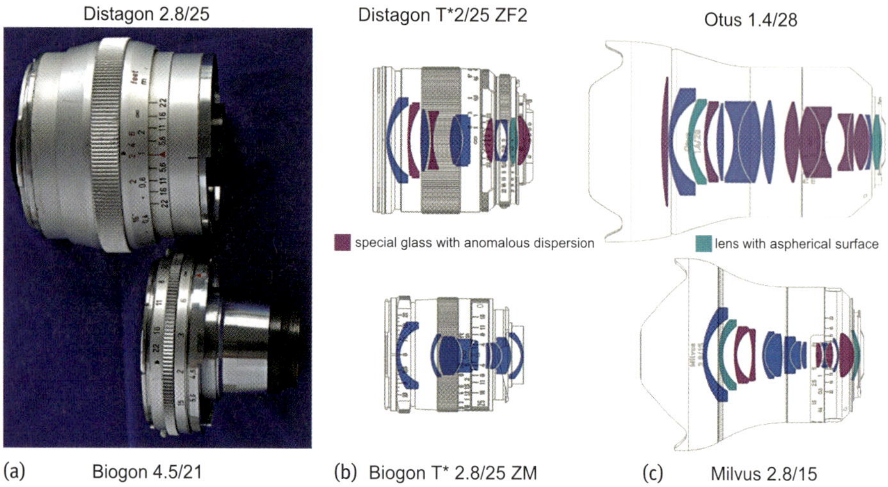

Fig. 6.24: Comparison of Zeiss wide-angle lenses. (a) Distagon 2.8/25 and Biogon 4.5/21, both with mounts for Contarex SLR cameras [Nas11c]; (b) construction details of Distagon T*2/25 ZF2 for DSLR cameras compared to details of Biogon T*2.8/25 ZM with M-mount for rangefinder cameras; (c) construction details of Otus 1.4/28 and Milvus 2.8/15, both designed for 35 mm DSLR cameras. (With kind permission of Carl Zeiss AG).

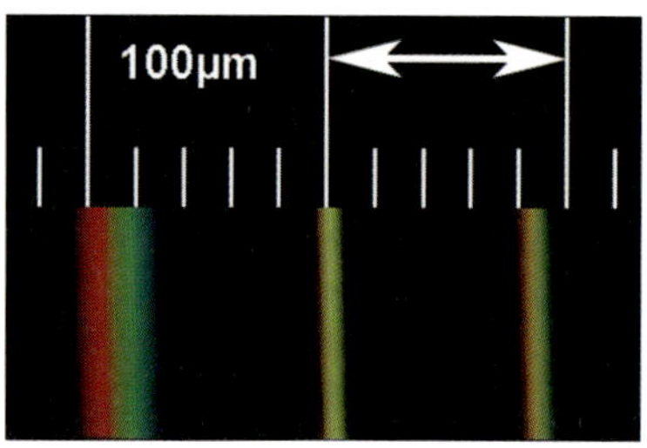

Fig. 6.25: Microscopic view of the tangential line images at 10 mm image height, $f_\# = 8$, for three different Zeiss lenses (from left to right): Distagon 4/24 (lens designed in the 1950s); Distagon T*2.8/21 for Contax SLR (center, lens designed in 1992); Biogon T* 2.8/21 ZM [Nas11c] (with kind permission of Carl Zeiss AG).

of Leica. The impressive improvement for the Distagon was achieved above all by the use of special glasses with anomalous partial dispersion whereas the design quality of the Biogon lens was mainly due to the more favorable symmetric construction conditions.

The recent design of the Zeiss Distagon T*2/25 ZF2 is also based on special glass types but also incorporates an element with two aspheric surfaces as next-to-last lens in the rear part. Details of the construction are given in Figure 6.24b. This element reduces further the spherical aberration and coma due to the large relative aperture of the lens and also corrects residual distortion. Like many lenses of high quality, the manual focusing is done using floating elements in order to maintain the quality over the whole range of object distances and not by unit focusing where the whole lens is displaced uniformly. The construction of the Zeiss Biogon T* 2.8/25 ZM lens of the same focal length and a one-stop less relative aperture is much more symmetric (Figure 6.24b) although a

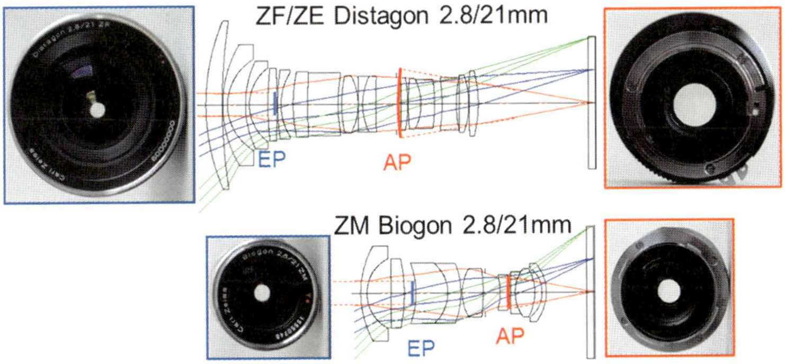

Fig. 6.26: Comparison of entrance pupils (left side) and exit pupils (right side) of the smaller Zeiss Biogon T*2.8/21 ZM and the bigger Distagon T*2.8/21; EP: entrance pupil, AP: exit pupil [Bla14] (with kind permission of Carl Zeiss AG).

certain asymmetry must be kept in order to not exceed a critical value of telecentricity, which is required for semiconductor image sensors.

The asymmetry of a retrofocus lens can be directly inspected by comparing its entrance respectively exit pupils. The exit pupil in this case is the virtual image of the stop as perceived through the rear lens groups from the image space and the corresponding image from the object side is the entrance pupil. Figure 6.26 shows the comparison of both pupils for a Zeiss Biogon and a Zeiss Distagon of identical focal length of 21 mm and identical f-number of 2.8. As the f-number is the ratio between the diameter of the entrance pupil and the focal length, it follows that the entrance pupil must be of the same size for both lenses as can be seen in the figure (left side). Likewise, the f-number is also the ratio between the diameter of the exit pupil and its distance to the image plane. This ratio is the same for both lenses. The comparison of both lenses shows that the Distagon has a larger exit pupil than the Biogon, and thus its distance from the image sensor is larger than that of the Biogon lens (right side). This can also be verified in the construction details of both lenses. Moreover, for symmetric lenses with the stop in its center, entrance and exit pupils have the same size. In the retrofocus design, however, due to the asymmetric distribution of the refractive powers with a negative front and a positive rear group the exit pupil is significantly larger than the entrance pupil. The comparison in Figure 6.26 illustrates this asymmetry for the Distagon with a pupil magnification of $M_p = 3.0$, and indicates a more symmetric distribution of the refractive powers for the Biogon with $M_p = 1.3$ [Bla14].

As the retrofocus construction of the Distagon with its larger back focal distance leads also to a larger distance of the aperture stop from the image plane, its telecentricity value θ_t is lower than that of the Biogon and makes this type of lens more appropriate for the use with semiconductor image sensors, which are sensitive for deviations from perpendicular ray incidence. The telecentricity value, which can be seen in the construction details by the deviations of the chief rays from the optical axis (see Section 3.4.3), is lower for the Distagon. Modern lenses, which are specially designed for use with mirrorless digital cameras must not fall below a certain back focal distance in order to guarantee a low θ_t. Thus, they cannot show the same symmetric design as the traditional lenses for rangefinder film cameras.

The most recent developments of high-quality DSLR camera lenses for the 35 mm format by Zeiss, for instance, the Zeiss Milvus 2.8/15 and Zeiss Otus 1.4/28 (Figure 6.24c), all feature the Distagon design with increasing complexity and also use special glass types and aspheric elements. It should be noted that even the normal lens Zeiss Otus 1.4/55 (Figure 6.19b) shows a Distagon and not a Planar lens design.

Due their large angular field of view, wide-angle lenses are more prone to vignetting than other types of lenses. Vignetting is the brightness fall-off at the peripheral parts of the image field and increases with the field angle. It can be roughly characterized by two parts, the mechanical vignetting and the natural vignetting (see Section 3.4.4). Mechanical vignetting is mainly due to the limited sizes of lens elements in the construction and can in general be reduced by stopping down. The natural vignetting varies with $\cos^4 \theta_t$ where $2\theta_t$ represents here the image space field angle. Since the field angles are related to each other by the pupil magnification M_p after Equation (3.97), lenses featuring a retrofocus design, having M_p significantly larger than 1, show much less shading at the corners due to natural vignetting than more symmetric lenses with M_p close to 1. Unlike mechanical vignetting, natural vignetting cannot be reduced by stopping down, but can be influenced by M_p, and thus by the lens construction. Therefore, the overall vignetting is less pronounced in retrofocus lenses, especially when stopping down, compared to more symmetric lenses with M_p close to 1 where the natural vignetting is dominant at large field angles.

Examples of other lenses for the 35 mm format are given in Figures 6.27 to 6.29. The Leica lenses are specially designed for the rangefinder cameras of the Leica M series with digital sensors. The Leica Summaron-M 28 mm $f/5.6$ is a modern replica of a traditional film camera lens from 1955. Consisting of only six lenses and featuring a nearly symmetrical design, it is very compact and offers all the advantages of symmetric lenses. The use with digital cameras delivers good images, however, with more vignetting inherent due to the construction. By comparison, the Leica Summilux-M 21 mm $f/1.4$ ASPH features a retrofocus design with aspheric lenses and floating elements for optimum manual focusing.

The Nikon lenses as well as the Canon lenses are all constructed for DSLR cameras, and thus require a larger back focal distance. Therefore, like all full format DSLR camera lenses with focal lengths of 35 mm and below, they feature a retrofocus design. All lenses operate in autofocus and manual mode. The AF Nikkor 28 mm 1:2.8 D is very compact and clearly shows the classical retrofocus setup with a large distance between the negative front group and the rear positive group. The AF Nikkor 20 mm 1:2.8 D has a more complex lens arrangement with significantly more lenses as the short focal length requires a higher effort for correction. In both lenses, the focusing is done using floating elements.

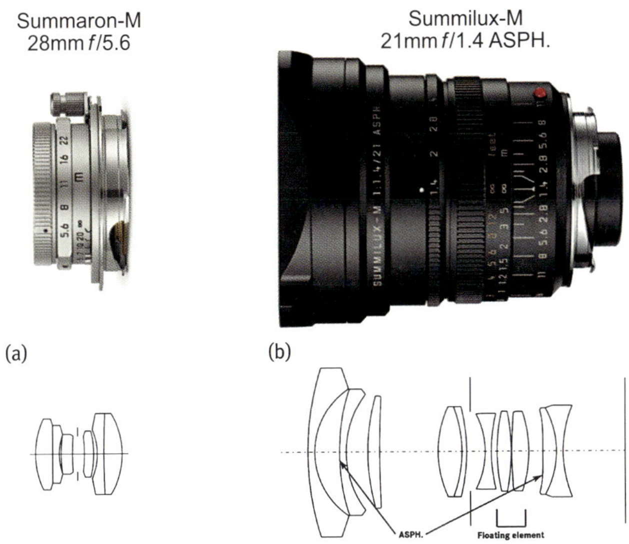

Fig. 6.27: Leica wide angle lenses for the 35 mm format (with kind permission of Leica Camera AG).

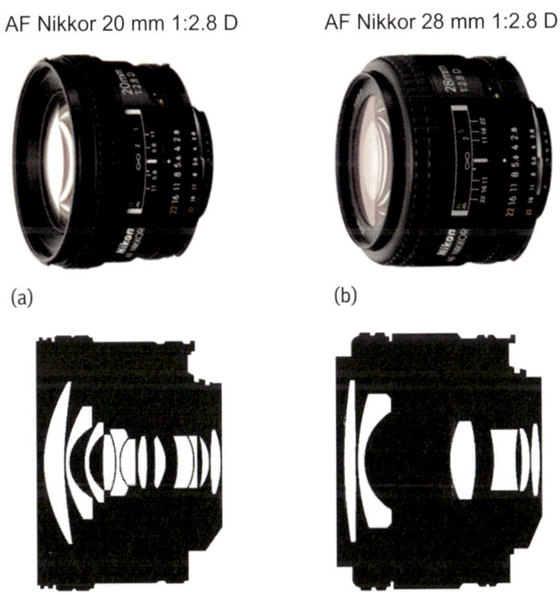

Fig. 6.28: Nikon wide angle lenses for the 35 mm format (with kind permission of Nikon).

The two wide-angle lenses of Canon shown in Figure 6.29 have both aspherical lens elements. The Canon EF 28/2.8 IS USM is equipped with an image stabilization mechanism (IS). The Canon EF 24/1.4 L II USM has a shorter focal length but is nevertheless longer and more complex than the 28 mm version. Its high relative aperture requires

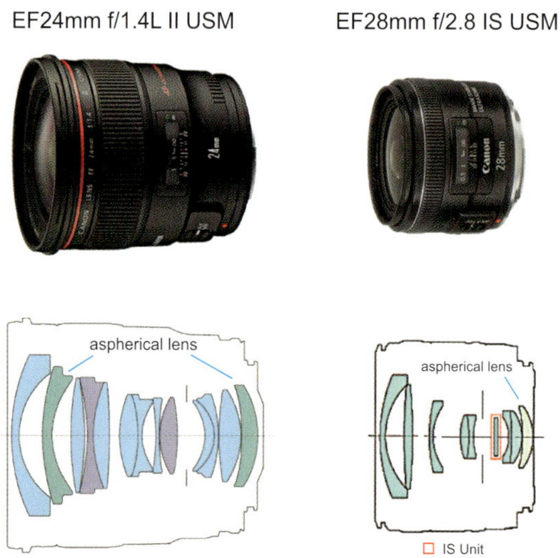

Fig. 6.29: Canon wide-angle lenses for the 35 mm format (with kind permission of Canon).

more effort for corrections. Besides the two aspheric lenses, it incorporates also two lenses of ultralow dispersion glasses for better correction of chromatic aberrations. The lens is focused using floating elements.

6.5.4 Fisheye lenses

Ultrawide angle lenses of 15 mm focal lengths, as described above, cover a total angular field of view of 110° across the diagonal. For wider angular of views, even exceeding 200° in extreme cases, special types of lenses are required, which are usually termed fisheye lenses. These lenses no longer render imaging according to the rectilinear perspective of central projection but show rather a curvilinear perspective according to an equidistant projection. The difference of these perspectives is illustrated in Figure 6.30.

The rectilinear perspective corresponds to the natural view as perceived by the human eye being the central point of projection (see also Sections 2.1 and 3.5.5). Objects perpendicular to the optical axis are imaged to the sensor true to scale. If the height of an object is doubled, then the corresponding image height h_i is doubled as well, thus we have a linear relationship (Figure 6.30a). Straight lines in the object space are imaged as straight lines on the sensor. According to Equation (3.130) the viewing angle β is related to the image height by the focal length f of the lens, namely $h_i = f \cdot \tan\beta$. This is the characteristic of the gnomonic projection (Equation (2.1)). It further implies that objects viewed under an angle that approaches 90° cannot be imaged as the image size becomes very large, and thus exceeds the sensor size. Moreover, the effective size of the entrance pupil decreases continuously with β, which means that the brightness in the image plane decreases with the image height. This is the effect of vignetting (see

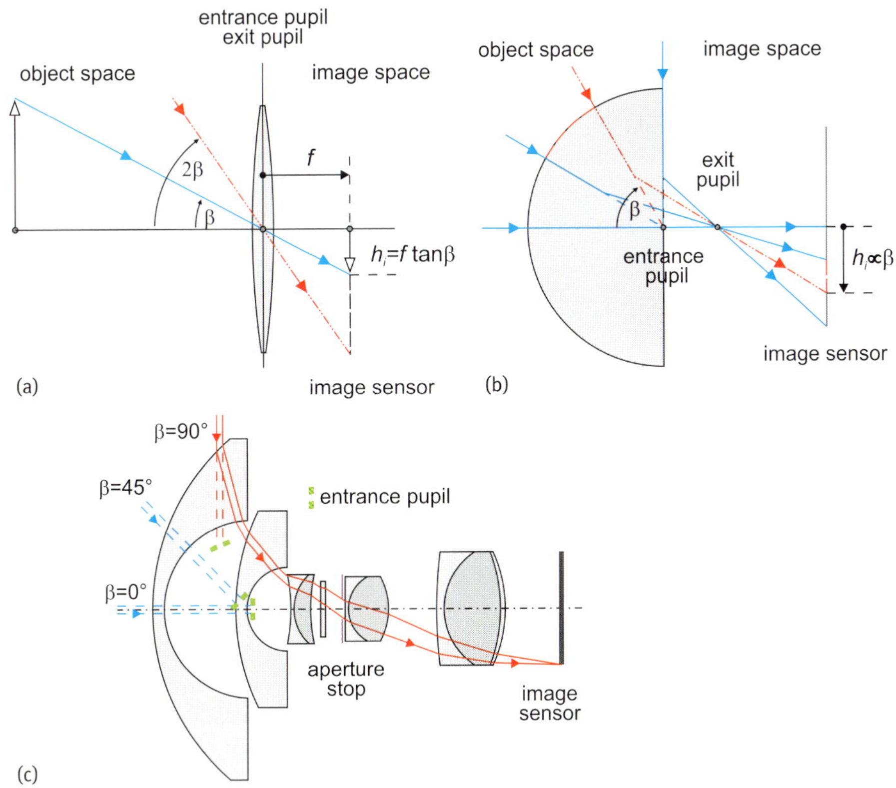

Fig. 6.30: (a) Rectilinear perspective due to central projection; (b) schematics of curvilinear perspective due to equidistant projection; (c) fisheye design by Miyamoto (1964).

also Section 3.4). Thus, it is impossible to image an object under an angle of 90° in this way. If, however, large angles are required, for instance for sky observations, a different projection is necessary. A feature of the equidistant projection is that distances in the object space that are perceived under the same angle are imaged with the same length in the image plane (Figure 6.30b). We get a linear relationship between image height and viewing angle, namely $h_i = C_c \cdot \beta$, where C_c is a proportionality constant, which in many cases is identical to f like in the central projection. The consequence is that we no longer get a linear relationship between the distances in the object and image space but instead, a deviation from the rectilinear perspective, which we designate as curvilinear. Whereas parallel lines in the rectilinear perspective tend to the vanishing point, straight lines in the curvilinear perspective are no longer imaged as straight lines but will be curved. This nonlinear characteristic and geometry distortion is inherent to the equidistant projection and must be distinguished from the third-order Seidel aberration termed distortion (Section 3.5.5). For very small angles of β, which means close to the optical axis, both perspectives are nearly identical. For larger angles of β, the deviations

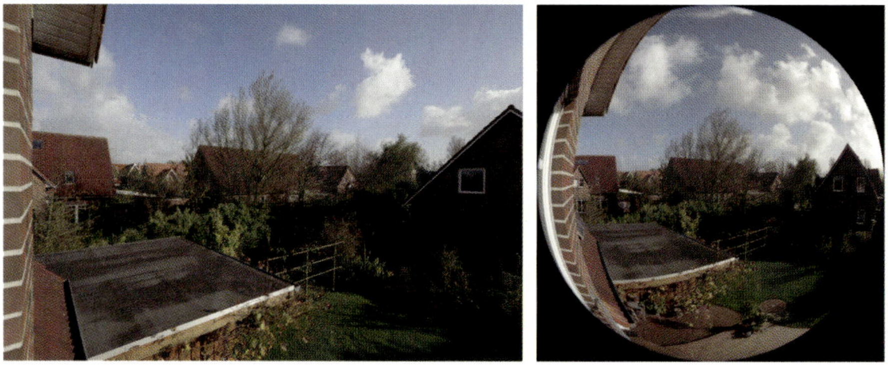

(a) (b)

Fig. 6.31: Comparison of wide-angle lenses for the 35 mm format. (a) wide-angle zoom lens at 18 mm focal length, total angle of view of 100°, exhibiting a slight barrel distortion; (b) typical strong barrel distortion of a fisheye lens at 11 mm focal length and a total angle of view of 150°. Straight lines remain rectilinear in (a) whereas they are imaged as curved lines in (b). The difference becomes more pronounced near the edge of the image circle.

become quite obvious showing a barrel-type distortion. Figure 6.31 illustrates the effect of the different perspectives. The image in part a) is captured using a wide-angle full format zoom lens at a focal length of 18 mm with a total angle of view of 100°. It reveals a slight barrel distortion, which can be best perceived at the left image side, and which is typical for the retrofocus design. Nevertheless, we have a rectilinear perspective. The part b) depicts the image of a full format fisheye lens at a focal length of 11 mm and a total angle of view of 150°. It features the typical strong barrel distortion due to the curvilinear perspective. Further lens distortions are difficult to separate from it. It can be seen that in the central projection of a) the lateral image parts seem to be expanded whereas in the equidistant projection the lateral image parts in b) are compressed and strongly distorted. In the center of the image, close to the optical axis, both perspectives are nearly identical.

The schematic construction of a fisheye lens is illustrated in Figure 6.30c after the patent of Miyamoto from 1964 who designed the first fisheye lens for series production by Nikon for 35 mm SLR film cameras. Very eye-catching are the large and strong negative meniscus lenses in the front group. They are followed by positive rear groups thus featuring a retrofocus design. This is needed to shift the principal planes to the image side since the focal length is very short and around 8–11 mm for the 35 mm format. The specific characteristic of this design is that the entrance pupil is tilted and shifted with increasing β and that its effective width slightly varies (Figure 6.30c). The changing location and orientation are important to ensure that light can enter the system even at extreme angles. A further particularity of fisheye lens is that the illumination at the peripheral parts of the image increases due to the barrel distortion, which can compensate other vignetting effects. The increase or decrease of the illumination also depends on the

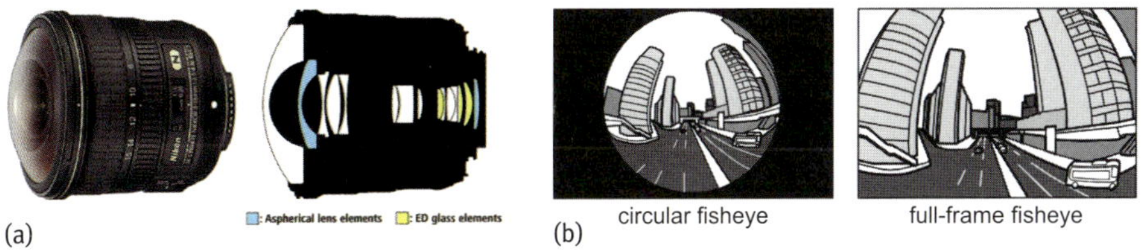

Fig. 6.32: Fisheye zoom lens. (a) AF-S FISHEYE–NIKKOR 8–15 mm 1:3.5–4.5E ED with construction details; (b) schematic image sections on the sensor for the shortest and longest focal length. (Reprinted with kind permission from Nikon.)

type of projection characteristics from the object space to the image space. For fisheye lenses, other types of projection are also possible, like the equisolid angle projection. In this case, areas in the object space perceived under the same solid angle are projected to areas of equal size in the image plane, which results in a different perspective than with the equidistant projection. However, the typical fisheye perspective distortion is still evident.

An example for a fisheye lens based on the equisolid projection is the Nikon lens AF-S FISHEYE–NIKKOR 8–15 mm 1:3.5–4.5E ED. Its construction is similar to the Miyamoto design but strongly modified in the rear part and also incorporating aspherical lenses and lenses made of special glass having anomalous dispersion. It should be mentioned that the more complex design is due to internal focusing mechanism and especially due to the zoom function where lens groups are moved separately and a higher amount of aberration corrections is required. By the zoom function, the focal length can be adjusted to the desired magnification, and thus the desired image section. The lens is designed to image the full image circle with a total angular field of view of 180° to the sensor format and the area outside the circle remains unexposed (Figure 6.32b, circular fisheye). By zooming to the longest focal length, the image is cropped and the whole sensor format is filled with the largest angular field of view of 175° along the image diagonal (Figure 6.32b, full-frame fisheye). As the focal length of the fisheye lens is very short, the depth of field is so high that the lens can be considered in many cases a fix-focus lens, especially at large f-numbers, and focusing is virtually unnecessary.

A similar fisheye lens for full-frame format (2× optical zoom, 8-15 mm focal length, f/2.8) and a total field of view of 180° is illustrated in Figure 6.33.[1] Part (a) shows the conventional lens design using spherical and aspherical lens elements for a flat image sensor. If the lens is redesigned for a curved image sensor, as given in Figure 6.33b, some interesting advantages can be achieved compared to the flat sensor design. Above all, the amount of image correction is reduced which has the consequence that the number of lens elements is reduced by more than 30 % and no aspheric lens is required. This leads to a more compact lens with lower weight. According to the manufacturer, distortions in

1 https://www.curve-onc.com/optical-design-fisheye-curve-one80/

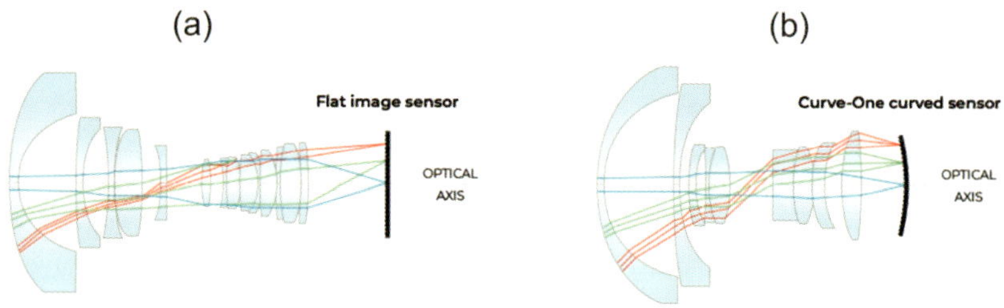

Fig. 6.33: Fisheye zoom lens with design optimized for flat image sensor (a) compared to a more relaxed design for curved sensors (b). Reprinted with kind permission from CURVE-ONE.[2] Curved CMOS sensor—company CURVE SAS—contact@curve-one.com— Curving all CMOS and IR sensor types.

the central of the image field up to 50° FOV are significantly reduced, and the irradiance can be kept high over the whole image field.[2] The disadvantage of that design, which is almost perfectly adapted to the curved sensor, is that the lens and sensor are combined as a module, and thus interchangeable optics for a given curved sensor may be difficult to realize. However, for smartphone cameras, consisting of multiple camera modules, which all require compact dimensions, curved sensors represent a very attractive option (see Chapter 7). Further details of curved sensors are discussed in Section 4.12.

6.6 Varifocal and zoom lenses

Camera lenses with variable focal length are generally termed varifocal lenses. The simplest way to achieve that goal is to use a combination of two lenses like in the above examples for the telephoto, respectively, retrofocus principle. If the separation between the lenses is varied, then the focal length of the combination varies but also the image position changes. This is not critical for projection lenses where a simple varifocal system may be used, and the image magnification is matched to the desired image size on the screen. However, the image might not yet be sharp on the screen and, therefore, the lens still needs to be shifted in a subsequent step in order to get a sharp image at the desired location. In other applications, for instance, with movie cameras, this is not practical because once the image position is fixed, changing the focal length while the camera is running must be possible without defocusing the image. This is possible with zoom lenses, sometimes also called true zoom lenses or parfocal zoom lenses, that have a special construction that allows for varying the image size without refocusing again. The first zoom lenses were developed for motion cameras and later adapted for other applications and have become the standard lenses in modern digital camera systems.

The zoom ratio or zoom factor is the ratio of the longest focal length to the shortest one. As the size and complexity of zoom lenses increases with the image format, typical

2 https://www.curve-one.com/optical-design-fisheye-curve-one80/

zoom ratios are in the range from 2 to 10, but also values up to 50 can be found for compact cameras with small image formats.

As for the design of zoom lenses, different principles are realized. At least two movable lens groups are required. They may be both positive or a positive/negative couple and can be arranged in the telephoto design with the positive group in the front and the negative group behind it or alternatively in the reversed way. In all cases, it is necessary that the back focal distance is positive in order to place the image sensor beyond the second lens. If the focal plane becomes virtual, then a third lens element is necessary to get a real image on the sensor. The third lens can be optionally split up in two parts again to form an afocal system with the front elements. If the aperture stop is located with the immobile rear element, this ensures a fixed relative aperture independently from the focal length. This will be discussed below in the example with three moving lens groups.

Let us start the consideration with a zoom lens consisting of two moving groups as depicted in Figure 6.34. If only one group is moved, the focal length changes, but also the image location. It can be shown that the image location remains unchanged if both lenses or lens groups move in a nonlinear relationship with each other [Flü55]. This nonlinear displacement is called mechanical compensation and is effectuated by turning or shifting the zoom ring of the lens barrel. Our example shows the most commonly used retrofocus lens arrangement for zoom lenses. Whereas the telephoto arrangement leads to construction lengths usually shorter than the focal length, the retrofocus design has the advantage of good corrections of both the relatively large aperture and large angle of view and to yield enough back focal clearance for system cameras with hinged mirrors.

Figure 6.34a shows the movement of both the negative front and positive rear element, represented as thin lenses, if the focal length is changed. We start the consideration with the image being sharp on the sensor in the image plane and both lenses being close together. In this case, both the focal length f and the back focal distance f_{Ei} from

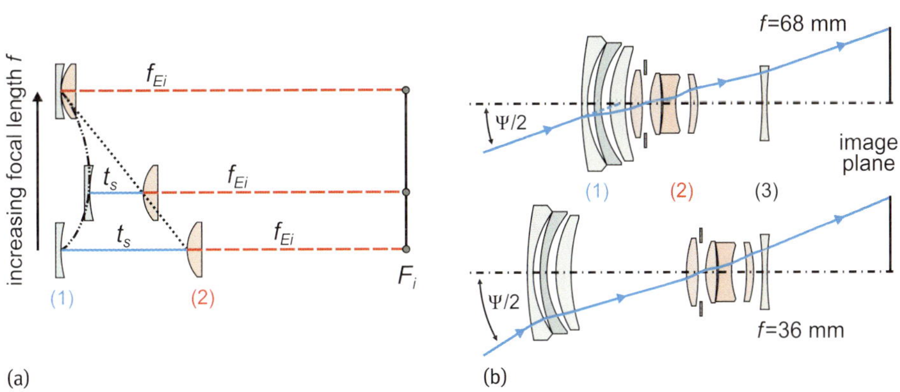

Fig. 6.34: Zoom lens of two lens groups with mechanical compensation. (a) Basic principle for a retrofocus lens arrangement; (b) construction details of a Pentax zoom lens with mechanical compensation.

vertex of the positive lens to the image focal point F_i have their maximum values as can be seen from the diagram in a). For parallel incident light, the image plane is located at F_i, which remains fixed during the lens movements we consider here in this example. Moving the positive lens toward the image plane decreases the overall focal length f linearly with the displacement of the positive lens. Simultaneously, the separation t_s between both lenses increases continuously in a way that the total length $l = t_s + f_{Ei}$ first of all decreases and then increases. We see that while f and f_{Ei} are linearly related to each other there is a nonlinear relationship between the length l of the lens combination and the focal length f with a minimum of l at some intermediate focal length.

It should be noted that this consideration is valid for a very distant object with the image located virtually in the focal plane of the lens. When the image is sharp on the sensor, zooming only changes the size of the image but not its sharpness. Focusing may be done by unit focusing where the whole lens barrel is shifted. When focusing to a nearer object, however, the image position remains not exactly fixed while zooming and a slight readjustment of the image position is necessary. For lenses used with DSLR cameras, this is no problem since the focus position is checked by the photographer either manually before the exposure or it is automatically adjusted by modern autofocus systems. The discussed principle works also when the lens is optimized for a nearer object distance. Zooming without refocusing is then possible only for that object distance whereas for other distances a slight correction is required to get the image sharp on the sensor. This means that the example for two moving lens groups is in the strict sense not a true zoom lens but merely a varifocal lens. The smaller the zoom factor and the higher the f-number, which may lead to a large depth of field, the lesser will be the necessary focus correction.

This principle of two moving lens groups is illustrated in Figure 6.34b for the example of a Pentax zoom lens designed for the focal length range from about 36 mm to 68 mm for the 35 mm format [Kin89]. The upper part of 6.33b shows the negative lens group (1) and positive lens group (2) as close together as possible thus yielding the longest focal length and the narrowest angle of view Ψ. The lower part is correspondingly related to the longest separation between the moving lens groups and the shortest focal length with the largest angle of view. Additionally, to part a) there is still a third lens element (3) in the setup, simply a biconcave lens, which is intended by the designer to shorten the overall length of the objective lens. It can also be understood as the amplifier lens in a projection setup to increase the image size (see Figure 6.21).

Another important point is the location of the aperture stop. It conventionally is fixed with the rear moving group. The size of the entrance pupil, as seen from the object space, changes with changing the focal length, but in general not in a linear way. Thus, the ratio between them, yielding the f-number and the relative aperture, respectively, changes with zooming. This also in general is not critical for still cameras as a changing relative aperture, and thus a changing illuminance in the image plane is compensated by a changing exposure time. For movie cameras, this is more critical as they operate with constant exposure times.

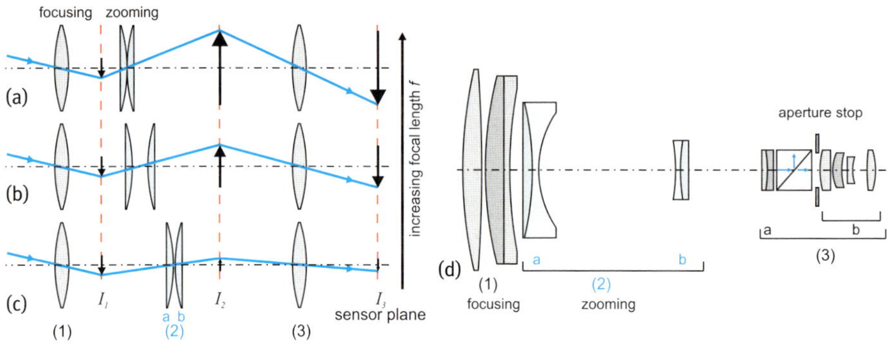

Fig. 6.35: Zoom lens of three lens groups with mechanical compensation. (a)–(c) basic principle; (d) construction details of Schneider Variogon 2.8/10–40 mm.

Let us now consider a different example for mechanical compensation where three lens groups are moved. This method is more complicated but has the advantage that unlike with two moving groups a true parfocal zoom lens for all object distances is achieved. Moreover, the f-number remains constant for all focal distances during zooming. The method is depicted in Figure 6.35. Its principle can be understood by the schematics in parts a) to c). We consider three lens groups (1), (2) and (3) with positive refractive powers. Group (2) is used to effectuate the change of the focal length and, therefore, consists of two individual lenses. The front group (1) renders an intermediate image of the object to the plane I_1 and is used for focusing. Its position must be adjusted for objects at different distances in order to keep the image at a fixed location in plane I_1.

Both lenses, 2a and 2b, act as one group (2) and have the highest refractive power when they are close together as in Figure 6.35a. They form the image of I_1 in the next fixed image plane located in I_2. Shifting (2) away from I_1 without changing the separation between 2a and 2b would change the position of the image generated by them. This can only be compensated by increasing the separation between them appropriately. As a consequence, the image is sharp in I_2 but of smaller size (Figure 6.35b). Shifting (2) further on and keeping the image sharp in I_2 requires that the separation between the lenses of (2) must be continuously increased in a nonlinear way to a maximum separation, which is given by the lens parameters. Shifting the lens group (2) further on necessitates that the separation is reduced again (Figure 6.35c). The maximum shift of group (2) is achieved when both lenses 2a and 2b are in contact again. The image in plane I_2 remains always sharp but its size is continuously reduced with the shift of (2) away from its initial position.

In a final imaging step, the image in I_2 is transferred by lens group (3) to the fixed image plane I_3, which is identical with the film or sensor plane. As lens (3) is fixed at its position like all image planes, the image on the sensor is always sharp once lens (1) is focused. The image size on the sensor at I_3 decreases with the shift of group (2) and the corresponding relative movements of 2a and 2b, which implies that the overall focal

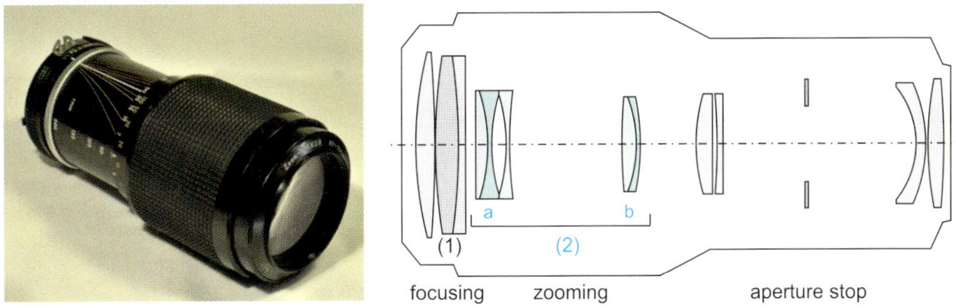

Fig. 6.36: Zoom–Nikkor 80–200 mm, f/4.5, with construction scheme.

length of the lens combination decreases continuously. We have the same situation like in the retrofocus zoom mechanism above that a nonlinear relative movement of two lenses is necessary to vary the focal length of the system without defocusing it. The difference is that focusing is done by lens group (1) alone. The positive refractive powers of the lenses assumed in this example are used for easier exemplification of the principle as we deal only with real images. However, this leads to an extended and large lens setup. The length can be effectively reduced if the lens group (2) is made of diverging lenses. In these cases, the intermediate images become virtual but the zoom principle is the same.

The described principle with a negative lens group (2) has been implemented in the Zoom–Nikkor 80–200 mm, $f/4.5$ (Figure 6.36). This is a long focus zoom lens for the 35 mm format featuring the typical telephoto design with a positive front group (1) and a negative group (2). It has a combined push-pull focus and zoom ring where focusing is done by turning the ring, and thus moving only group (1). Group (2) acts as a variator to adjust the focal length. By pushing and pulling the ring, the elements 2a and 2b move in a nonlinear way and the image on the sensor remains sharp. The aperture stop is located in the rear part and is fixed at its position like the remaining lenses. It is obvious that the exit pupil, being the image of the stop seen from the rear part, neither changes in size nor location when the lens is focused to other object distances or zoomed. Thus, the relative aperture always has the same value and is independent from the focal length.

The same principle can be found in the Variogon 2.8/10–40 mm by Schneider. The construction of this lens is shown in Figure 6.35d. The lens is a zoom lens for the super 8 movie film format with an image diagonal of 7.1 mm. The aperture stop is located in the third lens group and fixed with it. As in the Nikon zoom lens, the relative aperture is independent from the focal length or object distance. This is of high importance for movie shooting as the image brightness does not change during zooming or focusing. Another interesting point can be seen in the lens group (3). A telecentric ray path is achieved in this group by a splitting it up into a front lens 3a and a rear group 3b. In between, the beam is parallel to the optical axis. There is enough clearance to insert an optical beam splitter, which acts as an outcoupling element for an additional pathfinder in movie film

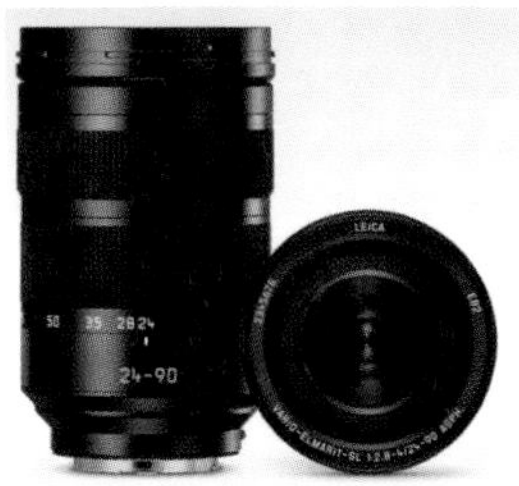

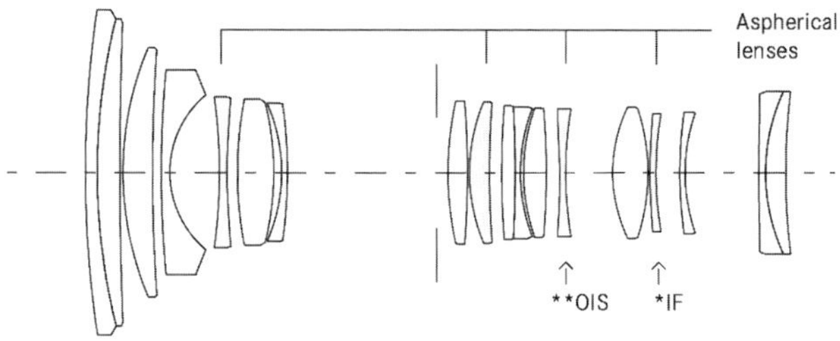

Fig. 6.37: Leica VARIO-ELMARIT-SL 24–90 f/2.8–4 ASPH. with construction scheme. **OIS: optical image stabilization, *IF: internal focus. (Reprinted with kind permission of Leica Camera AG.)

cameras. Similar modifications have been done for some types of TV cameras or cameras for movie film productions where the optical path is split up and directed to sensors for different colors.

A last example for a modern zoom lens for a mirrorless 35 mm format camera is depicted in Figure 6.37. The Leica VARIO-ELMARIT-SL 24–90 f/2.8–4 ASPH. consists of 18 lenses in six moving groups. Eleven lenses are made from glasses with anomalous partial dispersion. The lens incorporates multiple aspheric lenses and an optical image stabilization lens. Automatic focusing is based on an internal focusing mechanism where only one aspherical lens is shifted. Thus, the overall length of the lens barrel does not change on focusing. Like most zoom lenses with the aperture between moving lens groups, size and location of the entrance pupil vary and the relative aperture varies with the focal length.

6.7 Perspective control—tilt/shift lenses

Perspective control during image capturing can be achieved by special lenses that offer the possibility to incline the image plane and also to shift it with respect to the optical axis. This technique was originally used with large or medium size cameras in order to correct converging lines and also to select the plane of best sharpness without the conventional control of the depth of field by the aperture stop. The physical basics for the control are described by the Scheimpflug principle. The principle is named after the Austrian Scheimpflug who elaborated a method for aerial photography.

6.7.1 Scheimpflug principle

In order to understand the fundamentals of the Scheimpflug principle, let us consider the image formation illustrated in Figure 6.38a. We assume an inclined object plane, delineated by the line S_0 on which the points P_{01}, P_{02} and P_{03} are located. These points are imaged through the lens to the image space. The lens with the image focal length

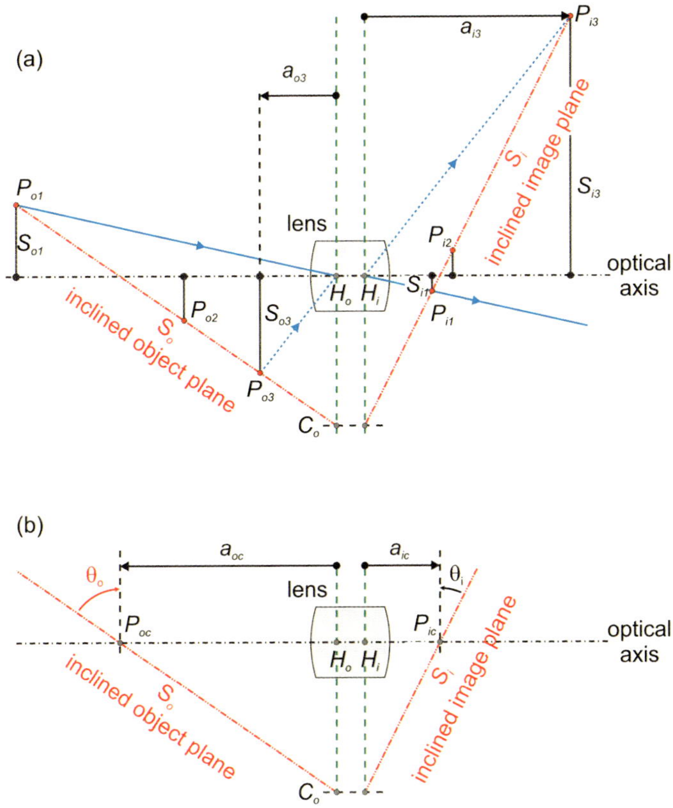

Fig. 6.38: Scheimpflug principle. (a) General relationship; (b) consideration relative to intersection with the optical axis.

f_i may be a complex lens arrangement in air and is characterized in the figure by its principal planes H_o and H_i in the object, respectively, image space. The location of the objects can be described by their object distances and heights in the object space, for instance, a_{o3} and S_{o3} for point P_{o3}. In the image space, their conjugated image points are also described by their image distances and image heights, for instance, a_{i3} and S_{i3} for point P_{i3}. The inclined object plane is described in the object space by the straight-line S_o.

The function S_i that describes the position of the imaged points in the image space can be calculated when we start with the line S_o as a linear function of the object distance a_o with m_o being the slope and C_o a constant in the object space:

$$S_o = m_o \cdot a_o + C_o. \tag{6.17}$$

It should be noted that the object and image distances a_o and a_i have opposite directions according to our convention. The relationship between them is given by the lens formula and we can write after Equation (2.7):

$$a_o = \frac{a_i \cdot f_i}{f_i - a_i}. \tag{6.18}$$

Substituting a_o in Equation (6.17) by this expression yields

$$S_o = m_o \cdot \frac{a_i \cdot f_i}{f_i - a_i} + C_o. \tag{6.19}$$

The lateral magnification M after Equations (2.6) and (6.18) can be rewritten as

$$M = \frac{f_i - a_i}{f_i}. \tag{6.20}$$

We may note that M is not a constant quantity for points on the object plane, respectively, their conjugated image points but locally dependent on the image distance a_i relative to the lens. The corresponding image size S_i is achieved by multiplying S_o by M:

$$S_i = M \cdot S_o = m_o \cdot \frac{a_i \cdot f_i}{f_i - a_i} \cdot \frac{f_i - a_i}{f_i} + C_o \cdot \frac{f_i - a_i}{f_i} = \left(m_o - \frac{C_o}{f_i}\right) \cdot a_i + C_o. \tag{6.21}$$

This result is very interesting. It shows that the images of all points located on the inclined object plane are also located on an inclined image plane. There is no need to argue with the depth of field since the object plane in its complete extension is imaged sharp to the inclined image plane. Like in the object space, the size of the images, relative to the optical axis, varies linearly with the image distance, however, with a different slope, but having the same vertical intercept C_o with the principal plane. If we assume a single thin lens with one principal plane instead of two characterizing the thick lens in Figure 6.38, then the Scheimpflug principle can be stated in the following way: an object plane is rendered sharp in an image plane if the object plane, image plane and the principal plane of the lens intersect in the same point, which in the figure is indicated by C_o. For the more general case of an optical system having two principal planes, the intercept points of the object and image planes with their corresponding principal planes are located on a straight line parallel to the optical axis.

Analogously to Equation (6.17), we can state, using Equation (6.20),

$$S_i = m_i \cdot a_i + C_o \tag{6.22}$$

where

$$m_i = m_o - \frac{C_o}{f_i}. \tag{6.23}$$

In case of an object plane S_o, which is nearly perpendicular to the optical axis, C_o can attain very large values. A relationship between the slopes m_i and m_o in the object and image space, respectively, can then be better expressed if we compute their values directly using the intersection points of the planes with the optical axis as given in Figure 6.38b.

The intersection points are P_{oc} and P_{ic} in the object and image plane, respectively. Their corresponding distances from the lens are a_{oc}, respectively, a_{ic}. The slope of the planes can be simply calculated using the triangles defined by P_{oc}, H_o, C_o and P_{ic}, H_i, C_o. Hence, we get

$$m_o = -\frac{C_o}{a_{oc}} \qquad m_i = -\frac{C_o}{a_{ic}}. \tag{6.24}$$

For the example represented in Figure 6.38, m_o is negative and m_i is positive, according to our sign convention where a_o is counted negative and a_i is positive. The ratio of both slopes then yields

$$\frac{m_i}{m_o} = \frac{C_o}{a_{ic}} \cdot \frac{a_{oc}}{C_o} = \frac{a_{oc}}{a_{ic}} = \frac{1}{M_c} \tag{6.25}$$

Here again, we made use of the relationship (2.6), which expresses the local image magnification M_c for the conjugated points P_{oc}, P_{ic} at the intersection of their planes with the optical axis by the ratio of their image to object distances. This relationship, in paraxial approximation, is generally valid for any slopes, distances or focal lengths. In case of standard photography, object and image planes are usually perpendicular to the optical axis and parallel to each other. Then the image magnification is the same for all points on the object plane, namely M_c. If object respectively image plane are slightly tilted, for instance, by the angles θ_o, respectively, θ_i relative to the perpendicular, the image magnification varies locally with their object distance from the lens. This effect is stronger the more the planes are tilted and the more they are extended. As a consequence, the images on the plane are sharp but distorted due to the varying magnification. This effect is known as Keystone distortion or Keystone effect.

According to Figure 6.38b, the slopes of the planes are related to their tilt angles by the following relations:

$$\tan\theta_o = -\frac{a_{oc}}{C_o} = \frac{1}{m_o} \qquad \tan\theta_i = -\frac{a_{ic}}{C_o} = \frac{1}{m_i}. \tag{6.26}$$

Hence, we get the relationship:

$$\frac{\tan\theta_i}{\tan\theta_o} = \frac{m_o}{m_i} = M_c. \tag{6.27}$$

It should be noted again that M_c is defined for the intersection points of the planes on the optical axis.

In order to meet the requirements of the Scheimpflug principle while taking photographs, it is necessary to tilt and shift the image plane, which is only possible for special lenses as the conventional lenses are always mounted with their principal planes oriented parallel to the film or sensor in the camera. The technical realization in camera lenses is discussed in the next sections.

6.7.2 Principal function of shift and tilt

6.7.2.1 Shift function

Central projection leads to images with rectilinear perspective (see Sections 2.1 and 6.5.4). In order to avoid converging lines in the image of an object with parallel lines, it is important that object and image planes are parallel to each other. If the viewing angle becomes very large, the image size becomes very large with the consequence that the image format is not large enough to capture the whole scenery. If, for instance, a high tower is to be imaged with its front area parallel to the camera, the optical axis points to the bottom part of the tower where the far distant horizon is (Figure 6.39a). Therefore, due to the limited size of the sensor, which is centered on the optical axis like the lens, the upper part of the image is cropped and the bottom part with the foreground area is also imaged but is of no interest. A solution to this problem would be to choose a camera with a different sensor size for the given focal length in order to capture the whole image, but there would be nevertheless the unwished foreground present. Another method would be to shift the camera sensor perpendicularly to the optical axis in vertical direction to where the top position of the image is located. As a consequence, the wanted section is fully imaged without crop and the perspective is not distorted due to the parallel image and sensor planes (Figure 6.39b). However, this can only be achieved in this case by a special lens where the complete lens arrangement is shifted

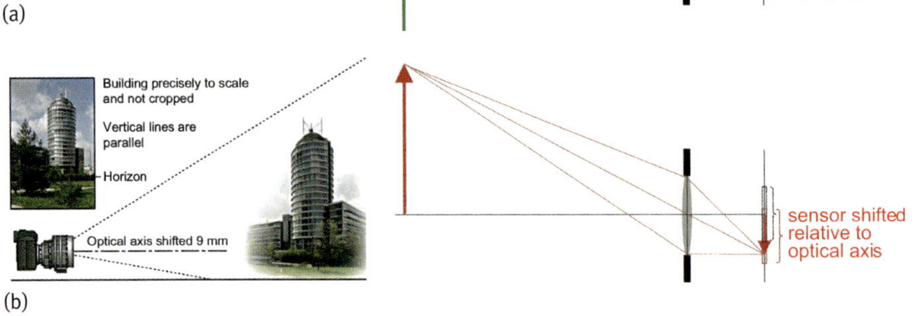

Fig. 6.39: Principle of a parallel shift between optical axis and sensor in the image plane. (a) No shift, optical axis centered on lens and image sensor; (b) vertical shift of optical axis centered on the lens relative to the image sensor (left side), respectively, shifted position of the sensor relative to the optical axis (right side). Inset on the left side reprinted with kind permission of Jos. Schneider Optische Werke.

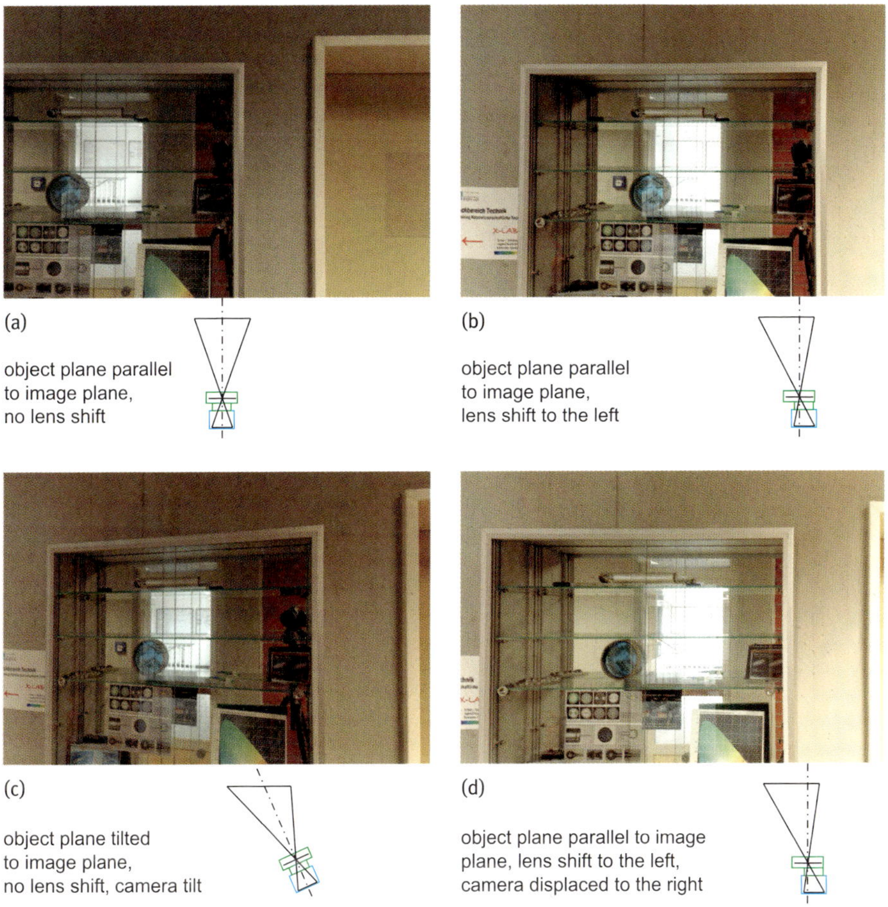

Fig. 6.40: Photographs of a showcase using a shift lens under different conditions.

in parallel to the sensor plane in the camera. The optical axis is fixed in the lens by the central line across the individual lenses that constitute the camera lens (Figure 6.44).

An example for this application is given in Figure 6.40 where different photographs are seen of a rectangular showcase with a mirror inside next to a doorframe. All photos have been taken by a tilt-shift lens for the 35 mm format with a focal length of 24 mm and $f_\# = 3.5$. Photos a) to c) are captured with the camera standing at the same position. In photo a), the camera is oriented perpendicularly to the showcase and its position can be seen in the mirror. The lens is exactly in the center of the image. For photo b), the principal plane of the lens has been shifted parallel to the image plane to the left side. There is no perspective distortion, as the camera is still at the same position, but it seems as if the center of the viewing perspective is located in front of the showcases center. If the shift function of the lens is not actuated, the entire lateral width of the showcase

can only be captured by turning the camera's orientation, and thus pointing under an oblique angle to the center of the case (photo c). As a consequence, the perspective distortion becomes obvious as the image plane and the front of the showcase are no longer parallel to each other. If the showcase should be imaged without perspective distortion but the camera should not be visible in the mirror, a further displacement of the camera to the right side of the case is necessary (photo d). Now the lens plane has to be shifted still more than for photo b) in order to image the whole width of the case.

6.7.2.2 Tilt function

The tilt function of the lens is the consequence of the Scheimpflug principle and is used to adjust the inclination of object and image planes to achieve the sharpness and depth of field. A practical example is given in Figure 6.41 for close-up imaging. The photographs were taken by a tilt-shift lens of $f = 24\,\mathrm{mm}$ for the 35 mm format with $f_\# = 3.5$. The picture in a) is taken without tilt, which means that the image plane, the principal plane of the lens and object plane are parallel to each other. The depth of field is conventionally controlled by the aperture stop of the lens (see Section 3.4.6). The best sharpness is seen near the middle of the photograph. Due to the relatively low f-number for that close-up imaging, the foreground as well as the background are no longer within the depth of field and are thus imaged blurred.

For comparison, picture b) is taken with the lens plane tilted to the image plane. Image distance, which implies focusing, and f-number were not changed and are identical to the settings in a). It can be seen in Figure 6.41b that the sharply imaged object plane in

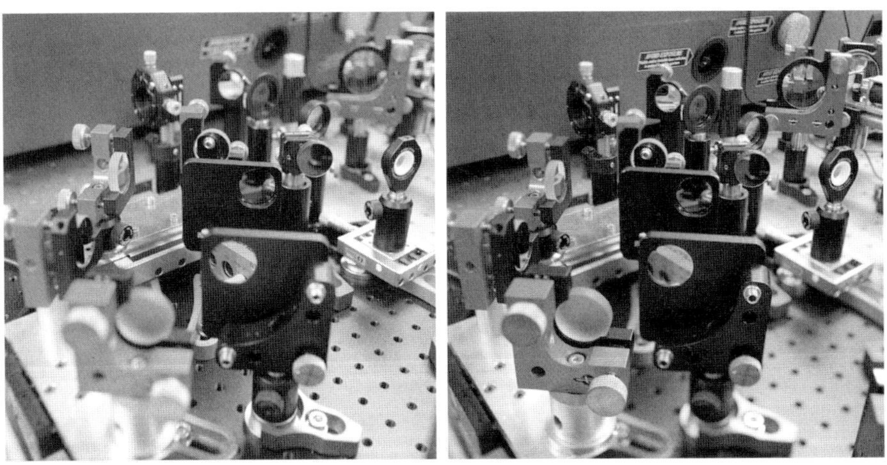

(a) (b)

Fig. 6.41: Close-up imaging of objects. (a) No tilt, the object plane is parallel to the image plane and the best sharpness is seen near the middle of the photograph; (b) using the tilt function of a lens to extend the depth of field to large distances (for details see text).

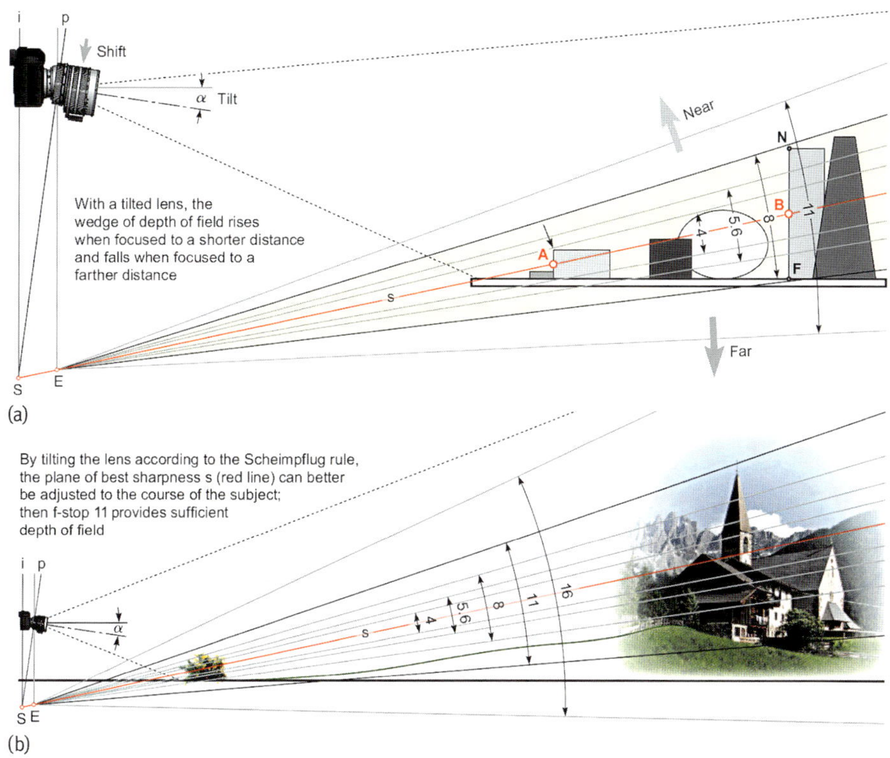

Fig. 6.42: Using the tilt function of a lens to define the plane of best sharpness; the wedge of depth of field is additionally adjusted by the f-number and focusing to nearer or farther distance. (a) Basic principle; (b) schematic application for landscape photography; i: image plane, p: principal lens plane, s: sharp imaged object plane. Reprinted with kind permission of Jos. Schneider Optische Werke.

this case is nearly parallel to the optical breadboard at a distance of some cm above it. Objects in the foreground, like the knurled screw, are sharply imaged as well as objects in the background, like the lower plate with characters. The sharpness above and below the image plane is restricted by the f-number. When further stopping down, the depth of field could be extended. This would lead to more sharpness in the foreground and background of image a). For image b), the extended depth of field would lead to increased sharpness seen for objects above and below the object plane.

This effect is illustrated in Figure 6.42a. Applying a tilt α between the image plane i and the lens plane p has the consequence that all objects in plane s are rendered sharp in the image plane if all three planes intersect in one line. The intersection according to the Scheimpflug principle is indicated by point S in the figure. In contrast to conventional imaging, all points at far and near distances on s are imaged sharp simultaneously, which leads to an obvious improvement compared to a lens without tilt (Figure 6.41). The extension of the sharpness in the plane using a lens tilt does not primarily depend

on the relative aperture of the lens and offers more flexibility for the exposure conditions like exposure time. However, the influence of the stop on the depth of field can be seen and exploited as mentioned above. Points above and below the plane s become blurred with increasing distance perpendicular to s. The wedge of depth of focus can be increased by stopping down and/or by focusing to nearer distance and decreased vice versa. The same principle can also be applied for long-distance landscape photography, especially with lenses at low f-numbers where the depth of field is usually very shallow. In many cases, a combination of both tilt and shift is done in order to fix the plane of best sharpness and to select the wanted image section.

6.7.3 Specifications and constructions of PC-lenses for 35 mm format

The principle of the tilt/shift lenses requires that the total image they generate must be sufficiently large so that the image sensor is still illuminated if the optical axis and the image plane are displaced relative to each other. In a lens without the tilt-shift-function, the diameter of the image circle is slightly larger than the image format diagonal in order to ensure that the sensor is fully illuminated. For the 35 mm format, the diagonal is 43.3 mm, thus the diameter of the image circle must be at least of 44 mm. Figure 6.43a depicts the image circle specifications of the Schneider PC-TS SUPER-ANGULON 2.8/50 HM for 35 mm format DSLR cameras. It has an image circle of 79.2 mm diameter, and thus offers the possibility to shift the optical axis by a maximum of 12 mm in lateral directions without taking into consideration a tilt. If the lens is additionally tilted by 8°, which is possible in every direction, a larger image circle than for shifting alone is necessary (Figure 6.43c). This area is indicated as reserve in the figure.

When the shift function is actuated, the whole lens arrangement is shifted parallel to its principal planes. This means that the mount of the lens fixed to the camera is moved

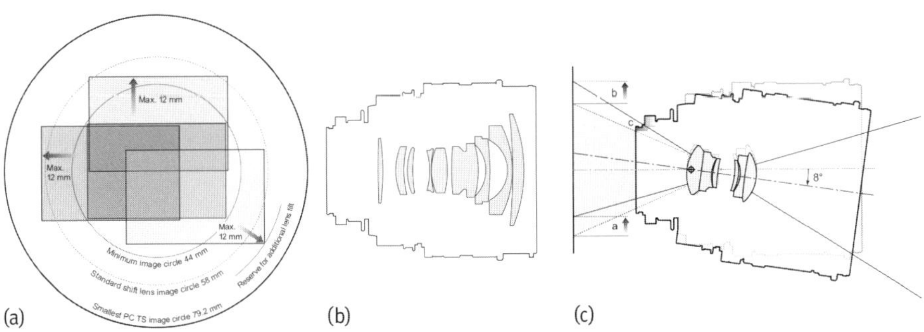

Fig. 6.43: (a) Image circle specifications of the Schneider PC-TS SUPER-ANGULON 2.8/50 HM; (b) construction details of PC-TS SUPER-ANGULON 2.8/50 HM; (c) tilt and construction details of Schneider PC-TS MAKRO-SYMMAR 4.5/90 HM; additional reserve in the image circle for tilt is indicated by b. Reprinted with kind permission of Jos. Schneider Optische Werke.

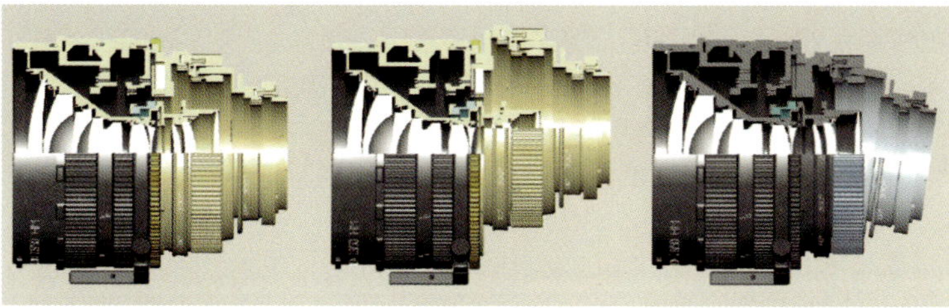

Fig. 6.44: Lateral cut of the Schneider PC-TS SUPER-ANGULON 2.8/50 HM for shift and tilt compared to its neutral position (reprinted with kind permission of Jos. Schneider Optische Werke).

into the opposite direction (Figure 6.44). When the lens is tilted, the sharp bend occurs at the rear part behind the lens arrangement close to the lens mount.

As for the lens construction, the Angulon lens has a significantly lower focal length than the diameter of its image circle. Their ratio is similar to that of a 30 mm focal length for the full format DSLR camera, which means that a typical retrofocus design can be expected. This is indeed seen in Figures 6.43b and 6.44. Although being a normal lens for the full format, it features the design of a retrofocus lens in order to ensure a good correction of all lens aberrations across the whole, enlarged image circle. Accordingly, the Schneider PC-TS MAKRO-SYMMAR 4.5/90 HM with a longer focal length of 90 mm and an image circle of 87.8 mm has a typical nearly 1:1 ratio between both quantities. Therefore, a nearly symmetrical lens construction for a relative aperture of $f/4.5$ is sufficient to achieve high quality. Due to the more simple setup, it has a lower weight and only 10 mm more length than the Angulon. Tilt shift lenses for DSLR full format cameras are available from about 17 mm focal lengths up to about 90 mm by all major lens manufacturers.

6.8 Antireflection coating and lens flares

As already mentioned above with the historical development of lenses, one of the main drivers for modern lens designs has been the invention of antireflection coating. This allowed for more sophisticated lens designs in order to reduce lens aberrations as the number of glass-air surfaces could be increased. Antireflection coating started after a patent to enhance the light transmission of optical components was issued to Smakula from Zeiss in 1935. When the light transmission of the lens is enhanced, there is not merely more light available for the image formation but it has a dramatic influence on the image quality. For instance, due to more light in the bright image parts, the contrast to dark parts is improved, whereas suppression of undesired lens flares may darken the black parts even more, and thus reduce the overall background brightness. In all these cases, the modulation transfer function MTF may be significantly improved, even at low spatial frequencies. The whole topic of reflections is quite complex and requires more consideration than we can do within the scope of the present book. Therefore, we will present here only some basic ideas for the reduction of light reflections in lenses and give some examples for their effects. A more comprehensive introduction to the topic can be found in the publication of Blahnik and Voelker [Bla16].

6.8.1 Antireflection coating

The path of an optical ray striking a surface is deflected if the refractive indices of the media on both sides of the interface are different. The principles are described by Snell's law (see Section 3.1.2) and constitute the basis for image formation using glass lenses. However, not only refraction occurs, but also reflection, and thus light intensity is scattered away and lost. This could be avoided if the refractive indices from one medium to the next one changes gradually. Then only a bending of the ray without reflection occurs. For the transition from air to glass, this is not possible as there is no solid bulk material having a refractive index below 1.3 to be matched to air. A different method is to use an intermediate layer between air and glass and exploit the interference properties of electromagnetic waves in order to minimize the reflectance of the light transition. Let us therefore consider the case depicted in Figure 6.45a where light is incident from air with the refractive index $n_0 = 1$ under an angle β to the surface of a solid coating layer (CL) with the refractive index n_{ar}. Part of the light is reflected at the air-CL interface, the remaining light is refracted to the CL and propagates to the CL-substrate interface where part of the light is back-reflected again, and the remaining part transmits to the substrate. Due to the limited thickness d_{ar} of the CL, the back-reflected part strikes the top CL-air interface and transmits to air while also a part of it is back-reflected again. The process of reflection and transmission goes on repeatedly with the light intensity of the ray decreasing after each interface strike. All partial rays leaving the CL to air at the top interface superpose each other and their electric fields interfere to establish a resulting reflected field of $\rho_E \cdot \mathcal{E}_0$ where ρ_E is the field reflection coefficient of the CL-substrate combination and $\mathcal{E}_0$ is the electric field of the incident ray. We can neglect in the first instance the lateral displacement of the partial rays as we generally have to deal with more extended beams and not too large values of β. Analogously, the rays leaving the CL to the substrate interfere to yield a total electric field of $\tau_E \cdot \mathcal{E}_0$ with τ_E being the corresponding transmission coefficient for the electric field.

For the computation of the interfering fields, it is important to differentiate between the different polarizations of light and to take into account the angle of incidence β. The reflection and transmission coefficient, considering also phase shifts at the interfaces, are given by the Fresnel equations, and the summation of the partial waves after multiple reflections and transmissions leads to a converging series. A destructive interference is observed if the waves associated with the reflected partial rays through the CL and exiting at the top boundary have a phase shift of 180° relative to the first reflected wave at the air-CL interface. This is equivalent to an optical path difference of $\lambda/2$ or an odd

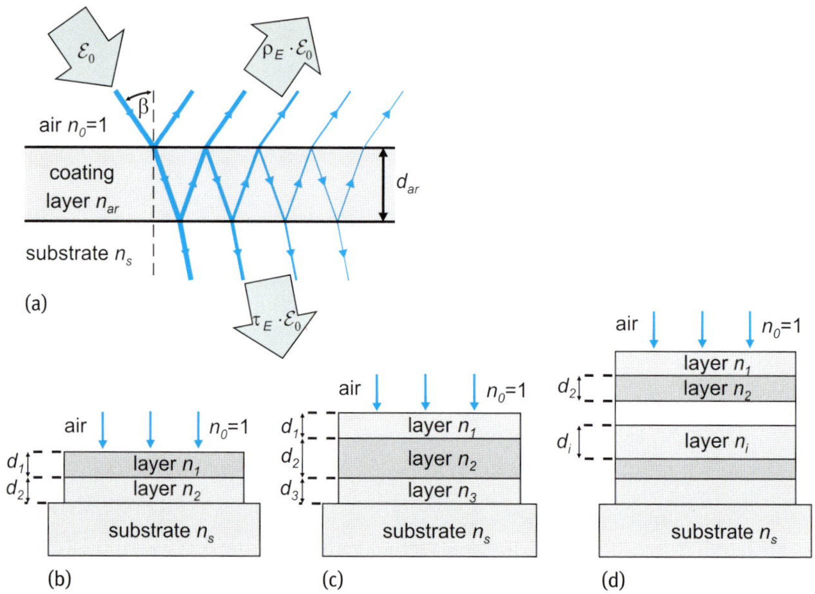

Fig. 6.45: Scheme of coating layer combinations on glass for light incidence from air. (a) Single layer coating; (b) double-layer coating; (c) triple-layer coating; (d) multilayer coating.

multiple of it between the interfering waves. As the CL is traversed two times or by an integer that is a multiple of two by the wavelets exiting at the top surface, an optical length of $\lambda/4$ for the CL leads to suppression of the reflected wave for perpendicular incidence. The mathematical treatment of the problem with multiple reflections for the different polarizations is quite complex and can be done by different methods. As we will see in the following, a single layer yields satisfying results only for some restricted conditions. Therefore, combinations of several layers of different thickness and material are more appropriate for most applications.

6.8.1.1 Single-layer coating

Let us consider the simple case of normal incidence with $\beta = 0$, where all light polarizations are equivalent, and take into account phase shifts at the interfaces. We find that a simple CL leads to minimum reflectance at a wavelength λ_0 in air if its optical thickness is equal to a quarter of the wavelength in the material, thus its thickness d_{ar} being [Ped08]:

$$d_{ar} = \frac{\lambda_0}{4 \cdot n_{ar}}. \tag{6.28}$$

Neglecting absorption losses in the CL, the power reflectance $\rho_P = |\rho_E|^2$ as well as the power transmittance $\tau_P = |\tau_E|^2$ can be calculated:

$$\rho_P = \left(\frac{n_0 \cdot n_s - n_{ar}^2}{n_0 \cdot n_s + n_{ar}^2}\right)^2 \quad \tau_P = 1 - \rho_P. \tag{6.29}$$

It is obvious that a zero reflectance is achieved if the refractive index n_{ar} is the geometric mean value of the surrounding media above and below it. Thus,

$$n_{ar} = \sqrt{n_0 \cdot n_s}. \tag{6.30}$$

For incidence from air, we get $n_{ar} = \sqrt{n_s}$. Assuming a typical value of $n_s = 1.52$ for Schott glass BK7, we need in this case a refractive index of $n_{ar} = 1.24$ for the CL in order to obtain zero reflectance. However, there is no bulk material with that index as can be seen from Table 6.2. For technical reasons, only a few materials are appropriate for a thin layer coating of glasses.

One of the most interesting materials is MgF_2 due to its high robustness and having a refractive index of as low as $n_{ar} = 1.38$ in the visible range. A zero reflectance cannot be achieved by a $\lambda/4$ layer of it on BK7, but according to Equation (6.29), a minimum reflectance of $\rho_P = 1.3\,\%$ compared to 4.3 % for a noncoated glass surface. Figure 6.46a shows the reflectance ρ_P as a function of the wavelength with the minimum centered at a wavelength of $\lambda_0 = 550$ nm in the green spectral range. The reflectance increases below and above λ_0 but remains below about 2.2 % within the whole visible range. The corresponding CL has a thickness of 100 nm and a variation of the layer thickness by Δd_{ar} results in a shift of $\Delta\lambda_0 = 4 \cdot n_{ar} \cdot \Delta d_{ar}$. Thus, a variation of the thickness by ±1 nm shifts λ_0 and simultaneously the whole curve by about ±5.5 nm.

For glasses of higher refractive indices, the same CL yields a lower minimum reflectance. A nearly perfect match is achieved with lanthanum dense flint glass LASF9

Tab. 6.2: Refractive index of materials for coatings in the visible spectral range [Bla16].

material	material name	refractive index n at 550 nm
Na_3AlF_6	cryolith	1.35
MgF_2	magnesium fluoride	1.38
SiO_2	silicon dioxide	1.45
Si_2O_3	disilicon trioxide	1.55
CeF_3	cerium fluoride	1.63
Al_2O_3	aluminium oxide	1.65
MgO	magnesium oxide	1.70
Nd_2O_3	neodymium oxide	1.95
ZrO_2	zirconium oxide	2.05
CeO_2	cerium oxide	2.22
ZnS	zinc sulfide	2.36
TiO_2	titanium dioxide	2.32
ZnSe	zinc selenide	2.65

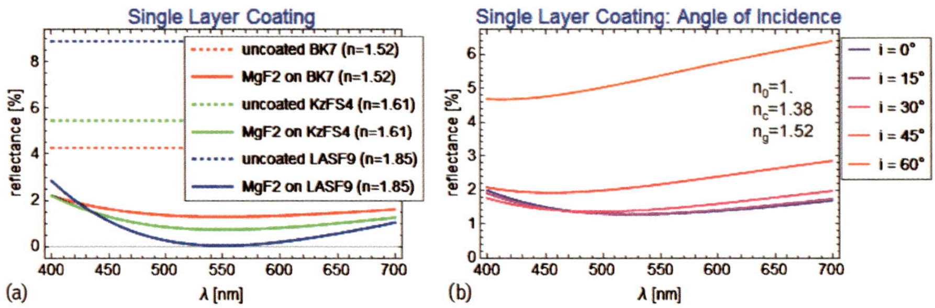

Fig. 6.46: Reflectance as a function of the wavelength in the visible range for a $\lambda/4$ single-layer coating. (a) Normal incidence on different glass substrates; (b) oblique incidence for MgF_2 ($n_c = 1.38$) coating on BK7 ($n_g = 1.52$) [Bla16]. Reprinted with kind permission of Carl Zeiss AG.

of $n_s = 1.85$ (Figure 6.46a). However, the reflectance curve is less flat than for lower index glasses and even exceeds the values of BK7 in the blue spectral range. For oblique incidence, the reflection at the interface increases according to the Fresnel equations and the optical path length in the CL increases. The consequence is that the overall reflectance at glass coated by a single layer increases with the angle of incidence and that wavelengths in the red spectral range are more strongly affected by the oblique incidence than in the blue-green range (Figure 6.46b). Thus, a red shift in the reflected light can be observed if the lens is viewed under large angles to its optical axis.

As the antireflection effect is due to light interference, not only $\lambda/4$ layers can be used but also odd integers of it, which generate the same phase difference between the interfering partial waves. With longer absolute lengths, however, the light paths become more sensitive to deviations from the perfect phase match caused for instance by variations of wavelength or angle of incidence. The practical consequence is that the spectral width of the antireflection coating decreases with increasing thickness, and thus the optimum thickness of a CL is that of $\lambda/4$.

6.8.1.2 Double-layer coating

The single-layer coating can be used to achieve a significant reduction of reflection at glass elements compared to noncoated elements. However, for high-quality lens arrangements with up to more than 20 individual reflecting surfaces the reduced reflectance is not sufficient, neither its magnitude nor its spectral width.

The remaining reflectance of a single layer coating can be further reduced by a subsequent layer combined with it (Figure 6.45b). It turns out that the best results will be obtained by a combination of two $\lambda/4$ layers or by one $\lambda/4$ and one $\lambda/2$ layer. The total reflectance of a $\lambda/4$–$\lambda/4$ layer combination can be calculated and yields for normal incidence ($\beta = 0$) [Ped08, Bla16]:

$$\rho_P = \left(\frac{n_0 \cdot n_2^2 - n_s n_1^2}{n_0 \cdot n_2^2 + n_s n_1^2}\right)^2 \qquad \tau_P = 1 - \rho_P. \tag{6.31}$$

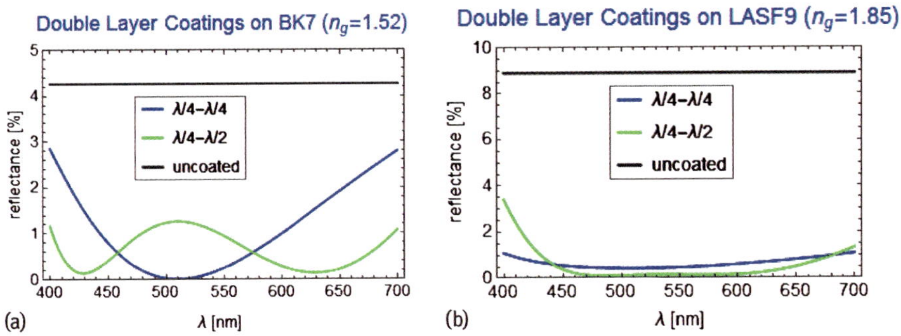

Fig. 6.47: Reflectance as a function of the wavelength in the visible range for different double-layer coatings, reference wavelength $\lambda_0 = 510$ nm. a) $\lambda/4$–$\lambda/4$ coating with $n_1 = 1.38$, $n_2 = 1.7$ and $\lambda/4$–$\lambda/2$ coating with $n_1 = 1.38$, $n_2 = 2.0$, both double layers on BK7 glass with $n_s = 1.52$; b) $\lambda/4$–$\lambda/4$ coating with $n_1 = 1.38$, $n_2 = 1.78$ and $\lambda/4$–$\lambda/2$ coating with $n_1 = 1.38$, $n_2 = 2.0$, both double layers on dense flint glass LaSF9 with $n_s = 1.85$ [Bla16]. Reprinted with kind permission of Carl Zeiss AG.

Thus, the reflectance is canceled when

$$\frac{n_2}{n_1} = \sqrt{\frac{n_s}{n_0}}. \tag{6.32}$$

This is more easily fulfilled than Equation (6.30) even for low-index glasses as only the ratio of the coating indices must be equal to the square root of the glass index for incidence from air. Figure 6.47a depicts the reflectance as a function of λ in the visible range for two $\lambda/4$ layers compared to a $\lambda/4$–$\lambda/2$ combination on BK7 glass. All layers are designed to have the corresponding thickness for $\lambda = 510$ nm. The $\lambda/4$–$\lambda/4$ curve features a V-type shape with the minimum reflectance at 510 nm and increasing at lower, respectively, higher wavelengths like a single CL. On the other hand, a $\lambda/4$–$\lambda/2$ combination does in general not have a zero reflectance on low index glasses. For the reference wavelength, it has the same reflectance like an individual single CL of material n_1 on the glass substrate as the $\lambda/2$ layer is not effective. At wavelengths above and below it, minima due to destructive interference are observed and lead to an overall broader spectral width than the $\lambda/4$–$\lambda/4$ combination. The curve is W-shaped with a maximum reflectance slightly above 1 %.

For high-index glass substrates, the overall reflectance is lower again as expected but may also strongly increase in the blue and red marginal spectral ranges (Figure 6.47b).

6.8.1.3 Triple-layer and multilayer coatings

It becomes obvious that a double-layer, especially for low-index glasses, yields better results than single-layer coating, but is also not yet sufficient for complex lens systems. Further improvements are achieved by triple-layer combinations, as again more param-

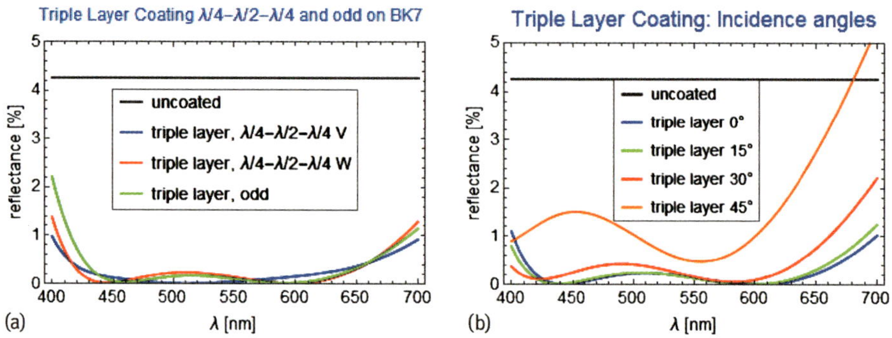

Fig. 6.48: Reflectance as a function of the wavelength for different triple-layer coatings on BK7 glass (n_s = 1.52), reference wavelength λ_0 = 510 nm. a) $\lambda/4$–$\lambda/4$–$\lambda/4$ coating with n_1 = 1.38, n_2 = 2.15, n_3 = 1.7, $\lambda/4$–$\lambda/2$–$\lambda/4$ coating with n_1 = 1.38, n_2 = 2.15, n_3 = 1.62, odd triple-layer with n_1 = 1.38, n_2 = 2.1, n_3 = 1.8, $d_1 = 567.2\,\text{nm}/(4 \cdot n_1)$, $d_2 = 212.3\,\text{nm}/(4 \cdot n_2)$ and $d_3 = 731.4\,\text{nm}/(4 \cdot n_3)$; b) $\lambda/4$–$\lambda/2$–$\lambda/4$ triple-layer coating with n_1 = 1.38, n_2 = 2.15, n_3 = 1.62 for different angles of incidence [Bla16]. Reprinted with kind permission of Carl Zeiss AG.

eters allow for better adjustment than for fewer layers (Figure 6.45c). Combinations of different layer thicknesses are possible, but the most important ones are $\lambda/4$–$\lambda/4$–$\lambda/4$ and $\lambda/4$–$\lambda/2$–$\lambda/4$ triple-layers. Minimum reflectance for a $\lambda/4$–$\lambda/4$–$\lambda/4$ combination is achieved for the index ratios:

$$\frac{n_1 \cdot n_3}{n_2} = \sqrt{n_0 \cdot n_s}. \tag{6.33}$$

As in the case of a $\lambda/4$–$\lambda/4$ double-layer coating, the spectral reflectance of this triple-layer coating also has a V-type profile, however, with lower values and a broader width (Figure 6.48a). A W-type shape is achieved with the $\lambda/4$–$\lambda/2$–$\lambda/4$ triple-layers under the condition for best index matching with light incident from air:

$$\frac{n_3}{n_1} = \sqrt{n_s}. \tag{6.34}$$

An example for this $\lambda/4$–$\lambda/2$–$\lambda/4$ triple layer is shown in Figure 6.48a in comparison to the quarter wave triple layer and a triple layer where the individual thicknesses have been optimized numerically and are neither of $\lambda/4$ nor $\lambda/2$ thickness. All these three triple layers are optimized for BK7 glass with its moderate refractive index. We can see from their spectral behavior that a low reflectance with $\rho_P \leq 0.4\,\%$ in the visible range from about 420 nm to 660 nm can be obtained. The reflectance values for higher index glasses tend to be still lower than for BK7 in that range.

Whereas the reflectance for normal incidence is low, oblique light incidence leads to more significant reflections as can be seen from Figure 6.48b. In this diagram, the spectral dependence of ρ_P is illustrated for a $\lambda/4$–$\lambda/2$–$\lambda/4$ triple-layer at different angles of incidence. The curves for normal incidence and for an angle of 15° are almost identical.

For larger angles, the W-profile of the reflectance curve shifts to shorter wavelengths, and thus a stronger increase is effective especially in the red visible light spectrum.

Although triple-layer coatings have already low reflectance, they may be not sufficient for superior quality of complex lens combinations. The reflectance can be reduced further by adding more layers according to the scheme presented in Figure 6.45d. With increasing number of layers, there is more freedom in the choice of the layer parameters to optimize the transmittance. The combination of layers then becomes more and more complex, and a simple calculation is no longer possible. However, a mathematical optimization before technical realization is mandatory and requires efficient methods. In general, the transmission and reflection characteristics of a layer can be formulated using transfer functions or matrices incorporating complex numbers to describe amplitudes and phases of the electromagnetic fields. For instance, the matrices establish a relation between input and output interfaces of a layer and can be applied in a sequential way to calculate the behavior of the total layer combination [Ped08, Hec16]. Due to the large number of parameters, all types of multilayers are computed numerically and require powerful computers. Two ways of designing the layer sequence turned out to be quite interesting: The first is to have a sequence of layers with alternating high- and low-refractive indices and in general also with varying thickness. The second one is to establish a nearly graded index coating consisting of a large number of very thin layers with continuously increasing index from the lowest index at the air side to the highest index at the glass side. The highest layer index, however, must not exceed the glass index. The ideal would be to achieve a kind of adiabatic tapering of the optical impedance for light waves. The optical impedance, which is the ratio of the electric by the magnetic field of the light wave, is inversely proportional to the refractive index of the medium in which light propagates. A graded index matching would thus ensure that no light is reflected back during propagation.

The reflectance of a multilayer coating consisting of seven single layers on BK7 glass is shown in Figure 6.49a. The reflectance is still inferior to that of a triple-layer coating and can be estimated with about $\rho_P \leq 0.2\,\%$ for the visible spectral range between about 400 nm and 650 nm for normal incidence. The curve is more flat and has a larger bandwidth especially to the blue/violet spectrum of light. As for the angle of incidence, the curve for 15° angle of incidence is nearly identical to that for normal incidence and all reflectance values at larger angles are significantly lower than those of triple-layer coatings. However, with increasing angle the reflectance increases in the red spectral range as for all other types of coatings. A drawback of coatings with increasing numbers of layers is that variations in the layer thickness influence much stronger the overall behavior than in the case of a single CL. Thus, a high precision control for the manufacturing process of multilayer coatings is required.

For so long, we only considered the visible spectral range for which the coatings have been optimized. It can be seen in Figure 6.49b that outside this range in the ultraviolet as well as in the near infrared range a drastic increase of ρ_P takes place. The reflectance becomes stronger outside the visible range when more layers have been

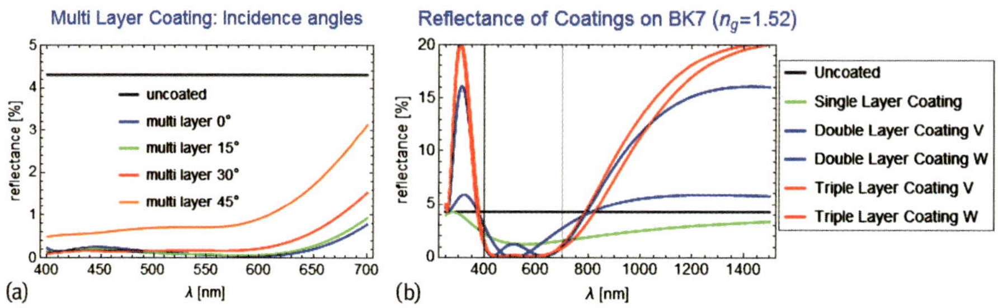

Fig. 6.49: Reflectance as a function of the wavelength for coatings on BK7 glass ($n_s = 1.52$). (a) Multilayer coating consisting of 7 single layers, curves for different angles of incidence; (b) reflectance outside the optimal visible range for different types of coatings [Bla16]. Reprinted with kind permission of Carl Zeiss AG.

combined. It achieves even a multiple of that of an uncoated substrate for triple-layer coatings. Therefore, care should be taken if glass substrates like lenses are to be used in other spectral ranges than for which they are specified. The use of photographic lenses in the UV-, respectively, IR-range may become very critical.

A modern technological development can be seen in the field of nanoporous films as coatings for mineral glasses as well as for polymers. Their principle is that due to very small solid particles of the order of some 10 nm sizes, fixed in a thin matrix, the overall refractive index can be significantly reduced to values of the order of 1.2 or even below. The refractive index can be controlled by the porosity of the material, for instance, the density and type of particles like nanorods or nanowedges.[3] By this approach, even single layer coatings for mineral glasses or graded index coatings may become possible. Some manufacturers of lenses and optical components use this technology and achieve, according to their own account, better results than using their multicoating layer technology.

6.8.2 Lens flares

Lens flares are the consequence of internal reflections between glass-air interfaces in lens combinations and between further surfaces of sensors and housing parts. They become manifest for instance as ghost images, straylight haze, background veiling, and are often termed generally as veiling glare. All these unintended effects falsify the images of a scenery since they would be absent if the lenses were perfectly transparent and without any internal reflections. They not only establish a kind of noisy background and,

3 J.-Q. Xi, Jong Kyu Kim, E. F. Schubert, Dexian Ye, T.-M. Lu, Shawn-Yu Lin, Jasbir S. Juneja: *Very low-refractive-index optical thin films consisting of an array of SiO_2 nanorods*, Optics Letters, **31**(5) (2006) 601–603.

therefore reduce the overall contrast but are the source for spurious ghost images as is demonstrated in some image examples below. Furthermore, they may be a cause that the modulation transfer function MTF degrades.

6.8.2.1 Double reflections

When light enters an optical system, the ray path splits up into a reflected and transmitted part each time it strikes the interface between two different media, which in our consideration is the air-glass surface. Figure 6.50a illustrates this for a single thick lens. The transmitted part of a ray emerging in the object space is imaged in the sensor plane and contributes to the desired illuminance of the image. The larger its part relative to an adjacent unexposed area, the higher the contrast. The first reflection at surface 1 is oriented away from the lens and does not enter the image plane. The reflected part at surface 2 is directed back to surface 1 and its transmitted part there is equally lost, but its second reflection at surface 1 is headed to surface 2. If it transmits the surface 2, it enters the image space as a 2-times reflected ray (2×). Following the paths of different rays, we see that only rays of even numbers of reflections, like 2×, 4×, 6×, etc., can contribute to the image formation whereas rays of odd numbers of reflection are scattered outside the optical system and need no longer be considered. With every reflection, the ray power decreases. A double reflected ray at a multilayer coated glass surface with a reflectance of $\rho_P = 0.2\,\%$ has only $4 \cdot 10^{-6}$ of the ray's initial power. If we assume that the reflected ray is imaged to a spot of the same size as the initial ray, we have a decrease in illuminance equally by a factor of $4 \cdot 10^{-6}$ compared to its initial value, which corresponds to a brightness change of −18 EV after Equation (2.22). For comparison, assuming $\rho_P = 5\,\%$ for a noncoated lenses, the double-reflected brightness reduction amounts to 0.25 % or −8.6 EV, whereas with $\rho_P = 1\,\%$ for a lens with single layer coating the corresponding reductions are 0.01 % or −13.3 EV.

The number of possible double reflections in lens combinations can be calculated if we consider Figure 6.50b. Here, we have two separate lenses with a total of four reflecting surfaces. Double reflected light in the image space can only originate from surface pairs. If we start with the 4th surface, double reflections originate from the surface pairs

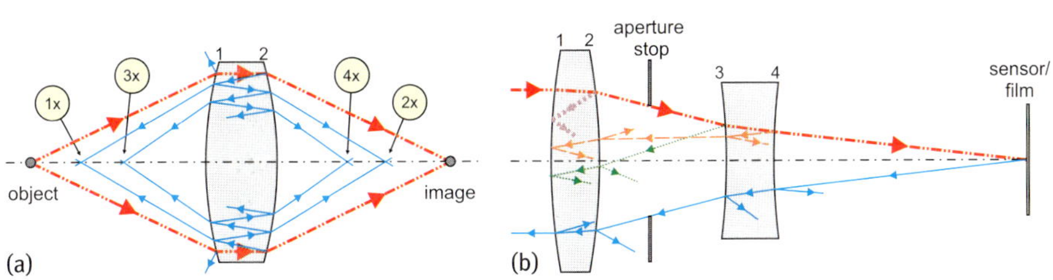

Fig. 6.50: Ray paths for multiple reflections in optical systems. (a) Odd and even reflections in a converging thick lens; (b) paths for double reflections between glass-glass (broken lines) and glass-sensor surface pairs (full lines, blue color) [Bla16].

(4|3), (4|2) and (4|1), which make 4 minus 1 reflections. Additional double reflections incorporating the 3rd surface come from the pairs (3|2), (3|1), which make 3 minus 1 reflections, and incorporating the 2nd surface we have only the additional contribution from the pair (2|1). The total sum $N_{2R,GG}$ of possible double reflections at lenses with m optical glass-air interfaces is given by

$$N_{2R,GG} = (m-1) + \cdots + 2 + 1 = \frac{m \cdot (m-1)}{2}. \tag{6.35}$$

A critical point when taking pictures are the internal reflections at the film and sensor surfaces. They are exploited in SLR or rangefinder cameras for measuring the total exposure during the imaging process by a nearby photo detector when the shutter is open (see Figure 2.21). These sensor reflections, however, may produce ghost images when they are back-reflected at the glass surfaces and strike the sensor again. If we include the sensor surface in our consideration, we get additionally m double reflections between the sensor and each of the m glass surfaces, the total sum of the double reflections $N_{2R,\text{total}}$ yields

$$N_{2R,\text{total}} = m + (m-1) + \cdots + 2 + 1 = \frac{(m+1) \cdot m}{2}. \tag{6.36}$$

The reflectance from the sensor surface is comparable to that of an uncoated glass surface. Its value depends on the wavelength and varies with the type of film or sensor. Typical values can be assumed to be of 5 % in the visible range [Bla16]. A double reflection between the sensor and a multilayer coated glass surface leads to a spurious ray of a brightness reduced by a factor 0.01 % or −13.3 EV compared to the initial ray. Double reflections between a sensor and an uncoated glass surface are comparable to those between uncoated glass surfaces and amount to 0.25 % or −8.6 EV. Also, double reflections between sensor and single-layer coated glasses are critical with values of 0.05 % or −11 EV.

As for fourfold reflections, their brightness is always about 10^{-4} lower than double reflections taking into consideration reflections at the sensor surface and coated glass. The number of fourfold reflections is proportional to m^4 with m being the number of glass surfaces. However, due to the low brightness values compared to double reflections and the original ray power, and also due to widening up, the overall intensity of fourfold reflections can be neglected when analyzing the structure of internal lens flares and the source is not extremely bright. Nevertheless, the higher-order reflections contribute to the overall stray light background, which is more or less continuous as compared to the more structured ghost flares due to the double reflections.

For the estimation of the multireflected intensities, we did not explicitly take into account the brightness change in the transmitted ray. When a ray traverses a glass surface its power decreases due to the transmittance $\tau_P = 1 - \rho_P$, which is larger the lower the reflectance is. After passing m glass surfaces and assuming the same reflectance for

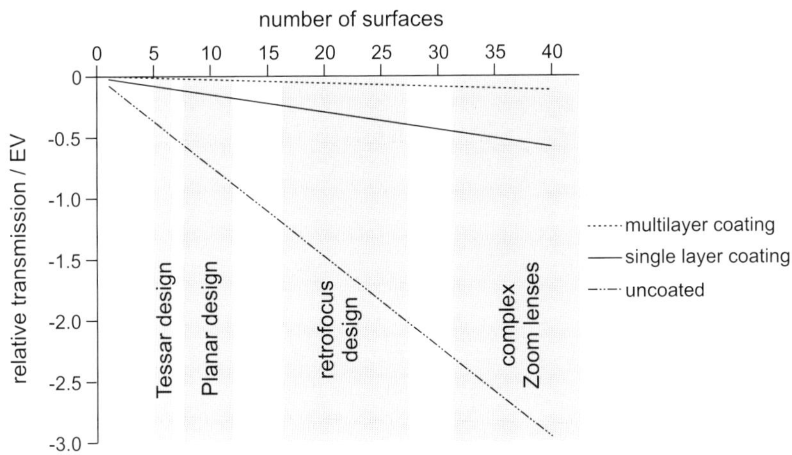

Fig. 6.51: Relative total lens transmission in EV as a function of the number of reflecting surfaces [Bla16].

all surfaces, the overall transmittance of a ray is proportional to $(1 - \rho_p)^m$. For complex lenses, thus the intensity of a ray in the image plane decreases with the number of glass-air surfaces. The brightness change of a beam relative to its initial value is shown in Figure 6.51 as a function of the number of glass surfaces. The change is expressed in EV and increases strongly for lenses without antireflection coating. Besides the internal lens flares, this brightness decrease in the image plane has been a major reason that only less complex lenses like the Tessar or Planar type were realized before 1930. Modern lenses with their retrofocus design, like the Zeiss Distagon type, or complex zoom lenses with more than about 30 glass-air surfaces require a multilayer coating for bright images of high contrast.

6.8.2.2 Structured ghost flares and stray light haze

As described above, the structured ghost flares in a lens are especially due to the double reflections between glass-glass and glass-sensor surfaces. In order to better demonstrate the need for antireflection treatment and the effect of lens flares, we will consider test photographs taken by a modern Zeiss Distagon 2.8/21 with its multilayer coating in comparison to a special variant of it manufactured without any antireflection coating of the glass surfaces (Figure 6.52) [Bla16]. The lens features a retrofocus design and consists of 29 refracting glass surfaces of which 26 are in contact with air (Figure 6.26).

In Figure 6.52a, we can see the ghost images produced by an intense light source in the object field. Whereas the ghost images are not visible in the coated lens (left side), they are clearly seen in the lens without coated glass surfaces. All ghost images of different sizes are lined up on a straight line from the nearly point-like flame of the candle as the primary light origin through the center of the image field. The ghost images in this example can be considered a good approximation of a point-spread function (PSF)

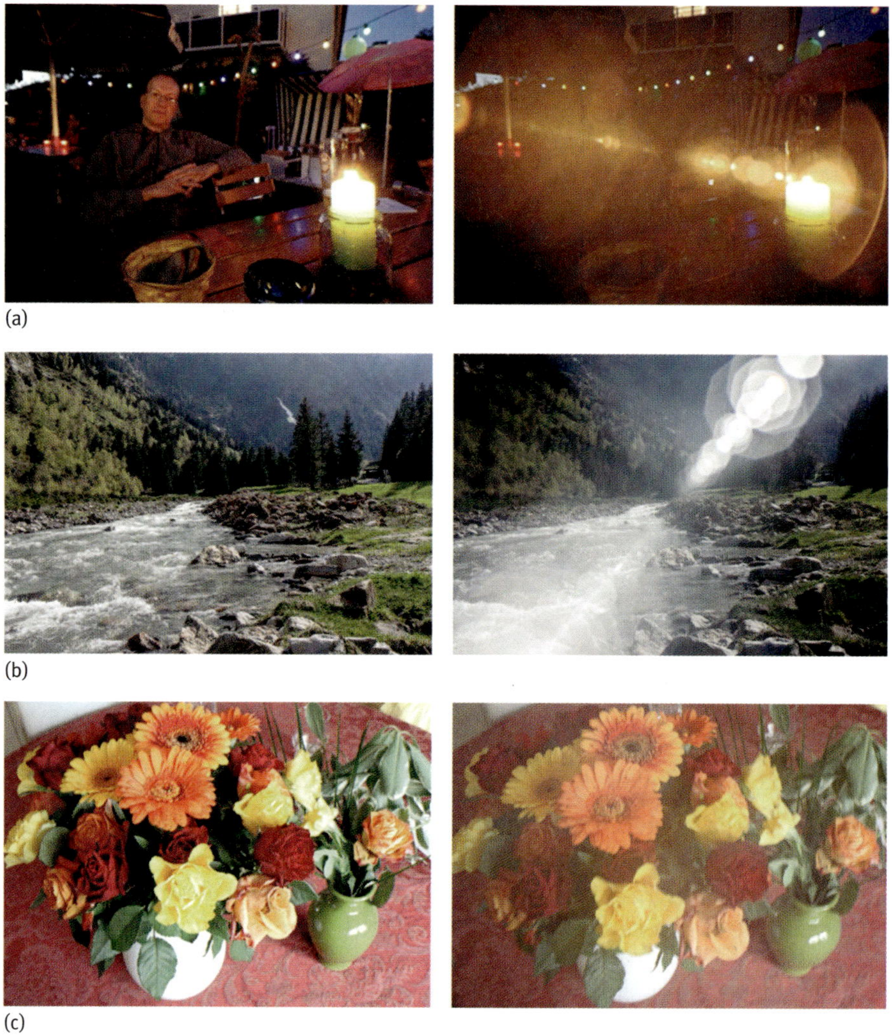

Fig. 6.52: Pictures taken by a Zeiss Distagon T*2.8/21 mm ZE lens with high-quality multilayer coating (left) as compared to pictures taken with the related uncoated demonstration lens [Bla16]. Note that with coating, the MTF is much better. This leads to better contrast, more concentrated PSF, etc. Reprinted with kind permission of Carl Zeiss AG.

of the lens *for that particular point* in the object space. The according PSF can be calculated based on the double reflections between all reflecting surfaces in the lens as expressed by Equation (6.36). In the case of defocusing, the corresponding PSF is discussed in Chapter 5 and depicted in Figure 5.25. The defocused images may also show up as blurred images of the iris stop (see below and Section 6.9.3).

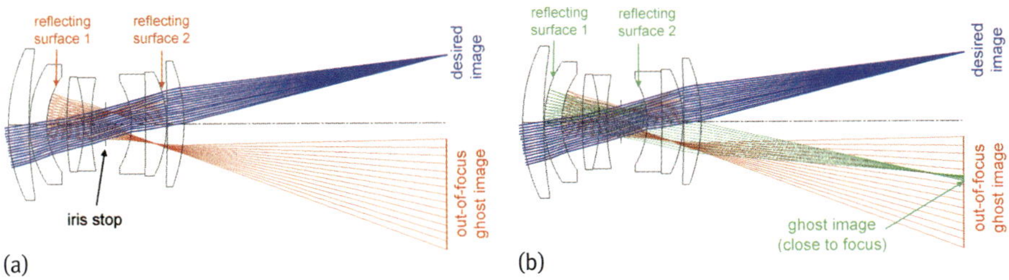

Fig. 6.53: Optical paths of rays after double reflections at surface pairs. (a) Formation of defocused ghost images shaped by the iris stop; (b) Formation of nearly sharp ghost images [Bla16]. Reprinted with kind permission of Carl Zeiss AG.

For the examples of the Distagon lens with $m = 26$, we thus expect 351 different ghost image reflections. The formation of one ghost reflection between two selected glass surfaces is illustrated in each of the Figures 6.53a,b for the example of a Zeiss Planar-type lens. The pair of reflecting surfaces is indicated as reflecting surface 1 or reflecting surface 2. The incoming light, represented by a bundle of individual rays from the object point, is imaged sharp to the desired image point. The first internal reflection occurs at surface 2 and the reflected partial rays are reflected for the second time at surface 1 back toward the image space. Considering the surface pair in Figure 6.53a, these double reflected rays converge in the center of the second lens group and then diverge in the image space. The corresponding ghost image on the sensor is out of focus and the cross-section of the blurred image is influenced by the iris stop, which has shaped the light bundle. Figure 6.53b represents the ghost image formation for the same incident ray bundle but after the double reflection at a different pair of surfaces, and thus for a different ray path. The partial rays after the reflection at surface 2 and then at surface 1 converge very close to the image sensor. After that, the bundle slightly diverges to strike the image plane forming a nearly sharp image spot.

All the different ray paths due to the different surface pairs can be calculated by numerical ray-tracing methods to establish the whole light distribution in the image plane generated by a bright light source. As we can see in Figure 6.53, the ray paths do not change when the lens is rotated along its optical axis and all ghost images are located with the desired image in a tangential or meridional plane, and thus containing the optical axis. The consequence is that all ghost images observed in the sensor plane, either sharp or out of focus, are lined up in a straight line across the center of the image as mentioned above.

The extent of the calculation for the light distribution in the image plane can be estimated for the example of the Distagon lens with its 26 glass to air surfaces and the sensor. Having 351 surface pair combinations and assuming a light bundle of 300×300 individual rays emerging from one object position of a point light source, then 32 million ray path calculations are necessary for one color of light or nearly 100 million calculations for three wavelengths simulating white light. Figure 6.54c shows the real image of

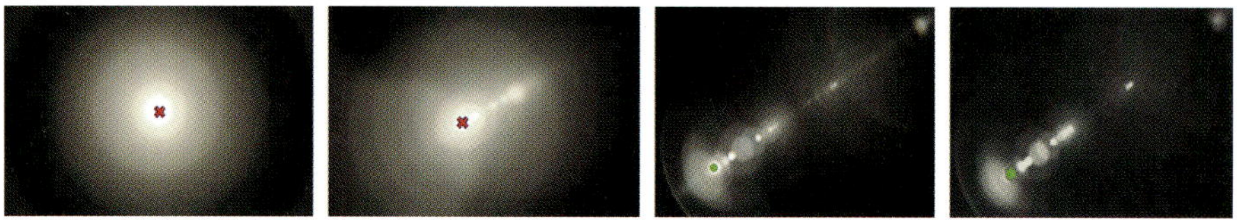

(a) (b) (c) (d)

Fig. 6.54: Ghost images of point light sources at different positions in the object space; position of the source marked by red crosses, respectively, green dots; (a) light source in the center; (b) light source slightly off center; (c) real sensor image with the light source far off center; (d) numerical simulation of the point spread function for the corresponding conditions as in image (c) [Bla16]. Reprinted with kind permission of Carl Zeiss AG.

all reflections on the sensor generated by the point light source seen down left in the image compared to the numerically calculated light distribution in Figure 6.54d [Bla16]. A good match for the most intense reflections between the experimental and calculated data can be observed. Thus, the numerical simulation of the light distribution in the image plane is a powerful method for optimizing the reflection behavior of a lens. The example is valid only for one location of the light source in the object space. In order to fully characterize a lens, more radial locations of the light source at different distances are required as well as a variation of the f-number. For instance, three radial positions and five object distances lead to about 1.4 billion computations of a prime lens with fixed focal length for one f-number. The computation of a zoom lens has to be multiplied by the number of focal lengths to be investigated. Therefore, powerful computer systems are necessary, which currently (2018) do such an analysis within approximately one day.

If we analyze the ghost images, we come to the conclusion that a bright image source within the object field tends to produce some sharp but also blurred images (Figures 6.52a, 6.54). Their number increases with the number of reflecting surfaces in the lens, and all reflections are located on a straight line from the source image through the center of the image field. The position of the light source is important. A similar effect can also be seen in Figure 4.34c.

If the light source is in the center, all ghost flares are rotational symmetric (Figure 6.54a). This may be critical in situations where lenses are used to determine the spot size of a light source. Since the superimposing reflections tend to deform its true shape and increase its size, the result may be falsified by these reflections. Moving the light source off center generates the distribution of ghost flares along a line and the shape of the ghost spots gets continuously deformed (Figure 6.54b).

Even if the light source is outside the object field and not imaged, the ghost flares exist in the image field (Figure 6.52b), although in a lower number. The position of the sun causing the flares in this image is indicated by the line from the image center along the chain of reflections. It should also be noted that in this figure the lens was stopped down and the shape of the iris stop can be observed in the defocused ghost images. In all images, the defocused ghost flares contribute to the stray light background and reduce the image contrast. Especially in Figure 6.52c, where no clear ghost reflection can be identified due to the lack of strong light sources, the stray light generates a nearly constant haze all over the image field.

Internal reflections also depend on the f-number of the lens as mentioned above. Unlike with many lens aberrations, where stopping down helps to improve the image quality, a general tendency cannot be given. There may be situations where stopping down is even more detrimental. Ghost images that are sharp near the location of the aperture stop may not be largely influenced then and remain while the overall illuminance in the image plane decreases. Hence, the contrast decreases. Conversely, stopping down could block reflections to reach the image plane. Ghost flares, apart from being disturbing, may be used in a beneficial way for artistic image design.

Another internal source for veiling glare might also be the diffraction of light at the blades of the iris aperture stop. Stopping down in connection with oversaturated imaging of the light source may lead to starburst effects, which become manifest in the image as stars with rays, the number of which depend on the structure of the aperture stop (see Section 6.9.3.2).

6.8.3 T-stop

The illuminance of light that is incident to the optical sensor in the image plane can be controlled by the aperture stop in the lens. According to Equation (2.15), the illuminance E_i is inversely proportional to the square of the f-number $f_\#$. The latter is defined as the ratio between the focal length and the diameter of the entrance pupil (Equation (2.14)) and, in a strict sense, is valid only for imaging of objects at infinite distance (see also Section 3.4.3). Nevertheless, it characterizes the brightness of a lens as for the exposure on the image sensor. It should be noted that $f_\#$ only takes into account the geometric dimensions of f and D_{en} but not the losses of light due to absorption and reflection when passing through the complete lens as described in the previous sections. In cinematographic applications, slight changes in exposure parameters for movies might become critical, which is usually not an issue for still-photography. Hence, there is a demand to characterize the transmitting light efficiency of a lens. This leads to the description by the T-stop parameter $T_\#$. The latter incorporates the power transmission coefficient τ_{PL} of the complete lens, which for high quality lenses without inbuilt filters is close to 1. We define the T-stop in relation to the f-number:

$$T_\# = \frac{f_\#}{\sqrt{\tau_{PL}}} \tag{6.37}$$

For a lens without internal losses, i. e., with $\tau_{PL} = 1$, $f_\#$ and $T_\#$ are identical. The T-stop number is always larger than the f-number, as the power transmission coefficient is inferior to 1. Let us, for instance, have a look at the Sony FE 100 mm F2.8 STF GM OSS.

This is a portrait lens for the Sony E-mount system with a special technology to produce high quality bokeh.[4] Its maximum aperture is given by the manufacturer as F/2.8 (T5.6), its minimum aperture as F/20 (T22). This means that the lens has the characteristics of one with $f_\# = 2.8$ at full aperture, but the light transmission is of only 25 % resulting in $T_\# = 5.6$. At its minimum aperture, the light transmission is of nearly 83 %. Usually, motion picture lenses are calibrated in T-stops rather than using f-numbers for still photography.[5]

6.9 Depth of focus, depth of field and bokeh

Depth of field and the related depth of focus are quantities that can be influenced by camera lens settings like f-number, focal length and object distance. There are different purposes to increase or reduce the depth, for instance, to get a sharp image over a large object distance and be independent on the exact focus setting. Conversely, the reduction of the depth may be intended to select a small range in the object distance and render foreground and background of a given scenery defocused. This is often done from an artistic point of view and is discussed under the topic of bokeh. The basic principles of depth of field and depth of focus have been described in Section 3.4.6. In the following sections, we present some practical examples and implications for setting the lens parameters. Moreover, the section of bokeh gives examples how details of the lens properties can be revealed by the defocused images shaped by the iris stop.

6.9.1 Depth of focus

Depth of focus is a parameter characterizing the range of sharp imaging in the image space around the sensor location (Figures 3.40, 6.55). According to Equation (3.120), this depth only depends on the f-number and the diameter of the circle of confusion u_i with the total depth of focus s_{DOFoc} being equal to $2 \cdot u_i \cdot f_\#$. When simply inspecting a printed image by the human eye with the print being at a distance of its diagonal, we assumed that the eye only resolves details larger than about 1/1500 of the diagonal. For instance, for a 5× magnification of the 35 mm format (24 mm × 36 mm) we get a print of 12 cm × 18 cm, which is viewed from 21.6 cm. Then details of about 0.14 mm can still be resolved by the eye, which is equivalent to a circle of confusion of about 30 μm on the sensor or film (see also examples given in Table 5.2 in Chapter 5). The same relationship is valid

4 https://electronics.sony.com/imaging/lenses/full-frame-e-mount/p/sel100f28gm; retrieved 02-Feb-2023.

5 "Kodak Motion Picture Camera Films" (https://web.archive.org/web/20021002095739/http://www.kodak.com/US/en/motion/support/h2/intro01P.shtml), archived from the original web page on 2002-10-02, retrieved 02-Feb-2023.

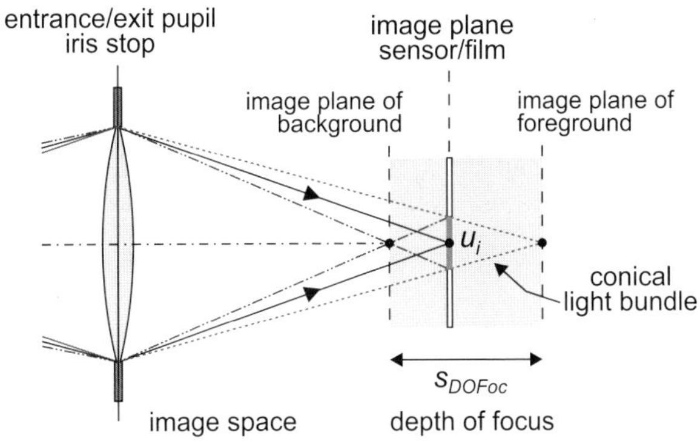

Fig. 6.55: Depth of focus in the image space with location of background, respectively, foreground images; u_i is the diameter of the circle of confusion in the image plane.

for larger images like poster formats, viewed from a correspondingly larger distance, when the natural viewing perspective should be preserved. If the photo is taken using a wide-angle lens, the image perspective has a larger angle than the natural viewing with a total angle of view of about 47°. Then, in order to have a nearly natural viewing impression, the observer has to step closer to the image, which requires a better resolution on the print respectively image sensor. Therefore, wide-angle lenses are designed by the manufacturers with higher requirements like 1/3000 of the diagonal or correspondingly 15 µm for lenses for the 35 mm format [Nas10]. From that consideration, it is clear that the circle of confusion does not have a fixed value but is always related to the image format and the type of application. The depth of focus influences the tolerance for the manufacturing of the camera lens and the lens should be mounted with a correspondingly adequate precision. For the 35 mm format with $f_\# = 1.4$ and $u_\text{i} = 15\,\mu\text{m}$, we get $s_\text{DOFoc} \approx 42\,\mu\text{m}$ for a high-quality lens and a more relaxed value of 168 µm for $f_\# = 2.8$ and $u_\text{i} = 30\,\mu\text{m}$. Lenses for the APS-C format have stricter requirements by a factor of about 1.5, which is equal to the crop factor *CF*. A mobile phone camera like that of Apple iPhone 7 Plus has a depth of focus of only 14 µm with $f_\# = 1.8$ and $u_\text{i} = 4\,\mu\text{m}$ as 1/1500 of its sensor diagonal. Thus, small format camera lenses need a higher manufacturing precision. It should be noted again that the depth of focus is virtually independent from the focal length of the lens whereas for the depth of field in the object space the situation is quite different.

6.9.2 Depth of field

According to the derivation in Section 3.4.6, the depth of field s_DOF, which characterizes the sharply imaged range in the object space, is a complex function of the focal length, the f-number, the circle of confusion and the object distance. If the lens is adjusted to a given object distance a_o, the near-point distance a_n and far-point distance a_f are not symmetrically located relative to a_o. The far point in general reaches to much longer distances than the near point. In order to investigate how the depth of field varies with the focal length or sensor format, we consider the hyperfocal distance a_{hf} given by Equation (3.117). This is the object distance to which the camera lens has to be set in order to have the far point located at infinity. The near point a_n is then at half of the distance from the lens center to a_{hf}. The depth of field then reaches from the near point to infinity. According to $|a_{hf}| \approx f_i^2/(u_i \cdot f_\#)$ and $a_n = a_{hf}/2$, we can assert that for given u_i and $f_\#$ the distances a_{hf} and a_n increase and the corresponding points are shifted further away from the lens if the focal length f_i of the lens increases. The far point remains at infinity. This implies that the perspective impression of an image changes as objects in the foreground become defocused in the image for longer focal lengths. As camera lenses with a given mount can be attached to camera bodies with different sensor formats, the question arises how the perspective changes when the same lens is used with different sensors. On the other hand, does the visual impression remain the same if for different image formats always the corresponding normal lens, and thus the same object field is imaged? To be more specific, do objects in the foreground become more or less blurred if a different sensor format is used but the camera settings like f-number or object distance are the same? For that purpose, we consider different situations.

6.9.2.1 Same lens used with different image formats

The natural viewing perspective is achieved when a camera of a given sensor format uses the corresponding normal lens. If the focal length f_i of the lens is longer than that of the normal lens, termed f_{norm}, the lens can be considered a long-focus lens for that format with a different perspective. Let us express the focal length f_i of a lens by its relative magnification factor M_{rel} to its normal lens according to Equation (2.11):

$$f_i = M_{rel} \cdot f_{norm}. \tag{6.38}$$

In many cases, the lenses of different formats are compared to the 35 mm format or full format as a reference. Therefore, we start our consideration for the full format. The hyperfocal distance for a full format lens with the normal focal length $f_{norm,FF}$ can then be expressed by

$$|a_{hf}| \approx \frac{f_i^2}{u_i \cdot f_\#} = \frac{M_{rel,FF}^2 \cdot f_{norm,FF}^2}{u_i \cdot f_\#}. \tag{6.39}$$

If the same lens will be used with a crop format sensor we should rewrite this equation using Equation (2.25) for the definition of the crop factor CF, which relates the diagonals or the normal lenses of any format to the full format:

$$|a_{hf}| = \frac{M_{rel,FF}^2 \cdot f_{norm,FF}}{f_\#} \cdot \frac{f_{norm} \cdot CF}{u_i}. \tag{6.40}$$

As discussed above, the diameter u_i of the circle of confusion should be determined relative to the diagonal d of the image format. If we compare images taken by different format cameras, the ratio d/u_i should always be same. Thus, the ratio f_{norm}/u_i should also be a constant for all formats. As a consequence, if we use the same lens for different formats, we can rewrite Equation (6.40) by subsuming $M_{rel,FF}$, $f_{norm,FF}$, and f_{norm}/u_i in a constant:

$$|a_{hf}| = \text{const} \cdot \frac{CF}{f_\#}. \tag{6.41}$$

This implies that the hyperfocal distance changes when the same lens is used by cameras of different formats, characterized by CF.

Let us consider the following example: A full format lens with $f_i = 50$ mm is used with a full format camera sensor at $f_\# = 8$. We assume for the circle of confusion $u_i = d/1500 = f_{norm,FF}/1500 \approx 30\ \mu$m. Then we get, using Equations (3.116) and (3.117), the hyperfocal distance $|a_{hf}| = 10.4$ m and the near distance $|a_n| = 5.2$ m. It should be noted that we use the absolute values as the object distances are counted negative due to our convention. If we view an image print at the distance of its diagonal, objects on the image are perceived sharp in the range from 5.4 m to infinity. As a next step, we mount the same lens to an APS-C camera, which has a crop factor of $CF \approx 1.5$. For the reduced format diagonal of the crop sensor, we have a smaller circle of confusion, namely $30\ \mu\text{m}/1.5 = 20\ \mu$m. Then we get $|a_{hf}| = 15.6$ m and the near distance $|a_n| = 7.8$ m. If we view the print of that image in the same size as that of the full format camera, we get a slightly magnified image as the 50 mm lens acts like a moderate portrait lens for the APS-C format and the scenery on the image is perceived sharp for objects in the range from 7.8 m to infinity (see also Figure 2.24). This is the same result as expressed by Equation (6.41), namely that the hyperfocal distance, and thus the near point of a given lens mounted on different formats increases by the crop factor CF if $f_\#$ is kept fixed.

A comparison between the images taken by a 50 mm full format lens with different sensor formats is given in Figure 6.56. Images a) and b) are taken with a Nikon FX full format sensor, images c) and d) with a Nikon DX crop format sensor with $CF = 1.5$. As parameters for the photographs, we used for the circle of confusion $u_i = 30\ \mu$m (FX) and $u_i = 20\ \mu$m (DX), object distance $a_o = -45$ cm and $M = -0.125$. The f-number is indicated in the images. All images are sections of a larger format and, therefore, the optimal viewing distance to assess the depth of field is three times the print diagonal for any format. The near and far points have been calculated according to Equations (3.107) and (3.108); their difference yields the depth of field s_{DOF} listed in the corresponding images. The cm ruler on the image is viewed under an angle of 45°, thus the sharp perceived range on its scale is related to the depth of field s_{DOF} by $s_{DOF} \cdot \sqrt{2}$. The near and far points in the images have been correspondingly converted to the ruler scale and are indicated in the images with the 90 cm mark as the focus point. In all images, the distance from the focus point to the far point is slightly larger than the distance from the focus point to the near point. In our examples with relatively close object distance, however, this

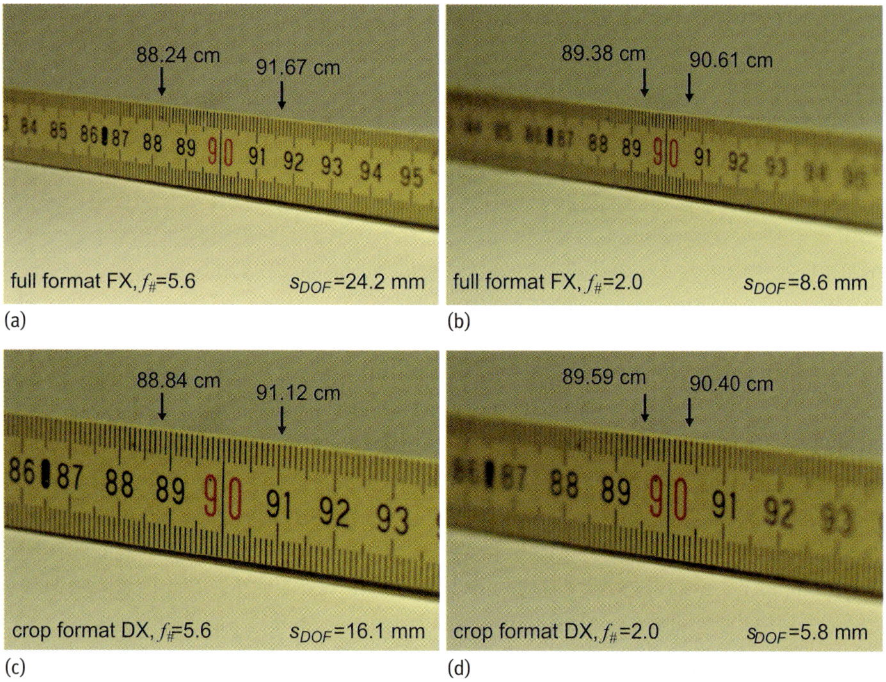

Fig. 6.56: Comparison of depth of fields for images taken with a full format 50 mm lens with different f-numbers and sensor formats. The near and far points have been calculated and converted to the ruler scale and are indicated by arrows. The values are based on a viewing distance of three times the image diagonal.

asymmetry is very small and virtually impossible to perceive visually. It can be clearly seen that stopping the aperture down from 2.0 to 5.6 by three stops leads to a significant increase of the depth of field. The 50 mm lens, being a normal lens for the FX format, acts as a moderate long-focus lens with the DX format. The resulting 1.5× magnification generates a shallower depth of field than with the full format. When viewing the image, this shallower depth is compensated by the image magnification for the crop format and, therefore, the corresponding far and near points are located at the same positions in the prints for the DX and FX formats if the f-number is not changed. It should be stressed, however, that the depth of field in the object space is different for the different formats, and thus the impression and the bokeh, for instance, in portrait photography, is also different.

6.9.2.2 Same object field with different image formats

Let us now consider the situation that images are taken by different cameras with each having its normal lens mounted to the camera body. For a normal lens, the total angle of view is about 47° and independent from the sensor diagonal. As a consequence, images taken by a full format camera and its normal lens show the same object field and have the same perspective as images taken by an APS-C camera with its normal lens. As for the hyperfocal distance using the normal lens, we get according to Equation (3.117) for the normal lens:

$$|a_{\mathrm{hf}}| \approx \frac{f_{\mathrm{norm}}^2}{u_{\mathrm{i}} \cdot f_{\#}} = \frac{f_{\mathrm{norm}}}{u_{\mathrm{i}}} \cdot \frac{f_{\mathrm{norm}}}{f_{\#}} = \mathrm{const} \cdot \frac{f_{\mathrm{norm}}}{f_{\#}}. \tag{6.42}$$

Like in the consideration above the ratio $f_{\mathrm{norm}}/u_{\mathrm{i}}$ is the same for all formats and can be substituted by a constant number in Equation (6.42). As the normal focal length for a full format camera is longer than for an APS-C camera, we see that the hyperfocal distance, and thus also the near point is farther away than for the APS-C camera. The depth of field using an APS-C camera is always larger than that of a full format camera if the same object field is imaged. This is very important, for instance, for portrait photography when only the facial part of a person should be imaged sharp. The larger the format of a camera is, the shorter the depth of field and the more selective the sharp section of the image. This is the reason why medium or full format cameras are very favorable for portrait photography. In order to achieve the same depth of field for cameras of different formats, the ratio between normal focal length and f-number, namely $f_{\mathrm{norm}}/f_{\#}$, must be the same. As a consequence, taking portrait photographs using an APS-C camera requires a further reduction of the f-number by the crop factor compared to the full format. This corresponds to approximately one stop value. Conversely, in order to achieve the same large depth of field of an APS-C camera when using a full format camera, the equivalent f-number must be larger by multiplication with the crop factor, correspondingly stopping down one value.

After these considerations, we can state that larger format cameras have larger depths of focus, and thus larger tolerances in the image space due to their large circles of confusion. Conversely, in the object space their depths of field are shallower than that of smaller format cameras. Therefore, miniature cameras require a high precision during manufacturing. Due to their generally large depth of field, it is possible to design them as fixed focus cameras without needing to adjust the focal position. A last point, which should be stressed here once again, is that the depth of field must always be seen in conjunction with the viewing condition, and thus with the circle of confusion. If for instance the image of a fixed focus camera appears sharp for objects at "infinity" when viewing the image print at the distance of its diagonal, infinitely distant objects may be rendered blurred when blowing up the print and viewing it at close proximity. In the latter case, the depth of field becomes clearly reduced as the perceivable circle of confusion has become smaller. Therefore, the purpose of the image, or how an image is presented, should be taken into account when calculating the best aperture settings for a desired depth of field for taking images.

6.9.3 Bokeh and starburst effect

The depth of field can be mainly influenced by the aperture stop but also by the focal length and object distance as described above in order to select the sharp imaged range in the object space. Objects outside the depth of field range are imaged with a blur that continuously increases with its distance to the optimum focus setting. There is no sharp transition between the focused and defocused areas, and the frontier between them is fixed by the definition of the circle of confusion and the resolving power of the human eye. The interplay between focused and defocused image parts can be intentionally deployed to draw the attention of an observer to certain areas with the defocused parts becoming a harmonious supplement to the scenery. This interplay often has an artistic character and is usually determined by the Japanese term in English transcription “bokeh.” Bokeh can be translated as “blurred” or “confused” [Nas10]. It is strongly influenced by the size as well as the type of the aperture stop. Another effect, which is strongly influenced by size and type of the aperture, is the starburst effect, which manifests itself in star-like appearances of bright light sources. Unlike bokeh, it is not restricted to unsharp image parts and becomes stronger with increasing f-numbers. Both effects will be discussed in the following sections.

6.9.3.1 Bokeh

Due to its often artistic or aesthetic properties, the classification of bokeh is of rather individual nature and cannot be assessed in an objective way. Bokeh is influenced by the quantity as well as the quality of the blurred image parts. As for the quality, the blurred image parts are shaped by the iris aperture stop of the lens as already discussed in the ghost images of the lens flares. Image points in the foreground of the depth of field are imaged sharp behind the sensor, and thus the corresponding conical light bundle intersects the sensor and generates a conical section with it (Figure 6.55). The blurred images of point-like light sources especially represent a kind of footprint of the iris stop and the lens construction and only have a round shape for fully opened iris stop and paraxial light bundles.

Figure 6.57 illustrates the strongly defocused images of bright point-like sources in the object space [Nas10]. Images a) to d) are taken with iris structures consisting of 5, 6, 8 and 9 blades. In image a), the aperture was stopped down only a half-stop, and thus short, curved sections of the full aperture between the points of the pentagon become visible. If parts of the incident beams are blocked by optical elements, the influence of this mechanical vignetting can be observed in the defocused iris images. Image e) depicts the defocused image of a fully open iris stop for a nearly paraxial beam whereas oblique beams at large angles of view are shaped by the vignetting cross-section as seen in image f). If the iris aperture is stepped down, its defocused image reflects the symmetry of the stop for paraxial beams (image g) whereas for an oblique beam the overall symmetry becomes distorted like in image h). The iris structure becomes less visible the

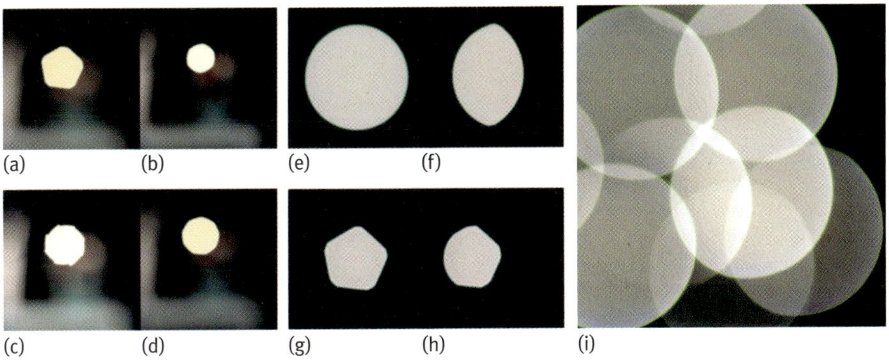

Fig. 6.57: Defocused images of different iris apertures [Nas10]. (a)–(d) Iris structures consisting of 5, 6, 8 and 9 blades; (e) fully opened circular aperture; (f) fully opened circular aperture impaired by mechanical vignetting at large angles of field; (g) pentagon-shaped aperture; (h) pentagon-shaped aperture impaired by mechanical vignetting at large angles of field; (i) superposition of defocused images of bright point sources; the inhomogeneous brightness distribution across the circular area is due to aspheric lens elements. Reprinted with kind permission of Carl Zeiss AG.

larger the bright objects are and then their defocused images reflect the shape of the large objects. This should be taken into consideration by the photographer when the bokeh is intentionally designed.

The defocused images of point light sources deliver also some technical information about the lens. The homogeneity of the brightness distribution in the defocused image is influenced by lens aberrations or imperfections. Figure 6.57i depicts the defocused image of a lens in which aspheric lens elements are implemented. The surface homogeneity of aspheric lenses is much more difficult to achieve than for spherical lenses. Thus, the rougher surface of an aspheric lens may generate brightness variations across the image as seen in the figure. Spherical lens surfaces are smoother and generate more homogeneous blur circles. This makes them more appropriate for portrait lenses with nice bokeh than aspheric lenses.

Before analyzing in more detail the defocused images, it is appropriate to differentiate between images from objects in the background and those from objects in the foreground. There is a certain asymmetry in the magnitude of the blur, which is due to the fact that the range in the depth of field is not symmetric with respect to the sharp imaged object plane. This asymmetry in the depth of field is expressed by the near and far point distances, which can be calculated. Moreover, there may be an additional asymmetry in the quality of the blur, due to lens aberrations.

When inspecting the blurred images of point-like sources in the foreground and background, some interesting differences can be observed. Figure 6.58a shows the photograph of a scenery where the focus is in the background, and thus renders the foreground defocused. The resulting blur circles of the defocused bright lights, which can be observed as reflections in the red ball ornament, are not homogeneously bright but

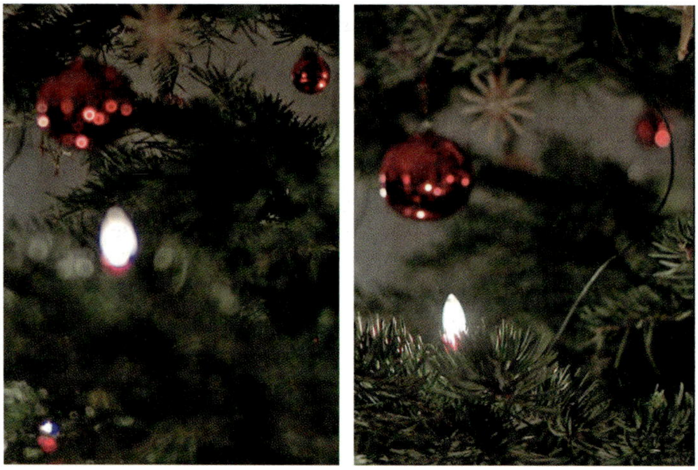

(a) (b)

Fig. 6.58: Photographs of the same scenery with focus in different planes in the object space, taken by a Zeiss Sonnar 1.5/50 ZM [Nas10]. a) The focus is in the background, thus bright objects in the foreground generate blurred foreground images; b) the focus is in the foreground rendering the background defocused. Reprinted with kind permission of Carl Zeiss AG.

reveal a clear circular ring structure. Conversely, the photograph of the same scenery with the focus in the foreground (Figure 6.58b) renders the background defocused and the corresponding blur images of the bright lights in the background balls show a nearly homogeneous structure.

The reason for this can be found when taking a closer look to the spherical aberration that has different influences to fore- and background images (see also Section 3.5.1). As for the location of these images, points in the foreground of the object space are imaged to planes behind the sensor (Figure 6.55) and correspondingly the image planes of background objects are in front of the sensor. As the foreground images are sharp behind the sensor, the cone of their light bundle intersects the sensor plane before their converging point. The sections of light bundles that are affected by an undercorrected spherical aberration are illustrated in Figure 6.59a. It can be seen that the cross-section of a bundle before reaching its converging point does not have a homogeneous brightness as the light rays have a higher density close to the envelope of the caustic than in the center. This leads to the observed blur image of a foreground point as depicted in Figure 6.59b with a bright ring structure at the edge. Conversely, the images of background objects are located in front of the sensor and their diverging conical light bundles intersect the sensor. We can see from Figure 6.59a that the section of the diverging bundle has a more continuously decaying brightness structure with the maximum brightness in the center. It appears more homogeneous than the foreground blur. This can also be directly perceived in the defocused image of a point-like background source in Figure 6.59c. While

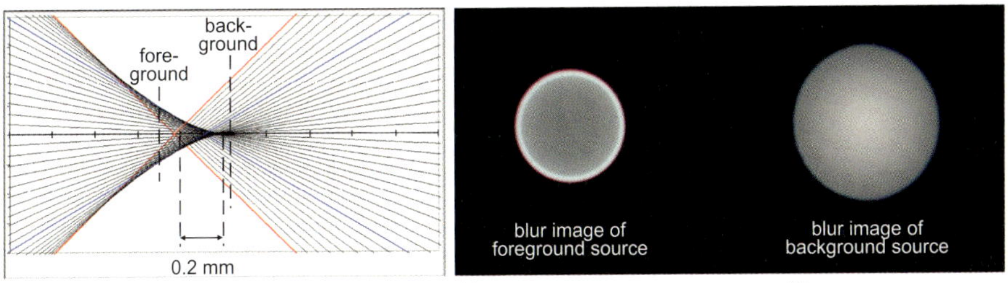

(a) (b) (c)

Fig. 6.59: Defocused images of bright light sources due to undercorrected spherical aberration [Nas10]. (a) Ray path simulation of a 50 mm lens with the red ray being the marginal ray for $f_\# = 1.4$ and the blue ray for $f_\# = 2.4$; (b) defocused image of a point-like foreground source with a bright circular peripheral ring red color fringe; (c) defocused image of a point-like background source with a nearly homogeneous bright circular peripheral ring red color fringe. Reprinted with kind permission of Carl Zeiss AG.

the examples are given for an undercorrected spherical aberration, other types of correction lead to different blur images with often less beautiful bokeh [Nas10].

After that consideration, we can intentionally look for similar asymmetries in other images. A close inspection of the images in Figure 6.56 reveals that the images b) and d) also show an asymmetry in the blurred black ciphers in the foreground and background. For instance, the ciphers at the 93 cm mark exhibit the typical structured foreground blur whereas the ciphers at the 87 cm mark are more homogeneously blurred. When stopping the iris down, not only the amount of blur is reduced but also its quality. This is due to the fact that stopping down significantly reduces the spherical aberration.

A last point that should be mentioned here is the color fringe structure that can be observed in the blur images. Like the spherical aberration the chromatic aberration is the reason that different rays from an object point do not converge in the same image point. In the case of normal dispersion, blue and green color rays are more strongly refracted than red ones. Here, we get a mixture of spherical and chromatic aberrations, the spherochromatism (see Section 3.5.6). The consequence is that the blue and green colors can be seen at the periphery of the background blur images whereas the foreground images have a rather reddish color fringe (Figure 6.59b,c). Colored fringes can be observed generally in blurred images parts like in Figure 6.60. Here, the foreground part also shows a rather reddish or purple color whereas the background fringes are more in the green and blue range. Spherochromatism can be reduced by stopping down as is the case for spherical aberration.

6.9.3.2 Starburst effect

Unlike bokeh, which is related to relatively large apertures and unsharp image sections, starburst effects can show up in parts of photographs with oversaturated images of bright light sources and when the lens aperture is relatively small, namely at high $f_\#$.

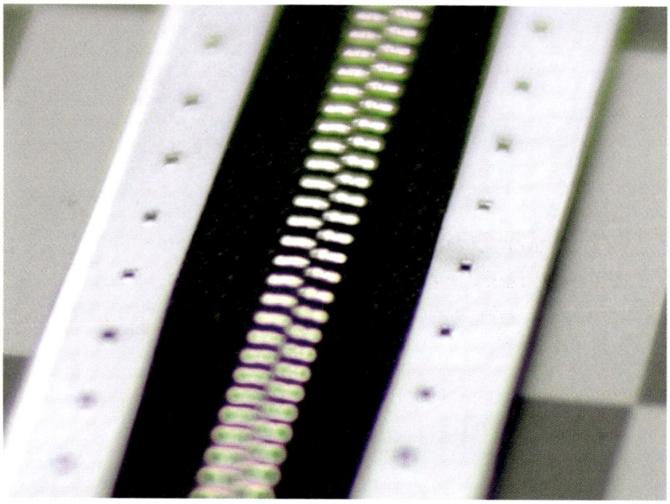

Fig. 6.60: Color bokeh around the sharp imaged plane; the foreground exhibits fringes in the purple range, the background rather green to blue fringes [Nas10] (reprinted with kind permission of Carl Zeiss AG).

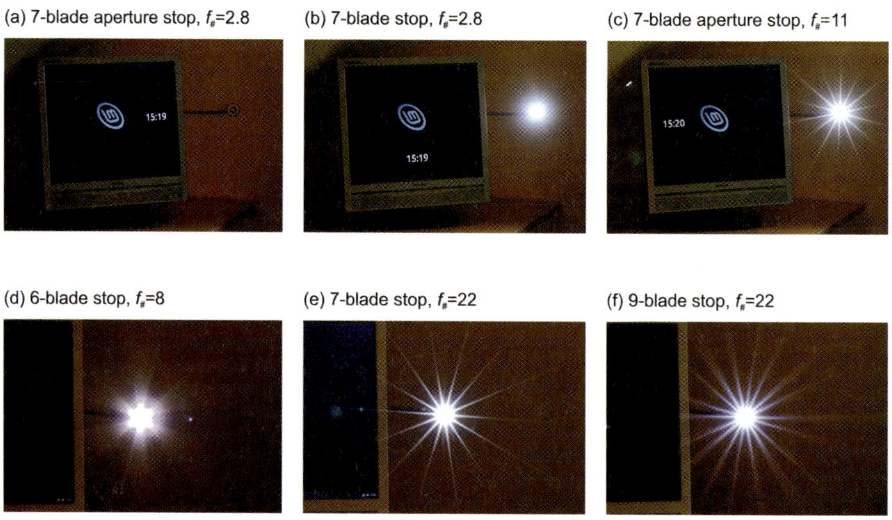

Fig. 6.61: Starburst effect in images taken with different lenses and different aperture stops. (a)–(c) photos taken by a full format lens with 7 blades; (a) $f_\# = 2.8$ without bright light, (b) bright light source at $f_\# = 2.8$; (c) bright light source at $f_\# = 11$, (d) compact camera at $f_\# = 8$ (6 blades, 1/1.7″ sensor); (e) full format lens, 7 blades and $f_\# = 22$; (f) APSC format lens, 9 blades and $f_\# = 22$.

This can be seen in the photographs shown in Figure 6.61. Part (a) shows the original scene with a computer screen and a switched-off lamp. Parts b), (c) and (e) show the same scene with the switched-on light source and increasing $f_\#$, captured by the same lens. It is evident that the star pattern of the oversaturated light image becomes more pronounced with higher $f_{\#}$. This is due to the fact that light is diffracted at the edges of the blades, which form the aperture stop. And diffraction increases with decreasing aperture, namely with increasing $f_{\#}$. If the bright light source was imaged correctly, i. e., not as an oversaturated exposure, then the diffracted part would be not bright enough to be perceived, similar to the light flare reduction in lenses with good antireflection coating.

Usually, aperture stops in camera lenses consist of multiple blades with straight or slightly curved edges. They are radially arranged to form a nearly circular iris diaphragm (Figure 2.8a). The more blades are used, the better is the approximation to a circle. Many system cameras have 7 or 9 blade aperture stops (Figure 2.12). Polygons with less than 7 corners can sometimes be found in the apertures of simpler compact cameras (see, e. g., Figures 2.8b and 2.11b).

In order to understand the star-like pattern, we will consider the diffraction of light at the edges of differently shaped apertures with polygonal form. For simplicity, we regard a rather bright point-like light source, which ideally is far away from the diffracting edge so that the Fraunhofer condition is fulfilled. Then the diffraction pattern is generated in the image space. In the ideal case with the stop being just in front of the lens, the pattern is imaged to the focal plane. However, for the considered conditions, this plane is quite close to the image plane so that in this plane the pattern is seen quite well as an only slightly defocused pattern. It can be calculated by the Fourier transformation of the incident light, which yields a straight line of high intensity along the direction of the edge. In transversal direction the spectrum can be described mathematically by relation no. 6 in Table A.2 in the Appendix. The resulting intensity function decays strongly with the transversal distance from the edge. The diffraction pattern is imaged close to the focal plane of the lens and can be perceived in the image plane of the photograph as a bright line across the image plane. The perception is better if the imaged light source is much brighter than the ambient space.

The diffraction at two parallel edges, e. g., formed by two parallel blades, yields nearly the same diffraction pattern, with the difference that a close inspection along the transversal direction shows interference maxima and minima. This is the classical single-slit diffraction pattern. However, the minima and maxima can virtually not be observed in practice. Thus, the diffraction pattern of two parallel blades is nearly identical to that of a single edge.

The diffraction pattern of an aperture with triangular structure can be understood as the superposition of the diffraction patterns of 3 individual edges that form an angle of 120° between each other (Figure 6.62). Hence, we get 3 straight lines across the image, which intersect at the center of the imaged bright light source. The diffraction image appears as a star with 6 rays. The images shown in the lower line of Figure 6.62 are the result of a numerical computation by a fast-Fourier-transformation algorithm.

A square aperture yields a diffraction pattern, which can be described by the superposition of the patterns of two single-slits with an angle of 90° between them. This

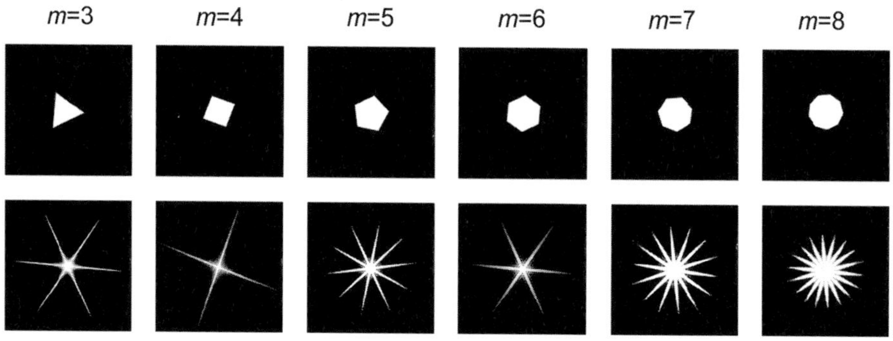

Fig. 6.62: The upper line shows the cross-sections of apertures formed by *m* blades with straight edges. The lower line shows their corresponding star patterns of a rather intense light source computed by a fast-Fourier-transformation algorithm.

shows the star pattern with 4 rays (Figure 6.62; also equivalent to 3D diagram of PSF in Figure 5.9a).

Now we can see the classification scheme for the diffraction patterns of apertures with polygonal cross-sections: If the aperture is formed by an even number m of blades, where 2 blades are in parallel at any one time, the resulting diffraction pattern can be approximated by a star pattern with m rays. If the aperture consists of an odd number m of blades, where no edge is parallel to any other edge, the star pattern is that of a star with $2m$ rays. In order to achieve a nearly perfect circular aperture made by individual blades, we should increase the number of blades to very high value. The consequence then is that we get the circular diffraction pattern of an Airy disk, which shows up as round spot with decaying intensity in radial direction from the center of the light source (see Figure 5.6 and 3D diagram of PSF in Figure 5.9b).

To compare this theoretical consideration with practical results, let us have a look at the lower section of Figure 6.61. Part (d) exhibits the starburst pattern of a compact camera with a 6-blades-stop and a 1/1.7″ sensor at $f_\# = 8$. According to the above-described scheme, a star with 6 ray can be identified. However, the rays are quite broad and also broad circular flares beyond the overexposed central part can be seen, which may be due to the fact that the stop consists of blades with rounded edges.

The same image taken by a full format lens with 7 blades yields a star with 14 rays (Figures 6.61c and e). The rays are finer than with the 6-star-pattern, confirming the straight edges of the blades. Also, according to the scheme, a 9-bladed aperture stop exhibits a star with 18 rays (Figure 6.61f). The rays are slightly fanned out, which seems due to the blades showing somewhat rounded edges.

As mentioned above, the starburst effect is due to diffraction of light at the aperture stop elements. Thus, the effect increases with increasing f-number. However, the f-number is a relative quantity while diffraction depends on the actual aperture size compared to the wavelength of light. As a consequence, the effect should be more pronounced at f-numbers in small-format cameras than at the same f-number in large-format cameras.

The starburst effect can be considered as an aberration if a true image from the object space is wanted. It may also be a source for contribution to the unwanted veiling glare, as described in Section 6.8.2. On the other hand, it is willingly exploited for artistic expression, in a similar way like bokeh.

7 Miniaturized imaging systems and smartphone cameras

Today miniaturized imaging systems take an important role in daily life, industrial applications and other fields of application. There is a continuously growing demand on them. This includes, in particular, cameras built in smartphones (smartphone camera modules, SPC), which nowadays even are the devices with which the vast majority of photos are taken. But there are a lot of other applications as well where miniaturized cameras are used. Those could be, e. g., found in the field of healthcare (micro endoscopy), automotive (dashboard camera, rear view camera), outdoor (action helmet cameras, cameras for motorcycles, cycling and/or sports), surveillance, industry and vision (professional pipe and wall inspection, drain, sewer and pipeline inspection, robotics), wildlife and nature observation and further mobile applications (including cameras for laptops and tablet computers). Consequently, today there is a huge market for miniaturized camera modules, which extremely surpasses that of classical cameras.

However, miniaturization of the optical and the sensor system in general is not straightforward and there are limitations. As a simple example may be the change of a length x to x'. Then the relative change is given by (x'/x). But area and volume scale differently to that, namely with $(x'/x)^2$ and $(x'/x)^3$, respectively. This has important consequences. Another example is the reduction of the focal length. As we have already seen in the previous chapters, this takes strong influence on the amount of light captured by the system.

The predominant part of world's population has a smartphone at their disposal, which implies the high-economic importance of SPC. However, there is also a big part of the remaining miniature cameras as well, for instance, cameras that are relevant for industrial applications. At least part of the benefits for those non-SPC miniaturized imaging systems originates to a significant extent from the SPC development. This has drawn our emphasis on SPC. However, we may mention as well that we have already discussed a miniaturized optical system. The subject of Section 3.3.6 has been an optical system based on ball lenses. Although the imaging quality of that system may be regarded as rather limited, on the other hand, it is rather simple, and thus may allow for application where image resolution is only required over small fields, such as for scanning probes in optical coherence tomography or fiber to fiber light transfer.

In modern cell or smart phones, SPC have to fulfill a multitude of tasks with high quality, just like a universal camera for anything. While they should be very small in size and operate at high speed, they should be at a low price as well. It should be noted that for high-end smartphones, SPC modules take approximately 14 to 20 % of the total cost. In order to satisfy these conflicting requirements, new technologies and trends in SPC have emerged during the last two decades. This may influence also the design of future cameras with larger formats. Therefore, our considerations in the following sections deal above all with SPC, always in relation to "classic" digital cameras. Nevertheless, to

https://doi.org/10.1515/9783110789966-007

some extent, other miniaturized imaging systems are discussed as well, in particular, modern developments that may play an important role in future.

Mobile phone cameras and miniature cameras have already been shortly presented in Section 2.6.4.2 as examples for camera systems in general. In the present chapter, we extend our discussion. Based on the contents of the preceding chapters, we start with the imaging optics. This is followed by sensor systems of smartphone and miniature cameras, which covers the special sensor issues of these systems. In that sense, this may be understood as an extension of Chapter 4. The subsequent section of camera system performance is about the overall quality of the imaging as in miniature systems where both optics and sensors are inseparably connected. In the following section, we consider new software-based technologies that have shown up to overcome the limitations of the systems inherent to their restricted dimensions. This is complemented with the last section of this chapter, which covers current modern developments for further advancements and miniaturization.

7.1 Imaging optics

7.1.1 Physical and optical properties

While mobile phones before the year 2000 were virtually only used for telephony, the first cameras integrated in the mobile phones showed up around the beginning of the current millennium. With the advancement of their electronic performance, more computational features have been integrated into mobile phones and transformed them into smartphones. In the years from about 2000 up to 2016, smartphones usually had one main camera module on their backside. In 2003, Sony Ericsson Z1010 was the first smartphone to feature an additional camera on the front side for "selfies" and video communication. This feature was adopted by nearly all smartphone manufacturers in the years to follow. It was only in 2016 that some manufacturers presented a dual camera system on the backside, in addition to the front camera. After that, high-end smartphones with multiple camera systems consisting of up to 5 rear cameras (e. g., Xiaomi Mi CC9 pro, Nokia 9 PureView, both 2019) showed up. This enabled more advanced features, some of which are discussed in Section 7.4, like depth profile evaluation or zoom properties. However, we may note that more recently many high-end SPC restrict to 3 or 4 main camera modules and potentially one camera for depth sensing. In addition, there is one or more front camera. For the discussion of the physical and optical properties of SPC, however, we mainly focus on single camera modules.

Imaging systems of SPC have to fulfill different requirements. Smartphones as an every-day companion for people should above all be small, compact and easy-to-manage. Simultaneously, they are expected to have a high-multimedia performance. Consequently, the integrated camera modules have to be significantly smaller in size than typical cameras with larger optical sensors, as for instance, compact cameras or

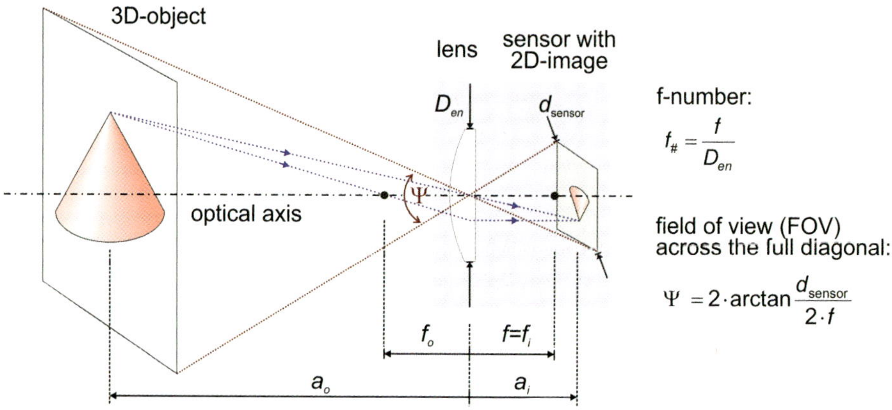

Fig. 7.1: Simple scheme of an optical system.

DSLR. In the same sense as in the previous chapters, the term DSLR should implicitly also include other high-quality system cameras without mirror as discussed in Chapter 2. A simple downsizing of all optical components in a typical system camera to achieve a good SPC-design, however, is not reasonable or even possible as the optical properties do not all scale in the same way. The reference for these properties is always the wavelength of visible light, thus in the range from about 0.4 µm to 0.7 µm. Reducing the geometrical dimension leads to increasing blur due to diffraction, less incident light power, and thus higher image noise.

For a better understanding, let us consider a simplified optical system as depicted in Figure 7.1. Here, we have summarized some properties that have been discussed in the preceding chapters in more detail. The system consists of the imaging optics, which in this scheme is the lens, the image sensor at the end of the optical ray path and the appropriate electronics to control the exposure and the processing of the captured image data. Usually, the principal planes of the lens and the image sensor plane are parallel to each other and perpendicular to the straight optical axis.

For our present discussion of the basics of optical imaging by SPC, it suffices to use the paraxial approximation for thin lenses given by Equation (2.5). We also make no differentiation between the object and image focal distances f_o, resp., f_i and, for simplification, we set the focal length $f = f_\mathrm{i}$. In photography, the distance a_o from the object to the lens is usually more than 10 times larger than the distance a_i from the image to the lens. As a consequence, the image distance is only slightly larger than the focal length f and becomes nearly identical to it when imaging far away objects. The total angular range in the object field, which is imaged to the sensor, is characterized by the field of view (FOV) and given by the full angle Ψ as measured across the diagonal of the sensor; see also Equation (2.10). It depends on the focal length f as well as on the size of the image sensor with its diagonal d_sensor. For images, which are captured by lenses with focal lengths that are nearly equal to the image sensor diagonal, we get $d_\mathrm{sensor}/f \approx 1$, and thus a resulting FOV ≈ 53° across the full diagonal (see Section 2.2). In that case, the images yield a perspective impression, which comes close to the natural perception of the human eye. For that reason, such a lens is termed "normal lens" and has a "normal focal length." Accordingly, wide-angle lenses have shorter focal lengths.

Current smartphones (2022) are usually equipped with multiple rear camera modules where the optical axis is predominantly perpendicular to the backside. Here, the thickness of the camera body with roughly 6–10 mm is the limiting quantity for the focal length f. For longer focal lengths, there are other designs, with folded optical paths for instance, as will be discussed below. While the sensor area should be as large as possible for high resolution, the limitation of the focal distance implies that the main camera modules have a shorter focal length than their image diagonal. Hence, usually the main camera has a wide-angle perspective with an equivalent focal length f_{eq}, defined by Equation (4.18), between approximately 26 mm and 28 mm. The latter one has been the standard f_{eq} of SPC between about 2010 and 2016 (see also Table 2.4 and Figure 7.8). This is different from the larger format cameras like DSLR where the normal focal length is $f_{eq} = 50$ mm. Thus, typical values of $f \approx 4$–6 mm in combination with larger image sensor diagonals of $d_{sensor} \approx 6$–8 mm result in a wide angle perspective with field angles FOV in the range from 70°–75° across the full diagonal according to Section 2.2. The consequence is a different perspective than with "normal" vision, which becomes more obvious at the edges of the image and a higher demand for correction of imaging errors.

The assessment of a lens perspective is only meaningful if the dimensions of the sensor are known for which the lens is designed. For a better comparability with other lenses and formats, we use the mentioned equivalent focal length f_{eq} and the crop factor CF from Equation (2.25). Both make reference to the full format which is usually taken as the standard in photographic issues. It should be recapitulated here that the full format with a sensor diagonal of $d_{FF} = 43.3$ mm has an aspect ratio of 3:2 and a normal lens of $f = 50$ mm, yielding a FOV = 47° while for most SPC sensor the aspect ratio is of 4:3 and CF mostly larger than 4 in case of wide angle camera modules (see Tables 2.4, 7.1, 7.4–7.6).

Tab. 7.1: Technical specifications of Apple iphone 10 (2017) smartphone with 3 camera modules.

Apple iphone 10 (2017)	front	Main	telephoto
equivalent focal length f_{eq}	32 mm	28	52 mm
f-number $f_{\#}$	2.2	1.8	2.4
field of view Ψ	68°	75°	45°
focal length f	4.4 mm	3.9 mm	6.4 mm
sensor diagonal d_{sensor}	6.0 mm	6.0 mm	5.3 mm
crop factor CF	7.2	7.2	8.1
pixel number	7 MP	12 MP	12 MP
pixel pitch p	1.2 µm	1.2 µm	1.0 mm
sensor format	1/3.0″	1/3.0″	1/3.4″
optical cutoff $N_{SB,cut}^{(2D)}$	$11.8 \cdot 10^6$	$17.7 \cdot 10^6$	$7.8 \cdot 10^6$

As an example for further discussions, let us choose a reference SPC with a 4:3 image sensor of a diagonal $d_{\text{sensor}} = 6\,\text{mm}$, as it is used in a main module of a SPC (Table 2.4, Table 7.1), and a lens of focal length $f = 3.9\,\text{mm}$ with $f_{\#} = 1.8$. We then get for the width of the sensor $PW = 4.8\,\text{mm}$ and $PH = 3.6\,\text{mm}$ for its height. The crop factor can be calculated to be $CF = 7.2$ and the true focal length $f = 3.9\,\text{mm}$ of the SPC corresponds to $f_{\text{eq}} = 28\,\text{mm}$, which is a typical wide-angle lens. If the sensor is designed for 12 MP, which means 4000 pixels in horizontal and 3000 pixels in vertical direction, we can calculate a square image pixel of $p = 1.2\,\mu\text{m}$ pitch, which is the distance between two adjacent cell centers. The pixel pitch of a 12 MP full-format sensor, having the same number of cells, yet on a bigger area, would be larger by the factor $CF = 7.2$ thus yielding $p = 8.6\,\mu\text{m}$. Typical image sensors of DSLR are currently ranging from about 24 up to 60 MP, and their pixel pitches are of 4–6 µm. As we will see below, some SPC even have a pixel pitch of only $p = 0.8\,\mu\text{m}$, which seems to be close to the least value, which still is reasonable for photographic quality.

The resolution of a camera system depends on many factors and has already been discussed in Section 5.3. It is usually characterized by the modulation transfer function of the system $\text{MTF}_{\text{system}}$. The two main factors in it are the optical part $\text{MTF}_{\text{optics}}$ and the sensor part $\text{MTF}_{\text{sensor}}$, which contribute in a multiplicative way. Somewhat differently, we get an idea of the optical quality of the camera when we calculate the limit value of image points that can be resolved in the system under the assumption that we have an ideal image sensor of unlimited resolution. Then the resolution of the complete system depends only on the optical part. The real resolution of the whole system will be considered later in Section 7.3. In the following, we only consider aspects of the optical resolution.

Let us start with a diffraction limited optical system without any lens aberrations, which has a circular entrance pupil and an aperture diameter D_{en}. The ideal lens produces a spot image of a light source at infinite distance, which is blurred due to diffraction. The spot can be described as an Airy-disk according to the Rayleigh criterion in 2D space with a diameter u_{d} after Equation (2.23) and δ_0 after Equation (5.43), respectively. Both are identical. Two image points, which are close together, can still be resolved as individual points in the image plane if the distance between their centers is equal to r_0, which is the radius of the Airy spot, and thus half of the blur diameter (see Figure 5.44). If the number of points, which still have the same distance of r_0 between their centers, increases, the contrast decreases and the visibility becomes very weak for an infinite number of points in a line. Based on the Abbe, respectively, Rayleigh criterion for the visibility of two isolated points, we get space bandwidth numbers $N_{\text{SB}}^{(1\text{D})}$ and $N_{\text{SB}}^{(2\text{D})}$. These are the number of points, each at a distance of r_0 from its neighbor point, along the picture height PH after Equation (5.52a), respectively, in the area $PH \cdot PW$ of the picture after Equation (5.52b). It should be noted that for these definitions the visibility is very low but not equal to zero. It seems that for this reason, $N_{\text{SB}}^{(2\text{D})}$ is in many cases defined as the optical resolution of a picture, as for instance, by Blahnik and Schindelbeck [Bla21]. As discussed in Section 5.3, a visible contrast, and thus a more realistic resolution of points is achieved when the distance between the image points is larger than about $1.06 \cdot r_0$. This is expressed by $N_{\mathrm{SB,points}}$ along PH, given by Equation (5.67).

For our following consideration, we choose the optical cutoff of vanishing contrast as the main criterion. When we consider the direction along PH, the limiting number of points is given by $N_{\mathrm{SB,cut}}$ (from Equation (5.69) with $\kappa = 1.22$ for a 2D circular aperture and $\alpha = 0$ for a lens without aberrations). Similarly, we may do the same for the picture width PW, and thus get for the optical cutoff $N_{\mathrm{SB,cut}}^{(2\mathrm{D})}$ in 2D:

$$N_{\mathrm{SB,cut}}^{(2\mathrm{D})} = 1.22 \cdot \frac{PH}{r_0} \cdot 1.22 \cdot \frac{PW}{r_0} = 1.49 \cdot \frac{PH \cdot PW}{r_0^2} \tag{7.1}$$

This is the upper limit of points in the image plane, which cannot be resolved any more due to diffraction and for which the contrast vanishes.

As mentioned above and discussed in Section 5.3, this number overestimates the number of image points, which can be regarded to be resolved according to Abbe's or Rayleigh's criterion. Nevertheless, $N_{\mathrm{SB,cut}}^{(2\mathrm{D})}$ serves as a reference for the comparison of different SPC modules. $N_{\mathrm{SB,cut}}$ and $N_{\mathrm{SB,cut}}^{(2\mathrm{D})}$ can be considered as optical quantities, which are analogous to the sensor cutoff expressed by the Nyquist frequency R_{N}. This is of importance for the resolution of the complete system, where equivalent conditions for both optical and sensor parts will be required.

Taking into account Equations (2.23) and (5.43), the radius of the Airy spot at a central wavelength in the visible spectral range ($\lambda = 0.55\,\mu\mathrm{m}$) for a lens without aberrations, i. e., $\alpha = 0$, is given by

$$r_0 = \frac{\delta_0}{2} = \frac{u_{\mathrm{d}}}{2} = 0.67\,\mu\mathrm{m} \cdot f_{\#} \tag{7.2}$$

For an image sensor with a 4:3 aspect ratio and an image sensor diameter d_{sensor}, the long and short sides of the sensor are $PH = 0.6 \cdot d_{\mathrm{sensor}}$ and $PW = 0.8 \cdot d_{\mathrm{sensor}}$, respectively. Substituting r_0 in Equation (7.1) by Equation (7.2), the upper limit of image points can then be written as

$$N_{\mathrm{SB,cut}}^{(2\mathrm{D})} - 1.59 \cdot \left(\frac{d_{\mathrm{sensor}}}{f_{\#} \cdot \mu\mathrm{m}}\right)^2 \tag{7.3}$$

For our reference camera with $d_{\mathrm{sensor}} = 6.0\,\mathrm{mm}$, $f_{\#} = 1.8$ the equations above yield $\delta_0 = 2.4\,\mu\mathrm{m}$, $N_{\mathrm{SB,cut}}^{(2\mathrm{D})} \approx 18 \cdot 10^6$ and $N_{\mathrm{SB}}^{(2\mathrm{D})} \approx 12 \cdot 10^6$. The design rules for high-quality lenses require that an allowable blur spot should be not larger than 1/3000 of the image diagonal (see Sections 1.5 and 6.9). The reference SPC with $d_{\mathrm{sensor}} = 6.0\,\mathrm{mm}$ thus should have a maximum blur spot 2.0 µm. In that case, the lens of our reference camera can be considered as aberration-free and only limited by diffraction, it nearly meets this requirement for a high-quality lens. As for the image sensor of 12 MP, its number of pixels is identical to $N_{\mathrm{SB}}^{(2\mathrm{D})}$ but inferior to the optical cutoff $N_{\mathrm{SB,cut}}^{(2\mathrm{D})}$. The size of the blur spot $\delta_0 = 2r_0$ is equal to $2p$, namely twice the pixel pitch, which means that $r_0 = p$. This is the usual method in order to match the sensor to the lens and to ensure that structures of the size of the blur spot can be unambiguously sampled according to the Nyquist theorem (see also Section 1.6 and Chapter 5). In that case, the 1D optical resolution, with $R_{\mathrm{cut}} = 1.22/r_0$, can be considered to be more than two times higher than that of the sensor with $R_{\mathrm{N}} = 1/(2p)$. The total resolution of the whole system is then dominated by the sensor, as shown in Figure 7.9c, and discussed in Section 7.3 for different SPC. Current SPC mostly fulfill this match between lens optics and image sensor for their standard/main modules. However, especially long-focus camera modules, and also many ultrawide angle modules do not meet this match and often have maximum f-numbers above 2.0 or even more (see Tables 2.4, 7.1, 7.4–7.6). Thus, they do not yield the overall resolution as expected by the sensor specifications because diffraction blur becomes dominant. The optical cutoff becomes inferior to the sensor cutoff and degrades the overall system resolution.

After Equation (2.16), the exposure in a still camera is directly proportional to $t_{\mathrm{x}}/f_{\#}^2$, and thus a combination of exposure time and f-number. That gives the photographer a certain flexibility for an appropriate choice of these parameters for his image layout. In the case of SPC modules, there is no variable iris diaphragm for aperture control as the lens is designed to operate close to its diffraction optimum with a fixed value. Stopping down, as discussed, would directly impair the optical resolution by blur whereas larger apertures are not reasonable while lens aberrations become more distinct and require a high correction effort and more space (see also Figure 7.19). Thus, the exposure in SPC is only controlled electronically by the exposure time. The sensitivity of the CIS can also be adapted to attain a larger span of exposure times (see also Section 7.2). Shorter exposure times, for instance, to reduce motion blur, require higher "ISO" settings which, however, may lead to less favorable SNR. This is a particular problem for systems of small pixel sizes. Moreover, the number of photons per pixel is inversely proportional to CF^2, as expressed by Equation (7.12) in Section 7.2. As a consequence, one pixel of our exemplary SPC above receives about 50 times less photons than that of a comparable full-format pixel, which results in a much lower signal-to-noise ratio with an increased noise manifestation.

7.1.1.1 Depth of field and bokeh

The fixed aperture in the SPC has a high impact on features like bokeh and depth of field (DOF). That takes strong influence on the aesthetic photographic artwork. "Bokeh" comprises the way in which the parts in front of or behind the sharply imaged object plane are rendered more or less blurred in the image (see also Section 6.9). A key parameter for its control is DOF (Section 3.4.6). DOF depends on the allowable circle of confusion of a diameter u_{i}, which is the limiting size of a spot that can be just detected by the human eye under normal viewing conditions. Furthermore, DOF increases linearly with $f_{\#}$ and nearly inversely with M^2 according to Equation (3.113). M is the image magnification, which is the ratio between image and object size. If we assume a minimum object distance of 10 cm for our exemplary SPC with $f = 3.9$ mm, the absolute value of the magnification is always below 0.05 according to Equation (2.8). Hence, we expect a significantly larger DOF in SPC, which have shorter focal lengths, when compared to system cameras with larger sensors and larger focal lengths. Note that this shorter DOF is in spite of the low f-numbers and small u_i of SPC.

To ease the comparison between different optical systems, let us first consider the hyperfocal distance a_{hf}, which is quite demonstrative for that purpose. If the optimum focus point of a camera is set to the hyperfocal distance, then the far-point a_f is approaching infinity whereas the near-point a_n is identical to half of a_{hf}, as given by Equations (3.116) and (3.117). All objects between a_n and infinity are rendered sharp on the image. For good quality imaging, the circle of confusion should be at least 1/1500 of the sensor diagonal or smaller, so we set $u_i = d_{sensor}/1500$ for our consideration. In order to compare different optical systems, we replace the sensor diagonal as well as the focal length by their corresponding values of the full format using Equations (2.25) and (4.18), taking into account the crop factor. As all distances on the object side are counted negative, due to sign convention of the present book, we take the absolute values for a_{hf} and a_n, respectively, and get

$$|a_{hf}| \approx \frac{f^2}{u_i \cdot f_\#} = \frac{1500}{d_{FF}} \cdot \frac{f_{eq}^2}{CF \cdot f_\#} \approx \frac{34.6}{\text{mm}} \cdot \frac{f_{eq}^2}{CF \cdot f_\#} \qquad |a_n| = \frac{|a_{hf}|}{2} \qquad |a_f| \to \infty \tag{7.4}$$

It becomes obvious from Equation (7.4) that the hyperfocal distance scales with the square of the focal length. Hence, wide-angle lenses with their short focal lengths are most important for achieving a large DOF, i. e., from a_n to infinity. Moreover, for lenses with the same equivalent focal length, $|a_{hf}|$ scales inversely with the product $CF \cdot f_\#$. That means that SPC with their large crop factor exhibit a very large DOF, which is intensified by the fact that the standard lens of SPC is a wide-angle one. Data calculated for some commercially available SPC modules in comparison to DSLR are listed in Table 7.2. Standard lenses in SPC with $f_{eq} = 28$ mm or shorter usually render images, which are sharp from about 1 m or less in the object space to infinity. Ultrawide angle lenses virtually

Tab. 7.2: Hyperfocal distance and near-point for different optical systems according to Equation (7.4).

	f_{eq}/mm	a_{hf}/m	a_n/m
$f_\# = 2.2$, SPC, $CF = 5.4$	14	0.6	0.3
$f_\# = 2.2$, DSLR	14	3.1	1.5
$f_\# = 1.8$, SPC, $CF = 7.2$	28	2.1	1.1
$f_\# = 1.8$, DSLR	28	15.1	7.5
$f_\# = 2.1$, SPC, $CF = 6.6$	50	6.3	3.1
$f_\# = 2.1$, DSLR	50	41.2	20.6
$f_\# = 2.4$, SPC, $CF = 8.5$	70	8.3	4.2
$f_\# = 2.4$, DSLR	70	70.7	35.4

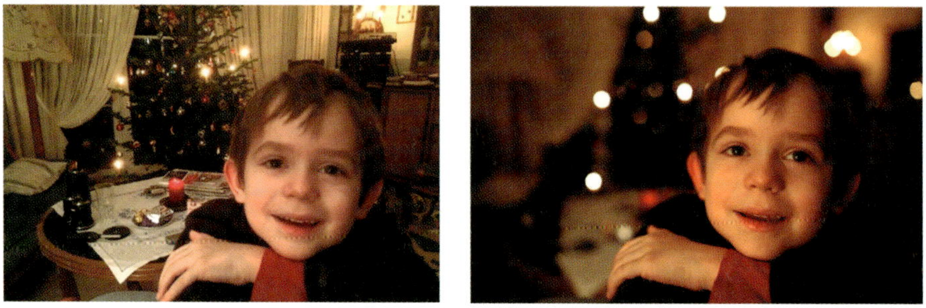

Fig. 7.2: Portrait photographs taken by different camera systems; left side: smartphone camera, $f_{\mathrm{eq}} = 29\,\mathrm{mm}$, right side: full format camera, $f_{\mathrm{eq}} = 28\,\mathrm{mm}$. Note that just for intended direct comparison the focal length of the full format camera was chosen to be nearly the same as f_{eq} of the SPC. However, good portrait pictures usually are made with longer focal length lenses, thus yielding an even better portrait mode [Bla21] (Author V. Blahnik, reprinted with kind permission).

do not need any adjustment of the focus position. This may be of advantage for quick snapshots, street or landscape photography.

For portrait photography or artistic artwork, however, where a narrow DOF with a blurred fore and background is wanted, this may be obstructive. Here, usually a moderate long focus lens with low $f_{\#}$ is required. Moreover, the perspective of normal or moderate long focus lenses is more appropriate for portrait photography as they have a more homogeneous magnification across the image field than wide-angle lenses with more DOF [Bla21]. As long focus lenses with a low f-number are difficult to implement in SPC due to their thin body, a good compromise is the use of a normal lens with $f_{\mathrm{eq}} = 50\,\mathrm{mm}$ for portrait photography. A closer look to the depth of field for a portrait photo is given in Figure 7.2 and in Table 7.3.

Figure 7.2 shows the comparison of two portrait photos, one taken by a SPC with $f_{\mathrm{eq}} = 29\,\mathrm{mm}$ and a second one taken by a full format lens of $f_{\mathrm{eq}} = 28\,\mathrm{mm}$ and the same $f_{\#} = 2.2$. Having nearly the same f_{eq}, they also have nearly the same FOV and the same perspective. The pictures were taken with a distance of the person at $|a_{\mathrm{o}}| \approx 0.45\,\mathrm{m}$ from the entrance pupil of the camera. The diagonal of the object field at 0.45 m is about 0.6 m–0.7 m for both lenses, which is a good range for portraits. While ranges and the perspectives are nearly identical, DOF is much larger in the picture of the SPC than of the full format lens. This is why the magnification on the SPC sensor is smaller by a factor of CF than for the full format. Using Equation (2.8) and substituting the focal length by the equivalent focal length and the crop factor (Equation (4.18)), we get

$$|M| = \frac{f}{|a_{\mathrm{o}}| - f} = \frac{f_{\mathrm{eq}}}{CF} \cdot \frac{1}{|a_{\mathrm{o}}| - \frac{f_{\mathrm{eq}}}{CF}} \approx \frac{f_{\mathrm{eq}}}{CF \cdot |a_{\mathrm{o}}|} \tag{7.5}$$

The simplification in this equation implies deviations from the exact result, which however in our considerations, are below 6 %, but help to see the relationship to the different parameters. DOF can also be estimated using Equation (3.113) under the conditions that the object distance a_o is not close to a_{hf}, the image magnification $|M| \ll 1$ and Equation (7.5) can be used. This yields

$$s_{DOF} \approx 2 \cdot u_i \cdot f_\# \cdot \frac{1}{M^2} \approx \frac{2 \cdot d_{FF}}{1500} \cdot CF \cdot f_\# \cdot \left(\frac{a_o}{f_{eq}}\right)^2 = CF \cdot f_\# \cdot \left(\frac{a_o}{f_{eq}}\right)^2 \cdot 0.058\,\text{mm} \quad (7.6)$$

The calculation of near- and far-point values after Equations (3.107) and (3.108) as well as DOF and $|M|$ for portrait photos taken at a_o by different lenses are compiled in Table 7.3. We now can state the following points:

- For portrait photography, where a well-defined object range is to be imaged with a shallow DOF, large format cameras are of advantage due to their larger magnification.
- DOF in SPC is almost one order of magnitude larger than in full-format cameras. Moreover, if there is automatic control of the focus point in SPC then DOF may be somewhat random. For the example in Table 7.3 with $f_{eq} = 29$ mm, the face is at a distance of 0.45 m from the camera lens. The autofocus may set $|a_o| = 0.45$ m or $|a_o| = 0.60$ m for the exposure. In both cases, the face is imaged sharp while DOF can vary between 23 cm and 43 cm. Thus, background sharpness is difficult to control in autofocus mode. But we may note that the situation may become somewhat better when the SPC allows for setting the focus spot manually.
- Choosing a longer focal length for portrait photography, like $f_{eq} = 50$ mm instead of $f_{eq} = 29$ mm in our example in Table 7.3, does not really change M neither DOF but ensures a more pleasant look in the image with less perspective distortions.

Hence, a nice bokeh as in the case of larger format cameras can virtually not be achieved in SPC optically by longer focal lengths and lower $f_\#$. Therefore, the only practical solution is an artificial generation of blurred fore and background by computational tricks. This will be shortly discussed below in Section 7.4.

Tab. 7.3: List of parameters when taking a photograph of the face of the person at a distance a_0 in the object space.

$f_\# = 2.2$	f_{eq}/mm	CF	$\|a_o\|$/m	$\|a_n\|$/m	$\|a_f\|$/m	DOF/cm	$\|M\|$
SPC	29	7.2	0.45	0.36	0.59	23	0.009
DSLR	28	1	0.45	0.44	0.47	3	0.064
SPC	29	7.2	0.60	0.45	0.89	43	0.007
$f_\# = 2.1$	f_{eq}/mm	CF	$\|a_o\|$/m	$\|a_n\|$/m	$\|a_f\|$/m	DOF/cm	$\|M\|$
SPC	50	6.6	0.9	0.79	1.05	26	0.008
DSLR	50	1	0.9	0.88	0.92	4	0.057

Finally, it should be noted again that, for all discussions on depth of focus, the range of sharpness depends on the conditions of how an image is viewed. We always assume that the image is viewed by the human eye from the distance of the image diagonal. If the size of the image is magnified and the viewing distance becomes significantly closer than the image diagonal, the range of visible sharpness of the image becomes narrower. For instance, if the camera was focused to a_{hf} while taking the photograph, a large magnification of the image may reveal that the image is no longer sharp at infinity but only in a limited range around a_{hf}. For further details, see Section 3.4.6.

7.1.1.2 Focus control

A simple SPC set-up without further technology for focus adjustment already facilitates sharp images beyond 1 m from the camera lens. Despite that fact, there are many situations when focus control is required. Close-up images below 1 m object distance are not possible without focus adjustment. And also when images are magnified to display crops, blurring may be detected in cropped parts, which are not visible under normal viewing conditions. Therefore, focus adjustment is enabled in all SPC of higher quality. For instance, if the above-mentioned reference SPC should be used for focusing to objects at 10 cm distance instead of 2 m, then according to Equation (2.7), a shift of the lens away from the sensor by only 150 µm is required. The image magnification for an object at $|a_o| = 0.1\,\mathrm{m}$ here is of 1:25. The comparison to a full-format lens with the same equivalent focal length of 28 mm shows that the bigger lens requires a shift by 9 mm to focus at objects at 0.1 m distance. But here the magnification of 1:2.7 is about one order of magnitude larger and has a narrower DOF than in case of the SPC. In order to focus from 0.1 m to infinity in SPC, a typical lens shift of a few 100 µm is sufficient. For further close-up imaging using SPC, special auxiliary close-up lenses are available.

According to Equation (7.6), DOF on the object side scales linearly with the product $f_\# \cdot CF$ and is therefore relatively large for SPC with small image sensors. On the other hand, the image-side depth of focus (DOFoc) scales with $f_\#/CF$, thus reciprocally with the crop factor. Using Equation (3.120), we get the following result:

$$s_{\mathrm{DOFoc}} \approx 2 \cdot u_i \cdot f_\# \approx \frac{2 \cdot d_{\mathrm{FF}}}{1500} \cdot \frac{f_\#}{CF} = \frac{f_\#}{CF} \cdot 0.058\,\mathrm{mm} \tag{7.7}$$

DOFoc is the range in which the position of the CIS can change without significantly impairing the image sharpness. Hence, it plays an important role for the manufacturing process of SPC. Considering a lens with $f_\# = 1.8$, we get a DOFoc in a full-format camera system of 104 µm whereas it is of only 14 µm for a SPC, and thus almost one order of magnitude smaller. The fact that SPC typically have low f-numbers and high-crop factors implies that SPC-modules have to be produced with a precision that is nearly one order of magnitude more stringent than for conventional DSLR. This also implies that interchangeable lenses, which should be used with the same sensor, are virtually impossible for SPC (an exception is a concept smartphone of Xiaomi which makes use of a mount for different Leica M camera lenses; however it is not on the market). This favors the integration of a lens with the sensor in a module.

Focus adjustment in SPC is achieved by a slight mechanical displacement of the complete lens assembly relative to the image sensor. Unlike in system cameras with larger sensors, where often a single lens or lens group is shifted along a distance of several mm, i. e., by internal focusing, the total shift range in SPC is of a few 100 μm. Conventionally, the lens barrel is electromagnetically displaced by a voice-coil motor (Figure 7.3) longitudinally, i. e., in the direction of the optical axis. The displacement is due to the magnetic interaction of a permanent magnet in the lens module with the electromagnetic fields produced by the electrical current through the coil wires. It can be precisely controlled by the electrical current but may be impaired by hysteresis effects, which require a control loop for focus adjustment [Bla21]. There are also recent developments aiming to shift an individual lens using a micro-electromechanical system (MEMS) integrated on a silicon substrate[1] (for illustration, see also[2]). Although that technology seems quite attractive as for its compactness and integrability with existing silicon technology, no commercial use of that approach has been reported up to the present (2022).

The focus position is controlled as in other cameras electronically by an autofocus mechanism. In its simplest form, the method is based on the evaluation of the image contrast, which is highest when the image is sharp on the sensor and low in a blurred image.

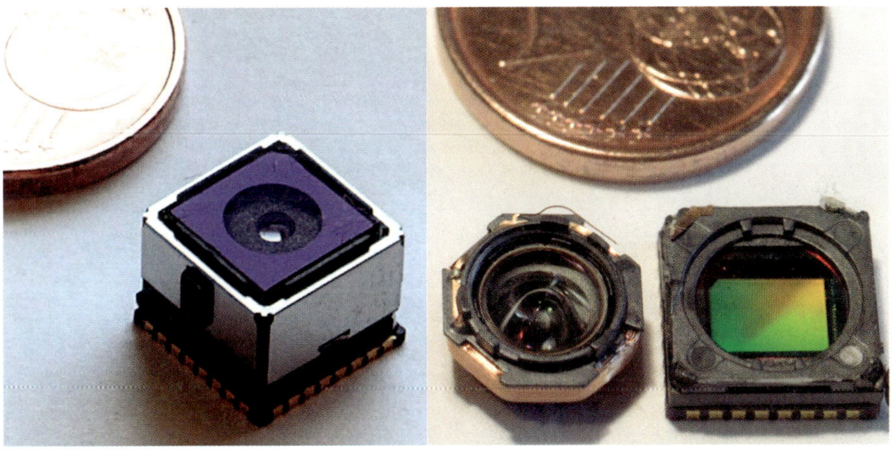

Fig. 7.3: Left side: Smartphone camera module; right side: lens barrel with voice-coil-motor for longitudinal focus adjustment and image sensor in comparison to a coin (Author: H. Brückner).[3]

1 https://patentscope.wipo.int/search/en/detail.jsf?docId=WO2017106955&tab=PCTBIBLIO

2 https://mechatronicsi.blogspot.com/2014/02/how-mobile-auto-focus-works-inandroid.html

3 U. Teubner, V. Blahnik, H. J. Brückner: *Smartphonekameras, Teil 2: Bildsensor und -verarbeitung - Tricksen für gute Bilder*, Physik in unserer Zeit, **51**(6) (2020) 290.

Unlike in DSLR where for the image analysis a part of the incoming light is mirrored to a dedicated sensor, in mirrorless cameras and SPC the evaluation has to be made using the image sensor signal directly before the image capture. Although the contrast autofocus mechanism yields very good results, it is a slow process and leads often to unacceptable time delays. A more advanced, quicker and today rather common mechanism is phase detection autofocus (PDAF; see Section 4.10.7.3). Due to lens adjustments, the position of their imaged points on the sensor is shifted. This shift is evaluated to determine the optimum lens position, which is then adjusted by the corresponding lens actuators. Autofocus mechanisms using contrast or phase detection are classified as passive mechanisms. Active mechanisms comprise a source, which emits a high-frequency signal, like ultrasound or infrared light, and detect the time delay of the reflected signal. A time of flight (TOF) sensor is based on that principle (see Section 4.10.7.4). In modern SPC, however, TOF sensors are usually not used for focus control but for artificial background manipulation in an image post-processing step (see also Section 7.4).

7.1.1.3 Image stabilization

Many modern high-quality cameras and lenses, respectively, have methods for image stabilization (IS) at their disposal. This helps to maintain the image quality with respect to resolution in case of a slight camera shake as, for instance, caused by hand tremor. Additionally, it allows for longer exposure times of handheld picture shootings corresponding to approximatively 3–5 EV. The motion blur impacted on the captured image is higher for cameras with lighter masses. Thus, as most pictures of lightweight SPC are taken in a handheld way, IS is also a crucial feature for high-quality SPC. As for the methods implemented in the cameras, we can roughly differentiate between electronic image stabilization (EIS) and optical image stabilization (OIS). EIS requires a large amount of electronic memory and computational image treatment as the captured images are analyzed and corrected after exposure. OIS, on the other hand, is more directly acting on the optical system during exposure where lens elements and/or the sensor are displaced physically. A quite detailed discussion of OIS technology and its implementation in SPC is given in a white paper by ST Microelectronics.[4] In the following, we will shortly present the basic principle of OIS and a possible implementation in SPC.

The image of an object at rest can be characterized as the entity of all signals of the pixels on the image sensor. Due to involuntary shake or hand tremor during the exposure time, the information contained in one pixel will be distributed over neighboring pixel as the light cone exposing one pixel is moved over the sensor surface. This leads to image blurring and can be counteracted by a relative movement between lens and sensor (Figure 7.4). In the simplest way, the whole lens barrel may be moved in transver-

4 *Optical Image Stabilization*, White Paper, ST Microelectronics https://www.st.com/content/ccc/resource/technical/document/white_paper/c9/a6/fd/e4/e6/4e/48/60/ois_white_paper.pdf/files/ois_white_paper.pdf/jcr:content/translations/en.ois_white_paper.pdf

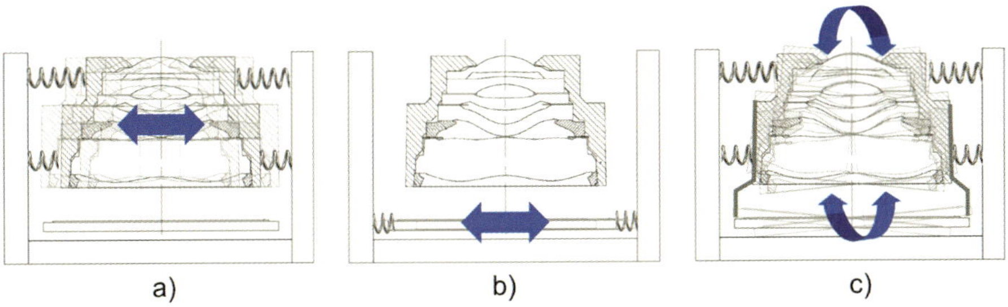

Fig. 7.4: Schematic camera module consisting of lens barrel (upper part), image sensor sensor (bottom) and return springs; other sensors and actuators are not shown. a) lateral movement of lens barrel relative to the fixed image sensor; b) lateral movement of image sensor relative to the fixed lens barrel, c) lateral and rotational movements of the lens barrel relative to the fixed image sensor [Bla21] (Author V. Blahnik, reprinted with kind permission).

sal direction relative to the image sensor fixed in the camera module. But also a shift of the image sensor with a fixed lens barrel has been implemented in some SPC (e. g., Apple iPhone 12 Pro and others). The sensor shift has the advantage of operating faster than the lens shift. A more advanced compensation method is realized by a miniaturized "gimbal" system (e. g., vivo X51 and others) where the lens barrel not only can be shifted transversally but also tilted around three rotational axes relative to a fixed sensor. For an efficient operation, a control system consisting of sensors, actuators and microcontroller is required. Sensors determine the location and tilt, respectively, rotation of the lens barrel relative to the image sensor. For instance, a Hall sensor is able to measure linear movements. Accelerometers, or better still gyroscopes, can determine rotational movements around three axes in space, caused by vibrations. All these sensors have the advantage that they can be integrated as MEMS in silicon technology alongside the microcontroller within the camera module. Their data are fed to the microcontroller to compute the required counteraction, which is then realized using actuators. As actuators, predominantly voice-coil motors are used for lateral and also for rotational movements. By this method, lateral or tilting movements of the camera can be compensated for vibrational frequency components up to about at least 20 Hz. EIS can additionally compensate for rotations around the optical axis.[5]

7.1.2 Lens design

The efficient design of a SPC-lens is predominantly governed by the limited space in smartphone housings. This leads to short focal lengths. Using the simple scheme pre-

5 *Optical Image Stabilization*, White Paper, ST Microelectronics https://www.st.com/content/ccc/resource/technical/document/white_paper/c9/a6/fd/e4/e6/4e/48/60/ois_white_paper.pdf/files/ois_white_paper.pdf/jcr:content/translations/en.ois_white_paper.pdf

sented in Figure 7.1, we can define a construction parameter V_c, which is the ratio between the construction length l_c of the SPC lens and the diameter d_{sensor} of the image sensor:

$$V_c = \frac{l_c}{d_{sensor}} \tag{7.8}$$

Under construction length, we understand in this case the extension from the front of the lens, which is virtually identical with the entrance pupil, to the image sensor. In this simple scheme, l_c is nearly the same as the focal length f, and thus we get a relation between l_c and the field of view given by Ψ:

$$V_c = \frac{l_c}{d_{sensor}} \approx \frac{f}{d_{sensor}} = \frac{1}{2 \cdot \tan \frac{\Psi}{2}} \tag{7.9}$$

A larger image sensor offers the potential of higher image resolution. As a consequence, larger sensors at a given l_c lead to larger FOV and shorter V_c, which is the reason that the standard lenses in SPC with optimized resolution have wide angle perspective as described above. But we may mention that this may change in future, when advanced optical elements such as space plates would be implemented (see Section 7.5.). On the other hand, wide-angle lenses require more effort for the correction of lens aberrations than normal lenses and have a slightly distorted perspective compared to them.

7.1.2.1 Standard wide-angle lens

The classical way to design a wide-angle lens with good corrections of all third- and most fifth-order aberrations is the combination of several spherical lenses. Let us here consider the Zeiss Biogon lens, which has already been discussed in Section 6.5. The Biogon (Figures 6.23b and 7.5a) is a wide-angle lens with a nearly symmetrical arrangement of lens elements around the aperture stop in the center. Due to the inherent symmetry, nearly all transversal aberrations can be avoided. Curvature of field can be corrected by the Petzval sum, chromatic aberrations by using achromatic lens combinations and spherical aberrations and partly astigmatism by choosing appropriate lens bendings (see Section 3.5). The lens elements close to the stop in the center are positive while they are negative at the outer parts of the lens. The Biogon can be downscaled to yield the same image size and focal length required for a smartphone. The $f_\#$ is maintained, but the size is still too large for a SPC [Ste12]. Nevertheless, this lens can serve as a reference to which a SPC lens is compared.

SPC lenses must have a large aperture to minimize diffraction as well as exposure noise. Hence, a good correction of aberrations is required. The fixed aperture stop, identical to the entrance pupil, is located at the front end of the lens. A correction in the classical way based on symmetry and adding more spherical lenses is no longer possible, given the overall size restrictions in SPC. Here, a different approach is required, which leads to a paradigm shift in the optical lens design [Bla21]. All elements in the

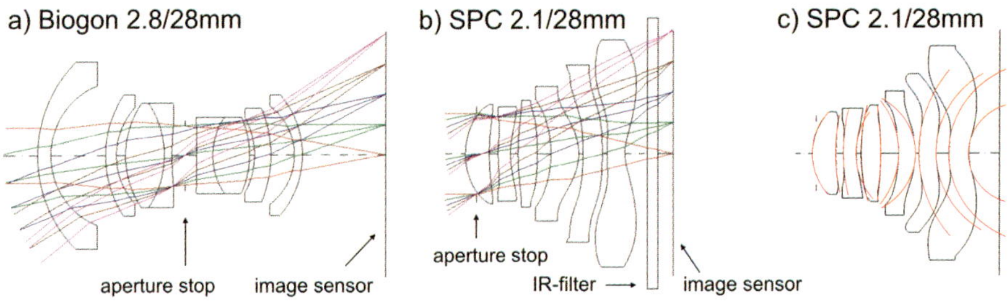

Fig. 7.5: a) Biogon 2.8/28 mm; b) SPC wide angle lens 2.1/28 mm, scaled to the same image size; c) the approximation in the center of the aspheric surfaces in the SPC lens by spheres is indicated by red lines. (Author: V. Blahnik).[6]

SPC have aspheric surfaces on their both sides (Figure 7.5). As each aspheric surface can be expressed mathematically by a power series with an arbitrary number of aspheric coefficients (see Section 3.5.7), there are enough parameters available for lens optimization. This, however, means a large number of possible solutions. A good start for the lens design can be a known lens based on spherical elements where low-order aberrations are well corrected. Then in subsequent steps and increasing FOV, the lens surfaces are modified to be aspheric.[7] While the first 2 to 3 aspheric surfaces behind the front end can be regarded as moderately aspheric with aspheric coefficients up to about the tenth order, the subsequent surfaces are rather extremely aspheric with coefficients up to the sixteenth order. In this design, it is not possible to eliminate third- and fifth-order aberrations by a number of spherical elements, but they are compensated by extremely high-order aberrations. Most of the lens elements can no longer be classified in the traditional sense of being positive or negative, but rather by having a complicated distribution of refractive power, which is a function of the position where the ray hits the lens. The basic lens form for paraxial rays, which travel close to the optical axis, still can be approximated by spherical surfaces, as shown in Figure 7.5c. Further away from the axis, the character of the element changes. This becomes most obvious for the rear elements. For the design of the complete lens, the path of all rays must be taken into consideration as they encounter position-dependent refraction. The difference between the classical Biogon design and the SPC lens becomes obvious from the ray paths shown in Figure 7.6.

Both lenses have nearly the same FOV and focal length. In the Biogon, the entering rays are bent in a way to reduce the divergence and achieve smaller angles relative to the optical axis in the center part of the lens (Figure 7.6a). All chief rays have a

6 H. J. Brückner, V. Blahnik, U. Teubner: *Smartphonekameras, Teil 1: Kameraoptik - Maximale Bildqualität aus Minikameras*, Physik in unserer Zeit, **51**(5) (2020) 236.

7 Dave Shafer: *A new theory of cell phone lens designs*, https://de.slideshare.net/operacrazy/a-new-theory-of-cell-phone-lenses

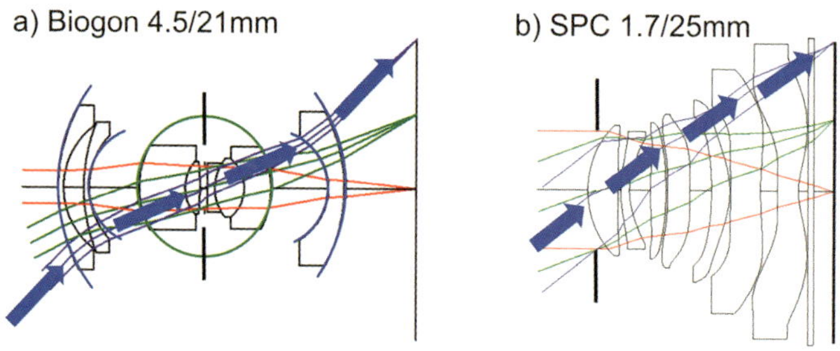

Fig. 7.6: Comparison of ray paths. (a) Biogon (4.5/21 mm); (b) SPC lens (1.7/25 mm); both lenses scaled to same image size [Bla21] (Author V. Blahnik, reprinted with kind permission).

nearly symmetrical path with respect to the aperture stop, which is located in the center of the lens, as mentioned above. In the SPC lens, being significantly shorter and having even a lower $f_{\#}$ than the Biogon design, the chief rays pass almost in a straight line through the lens (Figure 7.6b). The peripheral rays hit the rear lenses at high elevation and under large angles with the optical axis, which results in contributions to higher-order aberrations. It can be seen that the selected ray bundle in Figure 7.6b hits the last aspheric lens in a location where its refraction is that of a positive element whereas the bundle parallel to the optical axis encounters a rather negative, namely diverging section of the last lens element [Bla21]. The overall ray path is shorter than in the Biogon. As the various rays in the SPC lens have varying light path characteristics, their aberrations during propagation vary correspondingly, which is mirrored in the aberration curves [Bla21]. It turns out that the overall optical quality of the SPC lens is at least as good as that of the Biogon, and in some aspects even better [Ste12]. The optimization process for the aspheric lens design requires a large amount of computational power. The classical design rules based on spherical elements fail to work. Therefore, this method has only become possible with modern and fast computer systems in the last 1–2 decades.

The complicated wiggly aspheric shapes of most lenses cannot be realized in conventional silica-based glasses using the traditional polishing methods. Therefore, the lenses are prepared usually by molding techniques with plastic materials (see also Section 3.5.7). The drawback of the material is that it shows much higher dispersions and can be less controlled than conventional glass materials. Also, long-term stability may be an issue for system cameras with long-lasting lenses. Big advantages of the molding technique are that, besides the arbitrary shaping possibility, the lens elements are molded to have mounting sections at the edges. Thus, they can be easily stacked on top of each other, often with ring stops to prevent parasitic stray light, and no further mounting adjustment has to be made (Figure 7.7) [Bla21].

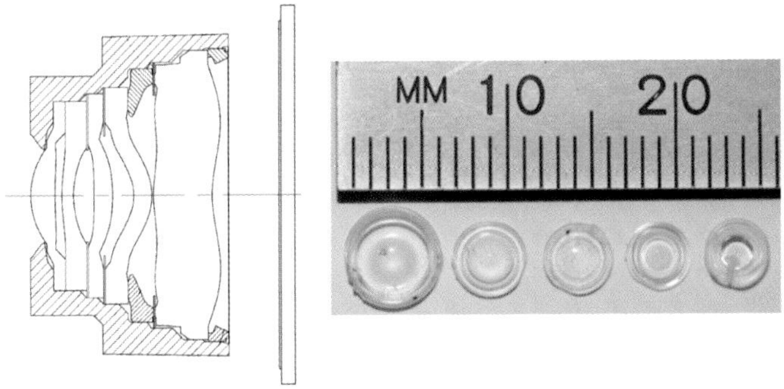

Fig. 7.7: SPC lens mount concept; molded lens elements can be directly stacked on top of each other (Author: V. Blahnik).[8]

7.1.2.2 Evolution of SPC lens modules

With the technological progress of the image sensors, the pixel size was continuously reduced to a common value of about 0.7 µm (see Section 7.2). This necessitates large lens apertures, i. e., low f-numbers, to diminish diffraction and also to improve the illuminance on the sensor chip. As discussed above, larger apertures lead to stronger lens aberrations that can be corrected by a more sophisticated lens design and using more elements. Figure 7.8 shows that from the years 2004 to 2016, f-numbers of standard wide-angle SPC modules could be improved from about 2.9 to 1.7. Simultaneously, the number of aspherical elements increased from 3 to 7, while the equivalent focal length of the standard lens settled down to 28 mm. The growing customer demand for adaptable FOV like in zoom lenses led to the development of multicamera systems. True zoom lenses of good quality, featuring a mechanical lens shift to adapt the focal length, were hardly compatible with the size limitations of compact smartphones. Around 2016, the first smartphones with a practical dual camera system became commercially available. Besides the standard wide-angle module with $f_{eq} \approx 28$ mm, they usually featured a moderate tele module with f_{eq} between 50 mm and 60 mm. It should be mentioned here that these modules are not strictly tele lenses in compliance with the usual classification of lenses. As tele lenses, we consider lenses with f_{eq} significantly larger than $f_{norm,FF} \approx 50$ mm or at least f significantly larger than its image sensor diagonal (see Sections 2.2 and 6.3). But compared to the standard modules with the characteristics of wide-angle lenses, they are conveniently termed tele lens modules. These modules are much more appropriate for portrait photography, as discussed above, and thus are sometimes called portrait lenses. In the subsequent years, SPC with three, four or even more lens modules emerged. These multicamera system are able to cover a FOV range

8 H. J. Brückner, V. Blahnik, U. Teubner: *Smartphonekameras, Teil 1: Kameraoptik - Maximale Bildqualität aus Minikameras*, Physik in unserer Zeit, **51**(5) (2020) 236.

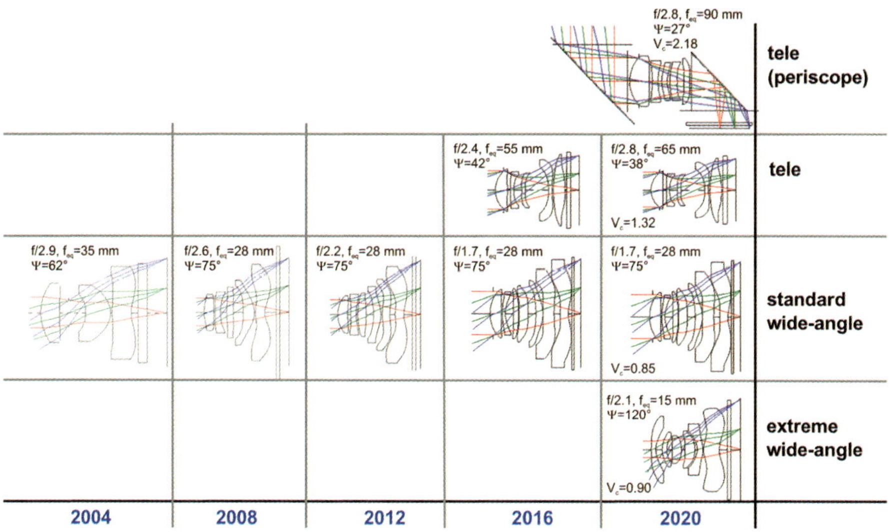

Fig. 7.8: Evolution of SPC lens design over time [Bla21] (Author V. Blahnik, reprinted with kind permission).

from extreme wide-angle with $\Psi \approx 120°$ down to $\Psi < 30°$, however, only by dedicated fixed lens modules (Figure 7.8).

Due to its relative long focal length compared to the image sensor diagonal, the implementation of a true tele lens for a SPC becomes still more difficult than for a standard lens due to the size restrictions. However, the construction length of a tele lens can be shortened by the application of the telephoto principle (see Sections 6.3.1 and 3.3.5). With positive lens groups at the front of the lens and negative groups behind, the total length becomes shorter than the focal length. But the smaller the telephoto ratio l_c/f, the stronger the refractive power of the lens groups, and thus the higher the effort for the correction of the lens aberrations. Here, the lens performance becomes limited. For tele lenses, a second difficulty arises, which is due to their reduced FOV. According to the simple consideration given by Equation (7.9), the construction parameter V_c increases with reduced field angle Ψ. As can be seen in Figure 7.8, the least V_c is achieved for the standard lenses and increases strongly for tele lenses. As a consequence, the useable diameter d_{sensor} of the image sensor becomes smaller while at same time the f-number of the tele lens, which is the ratio f/D_{en}, increases.

Figure 7.9 shows the cross-sectional view of the two lens modules of a dual SPC. Both image sensors have the same number of pixels, namely 12 MP but different image diagonals and pixel pitches. For the standard wide-angle lens with $f_\# = 1.8$, the blur diameter of the Airy disk is $u_d = 2.4\,\mu m$ (Equation (2.23)) while the pixel pitch is $p = 1.2\,\mu m$ on a sensor with $d_{sensor} = 6.0\,mm$. These are the same data as discussed for the reference SPC above. The sensor has the usual match with the pixel pitch being equal to half of the Airy diameter. This can be seen in Figure 7.9b where the distance between the zeroes of PSF_{optics} is identical to $2p = 2.4\,\mu m$. MTF_{optics} is over large ranges superior to MTF_{sensor}. The 1D cutoff of MTF_{optics} is achieved for $N_{SB,cut} = 3660\,lp/PH$ compared to the Nyquist limit of MTF_{sensor} at 1500 lp/PH (Figure 7.9c). The resolution of the whole system, expressed by MTF_{system}, is mainly dominated by MTF_{sensor}, which is the usual case for cameras with high-quality lenses.

In the tele lens with a smaller sensor and $f_{\#} = 2.8$, the Airy disk of 3.8 µm diameter is larger than in the standard module due to the higher $f_{\#}$ and is significantly larger than twice the pixel pitch (Figure 7.9b). The pitch in this module is of only 1.0 µm on a sensor with $d_{sensor} = 5.0\,mm$. The upper limit of image points yields $N^{(2D)}_{SB,cut} = 5.0 \cdot 10^6$ and is about a factor of 2.4 inferior than the number of pixels on the sensor. The 1D optical cutoff is at 1926 lp/PH according to Equation (5.69) and slightly above that of the sensor at 1500 lp/PH, using Equation (5.64b). It can be seen from Figure 7.9c that the sensor MTF is now almost over the whole range better than that of the optics. Nevertheless, the system cutoff is determined by the sensor. Despite this, the optical quality of the lens can now be seen as critical for the overall resolution of the complete system. This is typical for long focus camera modules. A more detailed discussion of system resolution will be given in Section 7.3 for different SPC.

The limitation of the focal length by the thickness of the smartphone body can be overcome when the optical axis is rotated by 90° using a folded mirror or a prism. This leads to the periscope set-up of the tele lens module as shown in Figures 7.8 and 7.10. With this transversal orientation of the lens module, the length of the optical axis is virtually not limited. However, the diameter of the lens elements, and thus the diameter D_{en} of the entrance pupil becomes the bottleneck as is limited by the thickness of the smartphone housing. If the smartphone should not become too thick, a realistic maximum diameter of $D_{en} \approx 4\,mm$ cannot be exceeded. This, however, determines the maximum useable focal length of the lens. From the definition of $f_{\#}$ according to Equation (2.14)

► **Fig. 7.9:** a) Cross-section photograph of a dual Smartphone camera with two 12 MP sensors; left side: module 1, $f_{eq} = 28\,mm$, $f_{\#} = 1.8$, $d_{sensor} = 6\,mm$, $p = 1.2\,\mu m$; right side: module 2, $f_{eq} = 56\,mm$, $f_{\#} = 2.8$, $d_{sensor} = 5\,mm$, $p = 1.0\,\mu m$. b) point-spread functions of the modules with Airy blur spot diameter u_d; left side: module 1, $u_d = 2.4\,\mu m$; right side: module 2, $u_d = 3.8\,\mu m$. c) modulation transfer functions MTF of the modules; left side: module 1, MTF_{sensor} dominates the system; right side: module 2; both MTFs have approximately the same importance. d) 2D false color images of the normalized PSF (true to scale); horizontal cut of these 2D plots through the center yield the curves plotted in (b); note the resolution of the sensor is $2p$ in horizontal and vertical direction, respectively. e) 2D false color images of the MTF (true to scale); horizontal cut of these 2D plots through the center yield the curves plotted in (c); note that PSF and MTF are directional; in diagonal direction they are worse than along the axes. Note that all curves in (b), (c), (d) and (e) show 2d computations for $\lambda = 550\,nm$ and for ideal conditions (in the sense of Figure 5.30b). Thus, they are expected to be better than in real systems. PSF is normalized to 1. (a) is reprinted from [9] (Author V. Blahnik).

9 H. J. Brückner, V. Blahnik, U. Teubner: *Smartphonekameras, Teil 1: Kameraoptik - Maximale Bildqualität aus Minikameras*, Physik in unserer Zeit, **51**(5) (2020) 236.

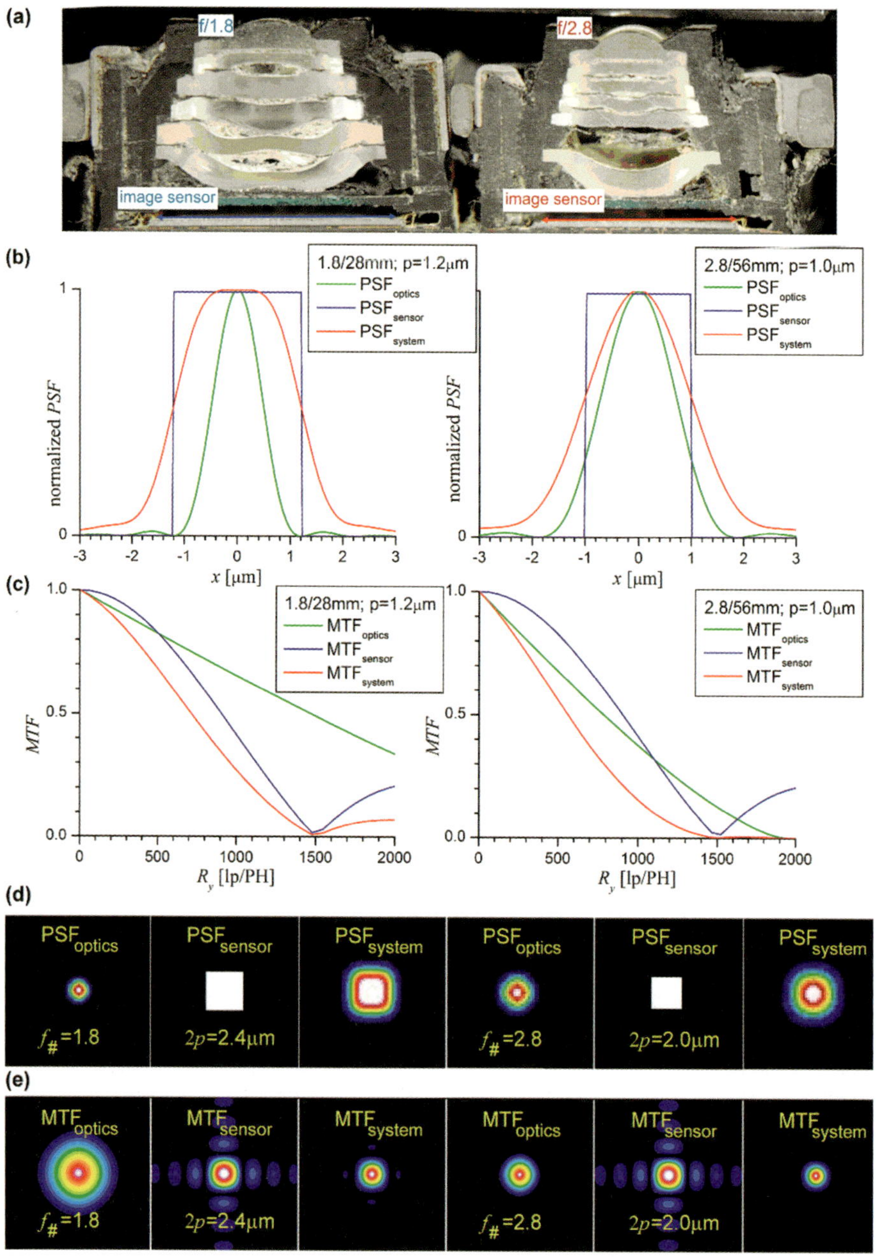
(a)
f/1.8
f/2.8
image sensor
image sensor
(b)
1.8/28mm; p=1.2μm
PSF optics
PSF sensor
PSF system
2.8/56mm; p=1.0μm
PSF optics
PSF sensor
PSF system
normalized PSF
x [μm]
(c)
1.8/28mm; p=1.2μm
MTF optics
MTF sensor
MTF system
2.8/56mm; p=1.0μm
MTF optics
MTF sensor
MTF system
MTF
R_y [lp/PH]
(d)
PSF optics
PSF sensor
PSF system
f# =1.8
2p=2.4μm
f# =2.8
2p=2.0μm
(e)
MTF optics
MTF sensor
MTF system
f# =1.8
2p=2.4μm
f# =2.8
2p=2.0μm

follows that $f = f_\# \cdot D_{en}$. It is difficult to achieve f-numbers as low as 2.8 for tele lenses. Assuming $f_\# \approx 2.8$ and $D_{en} \approx 4$ mm, a maximum useable focal length should not exceed 11 mm in a periscope design if a high resolution should be maintained. This focal length in combination with an image sensor of 4.3 mm diameter has the characteristics of a 110 mm equivalent lens of the full format. Here again, we may mention that length limitations may change in future, when advanced optical elements such as space plates would be implemented (see Section 7.5).

Some SPC with periscope modules have larger f-numbers, namely $f_\# = 3.4$ in combination with an image sensor diagonal of 4.5 mm (e. g., vivo X60 pro+, Huawei P30 pro). Their focal length of 13.0 mm corresponds to $f_{eq} = 125$ mm (see Table 7.4). The blur diameter of the Airy disk for $f_\# = 3.4$ is $u_d = 4.6$ µm after Equation (2.23), and thus more than 3 times larger than the pixel pitch. Hence, the overall image resolution, compared to the 8 MB sensor specification, is strongly reduced. In this example, the optical cutoff is $N^{(2D)}_{SB,cut} = 2.8 \cdot 10^6$ while the sensor of the modules has roughly 3 times more pixels. The 1D optical cutoff along the picture height according to Equation (5.69) yields 1445 lp/PH. In this case, it is slightly above the cutoff as given by the Nyquist limit $R_N = 1227$ lp/PH using Equation (5.64). In cases when the 1D optical cutoff is close to the Nyquist limit, the optical system may be the bottleneck for the whole camera resolution. This can be seen from the fact that $N^{(2D)}_{SB,cut}$ is significantly smaller than the number of pixels on the sensor (see also Equation (1.25)). Calculations of $N^{(2D)}_{SB,cut}$ are also done for other camera modules and listed in Tables 7.1, 7.4–7.6. If their $N^{(2D)}_{SB,cut}$ values are multiple times lower than the pixel number of the module, the overall camera resolution may be improved by a better optical system. However, in order to better understand the overall system resolution, the corresponding MTF curves must be discussed. This is done in Section 7.3 in comparison to other SPC modules.

Besides tele lens modules with a narrow FOV, extreme or ultrawide-angle lenses for an extended FOV up to 120° or more have been developed in the recent years [Bla21].

Tab. 7.4: Technical specifications of vivo X60 Pro + smartphone with 5 camera modules. (Source of some data: vivo.com, dxomark.com.)

vivo X60 Pro + (2021)	**ultra-wide**	**main**	**front**	**portrait**	**periscope**
equivalent focal length f_{eq}	14 mm	23 mm	26 mm	50 mm	125 mm
f-number $f_\#$	2.2	1.6	2.5	2.1	3.4
field of view Ψ	114°	87°	80°	47°	20°
focal length f	2.6 mm	6.4 mm	3.9 mm	7.5 mm	13.0 mm
sensor diagonal d_{sensor}	8.0 mm	12.1 mm	6.5 mm	6.5 mm	4.5 mm
crop factor CF	5.4	3.6	6.6	6.6	9.6
pixel number	48 MP	50 MP	32 MP	32 MP	8 MP*
pixel pitch p	0.8 µm	1.2 µm	0.8 µm	0.8 µm	1.1 µm
sensor format	1/2.0″	1/1.31″	1/2.8″	1/2.8″	1/4.0″
optical cut-off $N^{(2D)}_{SB,cut}$	$21.0 \cdot 10^6$	$90.9 \cdot 10^6$	$10.7 \cdot 10^6$	$15.2 \cdot 10^6$	$2.8 \cdot 10^6$
remarks	gimbal OIS				

Tab. 7.5: Technical specifications of Samsung Galaxy S22 ultra smartphone with 5 camera modules. For binning, see the discussion of multipixel cell technology in Section 4.10.7.1. (Source of some data: samsung.com, dxomark.com.)

Samsung Galaxy S22 ultra (2022)	ultra-wide	wide	front	telephoto	periscope
equivalent focal length f_{eq}	13 mm	23 mm	26 mm	70 mm	230 mm
f-number $f_\#$	2.2	1.8	2.2	2.4	4.9
field of view Ψ	120°	85°	80°	36°	11°
focal length f	2.1 mm	6.4 mm	3.8 mm	8.3 mm	27.2 mm
sensor diagonal d_{sensor}	7.0 mm	12.0 mm	6.4 mm	5.1 mm	5.1 mm
crop factor CF	6.2	3.6	6.8	8.5	8.5
pixel number	12 MP	108/12 MP	40/10 MP	10 MP	10 MP
pixel pitch p	1.4 μm	0.8 μm	0.7 μm	1.1 μm	1.1 μm
sensor format	1/2.55″	1/1.33″	1/2.82″	1/3.52″	1/3.52″
optical cut-off $N_{SB,cut}^{(2D)}$	$16.1 \cdot 10^6$	$70.7 \cdot 10^6$	$13.5 \cdot 10^6$	$7.2 \cdot 10^6$	$1.7 \cdot 10^6$
remarks		1:9 binning	1:4 binning		

Tab. 7.6: Technical specifications of Sony Xperia 5 III (2021) smartphone with 4 camera modules; the telephoto periscope operates as a varifocal zoom lens. (Source of some data: sony.com, gsmarena.com).

Sony Xperia 5 III (2021)	ultra-wide	wide	front	telephoto – periscope
equivalent focal length f_{eq}	16 mm	24 mm	27 mm	70–105 mm
f-number $f_\#$	2.2	1.7	2.0	2.3–2.8
field of view Ψ	107°	84°	78°	34°–23°
focal length f	2.6 mm	5.2 mm	2.8 mm	10–15 mm
sensor diagonal d_{sensor}	7.0 mm	9.5 mm	4.5 mm	6.23
crop factor CF	6.2	4.6	9.6	7.0
pixel number	12 MP	12 MP	8 MP	12 MP
pixel pitch p	1.4 μm	1.9 μm	1.0	1.2 μm
sensor format	1/2.55″	1/1.70″	1/4.0″	1/2.90″
optical cut-off $N_{SB,cut}^{(2D)}$	$16.1 \cdot 10^6$	$49.7 \cdot 10^6$	$8.0 \cdot 10^6$	$11.7–7.9 \cdot 10^6$
remarks	dual pix	dual pix, OIS		dual pix, OIS

Compared to standard wide-angle lenses, they typically have one or two “ray-bending” negative aspherical lenses in front of the aperture stop. These aspheric elements have curvatures that become stronger at their edges, and thus bend the incoming peripheral rays from about 60° to less than 40° when passing through the aperture stop (angles with respect to the optical axis). The ray path behind is similar to that of a standard wide-angle lens. The construction length can be kept relatively short, if a fisheye-like distortion of up to 20–30 % is permitted. The construction parameter V_c is somewhat larger than for standard lenses but can be kept below 1. Future developments with fisheye-like perspectives and larger FOV can be expected.

As for the overall quality of the SPC lenses, most of them come close to that of diffraction limited lenses if we consider the central part of the image field. Further away from the center, the quality decreases but not too much. Most lenses, except the ultrawide-angle ones, exhibit very low distortion of about 1 % or less and are virtually not visible [Bla21]. Wide-angle lenses tend to have a barrel-like distortion, which is allowed to be up to 20–30 % for ultrawide lenses with nearly fisheye characteristics, as mentioned above. In some cameras, these distortions can be corrected by software applications which, however, reduce the overall image resolution as image parts have to be cropped.

Finally, we may also mention the particular developments on image systems with curved sensors (see Section 4.12), which may lead to significant improvements (see Section 6.5.4 and[10]).

7.1.2.3 Zoom lenses, digital and hybrid zoom

The principle of zoom lenses has been discussed in Section 6.6. In order to continuously vary the focal length over a certain zoom range, which is usually a multiple of the shortest focal length, lens groups are displaced. This requires more space than is usually available in perpendicular direction to the smartphone back. Such a zoom lens was, for instance, realized in the Nokia smartphones N93 and N93i (2006–2007)[11] as a "bulky" transversal camera, but it was not successful on the market. Modern alternatives for a smartphone could be based on a telephoto lens with a folded light path as in the periscope design as claimed in a US patent from 2016.[12] A similar principle has been realized in the Sony Xperia 5 III (2021). It features a true varifocal lens, designed and made by Zeiss, with a zoom factor of 1.5 to continuously zoom from an equivalent focal length of 70 mm to 105 mm. Due to the increase of the focal length at fixed entrance pupil, the f-number rises from 2.3 at 70 mm to 2.8 at 105 mm. The schematic lens arrangement with the periscope prism mirror is similar to the periscope set-up shown in Figure 7.10. The lenses are combined in 3 groups: the first group behind the prism mirror, consisting of 3 lenses, and the last group near the image sensor, consisting of 2 lenses, are fixed in the entire barrel. All lenses are aspherical. If the center group of 2 lenses is close to the rear 3rd group near the sensor, we get the shortest zoom focal length with the lowest image magnification.[13] The longest focal length is achieved for the center group to be close to the 1st group. Besides the movement of the central group by a voice coil, the whole lens barrel is adjusted by another voice coil drive to achieve a sharp image on the sensor. The movement of the central group with the simultaneous barrel shift is usually done in a nonlinear way as described in Section 6.6. Additionally, voice coil actuators

10 D. Shafer: *Lens designs with extreme image quality features*, Adv. Opt. Techn. **2** (2013) 53–62.

11 https://en.wikipedia.org/wiki/Nokia_N93

12 https://patentimages.storage.googleapis.com/ab/aa/41/39443fc28d24f8/US9316810.pdf

13 https://www.techinsights.com/blog/sony-xperia-5-iii-uses-periscope-camera-two-zoomsettings?utm_medium=email&utm_source=techinsights&utm_campaign=2022_Q1_Image%20Sensor_Blog_Sony_Xperia_5_III

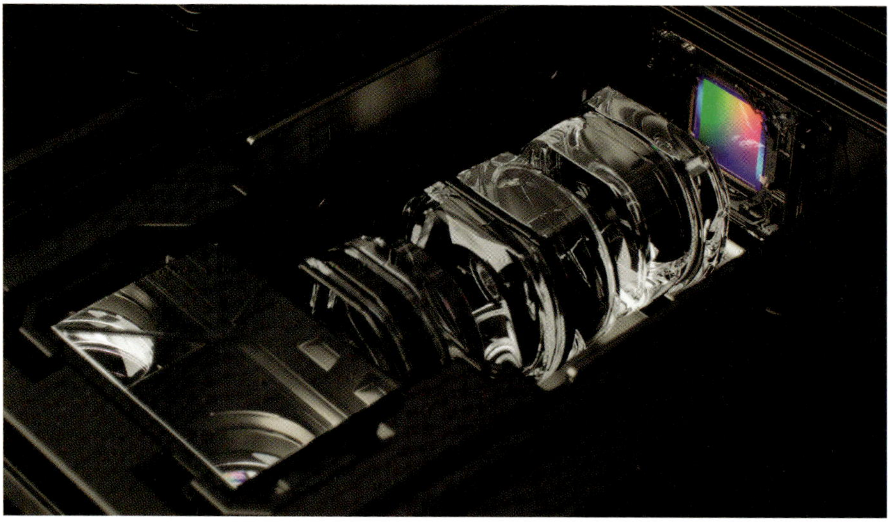

Fig. 7.10: Lens arrangement in the periscope lens of Sony Xperia 5 IV with fixed lens groups; the prism mirror on the left side rotates the optical axis by 90°; the lens arrangement in the zoom periscope of Sony Xperia 5 III is similar, but the central lens group is movable to achieve zooming. Image reprinted with kind permission of Sony.

connected to the periscope prism effectuate a tilt of the prism for optical image stabilization. It should be noted that the zoom factor of this module from 70 mm to 105 mm is 1.5 while the whole SPC is advertised as having a zoom factor of more than 4. The latter value can only be calculated if the longest available focal length of the tele module, namely $f_{eq} = 125$ mm, is related to that of the main wide-angle module with $f_{eq} = 24$ mm. This yields the zoom factor of the hybrid zoom of the total SPC system (see below).

Although there is more length available in the periscope design, the transversal aperture for the entrance pupil is still the bottleneck. This becomes more critical the longer the focal length is. Therefore, extreme long focal lengths like those found sometimes with system cameras, are not achievable for SPC. For an intermediate range of focal lengths, it seems sometimes more appropriate to use a so-called digital zoom, also termed software zoom, which is simply cropping an image and blow the content up. Unlike optical zooming, where zooming in the ideal case never deteriorates the optical image quality, the digital zooming is always associated with loss of resolution (see Section 1.4.3 and, in particular, Figure 1.14).

For instance, if an image is digitally zoomed by a given zoom factor in one direction, a part of the image is cropped and blown up to the full picture size. The remaining image part is discarded. This is the same effect as if in the camera module the original sensor would be replaced by a sensor of smaller size, and correspondingly, less pixels. For instance, a digital zoom by a factor of 2 reduces the picture height by 2, yielding a cropped value $PH' = PH/2$, and in the same way also its width. Thus, the total number of effective pixels is reduced by a factor of 4 for the cropped image, as already discussed in Section 1.4. But what does not change is the wavelength of light. This leads to the effect that diffraction blur relative to the picture height becomes more obvious when the picture is blown up to the usual size for inspecting it. This, for instance, is the case for zoomed pictures on the smartphone display or as paper print out. As a consequence, MTF_{optics} of the cropped image becomes worse due to the decreasing ratio of PH'/r_0, with r_0 being the radius of the Airy disk. Simultaneously, MTF_{sensor} degrades as the Nyquist cutoff decreases with the smaller number of pixels along PH'. Both MTF_{optics} and MTF_{sensor} are multiplied to yield the resolution of the system given by MTF_{system}.

Let us now consider cases in which digital zooming may be appropriate and in which not. If we take a look again to the example of the dual SPC in Figure 7.9, we can see that the upper limit of image points is $N_{SB,cut}^{(2D)} = 17.7 \cdot 10^6$ for the 28 mm lens, while for the 56 mm module it is of $N_{SB,cut}^{(2D)} = 5.1 \cdot 10^6$. If we take into account only the optical quality and disregard any sensor influence, then a digital zoom by a factor of 1.5 from the 28 mm lens to an equivalent focal length of 42 mm yields $N_{SB,cut}^{(2D)} = 7.9 \cdot 10^6$ (note that a factor of 1.5 in 1D yields a factor of 1.52 in 2D). This may be sufficient. But a digital zoom by a factor of 2 to 56 mm would yield $N_{SB,cut}^{(2D)} = 4.4 \cdot 10^6$, and thus has a lower optical resolution than that of the 56 mm module itself. Additionally, the Nyquist limit is reduced by digital zooming (see above). In this case, a switch to the module with the higher resolution and not a digital zoom is the better choice. This combination of a digital zoom in combination with a switch to a module with longer focal length and better resolution is termed hybrid zoom. Due to lack of high-quality optical zoom lenses, this is the conventional zooming strategy, which is applied in most modern smartphones with multicamera systems. It should be stressed again here that we have neglected for simplicity the sensor influence. Considering both influences of sensor and optical resolution is more appropriate. Some corresponding MTF-curves are discussed in Section 7.3.1.

As a different example, we may regard Sony's Xperia 5 III with its true optical zoom. The drop of the optical resolution when zoomed from 70 mm to 105 mm is indicated by a reduction of $N_{SB,cut}^{(2D)}$ from $11.7 \cdot 10^6$ to $7.9 \cdot 10^6$, namely by a factor of 1.4. This may be compared to a corresponding software zoom of the 70 mm module, which leads to $N_{SB,cut}^{(2D)} = 5.2 \cdot 10^6$. This is less than half of the original value. But what is more important is that the image size with the true zoom lens is not cropped. Hence, the sensor and its resolution are not impaired by zooming. The varifocal lens module of the Sony/Zeiss system has the advantage that a better optical resolution is achieved at all focal lengths when compared to any digital zoom. Hence, this justifies the real zoom lens technology. A more detailed discussion for the total resolution is given in Section 7.3.1.3, based on MTF-curves.

7.2 Sensor systems of smartphone and miniature cameras

Due to the small size, the sensors of miniature cameras and SPC modules, respectively, have to be rather small as well. Depending on the pixel pitch, this sets a limit to the total number of sensor pixels or vice versa, an intended number of pixels determines the pixel dimension. This actually depends on the application. For instance, cameras used for automotive or surveillance not necessarily need to have very high resolution, but sensitivity is an issue. Thus, less but larger pixels may be preferable.

On the other hand, for SPC this is partly different. Today many SPC have CIS with approximately 12 MP. For usual imaging conditions, this is regarded to be sufficient (see also the preceding sections). But there are also camera modules with tens of MP and even such ones exceeding 100 MP. Some selected examples are listed in Appendix A.5 in Table A.4. Moreover, CIS with a number of pixels $N_{pix} = N_h \cdot N_v$ of 150 MP and 200 MP have been announced by Samsung and potentially other suppliers, which then, of course, require an appropriate much advanced image signal processing and data handling. But once again, it should be noted that such a large number should not be mistaken for resolution. It rather may be announced due to other purposes or even advertising (see Section 1.6.4). Moreover, this large number poses a lot of critical questions as discussed later in this chapter (see also Section 5.2.7.2). Although in the following, again we mostly concentrate on SPC, the related discussion is more general and can be applied to other miniature camera systems as well.

7.2.1 CIS properties

Due to the compact construction of SPC, the applied CIS are very small. Table 7.7 provides data of several examples. When compared to full format sensors, the SPC sensor diagonal is about 4 to 10 times smaller, namely by the crop factor CF. As a consequence of the large pixel number within a small sensor area, the pixel size becomes very small. The typical width of a pixel pitch p is below 1.2 to 1.5 µm and the smallest pitch these days is slightly below 0.6 µm. $p = 0.7$ µm has become rather common. Comparison of CIS 1 and 2 to CIS 3 and 4, respectively, gives an impression of the progress in CIS development.

Thus, p has become smaller than the size of the Airy disk of red light (for the general discussions, we relate, e. g., to Sections 2.1 and 2.5). This results in several consequences.

Tab. 7.7: Example of CIS data of two somewhat older SPC (“CIS 1” and “CIS 2”) and two current high-end SPC (“CIS 3” and “CIS 4”).

	“CIS 1”	**“CIS 2”**	**“CIS 3”**	**“CIS 4”**
sensor diagonal	13.3 mm	7 mm	9.2 mm	8 mm
pixel number	40 MP	15 MP	64 MP	48 MP
pixel pitch p	1.4 µm	1.12 µm	0.8 µm	0.8 µm
FWC	2450 electrons	4570 electrons	6000 electrons	4500 electrons
read noise	2.0 electrons	1.8 electrons	1.4 to 1.5 electrons	1.2 electrons
η (peak)	0.41	0.3	>0.8	0.75
dynamic range	1225 (10.3 EV)	2540 (11.3 EV)	4280 (12 EV)	3750 (11.8 EV)

The first one has been discussed in Section 7.1. The second consequence is that small pixels are less sensitive to light when compared to larger ones and they have a significant smaller signal-to-noise ratio (SNR). Despite the advancements (see Section 4.10), this is still an issue. Here, we consider the following situation. A particular CIS, which we regard as the reference, is illuminated homogenously by light with a fixed amount of photons. Then we compare other CIS with other pixel pitches, other sensor diagonals and other total number of pixels, but we assume that all of them are illuminated exactly in the same way. From those values, one can easily calculate the amount of photons for one pixel of each of the other CIS.

To that purpose, we consider the illuminance E_i in the image sensor plane. Let us assume for simplicity a homogeneously radiating circular object at large distance from the camera. Then the illuminance on the sensor is directly related to the f-number of the lens used with the optical sensor, namely $E_i \propto f_\#^{-2}$ (see Equation (2.15)). We would like to remember that the illuminance is the light power density, shortly termed light intensity, and can be used to calculate the optical power incident to a pixel simply by multiplying it with the pixel area. As the number of photons is linearly related to the optical power, we get for the number of photons per pixel N_{ph} the following relation:

$$N_{ph} \propto E_i \cdot p^2 \propto \frac{p^2}{f_\#^2} \tag{7.10}$$

It should be noted that this is valid for a symmetrical pixel arrangement with square or circular geometry and for monochromatic light. Furthermore, for rectangular or circular sensor geometries the pixel area can be calculated by dividing the sensor area, which is proportional to the square of the sensor diagonal d_{sensor}, by the number N of pixels on that sensor:

$$N_{ph} \propto \frac{d_{sensor}^2}{N \cdot f_\#^2} \tag{7.11}$$

Let us consider the number of photons per pixel incident to any CIS with a diagonal d_{CIS} and the pixel number N_{CIS}. We compare this to the reference CIS with d_{ref} and N_{ref}. Both sensors are assumed to have the same aspect ratio. Then we define a relative photon efficiency η_p per pixel, related to the reference CIS:

$$\eta_p = \frac{d_{CIS}^2}{N_{CIS} \cdot f_{\#CIS}^2} \cdot \frac{N_{ref} \cdot f_{\#ref}^2}{d_{ref}^2} \tag{7.12}$$

In case that $f_\#$ is the same for both systems, η_p describes that more photons are collected per pixel with increasing sensor area and decreasing number of pixels per chip. Thus, a SPC captures less light than a full format DSLR, namely by a factor of $d_{CIS}^2/d_{FF}^2 = 1/CF^2$, even if their numbers of pixels are identical. For a SPC with a typical $CF = 7$, each pixel

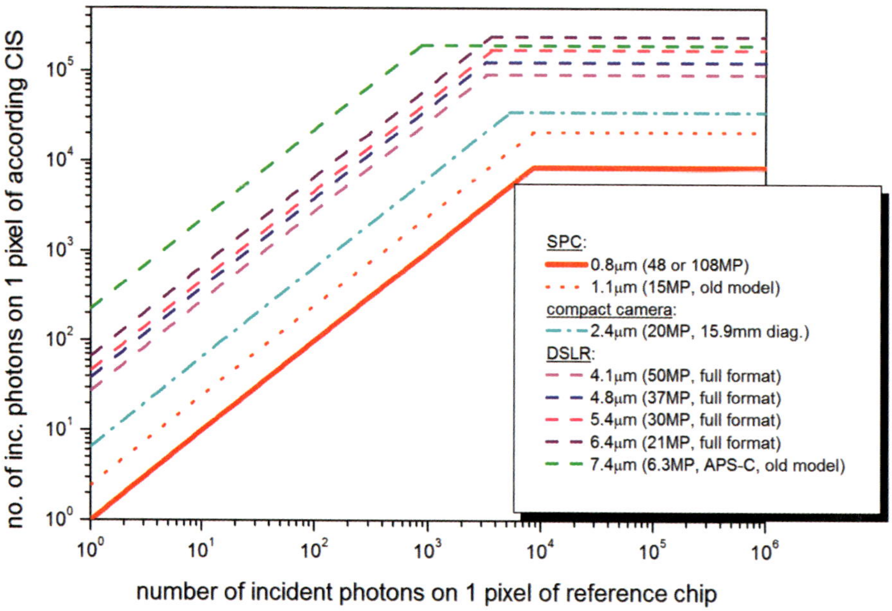

Fig. 7.11: Number of photons incident on a single pixel of each of various sensors when the illumination condition is exactly the same. The values of the x-axis provide the number of photons on one pixel of the $p = 0.8$ μm SPC-"CIS 4" of Table 7.7, which acts as the reference (see text; thick red line). For simplicity, we disregard fill factor, transmission losses, etc., $f_\#$ is the same for all cameras. Note that a change of $f_\#$ would lead to a parallel shift of the curves, but of course, this does not affect the saturation. η and FWC have been included in the calculations of each particular CIS.

gathers about 50 times less photons than a full-format camera of the same pixel number and same $f_\#$.

Equation (7.12) allows to compare different CIS with respect to their relative efficiency. This is illustrated in Figure 7.11. If we increase the number of photons incident on the reference CIS as displayed on the x-axis, then for the same illumination conditions, the amount of photons for one pixel of each of the other CIS increases as well (displayed on the y-axis). This continues until the number of photo electrons reaches FWC. Then the pixel signal becomes saturated. Although this diagram is trivial, it clearly illustrates the relative efficiency of different cameras and CIS, respectively.

For the presented example, we have chosen the 48MP SPC-CIS of Table 7.7 ("CIS 4") as the reference (thick red line) and compare it to an older SPC Table 7.7 ("CIS 2") and to CIS from a high-end compact camera, four full format DSLR and an older APS-C DSLR, which has quite large pixels. The corresponding pitches are provided in the legend. The lines show how much photons are received by one pixel of each CIS.

It is clear that a larger size of pixels and/or sensor surface leads to much more photons on each pixel. This provides a much better situation for DSLR when compared to SPC. Vice versa, the larger the number of pixels and/or the smaller the sensor, the less light is available to generate a signal within a pixel. This makes it more and more difficult to develop smaller sensors and to increase the number of mega pixels. Consequently, in situations with low light, illumination may be still sufficient for DSLR but not for SPC.

Light sensitivity is connected to "ISO-gain" (see S_{ISO} in Section 2.5, Section 4.8.8 and also the discussion on dual gain below). Hence, the "true ISO-values" according to ISO definition may be estimated to be 34 and 26 for the modern 48 MP and the older 15 MP SPC, respectively. For the modern DSLR, they are much higher (typically 80 for larger pixels; see Figure 4.54; note that doubling the ISO value means doubling the sensitivity). But in addition, larger sensors do receive more light in total (see Equation (7.12) and Figure 7.11). Thus, a typical SPC captures only between 1/40 and 1/20 of the light captured by a DSLR. This corresponds to 4.3 to 5.3 EV.

7.2.2 Further CIS properties: noise effects

Beside optical image quality and sensor sensitivity, noise is an important issue for cameras. As discussed in Sections 4.7 and 4.8, a minimum signal, namely a minimum number of photons, is necessary, in other words, the noise level has to be surpassed. Within an image, noise appears as a fluctuation of brightness. Thus, e. g., even a fully homogenous surface within an object leads to an observable random variation of image brightness within that area. Even more, coloured spots may appear and also wrong colors. This is illustrated in Figure 7.12, which shows an image and its crop taken with a somewhat older SPC to show the effect more clearly. However, albeit large progress has been made with SPC, similar noise effects are still present for today's SPC. In particular, there are situa-

(a) 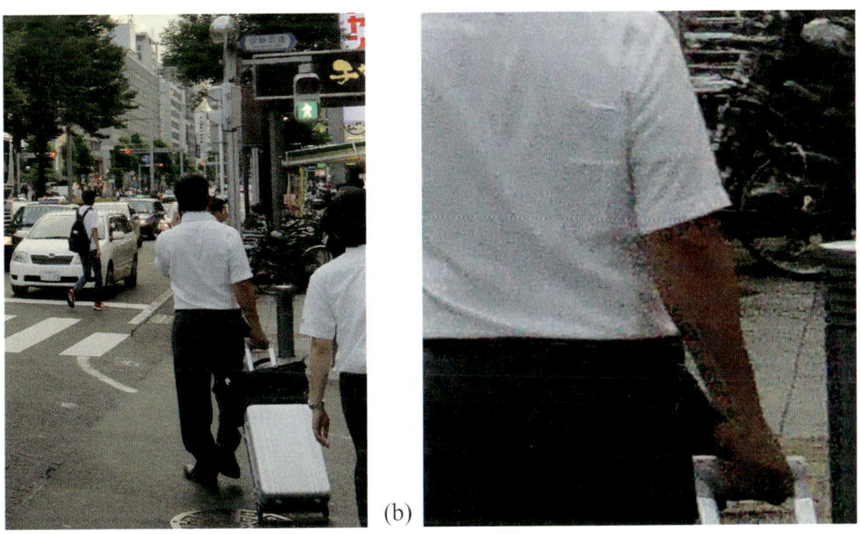 (b)

Fig. 7.12: (a) Illustration of noise in an SPC-image. (b) A detail of (a) shows this more clearly.

tions where even images taken with high-end SPC suffer from noise. Obviously, this degrades the image quality. We may mention that quite often noise, such as well seen in Figure 7.12, may not be seen in SPC-images due to denoising (see Sections 4.9 and 7.4). However, one has to be aware that this is always accompanied with loss of fine structures.

Noise has been treated in Section 4.7 in detail. Various sources of noise have been discussed and one may expect that some of them are of particular importance for SPC. An example may be thermal noise that leads to thermally activated electrons. Such temperature effects may be expected to be a particular problem with SPC, where in addition to the camera electronics there is a lot of other electronics within the smartphone and all that strongly concentrated in a small volume. However, due to the advanced chip architecture this kind of noise, and noise from the electronics in general, is rather low, even lower than for many DSLR (see, e. g., Table 7.7). This is for standard conditions. Nevertheless, one may note that any extensive use of electronics of a smartphone or of any camera will lead to thermal effects, which result in image degradation. A typical example is image capturing with high-framing rates (picture series or video), which moreover often is limited by the camera software or hardware to avoid too much heating of the device.

Another issue related to the small pixel size in SPC is blooming and cross talk (compare Section 4.7.5). However, CIS are generally less affected by blooming when compared to other image sensors such as CCD. Moreover, advanced CMOS technology makes use, e. g., of the introduction of additional potential wells (DTI; see Section 4.10.5.2) and overflow drains, CMOS stacking, sophisticated cross talk suppression techniques and new developed color filters (see Section 4.10.6). This has led to a further reduction of blooming and cross talk to less than 1 %.

The additionally rather large quantum efficiency of modern SPC-CIS ($\eta \approx 0.7$ to 0.8 or more; see Section 4.10.6) has led to an improved SNR. Thus, nowadays noise from electronics hardly plays a role under standard conditions of photography and may only become important for low light conditions. As a result, with exception of low light CIS illumination in general or in part of a captured image, the sensor itself and the electronics do not play much role for SPC (beside *PRNU*).

It may only become important for low light conditions and if exposure time is rather long, e. g., in case of astrophotography. Nevertheless, due to the fact that in a CIS the individual pixels are not perfectly the same, spatial noise such as FPN and PRNU (see Section 4.7.4) is still an issue, which may be compensated more or less well (one may obtain less than 1 % PRNU).

The noise situation is nearly fully determined by the photon nature of light, and thus by photon noise (see Section 4.7.2). This shows up also in the SNR (see Section 4.8.2). Figure 7.13 provides an example for SPC in comparison to DSLR and a compact camera.

From this diagram, it is obvious that SPC have a significant lower SNR when compared to compact cameras and DSLR. At large photon numbers, this is the result of the small pixel size (this is also related to Equation (7.12) because SPC have small sensors and large pixel numbers). This is the region where noise is dominated by σ_{ph}. Here, SNR

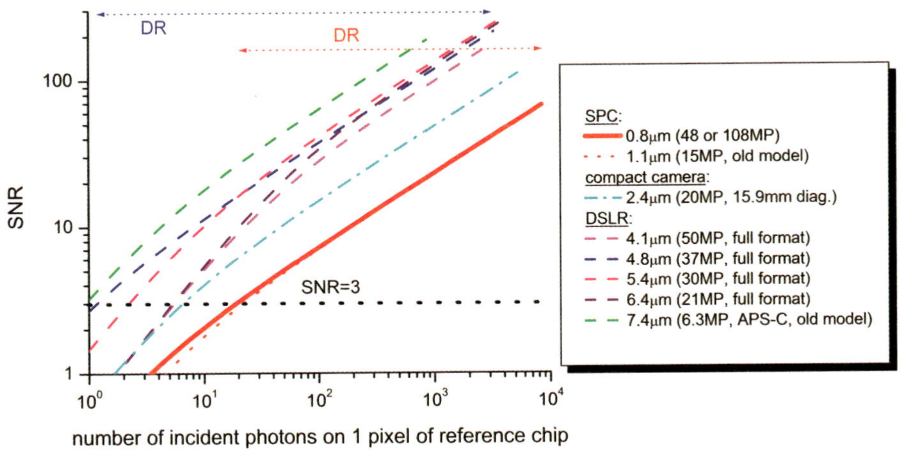

Fig. 7.13: Example of the SNR of SPC, compact cameras and DSLR. Note that within this double-logarithmic plot, a square root dependence of SNR as a function of incident photon numbers shows up as a straight line. The dotted horizontal lines with arrows indicate the limit of the useful dynamic range, here with respect to SNR ≥ 3 (note usually larger ranges are provided, in particular when the lower limits are set to noise level, i. e., to SNR = 1). This plot corresponds to Figure 7.11 and bases on the same data. Thus, similar to Figure 7.11, the values of the x-axis provide the number of photons on one pixel of SPC-"CIS 4" of Table 7.7, which again acts as the reference (thick red line). All lines are ended at the corresponding photon numbers when saturation, i. e., FWC is reached.

is given by the square root of the total number of photons within a pixel which is discernible from the straight line curve. This is valid for all image sensors as this kind of noise does not result from the sensor but from the light fluctuations emitted from or reflected from the object. It is the result of nature's laws and thus unavoidable even for the best cameras. Similar to Figure 7.11, for the same illumination condition much more photons are incident on a pixel of a DSLR, which leads to a better SNR value in Figure 7.13 when compared to a SPC, where the SNR is always below 100. At lower photon numbers, σ_{el} becomes important and thus the curves begin to bend. For DSLR, this occurs earlier than for SPC. Nevertheless, for the same amount of light within a scenery, their SNR is still larger.

A reasonable value when a signal still can be well discriminated from noise may be SNR = 3. This may mark the minimum of feature recognition within an image. In any case, SNR = 1 sets the limit which discriminates the regions where the signal becomes smaller and larger, respectively, than noise. Concerning image quality, a good SNR has a similar importance as a good MTF. Moreover, SNR has also an impact on the MTF and the resolution. This is subject of the following chapter.

7.3 Camera system performance

We may now discuss the MTF of SPC and compare it with the MTF of other cameras. In order to ease the discussion of the optical system, we restrict the consideration in this section to the influence of only the lens optics in conjunction with a sensor consisting of small pixels. We make also comparison to other cameras, such as DSLR. We start in Section 7.3.1 with disregarding noise. After that in Section 7.3.2, additionally sensor noise is taken into account. In the following, we assume a simple sensor design with MTF_{sensor} given by Equation (5.63).

7.3.1 MTF in absence of noise

7.3.1.1 Perfect systems

Figure 7.14 shows how the MTF of a complete SPC with a large number of pixels may look like, if the optics together with its wavelength dependence is taken into account (Equations (5.38) and (5.61)). Here, we assume the best case scenario for a 108 megapixel SPC with square shaped pixels with $p = 0.8\,\mu m$, a fill factor of 100 % and $PW: PH = 4:3$ (MTF_{sensor} displayed as dashed-dotted olive line). Furthermore, we assume a perfect lens system with $f_{\#} = 1.8$, namely one that is only limited by diffraction. The blue, green and red solid lines, respectively, in Figure 7.14a correspond to the MTF of that ideal lens (MTF_{optics}). The solid lines in Figure 7.14b indicate MTF_{system}, namely the product of MTF_{sensor} (according to Equation (5.63)) and MTF_{optics} for the same camera. Again, the

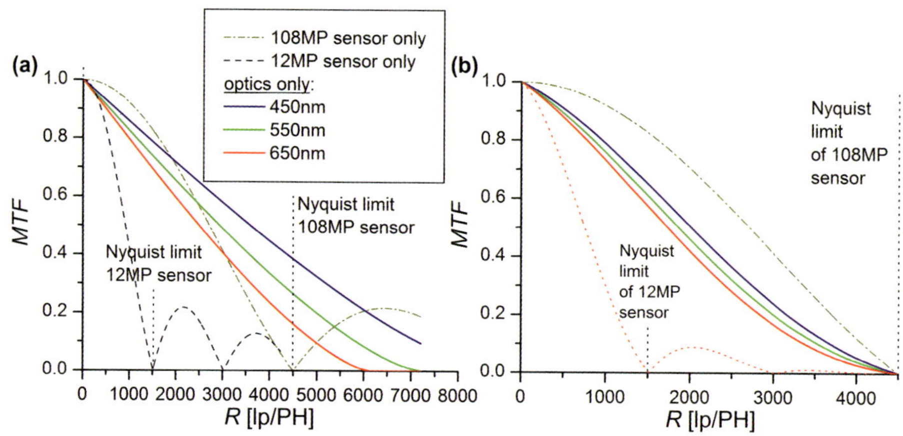

Fig. 7.14: (a) MTF_{sensor} of a 108 MP SPC (dashed-dotted olive line) and another one with 12 MP sensor (dashed black line). The solid colored lines correspond to MTF_{optics} of an ideal lens. (b) the same MTF_{sensor} of the 108 MP SPC is displayed again, now together with its MTF_{system} displayed in the corresponding colors. For comparison, the red dotted line shows MTF_{system} of the 12 MP SPC for red light. For detailed information, see the text.

colored solid lines correspond to the same wavelengths as in (a) (see legend). For comparison, MTF_{sensor} of the 108 MP sensor is shown as the dashed-dotted olive line and MTF_{sensor} of a 12 MP sensor with $p = 1.4\,\mu m$ as the dashed black line (with its Nyquist limit shown as black dotted line). Figure 7.14b also includes MTF_{system} of the 12 MP SPC with $f_\# = 1.5$ optics illuminated with red light as the dotted red line.

The diagrams clearly show that when the pixels are so small that this comes close to the wavelength of visible light, diffraction effects cannot be ignored. In particular, for red light, contrast and resolution are both lower when compared to light with a shorter wavelength (see Chapter 5). Moreover, due to the apparent differences for the different wavelengths, chromatic effects may play a negative role.

This results in $MTF_{total} = MTF_{system}$ curves that are significantly affected by both, MTF_{sensor} and MTF_{optics}, even in absence of aberrations. This is seen in the lower frequency range, where the contrast due to MTF_{optics} is below MTF_{sensor} and in the higher frequency range, where the contrast begins to vanish due to MTF_{sensor}. However, it is seen that the Nyquist frequency approaches the cut-off frequency of the optics, in particular, for red light. Thus, the diagrams clearly show that for an ideal sensor with such small pixels, MTF_{sensor} may be of such high-quality that the MTF of the whole camera system is significantly degraded even by a good lens. A real lens will lead to further degradation. MTF_{optics} also sets some limit for further improvement of MTF_{sensor} in the sense, that a larger number of pixels will not much improve MTF_{system} of the SPC, at least in the low and medium frequency range. But even with an ideal optics improved to $f_\# = 1$, the curve of $MTF_{system} = MTF_{sensor} \cdot MTF_{optics}$ of the SPC for red light is only just slightly above the present blue curve in Figure 7.14b (displayed also as purple dashed-dotted line in Figure 7.15b). Nevertheless, for the present example with ideal conditions, which disregards for instance the SNR related to pixel size, the MTF of the 108 MP SPC is very much better, when compared to a SPC with larger pixels and lower number of pixels even when illuminated with red light (see red dotted line in (b)).

Figure 7.15 shows a comparison of the same 108 MP SPC discussed before (dashed-dotted lines), now with a 50 MP full format DSLR with 50 MP and $p = 4.1\,\mu m$ (solid lines). In Figure 7.15a, both MTF_{sensor} curves are displayed in olive color. The red lines correspond to the MTF of an ideal optics set to $f_\# = 1.8$ (dashed-dotted line SPC, solid line DSLR; note the different *PH* as indicated in the insert). From Figure 7.15a, it is well seen that in contrast to the SPC, MTF_{optics} of the DSLR is much better than MTF_{sensor}. As a result, MTF_{system} magenta lines in Figure 7.15b is superior for low spatial frequencies and approximately similar in the intermediate range where the contrast sensitivity function of the human eye has its maximum. For large values of *R*, the high-end SPC is superior, but we may note that this will change, when noise is taken into account (see Section 7.3.2). It is only well surpassed by a rather expensive high-end professional medium format camera (dotted violet line). For comparison, the brown dashed line shows the artificial MTF of the SPC if one assumes an ideal $f_\# = 1$ optics. The corresponding curve for the DSLR changed to $f_\# = 1$ is rather similar to the solid magenta line.

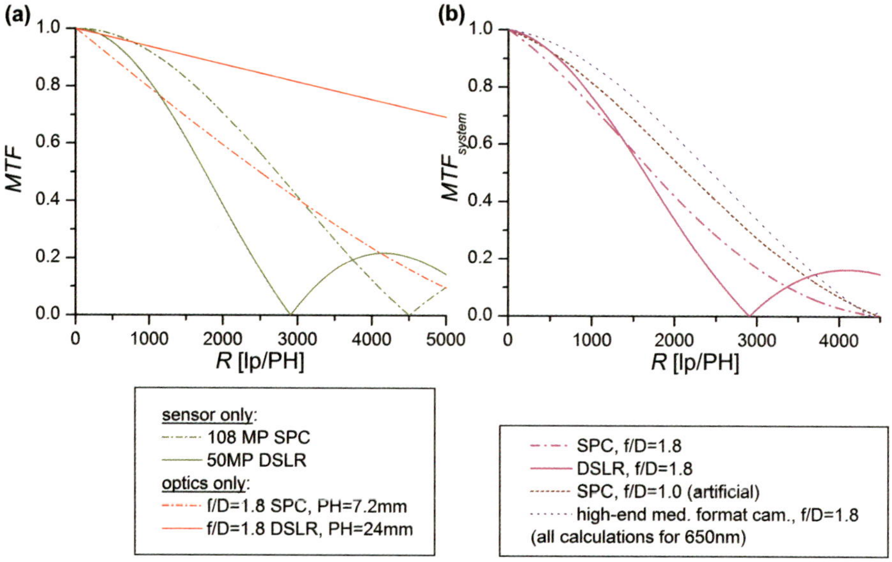

Fig. 7.15: (a) MTF_{sensor} and MTF_{optics} of a 108 MP SPC and of a full format 50 MP DSLR. (b) MTF_{system} for the same cameras and that of a high-end medium format camera. For comparison, an artificial MTF_{system} with a $f_{\#} = 1$ optics but otherwise same SPC is displayed as well. For detailed information, see the text.

Thus, huge resolution in photography becomes possible in theory. However, huge resolution, in particular, when expected for a camera with very small pixels, requires that camera shake is fully avoided. Otherwise, blur would become observable at high-spatial frequencies, which in consequence would foil the intended high resolution. At laboratory conditions, one may achieve good conditions. But in practice, in particular, with SPC this is an absolutely unusual situation because pictures are taken without tripod and also often objects move as well.

Nevertheless, the possibility of such a huge resolution is a new situation for consumer photography. For long time in history of cameras, the optics was superior when compared to the sensor. This has led to the fact, that the MTF of the whole camera system could be nearly fully determined by MTF_{sensor}. Another example is presented in Figure 7.16, which shows the MTF of a camera-lens combination. The camera is a professional full format DSLR with a $R_N = 1872\,\mathrm{lp}/PH$. It is equipped with a high-quality optics set to $f_{\#} = 4$. The thick lines are the results of a Siemens star measurement as published by a photo journal. The MTF_{sensor} and MTF_{optics} curves result from 1D calculations from Equations (5.63) and (5.39) and are displayed as dashed and dotted lines, respectively. Figure 7.16a shows the MTF obtained in horizontal direction in the center of the image. Figure 7.16b shows that one in diagonal direction measured in the corner. As the published data in Figure 7.16b seem not be measured correctly because MTF has to be 1 at $R = 0$, the dotted-dashed curve is a scaled version of the solid line. Furthermore, one has

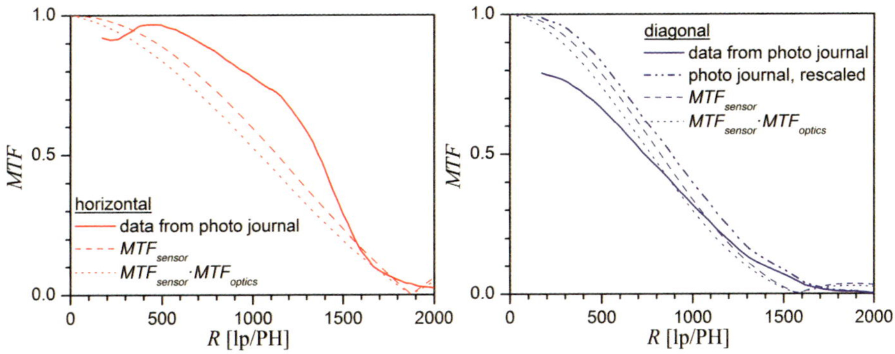

Fig. 7.16: MTF of a camera-lens combination (see text). (a) MTF obtained in horizontal direction in the center of the image. (b) MTF obtained in diagonal direction measured in the corner.

to note, that it is expected that in the corner some vignetting is present which degrades the MTF.

This diagram also shows that many tests published in photo journals or in the web may present just trivial information. Although it is claimed that the MTF shown in this figure characterizes the combination of the camera body and a particular lens, and although formally this is correct, conclusively this MTF is nearly fully determined by the MTF of a nearly ideal sensor. Moreover, the observed distinct differences, which are well seen as the hump at intermediate spatial frequencies, are neither the result of $\mathrm{MTF}_{\mathrm{optics}}$ nor of $\mathrm{MTF}_{\mathrm{sensor}}$. This can be purely attributed to image processing (see Sections 4.9 and 7.4). Neither the $\mathrm{MTF}_{\mathrm{optics}}$ curve nor the $\mathrm{MTF}_{\mathrm{sensor}}$ curve can exceed that of the corresponding curve for the best-case scenario as discussed before and, of course, not at all any real $\mathrm{MTF}_{\mathrm{system}}$ curve. At least, this is the case when we are interested to characterize the hardware, namely lens and sensor but not the performance of the image processing software.

We may also remark that the MTF and the resolution in diagonal direction are worse, when compared to horizontal or vertical direction, respectively. This is obvious, because for squared shaped pixels, the effective pitch of a pixel in diagonal direction is larger by $\sqrt{2}$, which results in less pixels per mm or per PH.

Furthermore, Figure 7.15 shows, that even a small amount of aberrations may not lead to a significant degradation of images captured by DSLR. For good DSLR, this is common. On the other hand, it is apparent that for SPC with a large pixel number, a rather small value of $f_{\#}$ is a strong demand and any optics which is not perfect will significantly impair image quality. To achieve the necessary high-image quality with real lenses is really a challenge (see Section 7.1).

Yet, nowadays, if well done, sensors with smaller and smaller pixels may lead to a spatial resolution of those sensors that significantly exceeds that of lenses, in particular, in SPC (in future that may change, see Section 7.5). But it must be noted that for very small pixels, in practice $\mathrm{MTF}_{\mathrm{electronics}}$ cannot be fully ignored. Consequently, $\mathrm{MTF}_{\mathrm{sensor}}$ may be degraded by wings, at least slightly. In addition, it must also be noted that there are further issues that are important for the total image quality and also that the received image quality may strongly depend on the conditions when photographs are taken. Some of these issues will be discussed in the following chapters.

7.3.1.2 Real systems

Figure 7.17 shows another comparison of the MTF of a SPC and an "upscaled version," which may be regarded as a full format camera. Both result from calculations, now not for perfect lens systems, but for a real one ("real lens"). Additionally, comparison is made to two "downscaled versions." Here, similar to Section 8.3.5, the MTF is provided as a function of the position distance h_i of the image center within the image field. This is displayed as a fraction of the full radius r_sensor, which is half of the sensor diagonal (Table 7.8).

The data from this example result from a simulation performed by V. Blahnik [Bla21]. The simulation has been made for the SPC-lens design shown in Figure 7.17a. This SPC lens is a $f_\# = 1.9$ wide-angle design (field of view $\Psi = 75°$) for a 1/3.3″ image sensor ($f = 3.52$ mm). The image diagonal is 5.4 mm, and thus for $PW/PH = 4/3$ one obtains $PH = 3.24$ mm. The MTF has been calculated as an average of the contributions of different wavelengths with the spectral relative weights 1 (405 nm), 3 (436 nm), 12 (486 nm), 24 (546 nm), 24(587 nm) and 11 (656 nm), respectively. The result is shown as the red curve for simulations for sagittal and tangential (= meridonal) rays in Figure 7.17b and c, respectively.

For comparison further simulations have been made, namely the mentioned upscaled and downscaled versions, respectively, both with the same aspect ratio as before (for better comparison we assumed also a 4:3-aspect ratio for the full format camera

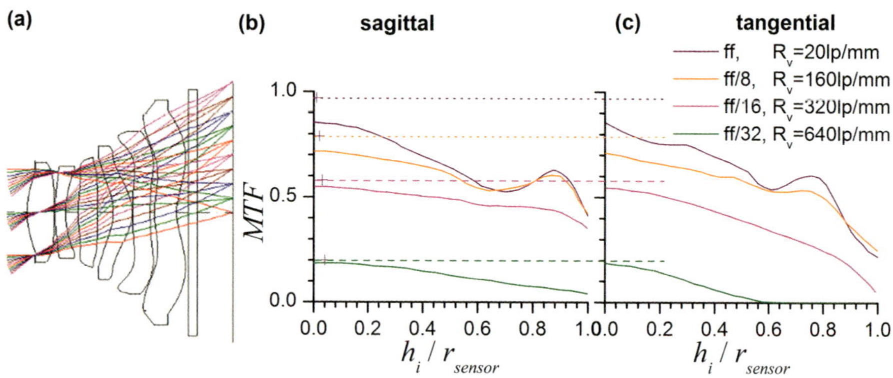

Fig. 7.17: (a) Lens design for a SPC used for the MTF simulation discussed in the main text [Bla21] (Author V. Blahnik, reprinted with kind permission). (b), (c): Simulated MTF of a SPC lens together with those of an upscaled and a downscaled version (see the text). The legend is valid for (b) and (c). The diagrams base on the data taken from [Bla21]. The dotted lines indicate the corresponding diffraction limit.

Tab. 7.8: Data and results for the example discussed in the main text. The Nyquist frequency results from Equation (5.64). For each of the R_v-curves in Figure 7.17, the MTF for the "real lens" is taken at $h_i = 0$. $MTF_{d\,ffr}$ is the MTF of a perfect lens and is calculated from Equation (5.39) for the same polychromatic case described in the text. $MTF_{sensor}(R_v)$ is calculated from Equation (5.63) for the indicated pixel pitch. It is calculated for the values of R_v provided in the corresponding row (see also legend of Figure 7.17). The yellow background marks the parameters for 22 MP cameras. The colors of the text match the colors of the curves in Figure 7.17.

notation	ff	ff/8	ff/16
sensor diagonal $2r_{sensor}$	43.2 mm	5.4 mm	2.7 mm
PH	25.92 mm	3.24 mm	1.62 mm
R_v	20 lp/mm = 518 lp/PH	160 lp/mm = 518 lp/PH	320 lp/mm = 518 lp/PH
pixel number for $p = 0.4$ µm	not realistic	10800 8100 = 87 MP	5400 · 4050 = 22 MP
pixel number for $p = 0.6$ µm	not realistic	7200 · 5400 = 39 MP	2700 · 2025 = 5.5 MP
pixel number for $p = 0.8$ µm	not realistic	5400 · 4050 = 22 MP	2250 · 2025 = 4.6 MP
pixel number for $p = 1.0$ µm	not realistic	4320 · 3240 = 14 MP	2160 · 1620 = 3.5 MP
pixel number for $p = 6.4$ µm	5400 · 4050 = 22 MP	not reasonable	not reasonable
Nyquist frequency for $p = 0.4$ µm	1250 lp/mm = 32400 lp/PH	1250 lp/mm = 4050 lp/PH	1250 lp/mm = 2025 lp/PH
Nyquist frequency for $p = 0.6$ µm	833 lp/mm = 21600 lp/PH	833 lp/mm = 2700 lp/PH	833 lp/mm = 1350 lp/PH
Nyquist frequency for $p = 0.8$ µm	625 lp/mm = 16200 lp/PH	625 lp/mm = 2025 lp/PH	625 lp/mm = 1013 lp/PH
Nyquist frequency for $p = 1.0$ µm	500 lp/mm = 12960 lp/PH	500 lp/mm = 1620 lp/PH	500 lp/mm = 810 lp/PH
Nyquist frequency for $p = 6.4$ µm	78 lp/mm = 2025 lp/PH	78 lp/mm = 253 lp/PH	78 lp/mm = 127 lp/PH
$MTF_{sensor}(R_v)$ for $p = 0.4$ µm	~1	0.97	0.90
$MTF_{sensor}(R_v)$ for $p = 0.6$ µm	~1	0.94	0.77
$MTF_{sensor}(R_v)$ for $p = 0.8$ µm	~1	0.90	0.62
$MTF_{sensor}(R_v)$ for $p = 1.0$ µm	~1	0.84	0.45
MTF_{sensor} for $p = 6.4$ µm	0.90	(>Nyquist limit)	(>Nyquist limit)
$MTF_{lens}(R_v)$	0.85	0.72	0.55
$MTF_{diffr}(R_v)$	0.97	0.79	0.58
$MTF_{lens}(R_v)/MTF_{diffr}(R_v)$	0.87	0.92	0.95
$MTF_{sensor}(R_v)/MTF_{lens}(R_v)$ for 22 MP	1.06	1.25	1.64
diffr. limited lens: $R_{MTF10} = 1000$ lp/mm	25920 lp/PH	3240 lp/PH	1620 lp/PH
for diffr. limited lens: $R_{MTF0} = 713$ lp/mm	18500 lp/PH	2310 lp/PH	1155 lp/PH
$N_{SB,cut}$ (related to the optics)	18500	2310	1155

ff and *PH* given in Table 7.8). Those scalings mean that the original lens and its sensor were increased in lateral size by a factor 8 or decreased by a factor 2 or 4. This then leads to a system close to a full format camera which is denoted by "ff" and to other ones, which are a factor of 16 or 32 smaller than "ff," and thus denoted by "ff/16" and "ff/32." The original design is one eighth of the size of "ff," and thus termed as "ff/8." For better comparability, we have recalculated the data from lp/mm to lp/*PH* with the corresponding sensor size. Thus, the (normalized) spatial frequency for all curves is the same, namely $R_V = 518\,\text{lp}/PH$. But one has to take into account for the different sensor sizes, which leads to different values of R_V when expressed in lp/mm. As discussed earlier in this book, this then directly allows comparison and the MTF curves are absolutely equivalent. The data are provided in Table 7.8.

Figure 7.17 and Table 7.8 may be discussed, though, with a partly different point of view when compared to [Bla21]. We will compare the 4 optical systems ff to ff/32 in combination with different sensors. In particular, we may compare 22 MP cameras (marked by yellow background), namely cameras where the sensor cut-offs given by the Nyquist frequency, are the same, namely $R_N = 2025\,\text{lp}/PH$, which corresponds to $N_{SB} = 2025$.

For further comparison, we add Figure 7.18, which shows the MTF of a diffraction limited lens, the sensor MTF and that of the system. First, we may have a look on the theoretical limitations provided by the optics. From Equation (5.39) and Figure 5.27 and with $f_\# = 1.9$, one obtains $R_{MTF0} = 1000\,\text{lp/mm}$. According to Equation (5.69), the corresponding $N_{SB,cut} = R_{MTF0} \cdot PH$ depends on *PH* and is displayed in the last row of Table 7.8. Furthermore, from Equation (5.68b) and with $\kappa = 1.22$ and $\lambda_0 = 550\,\text{nm}$, one can estimate $R_{MTF10} \approx 713\,\text{lp/mm}$. The MTF of all "real lenses" MTF_{lens} is close to $\text{MTF}_{\text{diffr}}$ when compared at the same intermediate value of $R_V = 518\,\text{lp}/PH$.

For the ff version, this apparently takes little influence on the MTF of the system. The lens has a high quality. The MTF across the image field has a high level and is rather flat, too. All this is consistent with the statement in [Bla21]: "The performance level of the full frame upscaled version of the SPC is excellent. The MTF performance is comparable with excellent full format lenses such as the ZEISS Otus 1.4/28 mm (when compared at SPC aperture f/1.9)" [Bla21]. Together with the large pixel size of 6.4 µm and thus expected good noise properties, an excellent camera may be set up.

The ff/8 version (this comes close to the iPhone with $CF = 7$), namely a good optical system for an SPC, nevertheless is worse. Although the "real lens" is even more close to the theoretical limit (also seen from the Strehl ratio, see Section 5.2.4), diffraction effects become important. This already shows up at the intermediate value of $R_V = 518\,\text{lp}/PH$. In addition, from Figure 7.18 it may be seen that for that reason even $\text{MTF}_{\text{system}}$ with a 56 MP sensor, namely one with very small pixels of $p = 0.5\,\mu\text{m}$, would be significantly worse when compared to the ff version with a 22 MP sensor. In particular, this is the case in the notified intermediate region which is of eminently importance for human observers (we remind of Section 5.2.9 and again we refer to the excellent articles by Nasse [Nas08, Nas09]). But the complete SPC system is still of good quality, which can also be seen from MTF as a function of h_i (see Figure 7.17).

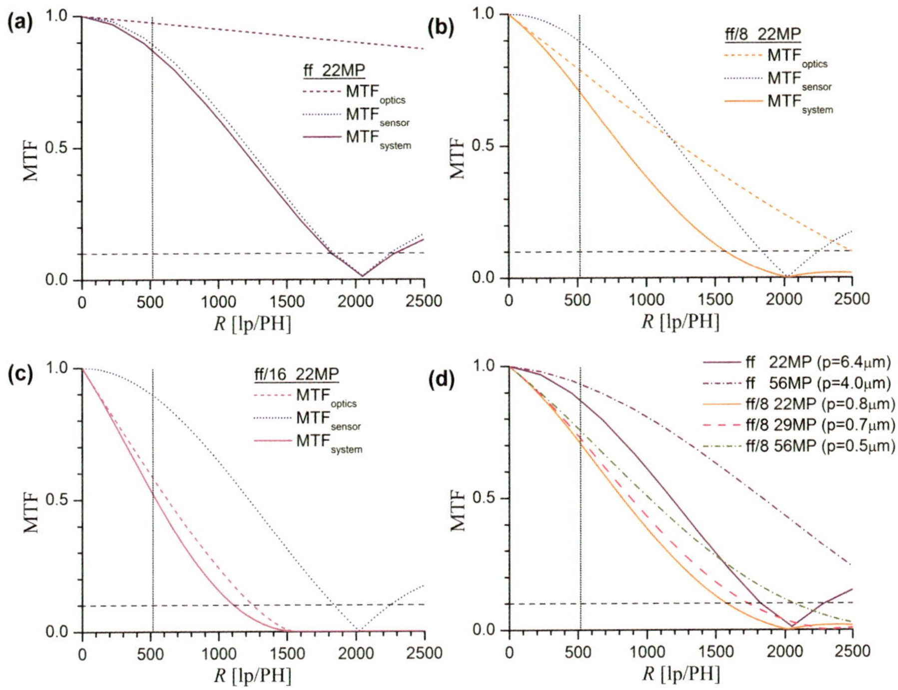

Fig. 7.18: 2D-MTF calculations for a diffraction limited spherical lens and a sensor with its MTF given by Equation (5.63). The dashed horizontal marks the resolution limit of $K = 10\,\%$. The dotted vertical lines mark $R_v = 518\ \mathrm{lp}/PH$ (see Figure 7.17 and Table 7.8). Note that Figure 7.17 shows $\mathrm{MTF_{optics}}$ of the "real lenses" and (a) to (c) $\mathrm{MTF_{diffr}}$, $\mathrm{MTF_{sensor}}$ and $\mathrm{MTF_{system}}$. (d) Comparison of $\mathrm{MTF_{system}}$ for the ff and the ff/8 system, respectively, with different sensors.

Even further downsizing leads again to more difficulties. Already pixels with $p = 0.6\ \mu\mathrm{m}$ would lead to a degradation of $\mathrm{MTF_{system}}$. If also combined with a 22 MP, the pixel pitch would be smaller than that of presently commercially available CIS and the rather small SNR would also deteriorate the image. The ff/16 version is strongly affected by the diffraction limit. Although the quality of the "real ff/16 lens" may be the best of all three compared lenses, nature laws set the limit. This trend is rather common for small optical systems because aberrations play less and less and diffraction effects more and more a role. If for instance the diameter of an aperture is a couple of wavelengths only, then on such a short distance no significant wavefront aberrations can occur (see Section 5.2.4.1). The negative trend for $\mathrm{MTF_{lens}}$ continues with the ff/32 version where $\mathrm{MTF_{lens}}(518\ \mathrm{lp}/PH)$ is always below 0.2. Further discussion is not needed here.

The discussion above shows that miniaturized cameras not only need lenses with extraordinary small aberrations, but also such ones with small f-numbers. This is in contrast to, e. g., DSLR and illustrated in Figure 7.19. For good DSLR, $\mathrm{MTF_{optics}}$ is much better than $\mathrm{MTF_{sensor}}$ and thus $\mathrm{MTF_{system}}$ is only little affected by $f_\#$. For the SPC with large pixel numbers this is vice versa, and thus $f_\#$ takes strong influence. Thus, lens design should

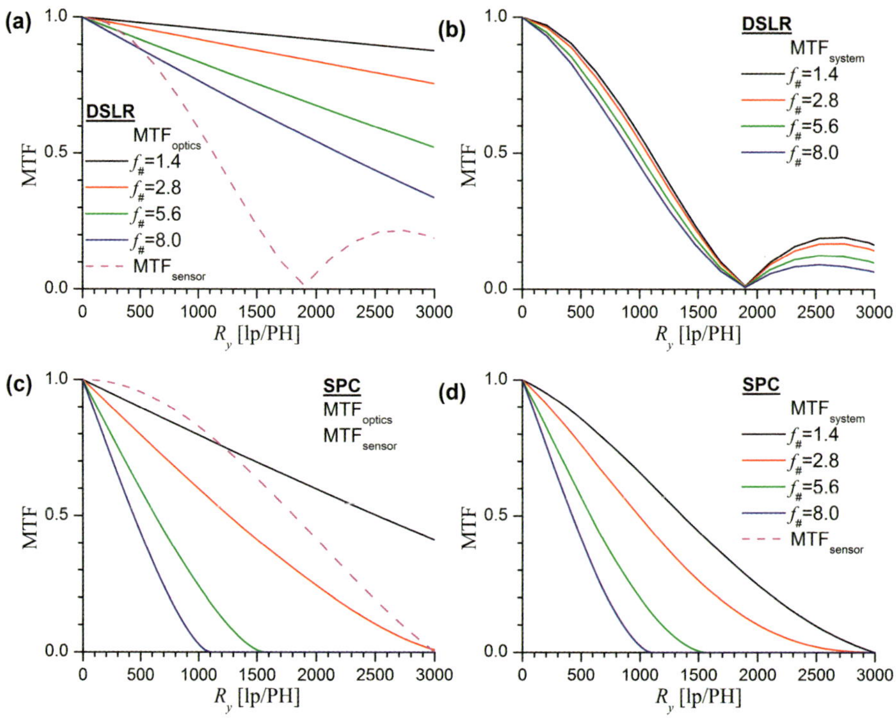

Fig. 7.19: Influence of the f-number on the image quality (MTF) for a full format DSLR with $p = 6.4\,\mu m$ ((a), (b)), $PH = 24\,mm$ and a SPC with an ideal sensor with $p = 0.8\,\mu m$ and an image diagonal of 8 mm (e. g., “CIS 4” in Table 7.7) ((c), (d)). Calculated curves have been made for 550 nm and an aberration free optics. For better comparison, we assumed the same aspect ratio of 4:3 for both cameras. The meaning of the lines in (c) is the same as in (a).

be made for as small $f_\#$ as possible but without significant aberrations. Nevertheless, there is also a small advantage of the significant diffraction effects. Since SPC usually are constructed without OLPF, and thus take a risk of Moiré effect (Section 1.6.2), the extended PSF limits MTF_{optics}, which takes the role of the low pass filter. Note that the OLPF is omitted to avoid too thick SPC modules and costs.

Finally, we may continue the discussion on the performance of SPC modules, which started in Section 7.1 and, in particular, compare two main cameras of different smartphones from different companies (see Table 7.4 and Table 7.5, resp.). Both cameras have an equivalent focal length of $f = 23\,mm$, but the concepts of the camera manufacturers strongly differ. In case of low aberrations, the main camera optics of vivo X60 Pro+ (Figure 7.20a) may be better. On the other hand, the corresponding module of Samsung Galaxy S22 ultra (Figure 7.20b) has a sensor with a superior MTF, at least theoretically. Moreover, the slightly larger value of $f_\#$ may allow slightly worse optical performance due to some aberrations. On the other hand, the significantly smaller pixel size will lead to a worse SNR, which is expected to take significant influence on the image quality (see

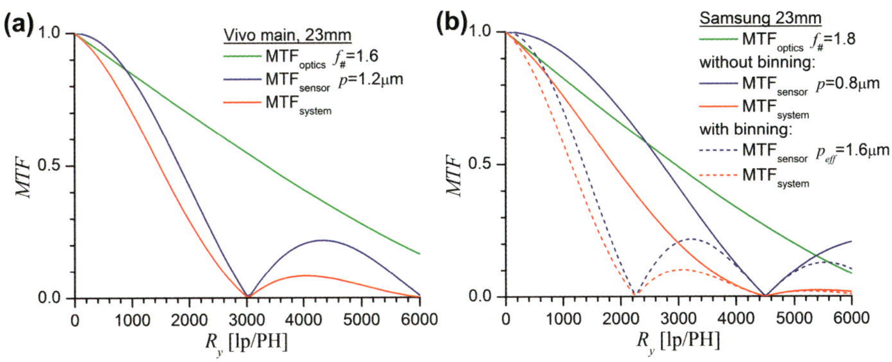

Fig. 7.20: MTF of (a) the vivo X60 Pro + (2021) and (b) the Samsung Galaxy S22 ultra (2022), respectively. Note that those MTF curves are not the real ones of those SPC. Instead, those are theoretical ones based in the parameters of the SPC and the assumption of ideal conditions (ideal sensor, no aberrations).

Section 7.3.2). But this may be even more complicate. The 108 MP CIS of the Samsung SPC may be operated in binning mode. In that case, noise properties may be not worse, but better when compared to the vivo camera. But the diagrams show that in case of binning MTF_{sensor} is much reduced when compared to the operation without it. Another issue is the possibility to apply PDAF, although there must not be an effect on image quality. This brief discussion clearly shows that it might be complicate to come to a stringent conclusion on the camera performance related to image quality. But the discussion also shows some of the important issues and what is relevant for setting up a proper camera system.

In conclusion, it is obvious that miniaturisation takes strong influence on image quality. Although still high image quality is possible, miniaturized cameras have to be well designed as a system with an adapted interplay of the optical and the sensor system. This is similar to larger cameras. But for miniaturized ones, diffraction sets severe limits with “classical lens systems.” In future, there might be other opportunities such as those based on nonclassical lens systems (see Section 7.5), which may reduce the diffractions effects. But this would then also require progress in sensor technology. This includes not only pixel pitch but also an improvement of the SNR. Further difficulties for shrinked pixels result, for instance, from the OMA. This is because the requirements for alignment increase and also here diffraction effects and resulting potential optical cross talk may become important. Also, the angular dependence of the microlens/pixel combination and the response (see Figure 4.21) set a challenge for the optical design of the optical systems.

7.3.1.3 Zoom lenses

In Section 7.1.2.3, we discussed the realization of zoom lenses for SPC. True zoom lenses, in which the focal length can be continuously varied from a starting value to its end value, are quite expensive to implement in SPC. This is why most SPC combine several single camera modules with a fixed focal length to act jointly as one camera. Its characteristics are that for a wanted image perspective the module with a larger FOV is chosen to capture an image, which in a next step is digitally zoomed to the wanted image crop. This method is termed hybrid zoom as compared to a true optical zoom (see Section 7.1.2.3). A digital zoom is always affected by a loss of resolution, namely by a degraded MTF_{optics} and simultaneously a degraded MTF_{sensor}. On the other hand, a true zoom in SPC based on a periscope technology, such as the module in Sony Xperia 5 III, suffers from increasing diffraction with the focal length, which however only impairs MTF_{optics} and not MTF_{sensor}. In order to understand the quality of zooming in SPC, it is important to consider the whole system, and thus MTF_{system}.

Let us start with two modern SPC, vivo X60 Pro + and Samsung Galaxy S 22 ultra (see Tables 7.4 and 7.5), which both dispose of 4 rear camera modules. The MTF of their main cameras are shown in Figure 7.20 and discussed in the preceding section. Their long focal lens module is of periscope type. The focal length of the Samsung periscope is nearly two times longer than that of the vivo periscope but the sensors are quite comparable. The overall resolution of their periscopes can be expressed by MTF_{system} as depicted in Figure 7.21. Their cut-off is slightly above 1000 lp/*PH* but clearly worse than the corresponding curves of the smartphones' main modules (see Figure 7.20) exhibiting cut-offs above or equal to 3000 lp/*PH*. Hence, the idea that a digital zoom from the high-quality lenses of shorter focal length could be a more appropriate solution. But it can be seen from Figure 7.21 that no digital zoom from any of the modules with shorter focal length attains the resolution of the periscope. In case of the vivo, a digital zoom from the 50 mm module to f_{eq} = 90 mm has a better resolution than images taken by the periscope. This may be no longer valid at 125 mm, so the strategy of the hybrid zoom in that SPC seems to

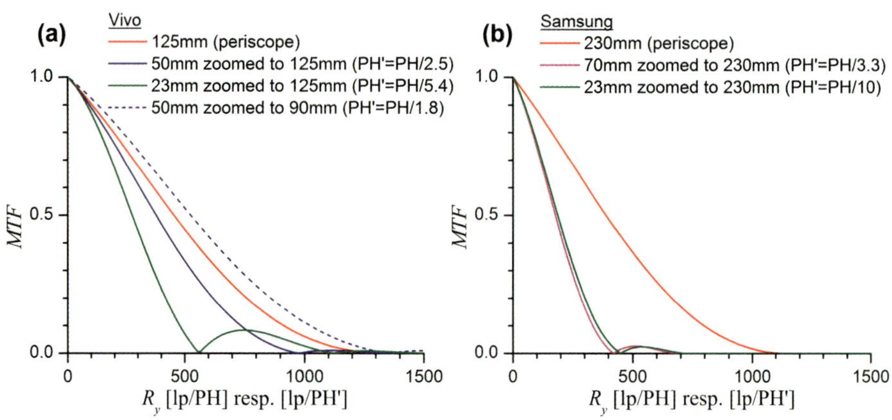

Fig. 7.21: MTF_{system} curves of (a) vivo X60 Pro + (2021); (b) the Samsung Galaxy S22 ultra (2022); the zoomed curves refer to digital zooming from the original picture height *PH* to a cropped value *PH'*. For details of lens specifications, see Tables 7.4 and 7.5.

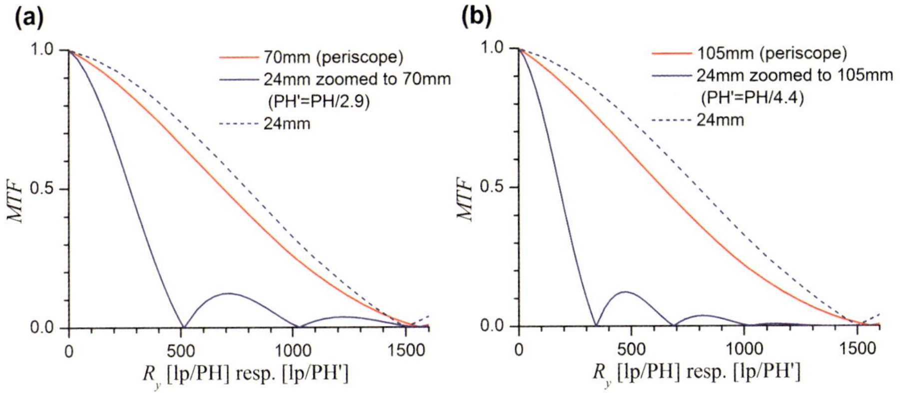

Fig. 7.22: MTF_{system} curves of Sony Xperia 5 III; comparison of the periscope to the 24 mm main module and its digital zoom to a) 70 mm and b) 105 mm, respectively; for detailed lens specifications see Table 7.6.

be well adapted. We write "may be" because the periscope would be superior if aberrations can be neglected. On the other hand, if aberrations cannot be neglected, the digital zoom from the 50 mm module may have a better quality. This is due to the fact that digital zooms are taken as a crop of the central part of the image, which usually suffers less from aberrations, in particular, when compared to the corners.

Now let us have a look to the Sony Xperia 5 III, which has a true optical zoom lens for the 70–105 mm focal length. It can be seen from Figure 7.22 that the MTF_{system}-curves of the zoom lens are at all focal lengths quite comparable to that of the main module at 24 mm. The nearly homogeneous behavior of the zoom lens on zooming is due to the fact that MTF_{sensor} remains unchanged and MTF_{optics} degrades with longer focal lengths but not dramatically. Overall, MTF_{system} of the 24 mm module is somewhat better than that of the zoom module. However, a digital zoom to some intermediate range, for instance, at 50 mm, should lead to a significant deterioration with an estimated cut-off around 750 lp/PH. The obvious quality difference between a digital zoom to 70 mm and the zoom lens itself at this focal length is given in Figure 7.22a). The difference becomes worse at 105 mm (Figure 7.22b). It should be mentioned that the main module of the Sony Xperia shows a cut-off around 1500 lp/PH, which is lower than those of the vivo or Samsung SPC considered above. This is a consequence of the larger pixel size, and thus lower number of pixels on the sensor. However, it can be expected that the main module of the Sony with larger pixels has a better SNR than the others and should thus be better adapted to low light situations.

To summarize these considerations for zoom lenses, we can state that a true optical zoom lens has the great advantage of having a more homogeneous quality over a larger range of focal length. Unlike digital zooming, the sensor MTF does not degrade. In most cases, a true zoom lens is superior to digital zooming. Digital zoom is only acceptable over a short range. We would like to stress again here that our considerations do not take into account any lens aberrations and are therefore best-case approximations, which cannot be achieved in reality.

As a criterion for all assessments of lens qualities, the highest importance should be given to the resolution in the medium range around 1000–1500 lp/*PH* and to a lesser extent to the maximum values.

7.3.2 Imaging in presence of noise

In the previous chapter, we have discussed the MTF and the resolution of SPC sensors in absence of noise. Now we would like to show how noise and low SNR take influence on image quality, which may be degraded even for a good MTF. For easier illustration, we will simplify the discussion somewhat: We will see that the influence of noise on image quality leads to unwanted fluctuations within the image and to a less clear representation. A discussion of MTF and noise issues will be subject of Sections 8.3.3 and 8.4.

This discussion relates on images of test gratings shown in Figure 7.23. We consider the situations with a spatial frequency in vertical direction $R_y = 300$, 900 and 1800 lp/*PH*, respectively, (i. e., SBN of 300, 900, 1800). Observation is made on a display which takes 1.6 % of PH. This is illustrated in Figure 7.23. We hint that plotting the y-coordinate as a fraction of PH yields that the diagrams become independent of any magnification of the image. We would like to note that in the image plane the grating is oriented in vertical direction, i. e., y-direction, as it is convenient to relate spatial frequency measurements to lp/*PH* as discussed in Chapter 5. However, for practical reasons of presentation, the y-direction is here chosen as the abscissa so that although calculations are done with respect to PH, here the grating is displayed by 90° rotation into horizontal direction. The red lines in the 3 diagrams show the brightness distribution along the y-axis of the grating structure, which is displayed above the plot under consideration. The direction of the x- and y-axis, respectively, shown exemplarily in Figure 7.23c is valid for all gratings displayed in Figure 7.23 to Figure 7.25.

Comparison is made for two cameras with the same illumination conditions. We will discuss that with an example, where we again compare the same high-end 108 MP SPC with $p = 0.8\,\mu\text{m}$ as before to the professional 30 MP full format DSLR with 5.4 µm pixels (see Section 7.2). For the SPC with $PH = 7.2\,\text{mm}$, one gets $N_v = 9024$ pixels in vertical, for the DSLR we get $PH = 24\,\text{mm}$ and $N_v = 4480$ pixels. This corresponds to $R_N = 4500$ lp/*PH* ($N_{SB,Nyquist} = 4500$) and $R_N = 2240$ lp/*PH* ($N_{SB,Nyquist} = 2240$), respectively. Thus, the spatial frequency of all gratings is well below the Nyquist limit. The present example has been chosen because the number of pixels per PH height of the SPC is in good approximation a factor of 2 larger when compared to the DSLR. In total, the SPC has approximately 4 times more pixels when compared to the DSLR. We would like to mention that the SPC of this example is a high-end camera which has taken profit of the advancements discussed before.

Now let us introduce some rather heavy noise. We consider the case of approximately 20 photons incident on one SPC-pixel in bright regions. From Figure 7.13, we see that then SNR = 3. For the same conditions, nearly 46 times more photons hit one pixel of the DSLR, thus roughly 900 photons (see Figure 7.11). SNR is approximately 17. Now we would like to image test objects, which similar to Figures 5.20 and 5.21 are the 3 bar gratings, but 2 of them together with an edge (transition to the large black bar on right-hand side; Figure 7.23). Although for mathematical reason sine-shaped gratings are preferential, here we have chosen bar gratings for clearer visibility of effects and because sharp edges are common in many pictures. Again, we chose the same conditions for both cameras in the sense that the image size of the gratings with respect to *PH* is always the same.

For the present conditions, in the bright regions noise is dominated by photon noise. In the regions which ideally should not receive light, read noise leads to limitations. In addition, due to contrast reduction by MTF, some signal occurs as well. Thus, any region in the image suffers from noise.

The calculated images of the test gratings captured in presence of noise with the SPC and the DSLR, respectively, are presented in Figure 7.24 and Figure 7.25.

Figure 7.24a to c show the images of the SPC with $R_y = 300\,\text{lp}/PH$, $900\,\text{lp}/PH$ and $1800\,\text{lp}/PH$. For the present camera, this corresponds to 30, 10 and 5 pixels per period. Figure 7.24d shows the brightness distribution along the y-axis, which would be the result of averaging a large amount of images. The ordinates of the diagrams are provided in arbitrary units. We would like to comment that the computed images above the diagrams are calculated for the assumption that the displayed brightness distribution is proportional to the sensor signal (including noise). However, in real images, due to the application of tone mapping the images may look different. But in no way, lacking structure information could be obtained back by any image processing.

Figure 7.25a to c show the images of the DSLR with $R_y = 300\,\text{lp}/PH$, $600\,\text{lp}/PH$ and $1800\,\text{lp}/PH$. For the present camera, this corresponds to 15, 5 and 2.5 pixels per period. Part (d) shows the brightness distribution along the y-axis, which would be the result of averaging a large amount of images. The ordinates of the diagrams are again provided in arbitrary units. Again, the computed images above the diagrams are calculated for the same assumption on the relation of the signal to the brightness. And again, we disregard tone mapping. Using raw data of DSLR, images such as those displayed, could be generated.

In spite of the high quality MTF in absence of noise, which similar to Figure 7.23, it is clearly seen that for the present illumination condition, noise leads to a strong degradation of the image quality for the SPC. For instance, even for a spatial frequency of 300 lp/mm, where the MTF values are larger than 90 % for both cameras, it is well seen that for the SPC the perceived contrast is reduced. For the DSLR, the effect is much less severe. In particular, when we observe the image in Figure 7.25c, with $R_y = 1800\,\text{lp}/PH$, which is the closest value to the Nyquist frequency for the DSLR ($R_N = 2240\,\text{lp/mm}$), the edge could be well recognized. This is in contrast to the SPC-image Figure 7.24c where

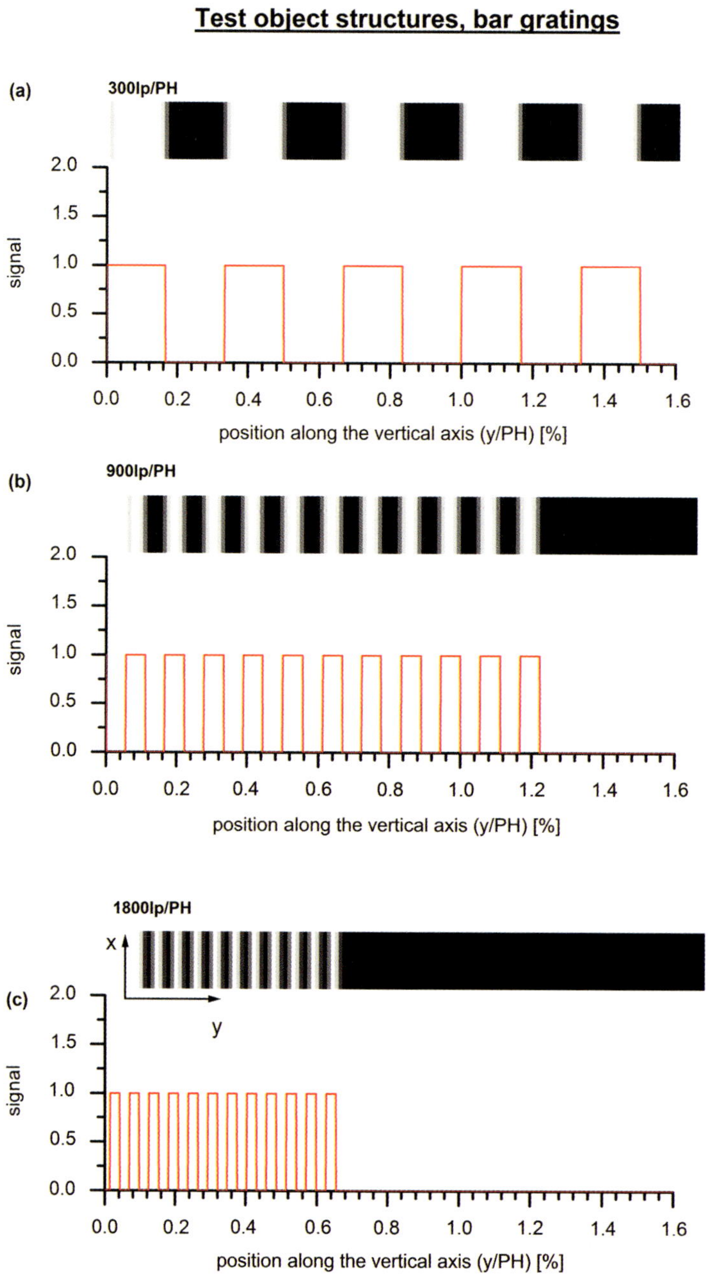

Fig. 7.23: Illustration of the test objects used for the example discussed in the text.

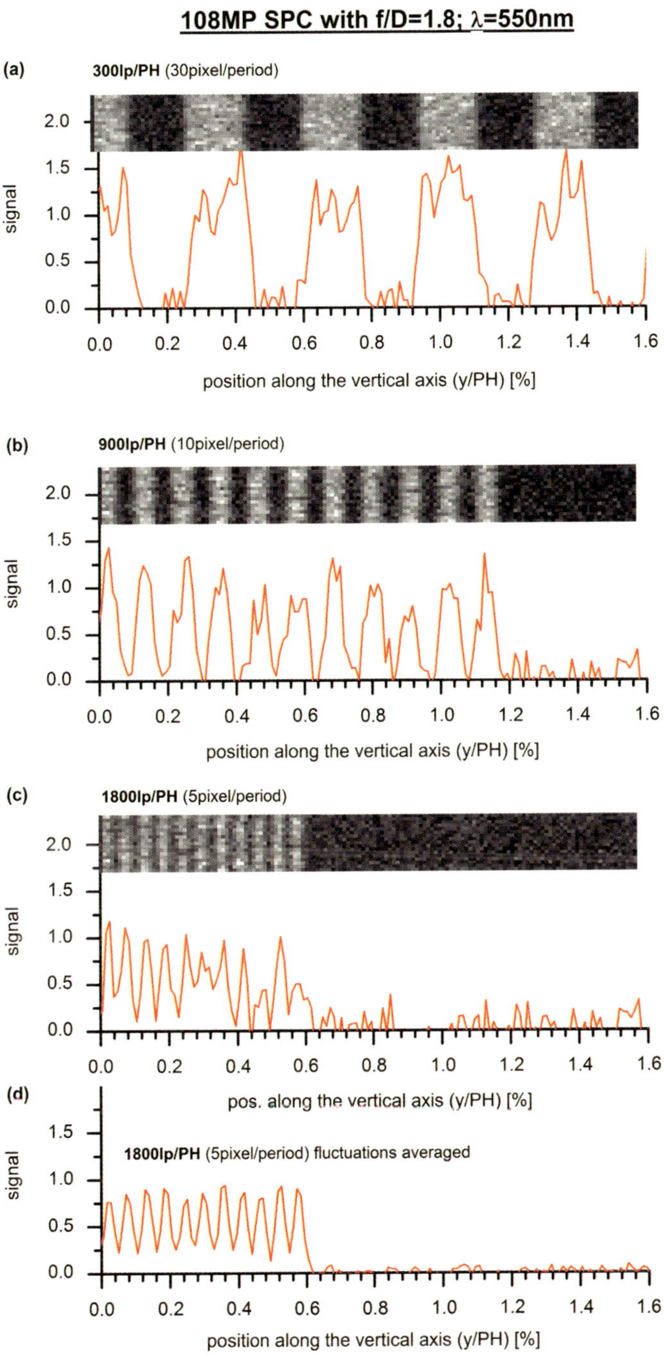

Fig. 7.24: Images of the test objects shown in Figure 7.23 captured with the 108 MP SPC in presence of noise (see the text).

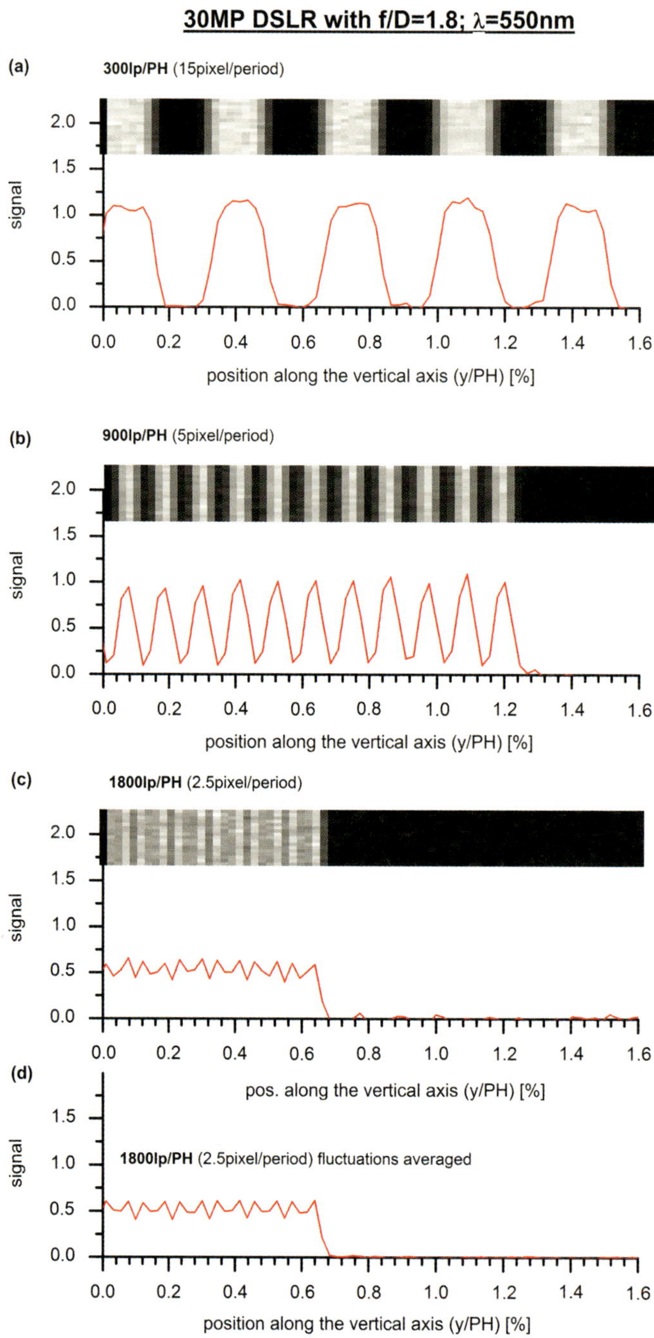

Fig. 7.25: Images of the test objects shown in Figure 7.23 captured with the 30 MP DSLR in presence of noise (see the text).

the sharp edge can be seen not very well, although here the Nyquist frequency of the SPC is a factor of 2 larger ($R_N = 4500$ lp/mm).

Before we continue, we would like to comment on the observation of the grating structure above the diagram in Figure 7.24c. From this “image,” one may get the impression of a still acceptable edge recognition. However, this is an artificial situation, because the reader already knows that this is the image of an edge. Thus, similar to Figure 7.24d where this is done numerically, the eye and brain of the observer automatically average all traces along the direction of the edge. Nevertheless, again such an object is somewhat artificial as such simple structures are not necessarily common in scenes, which are photographed. Furthermore, we may recall the discussion at the end of Section 1.6.1. Again, it is important to state that in imaging we do not prepare situations with a known configuration such as displayed in Figure 7.23 and usually we do not have objects made of periodic structures, which do allow more easily to remove noise within image processing without losing too much details.

Thus, from the single brightness distribution in diagram Figure 7.24c we conclude that resolution is close to be lost. Similar, Figure 7.25c provides still some structure information but also this is close to the resolution limit. Nonetheless, the edge is reproduced really sharp, even well seen from the red line in Figure 7.25c and even better from the image above. Thus, one can see that (strong) noise may degrade the image quality strongly, even for cameras with a high quality MTF curve. Thus, although the resolution limit, e. g., given by the Nyquist frequency, is not affected, practically in such a case of strong noise nevertheless resolution may be severely much reduced.

It is obvious that for less modulated structures, i. e., minima larger than zero and maxima well beyond saturation, i. e., a modulation between “dark gray” and “light gray,” the situation becomes even worse, in particular, for the SPC. On the other hand, in a real image captured by a SPC such strong noise effects are not necessarily well apparent, although image degradation at least in the sense of loss of details cannot be avoided. This is due to strong image processing as discussed in the next subchapter. An exception may be obtained, if a series of images of a static object could be made. In such a bit artificial situation, those images could be averaged, and thus noise reduced. Then fine structures could still be preserved and then, e. g., the 108 MP SPC would be superior when compared to the DSLR (Figure 7.24d and Figure 7.25d). We would additionally like to remark that imaging with such few photons as discussed in the present example is not restricted to low light conditions in general. But even normal sceneries often consist of regions with a lot of light, though there are also regions, which are rather dark, and thus affected as discussed. For less noise as in our discussed example, the situation is not as bad for the SPC. However, if illumination is not really bright, the degradation of image quality by noise may be still significant.

7.3.3 Summarizing remarks on SPC imaging quality

As we have seen, the image quality obtained with SPC may be very good. However, this is not generally the case. For good environmental conditions, such as enough light, no fast-moving objects etc., the image quality may be similar to that of DSLR and DSLM. However, this is by far not the case for other situations, e. g., for low light conditions. And this is not the case, when independent of the amount of available light, a particular view of a scenery is intended to be captured. In such a case, full format DSLR (again including DSLM) and most compact cameras allow to choose a particular lens, e. g. a "normal lens" with an equivalent focal length of 50 mm (see Section 2.2) or another particular one. Alternatively, those cameras allow to make use of a real zoom lens. For a smartphone camera, this is not possible (we exclude the very rare present exceptions). For good image quality, one is restricted to the focal length and the field of view of the main lens of the SPC, and potentially to that of the other built-in lenses. Nearly any zoom of the SPC is a software zoom, even so any for so-called hybrid zoom, which may lead to images of much less quality (see, e. g., Section 1.4.3 and Section 7.1). As selection of the field of view of a scenery is rather common in good photography, this is a severe restriction for photography with smartphones. Moreover, there are only very few SPC available which provide a long tele lens.

Further downsides of SPC have been discussed in the previous sections and chapters. These include noise and SNR issues (although this has been significantly improved in the last years), camera shake (in particular, for cameras with a lot of megapixels), the lack of raw data storage for many of them (note as well that DNG may not necessarily be regarded as raw) and restrictions with respect to high-speed imaging. The latter is a direct consequence of the rolling shutter effect. As rolling shutters are the common ones for SPC, they may take significant influence on the image even in situations, which are not regarded as high-speed imaging (see Figure 7.26; note that although rolling shutters are common for CIS of DSLR as well, those cameras are equipped additionally with a mechanical shutter which removes the rolling shutter effect). We will see in Section 7.4 that there are further issues, namely such ones that are related to the mostly automatic modus of the SPC.

On the other hand, there are a lot of advantages. Shooting photographs with SPC has led to a widespread usage. Nearly everybody can take pictures with at least moderate quality if we include both, technical issues and photographer qualities such as choice of motif. Here, one takes profit from the usual SPC mode that image capture, processing, storage and management is done automatically. Moreover, the user can directly "apply" the images, e. g., by mailing them directly with the smartphone. Moreover, smartphones are rather small and mostly with the owner.

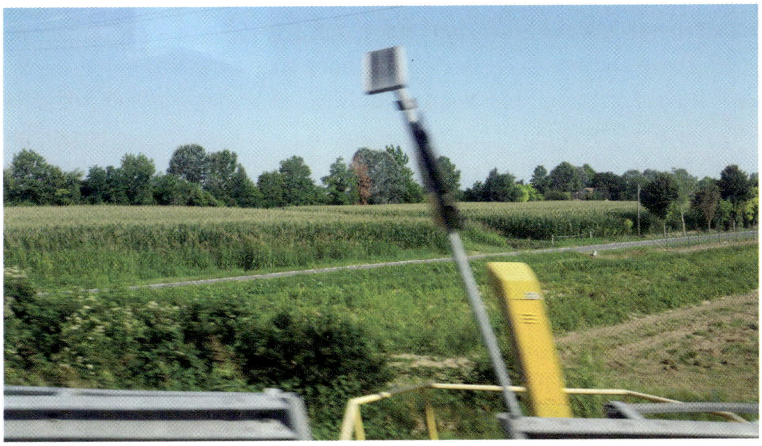

Fig. 7.26: Illustration of the rolling shutter effect from a rather simple picture captured with a smartphone camera. The picture has been taken out of the window of a moving car. Due to the readout procedure, the yellow tower and the gray pole are imaged as slanted objects (compare also Section 4.5.2.2 and Figure 2.22; compare also to Appendix A.4, although CIS and CCD readout are somewhat different). Picture from [14].

7.4 Computational imaging

Today, imaging with miniaturized cameras, and in particular SPC, has taken profit of the huge progress made in recent years. Now, this offers a large opportunity to take excellent photographs even without knowledge of photography by the user, which has led to a tremendous amount of pictures. Part of this progress is due to advances of the hardware, namely the optical and the sensor system, respectively. But another part of this progress is due to advances of computational imaging (CI, sometimes also called artificial intelligence, AI). The denotation CI may include software-based image processing and manipulation as discussed in Section 4.9 (see also Figure 4.1). Moreover, and based on advanced algorithms, it includes the compensation of shortcomings that are subject to the SPC modules and even to the conditions during image capture. However, within SPC modules image processing is carried out mostly by the smartphone system on chip. In such a way, CI is directly implemented in the overall image processing chain of an SPC.

Over time, new concepts based on these electronic image manipulations have shown up. Besides image capturing and standard electronic development, additional processes have become very important in order to achieve pleasing images as the result of the image capturing chain. In most cases, these processes do not enhance the technical quality of images but simply improve their general look and artistic impression.

14 U. Teubner, V. Blahnik, H. J. Brückner: *Smartphonekameras, Teil 2: Bildsensor und -verarbeitung - Tricksen für gute Bilder*, Physik in unserer Zeit, **51**(6) (2020) 290.

Usually, as part of the overall workflow when the raw images are developed to be converted to standard image formats like JPG, TIFF or others, some standard processes are carried through. These are also implemented in most external image editing software and can be used at will by the user. However, it requires some skill and time even for advanced users to efficiently exploit all the capacities of image manipulation. While for users of high-quality cameras of larger formats this is a typical post exposure process, which is done manually, in most SPC many interesting features are already available and optimize the image look in an automatic way, rather often without the possibility to take influence by the photographer. In the following, we will present some of these computational methods, which may also be integrated in modern system cameras. We will not go deeply into details as image processing in general is a complex and an advanced topic on its own which is treated in a lot of special books and not the subject of the present book (for a tutorial textbook see, e. g., [Jäh23]). We also refer to the previous chapters and just repeat several issues only shortly.

7.4.1 Sharpness control and tonal curve adaptation

One of the first steps after capturing the raw image is the sharpness adjustment of the images and tonal corrections. There are many ways to enhance the sharpness (see also Section 5.2.8). An image can be judged sharp, if the contrast difference between image parts is enhanced. The contrast can be increased globally for the total image but also locally to render details more visible. A typical method for increasing the local contrast and enhancing edges of structures is unsharp masking (see Section 5.1.9.2). This is a technique, which has been originally developed for analog darkroom photography, and was implemented for digital image processing. Moreover, there are other methods, which have been especially developed for digital image processing. Sharpening and adaptation of the tonal curve are usually done automatically after capturing the RAW image and converting it to JPG image data file. This can be seen in Figure 7.27 where the standard JPG image appears to be sharper than the original RAW image. Both images are based on the same original data, where the RAW format file is many times larger than the compressed JPG format file. It comprises a much larger contrast range and can in principle reveal more details. Due to sharpening and tonal adaptation, some details may be no longer resolved and the dynamic range decreases. Thus, the JPG image is of lower quality than the RAW image but may look more pleasant.

7.4.2 Noise reduction

A further unwanted effect is the emergence of noise in images when the captured scene is not well illuminated. Obviously, noise in general is a particular issue in imaging with SPC due to the small pixel sizes. However, the photographer will not necessarily notice

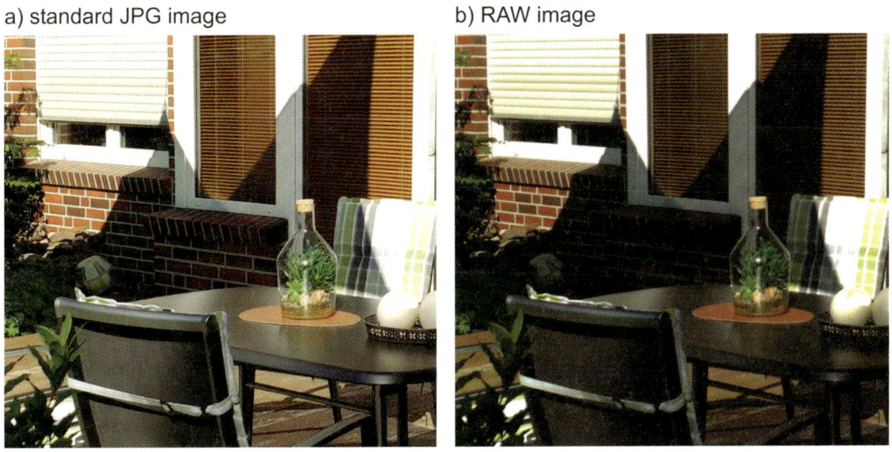

Fig. 7.27: Comparison of a standard JPG image (a) in comparison to a nonmanipulated RAW image (b) of a SPC. Both images are crops of the same original RAW image taken in automatic exposure mode.

increased noise because of the applied noise reduction by CI. This feature is done mostly automatically or it can be activated to be applied in some cameras. For that reason, denoising takes an important role during the image processing step after capturing (see also Sections 4.7.4, 4.9 and 5.2.8).

Noise reduction routines, also called denoising, are particularly applied to dark areas of the image or the image in total. If performed by a standard method, this is equivalent to the application of a suitable low pass filter, which acts like an additional MTF term in the product of Equation (5.53). This MTF reduces high frequencies in the Fourier space, and thus takes influence, in particular on the high frequency range of $MTF_{total} = MTF_{system}$. In other words, resolution is reduced. It depends on the automatically selected grade of denoising if this leads to a strong or only a weak effect. Modern noise reduction works especially well in smooth areas, where structure loss by denoising does not become apparent and the image will have a nice look. Also, fixed pattern noise may be well diminished. But in contrast to DSLR, where post-processing with raw converters can be performed under control of the user, for users of SPC with automatic application of this feature, the result is neither under their control nor is it predictable.

On the other hand, in regions with detailed structures denoising is not as easy. As it is the opponent of sharpening (see extended discussion in Section 5.2.8), it will lead to irrevocable smoothing of small details. This reduces resolution and, for instance, regions of low contrast like with human skin will be reproduced in an unnatural way. This can be seen in Figure 7.28, where the image taken by a SPC is compared to that of a DSLR. Not only details are lost in the smoothed SPC image, but also coarser structure blocs can be identified. The contrast difference is lowered. Further post-processing of the image could not compensate for that.

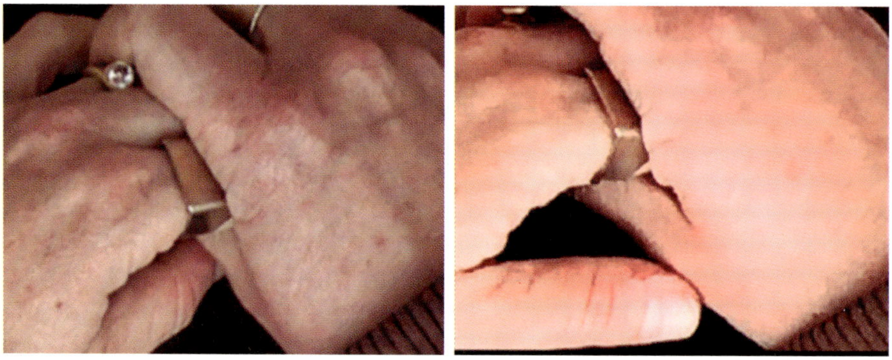

Fig. 7.28: Images of human skin captured at low ambient light by different cameras: DSLR (left) and SPC (right). Due to strong noise reduction by image treatment software, the structure of the skin is rendered in a somewhat unnatural way with less details and at lower contrast.[15] (Author: V. Blahnik).

A further unwanted effect is the emergence of noise in images when the captured scene is not well illuminated. The areal size of a pixel on the sensor determines the quantity of photons that strike the pixel per exposure time, and thus the number of electrons accumulated in its capacitor well. Especially at low light illumination, small pixels are strongly affected by noise. The signal-to-noise ratio (SNR) increases with the exposure, and thus it is lower for SPC than for cameras of larger formats. Noise manifests itself as a variation of brightness and of color in the near neighborhood of a pixel, even when the illumination in this neighborhood is of a constant value. This variation can be smoothed by mathematical algorithms averaging the brightness respectively color information around a pixel. Thus, a more homogeneous background in low light parts of the image is achieved. On the other hand, this is maybe at the detriment of resolution and contrast, if due to the averaging software significant natural variations in the original are smoothed away (see also discussion in the preceding sections).

CI tries to overcome the problem in regions with details, in particular, by interpreting the image content and reconstruction of the image based on that guess, whereas in the other parts of the image, standard operations of noise reduction may be sufficient. Locally different operations are applied. In other words, if advanced, or "intelligent" denoising is applied by modern CI, the ISP "makes assumptions" on the unknown object. It should be noted that the software is only artificially intelligent as it does not really know the photographer's intention. In some sense, this is similar to the discussed observation of the image displayed in Figure 7.24c. Due to the knowledge of Figure 7.23, we are well aware what is imaged, and thus it is much easier to see the edge and speculate to see the oscillation of the test grating. However, the situation would have been

15 U. Teubner, V. Blahnik, H. J. Brückner: *Smartphonekameras, Teil 2: Bildsensor und -verarbeitung - Tricksen für gute Bilder*, Physik in unserer Zeit, **51**(6) (2020) 290.

totally different if we had not known the object. It would have been even more difficult if there had been a structure in x-direction as well. But exactly that is the realistic situation, and consequently the ISP could only make a guess. This guess may be good or it may be wrong, but it allows denoising with pseudo "reduced loss of resolution." Thus, fine structures in the image may be "conserved." However, those do not result really from conservation of what was captured during the imaging process because after denoising these fine structures have been irrevocably removed. Then afterwards some kind of artificial intelligence based on the guess, tries to compensate and may introduce the eliminated fine structures artificially. If the guess has been good enough, the image may look as if resolution is still there. But if the guess has been wrong, the fine structures are really artificial. This is not an exceptional case but observed quite often. Many published tests with images taken from real objects confirm that statement. In any case, such kind of image processing means that the final image may be more likely generated by an image generation process than really captured. It is a challenge to avoid such impressions.

For that reason, several modern SPC offer the opportunity of noise reduction by capturing a series of images of the same scenery at the same conditions. Then the single frames can be fused or simply averaged, and thus the noise reduced (multiple frame stacking). However, this becomes difficult if the conditions are not stationary and ghosting has to be removed as well. Besides this, noise reduction is rather restricted as this scales typically with the square root of the number of frames.

Another possibility to reduce noise is binning (see Section 4.8.3). For the example of the $p = 0.8\,\mu\text{m}$ SPC in Figure 7.13, this increases the SNR by a factor of $\sqrt{4}$ (in 2 dimensions, see Section 4.8.3). For $N > 10$, the related SNR of the "macro pixel" is then almost identical with that of compact camera curve (compare dashed-dotted curve in cyan in Figure 7.13), but this is still significantly lower than the SNR of the DSLR (N is the number of photons displayed on the x-axis). But this is then at the expense of resolution: The Nyquist frequency is reduced by a factor of 2 (in one direction) and $\text{MTF}_{\text{system}}$ is significantly worse when compared to the full mode. Figure 7.29 illustrates that situation. Nevertheless, in situations without bright illumination, the overall image appearance may be improved. Binning can be considered as the hardware equivalent to the above discussed noise reduction by software.

7.4.3 High dynamic range

In many cases, when photographs are taken, the dynamic range of the scenery is much higher than the device can handle, on which the photograph is displayed. While in nature the dynamic range can be as high as 26 EV or more (see Section 4.8.4), the SPC sensor has a range of about 10-12 EV, depending on the technology (Table 7.7). The JPG image output has only 8 EV, a paper print has still less, and the smartphone display disposes of about the same range or slightly more. They can never reproduce the high dynamic

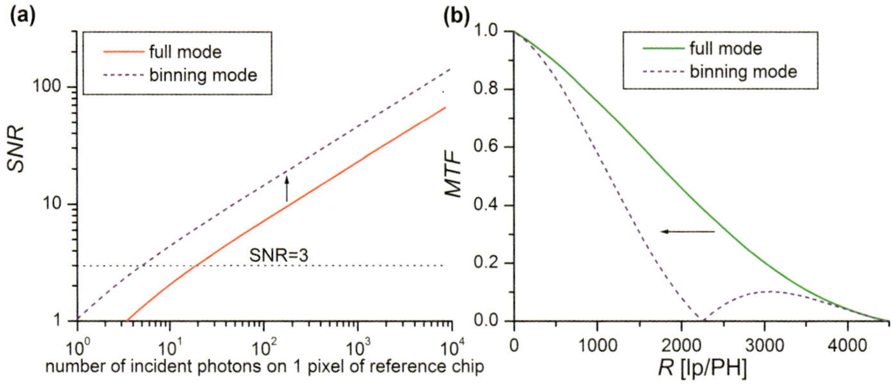

Fig. 7.29: (a) Example of the SNR of a 108 MP SPC where the red curve is exactly the same as that in Figure 7.13. The purple dotted one displays the improved SNR when operated in binning mode of 4 pixels. (b) MTF_{system} of a 108 MP sensor where the green curve is exactly the same as that in Figure 7.14. The purple dotted one displays the changed MTF_{system} when operated in binning mode of 4 pixels.

range (HDR) of the object. An appropriate way to cope with this situation is the generation of HDR images with a compressed dynamic range (see Sections 4.9.5 and, in particular, 4.9.5.1). This can be achieved by different methods: If the object's dynamic range is so high that it exceeds the dynamic range of the sensor, more than only one image is required for HDR generation. Taking for instance 3 images sequentially in a short time interval with different EV settings, a normal image exposure is achieved along with an overexposed and an underexposed image, respectively. These images can be numerically analyzed and overlaid after a tone mapping procedure to produce one final compressed HDR image. In this case, the bright light parts are reduced to exhibit their inherent structure whereas the low light parts are enhanced. Details in bright light and low light sections can be perceived. A similar effect can be achieved by blending for instance only 2 pictures taken at different EV conditions. In any of these cases, the camera may be controlled in a way that potentially on-chip dual conversion gain (DCG; see Section 4.10.6.3) is used when the 2 or more frames are taken.

Although these methods of combining different pictures are very efficient for high dynamic ranges, their disadvantage is that movements of objects during exposure or an unwanted camera shift may lead to "ghost images" in the blended final image. Alternatively, if the dynamic range of the sensor is sufficiently high, a simple tone mapping procedure, applied to a single image, may allow for expressing details in low and bright light parts of the image. Here, ghost images never show up, the method is quick, but the HDR compression is less efficient. An example for this is given in Figure 7.30 where images of a SPC are displayed. Figure 7.30a shows the standard JPG-image after automatic exposure where details in the dark parts can hardly be detected. In automatic HDR-mode, the shadow becomes brighter, details can be better perceived (Figure 7.30b). A much stronger effect is achieved by post-exposure raw image treatment using an external im-

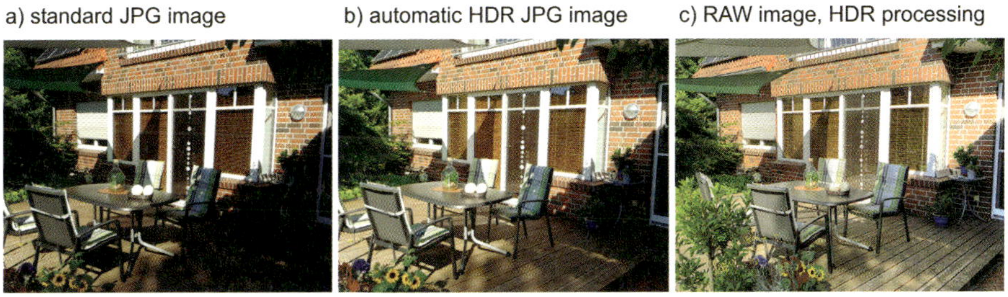

Fig. 7.30: Comparison of HDR images taken by a 16 MP SPC, $f/1.6$, f_{eq} = 30 mm; a) standard JPG image taken in automatic exposure mode; b) JPG image taken in automatic HDR-mode; c) raw image taken in manual mode after image processing by external converter.

age converter and tone mapping procedure (Figure 7.30c). However, when blowing such a manipulated image up, noise and other artifacts may become visible, above all in the shadowy parts.

7.4.4 Portrait mode

Another example of computational imaging in current SPC is the portrait mode. This setting is intended to isolate a sharp imaged person from a background and render the background blurred. In this way, a nice bokeh should be achieved, like in case of a full format camera with moderate tele lenses of large aperture. As discussed above in Section 7.1.1, a blurred background is hardly possible in a SPC image due to the large depth of field. However, a depth profile can be evaluated in modern SPC equipped with multiple camera modules and sensors in different ways [Bla21]. If at least 2 camera modules are available, common image sections can be used to determine by stereoscopic methods the distance to a given object. The methods work best if the object has a high contrast and is not occluded. A more recent technology is based on time-of-flight sensors (ToF, see Section 4.10.7.4) that allows for a depth profile measurement in the near range from about 0.1 m up to 1–2 m. The ToF-technology is independent from contrast and occlusion issues but has a relatively low resolution of about 200 x 200 pixels compared to SPC modules. In modern high-end smartphones, usually several methods are combined for optimum performance.

For the generation of the portrait image, the part of the person in the foreground is isolated based on the measured depth profile. The software renders the background artificially blurred. The quality of the result depends on the quality of the evaluated depth profile. If image parts are not well determined as for their depth, the computed image exhibits an unnatural look. An example for a portrait image is given in Figure 7.31. In comparison to the original image in Figure 7.31a, part b) exhibits a nice portrait look at first glance, similar to that taken by a larger format portrait lens at large aperture. By a closer inspection, however, it can be seen that peripheral sections of the head are not

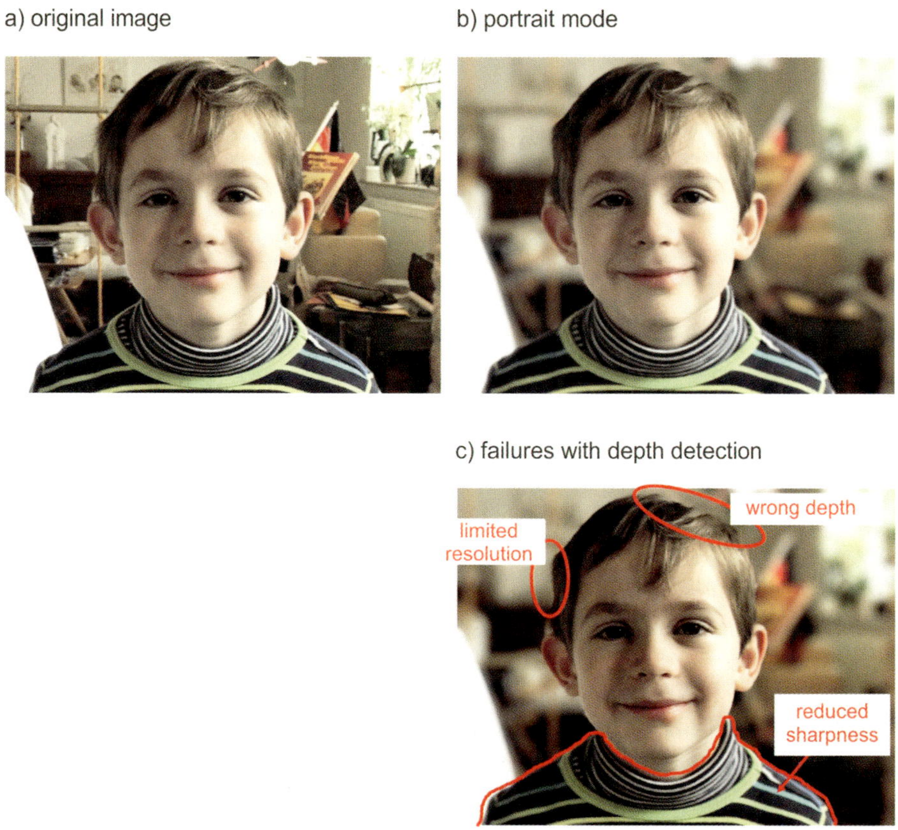

Fig. 7.31: Images taken with iPhone 7+. Comparison of original image with portrait mode application[16].

correctly rendered (Figure 7.31c). They are treated as background sections, and thus become blurred. Critical parts for portrait modes are always sections of fine structures like hairs. These details remain often unperceived when considering the images on a smartphone display. But magnified printouts usually do not yet achieve the quality of larger format "natural" portrait photographs. Further critical parts of the artificial bokeh are the nice, blurred image points like those of bright light sources in the background of night scenes (see also Section 6.9.3). There is an ongoing process to improve the sensors for 3D depth acquisition in conjunction with the required software to resolve these issues.

16 U. Teubner, V. Blahnik, H. J. Brückner: *Smartphonekameras, Teil 2: Bildsensor und -verarbeitung - Tricksen für gute Bilder*, Physik in unserer Zeit, **51**(6) (2020) 290.

7.4.5 Correction of lens aberrations

Most optical systems are usually not ideal and introduce aberrations. One of them is image distortion (Section 3.5). High-quality lenses made for DSLR are well corrected and minimize this “image error.” However, this may not be the case for modern lenses designed for many compact cameras and for almost all SPC. The philosophy has changed in order to simplify the lens design. The reason is not only to save costs, but mainly to make the lens more compact which is important for miniaturization of the camera system. For instance, a distortion as large as 20–30 % is regarded to be acceptable for wide-angle lenses in SPC [Bla21]. Then, to compensate for this negative impact, it is intended to apply software-based corrections for distortion as a post-processing step. In principle, this works well and leads to relatively high-quality images. We have added the word “relative,” because the presence of distortion in a captured image is already accompanied with a loss of information that cannot be compensated but only obscured (see Section 5.1.8). In total, image quality is less good when compared to an image obtained from a camera with a well-corrected optical system.

This can be shown by careful measurements similar to those discussed in Chapter 8. Yet we would like to provide a more illustrative example based on a typical photograph as given by Figure 7.32. Part (a) shows a raw data image taken by a modern compact camera, where no distortion corrections are performed. For better perception, we have added a horizontal and a vertical line through the image center. Due to image processing, distortion has been removed in part (b) by shifting of the hardware pixel signals to software generated new pixels. Obviously, this is not lossless in general and additional quantization errors have to be considered as well. Although this correction will take not much influence on regions close to the image center, losses are apparent closer to the borders. This may not be seen directly in the small reproduction scale in this book but it can be well understood from a comparison of the marked small restricted region located in a distance of 75 % from the image center to the corner. The red and blue circle, respectively, indicate the same region before and after distortion correction. Analysis of image brightness $\widetilde{B}_{\mathrm{im}}(k_x, k_y) = \widetilde{B}_{\mathrm{obj}}(k_x, k_y) \cdot \mathrm{MTF}(k_x, k_y)$ for the same edge formed by the border to the window (see circles) is displayed in Figure 7.32c. It is well seen that both curves differ and, in particular, there is a loss of approximately 7 to 8 % in lp/PW when the spatial frequency spectra are fit by straight lines. This may be seen from the different slopes and the different maximum frequencies of the extrapolations. Of course, this is a rather quick and rough qualitative analysis only. Nevertheless, it indicates the degradation. One has also to remind that this kind of degradation changes over the image plane. This corresponds to a change of resolution within the image, which is not intended for high quality images, at least not when this becomes observable (see also discussion in Section 5.2.9). On the other hand, similar corrections can be made by the raw converters applied to images of DSLR, but for high quality lenses the corrections are smaller. However, one also has to consider the MTF across the image field (see Section 8.3.5.3). We will not discuss that further and end the discussion of distortion details here.

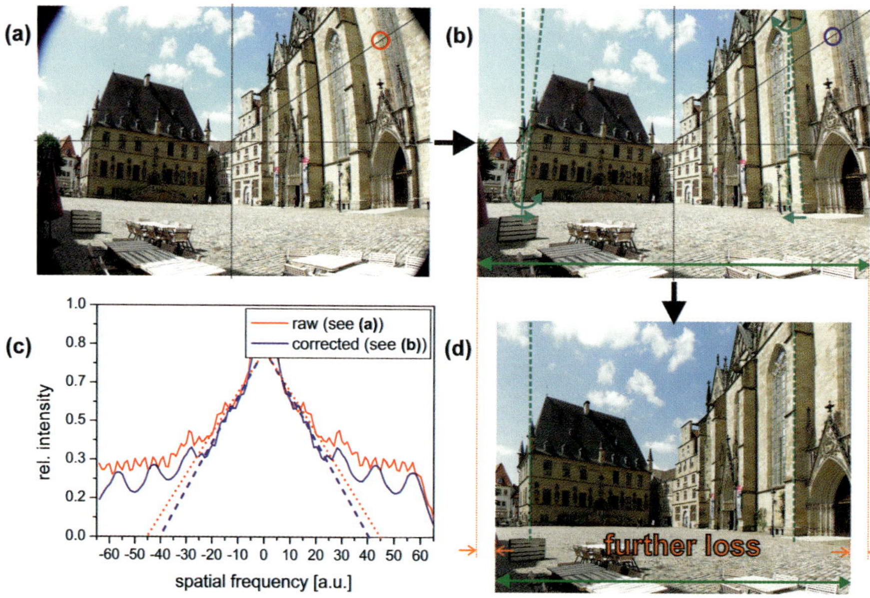

Fig. 7.32: (a) to (c) Example of corrections of distortion (see the text). (c) Shows $\widetilde{B}_{\mathrm{im}}(k_x) = \widetilde{B}_{\mathrm{obj}}(k_x) \cdot \mathrm{MTF}(k_x)$ before and after correction. As $\widetilde{B}_{\mathrm{obj}}(k_x)$ is identical for both cases, the difference of both curves reflect the degradation of the MTF due to distortion correction. Note that the superstructure of both curves is artificial because this is due to the rather small window where Fourier transformation is made ("diffraction" at this window). (d) Shows the further effect of the correction for converging lines and the resulting losses.

Nevertheless, a related correction should be mentioned, namely the straightening of converging lines also still present in picture Figure 7.32b. Such a correction is usually made by (special) software, whereas much superior tilt-shift lenses (see Section 6.7.2) are mostly used by specialists only. But similar to before, software-based solutions again introduce losses. This is illustrated in Figure 7.32d, where the losses result from stretching and shrinking of different parts of the picture. One also has to be aware of the additional crop, which may lead to a further loss when PW is changed to PW' (in the present example $PW' \approx 0.8 \cdot PW$). Thus, the SBN is decreased significantly. We may note that this may not be seen directly from the pixel number in the final image, because the reduced number after image processing may be blown up by suitable interpolation to the original one.

7.4.6 Final remarks to computational imaging and image processing in general

In the previous sections and chapters, we have seen that CI and image processing in general offer a lot of opportunities. In this section, we will not much discriminate except that CI is usually done automatically and image processing in general must not be necessarily automatic. Thus, images of high quality can be easily taken, at least in principle, by nearly everyone. Nevertheless, one has to differentiate the purpose of image capture. Purposes may be, e. g., scientific imaging, photo reportage, holiday and daily photography or artistic photography. Other examples may be found in medical and technical imaging, including vision. All those applications may set different boundary conditions for image manipulation and CI.

For instance, within scientific imaging, tone mapping has to be strictly avoided. As we will see in Section 8.3.1.4, a correct measurement in the scientific or technical sense is based on signals that are related and well-defined to the input signal. In other words, in must be possible to obtain a linear relation of the detected signal to the exposure reliably. Only this allows, for instance, to deduce ratios of the signals from different regions within the image correctly. Spectroscopy provides many examples. By no means this is possible if the image is taken by a camera, which does not allow for a well-defined control of image capture settings and when the camera does not allow for storage of raw images. And even then, it is required that those raw data provide a linear data set and maybe there are even more demands (see Section 4.9). For those applications, most cameras that are used for photography must be excluded. In particular, this excludes also nearly all smart phone cameras, although for other applications they may be interesting mobile mini-laboratories. Vice versa, usage of those cameras for the mentioned purpose will definitively lead to arbitrary results on which one must not rely. But we may note that not any technical or medical application has a need for linear data sets. In that case, other requirements may decide if an intended camera is suitable for an intended application.

On the other hand, and in contrast to before, tone mapping becomes essential if images are taken that should be perceived with at least good quality. In the following, we will concentrate on that. As we have discussed in Section 4.9.4, tone mapping is necessary for several reasons. One is the Weber–Fechner law, namely that the human eye has a logarithmic response characteristic and not a linear one. Another reason is the limited dynamic range of the common output devices, namely displays, prints, etc., with the consequence that the tone range has to be adapted appropriately. Even more, this becomes necessary if the dynamic range within the shooting scene is rather large (see HDR). In addition, appropriate tone curves may be applied that are, hopefully, well suited for the intended image (see, e. g., the examples in Figures 4.62 and 4.63). We will not recapitulate this here further, but we note that this kind of necessary image manipulation can be well done by image processing, including CI.

Other examples of necessary image manipulation are the sharpening and potentially noise reduction. Beside in Section 7.4.1, sharpening has been discussed extensively in Section 5.2.8. Sharpening intends to increase the contrast close to edges to improve the perception. In particular, for digital images based on the mosaic structure of the sensor (see, e. g., Figure 4.23) this is regarded to be important. The necessary de-mosaicing leads to some kind of smoothing, which, however, acts as the opponent of sharpening. Consequently, the resulting blur of an originally high-contrast region in the image then has to be increased again by a subsequent proper sharpening. If not well done, "oversharpened" images result (see below). Other issues of basic image processing and modification have been discussed in Section 4.6 and, e. g., in Section 7.4.2. In addition, image processing can be applied as well to compensate for drawbacks of the applied hardware (the optical/sensor system). Portrait mode of SPC or correction of lens aberrations of any camera are examples that have been discussed above together with their deficiencies.

Thus, part of image processing can be regarded as necessary, at least to some extent. Other parts may be optional. As usual, one has to weight advantages and disadvantages of that further image processing, in particular, when done automatically. Thus, e. g., despite of the advantage of the enormous simplification of photo shooting, CI has the disadvantage that this is done automatically. Thus, there is always the risk that artifacts are generated within the image, which may be well seen and in total degrade the image quality. Today this happens rather often. Artifacts result, e. g., from wrong calculation of hairs and more general quite often there is an unnatural rendering of textures. Although this may be improved in future, the general risk will remain. Examples of such degradations have been discussed for the portrait mode of SPC. Other examples may be found in Figures 5.40 and 8.11. Although these images display artificially "oversharpened" images, such images are quite often generated by the automatic mode of many cameras, including SPC. If instead, image post-processing is done "manually" via a raw converter and potentially further post-processing by a photo software such as Gimp, Photoshop® or another one, the user has a wide and much better control for image optimization.

Image processing in general and CI, in particular, has become very important for imaging. However, especially in a book concentrating on the physical and technical background, we may hint that not every image manipulation is desirable. Above all, this is the case when we consider the large amount of fake information and the potentially resulting consequences. In that sense, one has to bear in mind that image manipulation may be helpful, but one should be careful not to overreach.

In the above, we have discussed that image processing can be applied to compensate for drawbacks of the applied hardware. But it may be applied as well to compensate for "drawbacks of the scene" and then potentially turn photography to image generation rather than to image capture. A particularly grave example is the following, which today has been already implemented in some camera modules. An image is shot of a famous building, which during exposure, is partly covered by, e. g., a scaffold. Then CI looks up in a data base from the internet and replaces the partially obscured parts of the captured frame by blending with the image from the data base and saves this as the "captured image." This has not much to do with photography and opens a discussion.

In our opinion, unless done for pure artistic reasons, a photograph should at least represent the scene, which has been captured without too much manipulation. This shifts the discussion to "what is too much?" Obviously, as discussed above, a minimum of image manipulation is a must. This is similar to what has been done by default in the dark room when films have been developed. Today this procedure has been replaced by raw converters, which are external ones or such ones implemented in the camera. If restricted to the basic image processing, the processed image reflects more or less the scene the photographer has observed.

According to the taste of photographer, the next steps may be, e. g., stronger changes made on tonal curves, e. g., to change the mood of the image expression or retouching may be made on human skins and so on. Of course, each photographer may decide on that. But if done, image manipulation should be definitively communicated. Again, a bad example is a World-Press-Photo selected some time ago. Most of the light impression of that photography was thought to be computer generated. And if so, "image quality" could be regarded mainly to come from image manipulation done by an external company, rather than from the photographer's quality. There was a heavy discussion on that and even forensic investigations were made. For that reason, some award panels take the opportunity to cross-check the images with the captured raw data, which even more are watermark proofed. Already the introduction of watermarks shows that image manipulation is considered to be critical in general (watermarks will be removed when the image is manipulated). We will not judge that in detail. But our intention is to make you aware of that problem, in particular, for photographs that are regarded as trustworthy, such as press photos of journalists. And we would like to note that in that sense, SPC cannot be regarded as a trustworthy tool, because CI leads to automatic image manipulation interpreting the image content and reconstruction of the image based on a guess. Even for much advanced algorithms, CI mostly is not under control of the user. Yet it has been noticed in many reports of SPC analysis that even images that are expected to be quite similar look quite different. Again, this is in total contrast to images taken, e. g., with a DSLR, in particular, if the photographer is familiar with details of its handling.

And again, no post-processing, including CI, can shift physical limits. By no means missing signals and missing information can be really reconstructed. This includes deficiencies of the camera optics, the sensor, the system and those resulting from losses during image processing. Only locally applied tricky advanced methods may reduce loss of information in the processing chain. However, taking into account these limits, rather often image perception can be improved by image processing.

But finally, we would like to state, that albeit all those issues, really high-quality photographs take profit from "the good eye" and from the experience of the photographer as a human, and this is independent from the used camera.

7.5 Alternative concepts for miniature optics

In advance, we may note that a mixture of geometrical optics and wave optics, such as a mixture of rays and wave vectors, is physically not correct. Nevertheless, rays may be the interpretation as a line along the local direction of the related wave vector $\vec{k}$. For that reason, we make use of this mixture in this chapter because it simplifies the discussion.

Imaging bases on the tuning of the wavefronts of all individual rays or wavevectors properly. This tuning can be achieved by taking influence on the phase of each field at each local position of the imaging object as it is done in lenses, diffractive optical elements (see below) or phase plates. But as we have discussed in Chapter 5, instead of a description of the electric field $\mathcal{E}$ in terms of the amplitude $\mathcal{E}_0$ and the phase ϕ, any light field can be described in the Fourier space by its spectral field $\widetilde{\mathcal{E}} = \widetilde{\mathcal{E}}_0 \exp(i\varphi) + \text{c. c.}$, where $\widetilde{\mathcal{E}}_0$ is its spectral amplitude and φ its spectral phase where spectral may be related to ν or ω and/or to the spatial frequency $\vec{R}$ or $\vec{k}$. Hence, there is no physical reason to restrict phase tuning to the *xy*-space. Instead, it can be performed in the Fourier space, namely the spectral phase can be shifted in the $k_x k_y$-space of the spatial frequencies k_x and k_y. Then this operation becomes nonlocal because there is no direct relation in the sense of an individual effect of a local position within the imaging object onto a particular position in the image plane (cf. also Figure 5.3b). Any change at a particular position in the Fourier plane affects all positions in the image plane. This is described by Equation (5.36). In physics of ultrashort pulses or in laser plasma physics, such nonlocal effects are very common.

7.5.1 General description and diffractive optics

Stigmatic imaging means that any point in the object plane is imaged to a conjugated point in the image plane. Although in case of aberrations that may differ slightly, we make use of that idea for easier discussion in the following. Besides this, within that description the spread of the light distribution according to the PSF is not an issue at the moment. In that case, the electric field emitted at or reflected from an object point $\mathcal{E}_{\text{obj}}(x_{\text{obj}}, t)$ is propagated to the image point in a way that according to Fermat's principle, all *optical path* lengths are the same (see Section 3.1.4 and Section 5.1.1). There, the field that has propagated along all possible geometrical paths contributes to that measured in the image point $\mathcal{E}_{\text{im}}(x_{\text{im}}, t)$. Similarly, this is the case for $B_{\text{im}}(x_{\text{im}}, t)$.

The optical path length is given by the path integral along the propagation path (see Section 3.1.4 and Equation (3.13)). In the simplest case, propagation is along straight lines only and then the optical path length is given by the sum of all lengths within the straight-line regions multiplied with the local index of refraction n within those regions. Consequently, one can make use of a space dependence of the index of refraction $n(x, y)$, where here and in whole Chapter 7 we assume the optical axis in z-direction. In case of a normal lens, $n(x, y)$ can be considered to be binary (Figure 7.33a), namely

$$n(x,y) = \begin{cases} \text{approx. } 1 & \text{outside the lens} \\ n_{\text{lens}} & \text{inside the lens} \end{cases} \tag{7.13}$$

Thus, in total, of course, $n(x, y)$ depends on the shape of the lens. Although just the opposite of a miniaturized system, a special example is a lens for a light house, which clearly

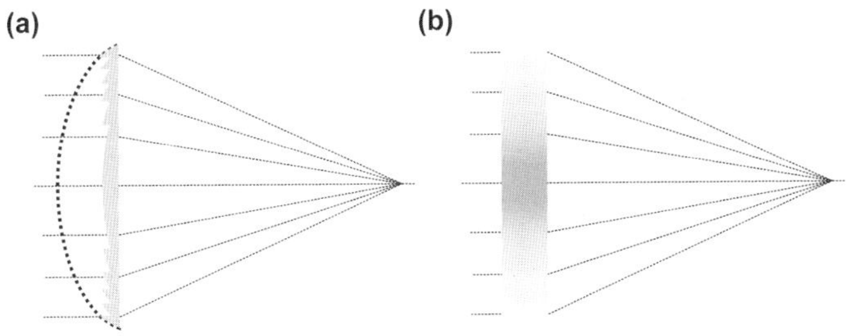

Fig. 7.33: a) focusing or imaging with a Fresnel lens shown in gray. For comparison, the dotted line indicates the shape of the corresponding normal lens. b) Same with a grin lens. The local index of refraction is indicated by the gray level.

has a large diameter. Thus, if shaped as a normal lens, it would be rather thick, and consequently heavy as well. To avoid that, such lenses are shaped differently, namely in a way that they become Fresnel lenses (Figure 7.33a). As it has become clear from Chapter 3 and from above, all optical lengths of relevant ray paths for a stigmatic imaging system have to be the same. However, this is only necessary with respect to their phases with a modulus of 2π. Any additional phase of multiples of $\pm 2\pi$ has no effect on the imaging properties. For that reason, at each height h_{lens} with respect to the optical axis, the thickness of the lens could be reduced in a way that the phase $\phi(h_{\text{lens}}) = 2\pi/\lambda_0 \cdot n_{\text{lens}} \cdot d(h_{\text{lens}})$ is suitably reduced by multiples of 2π. This can be regarded as unwrapping of the wavefront.

h_{lens} is the distance to the optical axis, measured perpendicularly to it, n_{lens} the index of refraction of the lens and λ_0 the wavelength in vacuum. The result is a Fresnel lens. Their application is not restricted to large lenses, but they are used, e. g., as rather flat lenses in overhead projectors. Application as miniaturized optical components or in microoptics is common as well because Fresnel lenses are rather flat devices with a minimum consumption of volume and mass. But we may note that careful manufacturing is necessary to avoid scattering and stray light generation in the regions where the lens shape is "folded back."

In contrast to geometrically shaped lenses, grin lenses may be made of plates with parallel end facets. Grin is the acronym for graded index, and a typical example for a grin lens is a short section of an optical multimode fiber [Sal19].[17] Here, the refractive index usually decreases slowly in radial direction from the center of the fiber core to the core-cladding interface of the fiber (Figure 7.33b). As a consequence, the path of rays in the lens oscillates around the optical axis.[17] The numerical aperture is a local function. It depends on the difference between the refractive index at a given position and that of the cladding. Due to the radial symmetry of the lens' index profile, the numerical aperture is maximum at the center of the lens and decays to zero at the core-cladding

17 H.-G. Unger: *Planar optical waveguides and fibers*, Clarendon Press Oxford 1977.

interface. The position of the focal point for incoming parallel light can be adjusted by the length of the fiber section. Imaging using grin lenses is typically done in special applications as, for instance, light coupling in fiber optical systems.

7.5.1.1 Diffractive optics

On the other hand, changes of optical path lengths may be achieved with $n(x,y) = \text{const}$, when one makes use of diffraction. This results in different geometrical path lengths. Such diffractive optical elements (DOE) or simply “diffractive lenses” can be applied directly or in the Fourier plane, mostly together with additional lenses within the system. This is similar to Figure 5.4a and b). Also, usage of DOE as optical components for beam shaping or as part of intraocular lenses is rather common. Examples of DOE are shown in Figure 7.34. Further examples are presented in the following three sections. Although further details of DOE will not be discussed further here, we may state that they are rather common in microoptics for many purposes. For an extended description we refer to the related literature (see, e. g., [18]).

Implicitly up to now, we have assumed a rather flat diffractive object, but there is no need for that in principle. In case of a volume grating with very different local object structures, those may be regarded as Huygens secondary wave emitters. This is rather general and very common within X-ray optics where the inner shell electrons of the atoms take this role. Consequently, $\mathcal{E}_{\text{obj}}$ after propagation and arrival at and subsequent reemission from a particular Huygens secondary emitter H, is given by $\mathcal{E}_H(x,t) = 1/2 \cdot (\mathcal{E}_{0H}(x,t) \cdot \exp(\text{i} \cdot \phi_H(x,t)) + \text{c.c.})$. Here, $\phi_H(x,t)$ is the local phase at that position. Again, c. c. denotes the complex conjugate. If the phase is shifted properly by this particular Huygens emitter (again with a modulus of 2π allowed) so that its phase front is correctly tuned, this electric field arrives at the image point. If the same is made for all Huygens emitters, or the other way round, if the Huygens emitters are made in such a way that this goal is achieved, the whole diffractive object may act as a device for stigmatic imaging. Here, we have neglected any losses. We may note that the phase shifts of optical paths must be arranged in a way that the integral over all contributions of $\mathcal{E}_H$ leads to the intended intensity maximum at the position of the image point. Below we will provide an example (Section 7.5.1.2). Another example is a phase radar, where the individual phases of the radio waves emitted from the particular antennas within a series of them can be adapted to emit the radar signal into an intended direction without mechanical movement of any of the antennas.

In the X-ray region, this well-known physical description is also discussed in terms of Bragg reflection and transmission. According to the Bragg condition, only in particular directions characterised by an angle $\theta_x(\lambda)$ constructive interference occurs (see

18 S. Sinzinger, J. Jahns: *Microoptics*, 2nd ed., Wiley-VCH, Weinheim, 2003.

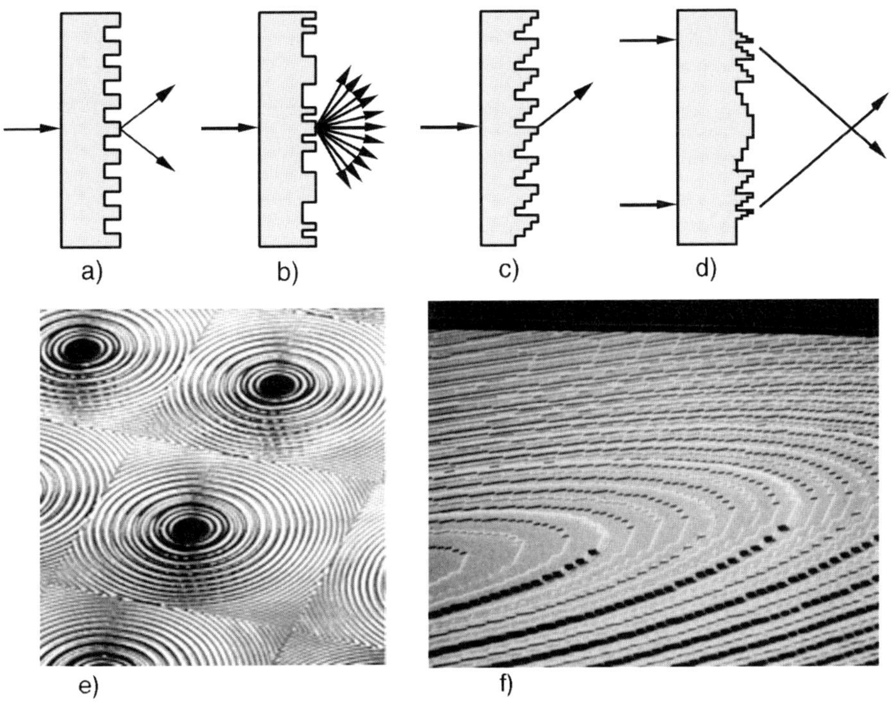

Fig. 7.34: Examples of DOEs: a) 1×2 beamsplitter; b) 1×N beamsplitter (e. g., Dammann grating); c) beam deflector; d) diffractive lens. e) SEM picture of a microlens array with 4 phase levels; f) SEM picture of a section of an 8-phase level DOE combining the functionality of a beam splitter and a diffractive microlens. From S. Sinzinger, J. Jahns: Microoptics, 2nd Edition, p. 138, 2003. Copyright Wiley-VCH GmbH. Reproduced with permission.

standard textbooks on physics, or particular ones on X-ray optics, such as[19]). Here, and in the following, for simplicity we restrict to a 2D geometry with X-ray propagation into z-direction. Extension to 3D geometry with $\theta_y(\lambda)$ etc., is straightforward. X-ray optics with bent crystals (see, Section 2.6.5 and, e. g.,[20]) makes use of that principle. For longer wavelengths, namely in the soft X-ray region, this method can be extended by making use of larger lattice spacings of artificial lattices, e. g., with HOPG (Highly Oriented Pyrolytic Graphite) crystals. In the visible region and close to it, equivalently, multilayer coatings, consisting of a large number of successive layers of low and high index of refraction materials and a thickness of d_{low} and d_{high}, respectively, can be applied as an artificial lattice to obtain highly reflective mirrors. One of the challenges is to avoid strong

19 D. Attwood, A. Sakdinawat: *X-Rays and Extreme Ultraviolet Radiation* (2nd ed.), Cambridge University Press, 2016/2017; see also D. Attwood: *Soft X-Rays and Extreme Ultraviolet Radiation*, Cambridge University Press, 1999.

20 T. Missalla, I. Uschmann, E. Förster: *Monochromatic focusing of subpicosecond x-ray pulses in the keV range*, Rev. Sci. Instrum. **70** (1999) 1288.

absorption, in particular, in the high index material. This can be regarded as the complement to the antireflection coatings discussed in Section 6.8.1.

7.5.1.2 Fresnel zone plates

Special DOEs are Fresnel zone plates and the somehow related photon sieves. Here, we will restrict to the first ones. These are flat elements made of concentric circular rings, which can be used as lenses that fulfill the lens Equation (1.18) and that of the magnification Equation (1.19) (Figure 7.35). Fresnel zone plates allow for very high-spatial resolution, but they are strongly chromatic (resolution down to 10 nm in the X-ray range).[21] In particular, they are used for soft and hard X-ray applications, such as soft X-ray microscopy around 4 nm, i. e., in the so-called water-window.

Transmission zone plates are circular diffraction gratings (for a detailed description, see standard textbooks of optics such as [Hec16] or special ones such as[21]). They consist of successive circular or elliptic rings that are either opaque or transparent. Transmission phase zone plates and reflective zone plates are available as well. In contrast to standard bar gratings (see Chapter 5), which lead to a 1D geometry with the LSF or PSF as the object point (in 1D, we do not have to discriminate), in 2D the circular geometry leads to a central spot where the PSF is the image of an infinitesimal small object point.

According to the discussion before, this requires that first, all phases are matched, and second, that the integral over all contributions of the Huygens emitters leads to the intended PSF, in other words the zone area has to be equal. For that reason, the radii of the zones have to vary so that for a given wavelength λ and an intended focal length f,

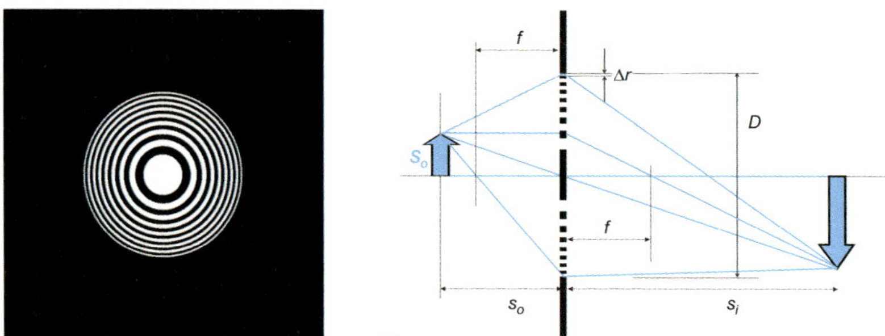

a) b)

Fig. 7.35: Fresnel zone plate a) Geometry of a transmission zone plate. White regions are fully transparent, black regions block the beam. b) Scheme of imaging of an object (corresponds to Figure 1.15, but here magnification >1). $f_{\#} = f/D$. f, D, Δr and the thickness of the zone plate are not in scale with respect to each other.

21 D. Attwood, A. Sakdinawat: *X-Rays and Extreme Ultraviolet Radiation* (2nd ed.), Cambridge University Press, 2016/2017; see also D. Attwood: *Soft X-Rays and Extreme Ultraviolet Radiation*, Cambridge University Press, 1999.

the radius r_i of the i-th zone is given by the relation,

$$r_i^2 = i\lambda f + \frac{i^2\lambda^2}{4} \tag{7.14}$$

It is well seen that r_i increases with zone number i, but the width of the rings decreases. Equation (7.14) is for the first diffraction order. Zone plates can be used in higher orders as well but then the efficiency becomes lower. The second term in Equation (7.14) results from spherical aberration. It can be ignored for a long focal length or a small $NA = \sin\theta = \lambda/(2\Delta r) \ll 1$, in particular, for very short wavelengths. In that case, $r_i \approx \sqrt{i \cdot \lambda \cdot f}$. With the width of the outermost zone (i. e., with largest i),

$$\Delta r = r_i - r_{i-1}, \tag{7.15}$$

NA and $f_\#$ can be estimated:[22]

$$NA \approx \frac{\lambda}{2\Delta r} \tag{7.16a}$$

$$f_\# \approx \frac{\Delta r}{\lambda} \tag{7.16b}$$

Together with Equations (5.44) and (5.48), this yields the spatial resolution rather similar to a normal lens.

An example is a Silicon zone plate used for X-ray microscopy of biological samples at a wavelength of 4 nm. A typical focal length and number of zones is 1.5 mm and 150, respectively. This yields a thickness of $\Delta r = 100$ nm for the outermost ring, and thus $f_\# \approx 25$. According to Tables 5.1 and 5.5, the 2D resolution is given by $2 \cdot r_0 = 1.22 \cdot \lambda \cdot f_\# = 1.22 \cdot \Delta r$, which for the present example is 122 nm.

7.5.1.3 Diffractive optics for camera lenses

Although zone plates can be very small (e. g., mm diameter) and although they are common lenses in the XUV and the X-ray range, in the optical range this is mostly not the case. On the other hand, modern lenses, e. g., such ones made for DSLR cameras, make use of that principle since more than a decade. Examples are Nikon PF-lenses and Canon DO-lenses, which combine one or two zone plates/Fresnel lenses together with a normal lens system. Figure 7.36 shows an example how such a combination can be used to correct for chromatic aberration by making use of the chromatic properties of a zone plate as discussed above and taking profit of the potentially high-spatial resolution. In total, the combination of lenses and diffractive optics may allow for better correction of chro-

22 D. Attwood, A. Sakdinawat: *X-Rays and Extreme Ultraviolet Radiation* (2nd ed.), Cambridge University Press, 2016/2017; see also D. Attwood: *Soft X-Rays and Extreme Ultraviolet Radiation*, Cambridge University Press, 1999.

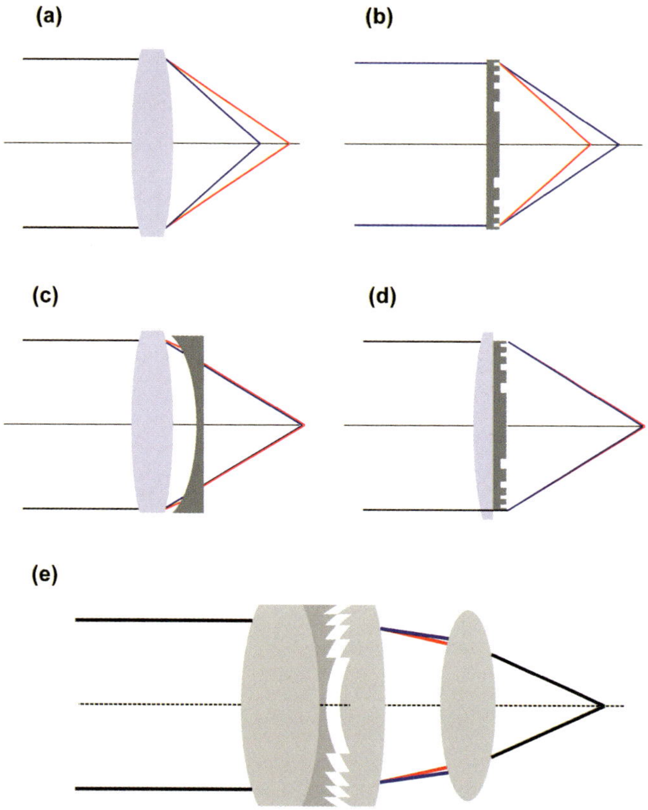

Fig. 7.36: a) A normal lens shows chromatic aberration, b) a Fresnel zone plate lens as well. c) reduction of chromatic aberration by combining lenses of different materials (see Section 3.5.6). d) reduction of chromatic aberration by a combination of a refractive and a diffractive lens. e) Example of a combination of standard lenses together with a twin DOE lens as it is used as part of a camera tele lens. Note that those are schemes only.

matic aberration and spherical aberration in parallel when compared to a combination of refractive lenses only. Moreover, that allows for a much more compact lens construction with less lenses and with much less weight.

7.5.1.4 Small diffractive lenses for a miniaturized systems

Diffractive optics with superwavelength features obtained in quasi 2D geometry, i. e., almost flat, can be generated as well. N. Mohammed et al. (and other groups) have shown that it is possible to generate achromatic diffractive lenses with relatively large NA, which can be used across the entire visible spectrum, e. g., with relatively large efficiency, in particular, when set up as multilevel diffractive lenses.[23,24] Furthermore,

23 N. Mohammad et al.: Sci. Rep. (2018) 8:2799.

24 S. Banerji et al.: *Imaging with flat optics: metalenses or diffractive lenses?*, Optica **6** (2019) 805.

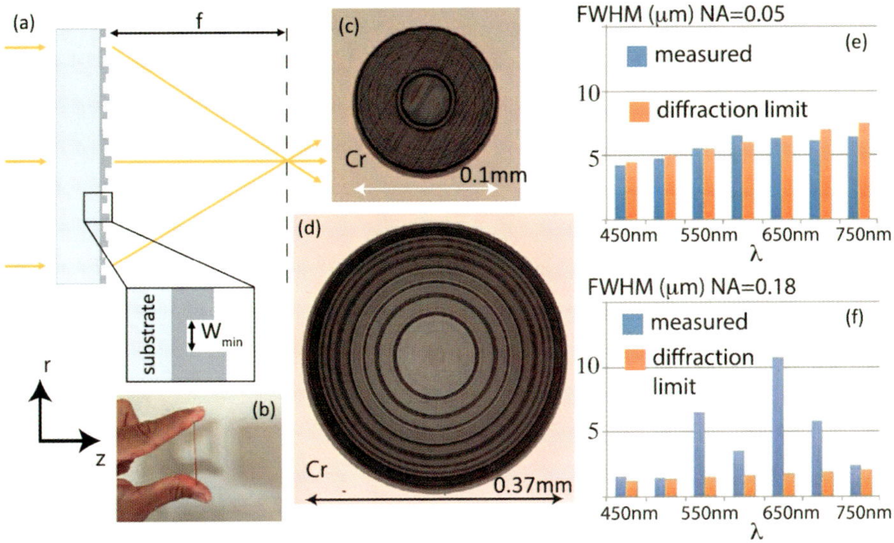

Fig. 7.37: (a) Schematic of a flat-lens design. (b) Photograph of a fabricated lens. Optical micrographs of a lens with $NA = 0.05$ (c) and with $NA = 0.18$ (d), respectively. Measured full width at half-maximum (FWHM) of the focal spot as a function of wavelength for (e) $NA = 0.05$ and (f) $NA = 0.18$ lenses. Reprinted from [26] according to Creative Commons Attribution 4.0 International License; http://creativecommons.org/licenses/by/4.0/.

those and other authors state that this kind of flat optics can be rather easily and cost-efficiently manufactured over large areas, which makes them superior to metalenses (see next Section and, e. g.,[25]). Figure 7.37 shows an example of such a design together with several experimental results of a lens made of concentric circular rings.

In (a), the scheme of the design which is a structure comprised of concentric rings of width, W_{min} and varying heights is presented. (b) shows a photography and (c) and (d) two variants of realized lenses: $NA = 0.05$ (3 µm and 2.4 µm ring width and maximum height, resp.) and $NA = 0.18$ (1.2 µm and 2.6 µm ring width and maximum height, resp.). For both lenses, the focal length is 1 mm and the number of gray levels 100.

7.5.2 Optics of metamaterials

Now we will see how the method of Huygens secondary emitters can be used for very special optics, which may allow for particularly small lenses.

25 S. Banerji et al.: *Imaging with flat optics: metalenses or diffractive lenses?*, Optica **6**, 805 (2019).

26 N. Mohammad et al.: Sci. Rep. (2018) 8:2799.

7.5.2.1 Basics of negative index of refraction

Within classical electrodynamics the propagation of waves and within geometrical optics the propagation of rays is described by Maxwell's equations, the material equations and the resulting wave equation etc. as summarized in Appendix A.12. All this has also been the basis of the discussion in the previous chapters. In particular, from Snell's law one has obtained the refraction conditions shown in Figure 3.2b and the process of imaging as sketched, e. g., in Figure 1.15b and Figure 3.7, which both are the result of, e. g., Equations (3.2) to (3.14). However, it must be noted that there, implicitly, it has not only been assumed that the involved materials such as lenses are transparent, but also that the permittivity ε and potentially the permeability μ are both larger than zero. According to Equation (A.8), ε relates $\vec{\mathscr{D}}$ and $\vec{\mathscr{E}}$ and μ relates $\vec{\mathscr{H}}$ and $\vec{\mathscr{B}}$; both, ε and μ are functions of the frequency ν. Consequently, according to Equation (A.12), n is real and positive as well. This leads to a propagation of the electromagnetic fields without damping or gain. If ε'' is a real positive, the situation will not change much, with the exception that absorption will occur, as, e. g., in an optical filter (see, e. g., Sections 4.2.1 and 4.6.3). All this is rather common and this is the situation of standard optics. Extended discussions on $\varepsilon(\nu)$, which are not restricted to optical materials can be found standard text books.

Nonetheless, there is no physical reason why the real parts of the complex functions of ε and μ have to be real and positive and that the square root of Equation (A.12) must be positive. For simplicity, here we write ε and μ instead of $\hat{\varepsilon}$ and $\hat{\mu}$. Again, the primed and double-primed variables indicate the real and imaginary part of the related complex variables or functions. And, indeed, it is possible to find materials or generate artificial ones with such very special values (see standard textbooks, e. g., [Sal19][27] or, e. g., our discussion in Section 4.10.2). Those materials include such ones where either ε' or μ' is negative, namely single negative media, or both of them, termed double negative media. For common materials, both of them are positive (double positive media), which then leads to the familiar wave and ray propagation. Single negative media are opaque and prevent propagation in the medium, but may lead to surface waves at the interface to double positive media. This is important, for instance, for the field of plasmonics and for special imaging on the microscopic level. But for the topic of optical imaging in the sense of this book, at present, this is not much an issue.

On the other hand, it can be shown that for double negative media n becomes negative, which leads to very unusual properties and wave propagation, which nonetheless, still is in agreement with the classical optics framework provided by Maxwell's equations and the material equations. Though in that case, in contrast to Equation (A.14a), both sides of the equation have a positive sign whereas in Equation (A.14b) the signs then have to differ. Hence, in contrast to double positive media where the electric field, the magnetic field and the wave vector form a set of vectors according to right-hand

---

27 J. Braat, P. Török: *Imaging Optics*, Cambridge University Press, Cambridge, 2019.

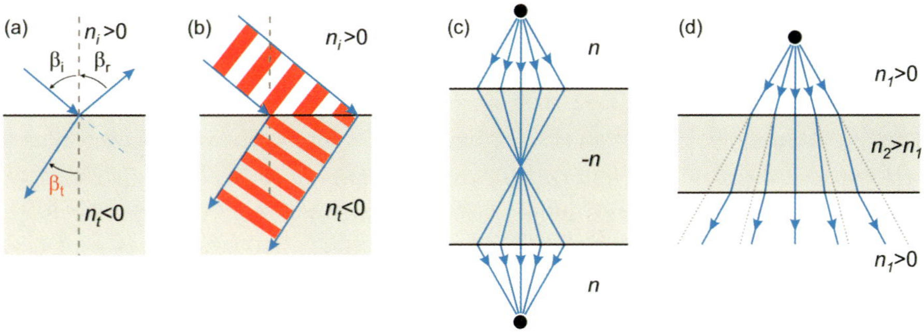

Fig. 7.38: (a), (b) Illustration of refraction at the interface between a double positive medium and a double negative medium, where the refractive indices have opposite signs; note that the angle printed in red is counted negative; (c) illustration of the application of a double negative medium for focusing; note that usually the thickness of the double negative medium has to be rather small; a converging beam becomes divergent after transmitting the interface and vice versa; (d) a refractive sheet with parallel faces acting as a space plate; in case of positive refractive indices, a diverging beam remains diverging after transmitting the interface, a converging beam remains converging.

rule, for double negative media the left-hand rule has to be applied. Like this, the corresponding media are so-called left-hand media. An important consequence is that if in Figure 3.2b the medium with n_t is replaced by a double negative one with n_{DN} one gets the beam path illustrated in Figure 7.38a,b.

While the transmitted beam at the interface between two media with refractive indices of equal signs has also the same sign of the angle as the incident one, it changes the sign at the interface between two media with indices of opposite signs. This feature can be used to focus a diverging beam in a positive refractive index, for instance air, by a metamaterial with the opposite refractive index as shown in Figure 7.38c. An incident diverging beam becomes initially converging in the metamaterial after transgressing the interface. Beyond the focal point in the metamaterial the beam becomes diverging and is converted into a converging beam again after transgressing the interface to the medium with the positive refractive index. Unlike that, in the classical case of a plane-parallel glass sheet, a diverging incident beam remains diverging (Figure 7.38d). We simply get a beam shift where the angles of the individual rays remain the same. The glass sheet acts as a space plate which changes the overall length of the light path.

So far, we have provided the basic issues. A deeper discussion is much beyond the scope of this book. Hence, we refer to the literature for this subject which is a topic of its own.

7.5.2.2 Realization of metamaterials, metasurfaces and metalenses

The preceding discussion provides the skeleton for the application of double negative media for optical imaging. But the problem is that nature does not provide such materials. On the other hand, it is possible to construct artificial ones, namely so-called metamaterials, which make use of sub-wavelength structures. As already introduced in Section 1.8, these allow to tune the index of refraction in a wide range, including negative values of n.

Before we discuss such materials, it is important to get an understanding how a macroscopic structure made of substructures with a lateral dimension smaller than the applied wavelength λ can take influence on wave propagation. Basically, this can be understood, e. g., by applying Huygens principle to the macroscopic structure made of very small features, which are separated laterally by a distance much smaller than λ. According to Huygens principle, they act as secondary point sources which emit spherical waves (Section 3.1). If we now consider a macroscopic volume, V_{mac} made of microscopic volumes each with a size V_{mic} smaller than λ^3, then the phase differences between all the emitted Huygens waves do not play an important role. The contribution of V_{mic} in total comprises an average over all the Huygens waves within that particular volume. Consequently, the refractive index n, which can be a function of the local coordinates x, y and z, namely $n = n(x, y, z)$, can only be considered to vary over distances larger than approximately λ, whereas it provides an average over V_{mic}. This has to be taken into account when applying, e. g., Equations (3.2) and (3.13). Then, if for instance a large amount of homogenously distributed tiny holes, each with a diameter $< \lambda$, is drilled into a macroscopic glass plate, the index of refraction is the weighted average of the index of refraction of the glass and the holes, respectively. Only, for much shorter wavelengths local variation of n may become observable (compare the scattering theory of X-rays at the atomic shell). Therefore, well-defined sub-λ structuring offers the opportunity to modify the index of refraction in a desired way.

Originally, metamaterials were realized for the microwave range, i. e., for wavelengths much larger than in the visible range. Consequently, even the sub-λ structures could be rather large, which offered the opportunity to set up those as resonant electronic impedance circuits or antennas. Note that the local frequency dependent impedance takes influence on properties such as reflection and transmission. Based on that, at least in principle, well-controlled artificial surfaces or volumes of metamaterials could be fabricated, which should have unique properties (2D metasurfaces and 3D metavolumes, resp.). Although there are a lot of difficulties, this allows to generate particular radiation enhancement or absorption, cloaking, the construction of super lenses and so on. We will not consider this wavelength range further and refer to the literature. Recent review papers are, e. g., [28] and[29].

But, although even more difficult, due to the large progress in micro and nanostructuring technology, today the same principle could be applied to electromagnetic waves

28 A. Li, S. Singh, D. Sievenpiper: *Metasurfaces and their applications*, Nanophotonics 7 (2018) 989–1011.

29 J. Hue et al.: *A Review on Metasurface: From Principle to Smart Metadevices*, Frontiers in Physics **8** (2021) 586087.

in the visible range. There is not only the possibility to generate a homogenous metasurface by a periodic set of nanoholes as described in the previous example but a particular well-designed arrangement of sub-λ structures offers additional opportunities. This arrangement can be made either with holes and/or with the complements, namely pillars or nanorods. An example of the first case is a distribution of millions of nanoholes in a 5 µm thick silicon plate with different diameters (between 170 and 310 nm) and interhole distances, which acts as a lens.[30] Another example is an array of different H-shaped gold nanorods that leads to an asymmetric reflection. Namely, the absolute value of the angle differences of the incident light and that of the reflected light in one direction differs from that in the other direction. We will see in Section 4.10.2 how one can make use of such properties.

Following the discussion above, more generally, a well-designed distribution of artificial Huygens emitters (i. e., a “Huygens’ surface”) allows the construction of a particular light-field distribution. Here, one takes advantage that there is a lot of degrees of freedom to tune amplitude, phase and polarization of the electromagnetic field, and thus, of course, also its wavefront. This includes also the application of metasurfaces with negative index of refraction, e. g., as very special lenses.

Thus, in principle a surface of such a Huygens’ surface may look similar to, e. g., Figure 7.34a or b or Figure 7.36b but the building blocks of those structures may be much smaller and more complicated as we will see below.

A specific example of a metalens is shown in Figure 7.39. Based on the idea of arranging metal elements of a special shape, the fabrication of a flat ultrathin lens without monochromatic aberrations has been reported by Aieta et al. in 2012 for the 1550 nm range.[31] The efficiency of the lens is very low as the design has not yet been optimized at the time of its publication. However, the basic principles of the metalens can be well understood in this example. The authors used different geometries of V-shaped antennas and combined them in a square array at a periodic distance of 750 nm (Figure 7.39a and c). The scattering amplitudes and phase shifts of the different V-antennas have been computed numerically. The sequence of the elements shown in Figure 7.39a exhibits a continuous phase shift by $\pi/4$. The overall arrangement of the metallic V-antennas on a planar SiO_2 wafer results in a circular lens of 0.9 mm diameter and a focal length of 3 cm (Figure 7.39c). Some detailed orientation and geometry of the antennas can be seen in the inserts (scanning electron microscope images) of this figure. The functionality of the lens was tested by an experimental set-up as depicted in Figure 7.39b and found to be free of aberrations. Besides its low efficiency and polarization dependency, there is an

30 S. W. D. Lim, M. L. Meretska, F. Capasso: *A High Aspect Ratio Inverse-Designed Holey Metalens*, Nano Letters **21**(20) (2021) 8642–8649.

31 F. Aieta et al.: *Aberration-Free Ultrathin Flat Lenses and Axicons at Telecom Wavelengths Based on Plasmonic Metasurfaces*, Nano Lett. **12** (2012) 4932–4936.

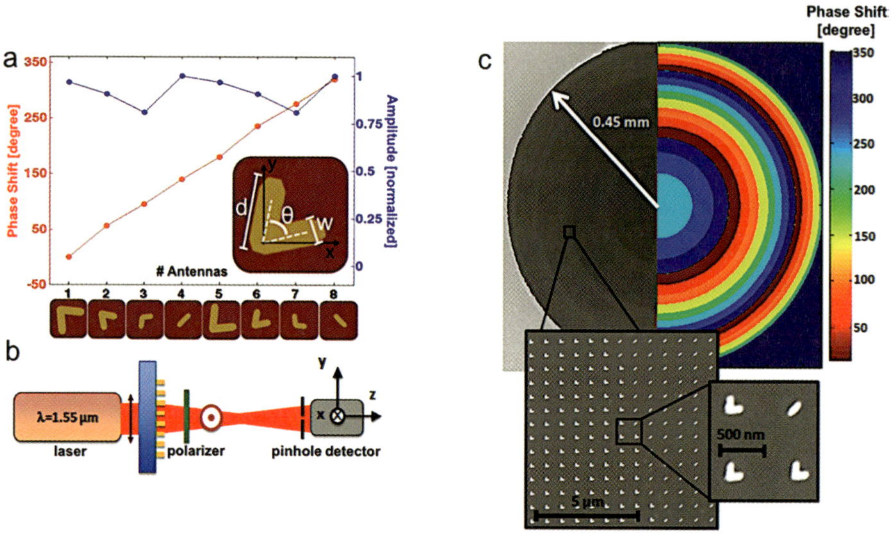

Fig. 7.39: a) Theoretical simulations used to obtain the phase shifts and scattering amplitudes for the eight elements used in the metasurfaces; the parameters characterizing the elements from 1 to 4 are $d = 180$, 140, 130 and 85 nm, and $\theta = 79$, 68, 104 and 175°; elements from 5 to 8 are obtained by rotating the first set of elements by an angle of 90° counterclockwise; the width of each antenna is fixed at $w = 50$ nm; b) experimental setup: a diode laser beam at $\lambda = 1.55$ µm is incident onto the sample with y-polarization; the light scattered by the metasurface in x-polarization is isolated with a polarizer; a detector mounted on a three-axis motorized translational stage collects the light passing through a pinhole, attached to the detector, with an aperture of 50 µm; c) SEM image of the fabricated lens with 3 cm focal distance; insets: close-up of patterned antennas; the distance between two neighboring antennas is fixed at $\Delta = 750$ nm in both directions. Reprinted/adapted with permission from [32]. Copyright 2012 American Chemical Society.

inherent wavelength dependence like in most metamaterials. More recent designs have overcome these early deficiencies.

A more broadband behavior can be achieved by a metamaterial of about 0.5 µm thickness based on a Polyimide film covered on both sides by very thin gold layers and patterned to exhibit a modified fishnet structure.[33] Due to the patterning the effective permittivity and permeability of the thin film are balanced to each other within a given wavelength range. This allows to match its impedance to that of the ambient medium, namely free space, and thus to nearly fully suppress reflection.

Metasurfaces containing metal elements are prone to ohmic losses. These can be reduced using dielectric elements such as nanopillars or fins. Dielectric pillars and fins, however, require much higher aspect ratios to achieve a full phase modulation as it can

32 F. Aieta et al.: *Aberration-Free Ultrathin Flat Lenses and Axicons at Telecom Wavelengths Based on Plasmonic Metasurfaces*, Nano Lett. **12** (2012) 4932–4936.

33 Z. H. Jiang et al.: *Tailoring Dispersion for Broadband Low-loss Optical Metamaterials Using Deep-subwavelength Inclusions*. Sci. Rep. **3** (2013) 1571; DOI:10.1038/srep01571

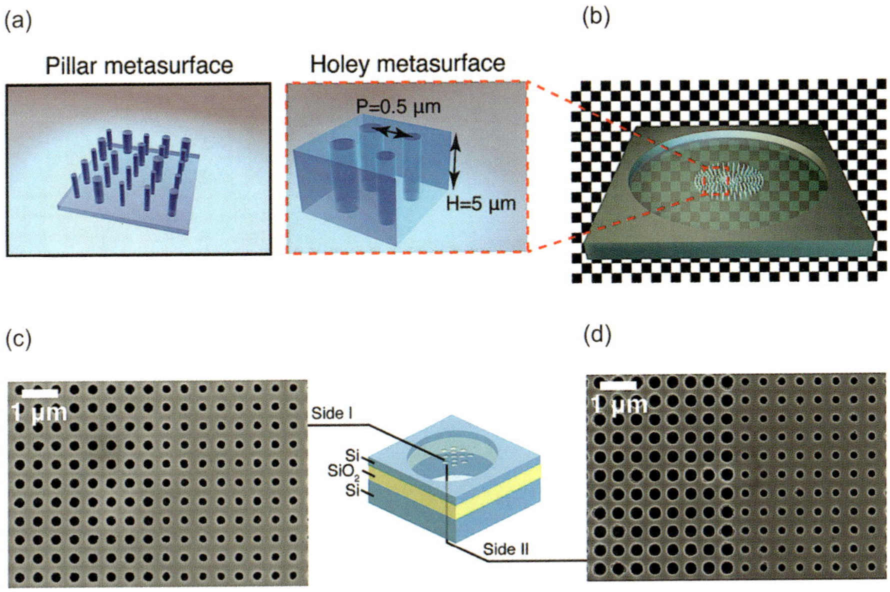

Fig. 7.40: Design of a holey metalens. (a) Comparison between a free-standing pillar metasurface and a holey metasurface. Holey metaatoms are more stable and robust compared to pillar metaatoms. (b) Artistic representation of a holey metalens. (c) SEM image of the fabricated holes on side I of the holey metalens. (d) SEM image of the fabricated holes on side II of the metalens. Reprinted with kind permission by the authors [34].

be done by metal elements. Free-standing pillars of high aspect ratios are more susceptible to mechanical damage. This can be overcome by an inverted design where via-holes are generated in a thin dielectric matrix which altogether constitute the metasurface.[34] Figure 7.40a,b shows the advanced design of a holey metalens for the near-infrared range. It features a converging lens of 2 mm diameter with a focal length of 4 mm and a numerical aperture of 0.24 at 1550 nm. The holes of varying diameters between 170 nm and 310 nm are arranged in a square pattern of 500 nm pitch (Figure 7.40c,d) in the substrate. It should be noted that the SiO_2 sacrificial layer that can be seen in this figure is only used to define the etch stop. Most of it as well as of the underlying Si-layer are removed after processing to yield a free-standing 5 µm thick silicon membrane, which is transparent for the application wavelength. When monochromatic light of $\lambda = 1.55$ µm is normally incident on this thin crystalline Si membrane with more than 12.5 million etched via nanoholes, the incident optical wavefront is modified by the holey structure and produces a diffraction-limited focal spot with a nearly Gaussian profile of approximately 3 µm diameter.

34 S. W. D. Lim, M. L. Meretska, F. Capasso: *A High Aspect Ratio Inverse-Designed Holey Metalens*, Nano Letters **21**(20) (2021) 8642–8649.

In general, in spite of the large potential of metaoptics, until recently the obtained image quality has mostly been far worse when compared to that of high-quality conventional lenses and the spatial resolution has been much less as well. But progress has been made. In principle, the application of metamaterial properties allows for the design of a "perfect lens," at least for a "perfect metamaterial" (see, e. g.,[35] and discussion above). Indeed, it can be shown that the optical resolution $\mathcal{R}$ of the optics can be much increased. For instance, for a wavelength of $\lambda = 400$ nm in free space, a conventional aberration-free optics with $f_\# = f/D = 1$ yields $\mathcal{R} = \lambda = 400$ nm (for Abbe's criterion, see Equation (1.20); alternatively $\mathcal{R} = 1/(2R_{max})$, see Chapter 5). Using a perfect metalens, this resolution could be much improved to approximately 10 nm.[35]

To the best of our knowledge, at present PSF with such small widths have not been achieved in the visible range. However, there is progress in that direction. For instance, F. Khorasaninejad et al. report on the experimental realisation of a metalens with a diameter of 240 µm and a focal length of $f = 90$ µm, which provides a subwavelength resolution.[36] That metalens consists of TiO_2 nanofins on a glass substrate. These are designed to tune the wavefront profile such that the lens functions like a spherical lens with an MTF0 of up to approximately 2000 lp/mm. As demonstrated in that work, focal spot sizes smaller than up to 1.5 times the diffraction limit were obtained for three design wavelengths between 660 and 405 nm (numerical aperture up to 0.8). Moreover, this work reports on the fabrication of a metalens for practical imaging.

As has been mentioned before, metalenses usually are highly chromatic. Nevertheless, dispersive phase compensation methods may be applicable to reduce or remove this negative property. A recent example of an achromatic (400 to 700 nm operation) polarization-insensitive high-quality "nanooptic imager" is shown in Figure 7.41. Tseng et al. present a large aperture (0.5 mm, $f_\# = 2$) wide field-of-view (FOV $= 40°$) "neural nanooptics" that consists of an optimized metasurface and an advanced deconvolution algorithm that allows for an order of magnitude lower image reconstruction error when compared to other existing works.[37] The authors claim that "no existing metaoptic demonstrated to date approaches a comparable combination of image quality, large aperture size, low f-number, wide fractional bandwidth, wide FOV, and polarization insensitivity ... and the proposed method could scale to mass production."

The discussed examples should not be considered as a review of the topic under discussion. Rather it should be considered as an illustration of metalens capabilities. There is much newer work as well, with a lot of advancements, in particular by F. Capasso's group. Altogether there are a lot of advances in the field of metaoptics by many groups

35 J. Braat, P. Török: *Imaging Optics*, Cambridge University Press, Cambridge, 2019.

36 M. Khorasaninejad et al.: *Metalenses at visible wavelengths: Diffraction-limited focusing and subwavelength resolution imaging*, Science **352** (2016) 1190.

37 E. Tseng, S. Colburn, J. Whitehead, L. Huang, S.-H. Baek, A. Majumdar, F. Heide: *Neural nano-optics for high-quality thin lens imaging*, Nature Comm. **12** (2021) 6493.

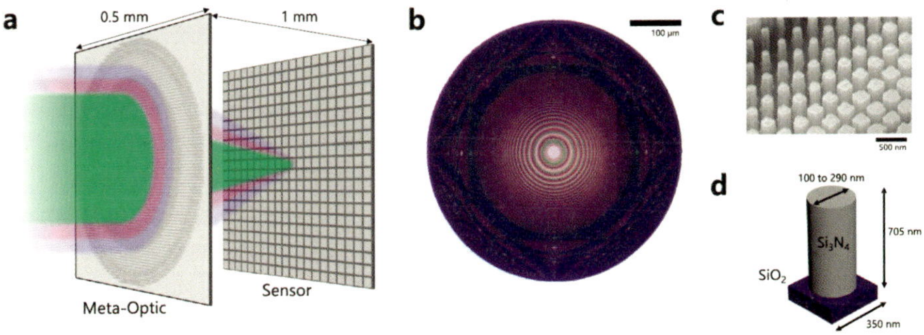

Fig. 7.41: Neural nanooptics. An ultrathin metaoptic as shown in (a) is 500 µm in thickness and diameter, allowing for the design of a miniature camera. The manufactured optic is shown in (b). A zoom-in is shown in (c) and nanopost dimensions are shown in (d). The design volume consists of a single metasurface that is 550,000× lower when compared to the six-element commercial compound lens used in the cited work. Figure and most of the figure caption reprinted from [38] according to Creative Commons Attribution 4.0 International License; http://creativecommons.org/licenses/by/4.0/.

and evolution still continues. This includes advancements in performance such as increased achromatic (broadband) imaging, polarization independence and, e. g., special properties such as chiral imaging as used in microscopy of biological compounds. Improvements have been made also on efficiency and NA, so that, e. g., wide-angle cameras can be realized. CMOS-compatible fabrication techniques and materials are another issue and even tuneable metalenses have been developed. But as it is not the goal of the present book to present a review, we omit citation of the large amount of the related work.

We have shown that metaoptics can be realized by a huge amount of very tiny structures such as nanorods or nanoholes. Usually all of them have a rather large height when compared to the lateral feature size. Even more, they may have a shape which is more complicated than just a simple cylindrical or rectangular pillar or its complement. Still even more, the design of the individual shapes and their distribution has to follow a sophisticated design according to the desired optical properties. This is by far not trivial and requires a sophisticated design with complex numerical simulations. Manufacturing is a challenge as well, especially due to the demand for very fine high-aspect ratio structures with very smooth sidewalls. For instance, atomic layer deposition or advanced lithographic methods and, in particular, electron beam lithography have to be applied. Potentially fabrication is also possible with femtosecond laser induced two-photon direct writing (see, e. g.,[39]). But although such methods are standardly available,

38 E. Tseng, S. Colburn, J. Whitehead, L. Huang, S.-H. Baek, A. Majumdar, F. Heide: *Neural nano-optics for high-quality thin lens imaging*, Nature Comm. **12** (2021) 6493.

39 S. Coelho et al.: *Direct-laser writing for subnanometer focusing and single-molecule imaging*, Nature Comm. **13** (2022) 647.

the effort is large, the process is time consuming and expensive. Thus, today metalens diameters are mostly restricted to sub-mm and commercial use is not yet reached.

Nevertheless, metaoptics, in principle, offer the path to very thin flat optical lenses, which in addition, may be achromatic and of similar high-optical quality properties as the up-to-date “normal” lenses. This makes them promising optical components for miniaturization of cameras. This includes both usage as main optics and usage as elements, e. g., to improve sensor performance (see Section 4.10.2). It has taken more than two decades of scientific research until today where there might be a breakthrough. For that reason, debates have been started on the performance of metalenses and that of microscale (multilevel) diffractive lenses (see Section 7.5.1.4). [40,41] provide such a discussion with detailed assessments.

7.5.3 Spaceplates

7.5.3.1 General issues on spaceplates

In the previous chapters, we have discussed how to miniaturize the optics of a system. In particular, metalenses (see previous section) may lead to the opportunity to create very small optics that may be used for, e. g., SPC or for scientific or medical applications. However, if we consider the whole system we may remember that in addition to optics and sensor, there is also a third component, which determines the size of the system, namely free space. From ray tracing as discussed in Section 3.3, it is obvious, that beside optical components such as lenses, that can be described by a transfer matrix, there is another crucial component with another matrix explicitly that accounts for free space propagation between the lenses and between the last lens and the sensor (see, for instance, Equation (3.31)). And indeed, the latter one takes large influence. In particular, if we consider for instance a telephoto lens with its long focal length, a lot of free space consumption is needed. Thus, as discussed for instance in Section 7.1.2, SPC suffer from this problem, which even cannot be removed by application of ultraflat optics. For that reason, new ideas to “compress” this space consumption are required. Then together with suitable miniaturized optics, such as metalenses, this may allow for ultraflat and compact camera systems.

A rather new idea to “compress space” is a so-called spaceplate, suggested by[42]. The idea is sketched in Figure 7.42. The goal of the spaceplate is to shrink the original distance to the image or focal plane. But there is the additional requirement that the optical path

---

40 J. Engelberg, U. Levy: *The advantages of metalenses over diffractive lenses*, Nature Comm. **11** (2020) 1991.

41 J. Engelberg, U. Levy: *Achromatic flat lens performance limits*, Optica **8** (2021) 834.

42 O. Reshef et al.: *An optic to replace space and its application towards ultra-thin imaging systems*, Nature Comm. **12** (2021) 3512.

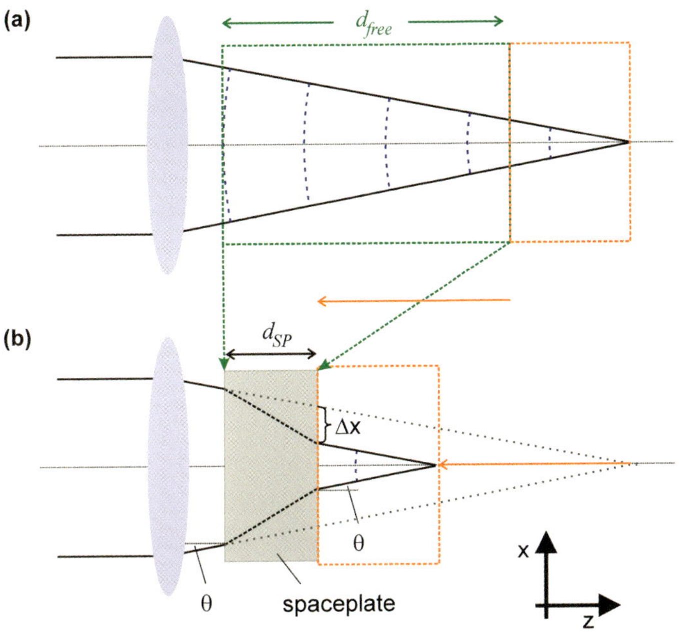

Fig. 7.42: Scheme of spaceplate operation. a) Usual beam path in an optical system, for instance when a collimated beam is focused by a lens. b) Similar situation when, in addition, a spaceplate is inserted. In that case, the propagation in the green marked free space region is shrunk to that one within the spaceplate marked in gray. Beam propagation within the orange marked region is identical in both cases. This includes also the wavefronts that are indicated by blue dashed lines. The amount of "saved space" is marked by the orange arrows. Δx indicates the shift in transversal direction.

in the region marked in green in Figure 7.42a should be identical to that in the spaceplate shown in Figure 7.42b (displayed in gray). In particular, the inclination of all rays should remain unchanged. That also requires that the length of the optical path of each ray j depends on the inclination angle θ_j with respect to the optical path at the position where ray j enters the plate. This can be described by a related wavevector (see below). Depending on θ_j, the spaceplate with thickness d_{sp} has to modify the optical path for each ray j properly. Then the length of the green region is reduced to the physical length d_{sp}, i. e., the thickness of the spaceplate (reduction by $d_{free} - d_{sp}$). The ratio of both lengths can be defined as compression factor $R_{compress} = d_{free}/d_{sp}$.

According to simple geometrical optics this can be achieved when a spaceplate is inserted that has a lower refractive index when compared to outside. This is the opposite case to the example shown in Figure 7.38d where the optical path is increased due to a spaceplate with a higher refractive index than outside. The problem in case of free space propagation, however, is that there is no classical material with a refractive index below 1 in order to satisfy that condition. Therefore, a different approach must be found for beam propagation in free space.

Physically, the spectral phase φ_{sp} of the spaceplate for a length d_{SP} must be the same as that for free space propagation according to d_{free}, where d_{SP} and d_{free} are defined in Figure 7.42. $\phi = \vec{k} \cdot \vec{r}$ where $\vec{r}$ is the spatial coordinate. If propagated the related shift is given by $\Delta\phi = \vec{k} \cdot \vec{d}$, where $\vec{d}$ is the distance vector between the coordinates. When we restrict to the xz-plane, then $k = (k_x^2 + k_z^2)^{1/2}$ where $k = 2\pi/\lambda$ and λ is the wavelength in the spaceplate. Due to spaceplate operation and in similarity to Figure 5.1, $k_z = |\vec{k}| \cdot \cos(\theta_x)$ (the component of $\vec{k}$ in propagation direction which here is the horizontal) and $k_x = |\vec{k}| \cdot \sin(\theta_x)$ in vertical direction. $|\vec{k}| = 2\pi/\lambda$. All this provides the tools for the calculation of the necessary φ_{sp}, which describes the phase shift and the corresponding OTF of the spaceplate as the optical element: $\mathrm{OTF}_{sp}(k_x) = \exp[\mathrm{i} \cdot \varphi_{sp}(k_x)]$ (for a lossless element). The intention is that this is achieved on a distance d_{SP} much shorter than d_{free}. Note that for all these considerations the factor $\exp(\mathrm{i}\omega t)$ in $\mathscr{E}$ (see Equation (5.1)) has been omitted because it does not take influence on the result.

From the previous description of a spaceplate, one can see that φ_{sp} is a function that is purely momentum dependent (note that momentum equals $\hbar|\vec{k}|$). Thus, in contrast to a position-dependent response, it cannot redistribute k-vectors and the related angles θ of light rays.[43] Consequently, it cannot redirect θ_j of any ray j within the optical system as usual lenses or mirrors do. “Therefore, a spaceplate is an optical element complementary to a lens” as Reshef et al. further state. As a further consequence, we see that the design of a suitable spaceplate requires $\varphi_{sp} = \varphi_{sp}(\theta)$, and hence effectively an index of refraction that depends on θ. We may mention that is according to a nonlocal property as discussed above. We may further mention that some metamaterials already include this, and thus lead to space compression. However, this is not the usual goal of metalenses and, in particular, at least ultrathin and almost planar (meta)optics definitively require additional free space propagation or a spaceplate. A third consequence is that spaceplates are translation invariant. This means that, for instance, in contrast to lenses, lateral shifts do not take influence on the spaceplate function.

7.5.3.2 Realisation of spaceplates and particular issues

Now the question may be how to realize a spaceplate. There are several possibilities. One of them bases on artificial multilayer systems as discussed in Section 7.5.1.1. But now multilayer technology has to be much extended, and more complicate arrangements have to be made. Also, there is some similarity to multilayer mirrors made for tuning of ultrashort pulses, where tuning is made in the spectral domain to change the pulse structure in the time domain, e. g., to generate or to compensate for temporal chirp of ultrashort laser pulses or to generate a more complicate pulse structure (see,

43 O. Reshef et al.: *An optic to replace space and its application towards ultra-thin imaging systems*, Nature Comm. **12** (2021) 3512.

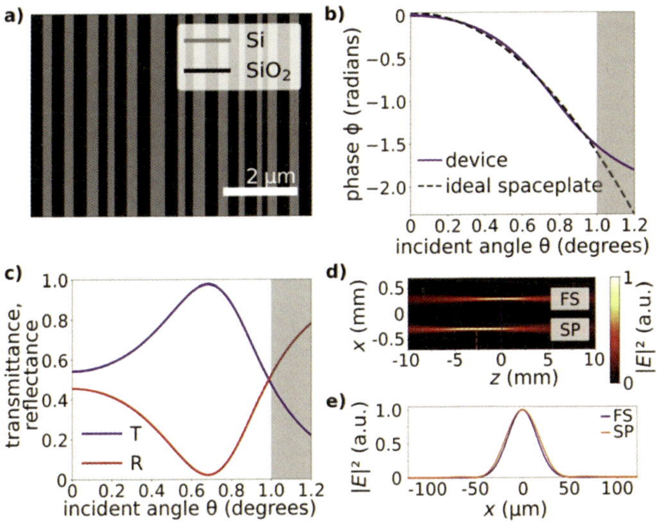

Fig. 7.43: Example of a spaceplate. (a) layer arrangement (b) phase after propagation through this spaceplate in comparison to an ideal one with the same $d_{free} = 2.6$ mm. (c) Normalized transmittance and reflectance of the spaceplate. Intensity distributions of a focusing Gaussian beam and cross-sections are displayed in d) and e), respectively (see text). The image is taken as Figure 2 from [47]. Reprinted with permission from Jordan T. R. Pagé et al., Opt. Expr. **30** (2022) 2197, https://doi.org/10.1364/OE.443067; #443067 Journal ©2022 Optica Publishing Group.

e. g.,[44,45]). Although not trivial, this is possible. For spaceplates, it is possible to design special multilayer structures that compress space by tuning in the spatial frequency domain. Figure 7.43a shows an example[46] of layer arrangement of a spaceplate designed for monochromatic operation with p-polarized light with $\lambda = 1550$ nm.

Here, compression is rather large. A free space length $d_{free} = 2.6$ mm has been compressed by a factor $R_{compress} = 340$ to a length of $d_{sp} = 7.6$ µm. Compression has been achieved by 27 alternating layers of Si (black) and SiO_2 (gray) surrounded by air (Figure 7.43a). Part b) displays the optical phase of the device together with a fit to the expected output phase of an ideal spaceplate. Obviously, φ_{sp} depends on θ (see also (c)). Part d) shows the intensity of a focusing Gaussian beam with a waist of 30 µm and a divergence of 0.94 (analytic Fourier optics propagation), both for free space propagating (FS) and propagating through the spaceplate (SP) from left to right. Part e) shows the

44 J.-C. Diels, W. Rudolph: *Ultrashort laser pulse phenomena*, Elsevier, 2006.

45 A. M. Weiner: *Ultrafast Optics*, Wiley, 2008.

46 J. Page et al.: *Designing high-performance propagation-compressing spaceplates using thin-film multilayer stacks*, Opt. Express **30** (2022) 2197.

47 J. Page et al.: *Designing high-performance propagation-compressing spaceplates using thin-film multilayer stacks*, Opt. Express **30** (2022) 2197.

PSF obtained without and with spaceplate, respectively (beam cross-sections measured at $z = 0$ mm (FS, shown in blue) and $z = -2.59$ mm (SP, shown in orange)). Both are rather identical. Thus, those experimental results indicate that the spaceplate works well, even for such a large compression factor. However, $NA = 0.017$ (at $\theta = 1°$) is rather low and that spaceplate was not yet lossless. Page et al. have investigated other configurations as well, e. g., such ones with $NA = 0.42$ but then $R_{compress} = 5.5$ was much lower. Even other configurations have made use of, e. g., layers of 101 layers each with 5 nm thickness together with various n tested, even $n < 0$ or $\varepsilon \to 0$ (see Section 7.5.2.1; but note that that increases large losses). For further details, we refer to[48].

Besides this example based on multilayer technology, there are several other configurations to realise a spaceplate. Examples are the usage of birefringent material, photonic crystals, low index materials, metamaterials and other more. Shastri et al.[49] provide a current brief collection and comparison of such designs together with their references. But although demonstrations have been successful, there are currently still a lot of drawbacks. In particular, spaceplates have been working well, e. g., with large compression factors, but with low NA and for small bandwidth only. Or broad bandwidth operation has been verified, e. g., by space compression with an absolute value of 3.4 mm for an entire color image,[50] or NA is relatively large but the compression factor and/or the bandwidth is small. Moreover, there are further issues that should be considered.

In total, that means that the demand on a large compression for a large bandwidth operation together with large NA, polarization independence, low losses and maybe more should be fulfilled. This demand is similar to that for metaoptics. Some discussion on that topic, e. g., on the compression factor and on bandwidth limits, is made, e. g., in[49]. Of course, as part of the whole optical system, a spaceplate should also have a good optical performance, e. g., as characterized by its MTF. But one has to state that a full understanding of the related optics and a proper design is very challenging. The topic of spaceplates is rather new and just at the beginning of its investigation. Beside a potential improvement of miniaturization of cameras (see Figure 7.44), in particular SPC, other applications such as photovoltaic systems, collimators for light sources, or applications within integrated optics may take profit. And indeed, at least one large manufacturer of smartphones intends to go into that direction. We may also note that due to their principle, spaceplates may offer the opportunity to use larger sensors with larger pixels even in miniaturized systems (remember that the crop factor is given by f/f_{eq}, see also Section 7.1). For miniaturized cameras, f is rather short, and thus is the applicable sensor

48 J. Page et al.: *Designing high-performance propagation-compressing spaceplates using thin-film multilayer stacks*, Opt. Express **30** (2022) 2197.

49 K. Shastri et al.: *To what extent can space be compressed? Bandwidth limits of spaceplates*, Optica **9** (2022) 738.

50 O. Reshef et al.: *An optic to replace space and its application towards ultra-thin imaging systems*, Nature Comm. **12** (2021) 3512.

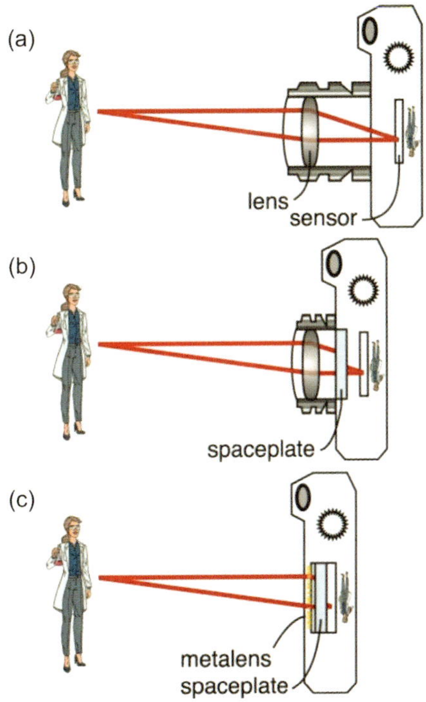

Fig. 7.44: Illustration how the size of a "normal" camera (a) can be reduced by the usage of a spaceplate (b). A stronger effect is achieved by a combination with a metalens (c). In principle, metalenses and spaceplates can be integrated monolithically with the sensor to form a single optical element. Reprinted from [51] according to Creative Commons Attribution 4.0 International License; http://creativecommons.org/licenses/by/4.0/.

size. With a spaceplate, however, f could be effectively lengthened without decreasing resolution. As discussed, this would lead to better performance, for instance due to improved noise properties. Finally, we may note that the combination of a metalens with a spaceplate in one optical element is possible as well.

51 O. Reshef et al.: *An optic to replace space and its application towards ultra-thin imaging systems*, Nature Comm. **12** (2021) 3512.

8 Characterization of imaging systems

8.1 General

Full characterization of a camera system may be rather complex. Besides characterization of the optical performance, this requires the analysis of the opto-electronic properties of the image sensor, e. g., dynamic range and noise, etc., which is a comprehensive topic on its own. There are even further properties that may have to be characterized, such as lens centering, shutter performance, the additional video performance of a camera, if available, and rather simple ones such as the handling properties of the camera. However, these further properties are not covered by the topics of the present book. Nevertheless, such discussion can be found in popular journals on photography, sometimes even in scientific journals. Table 8.1 provides a list of measurements that may be recommended for the evaluation of a camera system used for photography, in particular, for a DSLR, namely a system consisting of the camera body and the camera lens. With some restrictions, this may be also valid, e. g., for compact cameras. More information on tests may be found as well on the websites of test laboratories such as DXO Mark Lab.

Some of the recommendations in Table 8.1 may be useful for other cameras as well, e. g., ones used for scientific or technical purposes. Of course, as discussed in the pre-

Tab. 8.1: Recommended measurements for the evaluation of a camera system used for photography. The listed recommendations are partly taken from[1]. Further optional ones are provided in the same paper and also in other articles, in particular, the EMVA 1288 Standard for Characterization of Image Sensors and Cameras.[2]

mandatory	further recommended
Opto-electronic conversion function OECF	Distortion
White balancing	Shading/vignetting, sensor uniformity
Dynamic range (related scene contrast)	Flare
Used digital values	Chromatic aberration
Noise, SNR	Color reproduction quality
Resolution (limiting resolution center, corner)	Unsharp masking
Sharpness	Shutter lag
	Aliasing and other artifacts
	Compression rates
	Exposure and exposure time accuracy and constancy
	ISO speed

1 See the White Paper *Image Engineering digital camera tests* of D. Wueller from Image Engineering, 2006.

2 EMVA Standard 1288, Standard for Characterization of Image Sensors and Cameras, Issued by European Machine Vision Association www.emva.org; see, e. g., www.emva.org or www.standard1288.org

https://doi.org/10.1515/9783110789966-008

vious chapters, for such cameras, emphasis may be put on other properties than for photography, such as sensor linearity and so on.

Of course, the optical performance of a camera system has a major influence on the image quality. Topics are resolution, monochromatic and chromatic aberrations, vignetting, color reproduction and so on. Camera system characterization with respect to those properties is the subject of Section 8.2. Some other properties are not pure optical properties as they are not independent from those related to the sensor and image processing system. As an example, we have seen in Section 5.2 that resolution and aberrations are related to the MTF and here the sensor may also take an important role. Due to that reason, but even more because it is a really key issue for the characterization of an optical system, Section 8.3 puts emphasis on MTF measurements. Finally, there are other properties that may be discussed separately, namely opto-electronic ones (e. g., dynamic range and noise). Measurements of that matter are discussed in Section 8.4.

Although we cannot give a complete discussion on the extended topic of the characterization of imaging systems, within this chapter we will discuss issues that may be most important for the user of optical systems. But we do concentrate on the basics and exclude ISO-standards and other standards. However, the interested reader may be referred, e. g., to the EMVA Standard 1288, which is the Standard for Characterization of Image Sensors and Cameras.[3] In advance and before we begin with that discussion, we would like to remind you of an important issue, namely that for a good characterization that allows a reliable judgment, it is always important to do the characterization for the whole system under consideration. This means the combination of the lens with the camera body. Within compact cameras or mobile phone cameras, this is automatically the case. Analysis of the properties of one or several components alone, such as lens only or sensor only, does not necessarily lead to consistent results, although such investigations sometimes are necessary for other reasons (see later). There are several rather clear examples for this; other ones are less apparent. A first example, which is rather straightforward, is that the same lens investigated with a camera system using an APS sensor may behave quite differently when used in a full format camera. But even lenses used, e. g., on a different full format DSLR may differ in their performance. A second example was mentioned in Section 4.6.1. There it was explained that it cannot be expected that a high-quality lens designed for an analog SLR works well with a DSLR. Due to the optical microlens array, wide-angle lenses then may perform worse. For telephoto lenses, this may be less severe. This is important to note because sometimes people do such tests and then are rather disappointed and judge a high-quality lens as poor quality. However, this does not allow for a negative judgment on the lens itself. One also has to keep in mind that image processing done in the camera may play a role. This is even the

3 EMVA Standard 1288, Standard for Characterization of Image Sensors and Cameras, Issued by European Machine Vision Association www.emva.org; see, e. g., www.emva.org or www.standard1288.org

case if one uses raw data (see Section 4.9). It is also important that any post-processing of the image has to be avoided if the interest is the characterization of the hardware and not of the software.

Many of the evaluation procedures rely on the analysis of images taken of test charts (i. e., targets; see the following sections) at well-defined conditions. Certainly, then it is absolutely essential that the chart is illuminated very homogeneously. This is not an easy task at all. To achieve a good homogeneous diffuse illumination is difficult because virtually all light sources have special radiation characteristics. There are particular investigations on that subject and articles can be found in the literature (see also special books related to the characterization of camera systems).

Nevertheless, we would like to provide an important comment. It is known that Ulbricht spheres are optical devices that are used to achieve a very homogeneous radiation field. Its surface acts as a Lambertian surface (see Section 1.2.4). Anywhere on its inner surface the radiant properties, such as the radiant exposure (fluence), the radiant intensity, etc., are the same. This almost does not even change if there are openings in the sphere. However, it cannot be expected that the emission from one of the openings is homogeneous unless it is measured within a curved surface in it. As a result, the opening cannot necessarily be regarded as a source that can be used for homogeneous illumination of a test chart. Consequently, Ulbricht spheres at best may be used as approximately virtually homogeneous sources only if, e. g., the diameter of the sphere is rather large and the diameter of the opening is relatively small in comparison.

However, if this problem is solved, a good illumination should also lead to an image where white regions within the object are as bright as possible and black ones as dark as possible.

In the next two sections, we would like to discuss the evaluation of some of the most important optical properties of a camera system and its components.

8.2 Evaluation of the optical properties, part 1: vignetting aberrations and optical dynamics

8.2.1 Vignetting and aberrations

In order to evaluate the imaging properties of camera and lens systems, in many cases special test targets are required. One example is depicted in Figure 8.1, which can be used particularly for the determination of distortion as well as of vignetting of lenses. Since the target is used as a reference for distortion, it is of importance that the geometry of the pattern is highly rectilinear without any obvious distortion. If the target is used for the evaluation of vignetting effects, a nearly perfect homogeneous illumination is required, and with it a correspondingly high homogeneity of the reflected light all over the chart.

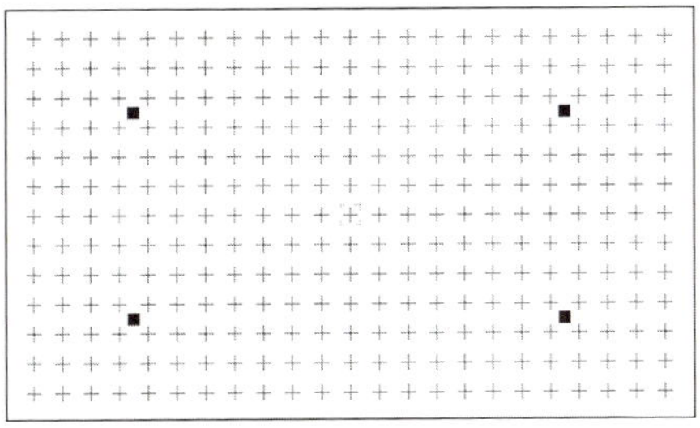

Fig. 8.1: Test target for evaluating distortion and vignetting (Courtesy of Image Engineering).

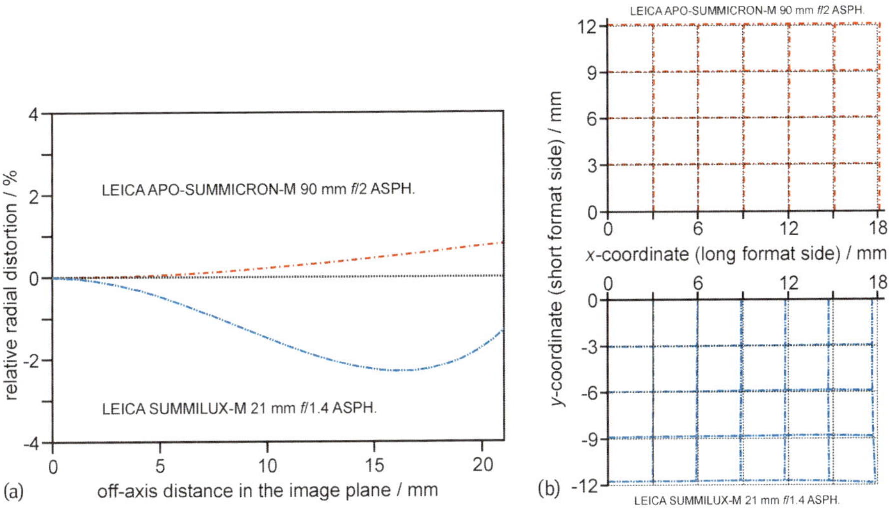

Fig. 8.2: Distortion properties of a long focus lens (LEICA APO-SUMMICRON-M 90 mm f/2 ASPH., upper part) and a super wide-angle lens (LEICA SUMMILUX-M 21 mm f/1.4 ASPH.). a) Relative radial distortion; b) distorted patterns in the image plane when compared to a rectilinear reference. $x = y = 0$ represents the center of the image (Diagrams redrawn after original datasheet from Leica Camera AG).

The distortion is evaluated by taking a photograph with the image plane in the camera oriented in parallel to the target and the optical axis fixed to the center of it. The positional data of the image points, namely those of the cross marks on the target, are given by the camera as coordinate pairs in pixel values and can be mapped to the x- and y-coordinate values of the image sensor format. Thus, a graphic presentation of the distortion can be reproduced as shown in Figure 8.2b for two different lenses. The corresponding relative radial distortion is shown in Figure 8.2a as a function of the off-axis distance from the center in the image plane (see also Section 3.5.5). The upper part of the figure presents the data from a LEICA long focus lens for the 35 mm format and exhibits a slight pincushion type distortion. The positive relative radial distortion is typical for long focus lenses. However, with the relative radial distortion being less than 1 % the lens can be considered free of distortion since it is virtually not perceptible. The lower part features the characteristics of a LEICA super wide-angle lens. The relative radial distortion is negative and typical for wide-angle lenses. Its maximum value is well below 3 % and even decreases at the corners of the image field. The type of distortion is a mixture of pincushion and barrel distortion. Also, for this type of super wide-angle lens the distortion is very low.

Vignetting of lenses can also be determined by the test target in Figure 8.1. As mentioned above, a nearly perfect, homogeneous illumination of the target is required. After taking a photograph of the test target, the relative illuminance in the image plane is given by the brightness data registered by the camera sensor for the homogeneously gray areas between the cross marks. Figure 8.3 shows the experimentally determined illuminance relative to the center as a function of the off-axis distance along the image diagonal for a Zeiss Otus lens for the 35 mm format. The vignetting is evaluated for two f-numbers, namely at full aperture of the lens with $f_\# = 1.4$ and with $f_\# = 4.0$. For comparison, the natural vignetting for this lens with a focal length of $f = 55$ mm and a corresponding angular field of view of $\Psi = 43°$ in the object space is indicated by the dotted line. As described in Section 3.4.4, the off-axis shading in the image plane can be understood as a consequence of two effects: the first is due to the natural vignetting yielding a maximum brightness fall-off proportional to $\cos^4(\beta_i)$ at the corners of the image field according to Equation (3.104). Here, β_i is the field angle in the image space. Natural vignetting is always present for diffuse illumination and is independent from the aperture stop. The second part, the mechanical vignetting, is due to the shading by lens elements and can usually be reduced by stopping down. It can be seen from the figure that for full aperture, the fall-off is the strongest. It decreases through stopping down. The lowest shading can be achieved at an f-number slightly above $f_\# = 4.0$ and is nearly indistinguishable from that seen in the figure at $f_\# = 4.0$. This implies that already by stopping three stops down, the mechanical vignetting is almost completely avoided. Here, it is also interesting to see that the relative illuminance is still higher than calculated on the base of the field angle in the object space. This is due to the fact that the Otus has an asymmetric retrofocus design, and thus the image space field angle is smaller than that in the object space. This effect can be seen with many modern high-

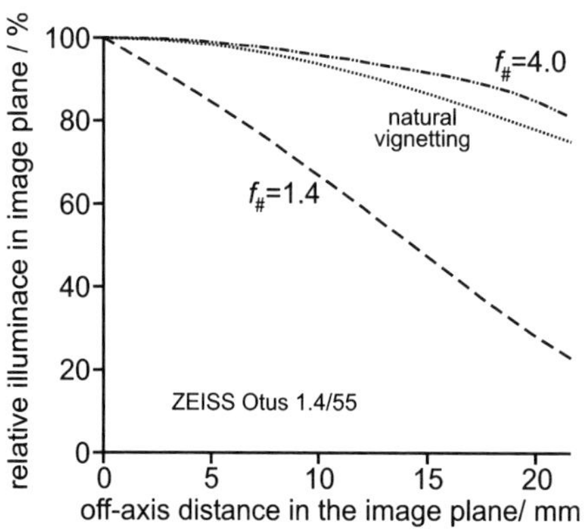

Fig. 8.3: Relative illuminance as a function of the off-axis distance in the image plane at different f-numbers (diagram redrawn after original datasheet from Zeiss).

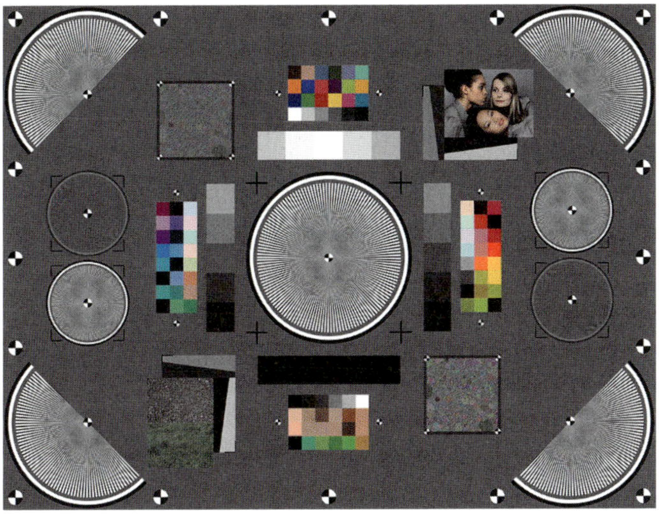

Fig. 8.4: "Multifunctional test chart" for the evaluation of various different tasks. Examples of possible measurements are those of shading/vignetting, distortion, chromatic aberration as discussed in the present section, resolution (based on Siemens stars, slanted edge, dead leafs; see Section 8.3), OECF (as this target is a reflective one, OECF evaluation here is much restricted; for a more advanced analysis, see Section 8.4) and color reproduction. Data analysis of those measurements can be performed with the corresponding software (iQ-Analyzer) of the same company that supplies the test chart (Courtesy of Image Engineering).

quality lenses, especially for the wide-angle domain, due to their complex construction and is very favorable for the reduction of vignetting.

A more complex test target is depicted in Figure 8.4 where besides color trueness many other characteristics of imaging can be checked. As for chromatic aberrations, they pose a severe problem that must be well compensated in order to ensure a high-image resolution, also for black and white imaging. The sharpness of a lens is highly improved by reducing the chromatic aberration as is shown in Figure 6.25 for an example of wide-angle lenses. The quality of a lens with respect to these aberrations can be tested by imaging black-and-white patterns with sharp contrast edges like the small circles with four sections in Figure 8.4 or from crosses such as are present in Figure 8.1 and Figure 8.4, when those consist of sharp contrast transitions. The narrower the transition between the wide and black section with a colored fringe in the captured image, the lower the chromatic aberration (see, e. g., Figure 6.25). For the quantitative determination of the longitudinal respectively transversal chromatic aberration, however, more advanced methods have to be used. We may note, that in many test charts a series of patches with different gray scales is included, so that at least a rough estimate of the tonal response, namely the tone curve, can be obtained. This then allows for signal linearization of the output signals with respect to the input signals, which is an absolute requirement for most MTF measurements. In case of linear output data, which may be raw data, this is not necessary but on the other hand this then allows for control of linearity. Furthermore, the patches allow for noise analysis (see Section 8.4), which may be important for MTF measurements as well (see below).

8.2.2 Optical dynamics and veiling glare

Optical systems or part of them may also suffer from unwanted illumination on the sensor, which is the result of scattering, reflections, etc. as discussed in Chapter 4 and Section 6.8 (see also the discussion of macro contrast in [Nas08]). Such effects are flares and veiling glare, which we will not discriminate here. They clearly may affect the optical properties and may lead to a loss of contrast and a reduction of the dynamic range of the optics and the system in total. Such spurious signals can be measured as illustrated in Figure 8.5. An Ulbricht sphere (integrating sphere) is used to generate a very homogenous object field, but with one exception, namely a spot, which ideally emits no photons (note that the ideally homogenous object field is only present on the inner surface of the sphere). Although in practice and due to Planck's law, this is possible only as an approximation, a rather good representation can be realized with a dark box together with a tiny whole in the half-sphere. As a point object, this hole is imaged onto a photo multiplier tube (see Section 4.11.2), which measures the signal strength with a dynamic range of typically 10^4. In such a way, the veiling glare index of a lens can be determined where a deeply black spot with an appropriate size is shifted across the image field. Veiling glare may be defined as the ratio of the signal within the dark area (ideally this is zero) to that of the homogenous bright background. The veiling glare index is the percentage value of the ratio (there are also ISO norms on that, e. g., ISO 1844 and IEC62676-5; note that the related measurements have to be done based on raw data; compare Section 8.3.1.4). Typical index values may be less than 1.5 % for high-quality DSLR lenses, approximately

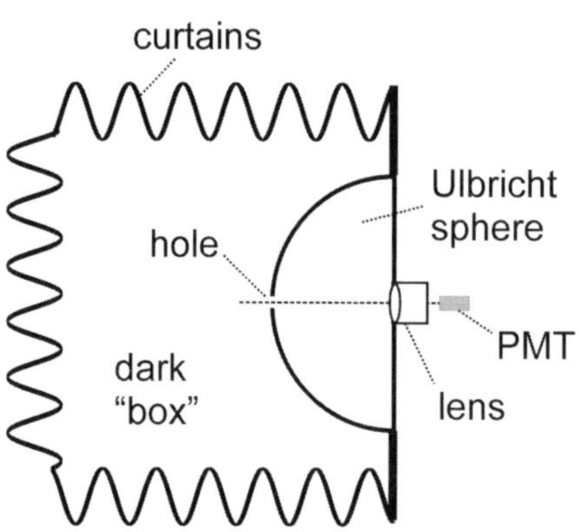

Fig. 8.5: Typical setup for the determination of the integrally measured straylight.

2 % for standard lenses, roughly 5 % for older SPC and less than that for newer models.[4] Other measurements may include spurious signals resulting from reflexes between sensor and lens.

8.3 Evaluation of the optical properties, part 2: MTF and SFR measurements

8.3.1 Grating-based methods

8.3.1.1 General

In principle, an MTF measurement relies on taking images of test objects as discussed in Section 5.2. If, for instance, the test object is a sine grating with a given period, and thus a given spatial frequency, the MTF is given by the ratio of the modulation within the image to that of the object (see Equation (5.41)). Due to the fact that the test object has a modulation of one (i. e., full contrast; otherwise, corrections have to be made; see below), the modulation directly yields the MTF. For nonsine gratings, see the discussion below.

As a direct consequence, one may expect that then the MTF could be obtained quite easily from Equation (5.41) or Equation (5.54). However, although this works well in theory, in practice in many cases this is not applicable. In particular, there will be large difficulties if the denominator becomes very small or even zero. This is apparent for a pure sine grating, where with the exception of its fundamental frequency, the spectral intensity is always zero (see Figure 5.19b) and also for the bar grating where besides the fundamental and its odd harmonics, the spectrum is zero, too (see Figure 5.20b). Due to manufacturing errors, for real gratings the nonzero regions may be slightly extended to the vicinity of the fundamental and the harmonics. As a result, the experimental determination of MTF curves usually is based on contrast measurements of test gratings analyzed in a different way.

In particular, if a lot of gratings such as those displayed in Figure 5.19a or Figure 5.20a, respectively, with different periods are photographed one after the other, one can obtain the contrast function (see Equation (5.50)), and thus the MTF (see Section 5.2.1). To simplify this procedure, there are test charts with all those gratings, or a selection, printed on the same test chart, sometimes even in different directions (see, e. g., Figure 8.6a).

Of course, this procedure requires a particular grating for each spatial frequency, and due to the restricted number of them, the number of different spatial frequency or sampling points in principle is limited. An easier way to get the same result with one grating only is to use a grating with variable line spacing (see, e. g., display in upper part

4 V. Blahnik: private communication.

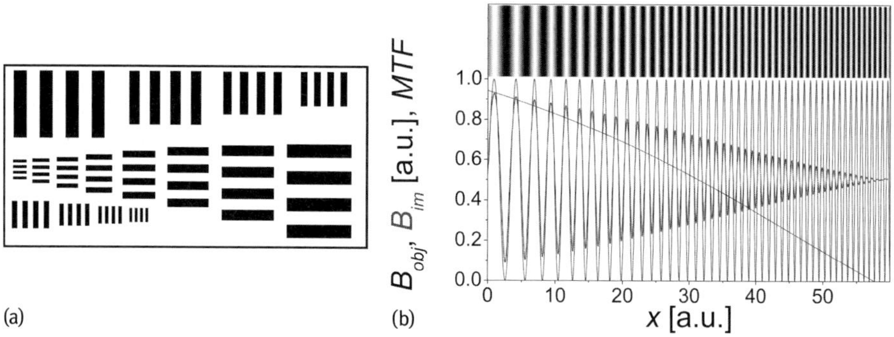

Fig. 8.6: Examples of test charts. (a) Target consisting of gratings of different periods and orientations. For possible images and corresponding MTF values, see Figure 5.21. (b) Grating with a changing period. Original object on top, profile measured along the horizontal line of the sine modulated brightness distribution (thin black line; ideal case for an MTF identical to unity everywhere). As an example, the thick line shows a brightness distribution of the image according to a real MTF. The related contrast (or MTF) is displayed as a solid line across the diagram.

of Figure 8.6b). In that case, depending on the chirp (the chirp describes the change of period, here along the horizontal axis), only a single grating is necessary and even more sampling points may be measured at once. Other test targets also make use of a spatially changing grating period, which results in the similar advantages or include more complex and/or different structures such as edges (see also Section 8.3.2). Some of them contain patches orientated in different directions, which allows for performing measurements of the MTF for the sagittal and meridional plane, respectively (see below).

A series of images of such test gratings with different periods gives access to the MTF curve. The principle is the same with test gratings with different given periods within the same image or a single variable line spacing grating where one can perform subsequent measurements at different positions. We will discuss that for a single grating with a particular period first. This period is either known or, via the demagnification factor, the period can be calculated for the sensor plane. However, it is very preferable to measure the period within the image in pixels. From the known pixel size, the period on the sensor can also be deduced in micrometers. For not too short periods, this procedure works quite well.

The corresponding spatial frequency is then the inverse of the period. The corresponding unit may be inverse pixel numbers (but this is not usual), μm^{-1} or mm^{-1}. However, more easily, we can divide the total number of pixels by the number of pixels within one period and then we directly obtain the spatial frequency in lp/*PH*, which is the usual unit. This is identical to a division of the sensor height in µm by the period in µm but avoids recalculation of period size in µm.

Next, we read out the contrast at that particular spatial frequency by measuring B_{max} and B_{min}, which are the maximum and minimum signal within the image of the grating (see, e. g., Figure 5.21a).

According to Equation (5.50), we then get the contrast $K(R_x)$. As a practical remark, we would like to hint that such contrast measurements require good illumination conditions. This includes a bright and homogenous illumination, which however, should not saturate, e. g., the detector because obviously, for a given value of B_{min}, any change of B_{max} changes K; see Equation (5.50). On the other hand, stray light has to be avoided because obviously, for a given value of B_{max}, any change of B_{min} changes K (see also Figure 8.5 and the related discussion). Here and in the following, for simplicity, we assume that the orientation is in the x-direction. Repeating this procedure for all the other gratings, where each of them has a different period, and thus a different R_x-value, we obtain the contrast transfer function CTF(R_x), which is equal to $K(R_x)$. For each period (or grating), the measurement yields a single data point of the CTF-curve or MTF-curve (see below).

8.3.1.2 Bar gratings

As discussed in Section 5.2.1, usually grating structures with sine modulated brightness distribution are preferable because the measurement of the contrast of a grating with a given period a_g then directly yields the MTF. If instead, one uses a bar grating oriented in the x-direction (see, e. g., Figure 5.20a), then one has to take into account that its Fourier transformation does not contain only a single frequency. The function can be described by a Fourier series (see Section 5.2.1 and Table A.2)

$$B_{obj}(x) = 1 - \frac{4}{\pi} \cdot \left(1 \cdot \sin\left(\frac{2\pi}{a_g}x\right) + \frac{1}{3}\sin\left(3\frac{2\pi}{a_g}x\right) + \frac{1}{5}\sin\left(5\frac{2\pi}{a_g}x\right) + \frac{1}{7}\sin\left(7\frac{2\pi}{a_g}x\right) + \cdots\right). \tag{8.1}$$

The Fourier components are all positive (see also Figure 5.20b). Here, we do have to comment. We define both the bar grating and the sine grating according to Figure 5.20a and Figure 5.19a, respectively, namely with an oscillation between 0 and 1. The resulting Fourier series is given by Equation (8.1) as it could be deduced directly or found tabulated, e. g., in [Bro79]. Although that description is straightforward, in several other articles and books that describe bar gratings and their MTF, a Fourier series based on cosine functions is used instead. This leads to terms with alternating signs. Besides the difference in phase (when compared to, e. g., Equation (8.1)), those articles often use a different amplitude as well, namely an amplitude of 2. Nevertheless, although the Fourier spectra of both descriptions differ, if amplitudes and phases are calculated properly, at the bottom one gets the same result.

In the following, we prefer the description given by Equation (8.1), because it may provide somewhat simpler description. As the sine terms lead to contributions for the fundamental and its harmonics and all that weighted according to the MTF at those frequencies (see Figure 5.19b), the not yet normalized contrast function is given by

$$\mathrm{CTF}(R_X) \approx \frac{4}{\pi} \cdot \left(\mathrm{MTF}(R_X) + \frac{\mathrm{MTF}(3R_X)}{3} + \frac{\mathrm{MTF}(5R_X)}{5} + \cdots \right). \tag{8.2}$$

As a rough approximation, one may restrict to the first term, and thus, $\mathrm{MTF}(R_X) \approx (\pi/4) \cdot \mathrm{CTF}(R_X)$.

The normalization that MTF is one at zero frequency has to also be done for sine gratings used in an experiment and thus normalization has to be done accordingly for bar gratings as well. This is also to get rid of nonoptimal illumination conditions and/or gamma values (see Chapter 4 and Section 8.3) and other related nonlinearities during image capture, etc. Altogether, it is typical to normalize the MTF to its low frequency value, so that $\mathrm{MTF}(0) = 1$ for a sine grating. We may use that also as a definition: $\mathrm{MTF}(R_X) = \mathrm{CTF}(R_X)/\mathrm{CTF}(0)$.

8.3.1.3 Siemens stars

A special kind of a test target is the so-called Siemens star (Figure 8.7). Similar to before, according to the sector structure, a profile measured along the horizontal line or in any other direction yields the contrast function. The brightness distribution of the sectors may either correspond to that of a bar grating or in the case of an advanced test chart as such depicted in Figure 8.7a, to that of a sine grating. The period depends on the distance of the line profile from the center of the star. It can be measured either as before, or it can

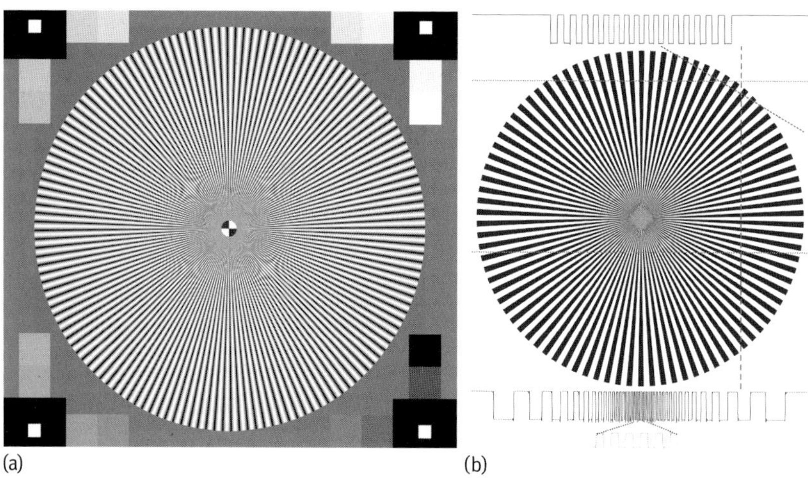

Fig. 8.7: (a) Test chart with a single Siemens star. Here, the gratings formed by the sectors are sine modulated (compare Figure 5.19a). The structures at the edges of the chart are necessary if the charts are used together with an image analyzing software, otherwise they can be omitted (Courtesy of Image Engineering). (b) Illustration of a Siemens star as a variable period grating. Here, the gratings formed by the sectors are modulated as a bar grating (compare Figure 5.20a). The profile measured along the horizontal line yields a grating of a specific period (at least in its center; for the upper horizontal dotted line, this is shown above the Siemens star, for the lower one below).

be deduced from geometry by taking into account the total number of sectors and the radial distance. We have to note that due to the tilt of the sectors the grating period is not constant along the line profile. In particular, this becomes apparent at larger distances from the line profile center (see Figure 8.7b). Thus, the contrast measurement has to be done only for the inner part of the line profile (see enlarged part of the line profile at the bottom of Figure 8.7b). The rest of the procedure is similar to before, in particular, if one measures the contrast as a function of radial distance of the center of the star, and thus as a function of the spatial frequency. One advantage of Siemens star targets is also that the line profiles can be done in any direction. In principle, this allows for deducing the MTF in different directions such as the sagittal or meridional direction.

An example of the analysis of a Siemens star measurement based on horizontal line profiles at different distances from the center of the star is shown in Figure 8.8. Each of the plots Figure 8.8a, b, c, etc. corresponds to one data point within the MTF curve. It is well seen that with decreasing distance from the star center the period becomes smaller, and thus R_X increases. At the same time, the contrast decreases as well, until in Figure 8.8g it becomes apparently zero. Of course, a good measurement of a MTF curve consists of more data points, and thus it makes sense to take more line profiles than presented here.

Figure 8.8h illustrates the measurement of R_X and contrast $K(R_X)$, respectively. Here, one period, i. e., 1 lp, corresponds to 12 pixels. The pixel width is 6.41 µm, and thus the period is 77 µm. The camera has a full format sensor, which means that $PH = 24$ mm, which corresponds to 3744 pixel. Therefore, here we obtain $R_X = 3744/12\,\text{lp}/PH = 312\,\text{lp}/PH$ (or 13 lp/mm). The Nyquist limit is given by the 2-pixel resolution, i. e., 1872 lp/PH. The contrast is indicated by the two horizontal lines ($K = 83\,\%$).

The result of this simple analysis is shown in Figure 8.9a, namely the MTF-curve (solid line). It is clearly seen that the MTF-curve of the system (solid line; for the moment, we do disregard the broken line, which is discussed below) decreases with R_X. As discussed before, in part, this is due to $\text{MTF}_{\text{optics}}$, but also $\text{MTF}_{\text{sensor}}$ does play a role. $\text{MTF}_{\text{sensor}}$ also includes also the effect of noise.

We would like to remark that here we have chosen a Siemens star with sectors of a bar shape for better illustration. But we would like to note that for advanced measurements also high-quality Siemens stars with sectors of sine modulated shape are available (see Figure 8.7a). It is not easy to produce such targets in high quality and bar targets are much more easily produced. Thus, the latter ones are often quite common. However, for a good measurement then corrections have to be made by taking into account the higher orders (see Equation (8.1)).

As another example, Figure 8.9b shows a measurement taken with the same camera but now equipped with a 250 µm pinhole as the optics (compare Section 2.1 and, in particular, Figure 2.4).

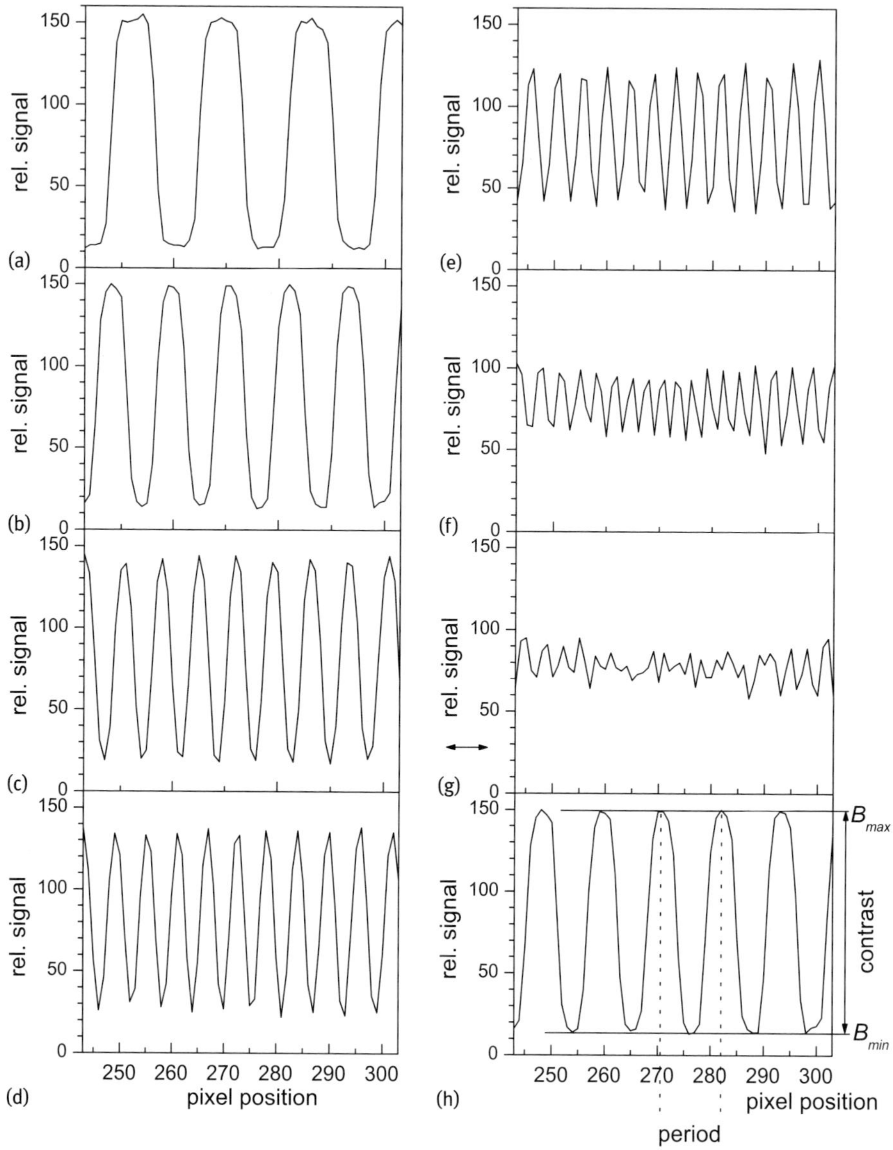

Fig. 8.8: Analysis of an image of a Siemens star (results from a real measurement of a DSLR with a high-quality lens). The test target was a Siemens star placed in the center of the object field. The profiles measured along the horizontal lines (a) to (h) have been measured at a radial distance of 99, 70, 46, 39, 31, 23 and 15 % from the center, respectively. The corresponding R_x are 210, 298, 447, 536, 670, 894 and 1341 lp/*PH*, respectively. (h) is the same profile as (b) and illustrates the example discussed in text.

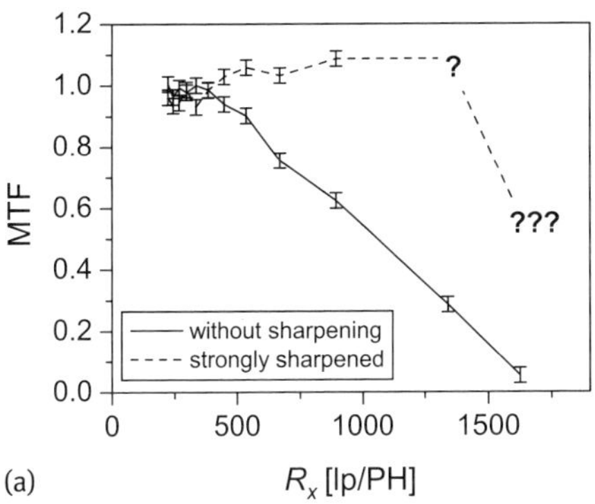

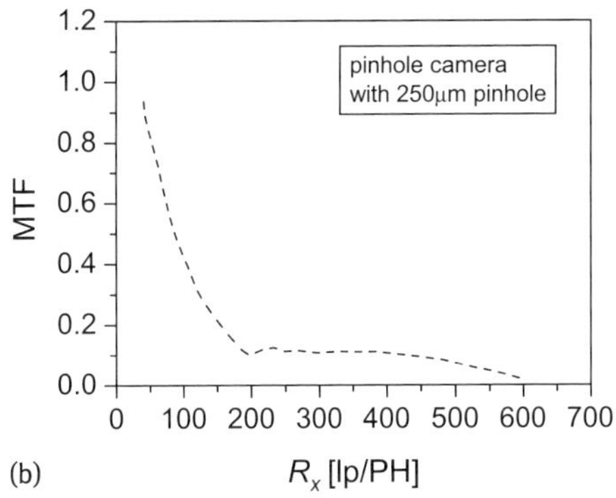

Fig. 8.9: (a) Experimentally deduced MTF curve of a DSLR with a high-quality lens (deduced from Figure 8.8). This curve is normalized to its maximum. (b) Measurement taken with the same camera but now equipped with a 250 μm pinhole instead of the lens.

8.3.1.4 Influence of tone curve

In the previous sections, we have discussed that the MTF can be deduced on the basis of images taken from Siemens stars or gratings. However, this is not necessarily straightforward. A correct measurement is based on signals that are related to the photon conversion curve, namely B_{pix}-values that are linearly related to the exposure. If then one analyzes line profiles such as shown in Figure 8.8, the maxima B_{max} and minima B_{min}, respectively, that are obtained are also linearly related to the corresponding exposure values, namely $B_{\text{max}} \propto H_{\text{max}}$ and $B_{\text{min}} \propto H_{\text{min}}$. Here, we disregard potential bias and offsets. Within the following example, this leads to the solid lines in Figure 8.10b, which show how the maxima decrease with R_{x} and how the minima increase (compare Figure 8.8a to Figure 8.8g). Accordingly, the calculated contrast $K(R_{\text{x}})$ can be directly and correctly calculated (black solid line in and Figure 8.10c).

In spite of this, usually images "suffer" from tonal corrections introduced by the image processor of the camera or performed by the raw converter. As an example, Figure 8.10a shows such a tone curve. In contrast to before, here B_{pix} is not proportional to H_{pix}. This leads to line profiles that may approximately look like those shown in Figure 8.8, but there are differences. The dotted lines in Figure 8.10b are based on the same exposure data as the solid lines but instead of a linear conversion of the maxima and minima, the conversion has been made in a nonlinear way according to the tonal curve shown in Figure 8.10a. From comparison of the blue and green curves, respectively, the difference is obvious. Unfortunately, just these curves are the result of the "measurement" on the basis of the tone-mapped data. Subsequently, they are used for the procedure described in the previous sections. As a result, the obtained MTF-curve, namely $K(R_{\text{x}})$, is not correct. It significantly differs from the theoretical one (thick solid red line in Figure 8.10c).

We may comment that the deviation of the experimental curve based on tonal corrections depends, of course, on the tone curve, but it depends also on test chart illumi-

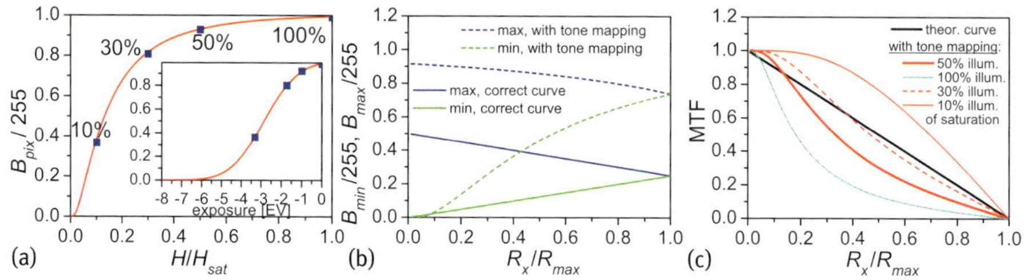

Fig. 8.10: Example of the influence of tonal corrections on the deduced MTF for a diffraction limited cylindrical lens. (a) 8-bit tone curve displayed on a lin-lin scale. The exposure is normalized to the exposure necessary to saturate the pixel and the brightness signal to the maximum possible number of counts, namely 255. The insert shows the same curve, but on a log-lin scale, namely with the abscissa provided in EV with the saturation value set to $EV = 0$. For marked points, see the text. (b) Maxima and minima taken from plots similar to those displayed in Figure 8.11 (solid lines). The same curves after tonal correction are shown as dotted lines. Note that both solid lines do converge at R_{max}, namely at $R_x/R_{max} = 1$. According to Equation (5.50), at this point the MTF becomes zero. (c) Theoretical MTF curve of a cylindrical lens (black solid line; compare Figure 5.8b). The thick red line is the MTF curve obtained from the tone mapped data in (b) and calculated from Equation (5.50). The thin red lines are MTF curves obtained for different illumination conditions (see the text).

nation and exposure. For the example discussed above, an exposure of $H_{pix} = 0.5 \cdot H_{sat}$ has been chosen, where H_{sat} is that value that leads to the saturation of the pixel (due to FWC). Thus, at $R_x = 0$ where we do have full contrast, $H_{max}/H_{sat} = 0.5$, and consequently, $B_{max}/255 = 0.5$ (blue solid curve in Figure 8.10b). But from observation of the tone-mapped image this has transformed to $B_{max}/255 = 0.92$ as can be seen from the point on the tone curve.

For other exposure conditions, namely an exposure that leads to 10, 30 or nearly 100 % of H_{sat} we obtain brightness values B_{pix} of 37, 80 and 98 % (these points are marked in Figure 8.10a). Even though the tone curve is still the same (Figure 8.10a), the deduced MTF-curves are all different (see thin red and cyan lines in Figure 8.10c). But in no case is the real MTF reproduced well enough. This shows that the result of the "measurement" significantly depends on exposure. Finding the best exposure is not easily predictable, but for the present example, an exposure that leads to $B_{pix} \approx 0.8 \cdot 255$ counts $\approx$ 200 counts may be the most suitable, even though the theoretical MTF is reproduced only absolutely roughly. The corresponding value of H_{pix} usually is not known, unless it is available from an independent measurement. Alternatively, one has to perform a sensor calibration, which may also be based on deduction of the tonal curve. At least an approximate calibration can be done if a series of patches with different gray scales is included in the image (see the discussion related to Figure 8.4).

As a result, we may conclude that if one intends to do measurements with a camera specified for photography, one has to always be very careful because of the strong influence of the usually unavoidable application of a tonal curve (see, e. g., example III in Appendix A.7). Again, photographic imaging cannot be considered as a measurement and in that sense it is more complicated than scientific imaging. Consequently, MTF measurements have to be done on the basis of data related to the photon conversion curve, for instance, on the basis of images taken as raw data that then have been converted linearly by a suitable raw converter such as *DCRaw*. This includes, e. g., also slanted edge measurements (see Section 8.3.2). Furthermore, this procedure has to also be applied for other measurements such as those related to vignetting (Section 8.2).

8.3.1.5 Postprocessing: the effect of sharpening and contrast enhancement

Figure 8.9a also shows the effect of strong sharpening and contrast enhancement due to post-processing based on the same raw data (dashed line). However, although MTF seems to be strongly improved, this is not the case. Of course, this is clear, because post-processing of the same image cannot improve the physical quality of the image (e. g., resolution). Figure 8.11 shows several line profiles. It can be seen that post-processing has led to much sharper boundaries, which becomes perceptible in the line profiles with 187 and 529 lp/*PH*. However, the line profile at 1251 lp/*PH* clearly shows that the resolution has been fully lost. The displayed structure does not resemble the grating structure, which still could be well seen in the line profile without post-processing. As a result, the MTF curve of the sharpened image displayed in Figure 8.9a is not at all realistic.

Evidently, this example shows that the evaluation of a camera system has to be made at best with data that are not affected by further processing. Consequently, raw data come close to that, at least if we consider the discussion in the previous section. Direct conversion into TIF or JPG images without post-processing then may also allow more or less reasonable evaluation. Nevertheless, data depth then is usually reduced (this is not the case for 16-bit TIF), and thus tone mapping also influences the MTF. Certainly,

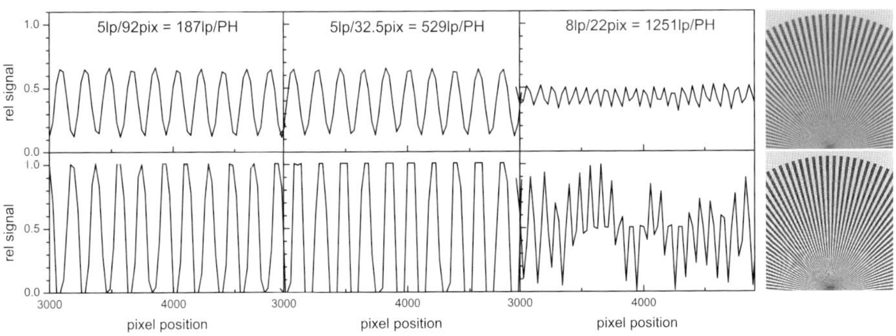

Fig. 8.11: Influence of sharpening. The upper row shows the image without any sharpening, contrast enhancement, etc. The lower row uses the same original raw data after they have been processed with strong sharpening and contrast enhancement. The right-hand side shows a part of the resulting images and the left-hand side three horizontal line profiles. Similar to the example displayed in Figure 5.39 and Figure 5.40, here again we applied “too much sharpening” to show effects more clearly. The negative effect of this can be seen in the image of the Siemens star.

if direct access to raw data is not possible, MTF analysis on the basis of preprocessed data, e. g., done by the image processor of the camera itself is the only way to get at least some information on the camera system performance. But again, then the MTF reflects a mixture of the optical performance and the influence of image processing, which makes it difficult to judge the camera system itself.

8.3.2 Edge gradient and sampling methods

8.3.2.1 Principle and knife edge method

The measurement of the MTF using a well-adapted edge-gradient method is rather simple. This also includes that the target itself is simple as long as the edge is sharp and straight. Bowing may lead to errors. However, alignment of the target is important and noise issues may play a role, though the latter may be a general problem for all methods.

Historically, edge-gradient methods have been applied to measure the resolution of scanning microdensitometers. On a larger scale, some kind of such methods today are still used as "standard methods," e. g., for laser beam profile analysis and known as the knife edge technique. Furthermore, it might be interesting that the same method can be applied for temporal characterization of femtosecond laser pulses from NIR down to XUV-range when it is applied in the time domain (see, e. g.,[5,6]).

Figure 8.12 illustrates the knife edge method, here for an illumination profile measurement in horizontal direction. The upper part of each plot shows the profile that should be measured. A knife edge is placed below the profile and a large planar detector is placed below indicated by the dotted box. Because this detector does not provide spatial resolution, it integrates over its surface. In other words, it just measures all of the energy of that part of the profile that is not blocked by the knife edge. In (a) to (d), the amount of collected energy is shown as the gray value of the detector shown as a box. In (a), the knife edge blocks most of the profile. Thus, the measured energy is low. This yields the first data point (cross indicated on the curve shown below; x-axis: position of the knife edge, y-axis, measured energy). In (b) and (c), more light could pass,, and consequently the measured energy is larger (second and third data point, again indicated by the crosses on the curve shown below). In (d), the situation for another data point is shown (here the profile is only slightly blocked).

In principle, a lot of data points may be measured by a lot of much smaller subsequent shifts Δx of the edge. This leads to an accurately measured curve as indicated

5 U. Teubner, U. Wagner, E. Förster: *Sub-Ten-Femtosecond Gating of Optical Pulses*, J. Phys. B **34** (2001) 2993–3002.

6 XUV-PUMA project, funding by Federal Ministry of Education and Research BMBF, funding code 05K16ME1; see also P. Finetti et al.: *Pulse duration of externally seeded free electron lasers*, Phys. Rev. X 7 (2017) 021043.

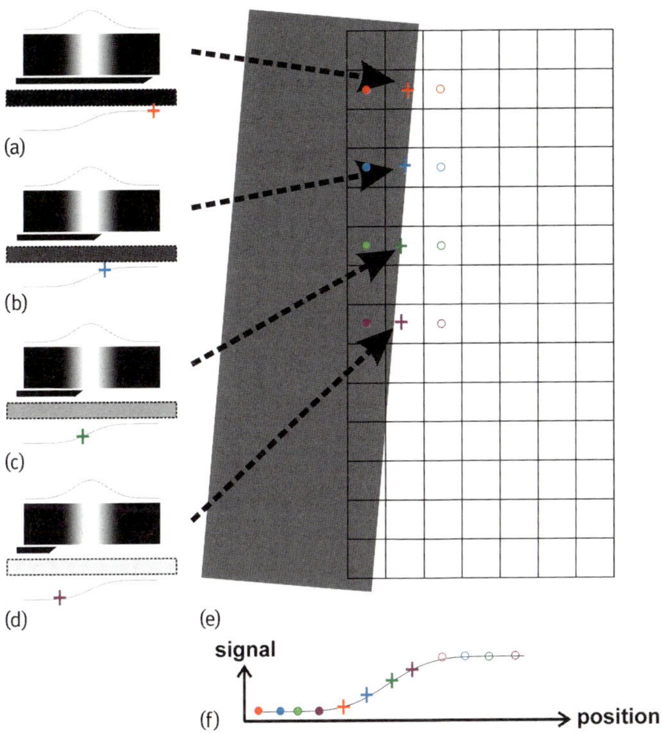

Fig. 8.12: (a) to (d) Illustration of the knife edge method to measure a spatial 1D profile within a multishot experiment. (e) Illustration of the slanted edge method to measure a spatial 1D profile within one shot. The knife edge or razor blade or something similar fully blocks the light on the detector surface directly behind. However, for better visibility of the pixel structure (white squares), here the blade is shown as a transparent gray box. The dots and crosses indicate the corresponding data points on the edge spread function ESF displayed in (f).

below each detector. From one data point to a neighbored one, the energy changes by an amount ΔE. The corresponding differential is $\Delta E/\Delta x$ (if infinitesimal, this is the first derivative). As a consequence, deriving the measured curve results in the spatial profile.

8.3.2.2 Edge spread function and line spread function

In a similar way, one may take a single image of a razor blade, which is illuminated from the rear side (Figure 8.13a). The relevant edge is indicated by the vertical arrow; the other part of the blade is irrelevant for the present measurement. It may be recognized that although the blade is very sharp, its brightness distribution in the image does extend over several pixels. Due to the large magnification, within the image the pixel structure can be seen in Figure 8.13a. Figure 8.13b shows the profile measured along the bright line shown in Figure 8.13a. This curve is the edge spread function, ESF. We would like to note that sometimes additional integration along the vertical line is performed. We

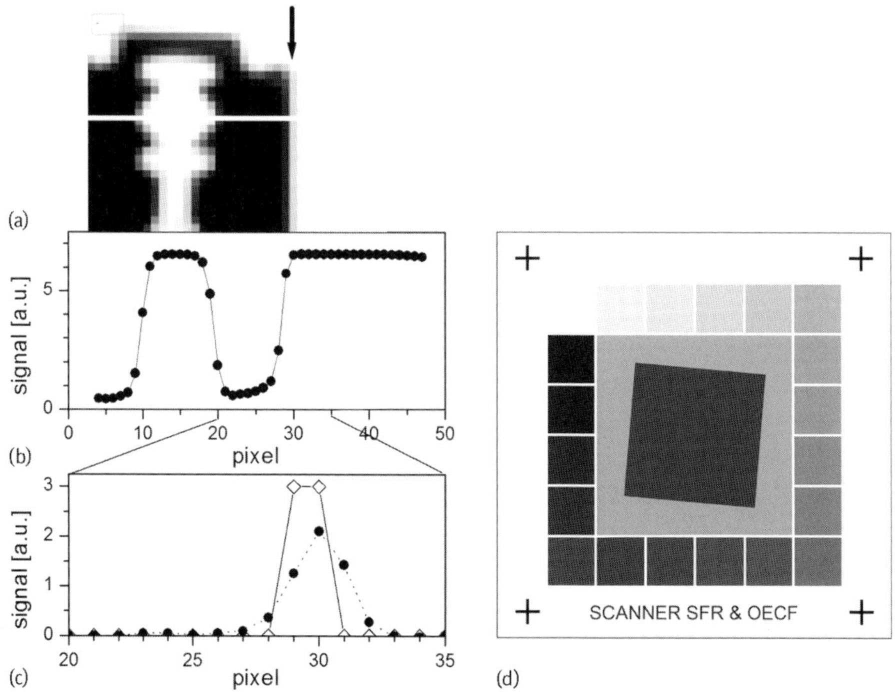

Fig. 8.13: (a) to (c) Illustration of a simple measurement using a razor blade (see the text); (d) typical slanted edge chart (Courtesy of Image Engineering).

would also like to remark that here the analysis is restricted to the vicinity of the edge and the center part of the razor blade has to be ignored. In the same way as described before, the first derivative of the ESF yields the spatial distribution of the corresponding resolution curve, namely the line spread function, LSF. This is the 1D equivalent to the 2D point-spread function PSF (Figure 8.13c, solid circles):

$$\mathrm{LSF}(x) = \frac{\mathrm{d}}{\mathrm{d}x}\mathrm{ESF}(x). \tag{8.3}$$

For the present example, we can see that the FWHM is approximately 3 pixels, which is slightly more than the 2-pixel limit according to the Nyquist limit. This may result from the fact that the sensor resolution is convoluted with that of the optics, which reduces the resolution. In the ideal case of ideal optics and sensors, one would obtain the 2-pixel resolution as indicated by the open symbols.

The described scheme is quite simple and works well if the image of the edge is spread over many pixels. In such a case, $\mathrm{MTF}_{\mathrm{system}}$ is determined mostly by $\mathrm{MTF}_{\mathrm{optics}}$ and the rather large sampling rate of the ESF may yield a good resolution. Alternatively, one could remove the sensor and replace it with a high-resolution microscope objective with another sensor behind. Then a scan along a line, which may yield a high sam-

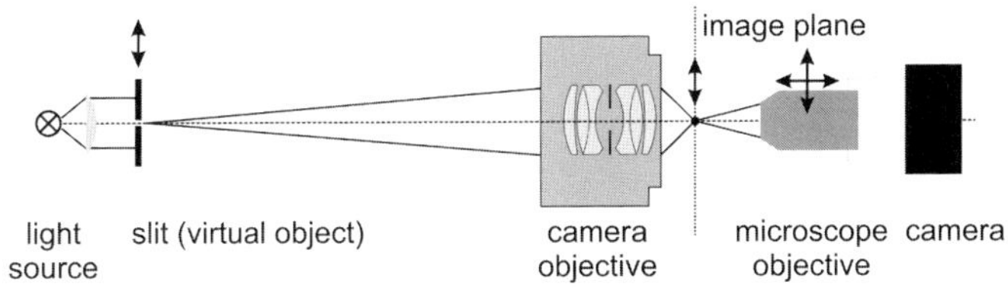

Fig. 8.14: Scheme of a setup to measure the MTF of a lens only. A slit with a small width is illuminated and acts as a virtual point-like object. This object is imaged by the lens that is tested. The PSF is observed in the image by means of a high-NA microscope objective equipped with a camera. The position of the "image spot" (i. e., the PSF or LSF, resp.) can be shifted as indicated.

pling rate as well, and then also allows for a well-resolved measurement of the ESF, and hence MTF_{optics}. This is also a typical setup for a measurement restricted just to MTF_{optics} (Figure 8.14). A rather similar setup is the MTF-Tester K8 of Carl Zeiss AG, Germany, where instead of a camera, a scanning slit system with PMT is used. There are also commercial systems available for high-precision MTF measurements, such as the ImageMaster series of Trioptics GmbH, Germany, which includes models for automatic and high-through-put and large serious inspections on industrial scale.

But such a situation with ESF distributed over many pixels is not the usual one for a digital sensor, namely a PDA. With digital devices, one has to be careful, in particular, because their spatial resolution may not be sufficiently large. In such cases, there is a strong risk of undersampling. This means that the sampling rate is below twice the Nyquist frequency, or somewhat simplified, and the sensor has less resolution than required to resolve the spread itself. This has also been illustrated in Figure 1.17, Figure 1.20 and Figure 1.21, and also in Figure 8.13; the feature on the sensor is smaller than the pixel, and thus the resulting signal also depends on the phase or offset.

8.3.2.3 Slanted edge method

Hence, for proper sampling also in such cases, a series of measurements should be carried out, with a shift of the phase in between. In principle, this can be done by either shifting the object or the camera subsequently by a small fraction of the pixel width. However, to realize this within a careful setup would be a rather difficult and also time-consuming. Consequently, one may apply a trick. Such a trick is illustrated in Figure 8.12e and Figure 8.12f. Instead of a realization of different phases by subsequent shifts, this can be performed at once, if the edge, e. g., of the razor blade, is slightly tilted (typically by 5 degrees). Then because different rows within the PDA detect different phases within the same image, the ESF can be obtained from a single exposure only (Figure 8.12f).

This procedure corresponds to an oversampling because the difference of the x-positions of the data points is much less than the pixel width (typically by a factor of 4). Then this "supersampled" ESF can be derived to obtain the LSF with good resolution (better than that in Figure 8.13c).

Fourier transformation then, in principle, allows to deduce the MTF. According to Equation (5.41) the OTF, and thus also the MTF can be obtained by division of $\mathrm{FT}[B_{\mathrm{im}}(x)] = \widetilde{B}_{\mathrm{im}}(k_x)$ by the spectrum $\mathrm{FT}[B_{\mathrm{obj}}(x)] = \widetilde{B}_{\mathrm{obj}}(k_x)$. In principle, one can make use of the ESF, which has a rather broad spectrum and no zero values. In particular, for $k_x > 0$ the spectrum is dominated by the $1/k_x$-term (see Appendix A.2, in particular, relation no. 6 in Table A.2) and only for large k_x-values the spectrum comes close to zero, which only then may lead to the difficulties mentioned at the beginning of Section 8.3.1. But note that due to the $1/k_x$-distribution, accuracy becomes worse for large k_x-values. Moreover, when the denominator becomes "too small," errors become huge. On the other hand, one can make use of the LSF and its spectrum, which ideally is a homogeneous distribution in the Fourier plane.

For realization of this method in practice, several aspects have to be included. First, as it is quite common to apply fast Fourier transformation, one has to take into account related problems in numerics. As an example, hard clippings lead to ringing effects (see, e. g., Figure 5.15; in Fourier mathematics this leads to Gibbs phenomenon) but this can be reduced by the application of appropriate filter functions (e. g., Hamming window). Second, data fluctuation has to be taken into account. This can be done by smoothing or just by fitting the data by a polynomial first, e. g., by a sixth-order polynomial. Third, the set of data points is discrete and oversampling has to be considered by down conversion of the data using an appropriate filter. This may be taken into account by an effective MTF, which may be just given by a sinc-weighting function. Fourth, the MTF deduced in this way is restricted to 1D in the direction perpendicular to the edge only or approximately to it, when we neglect the small slant.

All this can be implemented in a program for slanted edge analysis and if this is done properly the method works quite well and the results can be consistent with a Siemens star measurement. Nevertheless, although in principle the method may directly yield the MTF as discussed, deduction of the MTF is not so straightforward and several transformations have to be done. This is one reason why it is quite common to term the result spatial frequency response SFR (see Section 5.2.8) instead of MTF. If the measurement is carefully done and all calculations are correct, both functions are identical.

We may note that according to the ISO standard even further implementations have to be considered, e. g., detector response and OECF data. In any way, this and also the Siemens star method only works well, when the detector response is linear with respect to the incident light. If this is not the case, the measured brightness values have to be corrected in an appropriate way (see discussion of the influence of the ton curve in Section 8.3.1). We may mention, too, that also modern modifications of the method are available, which lead to further improvements.

Programs that implement all the discussed procedures are available, e. g., as commercial software or as a plugin for *ImageJ*. An example of such a measurement using *ImageJ* is displayed in Figure 8.15. We may remark that MTF measurements carried out for different tilt angles between 2° and 12° lead to almost the same results and that a slant of 5° is a good compromise between accuracy and (over)sampling rate.

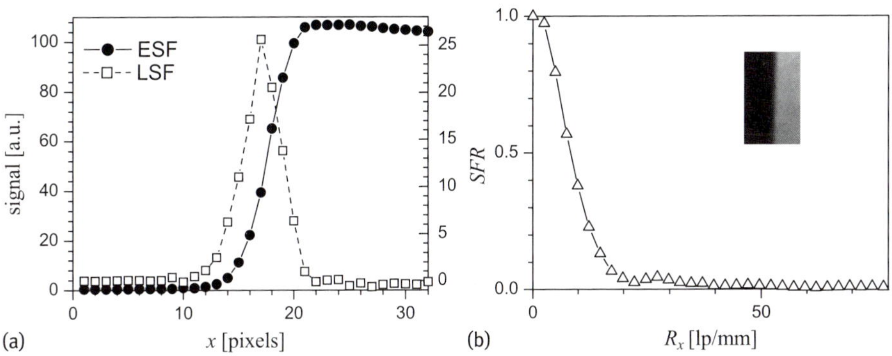

Fig. 8.15: Illustration of a slanted edge measurement: ESF (oversampled), LSF and SFR (i. e., MTF). The insert in (b) shows the slanted edge target (here cut of a razor blade; note the nonoptimized "white" area).

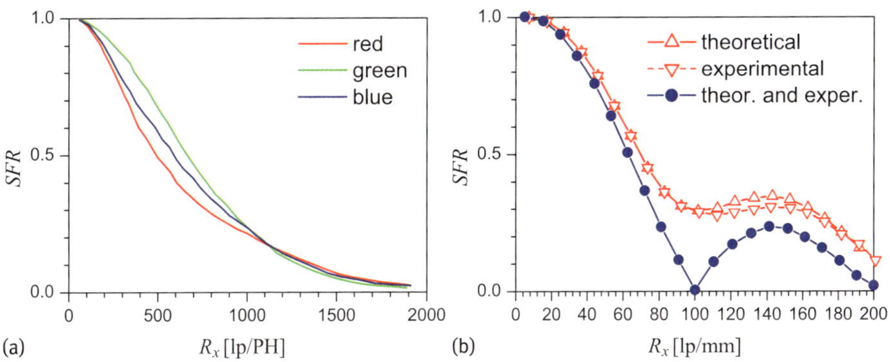

Fig. 8.16: Examples of slanted edge measurements (here SFR = MTF). (a) Measurement of the MTF_{system} of the three different color channels of a professional DSLR. (b) Measurement of the MTF_{sensor} of a CMOS sensor with 10 µm pixel pitch. Theoretical and experimental data, respectively, are shown for L-shaped pixels (red) and 100 % fill factor pixels (blue; theoretical and experimental values are almost the same) (data taken from[7]).

Another example is shown in Figure 8.16a. As discussed in Section 5.2.6, a sensor with a Bayer filter in front of it has an MTF that is affected by this mask. This is clearly seen in this plot.

Slanted edge measurements are not only useful to determine MTF_{system}, i. e., the MTF of a camera/lens combination, or MTF_{optics} but also to determine MTF_{sensor} only, even for a complex pixel topology (see the discussion in Section 5.2.4). This can be done, e. g., by placing the edge directly on the sensor surface or quite close to it. Figure 8.16b provides an example.

7 M. Estribeau, P. Magnan: *Fast MTF measurement of CMOS imagers using ISO 12233 slanted edge methodology*, Proc. SPIE **5251** (2004).

Finally, we may note that beside commercial packages, which are usable for automatic MTF measurements based on the slanted edge method, there is an open-source software package that is widely used also in science and industry: *MTF mapper*.[8]

8.3.3 Random and stochastic methods and noise problems

8.3.3.1 Dead leaves (and related) targets method

Direct MTF determination according to Equation (5.41) (or Equation (5.54)) is not possible when the denominator becomes fairly small or even zero. On the other hand, this method is very applicable, when this does not happen. Of course, for a real natural scenery it is not certain if this is the case. However, a synthetic object, i. e., a special target, where $\mathrm{FT}\{B_{\mathrm{obj}}(x,y)\} = \widetilde{B}_{\mathrm{obj}}(k_x, k_y)$ never comes too close to zero may be generated on a computer and then printed and photographed and analyzed. But we would like to note that sensor noise and other artefacts, e. g., from jpg data compression, do considerably affect the rapidly falling off spectrum, and thus $\widetilde{B}_{\mathrm{obj}}$ values close to zero cannot be completely avoided. On the other hand, for the dead leaves method described below, usually this problem is avoided by subtraction of a noise term from the experimental spectrum. Moreover, with a synthetic generation of such a target, there is the possibility of tuning $B_{\mathrm{obj}}(x,y)$ in such a way that it comes rather close to a typical scenery in photography. In such a typical scenery, usually there is no equal distribution of all spatial frequencies. Hence, it makes sense to compose the synthetic object from structures that reflect the frequency spectrum of a real scenery. Consequently, one may use the possibility to generate more structures within a particular spatial frequency range, which means that the respective frequencies will be given a larger weighting in the spectrum. Additionally, weighting may be influenced by the physiological response of the human eye so that usually midrange frequencies are given a higher influence, whereas high spatial frequencies appear less as it would be necessary for proper characterization for scientific or technical purposes. This then results also in the fact that high frequencies are more affected by noise and as a result the MTF measurement is not very sensitive to them (but note the above-mentioned correction by the noise term). Consequently, such kind of artificial targets may work well when used for photography, but they are less suitable for scientific and technical characterization of imaging systems. And even for photography, of course, there is no general artificial target that reflects the properties of all motifs.

Typical targets that take such properties are the so-called dead leaves or spilled coin targets. The basic idea of both targets is the same although some manufactures claim a better scale invariance for the latter ones. However, there is a variety of different dead leaves targets and special variants may just be named spilled coin targets (see also notes

8 https://sourceforge.net/projects/mtfmapper/

below). Dead leaves targets are made of a huge amount of disks with different diameters and different tonal grades, which all are superposed. Even more than 7 to 8 million different disks may have been generated within one target, but most of them are fully occluded so that only half a million or more could be seen. The targets may even be made from different colors if color information is relevant.

Target generation may be done in a stochastic process, e. g., the disk diameters d_{disk} are randomly distributed according to d_{disk}^{-3}. This leads to scale invariance (see below). The center positions are distributed according to a Poisson distribution. For details of targets and dead leaves measurements in general, see, e. g.,[9] and[10,11] and even more up-to-date articles (see also[12]). Here, we concentrate on the basics which are, among others, described in the mentioned articles. An example of such a target is shown in Figure 8.17a.

Due to the random distribution, the spectrum of a dead leaves pattern should be rather symmetric, which means that $\text{FT}\{B_{\text{obj}}(x,y)\} = \tilde{B}_{\text{obj}}(k_x,k_y) = \tilde{B}_{\text{obj}}(k_r)$ is almost a function of radial component k_r only. Rotation invariance is also a demand on such targets and scale invariance as well. The first demand means that the result of the measurement should not depend on the orientation of the target (if we do neglect the alignment structures, etc. around the dead leaves pattern itself. Note that rotation invariance is fulfilled in (k_x,k_y)-space only, but not in (x,y)-space). The second demand means that the analysis procedure should not much depend on the magnification when the image is captured. This is not always fulfilled and even at best conditions, there are limitations, due to a maximum and a minimum disk size. The texture of the dead leaves pattern should also look like a real structure.

Usually, the 2D power spectrum of the target almost follows a power law across a radial line. In other words, it is given by a function $a \cdot k_r^b$, with a typical value of $b \approx -1.93$ (for Figure 8.17, $b \approx -1.7$, though in (b) the modulus is plotted against the frequency) and a normalization coefficient a, which depends on the size of the image or its crop.

Because the generation of the chosen target pattern is known, of course, its spectrum $\text{FT}\{B_{\text{obj}}(x,y)\} = \tilde{B}_{\text{obj}}(k_x,k_y)$ is known exactly, and consequently, the MTF can simply deduced from power spectrum Equation (5.41) (or Equation (5.54)). An example of the Fourier transformation of the object brightness distribution in radial direction is shown in Figure 8.17b.

---

9 F. Cao, F. Guichard, H. Hornung: *Measuring texture sharpness of a digital camera*, Proc. SPIE **7250**, Digital Photography V (2009) 72500H.

10 F. Cao, F. Guichard, H. Hornung: *Dead leaves model for measuring texture quality on a digital camera*, Proc. SPIE **7537** (2010) 75370E.

11 J. McElvain et al.: *Texture-based measurement of spatial frequency response using the dead leaves target: extensions, and application to real camera systems*, Digital Photography VI, Proc. of SPIE-IS&T Electronic Imaging, SPIE **7537** (2010) 75370D.

12 IEEE Standard for Camera Phone Image Quality, IEEE Std 1858-2016 (or newer ones), IEEE Standards Association (ISBN 978-1-5044-2388-5 and ISBN 978-1-5044-2389-2).

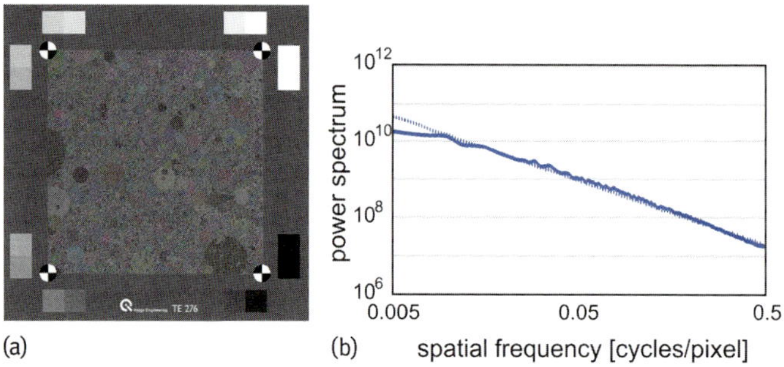

(a) (b)

Fig. 8.17: (a) Example of a colored dead leaves target (there are gray tone targets as well). The features outside the central part, i. e., the inner square with the "dead leaves," are used for alignment purposes and/or do serve for other measurements such as slanted measurements, that could be performed for comparison. (b) Spectrum of the target, namely the absolute value of the amplitude as a function of spatial frequency in radial direction (see the text; solid line spectrum of the target displayed in (a), dotted line exact power law); Source:[13] (Courtesy of Image Engineering).

8.3.3.2 Influence of image processing (i. e., image manipulation) and SFR

Dead leaves targets potentially may also make apparent that image processing has a large influence on the image generation in a way that cannot be controlled by the user at least if optical analysis is not made on the basis of raw data. In the following, we would like to discuss that briefly. We will recognize that image manipulation may be even more severe than would be expected from the previous discussion.

First,, one has to know that such a manipulation is influenced by the contents of the image, like brightness distribution within the object, its content of edge structures and the structure distribution itself, noise, etc. It is also influenced by the picture presets or styles (see Section 4.9 and, e. g., Figure 4.63) and often this is nonlinear. As an example, a preset for a portrait will result in a more smooth and less sharpened image, when compared to that obtained with a preset for sports photography. Clever tricks are applied to improve the perceived image quality. If this is well done, there is success in standard situations. But furthermore, interestingly, independent of all that also even a slight movement during exposure may have influence on the image manipulation process.

As discussed previously, perceived image quality depends on resolution, sharpness and noise (among other criteria). But again, sharpness and noise are opponents. Consequently, if the image processing reduces noise, this will affect sharpness and, most simply speaking, reduce resolution and so on. We would also like to remember that sensors with small pixels usually generate more noise and thus are affected stronger when compared to such ones with larger pixels. When the algorithm used for image process-

13 Images provided by U. Artmann, Image Engineering GmbH & Co. KG, Frechen, Germany; similar images are published by U. Artmann: *Measurement of Noise using the dead leaves pattern*, Electronic Imaging Conference 2018, Open Access: http://ist.publisher.ingentaconnect.com/content/ist/ei

ing cannot distinguish between fine real structures and noise, then real structures may be eliminated or at least set to poorer contrast (see also Section 8.4). This shows up as the so-called texture loss. In literature, this is discussed as an important issue within photography, and thus it is an intention to analyze this phenomenon through a suitable method. Measurements with dead leaves targets may fulfill this task and additionally provide information on noise (see above and below).

Vice versa to the above-described image manipulation, a sharpened image usually shows more noise although modern image processing algorithms try to avoid that, in particular, when they work locally (a post-processing software that does so is the software *Neat Image*). In both situations, the MTF produced by the hardware, i. e., the lens/sensor combination, is changed by the software, namely through image processing. Accordingly, one should not consider it an MTF in strict sense and it is better to use the more general term SFR (see Section 5.2.8).

If the described strong influence of the image processing on the observed SFR curve is not known, then a judgment based on SFR curves resulting from a dead leaves target analysis is likely to be interpreted in the wrong way. Unfortunately, it is not unusual that really good camera/lens combinations, and/or "good camera settings," are rated poor and poor quality combinations, and/or "worse camera settings," seem to reflect good performance. An example is shown in Figure 8.18.

To get rid of such misleading results, the SFR curves resulting from targets such as those under discussion have to be calculated in a way that incorporates the image manipulations. How to do so is extensively discussed in the community. One way to do so, is to correlate the input, namely the exitance distribution within the object and the output, namely the brightness distribution of the image, signals, and thus to include phase information and as well[14] (see example in Figure 8.18 and related discussion below). However, a further discussion of this quite special subject is beyond the scope of the present book.

From the red SFR curves in Figure 8.18, it seems that a higher ISO number does yield "better" SFR behavior, which would indicate better imaging performance. However, it is clear that an increased ISO value degrades image quality and that is also observed in the displayed images. On the other hand, properly corrected SFR curves seem to be quite consistent with the visual observation and seem to allow a better judgment (blue lines).

A similar observation can be made for a mobile phone camera (not shown here), where artefacts due to image processing, i. e., manipulation, are added to all frequencies. This results in a SFR curve that stays high even for frequencies close to the Nyquist limit, which is not reasonable. After correction, the SFR curves seem to allow a more reliable judgment.[14] It may be mentioned that even high-end cameras, although affected

14 L. Kirk et al.: *Description of texture loss using the dead leaves target: Current issues and a new intrinsic approach*, Proc. SPIE **9023**, Digital Photography X (2014) 90230C.

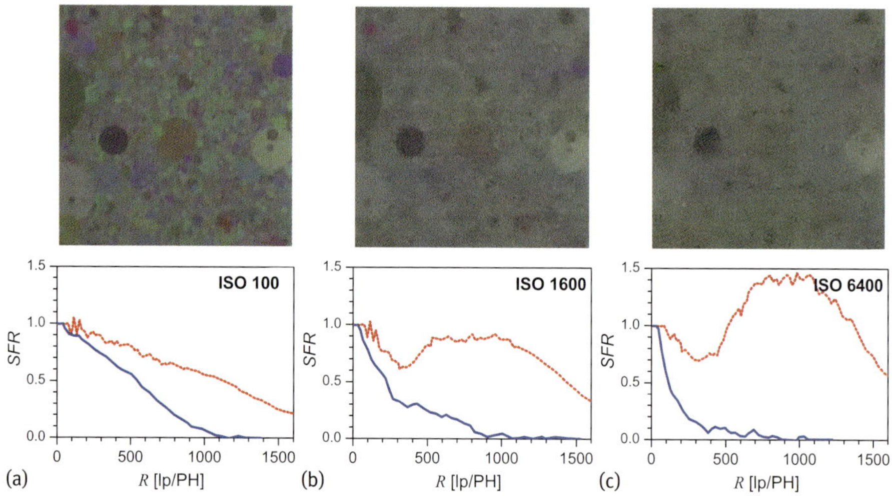

Fig. 8.18: Examples of dead leaves measurements using a colored target. Upper row:[15] images captured with a Panasonic Lumix DMC-TZ41 with different camera settings, namely ISO100 (a), ISO1600 (b), ISO6400 (c), respectively (from left to right). Lower row (Data taken from[14]): corresponding SFR plots. The red dotted curves show the "directly" obtained SFR and the blue solid lines the SFR that is based on the correlation of the input and output signals (Courtesy of Image Engineering).

less than compact or smart phone cameras, may be not free of image manipulations, and thus it makes sense to perform corrections to the SFR as well. Of course, unless raw data are recorded and analyzed, there is image manipulation by the image processor within the camera. However, dead leaves images on the basis of raw data may yield SFR curves that potentially do not much differ from MTF curves obtained from a measurement with Siemens stars and/or slanted edge targets, respectively (at least at not too high frequencies).

Thus, although the usage of dead leaves and spilled coin targets, etc. is established for the investigation of SFR of optical imaging systems, in particular, for such ones used for photography, there might be problems. Quite often noise and noise power spectrum (NPS, see Section 8.4) is investigated in addition and that suffers from the same problems. Particularly problems come up when the experimental conditions such as illumination, are not ideal and/or if nonlinear image processing steps are involved. The same problems are present in case of, e. g., MTF measurements as discussed in Section 8.3.1. In any case, such problems cannot be ignored when test chart analysis is made from images stored as jpg files (see also the discussion related to Figure 8.10) as it is the case, e. g., for many camera systems and most SPC. Thus, for instance, investigations of SPC systems

15 Images provided by U. Artmann, Image Engineering GmbH & Co. KG, Frechen, Germany; similar images are published by L. Kirk et al.: *Description of texture loss using the dead leaves target: Current issues and a new intrinsic approach*, Proc. SPIE **9023**, Digital Photography X (2014) 90230C.

becomes difficult rather often and special care has to be taken for the linearization of the output signals on base of special patches on the test charts (see end of Section 8.2.1; see also Section 8.4 and[16]). For that reason, several recent works concentrate on other methods to get the SFR (and NPS) from other test charts that again should be close representatives of natural scenes. It has been intended to provide robust measures that characterize the related system performance and the level of system scene dependency including noise power spectra[17] (note that in presence of significant noise Equation (5.41) has to be modified, namely usually NPS has to be subtracted from $\widetilde{B}_{\mathrm{im}}$). These measures should work well even in presence of nonideal conditions and potentially nonlinear image processing steps. However, we may note that such analysis may be extensive, and more details are beyond the scope of the present book.

8.3.4 Other methods and a brief comparison of the discussed methods

But once more, in any case SFR curves based on JPG-data do not reflect real MTF values (see also below). They allow judgments as well, but on a different level than physical MTF curves. And, if well done, they may allow comparison of the performance of different camera systems. In this case, comparison is made for the whole system consisting of optics, sensor and the image processing capability of the processor. Thus, this method may provide some useful tools for imaging performance characterization.

We also would like to mention that there are also other kinds of "random scale-invariant test charts" that may have other advantages or disadvantages, respectively, and in particular, speckle-based methods can be used for a measurement of the sensor-MTF. However, all that will not be discussed further here.

A comparison on the different methods to deduce MTF or SFR, respectively, which are discussed up to now is provided by Table 8.2.

The strong influence of image processing discussed before becomes clearly visibly in Figure 8.19.

In particular, it is obvious that the nonlinear tone curve that results from image processing will take a different influence on an MTF measurement than the linear photo response curve (see, e. g., Figure 4.62). Thus, if it is unavoidable to do analysis on the basis of the preprocessed raw data or even JPG-data, linearization of the image data, e. g., from a Siemens star measurement, has to be applied on the basis of OECF measurements (see Section 8.4) Linearization is not straightforward, and furthermore, depending on

16 IEEE Standard for Camera Phone Image Quality, IEEE Std 1858-2016 (or newer ones), IEEE Standards Association (ISBN 978-1-5044-2388-5 and ISBN 978-1-5044-2389-2).

17 E. W. S. Fry, S. Triantaphillidou, R. B. Jenkin, J. R. Jarvis, R. E. Jacobson: *Validation of Modulation Transfer Functions and Noise Power Spectra from Natural Scenes*, J. Imag. Sci. Tech. **63** (2019) 060406.

Tab. 8.2: Comparison of different methods to deduce MTF or SFR, respectively: slanted edge method (see Section 8.3.2), sine Siemens star, chirped grating (both Section 8.3.1), noise target and dead leaves target (present section). Information taken from[19].

method (target) ⟶	slanted edge	Siemens star	chirped grating	random scale-invariant	dead leaves
scale invariant	no	partly	yes	yes	yes
shift invariant	yes	no	no	yes	yes
exposure invariant	no	no	yes	yes	yes
rotation invariant	no	yes	no	yes	yes
texture like	no	no	no	no	yes
robust to denoising	no	yes	partly	partly	yes

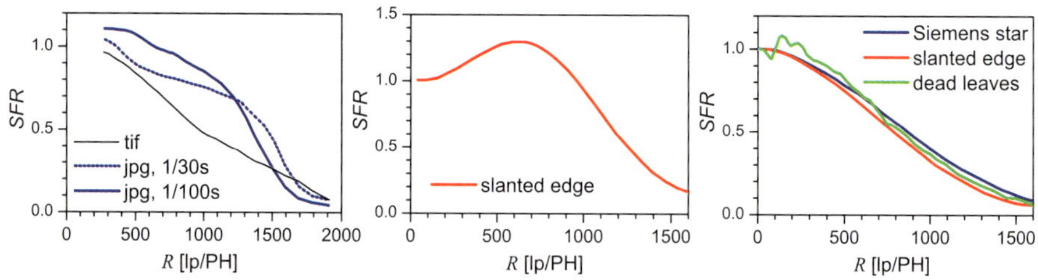

Fig. 8.19: Example of an MTF measurement performed with a Canon 5d III DSLR. (a) MTF curves obtained with the Siemens star method based on JPG-data (blue lines) and on TIFF-data obtained from raw data including a standard tone curve but no additional optimization (black line). Note also the dependence on exposure time indicated in the legend. (b) MTF curve obtained with the slanted edge method (with 60 % edge contrast) based on sRGB JPG-data. (c) Comparison of the Siemens star method, the slanted edge method and the dead leaves method, respectively. These measurements were performed with a Canon 5DMkII, RAW, ISO100, as low as possible image processing. The SFR from the Siemens star measurement is slightly higher when compared to that of the edge method as it includes the diagonal resolution. (data taken from[18] (a) and (b) and from[20], respectively). Here again, we would like to note, that MTF curves above the diffraction limited MTF, and in particular, MTF values larger than one, are physically not possible. Such curves require careful interpretation as discussed in both papers.

exposure time, image processing leads to different results (see, e. g., Figure 8.19a and[18]). Nevertheless, one may note that if data analysis is carefully done, and in particular, image processing is avoided as much as possible, the results obtained from the different methods may differ only slightly (Figure 8.19c).

18 U. Artmann: *Linearization and Normalization in Spatial Frequency Response Measurement*, Electronic Imaging **13** (2016) 1–6.

19 F. Cao, F. Guichard, H. Hornung: *Measuring texture sharpness of a digital camera*, Proc. SPIE **7250**, Digital Photography V (2009) 72500H.

20 U. Artmann: *Image quality assessment using the dead leaves target: experience with the latest approach and further investigations*, Proc. SPIE **9404**, Digital Photography XI (2015) 94040J.

Finally, we would like to mention that MTF or SFR measurements can also be used to investigate the performance of the autofocus of lenses and/or cameras. This is quite obvious because as discussed in Section 5.2.3, defocus leads to worse MTF and only at the best focus position, best resolution is obtained. An extensive investigation can be found, e. g., in[21]. As an example, this work shows as well, that quite a large amount of lenses are able to reach the best focal position.

8.3.5 MTF characterization across the image field

8.3.5.1 Measurements at different positions

Obviously, the MTF depends on $f_{\#}$, ISO-value, wavelength of target illumination, although usually one uses white light, etc. Up to now, we have assumed that for such a given parameters set there is a single MTF only. However, this is not the case. Even for a sensor that is homogeneous across its surface, due to the optical properties of the lens, such as aberrations and vignetting, the MTF changes within the image plane. A straightforward example of that is vignetting. Due to the $\cos^4$-law (see Section 3.4.4), brightness is reduced, in particular, in the corners of the image. This reduced brightness leads to a reduced contrast, and thus to smaller MTF values, even for a nearly aberration-free lens. For that reason, MTF evaluation has to be done at different places in the image field.

One of the simplest ways to measure the MTF at different locations is to position a single Siemens star or a slanted edge target at different places within the object field. If done correctly (see Appendix A.10), from the captured image, MTF curves can be generated for different positions within the image field and this for the sagittal and the meridional plane, respectively. We may point to Section 3.5.3 that often meridional is termed tangential and sagittal is termed radial. The discrimination between both planes is important because imaging performance usually differs for the sagittal and the meridional plane, respectively (see Figure 3.49).

The measurements can also be repeated to yield results for different conditions such as a different $f_{\#}$-values, different values of f, if, e. g., a zoom lens is investigated, different distances to the target and so on. For instance, it is obvious that the MTF may be different for objects that are far away or rather close. Thus, usually one obtains quite a lot of curves. Figure 8.20a shows an example but here restricted to two positions only.

The solid curves are related to a measurement at the center position of the image and object fields, respectively. Here, the measurements for both applied f-numbers do yield almost the same result. The other curves are obtained from a measurement in one corner. We may note that typical camera/lens test publications usually provide more

21 U. Artmann: *Auto Focus Performance—What can we expect from today's cameras?* Electronic Imaging **12** (2017) 219–226.

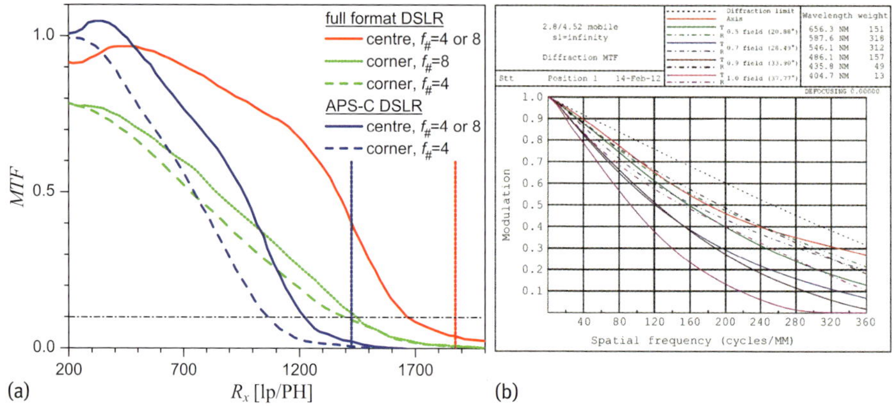

Fig. 8.20: (a) MTF curves resulting from a Siemens star measurement at two different field positions. Here, the same lens is mounted at two different camera bodies of the same company, namely a professional full format DSLR and a consumer APS-C DSLR. Measurements were made for two different f-numbers. (b) MTF charts of a mobile phone camera (with kind permission taken from [Ste12]). The curves in (a) serve for illustration only because such curves are usually found in many journals on photography or web sources. However, we would like to remind again that MTF curves above the diffraction limited MTF, and thus also MTF values larger than one, are physically not possible.

curves, namely curves for center, corners, middle top/bottom and middle left/right star. Moreover, sometimes there is an additional discrimination for sagittal and tangential values. In the present diagram, the vertical lines indicate the Nyquist limit of the sensor (red line full format camera, blue line APS-C camera). The black horizontal line indicates the MTF10 resolution limit. We may expect that the small hump arises from an influence of the image processor. It is generated within the camera or by the raw converter. Although a good measurement tries to avoid such an influence, this is not the subject of the present discussion.

Figure 8.20a clearly shows that although the same smooth MTF is intended for all positions, the MTF may be quite different within the image field. In particular, this is the case for large sensors such as the full format sensor. For smaller sensors, or equivalent to that, a restriction to the inner part only where the size is a small fraction of that of the full format sensor is expected to yield much better results for the same lens. Indeed, this may be observed here for the camera with the APS-C sensor. Also, the borders of the sensor, and, in particular, the corners, are located much closer to the optical axis and thus the image center. Therefore, one may simply expect that all the MTF curves come closer as well. Yet, this is not always the case and depends on the specific lens design. Thus, it is a challenge to provide good MTF and vignetting performance and over a large image field such as those present for large sensors.

Nevertheless, comparison of full format DSLR-lens combinations and APS-C DSLR-lens combinations are generally not straightforward, in particular, because quite often these cameras types are not necessarily equipped with the same lenses. Quite often

APS-C cameras make use of special lens constructions, which are not applicable to the full format. To take into account modern directions of camera technology, we would like to present MTF curves of a mobile phone camera as well. Just for comparison, but without further discussion, Figure 8.20b shows the MTF at different positions within the image field. Here, similar to before, the MTF is plotted for different positions within the image field, but now indicated by a corresponding angle (for details, see [Ste12]). Illumination is made with white light, which is simulated to consist of different wavelengths with appropriate weighting factors as displayed in the upper right corner of the diagram.

8.3.5.2 MTF across the image field

There are several possibilities for generating information on MTF values across the image field. The following discussion is related to such procedures. The related diagrams are usually displayed in catalogs and some examples will be discussed later.

As an example for the principle of a correctly performed MTF measurement, Figure 8.21a shows two times eight test gratings, which are located at eight different positions around a circle. This illustrates how the MTF could be measured in sagittal (with the green gratings) and meridional direction (with the red gratings), respectively. For a grating with a specific period, the corresponding MTF value can be measured at the position of a given radius of the circle, namely with respect to a given image height h_i (see Section 1.5.2 and, in particular, Figure 1.16). Similar measurements could be done for

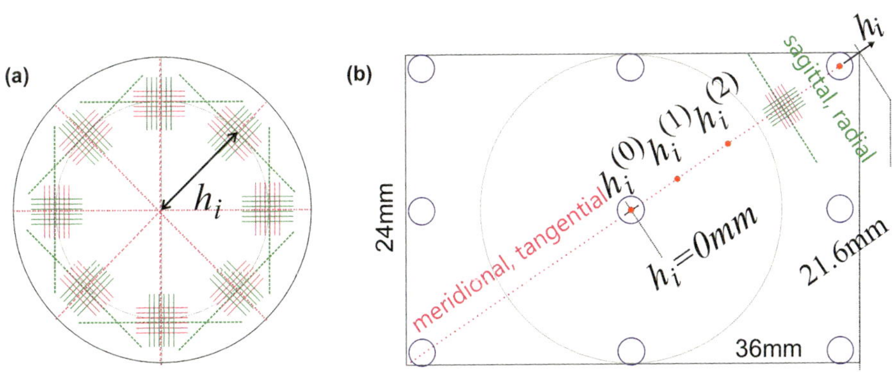

Fig. 8.21: (a) Illustration of MTF measurements with differently orientated gratings. The green gratings have line pairs that are oriented in parallel to the diagonal (radial lines = sagittal lines). The measurement results in a Fourier spectrum that is perpendicular to the diagonal (along green dashed line in (a) and (b)). This yields the sagittal MTF (= radial MTF). The purple gratings are oriented perpendicular to the diagonal (tangential lines = meridional lines) and perpendicular to the green gratings. The measurement results in a Fourier spectrum that is parallel to the diagonal (along purple dotted line in (a) and (b)). This yields the meridional MTF (= tangential MTF). (b) Illustration of a MTF measurement as a function of distance of the center for a full format sensor, which might be a digital one or a film. For further explanation, see the text.

other values of h_i. Here, the number of eight gratings positions on the circle is somewhat arbitrary and serves for illustration only.

Usually, lenses are symmetric and hopefully not decentered. But this has to be verified, e. g., by a measurement using two times eight test gratings as discussed before and also by other methods not discussed here. But to simplify the discussion, we restrict to one radial direction only, e. g., that along the diagonal shown in Figure 8.21b, as understanding of measurements in other directions then is straightforward. Then starting the measurement with the two test gratings positioned in the center and then with a subsequent continuation at different positions $h_i^{(j)}$ ($j = 0, 1, 2, \ldots$), i. e., at increased values of h_i until the corner is reached, yields the MTF(h_i)-dependence for a given spatial frequency R. Similar to before, R is given by the period of the selected grating, which means that grating periods have to be changed to obtain a different R. The procedure is illustrated in Figure 8.21b by four different measuring positions marked with red dots. The measurement should be done for the sagittal and meridional MTF, respectively, for instance, by shifting the displayed green and red grating properly.

Furthermore, at each position the measurement should be repeated with gratings of other periods. Typically, used gratings are chosen for spatial frequency values of R = 5, 10, 20, 30 and 40 lp/mm (sometimes other values are used as well). Sometimes those specific values are regarded to reflect the "contrast" or "resolution," respectively (for instance, Canon mostly displays their MTF(h_i)-curves for two R-values only, (but shows them in both orientations so that altogether four curves are displayed in their diagrams; see, e. g., Figure 8.21) and attributes the R = 10 lp/mm to "contrast" and R = 20 lp/mm to "resolution"). Several examples are shown below.

MTF analysis in the way just described may be rather time consuming and even more if one is interested in much more than just 2 or 5 R-values. Therefore, in principle, the set of gratings with different periods and oriented in the two directions, respectively, may be replaced by a single Siemens star. Instead, one may use differently oriented variable line spacing targets or slanted edge test charts, respectively. However, because then the further procedure is straightforward, we can skip further discussion on that and restrict to Siemens stars as the test targets.

Now, a measurement based on a sequence of images with a single Siemens star successively placed at different positions within the object field could be performed. Typically, the positions of interest include the center, middle top, middle bottom, middle left, middle right and the four corners of the object and image field, respectively, marked by the blue circles in Figure 8.21b. Further ones on the diagonal may be of interest as well, but here this is not a particular issue. At this moment, one has to be very careful because it is not necessarily correct to place the Siemens stars exactly within the positions marked by the blue circles. This is discussed in Appendix A.10.

A quite usual and simple method to avoid measurements based on a sequence of images with a single Siemens stars target successively placed at different positions as discussed before, is to use a single test chart consisting of multiple Siemens stars (Figure 8.22). In this case, one can expect that the analysis for the Siemens star positioned

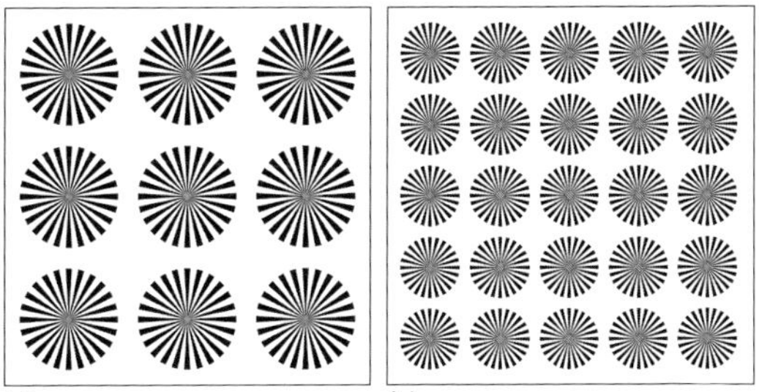

(a) (b)

Fig. 8.22: Illustration of a test charts with multiple Siemens stars. Note that this scheme should just give an idea of possible star arrangements. Neither the size of the sectors, nor the aspect ratio of the image field is chosen as suitable for a measurement, e. g., with a full frame camera.

in the center of the image field works well. However, for other positions this is not necessarily true, and thus more generally one has to take caution (see the discussion in Appendix A.10). In particular, one has to make sure that analysis is made properly, not only for the sagittal direction but also for the meridional direction. If this is not the case, then at best one could only expect rough estimates for the MTF values. This may be sufficient, but that should be verified.

8.3.5.3 Examples of MTF across the image field

In the following, we would like to discuss some examples of the MTF across the image field as provided from data sheets from different manufacturers. If measured, MTF curves of lenses only, usually are obtained from setups such as illustrated in Figure 8.14. However, we would like to note that most companies do not provide measured values but calculations only (see remark at the end of Section 5.2). The fact that the difference between calculated curves and real values of a particular lens may be severe could be of large importance for the user of a particular lens. However, the following discussion is related to a more general discussion, and insofar we may not have to discriminate between theoretical and experimental curves. We would further like to emphasize that with the following diagrams our intention is not to do a systematic comparison of objective lenses. Thus, e. g., we do not compare MTF curves measured at the same f-number, etc. for the same lens type offered by different manufacturers. And although from the plots the reader may identify good or poor properties, these plots do not serve as a base for a ranking in general or some kind of buyer advise. Instead, we would like to draw attention on some selected aspects and very briefly discuss them. An extensive discussion is beyond the scope of the present work. An example of discussion of Leica lenses can be found in[22].

22 See different chapters of "*Puts Kolumne*," e. g., document *Leica R-Objective by Erwin Puts*; there are different chapters (files) for different lenses, which can be found on the internet.

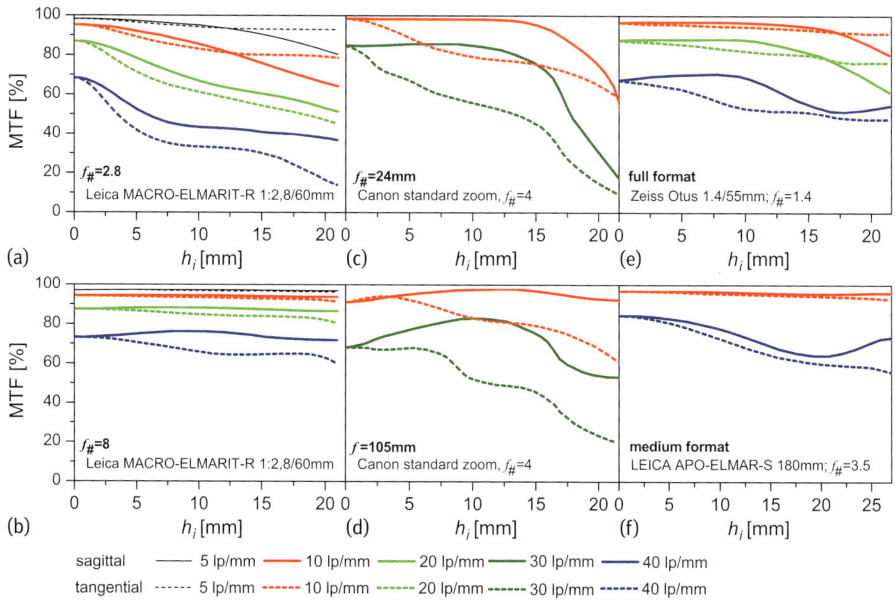

Fig. 8.23: MTF dependence across the image field for different objective lenses for SLR and DSLR. The ordinate displays the MTF, the abscissa the distance h_i from the image center across the diagonal. Curves are displayed for sagittal (i. e., radial, solid lines) and meridional (i. e., tangential, dashed lines) direction, respectively. The different colors of the lines correspond to spatial frequencies of 5, 10, 20, 30 and 40 lp/mm, respectively (120, 240, 480 and 960 lp/*PH* for full format) as assigned in the legend. For further information and discussion, see text. The data haven been taken from data sheets and other information on the lenses provided by the manufacturers.

As a first example, Figure 8.23a and b display the MTF charts of a famous and high-quality Leica MACRO-ELMARIT-R 1:2,8/60 mm lens that has been widely used together with analog camera bodies. Both for full aperture and when stopped down, it is apparent that there is very good contrast and acceptable sharpness in the center. These values are reduced with distance h_i as it is always a challenge to compensate for that. Here, we would like to remind of the natural decrease of illumination according to the $\cos^4$-law, which leads to a decrease of the MTF values as well. We would also like to remind the influence of aberrations for off-center parts within an image. Of course, the situation becomes much better when the lens is strongly stopped down (from $f_{\#} = 2.8$ to 8; see Figure 8.23a and b). In that case, we do get a quite flat response across the image field, extending even to the corners.

Figure 8.23c and d show another example, namely the MTF of the Canon "standard zoom lens" EF 24–105 mm 1:4L IS USM at its full aperture. Differences in the MTF for the wide angle and the telephoto region are apparent.

The next three examples are related to quite different high-quality lenses, namely one for the full format, one for the medium format and one for a mobile phone camera, respectively. An example for the first one is the Zeiss Otus 1.4/55 mm, which today may be one of the best lenses for full format cameras. It has superb image quality when used at an appropriate camera body. This can be seen from the MTF charts, where even at full aperture, the MTF values are excellent (Figure 8.23e). Moreover, the MTF curve is quite flat, and consequently, image quality is rather homogeneous over the whole image field. This is important for achieving an excellent image quality. At the same time, the contrast value is still very high so that this lens does outresolve most or even all available camera bodies. It is expected that this lens would support even 100 MP cameras.

An example for the MTF curves for a medium format lens of the Leica-S-lens program is displayed in Figure 8.23f (LEICA APO-ELMAR-S 180 mm f/3.5 (CS)). Here, it may be seen that this format may give access to a superb quality, at least for such a high-quality lens. Even more, the great quality of the presented lens is apparent. We would like to draw the attention to the very flat 10 lp/mm curve for the medium format lens and this even for the very large image field (i. e., 300 lp/*PH* for this format). Also, the high-spatial frequency of the 40 lp/mm curve (i. e., 1200 lp/*PH* for this format) with MTF-values between 60 and more than 80 % at $f_{\#} = 3.5$ indicates the high quality. It may be important to note that usually telephoto lenses show better MTF performance than lenses of shorter focal length. This is expected because their angle of acceptance is much smaller, and the light rays become relatively parallel to the optical axis. This obviously reduces aberrations, and thus also leads to a better MTF.

As the last example, Figure 8.24 shows the MTF of a mobile phone camera lens. Again, the reader may identify the corresponding values in the different curves of this diagram. This example shows that a well-constructed and well-manufactured mobile phone camera may perform quite well, too, but, of course, for the rather small area of the sensor.

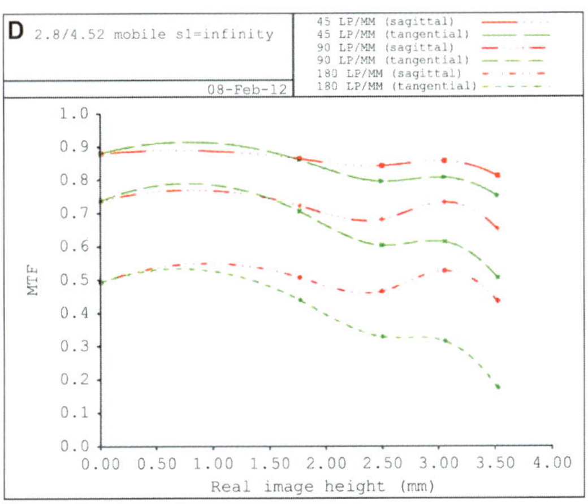

Fig. 8.24: MTF charts of a mobile phone camera (taken with kind permission from [Ste12]). These curves result from the same calculation as those displayed in Figure 8.20b.

Although we have not provided an extensive discussion, we have given some comments and some simple judgments. Some kind of rules for quality judgments are also given by several manufacturers or can be found in web, respectively. Just an example but without a further comment from our side, such rules are, e. g., that a good lens should have the 20 lp/mm line above 80 % (in center and >45 % in the corner), the 40 lp/mm line above 65 % (in center and >20 % in the corner) and the lines should extend at best straight into the corner regions and sagittal and meridional curves should not differ too much. But we would like to remark again that interpretation of MTF curves such as presented within this section is not easy. Our remark at the end of Section 5.2.9 is still valid. Nevertheless, there may be some simple hints.

MTF data are published by manufactures and independent institutions and in many journals, books or on the web. For instance, renowned ones are DXOMark or DPReview, but the reader may also have a look at our list of literature and web links. But one has to be careful with comparison because measurement conditions can vary greatly. Moreover, not all published measurements are done carefully or interpreted correctly and seriously. Remember that sometimes data are supplied that are in contradiction to physical laws, at least partly and we have also shown some such examples. Remember as well, e. g., calculated MTF curves are design curves. Such curves are always better than those measured for real lenses. Therefore, comparison of such curves with ones obtained from measurements is not honest.

8.4 Evaluation of the opto-electronic properties

Beside the optical properties and their characterization also the sensor properties are of much relevance for the performance of the whole system. In the following, we concentrate on related measurements that supplement those discussed in Chapter 4. Within this chapter, we concentrate on the opto-electronic conversion function OECF.

The method of measurement is rather simple and is based on the unmodified photon conversion curve. The sensor has to be illuminated with well-characterized intensities of different levels, which then are integrated over exposure time t_x. The absolute value of the intensities has not to be known but their relative values with respect to each other have to be supplied very accurately. The range of intensities should be large enough so that the lowest one is at maximum at the sensor noise level and the highest intensity must be at least large enough to drive the sensor into saturation. There should be enough different intensity levels between those limits so that the depth resolution (see Section 4.8.5 and Section 4.8.6) could be well resolved.

In principle, such a measurement could be done on a single pixel level, however, due to better statistics a sufficiently large number of pixels should be very homogeneously illuminated with the same intensity so that averaging becomes possible because then fluctuations due to noise will have less effect on the measurement. For measurements on a single pixel level, see also below. One possibility is a successive measurement where

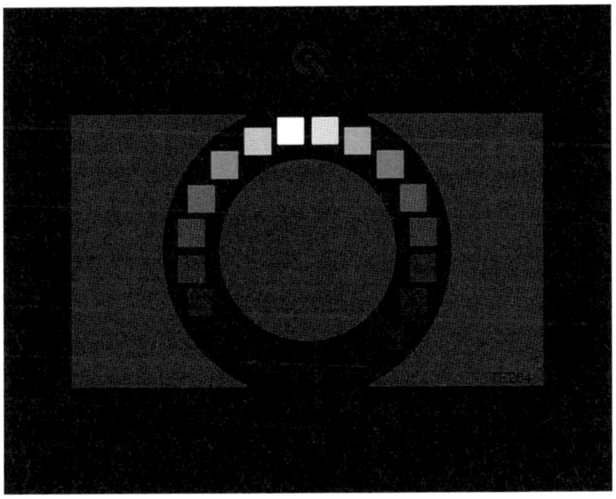

Fig. 8.25: OECF test target. The different patches have different OD. This is tabulated in the data sheet of the target. The transmission is 10^{-OD}. Here, it ranges from $OD = 0.03$ to 4.04, i. e., $DR = 13$ EV (Courtesy of Image Engineering).

images are acquired, each with a different intensity. Exposure may be varied either by a change of intensity on the sensor surface and fixed t_{x} or by a change of t_{x} at fixed intensity, although usually this is not recommended. The first way can be realized by a change of the intensity of the illumination, e. g., by application of very well-characterized transmission filters. Alternatively, for a fixed illumination, one could change the aperture, but this is not usual. Due to the large number of images that have to be carefully taken, this procedure is time consuming and needs very stable conditions and very reliable values of aperture, t_{x} and so on.

Another way is to measure the different intensities at once, i. e., within a single image. This can be achieved with a special slide-type target that is illuminated from the rear side and consists of well characterized transmission filters placed at different positions within the target. Well produced and characterized targets are on the market. Figure 8.25 shows an example.

Such a measurement has to be done in transmission geometry as just described. This allows for high dynamics because the dynamic range is only limited by the maximum and minimum transmission of the target, respectively. A measurement in reflection geometry is not suitable because the dynamic range of prints is rather limited (below 8 bit; see, e. g., Table 4.5). An essential requirement for this way is that the sensor has a spatially invariant pixel response (flat field), that the target is very homogeneously illuminated and that the transmission of the different patches in the target is very homogeneous as well. Altogether high accuracy is a must. Consequently, it is an advantage to do this measurement with the target well aligned and centered on the optical axis so that misalignment and, e. g., additional effects such as light fall-off at the edges, are not of importance. This usually requires also that the patches located within the center of the image field. But as a practical hint it should be mentioned as well that a slight defocusing is advantageous because then the signal of the individual patches are smoothed and the resulting averaged signal is more reliable. This avoids also artefacts due to a grain structure of the target. Additional spurious light, i. e., light that is not resulting from transmission through the target, has to be avoided.

It may also be important to consider the characteristics of the light source that illuminates the OECF test target. This is, in particular, because the quantum efficiency of the photodiodes depends on wavelength. Thus, it is clear that the spectral distribution has an influence on the result.

For a reasonable and reliable measurement, it is important to use the sensor signal directly, i. e., the photon transfer curves. Cameras applied for scientific or technical purposes usually allow this. Cameras used for photography usually do not as they supply only the tonal curves (see Section 4.9). However, if the camera allows storage of raw data and, if at best one can directly extract them from the stored file, then the sensors of those cameras can also be analyzed. In the case of color cameras, it may be important additionally that each color channel has to be accessed separately and de-mosaicing has to be avoided. Consequently, for cameras equipped with Bayer sensors, this leads to four channels: red, green, blue and a second green channel.

Data extraction from photon transfer curve can be done, e. g., by using knowledge how the data are stored or by linear data extraction modus of the program *DCraw* or by application of commercially available programs such as *RawDigger* or, e. g., the scanner software *SilverFast HDR-Studio*. After raw file reading, all those do allow for saving the data without further change, i. e., they can be stored such that the linear photon response curve is maintained. Again, to the best of our knowledge no raw converter besides *DCRaw* allows that (see Section 4.9.4). Alternatively, special software for image evaluation and optical system characterization can do this job. Even more, such software can also do the whole evaluation. Usually, this may be done also with respect to particular standards. In spite of that, it is still important to reiterate that in the case of a CMOS sensor the photon transfer curve stored in the raw data file may not be the originally measured nonlinear curve such as displayed as the dashed line in Figure 4.46b but a linearized version of it.

Because every patch covers a lot of pixels, noise analysis has to be made in parallel and taken into account. In principle, additional noise analysis can be made on the basis of the large patch within the smooth center of a test chart such as shown in Figure 8.25.

The input signal, i. e., the number of photons per pixel N_{ph} is not known in absolute values. However, for a homogeneous illumination of the test chart with N_0 photons per pixel (still unknown), we get for a patch with the optical density OD (see Figure 8.25):

$$N_{\mathrm{ph}} = N_0 \cdot 10^{-OD} \tag{8.4}$$

One can plot the signal S_{pix} (mean value in ADU, i. e., averaged over the patch) as a function of the optical density, and thus obtain curves such as shown in Figure 8.26 (see also Figure 4.51, Figure 4.55 and Figure 4.20a, resp.; note again that it is essential to use the

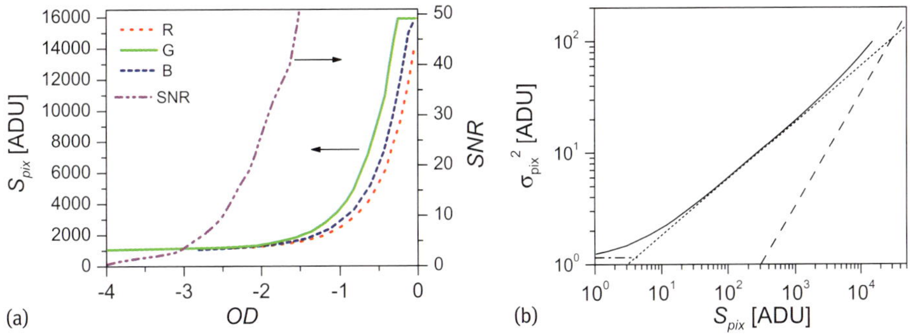

Fig. 8.26: (a) Measured photon conversion curve of a 11-bit DSLR for the red, blue and green channel, respectively, e. g., as displayed by commercial programs. The deduced SNR is displayed as well. Note that the red, green and blue curves are all linear although this cannot be seen from this diagram (but see Figure 4.51a and Figure 4.55a and d). (b) Measured photon transfer curve of a 14-bit DSLR. The range that the curve takes in the abscissa corresponds to the dynamic range. The horizontal line indicates read noise, the dotted line photon noise and the dashed line PRNU.

photon conversion curves; the tonal curves are not useful at all). This plot then allows for proof detector linearity and for deducing the dynamic range and SNR.

Furthermore, the noise signals or the SNR (see also Figure 4.33) can also be plotted as a function of the optical density. Then from both curves (see, e. g., Figure 4.38) calibration of the input signal is rather easy, because for signals measured close but still below the saturation value, which is given by the full well capacity FWC, photon noise is dominant. We will denote the equivalent optical density as OD_{sat}. Thus, SNR is given in Equation (4.39b). Consequently, the input value at that SNR value is just identical to SNR^2 (e. g., point marked as "$N_{\mathrm{ph,sat}}$" in Figure 4.38)

$$\mathrm{SNR}(OD_{\mathrm{sat}})^2 = N_0 \cdot 10^{-OD_{\mathrm{sat}}} \tag{8.5}$$

which yields N_0, and consequently, $N_{\mathrm{ph}}(OD)$.

Now one can deduce the conversion gain of the camera. If we restrict to the photoelectrons, the noise is given by Equation (4.11) and Equation (4.33), and thus

$$\sigma_{\mathrm{pe}}^2 = N_{\mathrm{pe}}. \tag{8.6}$$

According to Equation (4.44) $S_{\mathrm{pix}} = G_{\mathrm{c}} \cdot N_{\mathrm{pe}}$,and thus

$$\sigma_{\mathrm{pix}} = G_{\mathrm{c}} \cdot \sigma_{\mathrm{e,tot}}. \tag{8.7}$$

In similarity to Equation (4.34),

$$\sigma_{\mathrm{e,tot}} = \sqrt{\sigma_{\mathrm{pe}}^2 + \sigma_0^2 + \sigma_{\mathrm{read}}^2} = \sqrt{N_{\mathrm{pe}} + \sigma_0^2 + \sigma_{\mathrm{read}}^2} \tag{8.8}$$

where σ_0 is an unknown further contribution to noise. According to Equation (8.7) and Equation (8.8), this can be rewritten in a linear equation

$$\sigma_{\text{pix}}^2 = G_c S_{\text{pix}} + (G_c^2 \sigma_0^2 + G_c^2 \sigma_{\text{read}}^2) = a \cdot S_{\text{pix}} + b. \tag{8.9}$$

If now, for all measured optical densities the corresponding σ_{pix}^2 is plotted against S_{pix} (both in units ADU; see Figure 8.26b), then one obtains the photon transfer curve PTC. Then according to Equation (8.9) the slope directly yields the conversion gain. We would like to remark that, of course, for each color channel, or in general, each wavelength, one obtains a different PTC and also for different ISO settings.

However, although this method is rather simple, it is not fully correct. This can be seen from the PTC, which indeed is not exactly a straight line, but slightly bent. The PTC often can be better described by three different slopes, namely a constant, which corresponds to read noise, a line proportional to S_{pix} (according to Equation (8.9)) and a line proportional to S_{pix}^2 (see below). Thus, we must have a more careful look at the data. As we have discussed in Section 4.9.1, the images taken with a camera have a bias signal, they suffer from PRNU, etc. This has to be taken into account, which means, S_{pix} has to be the corrected signal in the way described in Section 4.9.1. The bias can be deduced as described in the mentioned chapter. Access to the bias signal, read noise and dark signal may also be obtained from special pixels outside the light sensitive region (see Section 4.5.2). The related signals can be accessed by special software such as *RawDigger*, otherwise usually these signals are not accessible, not even by raw converters.

The product $G_c \cdot \sigma_0$ may be identified with fixed pattern noise FPN. As the absolute value of the FPN is proportional to the absolute value of the signal, one can write

$$G_c \cdot \sigma_0 = k \cdot S_{\text{pix}}. \tag{8.10}$$

Consequently, one has to insert Equation (8.10) into Equation (8.9), which yields an equation with two unknown parameters. Alternatively and preferable, flat field correction (FFC) has to be done before, although one has to be aware that any image corrections may also lead to an increase of noise in general. We would like to note as well that determination of σ_{read} from the flat part of the curve is not very accurate. In the case of FFC, the linear Equation (8.9), but now without the second term (i. e., $\sigma_0 = 0$) allows to deduce G_c.

An alternative is a measurement on a single pixel level. This avoids fluctuations between different pixels, i. e., PRNU. In that case, of course statistics is obtained by a repeated measurement, i. e., a large number of images has to be taken at exactly the same conditions and then analyzed with respect to exactly the same pixel.

Based on the above discussion and that in Section 4.9.1, one can deduce photon number (i. e., the "real" input signal), electron number, conversion gain, dynamic range, SNR, quantum efficiency (see Equation (4.11)), FWC read noise, dark current, FPN, etc. For this or further analysis, the corresponding measurements can or even must be performed with different values of exposure time, ISO value, temperature and so on. All that is mostly straightforward, at least in principle. Measurements according to ISO standards, of course, do demand specific experimental conditions, but that is not an issue within the present book.

A more detailed description of the above procedure can be found in scientific and technical journal articles, in white papers of companies working on image and camera characterization such as Image Engineering GmbH, Germany or Imatest, USA, and others and also in the electronic measurement standard EMVA1288 developed and published by the European Machine Vision Association (see, e. g., web). An extended evaluation of image sensors is also provided in the book of Nakamura [Nak06].

As discussed several times, noise is another important parameter that has influence on image quality. For this reason, we would like to briefly comment on the noise measurements and give some remarks. From a physical point of view, a noise analysis based on measurements as described is quite useful. This allows some judgments on the camera as a diagnostic tool and allows us to deduce OECF parameters. However, when the camera should be used for photography, judgment of image quality of such noise measurements is not necessarily meaningful. The reason is again the response of the human eye together with its analyzing system, namely the brain. This response does not follow a reliable physical rule. The response even depends on the observed structure, e. g., if it is rather homogeneous or complex. It also depends on the specific conditions of observation and the noise pattern itself, e. g., its frequency spectrum. Furthermore, the response is different for fluctuations in brightness and color, respectively. The strong subjectivity is also illustrated in Figure 5.37, although this figure is related to another illusion.

To take into account the subjectivity, concepts have been developed to characterize the amount of perceived noise. One of these is the concept of so-called visual noise VN, which today this is also described in an ISO standard. Similar to SNR, VN is provided by a single number. A larger VN value indicates more perceived noise. The description of VN makes use of weighing the existing noise appropriately according to its visibility. For instance, from Figure 5.42 it becomes clear that noise structures occurring at particular spatial frequencies are hardly recognized (see also Figure 5.28 and the discussion in Section 5.2.8). Thus, it makes sense to give those frequencies adequate weight and, in particular, exclude contribution of that noise frequencies that are not observable. Further discussion would require a rather extended description of the complex relations. However, this does not give much more insight with respect to the basics behind it, and thus is omitted. The interested reader may be referred to the literature.

From the physical point of view, a characterisation of noise can be made by the noise power spectrum (NPS). The NPS, also called Wiener spectrum, is given by the Fourier transformation of the noise image $B_{noise}(x,y) = B_{im}(x,y) - \overline{B}_{im}(x,y)$. It describes the spatial frequency content of noise and characterizes the noise introduced by the imaging system including the illumination conditions. Here, B_{im} is the output image in presence of noise and $\overline{B}_{im}$ the average over a large amount of B_{im}, i. e., the mean. The NPS is a standard measure for image systems in general, and thus also plays an important role in medical imaging, in particular, in (X-ray) radiography. As we have seen in Section 8.3, NPS may also be necessary for SFR characterisation of imaging systems. For deeper discussion, we refer to the literature.

Finally, we would like to mention that with some similarity to the translation of the photon conversion curve into a tonal curve, usually the image processor or the raw converter performs the processing of noise. This shows up in noise measurements of the processed data. An interesting point is also that noise analysis allows us to get an idea of the extent of a potentially unwanted image manipulation. If, e. g., a region within an object (this may also be an appropriate test target) shows a brightness distribution that follows a normal distribution, in the ideal case this should be reproduced within the image, independent of the quality of the camera lens. However, image processing may interpret small structures within the regarded region as noise, and hence apply an "appropriate" smoothing process. As a result, a measurement of the fluctuations of the considered region will yield a distribution that deviates from the expected Gaussian. Usually, this shows up in a kurtosis value deviating from zero. The larger this value, the narrower the curve and this indicates an excess of smoothing. An additional measurement of the MTF should give more insight.

9 Outlook

Optical imaging is an ongoing success story with plenty developments within the last centuries. Large progress was made in imaging optics, but also in sensors and camera systems in general.

Of course, there are always improvements on camera lenses and cameras themselves. But although this is important, this is not an issue here as far as the type of imaging system and its basic principles are concerned. More dramatic is the change in the market within consumer cameras. Most recently, smartphones have shown tremendous technological advance. The advance still continues with excellent optics, usually with multiple camera modules in addition to a front camera, the possibility of taking images with raw data and so on. But we would like to remind also of the severe disadvantages of smart phone cameras discussed in the previous chapters. And we would like to emphasize that current developments such as complementation of missing parts of the image by "image processing" on base of available knowledge on the subject of the picture as presented, e. g., on the Photokina 2018 fair, has nothing to do with imaging. This is not capturing of a scenery and thus beyond our focus.

The change in the consumer market has contributed to a strong decrease in production and sale of compact cameras. With some kind of similarity there is a trend to system cameras. Due to their advantages, many of them, in particular mirrorless systems with more compactness and lower weight, may do replace DSLR. However, there are still advantages of DSLR as well, and it is short-sighted to predict their end. We will wait what the future brings. But also all that this is not an issue here, at least in the sense that there is no impact on the topic of the present book in that there is no new physics or technology. And anyway, for cameras used in science, industry, surveillance, etc., the situation presently remains unchanged in the same sense, even though there will be improvements in noise reduction, in dynamic range, in higher frame rates, and so on.

So what will the future bring? The answer is unknown, at least on long term. In the past, all over the world, most of the speculations on distant and far future and in all subjects have failed. Thus, we should not speculate. But there are new developments, which *may* become important in future. Within this outlook, we will just provide a few examples.

9.1 Sensors

The first set of examples is related to sensors. With exception of special applications, today films have been nearly fully replaced by electronic detectors based on semiconductor technology. We may expect that their usage will continue for some time, and thus also refer to the developments discussed in Chapter 4. If, for instance, one makes use of the individual control of each single pixel by its own "microcomputer," which may follow from the discussion in Section 4.10.5, this will have strong impact on optical imaging,

https://doi.org/10.1515/9783110789966-009

especially for photography. Of course, there are much more innovative ideas and some of them have started to go into the market as well.

9.1.1 Organic CIS, nano crystalline, quantum dots, graphene and other ones

Quite different advances would be the developments of new sensor types. One example is the MAPbX3-detector already discussed in Section 4.10.1. Another one still related to today's CMOS technology is a stacked organic CIS with electrically controllable near-infrared light sensitivity, e. g., developed by Panasonic. Even another one, namely one that has evolved from university research, is the Quantum 13 sensor or QuantumFilm sensor. It is not based on standard silicon technology but instead makes use of a specially designed thin film of nanocrystals. It is claimed that this sensor type has "a higher dynamic range, a dynamic pixel sizing, higher resolution and greater near-infrared and visible light sensitivity." Furthermore, it should have "a more accurate motion capture" for videos, however, we would like to emphasize again, that within the present book we do concentrate on still images. Topics of video and high-speed imaging are outside of our consideration.

Other sensor developments make use of nonsilicon materials and, in particular, of semiconductor compositions or, e. g., of graphen. Very attractive may be quantum dots (QD) or QD/graphene nanohybrids, which may have large potential for many sensor applications because they can be custom-tailored. Moreover, QD size ranges down to the nanometer scale, and thus follows the way of pixel shrink. Within these days, QD-sensors have become commercialized, e. g., such ones, which include QD for an improved sensitivity in NIR and SWIR region, respectively. Another example is an image sensor with integration of graphene developed by a group of the Barcelona Institute of Science and Technology, Spain, together with Graphenea SA, Spain.[1] This sensor makes use of graphene and quantum dots integrated in a CMOS circuit. Operation as a digital camera has been demonstrated. The idea is that next generation image sensor arrays based on graphene may be "designed to operate at higher resolution, in a broader wavelength range, and potentially even with a form factor that fits inside a smartphone or smartwatch" as the group states. Graphene may also be a well-adapted material for sensors used for light-field cameras, which we will briefly discuss below. We would also like to mention that instead of graphene, hexagonal Boron-Carbon-Nitrogen (h-BCN) has been investigated by a team of the University of Bayreuth, Germany, different US universities and the University of Krakow, Poland,[2] which may help for future improvements of image sensors.

---

1 S. Goossens et al.: *Broadband image sensor array based on graphene–CMOS integration*, Nature Photonics, **11** (2017) 366–371.

2 S. Beniwal et. al.: *Graphene-like Boron–Carbon–Nitrogen Monolayers*, ACS Nano (2017) DOI: 10.1021.

9.1.2 Single photon imaging and quanta image sensors

A fully different approach for new sensors, in particular, such ones which may support subdiffraction limited imaging is single photon imaging. Although somewhat older, a basic description can be found in[3]. This approach has some similarities to the single photon counting methods with a CCD as discussed at the end of Section 4.8.5.1 (see also Section 4.11 or single photon counting with EM-CCD) and can make use of various kinds of detectors. At the moment, beside SPAD-arrays the so-called quanta image sensors (QIS) are of much interest. The idea of QIS goes back until 2004/2005 but more practical research has begun just a decade ago with works of E. Fossum et al. at Dartmouth college, Hanover, USA, and a group of the École polytechnique fédérale de Lausanne, Switzerland.

QIS are also made of pixels, but of special ones, namely "binary pixels," SPADs (see Section 4.11.5) or "jots." Each jot (or SPAD) generates a binary signal, which either indicates the presence or the absence of a photon. The total light flux at its position is obtained by photon counting by repeated measurements (compare well-known photon counting from x-ray science and particle physics with Geiger-Müller counters and scintillator-based detectors, resp.). In a high repetition mode, a QIS takes a lot of "binary bit planes," i. e., frames, from which the image is calculated on the foundation of photon statistics (see Section 4.7.2). FWC is not really an issue, but very low read noise is important ("deep subelectron noise") and the possibility of really high-speed readout. HDR capabilities are an issue as well. Jots in modern multibit QIS with new sensor architecture can detect even more than a single photon. In any case, as single photon imaging with QIS requires that a huge amount of successive binary frames are captured, a well-suitable compensation of motion blur becomes necessary, at least if the scene is not still. This can be done computationally (quanta burst photography). Altogether, physics and technology of QIS is not trivial.

Current research and development is made for both, scientific and for commercial purpose, mainly with the goal to improve low light and high dynamic range capability. This includes also consumer photography and miniaturized cameras, and hence the new sensors should not only be very small, but are intended to have a low power consumption as well. Moreover, it would be advantageous if the fabrication process of QIS would be compatible to that of CIS. This leads to CIS-type QIS. Today, QIS is commercially available. A recent example is a sensor with 16.7 MP and a pixel pitch $p = 1.1\,\mu m$. Figure 9.1 shows an example of images taken with that sensor in comparison with high-quality CIS. For this particular example, despite the significantly smaller pixel size, the better image quality obtained with the QIS is obvious. Another recent development of a QIS based on CMOS technology is a device with 163 MP and $p = 1.1\,\mu m$ with very low noise

3 P. Seitz, A. J. P. Theuwissen (eds.): *Single-Photon Imaging*, Springer-Verlag, Berlin, Heidelberg 2011.

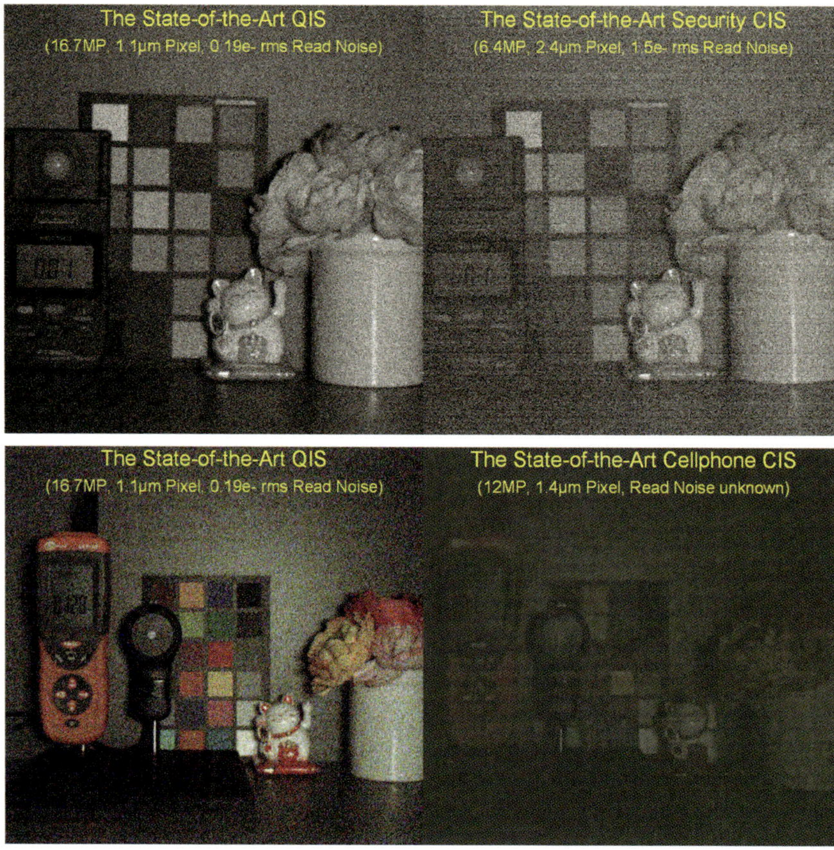

(a)

(b)

Fig. 9.1: (a) Example of images captured at ultralow light conditions with a QIS (left-hand side in (a) and (b), resp.) and two state-of-the-art CIS for security (right-hand side in (a); 4.8 times larger pixel size) and smartphone applications (left-hand side in (b); 1.8 times larger pixel size), respectively. Exposure times and lens configurations are the same for both images in (a) ($f_\# = 1.4$) and (b) ($f_\# = 1.6$), respectively. The QIS has 16.7 MP, a pixel pitch $p = 1.1$ µm and a read noise of 0.19 electrons (RMS). Its images are taken without advanced image post-processing for image quality enhancement such as denoising. For more details, see[5]. Images reprinted from J.Ma, S.Chan, E. R. Fossum: *Review of Quanta Image Sensors for Ultralow-Light Imaging*, IEEE Transactions On Electron Devices, **69** (2022) 2824–2839; work licensed under a Creative Commons Attribution 4.0 License. For more information, see https://creativecommons.org/licenses/by/4.0/.

(0.35 electrons RMS at room temperature), high dynamic range (16 EV) and a large FWC in extended mode (20000 with HGC).[4]

---

4 J. Ma, D. Zhang, D. Robledo, L. Anzagira, S. Masoodian: *Ultra-high-resolution quanta image sensor with reliable photon-number-resolving and high dynamic range capabilities*, Scientific Reports **12** (2022) 13869.

5 J. Ma, S. Chan, E. R. Fossum: *Review of Quanta Image Sensors for Ultralow-Light Imaging*, IEEE Transactions On Electron Devices, **69** (2022) 2824–2839.

There are developments for photon counting and/or photon number resolving sensors by other companies such as Sony, Panasonic, Canon and Hamamatsu ("qCMOS") as well. In particular, these are sensors that are based on CIS and SPAD, respectively, where both methods have their advantages and disadvantages. For a deeper discussion of QIS see, e. g., review articles such as[6] or[7].

Although well applicable cameras based on QIS or QD with "pixel size" well below the diffraction-limit are not yet available (e. g., pixel pitch of 200 nm or below), that may change in the future. It may be interesting if, when combined with, e. g., an optical system consisting of metalens generating subdiffraction PSF size, high-quality imaging may be possible. With these examples, we would like to finish the outlook on image sensors. Of course, there are many other important developments and ideas. For observation of the progress in sensor development, we may refer to[8].

9.2 Imaging optics

The second set of examples is related to optics, where in the last years large progress has been made as well. In Chapter 7, we have discussed miniaturization of cameras and, in particular, metalenses as very special and very small optics (Section 7.5). Currently, this kind of optics has entered the market. An example is a ToF sensor, which is the result of a conjoint development of a sensor company and a company specialized on metalenses. Here, the conventional lenses have been replaced by a metalens array. The application of metalenses may continue. SPC modules consisting of metalenses and of curved sensors can be easily thought of. Other camera modules may take profit as well. But there are other advances, too. We will briefly discuss two of them. Nonetheless, although the topics is interesting, this selection should not be a prediction of its future importance.

9.2.1 3D imaging

The first example is related to camera systems that are intended to capture more than just a 2D image. A particular case is the plenoptic camera, which bases on a concept developed at the beginning of the 20th century. This camera not only detects the light brightness distribution $B_{\mathrm{im}}(x,y)$ on the sensor, but also the direction where the light comes from. This is done by application of an additional microlens array in front of

6 E. R. Fossum, J. Ma, S. Masoodian, L. Anzagira, R. Zizza: *The Quanta Image Sensor: Every Photon Counts*, Sensors **16** (2016) 1260; doi:10.3390/s16081260

7 J. Ma, S. Chan, E. R. Fossum: *Review of Quanta Image Sensors for Ultralow-Light Imaging*, IEEE Transactions On Electron Devices, **69** (2022) 2824–2839.

8 http://image-sensors-world.blogspot.com

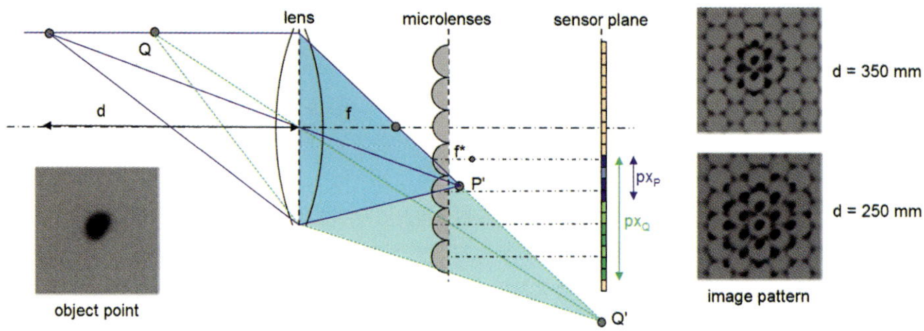

Fig. 9.2: Scheme of a light-field camera. It makes use of a microlens array in front of each pixel and an array of subpixels (see the text). The captured signal distribution on the array is made of all pixels, respectively, subpixels, and has a multifacet pattern. Its repetitions are used for distance-dependent parallax calculations from which the distances can be calculated. The right-hand side of this figure shows the image of an object recorded by the light-field camera at 2 different distances. The object point located at larger distance is imaged on fewer subpixel. Figure and part of the caption text by courtesy of T. Luhmann (see also [Luh19]).

each pixel (this has nothing to do with the OMA discussed in Section 4.6.1). Each pixel itself is made of a matrix of small photodiodes. An example is a sensor with 500x500 pixels where each pixel is made of a matrix consisting of 8x8 elements. This leads to 16 million light sensitive elements in total, i. e., a 16 MP sensor, which has 500x500 effective pixels.

Depending on the direction of the incident light within a pixel, the small microlens array directs the light to particular matrix elements where it is detected (see Figure 9.2). This allows correction of image sharpness within a post-capture process so that exact focusing is not an issue anymore. Indeed, it is possible to get a sharp image when it is captured such that the object is not focused. Here, we mean “focused” in the sense that the lens equation is fulfilled, but not in the sense that the object is really placed at the focal point (see Section 1.5). Of course, there is tremendous effort to get the necessary data and there is also an enormous amount of data when compared to standard sensors where a single pixel corresponds to a single photo diode.

Today, several institutions and companies, such as Raytrix, Kiel, Germany, are active in that field and offer so-called light-field cameras that make use of this method. As an example, since more than 10 years Raytrix GmbH, Kiel, Germany, offers 3D light-field cameras for industrial applications and research, where it claims worldwide leadership. On the other hand, in spite of the interesting opportunities, there has been only little progress within the last years in such cameras for the consumer market. Several companies that have intended to bring plenoptic cameras to the consumer market have not been successful. But there might be also exceptions, where, e. g., a recent work reports on the developments of a miniaturized 3D camera, that may lead to an implementation into smartphone cameras.[9]

A somewhat different approach has been done by K|lens GmbH, Saarbrücken, Germany, where standard imaging components such as lens and sensor are used. However, the optical system is modified in the way that there is a kaleidoscope inserted in between the Gaussian image plane and the standard image sensor. Thus, the original image is reflected and superposed multiple times when hitting the sensor yielding a complex light pattern. The final image is then reconstructed after intense computation. It should be possible to select the range of sharp imaged objects as well as the perspective after the exposure. Even a 3D reconstruction of the scenery should be possible with the newly developed "K-lens One," which has been intended to be usable for DSLR and/or DSLM.[10] But although this scheme is very interesting, there are a lot of difficulties, which have delayed the development of this lens. One may await its success and that of other technologies in future.

A simpler principle for 3D imaging is gated imaging (range gated camera) where a short flash illumination, e. g., in IR region, illuminates the scene. According to their distance dependent propagation time, the reflected light is detected by a gated sensor. Multiple gates (time slices) allow the light capture within a series of consecutive short time windows. Differences in signal brightness allow to get depth information in addition to the 2D scene images. Although straightforward, actually gated imaging is not yet advanced for broad usage. But this may change in future.

Of course, full 3D imaging would be even more interesting, especially of objects quite close to the observer. The ideal possibility to maintain full information (i. e., on amplitude and phase) is holography, which really delivers 3D images. However, this is rather special and a topic on its own. A lot of textbooks are available. Moreover, it is not straightforward applicable and not at all suitable for "simple" picture captures, even not with a much-advanced camera.

3D imaging finds important applications in industry, e. g., for machine vision (see, e. g., [Luh19]). But we would also like to comment that 3D images in general not always make sense. Even humans do not always see in "3D." Three-dimensional impression is only obtained if the objects are rather close to the observer, but not at all when they are far away. Consequently, 3D images of a scenery that is far away from the observer are much similar to 2D ones. A typical example for that is the image of a landscape. As 3D impression relies on an observation with two sensors placed in some lateral distance which usually is the distance between our eyes. The slight differences in the observation directions from the sensor to the object result in a 3D impression when those are considered appropriately by an image processor or the human brain.

---

9 H. M. Kim, M. S. Kim, G. J. Lee, H. J. Jan, Y. M. Song: *Miniaturized 3D Depth Sensing-Based Smartphone Light Field Camera*, **Sensors** 20 (2020) 2129.

10 K|lens GmbH, Saarbrücken, Germany, www.k-lens.de.

For technical applications, 3D information may be calculated from the different observation directions, for instance, obtained from two cameras. According to their distances and the distance to the object, respectively, depth information is obtained more or less accurately. For humans with their eyes as the cameras, the situation is similar. As the distance of the eyes is relatively small, let us say of the order of 0.1 m, one may easily estimate that a significant difference between the angles from one and the other eye, respectively, to the same object point requires that the distance to the object point is not many orders of magnitude larger than the distance between the eyes. As a result, for distances of, for instance, 100 m or more, detection of 3D information fails. In technology, the situation can be improved by setting two cameras far enough away from each other. One can make use of phase information as well, but then this does not compare with the discussed example of a human observer.

9.2.2 Liquid lens modules

The second example of a current development in optics relevant for miniaturized cameras is the application of liquid lens modules. As has been discussed in Chapter 7, focusing becomes an issue even for miniaturized systems such as SPC-modules. A solution may be the application of lenses with a focal length that could be set in a well-controlled way during camera operation. Liquid lenses may fulfil this requirement. Indeed, electrowetting, first explained by G. Lippmann in 1875, allows such a control. By using this method, e. g., the shape of a water droplet can be changed via the application of a voltage. The resulting electrical field leads to a change of surface tension, and thus to a change of the contact angle between the liquid and the substrate underneath. This was first demonstrated in the 1930s by A. Frumkin. Another type of liquid lens has been demonstrated rather recently by Xu et al.[11] It is made of a flat transparent cell, with a particular liquid (dibutyl adipate) enclosed by the housing and an ultrathin flexible membrane. The liquid is located between a ring electrode (copper wire) and an Indium Tin Oxide (ITO) covered glass substrate as the transparent rear electrode. Similar to before, an applied voltage generates an electrical field that changes the shape of the "liquid body." More general, the radius of curvature of a liquid lens can be well controlled with the consequence that such a lens gets an adjustable focal length. This offers flexibility, which makes such lenses interesting devices that are offered commercially.

A current example of a miniaturized camera, which applies the described technique is the smartphone Xiaomi Mi Mix Fold, where a 80 mm tele module is equipped with a liquid lens. It is stated that this allows easier focusing because shifts of a lens or a lens group can be avoided. Specifically, macro photography with that SPC module takes profit

11 M. Xu, Y. Liu, Y. Yuan, H. Lu, L. Qiu: *Variable-focus liquid lens based on electrically responsive fluid*, Opt. Lett. **47** (2022) 509–512.

due to the fact that objects with a distance of 3 cm only can be well imaged although the building length of the SPC module is moderate. On the other hand, of course, image quality is reduced because a particular lens design is optimized for particular shapes of all involved lenses. Any deviation of one or more lens shapes leads to an increased aberration of the system. Moreover, one may continue thinking about liquid lens technology, particularly its application for miniaturized true zoom lenses that up today are not easy to realize (see Section 7.1). This idea may be subject of near future.

9.3 Further developments and final statement

The discussed examples illustrate the continuing progress of developments of sensors, optics and systems, which finally may be applied for commercial products. Part of those developments make use of the evolution of computational imaging. It is fascinating how some of the current advanced developments can be combined. An example for this is a light field camera with an implementation of nanophotonics (metalens) together with an advanced neural network-based reconstruction algorithm.[12]

A more general remark should be done with respect to the driver for technological progress. Without doubt, the main driving force behind many developments in optics, respectively, electronics in the last two decades has been the widespread usage of smartphones in many applications of daily life. The requirement here has been to develop a device, which should be easy to use, relatively compact, with high performance and at low costs. The consequences in the electronics domain have been the development of new sensors, as described in the book chapters and mentioned in this outlook. However, the computational imaging (CI), as presented especially in Chapter 7, has become so powerful that even lens aberrations can be corrected "easily" in the post-exposure process without that the problem has been solved by the hardware. For smartphones, this was in many cases imperative as the hardware solution would lead to more bulky systems. As a consequence, the optic design of a lens could be simplified with an acceptance of aberrations, which could not be tolerated without computational correction. Another drastic change could be seen in the way how an efficient lens is designed. This could be even interpreted as paradigm shift (see Section 7.1.2). Unlike classical lenses, which traditionally evolved from the combination of spherical lens elements based on experienced design rules, SPC lenses consist of only aspheric elements with partially very weird shapes. Their combinations to form a powerful optical system can only be optimized by numerical computations, and moreover, in a relatively short time. As the current SPC lenses are made of plastic material and the operation of the whole system is

12 Q. Fan, W. Xu, X. Hu, W. Zhu, T. Yue, C. Zhang, F. Yan, L. Chen, H. J. Lezec, Y. Lu, A. Agrawal, T. Xu: *Trilobite-inspired neural nanophotonic light-field camera with extreme depth-of-field*, Nature Comm. **13** (2022) 2130, https://doi.org/10.1038/s41467-022-29568-y

computer-based, their life period is expected to be significantly shorter than traditional cameras. These aspects as well as the customer demands also have a large impact onto the development of new technologies in the optics domain in general. It can be expected that changes and technological trends seen with SPC will be seen in future also in the more traditional domain of optics.

Here, we would like to stop further discussion of special optics, sensors, cameras and modern developments in general. We are aware that this chapter is by far not complete but comprises a rather selected presentation. Nonetheless, some interesting aspects have been discussed within the current status of January 2023. This may also stimulate further reading of literature. As stated at the beginning of this chapter, we do not speculate. Easy speculations are trivial but speculations on important success in future usually fail.

A Appendix

A.1 Functions and relations

The following list summarizes some relations and functions that are used within the present book. The intention is not a full mathematical description but rather a presentation of the definitions that we use.

$$e^{i\varphi} = \cos(\varphi) + i \cdot \sin(\varphi)$$

$$\sin(\varphi) = \frac{1}{2i}(e^{i\varphi} - e^{-i\varphi})$$

$$\cos(\varphi) = \frac{1}{2}(e^{i\varphi} + e^{-i\varphi})$$

$\log(x)$	logarithm in general; may be with respect to any base
$\mathrm{ld}(x)$	logarithm with base 2: $\mathrm{ld}(x) = \log_2(x) = \ln(x)/\ln(2)$
$\lg(x)$	logarithm with base 10: $\lg(x) = \log_{10}(x) = \ln(x)/\ln(10)$
$\langle f(x)\rangle_x$	average of the function f with respect to x
$\mathrm{rect}(x) = \begin{cases} 1 & \lvert x\rvert < 1/2 \\ 1/2 & \lvert x\rvert = 1/2 \\ 0 & \text{otherwise} \end{cases}$	rectangle function
$\mathrm{rect}\left(\frac{x}{D}\right) = \begin{cases} 1 & \lvert x\rvert < D/2 \\ 1/2 & \lvert x\rvert = D/2 \\ 0 & \text{otherwise} \end{cases}$	rectangle function
$\mathrm{rect}\left(\frac{k_x}{2k_{x,\max}}\right) \begin{cases} 1 & \lvert k_x\rvert < k_{x,\max} \\ 1/2 & \lvert k_x\rvert = k_{x,\max} \\ 0 & \text{otherwise} \end{cases}$	rectangle function
$\mathrm{circ}(r) = \begin{cases} 1 & r < 1 \\ 1/2 & r = 1 \\ 0 & \text{otherwise} \end{cases}$	circle function (disk function may be a more appropriate name) for polar coordinates with $r = \sqrt{x^2 + y^2}$
$\mathrm{triang}(r) = \begin{cases} 1 - \lvert x\rvert & \lvert x\rvert \le 1 \\ 0 & \text{otherwise} \end{cases}$	triangle function

https://doi.org/10.1515/9783110789966-010

$\mathrm{sinc}(x) = \dfrac{\sin(x)}{x}$ — sinc function
Note: In some books sinc(x) is defined differently, namely $\sin(\pi x)/(\pi x)$

$\mathrm{jinc}(x) = \dfrac{J_1(x)}{(x)}$

$|\mathrm{jinc}(x)|^2 = \left|\dfrac{J_1(x)}{(x)}\right|^2$ — Airy pattern

$J_1(x)$ — Bessel function of first order

$\exp\left(-2\left(\dfrac{x}{x_{1e^2}}\right)^2\right) = \exp\left(-4\ln 2\left(\dfrac{x}{\mathrm{FWHM}}\right)^2\right)$ — Gaussian function

$\delta(x - x_0)$ — delta distribution (although mathematically not correct, for simplicity we may also use the term delta *function*)

properties:

$\delta(0) \to \infty \quad \int_{-\infty}^{\infty} \delta(x - x_0)\mathrm{d}x = 1 \quad \int_{-\infty}^{\infty} \delta(x - x_0) \cdot f(x)\mathrm{d}x = f(x_0) \quad \delta(x) = \dfrac{\mathrm{d}}{\mathrm{d}x}\Theta(x)$

the delta function may be expressed as

$\delta(x) = \lim_{\varepsilon \to 0}\left(\dfrac{1}{\sqrt{2\pi\varepsilon}} \exp\left(-\dfrac{x^2}{2\varepsilon}\right)\right)$ — (there are also other expressions) but for our purpose it may be just expressed by a very small needle

$\Theta(x)$ — Heavyside (or step) function

the Heavyside function may be expressed as

$\Theta(x) = \int_{-\infty}^{x} \delta(x')\mathrm{d}x' = \begin{cases} 1 & x > 0 \\ 1/2 & x = 0 \\ 0 & x < 0 \end{cases}$ — (sometimes alternatively it is defined as $\Theta(x \geq 0) = 1$)

$\mathrm{comb}(x) = \sum_{i:=-\infty}^{\infty} \delta(x - i)$ — comb function

$\mathrm{FT}[f]$ — Fourier transformation of the function f

$\mathrm{iFT}[\tilde{f}]$ — inverse Fourier transformation of the function $\tilde{f}$

$\mathfrak{Re}(x)$ — real part of the complex number x

$\mathfrak{Im}(x)$ — imaginary part of the complex number x

A.2 Fourier mathematics

Fourier formalism is a very fundamental and important method in almost every field of natural science and technology. Hence, many reviews, extensive discussions and also shorter descriptions of this standard method can be found in the literature. Thus, in the following we would like to briefly discuss only the most important aspects and relations, in particular, those that are of relevance to the present book. We would like to note that this is without the claim of a rigorous and mathematically absolutely correct formalism or a complete discussion. Furthermore, we would like to restrict to a discussion in one dimension and to quasimonochromatic light, namely a single color only, e. g., green light. Fourier transformation in two dimensions is then straightforward and also that for any other color. Just as a remark, quasimonochromatic means that the spectral bandwidth of the light is much smaller than its frequency.

Simply expressed, Fourier series or integral representation is a method for synthesizing a given function in terms of sine functions, where each of these functions is characterized by its amplitude and phase. In case of a periodic function, this function may be expressed as a series of Fourier terms, for instance, sine functions. As an example, Figure A.1 shows how a box-shaped structure may be described by a series of discrete sine terms. Each term is characterized by its harmonic frequency $k_{\mathrm{n}} = n \cdot k_1$ with an integer number $n = 1, 2, \ldots$ and k_1 being the fundamental frequency, or more correctly, the angular frequency. Each term has a particular amplitude and phase, which can each be calculated from Fourier analysis.

A single term (blue curve in Figure A.1) provides only a very rough approximation, whereas the original function (red) is significantly better represented, when two or more terms contribute (displayed in blue, green, etc.). In particular, the original function is well represented, by a sum of more than 50 terms (dark yellow line in (b)). For the present example, only odd terms contribute to the series; all even terms have zero amplitude. The different amplitudes can be easily recognized in Figure A.1a.

In case of an aperiodic function $f(x)$, instead of a Fourier series, a complex Fourier integral has to be applied, and thus the Fourier transformation (FT) yields the Fourier spectrum

$$\tilde{f}(k) \equiv \mathrm{FT}[f(x)] = N_1 \cdot \int_{-\infty}^{+\infty} f(x) \mathrm{e}^{-ikx} \mathrm{d}x. \tag{A.1}$$

In this case, instead of a discrete spectrum consisting of single lines at k_1, k_2, k_3 and so on, we now have a continuous spectrum in k-space, which is given by Equation (A.1). The original function $f(x)$ in x-space then is obtained from the inverse Fourier transformation (iFT)

$$f(x) = \mathrm{iFT}[\tilde{f}(k)] = N_2 \cdot \int_{-\infty}^{+\infty} \tilde{f}(k) \mathrm{e}^{ikx} \mathrm{d}k. \tag{A.2}$$

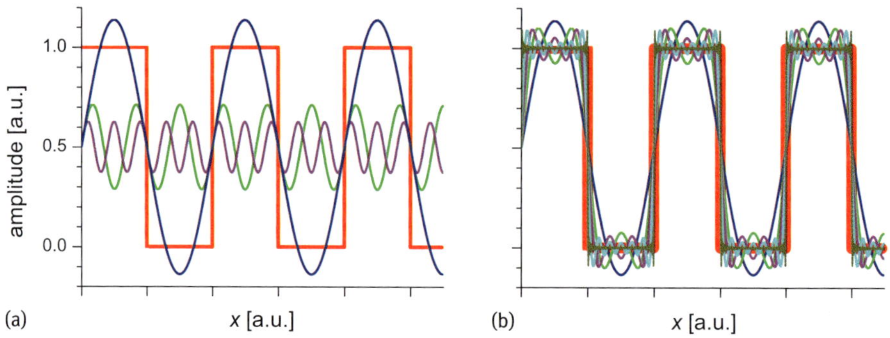

Fig. A.1: Synthesization of a periodic series of box functions by a series of sine functions. The individual terms are displayed in (a), whereas (b) shows the sum of those terms up to the first (blue curve), second (green), third (magenta) and sixth (cyan) term. Cut-out from a box-shaped structure, which contains more than two periods.

N_1 and N_2, respectively, are normalization factors that are required to make sure that inserting the calculated value $\tilde{f}(k)$ from Equation (A.2) into Equation (A.1) again yields exactly $f(x)$, i. e., that $f(x) = \mathrm{iFT}[\mathrm{FT}[f(x)]]$. This is fulfilled if $N_1 \cdot N_2 = (2\pi)^{-1}$. Usually, in mathematics, one chooses $N_1 = N_2 = (2\pi)^{-1/2}$ whereas in physics and engineering mostly $N_1 = 1$ and $N_2 = (2\pi)^{-1}$ are preferred. For this book, these normalization factors are mostly of minor importance, and thus often we do not take much care of them (we just set them to unity or ignore them). This is because we are mostly interested in shapes, structures, resolution, etc. and those do not depend on signal strength at least as long as we disregard, e. g., noise. However, if necessary, normalization factors may be always obtained quite simply from reasonable physical arguments. For example, if one identifies $f(x)$ with the light distribution emitted from an object or image structure, namely $f(x) = B_{\mathrm{obj}}(x)$, then an integration of $B_{\mathrm{obj}}(x)$ over x would yield the total amount of light emitted from the object, which is the total energy E_{tot}. On the other hand, integration can be performed in the Fourier plane instead (see, e. g., Figure 5.2 or Figure 5.3). But if we assume no additional losses, of course energy must be conserved, and thus the integration of $\tilde{f}(k)$, which represents the light distribution $= \tilde{B}_{\mathrm{obj}}(k)$; over k again must yield E_{tot}. Thus, from both integrals, N_1 and N_2 can be easily deduced. From a mathematical point of view, this physical argumentation of energy conservation is a result of Parseval's theorem (relation 5 in Table A.1) or its generalization (Plancherel theorem; relation 6). Table A.1 shows some important relations.

If we consider the spatial domain, x corresponds to the spatial coordinate and then k is the spatial frequency. If one is interested in the temporal domain, x has to be identified with time t, and consequently k has to be identified with the angular frequency ω. x and k and t and ω, respectively, are the conjugate variables.

An example of a synthesization of an aperiodic function is displayed in Figure A.2, namely a single rectangular or box function. In contrast to Figure A.1, here the box is not part of an extended periodic function, but the function is zero outside the displayed

Tab. A.1: Important relations for Fourier transformations. The asterisk denotes the conjugate complex of a function. Here, normalization is mostly ignored (see text).

no.	relation	Remark
1	$\mathrm{FT}[a \cdot f(x) + b \cdot g(x)] = a \cdot \tilde{f}(k) + b \cdot \tilde{g}(k)$	Linearity
2	$\mathrm{FT}[f(x - x_0)] = \exp(-ikx_0) \cdot \tilde{f}(k)$ $\mathrm{FT}[\exp(ikx_0) \cdot f(x)] = \tilde{f}(k - x_0)$	shift theorem (only spectral phase or phase, respectively, is changed)
3	$\mathrm{FT}[f(a \cdot x)] = \frac{1}{\lvert a \rvert}\tilde{f}\left(\frac{k}{a}\right)$	scale theorem
4	$\mathrm{FT}[(f \otimes g)(x)] = \sqrt{2\pi}\tilde{f}(k) \cdot \tilde{g}(k)$	convolution theorem
5	$\int_{-\infty}^{+\infty} \lvert f(x) \rvert^2 \mathrm{d}x = \int_{-\infty}^{+\infty} \lvert \tilde{f}(k) \rvert^2 \mathrm{d}k$	Parseval's theorem
6	$\int_{-\infty}^{+\infty} f(x) g(x)^* \mathrm{d}x = \int_{-\infty}^{+\infty} \tilde{f}(k) \tilde{g}(k)^* \mathrm{d}k$	Plancherel theorem
7	$\mathrm{FT}[f(x)^*] = \tilde{f}^*(-k)$	Fourier transformation of a complex conjugate function
8	$\mathrm{FT}\left[\frac{\mathrm{d}f(x)}{\mathrm{d}x}\right] = ik \cdot \tilde{f}(k)$	Fourier transformation of a derivative of a function

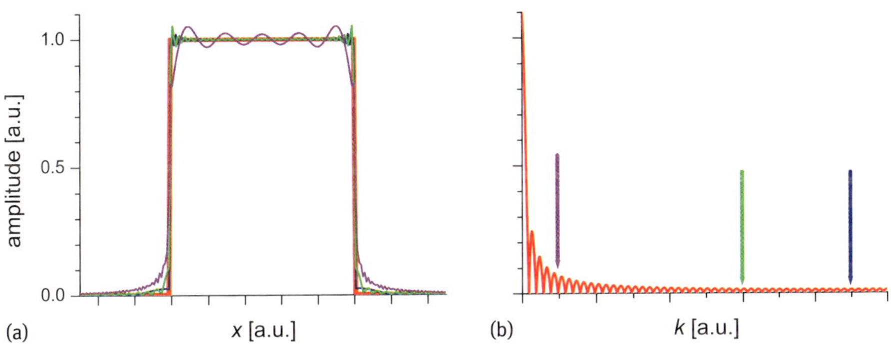

Fig. A.2: Synthesization of a single box function (red curve in (a)). (b) Spectrum.

region. Consequently, this function has to be synthesized by a Fourier integral instead of a Fourier series. As a result, the synthesization is made of a continuous frequency spectrum, which is in contrast to Figure A.1, which is based on a discrete frequency spectrum. In the case of no limitations on the contributing frequencies, the Fourier integral, and thus the spectrum ranges from minus to plus infinity and the original structure (red curve in (a)) is perfectly regenerated. Note, however, that Figure A.2b shows only the contributions of the positive frequency components. The contributions of the negative ones are symmetric to that of the positive ones and not shown there. If the bandwidth is limited, the spectrum may only extend from zero to the positions indicated by the ar-

Tab. A.2: Important relations for Fourier transformations. The asterisk denotes the conjugate complex of a function. Although factors that indicate scaling effects such as D are included, normalization is mostly ignored (see text; that means that more correctly "=" may have to be replaced by "∝").

no.	Relation	remark
1	$\mathrm{FT}[\exp(ik_0x)] = \int_{-\infty}^{+\infty} \exp(ik_0x)\exp(-ikx)\mathrm{d}k = \delta(k-k_0)$	exponential function
2	$\mathrm{FT}[1] = \delta(k)$	constant
3	$\mathrm{FT}[\delta(x)] = 1$	delta distribution
4	$\mathrm{FT}[\delta(x - x_0)] = \exp(-ikx_0)$	delta distribution (according to shift theorem)
5	$\mathrm{FT}[\mathrm{comb}(a \cdot x)] = \mathrm{comb}(k_x/(2\pi a))$	comb function
6	$\mathrm{FT}[\Theta(x)] = \frac{1}{\mathrm{i}\pi k} + \delta(k)$	Heaviside function
7	$\mathrm{FT}[\exp(-ax^2)] = \exp(-k^2/4a)$	Gaussian function
8	$\mathrm{FT}\left[\exp\left(-\frac{1}{2}\left(\frac{x}{a}\right)^2\right)\right] = \exp\left(-\frac{1}{2}a^2k_x^2\right)$	Gaussian function
9	$\mathrm{FT}[\exp(-ax)] = \frac{1}{a - \mathrm{i}k}$ (complex Lorentzian)	exponential function
10	$\mathrm{FT}\left[\mathrm{rect}\left(\frac{x}{D}\right)\right] = D \cdot \mathrm{sinc}(k\frac{D}{2})$	rectangle function (as defined in Appendix A.1)
11	$\mathrm{FT}\left[\mathrm{rect}\left(\frac{k}{2k_{\mathrm{max}}}\right)\right] = 2k_{\mathrm{max}} \cdot \mathrm{sinc}(k_{\mathrm{max}}x)$	rect function
12	$\mathrm{FT}[\mathrm{sinc}(k_{\mathrm{max}}x)] = \mathrm{rect}\left(\frac{k}{2k_{\mathrm{max}}}\right)$	sinc function
13	$\mathrm{FT}\left[D \cdot \mathrm{sinc}\left(k\frac{D}{2}\right)\right] = \mathrm{rect}\left(\frac{x}{D}\right)$	sinc function
14	$\mathrm{FT}\left[D \cdot \mathrm{sinc}^2\left(k\frac{D}{2}\right)\right] = \mathrm{triang}\left(\frac{x}{D}\right)$	sinc^2 function
15	$\mathrm{FT}[\mathrm{sinc}^2(k_{\mathrm{max}}x)] = \mathrm{triang}\left(\frac{k}{2k_{\mathrm{max}}}\right)$	
16	$\mathrm{FT}\left[\mathrm{triang}\left(\frac{x}{D}\right)\right] = D \cdot \mathrm{sinc}^2\left(k\frac{D}{2}\right)$	triangle function
17	$\mathrm{FT}[\sin(k_0 \cdot x)] = \frac{\delta(k-k_0) - \delta(k+k_0)}{2\mathrm{i}}$	sin function (the spectrums show only components at $\pm k_0$)
18	$\mathrm{FT}[\cos(k_0 \cdot x)] = \frac{\delta(k-k_0) + \delta(k+k_0)}{2}$	cos function (the spectrums shows only components at $\pm k_0$)
19	$\mathrm{FT}[\text{bar-grating}] = \frac{c_1 + c_2}{2} - 2\frac{c_1 - c_2}{\pi} \cdot \left(\sin kx + \frac{\sin 3kx}{3} + \frac{\sin 5kx}{5} + \cdots\right)$	the bar-grating has a value of c_1 in the range of $-\pi < x < 0$; it has a value of c_2 in the range of $0 < x < \pi$; c_1 and c_2 may be chosen; a bar grating that corresponds to a sine grating that oscillates between −1 and 1 has $c_1 = 1$ and $c_2 = -1$ (Fourier series)

rows. This yields to a representation that is worse than the original. The reconstructed "box functions" in (a) are displayed in the same color as the arrows of the corresponding cut-off frequencies marked in (b). The smaller the bandwidth, i. e., the extension until the cut-off, the stronger the deviation of the reconstructed signal to the original. This is the particular case for the spectrum clipped at the purple arrow.

Other helpful relations of Fourier transformations, in particular such ones that are relevant for Fourier optics are listed in Table A.2.

Sampling and Fourier analysis, practical notes

For numerical calculations in x-space, the coordinate is provided by a sequence of points with positions x_i, which can be calculated from $x_i = i \cdot \delta x$ with a usually equidistant step width δx and an integer number representing the counter index i. The zero position may be shifted; however, this is not important for the present discussion. The minimum value is x_0 and the maximum value is given by the number of data points N, namely by $x_N = N \cdot \delta x$.

A Fourier transformation transforms the x_i variables to the conjugate variables k_i, which can easily be obtained from $k_i = i \cdot \delta k$. The step width in k-space (frequency space or Fourier space) is given by $\delta k = (2\pi)/(N \cdot \delta x)$. The spectrum then extends from zero to positive and the negative direction. The minimum absolute value of k_i is δk, the maximum one is $(N/2) \cdot \delta k$, where the factor 1/2 comes from the Nyquist limit. Consequently, the spectrum extends from $(-N/2) \cdot \delta k$ to $+(N/2) \cdot \delta k$.

If now, in similarity to the optical bandwidth, we describe the bandwidth by the positive or negative range only, then the spectral bandwidth $\Delta k = (N/2) \cdot \delta k = \pi/\delta x$. Otherwise, of course, it is a factor of two larger.

If we use technical units for the frequency instead, then $R_i = k_i/(2\pi) = i \cdot \delta k$ with $\delta R = 1/(N \cdot \delta x)$. For a PDA, it is quite common that δx is identical to the pixel width. If, as an example, the PDA is made of $\delta x = 6\,\mu m$ pixels, then, e. g., we have $N = 4{,}000$ pixels within a full format PH. Then $\delta R = 1/(4{,}000 \cdot 6\,\mu m) = 0.042\,mm^{-1} = 1/PH$. The spectrum extends from zero to ± 83 lp/mm $= \pm 2{,}000$ lp/PH. But note that the minimum value in the spectrum is not zero but $0.042\,mm^{-1} = 1/PH$.

A.3 Convolution

This Appendix illustrates the basic principle of convolution. Let us assume several point or point-like signals, such as several object points, very small spectral lines or ultrashort laser pulses located at different spatial, spectral or temporal positions x_{j0}, respectively. j is an integer number, which indicates the corresponding line, etc. Each of the lines has a particular signal strength $y = f(x_{j0})$ given by its energy E_j. Due to its needle-like character, the height of the needle represents the energy, not the intensity. Seven of such points are shown in Figure A.3a. However, due to the apparatus that measures the spatial

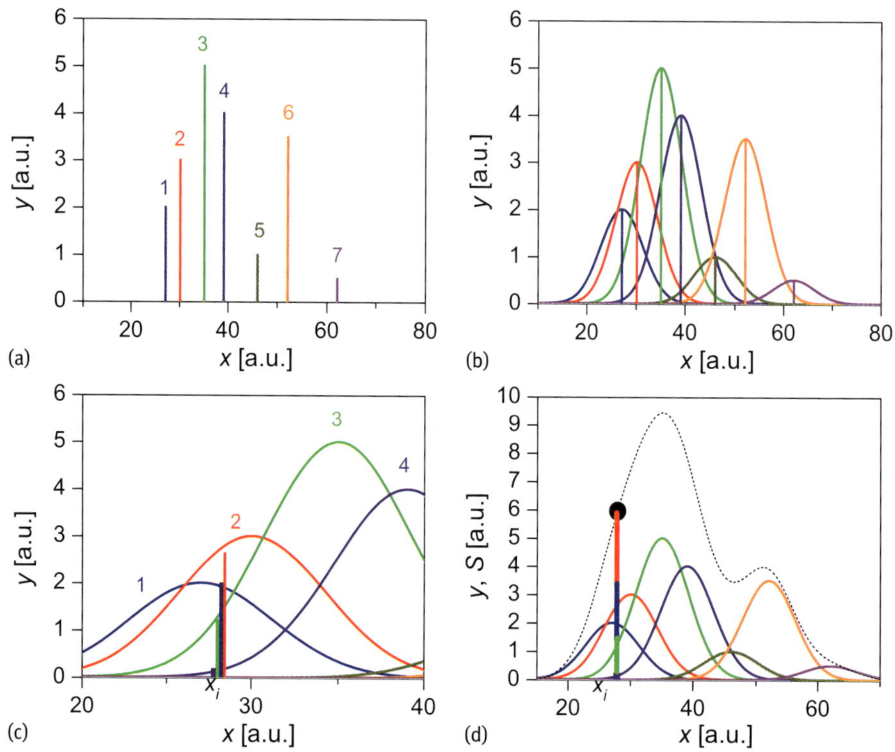

Fig. A.3: Principle of convolution (see text). The number at the lines indicates the value of j.

distribution of the points (i. e., the imaging system) or the spectral distribution (i. e., the spectrometer) or the temporal distribution (e. g., a photodiode), the signal is smeared. In the case of an imaging system, the apparatus function is the PSF, and in the other two cases, it is a corresponding function.

Here, for simplicity, we assume that the apparatus function is given by a Gaussian $G(x - x_{j0})$ (x_{j0} is the center position of line j). More general G may be another appropriate function. Of course, the maxima of the Gaussians reflect the corresponding signal strength, but the FWHM is the same for all smeared lines. This is shown in Figure A.3b.

Note that, in contrast to the single needles, now we do have distributions, provided here by the Gaussians. Such a distribution is given by $y = \mathrm{d}E/\mathrm{d}x$ and integration over x yields the total energy of the line. Due to energy conservation, this energy is the same as that of the original needle-like line. This means that for the very small needles y is an energy. For a 1D system $y = \mathrm{d}E(x)/\mathrm{d}x$, for a spectrometer this corresponds to $\mathrm{d}E(\lambda)/\mathrm{d}\lambda$ or $\mathrm{d}E(\nu)/\mathrm{d}\nu$ and for a temporal measurement y is the power $P(t) = \mathrm{d}E(t)/\mathrm{d}t$. For a 2D imaging system y is the fluence $F(x, y)$.

If one *would* measure the distribution of all points or lines separately one after another, one *would* measure, e. g., first the blue, then the red curve and so on. However, as we measure all points or lines at the same time, the distributions do overlap. Overlapping occurs at every position, e. g., at the position x_{i}. Consequently, at x_{i} we have to sum all the contributions of all the other points

$$S_{\mathrm{i}} = \sum_{\mathrm{j}} y_{\mathrm{j}}(x_{\mathrm{i}})$$

where the sum is over all lines that are present (here j = 1 to 7). $y_{\mathrm{j}}(x_{\mathrm{i}})$ is the contribution of line no. j at the specific position x_{i}. y_{j} is given by the center position of the corresponding line at $x_{\mathrm{j}0}$ and the strength of the signal $y_{\mathrm{j}0} = f(x_{\mathrm{j}})$ for the object point located at x_{j}.

$$y_{\mathrm{j}}(x_{\mathrm{i}}) = y_{\mathrm{j}0} \cdot G(x_{\mathrm{i}} - x_{\mathrm{j}0}) \Rightarrow S_{\mathrm{i}} = \sum_{\mathrm{j}} f(x_{\mathrm{j}}) \cdot G(x_{\mathrm{i}} - x_{\mathrm{j}})$$

These contributions of the blue (j = 1), red (j = 2), green (j = 3) and violet (j = 4) marked points, respectively, at an example point position x_{i} are shown as vertical bars in Figure A.3c in the corresponding colors (for better visibility the bars are slightly shifted in x-direction). This is the basic idea of convolution.

In Figure A.3d, the same contributions are shown as vertical bars, which are placed on top of each other, that corresponds to their sum at that position. Thus, we have obtained the final value (displayed as the black dotted line).

Of course, we have to repeat this procedure for all points within the interval of interest, e. g., for i = 0 to 100 points in between the range x = 10 to 80. This results in the black dotted curve, which is that which is measured (here this would be the image of the observed seven discrete tiny object points).

If we consider the imaging of a real object (here for simplicity in 1D geometry), we have an infinite number of object points instead of a finite sum of points x_{i}. Therefore, the sum in the above equations transforms into an integral. If the signal strength at a position x' in the object plane is given by $f(x')$, then the signal distribution in the image plane at a position x is given by $S(x)$,

$$S(x) = \int_{-\infty}^{\infty} f(x') \cdot G(x - x')\mathrm{d}x' = \int_{-\infty}^{\infty} f(x - x') \cdot G(x')\mathrm{d}x'$$

This convolution is often written as

$$S(x) = (f \otimes G)(x)$$

where the symbol "⊗" indicates the convolution.

A rather simple example is the convolution of Gaussian functions. If both $f(x')$ and $G(x-x')$ are Gaussians, namely $f(x') = \exp(-x^2/a)$ and $G(x-x') = \exp(-x^2/b)$, insertion of those functions in the above equation, after some calculation yields that the total width of the measured signal S is given by $(a^2 + b^2)^{1/2}$. This example also reflects that, e. g., the convolution of different noise contributions finally results in adding the squares of the individual noise contributions, which then leads to the total r. m. s. value.

A.4 CCD readout

This Appendix illustrates how an image may be affected when a global shutter is missing. The example shows images for a full frame slow-scan CCD used for scientific experiments. In contrast to the following examples, in the scientific experiments the object may be located in dark and then illuminated by an ultrashort flash. This situation is similar to the application of an ultrafast shutter. The principle of operation is that the nominal exposure time is set to a value t_{set}. The time interval t_{illum} during that the object is illuminated should be within the interval given by t_{set}. In the following, we assume that both times start at the same time.

Figure A.4 shows a scheme of image capturing in the presence of illumination during readout. The 6 × 6 matrix on the left-hand side is the CCD. Rows 1 to 6 are in the light sensitive region, row "A" is the readout register. When exposure begins the light distribution provided by the image, it leads to charge accumulation within t_{set} at the corresponding pixels. The red spot illustrates the image on the surface of the sensor.

In the following, we do restrict discussion to the column marked by the solid rectangle. On the right-hand side of the matrix the subsequent steps of readout of all 6 rows are shown according to the basic principle of CCD readout discussed in Section 4.4.1. This leads to charge generation not only in the pixel in row 4 but also to the charge genera-

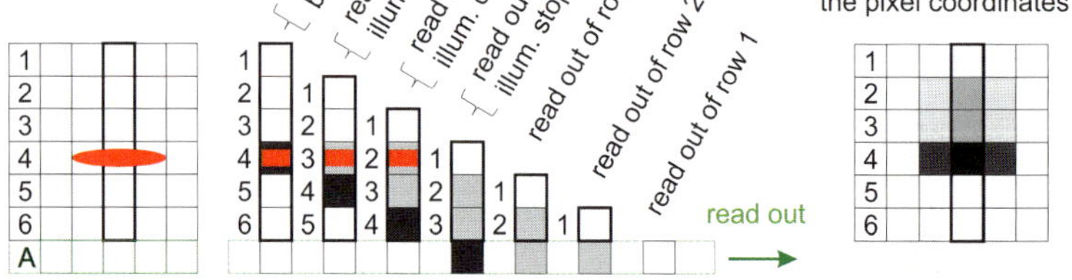

Fig. A.4: Scheme of image capture of a fixed light spot if illumination continues during readout (i. e., $t_{set} < t_{illum} < t_{read}$; see text). Note that after readout of row 5 illumination has stopped.

tion that *corresponds* to the signal in rows 3 and 2. But note that these charges are always generated at the position of row 4. The duration of this charge integration is identical to the duration of one shift t_{shift}. If there were also light at the position of row 5, within an integration time t_{shift} an additional charge would be added to that of the pixel of row 4. This happens in the next example discussed below.

Although the transfer leads to a transport of charges to other pixels, the center of Figure A.4 numerates the pixels according to the original row position. This is reasonable because after passage of the output amplifier the collected charges are attributed to their original pixel position. In the case of absent illumination after the end of the set exposure time, this attributed charge distribution or better the corresponding voltage signal or the corresponding ADU, would directly reflect the light field distribution of the image. However, due to the additional generation of further charges during charge transfer (each within t_{shift}), the finally recorded image shows a stripe above the "true image." If $t_{set} > t_{shift}$ the signal within the "true image" (here from the red spot) is larger, here indicated by more darkness than the signal from the stripes (see matrix on the right-hand side; in this figure, larger charge collection is shown by darker box).

The second example is mostly similar, but now in addition, the object is moving. The top row in Figure A.5 shows the illumination conditions, i. e., the light field distribution (in red) on the sensor. The object moves and as a result this shows up in the image as well. After some time, the movement stops, but illumination is still present. Even some time later, but before readout has finished, illumination is switched off. Total time for readout is t_{read}. The lower row of Figure A.5 shows the corresponding charge collection during transfer. First, the movement leads to a ghost image that is shifted, both in the horizontal (due to object movement) and in the vertical direction (due to readout). After the object comes to rest, the situation is similar to that in the previous example (namely generation of a horizontal stripe consisting of a smear to ghost image migration). The resulting image is shown in the upper right corner.

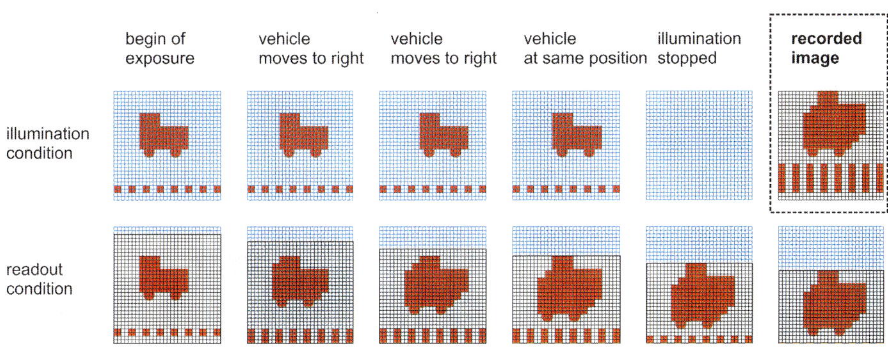

Fig. A.5: Scheme of capturing a moving object when illumination continues during readout ($t_{set} < t_{illum} < t_{read}$; see the text). The light sensitive region is marked in blue.

Figure A.6 shows real images according to the two discussed examples. In the first column, the images are taken with a snapshot illumination, i. e., $t_{illum} \le t_{set}$, the second column with $t_{set} < t_{illum} < t_{read}$ with a fixed object and the third column with $t_{set} < t_{illum} < t_{read}$ with a moving object. The first row displays the image of a light spot hitting a surface. The second and third row show the images of a model locomotive.

According to those conditions, Figure A.6a shows a sharp image without the artefacts discussed above. If the object stays at rest and $t_{set} < t_{illum} < t_{read}$, ghost images appear as vertical stripes (Figure A.6b, Figure A.6e and Figure A.6h; note that these artefacts have nothing to do with the "smear" discussed in Section 4.7.5). The limited time of artificial charge collection, i. e., $t_{illum} - t_{set}$, is indicated by the white arrows. Figure A.6c and Figure A.6f illustrate the situation of a moving object. This is very similar to the

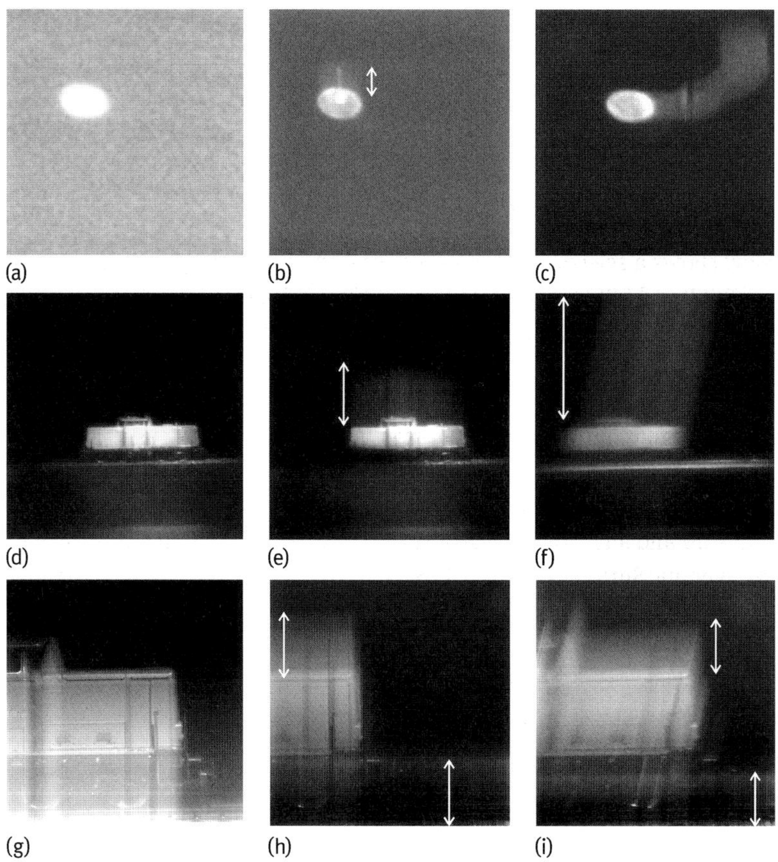

Fig. A.6: Image captured in the presence of illumination during readout. The position of the camera is unchanged in all cases. (a) to (c) show the image of a light spot and (d) to (f) a model locomotive from the side. (g) to (i) show images with details observed from closer distance. For further discussion, see the text.

situation shown in Figure A.5 and the related discussion. First, the object and image, respectively, move, which leads to slanted stripes and ghosts and then after the stop of light to spot movement. This continues in the vertical direction.

In Figure A.6c, the light movement within t_{set} even leads to a horizontal in-motion unsharpness first. The lower row in Figure A.6 shows a series comparable to the row in the middle. Here, the images are taken from a shorter distance, but the main difference is the setting of exposure time, which is a factor of five shorter, and thus now $t_{shift} \sim t_{set}$. Consequently, the number of charges accumulated during both times is not very different. Thus, the signal of the ghost has approximately the same strength as that of the original image. Hence, for the current situation the ghost image gives the impression of a 3D view with a look onto the engine hood as well (Figure A.6h and Figure A.6i). But this impression is wrong. The second row, and also Figure A.6g, clearly show that the perspective is restricted to a side view only. Altogether, for such a situation, the image is very strongly affected by ghosts.

Although readout is different when compared to CCD, also CMOS sensors operated without shutter show somehow similar effects (see also the rolling shutter effect discussed in the main part of the book).

A.5 Camera and sensor data

Table A.3 below shows a selection of typical camera and sensor chip data. They range from high-end DSLR and DSLM (summarized as DSL) with full format (FF) or other sizes, a view finder camera (Leica M11, see no. 7; this has currently the best image sensor of a full format camera according to DXOmark), to 3 consumer cameras, namely a bridge and 2 compact cameras. The list continues with several medium format (MF) cameras and a larger collection of cameras that are used for scientific, technical or industrial purposes (sci./tech.cameras). No. 15 is a large-scale scientific X-ray camera, which has rather large pixels with a huge FWC. No. 21 is an interline CCD. No. 27 is a giant sensor made of tiles of 189 scientific CCD and altogether 3.2 Giga-pixels. This sensor is set-up for the camera of 8.4 m Large Synoptic Survey Telescope (LSST).

Some of the listed cameras provide just examples of particular chips, which are used for other purposes as well (e. g., CIS no. 5 is used for industrial cameras as well). Most CMOS chips have a rolling shutter (no. 25 is just one example) but such ones with global shutter are available as well (see, e. g., no. 26). The list shows that today most consumer and industrial cameras are equipped with CIS. A lot of CMOS sensors are BSI, but there are still lot of FSI CIS. However, there is still a market for CCD, in particular, for very advanced or special applications. Examples for EM-CCD and iCCD are omitted here because some parameters depend on operation mode and gain. For related examples, see Section 4.11. A set of other examples is given by Table 5.4.

The data in the table include the internal quantum efficiency IQE (mostly for λ = 525 nm), the full well capacity FWC, read noise and corresponding dynamic range DR.

Tab. A.3: Typical sensor data (opto-electronic properties) of photo cameras.

no.	camera type (or chip)	chip type	bits (ADC)	IQE [%]	FWC [el.]	σ_{read} [el.]	DR	DR in EV	pixel pitch [µm]	no. pix. horiz.	no. pix. vert.	MP	chip size horiz. [mm]	chip size vert. [mm]	chip size diag. [mm]	asp.
1	DSL	LiveMOS	12	31	34000	7.2	4700	12	3.30	5184	3888	20.2	17.3	13	21.6	1.33
2	DSL, FF	CMOS	14	50	33000	16.5	2000	11	4.10	8688	5792	50.3	36	24	43.3	1.50
3	DSL, FF	CMOS	14	51	48800	3.6	13600	14	4.80	7424	4924	36.6	36	24	43.3	1.50
4	DSL, FF	CMOS	14	35	65000	27	2400	11	6.40	5616	3744	21.0	36	24	43.3	1.50
5	DSL, FF	CMOS	16	59	71586	1.1	65000	16		9568	6380	61.0	36	24	4333	
6	DSL, APS-C	CMOS	14	58	48000	3.8	12300	14	4.20	5568	3712	20.7	23.5	15.7	28.3	1.50
7	view finder, FF	CMOS	14	58	46060	0.9		15	3.76	9528	6328	60.3	36	24	43.3	1.50
8	bridge	MOS	12	65	14300	4	3600	12	2.40	5472	3648	20.0	13.2	8.8	15.9	1.50
9	compact	CMOS	12	72	15500	3.6	4300	12	2.40	5472	3648	20.0	13.2	8.8	15.9	1.50
10	compact	CCD	12	60	9700	4.7	2100	11	2.00	3744	2784	10.4	7.6	5.7	9.5	1.33
11	DSL, MF	CCD	14	34	32500	5.6	5800	13	6.00	7500	5000	37.5	45	30	54.1	1.50
12	DSL, MF	CMOS	14						5.30	8256	6192	51.4	43.80	32.90	55.0	1.33
13	DSL, MF	CMOS	16						3.76	11656	8742	101.9	48.64	40.38	55.0	1.33
14	scientif. cam.	CCD	16	85	100000	3	33000	15	13.50	2048	2048	4.2	27.6	27.6	39.0	1.00
15	scientif. cam.	CMOS			2.8 mio	<200	14000	14	48.00	512	1024	0.5	49.2	24.6	55.0	2.00
16	scientif. cam.	sCMOS		82	37500	1.3	37500	15	6.50	2048	2048	4.2	13.3	13.3	18.8	1.00
17	scientif. cam.	CCD		95	1000000	3.25	31000	15	20.00	1340	1300	1.7	26.8	26	37.3	1.03
IS	scientif. cam.	sCMOS	16	>95	80000	1.3	61500	16	11.0	2048	2048	4.2	22.53	22.53	31.9	1.00
19	chip	CCD		22	40300	12.5	3200	12	6.00	8176	6132	50.1	49.1	36.8	61.4	1.33
20	chip	CMOS		50	12000	14	860	9.7	4.50	4096	4096	16.8	18.4	18.4	26.0	1.00
21	sci./tech. cam.	CCD		50	20000	7	2900	1.5	4.65	1392	1040	1.4	12.7	12.7	18.0	1.00
22	sci./tech. cam.	CMOS		45	8100	17.2	470	8.9	5.50	2048	2048	4.2	11.3	11.3	16.0	1.00
23	chip	CMOS		53	10100	5.3	1900	11	2.50	2080	1552	3.2	5.2	3.9	6.5	1.33
24	chip	CCD			90000	4	22500	15	10.00	9216	9232	85.1	92.2	92.4	130.5	1.00
25	chip	CMOS	16						3.76	14192	10640	151.0	60.30	40.90	66.7	1.47
26	chip	CMOS	14						3.45	13400	9528	127.0			56.7	1.41
27	tiled	CCD			135000	5	27000	15	10.00			3200.0			640.0	

Those data are mostly provided for typical standard usage, e. g., ISO 100, and for slow read-out in case of scientific cameras. The number of bits in the column “ADC” is not equal to the dynamic range, this is just the ADC specification of the image processor. The dynamic range is given in column “DR” (see also Chapter 4 and Chapter 8). The data are obtained from product data sheets from the manufacturers and from internet sources such as photonstophotos.net by B. Claff.

The selection of cameras is quite arbitrary and should just reflect typical or some special examples. For instance, LiveMOS sensors are an outcome from a development by Panasonic. The intention was to include advantages of both CCD and CMOS in a new chip design.

Table A.3 illustrates that sensor data may be much different. For instance, even high-end DSLR and DSLM may differ strongly in their dynamic range, namely from slightly more than 11 EV to nearly 14 EV. Chips and cameras used for scientific or technical or industrial applications may have even larger dynamic ranges, some of them exceeding 15 EV.

Table A.4 presents selected examples of image sensors for mobile phones. This table should give an impression of camera sensors for various SPC models and camera modules and also an impression of SP-CIS evolution. For further data of CIS for smartphone cameras, we refer to Table 7.7. We may remark that this table has been made without any judgement on quality, etc. All listed camera modules are rear cameras only (note that some SPC have more than just one rear camera module). For explanation of Isocell, Teracell, Nonacell and PDAF, etc., see the discussion in the text. In addition, we would like to note that in 2022, Samsung and Omnivision have presented a 200 MP sensor with 0.56 μm pixels. But here we would like to comment that due to nature’s laws, with the present technology, in our opinion, such small pixels do not make sense in terms of camera performance but may be due to marketing reasons (see discussions in Chapter 7). If this will change, due progress in physics (see, e. g. Section 7.5) will be seen in the future. Currently, the largest sensors of smartphone cameras are the 1 inch CIS of Leitz Phone 1/Aquos R6 and their successors from Leica and Sharp (a Sony Xperia-series SPC has a similar sensor but makes use of only part of it). Such large sensors for consumer photography previously have been only used in compact cameras.

We would like to note that tables such as that shown in this Appendix may be rather soon out of date with respect to the presented cameras and chips. However, our intention is to present and discuss relations, and those do not depend on the actuality of camera or sensor models. Nevertheless, we may note that data of current cameras and sensors can be found online (see, e. g., the recommendable site[1] and also data sheets from manufacturers and vendors).

1 Photons to Photos by W. J. Claff, http://www.photonstophotos.net/

Tab. A.4: Selected examples of SPC CIS. Sources: Product data sheets from the companies and Wikipedia.

image sensor	smartphone (examples)	MP	diagonal [mm]	pixel pich	remarks
Sony IMX520	Huawei P30	8	5.0	1.22 µm	telephoto
Samsung S5K2P2 ISOCELL RGB	Samsung Galaxy S5	12	6.8	1.12 µm	
Sony IMX363	Asus ZenFone, Google Pixel 4, LG V40, Nokia X7	12	7.0	1.4 µm	
Samsung S5K2L1 ISOCELL RGB	Samsung Galaxy S7	12	7.1	1.4 µm	
Samsung S5K2L4 ISOCELL Plus RGB	Samsung Galaxy S10	12	7.1	1.4 µm	
Samsung S5K2LD	Samsung Galaxy S20 Ultra	12	9.1	1.8 µm	
Sony IMX555 RGB (dual pixel)	Sony Xperia 1 II	12	9.7	1.8 µm	including PDAF
Sony IMX503 RGB (dual pixel)	iPhone 11, 11 Pro, 11 Pro Max	12	14.0	1.4 µm	including PDAF
Sony IMX214	HTC One M8 Eye	13,5	5.9	1.1 µm	
OmniVision OV16885-4C 4-Cell	various	16	5.9	1.0 µm	
Sony IMX351 RGB	Huawei P30	16	5.8	1.1 µm	ultra-wide including PDAF
Sony IMX576 RGB	Huawei Nova 5i, Samsung Galaxy A50 and A7 and other SPC	25	6.5	0.9 µm	including PDAF
Samsung S5KGD1 Plus Tetracell	Huawei P30	32	6.4	0.8 µm	main
Sony IMX600 (RGB) and IMX600y (RYYB), Quad-Bayer	several Huawei SPC such as P20 and P30	40	9.2	1.01 µm	main (wide) including PDAF
Samsung S5KGH1 ISOCELL Plus Tetracell	Samsung Galaxy S20 Ultra	44	9.6	0.7 µm	front camera
Sony IMX586 Quad-Bayer 2x1 on-chip-lens (OCL)	a lot of SPC including such as OnePlus7 and Samsung Galaxy S20 Ultra	48	8.0	0.8 µm	including PDAF (IMX686: 64MP)
OmniVision OV48C	Xiaomi Mi10 ultra	48	12.0	1.2 µm	main; on-chip DCG HDR
Samsung S5KGW2 ISOCELL Plus RGB	Samsung Galaxy S20	64	9.2	0.8 µm	
Samsung S5KGW1 ISOCELL Plus Tetracell	Realme X50	64	9.2	0.8 µm	
Samsung Isocell Bright S5KHM1 Nonacell	Samsung Galaxy S20 Ultra	108	12.0	0.8 µm	
Samsung S5KHMX Isocell Bright Tetracell	Xiaomi Mi 10	108	12.0	0.8 µm	
Samsung S5KHP1 ISOCELL HP1		200	13.1	0.64 µm	
Samsung HP3, Omnivision		200	11.4	0.56 µm	
Sharp?	Leitz Phone 1, (Leica), Aquos R6 (Sharp)	20	"1 inch"		largest SPC sensor

A.6 Histograms

The x-axis of a histogram of an image shows the possible brightness values of a pixel on a scale from 0 to its maximum possible value. Thus, for physical quantities, or such ones that reflect the sensor signal without severe manipulation, like raw data images, this is B_{pix} (in counts) and for images with a tone-mapped brightness, this is B'_{pix}. Both B_{pix} and B'_{pix} are integer values. For simplicity, in the following, we will restrict our discussion to B_{pix}. The y-axis displays the number of pixels ΔN that show the corresponding brightness within this "channel" or "bin" of width ΔB_{pix}. Thus, these are integer values as well. Consequently, the y-values are given by $\Delta N/\Delta B_{pix}$ as a function of B_{pix}: the number of events with 0 counts, with 1 count, with 2 counts, etc.

The number of steps or "bins" depends on the chosen value of ΔB_{pix}, which is given by an integer number, namely 1, 2 and so on. For example, for an 8-bit scale and $\Delta B_{pix} = 1$, we have 256 bins, namely 0 to 255 channels. However, often it is better to group several bins. For instance, instead of increasing the pixel brightness always by one (i. e., $\Delta B_{pix} = 1$), a step or bin size of $\Delta B_{pix} = 4$ may be chosen, and thus of course within that bin much more pixels are counted. Within the following example, the number of pixels between 0 and 3 is collected by the first bin, those between 4 and 7 in the second one and so on. This corresponds to an averaging process, which may be reasonable sometimes, especially when pixels with specific values of B_{pix} are missing.

As an example, the following raw data image with linear conversion, namely without tone mapping, is shown together with the histogram (Figure A.7).

As has been discussed in Chapter 5, the response of the eye does not follow a linear scale of B_{pix} but is more adapted to a logarithmical scale, i. e., to $\mathrm{ld}(B_{pix})$. Thus, naively one could just take the above histogram and then plot the x-axis on a logarithmic scale instead of a linear one. However, this would lead to a totally wrong result. The reason for this is that in the above histogram the y-axis is given by $\Delta N/\Delta B_{pix}$ with the consequence that $\Delta B_{pix} = \text{constant}$ (e. g., always $\Delta B_{pix} = 4$ counts) all over the x-axis. For a logarithmic scale, this is no longer true, and thus the y-axis has to be rescaled as well, independently of the lin or log scaling of the y-axis. The rescaling is simply done by applying the derivative together with the chain rule:

$$\frac{\Delta N}{\Delta\,\mathrm{ld}(B_{pix})} = \frac{\Delta N}{\Delta B_{pix}} \cdot \frac{\mathrm{d}B_{pix}}{\mathrm{d}\,\mathrm{ld}(B_{pix})} = \frac{\Delta N}{\Delta B_{pix}} \cdot \left(\frac{\mathrm{d}\,\mathrm{ld}(B_{pix})}{\mathrm{d}B_{pix}}\right)^{-1} = \frac{\Delta N}{\Delta B_{pix}} \cdot B_{pix} \cdot \ln(2).$$

Now, for this log axis, the bin width is constant all over the x-axis as well, e. g., always $\mathrm{ld}(\Delta B_{pix}) = 1/10$ EV.

This rescaling is similar to a typical example from spectroscopy. The X-ray spectrum of an X-ray tube may be displayed (I) as the power per unit interval $\Delta\nu$, which is a function of frequency ν or (II) as the power per unit interval $\Delta\lambda$, which is a function of wavelength λ. According to $c = \lambda \cdot \nu$, $\lambda = c/\nu$, and consequently, $\mathrm{d}\nu/\mathrm{d}\lambda = |c/\lambda| = c/\lambda^2$. The absolute values are only used because power is always positive. The negative sign

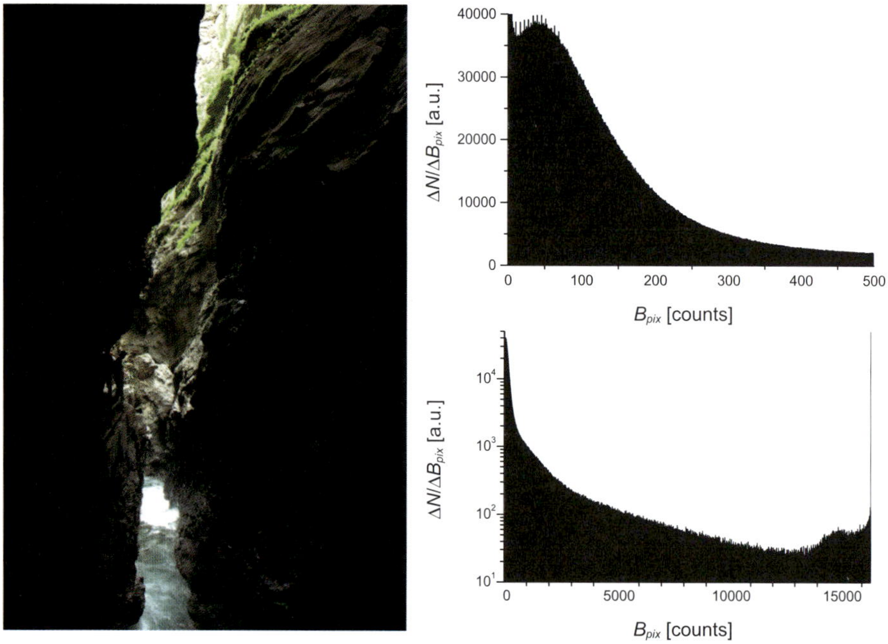

Fig. A.7: Example of an image and its histograms (see text). Histogram on top: linearly scaled y-axis, histogram; on bottom, same histogram, but for better visibility the y-axis is scaled logarithmically, and the x-axis is extended to a larger value.

from the derivative only indicates that an increase of ν corresponds to the decrease of λ. Thus, the units of the y-axis are simply rescaled to

$$\frac{\Delta P_x}{\Delta\lambda} = \frac{\Delta P_x}{\Delta\nu}\cdot\frac{\mathrm{d}\nu}{\mathrm{d}\lambda} = \frac{\Delta P_x}{\Delta\nu}\cdot\frac{c}{\lambda^2}.$$

Thus, from the spectrum Figure A.8a, Figure A.8b can be easily calculated. Although they look totally different, both spectra are absolutely identical.

Note that in any of the above cases, rescaling is important for the values of the y-axis, but if those are then further displayed on a linear or a logarithmic scale is not of importance.

Coming back to the rescaling of the histogram, appropriate rescaling leads to the following histograms, with one difference, namely that here the four channels within the Bayer filter mosaic are displayed separately (four channels within one 4×4 "macromatrix element"; Figure A.9).

Again, these are the histograms for the raw image shown above, but now on a log scale. This camera has a dynamic range of 11 bits (ranging from +3 EV down to −8 EV)

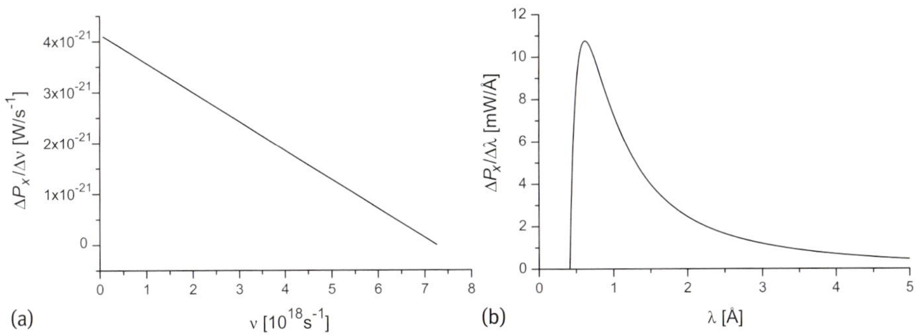

Fig. A.8: Example of a spectrum that illustrates proper recalculation of the ordinate (see the text).

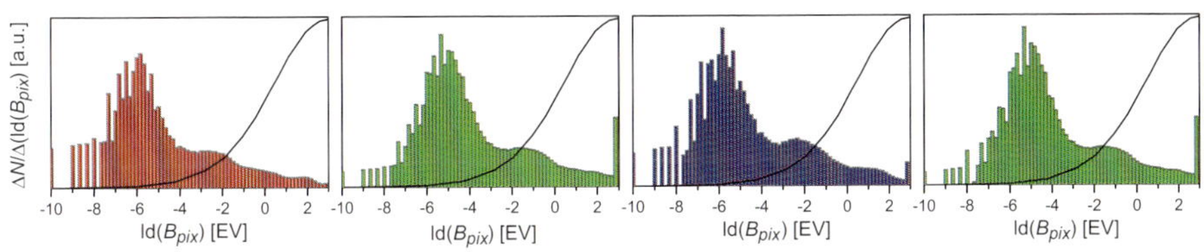

Fig. A.9: Example of histograms (see the text).

and an additional offset. The added curve in the histogram is a standard tone curve for that camera. Compared to our previously used calibrations of $EV = 0$ at the 18 % grey value or at the saturation point, here the zero point is somewhat arbitrary, but this is not of much relevance.

Besides the discussion of histogram rescaling, the present example also shows that this image extends over an exposure value that exceeds the dynamic range of the camera. This is seen well from the above histograms, which clearly show that B_{pix} is clipped at +3 EV and that below −8 EV noise overwhelms the signal.

Finally, we would like to note that sometimes there are differences between histograms that arise from different bin sizes, different offsets within the raw data or different scalings, when floating numbers are used, e. g., for the rescaling of 14-bit image data that are stored and then reloaded in/from 16-bit TIF files.

Poor image quality can also be seen directly from a histogram. As an example, Figure A.10 shows an image of good quality and the same one post-processed in a way that leads to posterization and banding. The image in (a) and the corresponding histograms show a smooth distribution with small grades between neighbored bins. The image in (b) shows banding effects. This is also seen in the histogram where a lot of bins are empty, i. e., some tonal grades are missing.

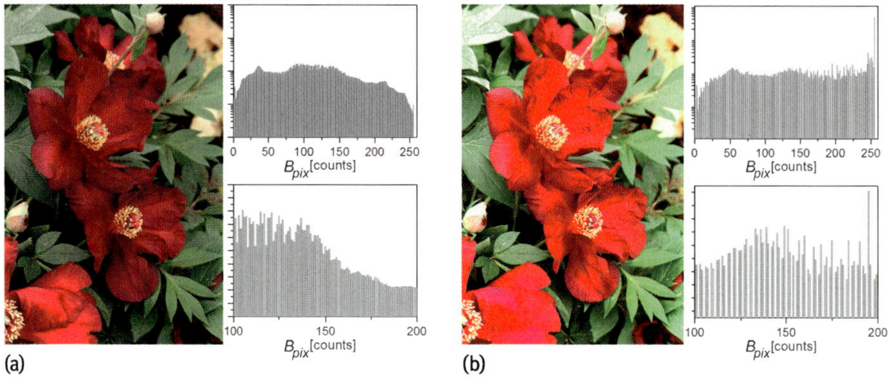

Fig. A.10: Example of banding (see the text). The ordinate of the histograms displays the number of pixels in the red color channel within the corresponding bin. The histograms are related to the image on their left-hand side. The upper histograms show bins between 0 and 255 counts with a log ordinate, the lower ones show a crop of the upper ones on a lin ordinate.

A.7 Tone mapping and tone curve discussion

To illustrate the effect of tone mapping for well-defined conditions, in the following different tone curves are discussed and also applied to simple artificial images. To do so, we make use of simulated curves. However, they are based on data of a real professional DSLR with 11-bit dynamic range and a realistic tone curve. The presented photon conversion and tone curves, respectively, are examples. Other tone curves may be chosen instead. We would like to emphasize that the present goal is just the visualization of the effects for better understanding, but not an optimization process.

Here, and in the following, $EV = 0$ is positioned at the saturation point and not at 18 % of the maximum (see photon conversion curve). B'_{pix} shows the rescaled signal ($B_{\text{pix}} \propto N_e$) on the usual 8-bit scale. For better visibility, here the tone curves are displayed as floating-point values.

I) Standard tone curve for this camera and tone curves for low light and a strong light, respectively

Diagram Figure A.11a (compare to Figure 4.62) displays both the photon conversion curve (dashed line) and the tone curve for standard conditions (solid red line). In addition, the same tone curve but shifted by −2 EV (green), which leads to an image that is +2 EV more bright is displayed as well and also a further one shifted by +2 EV (blue), which leads to an image that is less bright by −2 EV.

The standard tone curve covers the full range of the camera. At the lower end, it is limited by the noise, which is set at the end of the first channel (this extends from 0 to 1). Thus, the dynamic range of the camera of $2^0/2^{-11}$ is rescaled to 255/1.

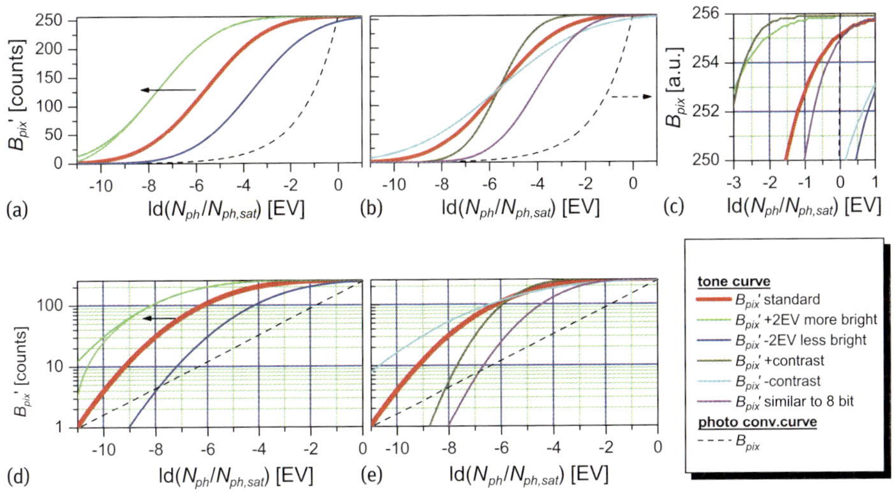

Fig. A.11: Examples of tone curves (B'_{pix}, ordinate on left-hand side; solid lines) and photon conversion curves (B_{pix}, ordinate on right-hand side; dashed lines). (a) shifted tone curves, (b) tone curves with different contrast and (c) details for the last 6 channels. (d) same as (a) but on a double log scale. (e) same as (b) but on a double log scale. $+x$ EV means that the image gets brighter by x aperture values (which corresponds to a shift of the tone curve to $-x$ EV).

This may be much better seen when the same diagram is plotted on a double logarithmic scale (Figure A.11d). Noise is reached at the lower end of the dynamic range of the camera where the SNR becomes one. Of course, it may be more reasonable to classify the reasonable range by SNR = 3, but this is not an issue of the present discussion. All exposure values up to the noise level converted via ADC may be *assigned*, e. g., to the first of channel, i. e., “1 count.”

At the upper end of the dynamic range, here at $EV = 0$, the tone curve may just reach the last channel, which extends from 255 to 256 so that values $> EV = 0$ are stored as saturated values and *assigned* to the last channel (Figure A.11c). By choosing such a tone curve, the full dynamic range of the camera is used and converted to the full range of the available 8-bit brightness scale of the image pixels.

If the image is too dark, one may apply the tone curve shown in green. Then signals from low exposures are *assigned* to upper channels.

For clearer understanding, we would like to discuss the curves briefly (see also Figure A.12). The color of the text should refer to the corresponding curve in Figure A.11. “Channel” corresponds to the number of counts assigned by the ADC.

The lowest value (the limit of the ordinate is one count). All signals yielding one count or less are put into the first channel, in particular all values below –11 EV (abszissa). Channels 2 to 4 (ordinate) contain data from –11 to –10 EV (abszissa).

Channels 1 to 11 can be considered as empty. They are not really empty because they are filled by signals with one to eleven counts resulting from exposures with less than −11 EV. However, for the present sensor all that is just noise so that for that the very first channel would have been sufficient to collect all those signals. Channels 12 to 30 contain the range −11 EV to −10 EV. Due to the rather large number of channels in that range of 1 EV width, this results in well-resolved low light regions. However, for high-light regions, resolution is poor (see Figure A.11c). Furthermore, saturation occurs not at 0 EV, but before.

In analogy, if a lot of light is available, signals occurring in high-light regions are reduced. Because in that region the blue curve is steeper than the red one, more channels are available, which results in better high-light resolution. But note that even at 0 EV only channels up to number 244 are occupied; higher ones stay empty. However, in the low-light region clipping now occurs at −9 EV, i. e., all signals resulting from illumination below −9 EV are put into channel 1 (note: in that region the red curve has 10 channels).

Of course, usually good image processing avoids empty channels because that reduces the dynamic range. In particular, for the rather limited 8-bit range this is an important issue. Thus, the green curve may be modified in the low light region (see, e. g., both green lines) or further rescaling by shifts, stretching and/or bending of the tone curve is applied within the 8-bit range, i. e., $B'_{\mathrm{pix}} \rightarrow B''_{\mathrm{pix}}$. This will not be discussed here because the present goal is not optimization, just explanation.

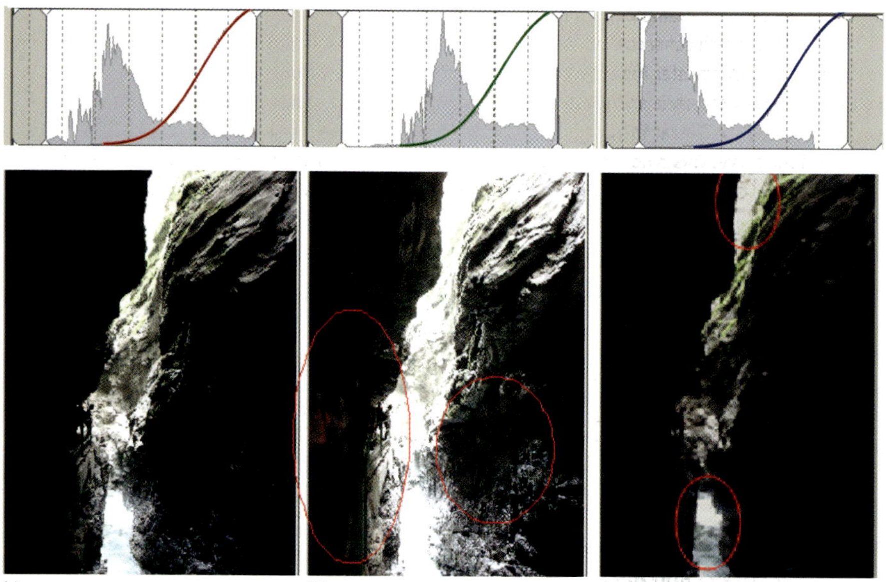

(a) (b) (c)

Fig. A.12: Examples of real pictures according to similar curves of same color as displayed in Figure A.11.

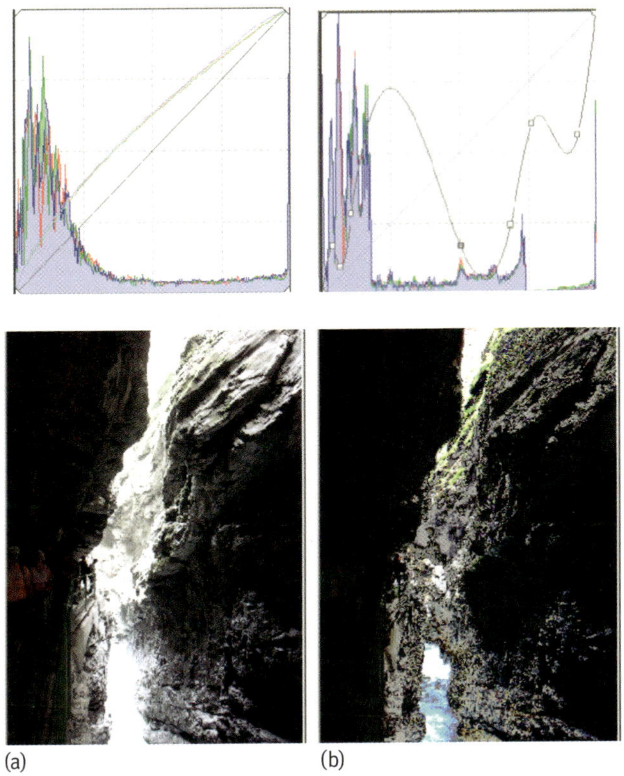

Fig. A.13: Real pictures examples of further tone mapping on the 8-bit level, i. e., an additional tone curve is applied to B'_{pix}, which consecutively leads to B''_{pix}. The original raw data image is shown in Figure A.7. (a) On top an *additional* transfer function for the tone curve given by a picture preset (i. e., $B'_{\mathrm{pix}} \rightarrow B''_{\mathrm{pix}}$). Here, in particular, pixel intensity is enhanced in low light regions, but in a nonlinear way. The result is shown in the picture below. (b) The *additional* transfer function here is somehow arbitrary and leads to a strongly deformed tone curve. This is not to improve image quality, but just to show that strange tone curves may result in strange images. Underneath the tone curves, the RGB histograms are displayed.

The discussed effects can be easily seen in a real photography. Here, we show again the picture shown in Figure A.7 but here first with the "standard curve" (red curve as before) and the two other ones. It is apparent that now more details are seen in the shadow region (b) or high-light region (c) as can be well seen in the marked regions of the pictures. Two other examples are shown in Figure A.13.

II) Comparison of the standard tone curve for the same 11-bit camera with different contrast curves and also with the standard tone curve of an 8-bit camera

Similar to before, we make use of colored text for easier discussion. The colors again refer to Figure A.11b, A.11c and A.11e.

The 11-bit curve covers the range of $DR = 2^{11}$ and the 8-bit curve that of $DR = 2^8$. Consequently, for >–8 EV, the 8-bit curve is steeper than the 11-bit curve, which leads to better resolution in that range. This is obvious because all 256 channels are used for the smaller range when compared to the 11-bit curve. Here, this makes use of 226 channels only. In contrast to that, the 256 channels of the 11-bit-curve cover the region below –8 EV as well. Thus, in the region below –8 EV there are 30 channels and no channels, respectively. The advantage is, of course, that there is resolution in that region as well, which is fully absent.

But an 11-bit curve is much more flexible. One could increase its contrast, which leads to the 11-bit-curve although the disadvantage then is clipping in the high-light region (see also Figure A.14). But it would be possible also to shift this curve, etc.

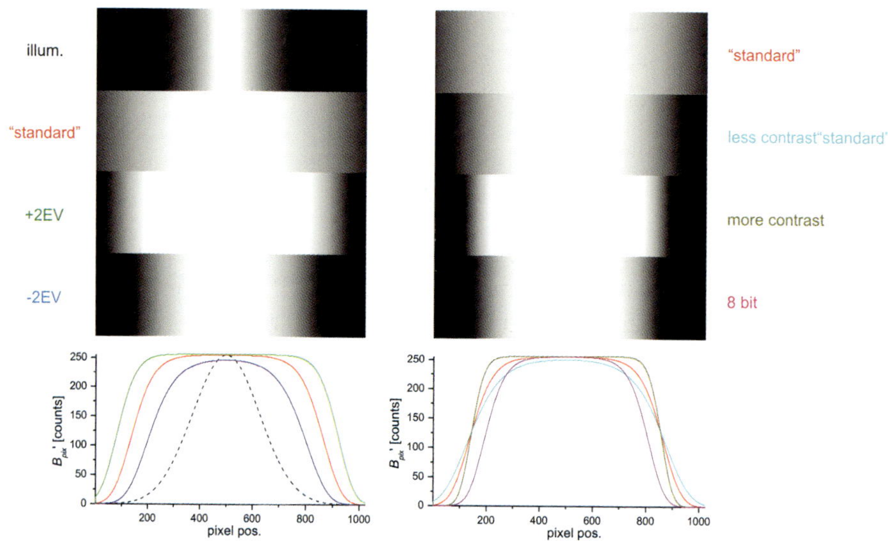

Fig. A.14: Calculated images of a Gaussian illumination. The images are generated with the example tone curves of Figure A.11. The use of colors is the same and explained in the legend in Figure A.11. The corresponding profiles measured along the horizontal lines are shown below. However, one has to be aware that this figure presents the image data. This may be seen differently when the data are displayed on a screen.

III) Artificial images of a well-known brightness distribution of the object

Figure A.14 shows another example. This is to illustrate that imaging within photography is far off from regarding it as a real scientific measurement of object light distributions. But, of course, the particular goals are both fulfilled; good perceived image in the first case, measurement of the real light distribution in the latter one.

For this example, we have chosen a Gaussian as the input signal (left column: image marked by "illum." and dashed curve in the profile along the horizontal line below).

A measurement with a linear detector without consecutive tone mapping would reproduce this quite well. However, the standard curve and the shifted curves with +2 EV and −2 EV, respectively, do obviously lead to significant differences when compared to the real input signal (see accordingly, images and line profiles in the left column). The difference is even more pronounced, when the contrast is enhanced, “the images become harder” (see accordingly, images and curves in the right column). The images and lineouts “standard” are the same in both columns.

As has been discussed in Chapter 4, the tone mapping process is unavoidable, even if raw data are processed by a raw converter. In that sense, raw data are never raw, and thus images processed by a raw converter can never be regarded as a measurement. Also, JPG images recorded by a camera are always far off any measurement. On the other hand, raw data are somehow raw when special programs such as *DCRaw* or *RawDigger* are used to extract the linear photon conversion curve. Therefore, the camera may be considered a measurement device, at least if one takes care of its linearity, which may not be trivial for CMOS based cameras. We may note that linearization of the CMOS is usually done by the image processor of the camera prior to data storage.

A.8 Summary of Fourier optics relations

Figure A.15 provides an illustration and a summary of the basic Fourier optics relations. The object is illuminated either by coherent light, described by $\mathcal{E}_{\mathrm{in}}$, or by incoherent light, described by I_{in}. Illumination may be in reflective geometry, but here for illustration we make use of transmission geometry, similar to a slide that is homogeneously illuminated. At best, $\mathcal{E}_{\mathrm{in}}$ and I_{in} both are constant. Here, in contrast to Chapter 5, we denote the variables in the object plane and image plane differently (namely $(x',y') \equiv (x_{\mathrm{o}},y_{\mathrm{o}})$ and $(x,y) \equiv (x_{\mathrm{i}},y_{\mathrm{i}})$). For a detailed discussion, see Chapter 5.

Figure A.16 shows the relations between PSF and OTF, etc. In case of a point source, in the Fourier plane there is no difference between coherent and incoherent light because a point source is per se spatially coherent (see Section 5.1.4) and we obtain a flat spatial frequency spectrum. Note that for our purpose we may regard the square of the delta function to be again a delta function. In the image, plane intensities are always observed. If, for instance, the system is dominated by a circular aperture only, OTF is given by the circle function, and thus the field and intensity by Equation (5.20b) and Equation (5.21b), respectively (diffraction of the field at a circular aperture). As a result, OTF is given by Equation (5.39) and Equation (5.40).

In the case of a source that is an infinite thin line in horizontal direction located at $y = 0$, $B_{\mathrm{obj}}(x,y) = \delta(y) \cdot 1$, and thus $\widetilde{B}_{\mathrm{obj}}(k_x,k_y) = 1 \cdot \delta(k_x)$. We do include “1” to illustrate

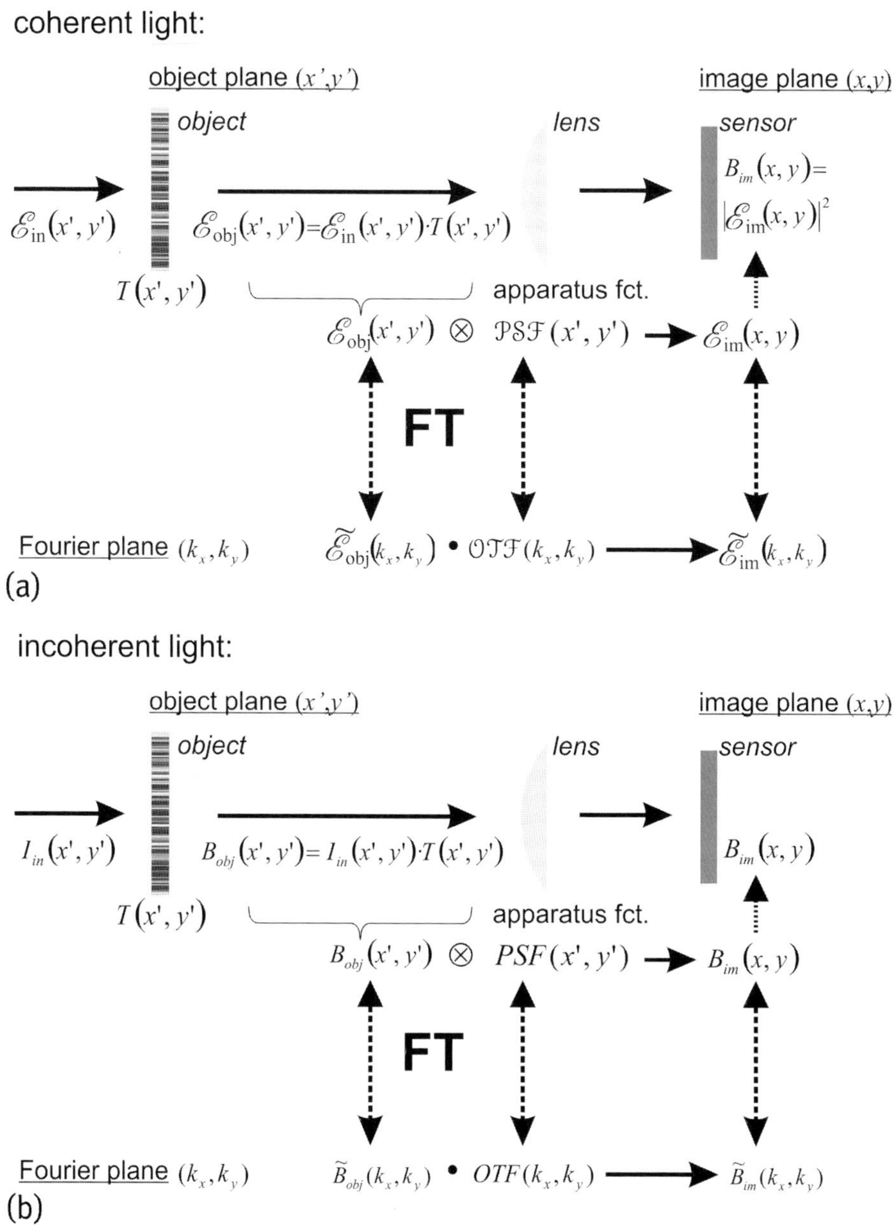

Fig. A.15: Fourier optics relations. "⊗" is the symbol used for convolution and a dot that for a simple multiplication. Fourier transformation is indicated by the broken arrows.

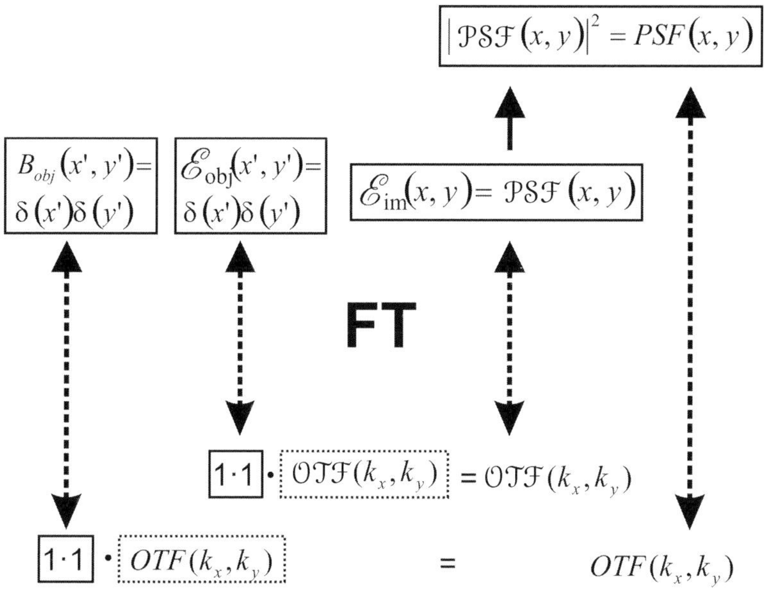

Fig. A.16: Relations for the pulse response. The dot is used for a simple multiplication. Fourier transformation is indicated by the broken arrows.

the corresponding Fourier pairs. The Fourier transformation is given by

$$\int_{-\infty}^{\infty}\int_{-\infty}^{\infty} dk_x\, dk_y \cdot \delta(k_x)\, \mathrm{MTF}(k_x, k_y) \exp(-ik_x \cdot x) \exp(-ik_y y)$$
$$= \int_{-\infty}^{\infty} dk_y \cdot \mathrm{MTF}(0, k_y) \exp(-ik_y y) = \mathrm{LSF}(y) \tag{A.3}$$

which defines the line spread function LSF. Here, for simplicity, we do restrict to the MTF. The LSF for the present case is

$$\mathrm{LSF}(y) = \int_{-\infty}^{\infty} \mathrm{PSF}(x, y) dx \tag{A.4}$$

Vice versa, a Fourier transformation of the LSF yields the MTF

$$\mathrm{FT}[\mathrm{LSF}(y)] = \int_{-\infty}^{\infty} dy \cdot e^{-ik_y y} \int_{-\infty}^{\infty} \mathrm{PSF}(x, y) dx = \int_{-\infty}^{\infty}\int_{-\infty}^{\infty} dx\, dy \cdot e^{-i0\cdot x} e^{-ik_y y}\, \mathrm{PSF}(x, y)$$
$$= \mathrm{FT}[\mathrm{PSF}(x, y)]_{k_x=0} = \mathrm{MTF}(0, k_y) \tag{A.5}$$

Thus, it is seen that the Fourier transformation of the LSF is identical to the profile along the horizontal line of the MTF through the center. This has been shown for LSF(y); for LSF(x) this is analogous. Further discussion is made in Section 8.3.2.

A.8.1 Remarks

Finally, there are two remarks. First, due to the consideration of the different planes, with the different coordinates $(x', y') \equiv (x_o, y_o)$ and $(x, y) \equiv (x_i, y_i)$ for the object plane and the image plane, respectively, the magnification of the optical system M is already included. For incoherent light and the condition that there would be no negative effect of the imaging system at all, i. e., without any limiting apertures and aberrations, $T(x', y')$ would be exactly reproduced by $T(x, y)$ with the only exceptions that the scale and its sign are changed by the magnification process. The negative sign of M indicates that the image is top-side down. In a realistic system, the "negative" effects introduced by the optical system and all relevant other components of the camera system are taken into account by the apparatus function, which is identical to the PSF. This apparatus function is convoluted with $T(x, y)$ (see Figure A.15b; note that the magnification may be regarded to be performed prior to the convolution process). Of course, for this reason, one also has to take care that the PSF and the related MTF have to be taken for exactly those conditions at which the image is captured. For the used camera that means that the PSF, respectively, MTF has to be taken for the selected lens (the sensor system is already fixed by the used camera), potentially the used (hardware) zoom, the used object distance, the used f-number, etc. because all that may have influence. For coherent light, this is considered analogous.

The second remark is related to the limiting case where a huge demagnification converts to focusing. An example is the "image" of a star. The star may have a huge size, but it is also located in a very far distance $|s_o|$. For this reason, the star can be considered as a virtual point object with $|S_o| \to 0$, and thus its "image" is given by the PSF. Hence, in the image plane there is no structure information of the star itself. As $|s_o| \to \infty$, $|s_i| \to f$.

On the other hand, all rays from such a distant object can be regarded as parallel, and consequently, the same situation can be described as focusing as discussed in Section 1.5.1. Hence, both considerations of the given situation with the star lead to the same result with a common light distribution provided by the PSF. The PSF itself is the result of the transfer function of the optical system, namely the MTF. Consequently, also the focal spot size δ_B and the width δ_0 of the PSF are identical (distance between the first zero positions in both cases). Therefore, Equation (1.17), Equation (5.43) and Equation (5.24) are equivalent (one has also to note the relation between $f_\#$ and NA (Equation (5.44))). In those equations, α may be regarded as a measure of the wavefront distortions introduced by the optical system (without aberrations $\alpha = 1$; note that in case of large aberrations a simple description by δ_B and δ_0 may not be sufficient).

A.8.2 Remark on focusing

The third remark is also related to focusing. One may ask what the focal spot size of a Gaussian laser beam with circular symmetry and a FWHM diameter D_{laser} may be when it is focused with a microscope objective of a given numerical aperture. In case of an ideal lens instead, the focal spot is given by Equation (5.43) where here δ_0 describes the FWHM or the $1/e^2$-width according to 2κ from Table 5.1. However, when a Gaussian beam enters the microscope objective, it is cut by its aperture D (see Equation (5.44)). But calculating the focal spot distribution is straightforward. The electric field $\mathscr{E}_{\mathrm{laser}}(r')$ corresponding to the intensity distribution of the Gaussian beam $I_{\mathrm{laser}}(r')$ (here r' is the radial coordinate in the object plane) is clipped by the aperture of the microscope lens, which can be described by $\mathscr{E}_{\mathrm{obj}}(r') = \mathscr{E}_{\mathrm{laser}}(r') \cdot \mathrm{rect}(r'/D)$ (for $\mathrm{rect}(r'/D)$ see Appendix A.1; note that the laser beam is coherent, which means that one has to apply the formalism displayed in Figure A.15a). Fourier transformation then yields the focal spot distribution $\widetilde{\mathscr{E}}(k_{\mathrm{r}})$ (compare Figure 5.1). The intensity distribution in the focal plane is obtained as $I(r) \propto |\widetilde{\mathscr{E}}(r)|^2$ where we make use of $k_r/k = \sin\theta_r$ (see Figure 5.1; here, r is the radial coordinate in the focal plane). With the θ_{r} as the corresponding angle and $\sin\theta_{\mathrm{r}} \approx \tan\theta_{\mathrm{r}} = r/f$, k_{r} is transferred to $r = k_{\mathrm{r}}/(2\pi) \cdot \lambda \cdot f$ (see Figure 5.12).

An example of this is shown in Figure A.17. The upper diagrams show $\mathscr{E}_{\mathrm{laser}}(r')$ (red), the $I_{\mathrm{laser}}(r')$ (purple) and the distribution when a plane wave enters the microscope objective (green curve). The product of the red and the green curve yields the "input field distribution" in the object plane $\mathscr{E}_{\mathrm{obj}}(r')$. Its Fourier transformation is the "output field distribution" in the focal plane (not shown). The square of the latter one is $I(r)$, displayed as the blue curve in the lower diagrams. For comparison, the lower diagrams show also the focal intensity distributions of a plane wave illumination of the microscope objective (green curves) and that obtained when the laser beam of a given diameter would have been focused by a lens with $D \gg D_{\mathrm{laser}}$ (purple). In the latter case, the focal spot size is always the smallest one, but then it must be calculated from Equation (5.43) with D_{laser} as the relevant aperture and not that of the optics.

From Figure A.17a, it is well seen that if the FWHM of the laser beam D_{laser} equals D, the focal spot size is nearly the same as expected from Equations (5.43) or (5.44) with NA or D given by the microscope objective. The same result is obtained for $D_{\mathrm{laser}} > D$ (not shown here), but due to the cut by the aperture of the microscope objective, this is always accompanied with a significant loss of energy. On the other hand, a smaller beam diameter D_{laser} leads to a larger focal spot because significant part of the microscope objective is not filled by laser light (Figure A.17b). This is in contrast to the usage of the microscope objective within a microscope, where automatically the full aperture is filled by the light from the observed object. But when used for the focusing of a Gausian beam, even more, due to the cut of it as aperture, the blue curve profile is more extended than that of the green and purple curves, respectively. It may be interesting, to see what happens if the beam is not well aligned. Figure A.17c shows that particular case. Finally, we may mention that in the same way, in a laser beam path, any cut by mirrors with

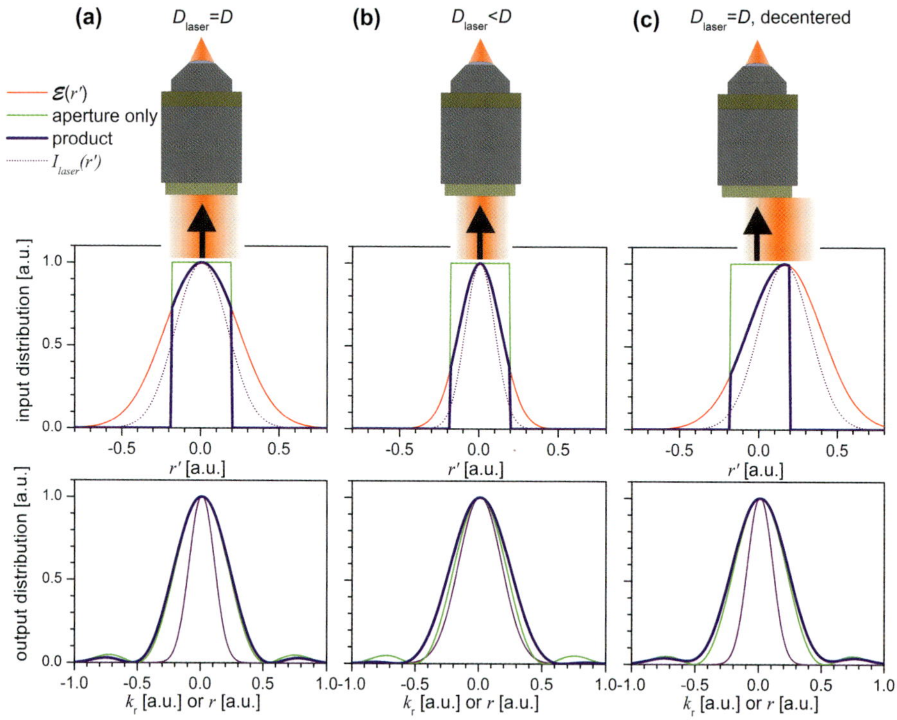

Fig. A.17: Illustration of focusing a Gaussian beam by a microscope objective with a given input aperture $D_{aperture}$ (see the text).

too small sizes will lead to clipping, and thus to larger focal spots (and other negative effects such as unwanted intensity peaks that may destroy optics, but this is not a subject here).

A.9 Examples of PSF and MTF in the presence of aberrations

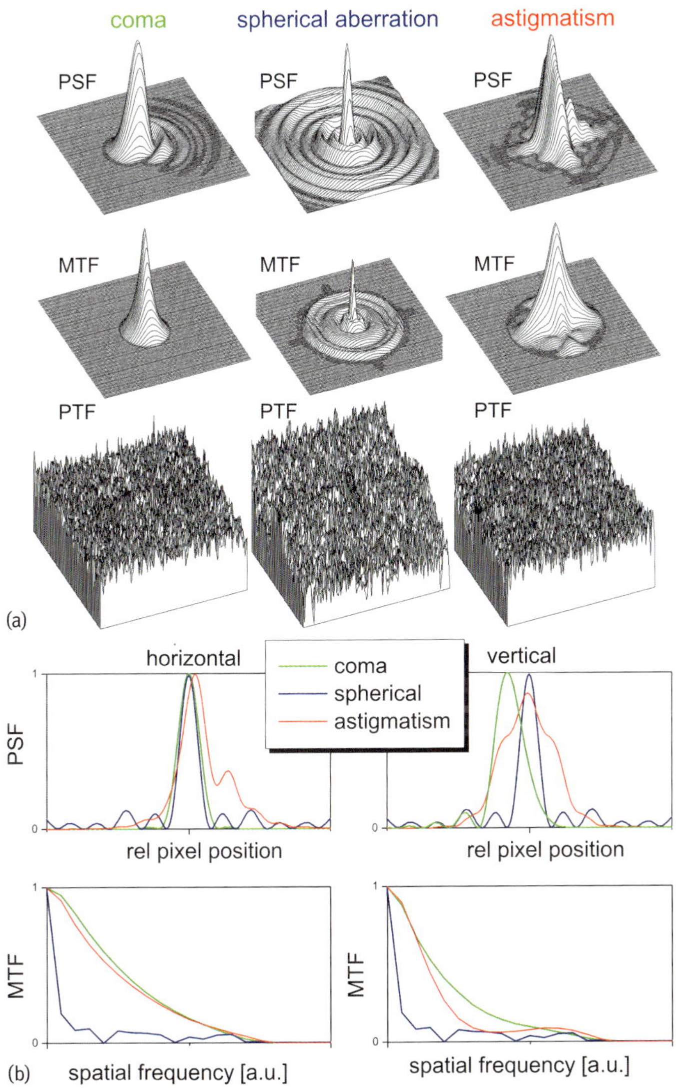

Fig. A.18: (a) PSF, MTF and PTF calculated for real lenses that are arranged to generate different aberrations. Examples are shown for coma, spherical aberration (here a rather strong one) and astigmatism. (b) Profiles measured along the lines in horizontal and vertical direction, respectively, through the center of the PSF and MTF, respectively. Note that the displayed profile of the MTF is equal to the Fourier transformation of the LSF (see Equation (A.4)). The example for coma is the same as displayed in Figure 5.10. The ordinate of the PTF ranges from $-\pi$ to $+\pi$. Calculations by courtesy of J. Napier.

A.10 MTF measurements with a Siemens star off-center

For MTF measurements with a Siemens star that is not located in the center of the object field, special care has to be taken for correct analysis. In particular, special attention has to be paid to measurements within the corners. In Figure A.19, this is illustrated for the top right corner. The field of the image on the sensor surface is indicated by the grey rectangle. Its center is marked by a cross. The yellow point marks a position h_i on the image field diagonal where the MTF should be measured. First, we will concentrate on the Siemens stars with black and white sectors. Before we continue, we would like to note that the displayed sector sizes are just chosen for proper illustration in this figure. But they are not suitable for a real measurement. For the latter one, a Siemens star as displayed in Figure 8.7a may be more appropriate.

A measurement along the green and red lines, respectively, displayed in Figure A.19 yields the pattern of the "grating" underneath formed by the sectors according to the local sector spacing. As usual, this yields the corresponding R_h (compare the line profile in Figure 8.7b; R_h is the frequency spectrum at the radial distance h from the center). The colors in Figure A.19 are the same as those in Figure 8.21; a contrast measurement along the green lines yields the sagittal MTF (= radial MTF). The grating bars, respec-

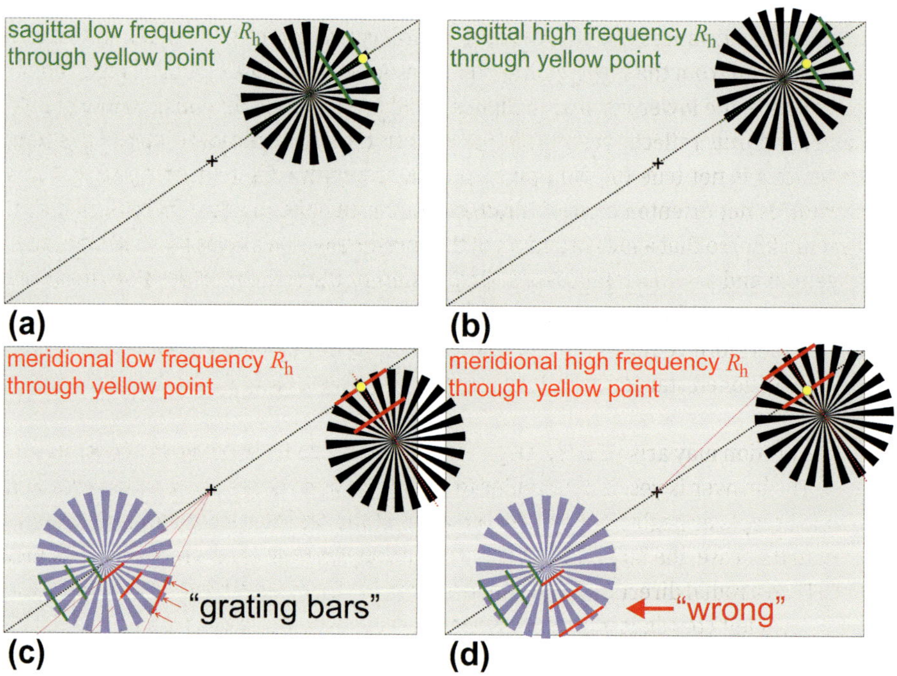

Fig. A.19: Illustration of a correct MTF measurement using a Siemens star, which is off-center. Beside the diagonal, other lines in the radial direction through the center are displayed in magenta. For a further explanation, see the text.

tively, sectors are oriented in radial direction, which results in a Fourier spectrum perpendicular to it as displayed in Figure A.19. A contrast measurement along the red lines yields the meridional MTF (= tangential MTF). For the intended position marked by the yellow point, for a specific R_h,-value, the Siemens star has to be placed as shown in Figure A.19a, for another sagittal R_h-value, the according position is shown in Figure A.19b. This means that the star has to be shifted along the image field diagonal.

If one assumes that the sagittal MTF-values do not differ significantly within the area provided by the Siemens star, one may even keep the star position fixed and perform the analysis for both green lines, e. g., those in Figure A.19a and *attribute* both R_h values to the same radial distance from the center r, e. g., to the position of the center of the star. In other words, one may neglect the effect of the shift. If that is accepted, the method may work well for the sagittal values because the green lines are always correctly oriented (i. e., perpendicular to the radial line).

Now, naïvely one may expect that a measurement of the meridional values with the same approximation may work well, too. However, this is not true because the situation is much different. Figure A.19c and Figure A.19d show that in those cases the center of the star is never located on the image field diagonal. But although the position of the center itself is not of importance, the direction of the measurement and the position where the measurement is performed matters. Thus, the measurement of the two different meridional R_h values in the example would not only lead to different spatial positions with different radial distances (for the moment we still concentrate on the black stars only). As it can be seen from the magenta line in Figure A.19d, the angle would change, too. As a consequence, the lower red line in Figure A.19d is well oriented along the image field diagonal, and thus reflects the meridional direction of the "spectrum" at this position. However, this is not true for the upper red line, which marks another "grating." This "spectrum" is not oriented along the radial line through the image field center (shown as magenta line) so that a measurement at that position would obviously yield a mixture of tangential and sagittal components. Only a shift of the star to the position displayed in Figure A.19c would yield a correct result. A correct measurement with Siemens stars off-center therefore requires a large sequence of measurements with a carefully and correctly positioned star for each R_h-value and this also separately for the two orientations.

The question may arise if it is also possible to use a single position for a Siemens star instead. The answer is yes, if it is similar to the discussion made for the different green lines, one accepts or neglects significant changes of the MTF within the area covered by the star. In that case, the blue star in Figure A.19c illustrates how a correct measurement of the MTF in a radial direction would have to be made for three different R_h-values close to the bottom left corner. It is apparent that the line profiles must be taken at much different positions. In contrast to that, a measurement according to the red lines of the blue star in Figure A.19d would be wrong. This example clearly shows that a correct measurement for the meridional direction (red and magenta lines) is not as straightforward as that for the sagittal direction (green lines).

A.11 Maxwell's equations, wave equation, etc.

The fundamental equations of (classical) electromagnetism are Maxwell's equations. Solving those together with the material equations and the appropriate boundary conditions for a specific situation, electrodynamics and related wave phenomena and, in particular, optics can be described theoretically, and predictions can be made. Hence, we regard it as reasonable to provide this fundament as well. On the other hand, this topic is well described in a huge number of books (see, e. g., [Lan85, Jac98]), including also advanced standard books of optics such as [Bor99]. For that reason, we would like to restrict and just to provide the most essential part of the theoretical framework with very rudimentary notes but almost without further discussion.

In SI system convention, Maxwell's four equations are

$$\nabla \cdot \vec{\mathscr{D}} = \rho_{el} \tag{A.6a}$$

$$\nabla \cdot \vec{\mathscr{B}} = 0 \tag{A.6b}$$

$$\nabla \times \vec{\mathscr{E}} = -\frac{\partial \vec{\mathscr{B}}}{\partial t} \tag{A.6c}$$

$$\nabla \times \vec{\mathscr{H}} = \vec{j} + \frac{\partial \vec{\mathscr{D}}}{\partial t} \tag{A.6d}$$

here written as differential equations, where ρ_{el} is the electric charge density and $\vec{j}$ the current density. In vacuum, both of them are zero. $\vec{\mathscr{E}}$ is the electric field, $\vec{\mathscr{D}}$ the electric displacement field, $\vec{\mathscr{H}}$ the magnetic field and $\vec{\mathscr{B}}$ is the magnetic induction, which often is simplified termed also as magnetic field.

In standard optics (linear optics, but not nonlinear optics), $\vec{\mathscr{D}}$ and $\vec{\mathscr{E}}$ on one side and $\vec{\mathscr{H}}$ and $\vec{\mathscr{B}}$ on the other side are related by the material equations

$$\vec{\mathscr{D}} = \varepsilon_0 \vec{\mathscr{E}} + \vec{\mathscr{P}} \tag{A.7a}$$

$$\vec{\mathscr{B}} = \mu_0 \vec{\mathscr{H}} + \vec{\mathscr{M}} \tag{A.7b}$$

where $\vec{\mathscr{P}}$ and $\vec{\mathscr{M}}$ are the polarisation vector and the magnetization vector, respectively. In the case of homogenous and isotropic matter, one further gets the linear relationships

$$\vec{\mathscr{D}} = \varepsilon_0 \varepsilon \vec{\mathscr{E}} \tag{A.8a}$$

$$\vec{\mathscr{B}} = \mu_0 \mu \vec{\mathscr{H}} \tag{A.8b}$$

where ε_0 is the dielectric permittivity of vacuum, ε the relative dielectric constant or the dielectric function, μ_0 the magnetic permeability and μ the relative (magnetic) permeability. Then one also gets

$$\vec{\mathscr{P}} = \varepsilon_0 \chi_e \vec{\mathscr{E}} \tag{A.9a}$$

$$\vec{\mathcal{M}} = \mu_0 \chi_m \vec{\mathcal{H}} \tag{A.9b}$$

and

$$\varepsilon = 1 + \chi_e \tag{A.10a}$$

$$\mu = 1 + \chi_m \tag{A.10b}$$

where χ_e and χ_m are the electric and magnetic susceptibility, respectively. In the case of anisotropic media, ε and μ and χ_e and χ_m become second-order tensors. It is important to comment that in several textbooks ε and μ are defined slightly differently. The notation of ε in those books is equal to our product $\varepsilon \cdot \varepsilon_0$, and similarly our $\mu \cdot \mu_0$ is identical with μ in those books. Thus, here ε and μ are dimensionless, which sometimes makes discussion easier.

Of further importance is Maxwell's relation

$$\varepsilon_0 \mu_0 c^2 = 1 \tag{A.11}$$

and the relation of the index of refraction with the dielectric function $\hat{\varepsilon}$ and the permeability $\hat{\mu}$ ($\hat{\varepsilon} \equiv \varepsilon$, $\hat{\mu} \equiv \mu$)

$$\hat{n}(\vec{r}, t) = \sqrt{\hat{\varepsilon}(\vec{r}, t)\hat{\mu}(\vec{r}, t)} \tag{A.12a}$$

here both expressed as a function of the spatial coordinate $\vec{r}$ and time t. In common optics $\hat{\mu} = \mu = 1$ (nonmagnetized materials), and thus Equation (A.12a) becomes

$$\hat{n}(\vec{r}, t) = \sqrt{\hat{\varepsilon}(\vec{r}, t)} \tag{A.12b}$$

Note that $\hat{n} = n' + in''$ and $\hat{\varepsilon} = \varepsilon' + i\varepsilon''$ in general are complex functions that also depend on frequency, where n' and ε' are the real parts, which describe the dispersion and n'' and ε'' the imaginary parts, which describe the absorption. n' is identical with the common index of refraction n. In transparent materials, the imaginary parts are zero.

From Equations (A.6a to d), one can derive the wave equation for electrodynamic waves

$$\nabla^2 \vec{\mathcal{E}}(\vec{r}, t) = \left(\frac{n}{c}\right)^2 \frac{\partial^2 \vec{\mathcal{E}}(\vec{r}, t)}{\partial t^2}, \tag{A.13}$$

here written for the electric field. c is the velocity of light in vacuum and c/n the phase velocity in the medium where the wave propagates.

We would like to remark that in case of nonrelativistic optics, such as optics of daily life, the electric forces dominate so that it is sufficient to restrict most calculations to the electric field (but in some exceptions it is easier to calculate $\vec{\mathcal{B}}$, and from that $\vec{\mathcal{E}}$). As $\vec{\mathcal{E}}$ and $\vec{\mathcal{B}}$ are directly related to each other by

$$\vec{\boldsymbol{k}} \times \vec{\mathscr{H}}_0 = -\omega \cdot \varepsilon \cdot \vec{\mathscr{E}}_0 \tag{A.14a}$$

$$\vec{\boldsymbol{k}} \times \vec{\mathscr{E}}_0 = \omega \cdot \mu \cdot \vec{\mathscr{H}}_0, \tag{A.14b}$$

and thus

$$|\vec{\mathscr{E}}| = \frac{c}{n}|\vec{\mathscr{B}}| \tag{A.15}$$

(for nonmagnetized materials $\mu = 1$) it is always possible to calculate one field from the other.

Equations (A.14) follow the familiar right-hand rule, which describes the orientations of $\vec{\mathscr{E}}$, $\vec{\mathscr{H}}$ (or $\vec{\mathscr{B}}$) and $\vec{\boldsymbol{k}}$ (for common optics; for double negative media the left-hand rule must be applied, see Section 7.5).

Now, for a specific situation with the according boundary conditions, Equation (A.13) may be solved. As discussed, e. g., in Section 5.1.1, a particular solution, namely the most simple one, is a plane monochromatic light wave.

If one now inserts the solution Equation (5.3) into Equation (A.13) and for simplicity omits the index and the arguments of the field one obtains

$$\nabla^2 \vec{\mathscr{E}} + k^2 \vec{\mathscr{E}} = 0 \tag{A.16}$$

where the absolute value of the wave vector $\vec{\boldsymbol{k}}$, namely the wave number is

$$k = |\vec{k}| = \frac{2\pi}{\lambda} \tag{A.17}$$

and

$$c/n = \lambda \cdot \nu = (\lambda_0/n) \cdot \nu \tag{A.18}$$

(see also Sections 3.1.1 and 5.1). Here, $k = k_0 \cdot n$ and $\lambda = \lambda_0/n$, respectively, are the wave number and wavelength in the medium where the light propagates. The index “0” indicates the corresponding quantities in vacuum. Equation (A.16) is the Helmholtz equation (see also Section 3.1.1). We may note that in paraxial approximation it has some similarity with Schrödinger’s equation.

A.12 Resolution and contrast sensitivity function of the human eye

The diagram below shows a chirped grating contrast chart that illustrates the resolution and the contrast sensitivity function CSF of the human eye. The book may be turned so that the bottom becomes the left-hand side. Then, by viewing the chart at various distances, one may observe the position where one perceives the best quality and one can see when the pattern vanishes. The contrast is varied along the image height direction with respect to the turned book. It is interesting that for a short viewing distance the left part of the test chart appears to have a rather low contrast, but at an increased distance this becomes better; at a rather short viewing distance, in the lower half and the left part of the test chart one may observe a rather smooth gray distribution. Small changes in this low contrast region could not well be recognized. This illustrates the reduced resolution at small values of R_φ, respectively. Within, e. g., 1°, this is less than one cycle, i. e., R_φ is much less than cycle/deg. On the far right, one still may observe the image of the grating lines (the viewing distance is short enough that the number of cycles per degree is not too large, and thus R_φ is approximately in the middle region displayed in Figure 5.42a). If the viewing distance is increased, the line structure in the left part becomes better visible. Due to the larger distance, now there are more cycles per degree visible, and thus R_φ shifts to the right direction, and thus the CSF increases, but the details of the fine structures at right are no longer visible (R_φ shifts from the middle to the right, and thus the CSF decreases).

By changing the viewing distance back and forth, one may observe that the region where resolution and contrast become optimum, changes. The reason for this observation is the CSF where a given grating period within the test chart, which is the object observed by the eye, translates into an R_φ-value (in lp/degree; see Figure 5.42a), which depends on viewing distance, or more general, on viewing conditions.

Bibliography

The following provides a list with selected literature. It is not the intention to offer a complete review with an extended compilation. But we would also like to remark that references can be found as footnotes, in particular, such ones, which do refer rather closely to the topics discussed on the corresponding pages.

Selected literature

[All11] E. Allen, S. Triantaphillidou (eds.): *The Manual of Photography*, Focal Press, Oxford 2011.

[Ber30] M. Berek: *Grundlagen der praktischen Optik*, de Gruyter Verlag, Berlin 1970.

[Bla14] V. Blahnik: *About the irradiance and apertures of camera lenses*, ZEISS Camera Lenses, July 2014, http://lenspire.zeiss.com/photo/app/uploads/2022/02/technical-article-about-the-irradiance-and-apertures-of-camera-lenses.pdf (visited January 2023).

[Bla16] V. Blahnik, B.Voelker: *About the reduction of reflections for camera lenses – How T^*-coating made glass invisible*, ZEISS Camera Lenses, March 2016, http://lenspire.zeiss.com/photo/app/uploads/2022/02/technical-article-about-the-reduction-of-reflections-for-camera-lenses.pdf (visited January 2023).

[Bla21] V. Blahnik, O.Schindelbeck: *Smartphone imaging technology and its applications*, Adv. Opt. Techn. **10**(3) (2021) 145–232.

[Bor99] M. Born, E. Wolf: *Principles of Optics*, 7th edn., Cambridge University Press, Cambridge 1999.

[Bro79] I. N. Bronstein, K. A. Semendjajew: *Taschenbuch der Mathematik*, 21 Aufl., BSB B.G. Teubner Verlagsgesellschaft, Leipzig und Verlag Nauka, Moskau, 1979.

[Eas10] R. L. Easton: *Fourier Methods in Imaging*, John Wiley & Sons, New York 2010.

[EMVA1288] www.emva.org or www.standard1288.org.

[Flü55] J. Flügge: Das Photographische Objektiv, in hrsg. v. K. Michel, *Die wissenschaftliche und angewandte Photographie, Band 1*, Springer Verlag, Wien 1955.

[Goo17] J. W. Goodman: *Introduction to Fourier Optics*, 4th edn., W. H. Freeman and Company, New York 2017.

[Gru02] S. M. Gruner, M. W. Tate, E. F. Eikenberry: *Charge-coupled device area x-ray detectors*, Rev. Sci. Instrum. **73** (2002) 2815–2842.

[Hec16] E. Hecht: *Optics*, 5th edn., Pearson Inc., 2016; see also: *Optik*, 7 Aufl., Walter de Gruyter, Berlin, 2018.

[Hön09] B. Hönlinger, H. H. Nasse: *Distortion*, Carl Zeiss, Camera and Lens Division, October 2009, https://lenspire.zeiss.com/photo/app/uploads/2018/04/Article-Distortion-2009-EN.pdf (visited January 2023).

[Hor06] A. Hornberg (ed.): *Handbook of Machine Vision*, Wiley-VCH Verlag, Weinheim 2006.

[Jac98] J. D. Jackson: *Klassische Elektrodynamik*, 4 Aufl., Walter de Gruyter, Berlin, 2006; *Classical Electrodynamics*, 3rd. edn., John Wiley & Sons, New York 1998.

[Jäh23] B. Jähne: *Digitale Bildverarbeitung und Bildgewinnung*, Springer Verlag Berlin, Heidelberg, 2023; see also previous eds. and english version: *Digital Image Processing*, Springer Verlag Berlin, Heidelberg, 2005.

[Kin10] R. Kingslake, R. B. Johnson: *Lens Design Fundamentals*, Academic Press, Burlington, Oxford 2010.

[Kin39] R. Kingslake: *The Optics of Photographic Lenses*, in: K. Henney, B. Dudley (eds.), *Handbook of Photography*, Whittlesey House, New York, London 1939.

[Kin89] R. Kingslake: *A History of the Photographic Lens*, Academic Press, San Diego 1989.

[Lan85] L. D. Landau, E. M. Lifschitz: *Lehrbuch der theoretischen Physik, Band II: Klassische Feldtheorie and Band VIII: Elektrodynamik der Kontinua*, Akademieverlag Berlin 1984 (Bd.II) and 1985 (Bd.VIII), resp.

https://doi.org/10.1515/9783110789966-011

[LdO03] *Lexikon der Optik*, Spektrum Akademischer Verlag GmbH, Heidelberg, 2003.

[Luh19] T. Luhmann, S. Robson, S. Kyle, J. Boehm: *Close-Range Photogrammetry and 3D Imaging*, De Gruyter, Berlin, Boston 2019.

[Nak06] J. Naklamura (ed.): *Image Sensors and Signal Processing for Digital Still Cameras*, CRC Press Taylor & Francis, Boca Raton 2006.

[Nas08] H. H. Nasse: *How to Read MTF Curves*, Carl Zeiss, Camera and Lens Division, December 2008, https://lenspire.zeiss.com/photo/app/uploads/2018/04/Article-MTF-2008-EN.pdf (visited January 2023).

[Nas09] H. H. Nasse: *How to Read MTF Curves – Part II*, Carl Zeiss, Camera and Lens Division, March 2009, https://lenspire.zeiss.com/photo/app/uploads/2018/04/CLN_MTF_Kurven_2_en.pdf (visited January 2023).

[Nas10] H. H. Nasse: *Depth of Field and Bokeh*, Carl Zeiss, Camera and Lens Division, March 2010, https://lenspire.zeiss.com/photo/app/uploads/2022/02/technical-article-depth-of-field-and-bokeh.pdf (visited January 2023).

[Nas11a] H. H. Nasse: *From the Series of Articles on Lens Names – Tessar*, Carl Zeiss, Camera and Lens Division, March 2011, https://lenspire.zeiss.com/photo/app/uploads/2022/02/technical-article-lens-names-tessar.pdf (visited January 2023).

[Nas11b] H. H. Nasse: *From the Series of Articles on Lens Names – Planar*, Carl Zeiss, Camera and Lens Division, July 2011, https://lenspire.zeiss.com/photo/app/uploads/2022/02/technical-article-lens-names-planar.pdf (visited January 2023).

[Nas11c] H. H. Nasse: *From the Series of Articles on Lens Names – Distagon, Biogon and Hologon*, Carl Zeiss, Camera and Lens Division, December 2011, https://lenspire.zeiss.com/photo/app/uploads/2022/02/technical-article-lens-names-distagon-biogon-hologon.pdf (visited January 2023).

[Ped08] F. Pedrotti, L. Pedrotti, W. Bausch, H. Schmidt: *Optik für Ingenieure*, Springer Verlag, Berlin, Heidelberg, 2008.

[Ped17] F. Pedrotti, L. Pedrotti, L. Pedrotti: *Introduction to Optics*, 3rd edn., Cambridge Press, Cambridge, 2017.

[Sal19] B. E. A. Saleh, M. C. Teich: *Optik und der Photonik*, 3. Aufl., Wiley-VCH, Weinheim 2020 see also: Fundamentals of Photonics, 3rd edn., John Wiley & Sons, 2019.

[Sch14] G. Schröder, H. Treiber: *Technische Optik*, Vogel Business Media GmbH, Würzburg 2014.

[Sch81] G. Schröder: *Technische Fotografie*, Vogel Verlag, Würzburg 1981.

[Smi08] W. J.Smith: *Modern Optical Engineering*, 4th edn., McGraw Hill, New York 2008.

[Ste12] T. Steinich, V. Blahnik: *Optical design of camera optics for mobile phones*, Adv. Opt. Techn. **1** (2012) 51–58, https://lenspire.zeiss.com/photo/app/uploads/2022/02/technical-article-optical-design-of-camera-optics-for-mobile-phones.pdf (visited January 2023).

Web links (visited January 2023)

Three recommendable websites (just an arbitrary rather short selection of a many good websites)

Image Sensors World by V. Koifman
http://image-sensors-world.blogspot.com

Photons to Photos by W. J. Claff
http://www.photonstophotos.net/

Lenspire by Zeiss
http://lenspire.zeiss.com/photo/en

Other recommendable sites can be found from companies such as Leica Camera AG, DXO, DXO Mark, Image Enginering, Imatest, PCO AG, etc.

Picture Credits

Thanks must go to those below in respect to the following illustrations and copyright images:

1.2:	Leica Camera AG, https://leica-camera.com
1.11:	https://commons.wikimedia.org/wiki/File:Eye_scheme_mulitlingual.svg (visited January 2023), by Talos, colorized by Jakov, CC BY-SA 3.0 (http://creativecommons.org/licenses/by-sa/3.0/), via Wikimedia Commons
1.29:	©Schott AG, http://www.schott.com/
2.1:	(a) https://commons.wikimedia.org/wiki/File:1646_Athanasius_Kircher_-_Camera_obscura.jpg – public domain; via Wikimedia Commons (b) [Hön09], Carl Zeiss AG
2.3:	https://commons.wikimedia.org/wiki/File:Rheda-Wiedenbr%C3%BCck,_stillgelegte_Eisenbahnbr%C3%BCcke,_Lochkamera.jpg (visited January 2023), by Joachim K. Löckener, CC BY 3.0 (https://creativecommons.org/licenses/by/3.0), via Wikimedia Commons
2.11(c):	Leica Camera AG, https://leica-camera.com
2.19(b):	Leica Camera AG, https://leica-camera.com
2.27(c):	[Nas11a], Carl Zeiss AG
3.12:	Carl Zeiss AG (visited January 2023): https://www.zeiss.com/content/dam/consumer-products/downloads/historical-products/photography/zm-lenses/en/datasheet-zeiss-zm-tele-tessar-485-en.pdf
3.27:	Portions Copyright ©2017 Synopsys, Inc. Used with permission. All rights reserved. Synopsys & Code V are registered trademarks of Synopsys, Inc.
3.47(b):	[Bla14], Carl Zeiss AG
3.55(a):	[Hön09], Carl Zeiss AG
3.62:	https://commons.wikimedia.org/wiki/File:Toric_lens_surface_2.png (visited January 2023), By HHahn, CC BY-SA 3.0 (https://creativecommons.org/licenses/by-sa/3.0), via Wikimedia Commons
4.79:	©Institut für Mikroelektronik Stuttgart
4.91, 4.86:	©ProxiVision GmbH, Bensheim, Germany
5.35(a):	Jan Holger Teubner
5.7:	[Nas08], Carl Zeiss AG
6.11:	[Bla14], Carl Zeiss AG
6.12, 6.13(b):	Nikon Corporation, https://www.nikon.de/de_DE
6.13(c):	Canon camera museum (visited January 2023): http://global.canon/en/c-museum/product/ef398.html
6.14(a):	Leica Camera AG, https://leica-camera.com (visited January 2023): https://de.leica-camera.com/Fotografie/Leica-M/M-Objektive/APO-Summicron-M-1-2-90-mm-ASPH/Downloads
6.14(b):	Carl Zeiss AG (visited January 2023): https://www.zeiss.de/consumer-products/fotografie/otus/otus-1455.html#data
6.14(c):	Carl Zeiss AG (visited January 2023): https://www.zeiss.de/consumer-products/fotografie/milvus/milvus-2135.html#data
6.15(c):	Nikon Corporation, https://www.nikon.de/de_DE
6.17:	Leica Camera AG, https://leica-camera.com (visited January 2023): https://de.leica-camera.com/Fotografie/Leica-M/M-Objektive/Leica-Summarit-M-1-2,4-50-mm/Downloads https://de.leica-camera.com/Fotografie/Leica-M/M-Objektive/APO-Summicron-M-1-2-50-mm-ASPH/Downloads

https://doi.org/10.1515/9783110789966-012

	https://de.leica-camera.com/Fotografie/Leica-M/M-Objektive/Noctilux-M-1-0,95-50-mm-ASPH/Downloads
	https://de.leica-camera.com/Fotografie/Leica-SL/SL-Objektive/Festbrennweiten-Objektive/SUMMILUX-SL-50
6.18:	Canon camera museum (visited January 2023):
	http://global.canon/en/c-museum/product/ef451.html
	http://global.canon/en/c-museum/product/ef392.html
	http://global.canon/en/c-museum/product/ef283.html
6.19:	Carl Zeiss AG (visited January 2023):
	https://www.zeiss.de/camera-lenses/fotografie/produkte/classic-objektive/planar-1450.html#daten
	https://www.zeiss.de/camera-lenses/fotografie/produkte/otus-objektive/otus-1455.html#daten
	[Nas11b], Carl Zeiss AG
	[Bla14], Carl Zeiss AG
6.20:	Nikon Corporation, https://www.nikon.de/de_DE
6.23(d):	[Nas11c], Carl Zeiss AG
6.24:	[Nas11c], Carl Zeiss AG (visited January 2023):
	https://www.zeiss.de/consumer-products/fotografie/otus/otus-1428.html#data
	https://www.zeiss.de/consumer-products/fotografie/milvus/milvus-2815.html#data
	https://www.zeiss.com/consumer-products/int/photography/zm/biogon-2825-zm.html#data
	https://www.zeiss.com/content/dam/consumer-products/downloads/historical-products/photography/classic-lenses/en/datasheet-zeiss-classic-distagon-225.pdf
6.25:	[Nas11c], Carl Zeiss AG
6.26:	[Bla14], Carl Zeiss AG
6.27:	Leica Camera AG, https://leica-camera.com (visited January 2023):
	https://de.leica-camera.com/Fotografie/Leica-M/M-Objektive/Summaron-M-1-5,6-28-mm/Downloads
	https://de.leica-camera.com/Fotografie/Leica-M/M-Objektive/SUMMILUX-M-1-1,4-21-mm-ASPH/Downloads
6.28:	Nikon Corporation, https://www.nikon.de/de_DE
6.29:	Canon camera museum (visited January 2023):
	http://global.canon/en/c-museum/product/ef421.html
	http://global.canon/en/c-museum/product/ef400.html
6.32:	Nikon Corporation, https://www.nikon.de/de_DE
6.37:	Leica Camera AG, https://leica-camera.com (visited January 2023):
	https://de.leica-camera.com/Fotografie/Leica-SL/SL-Objektive/Vario-Objektive/VARIO-ELMARIT-SL-24-90
6.39, 6.42, 6.43:	Jos. Schneider Optische Werke GmbH
6.44:	Jos. Schneider Optische Werke GmbH
6.46–6.49:	[Bla16], Carl Zeiss AG
6.52–6.54:	[Bla16], Carl Zeiss AG
6.57–6.60:	[Nas10], Carl Zeiss AG
8.1, 8.4:	Courtesy of Image Engineering GmbH & Co. KG, Frechen, Germany
8.7(a):	Courtesy of Image Engineering GmbH & Co. KG, Frechen, Germany
8.13(d):	Courtesy of Image Engineering GmbH & Co. KG, Frechen, Germany
8.17:	Courtesy of Image Engineering GmbH & Co. KG, Frechen, Germany
8.18, 8.25:	Courtesy of Image Engineering GmbH & Co. KG, Frechen, Germany (images in upper row)

8.20(b), 8.24: [Ste12], reprinted with kind permission

6.33: CURVE-ONE, https://www.curve-one.com/optical-design-fisheye-curve-one80/ (visited January 2023),
Curved CMOS sensor - company CURVE SAS - contact@curve-one.com - Curving all CMOS and IR sensors type

9.1: J. Ma, S. Chan, E. R. Fossum: *Review of Quanta Image Sensors for Ultralow-Light Imaging*, IEEE Transactions On Electron Devices, **69** (2022) 2824–2839; work licensed under a Creative Commons Attribution 4.0 License. For more information, see https://creativecommons.org/licenses/by/4.0/

9.2: [Luh19], reprinted with kind permission of the author

7.34: S. Sinzinger, J. Jahns: Microoptics, 2nd Edition, p. 138, 2003. Copyright Wiley-VCH GmbH; reproduced with permission

7.2: [Bla21], reprinted with kind permission

7.3: U. Teubner, V. Blahnik, H. J. Brückner: *Smartphonekameras, Teil 2: Bildsensor und -verarbeitung – Tricksen für gute Bilder*, Physik in unserer Zeit, **51**(6) (2020) 290

7.4: [Bla21], reprinted with kind permission

7.5: H. J. Brückner, V. Blahnik, U. Teubner: *Smartphonekameras, Teil 1: Kameraoptik – Maximale Bildqualität aus Minikameras*, Physik in unserer Zeit, **51**(5) (2020) 236

7.6: [Bla21], reprinted with kind permission

7.7: H. J. Brückner, V. Blahnik, U. Teubner: *Smartphonekameras, Teil 1: Kameraoptik – Maximale Bildqualität aus Minikameras*, Physik in unserer Zeit, **51**(5) (2020) 236

7.8: [Bla21], reprinted with kind permission

7.9a): H. J. Brückner, V. Blahnik, U. Teubner: *Smartphonekameras, Teil 1: Kameraoptik – Maximale Bildqualität aus Minikameras*, Physik in unserer Zeit, **51**(5) (2020) 236

7.10: Sony Europe

7.17a): [Bla21], reprinted with kind permission

7.26: U. Teubner, V. Blahnik, H. J. Brückner: *Smartphonekameras, Teil 2: Bildsensor und -verarbeitung – Tricksen für gute Bilder*, Physik in unserer Zeit, **51**(6) (2020) 290

7.28: U. Teubner, V. Blahnik, H. J. Brückner: *Smartphonekameras, Teil 2: Bildsensor und -verarbeitung – Tricksen für gute Bilder*, Physik in unserer Zeit, **51**(6) (2020) 290

7.31: U. Teubner, V. Blahnik, H. J. Brückner: *Smartphonekameras, Teil 2: Bildsensor und -verarbeitung – Tricksen für gute Bilder*, Physik in unserer Zeit, **51**(6) (2020) 290

7.37: N. Mohammad et al., Sci. Rep. **8** (2018) 2799; reprinted according to Creative Commons Attribution 4.0 International License; http://creativecommons.org/licenses/by/4.0/

7.39: F. Aieta et al.: *Aberration-Free Ultrathin Flat Lenses and Axicons at Telecom Wavelengths Based on Plasmonic Metasurfaces*, Nano Lett. **12** (2012) 4932–4936; reprinted/adapted with kind permission. Copyright 2012 American Chemical Society

7.40: S. W. D. Lim, M. L. Meretska, F. Capasso: *A High Aspect Ratio Inverse-Designed Holey Metalens*, Nano Letters **21**(20) (2021) 8642–8649; reprinted with kind permission by the authors

7.41: E. Tseng, S. Colburn, J. Whitehead, L. Huang, S.-H. Baek, A. Majumdar, F. Heide: *Neural nano-optics for high-quality thin lens imaging*, Nature Comm. **12** (2021) 6493; reprinted according to Creative Commons Attribution 4.0 International License; http://creativecommons.org/licenses/by/4.0/

7.43: Jordan T. R. Pagé et al.: *Designing high-performance propagation-compressing spaceplates using thin-film multilayer stacks*, Opt. Express **30** (2022) 2197; reprinted with permission; https://doi.org/10.1364/OE.443067; #443067 Journal ©2022 Optica Publishing Group

7.44: O. Reshef et al.: *An optic to replace space and its application towards ultra-thin imaging systems*, Nature Comm. **12** (2021) 3512; reprinted according to Creative Commons Attribution 4.0 International License; http://creativecommons.org/licenses/by/4.0/

4.70:	M. Miyata et al., Optica **8** (2021) 1596, ©2021 Optica Publishing Group; reprinted with permission
4.74b), c), 4.75b):	Z. Shukri: The State of the Art of CMOS Image Sensors | TechInsights Inc. Inc., reprinted with permission; https://www.techinsights.com/ebook/ebook-latest-development-trends-cmos-image-sensors (visited January 2023)
4.83, 4.84:	Z. Shukri: The State of the Art of CMOS Image Sensors | TechInsights Inc. Inc., reprinted with permission; *masked* https://www.techinsights.com/ebook/ebook-latest-development-trends-cmos-image-sensors (visited January 2023)
4.94 (a):	Courtesy Andanta GmbH, Olching, Germany
4.94 (b),(c):	Courtesy Curved CMOS sensor - company CURVE SAS - contact@curve-one.com - Curving all CMOS and IR sensors types
4.95:	B. Guenter et al., Opt. Expr. **25** (2017) 13010; #286903 Journal ©2017 Optica Publishing Group, reprinted with permission
4.96:	B. Guenter et al., Opt. Expr. **25** (2017) 13010; #286903 Journal ©2017 Optica Publishing Group, reprinted with permission
5.36:	B. Guenter et al., Opt. Expr. **25** (2017) 13010; #286903 Journal ©2017 Optica Publishing Group, reprinted with permission

Disclaimer

Although the authors have taken much care in writing the present book, mistakes cannot be excluded in general. The naming of companies, trademarks, products, etc. has to be considered for information only. The related names and designations are properties by their owners and may be protected by law.

The authors have made every effort to get the permission to reproduce pictures and figures from third parties. Those listed in the figure captions and Picture Credits are authorized by the right-holders. In case that the authors have overlooked a particular figure and missed to contact the right-holder, the right-holder is kindly asked to get in contact with the authors. They are kindly asked as well for subsequent approval.

Index

Note that in the following, keywords may be found as full terms or abbreviations or reference is made to both (but see also "Abbreviations" in the List of symbols at the beginning of the book). Moreover, we would like to note that terms such as CCD, CMOS, PDA, photodiode, DSLR, consumer camera, sensor, detector MTF, PTF, electron, imaging, etc. are so frequently used within this book that they are either not included in the index or they just refer to a few pages. Exceptions are special kinds of CCD and so on.

https://doi.org/10.1515/9783110789966-013